Y0-CAY-692

SECOND EDITION

# Wage Slave No More

## Law and Taxes for the Self-Employed

BY ATTORNEY STEPHEN FISHMAN

## Your Responsibility When Using a Self-Help Law Book

We've done our best to give you useful and accurate information in this book. But laws and procedures change frequently and are subject to differing interpretations. If you want legal advice backed by a guarantee, see a lawyer. If you use this book, it's your responsibility to make sure that the facts and general advice contained in it are applicable to your situation.

## Keeping Up to Date

To keep its books up to date, Nolo Press issues new printings and new editions periodically. New printings reflect minor legal changes and technical corrections. New editions contain major legal changes, major text additions or major reorganizations. To find out if a later printing or edition of any Nolo book is available, call Nolo Press at 510-549-1976 or check the catalog in the *Nolo News,* our quarterly publication. You can also contact Nolo Press on the Internet at www.nolo.com.

To stay current, follow the "Update" service in the *Nolo News*. You can get a free one-year subscription by sending us the registration card in the back of the book. In another effort to help you use Nolo's latest materials, we offer a 25% discount off the purchase of the new edition of your Nolo book if you turn in the cover of an earlier edition. (See the "Recycle Offer" in the back of this book.) This book was last revised in **February 1998**.


| | |
|---|---|
| Second Edition | FEBRUARY 1998 |
| Cover Design | TONI IHARA |
| Book Design | AMY IHARA |
| Production | SARAH TOLL |
| Index | NANCY MULVANY |
| Proofreading | PATRICIA WEST |
| Printing | CONSOLIDATED PRINTERS, INC. |

Fishman, Stephen.
Wage slave no more : law and taxes for the self-employed / by Stephen Fishman. -- 2nd ed.
p. cm.
Includes index.
ISBN 0-87337-458-4
1. Independent contractors--United States--Popular works. 2. Self-employed--Taxation--Law and Legislation--United States--Popular works. I. Title.
KF390.I54F57 1998
343.7305'26--dc21 97-44285
CIP





For information on bulk purchases or corporate premium sales, please contact the Special Sales Department. For academic sales or textbook adoptions, ask for Academic Sales. Call 800-955-4775 or write to Nolo Press, Inc., 950 Parker Street, Berkeley, CA 94710.

# Acknowledgments

Many thanks to:

Barbara Kate Repa for her superb editing.

Janet Portman for her editorial contributions.

Malcolm Roberts, CPA, for reviewing the tax materials.

Gary Gerard, of Gerard & Associates, Berkeley, California—a great landlord and brilliant translator, for sharing his experiences as an independent contractor.

The many independent contractors throughout the country who permitted me to interview them.

Nancy Mulvany for the helpful index.

Patricia West for thorough proofreading.

# Table of Contents

## 5 Obtaining Licenses, Permits and Numbers



## 6 Insuring Your Business and Yourself



## 7 Pricing Your Services and Getting Paid



## 8 Taxes and the Self-Employed



## 9 Reducing Your Income Taxes

CHAPTER

# 1

# Self-Employment: The Good, the Bad and the Ugly

This book is for people who are already self-employed or are thinking about becoming self-employed. Other terms used to describe the self-employed include independent contractor, consultant, freelancer, entrepreneur and business owner. All these terms describe the same thing—people who have left the ranks of the wage slaves to strike out on their own by going into business for themselves.

There are millions of self-employed people and their numbers are constantly growing as more people are going into business for themselves after their corporate employers make workforce cutbacks and businesses hire outside workers to do work that used to be performed by employees.

This book focuses on people who sell personal services. If your business involves selling goods to the public, see *The Legal Guide for Starting and Running a Small Business*, by Fred Steingold (Nolo Press).

Unfortunately, the lot of the self-employed is not always an easy one. You have to make what can be a difficult transition. You no longer have an employer around to take care of you. For example, you don't have a company payroll department to withhold and pay all your taxes for you. Many self-employed people—even those with plenty of clients—get into trouble because they don't run their operations in a businesslike manner. You don't have to start wearing a green eye shade, but you do need to learn a few rudiments of business and tax law. That's the focus of this book.

A few hours spent now learning the nuts and bolts of law and taxes for the self-employed can save you save you innumerable headaches and substantial time and money later on.

Before you delve into the details of law and taxes discussed in the following chapters, read this chapter for an overview of what's both good and bad about being self-employed as compared with being an employee. It may help you make an informed decision if you're thinking about becoming self-employed—or help confirm that you made the right decision if you're already working on your own.

## A. Self-Employment: The Good

Being self-employed can give you more freedom and privacy than employees have and result in tax benefits.

### 1. Independence

When you're self-employed, you're your own boss—with all the risks and rewards that entails. Most self-employed people bask in the freedom that comes from being in business for themselves. They would doubtless agree with the following sentiment expressed by one self-employed person: "I can choose how, when and where to work, for as much or little time as I want. In short, I enjoy working for myself."

The self-employed are masters of their economic fates. The amount of money they make is directly related to the quantity and quality of their work. This is not necessarily the case for employees. The self-employed don't have to ask their bosses for a raise; they go out and find more work.

Moreover, since you're normally not dependent upon a single company for your livelihood, the hiring or firing decisions of any one company don't have the impact on you they have on employees. One self-employed person explains: "I was laid off six years ago and chose to start my own company rather than sign on for another ride on someone else's rollercoaster. It's scary at first, but I'm now no longer at someone else's mercy."

### 2. Higher Earnings

You can often earn more when you're self-employed than as an employee in someone else's business. For example, an employee in a public relations firm decided to go out on her own when she learned that the firm billed her time out to clients at $125 per hour while only paying her $17 per hour. She now charges $75 per hour and makes a far better living than she ever did as an employee.

According to *The Wall Street Journal*, self-employed people who provide services are usually paid at least 20% to 40% more per hour than employees performing the same work. This is because hiring firms don't have to pay half of their Social Security taxes, pay unemployment compensation taxes, provide workers' compensation coverage or employee benefits like health insurance and sick leave. Of course, how much you're paid is a matter for negotiation between you and your clients. Self-employed people whose skills are in great demand may receive far more than employees doing similar work.

### 3. Tax Benefits

Being self-employed also provides you with many tax benefits that employees don't have. For example, no federal or state taxes are withheld from your paychecks as they must be for employees. Instead, the self-employed normally pay estimated taxes directly to the IRS four times a year. (See Chapter 10.) This means you can hold on to your hard-earned money longer without having to turn it over to the IRS. Moreover, it's up to you to decide how much estimated tax to pay, but there are penalties if you underpay. The lack of withholding and control over estimated tax payments can result in improved cash flow for the self-employed as compared with employees.

Even more important, you can take advantage of many business-related tax deductions that are limited or not available at all for employees. When you're self-employed, you can deduct from your income tax any necessary expenses related to your business as long as they are reasonable in amount and ordinarily incurred by businesses of your type. This may include, for example, office expenses including those for home offices, travel expenses, entertainment and meal expenses, equipment and insurance costs and more. (See Chapter 4.)

In contrast to the numerous deductions available to the self-employed, an employee's work-related deductions are severely limited. Some deductions available to the self-employed may not be taken by employees—for example, an employee may not deduct the cost of commuting to and from work, but a self-employed person who has a main office separate from that of a client may ordinarily deduct this expense. (See Chapter 9.) Even those expenses that are deductible may only be deducted to the extent they exceed 2% of the employee's adjusted gross income. This means that most expenses related to employment cannot be fully deducted.

The self-employed can also establish retirement plans such as SEP-IRAs and Keogh Plans that have tax advantages. These plans also allow them to shelter a substantial amount of their incomes until they retire. (See Chapter 16.)

Because of these tax benefits, the self-employed often pay less tax than employees who earn similar incomes.

### 4. More Privacy

If you're seeking to shield yourself from the prying eyes of the government, you'll have far more success if you're self-employed than if you work as an employee. The government uses employers to keep track of employees for a variety of purposes. For example, a federal law that took effect on October 1, 1997, requires all employers to report to the Department of Health and Human Services the name, address and Social Security number of each newly hired employee. This information will be placed in a huge database that is supposed to be used solely to aid in the collection of overdue child support.

Many states have similar requirements. Some mandate that employers provide them with even more information, such as telephone numbers, dates of birth, and details of insurance coverage provided to new employees.

When you're self-employed, such laws don't apply to you. It's far harder for the government to keep tabs on you or control you life.

## B. Self-Employment: The Bad

Despite the advantages, being self-employed may not be a bed of roses. Following are some of the major drawbacks and pitfalls.

### 1. No Job Security

As discussed above, one of the best things about being self-employed is that you're on your own. But this can be one of the worst things about it as well.

When you're an employee, you must be paid as long as you have your job, even if your employer's business is slow. This is not the case when you're self-employed. If you don't have business, you don't make any money. As one self-employed person says, "If I fail, I don't eat. I don't have the comfort of punching a timeclock and knowing the check will be there on payday."

However, many would argue that there is no such thing as job security for employees in modern America. According to a recent report by *The New York Times*, more than 43 million employee jobs have been erased from the U.S. economy since 1979—and the trend continues.

### 2. No Free Benefits

Although not required to by law, employers usually provide their employees with health insurance, paid vacations and paid sick leave. More generous employers may also provide retirement benefits, bonuses and even employee profit sharing.

When you're self-employed, you get no such benefits. You must pay for your own health insurance, often at higher rates than employers are able to pay. (See Chapter 6.) Time lost due to vacations and illness comes directly out of your bottom line. And you must fund your own retirement.

If you don't earn enough money to purchase these benefits yourself, you will have to forgo some or all of them.

### 3. No Unemployment Insurance Benefits

The self-employed also don't have the safety net provided by unemployment insurance. Hiring firms do not pay unemployment compensation taxes for the self-employed—and they can't collect unemployment when their work for a client ends.

### 4. No Workers' Compensation

Employers must generally provide workers' compensation coverage for their employees. Employees injured on the job are entitled to collect workers' compensation benefits even if the injury was their own fault.

Hiring firms do not provide workers' compensation coverage for the self-employed. If a work-related injury is a self-employed person's fault, he or she has no recourse against the hiring firm. (See Chapter 6.)

### 5. No Free Office Space or Equipment

Employers normally provide their employees with office space or other workplaces and whatever equipment they need to do the job. This is not necessary when a company hires a self-employed person, who must normally provide his or her own workplace and equipment.

### 6. Few or No Labor Law Protections

A wide array of federal and state laws protect employees from unfair exploitation and discrimination by employers. Among other things, these laws:

- impose a minimum wage and require many employees to be paid time and a half for overtime
- make it illegal for employers to discriminate against employees on the basis of race, color, religion, gender and national origin, and
- protect employees who wish to unionize.

Few such legal protections apply to the self-employed.

### 7. Complete Business Responsibility

When you're self-employed, you must run your own business. This means, for example, that you'll need to have at least a rudimentary recordkeeping system or hire someone to keep your records for you. (See Chapter 14.) You'll also likely have to file a far more complex tax return than you did when you were an employee. (See Chapter 8.)

### 8. Others May Discriminate

Because you don't have a guaranteed annual income as employees do, insurers, lenders and others may spurn your business or you may have to pay more than employees do for similar services. It can be particularly difficult, for example, for a self-employed person to obtain disability insurance, particularly if he or she works at home.

Health insurance is easier to get but could cost an arm and a leg. However, a federal pilot program allowing self-employed people to establish medical savings accounts may help you meet your health insurance needs. (See Chapter 6.)

Life will be more difficult if you want to buy a house because lenders are wary of self-employed borrowers. To prove you can afford a loan, you'll likely have to provide a prospective lender copies of your recent tax returns and a profit and loss statement for your business.

## C. Self-Employment: The Ugly

Unfortunately, the bad aspects of self-employment discussed above do not end this litany of woes. Being self-employed can get downright ugly.

### 1. Double Social Security Tax

For many, the ugliest and most unfair thing about being self-employed is that they must pay twice as much Social Security and Medicare taxes as employees. Employees pay a 7.65% on their salaries, up to the Social Security tax limit ($65,400 in 1997). Employers pay a matching amount. In contrast, self-employed people must pay the entire tax themselves—a whopping 15.3% up to the Social Security tax limit. This is in addition to federal and state income taxes. In practice, the Social Security tax is a little less than 15.3% because of certain deductions, but it's still a big bite of what you earn. (See Chapter 10.)

It is possible, however, to avoid paying at least part of this Social Security tax burden by incorporating your business. (See Chapter 2.)

### 2. Personal Liability for Debts

Employees are not liable for the debts incurred by their employers. An employee may lose his or her job when the employer's business fails, but will owe nothing to the employer's creditors.

This is not necessarily the case when you're self-employed. If you're a sole proprietor or partner in a partnership, you are personally liable for your business debts. You could lose most of what you own if your business fails. However, there are ways to decrease your personal exposure, such as incorporating your business and obtaining insurance. (See Chapters 2 and 6.)

### 3. Deadbeat Clients

Ugliest of all, you could do lots of business and still fail to earn a living. This is because many self-employed people have great difficulty getting their clients to pay them on time or at all. When you're self-employed, you bear the risk of loss from deadbeat clients. Neither the government nor anyone else is going to help you collect.

Clients who pay late or don't pay at all have driven many self-employed people back to the ranks of the wage slaves. However, there are many strategies you can use to help alleviate payment problems. (See Chapter 7.)

## D. How to Use This Book

This book helps you make what's good about self-employment even better, make the bad less bad and—hopefully—the ugly aspects a little more attractive.

Exactly which portions of the book you'll need to read now depends on whether you're already self-employed or are just starting out.

### 1. Starting Up Your Business

If you're just starting out, there are a number of tasks you'll need to complete before or soon after you start doing business. These include:

- choosing the legal form for your business (see Chapter 2)
- choosing a name for your business (see Chapter 3)
- deciding where to set up your office (see Chapter 4)
- obtaining business licenses and permits and a federal taxpayer ID number (see Chapter 5)
- obtain insurance for your business and yourself (see Chapter 6), and
- setting up at least a rudimentary bookkeeping system (see Chapter 14).

You should read the chapters discussing these tasks first.

### 2. Ongoing Legal and Tax Issues

Once your business is up and running, there are a number of ongoing legal and tax issues with which you may have to deal. These include:

- deciding how to price your services and taking steps to ensure you get paid (see Chapter 7)
- paying estimated taxes (see Chapter 11)
- keeping track of your tax deductible business expenses (see Chapters 9 and 14)
- taking steps to ensure the IRS doesn't view you as an employee if you're audited (see Chapter 15)
- using written client agreements (see Chapters 18, 19 and 20)
- dealing with ownership of copyrights, patents and trade secrets you create (see Chapter 17)
- deciding how to help fund your retirement (see Chapter 16), and
- dealing with taxes for employees or independent contractors you hire (see Chapter 13).

You can read the appropriate chapters when a problem arises or read them in advance to help avoid problems from the outset. ■

CHAPTER

# 2

# Choosing the Legal Form for Your Business

As a self-employed person, one of the most important decisions you have to make is what legal form your business will take. There are several alternatives—and the one you choose will have a big impact on how you're taxed, whether you'll be liable for your business's debts and how the IRS and state auditors will treat you.

There are four main forms in which to organize a business:

- sole proprietorship (see Section A)
- corporation (see Section B)
- partnership (see Section C), or
- limited liability company (see Section D).

If you own your business alone, you need not be concerned about partnerships; this business form requires two or more owners. If, like most self-employed workers, you're running a one-person business, your choice is between being a sole proprietor and forming a corporation or limited liability company.

Don't worry too much about making the wrong decision. Your initial choice about how to organize your business is not engraved in stone. You can always switch to another legal form later. It's common, for example, for self-employed people to start out as sole proprietors and then incorporate later when they become better established and start making a substantial income.

## WAYS TO ORGANIZE YOUR BUSINESS

| **Type of Organization** | **Main Advantages** | **Main Disadvantages** |
|---|---|---|
| **Sole Proprietorship** | • Simple and inexpensive to create and operate.<br>• Owner reports profits or loss on personal tax return. | • Owner personally liable for business debts.<br>• May make it difficult to establish sef-employed status. |
| **C Corporation** | • Clients have less risk from government audits.<br>• Owners have limited personal liability for business debts.<br>• Fringe benefits can be deducted as business expense.<br>• Owners can split corporate profit among owners and corporation, paying lower overall tax rate. | • More expensive to create and operate than sole proprietorship or partnership.<br>• Double taxation threat because a separate taxable entity.<br>• No beneficial employment tax treatment. |
| **S Corporation** | • Clients have less risk from government audits.<br>• Owners have limited personal liability for business debts.<br>• Owners can use corporate losses to offset income from other status.<br>• Owners can save on employment taxes by taking distributions instead of salary. | • More expensive to create and operate than sole proprietorship.<br>• Fringe benefits for shareholders limited. |
| **Partnership** | • Simple and inexpensive to create and operate.<br>• Owners report profit or loss on personal tax returns treatment. | • Owners personally liable for business debts.<br>• Two or more owners required.<br>• No beneficial employment tax. |
| **Limited Liability Company** | • Owners have limited liability from business debts if they participate in management.<br>• Profit and loss can be allocated differently than ownership interests. | • Tax treatment requires strict compliance with IRS guidelines.<br>• No beneficial employment tax treatment. |

Adapted from *The Legal Guide for Starting and Running a Small Business,* by Fred S. Steingold (Nolo Press).

## A. Sole Proprietorships

A sole proprietorship is simply a one-owner business. It is by far the cheapest and easiest way to legally organize your business. You don't have to get permission from the government to be a sole proprietor or pay any fees, except perhaps for a fictitious business name statement or business license. (See Chapter 5.) You just start doing business; if you don't incorporate or have a partner, you automatically are a sole proprietor. If you're already running a one-person business and haven't incorporated, you're a sole proprietor right now.

The vast majority of self-employed people—about 90%—are sole proprietors. Most sole proprietors run small operations, but a sole proprietor can hire employees and nonemployees, too. Indeed, some one-owner businesses are large operations with many employees.

**All in the Family—Almost**
You can share ownership of your business with your spouse and still maintain its status as a sole proprietorship. In the eyes of the law, you are both owners of the business—what the IRS calls co-sole proprietors. You can either split the profits from your business if you and your spouse file separate returns, or put them all into the family pot if you file a joint return. But only a spouse can be a co-sole proprietor. If any other family members share ownership with you, the business must be organized as a partnership, corporation or limited liability company.

### 1. Tax Concerns

When you're a sole proprietor, you and your business are one and the same for tax purposes. Sole proprietorships don't pay taxes or file tax returns. Instead, you must report the income you earn or losses you incur on your own personal tax return, IRS Form 1040. If you earn a profit, the money is added to any other income you have—for example, interest income or your spouse's income if you're married and file a joint tax return—and that total is taxed. If you incur a loss, you can use it to offset income from other sources.

Although you are taxed on your total income regardless of its source, the IRS does want to know about the profitability of your business. To show whether you have a profit or loss from your sole proprietorship, you must file IRS Schedule C, Profit or Loss From Business, with your tax return. On this form you list all your business income and deductible expenses. (See Chapter 9.) If you have more than one business, you must file a separate Schedule C for each one.

Sole proprietors are not employees of their proprietorships; they are business owners. Their businesses don't pay payroll taxes on their income or withhold income tax from their compensation. However, sole proprietors do have to pay self-employment taxes—that is, Social Security and Medicare taxes—on their net self-employment income. (See Chapter 10.) These taxes must be paid four times a year along with income taxes in the form of estimated taxes. (See Chapter 11.) Clients don't withhold any taxes from their compensation, but any client who pays a sole proprietor $600 or more in a year must file Form 1099-MISC with the IRS reporting the payment. (See Chapter 9, Section A.)

> **EXAMPLE:** Annie operates a computer consulting business as a sole proprietor—that is, she is the sole owner of the business. She must report all the income she receives from her clients on her individual tax return, IRS Form 1040, and file Schedule C. She need not file a separate tax return for her business. In

one recent year, she earned $50,000 from consulting and had $15,000 in business expenses, leaving a net business income of $35,000. She must add her $35,000 profit to any other income she has and report the total on her Form 1040. She must pay both income and self-employment taxes on her $35,000 profit.

## 2. Liability Concerns

One concern many business owners have is liability—that is, whether and to what extent they are responsible for paying their business's debts and business-related lawsuits.

However, this is often blown way out of proportion. You need not worry about liability for debts so long as you don't incur more debts than you're able to pay. Lawsuits are not as common as most people think and you can obtain insurance to protect you against most types of lawsuits. (See Chapter 6.)

### a. Business debts

When you're a sole proprietor, you are personally liable for all the debts of your business. This means that a business creditor—a person or company to whom you owe money for items you use in your business—can go after all your assets, both business and personal. This may include, for example, your personal bank accounts, your car and even your house. Similarly, a personal creditor—a person or company to whom you owe money for personal items—can go after your business assets, such as business bank accounts and equipment.

**EXAMPLE:** Arnie, a sole proprietor consultant, fails to pay $5,000 to a supplier. The supplier sues him in small claims court and obtains a $5,000 judgment. As a sole proprietor, Arnie is personally liable for this judgment. This means the supplier can not only tap Arnie's business bank account, but his personal accounts as well. And the supplier can also go after Arnie's personal assets such as his car and home.

Fortunately, however, some of your personal property is safe from creditors' reaches. How much of your property is protected depends on the state in which you live. For example, depending on how much they are worth, creditors may not be allowed to take your car, your business tools or your home and furnishings.

The fact that most of what you own is fair game to satisfy all debts is probably the main drawback of setting up your business as a sole proprietorship. However, bankruptcy is always an option if your debts get out of control.

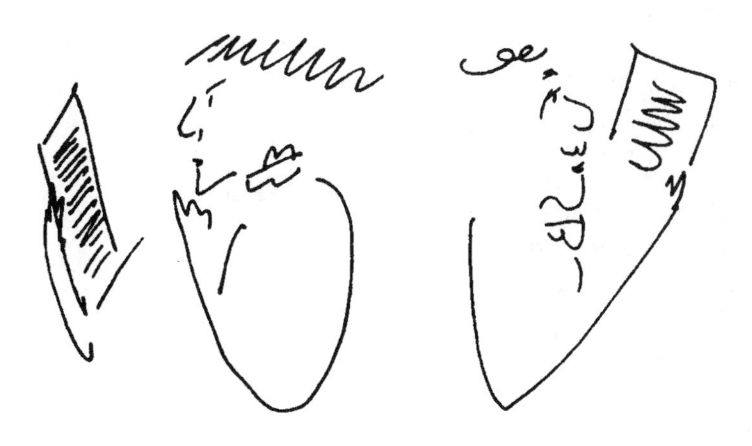

### A Mini-Course on Bankruptcy

By filing for bankruptcy, you can partly or wholly wipe out your debts and get a fresh financial start. There are two types of bankruptcy for individuals.

- Chapter 7 bankruptcy is the more familiar liquidation bankruptcy, in which many of your debts are wiped out completely without any further repayment. In exchange, you might have to surrender some of your property which can be sold to pay your creditors. Most people who file for Chapter 7 bankruptcy, however, don't have anything to turn over to their creditors. The whole process takes about three to six months and commonly requires only one trip to the courthouse. You can probably do it yourself, without a lawyer.
- Chapter 13 bankruptcy is a reorganization bankruptcy in which you rearrange your financial affairs, repay a portion of your debts and put yourself back on your financial feet. You repay your debts through a Chapter 13 plan. Under a typical plan, you make monthly payments to the bankruptcy court for three to five years. The money is distributed to your creditors. If you finish your repayment plan, any remaining unpaid balance on the unsecured debts is wiped out.

For a complete discussion of bankruptcy and the types and amounts of property that your creditors can't reach, see:

- *How to File for Bankruptcy,* by Stephen Elias, Albin Renauer & Robin Leonard, and
- *Chapter 13 Bankruptcy: Repay Your Debts,* by Robin Leonard.

(Both are published by Nolo Press.)

#### b. Lawsuits

If you're a sole proprietor, you'll also be personally liable for business-related lawsuits—for example, if someone slips and falls at your business location and sues for damages. Fortunately, you can obtain insurance to protect yourself against these types of risks. (See Chapter 6.)

### 3. IC Status

Companies that hire the self-employed always face the possibility of having to pay fines and penalties if the IRS or another government agency determines that the workers should have been classified as employees. (See Chapter 15.) For this reason, some firms are reluctant to hire self-employed people unless they are certain they can pass muster as independent contractors in case of a government audit.

A disadvantage of the sole proprietorship form is that it won't help you establish that you're self-employed in the eyes of the IRS or state auditors. Sole proprietors who provide services can look a lot like employees—especially if they work on their own without assistants and deposit their compensation in their personal bank account. After all, this is exactly what employees do. For this reason, some hiring firms prefer to hire self-employed people who have incorporated their businesses. (See Section B.) If you are reclassified as an employee by the IRS or state auditors, the consequences can be significant—for example, your client may decide to dispense with your services rather than pay the additional benefit and payroll tax costs of having an employee. The IRS could also disallow business deductions you took when you thought you were self employed. (See Chapter 9, Section B.)

### IRS Audit Rates are Higher for Sole Proprietors

Another factor to consider in choosing the legal form for your business is how it affects your chance of being audited. Sole proprietors have the greatest chance of being audited. Only corporations with assets over $1 million are audited more often. This undoubtedly reflects the IRS's belief that sole proprietors habitually underreport their incomes, take deductions to which they're not entitled and otherwise cheat on their taxes.

The IRS has grown particularly suspicious of sole proprietors who report very small incomes from their businesses. This is shown below by the jump in the audit rate of sole proprietors with incomes below $25,000.

However, the chances of being audited are still low and this factor alone should not dictate your choice of business form.

| | 1993 Audit Rate | 1994 Audit Rate |
|---|---|---|
| **Sole proprietors** | | |
| Income under $25,000 | 2.24% | 4.39% |
| $25,000 to $100,000 | 2.41% | 3.01% |
| $100,000 and over | 3.91% | 3.57% |
| **Partnerships** | 0.61% | 0.52% |
| **S Corporations** | 1.02% | 1.02% |
| **C Corporations** | | |
| Assets under $250,000 | 1.33% | 0.84% |
| $250,000 to $1 million | 3.94% | 2.47% |
| $1 million to $5 million | 9.35% | 7.11% |

## B. Corporations

The word corporation usually conjures up images of huge businesses such as General Motors or IBM. However, a business doesn't have to be large to be a corporation. Virtually any business can be a corporation, even if it has only one owner. Indeed, most corporations have only a few owners; such small corporations are often called closely held.

Relatively few self-employed people are incorporated, but this doesn't mean you shouldn't consider this form for your business. Incorporating your business can result in tax savings, limit your liability for business debts and even help you get clients.

### 1. What Is a Corporation

A corporation is simply a legal form in which you can organize and conduct a business and share in the profits or losses. Unlike a sole proprietorship, it has a legal existence distinct from its owners. It can hold title to property, sue and be sued, have bank accounts, borrow money, hire employees and do anything else in the business world that a human being can do.

In theory, every corporation consists of three groups:

- those who direct the overall business, called directors
- those who run the business day-to-day, called officers, and
- those who just invest in the business, called shareholders.

However, in the case of a small business corporation, these three groups can be and often are the same person—that is, a single person can direct and run the corporation and own all the corporate stock. So if you incorporate your one-person business, you don't have to go out and recruit and pay a board of directors or officers.

## 2. Your Employment Status

When you incorporate your business, you automatically become an employee of your corporation if you continue to work in the business, whether fulltime or parttime. This is so even if you're the only shareholder and are not subject to the direction and control of anybody else. In effect, you wear two hats—you're both an owner and an employee of the corporation.

> **EXAMPLE:** Ellen, an independent truck driver, forms a one-person trucking corporation, Ellen's Trucking, Inc. She owns all the stock and runs the business. The corporation hires her as an employee with the title of president.

When you have incorporated your business, clients hire your corporation, not you personally. You will sign any written agreement on behalf of your corporation. When you're paid, the client should issue the check to your corporation and you should deposit it in your corporate bank account, not your personal account. You can then pay the money to yourself in the form of salary, bonus or dividends. The method you choose to pay yourself can have important tax consequences. (See Section B5.)

Social Security and Medicare taxes must be withheld from any employee salary your corporation pays you and money must be paid to the IRS just as for any employee. (See Chapter 8, Section A.) However, your total Social Security and Medicare taxes are about the same as if you were a sole proprietor. They're just paid from two different accounts; half are paid by your corporation and half are withheld from your salary. Since all the money is yours, there is no practical difference from being a sole proprietor. Some additional state payroll taxes will be due, however—mostly state unemployment taxes. (See Chapter 8, Section A.)

You can also have your corporation provide you with employee fringe benefits such as health insurance and pension benefits. (See Section 6d.)

### Self-Employed By Any Other Name

Strictly speaking, when you incorporate your business, you are no longer self-employed: you are an employee of your corporation. Your corporation is neither legally self-employed nor an employee of the clients or customers for whom it provides services or produces products. Only individual human beings can be self-employed or employees.

However, people who own single-shareholder corporations and sell services to clients still often refer to themselves as ICs when they communicate with clients and customers and other self-employed people. This is understandable since their employee status is mainly a legal technicality.

## 3. Audit Risks

Many potential clients are fearful of hiring ICs because they are afraid they could get in trouble if the IRS audits them and claims the workers should have been treated as employees (See Chapter 15.) For years, tax experts have believed that firms that hire corporations have a much smaller chance of having worker classification problems with the IRS than if they hire sole proprietors to do the same work. This is because taking the time and trouble to incorporate is strong evidence that a worker is operating an independent business.

The IRS has recently confirmed this view in a manual training IRS auditors on how to determine the status of workers. (See Chapter 15.) The manual provides that an incorporated worker will usually not be treated as an employee of the hiring firm, but as an employee of the worker's corporation.

Because of this clear direction from the IRS, it's likely that more hiring firms will try to avoid hiring sole proprietors or partnerships and deal with incorporated businesses only. Others will give preference to a corporation if they have a choice between hiring a sole proprietor and corporation. The ability to get more business may alone justify the time and expense involved in incorporating.

Incorporating may be particularly helpful if you're a computer programmer, systems analyst, engineer, drafter or you perform similar technical services. Because special IRS rules make it harder for firms that hire such workers to win IRS worker classification audits, hiring firms generally classify them as employees. But they may make an exception if you're incorporated and they are able to hire your corporation instead of hiring you personally.

However, don't get the idea that you and your clients need not worry about the IRS at all if you incorporate. The IRS also directs that an incorporated worker may be reclassified as an employee of the hiring firm if the worker does not follow corporate formalities and in other cases of clear abuse. IRS auditors may disregard your corporate status and hold that you're a hiring firm's employee if you act like one—for example, if you:

- deposit your earnings directly into your personal bank account instead of putting them into a separate corporate account
- you fail to file tax returns for your corporation
- don't issue yourself stock, or
- fail to follow other corporate formalities such as holding an annual meeting or keeping corporate records. (See Section B8.)

**IRS DOCKS DOC, BUT NOT M.D., INC.**

One recent case shows why many clients prefer to hire corporations rather than sole proprietors. An outpatient surgery center hired two doctors to work as administrators. They both performed the same services. However, one of the doctors had formed a medical corporation of which he was an employee. The surgery center signed a written contract with the corporation, not the doctor. It also paid the doctor's corporation, not the doctor. The other doctor was a sole proprietor and had no written contract with the center.

The court concluded that the incorporated doctor was not an employee of the surgery center, but the unincorporated doctor was an employee. As a result, the center had to pay substantial back taxes and penalties for the unincorporated doctor, but not for the doctor who was incorporated. (*Idaho Ambucare Center v. U.S.*, 57 F.3d 752 (9th Cir. 1995).)

## 4. Liability Concerns

The main reason many business owners consider forming a corporation is to avoid personal liability for business debts and lawsuits. While incorporating your business can insulate you from liability to a certain extent, the protection is not nearly as great as most people think.

### a. Business debts

Corporations were created to enable people to invest in businesses without risking all their personal assets if the business failed or became unable to pay its debts. In theory, corporation owners are not personally liable for corporate debts or lawsuits. That is, they can lose what they invested in the corporation, but corporate creditors can't go after their personal assets such as their personal bank accounts or homes.

This theory holds true where large corporations are concerned. If you buy stock in IBM, for example, you don't have to worry about IBM's creditors suing you. But it often doesn't work that way for small corporations. Major creditors such as banks are probably not going to let you shield your personal assets by incorporating. Instead, they will likely demand that you personally guarantee business loans or extensions of credit—that is, sign a legally enforceable document pledging your personal assets to pay the debt if your business assets fall short. This means that you will be liable for the debt, just as if you were a sole proprietor. (See Section A.)

You can avoid having to pledge a personal guarantee for some business debts. These will most likely be routine and small debts. It's not likely, for example, that your office supply store will make you personally guarantee that your corporation will pay for its purchases. But, of course, once it gets wise to the fact that your business is not paying its bills, it won't extend you any more credit.

### b. Lawsuits

Being incorporated can also shield you from personal liability for some types of business-related lawsuits. For example, you can avoid being personally liable if an employee or self-employed person you hire accidentally harms someone while working for you.

**EXAMPLE:** Jane, a graphic designer, has incorporated her business. She is the sole shareholder and president of Jane's Graphics, Inc. Her corporation employs Bill as an assistant. He runs over and injures a pedestrian, Steve, while delivering some designs to a client. Steve sues the corporation and obtains a judgment against it. Since Jane's business is a corporation, she is not personally liable for the damages. Steve can collect against Jane's corporate bank accounts and even sell her corporation's assets, but can't go after her personal assets such as her personal bank accounts or house.

However, as a practical matter, incorporating is no substitute for a good insurance policy protecting you from the cost of lawsuits. As the above example shows, even if you incorporate, all the assets of your business will be available to satisfy any court judgment. Your corporate assets will probably amount to a large portion of your total net worth.

In addition, you can't use a corporation to shield yourself from personal liability for your own professional negligence. (See Section B10c—and see Chapter 6 for details on obtaining liability insurance.)

## 5. Basics of Corporate Taxation

When it comes to federal income tax, there are three very different types of corporations:

- C corporations, sometimes called regular corporations (see Section 6), and
- S corporations, also called small business corporations (see Section 7).

Basically, C corporations pay taxes and S corporations don't. You have the option of forming either type of corporation. Each has its benefits and drawbacks. Generally, S corporations are best for small businesses that have little income or suffer losses. C corporations are often advantageous for very successful businesses with substantial profits. You can start out as an S corporation and switch to a C corporation later, or vice versa.

If, after reading this chapter, you're not sure whether a C or S corporation is best for you, consult an accountant or other tax pro for help. (See Chapter 21.)

As explained below, you can save money on taxes by incorporating. There are some less tangible benefits as well. For one thing, small corporations are audited much less often than sole proprietorships. (See Section A.) And, even when they are audited, the IRS seems to take a less rigorous look at their tax deductions than it does for sole proprietors.

For additional information on corporate taxation, see:

- *Tax Savvy for Small Business,* by Frederick W. Daily (Nolo Press)
- IRS Publication 542, *Tax Information on Corporations,* and
- IRS Publication 589, *Tax Information on S Corporations.*

You can obtain these IRS publications free by calling the IRS at: 800-TAX-FORM; if you have a computer, you can also download them from the IRS Internet site at www.irs.ustreas.gov.

## 6. Taxes for C Corporations

When you form a corporation, it automatically becomes a C corporation for federal tax purposes. C corporations are treated separately from their owners for tax purposes. They must pay income taxes on their net income and file their own tax returns with the IRS using either Form 1120 or Form 1120-A. They also have their own income tax rates which are lower than individual rates at some income levels. C corporations generally take the same deductions as sole proprietorships to determine their net profits, but have some special deductions as well. This separate tax identity is a unique attribute of C corporations and can lead to tax savings.

### a. Income splitting

When you form a C corporation, you take charge of two separate taxpayers: your corporation and yourself. You don't pay personal income tax on income your incorporated business earns until it is distributed to you in the form of salary, bonuses or dividends. The nice thing about this is that you can split the income your business earns with your corporation. You can save on income tax this way because corporate tax rates can be lower than your personal tax rate. A C corporation pays less income tax than an individual on the first $75,000 of taxable income. (See the chart in section 6C below.)

You can safely keep up to $250,000 of your business earnings in your corporation—that is, let it stay in the corporate bank account. You can use the money, for example, to expand your business, buy equipment or to pay yourself employee benefits such as health insurance and pension benefits. However, if you keep more than $250,000, you'll become subject to an extra 39.6% tax called the accumulated earnings tax. This tax is intended to prevent you from sheltering too much money in your corporation.

There is another substantial tax benefit to income splitting: you don't have to pay Social Security and Medicare taxes, also called employment taxes, on profits you retain in your corporation. This is a 15.3% tax; so, for example, if you retain $10,000 in your corporation, you'll save $1,530 in taxes.

> **EXAMPLE:** Betty owns and runs her own incorporated construction contracting business. In one year, the corporation has a net profit of $20,000 after paying Betty a healthy $100,000 salary. Rather than pay herself the $20,000 in the form of additional salary or bonuses, Betty decides to leave the money in her corporation. She uses the money to buy equipment and pay salaries of her employees. The corporation pays only a 15% tax on these retained earnings. Had Betty taken the $20,000 as salary, she would have had to pay a 36% income tax and 15.3% employment tax on it.

### b. Consultants and other professionals

Income splitting is usually not a viable option if you're a consultant or engaged in certain other occupations involving professional services. The IRS calls corporations formed by such people personal service corporations. These corporations are required to pay corporate tax at a special flat 35% rate.

Corporations formed by self-employed consultants are personal service corporations if all the stock is owned by consultants who are corporate employees. Consulting means getting paid just to give a client your advice or counsel. You're not a consultant if you get paid only if the client buys something from you or from someone else through you. Unfortunately, huge numbers of self-employed people qualify as consultants.

> **EXAMPLE:** Acme Corporation hires Data Analysis, Inc., a C corporation solely owned by Tony, a data analyst, to determine its data processing needs. Tony, who is an employee of his corporation, studies Acme's business and recommends the type of data and information its employees need. Tony doesn't provide Acme with computer hardware or software, he just makes recommendations about how Acme's data processing system should be designed.
>
> Tony is considered a consultant and his corporation is a personal service corporation because all the stock is owned by consultant-employees—that is, by Tony. The corporation will be subject to a flat 35% tax rate.

Any corporation you form will also be a personal service corporation if all the stock is owned by corporate employees performing the following activities or professions:

- accounting
- engineering
- law
- health services—including doctors, nurses, dentists or other health care professionals
- performing arts, or
- actuarial science.

Because the 35% tax rate for personal service corporations is so high, their owners will usually end up paying more taxes if they leave any money in the corporation instead off taking it out of the business in the form of salary, bonuses and fringe benefits and paying taxes on it at their individual tax rates. Individual rates are lower than the 35% personal service corporation tax rate at income levels below $124,650 for individuals and $151,750 for married people filing joint returns. (See the chart below).

> **EXAMPLE:** Max, a self-employed engineer, forms his own corporation which qualifies as a personal service corporation because he owns all the stock and is the corporation's employee. His business earned $100,000 in one recent year. If Max takes all the earnings out of his corporation in the form of salary, he'll pay a maximum 31% individual income tax. But if he leaves any money in his corporation, it will be taxed at the 35% personal service corporation tax rate.

If your corporation qualifies as a personal service corporation and you think you might not be able to take all the profits out of the business, consider electing S corporation status. This way, all corporate profits will automatically pass to you each year and you'll avoid corporate taxes altogether. (See Section B7.)

### c. Comparison of tax rates

The following chart offers a comparison of the tax rates for individuals, corporations and personal service corporations.

The individual income tax brackets shown above are adjusted annually for inflation. This table shows the 1997 brackets. For the current brackets, see IRS Publication 505, Tax Withholding and Estimated Tax. You can obtain a copy free by calling the IRS at: 800-TAX-FORM; if you have a computer, you can download it from the IRS Internet site at www.irs.ustreas.gov.

**1997 INDIVIDUAL AND CORPORATE TAX RATES**

| Taxable Income | Individual Rate (Single) | Individual Rate (Married Filing Jointly) | Corporate Rate (Other than personal service corporations) | Personal Service Corporation Rate |
|---|---|---|---|---|
| Up to $24,650 | 15% | 15% | 15% | 35% |
| 24,651–40,200 | 28% | 15% | 15% | 35% |
| 40,201–50,000 | 28% | 28% | 15% | 35% |
| 50,001–59,750 | 28% | 28% | 25% | 35% |
| 59,751–75,000 | 31% | 28% | 25% | 35% |
| 75,001–99,600 | 31% | 28% | 34% | 35% |
| 99,601–100,000 | 31% | 31% | 34% | 35% |
| 100,001–124,650 | 31% | 31% | 39% | 35% |
| 124,651–151,750 | 36% | 31% | 39% | 35% |
| 151,751–271,050 | 36% | 36% | 39% | 35% |
| 271,051–335,000 | 39.6% | 39.6% | 39% | 35% |
| 335,001–10,000,000 | 39.6% | 39.6% | 34% | 35% |

## d. Fringe benefits

The other significant tax benefit of forming a C corporation is that your corporation can provide you—its employee—with fringe benefits and deduct the entire cost from the corporation's income as a business expense. No other form of business entity can do this.

Possible employee fringe benefits include:

- health insurance for you and your family members
- disability insurance
- reimbursement of medical expenses not covered by insurance
- deferred compensation plans
- group term life insurance
- retirement plans, and
- death benefit payments up to $5,000.

You need not include the value of premiums or other payments your corporation makes for these benefits in your personal income for income tax purposes. With health insurance costs skyrocketing, the ability to fully deduct these expenses is one of the best reasons to form a C corporation.

**EXAMPLE:** Marilyn incorporates her marketing business and is its only employee. Marilyn's corporation provides her with health insurance for her and her family at a cost of $6,000 per year. The entire cost can be deducted from the corporation's income for corporate income tax purposes, but is not included as income on Marilyn's personal tax return.

In contrast, if you're a sole proprietor, S corporation owner or partner in a partnership, you're allowed to deduct only a portion of health insurance premiums from your personal income tax. In an effort to help self-employed people pay for their health insurance, the deduction will gradually be increased from 40% in 1997 to 80% in 2006 and later, as follows:

**HEALTH INSURANCE DEDUCTABILITY**

| Year | Deduction Amount |
|---|---|
| 1997 | 40% |
| 1998 and 1999 | 45% |
| 2000 and 2001 | 50% |
| 2002 | 60% |
| 2003-2005 | 80% |
| 2006 | 90% |
| 2007 and later | 100% |

## The Bugaboo of Double Taxation

When you're a sole proprietor and you want to take money out of your business for personal use, you can simply write yourself a check. Such a transfer has no tax impact since all your sole proprietorship profits are taxed to you personally. It makes no difference whether you leave the money in the business or put it in your personal bank account.

Things are very different when you form a C corporation. Any direct payment of your corporation's profits to you will be considered a dividend by the IRS and taxed twice. First, the corporation will pay corporate income tax on the profit and then you'll pay personal income tax on it. This is called double taxation.

However, this problem rarely arises with small corporations. This is because you'll normally be an employee of your corporation and the salary, benefits and bonuses you receive are deductible expenses for corporate income tax purposes. If you handle things right, your employee compensation will eat up all the corporate profits so there's no taxable income left on which your corporation must pay income tax. You'll only pay income tax once on your employee compensation.

**EXAMPLE:** Al has incorporated his consulting business. He owns all the stock and is the company's president and sole employee. In one recent year, the corporation earned $100,000 in profits. During that year, Al's corporation paid him an $80,000 salary and a $20,000 Christmas bonus. The salary and bonus are tax deductible corporate business expenses, leaving the corporation with a net profit of zero. As a result, Al's corporation pays no income taxes. Al simply pays personal income tax on the income he received from the corporation, the same as any other employee.

The only time you might have a problem with double taxation is when your business profits are so great you can't reasonably pay them all to yourself in the form of employee compensation. The IRS only allows corporate owner-employees to pay themselves a reasonable salary for work they actually perform. Any amounts that are deemed unreasonable are treated as disguised dividends by the IRS and are subject to double taxation. One way to avoid this is to leave the excess profits in your corporation and distribute them to yourself as salary, bonus or benefits in future years.

**EXAMPLE:** Bob, a self-employed political consultant, spent $3,000 on health insurance for himself and his family in 1998. He may deduct 45% of this amount—or $1,350—from his income taxes for the year.

Note that by the year 2007, there won't be any point in forming a C Corporation to deduct your health insurance premiums; they'll be 100% deductible in any event.

Currently, you can deduct part of your remaining premiums plus other uncovered medical expenses as an itemized deduction. Itemized deductions are expenses you're allowed to deduct on your personal income tax return. But the deduction is limited to the amount your premiums and medical expenses exceed 7.5% of your adjusted gross income for the year. Your adjusted gross income is your total income minus deductions for IRA, Keogh and SEP-IRA contributions, the self-employed health insurance deduction, one-half of your self-employment tax and a few other items.

**EXAMPLE:** Bob had an adjusted gross income of $50,000 in 1998. He can deduct only those health insurance premiums and medical expenses that exceed $3,750 (7.5% x $50,000 adjusted gross income). Since his total medical expenses for the year were only $3,000, he gets no deduction.

### e. Interest free loans

Yet another benefit of forming a C corporation is that the shareholders can borrow up to $10,000 from the corporation free of interest. If you borrow any more than that, however, you must either pay interest or pay tax on the amount of interest you should have paid. The interest rate is determined by IRS tables. No other form of business entity offers this benefit.

Borrowing money from your corporation is very attractive tax-wise because the loan is not taxable income to you. However, shareholder loans must be true loans. As proof of the loan's veracity, you should sign a promissory note obligating you to repay it on a specific date or in regular installments. The loan should also be secured by your personal property such as your house or other property.

## 7. Taxes for S Corporations

When you incorporate, you have the option of having your corporation taxed as an S corporation for federal income tax purposes. An S corporation is taxed like a sole proprietorship. Unlike a C corporation, it is not a separate taxpaying entity. Instead, the corporate income and losses are passed through directly to the shareholders—that is, you and anyone else who owns your business along with you. The shareholders must split the taxable profit according to their shares of stock ownership and report that income on their individual tax returns.

An S corporation normally pays no taxes, but must file an information return with the IRS on Form 1120S telling the IRS how much the business earned or lost and indicating each shareholder's portion of the corporate income or loss.

**EXAMPLE:** Alice owns ABC Programming, Inc., an S corporation, and is its sole shareholder and sole employee. In one year, ABC earned $100,000 in gross income and had $90,000 in deductions, including a $80,000 salary for Alice. The corporation's $10,000 net profit is passed through directly to Alice, who must report it as income on her personal tax return. The S corporation files an information return with the IRS on Form 1120S, but pays no income taxes itself.

S corporations have become very popular with small business owners in recent years. Owning an S corporation can give you the best of both worlds. You're taxed as a sole proprietor, which is simpler than having a C corporation and is particularly helpful when you're first starting out and have little business income or may even have losses. At the same time, you still have the limited liability of a corporation owner. And there's one added benefit: You can save on self-employment taxes by setting up an S Corporation.

### a. Deducting business losses

You must report income or loss from an S corporation on your individual tax return. This means that if your business has a loss, you can deduct it from income from other sources including your spouse's income if you're married and file a joint return. You can't do this with a C corporation because it's a separate taxpaying entity; its losses must be subtracted from its income and can't be directly passed on to you. The ability to deduct business losses on your personal tax return may be particularly helpful when you're first starting out and have incurred losses which you can use to reduce your total taxable income.

> **EXAMPLE:** Jack and Johanna are a married couple who file a joint income tax return. Johanna earns $80,000 a year from her job. Jack quits his job as an employee-salesperson and becomes self-employed. He forms an S corporation with himself as the sole shareholder and only employee. In his first year in business his company earns $20,000 and has $40,000 in expenses. Jack and Johanna report this $20,000 loss on their joint tax return and subtract it from their total taxable income. Since Johanna's $80,000 salary puts them in the 31% marginal income tax bracket (see Section 6c), they've saved $6,200 in income tax (31% x $20,000).

### b. No income splitting

When you operate an S corporation, you can't split your income between two separate taxpaying entities as you can with a C corporation. If your business earns a substantial amount, income splitting can reduce your federal income taxes because C corporations pay less tax than individuals on profits up to $75,000. (See Section 6a.)

But the inability to split your income between yourself and your S corporation may be a drawback only in theory. Income splitting is a viable option only if your business earns enough money for you to leave some in your corporate bank account and not distribute it all to yourself in the form of salary, bonuses and benefits. Many self-employed people don't make enough money to even consider income splitting—particularly when they're first starting out. If you start making so much money that you want to split income, you can always convert your S corporation to a C corporation.

### c. Self-employment tax

A little-known tax benefit of forming an S corporation is that it can save you Social Security and Medicare tax. This is a flat 15.3% tax on your first $65,400 in income in 1997; the ceiling is adjusted annually for inflation. If you earn more than that amount, you pay a 2.9% Medicare tax on the excess.

If you're a sole proprietor, partner in a partnership or limited liability company member, all the income you receive from your business is subject to these taxes, called self-employment taxes. (See Chapter 10.) If you incorporate your business, you must be an employee of your corporation. The same 15.3% tax must be paid. You pay half out of your employee compensation and your corporation pays the other half.

Whether you are a sole proprietor, partner in a partnership, limited liability company member or employee of your C corporation, you must pay Social Security and Medicare taxes on all the income you take home. S corporations offer you a way to take home some money without paying these taxes. You report your corporation's earnings on your personal tax return and you must pay Social Security and Medicare taxes on any employee salary your S corporation pays you.

You do not, however, have to pay such tax on distributions from your S corporation—that is, on the net profits that pass through the corporation to you personally. The larger your distribution, the less Social Security and Medicare tax you'll pay.

**EXAMPLE:** Mel, a consultant, has formed an S corporation of which he's the sole shareholder and only employee. In one year, his corporation had a net income of $50,000. If Mel pays this entire amount to himself as employee salary, he and his corporation will have to pay Social Security and Medicare tax on all $50,000—a total tax of $7,650.

Instead, Mel decides to pay himself only a $30,000 salary. The remaining $20,000 is passed through the S corporation and reported as an S corporation distribution on Mel's personal income tax return, not employee salary. No Social Security or Medicare tax need be paid on this amount. Mel pays only $4,590 in Social Security and Medicare taxes instead of $7,650—a tax saving of $3,060.

Theoretically, if you took no salary at all, you would not owe any Social Security and Medicare taxes. As you might expect, however, this is not allowed. The IRS requires S corporation shareholder-employees to pay themselves a reasonable salary—at least what other businesses pay for similar services.

When you reduce your Social Security tax in this way you might receive smaller Social Security benefits when you retire, since benefits are based on your contributions. You can more than offset this, though, by putting the money you save into a tax-advantaged retirement plan such as an IRA, SEP-IRA or Keogh Plan. You'll probably earn more money from your contributions to such plans than you'd collect from Social Security. In addition, your contributions to such plans are usually tax deductible and you can start taking the money out when you reach age 59 1/2. In contrast, you can't collect Social Security until you're at least 62.

### d. S corporation rules

There are some IRS rules on who can establish an S corporation and how it's operated. For example:

- an S corporation can only have 75 shareholders
- none of the shareholders can be nonresident aliens—that is, noncitizens who don't live in the United States
- an S corporation can have only one class of stock—for example, you can't create preferred stock giving some shareholders special rights, and
- the shareholders can only be individuals, estates or certain trusts—for example, a corporation can't be an S corporation shareholder.

If you're running a one-person business, are a U.S. citizen or live in the U.S. and will be the only shareholder, these restrictions will not affect your S corporation operations in the least.

### e. How to elect S corporation status

To establish an S corporation, you first form a regular corporation under your state law. (See Section B9.) Then you file Form 2553 with the IRS. If you want your corporation to start off as an S corporation, you must file the form within 75 days of the start of your first tax year.

### f. State tax rules

Check with your state's corporation office to find out how an S corporation files and pays state taxes. Typically, states impose a minimum annual corporate tax or franchise fee. You may also face a state corporation tax on S corporation income—for example, California imposes a 1.5% tax on S corporation profits in addition to a minimum annual franchise tax of $800. However, you can deduct any state and local taxes as business expenses for your federal income taxes.

## 8. Disadvantages of the Corporate Form

There are some disadvantages to incorporating. You'll have to pay some taxes and fees other business entities don't pay. And you'll have to maintain some minimal corporate formalities that will take some time and effort.

### a. Corporate formalities

The IRS and state corporation laws require corporations to hold annual shareholder meetings and document important decisions such as choosing a federal or state tax election with corporate minutes, resolutions or written consents signed by the directors or shareholders. Fortunately, this is usually not a substantial burden for small businesses with only one or a few shareholders and directors. They usually dispense with holding real annual meetings. Instead, the secretary of the corporation prepares minutes for a meeting which takes place only on paper. There are also standard minute and consent forms you can use to ratify important corporate decisions.

If you're audited and the IRS discovers that you have failed to comply with corporate formalities, you may face drastic consequences. For example, if you fail to document important tax decisions and tax elections with corporate minutes or signed consents, you may lose crucial tax benefits and risk substantial penalties. Even worse, if you neglect these basic formalities, the IRS or a court may conclude that your corporation is a sham—and you may lose the limited liability afforded by your corporate status. This could leave you personally liable for corporate debts.

In addition, banks, trust, escrow and title companies, landlords and others often insist on a copy of a board or shareholder resolution approving a corporate transaction, such as borrowing money or renting property.

Two books provide a complete guide to handling corporate formalities in a streamlined manner and contain all the forms you'll need. They are:

- *Taking Care of Your Corporation, Vol. 1: Director and Shareholder Meetings Made Easy*, and
- *Taking Care of Your Corporation, Vol. 2: Key Corporate Decisions Made Easy.*

(Both are written by Anthony Mancuso and published by Nolo Press.)

### b. More complex bookkeeping

It is absolutely necessary that you maintain a separate corporate bank account if you incorporate. You'll need to keep a more complex set of books than if you're a sole proprietor. You'll also need to file a somewhat more complex tax return, or file two returns if you form a C corporation. And, since you'll be an employee of your corporation, you'll need to pay yourself a salary and file employment tax returns. (See Chapter 8.) All this costs time and money.

You'll probably want to use the services of an accountant or bookkeeper, at least when you first start out. Such a seasoned pro may be able to set up a bookkeeping system for you, make employment tax payments, provide guidance about tax deductions and prepare your tax returns.

If you have a computer, you can also use accounting software packages for small businesses such as *QuickBooks, Mind Your Own Business, Peachtree Accounting* and many others.

### c. Some increased taxes and fees

Finally, there are some fees and taxes you'll have to pay if you incorporate that are not required if you're a sole proprietor. For example, since you'll be an employee of your corporation, it will have to provide unemployment compensation for you. (See Chapter 9.) The cost varies from state to state, but is at least several hundred dollars per year.

You'll also have to pay a fee to your state to form your corporation and may have to pay additional fees throughout its existence. In most states, the fees are about $100 to $300. At least one state—California—imposes much higher fees. In California, you must pay the state $900 to incorporate and you must also pay a minimum $800 franchise tax to the state every year even if your corporation has no profits. However, you can deduct any state and local fees and taxes as business expenses for your federal income taxes.

## 9. Forming a Corporation

You create a corporation by filing the necessary forms and paying the required fees with your appropriate state agency—usually the Secretary of State or Corporations Commissioner. Each state specifies the forms to use and the filing cost.

You'll also need to choose a name for your corporation (see Chapter 3), adopt corporate bylaws, issue stock, and set up your corporate records. This all sounds complicated, but really isn't that difficult to do yourself. You can obtain preprinted articles, bylaws and stock certificates and simply fill in the blanks.

The following books explain how to form a corporation yourself in the four most populous states:

- *How to Form Your Own California Corporation*
- *How to Form Your Own Florida Corporation*
- *How to Form Your Own New York Corporation,* and
- *How to Form Your Own Texas Corporation.*

(All are written by Anthony Mancuso and published by Nolo Press.)

If you live in one of the other states, call your Secretary of State or Corporations Commissioner. They may have free forms you can use. If not, you can get inexpensive incorporation kits for most states; check with a bookstore or stationery store that carries legal forms.

If you live in one of the other 46 states, you have several options that are less expensive than hiring an attorney:

- Call your Secretary of State or Corporations Commissioner. That office probably provides free help in getting the required forms and instructions.
- Books by other publishers may also be helpful. Check with a bookstore or stationery store that carries legal forms.
- If you have access to the Internet, try contacting one or more of the several companies that will incorporate your business for you in any state. You provide all the necessary information online and pay a fee. A list of these incorporation services is contained in the Yahoo Internet directory at: www.yahoo.com/Business_and_Economy/Companies/Corporate_Services/. Contact several incorporation services, because their fees and services vary widely.

**There's No Place Like Home for Incorporating**

People who live in high tax states such as California and New York often hear that they can save money by incorporating in low tax states such as Nevada or Delaware.

This is a myth. You won't save a dime by incorporating outside your own state. If you form an out-of-state corporation, you will end up having to qualify to do business in your home state anyway.

This process is similar to incorporating in your state, and it costs the same. You also will have to pay any state corporate income taxes levied in your home state for income earned there. Even if another state has more modern or flexible corporation laws, these mostly favor large, publicly-held corporations, not the small closely-held corporations self-employed people form.

## 10. Professional Corporations

You may be required to form a special kind of corporation called a professional corporation or professional service corporation if you're involved in certain types of professions. The list of professionals who must form professional corporations varies from state to state, but usually includes:

- accountants
- engineers
- healthcare professionals such as doctors, dentists, nurses, physical therapists, optometrists, opticians and speech pathologists
- lawyers
- psychologists
- social workers, and
- veterinarians.

Call your state's corporate filing office, usually the Secretary of State or your Corporations Commissioner, to see who is covered in your state.

### a. Ownership requirements

Typically, a professional corporation must be organized for the sole purpose of performing professional services and all shareholders must be licensed to render that service. For example, in a medical corporation, all the shareholders must be licensed physicians.

### b. Formation requirements

Special forms and procedures must be used to establish a professional corporation—for example, you might be required to obtain a certificate of registration from the government agency regulating your profession, such as the state bar association. Special language will also have to be included in your articles of incorporation.

### c. Limits on limited liability

In most states, you can't use a professional corporation to avoid personal liability for your malpractice or negligence—that is, your failure to exercise your professional responsibilities with a reasonable amount of care.

> **EXAMPLE:** Janet, a civil engineer, forms a professional corporation of which she is the sole shareholder. She designs a bridge that collapses, killing dozens of commuters. Even though Janet is incorporated, she is personally liable along with her corporation for any damages caused by her negligence in designing the bridge. Janet's personal assets are at risk along with those of her corporation.

You can usually obtain additional business insurance to protect you against these types of risks, but it can be expensive. (See Chapter 6.)

However, if you're a professional involved in a group practice with other professionals, incorporating will shield you from personal liability for malpractice committed by other members of the group.

**EXAMPLE**: Marcus is a doctor involved in an incorporated medical practice with Susan, Florence and Louis. One of Florence's patients claims she committed malpractice and sues her personally and the group. While both the group and Florence are liable, Marcus is not personally liable for Florence's malpractice. This means his personal assets are not at risk.

**COMPARE AND CONTRAST: PROFESSIONAL SERVICE CORPORATIONS**

Workers in the fields of consulting, health, law, engineering, architecture, accounting or actuarial science constitute professional service corporations for federal tax purposes if all the stock is owned by employees who perform professional services for the corporation.

Such corporations are subject to a flat 35% corporate income tax. This tax rate will be applied by the IRS whether or not the individuals involved are organized as a professional corporation or a regular corporation under state law.

## C. Partnerships

If you are not the sole owner of your business, you can't be a sole proprietor. Instead, you automatically become a partner in a partnership unless you incorporate or form a limited liability company. However, if you co-own your business only with your spouse, you can both still be sole proprietors. (See Section A.)

A partnership is much the same as a sole proprietorship except there are two or more owners. Like a sole proprietorship, a partnership is legally inseparable from the owners: the partners. Unlike corporations, partnerships do not pay taxes, although they file an annual tax form. Instead, partnership income and losses are passed through the partnership directly to the partners and reported on the partners' individual federal tax returns. Partners must file IRS Schedule E with their returns showing their partnership income and deductions.

Like sole proprietors, partners are neither employees nor independent contractors of their partnership; they are self-employed business owners. A partnership does not pay payroll taxes on the partners' income or withhold income tax. Like sole proprietors, partners must pay income taxes (see Chapter 9) and self-employment taxes (see Chapter 10) on their partnership income.

For a detailed discussion of partnerships including how to write partnership agreements, see:

- *The Partnership Book,* by Denis Clifford and Ralph Warner (Nolo Press), and
- *Nolo's Partnership Maker,* a software program enabling you to write a partnership agreement on your computer.

### 1. Ownership

The main difference between a partnership and a sole proprietorship is that one or more people own the business along with you. This means that, among other things, you have to decide:

- how each partner will share in the partnership profits or losses
- how partnership decisions will be made
- what the duties of each partner are
- what happens if a partner leaves or dies, and
- how disputes are resolved.

Although not required by law, you should have a written partnership agreement answering these and other questions.

### 2. Personal Liability

Partners are personally liable for all partnership debts and lawsuits, the same as sole proprietors. (See Section A.) This means you'll be personally liable for business debts your partners incurred whether you knew about them or not.

**PARTNERSHIPS COMPARED WITH CORPORATIONS**

Partnerships are somewhat cheaper to form and operate than corporations and you don't have to comply with formalities such as holding annual meetings. Nor do you have to pay registration fees.

However, the fact that you open yourself up to greater liability than in a corporation is a huge drawback. Also, some prospective clients may prefer to hire corporations rather than partnerships because they think they'll have fewer problems with IRS and other government audits. (See Section B3.)

### 3. Limited Partnerships

A limited partnership is a special kind of partnership that has one or more general partners who run the partnership business and one or more partners who are called limited partners because they invest in the partnership but don't help run it. The limited partners are a lot like corporate shareholders: they aren't personally liable for the partnership's debts. The general partners are treated just like partners in normal partnerships: they are liable for all partnership debts and lawsuits.

Limited partnerships are most commonly used for real estate and similar investments. Self-employed people rarely form them. If there are people who want to invest in your business, but don't want to work in it or have any personal liability, you'd probably be better off forming a corporation and selling them shares. That way, you'll have the limited liability afforded by corporate status. (See Section B4.)

## D. Limited Liability Companies

The limited liability company, or LLC, is the newest type of business form in the United States. An LLC is taxed like a partnership but provides its owners with the same limited liability as a corporation.

For a complete discussion of limited liability companies, see *Form Your Own Limited Liability Company,* by Anthony Mancuso (Nolo Press).

### 1. LLC Owners

Until recently, there was a big legal roadblock that often made it difficult for people running one-person businesses to form LLCs: Most states required LLCs to have at least two owners, called members in LLC parlance. This was done to comply with complex IRS rules.

However, this has changed. The IRS threw out its old rules on January 1, 1997, and no longer cares how many members an LLC has. In response, a majority of states have amended their LLC laws to permit one-member LLCs. As of November, 1997, only the following states required that LLCs have at least two members:

Alabama
California
Connecticut
District of Columbia
Florida
Iowa
Kansas
Kentucky
Massachusetts
Nevada
New Jersey
North Carolina
Ohio
South Dakota
Tennessee
Wisconsin

It's likely that most of these states will eventually amend their LLC laws as well to permit one-member LLCs. If you live in a state on this list, contact your state's LLC office to doublecheck the status of your law.

Even if you live in a state that requires your LLC to have two members, you can have your spouse or a trusted friend, relative or business associate fill the second member slot. You can provide in your LLC operating agreement that this person has no voting power, leaving you in sole control.

## 2. Tax Treatment

IRS rules permit LLC owners to decide for themselves how they want their LLC to be taxed. An LLC can be taxed as a pass-through entity—like a sole proprietorship, partnership or S corporation—or as a regular C corporation.

### a. Pass-through entity

Ordinarily, LLCs are pass-through entities. They pay no taxes themselves. Instead, all profits or losses are passed through the LLC and reported on the LLC members' individual tax returns. This is the same as for a sole proprietorship, S corporation or partnership.

If the LLC has only one member, the IRS treats it as a sole proprietorship for tax purposes. The members' profits, losses and deductions are reported on his or her Schedule C, the same as for any sole proprietor. (See Section A1.)

If the LLC has two or more members, it must prepare and file each year the same tax from used by a partnership—IRS Form 1065, Partnership Return of Income—showing the allocation of profits, losses, credits and deductions passed through to the members. The LLC must also prepare and distribute to each member a Schedule K-1 form showing the member's allocations.

### b. Taxation as a C corporation

LLCs have the option of being taxed as a regular C corporation. This is very easily done by filing IRS Form 8832 and checking the appropriate box. If you wish to start your LLC out with this tax status, you must file the form within 75 days of the date your LLC Articles of Organization are filed. LLCs that elect corporate tax status must file the same corporate tax returns as a regular C corporation. The LLC pays income taxes on profits left in the business at corporate income tax rates, which can be lower than personal rates. (See Section B6.)

Of course, if you want to be taxed as a C corporation, you can form a C corporation instead of an LLC. The only reason to form an LLC instead of a C corporation in this instance is that an LLC is somewhat simpler and easier to run.

## 3. Liability Concerns

LLC company owners, called members, enjoy the same limited liability from business debts and lawsuits as corporation owners. You won't be liable for business debts unless you personally guarantee them. (See Section B4.)

## 4. Pros and Cons

LLCs appear to be a clear favorite over partnerships since they offer the same tax benefits, but also provide limited liability. They are also a serious alternative to forming a corporation, since they offer the same limited liability as a corporation along with some tax advantages.

Whether an LLC is right for you depends on whether you think the following advantages of an LLC outweigh the disadvantages.

### a. Advantages of LLCs

LLCs provide the same limited liability as a corporation; however, such "limited liability" is often more mythical than real. (See Sections B4 and B10.)

Setting up an LLC takes about the same time and money as a corporation, but thereafter a LLC is simpler and easier to run. With a corporation, you must hold and record regular and special shareholder meetings to transact important corporate business. Even if you're the only corporate owner, you need to document your decisions. This isn't required for an LLC.

LLCs allow more flexibility in allocating profits and losses among the business's owners than a corporation can. The owners of an S corporation must pay taxes on profits or get the benefits of losses in proportion to their stock ownership. For example, if there are two shareholders and each own 50% of the stock, they must each pay tax on 50% of the corporation's profits or get the benefits of 50% of the losses. In contrast, if you form an LLC, you have near total flexibility on how to allocate profits and losses among the owners—for example, one owner could get 75% of the profits and the other 25%. Of course, this will be useful only if two or more people own your business.

Moreover, an S corporation has less flexibility than an LLC to use borrowed money of the business to increase the tax deductions of the owners on their annual individual tax returns and lower the tax bite when the business is ultimately sold.

A C corporation cannot allocate profits and losses to shareholders at all. Shareholders get a financial return from the corporation by receiving corporate dividends or a share of the corporate assets when it is sold or liquidated.

LLCs don't have to comply with the rules limiting who can form and own an S corporation. (See Section B7d.)

And finally, LLCs members are not employees of the LLC, so the LLC doesn't have to pay federal and state unemployment tax for them. When you form a corporation you are an employee of the corporation and such taxes must be paid.

### b. Disadvantages of LLCs

Perhaps the biggest drawback for LLCs is that they offer no opportunity to save on self-employment taxes as you can when you form an S Corporation. LLC members who actively manage the business must pay self-employment tax on all the income they receive from the LLC—whether in the form of salary or distributions. In contrast, you can save on self-employment taxes by forming an S corporation because S corporation distributions—as opposed to salaries—are not subject to self-employment tax. (See Section B7.)

Moreover, money you retain in a S corporation or C corporation is not subject to self-employment taxes. This is not the case with an LLC that is treated as a pass-through entity. Whether or not you distribute your LLC profits to yourself or leave them in your company, you must pay self-employment taxes on your entire share of LLC profits.

Also, when you form an LLC, you will not have the ability to save on taxes by splitting your income between two tax entities. You can only do this by forming a C corporation. (See Section B6.)

LLC owners can't deduct the cost of employee benefits such a health insurance as can C corporations. (See Section B6.)

Several states still require that LLCs have two or more owners.

Finally, forming a corporation may give your business added credibility and help you get business. Many prospective clients will have no idea what an LLC is and may prefer to hire a corporation.

### Choice of Business Entity Confusion

There is no one best business form—and choosing which one will work best for you can be difficult. It all depends on your goals and preferences. The following chart may help you analyze which business form best furthers your own personal goals.

| Goal | Preferred Form of Business |
|---|---|
| Limit your personal liability | C or S corporation, LLC |
| Save on Social Security taxes | S corporation |
| Retain earnings in business | C corporation |
| Provide tax deductible benefits to employees, including yourself | C corporation |
| Deduct losses from your personal taxes | S corporation, LLC, Sole proprietorship |
| Easiest and cheapest to form and operate | Sole proprietorship |
| Avoid state and federal unemployment taxes | Sole proprietorship, LLC |
| Simplest tax returns | Sole proprietorship |
| Added credibility for your business | S or C corporation |
| Benefit from lower corporate tax rates | C corporation |
| Distribute high profits | S corporation, LLC, Sole proprietorship |

## 5. Forming an LLC

To form an LLC, you must file Articles of Organization with your state government. Your company's name will have to include the words "limited liability company" or "LLC" or a similar phrase as set forth in your state law. You should also create a written operating agreement, which is similar to a partnership agreement. (See Section C.) ■

CHAPTER

# 3

# Choosing and Protecting Your Business Name

This chapter is about choosing the name for your business. This is something you need to do right after you choose the legal form of your business. (See Chapter 2). You'll need to know what your business name will be to establish bank accounts, to print stationery and often, to market your business to others.

Business names are used in different contexts and the legal rules governing them differ depending on how they are used. A business name may be the name you use:

- to identify your business on your business checking account, invoices, business cards, contracts and letterhead, called a trade name or business name
- to market your services or goods, called a service mark or trademark, and
- to identify your company or yourself on the Internet, called an Internet domain name.

You may be able to use a different name in each of these contexts or use the same or similar name.

## A. Choosing a Trade Name

Your trade name is the name you use to identify your business when you open a business bank account, sign a contract or sue someone in the name of your business. If, like the vast majority of self-employed people, you're a sole proprietor (see Chapter 2, Section A), you can simply use your own name as your business name. For example, Roger Davis calls his consulting business Roger Davis Consulting.

However, you have the option of using a name other than your own as your business name. When a sole proprietor uses a name other than his or her own name, it's called a fictitious business name, assumed name or dba—short for doing business as.

> **EXAMPLE:** Roseann Zeiss quits her job with a public relations firm and sets up her own public relations business. Roseann is a sole proprietor. She could call her company Roseann Zeiss. She decides to call it AAA Publicity instead so she'll come first in her local business telephone directory.

Self-employed people often prefer to use assumed names instead of their personal names for their businesses to get a jumpstart on marketing. An assumed name can sound jazzier, help identify what your business does or make you seem more businesslike. But there is an added benefit that can be even more useful: It helps to establish your status as an independent contractor. Employees, obviously, don't use business names.

### 1. Choosing an Assumed Name

Use some care in choosing an assumed name. Avoid using a name that is very similar to that of a local or regional competitor or a large national company in your field. If it is so similar to a name that is already being used that it will confuse the public, you could be sued under state and federal unfair competition laws and may be required to change the name and even pay damages.

Also, make sure your name is not very similar to any famous business name, even if the name is used by a company in a different field—for example McDonald's, Proctor & Gamble, Honda and the like. Companies with famous names are usually fanatical about protecting them.

Check local business directories, Yellow Pages and other records to see if someone is already using the name you want. (See Section D1.)

### 2. Registering an Assumed Name

Any person who uses a name other than a surname to identify a business must register the name with the state or county as an assumed or fictitious name. If you fail to register, you'll have all sorts of problems. For example, you may not be able to open a bank account in your business name. You also may be barred from suing on a contract signed with the name.

To register, you usually file a certificate with the county clerk stating who is doing business under the name. In many states, you must publish the statement in a local newspaper. This is intended to help creditors identify the person behind an assumed business name. This makes it easier to track down those people who are in the habit of changing their business names to confuse and avoid creditors.

Some communities have newspapers that specialize in publishing such legal notices. Some states require, instead of or in addition to publication, that you file the statement with the state department of revenue or some other state agency.

Contact your county clerk and ask about the registration requirements in your locale. You'll have to fill out a simple form and pay a fee—usually between $15 and $50. The county clerk will normally check to see if any identical or very similar names have already been registered in the county. If so, you'll have to use another name. It's a good idea to think of several possible names before you attempt to register.

**Additions to a Surname May Require Registration**

You never have to file a fictitious business name statement if you use your full legal name as your business name—for example, if John Smith uses "John Smith" as his business name. However, some states require you to register your name if you use words in addition to your full name—for example, registration might be required if John Smith used "The Smith Group" as his business name. Ask your county clerk about your state's requirements.

## 3. Naming a Corporation or Limited Liability Company

If you form a corporation or limited liability company, you must get permission to use your corporate name by registering it with your Secretary of State or similar official.

### a. Registering a corporation name

Registering a corporate name involves three steps.

- **Step 1. Selecting a permissible name.** All but three states—Maine, Nevada and Wyoming—require you to include a word or its abbreviation indicating corporate status, such as "corporation," "incorporated," "company," "limited." Several states also require that the name be in English or Roman characters.
- **Step 2. Clearing your name.** Next you must make sure that your corporate name is distinguishable from any corporate name already registered in your state. Your state won't register a corporate name that too closely mimics a name already on file. The secretary of state or other corporate filing agency will do a search for you prior to authorizing the use of your name. In about half the states, you may phone to check on the availability of a name in advance. In the others you must write to request a search. Often you may request a search of more than one name at a time.
- **Step 3. Reserving your corporate name.** A corporation can usually reserve a name before incorporating if the name otherwise qualifies for registration. This freezes out other would-be registrants of that name or names that are deceptively similar during the period of reservation, usually 120 days. Most states permit you to extend the reservation for one or more additional 120-day periods for additional fees.

The reservation process involves sending an application for reservation to the Secretary of State, or the designated office, with a fee. Some states even permit you to reserve a corporate name over the telephone. You can find out the exact information by calling the secretary of state or corporate commissioner in your state.

### b. Registering a limited liability company name

Registering a name for a limited liability company (LLC) is very similar to registering a corporate name. You must choose a name conforming with your state's LLC requirements. Most states require you to use the words Limited Liability Company, Limited Company or abbreviations in your name.

You then call your state LLC filing office and ask personnel there to do a search to see if the name or names you've chosen are available. Most states allow you to reserve LLC names for 30 to 120 days by paying a small fee, usually no more than $50.

For a detailed discussion of LLC name requirements, see *Form Your Own Limited Liability Company*, by Anthony Mancuso (Nolo Press).

## 4. Legal Effect of Registering a Name

People often think that once they have complied with all the registration requirements for their trade name, they have the right to use it for all purposes. This isn't so. There are two very different contexts in which a business name may be used.

For bank accounts, creditors and potential lawsuits, the formal name of the business or trade name is used. A business may also use a name in a trademark or service mark to market its goods or services.

The registration requirements only control the formal business or trade name. Registering an assumed, corporate or LLC name gives you no ownership rights in the name in the sense of preventing others from using it. If someone else is the first to use your name as a mark, it doesn't make any difference whether you or they have previously registered it as an assumed or corporate or LLC name. They will still have the right to exclusive use of the name in the marketplace.

Simply put, if the name you have registered was already in use or federally registered as a trademark or service mark, you will have to limit your use of the name to your checkbook and bank account. The minute you try to use the name in connection with marketing your goods or services, you risk infringing the existing trademark or service mark.

If your business name figures in your future marketing plans, you must search for use of the name as a trademark in addition to complying with the name registration requirements. (See Section B.)

If you plan to market your goods or services on the Internet, then you'll also want to check to see whether someone else has already taken your proposed name as their domain name, which would mean, at the least, that you'd have to use a slightly modified name, since every domain name is unique. (See Section C.)

## B. Choosing a Trademark

A trademark is a distinctive word, phrase, logo or other graphic symbol that's used to distinguish one product from another—for example, Ford cars and trucks, Kellogg's cornflakes, IBM computers, Microsoft software.

A service mark is similar to a trademark, except that trademarks promote products while service marks promote services. Some familiar service marks include: McDonald's (fast food service), Kinko's (photocopying service), Blockbusters (video rental service), CBS's stylized eye in a circle (television network service), the Olympic

Games' multi-colored interlocking circles (international sporting event).

The word trademark is also a generic term used to describe the entire broad body of state and federal law that covers how businesses distinguish their products and services from the competition. Each state has it own set of laws establishing when and how trademarks can be protected. There is also a federal trademark law called the Lanham Act (15 U.S.C. 1050 and following), which applies in all 50 states. Generally, state trademark laws are relied upon for marks used only within one particular state, while the Lanham Act is used to protect marks for products that are sold in more than one state or across territorial or national borders.

If, like many self-employed people, you operate within a single state, you'll have to rely primarily on your state trademark law. But if you do business in more than one state, you can use either the federal or state law. The various trademark laws don't differ greatly except that some state laws allow trademark owners to collect greater damages against infringers than does federal law.

For a detailed discussion of trademarks, including all the forms and instructions you need to register a trademark with the U.S. Patent and Trademark Office, see *Trademark: How to Name a Business & Product*, by Kate McGrath and Stephen Elias (Nolo Press).

## 1. Trade Names Are Not Trademarks

Your trade name is neither a trademark nor a service mark and is not entitled to trademark protection unless it is used to identify a particular product or service you produce and sell to the public. Businesses often use shortened versions of their trade names as trademarks—for example, Apple Computer Corporation uses the name Apple as a trademark on its line of computer products, and Nolo Press, Inc. uses Nolo as a service mark for its online law services.

A trade name acts like a trademark when it is used in such a way that it creates a separate commercial impression. In other words, when it acts to identify a product or service. This can sometimes be a tricky determination, especially comparing trade names and service marks, because they often both appear in similar places—on letterheads, advertising copy, signs and displays. But some general principles apply:

- If the trade name is used with its full name, address and phone, it's probably a trade name.
- If a shortened version of the trade name is used, especially with a design or logo beside or incorporating it, the trade name becomes a trademark.

**EXAMPLE:** Joe, a self-employed computer programmer, calls his unincorporated business Acme Software Development and files a fictitious business name statement with his county clerk. When he uses the name Acme Software Development along with his office address on his stationery, it is just a trade name, not a trademark. However, Joe develops a software utility program he calls Acme Tools and markets over the Internet. Acme Tools is a trademark.

## 2. Selecting a Trademark

Not all trademarks are treated equally by the law. The best trademarks are "distinctive"—that is, they stand out in a customer's mind because they are inherently memorable. The more distinctive a trademark is, the more legal protection it will receive. Less distinctive marks may be entitled to little or no legal protection. Obviously, it is much better to have a strong trademark than a weak one.

Good examples of distinctive marks are arbitrary, fanciful or coined names such as Kodak and Xerox. Examples of poorly chosen marks include:

- personal names, including nicknames, first names, surnames and initials

- marks that describe the attributes of the product or service or its geographic locations—for example, marks such as Quick Printing and Oregon Marketing Research are initially legally weak and not extensively protectible until they have been in use long enough to be easily recognized by customers, and
- names with bad translations, unfortunate homonyms (sound-alikes) or unintended connotations—for example, the French soft drink called Pschitt had to be renamed by the U.S. market.

Generally, selecting a mark begins with brainstorming for general ideas. After several possible marks have been selected, the next step is often to use formal or informal market research techniques to see how the potential marks will be accepted by consumers. Next, a trademark search is conducted. This means that an attempt is made to discover whether the same or similar marks are already being used.

## 3. Registering a Trademark

If you use all or part of your business name or any other name as a trademark or service mark, consider registering it. Registration is not mandatory, but makes it easier for you to protect your mark against would-be copiers and puts others on notice that the mark is already taken.

If you do business only in one state, register your mark with your state trademark office. This is the case for most local service businesses that don't do business across state lines or sell to interstate travelers.

If you do business in more than one state, register with the U.S. Patent and Trademark Office in Washington, DC.

To register, you must fill out an application and pay a fee. Be prepared to work with your state or federal trademark officials to get your registration approved. (See the Appendix for a list of state trademark agencies.)

The Trademark Assistance Center of the U.S. Patent and Trademark Office can be contacted at: 703-308-9000. You can obtain a free booklet on trademark registration entitled *Basic Facts About Registering a Trademark* by calling 703-308-900 or writing to: Assistant Commissioner for Trademarks, Box New App/Fee, 2900 Crystal Drive, Arlington, VA 22202. The Assistance Center also has an excellent Website from which you can download application forms and obtain much other information. The URL is www.uspto.gov.

### Intent to Use Registration

If you seriously intend to use a trademark on a product or for a service sold in more than one state in the near future, you can reserve the right to use the mark by filing an intent to use registration with the U.S. Patent and Trademark Office (PTO).

If the mark is approved, you have six months to actually use the mark on a product sold to the public and file papers with the PTO describing the use, with an accompanying $100 fee. If necessary, this period may be increased by five additional six-month periods if you have a good explanation for each extension

The ownership becomes effective when the mark is put in use and the application process is complete, but ownership will be deemed to have begun on the date the application was filed.

You should promptly file an intent to use registration as soon as you have definitely selected a trademark for a forthcoming product. Your competitors are also trying to come up with good trademarks, and they may be considering using a mark similar to the one you want.

For step-by-step guidance on how to register a trademark, see *Trademark: How to Name a Business and Product*, by Kate McGrath & Stephen Elias (Nolo Press).

### 4. Trademark Notice

The owner of a trademark that has been registered with the U.S. Patent and Trademark Office (PTO) is entitled to use a special symbol along with the trademark. This symbol notifies the world of the registration. Use of trademark notices is not mandatory, but makes it much easier for the trademark owner to collect damages in case of infringement. It also deters others from using the mark.

The most commonly used notice for trademarks registered with the PTO is an "R" in a circle—®—but "Reg. U.S. Pat. & T.M. Off." may also be used. The "TM" superscript—™—may be used to denote marks that have been registered on a state basis only or marks that are in use but which have not yet officially been registered by the PTO.

Do not use the copyright symbol, or ©, with a mark. It has absolutely nothing to do with trademarks.

### 5. Enforcing Trademark Rights

Depending on the strength of the mark and whether and where it has been registered, a trademark owner may be able to bring a court action to prevent others from using the same or similar marks on competing or related products.

Trademark infringement occurs when an alleged infringer uses a mark that is likely to cause consumers to confuse the infringer's products with the trademark owner's products. A mark need not be identical to one already in use to infringe upon the owner's rights. If the proposed mark is similar enough to the earlier mark to risk confusing the average consumer, its use will likely be infringement.

**COURT CLAMPS DOWN ON COPYCATS**

A Sandusky, Ohio, insurance agent invented the term "securance"—a contraction of the words security and insurance—and used it as a service mark to market his insurance services on his stationery, billboard and newspaper advertising and on giveaway items such as calendars and pencils. He registered the mark with the State of Ohio and the U.S. Patent & Trademark Office.

Four years later, the Nationwide Mutual Insurance Co. began using the word as its own service mark in its national advertising.

The agent sued the insurer for trademark infringement and won. The court held that the agent had the exclusive right to use the word securance to identify insurance services in the geographic areas in which he did business since he had been the first to use it. The court ordered the insurer to stop using securance in its advertising in the entire state of Ohio. (*Younker v. Nationwide Mut. Ins. Co.*, 191 N.E.2d 145 (1963).)

## C. Choosing an Internet Domain Name

The Internet is a computer network that allows near-instantaneous communication between computers anywhere in the world. Unlike the telephone, or fax, the Internet allows the transmission of digital files which can be read by computers to produce software, text, photographs, graphics, sounds, videos and movies. The World Wide Web—a part of the Internet—adds to the Internet by providing an appealing graphical interface which makes it much easier to get around and that, for businesses and consumers, has become a powerful new type of yellow pages extending to all parts of the country and many parts of the world.

There are more than 600,000 business-related sites on the Web. Most national businesses have a Website, as do a huge number of small and medium sized businesses offering all types of products and services. More and more self-employed people have Websites as well—particularly those involved in hi-tech fields such as computer programming.

If you want to jump on the Web bandwagon, you'll need to create a Web page—your own site on the Internet containing the information you want to make available to Web users. An important issue that arises when you create a Web page has to do with the name you give your Website—called an Internet domain name.

For a detailed discussion of all the legal issues involving Internet domain names, see *Trademark: How to Name Your Business & Product*, by Stephen Elias and Kate McGrath (Nolo Press).

### 1. What Is a Domain Name

Every business on the Web has a domain name—a unique address that computers understand. If you enter a particular domain name in a Web browser, the computer will link your computer with the Website connected with the domain name you entered.

Most World Wide Web business addresses consist of two main sections: a beginning section containing the letters http://www and a section containing the domain name itself. For example, the domain name for the Nolo Press Website is http://www.nolo.com.

### 2. Clearing a Domain Name

Because each domain name must be unique—so that all the computers attached to the Internet can find it—it is impossible for two different businesses to have the same domain name. If somebody is already using a name you want, you likely won't be able to use it.

There are two ways to find out whether you will be able to use your proposed mark as a domain name. One is to buy a domain name search from Thompson and Thompson [http://ttdomino.thomson-thomson.com/]. The other is to do it yourself by visiting Tabnet on the World Wide Web [www.tabnet.com]. Tabnet not only lets you search to see whether your name is available, but also provides information about the registrant if it turns out that someone got there ahead of you. Selling domain names is a cottage industry; they go for about $2,000. So if you are set on a particular name but find that you are blocked, you may want to try to negotiate a sale.

### 3. Registering a Domain Name

Domain names are reserved for your use alone by registering the name so that no one else can use it for the same purpose on the Internet.

There are three ways to register your name:

- use Tabnet (www.tabnet.com) to do it for you for $50 plus the registration fees,
- get help from your Internet service provider, or
- do it yourself.

If you choose to do it yourself, the first step to registering a domain name is to acquire operational name service. This is usually done by signing on with an Internet service provider (ISP). These providers are companies that connect to the Internet and can be located in the Yellow Pages or in computer magazines.

## D. Conducting a Name Search

Before choosing a trade name, trademark or domain name, conduct a name search to see if someone in a related business is already using the same or a very similar name. If so, choose a different name. Obviously, you don't want to spend money on marketing and advertising a name or mark only to discover it infringes another name or mark and must be changed.

The amount of time and money you put into such a search depends on the nature of your business and your plans for expansion.

### 1. Local Searches

If you only do business locally, you can feel reasonably safe if you search for conflicting names at the state and local levels. Check telephone books and business directories for the cities in which you plan do business, as well as for surrounding areas. Many public libraries have excellent phone book collections. While at the library, look at trade magazines for your industry and compilations of names of related businesses. The reference librarian should be able to help you.

Also, call your local county clerk's office to ask how you can check assumed business name filings. In most states, assumed or fictitious business name statements are filed with the county clerk's office. Generally, you must go in and check the files in person. It takes just a few minutes to do this.

To make sure no corporation registered in your state is already using the name, call the state office where corporations are registered—usually the Secretary of State or Corporations. Workers there may do a name search for you over the phone or may require that you send a letter requesting a search.

Finally, check the records of your state trademark agency to see if a similar name has already been registered. Many agencies will do a search for you by phone, either free or for a small charge; some require you to request a search by mail. (See Appendix 1 for a list of state trademark agencies.)

### 2. National Searches

If you intend to do business regionally or nationally, you'll need to do a more widespread national search, generally called a trademark search. A national search is also necessary if you plan to establish a presence on the Internet because you'll be sending your business name all over the country.

You can hire a search firm to do a trademark search for you, or do one yourself.

#### a. Hiring a search firm

Traditionally, most trademark searches were conducted by specialized trademark search firms at the behest of trademark attorneys who were handling the trademark registration process. Even today, some of the largest trademark search firms refuse to conduct searches for anyone but a lawyer. But most search firms aren't so choosy and will conduct a search for anyone willing to pay them.

The services provided by various trademark search firms, and the fees they charge for different types of searches, vary considerably. For this reason, it pays to shop around and compare the services and prices being offered.

You can find a trademark search firm by:

- looking in the Yellow Pages under trademark consultants or information brokers
- consulting a legal journal or magazine which will usually contain ads for search firms, or
- doing an Internet search.

**THE ROLE OF ATTORNEYS IN TRADEMARK SEARCHES**

If you decide to hire a trademark attorney to advise you on the choice and registration of a trademark or service mark, the attorney will be able to arrange for the trademark search. Some attorneys do it themselves, but most farm the search out to a search firm.

Once the report comes back from the search firm, the attorney will interpret it for you and advise you on whether to go ahead with your proposed mark. Although you are getting considerably more in this attorney package than you'll get from a search service, it will cost you.

### b. Doing a search on your own

Many resources are available to enable you to do all or part of a trademark search yourself, including:

- numerous search services on the Internet
- trademark search services offered by two private subscription-based companies, Dialog and CompuServe
- Patent and Trademark Depository Libraries (PTDLs) located throughout the country. Using a Patent and Trademark Depository Library (PTDL) to do your own federal trademark search involves the least cash outlay, but will cost you in time and transportation expenses unless you live or work close to one. (See Appendix 1 for contact information.)

For more on trademark searches, see *Trademark: How to Name a Business and Product*, by Kate McGrath & Stephen Elias (Nolo Press). ■

CHAPTER

# Home Alone or Outside Office?

When you're self-employed, you have the option of working from home or from an outside office. This chapter covers the pros and cons of working from home, with emphasis on zoning and other restrictions and the home office tax deduction.

## A. Pros and Cons of Working at Home

Legally, it makes little difference whether you work at home or in an outside office. The basic legal issues discussed in this book such as deciding whether to operate as a sole proprietor or corporation, picking a name for your business and collecting clients are the same whether you run your business from home or from the top floor of a high-rise office building.

There may be a difference as to whether you can deduct your office expenses (see Section D), but otherwise, your taxes are the same whether you work at home or in an office.

The issues to consider in deciding whether to work at home are more practical than legal: whether you can afford to pay office rent, the benefits of less commuting and other factors that may make a home office more convenient and whether working at home will disrupt your home or neighborhood.

### 1. Benefits of Working at Home

According to Find/SVP, a New York marketing research firm, 12.3 million self-employed home-based businesses are now in operation. Of the 4.5 million new business start-ups in 1996, it's estimated that more than two out of three were home-based, according to the National Federation of Independent Businesses.

Working at home is popular for the self-employed because it can save time and money and improve productivity.

#### a. No office rent expenses

For many self-employed people, the greatest benefit of working at home is that you don't have to pay rent for an office. Office rents vary enormously depending upon the area, but you'll likely have to pay a few hundred dollars for even a small office. In large cities, you may have to spend much more. Look at the commercial real estate advertising section of your Sunday newspaper to get an idea of the going rates.

Working at home will save you several thousand dollars a year in rent. This is money you can use to expand your business or pay your living expenses. It's true that you can deduct your office rent as a business expense (see Chapter 8), but you may be able to deduct home office expenses as well (see Section D). Keep in mind, however, that you can't deduct all your rent—and the amount you'd save in taxes depends on your tax bracket. (See Chapter 8.)

One way to reduce the costs of obtaining office space is to obtain it from a client. Many clients are willing to provide outside workers with desk space. This is particularly likely if having you around will make life easier for them. Some clients may even offer to provide you with office space at no cost to you. However, to safeguard your self-employed status, it's best that you pay something for the space. It doesn't have to be much. You can charge the client slightly more for your services to cover the cost. The client shouldn't mind because it's getting the money back and this procedure will help the client if the IRS conducts an audit and questions your status.

#### b. No commuting time or expenses

Working at home means you don't have to commute to an outside office every day. In 1997, the IRS allowed a commuting expense deduction of 31.5 cents per mile. Using this figure, if working at home allows you to drive 6,000 fewer miles per year (500 miles per month) you'd save $1,890 per year.

Not having to commute saves you not only money, but time. If you commute just 30 minutes each day, you're spending 120 hours each year behind the wheel of your car. That's three full 40 hour weeks that you could use earning money in your home office.

#### c. You can deduct home office expenses

If you arrange things right, you can deduct your home office expenses—including a portion of your rent or mortgage payment, utilities and other expenses. Changes in the tax law slated to take effect in 1999 will make the home office deduction available to more self-employed people than in past years. (See Section D.)

The home office deduction is particularly valuable if you rent your home. It enables you to deduct a portion of what is likely your largest single expense—your rent—an item that is ordinarily not deductible.

#### d. You can deduct some commuting costs

When you have an outside office, you can't deduct your commuting expenses—that is, the cost involved in going from home to your office. However, if your main office is at home, you may deduct the cost of driving from home to meet clients or to other locations to conduct business.

#### e. Benefits other than money

Of course, the benefits of working at home are not just monetary. For many self-employed people, they are outweighed by other factors such as the increased flexibility they have over their daily schedule. You can, if you wish, work in the evenings or late at night in your pajamas—something that can be difficult to do if you're renting an office.

When you work at home, it's also easier to deal with childcare problems and household chores and errands. You may also have more contact with your family.

### 2. Drawbacks of Working at Home

There are potential drawbacks to working from home—or example, it may not help you project a professional image. Even worse, working at home might be against the law where you live. However, there are usually things you can do to avoid or ameliorate the problems.

#### a. Clients don't take you seriously

The major problem many self-employed people who work at home say they have is that clients don't take them seriously. Some clients may be reluctant to deal with a home-based businessperson. This can make it harder for you to get your business established.

However, there are many things you can do to help create and maintain a professional image. For example:

- obtain a separate telephone line for your business which you use only for business calls
- use an answering service to answer your business phone when you're not home
- obtain and use professional looking business cards, envelopes and stationery
- hold meetings at clients' offices instead of at your home
- rent a mailbox to receive your business mail instead of using your home address
- use an assumed name for your business rather than your own name (see Chapter 3)
- use a federal taxpayer identification number rather than your personal Social Security number (see Chapter 5), and
- consider incorporating or forming a limited liability company so that clients will be hiring a corporation or company, not you personally (see Chapter 2).

### b. Restrictions on home-based businesses

Another major problem for the home-based self-employed are restrictions on home businesses imposed by cities, condominium associations and deed restrictions. It may actually be illegal for you to work at home. (See Section D.)

### c. Obtaining services can be difficult

Businesses that provide services to businesses sometimes discriminate against those who work at home. For example, UPS charges more for deliveries to a home business than to one at an outside business office. And many temporary agencies won't deal with a home based business because they're afraid they won't get paid.

### d. Lack of security

Your home is likely not as secure an environment as an office building that is filled with people, has burglar alarms, employs security guards and even has hidden security cameras. If you're handling large amounts of cash or other highly valuable items, you may prefer to work in a more secure outside office than at home.

However, there are many commonsense precautions you can take to make your home office more secure—for example:

- rent a post office box to receive your mail instead of having it delivered at home
- don't let equipment servicers or vendors visit without an appointment
- obtain good locks and use them, and
- if you have a separate business phone line, don't alert people to your absence with a message on the answering machine such as "I won't be here for a week."

### e. Isolation, interruptions and other factors

Finally, some people have trouble adapting to working at home because of the isolation. They miss the social interaction of a formal office setting. However, renting an outside office where you'll be all by yourself won't necessarily end the isolation problem.

In contrast, other self-employed people find it difficult to get any work done at home because of a lack of privacy and because of interruptions from children and other family members. Others gain weight because the refrigerator is always nearby or end up watching television instead of working.

However, most of the millions of self-employed people who work at home are not fazed by these problems. Nearly 98% of the 4,200 home workers who responded to a 1992 survey by the magazine *Home Office Computing* said they were happier at home than in a corporate office.

## 3. Businesses Well-Suited to Home Offices

Home offices can work well for any business that is normally done in a simple office setting. This includes a multitude of service businesses—for example:

- desktop publishing
- accounting and bookkeeping
- computer programming
- consulting
- writing
- telemarketing
- graphic artwork
- information brokering, and
- financial planning.

A home office is also an ideal choice for businesses in which most of the work is done at clients' offices or other outside locations—for example:

- building contracting
- traveling sales
- house and carpet cleaning

- home repair work
- courier or limousine service
- piano tuning
- pool cleaning
- hazardous waste inspection, and
- catering service

Under tax law changes that will go into effect in 1999, you'll be able to spend most of your time working away from home and still qualify for the home office deduction. (See Section D.)

### 4. Businesses Poorly Suited to Home Offices

Any business that will disrupt your home or neighborhood is not well suited for the home. These include businesses that generate substantial amounts of noise, pollution or waste.

A home office is not your best choice if substantial numbers of clients or customers must visit you in your office. This could cause traffic and parking problems in your neighborhood and cause neighbors to complain. One possible solution to this problem is to rent a office part-time or by the hour just to meet clients. Such rentals are available in many cities and will be cheaper than renting a full-time office. Look in your Yellow Pages under office rentals for "business identity programs."

You may also have problems if your business requires you to store a substantial amount of inventory. However, you can get around this and still spend most of your time at home by renting a separate storage space for your inventory.

Finally, a home office may not work well if you need to have several employees working with you. This could cause parking problems in your neighborhood and space problems in your home. Moreover, many local zoning laws prevent home businesses from having more than one or two employees. One way around this problem is to allow your employees to work in their own homes.

## B. Pros and Cons of an Outside Office

Having an outside office can avoid the problems discussed in the previous section. It can help establish your credibility and provide a better setting for meeting clients or customers than a home office.

Renting an outside office will also help establish that you are self-employed if you're audited by the IRS or your state tax department. (See Chapter 15.)

An outside office also enables you to help keep your home and work lives separate and may enable you to work more efficiently.

The drawbacks of an outside office are the flipside of the benefits of having a home office. (See Section A.) You must pay rent for your office and drive to and from it every day. You won't be entitled to a home office deduction, although you can deduct your outside office rent, utilities and other expenses. You also lose much of the flexibility afforded by a home office.

## C. Restrictions on Home-Based Businesses

If, like millions of self-employed people, you plan to work at home, you may have potential problems with your local zoning laws or with land use restrictions in your lease or condominium rules. You should investigate to determine whether you may have problems before you open your home office. Even if your community is unfriendly to home offices, there are many things you can do to avoid difficulties.

### 1. Zoning Restrictions

Municipalities have the legal right to establish rules about what types of activities can be carried out in different geographical areas. For example, they often establish commercial zones for stores and offices, industrial zones for factories and residential zones for houses and apartments.

Some communities—Houston, for example—have no zoning restrictions at all. However, most do and most have laws limiting the kinds of businesses you can conduct in a residential zone. The purpose of these restrictions is to help maintain the peace and quiet of residential neighborhoods.

Fortunately, the growing trend across the country is to permit home businesses. Many cities—Los Angeles and Phoenix, for example—have recently updated their zoning laws that permit many home businesses. However, some communities remain hostile to home businesses.

### a. Research your local zoning ordinance

Your first step to determine whether you might have a problem working at home is to carefully read your local zoning ordinance. Get a copy from your city or county clerk's office or your public library.

Zoning ordinances are worded in many different ways to limit businesses in residential areas. Some are extremely vague, allowing "customary home-based occupations." Others allow homeowners to use their houses for a broad but, unfortunately, not very specific list of business purposes—for example, "professions and domestic occupations, crafts and services." Still others contain a detailed list of approved occupations, such as "law, dentistry, medicine, music lessons, photography, cabinetmaking."

Ordinances that permit home-based businesses typically include detailed regulations on how you can carry out your business activities. These regulations vary widely, but the most common type limit your use of on-street signs, limit car and truck traffic and restrict the number of employees who can work at your house on a regular basis; some prohibit employees altogether. Some ordinances also limit the percentage of your home's floorspace that can be devoted to your business. Again, study your ordinance carefully to see how these rules apply to you.

If you read your ordinance and still aren't sure whether your business is allowed or don't understand the local regulations, you may be tempted to discuss the matter with zoning or planning officials. However, until you figure out what the rules and politics of your locality are, it may be best to do this without identifying and calling attention to yourself. For example, have a friend who lives nearby make inquiries.

#### Hard Lobbying Can Pay off

If your town has an unduly restrictive zoning ordinance, you can try to get it changed. For example, a self-employed person in the town of Melbourne, Florida, was surprised to discover that his local zoning ordinance barred home-based businesses and decided to try to change the law.

He sent letters to his local public officials, but got no response.

He then reviewed the zoning ordinances favoring home offices from nearby communities and drafted an ordinance of his own that he presented to the city council. He enlisted support from a local home business association and got a major story about his battle printed in the local newspaper.

After several hearings, the city council voted unanimously to amend the zoning ordinance to allow home offices.

### b. Determine the attitude toward enforcement

Even if your locality has a restrictive zoning law on the books, you won't necessarily have problems. In most communities, such laws are rarely enforced unless one of your neighbors complains to local officials. Complaints usually occur because you make lots of noise or have large numbers of clients, employees or delivery people coming and going, causing parking or traffic problems. If you're unobtrusive—for example, you work quietly in your home office all day and rarely receive business visitors—it's not likely your neighbors will complain.

Unfortunately, some communities are extremely hostile toward home businesses and actively try to prevent them. This is most likely to be the case if you live in an upscale purely residential community. Even if you're unobtrusive, these communities may bar your from working at home if they discover your presence. If you live in such a community, you may want to consider moving—or you'll really need to keep your head down to avoid discovery.

To determine your community's enforcement style, try talking with your local chamber of commerce and other self-employed people you know in your town. Friends or neighbors who are actively involved with your local government may also be knowledgeable.

**Keeping Your Home Business Unobtrusive**

There are many ways to help keep your home business as unobtrusive as possible in the interest of warding off neighbor complaints. For example, if you get a lot of deliveries, arrange for mail and packages to be received by a private mailbox service such as Mail Boxes, Etc. Don't put your home address on your stationery and business cards.

Also, try to visit your clients in their offices instead of having them come to your home office.

A husband and wife team of psychiatrists who ran a 24-hour group therapy practice in a quiet neighborhood of Victorian homes provide a perfect example of how to get neighbors to complain about a home office: they paved every inch of their yard for parking and installed huge lights to illuminate the entire property.

### c. Inform your neighbors

Good neighbor relations are the key to avoiding problems with zoning. Tell your neighbors about your plans to start a home business so they'll know what to expect and will have the chance to air their concerns. Explain that there are advantages to you working at home—for example, having someone home during the day should improve security for the neighborhood. You might even offer to meet a neighbor's repair people or accept his or her packages.

If any of your neighbors are retired, try to be particularly helpful to them. Retired people who stay at home all day are more likely to complain about a home office than neighbors who work during the day.

On the other hand, if your relations with your neighbors are already shaky or you happen to be surrounded by unreasonable people, you may be better off not telling them you work at home. If you're inconspicuous, they may never know what you're doing.

### d. If your neighbors complain

If your neighbors complain about your home office, you'll probably have to deal with your local zoning bureaucracy. If local zoning officials decide you should close your home business, they'll first send you a letter ordering you to do so. If you ignore this and any subsequent letters, they may file a civil lawsuit against you seeking an injunction—that is, a court order prohibiting you from violating the zoning ordinance by operating your home business. If you violate such an injunction, a judge can fine you or even put you in jail.

Don't ignore the problem; it won't go away. Immediately after receiving the first letter from zoning officials, talk with the person at city hall who administers the zoning law—usually someone in the zoning or planning department. City officials may drop the matter if you'll agree to make your home business less obtrusive.

If this doesn't work, you can apply to your planning or zoning board to grant a variance allowing you to violate the zoning ordinance. To obtain such a variance, you'll need to show that your business does no harm to your neighborhood and that relocation would deprive you of your livelihood. Be prepared to answer the objections of unhappy neighbors who may be at the planning commission meeting loudly objecting to the proposed variance.

You can also try to get your city council or zoning board to change the local zoning ordinance. You'll probably have to lobby some city council members or planning commissioners. It will be useful to enlist the support of the local chamber of commerce and other business groups. Try to enlist the support of your neighbors as well—for example, have as many of them as possible sign a petition favoring the zoning change. Many people with home offices are organizing on local, state and national levels to lobby for new zoning laws permitting home offices.

Two national membership organizations pushing for changes that may be able to give you advice.

- American Association of Home-Based Businesses, P.O. Box 10023, Rockville, MD 20849-0023; 800-447-9710, fax: 301-963-7042, Internet: www.aahbb.org, and
- Home Office Association of America, 909 Third Avenue, Suite 990, New York, NY 10022; 800-809-4622, fax: 800-315-4622, Internet: www.hoaa.com.

Finally, you can take the matter to court, claiming that the local zoning ordinance is invalid or that the city has misinterpreted it. You'll probably need the help of a lawyer familiar with zoning matters to do this. (See Chapter 21.)

### Home-Based Entrepreneur Fights City Hall—And Wins

Many self-employed people have successfully fought decisions that their home businesses violate local zoning laws. For example, Judy ran a theatrical costume business from her home in Los Angeles for several years. She never received any complaints from her neighbors about her business, but someone with a grudge against her alerted the Los Angeles Building Department that Judy was operating a home business in violation of the Los Angeles zoning ordinance.

The Building Department sent an inspector to look at her property. He handed her an order on the spot requiring her to stop doing business at home. Judy decided to fight rather than switch her office location. She applied for a zoning variance from the Los Angeles Planning Commission, but was turned down after a hearing.

Judy refused to take no for an answer and appealed to the Los Angeles Zoning Appeals Board. She went to work to make sure she was prepared for the appeal. She enlisted the support of her neighbors. She was head of the Neighborhood Watch—a group of volunteers dedicated to keeping the locale safe—so she already knew most of them. Nearly 140 neighbors signed a petition urging the board to permit her home office; several also wrote letters to the board.

Judy drew a map of her property and took pictures which helped show that her home business did not disrupt the neighborhood. She also conducted a neighborhood traffic survey which showed that her business did not increase traffic in the neighborhood substantially.

The appeals board was so impressed by her evidence that it granted one of the first variances ever given to a home office in Los Angeles.

In the most significant victory for home-based business people to date, the City of Los Angeles has since amended its local zoning laws to permit most home businesses.

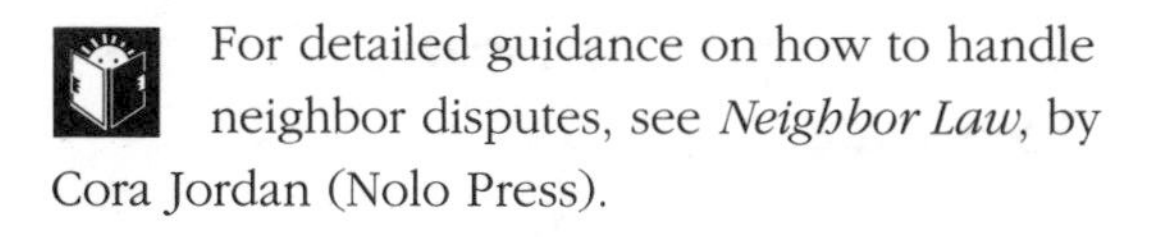
For detailed guidance on how to handle neighbor disputes, see *Neighbor Law*, by Cora Jordan (Nolo Press).

## 2. Private Land Use Restrictions

The government uses zoning laws to restrict how you can use your property. There can also be private restrictions on how you can use your home.

Depending on the part of the country in which you live, use restrictions are commonly found in:

- property deeds
- homeowner association rules, and
- leases.

### a. Property deed restrictions

Property deeds often contain restrictions, called restrictive covenants, limiting how you can use your property. Restrictive covenants often bar or limit the use of home offices.

You can find out if your property is subject to such restrictions by reading your title insurance policy or checking your deed. If your neighbors believe you're violating these restrictions, they can take court action to stop you. Such restrictions are usually enforced by the courts unless they are unreasonable or the character of the neighborhood has changed so much since they were written that it makes no sense to enforce them.

**RESTRICTIONS STRICTLY ENFORCED**

Sheldon and Raye Isenberg, both psychiatrists, purchased a home in an Illinois subdivision. Their deed contained a restrictive covenant providing that: "No lot shall be used except for single residential purposes."

Sheldon and Raye together saw about 30 patients per week at their home. However, there seemed to be little or no disruption of the neighborhood. The patients never came late at night or on the weekends and parked in the Isenbergs' driveway, not on the street. No one with a criminal or drug use record or who might endanger others was permitted to come to the Isenbergs' home.

The only problem that ever occurred was when two patients mistook a neighbor's house for the Isenbergs' and the neighbor spotted them in his yard. Nevertheless, several of the Isenbergs' neighbors took them to court, claiming they violated the express words of the restrictive covenant. The judge found that the restrictive covenant was valid and ordered the Isenbergs to discontinue using their home as an office. (*Wier v. Isenberg*, 420 N.E.2d 790 (Ill. App. 1981).)

### b. Homeowners' association rules

One in six Americans lives in a planned community with a homeowners' association. When you buy property in such a development, you automatically become a member of the homeowners' association and become subject to its rules, which are usually set forth in a lengthy document called Covenants, Conditions and Restrictions, or CC&Rs for short. CC&Rs often regulate what you can do on, in and to your property in minute detail. The homeowners' association is in charge of modifying and enforcing these rules.

The CC&Rs for many developments specifically bar home offices. The homeowners' association may be able to impose fines and other penalties against you if your business violates the rules. It could also sue you in court to get money damages or other penalties. Some homeowners' associations are very strict about enforcing their rules against home businesses, others are much less so.

Carefully study the CC&Rs before you buy into a condominium, planned developed or cooperative to see if home offices are prohibited. If so, you may want to buy somewhere else.

If you're already in a development that bars home offices, you may be able to avoid problems if you're unobtrusive and your neighbors are unaware you have a home office. However, the best course may be to seek to change the CC&Rs. Most homeowner associations rule through a board of directors whose members are elected by all the members of the association. Lobby members of the board about changing the rules to permit home offices. If that fails, you and like-minded neighbors could try to get seats on the board and gain a voice in policymaking.

#### c. Lease restrictions

If you're a renter, check your lease before you start your home business. Many standard lease forms prohibit tenants from conducting a business on the premises—or prohibit certain types of businesses. Your landlord could evict you if you violate such a lease provision. However, most landlords don't want to evict their tenants. Most don't care what you do on your premises as long as it doesn't disturb your neighbors or cause damage. Keep up good neighbor relations to prevent complaints. (See Section C.)

However, if you have business visitors, your landlord may require you to obtain liability insurance in case a visitor has an accident such as a trip or fall on the premises. (See Chapter 6.)

## D. Deducting Your Home Office Expenses

If you elect to work from home, the federal government is prepared to help you out by allowing you to deduct your home office expenses from your taxes. This is so whether you own your home or apartment or are a renter. Although this tax deduction is commonly called the home office deduction, it is not limited to home offices. You can also take it if, for example, you have a workshop or studio at home.

If you've heard stories about how difficult it is to qualify for the home office deduction and felt it wasn't worth the trouble involved, you can breathe more easily. Changes in the tax law that take effect in 1999 make it much easier for many self-employed people to qualify for the deduction. Even if you haven't qualified for the deduction in the past, you may be entitled to it in 1999 and future years.

Because some people still believe the home office deduction is an audit flag for the IRS and preach that fear loudly, many self-employed people who may qualify for it are afraid to take it. Although taking the home office deduction might increase your chance of being audited, the chances are still relatively small. Also, you have nothing to fear from an audit if you're entitled to the deduction.

However, if you intend to take the deduction, you should also take the effort to understand the requirements and set up your home office so as to satisfy them. Before you start moving your furniture around, read this section.

**SIDELINE BUSINESS MAY QUALIFY FOR OFFICE DEDUCTION**

You don't have to work full-time in a business to qualify for the home office deduction. If you satisfy the requirements, you can take the deduction for a sideline business you run from a home office. However, the total amount you may deduct cannot exceed your income from the business. (See Section D.)

**EXAMPLE:** Barbara works full-time as an editor for a publishing company. An avid bowler, she also spends about 15 hours a week writing and publishing a bowling newsletter. She does all the work on the newsletter from an office in her apartment. Barbara may take the home office deduction. But she can't deduct more than she earns as income from the newsletter.

### 1. Regular and Exclusive Business Use

You can't take the home office deduction unless you regularly use part of your home exclusively for a trade or business.

Unfortunately, the IRS doesn't offer a satisfactory definition of regular use to guide you. The agency has decreed only that you must use a portion of your home for business on a continuing basis—not just for occasional or incidental business. You'll likely satisfy this test if you use your home office a few hours each day.

Exclusive use means that you use a portion of your home *only* for business. If you use part of your home as your business office and also use that part for personal purposes, you cannot meet the exclusive use test and cannot take the home office deduction.

**EXAMPLE:** Johnny, an accountant, has a den at home furnished with a desk, chair, bookshelf, filing cabinet and bed. He uses the desk and chair for both business and personal reasons. The bookshelf contains both personal and business books. And the filing cabinet contains both personal and business files. Johnny can't claim a business deduction for the den since it is not used exclusively for business purposes.

You needn't devote an entire separate room in your home to your business. But some part of the room must be used exclusively for business.

**EXAMPLE:** Johnny the accountant keeps his desk, chair, bookshelf and filing cabinet in one part of his den and uses them exclusively for business. The remainder of the room—one-third of the space—is used to store a bed for house guests. Johnny can take a home office deduction for the two-thirds of the room used exclusively as an office.

As a practical matter, the IRS isn't going to make a surprise inspection of your home to see whether you're complying with these requirements. You can always shift your furniture around if you get audited. However, complying with the rules from the beginning avoids having to lie to the IRS if you are audited.

This means, simply, that you'll have arrange your furniture and belongings so as to devote a portion of your home exclusively to your home office. The more space you use exclusively for business, the more your home office deduction will be worth. (See Section D.)

Although not explicitly required by law, it's a good idea to physically separate the space you use for business from the rest of the room. For example, if you use part of your living room as an office, separate it from the rest of the room with room dividers or bookcases.

### 2. Qualifying for the Deduction

Unfortunately, satisfying the requirement of using your home office regularly and exclusively for business is only half the battle.

It must also be true that:

- your home office is your principal place of business

- you meet clients or customers at home, or
- you use a separate structure on your property exclusively for business purposes.

**Ways to Solidify Home Office Deduction**

Here are some ways to help convince the IRS you qualify for the home office deduction.

- Take a picture of your home office and draw up a diagram showing your home office as a portion of your home.
- Have all your business mail sent to your home office.
- Use your home office address on all your business cards, stationery and advertising.
- Obtain a separate phoneline for your business and keep that phone in your home office. The tax law helps you do this by allowing you to deduct the monthly fee for a second phone line in your home if you use it for business. You can't deduct the monthly fee for a single phone line, even if you use it partly for business; however, you can deduct the cost of business calls you place from that line. Having a separate business phone will also make it easier for you to keep track of your business phone expenses.
- Encourage clients or customers to regularly visit your home office and keep a log of their visits.
- To make the most of the time you spend in your home office, communicate with clients by phone, fax or electronic mail instead of going to their offices. Use a mail or messenger service to deliver your work to customers.
- Keep a log of the time you spend working in your home office. This doesn't have to be fancy; notes on your calendar will do.

## 3. Home as Principal Place of Business

The most common way to qualify for the home office deduction is to use your home as your principal place of business. If, like many self-employed people, you do all or most of your work in your home office, it is clearly your principal place of business—for example, you're a writer and you do most of your writing in your home office, or you sell by phone and make most of your sales calls from home.

Things can get more complicated, however, if you spend a substantial amount of time working outside your home. Fortunately, the rules have been changed to make it much easier for your home office to qualify as your principal place of business. However, these changes don't take effect until 1999. So, if you're using this book in 1998, be sure to read the next section. You might not qualify for the deduction in 1998, but you may very well qualify for it in 1999 and later years.

### a. Rules for 1998 and earlier

The rules for 1998 and earlier are quite restrictive if you work a lot outside your home office. If this is true, your home office will be your principal place of business only if it is where you do most of your important business or spend most of your working hours—that is, both the amount of time you spend in your home office and the importance of the work you do there are important concerns.

There is some help for you in evaluating whether a workspace you use in this way can qualify for a deduction.

First, compare the business activities you perform at your home office with those you perform outside your home office to determine which ones are more important for your business. Activities that directly help you make money are most important; those that don't directly lead to earning income are less important. For example, if you're a salesperson, calling on customers and making sales are your most important activities. Without sales, you earn no income. Recordkeeping is less important for a salesperson since it doesn't directly lead to earning income.

If you must meet with clients or patients outside your office or if you deliver services or goods to customers, your home office will likely not qualify as your principal place of business. For example, if you're a salesperson and make your sales by calling on customers in their offices, you probably cannot claim that your home office is your principal place of business.

It may not be clear which location is most important for your business—if, for example, you perform income-producing services for your clients both at your home office and their premises. Compare the amount of time you spend on business in your home office with the amount of time you spend elsewhere. If you spend the majority of your time in your home office, it may qualify as your principal place of business.

**EXAMPLE:** Joe, a self-employed salesperson, makes sales both by phone from his home office and by visiting customers at their premises. As a result, he does equally important work both in his home office and outside it. However, Joe spends 80% of his time making sales from home and only 20% on the road. Since Joe spends the majority of his time in his home office, it qualifies as his principal place of business.

### b. Rules for 1999 and later

Starting in 1999, it will be far easier to satisfy the principal place of business requirement. Indeed, it's likely that any self-employed person with a home office can satisfy it.

Under the new rules, a home office qualifies as your principal place of business if:

- you use the office to conduct administrative or management activities for your business, and
- there is no other fixed location where you conduct such activities.

What this means is that, to qualify for the home office deduction, your home office need no longer be the place where your generate most of your business income. It's sufficient that you regularly use it to administer or manage your business—for example, keep your books, schedule appointments, do research and order supplies. As long as you have no other fixed location where you do such things—for example, an outside office—you'll get the deduction.

**EXAMPLE:** Sally, a handyperson, performs home repair work for clients in their homes. She also has a home office where she uses regularly and exclusively to keep her books, arrange appointments and order supplies. Beginning in 1999, Sally will be entitled to a home office deduction.

This legal change will enable the huge numbers of self-employed people who work primarily outside of their homes to take the home office deduction—for example:

- construction workers who work primarily on construction sites
- doctors who work primarily in hospitals
- traveling salespeople who visit clients at their place of business, and
- house painters, gardeners and home repair people who work primarily in their customers' homes.

Because of these new rules, most self-employed people will qualify for the home office deduction. All you have to do is set up a home office that you regularly use to manage or administer your business. Even if you spend most of your work time away from home, you can still find plenty of business-related work to do in your home office.

## 4. Meeting Clients or Customers at Home

Even if your home office is not your principal place of business, you may deduct your expenses for the part of your home used exclusively to meet with clients, customers or patients. You must physically meet with others at home; phoning them from home is not sufficient. And the meetings must be a regular part of your business; occasional meetings don't qualify.

There is no numerical standard for how often you must meet clients at home for those meetings to be considered regular. However, the IRS has indicated that meeting clients one or two days a week is sufficient. Exclusive use means you use the space where you meet clients only for business. You are free to use the space for business purposes other than meeting clients—for example, doing your business bookkeeping or other paperwork. But you cannot use the space for personal purposes such as watching television.

**EXAMPLE:** June, an attorney, works three days a week in her city office and two days in her home office which she uses only for business. She meets clients at her home office at least once a week. Since she regularly meets clients at her home office, it qualifies for the home office deduction. This is so even though her city office is her principal place of business.

If you want to qualify for this deduction, encourage clients or customers to visit you at home. Keep a log or appointment book showing all their visits.

## 5. Using a Separate Structure for Business

You can also deduct expenses for a separate free-standing structure such as a studio, garage, or barn if you use it exclusively and regularly for your business. The structure does not have to be your principal place of business or a place where you meet patients, clients or customers.

As always where the home office deduction is involved, exclusive use means you use the structure only for business—for example, you can't use it to store gardening equipment or as a guest house. Regular use is not precisely defined, but it's probably sufficient for you to use the structure 10 or 15 hours a week.

**EXAMPLE:** Deborah is a freelance graphic designer. She has her main office in an industrial park, but also works every weekend in a small studio in her back yard. Since she uses the studio regularly and exclusively for her design work, it qualifies for the home office deduction.

### Storing Inventory or Product Samples at Home

You can take the home office deduction if you're in the business of selling retail or wholesale products and you store inventory or product samples at home.

To qualify, you can't have an office or other business location outside your home. And you must store your inventory at a particular place in your home—for example, a garage, closet or bedroom. You can't move your inventory from one room to the other. You don't have to use the storage space exclusively to store your inventory to take the deduction. It's sufficient that you regularly use it for that purpose.

**EXAMPLE:** Janet sells costume jewelry door to door. She rents a home and regularly uses half of her attached garage to store her jewelry inventory and also uses it park her Harley Davidson motorcycle. Janet can deduct the expenses for the storage space even though she does not use her garage exclusively to store inventory. Her garage accounts for 20% of the total floor space of her house. Since she uses only half of the garage for storing inventory, she may deduct one half of this, or 10%, of her rent and certain other expenses. (See Section D.)

## 6. Amount of Deduction

To figure out the amount of the home office deduction, you need to determine what percentage of your home is used for business. To do this, divide the square footage of your home office by the total square footage of your home. For example, if your home is 1,600 square feet and you use 400 square feet for your home office, 25% of the total area is used for business.

Or if all the rooms in your home are about the same size, figure the business portion by dividing the number of rooms used for business by the number of rooms in the home. For example, if you use one room in a five-room house for business, 20% of the area is used for business.

The home office deduction is not one deduction, but many. You are entitled to deduct from your gross income your home office use percentage of:

- your rent if you rent your home, or
- depreciation, mortgage interest and property taxes if you own your home.

In addition, owners and renters may deduct this same percentage of other expenses for keeping up and running an entire home. The IRS calls these indirect expenses.

They include:

- utility expenses for electricity, gas, heating oil and trash removal
- homeowner's or renter's insurance
- home maintenance expenses that benefit your entire home including your home office, such as roof and furnace repairs and exterior painting
- condominium association fees
- snow removal expenses
- casualty losses if your home is damaged—for example, in a storm, and
- security system costs.

You may also deduct the entire cost of expenses just for your home office. The IRS calls these direct expenses. They include, for example, painting your home office or paying someone to clean it. If you pay a housekeeper to clean your entire house, you may deduct your business use percentage of the expense.

> **EXAMPLE:** Jean rents a 1,600 square foot apartment and uses a 400 square foot bedroom as a home office for her consulting business. Her percentage of business use is 25% (400 divided by 1,600). She pays $12,000 in annual rent and has a $1,200 utility bill for the year. She also spent $200 to paint her home office. She is entitled to deduct 25% of her rent and utilities and the entire painting expense for a total home office deduction of $3,500.

Be sure to keep copies of all your bills and receipts for home office expenses—for example, keep:

- IRS Form 1098 sent by whoever holds your mortgage showing the interest you paid on your mortgage for the year
- property tax bills and your canceled checks
- utility bills, insurance bills, and receipts for payments for repairs to your office area along with your canceled checks paying for these items, and
- a copy of your lease and your canceled rent checks if you're a renter.

The home office deduction can be very valuable if you're a renter because you get to deduct part of your rent—a substantial expense that is ordinarily not deductible. If you own your home, the home office deduction is worth less because you're already allowed to deduct your mortgage interest and property taxes just because you're a homeowner.

Taking the home office deduction won't increase your income tax deductions for these items, but it will allow you to deduct them from your self-employment taxes. You'll save $153 in self-employment taxes for every $1,000 in mortgage interest and property taxes you deduct. You'll also be able to deduct a portion of repairs, utility bills, cleaning and maintenance costs and depreciation.

**Depreciating Office Furniture and Other Personal Property**

Whether or not you qualify for or take the home office deduction, you can depreciate or expense under Section 179 the cost of office furniture, computers, copiers, fax machines, and other personal property you use for your business and keep at home. These costs are deducted directly on your Schedule C, Profit or Loss From Business. They don't have to be listed on the special tax form used for the home office deduction. (See Section D.)

If you use the property for both business and personal reasons, the IRS requires you to keep records showing when the item was used for business and when for personal reasons—for example, a diary or log with the dates, times and reason the item was used. (See Chapter 14, Section A.)

### a. Profit limit for deductions

There is an important limitation on the home office deduction: it may not exceed the net profit you earn from your home office.

Determining this amount can be complex.

First, you have to figure out how much money you earn from using your home office. If you do all your work at home, this will be 100% of your business income. But if you work in several locations, you must determine the part of your gross income that came from working in your home office. To do this, consider the time you spend in your home office and elsewhere and the type of work you do in each location.

You then subtract from this amount:

- the business percentage of your otherwise deductible mortgage interest and real estate taxes; you'll only have these expenses if you own your home, and
- business expenses that are not attributable to the business use of your home—for example, supplies, depreciation of business equipment, business phone, advertising, salaries.

You can carry over any excess in home office deductibles and deduct them in the first year in which your business earns a profit. However, whether or not your business incurs a loss, you can still deduct all your home mortgage interest and property taxes because you're a homeowner.

**EXAMPLE:** Sam runs a part-time consulting business out of his home office, which occupies 20% of his home. In one recent year, his gross income from the business was $6,000. He does all his consulting work at home, so this entire amount is attributable to his home office. He determines how much of his home office expenses he may deduct as follows:

- First, he subtracts 20% of his home mortgage interest and property taxes from his $6,000 gross income. This is $3,000, so he has $3,000 left.
- Next, he subtracts his business expenses other than for the use of his home office. These amount to $2,000, so he is left with $1,000.

Sam may only deduct $1,000 of home office expenses. These expenses totaled $2,000 for the year, so Sam has $1,000 left over that he may not deduct for the year. He may deduct this amount next year if he has sufficient income from his business.

### b. Special concerns for homeowners

If you're not careful when you take the home office deduction, you may have to pay extra taxes when you sell your home.

You normally pay no tax on any profit you earn when you sell your home if you buy or build a more expensive replacement residence within two years. However, if you take the home office deduction, this rule does not apply to the portion of your house you use for business. Instead, your old house is treated as two separate properties for tax purposes.

You'll have to pay income tax on the profit you earn from the portion of your house used as a home office. For example, if 20% of your house was used as a home office, you'd have to pay tax on 20% of the profit you earn when you sell your home. If your home has gone up in value dramatically since you bought it, you'll have a huge tax bill.

To avoid this tax trap, before January 1 of the year you plan to sell your house, you must either:

- move out of your home office and rent an outside office, or
- stop using your home office exclusively for business—for example, move in a couch, bed, television or exercise machine and even let your spouse or children use the space.

If you want to be able to move on short notice without worrying about taxes, don't take the home office deduction.

### 7. IRS Reporting Requirements

All unincorporated taxpayers who take the home office deduction must file IRS Form 8829 with their tax returns. Renters who take the deduction must also file Form 1099-MISC with the IRS.

#### a. IRS Form 8829

If you qualify for the home office deduction and are a sole proprietor or partner in a partnership, you must file IRS Form 8829, Expenses for Business Use of Your Home, along with your personal tax return. The form alerts the IRS that you're taking the deduction and shows how you calculated it. You should file this form even if you're not allowed to deduct your home office expenses because your business has no profits. (See Section D.) By filing, you can apply the deduction to a future year in which you earn a profit.

If you organize your business as an S corporation instead of a sole proprietorship or partnership, you don't have to file Form 8829. (See Chapter 2, Section B.)

For additional information, see IRS Publication 587, *Business Use of Your Home*. You can obtain this and all other IRS publications by calling the IRS at 800-TAX-FORM, visiting your local IRS office or downloading the publications from the IRS Internet site at www.irs.ustreas.gov.

#### b. Filing requirement for renters

If you're a renter and take the home office deduction, you should file an IRS Form 1099-MISC each year reporting the amount of your rental payments attributable to your home office.

> **EXAMPLE:** Bill rents a house and takes the home office deduction. He spends $12,000 per year on rent and uses 25% of his house as a home office. He should file a Form 1099 reporting $3,000 of his rental payments.

You file three copies of the 1099.

- File one copy with the IRS by February 28.
- Give one copy to your landlord by January 31.
- File one copy with your state tax department if your state has income taxes. (See Chapter 9.)

Your landlord may not appreciate receiving a 1099 from you, but it will definitely be helpful if you're audited by the IRS and your home office deduction questioned. It helps to show that you were really conducting a business out of your home.

The only time a 1099 is not required is where your landlord is a corporation. Form 1099 need not be filed for payments to corporations. A 1099 is also not required in the unlikely event that your rental payments for your home office total less than $600 for the year. ■

CHAPTER

5

# Obtaining Licenses, Permits and Numbers

Once you've decided how to organize and name your business, you'll need to obtain any necessary licenses, permits and numbers. This can be a bit of a pain and require paperwork and paying some fees, but it's worth it. You can suffer fines and other penalties if you don't satisfy the government requirements.

Also, having all required business licenses helps you look like an independent businessperson instead of an employee. Some potential clients or customers may even ask for copies of your licenses, permits or numbers before agreeing to hire you because they know this will help them if they're audited by the IRS or other government agencies.

## A. Business Licenses

Whether and what business licenses or permits are required depends on the kind of work you do and where you do it. Licenses or permits may be required from the federal, state and local governments. Professional organizations, other self-employed people and your local chamber of commerce may all be able to give you information on licensing requirements for your business.

### 1. Federal Licenses and Permits

The federal government doesn't require licenses or permits for most small businesses. One notable exception, however, is trucking. Trucking companies must be licensed by the Interstate Commerce Commission. License requirements are also imposed on investment advisors by the Securities and Exchange Commission.

### 2. State Requirements

A few states, for example, Alaska and Washington, require all businesses to obtain state business licenses. These are required in addition to local licenses.

Most states don't issue or require general business licenses. However, all states require special licenses for people who work in certain occupations. Doctors, lawyers, architects, nurses and engineers must be licensed in every state. Most states require licenses for other occupations that require extensive training or where consumers need to be protected from fraud or potential hazards—for example, most states license barbers, bill collectors, building contractors, tax preparers, insurance agents, cosmetologists, real estate agents or brokers and auto mechanics. Your state may require licenses for other occupations, too.

Procedures for obtaining a license vary from state to state and occupation to occupation. You may have to meet specific educational requirements or have training or experience in the field. You may even have to pass a written examination. Of course, you'll have to pay a license fee. Some states may also require that you have liability insurance before you can be issued a license. (See Chapter 6.)

If your state government discovers that you're doing business without a required license, various bad things can happen to you. You'll undoubtedly be ordered to stop doing business. You may also be fined. And depending on your occupation, failure to obtain a license may be a crime—a misdemeanor or even a felony.

Many states have agencies designated to help businesses get started. This is the first place to call to obtain information on your state's license requirements. These agencies often have free or inexpensive publications that discuss licensing requirements. For example, the California Department of Commerce publishes *The California License Handbook* and *California Permit Handbook.* (Appendix 1 contains a list of these state agencies.)

Also, many state agencies now have sites on the Internet which may contain information about licensing requirements. A good place to start an Internet search for such information is The State Web Locator maintained by Villanova University.

The Internet address is www.law.vill.edu/State-Agency/index.html.

### 3. Local Requirements

Many cities, counties and municipalities require business licenses or permits for all businesses—even one-person home-based operations. Usually, you just have to pay a fee to get such licenses; they are simply a tax in disguise. Other cities have no license requirements at all, or exempt very small businesses.

If you're doing business within a city limits, you'll need to get a license from the city. If you're in an unincorporated area, you'll need to contact your county government. If you're doing business in more than one city or county, you may have to get a license in each city or county.

To find out what to do, call the appropriate local official in charge of business licensing. This is often the city or county clerk, planning or zoning department, city tax office, building and safety department or public works department. More than one local license may be needed, so you may have to deal with more than one local agency. Your local chamber of commerce may be able to direct you to the agency or person to contact.

To obtain a license, you'll be required to fill out an application and pay a fee. Fees vary from locality to locality—from as little as $15 or several hundred dollars. Fees are often based on your projected gross revenues—for example, 10 cents per $1,000 of revenue. You'll be required to renew your license and pay a new fee, usually every year. It's also likely that you'll be required to post your license at your place of business.

Many self-employed people, particularly those who work at home, never bother to get a local business license. If your local government discovers you're running an unlicensed business, it may fine you and bar you from doing business until you obtain a license.

**Problems for the Home-Based Self-Employed**

If you work at home, be careful about applying for a local business license. You'll have to provide your business address to obtain one. Before granting a license, many cities first check to see whether the area in which your business is located is zoned for business. If your local zoning ordinance bars home offices in your neighborhood, you could be in for trouble. (See Chapter 4.)

## B. Obtaining an Employer Identification Number

A federal employer identification number, or EIN, is a nine-digit number the IRS assigns to businesses for tax filing and reporting purposes. The IRS uses the EIN to identify the taxpayer.

### 1. When an EIN Is Required

EINs are free and easy to obtain. Use your EIN on all business tax returns, checks and other documents you send to the IRS. Your state tax authority may also require your EIN on state tax forms.

#### a. Sole proprietors

If you're a sole proprietor, you must have an EIN if you hire any employees, including household employees. Even if you have no employees, you must get an EIN if you have a Keogh retirement plan. (See Chapter 16.) Also, some banks require you to have an EIN before they'll set up a bank account for your business.

However, if, like most self-employed people, you're a sole proprietor and don't have any employees or a Keogh plan, you don't have to obtain an EIN; you can use your Social Security number instead. Note that sole proprietors without employees are permitted to obtain EINs and you should do so. Using an EIN instead of your

Social Security number on your tax returns and payments helps to show that you're an independent businessperson—in other words, an independent contractor and not an employee.

#### b. Corporations, partnerships and limited liability companies

You must have an EIN if you form a corporation, partnership or limited liability company. An EIN is also required if you were formerly a sole proprietor and form any of these entities.

### 2. Obtaining an EIN

You can obtain an EIN by filing IRS Form SS-4, Application for Employer Identification Number, with the IRS. Filling out the form is simple and the SS-4 form has detailed instructions.

Note these possible trouble spots.

- **Space 1:** List your full legal name if you're a sole proprietor. If you've incorporated, list the corporation's name—the name on your articles of incorporation or similar document establishing your corporation.
- **Space 7:** Leave this space blank if you're a sole proprietor.
- **Space 11:** For most self-employed people, the closing month of the tax year is December. (See Chapter 8, Section A.)
- **Space 12:** If you don't plan to hire any employees, enter "N/A" in this space.

#### a. Applying by mail

You can obtain your EIN by mailing the completed SS-4 to the appropriate IRS service center listed in the form's instructions. The IRS will mail the EIN to you in about a month.

#### b. Applying by phone

If you need an EIN right away, you can get it over the phone by using the IRS's Tele-TIN program. Here's what to do.

- Fill out the SS-4 form.
- Call the IRS at the number shown in the SS-4 instructions; an IRS representative will take the information off your SS-4 and assign you an EIN which you can start using immediately.
- Write your EIN in the upper right-hand corner of the SS-4 and sign and date the form.
- Mail or fax the signed SS-4 within 24 hours to the Tele-TIN unit at the IRS service center address for your state; the addresses are provided in the SS-4 instructions or the IRS representative with whom you speak will give you the fax number.

## C. Sales Tax Permits

Almost all states and many municipalities impose sales taxes of some kind. The only states without sales tax are Alaska, Delaware, Montana, New Hampshire and Oregon.

In some states, sales tax is imposed on sellers, who then have the option of passing the tax along to their purchasers. In other states, the tax is imposed directly on the purchaser, and the seller is responsible for collecting the tax and remitting it to the state. In a few of the states, sellers and purchasers share sales tax.

### 1. Selling Products or Services

If you sell tangible personal property to the public—things you can hold in your hand—you'll undoubtedly have to pay sales taxes. All states that have sales taxes impose them on sales of goods or products to end users.

On the other hand, if you only provide services to clients or customers—that is, you don't sell or transfer any type or personal property—you probably don't have to worry about sales taxes because most states don't tax services at all or only tax certain specified services. Notable exceptions are Hawaii, New Mexico and South Da-

## IRS FORM SS-4

Form **SS-4** (Rev. December 1995) Department of the Treasury Internal Revenue Service

**Application for Employer Identification Number**

**(For use by employers, corporations, partnerships, trusts, estates, churches, government agencies, certain individuals, and others. See instructions.)**

**▶ Keep a copy for your records.**

EIN

OMB No. 1545-0003

Please type or print clearly.

1 Name of applicant (Legal name) (See instructions.)
Ted Anderson

2 Trade name of business (if different from name on line 1)
The Poster Warehouse

3 Executor, trustee, "care of" name

4a Mailing address (street address) (room, apt., or suite no.)
555 Main Street

5a Business address (if different from address on lines 4a and 4b)

4b City, state, and ZIP code
Ann Arbor, MI 48104

5b City, state, and ZIP code

6 County and state where principal business is located
Washtenaw

7 Name of principal officer, general partner, grantor, owner, or trustor—SSN required (See instructions.) ▶ 555-55-5555
Ted Anderson

8a Type of entity (Check only one box.) (See instructions.)
[X] Sole proprietor (SSN) 555 : 55 : 5555
[ ] Partnership
[ ] REMIC
[ ] State/local government
[ ] Other nonprofit organization (specify) ▶ ______ (enter GEN if applicable) ______
[ ] Other (specify) ▶
[ ] Personal service corp.
[ ] Limited liability co.
[ ] National Guard
[ ] Estate (SSN of decedent) ______
[ ] Plan administrator-SSN ______
[ ] Other corporation (specify) ▶
[ ] Trust
[ ] Federal Government/military
[ ] Farmers' cooperative
[ ] Church or church-controlled organization

8b If a corporation, name the state or foreign country (if applicable) where incorporated | State | Foreign country

9 Reason for applying (Check only one box.)
[X] Started new business (specify) ▶ Consulting
[ ] Hired employees
[ ] Created a pension plan (specify type) ▶
[ ] Banking purpose (specify) ▶
[ ] Changed type of organization (specify) ▶
[ ] Purchased going business
[ ] Created a trust (specify) ▶
[ ] Other (specify) ▶

10 Date business started or acquired (Mo., day, year) (See instructions.)
August 1, 19xx

11 Closing month of accounting year (See instructions.)
December

12 First date wages or annuities were paid or will be paid (Mo., day, year). **Note:** *If applicant is a withholding agent, enter date income will first be paid to nonresident alien. (Mo., day, year)* . . . . ▶ September 1, 19xx

13 Highest number of employees expected in the next 12 months. **Note:** *If the applicant does not expect to have any employees during the period, enter -0-. (See instructions.)* . . . ▶

| Nonagricultural | Agricultural | Household |
|---|---|---|
| 1 | 0 | 0 |

14 Principal activity (See instructions.) ▶ Consulting

15 Is the principal business activity manufacturing? . . . . [ ] Yes [X] No
If "Yes," principal product and raw material used ▶

16 To whom are most of the products or services sold? Please check the appropriate box. [ ] Business (wholesale)
[ ] Public (retail) [ ] Other (specify) ▶ [ ] N/A

17a Has the applicant ever applied for an identification number for this or any other business? . . . . [ ] Yes [X] No
**Note:** *If "Yes," please complete lines 17b and 17c.*

17b If you checked "Yes" on line 17a, give applicant's legal name and trade name shown on prior application, if different from line 1 or 2 above.
Legal name ▶ Trade name ▶

17c Approximate date when and city and state where the application was filed. Enter previous employer identification number if known.
Approximate date when filed (Mo., day, year) | City and state where filed | Previous EIN

Under penalties of perjury, I declare that I have examined this application, and to the best of my knowledge and belief, it is true, correct, and complete.

**Business telephone number (include area code)**

**Fax telephone number (include area code)**
(313) 555-5555

Name and title (Please type or print clearly.) ▶ Ted Anderson, Owner

Signature ▶ Ted Anderson Date ▶ July 14, 19xx

**Note:** *Do not write below this line. For official use only.*

| Please leave blank ▶ | Geo. | Ind. | Class | Size | Reason for applying |
|---|---|---|---|---|---|

**For Paperwork Reduction Act Notice, see page 4.** Cat. No. 16055N Form **SS-4** (Rev. 12-95)

kota—all of which impose sales taxes on all services, subject to certain exceptions.

Determining whether you're selling property or providing a service can be difficult because the two are often involved in the same transaction. For example, a piano tuner may have to replace some piano wire to tune a piano, or a dentist may provide a patient with a gold filling in the process of filling a tooth. In these instances, many state taxing authorities look at the true object of the transaction to determine if sales tax will be assessed. That is, they look at whether the main purpose of the transaction is to provide the consumer with a service or sell property. It seems clear that the main purpose of hiring a piano tuner or dentist is to obtain a service—that is, a tuned piano or filled tooth. The property used to provide the service is incidental.

## 2. Contacting Your State Sales Tax Department

Each state's sales tax requirements are unique. A product or service taxable in one state may be tax-free in another. The only way to know is to contact your state sales tax department to find out if the products or services you provide are subject to sales taxes. (See Appendix 1 for a list of state sales tax offices.) If you don't understand the requirements, seek help from a tax pro. (See Chapter 21.)

## 3. Obtaining a State Sales Tax Permit

If the products or services you provide are subject to sales tax, you'll have to fill out an application to obtain a state sales tax permit. Complete and mail the application before you make a taxable sale. Many states impose penalties if you make a sale before you obtain a sales tax permit. Generally, you pay sales taxes four times a year, but you might have to pay them monthly if you make lots of sales. Be sure to collect all the taxes due because if you fail to do so, you can be held personally liable for the full amount of uncollected tax.

**Keep a Stern Lookout**
States constantly change their sales tax laws, so be on the lookout for changes affecting your business. Professional organizations and your state chamber of commerce can be good sources of information on your state sales taxes. ■

CHAPTER

6

# Insuring Your Business and Yourself

Most employees don't need to worry much about health or liability insurance or insurance for business property; their employers take care of their insurance needs. Unfortunately, this is not the case when you're self-employed. Self-employed people must purchase all their insurance themselves, and they usually need more coverage than employees. Insurance is the single greatest expense for many of the self-employed.

The best time to obtain insurance is when you first become self-employed or even before you quit your job to do so. Insurance is cheapest and easiest to obtain before you have a problem. Don't wait until you become ill, are being sued or have business property damaged or stolen to start thinking about insurance. By then it may be too late. And even if you're able to obtain insurance, it will likely not provide coverage for your pre-existing condition or problem.

For more information on insurance, see:

- *The Buyer's Guide to Business Insurance*, by Don Bury and Larry Heischman (Oasis Press)
- *Health Insurance: A Guide for Artists, Consultants, Entrepreneurs & Other Self-Employed*, by Lenore Janecek (ACA Books)
- *Insuring the Bottom Line*, by David Russell (Merritt Publishing), and
- *Insuring Your Business*, by Sean Mooney (Insurance Information Institute Press).

## A. Health Insurance

Health insurance pays at least part of your doctor, hospital and prescription expenses if you or a family member get sick. When you're self-employed, you have to obtain your own health insurance. Your clients or customers need not and will not provide it for you. Even if you're in perfect health, you should obtain health insurance. Medical costs for even relatively minor illnesses or injuries can be huge.

And if you're uninsured, the reality is that you may have difficulty finding a doctor or hospital willing to treat you. A 1996 study by the Harvard School of Public Health found that 45% of uninsured people had trouble getting adequate healthcare.

Although the laws of most states prevent an insurer from denying you coverage because you or a family member have a pre-existing medical problem, it can still be very difficult for self-employed people to obtain affordable health insurance if you or a family member have a chronic or serious illness. Some people who would like to quit their jobs and become self-employed refrain from doing so because they're afraid they won't be able to get health insurance.

The availability and cost of health insurance depends on many factors, some of which are within your control and others over which you have no control. Among the factors you have no control over are:

- your age and gender
- your health history, and
- where you live.

The most important factors over which you do have control are whether you obtain a group or individual policy and the type of plan you purchase.

### 1. Group and Individual Policies

You can obtain health insurance either through a group or as an individual. Group policies insure all members of the group. Insurers prefer to cover large groups because the risks and administrative costs are spread over many people.

Industry trade associations, professional groups and similar membership organizations can act as groups, so you can often obtain a group policy by joining one of them. There is a professional group or trade association for virtually every occupation. If you don't know of a group you can join, ask other self-employed people in your field. Many organizations have Websites, so if you have Internet access, you may be able to find one by doing a search with a search engine such as Alta Vista at www.altavista.digital.com. Thousands of membership organizations are also listed in the *Directory of Organizations* (Gale Research), which may be available in the reference section of your local library.

You may also be able to obtain health insurance by joining your local chamber of commerce.

Finally, there are national membership organizations for the self-employed that provide insurance, such as the National Association of the Self-Employed, which you can reach at: 800-232-6273, www.nase.org. Other organizations set up specifically to provide health and other insurance benefits to members include the Support Services Alliance, 800-322-3920, www.ssainfo.com; and the Small Business Service Bureau, 800-222-5678.

In addition, several states have formed cooperatives that small business owners may join to obtain group health insurance coverage. These are called Cooperatives for Health Insurance Purchasing or CHIPs for short. To find out if your state offers a CHIP, call the Institute for Health Policy Solutions at: 202-857-0810.

Individual coverage is typically 25% to 30% more expensive than group coverage, and the coverage limits are usually lower than those offered under group coverage. For example, a group health insurance policy often will not have any limit on the total benefits paid during your lifetime, while individual coverage often limits total lifetime benefits to one or two million dollars. With the skyrocketing costs of medical care, you can reach the limit surprisingly quickly if you have a chronic illness.

**PORTABILITY LAW DOESN'T HELP SELF-EMPLOYED**

In late 1996, Congress passed a law making employee health insurance portable—that is, employees who go from one job where they had group health coverage to another company that has health insurance can't be denied coverage solely because of their health status, claims history or medical condition.

The employer's health insurer can now exclude coverage for an employee's pre-existing medical condition for 12 months at most. A pre-existing condition is defined as one for which the employee received treatment within six months before the enrollment date in the new plan. The law took effect July 1, 1997. Unfortunately, the law does nothing for the millions of uninsured Americans or for people in business for themselves—that is, the self-employed.

## 2. Types of Plans

A bewildering array of health insurance plans are available, with a bewildering amount of lingo in their policies. As a self-employed person, you will most likely be able to choose among traditional and managed care plans.

Which plan is best for you depends on how much money you can afford to spend for health insurance, your prior health history, which plans are available in your area and will accept you and how important choice of doctors is to you. It's best to shop around and investigate as many different plans as possible since costs and benefits vary widely.

### a. Traditional plans

The traditional form of health insurance is now becoming increasingly rare. In this type of fee for service plan, you're allowed to go to any doctor or hospital you choose. Either you or your doctor submits a claim to the insurer for reimbursement of the cost. However, the plan will pay only for care that is medically necessary and covered by the plan.

These plans typically require you to pay a deductible and a co-payment—that is, pay a portion of your medical bills out of your own pocket before your insurance kicks in to cover expenses.

The deductible can be anywhere from $100 to several thousand dollars. Ordinarily, the deductible accumulates throughout the calendar year. This means any medical bills you pay from January 1 to December 31 count toward your deductible. Once you have met your deductible, your insurer starts paying benefits.

The co-payment is usually 20%, although 10% or 30% co-payments are not uncommon.

The higher the deductible and co-payment, the lower your premiums will be; but if you get sick, you'll have to pay a substantial amount out of your own pocket. For example, if you have a policy with a $1,000 deductible and a 20% co-payment, you'll have to pay the first $1,000 of your yearly medical expenses yourself; thereafter, the insurer will pay 80% of the cost and you'll have to pay the other 20%. If you incur $5,000 in medical expenses under such a plan, you'd have to pay $1,800 of the cost yourself.

If you have a substantial medical bill—and a serious illness could cost several hundred thousand dollars—making a 20% co-payment will be a real hardship. To avoid this, most insurance plans change their co-insurance percentages to 100% after you've incurred a certain amount of paid expenses for the year. This is called the out of pocket maximum.

### b. Managed care

Managed care has become the norm in the United States. Under a managed care plan, instead of paying for each service you receive separately, your coverage is paid for in advance. Managed care is usually cheaper than a traditional plan, but you get a more limited choice of doctors and hospitals.

- **Health maintenance organizations.** Often called HMOs, these are pre-paid health programs that require you to use doctors and hospitals that are part of the organization. Some HMOs employ their own doctors and run their own hospitals, while others are affiliated with private physicians. With an HMO, you get your medical care for a fixed price. Ordinarily you don't pay a deductible, but you may be charged a small co-payment—$10 or $15—for certain services. However, you can't go to a doctor or hospital outside the plan except in a medical emergency.

  HMOs often offer the lowest premiums around, but these savings come at a price. When you're in an HMO, your primary care doctor is in complete charge of your care. You can't visit a specialist without a referral from another doctor. HMOs often discourage referrals to expensive specialists and your primary doctor's compensation from the HMO may even be reduced if he or she refers you to someone else.

  And many HMOs require you to get prior approval for any treatment you get. They also have detailed guidelines governing your care. For example, HMO rules may not allow you to be given experimental treatments, such as bone marrow transplants, if you get cancer.
- **Preferred provider organization plans.** Also called PPOs, these plans are a cross between traditional fee for service plans and HMOs. PPOs establish networks of doctors and hospitals that agree to provide care at a price that is usually lower than that available outside the plan. You obtain full benefits only if you go to a doctor or hospital within the network. You are permitted go to a doctor or hospital outside

the network, but you'll usually be required to pay a deductible and co-payment.

- **Point of service plans.** Point of service plans are similar to PPOs, except that you ordinarily have a primary care physician who is in charge of your medical care. PPOs ordinarily don't require this.

## 3. Comparing Plans

Comparing health insurance plans can be confusing. Be sure to read the plan literature carefully. If you don't understand something, ask for an explanation. Compare the plans' coverage and costs.

No plan covers everything. HMOs typically provide broader coverage than fee for service plans. Most plans don't cover eyeglasses, hearing aids or cosmetic surgery. However, you may pay for such items with funds from a Medical Savings Account. (See Section 4.) Look carefully to see what medical expenses are covered. Typical expenses to check include:

- inpatient hospital services
- outpatient surgery
- physician visits in the hospital
- office visits
- skilled nursing care
- medical tests and x-rays
- prescription drugs
- mental health care
- drug and alcohol abuse treatment
- home health care visits
- rehabilitation facility care
- physical therapy
- hospice care
- maternity care
- preventive care and checkups, and
- well-baby care

See if the plan excludes pre-existing conditions or specific illnesses.

Next, compare the amount of the premium and the deductible or co-payment for each plan you look at. Just as important as the premium, however, is the maximum amount the plan will pay. You should seek a benefit limit of at least $1 million. Beware of plans that advertise a high benefit ceiling but have a much lower benefit limit per claim or a lower maximum benefit per year.

Finally, see what type of hoops the plan requires you to jump through to get treatment. You may be required to get advance authorization for treatment or obtain a second opinion before you're allowed to have surgery. To save money, more and more traditional plans are imposing these types of restrictions usually associated with managed care.

If you're looking for a new plan, but want to continue with your current doctor, make sure he or she belongs to the plan you're considering. Ask for your doctor's opinion about the various plans before you sign up. Most doctors belong to several plans and can tell you which are easy to deal with and which are Byzantine bureaucracies.

## 4. Medical Savings Accounts

The federal government has instituted a pilot program allowing medical savings accounts, or MSAs. MSAs, which became available in 1997, are designed to be used by self-employed people who purchase health insurance with a high deductible—that is, the amount they must pay themselves before the insurance starts paying benefits. If you have individual coverage, your deductible must be between $1,500 to $2,250. If you have family coverage, the deductible must be between $3,000 to $4,500. Employees who work for small companies—those with 50 or fewer employees—may also establish MSAs.

When you purchase health insurance with such a high deductible, you pay lower premiums. You can use the money you save on premiums or any other money you have to establish an MSA, which is similar to an IRA. You set up an MSA with a bank or other financial institution. Your contributions to the account are 100% tax deductible and you pay no tax on the interest you earn on the money in your account.

### a. Withdrawing funds

If you or a family member becomes sick, you can withdraw your money from the MSA to pay your deductible or any other expenses that the IRS says are deductible as medical expenses. This means you can use your MSA funds not only to pay for doctor's office visits, hospital care and prescriptions, but for many other expenses ordinarily not covered by health insurance such as chiropractic care, dental and orthodontic care, hearing aids and eyeglasses.

Ordinarily, you may not use MSA funds to pay health insurance premiums. However, MSA funds can be used to pay premiums for long-term care insurance and for continuation coverage, such as COBRA coverage. (See Section 5.)

Importantly, you pay no federal tax at all on MSA withdrawals you use to pay medical expenses. Most states recognize MSAs under their own tax laws—that is, you'll likely not have to pay state income tax on MSA contributions or withdrawals.

> **EXAMPLE:** Jane, a self-employed consultant, obtains health insurance coverage with a $2,000 deductible. She sets up an MSA with a bank and deposits $1,000 every year for three years. She deducts each $1,000 contribution from her gross income for the year for income tax purposes. Jane pays no taxes on the interest she earns on the money in her account, which is invested in a money market fund. By the end of three years, she has $4,000 in the account. When Jane becomes ill after the third year and is hospitalized, she withdraws $2,000 from her MSA to pay her deductible. She pays no federal tax on the withdrawal.

However, if you withdraw funds from your MSA to use for something other than medical expenses, you must pay the regular income tax on the withdrawals plus a 15% penalty. For example, if you were in the 28% income tax bracket, you'd have to pay a 43% tax on your withdrawals.

But after you reach age 65, or become disabled, you can withdraw your MSA funds for any reason without penalty. You only pay the regular income tax on the withdrawals if they are used for other than medical expenses.

If you enjoy good health while you have your MSA and don't have to make withdrawals, you may end up having a substantial amount in your account you can withdraw without penalty for any purpose after age 65. In effect, then, an MSA can serve as a supplemental retirement account, much like an additional IRA.

### b. Limit on contributions

Contributions to an MSA are subject to an annual limitation, which is a percentage of your health plan deductible. If you have individual coverage, you may contribute up to 65% of your deductible—a maximum of $1,464 per year. If you have family coverage, you may contribute 75% of your deductible—a maximum of $3,375 each year. This maximum is prorated during the initial year based on the number of months the MSA is effective.

In addition, self-employed people may not contribute more than they earn from their businesses each year.

> **EXAMPLE:** Barry is self-employed, and obtains individual health coverage with a $2,250 deductible, the maximum allowed. He opens an MSA on July 1, 1998. He can contribute $122 per month for the last six months of 1998, or $732. In 1999, he can contribute $122 every month up to $1,464 for the entire year. However, Barry must earn at least as much from his business as he contributes to his MSA each year.

You have until April 15 of the current year to make a tax deductible MSA contribution for the previous year. You can contribute monthly, or make a lump sum contribution. You need not contribute to your MSA every year. It's up to you to decide how much to put in, up to the annual limit.

## c. How to open an MSA

First, you can't have an MSA if you're covered by other health insurance—for example, if your spouse has family coverage for you from his or her job. So you may have to change your existing coverage.

To participate in the MSA program, you need two things:

- a high-deductible health plan, and
- an MSA account.

If you already have a high-deductible health plan that meets the criteria, you can immediately establish an MSA with any bank, insurance company, or financial institution that offers an MSA product. However, most traditional high-deductible health plans don't meet the criteria: either the deductible is too low or the cap on out of pocket expenses is too high. To qualify, the policy must have a cap on the out of pocket expenses you must pay of $5,000 for families and $3,000 for singles. Check your policy carefully to see it meets these requirements.

An MSA must be established with a trustee. Any person, insurance company, bank or financial institution already approved by the IRS to be a trustee or custodian of an IRA is approved automatically as an MSA trustee. Others have applied and been approved under IRS procedures for MSAs.

If you're not insured by a high-deductible health plan, you must enroll in such a health plan before setting up an MSA. You may obtain coverage from an HMO, PPO or traditional plan. Some insurers administer both the health plan and the MSA; others have a bank or other financial institution handle the MSA.

Whoever administers your account will usually give you a checkbook or debit card to use to withdraw your funds from the account.

Over 50 insurers are now offering MSAs—including Blue Cross and Blue Shield in many states. One company, Golden Rule Insurance Co., presently controls over half the market; it can be reached at: 800-589-8911; on the Internet at: www.medicalsavings.com. Other companies offering MSAs include:

- American Health Value, 800-914-3248
- Medical Savings of America, 800-853-7321, and
- Time Insurance, 800-800-1212.

You can find a list of many other insurers offering MSA coverage on the American Medical Association Internet site at www.ama-assn.org/adcom/msas.htm. Another list can be found at the Health Insurance Association of America site at www.hiaa.org/consumerinfo/medical.html.

Look at the plans offered by several companies to see which offers the best deal. Compare how much the set-up and service fees are, what the account is invested in and whether you're allowed to move your MSA in the future to another trustee without losing your coverage.

## d. Advantages for the self-employed

MSAs represent the most radical change in healthcare financing since World War II and are a real boon for the self-employed. You can purchase a health plan with a high deductible, pay significantly lower premiums and have the security of knowing you can dip into your MSA if you get sick and must pay the deductible or other uncovered medical expenses.

But if you don't tap into the money, it will keep on accumulating free of taxes. You also get the benefit of deducting 100% of your MSA contributions from your taxes. Currently, only a portion of health insurance premiums are deductible. (See Chapter 9.)

MSAs are particularly well-suited for self-employed people who are young or in good health and don't go to the doctor often.

However, the MSA program is a four-year pilot program. Congress has authorized only 750,000 MSAs to be established nationwide on a first-come, first-served basis by the end of the year 2000. The IRS will announce when the limit is reached and no new MSAs will be allowed unless Congress expands the program. Even if the limit is not reached, no new MSAs may be established after 2000 unless Congress amends the law. People who meet the deadline may continue to contribute to their MSAs after 2000.

Thus far, MSAs have been catching on very slowly. The IRS announced that as of June 30, 1997, only 22,051 MSAs had been purchased, many fewer than expected. It's likely that as self-employed people become aware of the benefits of MSAs, the numbers of people obtaining them will grow. Therefore, you should act to establish your own MSA as soon as possible.

## 5. Insurance From a Former Job

If you're laid off or quit, and your employer provided you with group health insurance coverage, you may be able to keep your old health insurance coverage. A federal workplace law called COBRA—short for Consolidated Omnibus Budget Reconciliation Act—requires your former employer to offer you and your spouse and dependents continuing insurance coverage if you lose your job for any reason other than being fired or resigning on account of gross misconduct.

### Employers Trying to Snake Away From COBRA

Because COBRA applies to employees who are terminated from their jobs, some employers claim that it doesn't apply to employees who quit their jobs—for example, to become ICs. This legal issue is unresolved. However, you should assume that COBRA applies to you even if you quit and demand coverage if you want it and otherwise qualify for it.

The law applies to all employers with 20 or more employees. Your employer's health plan administrator is supposed to inform you within 14 days after you leave your job that you can continue your coverage. Coverage must be offered regardless of any pre-existing medical conditions you have. You have 60 days after receiving the notice to decide whether to obtain the continuing coverage. If you elect to obtain this coverage, it's retroactive to the date you left your job.

You usually have to pay for this coverage yourself. Your employer may charge you up to 102% of what it pays for the coverage; the extra 2% is for administrative costs. However, some employees who are laid off are able to negotiate free coverage for a time as part of a severance package.

Your coverage can last for up to 18 months. At the end of that time, you have the right to convert to an individual policy. However, such a policy will likely be much more expensive than your employer's group policy.

Employers and health insurance plan administrators who violate COBRA can be fined. Unfortunately, the law generally cannot be enforced by any means other than a complex and expensive lawsuit. Such lawsuits are usually brought only by large groups of former employees who can afford to share the legal fees involved. If you run into problems claiming COBRA benefits, you can try calling your local Labor Department or IRS office; both agencies administer COBRA. But you'll likely get better results by calling the Older Women's League; 666 11th Street, NW; Suite 700; Washington, DC 20001; 202-783-6686. This organization also assists younger men and women.

Many states have laws similar to COBRA that are easier for you to enforce and, more importantly, usually provide broader benefits and apply to smaller employers than does COBRA. These laws vary greatly from state to state. Contact your state insurance department for more information.

You can download from the Internet a summary of COBRA prepared by the U.S. Department of Labor Pension and Welfare Benefits Administration at http://gatekeeper.dol.gov/dol/pwba/public/pubs/COBRA/cobra95.htm.

You may also obtain a report on COBRA called "COBRA Continuation Coverage" from the International Foundation of Employee Benefit Plans; 18700 West Bluemound Road; P.O. Box 69; Brookfield, WI 53008; 414-786-6700.

## B. Disability Insurance

Disability insurance is designed to replace the income you lose if you become so sick or injured you're unable to work for a lengthy period or never able to work again. While disability insurance is often overlooked, as a self-employed worker, it's one of the most useful types of insurance you can have. Disability coverage is much more important than life insurance, for example, since most workers are far more likely to become disabled than die. Indeed, one out of four workers become disabled during some part of their lifetimes.

Disability insurance pays you a monthly benefit if you're unable to work. The cost of disability insurance depends on many factors.

- **The amount of coverage you obtain.** The maximum benefit you can obtain is usually two-thirds of your income. You can obtain a smaller benefit and pay a smaller premium. At a minimum, try to obtain a benefit large enough to pay your monthly mortgage costs or rent and other fixed expenses.
- **The term of your coverage.** Some disability insurance plans offer only short-term benefits; the periods range from 13 weeks to five years. More expensive long-term plans pay you until you reach 65 or pay indefinitely. If you can afford it, a long-term policy is best.
- **The elimination period** This is how long you have to wait after you become disabled before you start getting benefits. Periods range from 30 to 730 days. A 90-day period is most common.
- **The nature of your work.** The amount of your premiums will depend also on the nature of your work. People in hazardous occupations—construction, for example—pay more than people with relatively safe jobs.
- **How your policy defines disability.** More expensive plans pay you full benefits if you can't work in your particular occupation, even if you may be able to do other types of work. Less expensive plans pay you only if you are unable to work in any occupation for which you're suited.
- **Your health.** Your current health is also an important factor. Usually, some type of physical exam will be required. If you smoke or suffer from a pre-existing medical condition, be prepared to pay more and search harder for coverage.

Unfortunately, it can be very difficult for self-employed people to obtain disability insurance. Many disability insurers don't like to issue policies to the self-employed because their incomes often fluctuate dramatically and they may not be able to pay their premiums. Also, because the self-employed don't have employers to supervise them and verify they're disabled, it can be difficult for an insurer to know for sure whether they're really unable to work. This is a particular problem if you work at home. Some insurers won't issue a disability policy to anyone who spends more than half the time working at home.

Many insurers will not issue you a policy until you've been self-employed for at least six months. They want to see how much money you've earned during this period so they'll know whether you can afford the cost of premiums.

You'll have an easier time obtaining disability coverage if you can show an insurer that you're operating a successful, established business—for example, you have:

- employees
- long-term contracts with clients you can show the insurer
- a detailed financial forecast statement showing how much money you expect your business to earn in future years, and
- good credit references.

If you're still employed, try to obtain an individual disability policy before you quit your job and become self-employed.

If already self-employed, try to obtain group coverage though a professional organization or trade group. If this doesn't work, you'll have to obtain a individual policy.

There are five main disability insurers: Chubb Group of Insurance Companies, Northwestern Mutual Life Insurance Company, Paul Revere Life Insurance Company, Provident Life & Accident Insurance Company and UNUM Life Insurance Company of America. Try to get quotes from them all. You can likely find these listed in your local Yellow Pages under insurance. If not, find local insurance agents who represent them.

You can find a directory of many other companies that offer disability coverage on the Internet site maintained by the Health Insurance Association of America, an industry trade group. The Internet address is: www.hiaa.org/consumerinfo/disability_dir.html.

## C. Business Property Insurance

Business property insurance helps compensate for loss to your business assets: computers, office furniture, equipment and supplies. If, for example, your office burns down or is burglarized and all your business equipment lost, your business property insurance will pay you a sum of money. Three main factors determine the cost of such insurance: the policy limits, type and scope of coverage.

### 1. Policy Limits

All policies have a maximum limit on how much you will be paid, no matter how great your loss. The greater your policy limit, the more expensive the insurance will be.

### 2. Replacement or Cash Value Coverage

Property insurance can be for your property's replacement cost or its actual present cash value. A replacement cost policy will replace your property at current prices regardless of what you paid for it. An actual cash value policy will only pay you what your property was worth when it was lost or destroyed. If the item has depreciated in value, you may obtain far less than the amount needed to replace it. A replacement cost policy is always preferable, but costs more than a cash value policy.

### 3. Scope of Coverage

Business property insurance comes in one of two forms: Named Peril and Special Form.

Named Peril policies only cover you for perils listed in the policy. For example, the cheapest type of named peril policy only covers losses caused by fire, lightning, explosion, windstorm, hail, smoke, aircraft, vehicles, riot, vandalism, sprinkler leaks, sinkholes and volcanoes.

In contrast, a Special Form policy will cover you for anything except for certain perils that are specifically excluded—for example, earthquakes. Special Form policies cost somewhat more, but may be worth it.

Before you purchase business property insurance, take an inventory of all your business property and estimate how much it would cost you to replace it if it was lost, destroyed, damaged or stolen. Obtain replacement value business property coverage with a policy limit equal to this amount. If you can't afford that much coverage, consider a policy with a higher deductible. This is usually much wiser than obtaining coverage with a lower policy limit. If you insure your property for less than its full value, you won't be covered if you suffer a total loss.

Note that losses from earthquakes and floods normally aren't covered by business property policies. Earthquake insurance can be obtained through a separate policy or as an endorsement to your business property coverage. Flood insurance is usually handled through a separate policy called Difference in Conditions. Unfortunately, if you live in a part of the country where such hazards are common, such insurance can be expensive.

### Cheap Insurance for Your Computer

If the only valuable business equipment you have is a computer, you may only need computer insurance. A company called Safeware will insure your computer equipment against any type of loss except theft of computer equipment left in an unattended car. The rates are based on the replacement cost of your computers—not their present cash value—and are quite reasonable. For example, in 1996 you could purchase $5,000 in coverage for $69 per year. You can contact Safeware at: 800-848-3469.

## 4. If You Work at Home

If you work at home, there are several ways to obtain insurance coverage for your business property.

- **Homeowner's policies.** If you have homeowner's insurance, take a careful look at your policy. It may provide you with a limited amount of insurance for business property—usually no more than $2,500 for property damaged or lost in your home and $250 away from your home. Computer equipment may not be covered at all. If you have very little business property, this might be enough coverage for you.
- **Homeowner's insurance endorsements.** If your homeowner's policy doesn't provide any coverage for business property or you need more coverage than it provides, you may be able to purchase an endorsement—that is, an add-on page—to your policy providing additional coverage for your business property. This coverage is often inexpensive—typically less than $50 per year. Ask your homeowner's insurer or insurance agent about it.
- **BOP policies.** You can also buy an insurance policy especially designed for small businesses called a business owner's packaged policy or BOP. Such policies combine both property and liability insurance coverage in a single policy.
- **Business property policies.** Some policies just cover your business property. This might be a good idea if you have extremely valuable business equipment.

## 5. If You Rent an Office

If you rent an office outside your home, read your lease carefully to see if it requires you to carry insurance. Many commercial landlords require their tenants to carry insurance to cover damage the tenant does to the premises and injuries suffered by clients and other visitors.

The lease may specify how much insurance you must carry. Your best bet will probably be to get a BOP policy providing both property and liability coverage. Your landlord will probably require you to submit proof that you have insurance—for example, a photocopy of the first page of your policy.

## D. Liability Insurance

Liability insurance protects you when you're sued for something you did or failed to do that injured another person or damaged some property. It pays the legal fees for defending such lawsuits and any settlement or judgment obtained against you up to the amount of the policy, as well as the injured person's medical bills. In our lawsuit-happy society, such insurance is often a must. One recent survey by the Gallup Organization found that 24% of small business owners had either been sued or were threatened with court action during the previous five years.

There are two very different types of liability insurance:

- general liability insurance, and
- professional liability insurance.

You may need both types of coverage.

**INCORPORATING PROVIDES SOME LAWSUIT PROTECTION**

Incorporating your business gives you some protection from lawsuits, but not as much as you may think. For example, incorporating may protect your personal assets from lawsuits by people who are injured on your premises, but it won't protect you from personal liability if someone is injured or damaged because of your malpractice or negligence—that is, your failure to exercise your professional responsibilities with a reasonable amount of care.

Also, unless you have a decent insurance policy, all the assets of your incorporated business—which will probably amount to a large portion of your net worth—can be taken to satisfy a court judgment obtained by an injured person. (See Chapter 2.)

### 1. General Liability Insurance

General liability insurance provides coverage for the types of lawsuits any business owner could face. For example, this type of insurance protects you if: a client visiting your home office slips on the newly washed floor and shatters her elbow, or you knock over and shatter an heirloom vase while visiting a client in his home.

You definitely need this coverage if clients or customers visit your office. If you already have a homeowners' or renters' insurance policy, don't assume you're covered for these types of claims. Such policies ordinarily don't provide coverage for injuries to business visitors unless you obtain and pay for a special endorsement.

You also need general liability insurance if you do any part of your work away from your office—for example, in clients' offices or homes. You could injure someone or damage property while working there.

On the other hand, if you have little or no contact with the public, you may not need such insurance. For example, a freelance writer who works at home and never receives business visitors probably wouldn't need general liability coverage.

However, whether you want it or not, some clients may require you to carry liability insurance as a condition of doing business with you. For example, a software tester reports that her clients require that she carry at least $500,000 in general liability insurance. Many clients are afraid that if you don't have insurance and injure someone while working for them, the injured person will sue them instead. This fear is well-founded: lawyers always go after the deepest pockets—the person with insurance or money to pay a judgment. You might think this means you'd be better off with no insurance at all because people won't sue you, but this is not necessarily the case. If you have money or property, there's a good chance you'll get sued. Liability insurance will protect you from losing everything you own.

Luckily, general liability insurance is not terribly expensive; you can usually obtain it for a few hundred dollars per year. You can purchase such coverage:

- as part of a package policy such as a business owner's package or BOP policy, or
- by obtaining a separate general liability insurance policy known as a commercial general liability or CGL policy; this will probably cost the most, but give you higher policy limits—that is, more coverage.

If you work at home, you may also be able to add an endorsement to your homeowner's policy covering injuries to business visitors.

## 2. Professional Liability Insurance

General liability insurance does not cover professional negligence—that is, claims for damages caused because of an alleged error or omission in the way you perform your services. You need a separate professional liability insurance policy, also known as errors and omissions or E & O coverage. Some types of workers—doctors and lawyers, for example—are required by state law to obtain such insurance.

> **EXAMPLE:** Janet, an architect, designs a factory building that collapses, costing her client a fortune in damages and lost business. The client claims that Janet's design for the building was faulty and sues her for the economic losses it suffered. Janet's general liability policy won't cover such a claim. She needs a special E & O policy for architects.

Common types of professional liability insurance policies cover the following types of workers:

- accountants
- architects
- attorneys
- doctors
- engineers
- insurance agents and brokers
- pension plan fiduciaries, and
- stockbrokers.

You can obtain E & O coverage for many other occupations as well if you're willing to pay the price. Because of the growing number of professional negligence suits and the huge costs of litigation, such insurance tends to be expensive, ranging from several hundred to several thousand dollars per year. The premiums you'll have to pay depend on many factors, including:

- the claims history for your type of business; insurance costs more for businesses that generate lots of lawsuits
- the size of your business; the more work you do, the more opportunity there is for you to make a mistake resulting in a lawsuit
- your knowledge and experience in your field; less experienced self-employed people are more likely to make mistakes, and
- the size of your clients' businesses; mistakes involving large businesses will likely result in larger lawsuits than those involving small businesses.

If you need E & O insurance, the first place to look is a professional association. Many of them arrange for special deals with insurers offering lower rates or can at least steer you to a good insurer.

**Home-Based Architect Gets Liability Insurance**

Mel, an architect, recently left a job with a large architectural firm in San Francisco and set up his own architecture business, designing homes and small commercial offices. He works out of an office in a detached studio in his backyard. It soon dawned on Mel that he needed liability insurance.

First, Mel needed general liability insurance because clients, delivery people and other business visitors come and go from his home office every week. Mel could be subject to a huge lawsuit, for example, if a client was injured after slipping on a roller skate left by Mel's son. Mel called his homeowner's insurer and obtained an endorsement to his existing homeowner's policy covering injuries to business visitors and insuring up to $25,000 worth of his business equipment well. He had to pay an additional $150 annual premium for $500,000 in liability coverage.

Mel also needed E & O insurance because he could also be subject to a lawsuit for professional negligence if a problem occurred with one of his buildings. He shopped around and decided to purchase coverage through the American Institute of Architects in Washington, DC—a leading membership organization for architects. Mel obtained a one million dollar architect liability policy for $3,300 per year.

## E. Car Insurance

If, like most self-employed people, you use your automobile for business as well as personal use—for example, visiting clients or transporting supplies in addition to grocery shopping—you need to make certain that your automobile insurance will protect you from accidents that may occur while on business. The personal automobile policy you already have may also cover your business use of your car. On the other hand, it may specifically exclude coverage if you use your car on business.

Review your policy and discuss the matter with your insurance agent or auto insurer. You may need to purchase a separate business auto insurance policy or obtain a special endorsement covering your business use. Whatever you do, make sure your insurer knows you use your car for business, not just for private uses or driving to and from your office. If you do not inform your company about this, it may cancel your coverage if a claim occurs that reflects a business use—for example, you get into an accident while on a business trip.

If you keep one or more cars strictly for business use, you will definitely need a separate business automobile policy. You may be able to purchase such a policy from your personal auto insurer.

## F. Workers' Compensation Insurance

Each state has its own workers' compensation system that is designed to provide replacement income and cover medical expenses for employees who suffer work-related injuries or illnesses. To pay for this, employers are required to pay for workers' compensation insurance for their employees, either though a state fund or a private insurance company.

Before the first workers' compensation laws were adopted about 80 years ago, an employee injured on the job had only one recourse: to sue the employer in court for negligence—a difficult, time-consuming and expensive process. The workers' compensation laws changed all this by establishing a no-fault system. Injured employees gave up their rights to sue in court. In return, employees became entitled to receive compensation without having to prove that the employer caused the injury. In exchange for paying for workers' compensation insurance, employers were spared from having to defend lawsuits by injured employees and paying out damages.

### 1. Restricted to Employees

Workers' comp is for employees, not self-employed people, who have the legal status of independent contractors. If you qualify as an independent contractor under your state's workers' compensation insurance law, your clients or customers need not provide you with workers' comp coverage. Each state has its own test to determine if a worker qualifies as an employee or independent contractor for workers' comp purposes.

For detailed information on how states classify workers for workers comp purposes, see *Hiring Independent Contractors: The Employer's Legal Guide*, by Stephen Fishman (Nolo Press).

You should satisfy your state's test if you act to preserve your status as an independent contractor. (See Chapter 15.) However, whether you're an independent contractor or employee for workers' comp purposes is the client's determination to make, not yours.

### 2. Your Worker Status

Not having to provide you with workers' comp coverage saves your clients a lot of money, but also presents them with a problem: If you're injured while working on a client's behalf, you could file a workers' compensation claim and allege that you're really the client's employee. If you prevail on your claim, you can collect workers' comp benefits even if your injuries were completely your own fault. Fines and penalties can also be imposed against your client by the state workers' compensation agency if it determines the client misclassified you as an independent contractor.

The response of many hiring firms to these fears is to require you to obtain your own workers' compensation coverage, even if you don't want it. They're afraid that if you don't have your own coverage, you'll file a workers' comp claim against them if you're injured on the job. Also, many workers' comp insurers require hiring firms to pay additional premiums for any independent contractors they hire that don't have their own workers' comp coverage.

### 3. If You Have Employees

Even if your clients don't require you to have it, you must obtain workers' comp coverage if you have employees. However, if you have only a few employees, you might not need to obtain workers' comp coverage. The workers' compensation laws of about one-third of the states exclude many small employers. (See the chart below.)

Many knowledgeable clients will want to see proof that you have workers' compensation insurance for your employees before they hire you. The reason for their adamance is because, if you don't have insurance, your state law will probably require your client to provide it. The purpose of these laws is to prevent employers from avoiding paying for workers' comp insurance by subcontracting work out to independent contractors who don't insure their employees.

**State Employee Minimums for Workers' Compensation**

| Workers' Comp not required if you have two or fewer employees | Workers' Comp not required if you have three or fewer employees | Workers' Comp not required if you have four or fewer employees |
|---|---|---|
| Arkansas | Pennsylvania | Alabama |
| Florida | Rhode Island | Minnesota |
| Michigan | | Mississippi |
| New Mexico | | Tennessee |
| North Carolina | | |
| Virginia | | |
| Wisconsin | | |

### 4. Obtaining Coverage

Most small businesses buy insurance through a state fund or from a private insurance carrier.

In the following states, you must purchase coverage from the state fund: Nevada, North Dakota, Ohio, Washington and West Virginia.

In a number of states, you have a choice of buying coverage from either the state or a private insurance company. They include: Arizona, California, Colorado, Hawaii, Idaho, Maryland, Michigan, Minnesota, Montana, New York, Oklahoma, Oregon, Pennsylvania and Utah.

If private insurance is an option in your state, you may be able to save money on premiums by coordinating workers' comp coverage with property damage and liability insurance. (See Sections C and D.)

### 5. Cost of Coverage

The cost of workers' compensation varies from state to state and depends upon a number of factors including:

- the size of your payroll
- the nature of your work, and
- how many claims have been filed in the past by your employees.

As you might expect, it costs far more to insure employees in hazardous occupations such as construction than to insure those who work in relatively safe jobs such as clerical work. It might cost $200 to $300 a year to insure a clerical worker and perhaps ten times as much to insure a construction worker or roofer.

## G. Other Types of Insurance

There are several other types of insurance policies that may be useful for self-employed people:

- business interruption insurance, designed to replace the income you lose if your business property is damaged or destroyed due to fire or other disasters and you're forced to close, relocate or cut your business back when you recover
- electronic data processing or EDP insurance, which compensates you for the cost

### Suing Your Clients for Negligence

Even if you have your own workers' comp insurance, you can still sue a hiring firm for damages if its negligence caused or contributed to a work-related injury. Since you're not the hiring firm's employee, the workers' comp provisions barring lawsuits by injured employees won't apply to you. The damages you can obtain through a lawsuit can far exceed the modest workers' compensation benefits to which you may be entitled.

**EXAMPLE:** Trish, a self-employed trucker, contracts to haul produce for the Acme Produce Co. Trish is self-employed and Acme does not provide her with workers' comp insurance. At Acme's insistence, however, Trish obtains her own workers' comp coverage. Trish loses her little finger when an Acme employee negligently drops a load of asparagus on her hand. Since Trish is self-employed, she can sue Acme in court for negligence even though she has workers' comp insurance. If she can prove Acme's negligence, Trish can collect damages for her lost wages, medical expenses and her pain and suffering as well. These damages could far exceed the modest workers' comp benefits Trish may be entitled to for losing her finger.

However, if you receive workers' compensation benefits and also obtain damages from the person that caused the injury, your workers' comp insurer may be entitled to be reimbursed by you for any amounts it paid for your medical care. Your insurer might also be able to bring its own lawsuit against your client.

of reconstructing the data you lose when your computer equipment is damaged or destroyed, and

- product liability insurance, which covers liability for injuries caused by products you design, manufacture or sell.

If you're interested in such coverage, talk to several agents who have experience dealing with self-employed people in your field. Professional and trade organizations and may also be able to give you helpful advice.

## H. Ways to Find and Save Money on Insurance

There are a number of things you can do to make it easier to find and pay for insurance.

### 1. Seek Out Group Plans

For many self-employed people, the cheapest and easiest way to obtain insurance is through a professional organization, trade association or similar membership organization.

There are hundreds of such organizations representing every conceivable occupation—for example, the American Society of Home Inspectors, the Association of Independent Video and Filmmakers and the Graphic Artists Guild.

There are also national membership organizations that allow all types of self-employed people to join—for example, the National Association of the Self Employed and the Home Office Association of America. Many of these organizations give their members access to group health and business insurance. Because these organizations have many members, they can often negotiate cheaper rates with insurers than you can yourself. Your local chamber of commerce or alumni association may also offer insurance benefits.

If you don't know the name and address of an organization you may be eligible to join, ask other self-employed people or check out the *Encyclopedia of Associations* (Gale Research); it should be available in your public library. Also, many of these organizations have Websites on the Internet, so you may be able to find the one you want by doing an Internet search; you'll need a computer and modem for this. (See Chapter 21.)

### 2. Buying From an Insurance Company

If you're unable to arrange coverage through a group, try to purchase insurance from one of the growing number of companies that sell policies directly to the public rather than using insurance agents or brokers. These companies can usually offer you lower rates because they don't have to pay commissions to insurance agents. Other self-employed people may be able to recommend a company to you, or you can find them listed in the Yellow Pages under insurance.

### Be Wary of Insurance Agents

Insurance agents or brokers can be a useful source of information. The terms agent and broker mean different things in different parts of the country. In some states, an agent is a person who represents a specific insurance company and a broker is a person who is free to sell insurance offered by various companies. Elsewhere, the term insurance agent is used more broadly to cover both types of representatives. If you want to use an agent, find one who is familiar with businesses such as yours and who represents more than one insurer.

However, be wary about what any agent tells you. Insurance agents are salespeople who earn their livings by selling insurance to the public, for which they are paid commissions by insurance companies. The more insurance they sell you, the more money they make. Agents often recommend insurance from companies paying the highest commissions, whether or not it's the cheapest or best policy for you. If you use an agent, try to get quotes from more than one and compare them with the coverage you can obtain through a professional organization or by dealing directly with an insurer.

## 3. Comparison Shop

Insurance costs vary widely from company to company. You may be able to save a lot by shopping around. Also, review your coverage and rates periodically. Insurance costs go up and down periodically. If you're shopping for insurance during a time when prices are low, try locking in a low rate by signing up for a contract for three or more years.

## 4. Increase Your Deductibles

Your premiums will be lower if you obtain policies with high deductibles. For example, the difference between a $250 and $500 deductible may be 10% in premium costs, and the difference between a $500 and $1,000 deductible may save you an additional 3% to 5%.

## 5. Find a Comprehensive Package

It's often cheaper to purchase a comprehensive insurance package that contains many types of coverage than buying coverage piecemeal from several companies. Many insurers offer special policies for small business owners, also called BOP policies, that provide liability coverage against injuries to clients or customers or damage to their property while on your premises, fire and theft coverage and business interruption insurance.

## 6. Use the Internet

If, like millions of Americans, you have access to the Internet, you can obtain a great deal of information about insurance from your computer. Insurance companies, agents and organizations all have their own sites. Some good places to start an Internet search about insurance are:

- **Health Insurance Association of America.** This is a nationwide trade association of over 250 health insurers. Its Internet site contains useful articles on all aspects of health insurance and directories of disability and long-term health insurance providers, and companies providing Medical Savings Accounts. The Internet address is www.hiaa.org.
- **Quotesmith.** This company provides free quotes from hundreds of insurance companies. The Internet address is www.quotesmith.com.
- **Insurance InLinea.** This is an informational site created and maintained by a nationwide group of insurance agents. It provides useful information on all forms of insurance

**CHECK ON AN INSURER'S FINANCIAL HEALTH**

Several insurance companies have gone broke in recent years. If this happens and you have a loss covered by a policy, you may only receive a small part of the coverage you paid for or none at all. The best way to avoid this is to obtain coverage from an insurer that is in good financial health.

You can check out insurers' financial status in these standard reference works which rate insurance companies for financial solvency:

- *Best's Insurance Reports* (Property-Casualty Insurance Section)
- *Moody's Bank and Financial Manual* (Volume 2)
- *Duff & Phelps* (Insurance Company Claims-Paying Ability Rating Guide), and
- *Standard & Poor's.*

An insurance agent should be able to give you the latest rating from these publications. They may also be available in your public library.

If you have access to the Internet, you can obtain insurance company ratings from Standard & Poor's and Duff & Phelps for free on the Insurance News Network, an Internet site maintained by the Independent Insurance Agents of America. The Internet address is www.insure.com.

for small businesses. The Internet address is www.inlinea.com/.

- **Insurance Company Links.** This is a directory of private insurance companies on the Internet. The URL is http://connectyou.com/ins/comp.htm.
- **Insurance Yellow Pages.** This is an Internet search engine that allows browsers to do key word searches for insurance-related Web sites. The URL is http://connectyou.com/ins/yellow.htm.
- **AA Insurance World Network.** This is a state-by-state directory of insurance agents and their specialty areas. The URL is http://ns.nomius.com/~I-World/.

## 7. Deduct Your Business Insurance Costs

You can deduct the premiums for any type of insurance you obtain for your business from your income taxes. This includes business property insurance, liability insurance, insurance for business vehicles and workers' compensation insurance.

The premiums for health insurance you obtain for yourself are partly deductible if you're a sole proprietor, partner in a partnership or S corporation owner. They are completely deductible if you form a C corporation and it provides insurance for you as its employee. (See Chapter 2.)

Car insurance and homeowners' or renters' insurance premiums are deductible to the extent you use your car or home for business. (See Chapter 9.)

However, you may not deduct premiums for life or disability insurance for yourself. But, if you become disabled, the disability insurance benefits you receive are not taxable. ■

CHAPTER

# 7

# Pricing Your Services and Getting Paid

Two difficult problems self-employed workers face are deciding how much to charge clients and getting paid for their services. This chapter sets forth a rational way to go about setting your fees and gives you ideas about what to do when clients or customers don't pay when they're supposed to.

## A. Pricing Your Services

New and experienced self-employed people alike are often perplexed about how to determine their fees. No book can tell you how much your services are worth, but this section guides you in making this determination.

### 1. How the Self-Employed Are Paid

There are no legal rules controlling how you are paid. It is entirely a matter for negotiation between you and your clients. The method you choose will be determined by the customs and practices in your particular field, your personal preferences and those of your clients. Some clients insist on paying all self-employed people they hire a particular way—for example, a fixed fee; others are more flexible. Many self-employed people also have strong preferences for particular payment methods—for example, some insist on always being paid by the hour.

When you're first starting out, you may wish to try several different payment methods with different clients to see which works best for you. However, if the customary practices in your field dictate a particular payment method, you may have no choice in the matter. Other self-employed workers and professional organizations can give you information on the practices in your particular field.

This section provides an overview of the more common payment methods for the self-employed. However, these are by no means the only ways you can be paid. Other methods are used in some fields—for example, freelance writers are often paid a fixed amount for each word they write.

For more information on payment methods and setting fees, see:

- *How to Set Your Fees and Get Them*, by Kate Kelly (Visibility Enterprises)
- *Selling Your Services*, by Robert Bly (Henry Holt), and
- *The Contract and Fee Setting Guide for Consultants and Professionals,* by Howard Shenson (John Wiley & Sons).

#### a. Fixed fee

In a fixed fee agreement, you charge a set amount for an entire project. Your fixed fee can include all your expenses—for example, materials costs, travel expenses, phone and fax expenses, photocopying charges—or they can be billed separately to the client.

Most clients like fixed fee agreements because they know exactly what they'll have to pay for your services. However, fixed fees can be risky for you. If you underestimate the time and expense required to complete the project, you could earn much less than your work was worth or even lose money. Many self-employed people refuse to use fixed fees for this reason. For example, one self-employed technical writer always charges by the hour because she says she's never had a project that didn't last longer than both she and the client thought it would.

**EXAMPLE:** Ellen, a graphic artist, agrees to design a series of book covers for the Scrivener & Sons Publishing Co. Her fixed price contract provides that she'll be paid $5,000 for all the covers. Ellen estimates that the project would take 75 hours at most, so she would earn at least $66 per hour, more than her normal hourly rate of $50 per hour.

However, due to the publisher's exacting standards and demands for revisions, the project ends up taking Ellen 125 hours. As a result, she earns only $40 per hour for the project—far less than what she would have charged had she billed by the hour.

Although fixed fee agreements can be risky, they can also be very rewarding if you work efficiently and if you accurately estimate the time and expense involved in completing a project. Surveys of the self-employed have consistently found that fixed fee agreements are more profitable than other types of contracts. For example, a study recently conducted by the hosts of the consultant's forum on the America Online service found that self-employed people who charged fixed fees earned on average 150% more than those who charged by the hour for the same services. A similar survey conducted by a trade journal called *The Professional Consultant* found that self-employed people charging fixed fees earned 95% more than their colleagues who charged by the hour or day.

**REDUCING THE RISKS OF FIXED FEES**

There are several ways to reduce the risks involved in charging a fixed fee.

- First and foremost, carefully and thoroughly define the scope of the project in writing before determining your fee. If this will take a substantial amount of time, you may wish to charge the client a flat fee or hourly rate to compensate you for the work involved in this assessment process.
- Leave some room for error or surprises when you calculate your fee—that is, charge the client as if the project will take a bit longer than you think it will.
- Consider placing a cap on the total number of hours you'll work on the project. Once the cap is reached, you stop work and you and the client must negotiate new payment terms. For example, the client might increase your fixed fee or agree to pay you by the hour until the project is finished.
- Make sure your agreement with the client contains a provision allowing you to renegotiate your price if the client makes changes or the project takes longer than you estimated. (See Chapter 19.)

### b. Unit of time

It's safer for you to be paid by unit of time—that is, by the number of hours or days you spend on a project—rather than a fixed fee. This is especially true where you are unsure how long or difficult the project will be or the client is likely to demand substantial changes midstream. Many self-employed people refuse to work any other way. This method of payment is customary in many fields such as law and accounting.

However, clients are often nervous about paying you by the hour. They're afraid you'll spread out the project as long as possible to earn more money. Clients will often seek to place a limit on the total number of hours you can spend on the project. This way, they limit the total amount they'll have to spend. Others will require you to provide a time estimate. If you do this, be sure to call the client before spending more time than you estimated.

As with fixed fees, it's a good idea to leave some margin for error when you provide a time estimate. One self-employed person says she determines how many hours a job will take by first deciding how long it should take, doubling that number and then adding 25%. You may not need to go to this extreme, but it's wise to be conservative when estimating the time involved in any project.

### c. Fixed and hourly fee combinations

You can also combine a fixed fee with hourly payment. This arrangement allows you to reduce the risk that you'll be underpaid by charging by the hour for tasks for which it is difficult to estimate the time and work involved and a fixed fee for those tasks where such an estimate is easy. For example, if your work involves some tasks that are essentially mechanical and others that are highly creative, you can probably accurately estimate how long the mechanical work will take but may have great difficulty estimating how much time the creative work will require. You can reduce the risk you won't be paid enough by charging a fixed fee for the mechanical work and billing by the hour for the creative work.

**EXAMPLE:** Bruno, a freelance graphic artist, is hired by Scrivener & Sons Publishing Co. to produce the cover for its new detective thriller, *And Then You Die.* Bruno has absolutely no idea how long it will take him to come up with an acceptable design for the cover. He charges Scrivener $75 per hour for this design work. Once a design is accepted, however, Bruno knows exactly how long it will take him to produce a camera-ready version. He charges Scrivener a fixed fee of $1,000 for this humdrum production work.

### d. Retainer agreements

With a retainer agreement, you receive a fixed fee upfront in return for promising to be available to work a certain number of hours for the client each month or to perform a specified task. Often, the client pays a lump sum retainer fee at the outset of the agreement; or you can be paid on a regular schedule—for example, monthly, quarterly or annually.

**EXAMPLE:** Jean, an accountant, agrees to perform up to 20 hours of accounting services for Acme Corp. every month, for which Acme pays her $1,500 per month.

Many self-employed people like retainer agreements because they provide a guaranteed source of income. But in return for this security, you usually have to charge somewhat less than you do when paid on a per project basis. Also, retainer agreements can contradict your work status. If you spend most of your time working for a single client, the IRS may view you as that client's employee. (See Chapter 15.)

### e. Performance billing

Perhaps the most risky form of billing of all is performance billing, also known as charging a contingency fee. Basically, this means you get paid according to the value of the results you achieve for a client. If you get poor results, you may receive little or nothing. Clients generally favor this type of arrangement because they don't have to pay you if your services don't benefit them. Using this type of fee arrangement can help you get business if a client is skeptical that you'll perform as promised or you're providing a new service with benefits that are not generally understood.

This type of fee arrangement is used most often for sales or marketing projects in which the fee is based on a percentage of the increased business.

> **EXAMPLE:** Alice, a marketing consultant, contracts to perform marketing services for Acme Corp. to help increase its sales. Acme agrees to pay her 25% of the total increase in gross sales over the next 12 months. If sales don't go up, Alice gets nothing.

Some self-employed people reduce the risks involved in performance billing by requiring their clients to pay them a minimum amount regardless of the results they achieve. For example, a contract might provide that a sales trainer would receive $5,000 for providing training services plus 10% of the increase in the client's sales for a specified number of months after the training program.

If you use a performance contract, don't tie your compensation to the client's profits. Clients can easily manipulate their profits through accounting gimmicks—and also reduce your compensation. Use something that is easy to measure and harder to manipulate such as the client's gross sales or some measurable cost saving.

### f. Commissions

Self-employed people who sell products or services are often paid by commission—that is, an amount for each sale they make. This includes, for example, many independent sales representatives, brokers, distributors and agents.

> **EXAMPLE:** Mark is a self-employed salesperson who sells industrial filters. He receives a commission from the filter manufacturer for each filter he sells. The commission is equal to 20% of the price of the filter.

If you're a good salesperson, you can earn far more by being paid by commission than with any other payment method. But if business is poor, your earnings will suffer.

## 2. Determining Your Hourly Rate

However you're paid, you need to determine how much to charge per hour. This is so even if you're paid a fixed fee for an entire project. To determine the amount of a fixed fee, you must estimate how many hours the job will take and multiply the total by your hourly rate; then add the amount of your expenses. Knowing how much you should earn per hour will also help you know whether a retainer agreement or performance billing is cost effective or a sales commission is fair.

If you're experienced in your field, you probably already know what to charge because you are familiar with market conditions. However, if you're just starting out, you may have no idea what you can or should charge. If you're in this boat, try using a two-step approach to determine your hourly rate:

- calculate what your rate should be based on your expenses, and
- then investigate the marketplace to see if you should adjust your rate up or down.

### a. Hourly rate based on expenses

A standard formula for determining an hourly rate requires you to add together your labor and overhead costs, then add the profit you want to earn and divide the total by your hours worked. This is the absolute minimum you must charge to pay your expenses, pay yourself a salary and earn a profit. Depending on market conditions, you may be able to charge more for your services or you might have to charge less. (See Section A2c.)

To determine how much your labor is worth, pick a figure for your annual salary. This can be what you earned for doing similar work when you were an employee, what other employees earn for similar work or how much you'd like to earn.

Next, compute your annual overhead. Overhead includes all the costs you incur to do business—for example:

- rent and utilities
- business insurance
- stationery and supplies
- postage and delivery costs
- office equipment and furniture
- clerical help
- travel expenses
- professional association memberships
- legal and accounting fees
- telephone expenses
- business-related meals and entertainment, and
- advertising and marketing costs—for example, the cost of a Yellow Pages ad or brochure.

Overhead also includes the cost of your fringe benefits, such as medical insurance, disability insurance and retirement fund money. Also include your income and self-employment taxes.

If you're just starting out, you'll have to estimate these expenses or ask other self-employed people what their overhead is and use that amount.

You're also entitled to earn a profit over and above your labor and overhead expenses. Your salary does not include profit; it's part of your costs. Profit is the reward you get for taking the risks involved in being in business for yourself. It also provides money to expand and develop your business. Profit is usually expressed as a percentage of total costs. There is no standard profit percentage, but a 10% to 20% profit is common.

Finally, you must determine how many hours you'll work and get paid for during the year. Assume you'll work a 40-hour week for purposes of this calculation, although you may end up working more than this. If you want to take a two-week vacation, you'll have a maximum of 2,000 billable hours (50 weeks x 40 hours). If you want to take a longer vacation, you'll have fewer billable hours.

However, you'll probably spend at least 25% to 35% of your time on tasks such as bookkeeping and billing, marketing your services, upgrading your skills and doing other things you can't bill to clients. This means you'll likely have at most 1,300 to 1,500 hours for which you can get paid each year—if you factor in two vacation weeks away from work.

**EXAMPLE:** Sam, a self-employed computer programmer, earned a $50,000 salary as an employee and wants to receive at least the same salary. He estimates that his annual overhead amounts to $20,000 per year. He wants to earn a 10% profit and estimates he'll have 1,500 billable hours each year. To determine his hourly rate, Sam must:

- add his salary and overhead together: $50,000 + $20,000 = $70,000
- multiply this total by his 10% profit margin and add the amount to his salary and overhead: $70,000 x 10% = $7,000; $70,000 + $7,000 = $77,000, and
- divide the total by his annual billable hours: $77,000 ÷ 1,500 = $51.33.

Sam determines that his hourly rate should be $51.33. He rounds this off to $50. However, depending on market conditions, Sam might realistically charge more or less.

### b. Hourly rate worksheet

You can use the following worksheet to calculate your hourly rate.

**HOURLY RATE WORKSHEET**

Annual salary you want to earn ____________

Desired Profit % ____________

Billable hours per year ____________

**Yearly Expenses**

- Marketing ____________
- Travel ____________
- Legal and accounting costs ____________
- Insurance ____________
- Supplies ____________
- Rent ____________
- Utilities ____________
- Telephone ____________
- Professional association memberships ____________
- Business meals and entertainment ____________
- Other ____________

**Total Expenses** ____________

**Calculation:**

- Step 1: Salary + Expenses = X ____________
- Step 2: X x Profit % = Y ____________
- Step 3: X + Y = Z ____________
- Step 4: Z ÷ Billable Hours = Hourly Rate ____________

### A Calculating Shortcut

An easier but less accurate way to figure your hourly rate is to find out what hourly salary you'd likely receive if you were to provide your services as an employee in someone else's business and multiply this by 2.5 or 3. This is a much cruder way business management experts have developed to calculate how much money you need to earn to pay your expenses, salary and earn a profit.

> **EXAMPLE:** Betty, a freelance word processor, knows that employees performing the same work receive $15 per hour. She should charge $37.50 to $45 per hour.

If you don't know what employees receive for doing work similar to yours, try calling several employment agencies in your area and ask what you'd earn per hour as an employee.

You can also obtain salary information for virtually every conceivable occupation from *The Occupational Outlook Handbook*, published by the U.S. Department of Labor. You should be able to find it in your public library.

#### c. Investigate the marketplace

It's not enough to calculate how much you'd like to earn per hour. You have to determine whether this figure is realistic. This requires that you do a little sleuthing to find out what other self-employed people are charging for similar services and what the clients you'd like to work for are willing to pay. There are many ways to gather this information.

- Contact a professional organization or trade association for your field. It may be able to give you good information on what other self-employed people are charging in your area.
- Some professional organizations even publish pricing guides. For example, the National Writers Union publishes the *National Writers Union Guide to Freelance Rates & Standard Practice* (Writer's Digest Books), which lists rates for all types of freelance writing assignments. And the Graphic Artists Guild puts out a *Handbook of Pricing & Ethical Guidelines* for freelance graphic artists.
- Ask other self-employed people what they charge. If you have a computer and modem, you can communicate pricing concerns with other self-employed people on commercial online services such as Compuserve and America Online or on the World Wide Web. (See Chapter 21, Section E.)
- Talk with potential clients and customers—for example, attend trade shows and business conventions.

### 3. Experimenting With Charging

Pricing is an art, not an exact science. There are no magic formulas. And nostrums you often hear, such as "charge whatever you can get," are not very helpful. The best way—indeed, the only way—to discover how to charge and how much to charge is to experiment. Try out different payment methods and fee structures with different clients and see which work best for you.

You may discover that you will not likely get your ideal hourly rate because other self-employed people are charging less in your area. However, if you're highly skilled and performing work of unusually high quality, don't be afraid to ask for more than other self-employed people with lesser skills charge. Lowballing your fees won't necessarily get you business. Many potential clients believe they get what they pay for and are willing to pay more for quality.

One approach is to start out charging a fee that is at the lower end of the spectrum for self-employed people performing similar services and then gradually increase it until you start meeting price resistance. Over time, you should be able to find a payment method and fee structure that enables you to get enough work and that adequately compensates for your services.

**THE SELF-EMPLOYED SHOULD BE PAID MORE THAN EMPLOYEES**

Don't be afraid to ask for more per hour than employees earn for doing similar work. Self-employed people should be paid more than an employee. Unlike employees, the self-employed are not provided with employee benefits such as health insurance, vacations, sick leave or retirement plans; nor do hiring firms have to pay payroll taxes for them. This saves a hiring firm a bundle—employee benefits and payroll taxes add at least 20% to 40% to employers' payroll costs. You also have many business expenses employees don't have, such as office rent, supplies and marketing costs.

In addition, in our economic system, people in business for themselves are supposed to earn more than employees because they take much greater risks. Unlike most employees, self-employed workers don't get paid if business is bad. It's only fair, then, that they should be paid more than employees when business is good.

## B. Getting Paid

Hiring firms normally pay their employees like clockwork. Employers know that if they don't pay on time, their employees can get the state labor department to investigate and fine them. Also, employers usually depend upon their employees for the daily operation of their businesses and need to keep the workforce as content as possible.

Unfortunately, there are no similar incentives working on hiring firms when it comes to paying the self-employed. Many self-employed people have trouble getting paid by their clients. Some hiring firms feel free to pay outside workers late; some never pay at all. Sometimes this is because of cashflow problems; but often it's because hiring firms know that the self-employed often don't have the time, money or will to force them to pay on time. One computer consultant complains that delaying payment is an almost automatic response for a lot of companies. They seem to figure that if you don't nag, you don't really want to get paid.

As an independent businessperson, it's entirely up to you to take whatever steps are appropriate and necessary to get paid. No government agency will help you. Following are some strategies you can use to get clients to pay on time—or at least pay you eventually.

### 1. Avoiding Payment Problems

Taking a healthy dose of preventative medicine before you sign on with a client can help you eliminate, or at least reduce, payment problems.

#### a. Use written agreements

If you only have an oral agreement with a client who fails to pay you, it can be very hard to collect what you're owed. Without a writing to contradict him or her, the client can claim you didn't perform as agreed or can easily dispute the amount due. Unless you have witnesses to support your version of the oral agreement, it will be just your word against the client's. At the very least, you should have a writing containing: a description of the services you agree to perform, the deadline for performance and the payment terms. (See Chapter 19.)

### b. Find out if a purchase order is required

A purchase order is a document used by a client authorizing you to be paid for your services. (See Chapter 20, Section D.) Some clients will not pay you unless you have a signed purchase order, even if you already have a signed contract. Find out whether your client uses purchase orders and obtain one before you start work to avoid payment problems later on.

### c. Ask for a down payment

If you're dealing with a new client or one who has money problems, ask for a down payment before you begin work. Some self-employed people refer to such a payment as a retainer. This will show that the client is serious about paying you. And even if the client doesn't pay you in full, you'll at least have obtained something. Some self-employed people ask for as much as one-third to one-half of their fees in advance.

### d. Use periodic payment schedules

For projects lasting more than a couple of months, try using payment schedules requiring the client to pay you in stages—for example, one-third when you begin work, one-third when you complete half your work and one-third when you finish the entire project. Complex projects can be divided into phases or milestones with a payment due when you complete each phase.

If a client misses a payment, you can stop work. If you're never paid in full, you'll at least have obtained partial payment, so the entire project won't be a dead loss. A staged payment schedule will also improve your cashflow.

### e. Check clients' credit

The most effective way to avoid payment problems is to not deal with clients who have bad credit histories. A company that habitually fails to pay other creditors will likely give you payment problems as well. If you're dealing with a well-established company or government agency that is clearly solvent, you may forgo a credit check. But if you've never heard of the company, a credit check is prudent.

The most effective way to check a client's credit is to obtain a credit report from a credit reporting agency. Dun & Bradstreet, the premier credit reporting agency for businesses, maintains a database containing credit information on millions of companies. You can obtain a credit report on any company in Dun & Bradstreet's database by calling: 800-552-3867. The report will be faxed to you the same day or mailed. You can also obtain reports via your computer from Dun & Bradstreet's Internet site at: www.dbisna.com, or from the Compuserve online service.

A basic Dun & Bradstreet credit report, called a business information report, costs $45. It contains information on the company's payment history, financial condition and business history. It will also tell you whether the company has had any lawsuits, judgments or liens filed against it and whether it has ever filed for bankruptcy. The name of the company's bank is also listed. Dun & Bradstreet also assigns companies a credit rating which is intended to help you predict which ones will pay slowly or not at all.

A cheaper but more time-consuming way to check a potential client's credit is to ask the client to provide you with credit information and references. This is better than nothing, but may not give you an accurate picture because potential clients usually try to avoid telling you about their financial problems and give you the names of references who have not had problems with them.

Credit checks are routine these days so your request for credit credentials is not likely to drive away business. Be wary of any potential client that refuses to give you credit information. Provide the client with a request for credit information such as the following.

## REQUEST FOR CREDIT INFORMATION

**Andre Bocuse Consulting Services**
**123 4th Street**
**Marred Vista, CA 90000**
**555-1234**

Please provide the following information so we can extend credit for our services. All responses will be held in confidence. Please mail this form to the address shown below, or fax it to 100-555-1222. Thank you.

1. Company Name ______________________
Address ______________________
______________________
2. Contact person ______________________
3. Federal Tax ID No. ______________________
4. Type of Business ______________________
5. Number of employees ______________________
6. Date business established ______________________
7. Check one of the following forms of business.

☐ **Corporation**

☐ **State of incorporation**

Names, titles and addresses of your three chief corporate officers:

Name: ______________________
Address: ______________________
______________________
______________________

Name: ______________________
Address: ______________________
______________________
______________________

Name: ______________________
Address: ______________________
______________________
______________________

☐ **Partnership**

Names and addresses of the partners:

Name: ______________________
Address: ______________________
______________________
______________________

Name: ______________________
Address: ______________________
______________________
______________________

Name:

Address:

☐ **Limited Liability Company**

Name:

Address:

Name:

Address:

**Names and addresses of the owners:**

Name:

Address:

Name:

Address:

Name:

Address:

☐ **Sole Proprietorship**

8. Purchase order required? Yes ☐ No ☐

9. Bank References

Bank #1

Account # Phone:

Contact Person:

Address:

Bank #2

Account # Phone:

Contact Person:

Address:

10. Please provide the following information for three vendors you use regularly:

Reference #1

Name: ______________________________

Address: ______________________________

______________________________

Phone: ______________

Reference #2

Name: ______________

Address: ______________________________

______________________________

Phone: ______________

Reference #3

Name: ______________________________

Address: ______________________________

______________________________

Phone: ______________________________

My company and I authorize the disclosure and release of any credit-related information based on this document to Andre Bocuse Consulting.

Authorized Signature:

______________________________

Printed Name:

______________________________

Title: ______________________________

Date: ______________

Call the accounting department of the credit references listed and ask if the company has experienced any payment problems with the client. Accounting departments are typically asked for this information and will usually provide it freely.

If a credit report or your own investigation reveals that a potential client has a bad credit history or is in financial trouble, you may prefer not to do business with it. However, you may not be able to afford to work only for clients with perfect credit records. If you want to go ahead and do the work, obtain as much money upfront as possible and be on the lookout for payment problems. If the client is a corporation or limited liability company, you may seek to have its owners sign a personal guaranty. (See Section B1g.)

### f. Checking form of ownership

Your investigation of a potential client should include determining how its business is legally organized. This could have a big impact on your ability to collect a judgment against if it fails to pay you.

If you sue a client that fails to pay you and win, the court will order the client to pay you a specified sum of money. This is known as a court judgment. Unfortunately, if the client fails or refuses to pay the judgment, the court will not help you collect it. You've got to do it yourself or hire someone to help you. However, having a court judgment gives you many legal tools that can help you collect. For example, you can file liens on the client's property that make it impossible for the client to sell the property without paying you, get hold of the client's bank accounts and even have business or personal property such as the client's car seized by local law enforcement and sold.

Your ability to collect a court judgment may be helped or severely hindered depending on the way the client's business is legally organized.

- If the client is a sole proprietorship—that is, individually owns the business—he or she is personally liable for any debts the business owes you. This means the proprietor's own personal assets—as well as those of the business—are available to satisfy the debt. For example, both the proprietor's business and personal bank accounts may be tapped to pay you.
- If the client is a partner in a partnership, you can go after the personal assets of all the general partners. Be sure to get all their names before you start work. If the partnership is a limited partnership, you can't touch the assets of the limited partners, so don't worry about getting their names.
- If the client is a corporation, you could have big problems collecting a judgment. You normally can't go after the personal assets of a corporation's owners, such as the personal bank accounts of the shareholders and officers. Instead, you're limited to collecting from the corporation's assets. If the corporation is insolvent or goes out of business, there may be no assets to collect.
- If the client is a limited liability company, its owners will normally not be personally liable for any debts the business incurred, just as if it were a corporation.

**WHAT'S IN A NAME? A LOT**

Often, you can tell how a potential client's business is legally organized just by looking at its name.

If it's a corporation, its name will normally be followed by the words Incorporated, Corporation, Company or Limited; or the abbreviations Inc., Corp., Co. or Ltd.

Partnerships often have the words Partnership or Partners in their name, but not always. A limited liability company will usually have the words Limited Liability Company, Limited Company or the abbreviations L.C., LLC or Ltd. Co. in its name. Sole proprietors often use their own names, but they don't have to do so. They may use fictitious business names or dbas that are completely different from their own names.

### g. Obtaining personal guarantees

If you're worried about the credit-worthiness of a new or small incorporated client or limited liability company, you may seek to have its owners sign a personal guarantee. A person who signs a personal guarantee, known as a guarantor, promises to pay someone else's debt. This is the same as co-signing a loan. A guarantor who doesn't pay can be sued for the amount of the debt by the person to whom the money is owed.

You can obtain a personal guaranty from the officers or owners of a company you're afraid will not pay you. The guarantee legally obligates them to pay your fee if the company doesn't pay it. This means that if the client fails to pay you, you can sue not only the client, but the guarantors as well and go after their personal assets if you obtain a court judgment.

> **EXAMPLE:** Albert, a self-employed consultant, contracts to perform services for Melt, Inc., a company involved in the highly competitive ice cream business. Melt is a corporation owned primarily by Barbara, a multi-millionaire. Albert's contract contains a personal guaranty requiring Barbara to pay him if the corporation doesn't.
>
> Melt goes broke when botulism is discovered in its ice cream and the company fails to pay Albert. Albert files a lawsuit against Barbara, and easily obtains a judgment on the basis of the personal guaranty. When Barbara refuses to pay, Albert gets a court order enabling him to tap into one of her fat bank accounts and is paid in full.

Having a personal guaranty will not only help you collect a judgment, it will help prevent payment problems because the guarantors will have a strong incentive to make sure you're paid in full and on time. By doing so, they safeguard their personal assets from being snatched away.

Not many self-employed people ever think about asking for personal guarantees, so some clients may be taken aback if you do so. Explain that you need the added protection so you can extend the credit the client seeks. Also, note that signing a guaranty presents no risk at all to the business's owners so long as you're paid on time. You might also give the client a choice: the business's owners can either provide you with a personal guaranty or give you a substantial down payment upfront.

A personal guaranty can be a separate document, but the easiest way to draft one is to include a guaranty clause at the end of your contract with the client.

> **EXAMPLE:** Andre Bocuse, a self-employed consultant, agrees to perform consulting services for Acme Corporation, a one-person corporation owned by Joe Jones. Because Andre has never worked for Acme before and is worried about being paid, he asked Joe Jones to sign the following personal guaranty. This way he knows he can go after Jones' personal assets if Acme doesn't pay up. He adds the following clause to the end of his contract with Acme and has Jones sign it:
>
> *In consideration of Andre Bocuse entering into this Agreement with Acme Corporation, I personally guarantee the performance of all of the contractual obligations undertaken by Acme Corporation, including complete and timely payment of all sums due Andre Bocuse under the Agreement.*
>
> ____________________________
>
> Joe Jones

## 2. Sending Invoices to Your Clients

Send invoices to your clients as soon as you complete work. You don't have to wait until the end of the month. Create a standard invoice to use with all your clients. If you have a computer, accounting or invoice programs can create invoices for you. You can also choose to have your own invoices printed.

Your invoice should contain:

- your company name, address and phone number
- the client's name
- an invoice number
- the date
- the client's purchase order number or contract number, if any
- the terms of payment (see Section A1)
- the time period covered by the invoice

- a brief description of the services you performed; if you're billing by the hour, list number of hours expended and the hourly rate
- if you're billing separately for expenses or materials, the amounts of these items
- the total amount due, and
- your signature.

Include a self-addressed return envelope with your invoice. This tiny investment can help speed up payment.

Make at least two copies of each invoice: one for the client and one for your records. You may also want to may a third copy to keep in an unpaid invoices folder so you can keep track of when payments are overdue. (See Section B3.)

Following is an example of a self-employed worker's invoice.

### a. Terms of payment

The terms of payment is one of the most important items in your invoice. It sets the ultimate deadline which the client must pay you. This varies from industry to industry and will also vary from client to client. Thirty days is common, but some clients will want longer—45, 60 or even 90 days.

Obviously, the shorter the payment period, the better off you'll be. This is something you should discuss with the client before you agree to take a job. Some self-employed people ask for payment within 15 days or immediately after the services are completed. However, some clients' accounting departments aren't set up to meet such short deadlines.

**INVOICE**

JOHN SMITH
1000 GRUB STREET
MARRED VISTA, CA 90000
510-555-5555

Date: 4/30/97
Invoice Number: 103
Your Order Number: A62034
Terms: Net 30
Time period of: 4/1/97-4/30/97
To: Susan Elroy
Accounting Department
Acme Widget Company
10400 Long Highway
Marred Vista, CA 90000

**Services:**
Consulting services of John Smith on thermal analysis of Zotz 650 control unit.
50 hours @ $100.00 per hour.

**Subtotal:** $5,000
**Material Costs:** None
**Expenses:** 0

**TOTAL AMOUNT OF THIS INVOICE:** $5,000

Signed by: *John Smith*

If you have a written client agreement, it should provide how long the client has to pay you. (See Chapter 19, Section A3.)

The standard way to indicate the payment terms in your invoice is to use the word Net followed by the number of days the client has to pay after receipt of the invoice. For example, Net 30 means you want full payment in 30 days.

Some self-employed people offer discounts to clients or customers that pay quickly. A common discount is 2% for payment within ten days after the invoice is received. If such a discount can get a slow paying client to pay you quickly, it's worth it. You're always better off getting 98% of what you're owed right away rather than having to wait months for payment in full.

If you decide to offer a discount, indicate on your invoice the percentage followed by the number of days the client has to pay to receive the discount—for example, 2% 10 means the client can deduct 2% of the total due if it pays you within ten days. State your discount before the normal payment terms—for example, 2% 10 Net 30 means the client gets a 10% discount if it pays within ten days, but the full amount must be paid within 30 days.

### b. Charging late fees

One way to get clients to pay on time is to charge late fees for overdue payments. One consultant was experiencing major problems with late paying clients—40% of his clients were over 30 days late in paying him. He began charging a late fee—and the number of those paying late dropped to 5%.

However, late fees don't always work; some clients simply refuse to pay them. Not paying a late fee when required in your invoice is a breach of contract by the client, but it's usually not worth the trouble to go to court to collect a late fee.

If you wish to charge a late fee, make sure it's mentioned in your agreement. You should also clearly state the amount of your late fee on all your invoices—for example, your invoices should include the phrase: "Accounts not paid within terms are subject to a ____% monthly finance charge."

The late fee is normally expressed as a monthly interest charge.

**State Restrictions on Late Fees**

Your state might have restrictions on how much you can charge as a late fee. You'll have to investigate your state laws to find out. Check the index to the annotated statutes for your state—sometimes called a code—available in any law library. (See Chapter 21, Section D.) Look under the terms interest, usury or finance charges. Also, your professional or trade organization may have helpful information.

No matter what state you live in, you can safely charge as a late fee at least as much as banks charge businesses to borrow money. Find out the current bank interest rate by calling your bank or looking in the business section of your local newspaper.

The math takes two parts. First, divide the annual interest rate by 12 to determine your monthly interest rate.

**EXAMPLE:** Sam, a self-employed consultant, decides to start charging clients a late fee for overdue payments. He knows banks are charging 12% interest per year on borrowed money and decides to charge the same. He divides this rate by 12 to determine his monthly interest rate: 1%.

Then, multiply the monthly rate by the amount due to determine the amount of the monthly late fee.

**EXAMPLE:** Acme Corp. is 30 days late paying Sam a $10,000 fee. Sam multiples this amount by his 1% finance charge to determine his late fee: $100 (.01 x $10,000). He adds this amount to Acme's account balance. He does this every month the payment is late.

## 3. Collecting Overdue Accounts

If your invoice isn't paid on time, act quickly to collect. Clients who consistently pay late or do not pay at all can put you out of business. Moreover, experience shows that the longer a client fails to pay, the less likely it is you'll ever be paid.

### a. When an account is overdue

Money that your clients or customers owe you is called an account receivable. Keep track of the age of your accounts receivable so you know when a client's payment is late and how late it is.

Many computer accounting programs can keep track of the age of your accounts receivable. However, there is a simple way to do this without a computer. Make an extra copy of each of your invoices and keep them in a folder or notebook marked Unpaid Invoices. When a client pays you, throw away the applicable invoice. By looking in this notebook, you can tell exactly which clients haven't paid and how late their payments are.

Consider a client late if it fails to pay you within ten days after the due date on your invoice. At this point, you should contact the client and find out why you haven't been paid.

**Note Where Clients Bank**
When a client pays you by check, make a note of the name and address of the bank and the account number and place it in your client file. This information will come in very handy if you ever have to collect a judgment against the client.

### b. Contacting the client

Don't rely on collection letters. A phone call will have far more impact. Unfortunately, it can often be hard to get clients to return your collection calls. To reduce phone tag, ask what time the client will be in and call back then. If you leave a phone message, state the time of day you receive return calls—for example, every afternoon from 1 p.m. to 5 p.m. If you can't reach the client by phone, try sending faxes. One self-employed worker even reports good success from sending telegrams to nonpaying clients.

### c. Your first collection call

During your first collection call, you want to either solve a problem that has arisen or handle a stalled payment. Write down who said what during this and each subsequent phone call.

Prepare before you make your call. You should know exactly how much the client owes you and have a copy of your invoice and any purchase order in front of you when you dial the phone. Also, make sure you speak with the appropriate person. In large companies, this may be someone in the accounts payable department or purchasing department; in small companies, it could be the owner.

Politely inform the client that its payment to you is past due. About 80% of the time, late payments are caused by problems with invoices—for example, your client didn't receive an invoice or misplaced or didn't understand it. You may simply have to send another invoice or provide a brief explanation of why you charged as you did.

On the other hand, some clients may refuse to pay you because they are dissatisfied with your services or charges. In this event, schedule a meeting with the client as soon as possible to work out the problem. If the client is dissatisfied with only part of your work, ask for partial payment immediately.

Other clients may be satisfied with your services but don't have the money to pay you. They may want an extension of time to pay, or ask to work out a payment plan with you—for example, pay a certain amount every two weeks until the balance is paid. If such a client seems sincere about paying you, try to work out a reasonable payment plan. At least you are likely to get paid eventually this way and may also get repeat business from the client. If you agree to any new payment terms, set them forth in a confirming letter and send it to the client. The letter should state when and how much you'll be paid. Keep a copy in your files.

**Sample Confirming Letter**

April 15, 199X
Sue Jones, President
Acme Corporation
123 Main Street
Marred Vista, CA 90000

Re: Your contract # 1234
Invoice # 102

Dear Sue:

Thank you for your offer to submit $500 per month to pay off your company's outstanding balance on the above account.

As agreed, I am willing to accept $500 monthly payments for four months until this debt is satisfied. The payments are due on the first of each month, beginning May 1, 199X and continuing monthly through August 1, 199X.

As long as the payments are timely made, I will withhold all further action.

Thank you for your cooperation.

Very truly yours,

Andre Bocuse

Some clients may offer to pay a part of what they owe if you'll accept it as full payment. Although it may be galling to agree to this, it may make more economic sense than fighting with the client for full payment. If you orally agree to this, send the client a confirming letter setting forth the new payment terms. Keep a copy for yourself.

**Sample Confirming Letter**

August 1, 199X
John Anderson
200 Grub Street
Albany, NY 10000

Re: My Invoice # 102

Dear John:

As we orally agreed, I'm willing to accept $4,500 as a full and complete settlement of your account.

This sum must be paid by September 1, 199X or this offer will become void.

Thank you for your cooperation and I look forward to receiving payment.

Very truly yours,

Yolanda Allende

### Beware of Payment in Full Checks

Be careful about accepting and depositing checks from a client that have the words Payment in Full or something similar written on them. If the client owes you more than the face value of the check, you may be barred from collecting the additional amount.

Where there's a dispute about how much the client owes you, depositing a full payment check usually means that you accept the check in complete satisfaction of the debt. Crossing out the words Payment in Full generally won't help you. You'll still be prevented from suing for the balance once you deposit the check.

However, several states have changed this rule to help creditors. In these states, you normally can cash a full payment check and still preserve your right to sue for the balance by writing the words Under Protest or Without Prejudice on your endorsement. These states include Alabama, Delaware, Massachusetts, Minnesota, Missouri, New Hampshire, New York, Ohio, Rhode Island, South Carolina, South Dakota, West Virginia and Wisconsin.

Californians may cross out the full payment language, cash the check and sue for the balance. However, the client may be able to get around this by sending a written notice that cashing the check means it was accepted as payment in full. (Calif. Civil Code Section 1526). Luckily, few clients are aware of this rule, so crossing out the full payment language usually works just fine.

## d. Subsequent collection efforts

If you haven't received payment after more than a month and a first reminder, send the client another invoice marked Second Notice. Call the client and send invoices monthly. If you've been dealing with someone other than the owner of the company, don't hesitate to call the owner. Explain that cashflow is important to your company, and you can't afford to carry this receivable any longer.

Be persistent. When it comes to collecting debts, the squeaky wheel usually gets the money. A client with a faltering business and many creditors who has the money to pay just one debt will likely pay the creditor who has made the most fuss.

At this point, you should feel free to stop all work for the client and not deliver any work you've completed but not yet delivered. You'll go broke fast if you keep working for people who don't pay you.

However, don't allow your insistent behavior to cross the line so that you harass the client; that could get you into legal trouble.

### DON'T HARASS DEADBEAT CLIENTS

No matter how angry you are at a client who fails to pay you, don't harass him or her.

Harassment includes:

- threatening or using physical force if the client doesn't pay
- using obscene or profane language
- threatening to sue the client when you really don't intend to do so
- threatening to have the client arrested
- phoning the client early in the morning or late at night
- causing a phone to ring repeatedly or continuously to annoy the client, or
- communicating with the client unreasonably often.

Many states have laws prohibiting these types of collection practices. And even in states that don't have specific laws against it, court decisions often penalize businesses that harass debtors. A client could sue you for engaging in this kind of activity. Use your common sense and deal with the client in a businesslike manner, however much he or she owes you—and however agitated that debt makes you feel.

## 4. If a Client Won't Pay

If a client refuses to pay or keeps breaking promises to pay, you must decide whether to write off the debt or take further action. If the client has gone out of business or is unable to pay you anything, either now or in the future, your best option may be to write off the debt. There's no point in spending time and money trying to get blood out of a turnip.

But if the client is solvent, you should seriously consider:

- taking legal action against the client yourself
- hiring an attorney to take legal action against the client, or
- hiring a collection agency.

### a. Sending a final demand letter

Before you start any type of legal action against a client, send a final demand letter to the client informing him or her that you will sue if you don't receive payment by a certain date. Many clients will pay you voluntarily after receiving such a letter because they don't want to have to go to court and don't want their credit ratings damaged if you obtain a judgment against them.

The letter should state how much the client owes you and inform the client that you'll take court action if full payment isn't received by a specific date. Following is an example of such a letter.

#### SAMPLE FINAL DEMAND LETTER

April 24, 199X
Dick Denius
123 Grub Street
Anytown, AK 12345

Re: Your account number: 678

Dear Mr. Denius:

Your outstanding balance of $6,000 is over 120 days old.

If you do not make full payment by 5/15/9X, a lawsuit will be filed against you. A recorded judgment will be a lien against your property and can only have an adverse effect on your credit rating.

I hope to hear from you immediately so that this matter can be resolved without filing a lawsuit.

Very truly yours,

Natalie Kalmus

### b. Suing in small claims court

All states have a wonderful mechanism that helps businesses collect small debts: small claims court. Small claims courts are set up to resolve disputes involving relatively modest amounts of money. The limit is normally between $2,000 and $7,500, depending on the state in which you file your lawsuit. If you're owed more than the limit, you can still sue in small claims court and waive the excess.

Small claims court is particularly well-suited to collecting small debts because it's inexpensive and usually fairly quick. In fact, debt collection cases are by far the most common type of cases heard by small claims court.

You don't need a lawyer to go to small claims court. Indeed, a few states—including California, New York and Michigan—bar lawyers from small claims court.

For detailed advice about how to handle a small claims court suit, see *Everybody's Guide to Small Claims Court* (National and California Editions), by Ralph Warner (Nolo Press).

You begin a small claims lawsuit by filing a document called a complaint or statement of claim. These forms are available from your local small claims court clerk and are easy to fill out. You may also be asked to attach a copy of your written agreement, if you have one. You then notify the client, now known as the defendant, of your lawsuit. Depending on your state, the notice can be delivered by certified mail or by a process server. Many clients pay up when they receive a complaint because they don't want to go to court.

A hearing date is then set. If the client doesn't show up in court, you'll win by default. A substantial percentage of clients don't contest claims for unpaid fees in court because they know they owe the money and can't win. If the client does attend the court session, you present your case to a judge or court commissioner under rules that encourage a minimum of legal and procedural formality. Be sure to bring all your documentation to court, including your invoices, client agreement and correspondence with the client.

Unfortunately, getting a small claims judgment against a client doesn't guarantee you'll be paid. Many clients will automatically pay a judgment you obtain against them, but others will refuse to pay. The court will not collect your judgment for you. You've got to do it yourself or hire someone to help you.

**BIG FEE, SMALL CLAIM**

Gary, a freelance translator, recently contracted with the San Francisco office of a national brokerage firm to perform translating services on a rush basis. He completed the work on time and faxed it to the company with his invoice. The client failed to pay the invoice within 30 days.

Over the next three months, Gary sent the company a stream of collection letters demanding payment, but never heard a word. He finally got sick of waiting and decided to sue the client. He was owed $3,000, well within the $5,000 California small claims court limit, so filed his suit in the San Francisco small claims court.

He then had the San Francisco County Sheriff's Department serve his complaint on the client at its office. The next day he received a fax from the company's legal department at its New York headquarters apologizing for the delay in payment and promising to pay at once. He received a check within a few days.

### c. Suing in other courts

If the client owes you substantially more than the small claims court limit for your state, you may wish to sue in a formal state trial court, usually called the municipal court or superior court. Debt collection cases are usually very simple, so you can often handle them yourself or hire a lawyer for the very limited purpose of giving you advice on legal points or helping with strategy. In truth, few collection cases ever go to trial. Usually, the defendant either reaches a settlement with you before trial or fails to show up in court and you get a default judgment.

For detailed guidance on how to represent yourself in courts other than small claims courts, see:

- *Represent Yourself in Court: How to Prepare and Try a Winning Case*, by Paul Bergman and Sara Berman-Barrett (Nolo Press). This book explains how to handle a civil case yourself, without a lawyer, from start to finish.
- *How to Sue for Up to $25,000 . . . and Win!,* by Judge Roderic Duncan (Nolo Press). This book shows you how to handle claims in California courts.

### d. Arbitration

Before you think about suing the client in court, look at your contract to see whether it contains an arbitration clause. If your contract has such a clause, you'll be barred from suing the client in small claims or any other court. This is not necessarily a bad thing. Arbitration is similar to small claims court in that it's intended to be speedy, inexpensive and informal. The main difference is that an arbitrator, not a judge, rules on the case. An arbitrator's judgment can be entered with a court and enforced the same as a regular court judgment. (See Chapter 21, Section A1.)

### e. Hiring an attorney

Hiring an attorney to sue a client for an unpaid bill is usually not worth the expense involved unless the debt is very large and you know the client can pay. However, it can be effective to have a lawyer send a dunning letter to a client. Some clients take communications from lawyers more seriously than they take a letter you write on your own. Some lawyers are willing to do this for a nominal charge. (See Chapter 21, Section B.)

### f. Hiring a collection agency

Collection agencies specialize in collecting debts. You don't pay them anything. Instead, they take a slice of the money they collect. This can range from 15% to 50% depending on the size, age and type of debts involved. Collection agencies can be particularly good at tracking down skips—people who hide from their creditors.

Sicking a collection agency on a client will likely alienate the client and mean that you will not get any repeat business from him or her. Use this alternative only if you don't want to work for a particular client again.

You may have trouble finding an agency to deal with you if you only have a few debts, particularly if they're small. Try to get a referral to a good collection agency from colleagues or a professional organization or trade group. Ask for references before hiring any agency and call them. It's also advisable to get a fee agreement in writing.

## 5. Deducting Bad Debts From Income Taxes

In a few situations, you can deduct, the value of an unpaid debt from your income taxes. This is called a bad debt deduction. Unfortunately, if you're like the avast majority of self-employed people—a cash basis taxpayer who sells services to your clients—you can't claim a bad debt deduction if a client fails to pay you. Since you don't report income until it is actually received, you aren't considered to have an economic loss when a client fails to pay.

> **EXAMPLE:** Bill, a self-employed consultant, works 50 hours for a client and bills it $2,500. The client never pays. Bill cannot deduct the $2,500 loss from his income taxes. Since Bill is a cash basis taxpayer, he never reported the $2,500 as income because he never received it. As far as the IRS is concerned, this means Bill has no economic loss.

This rule seems absurd since you've lost the value of your time and energy when a client fails to pay you for your services, but it's strictly enforced by the IRS.

The only time a business can deduct a bad debt is if it actually lost cash on the account or it previously reported income from sale of the item. Few self-employed people give out cash, and only businesses using the complex accrual method of accounting report income from a sale for which no payment is received. (See Chapter 14, Section C.) There's no point in trying to switch to the accrual method to deduct bad debts. You won't reduce your taxes since the bad debt deduction merely wipes out a sale you previously reported as income. ■

CHAPTER

8

# Taxes and the Self-Employed

Employees typically don't need to worry much about taxes. All or most of their taxes are withheld from their paychecks by their employers and paid to the IRS and state tax department. The employer calculates how much to withhold. The employee's only responsibility is to file a tax return with the IRS and state tax department each year.

But when you become self-employed, your tax life changes dramatically. You have no employer to pay your taxes for you; you must pay them directly to the IRS and your state. This requires periodic tax filings you probably never made before—and it will be up to you to calculate how much you owe. For this reason, you'll also need to keep accurate records of your business income and expenses. And the tax return you must file each year will likely be more complicated than when you were an employee.

This chapter provides an overview of the brave new world of taxation you are about to enter as a self-employed worker and explains some ways to deal with it efficiently.

## A. Tax Basics for ICs

All levels of government—federal, state and local—impose taxes. You need to be familiar with the requirements for each.

### 1. Federal Taxes

The federal government puts the biggest tax bite on the self-employed. When you're in business for yourself, the federal government may impose a number of taxes on you.

These federal taxes include:

- income taxes
- self-employment taxes
- estimated taxes, and
- employment taxes.

#### a. Income taxes

Everyone who earns above a minimum amount of dollars must pay income taxes. Unless you're one of the few self-employed people who have formed a C corporation, you'll have to pay personal income tax on the profits your business earns. Fortunately, you may be able to take advantage of a number of business-related deductions to reduce your taxable income when you're self-employed. (See Chapter 9.)

By April 15 of each year, you'll have to file an annual income tax return with the IRS showing your income and deductions for the year and how much estimated tax you've paid. You file IRS Form 1040 and must include a special tax form in which you list all your business income and deductible expenses. Most self-employed people use IRS Schedule C, Profit or Loss From Business.

Tax matters are more complicated if you incorporate your business. If you form a C corporation, it will have to file its own tax return and pay taxes on its profits. You'll be an employee of your corporation and must file a personal tax return and pay income tax on the salary your corporation paid you. (See Chapter 2, Section B.)

#### b. Self-employment taxes

Self-employed people are entitled to Social Security and Medicare benefits when they retire, just like employees. And just like employees, they have to pay Social Security and Medicare taxes to help fund these programs. These taxes are called self-employment taxes, or SE taxes. You must pay SE taxes if your net yearly earnings from self-employment are $400 or more. When you file your annual tax return, you include IRS Form SE, showing how much SE tax you were required to pay. (See Chapter 10, Section E.)

### c. Estimated taxes

Federal income and self-employment taxes are pay-as-you-go taxes. You must pay these taxes as you earn or receive income during the year. Unlike employees, who usually have their income and Social Security and Medicare tax withheld from their pay by their employers, self-employed people normally pay their income and Social Security and Medicare taxes directly to the IRS. These tax payments are called estimated taxes and are usually made four times every year on IRS Form 1040-ES—on April 15, June 15, September 15 and January 15. You have to figure out how much to pay; the IRS won't do it for you. (See Chapter 11, Section B.)

### d. Employment taxes

Finally, if you hire employees to help you in your business, you'll have to pay federal employment taxes. These consist of half your employees' Social Security and Medicare taxes and all of the federal unemployment tax. You must also withhold half your employees' Social Security and Medicare taxes and all their income taxes from their paychecks. You must pay these taxes monthly, by making federal tax deposits at specified banks. However, if the total owed is $500 or less, the deposits are due quarterly.

When you have employees, you'll have to keep lots of records and file quarterly and annual employment tax returns with the IRS. (See Chapter 13, Section B.)

When you hire other self-employed people, however, you don't have to pay any employment taxes. You need only report payments over $600 for business-related services to the IRS and to your state tax department if your state has income taxes. (See Chapter 13, Section C.)

## 2. State Taxes

Life would be too simple if you were only required to pay federal taxes. States get into the act as well, imposing their own taxes to fund their governments.

### a. Income taxes

All states except Alaska, Florida, Nevada, South Dakota, Texas, Washington and Wyoming impose income taxes on the self-employed. Some states simply charge you a percentage of the federal income tax you have to pay for the year—for example, Vermont residents must pay 25% of the amount of their federal income taxes. Most other states charge a percentage of the income shown on your federal income return. Depending on the state in which you live, these percentages range anywhere from 3% to 11%.

In most states, you have to pay your state income taxes during the year in the form of estimated taxes. These are usually paid at the same time you pay your federal estimated taxes.

You'll also have to file an annual state income tax return with your state tax department. In all but six states—Arkansas, Delaware, Hawaii, Iowa, Louisiana and Virginia—the return must be filed by April 15, the same deadline as for your federal tax return. (See Section A4.)

If you're incorporated, your corporation will likely have to pay state income taxes and file its own state income tax return

Each state has its own income tax forms and procedures. Contact your state tax department to learn about your state's requirements and obtain the forms. (See the Appendix for a list of state tax offices.)

### b. Employment taxes

If you live in a state with income taxes and have employees, you'll likely have to withhold state income taxes from their paychecks and pay the money over to your state tax department. You'll also have to provide your employees with unemployment compensation insurance by paying state unemployment taxes to your state unemployment compensation agency. (See Chapter 13, Section B2.)

### c. Sales taxes

Almost all states and many municipalities impose sales taxes of some kind. The only states without sales tax are Alaska, Delaware, Montana, New Hampshire and Oregon.

All states that have sales taxes impose them on sales of goods or products to end users. If you only provide services to clients or customers, you probably don't have to worry about sales taxes because most states don't tax services at all or only tax certain specified services. Notable exceptions are Hawaii, New Mexico, and South Dakota—all of which impose sales taxes on all services, subject to certain exceptions.

If the products or services you provide are subject to sales tax, you'll have to fill out an application to obtain a state sales tax number. (See Chapter 8.) Many states impose penalties if you make a sale before you obtain a sales tax number. Generally, you pay sales taxes four times a year, but you might have to pay them monthly if you make lots of sales.

### d. Other state taxes

A hodgepodge of other taxes on businesses are imposed by various states. These are too numerous and diverse to summarize here.

For example:

- Connecticut has a personal property tax on business equipment and materials that is levied at the same rate as the property tax.
- Nevada imposes a Business Privilege Tax of $25 per employee.
- Hawaii imposes a general excise tax on businesses ranging from .5% to 4% of the gross receipts businesses earn.
- Michigan has a Single Business Tax of 2.3% of net business profits over $45,000.

Contact your state tax department for information on these and other similar taxes your state might impose.

## 3. Local Taxes

You might have to pay local business taxes in addition to federal and state taxes. For example, many municipalities have their own sales taxes which you may have to pay to a local tax agency.

Some cities and counties also impose property taxes on business equipment or furniture. You may be required to file a list of such property with local tax officials, along with cost and depreciation information. Some cities also have a tax on business inventory. This is why many retail businesses have inventory sales: they want to reduce their stock on hand before the inventory tax date.

A few large cities—for example, New York City—impose their own income taxes. Some also charge annual business registration fees or business taxes.

Your local chamber of commerce should be able to give you good information on your local taxes, or contact your local tax department.

## 4. Calendar of Important Tax Dates

The following calendar shows you important tax dates during the year. If you're one of the few self-employed people who uses a fiscal year instead of a calendar year as your tax year, these dates will be different. (See Chapter 14, Section C.) If you have employees, you must make additional tax filings during the year. (See Chapter 13, Section B.)

The dates listed below represent the last day you have to take the action described. If any of the dates fall on a holiday or weekend, you have until the next business day to take the action.

## B. Handling Your Taxes

Many self-employed people take care of their taxes completely by themselves. Others hire tax professionals to take care of everything for them. Many more are somewhere in between. Your approach depends on how complex your tax affairs are and whether you have the time, energy and desire to do some or all of the work yourself.

Following are three different approaches self-employed workers commonly use in preparing their taxes.

**TAX CALENDAR**

| Date | Action |
|---|---|
| January 15 | Your last estimated tax payment for the previous year is due. |
| January 31 | If you file your tax return by now, you don't have to make the January 15 estimated tax payment. (See Chapter 11.)<br>If you hired independent contractors last year, you must provide them with Form 1099-MISC. |
| February 28 | If you hired independent contractors last year, you must file all your 1099s with the IRS. |
| March 15 | Corporations must file federal income tax returns. |
| April 15 | Your individual tax return must be filed with the IRS and any tax due paid. Or, you can pay the tax due and file for a four-month extension of time to file your return.<br>You must make your first estimated tax payment for the year.<br>Partnerships must file information tax return.<br>State individual income tax returns due in all states except Arkansas, Delaware, Hawaii, Iowa, Louisiana and Virginia. |
| April 20 | Individual income tax returns due in Hawaii. |
| April 30 | Individual income tax returns due in Delaware, Iowa and Virginia. |
| May 15 | Individual income tax returns due in Arkansas and Louisiana. |
| June 15 | Make your second estimated tax payment for the year. |
| September 15 | Make your third estimated tax payment for the year. |

Self-employed people whose tax affairs are relatively simple can easily handle all their tax work themselves, particularly if they're comfortable using computers and use accounting and tax preparation computer programs.

**EXAMPLE:** Steve, a freelance writer, does all his taxes himself. Like most self-employed workers, he is a sole proprietor and does all the work in his writing business himself—that is, he does not hire employees or independent contractors to help him. As a writer, he works at home and doesn't need much in the way of equipment or supplies and has few business expenses to track other than his home office expenses.

He has a computer and easily keeps track of his income and expenses by using a simple computer accounting program. He also uses a computer tax preparation program to prepare his tax returns. He estimates he spends no more than one hour per month on bookkeeping and perhaps five hours to prepare his annual tax return.

On the other hand, self-employed people with larger businesses often hire tax pros to do all the work for them. This may be a particularly good idea if you incorporate your business or have employees.

**EXAMPLE:** Carol, a software tester, has formed a C corporation and has two employees. Her tax affairs are much more complicated than Steve's. She must file tax returns both for herself and her corporation. She must also withhold employment taxes from her own pay and her employees' pay and file quarterly and annual employment tax returns with the IRS and state of California.

Her bookkeeping requirements are also more complex than those of a sole proprietor like Steve. Carol does none of her tax work herself. She hires an accountant to do her bookkeeping and prepare her tax returns and uses a payroll service to calculate and pay the employment taxes for herself and her employees.

Even if you have a fairly complex tax return and want a tax pro to prepare it for you, you can still save money by doing some work yourself such as routine bookkeeping.

**EXAMPLE:** Gary, a self-employed translator, is a sole proprietor like Steve; but, unlike Steve, he hires independent contractor translators to work for him. He has to keep track of his payments to the independent contractors and report them to the IRS and his state tax department. He also rents an outside office and must keep track of this and other business expenses. A trained engineer with a mathematical bent, he does all his bookkeeping himself using a manual system. However, he hires an accountant to prepare his annual tax return which includes some rather complex business deductions such as depreciation of business equipment.

### 1. Doing the Work Yourself

The more tax work you do yourself, the less money you'll have to pay a tax pro such as an enrolled agent or certified public accountant—CPA—to help you through. This will not only save you cash, but give you more personal control over your financial life. The subject of taxes might seem daunting, but you don't need to become a CPA to take care of your own tax needs. Nor do you have to do everything yourself—for example, you can hire a tax pro to prepare your annual tax return or to advise you if you encounter a particularly difficult problem.

You can do some or all of the following tasks yourself.

### a. Bookkeeping

Even if you do hire a tax pro to prepare your tax returns, you'll save money if you keep good records. Tax pros have many horror stories about clients who come in with plastic bags or shoe boxes filled with a jumble of receipts and canceled checks. As you might expect, these people end up paying much more than those who have a complete and accurate set of income and expense records. It is not difficult to set up and maintain a bookkeeping system for your business. You should do so when you first start business. (Chapter 14, Section A describes a simple bookkeeping system adequate for most self-employed people. Chapter 9 provides an overview of the business tax deductions you'll need to track.)

### b. Paying estimated taxes

If your business makes money, you'll need to pay estimated taxes during the year. It's usually not too difficult to calculate what you owe and the mechanics of sending in your money are simple. You may have to make estimated tax payments soon after you start doing business, so don't delay this task. (See Chapter 11.)

If you live in one of the 43 states with income taxes—that is, all states but Alaska, Florida, Nevada, South Dakota, Texas, Washington and Wyoming—call your state tax department to find out whether you must make estimated tax payments. (See Appendix 1 for contact details.)

### c. Paying state and local taxes

You may have to pay various state and local taxes, such as a gross receipts tax, sales tax or personal property tax. (See Section A2.) This is an area where seeking guidance from a tax pro can be very helpful, particularly if the expert is familiar with businesses similar to yours. He or she can advise you as to whether you must pay any of these taxes. Otherwise, you'll need to contact your state tax department and local tax office for information. However, once you learn about the requirements and obtain the proper forms, it's usually not difficult to compute these taxes on your own.

### d. Filing your annual tax returns

The most difficult and time-consuming tax-related task you'll face is filing your annual tax return. This involves determining and adding up all your deductible expenses for the year and subtracting them from your gross income to determine your taxable income.

As you probably know, your federal tax return is due by April 15. If you live in one of the 43 states that have income taxes, you'll have to file a state income tax return as well. These are also due by April 15 except in six states that have later dates. (See the chart in Section A4.)

One way to make your life easier is to hire a tax pro to prepare your returns the first year you're in business. You can then use those returns as a guide when you do your own returns in future years.

### Taxes at the Touch of a Button

Many self-employed people do their tax returns themselves. If your business is small, it's usually not that difficult. This task is made much easier by the fact that there are many excellent publications and computer programs available to help you.

Having a computer will make preparing your tax returns much easier. Several tax preparation programs are available that contain all the forms you'll need. These programs not only do all your tax calculations for you, they also contain online tax help and questionnaires you can complete to figure out what forms to use.

The programs automatically put the information and numbers you type into the proper blanks on the forms. When you're done, the program prints out your completed tax forms. All you have to do by hand is sign the forms and put them in the mail. If you don't want to use the mail, you can even file your taxes electronically. Most tax packages also have additional programs you can purchase for your state taxes.

Three of the most highly regarded tax preparation programs are *TurboTax* (for Windows and DOS), *MacInTax* (for the Macintosh) and *Kiplinger Taxcut* (For Windows, DOS and Macintosh). There are versions of *TurboTax* and *MacInTax* specially designed for small business owners.

If you don't have a computer, there are several tax preparation guides that are published each year that provide much better guidance than the IRS instructions that come with your tax forms. These include *Consumer Reports' Guide to Income Tax* (Consumer Reports), and *The Ernst and Young Tax Guide* (John Wiley & Sons).

#### e. Paying employment taxes

If you have employees, your tax life becomes much more complicated than if you work alone or just hire independent contractors. You'll need to file both annual and quarterly employment tax returns. You're also required to withhold part of your employees' pay and send it to the IRS along with a contribution of your own. (See Chapter 13, Section B.)

This is an area where many business owners seek outside help, since calculating withholding can be complex. Many use an accountant or payroll tax service to perform these tasks. However, if you have a computer, accounting programs such as *QuickBooks* and *PeachTree Accounting* can calculate your employee withholding and prepare employment tax returns.

If you handle your taxes yourself, you'll likely want to eventually obtain a more detailed book specifically on taxation. Many excellent books are available. One of the best of these tax guides is *Tax Savvy for Small Business*, by Frederick W. Daily (Nolo Press).

The IRS also has publications on every conceivable tax topic. These are free, but can be difficult to understand. IRS Publication 910, *Guide to Free Tax Services*, contains a list of these publications and many of the most useful ones are cited in the following chapters. One that you should obtain is Publication 334, *Tax Guide for Small Business.* You can obtain these and all other IRS publications by calling the IRS at 800-TAX-FORM, visiting your local IRS or downloading them from the IRS's Internet site at www.irs.ustreas.gov.

### 2. Hiring a Tax Pro

Instead of doing the work yourself, you can hire a tax professional to perform some or all of it for you. A tax pro can also provide guidance to help you make key tax decisions, such as choosing the best set-up for your business and helping you deal with the IRS if you get into tax trouble. (See Section C.)

### a. Types of tax pros

There are several different types of tax pros. They differ widely in training, experience and cost.

**Tax preparers.** As the name implies, tax preparers prepare tax returns. The largest tax preparation firm is H & R Block, but many mom and pop operations open for business in storefront offices during tax time. In most states, anybody can be a tax preparer and no licensing is required. Most tax preparers don't have the training or experience to handle taxes for businesses and so are probably not a wise choice.

**Enrolled agents.** Enrolled agents or EAs are tax advisors and preparers who are licensed by the IRS. They have to have at least five years of experience and pass a difficult test. EAs are often the best choice for self-employed workers. They can often do as good a job as a certified public accountant, but charge less. Many also offer bookkeeping and accounting assistance.

**Certified Public Accountants.** Certified public accountants, or CPAs, are licensed and regulated by each state. They undergo lengthy training and must pass a comprehensive exam. CPAs represent the high end of the tax pro spectrum. In addition to preparing tax returns, they perform sophisticated accounting and tax work. Large businesses routinely hire CPAs for tax help. However, if you're running a one-person business, you may not require a CPA's expertise and will do just as well with a less expensive EA.

**Tax attorneys.** Tax attorneys are lawyers who specialize in tax matters. The only time you'll ever need a tax attorney is if you get into serious trouble with the IRS or other tax agency and need legal representation before the IRS or in court. Some tax attorneys also give tax advice, but they are usually too expensive for small businesses. You're probably better off hiring a CPA if you need specialized tax help.

### b. Finding a tax pro

The best way to find a tax pro is to obtain referrals from business associates, friends or professional associations. If none of these sources can give you a suitable lead, the National Association of Enrolled Agents at 800-434-4339 can help you locate an EA in your area. Local CPA societies can give you referrals to local CPAs. You can also find tax pros in the telephone book under "Accountants, Tax Return."

Your relationship with your tax pro will be one of your most important business relationships. Be picky about the person you choose. Talk with at least three tax pros before hiring one. You want a tax pro who takes the time to listen to you, answers your questions fully and in plain English, seems knowledgeable and makes you feel comfortable. Make sure the tax pro works frequently with small businesses. It can also be helpful if the tax pro already has clients in businesses similar to yours. A tax pro already familiar with the tax problems posed by your type of business can often give you the best advice for the least money.

### c. Tax pros' fees

Ask about a tax pro's fees before hiring him or her and, to avoid misunderstandings, obtain a written fee agreement before any work begins.

Most tax pros charge by the hour. Hourly rates vary widely depending on where you live and on the type of tax pro you hire. Enrolled agents often charge $25 to $50 per hour. CPAs typically charge about $100 per hour. Some tax pros charge a flat fee for specific services—for example, preparing a tax return.

These fees are rarely set in stone; you can usually negotiate. You'll be able to get the best possible deal if you hire a tax pro after the tax season when he or she is less busy—that is, during the summer or fall.

## C. IRS Audits

In an audit, the IRS examines you, your business, your tax returns and the records used to create the returns. If an IRS auditor determines you didn't pay enough tax, you'll have to pay the amount due plus interest and penalties.

Unfortunately, the self-employed have a much higher chance of being audited than employees. If you're a sole proprietor, the odds are about 50-50 that you'll be audited at least once if you're in business more than 11 years. Some self-employed workers are audited far more often. Repeat audits are especially likely if past audits turned up serious problems, such as your failure to report income.

### 1. Audit Time Limit

As a general rule, the law allows the IRS to audit a tax return up to 36 months after it's filed. This means you normally don't have worry about audits for tax returns you filed more than three years ago. The IRS calls the years it can audit you "open years."

### 2. Type of Audits

The IRS auditor will seek to interview you. If, like most self-employed people, you're a sole proprietor and gross less than $100,000 per year, the auditor will probably request that you come to the local IRS office for the audit. Office audits usually last from two to four hours. In an office audit, a typical business taxpayer is typically hit for additional taxes averaging $4,000.

However, if your business is a corporation, partnership or sole proprietorship grossing more than $100,000 per year, the IRS will usually seek to conduct the audit at your place of business. This is known as a field audit. You're entitled to request that the audit be conducted elsewhere—for example, the IRS office or the office of your attorney or accountant. Explain that your business will be disrupted if the auditor comes there. This is usually a good idea because you don't want an IRS auditor snooping around your business premises.

Field audits are conducted by IRS revenue agents, who are much better trained than IRS office auditors. They are more thorough than office audits—that is, a field auditor will spend more time on the audit and look at more records than an office auditor. As a result, these audits usually result in a larger tax bite. The typical field audit costs a business about $17,000.

### 3. What the Auditor Does

IRS auditors look primarily at two issues: whether you've under-reported your income and whether you've claimed tax deductions to which you're not entitled—for example, claimed that nondeductible personal expenses were deductible business expenses. Using a separate bank account for your business expenses will help convince the auditor that you're not mixing your personal bills with your business expenses.

The auditor will want to see the business records you used to prepare your tax returns, including your books, check registers, canceled checks and receipts. The auditor will also ask to see your records supporting your business tax deductions—for example, a record of the miles you've driven your car if you took a deduction for business use of your car.

IRS auditors can also obtain your bank records, either from you or your bank and check them to see if your deposits match the income you reported on your tax return. If the total of all your deposits is larger than your reported income, the auditor will assume you failed to report all your income unless you can show the deposits you didn't include in your tax return weren't income—for example, they were loans, inheritances, or transfers from other accounts. This is why you need to keep good records of your income. (See Chapter 14, Section A.)

An IRS auditor may also investigate whether you really qualify as an independent contractor or should have been classified as an employee by your clients or customers. (See Chapter 15.)

## 4. Handling Audits

You have the legal right to take anyone along with you to help during an audit—a bookkeeper, tax pro or even an attorney. If you've hired a tax pro to prepare your returns, it can be helpful for him or her to attend the audit to help explain your business receipts and records and to explain how the returns were prepared. Some tax pros include free audit services as part of a tax preparation package.

However, if you prepared your tax returns yourself, you can probably deal with an office audit yourself. It could cost more to hire a tax pro to represent you in an office audit than the IRS is likely to bill you. If you're worried that some serious irregularity will come to light—for example, you've taken a huge deduction and can't produce a receipt or canceled check to verify it—consult with a tax pro before the audit.

No matter who prepared your tax returns, it usually makes sense to have a tax pro to represent you in a field audit, since these can result in very substantial assessments.

For a detailed discussion of IRS small business audits, see *Tax Savvy for Small Business*, by Frederick W. Daily (Nolo Press). ■

CHAPTER

# 9

# Reducing Your Income Taxes

If you're like the vast majority of self-employed people, you must pay personal federal income tax on the net profit you earn from your business activities. This is true whether you're legally organized as a sole proprietor, S corporation, partnership or limited liability company. The only exception is if you've formed a C corporation. (See Chapter 2, Section B.)

The key phrase here is "net profit." You are entitled to deduct the total amount of your business-related expenses from your gross income—that is, all the money or the value of other items you receive from your clients or customers. You pay income tax on your resulting net profit, not all your IC income.

> **EXAMPLE:** Karen, a sole proprietor, earned $50,000 this year from her consulting business. Fortunately, she doesn't have to pay income tax on the entire $50,000. This is because she qualifies for several business-related tax deductions, including a $5,000 home office deduction (see Chapter 4.) and a $10,000 deduction for equipment expenses (see Section D). She deducts these amounts from her $50,000 gross income to arrive at her net profit: $35,000. She only pays income tax on that amount.

This chapter provides an overview of the many business-related federal income tax deductions that are available to reduce your net profits and so reduce the amount of income tax you have to pay. It is a starting point for understanding income taxes, but by no means a complete discussion of this complex subject.

**Most States Have Income Taxes, Too**

All states except Alaska, Florida, Nevada, South Dakota, Texas, Washington and Wyoming also impose personal income taxes on the self-employed. If you're incorporated, your corporation will likely have to pay state income taxes as well. Contact your state tax department for income tax information and the appropriate forms. (See the Appendix for a list of state income tax offices.)

For a more detailed treatment of income taxes, see:

- *Tax Savvy for Small Business*, by Frederick W. Daily (Nolo Press), and
- *Tax Planning and Preparation Made Easy for the Self-Employed*, by Gregory L. Dent and Jeffrey E. Johnson (John Wiley & Sons).

In addition, many IRS publications dealing with income tax issues are mentioned below. You can obtain a free copy of any IRS publication by calling 800-TAX FORM or contacting your local IRS office. You can also download the publications from the IRS Internet site at www.irs.ustreas.gov. (See Chapter 21, Section E.)

## A. Income Reporting

Employers deduct income tax from their employees' paychecks and remit and report it to the IRS. They give all their employees an IRS Form W-2, Wage and Tax Statement, showing wages and withholding for the year. Employees must file a copy of the W-2 with their income tax returns so that the IRS can compare the amount of income they report with the amounts their employers claim they were paid.

When you're self employed, no income tax is withheld from your compensation and you don't receive a W-2 form. However, this does not mean that the IRS doesn't have at least some idea of how much money you've made. If the total of all the payments you receive from a client over the course of a year is $600 or more, the client must complete and file IRS Form 1099-MISC reporting the payments.

The client must complete and file a copy of the 1099 with:

- the IRS
- your state tax office if your state has income tax, and
- you.

To make sure you're not underreporting your income, IRS computers check the amounts listed on your 1099s against the amount of income you report on your tax return. If the amounts don't match, you have a good chance of being flagged for an audit.

### 1. When 1099s Are Not Required

Your clients need not file any 1099 form if you've incorporated your business and the client hires your corporation, not you personally. This is one reason clients often prefer to hire incorporated businesses. The IRS uses 1099s as an important audit lead. If a company files more 1099s than average for its type of business, the IRS often concludes that it must be misclassifying employees as independent contractors and may conduct an audit.

The only exception to the rule that you need not file a 1099 for a corporation is if you're a medical doctor and have formed a medical corporation. In this set-up, clients who retain your services for their businesses must file 1099s just as if you were a sole proprietor.

In addition, 1099s are not required when you perform services for a person who is not engaged in a business, such as a homeowner or patient. Nor are they required to report payments to you solely for merchandise or inventory. (See Chapter 13, Section C1.)

### 2. What to Do With Your 1099s

You should receive all your 1099s for the previous year by January 31 of the current year. Check the amount of compensation your clients say they paid you in each 1099 against your own records, to make sure they are consistent. If there is a mistake, call the client immediately and request a corrected 1099. Insist that a corrected 1099 be filed with the IRS. You don't want the IRS to think you've been paid more than you really were.

You don't have to file your 1099s with your tax returns. Just keep them in your records.

### 3. If You Don't Receive a 1099

It's not unusual for clients to fail to file required 1099s. This may be unintentional—for example, because the client doesn't understand the rules or is negligent. On the other hand, some clients purposefully fail to file 1099s because they don't want the IRS to know they're hiring independent contractors.

If, by January 31, you don't receive a 1099 from a client who paid you more than $600 for business-related services the prior year, call the client and ask for it. If the client still does not produce the form, don't worry about it. It's not your duty to see that 1099s are filed. This is your client's responsibility. The IRS will not impose any fines or penalties on you if a client fails to file a 1099. It may, however, impose a $50 fine on the client—and exact far more severe penalties if an IRS audit reveals that the client should have classified you as an employee.

However, whether or not you receive a Form 1099, it is your duty to report all the self-employed income you earn each year to the IRS. If you're audited by the IRS, it will, among other things, examine your bank records to make sure you haven't underreported your income. If you have underreported, you'll have to pay back taxes, fines and penalties.

#### Your Reported Income Must Jibe with 1099s

It's very important that the self-employment income you report in your tax return at least be equal to that reported to the IRS on the 1099s your clients send in. If a client reimbursed you for expenses such as travel, be sure to check and see if the 1099 the client provides you includes this amount. Some clients routinely include expense reimbursements on their 1099s, others do not.

If a 1099 includes expenses, you must report the entire amount as income on your tax return. You then deduct the amount of the expense reimbursement as your own business expense on your Schedule C. This way, your net self-employment income will come out right.

## B. Introduction to Income Tax Deductions

A deduction is an expense or the value of an item that you can subtract from your gross income to determine your taxable income—that is, the amount you earn that is subject to taxation. The more deductions you have, the less income tax you pay. When people speak of taking a deduction or deducting an expense from their income taxes, they mean they subtract it from their gross incomes.

Most of the work involved in doing your taxes is determining what deductions you can take, how much you can take and when you can take them. You don't have to become an income tax expert. But even if you have a tax pro prepare your tax returns, you need to have a basic understanding of what expenses are deductible so that you can keep proper records. This takes some time, but it's worth it. There's no point in working hard to earn a good income only to turn most of the money over to the government.

### 1. What You Can Deduct

Virtually any expense is deductible as long as it is:

- ordinary and necessary
- directly related to your business, and
- for a reasonable amount.

#### a. Ordinary and necessary expenses

An expense qualifies as ordinary and necessary if it is common, accepted, helpful and appropriate for your business or profession. An expense doesn't have to be indispensable to be necessary; it need only help your business in some way, even in a minor way. It's usually fairly easy to tell if an expense passes this test.

**EXAMPLE:** Bill, a freelance writer, hires a research assistant for a new book he's writing about ancient Athens and pays her $15 an hour. This is clearly a deductible business expense. Hiring research assistants is a common and accepted practice among professional writers. The assistant's fee is an ordinary and necessary expense for Bill's writing business.

**EXAMPLE:** Bill, the freelance writer, visits a masseuse every week to work on his bad back. Bill claims the cost as a business expense, reasoning that avoiding back pain helps him concentrate on his writing. This is clearly not an ordinary or customary expense for a freelance writer and the IRS would not likely allow it as a business expense.

#### b. Expense must be related to your business

An expense must be related to your business to be deductible. That is, you must use the item you buy for your business in some way. For example, the cost of a personal computer is a deductible business expense if you use the computer to write business reports.

You cannot deduct purely personal expenses as business expenses. The cost of a personal computer is not deductible if you use it just to play computer games. If you buy something for both personal and business reasons, you may deduct the business portion of the expense. For example, if you buy a cellular phone and use it half the time for business calls and half the time for personal calls, you can deduct half the cost of the phone as a business expense.

However, the IRS requires you to keep records showing when the item was used for business and when for personal reasons. One acceptable form of record would be a diary or log with the dates, times and reason the item was used. (See Chapter 14, Section A.) This kind of recordkeeping can be burdensome and may not be worth the trouble if the item isn't very valuable.

To avoid having to keep such records, try to use items either only for business or only for personal use. For example, if you can afford it, purchase two computers and use one solely for your business and one for playing games and other personal uses.

### c. Deductions must be reasonable

There is usually no limit on how much you can deduct so long as it's not more than you actually spend and the amount is reasonable. Certain areas are hot buttons for the IRS—especially entertainment, travel and meal expenses. The IRS won't allow such expenses to the extent it considers them lavish. (See Section G.)

Also, if the amount of your deductions is very large relative to your income, your chance of being audited goes up dramatically. One recent analysis of almost 1,300 tax returns found that you are at high risk for an audit if your business deductions exceed 63% of your revenues. You're relatively safe so long as your deductions are less than 52% of your revenue. If you have extremely large deductions, make sure you can document them in case you're audited. (See Chapter 14, Section A3.)

### d. Common deductions for the self-employed

Self-employed workers typically are entitled to take a number of income tax deductions. The most common include:

- advertising costs—for example, the cost of a Yellow Pages advertisement or brochure
- attorney and accounting fees for your business
- bank fees for your business bank account
- business start-up costs
- car and truck expenses (see Section E)
- costs of renting or leasing vehicles, machinery, equipment and other property used in your business
- depreciation of business assets (see Section D2)
- education expenses—for example, the cost of attending professional seminars or classes required to keep up a professional license
- expenses for the business use of your home (see Section C)
- fees you pay to other self-employed workers you hire to help your business—for example, the cost of paying a marketing consultant to advise you on how to get more clients
- health insurance for yourself and your family (see Section H)
- insurance for your business—for example, liability, workers' compensation and business property insurance (see Chapter 6)
- interest on business loans and debts—for example, interest you pay for a bank loan you use to expand your business
- license fees—for example, fees for a local business license or occupational license (see Chapter 5)
- office expenses, such as office supplies
- office utilities
- postage
- professional association dues
- professional or business books you need for your business
- repairs and maintenance for business equipment such as a photocopier or fax machine
- retirement plan contributions (see Section H3)
- software you buy for your business (see Section D3)
- subscriptions for professional or business publications
- travel, meals and entertainment (see Sections F and G), and
- wages and benefits you provide your employees.

**INVENTORY IS NOT A BUSINESS EXPENSE**

If you make or buy goods to sell, you are entitled to deduct the cost of those goods actually sold on your tax return. This is what you spent for the goods or their actual market value if they've declined in value since you bought them.

However, money spent for goods to sell is not treated as a business expense. Instead, you deduct the cost of goods you've sold from your business receipts to determine your gross profit from the business. Your business expenses are then deducted from your gross profit to determine your net profit, which is taxed.

Businesses that make, buy or sell goods must determine the value of their inventories at the beginning and the end of each tax year using an IRS-approved accounting method. Conducting inventories can be burdensome. Many self-employed people don't have to worry about inventories because they sell personal services to their clients or customers, not goods or products. For example, doctors, lawyers, carpenters and painters usually do not keep inventories of goods. However, if in addition to selling your services, you also regularly charge clients for materials and supplies, you must maintain an inventory for tax purposes.

For more information on inventories, see the Cost of Goods Sold section in Chapter 7 of IRS Publication 334, Tax Guide for Small Businesses; and IRS Publication 538, Accounting Periods and Methods, and Publication 970, Application to Use LIFO Inventory Method.

You can obtain these and all other IRS publications by calling the IRS at 800-TAX-FORM, visiting your local IRS office or downloading the publications from the IRS Internet site at www.irs.ustreas.gov.

## 2. When to Deduct

Some expenses can be deducted all at once; others have to be deducted over a number of years. It all depends on how long the item you purchase can reasonably be expected to last—what the IRS calls its useful life.

### a. Current expenses

The cost of anything you buy for your business that has a useful life of less than one year must be fully deducted in the year it is purchased. This includes, for example, rent, telephone and utility bills, photocopying costs and postage and other ordinary business operating costs. Such items are called current expenses.

**EXAMPLE:** Max, a self-employed telemarketer, spends $5,000 on phone bills in one year. He can deduct this entire expense from his income the year he incurs it.

### b. Capital expenses

Certain types of costs are considered to be part of your investment in your business instead of operating costs. These are called capital expenses. Subject to an important exception for a certain amount of personal property (see Section D1), you cannot deduct the full value of such expenses in the year you incur them. Instead, you must spread the cost over several years and deduct part of it each year.

There are two main categories of capital expenses. They include:

- the cost of any asset you will use in your business that has a useful life of more than one year—for example, equipment, vehicles, books, furniture, machinery, patents (see Section D2), and
- business start-up costs such as fees for doing market research or attorney and accounting fees paid to set up your business.

## 3. Businesses That Lose Money

If the money you spend on your business exceeds your business income for the year, your business incurs a loss. This isn't as bad as it sounds because you can use a business loss to offset other income you may have—for example, interest income or your spouse's income if you file jointly. You can even accumulate your losses and apply them to reduce your income taxes in future or past years.

For detailed information on deducting business losses, see IRS Publication 536, *Net Operating Losses*. You can obtain this and all other IRS publications by calling the IRS at 800-TAX-FORM, visiting your local IRS office or downloading the publications from the IRS Internet site at www.irs.ustreas.gov.

### a. Recurring losses

If you keep incurring losses year after year, you need to be very concerned about running afoul of the hobby loss rule. This tax law could cost you a fortune in additional income taxes.

The IRS created the hobby loss rule to prevent taxpayers from entering into ventures primarily to incur expenses they could deduct from their other incomes. The rule halts this form of tax avoidance by allowing you to take a business expense deduction only if your venture qualifies as a business. Ventures that don't qualify as businesses are called hobbies. If the IRS views what you do as a hobby, there will be severe limits on what expenses you can deduct.

A venture is a business if you engage in it to make a profit. It's not necessary that you earn a profit every year. All that is required is that your main reason for doing what you do is to make a profit. A hobby is any activity you engage in mainly for a reason other than making a profit—for example, to incur deductible expenses or just to have fun.

The IRS can't read your mind to determine whether you want to earn a profit. And it certainly isn't going to take your word for it. Instead, it looks to see whether you do actually earn a profit, or whether you behave as if you want to earn a profit.

### b. Profit test

You don't have to worry about the IRS labeling your business as a hobby if you earn a profit from it in any three of five consecutive years. If your venture passes this test, the IRS presumes it is carried on for profit.

You have a profit when the gross income from an activity is more than the deductions for it. You don't have to earn a big profit to satisfy this test. Careful year-end planning can help your business show a profit for the year. For example, if clients owe you money, press for payment before the end of the year. Also, put off paying expenses or buying new equipment until the new year.

A loss for one or two years isn't fatal. But do everything possible to avoid showing losses for three consecutive years. This will definitely set the IRS computers to whirring and increase your chance of being audited. If you are audited, the hobby loss issue will definitely be raised.

### c. Behavior test

If you keep incurring losses and can't satisfy the profit test, you by no means have to throw in the towel and decide your venture is a hobby. You can continue to treat it as a business and fully deduct your losses. However, you must take steps that will convince the IRS your business isn't a hobby if you're audited.

You must be able to convince the IRS that earning a profit—not having fun or accumulating tax deductions—is the primary motive for what you do. This can be particularly difficult if you're engaged in an activity that could objectively be considered fun—for example, creating artwork, photography and writing—but it can still be done. People who have incurred losses for seven, eight or nine years in a row have convinced the IRS they were running a business.

### Losing Golfer Scores Hole-in-One in Tax Court

Donald, a Chicago high school gym teacher, decided to become a golf pro when he turned 40. He became a member of the Professional Golfers of America, which entitled him to compete in certain professional tournaments. Donald kept his teaching job, but played in various professional tournaments during the summer. His expenses exceeded his income from golfing for five straight years.

| Year | Golf Earnings | Golf Expenses | Losses |
|---|---|---|---|
| 1978 | $ 0 | $2,538 | $2,538 |
| 1979 | 148 | 1,332 | 1,184 |
| 1980 | 400 | 4,672 | 4,272 |
| 1981 | 904 | 4,167 | 3,263 |
| 1982 | 1,458 | 8,061 | 6,603 |

The IRS sought to disallow the losses for 1981 and 1982, claiming that golf was a hobby for Donald. Donald appealed to the Tax Court and won.

The court held that Donald played golf to make a profit, not just to have fun. He carefully detailed the expenses he incurred for each tournament he entered and recorded the prize money available. He attended a business course for golfers, and assisted a professional golfer from whom he also took lessons. He practiced every day, up to 12 hours during the summer; he also traveled frequently to Florida during the winter to play. Although his costs increased over the years, his winnings steadily increased each year as well. The court concluded that, although Donald obviously enjoyed golfing, he honestly wanted to earn a profit from it. (*Kimbrough v. Commissioner,* 55 TCM. 730 (1988).)

You must show the IRS that your behavior is consistent with that of a person who really wants to make money. There are many ways to accomplish this.

First and foremost, you must show that you carry on your enterprise in a businesslike manner—for example, you:

- maintain a separate checking account for your business (see Chapter 14, Section A)
- keep good business records (see Chapter 14, Section A2)
- make some effort to market your services—for example, have business cards and, if appropriate, a Yellow Pages or similar advertisement
- have business stationery and cards printed
- obtain a federal employer identification number (see Chapter 5)
- secure all necessary business licenses and permits (see Chapter 5)
- have a separate phone line for your business if you work at home
- join professional organizations and associations, and
- develop expertise in your field by attending educational seminars and similar activities.

It is also helpful to draw up a business plan with forecasts of revenue and expenses. This will also be a big help if you try to borrow money for your business.

For detailed guidance on how to create a business plan, see *How to Write a Business Plan*, by Mike McKeever (Nolo Press).

The more time and effort you put into the activity, the more it will look like you want to make money. So try to devote as much time as possible to your business and keep a log showing the time you spend on it.

It's also helpful to consult with experts in your field and follow their advice about how to modify your operations to increase sales and cut costs. Be sure to document your efforts.

**EXAMPLE:** Otto, a professional artist, has incurred losses from his business for the past three years. He consults with Cindy, a prominent art gallery owner, about how he can sell more of his work. He writes down her recommendations and then documents his efforts to follow them—for example, he visits art shows around the country and talks with a number of gallery owners about representing his work.

You'll have an easier time convincing the IRS your venture is a business if you earn a profit in at least some years. It's also very helpful if you've earned profits from similar businesses in the past.

#### d. Tax effect

If the IRS determines your venture is a hobby, you'll lose valuable deductions and your income tax burden will increase. Unlike business expenses, expenses for a hobby are personal expenses that you can deduct only from income from the hobby. They can't be applied to your other income such as your or your spouse's salary or interest income.

**EXAMPLE:** Bill holds a fulltime job as a college geology teacher and also paints parttime and shows his work in art galleries. The IRS has decided that painting is a hobby for Bill, since he's never earned a profit from it. In one year, Bill spent $2,000 on the painting hobby, but earned only $500 from the sale of one painting. His expenses can only be deducted from the $500 income derived from painting. This wipes out the $500, but Bill cannot apply the remainder to his other income. Because of the hobby loss rule, Bill has lost $1,500 worth of tax deductions.

### 4. Tax Savings From Deductions

Only part of any deduction will end up as an income tax saving—for example, a $5,000 tax deduction will not result in a $5,000 income tax saving.

How much you'll save depends on your tax rate. The tax law assigns a percentage income tax rate to specified income levels. People with high incomes pay income tax at a higher rate than those with lower incomes. These percentage rates are called tax brackets.

To determine how much income tax a deduction will save you, you need to know your marginal tax bracket. This is the tax bracket in which the last dollar you earn falls. It's the rate at which any additional income you earn would be taxed. The income tax brackets are adjusted each year for inflation.

For the current brackets, see IRS Publication 505, *Tax Withholding and Estimated Tax*. You can obtain this and all other IRS publications by calling the IRS at 800-TAX-FORM, visiting your local IRS office or downloading the publications from the IRS Internet site at www.irs.ustreas.gov.

The following table shows the tax brackets for 1997. For example, if you were single and earned $50,000 in 1997, your marginal tax bracket was 28%—that is, you had to pay 28 cents in income tax for every additional dollar you earned.

To determine how much tax a deduction will save you, multiply the amount of the deduction by your marginal tax bracket. If your marginal tax bracket is 28%, you will save 28 cents in income taxes for every dollar you are able to claim as a deductible business expense.

**1997 TAX BRACKETS**

| Tax Bracket | Income If Married Filing Joint Return | Income If Single |
|---|---|---|
| 15% | Up to $41,200 | Up to $24,650 |
| 28% | From $41,201 to $99,600 | $24,650 to $59,750 |
| 31% | $99,600 to $151,750 | $59,751 to $124,650 |
| 36% | $151,751 to $271,050 | $124,651 to $271,050 |
| 39.6% | On all over $271,050 | On all over $271,050 |

**EXAMPLE:** Barry, a single self-employed consultant, earned $50,000 in 1997 and was therefore in the 28% marginal tax bracket. He was able to take a $5,000 home office deduction. (See Chapter 4.) His actual income tax saving was 28% of the $5,000 deduction, or $1,400.

You can also deduct most business-related expenses from your income for self-employment tax purposes. (See Chapter 10.)

## C. Business Use of Your Home

Many self-employed people work from home, particularly when they're starting out. If you can meet some strict requirements, you're allowed to deduct your expenses for the business use of part of your home. This is commonly called the home office deduction. (See Chapter 4)

However, this deduction is not limited to home offices. You can also take it if, for example, you have a workshop or studio at home.

The deduction can be taken for a house, apartment, condominium, mobile home, boat or separate unattached structure on your property.

You can take it whether you are an owner or renter.

And you can take it for a business that is a sideline, not only for your primary means of income.

You can also take it if you regularly store business inventory at home.

Unfortunately, it isn't nearly as easy as it used to be to qualify for the home office deduction because the IRS and U.S. Supreme Court have imposed strict requirements on who can take it.

### Home Alone or Outside Office?

Your decision about whether to work at home or in an outside office should not be based primarily on tax considerations; it should be based on what is best for you and your business. If clients regularly need to meet with you in your office, an outside office is desirable if you can afford it. It will also help establish that you're an IC. (See Chapter 15.)

Office rent and utilities are deductible business expenses. On the other hand, if you never see clients or only deal with them over the phone or fax machine, a home office may be just fine provided you have adequate space and enough peace and quiet to get your work done.

### 2. Qualifying for the Deduction

Unfortunately, satisfying the requirement of using your home office regularly and exclusively for business is only half the battle.

It must also be true that:

- your home office is your principal place of business
- you meet clients or customers at home, or
- you use a separate structure on your property exclusively for business purposes.

### Ways to Solidify Home Office Deduction

Here are some ways to help convince the IRS you qualify for the home office deduction.

- Have all your business mail sent to your home office.
- Use your home office address on all your business cards, stationery and advertising.
- Obtain a separate phone line for your business and keep that phone in your home office. The tax law helps you do this by allowing you to deduct the monthly fee for a second phone line in your home if you use it for business. You can't deduct the monthly fee for a single phone line, even if you use it partly for business; however, you can deduct the cost of business calls you place from that line. Having a separate business phone will also make it easier for you to keep track of your business phone expenses.
- Encourage clients or customers to regularly visit your home office and keep a log of their visits.
- To make the most of the time you spend in your home office, communicate with clients by phone, fax or electronic mail instead of going to their offices. Use a mail or messenger service to deliver your work to customers.
- Keep a log of the time you spend working in your home office. This doesn't have to be fancy; notes on your calendar will do.

### a. Home as principal place of business

The most common way to qualify for the home office deduction is to use your home as your principal place of business. If, like many ICs, you do all or most of your work in your home office, it is clearly your principal place of business—for example, you're a writer and you do most of your writing in your home office, or you sell by phone and make most of your sales calls from home.

Things get much more complicated, however, if you spend a substantial amount of time working outside your home. In this event, your home office will be your principal place of business only if it is where you do most of your important business or spend most of your working hours—that is, both the amount of time you spend in your home office and the importance of the work you do there are important concerns.

There is some help for you in evaluating whether a workspace you use in this way can qualify for a deduction.

First, compare the business activities you perform at your home office with those you perform outside your home office to determine which ones are more important for your business. Activities that directly help you make money are most important; those that don't directly lead to earning income are less important. For example, if you're a salesperson, calling on customers and making sales are your most important activities. Without sales, you earn no income. Recordkeeping is less important for a salesperson since it doesn't directly lead to earning income.

If you must meet with clients or patients outside your office or if you deliver services or goods to customers, your home office will likely not qualify as your principal place of business. For example, if you're a salesperson and make your sales by calling on customers in their offices, you probably cannot claim that your home office is your principal place of business.

It may not be clear which location is most important for your business—if, for example, you perform income-producing services for your clients both at your home office and their premises. Compare the amount of time you spend on business in your home office with the amount of time you spend elsewhere. If you spend the majority of your time in your home office, it may qualify as your principal place of business.

> **EXAMPLE:** Joe, a self-employed salesperson, makes sales both by phone from his home office and by visiting customers at their premises. As a result, he does equally important work both in his home office and outside it. However, Joe spends 80% of his time making sales from home and only 20% on the road. Since Joe spends the majority of his time in his home office, it qualifies as his principal place of business.

## D. Deducting the Cost of Business Assets

One of the nice things about being self-employed is that you can deduct from your income taxes what you spend for things you use to help produce income for your business—for example, computers, calculators and office furniture. You can take a full deduction whether you pay cash for an asset or buy on credit.

If you qualify for the Section 179 deduction discussed below, you can deduct the entire cost of these items in the year you pay for them. Otherwise, you have to deduct the cost over a period of years—a process called depreciation.

The rules for deducting business assets can be complex, but it's worth spending the time to understand them. After all, by allowing these deductions, the U.S. government is in effect offering to help you pay for your equipment and other business assets. All you have to do is take advantage of the offer.

### 1. Section 179 Deduction

If you learn only one section number in the tax code, it should be Section 179. It is one of the greatest tax boons for small businesspeople. Section 179 permits you to deduct a large amount of your business asset purchases in the year you make them, rather than having to depreciate them over several years. (See Section D2.) This is called first year expensing or Section 179 expensing. It allows you to get a big tax deduction all at once, rather than having to mete it out a little at a time.

> **EXAMPLE:** Ginger buys a $8,000 photocopy machine for her business. She can use Section 179 to deduct the entire $8,000 expense from her income taxes for the year.

It's up to you to decide whether to use Section 179. It may not always be in your best interests to do so. (See Section D2.) If you do use it, you can't change your mind later and decide to use depreciation instead.

### a. Property that can be deducted

You can use Section 179 to deduct the cost of any tangible personal property you use for your business that the IRS has determined will last more than one year—for example, computers, business equipment and office furniture. (See Section D3.) Special rules apply to cars. (See Section D3.) You can't use Section 179 for land, buildings or intangible personal property such as patents, copyrights and trademarks.

If you use property both for business and personal purposes, you may deduct under Section 179 only if you use it for business purposes more than half the time. The amount of your deduction is reduced by the percentage of personal use. You'll need to keep records showing your business use of such property. (See Chapter 14.)

If you use an item for business less than half the time, you must depreciate it. (See Section D2.)

### b. Deduction limit

There is a limit on the total amount of business expenses you can deduct each year using Section 179. Over the next several years, the limit will gradually go up from $17,500 in 1996 to $25,000 in 2003 and later as noted on the following chart.

**LIMITS ON SECTION 179 DEDUCTION**

| Year | Section 179 Deduction |
|---|---|
| 1996 | $17,500 |
| 1997 | $18,000 |
| 1998 | $18,500 |
| 1999 | $19,000 |
| 2000 | $20,000 |
| 2001 | $24,000 |
| 2002 | $24,000 |
| 2003 and later | $25,000 |

This dollar limit applies to all your businesses together, not to each business you own and run. You do not have to claim the full amount. It's up to you decide how much of the cost of property you want to deduct. But you don't lose out on the remainder; you can depreciate any cost you do not deduct under Section 179. (See Section D2.)

If you purchase more than one item of Section 179 property during the year, you can divide the deduction among all the items in any way, as long as the total deduction is not more than the Section 179 limit. It's usually best to apply Section 179 to property that has the longest useful life and therefore the longest depreciation period. This reduces the total time you have to wait to get your deductions. (See Section D2c.)

**EXAMPLE:** In 1997, Ben, a self-employed consultant, buys for his business a $10,000 computer, an $10,000 copier and $10,000 in office furniture. The total for all these purchases is $30,000—more than the $18,000 Section 179 limit for 1997. He can divide his Section 179 deduction among these items any way he wants. The copier and computer would have to be depreciated over five years, but the furniture over seven years. He should apply Section 179 to the furniture first and then to the computer or copier. Any portion of the cost of the copier or computer that exceeds the Section 179 limit can be depreciated over five years. This way, he avoids having to wait seven years to get his full deduction for the furniture.

### c. Limit on Section 179 deduction

You can't use Section 179 to deduct more in one year than your total profit from all of your businesses and your salary if you have a job in addition to your business. If you're married and file a joint tax return, your spouse's salary and business income is included as well. You can't count investment income—for example, interest you earn on your savings.

You can't use Section 179 to reduce your taxable income below zero. But any amount you cannot use as a Section 179 deduction you can carry to the next tax year and possibly deduct it then.

> **EXAMPLE:** In 1997, Amelia earned a $5,000 profit from her engineering consulting business and $10,000 from a parttime job. She spent $17,000 for computer equipment. She can use Section 179 to deduct $15,000 of this expense for 1997 and deduct the remaining $2,000 the next year.

In the unlikely event that you buy over $200,000 of Section 179 property in a year, your deduction is reduced by one dollar for every dollar you spend over that amount.

### d. Minimum period of business use

When you deduct an asset under Section 179, you must continue to use it for business at least 50% of the time for as many years it would have been depreciated. (See Section D2c.) For example, if you use Section 179 for a computer, you must use it for business at least 50% of the time for five years since computers have a five-year depreciation period.

If you don't meet these rules, you'll have to report as income part of the deduction you took under Section 179 in the prior year. This is called recapture.

For more information, see IRS Publication 334, *Tax Guide for Small Business, Chapter 12, Depreciation*. You can obtain this and all other IRS publications by calling the IRS at 800-TAX-FORM, visiting your local IRS office or downloading the publications from the IRS Internet site at www.irs.ustreas.gov.

## 2. Depreciation

Because it provides a big tax deduction immediately, most small business owners look first to Section 179 to deduct asset costs.

However, you must use depreciation instead if you:

- don't qualify to use Section 179—for example, you want to deduct the cost of a building or a patent, or
- use up your Section 179 deduction for the year.

Depreciation involves deducting the cost of a business asset a little at time over a period of years. This means it will take you much longer to get your full deduction than under Section 179. However, this isn't always a bad thing. Indeed, you may be better off in the long run using depreciation instead of Section 179 if you expect to earn more in future years than you will in the current year. Remember that the value of a deduction depends on your income tax bracket. If you're in the 15% bracket, a $1,000 deduction is worth only $150. If you're in the 31% bracket, it's worth $310. (See Section B4.) So spreading out a deduction until you're in a higher tax bracket can make sense.

**EXAMPLE:** Marie, a self-employed consultant, buys a $5,000 photocopier for her business in 1997. She elects to depreciate the copier instead of using the Section 179 deduction. This way, she can deduct a portion of the cost from her gross income each year for the next six years. Marie is only in the 15% tax bracket in 1997, but expects to be in the 31% bracket in 1998.

The following chart shows how much more money Marie saves by taking depreciation and deducting a portion of the copier's cost over six years rather than taking a Section 179 deduction for the entire cost in one year. Marie is using the straight line depreciation method which produces the same deduction every year, except the first and last. (See Section D2d.)

Depreciation may also be preferable to using Section 179 if you want to puff up your business income for the year. This can help you get a bank loan or help your business show a profit instead of incurring a loss and avoid running afoul of the hobby loss limitations. (See Section B3.)

### a. What must be depreciated

Whether you must depreciate an item depends on how long it can reasonably be expected to last—what the IRS calls its useful life. Depreciation is used to deduct the cost of any asset you buy for your business that has a useful life of more than one year—for example, buildings, equipment, machinery, patents, trademarks, copyrights and furniture. Land cannot be depreciated because it doesn't wear out. The IRS, not you, decides the useful life of your assets. (See Section D2c.)

You can also depreciate the cost of major repairs that increase the value or extend the life of an asset—for example, the cost of a major upgrade to make your computer run faster. However, you deduct normal repairs or maintenance in the year they're incurred as a business expense.

### b. Mixed use property

If you use property both for business and personal purposes, you can take depreciation only for the business use of the asset. Unlike for the Section 179 deduction, you don't have to use an item over half the time for business to depreciate it.

**COMPARISON OF DEDUCTIONS: SECTION 179 AND DEPRECIATION**

| Year | Marie's Marginal Tax Rate | Tax Savings Using Section 179 Deduction | Tax Savings Using Depreciation Deduction |
|---|---|---|---|
| 1997 | 15% | $750 (15% x $5,000, the cost of the copier) | $75 (15% x $500 of copier's $5,000 cost) |
| 1998 | 31% | 0 | $310 (31% x $1,000) |
| 1999 | 31% | 0 | $310 (31% x $1,000) |
| 2000 | 31% | 0 | $310 (31% x $1,000) |
| 2001 | 31% | 0 | $310 (31% x $1,000) |
| 2002 | 31% | 0 | $155 (31% x $500) |
| TOTAL | | $750 | $1,470 |

**EXAMPLE:** Carl uses his photocopier 75% of the time for personal reasons and 25% for business. He can depreciate 25% of the cost of the copier.

Keep a diary or log with the dates, times and reason the property was used to distinguish between the two uses. (See Chapter 14, Section A.)

### c. Depreciation period

The depreciation period—called the recovery period by the IRS—begins when you start using the asset and lasts for the entire estimated useful life of the asset. The tax code has assigned an estimated useful life for all types of business assets, ranging from 3 to 39 years. Most of the assets you buy for your business will probably have an estimated useful life of five or seven years .

If you need to know the depreciation period for an asset not included in this table, see IRS Publication 534, *Depreciation,* for a complete listing. You can obtain this and all other IRS publications by calling the IRS at 800-TAX-FORM, visiting your local IRS office or downloading the publications from the IRS Internet site at www.irs.ustreas.gov.

You are free to continue using property after its estimated useful life expires, but you can't deduct any more depreciation.

### d. Calculating depreciation

You can use three different methods to calculate the depreciation deduction: straight line or one of two accelerated depreciation methods. Once you choose your method, you're stuck with it for the entire life of the asset.

In addition, you must use the same method for all property of the same kind purchased during the year. For example, if you use the straight line method to depreciate a computer, you must use that method to depreciate all other computers you purchase during the year for your business.

The straight line method requires you to deduct an equal amount each year over the useful life of an asset. However, you ordinarily deduct only a half-year's worth of depreciation in the first year. You make up for this by adding an extra year of depreciation at the end.

**EXAMPLE:** Sally buys a $1,000 fax machine for her business in 1996. It has a useful life of five years. (See the chart in Section D2c.) Using the straight line method, she would

**ASSET DEPRECIATION PERIODS**

| Type of Property | Recovery Period |
|---|---|
| Computer software (software that comes with your computer is not separately depreciable unless you're separately billed for it) | 3 years |
| Office machinery (computers and peripherals, calculators, copiers, typewriters) | 5 years |
| Autos and light trucks | 5 years |
| Construction and research equipment | 5 years |
| Office furniture | 7 years |
| Residential buildings | 27.5 years |
| Nonresidential buildings purchased before 5/12/93 | 31.5 years |
| Nonresidential buildings purchased after 5/12/93 | 39 years |

depreciate the asset over six years. Her annual depreciation deductions are as follows:

| | |
|---|---|
| 1996 | $100 |
| 1997-2000 | $800 ($200 each year) |
| 2001 | $100 |
| TOTAL | $1,000 |

Most small businesses use one of two types of accelerated depreciation: the double declining balance method and the 150% declining balance method. The advantage to these methods is that they provide larger depreciation deductions in the earlier years and smaller ones later on. The double declining balance method starts out by giving you double the deduction you'd get for the first full year with the straight line method. The 150% declining balance method gives you one and one-half times the straight line deduction.

**EXAMPLE:** Sally decides to use the fastest accelerated depreciation method to depreciate her $1,000 fax machine—the double declining balance method. Her annual depreciation deductions are as follows:

| | |
|---|---|
| 1996 | $400 |
| 1997 | $240 |
| 1998 | $144 |
| 1999 | $108 |
| 2000 | $108 |
| TOTAL | $1,000 |

However, accelerated depreciation is not necessarily best if you expect your income to go up in future years. There are also some restrictions on when you can use accelerated depreciation. For example, you can't use it for cars, computers and certain other property that is used for business less than 50% of the time. (See Section D3.)

Determining which depreciation method is best for you and calculating how much depreciation you can deduct is a complex task usually best left to an accountant. If you want to do it yourself, seriously consider obtaining a tax preparation computer program that can help you do the calculations. (See Chapter 8.)

For more information, see IRS Publication 534, *Depreciation*. You can obtain this and all other IRS publications by calling the IRS at 800-TAX-FORM, visiting your local IRS office or downloading the publications from the IRS Internet site at www.irs.ustreas.gov.

### 3. Cars, Computers and Cellular Phones

There are special rules for certain items that can easily be used for personal as well as business purposes. These items are called listed property and include:

- cars, boats, airplanes and other vehicles (see also Section E for rules on mileage and vehicle expenses)
- computers
- cellular phones, and
- any other property generally used for entertainment, recreation or amusement—for example, VCRs, cameras and camcorders.

The IRS fears that taxpayers might use listed property items such as computers for personal reasons but claim business deductions for them. For this reason, you're required to document your business use of listed property. You can satisfy this requirement by keeping a logbook showing when and how the property is used. (See Chapter 14, Section A.)

#### a. Exception to recordkeeping rule

You normally have to document your use of listed property even if you use it 100% for business. However, there is an exception to this rule: If you use listed property only for business and keep it at your business location, you need not comply with the recordkeeping requirement. This includes listed property you keep at your home office if the office qualifies for the home office deduction. (See Section C.)

**EXAMPLE:** John, a freelance writer, works fulltime in his home office which he uses exclusively for writing. The office is clearly his principal place of business and qualifies for the home office deduction. He buys a $4,000 computer for his office and uses it 100% for his writing business. He does not have to keep records showing how he uses the computer.

### b. Depreciating listed property

If you use listed property for business more than 50% of the time, you can depreciate it just like any other property. However, if you use it 50% or less of the time for business, you must use the straight line depreciation method and an especially long recovery period. If you start out using accelerated depreciation and your business use drops to 50% or less, you have to switch to the straight line method and pay taxes on the benefits of the prior years of accelerated depreciation.

## E. Car Expenses

Most self-employed people do at least some driving related to business—for example, to visit clients or customers, to pick up or deliver work, to obtain business supplies or to attend seminars. Of course, driving costs money—and you are allowed to deduct your driving expenses when you use your car, van, pickup or panel truck for business.

There are two ways to calculate the car expense deduction. You can:

- use the standard mileage rate—which requires relatively little recordkeeping, or
- deduct your actual expenses—which requires much more recordkeeping but might give you a larger deduction.

If you own a late model car worth more than $15,000, you'll usually get a larger deduction by using the actual expense method because the standard mileage rate doesn't include enough for depreciation of new cars. On the other hand, the standard mileage rate will be better if you have an inexpensive or old car and put in a lot of business mileage.

Either way, you'll need to have records showing how many miles you drive your car for business during the year—also called business miles. Keep a mileage logbook for this purpose. (See Chapter 14, Section A2c.)

**Commuting Expenses Are Not Deductible**

You usually cannot deduct commuting expenses—that is, the cost involved in getting to and from work. However, if your main office is at home, you may deduct the cost of driving to meet clients. This is one of the advantages of having a home office. (See Chapter 4.)

### 1. Standard Mileage Rate

The easiest way to deduct car expenses is to take the standard mileage rate. When you use this method, you need only keep track of how many business miles you drive, not the actual expenses for your car such as gas or repairs.

You can use the standard mileage rate only for a car that you own. You must choose to use it in the first year you start using your car for your business. In later years, you can choose to use the standard mileage rate or actual expenses.

Each year, the IRS sets the standard mileage rate—a specified amount of money you can deduct for each business mile you drive. In 1997, for example, the rate was 31.5 cents per mile. To figure your deduction, multiply your business miles by the standard mileage rate for the year.

**EXAMPLE:** Ed, a self-employed salesperson, drove his car 10,000 miles for business in 1997. To determine his car expense deduction, he simply multiplies the total business miles he drove by 31.5 cents. This gives him a $3,150 deduction (31.5 cents x 10,000 = $3,150).

If you choose to take the standard mileage rate, you cannot deduct actual operating expenses—for example, depreciation or Section 179 deduction, maintenance and repairs, gasoline and its taxes, oil, insurance and vehicle registration fees. These costs are already factored into the standard mileage rate.

You can deduct any business-related parking fees and tolls—for example, a parking fee you have to pay when you visit a client's office. But you cannot deduct fees you pay to park your car at your place of work.

## 2. Actual Expenses

Instead of taking the standard mileage rate, you can elect to deduct the actual expenses of using your car for business. To do this, deduct the actual cost of depreciation for your car subject to limitations, interest payments on a car loan, lease fees, rental fees, license fees, garage rent, repairs, gas, oil, tires and insurance. The total deductible amount is based on the percentage of time you use your car for business. You can also deduct the full amount of any business-related parking fees and tolls.

Deducting all these items will take more time and effort than using the standard mileage rate because you'll need to keep records of all your expenses. However, it may provide you with a larger deduction than the standard rate.

**EXAMPLE:** Sam drives his $20,000 car 15,000 miles for business in 1997, and doesn't drive it at all for personal use. If he took the standard mileage deduction, he could deduct $4,725 from his income taxes (31.5 cents x 15,000 = $4,725). Instead, however, he takes the actual expense deduction. He keeps careful records of all his costs for gas, oil, repairs, parking, insurance and depreciation. These amount to $7,000 for the year. He gets an extra $2,275 deduction by using the actual expense method.

### a. Mixed uses

If you use your car for both business and personal purposes, you must also divide your expenses between business and personal use.

**EXAMPLE:** In one recent year Laura, a salesperson, drove her car 10,000 miles for her business and 10,000 miles for personal purposes. She can deduct 50% of the actual costs of operating her car.

If you only own one car, you normally can't claim it's used only for business. An IRS auditor is not likely to believe that you walk or take public transportation everywhere except when you're on business. The only exception might be if you live in a place with developed transportation systems, such as Chicago, New York City or San Francisco, and drive your car only when you go out of town on business.

### b. Expense records required

When you deduct actual car expenses, you must keep records of the costs of operating your car. This includes not only the number of business miles and total miles you drive, but also gas, repair, parking, insurance and similar costs. (See Chapter 14) If this seems to be too much trouble, use the standard mileage rate. That way, you'll only have to keep track of how many business miles you drive, not what you spend for gas and similar expenses.

### c. Limits on depreciation deductions

Regardless of how much you spend for an automobile, your depreciation deduction is strictly limited. For example, for cars purchased in 1996, the annual depreciation deduction is limited to a maximum of $3,060 the first year, $4,900 the second year, $2,950 the third year and $1,775 thereafter. These amounts change each year.

There is no point in using Section 179 to write off a car because the limit is the same as under the regular depreciation rules. For example, if you purchased a $17,500 car in 1996, you'd only be allowed to write off $3,060 of the cost for the first year. Writing the car off under Section 179 instead of depreciating the cost will not increase your deduction. It will only use up $3,060 of the $17,500 Section 179 deduction and mean you can't use Section 179 to write off that amount of other personal property. Save Section 179 for other items not subject to these restrictions.

### d. Leasing a car

If you lease a car that you use in your business, you can deduct the part of each lease payment that is for the business use of the car. However, you cannot deduct any part of a lease payment that is for commuting or for any personal use of the car.

Leasing companies typically require you to make an advance or down payment to lease a car. You must spread such payments over the entire lease period. You cannot deduct any payments you make to buy a car even if the payments are called lease payments.

When you lease a car, you cannot use the standard rate method to deduct your expenses for gas, oil, maintenance and repairs, license fees and insurance. You must use the actual expense method and keep records of your expenses.

**Tax Whammy for Luxury Cars**

If you lease what the IRS considers to be a luxury car for more than 30 days, you may have to add to your income an inclusion amount.

This is to make up for the fact that the lease payments on a luxury car are higher than on a lower priced car. The inclusion amount is listed in IRS leasing tables and is usually quite small. For example, if you leased a $40,000 car in 1996 and used it 100% for business, you'd have to add $399 to your income.

The government doesn't want to subsidize people who lease expensive cars. A luxury car is currently defined as one with a fair market value of more than $15,800.

For more information about the rules for claiming car expenses, see IRS Publication 917, *Business Use of a Car.* You can obtain this and all other IRS publications by calling the IRS at 800-TAX-FORM, visiting your local IRS office or downloading the publications from the IRS Internet site at www.irs.ustreas.gov.

## F. Travel Expenses

If you travel for your business, you can deduct your airfare, hotel bills and other expenses. If you plan your trip right, you can even mix business with pleasure and still get a deduction for your airfare. However, IRS auditors closely scrutinize these deductions because many taxpayers claim them without complying with the copious rules attached to them. This is why you need to understand the limitations on this deduction and keep proper records.

## 1. Travel Within the United States

Some business people seems to think they have the right to deduct the cost of any trip they take. This is not the case. You can deduct a trip within the United States only if:

- it's primarily for business
- you travel outside your city limits, and
- you're away at least overnight or long enough to require a stop for sleep or rest.

### a. Business purpose of trip

For your trip to be deductible, you must spend more than half of your time on activities that can reasonably be expected to help advance your business.

Acceptable activities include:

- visiting or working with clients or customers
- attending trade shows, or
- attending professional seminars or business conventions where the agenda is clearly connected to your business.

Business does not include sightseeing or recreation that you attend by yourself or with family or friends, nor does it include personal investment seminars or political events.

Use common sense before claiming a trip is for business. The IRS will likely question any trip that doesn't have some logical connection to your business. For example, if you build houses in Alaska, an IRS auditor would probably be skeptical about a deduction for a trip you took to Florida to learn about new home air conditioning techniques.

To repeat, if your trip within the United States is not primarily for business, none of your travel expenses are deductible. But you can still deduct expenses you have while at your destination that are directly related to your business—for example, the cost of making long distance phone calls to your office or clients while on vacation.

### b. Travel outside city limits

You don't have to travel any set distance to get a travel expense deduction. However, you can't take this deduction if you just spend the night in a motel across town. You must travel outside your city limits. If you don't live in a city, you must go outside the general area where your business is located.

### c. Sleep or rest

Finally, you must stay away overnight or at least long enough to require a stop for sleep or rest. You cannot satisfy the rest requirement by merely napping in your car.

**EXAMPLE:** Phyllis, a self-employed salesperson based in Los Angeles, flies to San Francisco to meet potential clients, spends the night in a hotel and returns home the following day. Her trip is a deductible travel expense.

**EXAMPLE:** Andre, a self-employed truck driver, leaves his workplace on a regularly scheduled roundtrip between San Francisco and Los Angeles and returns home 18 hours later. During the run, he has six hours off at a turnaround point where he eats two meals and rents a hotel room to get some sleep before starting the return trip. Andre can deduct his meal and hotel expenses as travel expenses.

### d. Combining business with pleasure

Provided that your trip is primarily for business, you can tack on a vacation to the end of the trip, make a side trip purely for fun or go to the theater and still deduct your entire airfare. What you spend while having fun is not deductible, but you can deduct your expenses while on business.

**EXAMPLE:** Bill flies to Miami for a four-day business meeting. He then spends three days in Miami swimming and enjoying the sights. Since Bill spent over half his time on business—four days out of seven—the cost of his flight is entirely deductible, as are his hotel and meal costs during the business meeting. He may not deduct his hotel, meal or other expenses during his vacation days.

## 2. Foreign Travel

The rules differ if you travel outside the United States, and are in some ways more lenient. However, you must have a legitimate business reason for your foreign trip. A sudden desire to investigate a foreign business won't qualify because you can't deduct business expenses for a business if you're not already in it.

### a. Trips lasting no more than seven days

If you're away no more than seven days, and you spend the majority of your time on business, you can deduct all of your travel costs.

However, even if your trip was primarily a vacation, you can deduct your airfare and other transportation costs as long as at least part of the trip was for business. You can also deduct your expenses while on business. For this reason, it's often best to limit business-related foreign travel to seven days.

**EXAMPLE:** Jennifer flies to London for a two-day business meeting. She then spends five days sightseeing. She can deduct the entire cost of her airfare, and the portion of her hotel and meals she spent while attending the meeting.

### b. Trips lasting more than seven days

More stringent rules apply if your foreign trip lasts more than one week. To get a full deduction for your expenses, you must spend at least 75% of your time away on business.

If you spend less than 75% of your time on business, you must determine the percentage of your time spent on business by counting the number of business days and the number of personal days. You can only deduct the percentage of your travel costs that relates to business days. A business day is any day you have to be at a particular place on business or in which you spend four or more hours on business matters. Days spent traveling to and from your destination also count as business days.

**EXAMPLE:** Sam flies to London and stays 14 days. He spends seven days on business and seven days sightseeing. He therefore spent 50% of his time on business. He can deduct half of his travel costs.

### c. Foreign conventions

Different rules apply if you attend a convention outside North America. In such cases, a deduction is allowed only if:

- the meeting has a definite, clear connection to your business, and
- it's reasonable for the convention to be held outside North America—for example, if all those attending are plumbers from New York, it would be hard to justify a convention in Tahiti.

## 3. Travel on Cruise Ships

Forget about getting a tax deduction for a pleasure cruise. However, you may be able to deduct part of the cost of a cruise if you attend a business convention, seminars or similar meetings directly related to your business while onboard.

Personal investment or financial planning seminars don't qualify.

But there is a major restriction: you must travel on a U.S. registered ship that stops only in ports in the United States or its possessions such as Puerto Rico or the U.S. Virgin Islands. Not many cruise ships are registered in the United States, so you'll likely have trouble finding a cruise that qualifies. If a cruise sponsor promises you'll be able to deduct your trip, investigate carefully to make sure it meets these requirements.

If you go on a cruise that is deductible, you must file with your tax return a signed note from the meeting or seminar sponsor listing the business meetings scheduled each day aboard ship and certifying how many hours you spent in attendance. Your annual deduction for cruising is limited to $2,000.

### 4. Taking Your Family With You

You generally can't deduct the expense of taking your spouse, children or others along with you on a business trip or to a business convention. The only deductions allowed are for expenses of a spouse or other person who is your employee and has a genuine business reason for going on a trip with you. Typing notes or assisting in entertaining customers are not enough to warrant a deduction; the work must be essential. For example, if you hire your son as a salesperson for your product or service and he calls on prospective customers during the trip, both your expenses and his expenses are deductible.

When you travel with your family, you deduct your business expenses as if you were traveling alone. However, the fact that your family is with you doesn't mean you have to reduce your deductions. For example, if you drive to your destination, you can deduct the entire cost even if your family rides along with you. Similarly, you can deduct the full cost of a single hotel room even if you obtain a larger, more expensive room for your whole family.

### 5. Deductible Expenses

You can deduct virtually all of your expenses when you travel on business, including:

- airfare to and from your destination
- hotel or other lodging expenses
- taxi, public transportation and car rental expenses
- telephone and fax expenses
- the cost of shipping your personal luggage or samples, displays or other things you need for your business
- computer rental fees
- laundry and dry cleaning expenses, and
- tips you pay on any of the other costs.

However, only 50% of the cost of meals are deductible. The IRS imposes this limitation based on the reasoning that you would have eaten had you stayed home.

You must keep good records of your expenses. (See Chapter 14, Section A.) You cannot deduct expenses for personal sightseeing or recreation.

## G. Entertainment and Meal Expenses

Depending on the nature of your business, you may find it helpful or even necessary to entertain clients, customers, suppliers, employees, other self-employed people, professional advisors, investors and other business associates. It's often easier to do business in a nonbusiness setting. Entertainment includes, for example, going to restaurants, the theater, concerts, sporting events and nightclubs, throwing parties and boating, hunting or fishing outings.

In the past, you could deduct entertainment expenses even if business was never discussed. For example, if you took a client to a restaurant, you could deduct the cost even if you spent the whole time drinking martinis and talking about sports. This is no longer the case. To deduct an entertainment expense, you must discuss business either before, during or after the entertainment.

The IRS doesn't have spies lurking about in restaurants, theaters or other places of entertainment, so it has no way of knowing whether you really discussed business with a client or other business associate. You're pretty much on your honor here. However, be aware that the IRS closely scrutinizes this deduction because many taxpayers cheat when taking it and you'll have to comply with stringent recordkeeping requirements. (See Chapter 14, Section 3A.)

### 1. Discussing Business During Entertainment

You're entitled to deduct part of the cost of entertaining a client or other business associate if you have an active business discussion during the entertainment aimed at obtaining income or other benefits. You don't have to spend the entire time talking business, but the main character of the meal or other event must be business.

> **EXAMPLE:** Ivan, a self-employed consultant, takes a prospective client to a restaurant where they discuss and finalize the terms of a contract for Ivan's consulting services. Ivan can deduct the cost of the meal as an entertainment expense.

The IRS will not believe you discussed business if the entertainment occurred in a place where it is difficult or impossible to talk business because of distractions—for example, at a nightclub, theater, or sporting event; or at an essentially social gathering such as a cocktail party.

On the other hand, the IRS will presume you discussed business if a meal or entertainment took place in a clear business setting—for example, a catered lunch at your office.

### 2. Discussing Business Before or After Entertainment

You are also entitled to deduct the full expense of an entertainment event if you have a substantial business discussion with a client or other business associate before or after it. This requires that you have a meeting, negotiation or other business transaction designed to help you get income or some other specific business benefit.

Generally, the entertainment should occur on the same day as the business discussion. However, if your business guests are from out of town, the entertainment can occur the day before or the day after.

The entertainment doesn't have to be shorter than your business discussions, but you can't spend only a small fraction of your total time on business. You can deduct entertainment expenses at places such as nightclubs, sporting events or theaters.

> **EXAMPLE:** Following lengthy contract negotiations at a prospective client's office, you take the client to a baseball game to unwind. The cost of the tickets is a deductible business expense.

### 3. 50% Deduction Limit

You can deduct only entertainment expenses you have paid. If a client picks up the tab, you obviously get no deduction. If you split the expense, you may deduct only what you paid.

Moreover, you're allowed to deduct only 50% of your expenses—for example, if you spend $50 for a meal in a restaurant, you can only deduct $25. However, you must keep track of all you spend and report the entire amount on your tax return. The cost of transportation to and from a business meal or other entertainment is not subject to the 50% limit.

You can deduct the cost of entertaining your spouse and the client's spouse only if it's impractical to entertain the client without a spouse and your spouse joins the party because the client's spouse is attending.

If you entertain a client or other business associate while away from home on business, you can deduct the cost either as a travel or entertainment expense, but not both.

**Champagne and Caviar Might Not Be Deductible**

Your entertainment expenses must be reasonable to be fully deductible. You can't deduct entertainment expenses if the IRS considers them lavish or extravagant. There is no dollar limit on what is reasonable. Nor are you necessarily barred from entertaining at deluxe restaurants, hotels, nightclubs or resorts.

Whether your expenses will be considered reasonable depends on the particular facts and circumstances—for example, a $250 expense for dinner with a client and two business associates at a fancy restaurant would likely be considered reasonable if you closed a substantial business deal during the meal. Since there are no concrete guidelines, you have to use common sense.

For additional information, see IRS Publication 463, *Travel, Entertainment and Gift Expenses*. You can obtain this and all other IRS publications by calling the IRS at 800-TAX-FORM, visiting your local IRS office or downloading the publications from the IRS Internet site at www.irs.ustreas.gov.

## H. Health Insurance Deductions

Self-employed people must provide their own health insurance and fund their own retirements. If you don't make a lot of money, this can be tough. Fortunately, there are some specific tax deductions designed to help you.

### 1. Deducting Health Insurance Premiums

If you're a sole proprietor, partner in a partnership, owner of an S corporation or member of a limited liability company, you may deduct a portion of the cost of health insurance covering you, your spouse and your dependents. However, this deduction can't exceed the net profit from your business. The amount of this deduction will be gradually increased from 30% to 100% of insurance costs over 11 years as noted in the following chart.

**HEALTH INSURANCE DEDUCTION LIMITS**

| Year | Deduction Amount |
|---|---|
| 1996 | 30% |
| 1997 | 40% |
| 1998 - 1999 | 45% |
| 2000 - 2001 | 50% |
| 2002 | 60% |
| 2003 - 2005 | 80% |
| 2006 | 90% |
| 2007 and later | 100% |

**EXAMPLE:** Bob, a self-employed political consultant, earned $50,000 in net profits from his business during 1998. He spent $3,000 on health insurance for himself and his family. He may deduct 45% of this amount, or $1,350, from his income taxes for the year.

You can't take this deduction if you're an employee and are eligible for health insurance through your employer, or your spouse is employed and you're eligible for coverage through his or her employer.

If you form a C corporation, it may deduct the entire cost of health insurance it provides you and any other employees. (See Chapter 2, Section B.)

### 2. Medical Savings Accounts

Starting in 1997, the federal government instituted a pilot program allowing medical savings accounts, or MSAs. MSAs are designed to be used by self-employed people who purchase health insurance with a high deductible. (See Chapter 6.)

## I. Deducting Start-Up Costs

Expenses you incur before you actually start your business—for example, license fees, fictitious business name registration fees, advertising costs, attorney and accounting fees, travel expenses, market research and office supplies expenses—are deductible from your federal income taxes. These expenses are called business start-up costs and must be deducted in equal amounts over the first 60 months you're in business, a process called amortization.

**EXAMPLE:** Bill decides to start a freelance public relations business. Before he opens his office and takes on any clients, he spends $6,000 on license fees, advertising, attorney and accounting fees and office supply expenses. He can't deduct all $6,000 in start-up costs at once. Instead, he can deduct only $100 per month for the first 60 months he's in business—that is, a maximum of $1,200 per year.

You can avoid having to stretch out these deductions over 60 months and instead deduct them all in the first year you're in business if you:

- delay paying start-up costs until you open your doors and start serving clients or customers, or
- start your business on a very small scale and avoid incurring start-up expenses until you've made some money; it doesn't have to be a lot.

For more information on business start-up costs, see IRS Publication 535, Business Expenses. You can obtain a free copy by calling the IRS at 800-TAX-FORM or by calling or visiting your local IRS office. You can also download a copy from the IRS Internet site at www.irs.ustreas.gov. ■

CHAPTER

# The Bane of Self-Employment Taxes

All Americans who work in the private sector are required to pay taxes to help support the Social Security and Medicare systems. Although these taxes are paid to the IRS, they are entirely separate from federal income taxes.

Employees have their Social Security and Medicare taxes directly deducted from their paychecks by their employers, who must make matching contributions. Such taxes are usually referred to as FICA taxes.

But if you're self-employed, your clients or customers will not pay or withhold your Social Security and Medicare taxes. You must pay them to the IRS yourself. When self-employed workers pay these taxes, they are called self-employment taxes or SE taxes. This chapter shows you how to determine how much SE tax you must pay.

**Very Low Income ICs Are Exempt**
If your net income from your business for the year is less than $400, you don't have to pay any self-employment taxes and you can skip this chapter.

For additional information on self-employment taxes, see IRS Publication 533, *Self-Employment Tax*. You can obtain this and all other IRS publications by calling the IRS at 800-TAX-FORM, visiting your local IRS office or downloading them from the IRS Internet site at www.irs.ustreas.gov.

## A. Who Must Pay

Sole proprietors, partners in partnerships and members of limited liability companies must all pay SE taxes if their net income for the year is $400 or more.

Corporations do not pay SE taxes. However, if you're incorporated and work in your business, you are an employee of your corporation and will ordinarily be paid a salary. Instead of paying SE taxes, you must pay FICA taxes on your salary just like any other employee. Half of your Social Security and Medicare taxes must be withheld from your salary and half paid by your corporation. (See Chapter 6.)

## B. SE Tax Rates

The self-employment tax consists of a 12.4% Social Security tax and a 2.9% Medicare tax for a total tax of 15.3%. But in practice, the bite it takes is smaller because of certain deductions. (See Section C.)

The SE tax is a flat tax—that is, the tax rate is the same no matter what your income level. However, there is an income ceiling on the Social Security portion of the tax. You need not pay the 12.4% Social Security tax on your net self-employment earnings that exceed the ceiling amount. If the ceiling didn't exist, people with higher incomes would end up paying far more than they could ever get back as Social Security benefits. The Social Security tax ceiling is adjusted annually for inflation. In 1997, the ceiling was $65,400.

However, there is no similar limit for Medicare: you must pay the 2.9% Medicare tax on your entire net self-employment income, no matter how large. Congress enacted this rule a few years ago to save Medicare from bankruptcy.

> **EXAMPLE:** Mona, a self-employed consultant, earned $90,000 in net self-employment income in 1997. She must pay both Social Security and Medicare taxes on the first $65,400 of her income—a 15.3% tax. She only pays the 2.9% Medicare tax on her remaining $24,600 in income. Mona owes a total of $10,719 in SE taxes.

## C. Earnings Subject to Self-Employment Tax

You pay self-employment taxes on your net self-employment income, not your entire income. To determine your net self-employment income, you first figure the net income you've earned from your business. Your net business income includes all your income from your business, minus all business deductions allowed for income tax purposes. However, you can't deduct retirement contributions you make for yourself to a Keogh or SEP plan or the self-employed health insurance deduction. (See Chapter 9, Section H.) If you're a sole proprietor, as are most self-employed people, use IRS Schedule C, Profit or Loss From Business, to determine your net business income.

If you have more than one business, combine the net income or loss from them all. If you have a job in addition to your business, your employee income is not included in your SE income. (See Section F.) Nor do you include investment income, such as interest you earn on your savings.

You then get one more valuable deduction before finally determining your net self-employment income. You're allowed to deduct 7.65% from your total net business income. This is intended to help ease the SE tax burden on the self-employed. To do this, multiply your net business income by 92.35% or .9235.

> **EXAMPLE:** Billie, a self-employed consultant, earned $70,000 from her business and had $20,000 in business expenses, leaving a net business income of $50,000. She multiplies this amount by .9235 to determine her net self-employment income, which is $46,175. This is the amount on which Billie must pay SE tax.

The fact that you can deduct business expenses from your SE income makes them doubly valuable: They will not only reduce your income taxes, but your SE taxes as well. Your actual SE tax savings will be 15.3% of the amount of such deductions—for example, a $1,000 home office deduction will save you $153 in SE taxes. (See Chapter 9, Section B4.)

**S Corporation Status: A Way Around the SE Tax Thicket**

As a person in business for yourself, you may be able to take advantage of an important wrinkle in the SE tax rules: Distributions from S corporations to their owners are not subject to SE taxes. This is so even though such distributions are included in your income for income tax purposes.

If you incorporate your business and elect to become an S corporation, you may distribute part of your corporation's earnings to yourself without paying SE taxes on them. You can't distribute all your earnings to yourself this way, however, because your S corporation must pay you a reasonable salary on which FICA taxes must be paid. (See Chapter 2, Section B7.)

## D. Computing the SE Tax

It's easy to compute the amount of your SE tax. First, determine your net self-employment income as described above. If your net self-employment income is below the Social Security tax ceiling—$65,400 in 1997—multiply it by 15.3% or .153.

> **EXAMPLE:** Mark, a self-employed consultant, had $50,000 in net self-employment income in 1997. He must multiply this by .153 to determine his SE tax, which is $7,650.

If your net self-employment income is more than the Social Security tax ceiling, things are a bit more complicated. Multiply your income up to the ceiling by 12.4% and all of your income by the 2.9% Medicare tax; then add both amounts together to determine your total SE tax.

> **EXAMPLE:** Martha had $100,000 in net self-employment income in 1997. She multiplies the first $65,400 of this amount by the 12.4% Social Security tax, resulting in a tax of $8,109.60. She then multiplies her entire $100,000 income by the 2.9% Medicare tax, resulting in a $2,900 tax. She adds these amounts together to determine her total SE tax, which is $11,009.60.

In another effort to make the SE tax burden a little easier for the self-employed, you're allowed to deduct half of the amount of your SE taxes from your business income for income tax purposes. For example, if you pay $10,000 in SE taxes, you can deduct $5,000 from your gross income when you determine your taxable income.

## E. Paying and Reporting SE Taxes

Pay SE taxes directly to the IRS during the year as part of your estimated taxes. You have the option of either:

- paying the same amount in tax as you paid the previous year, or
- estimating what your income will be this year and base your estimated tax payments on that. (See Chapter 11, Section B.)

When you file your annual tax return, you must include IRS Form SE, Self-Employment Tax, along with your income tax return. This form shows the IRS how much SE tax you were required to pay for the year. You file only one Form SE no matter how many unincorporated businesses you own. Add the SE tax to your income taxes on your income tax return, Form 1040, to determine your total tax.

Even if you do not owe any income tax, you must still complete Form 1040 and Schedule SE if you owe $400 or more in SE taxes.

## F. Outside Employment

If, in addition to being self-employed, you have an outside job in which you're classified as an employee and have Social Security and Medicare taxes withheld from your wages, you must pay the Social Security tax on your wages first. If your wages are at least equal to the Social Security tax ceiling, you won't have to pay the 12.4% Social Security tax on your SE income. But no matter how much you earn from your job, you'll have to pay the 2.9% Medicare tax on all your SE income.

> **EXAMPLE:** Anne earned $70,000 in employee wages and $10,000 in self-employment income from a business in 1997. She did not have to pay Social Security taxes on her earnings above the $65,400 Social Security tax ceiling for the year. Her employer withheld 7.65% in Social Security taxes up to $65,400 of her wages and 1.45% (the Medicare portion of an employee's FICA taxes) on her earnings between $65,400 and $70,000. Anne also had to pay the 2.9% Medicare portion of the SE tax—but not the 12.4% Social Security tax—on her $10,000 in self-employment earnings.

However, if your employee wages are lower than the Social Security tax ceiling, you'll have to pay Social Security taxes on your SE income until your wages and SE income combined exceed the ceiling amount.

**EXAMPLE:** Bill earned $20,000 in employee wages and $50,000 in self-employment income in 1997. His wages were lower than the $65,400 Social Security tax ceiling for the year. His employer withheld a 7.65% FICA tax on his wages and he had to pay a 12.4% Social Security tax on $45,400 of his SE income. He stopped paying the Social Security tax after his wages and income combined equaled $65,400. This meant he didn't have to pay the Social Security tax on $4,600 of the $50,000 he earned as an IC. However, he had to pay the 2.9% Medicare tax on all his SE income. ■

CHAPTER

# 11

# Paying Estimated Tax

What many ICs like best about their employment status is that it allows freedom in planning and handling their own finances. Unlike employees, they don't have taxes withheld from their compensation by their clients or customers. As a result, many self-employed people have higher take-home pay than employees earning similar amounts.

Unfortunately, however, self-employed workers do not have the luxury of waiting until April 15 to pay all their taxes for the previous year. The IRS wants to get its money a lot faster than that, so the self-employed are required to pay taxes on their estimated annual incomes in four payments spread out over each year. These are called estimated taxes and are used to pay both income taxes (see Chapter 9) and self-employment taxes (see Chapter 10).

Because of estimated taxes, self-empoyed people need to carefully budget their money. If you to fail set aside enough of your earnings to pay your estimated taxes, you could face a huge tax bill on April 15—and have a tough time coming up with the money to cover it.

**Most States Have Estimated Taxes, Too**
If your state has income taxes, it probably requires the self-employed to pay estimated taxes. The due dates are generally the same as for federal estimated tax. State income tax rates are lower than federal income taxes. The exact rate depends on the state in which you live. Contact your state tax office for information and the required forms. (See Appendix 1 for contact details.)

## A. Who Must Pay

You must pay estimated taxes if you are a sole proprietor, partner in a partnership or member of a limited liability company and you expect to owe at least $1,000 in federal tax for the year. You'll owe $1,000 if you have $3,000 in net taxable income. If you've formed a C corporation, it may also have to pay estimated taxes.

However, if you paid no taxes last year—for example, because your business made no profit or you weren't working—you don't have to pay any estimated tax this year no matter what your tax tally for the year. But this is true only if you were a U.S. citizen or resident for the year and your tax return for the previous year covered the whole 12 months.

### 1. Sole Proprietors

Most ICs are sole proprietors. A sole proprietor and his or her business are one and the same for tax purposes, so you simply pay your estimated taxes out of your own pocket. You, not your business, pays the taxes.

### 2. Partners and Limited Liability Companies

Partnerships and limited liability companies, or LLCs, are similar to sole proprietorships. They don't pay any taxes; instead, all partnership and LLC income passes through to the partners or LLC members. (See Chapter 2, Sections C and D.) The partners or LLC members must pay individual estimated tax on their shares of partnership or LLC income. This is so whether it's actually paid to them or not. The partnership or LLC itself pays no tax.

### 3. Corporations

A corporation is separate from you for tax purposes. Both you and your corporation might have to pay estimated taxes.

You will ordinarily be an employee of your corporation and receive a salary directly from it. Income and employment taxes must be withheld from your salary just as for any employee. (See Chapter 13.) You won't need to pay any estimated tax on your salary. But if you receive dividends or distributions from your corporation, you'll need to pay tax on them during the year—unless the total tax due on the amounts received is less than $500. You can cover the taxes due either by paying estimated tax or increasing the tax withheld from your salary; it doesn't make much practical difference which you choose.

If you've formed a C corporation, it must pay quarterly estimated taxes if it will owe $500 or more in corporate tax on its profits for the year. These taxes are deposited with a bank, not paid directly to the IRS. However, most small C corporations don't have to pay any income taxes or estimated taxes because all the profits are taken out of the corporation by the owners in the form of salaries, bonuses and benefits. (See Chapter 2.)

S Corporations ordinarily don't have to pay estimated taxes because all profits are passed through to the shareholders, as in a partnership. (See Chapter 2.)

For detailed guidance on C corporation estimated taxes, see IRS Publication 542, *Tax Information for Corporations*. You can obtain this and all other IRS publications by calling the IRS at 800-TAX-FORM, visiting your local IRS office or downloading them from the IRS Internet site at www.irs.ustreas.gov.

## B. How Much You Must Pay

You should normally determine how much estimated tax to pay after completing your tax return for the previous year. Most people want to pay as little estimated tax as possible during the year so they can earn interest on their money instead of handing it over to the IRS. However, the IRS imposes penalties if you don't pay enough estimated tax. (See Section E1.) There's no need to get excessively concerned about these penalties. They aren't terribly large in the first place and it's easy to avoid having to pay them. All you have to do is pay at least the smaller of:

- 90% of your total tax due for the current year, or
- 100% of the tax you paid the previous year or more in 1999 or later if you're a high income taxpayer (see below).

You normally make four estimated tax payments each year. There are three different ways you can calculate your payments. You can use any one the three methods and you won't have to pay a penalty as long as you pay the minimum total the IRS requires as explained above. One of the methods—basing your payments on last year's tax—is extremely easy to use. The other two are more complex to figure out, but might permit you to make smaller payments.

### 1. Payments Based on Last Year's Tax

The easiest and safest way to calculate your estimated taxes is to simply pay 100% of the total federal taxes you paid last year, or more in 1999 or later if you're a high income taxpayer as described below. You can base your estimated tax on the amount you paid the prior year even if you weren't in business that year, but your return for the year must have been for a full 12-month period.

You should determine how much estimated tax to pay for the current year at the same time as you file your tax return for the previous year—no later than April 15. Take the total amount of tax you had to pay for the year and divide by four. If this comes out to an odd number, round up to get an even number. These are the amounts you'll have to pay in estimated tax. You'll make four equal payments throughout the year and the following year. (See the chart in Section C for when you must make your payments).

**EXAMPLE:** Gary, a self-employed consultant, earned $50,000 last year. He figures his taxes for the prior year on April 1 of this year and determines he owed $9,989.32 for the year.

To determine his estimated tax for the current year he divides this amount by four. $9,989.32 divided by four equals $2497.33. He rounds this up to $2,500. He'll make four $2,500 estimated tax payments to the IRS. So long as he pays this much, Gary won't have to pay a penalty even if he ends up owing more than $10,000 in tax to the IRS for the year because his income goes up, his deductions go down or both.

### a. High income taxpayers must pay more

Beginning in 1999, high income taxpayers—those with adjusted gross incomes of more than $150,000 or $75,000 for married couples filing separate returns must pay more than 100% of their prior year's tax.

Your adjusted gross income or AGI is your total income minus deductions for:

- IRA, Keogh and SEP-IRA contributions
- the self-employed health insurance deduction
- one-half of your self-employment tax, and
- alimony, deductible moving expenses and penalties you pay for early withdrawals from a savings account before maturity or early redemption of certificates of deposit.

To find out your AGI, look at line 31 on your last year's tax return, Form 1040.

As the following chart shows, the estimated tax amounts that must be paid will vary over the next several years until 2003.

| Current Year | Percentage of Prior Year's Tax That Must be Paid as Estimated Tax |
|---|---|
| 1998 | 100% |
| 1999-2001 | 105% |
| 2002 | 112% |
| 2003 and later | 110% |

**EXAMPLE:** Mary, a self-employed consultant, earned $250,000 in gross income in 1998. Her adjusted gross income was $200,000 after subtracting the value of her Keogh Plan contributions, health insurance deduction and half her self-employment taxes. Mary paid $50,000 in income and self-employment taxes in 1998. In 1999, Mary must pay 105% of the tax she paid in 1998—$52,500 in estimated tax. As long as she pays this amount she wont' have to pay a penalty to the IRS even if she earns more than she did in 1998.

### b. Mid-course correction

Your third estimated tax payment is due on September 15. By this time you should have a pretty good idea of what your income for the year will be. If you're reasonably sure that your income for the year will be at least 25% less than what you earned last year, you can forgo the last estimated tax payment due on January 15 of next year. You would have already paid enough estimated tax for the year.

If it looks as if your income will be greater than last year, it is not necessary that you pay more estimated tax. The IRS cannot penalize you so long as you pay 100% of what you paid last year—more if you're a high income taxpayer in 1999 and later.

### c. You may owe tax on April 15

Basing your estimated tax on last year's income is generally the best method to use if you expect your income to be higher this year than last year. You'll be paying the minimum possible without incurring a penalty. However, if you do end up earning more than last year, using this method will cause you to underpay your taxes. You still won't have to pay a penalty, but you'll have to make up the underpayment when you file your tax return for the year. This could present you with a big tax bill if your income rose substantially from last year. To make sure you have enough money for this, it's a good idea to sock away a portion of your income in a separate bank account just for taxes. (See Section C.)

## 2. Payments Based on Estimated Taxable Income

If you're absolutely certain your net income will be less this year than last year, you'll pay less estimated tax if you base your tax on your taxable income for the current year instead of basing it on last year's tax. This is not worth the time and trouble, however, unless you'll earn at least 30% less this year than last.

The problem with using this method is that you must estimate your total income and deductions for the year to figure out how much to pay. Obviously, this can be difficult or impossible to compute accurately. And there are no magic formulas to look to for guidance. The best way to proceed is to sit down with your tax return for the previous year.

Try to figure out whether your income this year will be less than you had last year. You'll find all your income for last year listed on lines 7 through 22 of your Form 1040. Also, determine whether your deductions will be greater than last year. You'll find these listed in the Adjustments to Income and Computation sections of your Form 1040, lines 23 through 37.

Pay special attention to your business income and expense figures in Parts I and II of your Schedule C, Profit or Loss From Business. Decide whether it's likely you'll earn less business income this year than last. You'll probably earn less, for example, if you plan to work fewer hours than last year, you've lost important clients or business conditions are generally poor. Also, determine whether your deductible business expenses will be greater this year than last—for example, because you plan to purchase expensive business equipment and deduct the cost. (See Chapter 9, Section B.)

Take comfort in knowing that you need not make an exact estimate of your taxable income. You won't have to pay a penalty if you pay at least 90% of your tax due for the year.

> **EXAMPLE:** Larry, a self-employed consultant, earned $45,000 last year and paid $10,000 in income and self-employment taxes. Larry expects to earn much less this year because a key client has gone out of business. The lost client accounted for more than one-third of Larry's income last year, so Larry estimates he'll earn about $30,000 this year. The minimum estimated tax Larry must pay is 90% of the tax he will owe on his $30,000 income, which he estimates to be $6,000.

IRS Form 1040-ES contains a worksheet to use to calculate your estimated tax. You can obtain the form by calling the IRS at 800-TAX-FORM, visiting your local IRS office or downloading it from the IRS Internet site at www.irs.ustreas.gov.

Or if you have a tax preparation computer program, it can help you with the calculations.

If you have your taxes prepared by an accountant, he or she should determine what estimated tax to pay. If your income changes greatly during the year, ask your accountant to help you prepare a revised estimated tax payment schedule.

## 3. Payments Based on Quarterly Income

A much more complicated way to calculate your estimated taxes is to use the annualized income installment method. It requires that you separately calculate your tax liability at four points during the year—March 31, May 31, August 31 and December 31—prorating your deductions and personal exemptions. You base your estimated tax payments on your actual tax liability for each quarter. (See the chart in Section C.)

This method is often the best choice for people who receive income very unevenly throughout the year—for example, those who work in seasonal businesses. Using this method, they can pay little or no estimated tax for the quarters in which they earned little or no income.

> **EXAMPLE:** Ernie's income from his air conditioning repair business is much higher in the summer than it is during the rest of the year. By using the annualized income installment method, he makes one large estimated tax payment on September 15, after the quarter in which he earns most of his income. His other three annual payments are quite small.

If you use this method, you must file IRS Form 2210 with your tax return; this form shows your calculations.

**You'll Need Help With This Math**
You really need a good grasp of tax law and mathematics to use the annualized income installment method. The IRS worksheet used to calculate your payments using this method contains 43 separate steps. If you want to use this method, give yourself a break and hire an accountant or at least use a tax preparation computer program to help with the calculations.

See IRS Publication 505, *Tax Withholding and Estimated Tax,* for a detailed explanation of the annualized income method. You can obtain the form by calling the IRS at: 800-TAX-FORM, visiting your local IRS office or downloading it from the IRS Internet site at www.irs.ustreas.gov.

## C. When to Pay Estimated Tax

Estimated tax must ordinarily be paid in four installments, with the first one due on April 15. However, you don't have to start making payments until you actually earn income. If you don't receive any income by March 31, you can skip the April 15 payment. In this event, you'd ordinarily make three payments for the year starting on June 15. If you don't receive any income by May 31, you can skip the June 15 payment as well and so on.

The following chart shows the due dates and the periods each installment covers

**ESTIMATED TAX DUE**

| Income received for the period | Estimated tax due |
|---|---|
| January 1 through March 31 | April 15 |
| April 1 through May 31 | June 15 |
| June 1 through August 31 | September 15 |
| September 1 through December 31 | January 15 of next year |

**SPECIAL RULE FOR FARMERS AND FISHERMEN**

If at least two-thirds of your annual income comes from farming or fishing, you need make only one estimated tax payment on January 15. The first three payment periods in the chart above don't apply.

Also, you can skip the January 15 payment if you file your tax return and pay all taxes due for the previous year by January 31 of the current year. This is a little reward the IRS gives you for filing your tax return early.

However, it's rarely advantageous to file early because you'll have to pay any tax due on January 15 instead of waiting until April 15—meaning you'll lose three months of interest on your hard-earned money.

Your estimated tax payment must be postmarked by the dates noted above, but the IRS need not actually receive them then. If any of these days falls on a weekend or legal holiday, the due date is the next business day.

**The Year May Not Begin in January**
Don't get confused by the fact that the January 15 payment is the fourth estimated tax payment for the previous year, not the first payment for the current year. The April 15 payment is the first payment for the current year.

**Beware the Ides of April**
April 15 can be a financial killer for ICs because there are two separate financial obligations due on that date. By then, you not only have to pay any income and self-employment taxes that are due for the previous year, you also usually have to make your first estimated tax payment for the current year. If you've underpaid your estimated taxes by a substantial amount, you could have a whopping tax bill.

Many self-employed people establish separate bank accounts into which they deposit a portion of each payment they receive from their clients. This way they have some assurance that they'll have enough money to pay their taxes. The amount you should deposit depends on your federal and state income tax brackets and the amount of your tax deductions. Depending on your income, you'll probably need to deposit 25% to 50% of your pay. If you deposit too much, of course, you can always spend the money later on other things.

**EXAMPLE:** Wilma, a self-employed worker who lives in Massachusetts, is in the 28% marginal federal income tax bracket and must pay a 6% state income tax. She must also pay a 15.3% self-employment tax. All these taxes amount to 49.3% of her pay. But she doesn't have to set aside this much because her deductions will reduce her actual tax liability. Using the amount of her deductions from last year as a guide, Wilma determines she needs to set aside 35% to 40% of her income for estimated taxes.

## D. How to Pay

The IRS wants to make it easy for you to send in your money, so the mechanics of paying estimated taxes are very simple. You file federal estimated taxes using IRS Form 1040-ES. This form contains instructions and four numbered payment vouchers for you to send in with your payments. You must provide your name, address, Social Security number and amount of the payment on each voucher. You file only one payment voucher with each payment, no matter how many unincorporated businesses you have.

If you're married and file a joint return, the names on your estimated tax vouchers should be exactly the same as those on your income tax return. Even if your spouse isn't self-employed, he or she should be listed on the vouchers so that the money gets credited to the right account.

If you made estimated tax payments last year, you should receive a copy of the current year's Form 1040-ES in the mail. It will have payment vouchers preprinted with your name, address and Social Security number.

If you did not pay estimated taxes last year, get a copy of Form 1040-ES from the IRS. Do so by calling the IRS at 800-TAX-FORM, visiting your local IRS office or downloading it from the IRS Internet site at www.irs.ustreas.gov. After you make your first payment, the IRS should mail you a Form 1040-ES package with the preprinted vouchers.

Use the addressed envelopes that come with your Form 1040-ES package. If you use your own envelopes, make sure you mail your payment vouchers to the address shown in the Form 1040-ES instructions for the place where you live. Do not mail your estimated tax payments to the same place you sent your Form 1040.

**Keep Your Canceled Checks**

It's not unheard of for the IRS to make a bookkeeping error and then claim that you paid less estimated tax than you did or to apply your payment to the wrong year. If this happens, provide the IRS with a copy of the front and back of your canceled estimated tax checks. The agency encodes a series of tracking numbers on the endorsement side of any check that enable it to locate where payments were applied in its system. This points up the importance of keeping your canceled estimated tax checks.

Don't use money market checks to pay estimated taxes or checks from an account in which the bank doesn't return the original checks to you. Even if your bank promises to give you free copies of your checks, the copies may be so poor the IRS can't read them.

**Tell the IRS If You Move**

You're supposed to notify the IRS if you are making estimated tax payments and you change your address during the year. You can use IRS Form 8822, Change of Address, for this purpose. It's a simple change of address form. Or you may send a signed letter to the IRS Center where you filed your last return stating:

- your full name and your spouse's full name
- your old address and spouse's old address if different
- your new address, and
- Social Security numbers for you and your spouse.

## E. Paying the Wrong Amount

If you pay too little estimated tax, the IRS will make you pay a penalty. If you pay too much, you can get the money refunded or apply it to your current year's estimated taxes.

### 1. Paying Too Little

The IRS imposes a money penalty if you underpay your estimated taxes. Fortunately, the penalty is not very onerous. You have to pay the taxes due plus a percentage penalty for each day your estimated tax payments were unpaid. The percentage is set by the IRS each year. The penalty was 9% in 1995 and 9% in 1996.

The penalty has ranged between 8% and 10% in recent years. This is the mildest of all IRS interest penalties. Even if you paid no estimated tax at all during the year, the underpayment penalty you'd have to pay would be no more than 5% to 6% of your total taxes due for the year.

You can find out what the current penalty is in the most recent version of IRS Publication 505, *Tax Withholding and Estimated Tax*. You can obtain the form by calling the IRS at 800-TAX-FORM, visiting your local IRS office or downloading it from the IRS Internet site at www.irs.ustreas.gov.

The penalty is comparable to the interest you'd pay on borrowed money. Many self-employed people decide to pay the penalty at the end of the tax year rather than take money out of their businesses during the year to pay estimated taxes. If you do this, though, make sure you pay all the taxes you owe for the year by April 15 of the following year. If you don't, the IRS will tack on additional interest and penalties. The IRS usually adds a penalty of 1/2% to 1% per month to a tax bill that's not paid when due.

Because the penalty is figured separately for each payment period, you can't avoid having to pay it by increasing the amount of a later payment. For example, you can't avoid a penalty by doubling your June 15 payment to make up for failing to make your April 15 payment. If you miss a payment, the IRS suggests that you divide the amount equally among your remaining payments for the year. But this won't avoid a penalty on payments you missed or underpaid.

Since the penalty must be paid for each day your estimated taxes remain unpaid, you'll have to pay more if you miss a payment early in the tax year rather than later. For this reason, try to pay your first three estimated tax payments on time. You can let the fourth payment (due on January 15) go. The penalty you'll have to pay for missing this payment will likely be very small.

The IRS will assume you've underpaid your estimated taxes if you file a tax return showing that you owe $500 or more in additional tax, and the amount due is more than 10% of your total tax bill for the year.

If you have underpaid, you can determine the amount of the underpayment penalty by completing IRS Form 2210, Underpayment of Estimated Tax by Individuals, and pay the penalty when you send in your return. Tax preparation programs can do this for you. However, it is not necessary for you to compute the penalty you owe. You can leave it to the IRS to determine the penalty and send you a bill. If you receive a bill, you may wish to complete Form 2210 anyway to make sure you aren't overcharged.

## 2. Paying Too Much

If you pay too much estimated tax, you have two options: you may have the IRS refund the overpayment to you, or you can credit all or part of the money to your current year's estimated taxes. Unfortunately, you can't get back the interest your overpayment earned while sitting in the IRS coffers; that belongs to the government.

To take the credit, write in the amount you want credited instead of refunded to you on line 64 of your Form 1040. The payment is considered to have been made on April 15. You can use all the credited amount toward your first payment, or you can spread it out in any way you choose among any or all of your payments. Be sure to take the amount you have credited into account when figuring your estimated tax payments.

It doesn't make much practical difference which option you choose. Most people take the credit so they don't have to wait for the IRS to send them a refund check. ■

CHAPTER

# 12

# Rules for Salespeople, Drivers and Clothing Producers

You need to read this chapter only if you work as a:

- business-to-business salesperson
- fulltime life insurance salesperson
- clothing or needlecraft producer who works at home
- driver who distributes food products, beverages or laundry
- direct seller, or
- licensed real estate agent.

If you fall within the first four categories, you may be a statutory employee; read Section A, below. If you fall within the last two categories, you may be a statutory independent contractor (IC); skip directly to Section B.

## A. Statutory Employees

If you are a business-to-business or life insurance salesperson, clothing producer who works at home or driver who distributes food products, beverages or laundry and you also meet several other primary requirements (see Section A1), you are a statutory employee. This label derives from the fact that your employment status for certain purposes is defined and explained by statutes passed by Congress.

Practically speaking, being a statutory employee only has two consequences for you, but neither of them are particularly good. Fortunately, it's easy to avoid being classified as a statutory employee and continue being an independent contractor. (See Section A1.)

The first and worst consequence of statutory employee status is that the hiring firms for which you work must pay half of your Social Security and Medicare taxes themselves and withhold the other half from your pay and send it to the IRS, just as for any other employee. (See Chapter 13.) As a result, you'll receive less take-home pay. This is not only because a whopping 7.65% must be deducted from your paychecks, but also because hiring firms will probably insist on paying you less compensation than they would if you were an IC for employment tax purposes. This is to make up for the fact that they have to pay a 7.65% Social Security and Medicare tax for you out of their own pockets. You won't need to include these Social Security and Medicare taxes in your estimated tax payments. (See Chapter 11.)

You'll also receive a Form W-2, Wage and Tax Statement, from the hiring firms for which you work instead of a Form 1099-MISC, the form used to report ICs' income to the IRS. (See Chapter 13, Section B1d.) The W-2 will show the Social Security and Medicare tax withheld from your pay and your Social Security and Medicare income. You'll have to file your W-2s with your tax returns, so the IRS will know exactly how much income you've earned as a statutory employee. The hiring firms will also file a copy with the Social Security Administration.

As a statutory employee, you report income and earnings on Schedule C, Profit or Loss From Business, the same form used by self-employed people who are sole proprietors. Unlike regular employees whose business deductions are strictly limited, you can deduct the full amount of your business expenses. (See Chapter 9.)

You are an IC for income tax purposes provided you qualify as one under the regular IRS rules. (See Chapter 15.) If you qualify, no income tax need be withheld from your compensation and you'll have higher take-home pay as a result. But you will need to pay your income taxes four times during the year in the form of estimated taxes. (See Chapter 11.)

Most hiring firms would prefer that you be classified as an IC for employment tax purposes instead of as a statutory employee. That would allow them to avoid the burden of paying half your FICA taxes themselves and withholding your share from your pay. This chapter describes some simple ways you can avoid this status. Doing so may help you get work you might be denied if classified as a statutory employee.

**If You Have Multiple Businesses**

If you have another business in which you are not classified as a statutory employee, you must file a separate Schedule C for that business. You aren't allowed to use expenses from your statutory employment to offset earnings from your self-employment.

> **EXAMPLE:** Margaret works as a business-to-business salesperson in which she qualifies as a statutory employee. She also works parttime as a self-employed marketing consultant. This year she had $5,000 in travel expenses from her selling job. She can't deduct any of this expense from her earnings as a consultant. She can only deduct it from her income as a salesperson. When Margaret does her taxes, she must file a separate Schedule C for both of her occupations.

## 1. Requirements for Statutory Employee Status

You're a statutory employee only if you satisfy all of these requirements:

- you perform services personally for the hiring firm
- you make no substantial investment in the equipment or facilities you use to work, and
- you have a continuing relationship with the hiring firm.

### a. Personal service

You're a statutory employee only if your oral or written agreement with the hiring firm requires you to do substantially all the work yourself. In other words, you can't hire helpers or subcontract the work out to others.

### b. No substantial investment

Statutory employees must not have a substantial investment in equipment or premises used to perform their work—for example, office space, machinery and office furniture. An investment is substantial if it is more than an employee would be expected to provide—for example, paying office rent or expensive computer equipment. Vehicles you use on the job do not count as substantial investments.

### c. Continuing relationship

A continuing relationship means you work for the hiring firm on a regular or recurring basis. A single job is not considered a continuing relationship. But regular parttime or regular seasonal employment qualifies.

### Avoiding Statutory Employee Status

Even if you do the type of work normally performed by a statutory employee, you can avoid being classified as one. To do this, set up your work relationship so that it does not satisfy one or more of the three requirements discussed in Section 1. For example, you can:

- Sign a written agreement with the person or firm that hires you stating that you have the right to subcontract or delegate the work out to others. This way, you make clear that you don't have to do the work personally.
- Avoid having a continuing relationship with any one hiring firm by working on single projects, not ongoing tasks.
- Invest in outside facilities, such as your own office.

However, even if you do these things, some hiring firms may want to classify you as a statutory employee because they don't understand the law. You may have to educate the people or firms you work for about these rules.

## 2. Types of Statutory Employees

You won't be a statutory employee just because you satisfy the three threshold requirements explained above. You must also satisfy additional requirements for each type of statutory employee occupation. These rules present you with yet more ways to avoid statutory employee status. If the type of work you do and how you do it don't fall squarely within the additional rules below, you won't be classified as a statutory employee.

### a. Business-to-business salespeople

If you're a business-to-business salesperson, you're a statutory employee only if you satisfy the three threshold requirements discussed above (see Section A1) and you also:

- work at least 80% of the time for one person or company, except, possibly, for sideline sales on behalf of someone else
- sell on behalf of, or turn your orders over to, the hiring firm
- sell merchandise for resale or supplies for use in the buyer's business operations, as opposed to goods purchased for personal consumption at home, and
- sell only to wholesalers, retailers, contractors or those who operate hotels, restaurants or similar establishments; this does not include manufacturers, schools, hospitals, churches, municipalities or state and federal governments.

**EXAMPLE:** Linda sells books to retail bookstores for Scrivener & Sons Publishing Company. Her territory covers the entire midwest. She works only for Scrivener and is paid a commission based on the amount of each sale. She turns her orders over to Scrivener's, which ships the books to each bookstore customer. Linda is a Scrivener's statutory employee.

### b. Life insurance salespeople

If your fulltime occupation is soliciting life insurance applications or annuity contracts, you're a statutory employee only if you satisfy the threshold requirements explained in Section A1, and:

- you work primarily for one life insurance company, and
- the company provides you with work necessities such as office space, secretarial help, forms, rate books and advertising material.

**EXAMPLE:** Walter sells life insurance fulltime for the Old Reliable Life Insurance Company. He works out of Old Reliable's Omaha office where he is provided with a desk, clerical help and rate books and insurance applications. Walter is Old Reliable's statutory employee.

### c. Clothing or needlecraft producers

If you make or sew buttons, quilts, gloves, bedspreads, clothing, needlecraft products or similar products, you're a statutory employee only if you satisfy the threshold requirements explained in Section A1, and you:

- work away from the hiring firm's place of business—usually in your own home or workshop, or in another person's home
- work only on goods or materials the hiring firm furnishes
- work according to the hiring firm's specifications; generally, such specifications are simple and consist of patterns or samples, and
- are required to return the processed material to the hiring firm or person designated by it.

If your work set-up meets all these requirements, the hiring firm must pay the employer's share of FICA on your compensation and withhold your share of FICA taxes from your pay. However, no FICA tax is imposed if the hiring firm pays you less than $100 for a calendar year.

**EXAMPLE:** Rosa sews buttons on shirts and dresses. She works at home. She does work for various companies, including Upscale Fashions, Inc. Upscale provides Rosa with all the clothing and the buttons she must sew. The only equipment Rosa provides is a needle. Upscale gives Rosa a sample of each outfit showing where the buttons are supposed to go. When Rosa finishes each batch of clothing, she returns it to Upscale. Rosa is a statutory employee.

### d. Food, beverage and laundry distributors

You're also a statutory employee if you work as a driver and distribute meat or meat products, vegetables or vegetable products, fruits or fruit products, bakery products, beverages other than milk, or laundry or dry cleaning to customers designated by the hiring firm as well as those you solicit. It makes no difference whether you operate from your own truck or trucks belonging to the person or firm that hired you.

**EXAMPLE:** Alder Laundry and Dry Cleaning enters into an agreement with Sharon to pick up and deliver clothing for its customers. Sharon is a statutory employee because she meets all three threshold requirements: her agreement with Alder acknowledges that she will do the work personally, she has no substantial investment in facilities (her truck doesn't count since it's used to deliver the product) and she has a continuing relationship with Alder.

## B. Statutory Independent Contractors

If you're a direct seller or licensed real estate agent, you are automatically considered an independant contractor for Social Security, Medicare and federal unemployment tax purposes provided that:

- your pay is based on sales commissions, not on the number of hours you work, and
- you have a written contract with the hiring firm providing that you will not be treated as an employee for federal tax purposes.

Consider yourself lucky and be thankful your industry lobbyists were able to get these special rules adopted by Congress. Since your worker status is automatically determined by law, hiring firms need not worry about the IRS. This should make it very easy for you to get work as an independent contractor in your chosen field.

You undoubtedly know whether or not you're a licensed real estate agent whom the IRS will automatically classify as a statutory independent contractor.

You're a direct seller for IRS purposes if you sell consumer products to people in their homes or at a place other than an established retail store—for example, at swap meets. Consumer products include tangible personal property that is used for personal, family or household purposes—for example, vacuum cleaners, cosmetics, encyclopedias and gardening equipment. It also includes intangible products such as cable services and home study educational courses.

> **EXAMPLE:** Larry is a Mavon Guy. He sells men's toiletries door-to-door. He is paid a 20% commission on all his sales. This is his only remuneration from Mavon. He has a written contract with Mavon that provides that he will not be treated as an employee for federal tax purposes. Larry is a statutory nonemployee—that is, an independent contractor, for federal employment tax purposes.

If they also satisfy the requirements outlined above, people who sell or distribute newspapers or shopping news are also considered to be direct sellers. This is true whether they are paid by the publisher based on the number of papers delivered, or they purchase newspapers from the publisher and then sell them and keep the money. ■

CHAPTER

# 13

# Taxes for Workers You Hire

If you do all the work in your unincorporated business yourself, you don't need to read this chapter. However, if, like many self-employed people, you hire others to assist you or incorporate your business, it's wise to learn a little about federal and state tax requirements that apply to you and them.

## A. Hiring People to Help You

Sooner or later, most self-employed people need to hire people to help them—for example:

> Agnes, a freelance graphic designer, hires a parttime assistant to help her meet a pressing deadline from her biggest client.
>
> Arthur, an architect, hires a computer consultant to help him choose and install a new computer system and show him how to use new design software.
>
> Amy sells books from several publishers to New England bookstores; she hires Bill to cover her sales territory while she's on vacation.

Whenever you hire a helper, you need to be concerned about obeying federal and state tax laws.

### 1. Different Tax Rules for Independent Contractors and Employees

The tax rules you have to follow when you hire helpers differ depending upon whether they qualify as employees or are self-employed—also called independent contractors (ICs) by the IRS and other government agencies.

If you hire an employee to help you in your business, you become subject to a wide array of state and federal tax requirements. You must withhold taxes from your employees' pay and pay other taxes yourself. You must also comply with complex and burdensome bookkeeping and reporting requirements.

If you hire an IC, you need not comply with these requirements. All you have to do is report the amount you pay the IC to the IRS and your state tax department. However, hiring an IC is not necessarily cheaper than hiring an employee. Some ICs charge far more than what you'd pay an employee to do similar work. Nevertheless, many self-employed still prefer to hire ICs instead of employees because of the smaller tax and bookkeeping burdens.

### 2. Determining Worker Status

Initially, it's up to you to determine whether any person you hire is an employee or an IC. If you decide that a worker is an employee, you must comply with the federal and state tax requirements discussed in Section B. If you decide the worker is an IC, you need only comply with the income reporting and tax identification number requirements covered in Section C.

However, your decision about how to classify a worker is subject to review by various government agencies, including:

- the IRS
- your state's tax department
- your state's unemployment compensation insurance agency, and
- your state's workers' compensation insurance agency.

Any agency that determines that you misclassified an employee as an IC may impose back taxes, fines and penalties.

Scrutinizing agencies use various tests to determine whether a worker is an IC or an employee. The determining factor is usually whether you have the right to control the worker. If you have the right to direct and control the way a worker performs—both as to the final results and the details of when, where and how the work is done—then the worker is your employee. On the other hand, if your control is limited to accepting or rejecting the final results the worker achieves, he or she is an IC. (See Chapter 15 for a detailed discussion of how to determine whether a worker is an IC or an employee.)

### Safe Harbor Makes it Easier to Win Audits

Recent changes in rules known as the IRS Safe Harbor or Section 530 have made it easier for firms that hire ICs to win IRS employment tax audits. If you're audited for employment tax purposes, the auditor must first determine whether you qualify for Safe Harbor protection. If you do, no assessments or penalties may be imposed. You may continue treating the workers involved as ICs for these purposes and the IRS will not question their status.

This means you need not pay the employer's share of the workers' Social Security and Medicare taxes or withhold income or Social Security taxes from their pay. However, the workers could still be considered employees for other purposes, such as pension plan rules and state unemployment compensation taxes.

To qualify for Safe Harbor protection, you must satisfy three requirements. You must have:

- filed all required 1099s for the workers in question
- consistently treated the workers involved and others doing substantially similar work as ICs, and
- had a reasonable basis for treating the workers as ICs—for example, treating such workers as ICs is a common practice in your industry or an attorney or accountant told you they qualified as ICs.

**Parttimers and Temps Can Be Employees**

Don't think that a person you hire to work parttime or for a short period must be an IC. People who work for you only temporarily or parttime are your employees if you have the right to control the way they work.

For a detailed discussion of the practical and legal issues involved when hiring ICs from the employer's point of view, see *Hiring Independent Contractors: The Employer's Legal Guide*, by Stephen Fishman (Nolo Press).

## B. Tax Concerns When Hiring Employees

Whenever you hire an employee, you become an unpaid tax collector for the federal government. You are required to withhold and pay both federal and state taxes for the worker. These taxes are called payroll taxes or employment taxes.

You must also satisfy these requirements if you incorporate your business and continue to actively work in it. In this event, you will be an employee of your corporation.

### 1. Federal Payroll Taxes

The IRS regulates federal payroll taxes, which include:

- Social Security and Medicare taxes—also known as FICA
- unemployment taxes—also known as FUTA, and
- income taxes—also known as FITW.

You must also pay these payroll taxes for yourself if you incorporate your business and work as an employee of your corporation. The taxes must be paid by your corporation, not by you personally. (See Chapter 2, Section B.)

## Pros and Cons of Hiring Employees and ICs

There are advantages and disadvantages to hiring both employees and ICs.

### HIRING EMPLOYEES

| Pros | Cons |
|---|---|
| You can closely supervise employees. | You must pay federal and state payroll taxes for them. |
| You can give employees extensive training. | You must usually provide them with workers' compensation coverage. |
| You automatically own any intellectual property employees create on the job. | You must provide them with office space and equipment. |
| You don't need to worry about government auditors claiming you misclassified employees | You ordinarily provide them with employee benefits such as vacations and sick leave. |
| Employees can't sue you for damages if they are injured on the job provided you have workers' comp insurance. | You're liable for their actions. |
| Employees can generally be fired at any time. | You can be sued for labor law violations. |

### HIRING ICS

| Pros | Cons |
|---|---|
| No need to pay federal and state payroll taxes for ICs. | You risk exposure to audits by the IRS and other agencies. |
| No need to provide workers' comp insurance for ICs. | You can't closely supervise or train them. |
| No need to provide office space or equipment for ICs. | ICs usually can't be terminated unless they violate their contract. |
| No employee benefits need be provided ICs. | ICs can sue you for damages if they are injured on the job. |
| You're generally not liable for ICs' actions. | Possible loss of copyright ownership if you don't obtain an assignment of rights. |
| Reduced exposure to lawsuits for labor law violations. | ICs may usually work for your competitors as well as you. |

IRS *Circular E, Employer's Tax Guide,* provides detailed information on federal payroll taxes. It is an outstanding resource that you should have if you hire employees. You can get a free copy by calling the IRS at 800-TAX-FORM, by calling or visiting your local IRS office or by downloading it from the IRS Web site at www.irs.ustreas.gov.

### a. FICA

FICA is an acronym for Federal Income Contributions Act, the law requiring employers and employees to pay Social Security and Medicare taxes. The IRS imposes FICA taxes on both employers and employees. If you hire an employee, you must collect and remit his or her part of the taxes by withholding it from paycheck amounts and also pay a matching amount.

The amounts you must withhold and pay are listed in the current edition of IRS Circular E. For 1997, for example, employers and employees were each required to pay 7.65% on the first $65,400 of employee's annual wages. The 7.65% figure is the sum of the 6.2% Social Security tax and the 1.45% Medicare tax.

There is no Social Security tax on the portion of an employee's annual wages that exceed the $65,400 ceiling. However, the Medicare tax marches on: both you and the employee must pay the 1.45% Medicare tax on any wages over $65,400. The ceiling for the Social Security tax changes annually. You can find out what the Social Security tax ceiling is for the current year from IRS Circular E, *Employer's Tax Guide*; the amount is printed right on the first page.

## b. FUTA

FUTA is an acronym for the Federal Unemployment Tax Act; the law establishes federal unemployment taxes. Most employers must pay both state and federal unemployment taxes. But even if you're exempt from the state tax, you may still have to pay the federal tax. Employers alone are responsible for FUTA. You may not collect or deduct it from employees' wages.

You must pay FUTA taxes if:

- you pay $1,500 or more to employees during any calendar quarter—that is, any three-month period beginning with January, April, July or October, or
- in each of 20 different calendar weeks during the year, there was at least a part of the day in which you had an employee to whom you paid $1,500 or more during a calendar quarter. The weeks don't have to be consecutive, nor does it have to be the same employee each week.

Technically, the FUTA tax rate is 6.2%, but in practice, you rarely pay this much. You are given a credit of 5.4% if you pay the applicable state unemployment tax in full and on time. This means that the actual FUTA tax rate is usually 0.8%. In 1997, the FUTA tax was assessed on the first $7,000 of an employee's annual wages. The FUTA tax, then, usually is $56 per year per employee.

## c. FITW

FITW is an acronym for federal income tax withholding. When you hire an employee, you're not only a tax collector for the government, but you are a manager of sorts of your employee's income. The IRS fears that employees will not save enough from their wages for their tax bill on April 15 and wants, of course, to speed up tax collections. So the IRS tells you, the employer, not to pay the employees their entire wages but to send the money to the IRS—the employee version of the estimated taxes ICs must pay. (See Chapter 11.)

You must calculate and withhold federal income taxes from all your employees' paychecks. You normally deposit the funds in a bank, which transmits the money to the IRS. Employees are solely responsible for paying federal income taxes. Your only responsibility is to withhold the funds and remit them to the government.

You must ask each employee you hire to fill out IRS Form W-4, Employee's Withholding Allowance Certificate. The information on this form is used to help determine how much tax must be withheld from the employee's pay.

By January 31 of each year, you must give each employee you hired the previous year a copy of IRS Form W-2, Wage and Tax Statement, showing how much he or she was paid and how much tax was withheld for the year. You must also send copies to the Social Security Administration.

You can obtain copies of these forms by calling the IRS at 800-TAX-FORM, by visiting your local IRS office or by downloading them from the IRS Internet site at www.irs.ustreas.gov.

For detailed information on FITW, see IRS Publication 505, *Tax Withholding and Estimated Tax.* You can obtain a copy by calling the IRS at 800-TAX-FORM, by visiting your local IRS office or from the IRS Internet site at www.irs.ustreas.gov.

## d. Paying payroll taxes

You pay FICA, FUTA and FITW monthly—by making federal tax deposits at specified banks. But if you owe a total of $500 or less, the deposits are due quarterly. You must submit an IRS Federal Tax Deposit coupon (Form 8109-B) with each payroll tax payment. If you have employees, you must also report these payments to the IRS on Form 941, Employers Quarterly Federal Tax Return, after each calendar quarter that you have employees. Form 941 shows how many employees you had, how much they were paid and the amount of FICA and income tax withheld.

Once each year you must also file IRS Form 940, Employer's Annual Federal Unemployment Tax Return or the simpler Form 940-EZ. This form shows the IRS how much federal unemployment tax you owe.

Starting in 1998, the IRS requires all businesses with more than $50,000 of federal employment tax deposits to deposit their employment taxes electronically rather than using paper federal tax deposit coupons. With the IRS's new Electronic Federal Tax Payment System (EFTPS), deposits may be made by telephone or personal computer. For information on EFTPS or to get an enrollment form, call EFTPS Customer Service at 800-555-4477 or 800-945-8400.

Figuring out how much to withhold, doing the necessary recordkeeping and filling out the required forms can be complicated. If you have a computer, computer accounting programs such as QuickBooks can help with all the calculations and print out your employees' checks and IRS forms.

You can also hire a bookkeeper or payroll tax service to do the work. Payroll tax services are usually not expensive; if you only have one or two employees, such a service could cost less than $50 per month.

## e. Penalties for failing to pay FICA and FITW

As far as the IRS is concerned, an employer's most important duty is to withhold and pay over Social Security and income taxes. Employee FICA and FITW are also known as trust fund taxes because the employer is deemed to hold the withheld funds in trust for the U.S. government.

If you fail to pay trust fund taxes, you can get you into the worst tax trouble there is. The IRS can—and often does—seize a business's assets and force it to close down if it owes back payroll taxes. You can also get thrown in jail, but this rarely happens.

At the very least, you'll have to pay all the taxes due plus interest. The IRS may also impose a penalty known as the trust fund recovery penalty if it determines that you willfully failed to pay the taxes. The agency can claim you willfully failed to pay taxes if you knew the taxes were due and didn't pay them. Good evidence that you knew such taxes were due is that you paid them in the past, but stopped.

The trust fund recovery penalty is also known as the 100% penalty because the amount of the penalty is equal to 100% of the total amount of employee FICA and FITW taxes the employer failed to withhold and pay to the IRS. This can be a staggering sum. As a business owner, you'll be personally liable for the 100% penalty—that is, you will have to pay it out of your own pocket. This is so even if you've incorporated your business.

**THEIR DOGS ATE THEIR HOMEWORK**

You have to make sure these taxes are paid. The IRS does not take kindly to excuses. For example, the IRS assessed a $40,000 penalty against two brothers in the floor covering business when their company failed to pay FITW and FICA for its employees for over two years. The brothers had the money to pay the taxes and had entrusted the task to an office manager. The manager failed to make the payments. The brothers pleaded ignorance about it, claiming that the manager intercepted and screened the mail and altered check descriptions and quarterly reports.

Both the IRS and court were unmoved. Although the court stated it was not unreasonable for the brothers to entrust the payments to their office manager, they were still ultimately responsible to make sure they were paid and were therefore liable for the penalty. (*Conklin Bros. v. United States*, 986 F.2d 315 (9th Cir. 1993).)

For guidance on how to deal with the IRS if you are having trouble meeting your payroll tax obligations, see *Tax Savvy for Small Business*, by Fred Daily (Nolo Press).

### f. Rules for family members

Self-employed people often hire family members to help them. What argues in favor of this family togetherness is that if you hire your child, spouse or parent as an employee, you may not have to pay FICA and FUTA taxes.

**Employing your child.** You need not pay FUTA taxes for services performed by your child who is under 21 years old. You need not pay FICA taxes for your child under 18 who works in your trade or business, or your partnership if it's owned solely by you and your spouse.

> **EXAMPLE:**: Lisa, a 16-year-old, makes deliveries for her mother's mail order business, which is operated as a sole proprietorship. Although Lisa is her mother's employee, her mother need not pay FUTA until Lisa reaches 21 and need not pay FICA taxes until she reaches 18.

However, these rules do not apply—and you must pay both FICA and FUTA—if you hire your child to work for:

- your corporation, or
- your partnership unless all the partners are parents of the child.

> **EXAMPLE:** Ron works in a computer repair business that is half owned by his mother and half owned by her partner, Ralph, who is no relation to the family. FICA and FUTA taxes must be paid for Ron because he is working for a partnership and not all the partners are his parents.

### Income Tax Break for Child-Employees

You must withhold income taxes from your child's pay only if it exceeds the standard deduction for the year. The standard deduction was projected to be $4,250 in 1998 and is adjusted every year for inflation. A child who is paid less than this amount need not pay any income taxes on his or her salary.

You might consider getting your child to do some work around the office instead of paying him or her an allowance for doing nothing. If your child's pay is below the standard deduction amount, it is not only tax-free, but you can also deduct the amount from your own taxes as a business expense if the child's work is business-related—for example, cleaning your office, answering the phone or making deliveries. However, you can only deduct your child's wages if they are reasonable—that is, what you'd pay a stranger for the same work. Don't try paying your child $100 per hour for office cleaning so you can get a big tax deduction.

If you pay your child $600 or more during the year, you must file Form W-2 reporting the earnings to the IRS.

**Employing your spouse.** If you pay your spouse to work in your trade or business, the payments are subject to FICA taxes and federal income tax withholding, but not to FUTA taxes.

> **EXAMPLE:** Kay's husband, Simon, is a sole proprietor computer programmer. Kay works as his assistant and is paid $1,500 per month. Simon must pay the employer's share of FICA taxes for Kay and withhold employee FICA and federal income taxes from her pay.

But this rule does not apply—and FICA, FUTA and FITW must all be paid—if your spouse works for:

- a corporation, even if you control it, or
- a partnership, even if your spouse is a partner along with you.

> **EXAMPLE:** Laura's husband, Rob, works as a draftsperson in Laura's architectural consulting firm, a corporation of which she is the sole owner. The corporation must pay FICA, FUTA and FITW for Rob.

**Employing a parent.** The wages of a parent you employ in your trade or business are subject to income tax withholding and FICA taxes.

> **EXAMPLE:** Don owns and operates a graphic design firm and employs Art, his father, as a parttime designer. Since the firm is a business, Don must pay the employer's share of FICA taxes for Art and withhold employee FICA and federal income taxes from his pay.

## 2. State Payroll Taxes

Employers in all states are required to pay and withhold state payroll taxes for employees. These taxes include:

- state unemployment compensation taxes in all states
- state income tax withholding in most states, and
- state disability taxes in a few states.

### a. Unemployment compensation

Federal law requires that all states provide most types of employees with unemployment compensation, also called UC, or unemployment insurance.

Employers are required to contribute to a state unemployment insurance fund. Employees make no contributions, except in Alaska, New Jersey, Pennsylvania and Rhode Island where employers must withhold small employee contributions from employees' paychecks. An employee who is laid off or fired for other than serious misconduct is entitled to receive unemployment benefits from the state fund. You need not provide unemployment for ICs.

If your payroll is very small—below $1,500 per calendar quarter—you probably won't have to pay UC taxes. In most states, you must pay state UC taxes for employees if you're paying federal UC taxes, also called FUTA taxes. (See Section B1b.) However, some states have more strict requirements. Contact your state labor department for the exact service and payroll amounts.

#### b. State income tax withholding

All states except Alaska, Florida, Nevada, South Dakota, Texas, Washington and Wyoming have income taxation. If you do business in a state that imposes state income taxes, you must withhold the applicable tax from your employees' paychecks and pay it to the state taxing authority. No state income tax withholding is required for workers who qualify as ICs.

It's easy to determine whether you need to withhold state income taxes for a worker: if you are withholding federal income taxes, then you must withhold state income taxes as well. Each state has its own income tax withholding forms and procedures. Contact your state tax department for information. (See Appendix 1 for contact details.)

#### c. State disability insurance

Five states have state disability insurance that provides employees with coverage for injuries or illnesses that are not related to work. Injuries that are job-related are covered by workers' compensation. (See Section B3.) The states with disability insurance are: California, Hawaii, New Jersey, New York and Rhode Island. Puerto Rico also has a disability insurance program.

In these states, employees contribute to disability insurance in amounts their employers withhold from their paychecks. Employers must also make contributions in Hawaii, New Jersey and New York.

Except in New York, the disability insurance coverage requirements are the same as for UC insurance. If you pay UC for a worker, you must withhold and pay disability insurance premiums as well. You need not provide disability for ICs.

### 3. Workers' Compensation Insurance

Subject to some important exceptions, employers in all states must provide their employees with workers' compensation insurance to cover work-related injuries. Workers' compensation is not a payroll tax. You must purchase a workers' compensation policy from a private insurer or state workers' compensation fund. (See Chapter 6, Section F.)

## C. Tax Reporting for Independent Contractors

If you hire an IC, you don't have to worry about withholding and paying state or federal payroll taxes or filling out lots of government forms. This is one reason ICs generally prefer hiring other ICs rather than employees.

However, if you pay an unincorporated IC $600 or more during the year for business-related services, you must:

- file IRS Form 1099-MISC telling the IRS how much you paid the worker, and
- obtain the IC's taxpayer identification number.

The IRS imposes these requirements because it is very concerned that to avoid paying taxes many ICs don't report all the income they earn. To help prevent this, the IRS wants to find out how much you pay ICs you hire and make sure it has their correct tax ID numbers.

The filing and ID requirements apply to all ICs you hire who are sole proprietors or partners in partnerships, which is the vast majority of ICs. However, they don't apply to corporations, probably because large businesses have a strong legislative lobby. The IRS has attempted to change the law to include corporations, but so far it hasn't succeeded.

This means that if you hire an incorporated IC, you don't have to file anything with the IRS.

**EXAMPLE:** Bob, a self-employed consultant, pays $5,000 to Yvonne, a CPA, to perform accounting services. Yvonne has formed her own one-person corporation called Yvonne's Accounting Services, Inc. Bob pays the corporation, not Yvonne personally. Since Bob is paying a corporation, he need not report the payment on Form 1099-MISC or obtain Yvonne's tax ID number.

This is one of the main advantages of hiring incorporated ICs, because the IRS uses 1099s as a lead to find people and companies to audit.

However, it's wise to make sure you have the corporation's full legal name and obtain its federal employer identification number. Without this information, you may not be able to prove to the IRS that the payee was incorporated. An easy way to do this is to have it fill out IRS Form W-9, Request for Taxpayer Identification Number, and keep it in your files. This simple form merely requires the corporation to provide its name, address and EIN. (See Appendix 1 for a copy of the form.)

**When in Doubt, File a 1099**

The IRS may impose a $50 fine if you fail to file a Form 1099 when required. But, far more serious, you'll be subject to severe penalties if the IRS later audits you and determines you misclassified the worker.

If you're not sure whether you must file a Form 1099-MISC for a worker, go ahead and file one anyway. You lose nothing by doing so and will save yourself the severe consequences of not filing if you were legally required to do so.

For a detailed discussion of the consequences of not filing a 1099 Form, see *Hiring Independent Contractors: The Employer's Legal Guide*, by Stephen Fishman (Nolo Press).

The only exception to this rule is for payments to medical corporations; you must report such payments on Form 1099-MISC if they are made for your business.

## 1. $600 Threshold for IC Income Reporting

You need to obtain an unincorporated IC's taxpayer ID number and file a 1099 with the IRS only if you pay the IC $600 or more during a year for business-related services. It makes no difference whether the sum was one payment for a single job or the total of many small payments for multiple jobs.

**EXAMPLE:** Andre, a computer consultant, hires Thomas, a self-employed programmer, to help create a computer program. Andre classifies Thomas as an IC and pays him $2,000 during the year. Thomas is a sole proprietor. Since Andre paid Thomas more than $599 for business-related services, Andre must file Form 1099 with the IRS reporting the payment and obtain Thomas's taxpayer ID number.

In calculating whether the payments made to an IC total $600 or more during a year, you must include payments for parts or materials the IC used in performing the services. For example, if you hire a painter to paint your home office, the cost of the paint would be included in the tally.

However, not all payments you make to ICs are counted towards the $600 threshold.

### a. Payments for merchandise

You don't need to count payments solely for merchandise or inventory. This includes raw materials and supplies that will become a part of merchandise you intend to sell.

> **EXAMPLE:** Betty pays $5,000 to purchase 100 used widgets from Joe's Widgets, a sole proprietorship he owns. Betty intends to repair and resell the widgets. The payment need not be counted toward the $600 threshold because Betty is purchasing merchandise from Joe, not services.

### b. Payments for personal services

You need only count payments you make to ICs for services they perform in the course of your trade or business. A trade or business is an activity carried on for gain or profit. You don't count payments for services that are not related to your business, including payments you make to ICs for personal or household services or repairs—for example, payments to babysitters, gardeners and housekeepers. Running your home is not a profit-making activity.

> **EXAMPLE:** Joe, a self-employed designer, pays Mary a total of $1,000 during the year for gardening services for his residence. None of the payments count toward the $600 threshold because they don't relate to Joe's design business. Joe need not obtain Mary's taxpayer ID number or file a 1099 reporting the payments to the IRS.

## 2. Obtaining Taxpayer Identification Numbers

Some ICs work in the underground economy—that is, they're paid in cash and never pay any taxes or file tax returns. The IRS may not even know they exist. The IRS wants you to help it find these people by supplying the taxpayer ID numbers from all ICs who meet the requirements explained above.

If an IC won't give you his or her number or the IRS informs you that the number the IC gave you is incorrect, the IRS assumes the person isn't going to voluntarily pay taxes. So it requires you to withhold taxes from the compensation you pay the IC and remit them to the IRS. This is called backup withholding. If you fail to backup withhold, the IRS will impose an assessment against you equal to 31% of what you paid the IC.

### a. Avoiding backup withholding

Backup withholding can be a bookkeeping burden for you. Fortunately, it's very easy to avoid it. Have the IC fill out and sign IRS Form W-9, Request for Taxpayer Identification Number, and retain it in your files. (See Appendix 1 for a copy of the form.) You don't have to file the W-9 with the IRS. This simple form merely requires the IC to list his or her name and address and taxpayer ID number. Partnerships and sole proprietors with employees must have a federal employer identification number (EIN), which they obtain from the IRS. In the case of sole proprietors without employees, the taxpayer ID number is either the IC's Social Security number or an EIN if the IC has one.

If the IC doesn't already have an EIN, but promises to obtain one, you don't have to backup withhold for 60 days after he or she applies for one. Have the IC fill out and sign the W-9 form, stating "Applied For" in the space where the ID number is supposed to be listed. If you don't receive the IC's ID number within 60 days, start backup withholding.

### b. Backup withholding procedure

If you are unable to obtain an IC's taxpayer ID number or the IRS informs you that the number the IC gave you is incorrect, you'll have to do backup withholding. You must begin doing so after you pay an IC $600 or more during the year. You need not backup withhold on payments totaling less than $600.

To backup withhold, deposit with your bank 31% of the IC's compensation every quarter. You must make these deposits separately from the payroll tax deposits you make for employees. You report the amounts withheld on IRS Form 945, Annual Return of Withheld Federal Income Tax. This is an annual return you must file by January 31 of the following year. See the instructions to Form 945 for details. You can obtain a copy of the form by calling the IRS at 800-TAX-FORM, by contacting your local IRS office or by downloading it from the IRS Internet site at www.irs.ustreas.gov.

## 3. Filling Out Your 1099 Form

One 1099-MISC form must be filed for each IC to whom you paid $600 or more during the year. You must obtain original 1099 forms from the IRS. You cannot photocopy this form because it contains several pressure-sensitive copies. Each 1099 form contains three parts and can be used for three different workers. All your 1099s must be submitted together along with one copy of Form 1096, which is a transmittal form—the IRS equivalent of a cover letter. You must obtain an original Form 1096 from the IRS; you cannot submit a photocopy. Obtain these forms by calling the IRS at 800-TAX-FORM or by contacting your local IRS office.

Filling out Form 1099-MISC is easy. Follow this step-by-step approach.

- List your name and address in the first box titled Payer's name.
- Enter your taxpayer identification number in the box entitled Payer's Federal identification number.
- The IC you have paid is called the "Recipient" on this form, meaning the person who received the money. You must provide the IC's taxpayer identification number, name

9595 ☐ VOID ☐ CORRECTED

| PAYER'S name, street address, city, state, and ZIP code | | | 1 Rents $ | OMB No. 1545-0115 | Miscellaneous Income |
|---|---|---|---|---|---|
| | | | 2 Royalties $ | 1997 | |
| | | | 3 Other income $ | Form 1099-MISC | |
| PAYER'S Federal identification number | RECIPIENT'S identification number | | 4 Federal income tax withheld $ | 5 Fishing boat proceeds $ | Copy A For Internal Revenue Service Center |
| RECIPIENT'S name | | | 6 Medical and health care payments $ | 7 Nonemployee compensation $ | File with Form 1096. |
| Street address (including apt. no.) | | | 8 Substitute payments in lieu of dividends or interest $ | 9 Payer made direct sales of $5,000 or more of consumer products to a buyer (recipient) for resale ▶☐ | For Paperwork Reduction Act Notice and instructions for completing this form, see Instructions for Forms 1099, 1098, 5498, and W-2G. |
| City, state, and ZIP code | | | 10 Crop insurance proceeds $ | 11 State income tax withheld $ | |
| Account number (optional) | | 2nd TIN Not. ☐ | 12 State/Payer's state number | | |

Form 1099-MISC 36-2515832 Do NOT Cut or Separate Forms on This Page Department of the Treasury - Internal Revenue Service and address in the boxes indicated. For sole proprietors, you must list the individual's name first, and then may list a different business name, though this is not required. You may not enter only a business name for a sole proprietor.

- Enter the amount of your payments to the IC in Box 7, entitled "Nonemployee compensation." Be sure to fill in the right box or the 1099-MISC will be deemed invalid by the IRS.
- Finally, if you've done backup withholding for an IC who has not provided you with a taxpayer ID number, enter the amount withheld in Box 4.

The 1099-MISC form contains five copies. These must be filed as follows:

- Copy A, the top copy, must be filed with the IRS no later than February 28 of the year after payment was made to the IC. If you don't use the remaining two spaces for other ICs, leave those spaces blank. Don't cut the page.
- Copy 1 must be filed with your state taxing authority if your state has a state income tax. The filing deadline is probably February 28, but check with your state tax department to make sure. Your state may also have a specific transmittal form or cover letter you must obtain.
- Copy B and Copy 2 must be given to the worker no later than January 31 of the year after payment was made.
- Copy C is for you to retain for your files.

All the IRS copies of each 1099 are filed together with Form 1096, a simple transmittal form. You must add up all the payments reported on all the 1099s and list the total in the box indicated on Form 1096. File the forms with the IRS Service Center listed on the reverse of Form 1096.

DO NOT STAPLE 6969

Form **1096** Department of the Treasury Internal Revenue Service

**Annual Summary and Transmittal of U.S. Information Returns**

OMB No. 1545-0108 **1997**

ATTACH IRS LABEL HERE

FILER'S name

Street address (including room or suite number)

City, state, and ZIP code

If you are not using a preprinted label, enter in box 1 or 2 below the identification number you used as the filer on the information returns being transmitted. Do not fill in both boxes 1 and 2.

Name of person to contact if the IRS needs more information

Telephone number
( )

**For Official Use Only**

| 1 Employer identification number | 2 Social security number | 3 Total number of forms | 4 Federal income tax withheld | 5 Total amount reported with this Form 1096 |
|---|---|---|---|---|
| | | | $ | $ |

Enter an "X" in only one box below to indicate the type of form being filed. If this is your FINAL return, enter an "X" here . . ▶ ☐

| W-2G 32 | 1098 81 | 1099-A 80 | 1099-B 79 | 1099-C 85 | 1099-DIV 91 | 1099-G 86 | 1099-INT 92 | 1099-MISC 95 | 1099-OID 96 | 1099-PATR 97 | 1099-R 98 | 1099-S 75 | 5498 28 |
|---|---|---|---|---|---|---|---|---|---|---|---|---|---|
| ☐ | ☐ | ☐ | ☐ | ☐ | ☐ | ☐ | ☐ | ☐ | ☐ | ☐ | ☐ | ☐ | ☐ |

CHAPTER

14

# Recordkeeping and Accounting Made Easy

You probably didn't become self-employed so you could turn into a bookkeeper or accountant. Although it can be a bit of a pain, all self-employed people need to keep records of their income and expenses. Among other things, keeping good records will enable you to reap a rich harvest in tax deductions. So time spent on recordkeeping is usually time well spent.

## A. Simple Bookkeeping

Except in a few cases, the IRS does not require any special kind of records. You may choose any system suited to your business that clearly shows your income and expenses. If you are in more than one business, keep a separate set of books for each business.

If, like most self-employed people, you run a one-person service business and are a sole proprietor, you don't need a fancy or complex set of books. You can get along very nicely with just a few items. They include:

- a business checking account
- income and expense journals
- files for supporting documents such as receipts and canceled checks, and
- if you buy equipment such as computers or copiers to use in your business, an asset log to support your depreciation deductions.

**Special Concerns If You Hire Employees**

If you have employees, you must create and keep a number of records, including payroll tax records, withholding records and employment

### Benefits of Keeping Records

Keeping good records will help you in a number of ways.

**Monitor the progress of your business.** Without records, you'll never have an accurate idea how your business is doing. You may think you're making money when you're really not. Records can show whether your business is improving or what changes you need to make to increase the likelihood of success.

**Prepare financial statements.** You need good records to prepare accurate financial statements. These include income (profit and loss) statements and balance sheets. These statements can be essential in dealing with your bank or creditors.

**Keep track of deductible expenses.** You may forget expenses when you prepare your tax return unless you record them when they occur. Every $100 in expenses you forget to deduct will cost you $43.30 in additional income and self-employment taxes if you earn a mid-level income, putting you in the 28% marginal tax bracket.

**Prepare your tax returns.** You need good records to prepare your tax return or to enable an accountant to prepare your return for you in a reasonable amount of time. These records should show the income, expenses and credits you report. Generally, these are the same records you use to monitor your business and prepare your financial statements.

**Win IRS audits.** If you're audited by the IRS, it will be up to you to prove that you have accurately reported your income and expenses on your tax returns. An IRS auditor will not simply take your word that your return is accurate. You need accurate records and receipts to back up your returns.

tax returns. And you must keep these records for four years.

For detailed information, see IRS Circular E, *Employer's Tax Guide*. You can get a free copy by calling the IRS at 800-TAX-FORM, by calling or visiting your local IRS office or by downloading it from the IRS Web site at www.irs.ustreas.gov.

Also, contact your state tax agency for your state's requirements. (See Appendix 1 for contact details.)

## 1. Business Checkbook

One of the first things you should do when you become self-employed is to set up a separate checking account for your business. Your business checkbook will serve as your basic source of information for recording your business expenses and income. Deposit all your self-employment compensation, such as the checks you receive from clients, into the account and make all business-related payments by check from the account. Don't use your business account for personal expenses or your personal account for business.

Keeping a separate business account is not required by law if you're a sole proprietor, but it will provide many important benefits.

- It will be much easier for you to keep track of your business income and expenses if they're paid from a separate account.
- Your business account will also clearly separate your personal expenses and business income and expenses; this will prove very helpful if you're audited by the IRS.
- Your business account will help convince the IRS that you are running a business and not engaged in a hobby. Hobbyists don't generally have separate bank accounts for their hobbies. This is a huge benefit if you incur losses from your business because losses from hobbies are not fully deductible. (See Chapter 9.)
- Perhaps most important, having a separate business bank account will help to establish that you're an independent contractor, not an employee. Employees don't have separate business accounts. Your business account helps show you're in business and are therefore an independent contractor.

### a. Setting up your bank account

At a minimum, you'll need to open a business checking account in which you will deposit all your self-employment income and from which you will pay all your business expenses. This account should be in your business name. If you're a sole proprietor, this can be your own name. If you've formed a corporation or limited liability company, the account should be in your corporate or company name.

There is no need to open your business checking account at the same bank where you have your personal checking account. Shop around and open your account with the bank that offers you the best services at the lowest price.

If you're doing business under your own name, consider opening up a second account in that name and using it solely for your business instead of a separate business account. You'll usually pay less for a personal account than for a business account.

If you're a sole proprietor and doing business under an assumed name, you'll likely have to give your bank a copy of your fictitious business name statement. (See Chapter 3.)

**USE A SEPARATE CREDIT CARD FOR BUSINESS**

Use a separate credit card for business expenses instead of using one card for both personal and business items. Credit card interest for business purchases is 100% deductible while interest for personal purchases is not deductible at all. Using a separate card for business purchases will make it much easier for you to keep track of how much interest you've paid for business purchases. The card doesn't have to be in your business name. It can just be one of your personal credit cards.

If you've incorporated, call your bank and ask what documentation is required to open the account. You will probably need to show the bank a corporate resolution authorizing the opening of a bank account and showing the names of the people authorized to sign checks.

Typically, you will also have to fill out, and impress your corporate seal on, a separate bank account authorization form provided by your bank. You will also need to have a federal employer identification number. (See Chapter 5.)

Similarly, if you've established a limited liability company (see Chapter 2), you'll likely have to show the bank a company resolution authorizing the account.

You may also want to establish interest-bearing accounts for your business in which you place cash you don't immediately need. For example, you may set up a business savings account or a money market mutual fund in your business name.

### a. Paying yourself

To pay yourself when you're a sole proprietor, write a business check to yourself and deposit the money in your personal account. This is known as a withdrawal or personal draw. Use your personal account to pay all your non-business or personal expenses.

### b. When you write checks

If you already keep an accurate, updated personal checkbook, do the same for your business checkbook. If, however, like many people you tend to be lax in keeping up your checkbook, you're going to have to change your habits. Now that you're in business, you can't afford this kind of carelessness. Unless you write large numbers of business checks, maintaining your checkbook won't take much time.

When you write business checks, you may have to make some extra notations besides the date, number, amount of the check, and the name of the person or company to which the check is written. If it's not clear from the name of the payee what a check is for, describe the business reason for the check—for example, the equipment or service you purchased.

You can use the register that comes with your checkbook and write in all this information manually, or you can use a computerized register. (See Section B.)

**Don't Write Checks for Cash**
Avoid writing checks payable to cash, since it's unclear what specific business purpose such checks have. Writing cash checks might lead to questions from the IRS if you're audited. If you must write a check for cash to pay a business expense, be sure include the receipt for the cash payment in your records.

### c. Making deposits

When you make deposits into your business checking account, record in your check register:

- the date and amount, and
- a description of the source of the funds—for example, the client's name.

## 2. Income and Expense Records

In addition to a business checkbook, you should maintain income and expense records. Create what accountants call a chart of accounts—a listing by category of all your expenses and income.

These records, which should be updated at least monthly, will show you how much you're spending and for what, and will also clearly indicate how much money you're making. They will also make it much easier for you or a tax pro to prepare your tax returns. Instead of having to locate, categorize and add up the amount of each bill or canceled check at tax time, you can simply use the figures in your chart of accounts.

You can either create a chart of accounts on paper or use a computerized system. (See Section B.)

### a. Expense journal

Your expense journal will show what you buy for your business. You can easily create it by using ledger sheets you can get from any stationery or office supply store. Get ledger sheets with at least 12 or 14 columns. Devote a separate column to each major category of expenses you have. Alternatively, you can purchase accounting record books with the expense categories already printed on them. These cost more, however, and may not offer categories that meet your needs.

To decide what your expense categories should be, sit down with your first month's bills and receipts and divide them up into categorized piles. Some common expense categories many self-employed people have include:

- business meals and entertainment
- travel
- telephone
- supplies and postage
- office rent
- utilities for an outside office
- professional dues, publications and books
- business insurance
- payments to other self-employed people
- advertising costs
- equipment, and
- license fees.

You should always include a final category called Miscellaneous for various and sundry expenses that are not easily pigeon-holed.

Depending on the nature of your business, you may not need all these categories or you might have others. For example, a graphic designer might have categories for printing and typesetting expenses, or a writer might have a category for agent fees.

You can add or delete expense categories as you go along—for example, if you find your Misc. category contains many items for a particular type of expense, add it as an expense category.

You don't need a category for automobile expenses, since these expenses require a different kind of documentation for tax purposes.

In separate columns, list the check number for each payment, date and name of the person or company paid. If you pay by credit card or check, indicate it in the check number column.

Once a month, go through your check register, credit card slips, receipts and other expense records and record the required information for each transaction. Also, total the amounts for each category when you come to the end of the page and keep a running total of what you've spent for each category for the year to date.

The following example shows a portion of an expense journal.

(See Appendix 1 for a sample of an expense journal that you can copy or adapt and use.)

**EXPENSE JOURNAL**

| Date | Check No. | Transaction | Amount | Adver-tising | Outside Contrac-tors | Utilities | Supplies | Rent | Travel | Equip-ment | Meals & Entertain-ment | Misc. |
|---|---|---|---|---|---|---|---|---|---|---|---|---|
| 5/1 | 123 | ABC Properties | 500 | | | | | 500 | | | | |
| 5/1 | 124 | Office Warehouse | 150 | | | | 150 | | | | | |
| 5/10 | VISA | Computer World | 1000 | | | | | | | 1000 | | |
| 5/15 | VISA | Cafe Ole' | 50 | | | | | | | | 50 | |
| 5/16 | CASH | Sam's Stationery | 50 | | | | 50 | | | | | |
| 5/18 | 125 | Electric Co. | 50 | | | 50 | | | | | | |
| 5/30 | 126 | Bill Carter | 500 | | 500 | | | | | | | |
| **TOTAL This Page** | | | 2300 | | 500 | 50 | 200 | 500 | | 1000 | 50 | |
| **TOTAL Year to Date** | | | 7900 | 200 | 2000 | 250 | 400 | 2500 | 300 | 1500 | 250 | 500 |

## b. Income journal

The income journal shows you how much money you're earning, and also keeps track of its sources. At a minimum, your income ledger should have columns for the source of the funds—for example, the client's name, your invoice number if there is one, the amount and the date the payment was received. If you have lots of different sources of income, you can create different categories for each source and devote separate columns to them in your journal.

(See Appendix 1 for a sample of an income journal that you can copy or adapt and use.)

**INCOME JOURNAL**

| Source | Invoice | Amount | Date Received |
|---|---|---|---|
| Joe Smith | 123 | $2,500 | 5/5 |
| Acme Inc. | 124 | $1,250 | 5/15 |
| Sue Jones | Personal Loan | $2,000 | 5/20 |
| **Total** | | $5,750 | |

### c. Automobile mileage and expense records

If you use a car or other vehicle for business purposes other than just commuting to and from work, you're entitled to take a deduction for gas and other auto expenses. You can either deduct the actual cost of your gas and other expenses or take the standard rate deduction based on the number of business miles you drive. In 1997, the standard rate was 31.5 cents per mile. (See Chapter 9, Section E.)

Either way, you must keep a record of the total miles you drive during the year. And if you use your car for both business and personal use, you must record your business and personal mileage. Obtain a mileage log book from a stationery or office supply store; you can get one for a few dollars. Keep it in your car with a pen attached. Note your odometer reading in the logbook on the day you start using your car for business. Record your mileage every time you use your car for business and note the business purpose for the trip. Add up your business mileage when you get to the end of each page in the logbook. This way you'll only have to add the page totals at the end of the year instead of all the individual entries.

Here's an example of a portion of a page from a mileage logbook.

**CAR MILEAGE AND EXPENSE LOGBOOK**

| Date | Destination | Business Purpose | Mileage at Beginning of Trip | Mileage at End of Trip | Business Miles | User's Name |
|---|---|---|---|---|---|---|
| 5/1/98 | Fresno | See Art Andrews—potential client | 50,000 | 50,100 | 100 | Jack S. |
| 5/5/98 | Stockton | Delivered documents to Bill James | 50,500 | 50,550 | 50 | Jack S. |
| 5/15/98 | Sacramento | Meeting at Acme Corp. | 51,000 | 51,100 | 100 | Jack S. |
| **TOTAL Business Miles** | | | | | 250 | |

If you think you may want to take the deduction for your actual auto expenses instead of the standard rate, keep receipts for all your auto-related expenses including gasoline, oil, tires, repairs and insurance. You don't need to include these expenses in your ledger sheets; just keep them in a folder or envelope. At tax time, add them to determine how big a deduction you'll get using the actual expense method. Also add in the amount you're entitled to deduct for depreciation of your auto. (See Chapter 9, Section D3.) Your total deduction using your actual auto expenses may or may not be larger than the deduction you'll get using the standard rate. You're generally allowed to use whatever method gives you the largest deduction.

**Use a Credit Card for Gas**

If you use the actual expense method for car expenses, use a credit card when you buy gas. It's best that this be a separate card—either a gas company card or a separate bank card. The monthly statements you receive will serve as your gas receipts. If you pay cash for gas, you must either get a receipt or make a note of the amount in your mileage logbook.

Costs for business-related parking other than at your office and for tolls are separately deductible whether you use the standard rate or the actual expense method. Get and keep receipts for these expenses.

## 3. Supporting Documents

The IRS lives by the maxim that Figures Lie and Liars Figure. It knows very well that you can claim anything in your books, since you create them yourself. For this reason, the IRS requires that you have documents to support the entries in your books and on your tax returns. You don't have to file any of these documents with your tax returns, but you must have them available to back your returns if you're audited.

### a. Income documents

When the IRS audits a small business, it usually asks for both your business and personal bank statements. If you don't have them, the IRS may subpoena them from your bank. If your bank deposits are greater than your reported income on your tax return, the IRS auditor will assume you've underreported your income and impose additional tax, interest and penalties.

To avoid this, you need to be able to prove the source of all your income. Keep supporting documents showing the source and amounts of all the income you receive as an independent contractor. This includes bank deposit slips, invoices and the 1099-MISC forms your clients give you. Keep your bank statements as well.

### b. Expense documents

You also need supporting documents for business expenses. In the absence of a supporting document, an IRS auditor will likely conclude that an item you claim as a business expense is really a personal expense and refuse to allow the deduction. If you're in the mid-level income, 28% marginal tax bracket, every $100 in disallowed deductions costs you $28, plus interest and penalties.

The best supporting document for an expense is a paid receipt that shows who you paid, how much, the date and the item or service purchased.

If you don't have a receipt, keep your canceled check. A canceled check isn't as good as a receipt because it doesn't show what you bought. You can prove this, however, by matching the check with your bill or invoice.

If you buy by credit card, keep your credit card slips and save your monthly billing statements. The best approach is to set aside a separate credit card just for business expenses. If you pay an expense with an ATM card or another electronic funds transfer method, keep your receipt and bank statement.

The reason for saving supporting documents is to prove to the IRS that an expense was related to your business. Sometimes it will be clear from the face of a receipt, sales slip or the payee's name on your canceled check that the item you purchased was for your business. But if it's not clear, note what the purchase was for on the document.

### c. Entertainment, meal and travel expense records

Deductions for business-related entertainment, meals and travel are a hot button item for the IRS because they have been greatly abused by many taxpayers. You need to have more records for these expenses than for almost any others and they will be closely scrutinized if you're audited.

Whenever you incur an expense for business-related entertainment, meals or travel, you must document:

- the date the expense was incurred
- the amount
- the place
- the business purpose for the expense, and
- if entertainment or meals are involved, the business relationship of the people at the event—for example, their names and occupations and any other information needed to establish their business relation to you.

## Recordkeeping Made Easy When You Travel

Here's an easy way to keep your travel expense records. When you go on a business trip, keep all your receipts in an envelope. When you get back from your trip, sort all your receipts by category—for example, airfare, hotel bills, meals and tips, entertainment, taxis and car rental.

Total your expenses for each category and write the totals on the outside of the envelope. Then list the total for all your travel expenses excluding meals and entertainment.

Note on the envelope the dates and place of your trip and the percentage of your time devoted to business. Also note the business reason for the trip. If you have correspondence or other records documenting the business purpose, keep them in the envelope as well. Keep a separate envelope for each trip you take.

To make things even more simple, you can avoid having to record the actual cost of your own meals and other incidentals while traveling on business if you use the IRS's standard allowance for these items. The standard allowance ranges between $26 to $38 per day based on your destination and the number of days you were away from home.

To determine the standard allowance for a trip, see IRS Publication 463, *Travel, Entertainment and Gift Expenses*. You can get a free copy by calling the IRS at 800-TAX-FORM, by calling or visiting your local IRS office or by downloading it from the IRS Web site at www.irs.ustreas.gov.

Your final step is to transfer the figures from your envelope to your monthly expense journal. (See Section A2a.) Enter in your journal the total for all your expenses excluding meals and entertainment under Travel, or Miscellaneous if you don't have a separate travel category.

If your trip was less than 100% for business, first multiply the total by your business percentage and instead enter that in your journal. For example, if you spent 75% of your time on business, multiply your expense total by that amount. If your trip was less then 50% for business, you can't deduct airfare or other transportation costs to your destination (see Chapter 9, Section F), so don't include this amount in your expense journal.

If your trip was 100% for business, list the total amount of your meals and entertainment separately in a Business Meals Entertainment category if you have one, or under Miscellaneous. If your trip was not solely for business, only list your meals and entertainment on days devoted to business. Make a note in your journal if you use the IRS standard allowance for meals.

All this recordkeeping is not as hard as it sounds. Your receipts will ordinarily indicate the date, amount and place in which you incurred the expense. You just need to describe the business purpose and business relationship if entertainment or meals are involved. You can write this directly on your receipt.

**EXAMPLE:** Mary, a freelance computer programmer, has lunch with Harold, president of Acme Technologies, Inc., to discuss performing programming work. Her restaurant receipt shows the date, the name and location of the restaurant, the number of people served and the amount of the expense. Since Mary paid by credit card, the receipt even shows the amount of the tip. Mary just has to document the business purpose for the lunch. She writes on the receipt: "Lunch with Harold Lipshitz, President Acme Technologies, Inc. Discussed signing contract for programming services."

You must keep supporting documents for expenses other than lodging that tally more than $75. Keep your receipts or credit card slips for such expenses. Canceled checks alone are not sufficient; you must have the bill for the expense as well as proof.

### d. Filing supporting documents

If you don't have a lot of receipts and other documents to save, you can simply keep them all in a single folder, shoebox or other safe place.

If you have a lot of supporting documents to save or are the type of person who likes to be extremely well-organized, separate your documents by category—for example, income, travel expenses, equipment purchases. You can use a separate file folder for each category or get an accordion file with multiple pockets.

## 4. Asset Records

When you purchase property such as computers, office furniture, copiers or cellular telephones to use in your business, you must keep records to verify:

- when and how you acquired the asset
- the purchase price
- cost of any improvements—for example, a major upgrade for your computer
- Section 179 deduction taken (see Chapter 9, Section D1)
- deductions taken for depreciation (see Chapter 6, Section D2)
- how you used the asset
- when and how you disposed of the asset
- selling price, and
- expenses of sale.

Set up an asset log showing this information for each item you purchase.

**ASSET LOG**

| Description of Property | Date Placed in Service | Cost or Other Basis | Business/ Investment Use % | Section 179 Deduction | Depreciation Prior Years | Basis for Depreciation | Method/ Convention | Recovery Period | Rate or Table % | Depreciation Deduction |
|---|---|---|---|---|---|---|---|---|---|---|
| Computer | 1/3 | 3,200 | 100% | 3,200 | 0 | 0 | N/A | N/A | N/A | 0 |
| **TOTAL** | | | | 3,200 | | | | | | 0 |

**EXAMPLE:** Patty purchases a $3,200 computer for her business on January 3. She uses it 100% for business and decides to deduct the entire purchase price in a single year using Section 179. She prepares the following asset log for the computer.

You can purchase asset logs from stationery or office supply stores, or set one up yourself using ledger paper. (See Appendix 1 for a sample log you can copy and use.) You can also use a computer accounting program such as *QuickBooks* to do so. If you have an accountant prepare your tax returns, he or she can create an asset log for you.

Also be sure to keep your receipts for each asset purchase, since they'll usually verify what you purchased, when and how much you paid.

### a. Listed property

Listed property means certain type of business assets that can easily be used for personal as well as business purposes. Listed property includes:

- cars, boats, airplanes and other vehicles
- computers
- cellular phones, and
- any other property generally used for entertainment, recreation or amusement—for example, VCRs, cameras and camcorders.

Unless you use listed property 100% for business and keep it at your business location, the IRS imposes special recordkeeping requirements to depreciate or take a Section 179 deduction for it.

**EXAMPLE:** Mary, a freelance writer, does all of her work in her home office, which is clearly her business location. She purchases a computer for her office which she uses 100% for writing. She doesn't need to keep a log of her business use.

If you use listed property both for business and personal uses, you must document your usage—both business and personal. Keep a log book, business diary or calendar showing the dates, times and reasons for which the property is used. (See Appendix 1 for a log sample you can copy and use.) You also can purchase log books for this purpose at stationery or office supply stores.

**EXAMPLE:** Bill, an accountant, purchases a computer he uses 50% for business and 50% to play games. He must keep a log showing his business use of the computer. Following is a sample from one week in his log.

**USAGE LOG FOR PERSONAL COMPUTER**

| Date | Time of Business Use | Reason for Business Use | Time of Personal Use |
|---|---|---|---|
| 5/1/97 | 4.5 hours | Prepared client tax returns | 1.5 hours |
| 5/2/97 | | | 3 hours |
| 5/3/97 | 2 hours | Prepared client tax returns | |
| 5/4/97 | | | 2 hours |

## B. How Long to Keep Records

Keep all your supporting documents for at least three years after you file your tax returns. If you've failed to file tax returns, underreported your income or taken deductions to which you're not entitled, keep your supporting documents indefinitely. They may help you if you're audited.

Keep your asset records for three years after the depreciable life of the asset ends. For example, keep records for five-year property such as computers for eight years.

You should keep your ledger sheets for as long as you're in business, since a potential buyer of your business might want to see them.

### Using Your Computer for Recordkeeping

If you have a personal computer, there are many inexpensive accounting and recordkeeping programs you can use. The most popular of these is *Quicken.* With *Quicken,* you enter all your withdrawals from and deposits to your business checking account in a computerized check register. You note the category of the expense in the register. *Quicken* can then print your checks on your computer printer. You can also set up registers for your credit card accounts.

The great thing about *Quicken* is that it automatically creates the income and expense reports you're required to keep. (See Section A2.) That is, it will show you the amounts you've spent for each category, such as rent or taxes. You can even import these amounts into income tax preparation programs such as *TurboTax* and *MacInTax* when you do your taxes.

Far more sophisticated accounting programs are also available. Programs such as *QuickBooks, Peachtree Accounting* and *Mind Your Own Business* can help you prepare all the books and financial statements that accountants create for small businesses. They can also calculate payroll taxes and print out checks for employees.

Using your computer for recordkeeping can save you time, but it does not relieve you of your obligation to keep the supporting documents. You'll need receipts to back up your computerized records.

## C. Accounting Methods and Tax Years

To put it mildly, accounting methods and tax years are rather dry subjects, but it's essential that you understand the basics about them. The vast majority of self-employed people use the cash method and calendar tax year—the simplest of the methods available.

### 1. Methods of Accounting

There are two basic methods of accounting: cash basis and accrual basis. Using the cash basis method is like maintaining a checkbook. You record income only when the money is received, and expenses when they are actually paid.

> **EXAMPLE:** You sign a contract to perform $25,000 worth of services for a client. Under the cash basis accounting method, you report as income only that portion of the promised $25,000 the client actually paid.

In contrast, in accrual basis accounting, you report income or expenses as they are earned or incurred rather than when they are actually collected or paid. Under this method, you report the money promised by a contract as income for the current year, even if the client had not yet paid you any of the money.

The cash method is by far the simplest and is used by most self-employed people who provide services and do not maintain inventory or offer credit. The accrual method can be difficult to use because there are complex rules to determine when income or expenses are accrued. The accrual method is used by businesses that provide for credit sales or maintain an inventory. C corporations are generally required to use the accrual method.

**It's Not So Easy to Change Your Mind** You choose your accounting method by checking a box on your tax form when you file your tax return. Once you choose a method, you can't change it without getting permission from the IRS. IRS Form 3115, Application for Change in Accounting Method, must be filed 180 days before the end of the year in which you want to make the change. Changing your accounting method can have serious consequences, so consult a tax pro before filing this form.

### 2. Tax Years

You are required to pay taxes for a 12-month period, also known as the tax year. Sole proprietors, partnerships, limited liability companies, S corporations and personal service corporations (see Chapter 2) are required to use the calendar year as their tax years—that is, January 1 through December 31.

However, there are exceptions that permit some small businesses to use a tax year that does not end in December, also known as a fiscal year. You need to get the IRS's permission to use a fiscal year. The IRS doesn't like businesses to use fiscal years, but it might grant you permission if you can show a good business reason for it.

One good reason to use a fiscal year is that your business is seasonal. For example, if you earn most of your income in the spring and incur most of your expenses in the fall, a tax year ending in July or August might be better than a calendar tax year ending in December because the income and expenses on each tax return will be more closely related. To get permission, you must file IRS Form 8716, Election to Have a Tax Year Other Than a Required Tax Year. ■

CHAPTER

# 15

# Safeguarding Your Self-Employed Status

The IRS and many other government agencies would prefer you to be an employee rather than self-employed. This chapter shows you how to avoid being viewed as an employee by the powers that be.

The term generally used to describe self-employed people for tax purposes is independent contractor. Independent contractor is therefore used to identify the self-employed throughout this chapter. Don't get confused, however: independent contractor and self-employed mean the same thing.

## A. Who Decides Your Work Status

Initially, it's up to you and each hiring firm you deal with to decide whether you should be classified as an independent contractor or employee. But the decision about how you should be classified is subject to review by various government agencies, including:

- the IRS
- your state's tax department
- your state's unemployment compensation insurance agency
- your state's workers' compensation insurance agency, and
- the United States Labor Department and National Labor Relations Board.

The IRS considers worker misclassification to be a serious problem costing the U.S. government billions of dollars in taxes that would otherwise be paid if the workers involved were classified as employees and taxes automatically withheld from their paychecks. Most state agencies live by the same theory.

The IRS or your state tax department might question your status in a routine audit of your tax returns. More commonly, however, you'll come to the government's attention when it investigates the classification practices of a firm that hired you. Government auditors may question you and examine your records and the hiring firm's records, too. Because the rules for determining whether you're an independent contractor or employee are rather vague and subjective, it's often easy for the government to claim that you're an employee even though both you and the hiring firm sincerely believed you qualified as an independent contractor. (See Section D.)

## B. If the Government Reclassifies You

If you're like most independent contractors, you probably think that a government agency determination that one or more of your clients should classify you as an employee is solely the client's problem. Unfortunately, this is not the case. What's bad for your clients can also be very bad for you.

If the IRS or other government agency audits you or a hiring firm you've worked for and determines that you should have been classified as an employee instead of an independent contractor, it can and probably will impose assessments and penalties on the firm. Some companies have gone bankrupt because of such assessments.

Rest assured that the government will not penalize or fine you if you've been misclassified as an independent contractor. However, this does not mean being reclassified as an employee won't adversely affect you. For example, the hiring firm may dispense with your services because it doesn't want to pay the additional expenses involved in treating you as an employee. It is not unusual for IRS settlement agreements with hiring firms to require that the firms terminate contracts with independent contractors—with no input from the independent contractors. Or the hiring firm may insist on reducing your compensation to make up for the extra employee expenses. But even if none of these things happen, it's likely you'll be treated very differently on the job. For example, the hiring firm—now your employer—will probably expect you to follow its orders and may attempt to restrict you from working for other companies.

### Worker Gets the Ax When the IRS Calls

Dave, a financial analyst, was hired by a large New York bank and classified as an independent contractor for IRS purposes. He signed an independent contractor agreement and submitted invoices to the bank's accounting department to be paid. The bank withheld no taxes from his pay, paid no Social Security or Medicare taxes for him and provided him with no employee benefits.

However, Dave was otherwise treated largely as an employee. He worked on a team along with regular bank employees and shared their supervisor. He performed the same functions as the employees and worked the same core hours. Because the bank required him to work at its headquarters, he received an admittance card key, office equipment and supplies from the bank.

Dave's happy worklife changed abruptly when the IRS notified the bank that it wanted to examine its employment records to determine whether the company was complying with the tax laws. Fearing that it should have classified Dave as an employee instead of an independent contractor, the bank summarily fired him in the hope this would lessen potential problems with IRS auditors. It didn't work. The bank was required to reclassify Dave as an employee for the two years he had worked for it. Dave was entitled to a refund of half his Social Security and Medicare taxes for those years and was able to collect unemployment benefits, but he was still out of a job.

## 1. Tax Consequences

An IRS determination that you should be classified as an employee can also have adverse tax consequences for you. You're supposed to file amended tax returns for the years involved. On the plus side, you'll be entitled to claim a refund for half the self-employment taxes you paid on the compensation you received as a misclassified independent contractor.

But you might lose valuable business deductions because business expenses such as home offices and health insurance premiums are either not deductible or limited for employees. You'll end up owing more taxes if the value of these lost deductions exceeds the amount of your self-employment tax refund. Your employer will also have to start withholding your income and Social Security taxes from your pay.

From a tax perspective, however, by far the worst thing that can happen to you if you're reclassified as an employee by the IRS is that your Keogh retirement plan will lose its tax qualified status. Of course, if you don't have a Keogh plan, you will not be affected by this threat.

Generally, in a Keogh plan, your contributions are tax deductible and you don't pay any tax on the interest your investment earns until you retire. (See Chapter 16.) But if your Keogh is disqualified, you'll have to pay tax on your contributions and on the interest you've earned from your investments. If you have a substantial amount invested in a Keogh plan, you could face a staggering tax bill.

## 2. Qualifying for Employee Benefits

One good thing that can happen if you're reclassified as an employee is that you may qualify for benefits your employer gives to its other employees, such as health insurance, pension benefits and unemployment insurance.

If you've incurred out-of-pocket expenses for medical care, you may be entitled to be reimbursed. However, these benefits may be short-lived if the hiring firm decides it can't afford to keep you on as an employee.

**IRS Audit Ruins Independent Contractor's Life**

John, a New Hampshire-based narrator of corporate videos and TV commercials, thought he was an independent contractor and was characterized as one by hundreds of clients for which he usually worked for only a few hours or days.

However, when the IRS audited him, it determined that he was really his clients' temporary employee because some paid his union dues and he worked on the clients' premises. When John told his clients that they had to classify him as an employee, issue a W-2 and pay payroll taxes, some of them told him they couldn't afford to hire him any more because of the added expense.

In addition, John had to refile his tax returns for the years in question and recharacterize the compensation he received as wages instead of self-employment income. John was entitled to a refund for half of the self-employment taxes he paid on this compensation. But he lost some substantial business deductions—for example, a $10,000 deduction he took in one year for mileage and auto expenses. The loss of these deductions more than outweighed the refund of self-employment taxes. The net result of being an employee for John is that he lost work for future years and had to pay more taxes for past years.

## C. Determining Worker Status

Various government agencies and courts use slightly different tests to determine how workers should be classified. Unfortunately, these tests are confusing, subjective and often don't lead to a conclusive answer about whether you're an independent contractor or employee. This is why some hiring firms are afraid to hire independent contractors.

This section provides an overview of the most important test for independent contractor status. However, it's practically impossible for anyone to learn and follow all the different tests various government agencies use to determine worker status. Instead, follow the guidelines for preserving your independent contractor status. (See Section E.)

Most, but not all, government agencies use the right of control test to determine whether you're an employee or independent contractor. You're an employee under this test if a hiring firm has the right to direct and control you in how you work—both as to the final results and as to the details of when, where and how you perform the work.

The employer may not always exercise this right—for example, if you're experienced and well trained, your employer may not feel the need to closely supervise you; but the employer has the right to do so at any time.

**EXAMPLE:** Mary takes a job as a hamburger cook at the local AcmeBurger. AcmeBurger personnel carefully train her in how to make an AcmeBurger hamburger—including the type and amount of ingredients to use, the temperature at which the hamburger should be cooked and so forth.

Once Mary starts work, AcmeBurger managers closely supervise how she does her job. Virtually every aspect of Mary's behavior on the job is under AcmeBurger control—including what time she arrives at and leaves work, when she takes her lunch break, what she wears and the sequence of the tasks she must perform. If Mary proves to be an able and conscientious worker, her supervisors may not look over her shoulder very often. But they have the right to do so at any time. Mary is AcmeBurger's employee.

In contrast, you're an independent contractor if the hiring firm does not have the right to control you on the job. Because you're an independent businessperson not solely dependent on the firm for your livelihood, its control is limited to accepting or rejecting the final results you achieve. Or if a project is broken down into stages or phases, the firm's input is limited to approving the work you perform at each stage. Unlike an employee, you are not supervised daily.

**EXAMPLE:** AcmeBurger develops a serious plumbing problem. AcmeBurger does not have any plumbers on its staff, so it hires Plumbing by Jake, an independent plumbing repair business owned by Jake. Jake looks at the problem and gives an estimate of how much to will cost to fix. The manager agrees to hire him and Jake and his assistant commence work.

In a relationship of this kind where Jake is clearly running his own business, it's virtually certain that AcmeBurger does not have the right to control the way Jake performs his plumbing services. Its control is limited to accepting or rejecting the final result. If AcmeBurger doesn't like the work Jake has done, it can refuse to pay him.

The difficulty in applying the right of control test is deciding whether a hiring firm has the right to control you. Government auditors can't look into your mind to see if you are controlled by a hiring firm. They have to rely instead on indirect or circumstantial evidence indicating control or lack of it—for example, whether a hiring firm provides you with tools and equipment, where you do the work, how you're paid and whether you can be fired.

The factors each agency relies upon to measure control vary. Some agencies look at 14 factors to see if you're an employee or independent contractor; some look at 11; some consider only three. Which of these factors is of the greatest or least importance is anyone's guess. This can make it very difficult to know whether you pass muster as an independent contractor.

## D. IRS Approach to Worker Status

The IRS uses the right of control test just mentioned to determine whether you're an independent contractor or employee for tax purposes. The agency developed a list of 20 factors its auditors were supposed to use to measure how much control a hiring firm had over you. This test has become very well known—discussed in countless magazine and journal articles, posted all over the Internet and practically memorized by many independent contractors. Many of your prospective clients will be aware of the test and may even ask you about it.

**Special IRS Rules for Some Workers**

If you're a licensed real estate agent, direct seller, business-to-business salesperson, home worker who makes clothing items, fulltime life insurance salesperson or driver who distributes food products, beverages or laundry, your status may be predetermined for IRS purposes under special rules. (See Chapter 12.)

It's important to understand, however, that the 20 Factor Test is only an analytical tool IRS auditors use to measure control. It is not the legal test the IRS uses for determining worker status. IRS auditors have never been restricted to considering only the 20 factors on the test, nor are they required to consider them all. In short, the test is not an end in and of itself.

Unfortunately, the 20 Factor Test left much to be desired as an analytical tool. It proved to be so subjective and complex that trying to apply it was often a waste of time. Even the IRS found it difficult to apply the test consistently. People performing the same services have been found to be employees in some IRS districts and independent contractors in others. In one case, for example, a Methodist minister was found the be an employee of his church; a Pentecostal pastor was found to be an independent contractor in another case the same year. (*Weber vs. Commissioner*, 103 TC 378 (1994); *Shelley v. Commissioner*, TCM 1994-432 (1994).)

Hiring firms, independent contractors, attorneys and tax experts complained bitterly for years that it was impossible to know who was and was not an independent contractor under the 20 Factor Test and that aggressive IRS auditors automatically classified every worker as an employee. Legislation was introduced in Congress to clarify the issue by establishing a simple test for determining worker status—a test that would probably have led to far more workers qualifying as independent contractors than under current law.

In an obvious attempt to block such legislation and to set the rules itself, the IRS recently issued a training manual for its auditors setting forth a somewhat simpler approach to measure control. It should be easier for you to qualify as an independent contractor for IRS purposes under this manual than under the 20 Factor Test. Although the manual doesn't have the force of law, the IRS apparently expects its auditors to follow it and began training them to do so in late 1996.

The manual provides the best guidance the IRS has ever made public on how to classify workers. Even more important, it evinces a much kinder, gentler attitude by the IRS on worker classification issues. Declaring that classifying a worker as an independent contractor instead of an employee "can be a valid and appropriate business choice," the manual admonishes IRS auditors to "approach the issue of worker classification in a fair and impartial manner."

## 1. How the IRS Measures Control

The IRS has instructed its auditors to look at three areas to determine whether a hiring firm has the right to control a worker. These are:

- your behavior on the job
- your finances, and
- your relationship with the hiring firm.

The following chart shows the primary factors the IRS looks at for each area.

The current IRS approach is somewhat simpler than the old 20 Factor Test, but is no panacea. It still remains unclear how important each factor is and how many factors must indicate independent contractor status for you to be classified as an independent contractor. The IRS says there is no magic number of factors. Rather, the factors showing lack of control must outweigh those that indicate control. No one factor alone is enough to make you an employee or an independent contractor.

To make your life easier, this chapter offers you a list of eight guidelines for you to follow during your worklife. If you do, it's likely that you will be viewed as an independent contractor by the IRS and any other government agency.

What's really helpful about the IRS audit manual is that it recognizes that, because of the changing nature of work today, several factors that used to be considered crucial in determining worker status are no longer so important. It also takes a more liberal approach to other factors that are still considered important. Altogether, these changes should give you and your clients more leeway in how you deal with each other.

## IRS TEST FOR WORKER STATUS

| | | |
|---|---|---|
| **Behavioral Control**<br>Factors showing whether a hiring firm has the right to control how you perform the specific tasks you've been hired to do. | **You will more likely be considered self-employed if you:**<br>• are not given instructions by the hiring firm<br>• provide your own training | **You will more likely be considered an employee if you:**<br>• receive instructions you must follow about how to do your work<br>• receive detailed training from the hiring firm |
| **Financial Control**<br>Factors showing whether a hiring firm has a right to control your financial life. | **You will more likely be considered self-employed if you:**<br>• have a significant investment in equipment and facilities<br>• pay business or travel expenses yourself<br>• make your services available to the public<br>• are paid by the job<br>• have opportunity for profit or loss | **You will more likely be considered an employee if you:**<br>• have equipment and facilities provide by the hiring firm free of charge<br>• have your business or traveling expenses reimbursed<br>• make no effort to market your services to the public<br>• are paid by the hour or other unit of time<br>• have no opportunity for profit or loss—for example, because you're paid by the hour and have all expenses reimbursed |
| **Relationship Between Worker and Hiring Firm**<br>Factors showing whether your and the hiring firm believe you're self-employed or an employee | **You will more likely be considered self-employed if you:**<br>• don't receive employee benefits such as health insurance<br>• sign a client agreement with the hiring firm<br>• can't quit or be fired at will<br>• are performing services that are not a regular part of the hiring firm's regular business activities | **You will more likely be considered an employee if you:**<br>• receive employee benefits<br>• have no written client agreement<br>• can quit at any time without incurring any liability to the hiring firm<br>• can be fired at any time<br>• are performing services that are part of the hiring firm's core business |

### a. Payment by the hour

One of the most important changes in the manual is the recognition by the IRS that some independent contractors—lawyers, for example—are usually paid by the hour instead of being paid a flat fee. If you want to be classified as an independent contractor, it's still better not to be paid an hourly wage, but it seems clear you can get away with it if it's a common practice in your line of business.

### b. Reimbursement of expenses

The IRS now advises auditors that they should not focus on whether a worker has business or travel expenses reimbursed by a hiring firm. Instead, they should determine whether a worker has unreimbursed expenses such as fixed ongoing costs—office rent, for example.

### c. Advertising

Failure to offer your services to the public is a sign of employee status. In the past, the IRS usually only considered advertising in telephone books and newspapers as evidence that an independent contractor offered services to the public. Of course, many independent contractors don't get business this way; they rely primarily on word of mouth. The IRS now recognizes this fact of life and provides that advertising is not essential for independent contractor status.

### d. Form of direction

For the first time, the manual makes a distinction between receiving instructions on how to work—a very strong indicator of employee status—and being given suggestions. A suggestion about how work is to be performed does not constitute the right to control. However, if you must comply with suggestions or you would suffer adverse consequences if you don't comply, such as being fired or not assigned more work, then the suggestions are, in fact, instructions.

### e. Training

Periodic or ongoing training about how to do your work is strong evidence of an employment relationship. However, a client may provide you with a short orientation or information session about the company's policies, new product line or new government regulations without jeopardizing your independent contractor status. Training programs that are voluntary and that you attend without receiving pay do not disqualify you from being classified as an independent contractor.

### f. Investment

A significant investment in equipment and facilities is not necessary for independent contractor status. The manual notes that some types of work simply do not require expensive equipment—for example, writing and certain types of consulting. But even if expensive equipment is needed to do a particular type of work, an independent contractor can always rent it.

### g. Performing key services

One of the most important factors IRS auditors look at is whether the services performed by a worker are key to the hiring firm's regular business. The IRS figures that if the services you perform are vital to a company's regular business, the company will be more likely to control how you perform. For example, a law firm is less likely to supervise and control a painter it hires to paint its offices than a paralegal it hires to work on its regular legal business.

However, IRS auditors are required to examine all the facts and circumstances. For example, a paralegal hired by a law firm could very well be an independent contractor if he or she was a specialist hired to help with especially difficult or unusual legal work.

**Written Agreements Are More Important Than Ever**

A written independent contractor agreement can never make you an independent contractor by itself. However, if the evidence is so evenly balanced that it is difficult or impossible for an IRS auditor to decide whether you're an independent contractor or employee, the existence of a written independent contractor agreement can tip the balance in favor of independent contractor status. This makes using written independent contractor agreements more important than ever before. (See Chapter 18.)

### h. Fulltime work

It used to be considered the kiss of death for an independent contractor to work fulltime for a client. IRS auditors saw this as very strong evidence of employee status. The manual now recognizes that: "An independent contractor may work fulltime for one business either because other contracts are lacking, because this is the way time is allocated, or because the contract by its terms requires fulltime exclusive effort."

### i. Longterm work for a single client

Performing services for the same client year after year used to be seen as a sign you were an employee. No longer. The IRS now recognizes that independent contractors may work for a client on a longterm basis, either because they sign longterm contracts or because their contracts are regularly renewed by the client because they do a good job, price their services reasonably or no one else is readily available to do the work.

### j. Time and place of work

It is not important to the IRS where or when you work. For example, the fact that you work at a client's offices during regular business hours is not considered evidence of employee status. On the other hand, the fact that you work at home at hours of your own choosing is not strong evidence that you're an independent contractor, since many employees now work at home as well.

## 2. Rules for Technical Services Workers

If you're a computer programmer, systems analyst, engineer or provide similar technical services and obtain work through brokers, special tax rules may make it very difficult for you to work as an independent contractor even if you may qualify as one under the IRS test.

This is the result of a Tax Code provision called Section 1706. The law prevents brokers who provide certain types of technical services workers to others from using a defense against the IRS called Section 530. By using Section 530, an employer who misclassifies a worker as an independent contractor can avoid paying any fines or penalties for failing to pay employment taxes if it can show it had a reasonable basis for treating the worker as an independent contractor. Section 530 has made hiring firms' lives a little easier by giving them an additional defense against the IRS.

Section 1706 provides that the Section 530 defense may not be used by brokers that contract to provide their clients with:

- engineers
- designers
- drafters
- computer programmers
- systems analysts, or
- other skilled workers in similar technical occupations.

Section 1706 doesn't make anyone an employee. But it does make it harder for brokers to avoid paying assessments and penalties if the IRS claims they've misclassified a technical services worker as an independent contractor. Because of Section 1706, brokers who contract to provide companies with technical services workers usually classify the workers as their employees and issue them W-2s. One result of this is that you'll likely receive less pay from the broker than if you were classified an independent contractor because it has to pay payroll taxes for you and provide workers' compensation.

Some brokers may make an exception and treat you as an independent contractor if you are clearly running an independent business—for example, you deal directly with many clients and are incorporated. As one case in point, one Silicon Valley software tester occasionally obtains work through brokers and is never classified as an employee of the broker. Brokers feel safe treating her as an independent contractor because she has incorporated her business, has employees and has many clients, including several Fortune 500 companies. The brokers sign a contract with the tester's corporation, not with her personally.

Section 1706 has no application at all if you contract directly with a client instead of going through a broker. But many high tech firms are still wary of hiring independent contractors.

## E. Preserving Your Status

If you consistently follow the guidelines discussed below, it's likely that any government agency or court would determine that you qualify as an independent contractor. However, there are no guarantees.

Some of these guidelines may be a bit more strict than those now followed by the IRS. This is because not all government agencies follow the IRS standards. Various state agencies, such as state tax departments and unemployment compensation and workers' compensation agencies, may have tougher classification standards than the IRS. This means you can't rely solely on the IRS rules.

> **EXAMPLE:** Debbi, an independent contractor accountant based in Rhode Island, clearly qualifies as an independent contractor under the IRS test. However, very restrictive Rhode Island employment laws require that she be classified as an employee for state purposes if she performs services for other accounting firms. This means they must withhold state income taxes from her pay, pay unemployment taxes and provide her with workers' compensation insurance. In addition, if Debbi has employees of her own, they may become the firm's employees as well under Rhode Island law.

Some hiring firms are terrified of government audits and are very nervous about hiring independent contractors. These may include companies that have had problems with government audits in the past or are in industries that are targeted by government auditors. Recent targets include trucking firms, courier services, securities dealers, high technology firms, nurse registries, building contractors and manufacturer representatives. Such hiring firms may be more willing to hire you as an independent contractor if you can show that you follow the guidelines. Document your efforts and be ready to show the documentation to nervous clients.

Many other companies don't give government audits a second thought and are more than happy to classify you as an independent contractor and save the money and headaches involved in treating you as an employee. These companies may not worry whether you qualify as an independent contractor, but you should. You could still end up getting fired, taking a pay cut or having to pay extra taxes if some government bureaucrat decides you're really an employee. So even though your client may not appreciate your efforts, continue to follow the guidelines discussed here; both you and your client will be glad you did if the government comes calling.

### 1. Retain Control of Your Work

The most fundamental difference between employees and independent contractors is that employers have the right to tell their employees what to do. (See Section D.) Never permit a hiring firm to supervise or control you the way it does its employees. It's perfectly all right for the hiring firm to give you detailed guidelines or specifications for the results you're to achieve. But how you go about achieving those results should be entirely up to you.

A few guidelines will help emphasize that you are the one who is responsible.

- Do not ask for or accept instructions or orders from the hiring firm about how to do your job. For example, you should decide what equipment or tools to use, where to purchase supplies or services, who will perform what tasks and what routines or work patterns must be used. It's fine for a hiring firm to give you suggestions about these things, but you always have the option of accepting or rejecting such suggestions.
- Do not ask for or receive training on how to do your work from a hiring firm. If you need additional training, seek it elsewhere.
- A hiring firm may give you a deadline for when your work should be completed, but you should generally establish your own working hours. For example, if you want, you could work 20 hours two days a week and take the rest of the week off. In some cases, however, it may be necessary to coordinate your working hours with the client's schedule—for example, where you must perform work on the client's premises or work with its employees.
- Decide on your own where to perform the work—that is, a client should not require you to work at a particular location. Of course, some services must be performed at a client's premises or other particular place so you won't have a choice about where to work.
- Decide whether to hire assistants to help you and, if you do, pay and supervise them yourself. Only you should have the right to fire your assistants.
- Do not attend regular employee meetings or functions such as employee picnics.
- Avoid providing frequent formal written reports about the progress of your work—for example, daily phone calls to the client. It is permissible, however, to give reports when you complete various stages of a project.
- Do not obtain, read or pay any attention to a hiring firm's employee manuals or other rules for employees. The rules governing your relationship with the hiring firm are contained solely in your independent contractor agreement, whether written or oral. (See Chapter 18.)

If you work outside the client's premises, it's usually not difficult to avoid being controlled. Neither the client nor its employees will have much opportunity to try to supervise you. For example, Katherine, a freelance legal writer, never has any problems being controlled by a large legal publisher for whom she performs freelance assignments. She says anonymity on the job helps with this: "I get my freelance assignments by phone, do all the work at home and in the local law library and then transmit the project from my computer to the publisher by modem. I've hardly ever been in the publisher's office."

On the other hand, you could have problems if you work in a client's workplace. The client's supervisors or managers may try to treat you as if you were an employee. Before you start work, make clear to the client that you do not fall within its regular personnel hierarchy; you are an outsider. You might ask the client to designate one person with whom you will deal. If anyone in the company gives you problems, you can explain that you deal only with your contact person and refer the person to your contact.

Note, however, that it's fine for a client to require you to comply with government regulations about how to perform your services. For example, a client may require a construction contractor to comply with municipal building codes that impose detailed rules on how a building is constructed. The IRS and most other government agencies would not consider this to be an exercise of control over the contractor by the client.

## 2. Opportunity for Profit or Loss

Because they are in business for themselves, independent contractors have the opportunity to earn profits or suffer losses. If you have absolutely no risk of loss, you're probably not an independent contractor.

### a. Business expenses

The best way to show an opportunity to realize profit or loss is to have recurring business expenses. If receipts do not match expenses, you lose money and may go into debt. If receipts exceed expenses, you earn a profit.

Good examples of independent contractor expenses include:

- salaries for assistants
- travel and other similar expenses incurred in performing your services
- substantial investment in equipment and materials
- rent for an office or workplace
- training and educational expenses
- advertising
- licensing, certification and professional dues
- insurance
- leasing of equipment
- supplies, and
- repairs and maintenance of business equipment.

Don't go out and buy things you don't really need. But if you've been thinking about buying equipment or supplies to use in your business, go ahead and take the plunge. You'll not only solidify your independent contractor status, but get a tax deduction as well. (See Chapter 9.)

In addition, it's best that you don't ask clients to reimburse you for expenses such as travel, photocopying and postage. It's a better practice to bill clients enough to pay for these items yourself. Setting your compensation at a level that covers your expenses also frees you from having to keep records of your expenses. Keeping track of the cost of every phone call or photocopy you make for a client can be a real chore and may be more trouble than it's worth.

### b. Get paid by the project

Another excellent way to have opportunity for profit or loss is to be paid an agreed price for a specific project, rather than to bill by unit of time such as hourly payment. If the project price is higher than the expenses, you'll make money; if not, you'll lose money. However, this form of billing is too risky for many independent contractors.

Many independent contractors—for example, attorneys and accountants—are typically paid by the hour. If hourly payment is customary in your field, this factor should not affect your independent contractor status.

## 3. Look Like an Independent Business

Take steps to make yourself look like an independent business person. There are several things you can do to cultivate this image.

- Don't obtain employee-type benefits from your clients such as health insurance, paid vacation, sick days, pension benefits or life or disability insurance; instead, charge your clients enough to purchase these items yourself.
- Incorporate your business instead of operating as a sole proprietor. (See Chapter 2.)
- Obtain a fictitious business name instead of using your own name for your business. (See Chapter 3.)
- Obtain all necessary business licenses and permits. (See Chapter 5.)
- Obtain business insurance. (See Chapter 6.)

- Maintain a separate bank account for your business. (See Chapter 14.)
- Obtain a federal employer identification number. (See Chapter 5.)

You may have an easier time getting work if you do these things. For example, a large corporation that regularly used the services of one independent contractor asked her to incorporate or at least obtain a business license because it was worried she might otherwise be viewed as the company's employee.

## 4. Work Outside Hiring Firms' Premises

The IRS no longer considers working at a hiring firm's place of business to be an important factor in determining whether a worker is an independent contractor or employee, but many state agencies still do. For example, in half the states you may be considered an employee for unemployment compensation purposes if you work at the hiring firm's place of business or another place it designates.

Working at a location specified by a hiring firm implies that the firm has control, especially where the work could be done elsewhere. If you work at a hiring firm's place of business, you're physically within the firm's direction and supervision. If you can choose to work off the premises, the firm obviously has less control.

Unless the nature of the services you're performing requires it, don't work at the hiring firm's office or other business premises. An independent contractor hired to lay a carpet or paint an office must obviously work at the hiring firm's premises. But if your work can by done anywhere, do it anywhere but at the client's premises.

Working at a home office will not show you're an independent contractor as far as the IRS is concerned since many employees are now doing so. Renting an office outside your home definitely will show independent contractor status. This really shows you're operating your own business and gives you a recurring business expense to help establish risk of loss.

## 5. Make Your Services Widely Available

Independent contractors normally offer their services to the general public, not just to one person or entity. The IRS recognizes that many independent contractors rely on word of mouth to get clients and don't do any active marketing. (See Section E1.) However, nervous clients and other government auditors will be impressed if you market your services to the public.

There are many relatively inexpensive ways you can do so—for example:

- obtain a business card and stationery
- hang out a shingle in front of your home or office advertising your services
- maintain listings in business and telephone directories
- attend trade shows and similar events
- join professional organizations
- advertise in newspapers, trade journals and magazines
- mail brochures or other promotional materials to prospective clients, and
- phone potential clients to drum up business.

Keep copies of advertisements, promotional materials and similar items to show prospective clients and government auditors.

## 6. Have Multiple Clients

IRS guidelines provide that you can work fulltime for a single client on a longterm basis and still be an independent contractor. (See Section D.) Nevertheless, having multiple clients shows that you're running an independent business because you are not dependent on any one firm for your livelihood. Some potential clients may be afraid to hire you as an independent contractor unless you perform services for others besides them.

It's best to have more than one client at a time. Government auditors will rarely question the status of an independent contractor who works for three or four clients simultaneously. However, the nature of your work may require that you work fulltime for one client at a time. In this event, at least try to work for more than one client over the course of a year—for example, work fulltime for one client for six months and fulltime for another client for the other six months.

Having multiple clients also gives you increased economic security. One independent contractor thinks of herself as a eight-legged spider, with each client a separate leg. If she loses one client, things are still economically stable because she has her other legs to stand on.

If you seem to be locked into having just one client, you may be able to drum up new business by offering special rates for small jobs you would otherwise lose.

**Don't Sign Noncompetition Agreements** Some clients may ask or require you to sign noncompetition agreements restricting your ability to work for the client's competitors while working for the client, afterwards or both. You should avoid such restrictions like the plague. Not only will they make you look like an employee, they may make it impossible for you to earn a living.

## 7. Use Written Agreements

Use written independent contractor agreements for all but the briefest, smallest projects. Among other things, the agreement should make clear that you are an independent contractor and the hiring firm does not have the right to control the way you work. A written agreement won't make you an independent contractor by itself, but it is helpful—particularly if you draft it yourself.

Also, don't accept new projects after the original project is completed without signing a new independent contractor agreement. You can easily be converted from an independent contractor to an employee if you perform assignment after assignment for a client without negotiating new contracts.

## 8. Avoid Accepting Employee Status

Some clients will refuse to hire you as an independent contractor and insist on classifying you as an employee. This is a particularly common problem for independent contractors who work for high technology companies. Because the government closely scrutinizes such firms' hiring practices and because of special IRS rules, high tech companies are often wary of using independent contractors. (See Section D2.)

Some companies may hire you themselves as an employee. Others may insist that you contract with a broker or employment agency who treats you as its employee. The firm then hires you through the broker.

It's best to avoid performing the exact same services as an independent contractor and employee because:

- being classified as both an independent contractor and employee on your tax returns may make an IRS audit more likely
- it could lead government auditors to conclude you're an employee for all purposes
- you'll generally be paid less than if you were an independent contractor because the hiring firm or broker will have to provide you with workers' compensation and unemployment insurance; on the plus side, however, you may be able to collect unemployment when your services end
- you may not be able to deduct unreimbursed expenses incurred while you were an employee or the deductions may be limited (see Chapter 9), and
- you can't apply your employee income to your independent contractor business to help it show a profit; if your business keeps showing losses, the IRS might conclude it's a hobby and disallow your business deductions. (See Chapter 9.) ■

CHAPTER

# 16

# Retirement Options for the Self-Employed

When you're self-employed, you don't have an employer to provide you with a pension. It's up to you to establish and fund your own pension plan to supplement any Social Security benefits you'll receive. This chapter provides an overview of the main retirement options for the self-employed.

This chapter is not a guide about where to invest your money. There are hundreds of investment choices—including stocks, bonds, mutual funds, money market accounts and certificates of deposit. And there are a seemingly equal number of books, magazines, Websites and other resources devoted to investment strategy.

Deciding what to invest in is not the first decision you must make when planning for your retirement. Rather, you must first decide what type of retirement account or accounts to establish. There are several types of accounts specifically designed for the self-employed that provide terrific tax benefits. These will help you save for retirement and may reduce your tax burden when you do retire.

Choosing what type of account to establish is just as important as deciding what to invest in once you open your account, if not more so. Once you establish your account, you can always change your investments within it with little or no difficulty. But changing the type of accounts you have may prove difficult and costly. This chapter focuses on this initial crucial retirement decision.

Two easy-to-understand guides on retirement investing are:

- *Get a Life: You Don't Need a Million to Retire Well,* by Ralph Warner (Nolo Press), and
- *Investing for Dummies,* by Eric Tyson (IDG Books).

For additional information on the tax aspects of retirement, see:

- IRS Publication 560, *Retirement Plans for the Self-Employed,* and
- IRS Publication 590, *Individual Retirement Accounts.*

You can obtain these and all other IRS publications by calling the IRS at 800-TAX-FORM, visiting your local IRS office or downloading the publications from the IRS Internet site at: www.irs.ustreas.gov.

## A. Retirement Plans

In all likelihood you will receive Social Security benefits when you retire. However, Social Security will likely provide you with no more than half of your needs when you retire, possibly less depending upon your retirement lifestyle. You'll need to make up the shortfall with your own retirement investments.

Luckily, when it comes to saving for retirement, the self-employed are actually better off than most employees. This is because the federal government allows you to set up retirement accounts specifically designed for small business people that provide some terrific income tax benefits. These are called IRAs, SEP-IRAs and Keogh Plans.

If, like most self-employed people, you are a sole proprietor, a partner in a partnership or an owner of a limited liability company, you can establish such an account through a bank, savings and loan, credit union, insurance company, brokerage house, mutual fund company or other financial institution.

Retirement accounts are simply shells that protect you from being taxed by the government. After deciding what type of account you want you must then decide what to invest in. You can invest in almost anything—for example, stocks, bonds, mutual funds, money market funds or certificates of deposit. You can transfer your money from one type of investment to another within your account as market conditions change.

These accounts provide you with two enormous tax benefits in that you:

- pay no taxes on the income your retirement investments earn until you withdraw the funds upon retirement, and

- deduct the amount you contribute to your retirement account from your income taxes for the year, subject to certain limits.

### 1. Tax Deferral

The money you earn on an investment is ordinarily taxed the year you earn it. For example, you must pay taxes on the interest you earn on a savings account or certificate of deposit when that interest accrues. When you sell an investment at a profit, you must pay income tax on this amount as well—for example, you must pay tax on the profit you earn from selling stock.

But this is not the case when you invest in a tax qualified retirement account such as an IRA, SEP-IRA or Keogh plan. The money your investment earns is not taxable to you until you withdraw the funds when you retire—when you will usually be in a lower income tax bracket than you were during your working years.

You're not supposed to make withdrawals until you reach age 59-1/2, subject to certain exceptions. If you make early withdrawals, you must ordinarily pay regular income tax on the amount you take out, plus a 10% federal tax penalty. For example, if you're in the 28% income tax bracket, you'll have to pay a 38% federal tax on your withdrawal. Your state may impose additional penalties for early withdrawal.

This means you must be prepared to leave your money in your retirement accounts for many years, perhaps decades. The following chart shows how long you must leave your money in a tax-deferred retirement account to earn as you would have done had you put the money in a regular nonretirement account not subject to a penalty upon withdrawal. For example, if you earn 10% a year on your investment, and you withdraw your funds after 15 years, you would have been better off not investing in the retirement account at all. This calculation assumes that $2,000 is invested yearly and that your tax bracket is 30%.

**INVESTMENT IN RETIREMENT ACCOUNT**

| Return on investment | Years to break even |
|---|---|
| 6% | 24 |
| 7% | 21 |
| 8% | 18 |
| 9% | 16 |
| 10% | 15 |
| 11% | 14 |
| 12% | 13 |
| 13% | 12 |
| 14% | 12 |
| 15% | 11 |

### 2. Tax Deduction

Avoiding income tax on your retirement investments until you retire is a good deal in and of itself. But there is an additional outstanding tax benefit to retirement accounts: You can usually deduct the amount you contribute to your retirement account from your income taxes for the year. This can give you a substantial income tax saving.

> **EXAMPLE:** Art, a self-employed sole proprietor, establishes a retirement account at his local bank and contributes $10,000 this year. He can deduct the entire amount from his income taxes. Since Art is in the 28% tax bracket, he saves $2,800 in income taxes for the year (28% x $10,000) and has saved $10,000 toward his retirement.

How much you can contribute each year depends on the type of account you establish and how much money you earn. However, you won't be able to make any contributions if you have a loss from your business. You must have self-employment income to fund these retirement accounts.

The combination of tax deferral and a current tax deduction makes a huge difference on how much you'll save for retirement. The following chart shows the difference in growth between a tax-deferred and taxable account over 30 years.

## GROWTH OF A TAXABLE VS. TAX-DEFERRED ACCOUNT

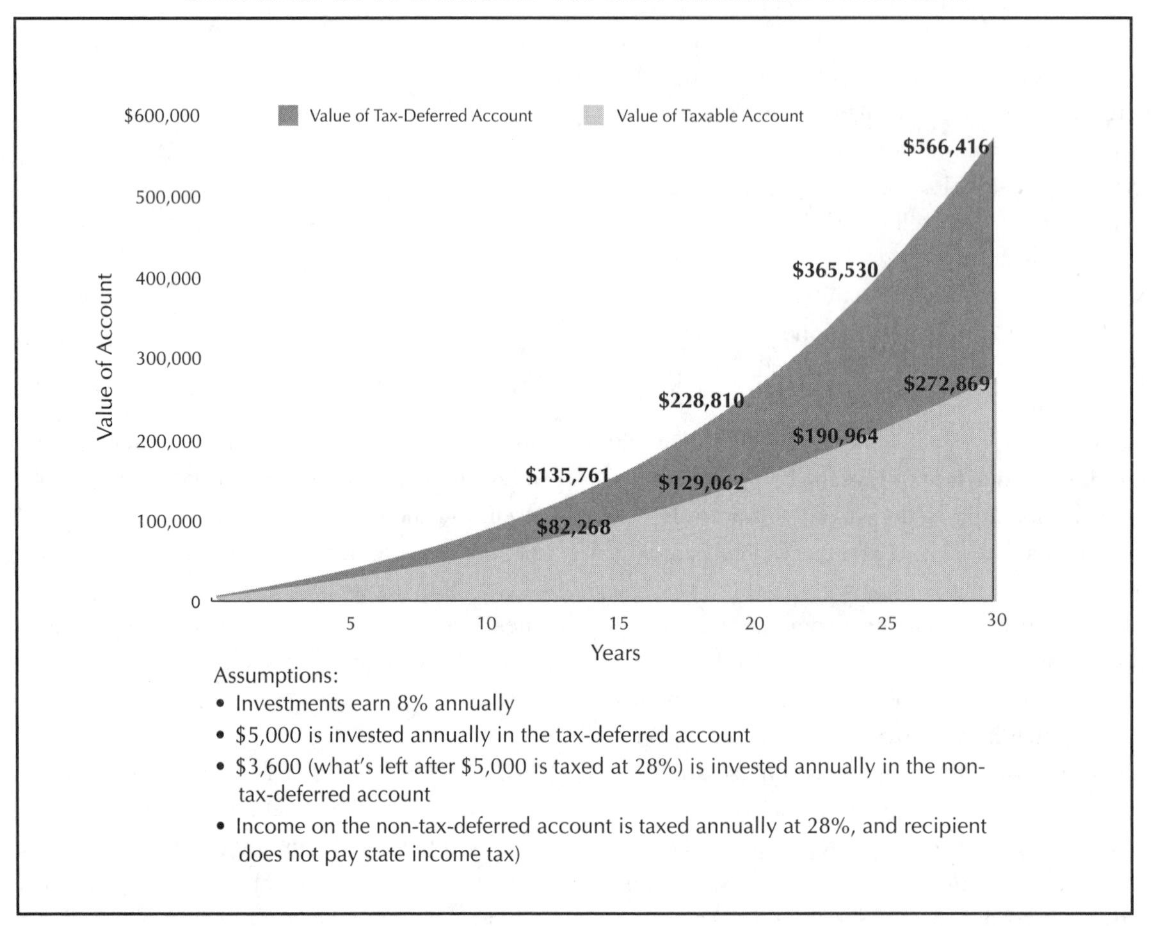


Assumptions:

- Investments earn 8% annually
- $5,000 is invested annually in the tax-deferred account
- $3,600 (what's left after $5,000 is taxed at 28%) is invested annually in the non-tax-deferred account
- Income on the non-tax-deferred account is taxed annually at 28%, and recipient does not pay state income tax)

From *Get a Life: You Don't Need a Million to Retire Well,* By Ralph Warner (Nolo Press).

**HOW MUCH MONEY WILL YOU NEED WHEN YOU RETIRE?**

How much money you'll need when you retire depends on many factors, including your lifestyle. You could need to have anywhere from 50% to 100% of the amount you earned while employed. The average is about 70% to 80% of pre-retirement earnings. But this is only an average, and may not apply to you.

For a detailed explanation of how to arrive at a reasonably accurate estimate of your annual retirement financial needs and how much you'll need to save to meet them, see *Get a Life: You Don't Need a Million to Retire Well*, by Ralph Warner (Nolo Press).

## B. IRAs

The simplest type of tax-deferred retirement account is the individual retirement account or IRA. If you can afford to invest no more than $2,000 per year in a retirement plan, an IRA is a good choice.

An IRA is a trust or custodial account set up for the benefit of an individual or his or her beneficiaries. The trustee or custodian administers the account. The trustee can be a bank, mutual fund, brokerage firm or other financial institution such as an insurance company which has been approved by the IRS. The custodian must meet strict IRS requirements regarding safekeeping of your account.

IRAs are extremely easy to set up and administer. You need not file any special tax forms with the IRS. The financial institution you use to set up your account will ordinarily request that you complete IRS Form 5305, Individual Retirement Trust Account, which serves as a preapproved IRA agreement. Keep the form in your records.

Most financial institutions offer an array of IRA accounts that provide for different types of investments. You can invest your IRA money in just about anything—stocks, bonds, mutual funds, Treasury bills and notes and bank certificates of deposit. However, you can't invest in collectibles such as art, antiques, stamps or other personal property.

Recent changes in the tax laws have complicated the IRA picture. Beginning in 1998, there are two different retirement IRAs to choose from:

- traditional IRAs, and
- Roth IRAs.

You can have as many IRA accounts as you want. But the maximum amount you can contribute each year to all your IRA accounts is $2,000—for example, you could contribute $1,000 to a traditional IRA and $1,000 to a Roth IRA.

The IRA contribution limit is higher, however, if you're married. A married couple may now put up to $2,000 per spouse into an IRA, for a total of $4,000 per year. This is so even if one spouse isn't working. But you must file a joint return and the working spouse's earnings must be at least as much as the IRA contribution.

### 1. Traditional IRAs

Traditional IRAs have been around since 1974. The principal feature of these IRAs is that you receive an income tax deduction for the amounts you contribute each year to your account. For example, if you contribute $2,000, you may deduct $2,000 from your gross income for that year for income tax purposes. If you're in the 28% income tax bracket, this will save you $560 in income taxes. Thereafter, your earnings accumulate in the account tax free until you withdraw them.

#### a. Deducting your IRA contributions

You can deduct the amount of your annual IRA contributions from your income taxes so long as:

- you have at least as much earned income as your contribution, and
- you're not covered by another tax-favored retirement plan such as a SEP-IRA or Keogh Plan.

In the past, you couldn't deduct an IRA contribution if your spouse worked as an employee and was covered by an employer-sponsored pension plan. This is no longer the case. Beginning in 1998, you can take a full $2,000 deduction for an IRA contribution regardless of whether your spouse is covered by a retirement plan at work. However, the deduction is phased out if you and your spouse together earn over $150,000 in adjusted gross income.

Your adjusted gross income or AGI is your total income minus deductions for:

- IRA, Keogh and SEP-IRA contributions
- the self-employed health insurance deduction
- one-half of your self-employment tax, and
- alimony, deductible moving expenses and penalties you pay for early withdrawals from a savings account before maturity or early redemption of a certificate of deposit.

To find out what your AGI is, look at line 31 on your last year's tax return, Form 1040.

Even if your contribution isn't deductible, you can still set up and contribute to a nondeductible IRA. In this case, your contribution won't be deductible from your income taxes, but you won't have to pay taxes on the interest those IRA investments earn until you withdraw the money.

### b. When can you withdraw your money?

The money in an IRA is not supposed to be withdrawn until you reach age 59-1/2 unless you die or become disabled. The amounts you withdraw are then included in your regular income for income tax purposes. You must begin withdrawing your money from your IRA by April 1 of the year after the year you turn 70.

As a general rule, if you make early withdrawals, you must pay regular income tax on the amount you take out plus a 10% federal tax penalty. For example, if you're in the 28% income tax bracket, you'll have to pay a 38% federal tax on your withdrawal. Your state may impose additional penalties for early withdrawal.

However, you don't have to pay the 10% penalty for early withdrawals under a number of circumstances.

- **Paying medical bills.** You are now allowed to withdraw funds from an IRA before you reach age 59 1/2 without a 10% penalty tax if you use the money to pay medical expenses that exceed 7.5% of your adjusted gross income. For example, if your adjusted gross income is $20,000, you may use IRA withdrawals to pay medical expenses in excess of $1,500 (7.5% x $20,000). Qualifying expenses include doctor or hospital bills, nursing care, prescription drugs, eyeglasses, dentures and hearing aids.

  Also, no penalties are exacted if you use an IRA withdrawal to pay health insurance premiums, provided that your business is so poor that you would be entitled to receive unemployment compensation if you were an employee.
- **Purchasing a home.** Starting in 1998, you may withdraw up to $10,000 from your IRAs during your entire life to help buy or build a first home. You can use the money as a down payment or to help pay other home buying expenses such as closing costs. You're considered a first-time homebuyer as long as you and your spouse didn't own a home two years before the date your new home is acquired.
- **Paying education expenses.** The 10% penalty tax for early IRA withdrawals is also eliminated if you use the money to pay higher education expenses for yourself, your spouse or any child or grandchild of yourself or your spouse. Such expenses include tuition, fees, books and supplies required to attend any college or university.

### c. Establishing your account

You have lots of time to set up an IRA. Indeed, you may be able to do so this year and receive a deduction on last year's taxes. To receive a deduction for any tax year, you need only set up a new IRA or contribute to an existing IRA by the time your income tax return is due for the year—that is, by April 15 of the following year. For example, you have until April 15, 1999 to make an IRA contribution for 1998. Penalties may be imposed if your contribution is not made by the deadline.

## 2. Roth IRAs

Roth IRAs are the new kid on the block. They are essentially the same as traditional IRAs except for one crucial difference: your contributions are not deductible, but you don't have to pay any taxes on withdrawals after you reach age 59 1/2 or you make withdrawals because you have become disabled or use the money for a first-time home purchase. (See Section 1 above.) This means you don't have to pay any income taxes on the interest your Roth IRA investments earn.

However, to receive this favorable tax treatment, you must keep your money in your Roth IRA for at least five years, even if you are over age 59 1/2. You must pay income taxes plus a 10% penalty on early distributions.

> • **EXAMPLE:** Andy establishes a Roth IRA in 1998 and contributes a total of $20,000 over the next 10 years. He receives no deductions from his income taxes for these contributions. When he retires in 2018 at age 60, his Roth IRA is worth $100,000. He pays no income taxes on the money he withdraws from the account.

### a. Evaluating Roth IRAs

You'll save money on income taxes by establishing a Roth IRA rather than a traditional IRA if you will be in a higher tax bracket when you retire than you are now. This is not the case with most retired people, but it might be the case for you if you're relatively young and you plan to save a substantial amount for your retirement or you'll have substantial income from other sources such as an inheritance.

But if you expect your tax rate to be lower when you retire, you're better of with a traditional deductible IRA. If you expect your income to be about the same when you retire as it is now, there is little or no tax difference between a traditional and Roth IRA.

### b. Income limit

Roth IRAs are subject to income limits. The maximum yearly contribution that can be made to a Roth IRA is phased out for single taxpayers with adjusted gross income of between $95,000 and $110,000 and for married people filing jointly between $150,000 and $160,000.

### c. Rolling over traditional IRAs into Roth IRAs

If your adjusted gross income for the tax year is less than $100,000, you may transfer your existing traditional IRA to a Roth IRA. This is called a rollover. However, when you do such a rollover, you have to pay income taxes on all the amounts you contributed to your IRA in prior years plus all the interest your IRA earned. If you roll over a traditional IRA into a Roth IRA, you may spread the taxes you'll have to pay over the next four years, paying one-fourth of the tax due each year.

Determining whether doing a rollover will save you money in the long run can be complex. Before you do a rollover, consult with a tax pro to see if it makes economic sense.

## C. SEP-IRAs

IRAs are all well and good, but the $2,000 annual limit on contributions is likely too small to adequately fund your retirement. For most self-employed people, the best and simplest alternative to IRAs are simplified employee pensions or SEP-IRAs.

A simplified employee pension is called a SEP-IRA because it's very similar to an IRA except you can contribute much more money each year. Instead of being limited to a $2,000 annual contribution, you can invest up to 13.04% of your net earnings every year—up to a maximum of $24,000 a year in 1997.

You can deduct your contributions to SEP-IRAs from your income taxes up to this limit and the interest on your SEP-IRA investments accrues free until you withdraw the money upon retirement.

### 1. Who Can Establish a SEP-IRA

SEP-IRAs are specifically designed for the self-employed. Any person who receives any self-employment income by providing a service can establish a SEP-IRA. This is so whether you work fulltime or parttime. You can even have a SEP-IRA if you are covered by a retirement plan at a fulltime employee job.

It makes no difference how you organize your business; you can be a sole proprietor, partner in a partnership, member of a limited liability company or owner of a regular or S corporation and establish a SEP-IRA. (See Chapter 2.)

### 2. Setting Up Your Account

As with IRAs, SEP-IRAs are self-directed accounts you can set up with a bank, mutual fund, brokerage firm or other financial institution. There are no complicated forms to fill out and you need not make annual filings with the IRS. The financial institution with which you open your account will likely have its own pre-approved SEP-IRA plan or you can use the model SEP-IRA plan contained on IRS form 5205-SEP.

You can set up a SEP-IRA any time up to the date your tax return is due for the prior year—that is, April 15 of the current year. You must make your annual contribution by the due date for your tax return for the year; this ordinarily means April 15 as well. However, if you wish, you can file for an automatic extension of time to file your tax return until August 15. This will give you an additional three months to make your SEP-IRA contribution.

You don't have to make contributions every year—and your contributions can vary from year to year. As with IRAs, the range of investments available is nearly limitless.

### 3. Withdrawing Your Money

Withdrawals from SEP-IRAs are subject the same rules as IRAs. This means that if you withdraw your money from your SEP-IRA before you reach age 59 1/2, you'll have to pay a 10% tax penalty plus regular income taxes on the amount unless an exception applies. (See Section B1.)

#### SEP-IRAs and Employees

SEP-IRAs are not a good option if you have employees or ever intend to have them. This is because you must provide a SEP-IRA for all employees over 20 years of age who have been employed by you for three out of the last five years and who have earned at least $400 per year. Even parttime and occasional employees must be given SEP-IRAs; no particular number of hours must be worked to be eligible.

Your employees' SEP-IRAs can be funded from voluntary contributions from their paychecks. But, if they refuse to do so, you must make the contributions. If your employees' SEP-IRAs aren't funded, you can't contribute to your own SEP-IRA. Your employees become vested in their accounts immediately—meaning they are entitled to the money right away.

If you have more than 25 employees, you can't have a SEP-IRA. You must adopt a Keogh plan instead or use another plan designed for employers.

## D. Keogh Plans

Keogh Plans—named after the Congressman who sponsored the legislation that created them—are just for self-employed people.

You can't have a Keogh if you incorporate your business. Keoghs require more paperwork to set up than SEP-IRAs, but offer more options. You can contribute more and still take an income tax deduction.

### 1. Types of Keogh Plans

There are two basic types of Keogh Plans:

- defined contribution plans, in which benefits are based on the amount contributed to and accumulated in the plan, and
- defined benefit plans, which provide for a set benefit upon retirement; you must be willing and able to contribute more than $30,000 per year to have such a plan, so they are rarely used.

There are two types of defined contribution plans: profit-sharing plans and money purchase plans. These plans can be used separately or in tandem with one other.

#### a. Profit-sharing plans

You can contribute up to 13.04% of your net self-employment income to a Keogh Plan, but the upper limit for contributions is $30,000. You can contribute as much as you want each year up to the limit or not contribute at all.

#### b. Money purchase plans

In a money purchase plan, you contribute a fixed percentage of your net earnings every year; the amount is up to you. Make sure you can afford to make the contributions each year because you can't skip making them. If you do, the IRS will fine you 5% of the amount of the underfunding and tell you to make up the difference within 90 days or be assessed a penalty of 100% of the shortfall.

In return for giving up flexibility, you can contribute more to a money purchase plan—20% of your net earnings. But the maximum allowed is $30,000 per year.

#### c. Paired plans

You may, if you wish, have both a profit-sharing plan and a money purchase plan, as long as your combined contributions total no more than 25% of your self-employment income up to $30,000. This enables you to make the maximum possible Keogh contribution while giving you the ability to vary your contributions somewhat each year.

For example, you could set up a money purchase plan and contribute 10% of your self-employment income each year and also have a profit-sharing plan into which you can contribute up to 15% of your income. But the contributions to your profit-sharing plan would be completly discretionary—that is, you can contribute as much or little as you want up to the limit or make no contribution at all.

### 2. Setting Up a Keogh Plan

You must establish the plan by the end of your tax year—but you still have until your tax return is due the following year to make contributions. Ordinarily, your return is due by April 15; but you may file a request for an extension until August 15. By doing so, you'll have three extra months to contribute to your plan.

If you wish, you can deposit a small advance to set up your account and then wait until you file your tax return to make your full contribution for the year.

As with IRAs and SEP-IRAs, you can set up a Keogh Plan at most banks, brokerage houses, mutual funds and other financial institutions and trade or professional organizations and choose among a huge array of investments.

To set up your plan, you must adopt a written Keogh plan and set up a trust or custodial account with our plan provider to invest your funds. Your plan provider will ordinarily have an IRS-approved master or prototype Keogh plan for you to sign. You can also have a special plan drawn up for you, but this is expensive and unnecessary for most self-employed people.

**KEOGHS AND EMPLOYEES**

If you have employees, having a Keogh plan can get expensive because you are required to match your personal contributions to your own Keogh account with contributions for all eligible employees. All employees over 20 years old with at least one year of service must be allowed to participate in the Keogh plan. An employee is considered to have one year of service if he or she works for you 1,000 or more hours during the year.

However, unlike SEP-IRAs, you can establish a vesting schedule requiring your employees to work for you for a specified number of years before they earn the right to all the money in their retirement account.

## 3. IRS Reporting Requirements

If your Keogh assets exceed $100,000 in value, you'll need to make an annual filing with the IRS. If you're the only participant in your plan, you can file on IRS Form 5500-EZ, which is a simplified version of the regular Form 5500. The form shouldn't take much time to prepare, particularly after you complete the first one. The report is due on the last day of the seventh month after your plan year ends.

## 4. Withdrawing Your Money

You may begin to withdraw money from your Keogh plan after you reach age 59 1/2. Early withdrawals are permitted without penalty if you become disabled or to pay health expenses in excess of 7.5% of your adjusted gross income. Otherwise, early withdrawals are subject to a 10% penalty.

## 5. Comparing Retirement Plans

The following chart allows you to compare the attributes of IRAs, ROTH IRAs, SEP-IRAs and Keogh Plans. Make careful note of the deadlines for establishing the plans and contributing to them. If you miss the deadline, you'll have to forgo your tax deduction for the year.

## COMPARISON OF RETIREMENT PLANS FOR THE SELF-EMPLOYED

| | **IRA** | **ROTH IRA** | **SEP-IRA Sharing Plan** | **Keogh Profit Purchase Plan** | **Keogh Money** |
|---|---|---|---|---|---|
| **Maximum contribution allowed** | $2,000 per year | 2,000 per year | Lesser of 13.04% of self-employ- ment in- come or $24,000 | Lesser of 13.04% of self-employ- ment in- come or $30,000 | Lesser of 20% of self- employment income or $30,000 |
| **Annual contribution requirements** | None | None | None | None | Each year you must contribute amount desig- nated when plan begun |
| **Set-up and contribution deadlines** | Plan can be adopted and contribution made up until April 15 | Plan can be adopted and contribution made up until April 15 | Plan must be adopted by April 15, contributions can be made until tax return due | Plan must be adopted by Dec. 31, contributions can be made until tax return due | Plan must be adopted by Dec. 31, contributions can be made until tax return due |
| **Access to assets** | Withdrawals before age 59 1/2 sub- ject to 10% penalty with certain excep- tions | Withdrawals before age 59 1/2 sub- ject to 10% penalty with certain excep- tions | Withdrawals before age 59 1/2 sub- ject to 10% penalty with certain excep- tions | Withdrawals before age 59 1/2 sub- ject to 10% penalty with certain excep- tions | Withdrawals before age 59 1/2 sub- ject to 10% penalty with certain excep- tions |
| **IRS reporting requirements** | None | None | None | IRS Form 5500 must be filed annually unless less than $100,000 in assets | IRS Form 5500 must be filed annually unless less than $100,000 in assets |
| **Key advan- tage** | Simple, simple, simple | No taxes on withdrawals | Can contrib- ute more than to IRA | Can contrib- ute more than to SEP-IRA | Highest contri- bution maxi- mum |

## E. Retirement Plans If You Have Employees

The following plans are designed for companies that have employees. If you have employees and are interested in setting up such plans, talk with a pension professional.

### 1. 401(k) Plans

This type of plan is typically used by companies with 25 or more employees. It permits employees to open retirement accounts and contribute a portion of their salaries. They pay no income tax on their contributions. Employers may make matching contributions to help fund their employees' retirement, but this is not required. These employer contributions can be made subject to a vesting schedule—that is, an employee can be required to work a set number of years to be entitled to the money.

You can't have a 401(k) plan if you work alone in your business unless you form a C corporation. This eliminates the vast majority of self-employed.

### 2. SIMPLE Retirement Plans

A SIMPLE plan is specifically designed for businesses with 100 or fewer employees and is intended to involve less administrative complexity than regular 401(k) plans. A SIMPLE plan can take the form of a 401(k) account or an IRA account established for each eligible employee. Employees contribute a portion of their salaries into their retirement accounts. Employers are required to make matching contributions up to 3% of pay. SIMPLE plans have not proved very popular with small businesses because employee contributions are mandatory.

## F. Social Security Income

You're not going to have to rely solely on your own financial resources when you retire. You'll receive Social Security benefits as well. Despite what you may have heard, Social Security is not going to disappear any time soon.

This section is based on Chapter 8 of *Get a Life: You Don't Need a Million to Retire Well*, by Ralph Warner (Nolo Press).

However, there will be some changes. To reflect the fact that the average person now lives considerably longer than when the Social Security system was established in the 1930s, the age at which workers will become eligible for benefits will creep up. It's already scheduled to go to 66 for people born after 1943, and may increase to 67 or even 68. And there will be increasing pressure to pass legislation allowing people to invest at least a portion of their Social Security contributions in mutual funds and other investments calculated to increase their eventual benefits.

But no matter how much tinkering politicians do, benefit levels, which are currently indexed to the rate of inflation, are unlikely to decrease. The political power of those over 65 is just too great for this to occur.

### 1. How Much You Will Receive

It's impossible to determine the exact amount you'll receive in Social Security benefits, because the amount of your Social Security check will depend on:

- how much you earn over all the years you work
- how much benefits increase in future years as a result of inflation, and
- the age at which you retire.

For example, you will get a substantially larger Social Security payment if you retire at 70 than if you retire earlier, in part because you will have been paying into the system longer.

Typically, you need to work about ten years in jobs covered by Social Security to be eligible for benefits. All your time spent as a self-employed person counts toward this total since you pay Social Security taxes when you're self-employed. Also count your time spent as an employee in jobs covered by Social Security, which includes virtually all non-government jobs.

**PRE-1951 SE EARNINGS NOT COUNTED FOR SOCIAL SECURITY**

Self-employed people have been required to pay Social Security taxes only since 1951. Self-employment earnings before then had no Social Security tax obligations and so were not applied to a worker's earnings record.

If you had self-employment earnings before 1951, they will not help you qualify for Social Security benefits. If you do qualify for benefits, the amounts you earned in self-employment before 1951 will not be counted in figuring how much those benefits will be.

If you meet this requirement, the amount of your benefits will depend on many factors, including your date of birth and the type of benefit—such as retirement or survivors' benefits—for which you are you are eligible. It will also be influenced by your age at retirement. Currently, you can claim some retirement benefits as early as age 62, but normal retirement age is 65, and is scheduled to increase to 66 for retirees born after 1943.

Average benefits for a person who first claims them in 1997 at age 65 are about $750 per month. For a couple, the monthly average is about $1,250. For most people, this amounts to no more than one-half to two-thirds of how much they'll need to live on when they retire.

## 2. Obtaining an Estimate of Your Benefits

You can get an estimate of your benefits by requesting one from Social Security. Contact the Social Security Administration (800-772-1213) and ask for a Personal Earnings and Benefit Estimate Statement (Form SSA 7004).

To fill out the form, you answer a few simple questions: your name, approximate earnings and age at which you plan to retire. Social Security then sends you an estimate of how much your benefits would be if you retired at age 62, 65 or 70.

You can also fill out Form SSA 7004 at the Social Security Administration's Internet site: www.ssa.gov. While this online service is relatively new, you are likely to receive your estimate more quickly than if you submitted the paper version.

## 3. Social Security and Working After Retirement

Many retirees think about becoming self-employed but are fearful that working after they retire will savage their Social Security benefits. This is no longer true. In 1996, legislation was passed to more than double the earnings limit before Social Security retirement benefits are reduced. Generally increasing each year, by 2002 Social Security recipients ages 65 to 69 will be able to earn $30,000 before losing any Social Security benefits. Once you turn 70, there is no limit on the amount you can earn and still receive full Social Security benefits.

However, if you earn a large income during your 60s, you won't necessarily lose out on Social Security entitlements for good. This is because each year you delay applying for benefits, you receive a Delayed Retirement Credit, which raises future benefits. If you wait until your full retirement age to claim Social Security retirement benefits, your benefit amounts will be permanently higher. Your benefit amount is increased by a certain percentage each year you wait, up until age 70. After you reach 70, there is no longer any increase, and so no reason to further delay claiming.

For more detailed information on all aspects of Social Security, see *Social Security, Medicare and Pensions*, by Joseph L. Matthews (Nolo Press). ■

CHAPTER

# 17

# Copyrights, Patents and Trade Secrets

Self-employed people are often hired to create or contribute to the creation of copyrights, patents or trade secrets—for example, writings, photos, graphics, music, software programs, designs or inventions. This chapter explains the legalities surrounding these creative works—who owns them and how those who create them can protect their rights in them.

## A. Intellectual Property

Products of the human intellect that have economic value—that is, ideas or creations that are worth money—are tagged with the lofty name of intellectual property. This includes work of which you are the author, such as writings, computer software, films, music and inventions and information or know-how not generally known.

A body of federal and state laws has been created giving the owners of intellectual property ownership rights similar to those enjoyed by owners of tangible personal property such as automobiles. Intellectual property may be owned, bought and sold the same as other property.

There are three separate bodies of law that protect intellectual property. Copyrights and patents are governed solely by federal law. Trade secrets are protected under state laws.

### 1. Copyrights

The federal copyright law (17 U.S.C. §§101 and following) protects all original works of authorship. A work of authorship is any work created by a person that other people can understand or perceive, either by themselves or with the help of a machine such as a computer or television. Authorship can include all kinds of written works, plays, music, artwork, graphics, photos, films and videos, computer software, architectural blueprints and designs, choreography and pantomimes.

The owner of a copyright has a bundle of rights that enable him or her to control how the work may be used. These include the exclusive right to copy and distribute the protected work, to create works derived from it—updated editions of a book, for example—and to display and perform it. These rights come into existence automatically the moment a work of authorship is created. The owner need not take any additional steps or file legal documents to secure a copyright.

Copyright owners typically profit from their works by selling or licensing all or some of these rights to others—publishers, for example. (See Section B for more on establishing and transferring copyrights.)

Many self-employed people earn their livings by creating works of authorship for their clients—for example, freelance writers and graphic artists, self-employed computer programmers and photographers.

For a detailed discussion of copyright, see *The Copyright Handbook: How to Protect and Use Written Works,* by Stephen Fishman and *Software Development: A Legal Guide,* by Stephen Fishman (both published by Nolo Press).

### 2. Patents

The federal patent law (35 U.S.C. §§100 and following) protects inventions that are new, useful and not obvious to someone versed in the relevant technology. To obtain a patent, an inventor must file an application with the U.S. Patent and Trademark Office in Washington, DC, and pay a fee. If the Patent Office determines that the invention is sufficiently new, useful and unobvious, it will issue the inventor a patent.

A patent gives an inventor a monopoly on the use and commercial exploitation of the invention. A patent lasts 20 years from the application date. Anyone who wants to use or sell a patented invention during a patent's term must obtain the patent owner's permission. A patent may protect articles—for example, machines, chemicals, manufactures and biological creations; and processes—that is, methods of accomplishing things.

Self-employed workers with technical expertise are often called on by their clients to help develop new inventions that may end up being patentable. These people need to understand the rules governing who will own the patent. (See Section C.)

For a detailed discussion of patents, see *Patent it Yourself*, by David Pressman (Nolo Press).

### 3. Trade Secrets

A trade secret is information that other people do not generally know and that provides its owner with a competitive advantage in the marketplace. The information can be an idea, written words, formula, process or procedure, technical design, customer list, marketing plan or any other secret that gives the owner an economic advantage.

A person who takes reasonable steps to keep the confidential information or know-how secret—for example, does not publish it or otherwise make it freely available to the public—becomes the owner of the trade secret. The laws of most states will protect the owner from disclosures of the secret by:

- the owner's employees
- people who agree not to disclose it, such as ICs the owner hires
- industrial spies, and
- competitors who wrongfully acquire the information.

In the course of your work, you may be exposed to your client's most valuable trade secrets—for example, top secret marketing plans, new products under development, manufacturing techniques or customer lists. Understandably, your client probably doesn't want you blabbing its trade secrets to others, particularly to its competitors.

To make sure you'll keep such information confidential, many clients will ask you to sign a nondisclosure agreement stating that you may not reveal the client's trade secrets to others without permission. A nondisclosure provision can be included in a client agreement or can be a separate document. Carefully review any nondisclosure provision a client asks you to sign. (See Chapter 20, Section B.)

For detailed information on trade secrets, see *Trade Secrets*, by James H.A. Pooley (Amacom).

## B. Copyright Ownership

Any work of authorship you produce for a client is automatically protected by copyright the moment it is created. At that same moment, somebody becomes the owner of the copyright. Who owns the copyright in a work you create is important because the owner alone has the right to copy, distribute or otherwise commercially exploit the work—that is, earn money from it. Unfortunately, self-employed people and hiring firms can get into disputes about who owns the copyright in the work product.

To avoid disputes over ownership, you need to have a basic understanding of some of the basics of copyright law.

**YOU DON'T NEED A C IN A CIRCLE TO HAVE A COPYRIGHT**

Many self-employed people are unaware that a work need not contain a copyright notice—the familiar © symbol followed by the publication date and copyright owner's name—to be protected by copyright laws. Likewise, you are not required to register your copyright. This involves filling out a short registration form and sending it to the Copyright Office in Washington, DC, along with one or two copies of the work and paying a fee.

Giving others notice of the copyright and registering it are both optional. However, both are highly desirable if the work is published. They enable the copyright owner to receive the maximum damages possible if someone copies and uses the work without permission and the owner files and wins a copyright infringement suit.

For detailed discussion, see *The Copyright Handbook: How to Protect and Use Written Works*, by Stephen Fishman (Nolo Press).

## 1. Works Created By Independent Contractors

Self-employed workers ordinarily qualify as independent contractors for copyright purposes. The fundamental rule is that, as an independent contractor, you own the copyright in a work of authorship you create for a client unless you sign a piece of paper transferring your copyright rights to the client. Without your signature on such a document, the client—the person who paid you to create the work—may have no copyright rights at all or at most may share copyright ownership with you.

As the owner of the copyright, you have the right to resell your work to others or copy, distribute and create new works based on the work.

> **EXAMPLE:** Tom hires Jane, an independent contractor computer programmer, to create a computer program. Tom and Jane have an oral work agreement. Jane creates the program. Since Jane never signed an agreement transferring any of her copyright rights to Tom, Jane still owns all the copyright rights in the program. Jane has the exclusive right to sell the program to others or permit them to use it. Even though Tom paid Jane to create the program, he doesn't own it and can't sell or license it to others.

In the real world, copyrights are rarely left to fate. Independent contractors are normally asked to sign written agreements transferring all or some of their copyright rights to the client who hires them. But if you find yourself working with an inexperienced client who doesn't understand the ownership rules, be sure to take the initiative and set forth in writing who will own the copyright in your work. (See Chapter 19, Section B8.)

Copyright transfers can take one of three forms. You can:

- transfer some of your rights
- transfer all of your rights, or
- sign a work made for hire agreement which transfers all your copyright rights and then some.

Which rights you transfer to clients and which you keep for yourself is a matter for negotiation. After you and the client reach an agreement on copyright ownership, one of you must write a copyright transfer agreement and you must sign it to make it legally valid. You may include the understanding as a clause in a client agreement or negotiate it as a separate freestanding agreement.

**Beware of Conflicting Copyright Transfers**

If you're performing similar work for two or more clients simultaneously and agree to assign the intellectual property rights in your work to both clients, you could end up transferring the same rights twice. Before you agree to sell a client anything you create, review your existing agreements to make sure you haven't already sold it.

**PHOTOGRAPHER WINS PYRRHIC VICTORY AGAINST MAGAZINE**

Resolving copyright ownership disputes can take a great deal of time, angst and money if you have to go to court. Even if you eventually win your case, you may feel like a loser. Consider the case of Marco, a professional photographer who took photographs for several issues of *Accent Magazine,* a trade journal for the jewelry industry, over a six-month period. Marco had an oral agreement with the magazine and was paid a fee of about $150 per photograph. *Accent,* claiming it owned the copyright in the photos, wanted to re-use them without paying Marco an additional fee. Marco claimed that he owned the photos and that *Accent* had to pay him for permission to use them again. Marco and the magazine were never able to reach an accord on who owned the photos.

When the magazine tried to use the photos without Marco's permission, he asked a court to block publication. The court refused and Marco filed an appeal with the federal appeals court in Philadelphia. After about two years, Marco won his appeal. But he probably ended up spending far more on attorney fees than the photos were worth. (*Marco v. Accent Publishing Co., Inc.*, 969 F.2d 1547 (3d Cir. 1992).)

## 2. Transferring Some Rights

You are not legally required to give a client all your copyright rights. You can transfer some rights and retain others. When you do this, you may sell the rights you retain to people other than the client.

As discussed above, a copyright is really a bundle of rights including the exclusive rights to copy, distribute, perform, display and create derivative works such as adaptations or new editions from a work. Each of these rights can be transferred together or separately. They can also be divided and subdivided by media, geography, time, market segment or any other way you and a client can think up. You can often make more money by dividing up your copyright rights and selling them piecemeal to several different purchasers than you could selling them all to a single client.

The bundle of copyright rights in any work of authorship can be divided and sold separately. Exactly how you can profitably divide your copyright rights depends on the nature of the work and the market for it. For example, the copyright in a computer program is often divided by geography or type of computer system.

> **EXAMPLE:** Bill, a famous freelance videogame designer, is hired by Fun & Sun Gameware to create a new videogame. He signs an agreement transferring to Gameware only the right to distribute the game for the Nintendo videogame system in the United States. Bill retains all his other copyright rights. He sells the right to publish the game in Japan to Nippon Games and sells the right to create a film based on the game to Repulsive Pictures.

The copyright in a magazine article may be divided by priority of publication—that is, a writer may grant a magazine the right to publish an article for the first time and retain the rights to sell it to others later. Freelance writers often earn substantial income by selling reprint rights to their work. Similarly, graphic artists often grant a client only the right to use an image or design in a certain publishing category and keep the right to resell their work for use in other categories.

> **EXAMPLE:** Sally, a self-employed graphic designer, creates and sells 25 spot illustrations for use in a school textbook. She grants the

textbook publisher the exclusive right to use the images in textbooks, but Sally may sell them to others to use for different purposes—for example, in magazine articles.

When you divide up your copyright rights this way, the transfer is often called a license. Licenses fall into two broad categories: exclusive and non-exclusive.

**DIVIDING UP YOUR COPYRIGHT BUNDLE**

It's often wise to keep as many rights for yourself as possible since you may be able to sell them to others and make additional money from your work.

However, you may not be able to keep many copyright rights. Many clients want to have all the copyright rights in works they pay self-employed people to create. Other clients will pay you substantially less for some rights than if you give them all your rights. If the market for your work is limited, it may make more economic sense to sell all your rights and get as much money as possible from the client instead of taking less and then trying to sell your work to others.

The normal practices in your particular field will usually have a big impact on the negotiations. These traditions and the terminology used to describe various types of copyright transfers vary widely. If you're not familiar with them, ask others in your field or contact professional organizations or trade groups for information.

### a. Exclusive licenses

When a copyright owner grants someone an exclusive license, he or she gives that person, called the licensee, the exclusive right to one or more, but not all, of the copyright rights. An exclusive license is a transfer of copyright ownership and must be in writing to be valid.

**EXAMPLE:** Jane writes an article on economics and grants *The Economist's Journal* the exclusive right to publish it for the first time in the United States and Canada. Jane has granted the *Journal* an exclusive license. Only the *Journal* may publish the article for the first time in the U.S. and Canada. The magazine owns this right. But Jane retains all her other copyright rights. This means she has the right to republish her article after it appears in the *Journal* and to include it in a book. She also retains the right to create derivative works from it—for example, to expand it into a book-length work.

### b. Nonexclusive licenses

In contrast, a nonexclusive license gives a person the right to exercise one or more of a copyright owner's rights, but does not prevent the copyright owner from giving other people permission to exercise the same right or rights at the same time. A nonexclusive license is the most limited form of rights transfer you can grant a client.

As with exclusive licenses, nonexclusive licenses may be limited as to time, geography, media or in any other way. They can be granted orally or in writing. The much better practice, however, is to use some sort of writing; this can avoid possible misunderstandings.

**EXAMPLE:** Bill, a freelance computer programmer, agrees to create an accounting program for AcmePool, a swimming pool company. Bill thinks other swimming pool companies might be interested in buying the program as well, so he grants AcmePool only a nonexclusive right to use the program. This means he can sell it to others, not just Acme-

Pool. AcmePool has no right to sell the program. AcmePool agreed to the deal because Bill charged it much less than he would have charged had AcmePool acquired ownership of the program.

## 3. Transferring All Rights

When people transfer all the copyright rights they own in a work of authorship, the transaction is called an assignment or sometimes an all rights transfer. When such a transaction is completed, the original copyright owner no longer has any ownership rights at all. The new owner—the assignee—has all the copyright rights the transferor formerly held. The new owner is then free to sell licenses or to assign the copyright to someone else.

An assignment can be made either before or after a work is created, but must be in writing to be valid. An assignment can be a separate document or it can be included in a client agreement.

**EXAMPLE:** Tom hires Jane, a self-employed programmer, to create a computer program. Before Jane starts work, Tom has her sign an independent contractor agreement providing, among other things, that she transfers all her copyright rights in the program to Tom. Jane completes her work, delivers the program and Tom pays her. Tom owns all the copyright rights in the program.

### Assignments Can Be Revoked—Eventually

There are many sad stories about authors and artists who were paid a pittance when they were young or unknown, only to have their work become extremely valuable later in their lives or after their deaths. For example, the creators of Superman sold all their rights to a comic book company for a mere $10,000 when they were in their early twenties. They lived to see the company earn millions from their creation, but they shared in none of this money.

To protect copyright owners and their families from unfair exploitation, the Copyright Act gives authors or their heirs the right to get back full copyright rights 35 years after they were assigned. Authors don't have to pay anything to get back their rights. They simply need to file the appropriate documentation with the Copyright Office and former copyright owner.

**EXAMPLE:** Art, a teenage videogame enthusiast, is hired as an independent contractor in 1980 to create a new computer arcade game by Fun & Sun Gameware. He assigns all his copyright rights in the game to Fun & Sun Gameware for $1,500. The game becomes a bestseller and earns Fun & Sun Gameware millions. Art is entitled to none of this money, but he or his heirs can terminate the transfer to Fun & Gameware in the year 2015 and get back all rights in the game without paying Fun & Gameware anything.

This termination right may be exercised only by individual authors or their heirs, and only as to copyright transfers made after 1977. This means the earliest any assignment can be revoked is the year 2013 for works created in 1978. However, the creator of a work made for hire has no such termination rights. (See Section B4.)

## 4. Works Made for Hire

When employees create works of authorship as part of their jobs, their employers automatically own all the copyright rights in the work. Works created on the job are called works made for hire. Certain types of works created by independent contractors can also qualify as works made for hire. When a work is a work made for hire, the person who ordered or commissioned it and paid for it is considered to be the author for copyright purposes, not the person who created it. This commissioning party—not the actual creator—automatically owns all the copyright rights.

### a. Types of works made for hire

Nine categories of works created by independent contractors can be works made for hire:

- a contribution to a collective work—for example, a work created by more than one author, such as a newspaper magazine, anthology or encyclopedia
- a part of an audiovisual work—for example, a motion picture screenplay
- a translation
- supplementary works—for example, forewords, afterwords, supplemental pictorial illustrations, maps, charts, editorial notes, bibliographies, appendixes and indexes
- a compilation—for example, an electronic database
- an instructional text
- a test
- answer material for a test, and
- an atlas.

### b. Written agreement requirement

It's not enough that your work falls into one of these categories. You and the client must sign a written agreement stating that the work you create shall be a work made for hire. The agreement must be signed before you begin work to be effective. The client cannot wait until after your work is completed and delivered and then decide that it should be a work made for hire.

Think of a work for hire agreement as the hydrogen bomb of copyright transfers. When you sign such an agreement, you not only give up all your ownership rights in the work for the rest of time, you're not even considered the work's author. You won't be legally entitled to any credit for your work, such as having your name attached if the work is published unless your agreement with the client requires that you be given credit.

**EXAMPLE:** The editor of *The Egoist Magazine* asks Gloria, a freelance writer, if she would be interested in writing an article for the magazine on night life in Palm Beach. Gloria agrees and the editor sends her a letter agreement to sign setting forth such terms as compensation, the deadline for the article and its length. The letter also specifies that the article "shall be a work made for hire."

Gloria signs the agreement, writes the article and is paid by the magazine. Since the article qualifies as a work made for hire, the magazine is the initial owner of all the copyright rights in the article. Gloria owns no copyright rights. As the copyright owner, the magazine is free to sell reprint rights in the article, to sell film and TV rights, translation rights and any other rights anyone wants to buy. Gloria is not entitled to license or sell any rights in the article because she doesn't own any. She gave up all her copyright rights by signing the work for hire agreement.

Many writer and artist organizations strongly advise their members to refuse to sign work for hire agreements. However, an increasing number of clients insist on such agreements. In some cases, signing a work for hire agreement is a take it or leave it proposition: if you don't agree to sign, the client will find someone else who will. For example, *The New York Times* now requires freelance contributors to sign work for hire agreements; the paper will simply refuse to publish a freelance article unless the author first signs such an agreement.

On the other hand, you can often get a client to agree to something less than a work for hire agreement if you ask. If the client refuses, you can still sign the work for hire agreement, but you would have lost nothing by asking.

### 5. Sharing Ownership With Clients

If a client does not obtain a copyright transfer from you, you will own the work you create for him or her. However, a client who qualifies as a co-author might be considered a co-owner of the work along with you. For this to occur, the client must actually help you create the work. Giving suggestions or supervision is not enough.

A client who qualifies as a co-author will jointly share copyright ownership in the work with you. As co-authors, you're each entitled to use or let other people use the work without obtaining approval of the other co-author. But any profits you make must be shared with the other co-authors. This could cause problems—for example, if co-authors sold the same work to competing publishers, the value of the work could be diluted. It's usually in co-authors' interests to work together to avoid this.

> **EXAMPLE:** Marlon, a legendary actor, hires Tom, a freelance writer, to ghostwrite his autobiography. Tom is an IC, not Marlon's employee. Marlon fails to have Tom sign a written agreement transferring to Marlon his copyright rights in his work. However, Marlon worked closely with Tom in writing the autobiography, contributing not only ideas, but also writing portions of the book.
>
> As a result, Marlon and Tom would probably be considered co-authors and joint owners of the autobiography. Both would have the right to sell the autobiography to a publisher, to serialize it in magazines, sell it to movie producers or otherwise commercially exploit the work. However, Marlon and Tom would have to share any profits earned.

If a client doesn't qualify as a co-author, at most it will have a nonexclusive right to use the work. But it wouldn't be allowed to sell or license any copyright rights in the work because it wouldn't own any. The independent contractor would own all the rights and be able to sell or license them without the client's permission and without sharing the profits with the client.

> **EXAMPLE:** Mark pays Sally, a freelance photographer, to take some pictures of toxic waste dumps to supplement his treatise on toxic waste management. Sally failed to sign an agreement transferring her copyright rights in the photos to Mark. Mark does not qualify as a co-author of the photos because he didn't help take them. As a result, Sally owns the copyright in the photos. However, Mark has a nonexclusive license to use the photos in his treatise. But this doesn't prevent Sally from selling the photos to others or otherwise exploiting her copyright rights.

## C. Patent Ownership

Patent rights initially belong to the person who develops an invention. However, patent rights can be assigned—that is, transferred—to others just as copyrights can. Firms that hire independent contractors to help create new technology or anything else that might qualify for a patent normally have the independent contractors assign to them in advance all patent rights in the work. Such an assignment may be included in an independent contractor agreement or in a separate document.

If you have no signed assignment, it's far from clear who will own inventions you develop. Unlike copyrights, where silence indicates independent contractor ownership, there is no such presumption in the world of patents. The law regarding ownership of inventions by employees is very clear—an employer owns any inventions an employee was hired to create. But courts have barely addressed ownership of inventions by independent contractors in the absence of a written ownership agreement.

You can assert ownership over your invention or other work, but be prepared for a costly legal dispute if you do. The client could claim you had a duty to assign your patent rights to it even though there was no written assignment agreement. You would probably have a good chance of winning such a case, but it would be no sure thing. It's best to avoid such disputes in advance by clearly defining in writing who will own the work.

## D. Trade Secret Ownership

Trade secret ownership rules are similar to those for patents. You are the initial owner of any trade secrets you develop while working for a client. But ownership rights in trade secrets can be assigned to others just as patents can be. Hiring firms typically require independent contractors to sign written agreements assigning their trade secret ownership rights to the client in advance.

In the absence of such an agreement, you probably own your trade secrets and can sell them to others, including the client's competitors. However, if you used the client's resources to develop the trade secret information, the client might have a nonexclusive license to use it without your permission. This is called a shop right. ■

CHAPTER

# 18

# Using Written Client Agreements

A contract—also called an agreement—is a legally binding promise. Whenever you agree to perform services for a client, you enter into a contract; you promise to do work and the client promises to pay you for it.

The word contract often intimidates people who conjure up visions of voluminous legal documents laden with legalese. However, a contract need not be long or complex. Many contracts consist of only a few simple paragraphs. Indeed, most contracts don't even have to be in writing. Even so, it's never a good idea to rely on an oral agreement with a client.

**CLIENT AGREEMENTS DON'T HAVE TO BE INTIMIDATING**

Some self-employed workers shy away from using written agreements because they're afraid they will intimidate their clients or make them think they don't trust them. This could happen if you're not careful. For example, a prospective client might think twice about hiring you if you present it with a 20-page contract for a simple one-day project. The client might conclude that you're either paranoid, will be hard to deal with or both.

You can avoid this problem, however, if you calibrate your agreements to your assignments. Use simple contracts or short letter agreements for simple projects and save the longer, more complex agreements for long or complex projects.

## A. Reasons to Use Written Agreements

Most contracts don't have to be in writing to be legally binding. For example, you and a client can enter into a contract over the phone or during a lunch meeting at a restaurant. No magic words need be spoken. You just have to agree to perform services for the client in exchange for something of value—usually money. Theoretically, an oral agreement is as valid as a 50-page contract drafted by a high-powered law firm.

**EXAMPLE:** Gary, a freelance translator, receives a phone call from a vice-president of Acme Oil Co. The VP asks Gary to translate some Russian oil industry documents for $2,000. Gary says he'll do the work for the price. Gary and Acme have a valid oral contract.

**SOME AGREEMENTS MUST BE IN WRITING**

Some types of agreements must be in writing to be legally enforceable. Each state has a law, usually called the Statute of Frauds, listing the types of contracts that must be in writing to be valid.

A typical list includes:

- any contract that cannot possibly be performed in less than one year. Example: John agrees to perform consulting services for Acme Corp. for the next two years for $2,000 per month. Since the agreement cannot be performed in less than one year, it must be in writing to be legally enforceable.
- contracts for the sales of goods—that is, tangible personal property such as a computer or car—worth $500 or more.
- a promise to pay someone else's debt. Example: The president of a corporation personally guarantees to pay for the services you sell the corporation. The guarantee must be in writing to be legally enforceable.

- contracts involving the sale of real estate, or real estate leases lasting more than one year.

Any transfer of copyright ownership must also be in writing to be valid. (See Chapter 17, Section B.)

In the real world, however, using oral agreements is like driving without a seatbelt. Things will work out fine as long as you don't have an accident; but, if you do have an accident, you'll wish you had buckled up. An oral agreement can work if you and your client agree completely about its terms and obey them. Unfortunately, things don't always work so perfectly.

Following are some of the most important reasons why you should always sign a written agreement with a client before starting work.

### 1. Avoiding Misunderstandings

Courts are crowded with lawsuits filed by people who enter into oral agreements with one another and later disagree over what was said. Costly misunderstandings can develop if you perform services for a client without a writing clearly stating what you're supposed to do. Such misunderstandings may be innocent; you and the client may have simply misinterpreted one another. Or they may be purposeful; without a writing to contradict him or her, a client can claim that you orally agreed to anything.

Consider a good written client agreement to be your legal lifeline. If disputes develop, the agreement will provide ways to solve them. If you and the client end up in court, a written agreement will establish your legal duties to one another.

For these same reasons, your clients should be happy to sign a well-drafted contract. Be wary of any client who refuses to put your agreement in writing. Such a client might be a bad credit risk. If a prospective client balks at signing an agreement, you may wish to obtain a credit report on him or her, or talk with others who have worked for that client to see if they had problems. If you think you might have payment problems, ask for a substantial downpayment upfront and for progress payments if the project is lengthy. (See Chapter 7, Section B.)

**Written Agreements Preempt Cheap Talk**

When you put your agreement in writing, it is usually treated as the final word on the areas covered. That is, it takes precedence over anything you and the client said to each other but did not include in your written agreement.

However, if you and the client end up in court or arbitration because you disagree over the terms or meaning of your contract, things you and the client said to each other during the negotiating process but didn't write down can be used to explain unclear terms in the written contract or to prove additional terms where the writing is incomplete. Since you and the client may disagree about what was said during negotiations, it's best to make the written contract as clear and complete as possible.

### 2. Assuring That You Get Paid

A written agreement clearly setting out your fees will help ward off disputes about how much the client agreed to pay you. If a client fails to pay and you have to negotiate or eventually even sue for your money, the written agreement will be proof of how much you're owed. Relying on an oral agreement with a client can make it very difficult for you to get paid in full or to get paid at all.

**Oral Agreement Costs IC $600**

One self-employed worker recently learned the hard way that an oral agreement isn't worth the paper on which it's printed. Jane, a commercial illustrator who works as a freelancer for a variety of ad agencies and other clients, orally agreed to do a series of drawings for a dress designer. Jane did the drawings and submitted her bill for $2,000. The designer refused to pay, alternately claiming that payment was conditional on the drawings being published in a fashion magazine and that Jane was charging too much.

Jane filed a lawsuit against the designer in small claims court claiming that the client owed her the $2,000. The judge had no trouble finding that Jane had an enforceable oral contract with the designer who admitted that he had asked Jane to do the work. However, the judge awarded Jane only $1,400 because she could not document her claim that she was to be paid $100 per hour and the designer made a convincing presentation that designers usually charge no more than $70 per hour.

### 3. Defining Projects

The process of deciding what to include in an agreement helps both you and the client. It forces you both to think carefully, perhaps for the first time, about exactly what you're supposed to do. Hazy and ill-defined ideas get reduced to a concrete contract specification or description of the work that you will perform. This gives you and the client a yardstick by which to measure your performance and is the best way to avoid later disputes that you haven't performed adequately.

### 4. Establishing IC Status

A well-drafted client agreement will also help establish that you are an independent contractor, not the client's employee. You can suffer severe consequences if the IRS or another government agency decides you're an employee instead of an independent contractor. (See Chapter 15.)

## B. Reviewing a Client's Agreement

Many clients have their own agreements they'll ask you to sign. This may be a copy of an agreement they've used in the past with other people they've hired, a standard agreement prepared by an attorney or a letter summarizing the terms to which you've agreed. You'll almost always be better off if you use your own agreement, not one provided by your client, because it gives you the greatest control over the contract terms. Take the initiative and send the client an agreement to sign immediately after you accept an assignment. Don't wait for the client to provide you with an agreement of its own.

If a client insists on using its own agreement, read that document carefully because it may contain provisions that are unfair to you—for example, provisions requiring you to repay the client if an IRS auditor determines you're an employee and imposes fines and penalties against the client. Pay special attention to noncompetition provisions restricting your right to work for other companies and nondisclosure provisions that prevent you from using information you learn while working for the client.

Remember that any contract can be rewritten, even if the client claims it's a standard agreement that all outside workers sign. This is a matter for negotiation. Seek to delete or rewrite unfair provisions. There may also be provisions you may wish to add. (See Chapter 20, Section D.)

For practical guidance on how to conduct contract negotiations, see:

- *Field Guide to Negotiation,* by Gavin Kennedy (Harvard Business School Press)
- *Getting to Yes: Negotiating Agreement Without Giving In,* by Roger Fisher and William Ury (Penguin Books)
- *Negotiating Rationally,* by Max H. Bazerman and Margaret A. Neale (Free Press), and
- *Win-Win Negotiating,* by Fred E. Jandt (John Wiley & Sons).

## C. Creating Your Own Client Agreements

You don't need to hire a lawyer to draft an independent contractor agreement. All you need is a brain and a little common sense. This book contains a number of sample forms you can use and copious guidance on how to custom tailor them to fit your particular needs. This may take a little work, but when you're done, you'll have an agreement you can use over and over again with minor alterations.

### 1. Types of Agreements

There are two main types of client agreements. Use the type of agreement that best suits your own needs and the needs of your clients.

#### a. Letter agreements

A letter agreement is usually a short contract written in the form of a letter. After you and the client reach a tentative agreement—over the phone, in a meeting, by fax or electronic mail—you set forth the contract terms in a letter written on your stationery, sign two copies and send them to the client. If the client agrees to the terms in the letter, he or she signs both copies at the bottom and returns one signed copy to you. At that point, you and the client have a fully enforceable, valid contract. Many self-employed people use letter agreements because they seem less intimidating to clients than more formal looking contracts. Others use them because of business custom. (See Chapter 19, Section D.)

#### b. Standard contracts

A standard contract usually contains a number of paragraphs and captions. It is usually longer and more comprehensive than a letter agreement. It's a good idea to use a standard contract if:

- the project is long and complex or involves a substantial amount of money, or
- you're dealing with a client you don't trust, either because it's a new client or your past dealings indicate the client is untrustworthy. (See Chapter 19.)

**PROPOSALS ARE NOT CLIENT AGREEMENTS**

Clients often choose the outside workers they hire through a competitive bidding process. They ask a number of people to submit written proposals describing how they'll perform the work and the price they'll charge. The client chooses the IC who submits the best proposal.

Some self-employed people write quite lengthy and detailed proposals and depend on them instead of standard contracts. This is a mistake. Proposals normally do not address many important issues that should be covered in a contract, such as the term of the agreement, how it can be terminated and how disputes will be resolved. If a problem arises that is not covered by the proposal, you and the client will have to negotiate a solution in the middle of the project. If you can't reach a solution, either you or the client may end up taking the other to court to resolve the battle.

The best course is to have a signed contract in hand before starting work. If you submit a proposal to a client, it should state that if the client agrees to proceed, you will forward a contract for signature within a specific number of days.

## 2. The Drafting Process

Draft contract forms you can use over and over again with various clients. You can use the forms as the basis for your agreement and make alterations you need to suit the particular situation. If you perform the same type of work for every client, you might need just one or two form agreements. This might include a brief letter agreement for small jobs, and a longer standard contract for larger projects. However, if your work varies, you may need several different agreements. If you hire self-employed people to work for you, you'll need an agreement for them as well. (See Chapter 19.)

**EXAMPLE:** Ellen, a freelance publicist, usually uses a letter agreement with her clients. However, when she is hired to do a particularly long project, she prefers to use a longer standard contract because it affords her added protection if something goes wrong.

She also occasionally hires other free–lance publicists to work for her as ICs when she gets very busy. Ellen uses a lengthy subcontractor agreement with them because it helps to establish that they're ICs instead of her employees and makes their duties as clear as possible.

Ellen uses the forms on the computer disk contained in this book to draft three different IC agreements she keeps on her computer and uses over and over again: a letter agreement, a standard IC agreement and a subcontractor agreement.

## 3. Using the Forms in This Book

This book contains a sample client agreement that almost any self-employed person who sells services can adapt to meet his or her needs. Each of the clauses in the agreement is discussed in detail and a sample briefer letter agreement is also provided. (See Chapter 19.)

There are two ways to use these clauses. You can retype the language suggested in this book to assemble your own final agreement. This will allow you to tailor your document to your specific needs. Alternatively, you can complete and use the tear-out sample agreement in the Appendix. This is easier than creating your own agreement, but gives you less flexibility regarding the content of your agreement.

## D. Putting Your Agreement Together

Whether you use a simple letter agreement or formal standard contract, make sure it's properly signed and put together. This is not difficult if you know what to do. And this section provides all the instruction you should need to get it right.

### 1. Signatures

It's best for both you and the client to sign the agreement and to do it in ink. You need not be together when you sign and it isn't necessary for you to sign at the same time. There's no legal requirement that the signatures be located in any specific place in a business contract, but they are customarily placed at the end of the agreement; that helps signify that you both have read and agreed to the entire document.

It's very important that both you and the client sign the agreement properly. Failure to do so can have drastic consequences. How to sign depends on the legal form of your business and the client's business.

#### a. Sole proprietors

If you or the client are sole proprietors, you can simply sign your own names because a sole proprietorship is not a separate legal entity.

However, if you use a fictitious business name (see Chapter 2), it's best for you to sign on behalf of your business. This will help show you're an IC, not an employee.

**EXAMPLE:** Chris Kraft is a sole proprietor who runs a marketing research business. Instead of using his own name for the business, he calls it AAA Marketing Research. He should sign his contracts like this:

AAA Marketing Research

By: ______________________________

Chris Kraft

#### b. Partnerships

If either you or the client is a partnership, a general partner should sign on the partnership's behalf. Only one partner needs to sign. The signature block for the partnership should state the partnership's name and the name and title of the person signing on the partnership's behalf. If a partner signs only his or her name without mentioning the partnership, the partnership is not bound by the agreement.

**EXAMPLE:** Chris, a self-employed marketing consultant, contracts to perform marketing research for a Michigan partnership called The Argus Partnership. Randy Argus is one of the general partners. He signs the contract on the partnership's behalf like this:

The Argus Partnership
A Michigan Partnership

By: ______________________________

Randy Argus, a General Partner

If the client is a partnership and a person who is not a partner signs the agreement, the signature should be accompanied by a partnership resolution stating that the person signing the agreement has the authority to do so. The partnership resolution is a document signed by one or more of the general partners stating that the person named has the authority to sign contracts on the partnership's behalf.

#### c. Corporations

If either you or the client is a corporation, the agreement must be signed by someone who has authority to sign contracts on the corporation's behalf. The corporation's president or chief executive officer (CEO) is presumed to have this authority.

If someone other than the president of an incorporated client signs—for example, the vice-president, treasurer or other corporate officer—ask to see a board of directors resolution or corporate bylaws authorizing him or her to sign. If the person signing doesn't have authority, the corporation won't be legally bound by the contract.

Keep in mind that if you sign personally instead of on your corporation's behalf, you'll be personally liable for the contract.

The signature block for a corporation should state the name of the corporation and indicate by name and title the person signing on the corporation's behalf.

> **EXAMPLE:** Chris, a self-employed marketing consultant, contracts to perform marketing research with a corporation called Kiddie Krafts, Inc. The contract is signed by the corporation's president, Susan Ericson, so Chris doesn't need a corporate resolution showing she has authority to bind the corporation. The signature block should appear in the contract like this:
>
> Kiddie Krafts, Inc.
> A California Corporation
>
> By: ______________________________
> Susan Ericson, President

## 2. Dates

When you sign an agreement, include the date and make sure the client does, too. You can simply put a date line next to the place where each person signs—for example:

Date: _______, 199X

You and the client don't have to sign on the same day. Indeed, you can sign weeks apart.

## 3. Attachments or Exhibits

An easy way to keep a letter agreement or standard contract as short as possible is to use attachments, also called exhibits. You can use them to list lengthy details such as performance specifications. This makes the main body of the agreement shorter and easier to read.

If you have more than one attachment or exhibit, they should be numbered or lettered—for example, Attachment 1 or Exhibit A. Be sure that the main body of the agreement mentions that the attachments or exhibits are included as part of the contract.

## 4. Altering the Contract

Sometimes it's necessary to make last minute changes to a contract just before it's signed. If you use a computer to prepare the agreement, it's usually easy to make the changes and print out a new agreement.

However, it's not necessary to prepare a new contract. Instead, the changes may be handwritten or typed onto all existing copies of the agreement. If you use this approach, be sure that all those signing the agreement also sign their initials as close as possible to the place where the change is made. If both people who sign don't initial each change, questions might arise as to whether the change was part of the agreement.

## 5. Copies of the Contract

Prepare at least two copies of your letter agreement or standard contract. Make sure that each copy contains all the needed exhibits and attachments. Both you and the client should sign both copies—and both of you should keep one signed copy of the agreement.

## 6. Faxing Contracts

It has become very common for self-employed people and their clients to communicate by fax machine. People often send drafts of their proposed agreement back and forth to each other by fax. When a final agreement is reached, one signs a copy of the contract and faxes it to the other who signs it and faxes it back.

A faxed signature is probably legally sufficient if neither you nor the client dispute that it is a fax of an original signature. However, if a client claims that a faxed signature was forged, it could be difficult or impossible to prove it's genuine, since it is very easy to forge a faxed signature with modern computer technology. Forgery claims are rare, however, so this is usually not a problem. Even so, it's a good practice for you and the client to follow up the fax with signed originals exchanged by mail or air express.

## E. Changing the Agreement After It's Signed

No contract is engraved in stone. You and the client can always modify or amend your contract if circumstances change. You can even agree to call the whole thing off and cancel your agreement.

> **EXAMPLE:** Barbara, a self-employed well digger, agrees to dig a 50-foot deep well on property owned by Kate for $2,000. After digging 10 feet, Barbara hits solid rock that no one knew was there. To complete the well, she'll have to lease expensive heavy equipment. To defray the added expense, she asks Kate to pay her $4,000 instead of $2,000 for the work. Kate agrees. Barbara and Kate have amended their original agreement.

Neither you nor the client is ever obligated to accept a proposed modification to your contract. Either of you can always say no and accept the consequences. At its most dire, this may mean a court battle over breaking the original contract. However, you're usually better off reaching some sort of accommodation with the client, unless he or she is totally unreasonable.

Unless your contract is one that must be in writing to be legally valid—for example, an agreement that can't be performed in less than one year (see Section A), it can usually be modified by an oral agreement. In other words, you need not write down the changes.

> **EXAMPLE:** Art signs a contract with Zeno to build an addition to his house. Halfway through the project, Art decides that he wants Zeno to do some extra work not covered by their original agreement. Art and Zeno have a telephone conversation in which Zeno agrees to do the extra work for extra money. Although nothing is put in writing, their change to their original agreement is legally enforceable.

Self-employed workers and their clients change their contracts all the time and never write down the changes. The flexibility afforded by such an informal approach to contract amendments might be just what you want. However, misunderstandings and disputes can arise from this approach. It's always best to have some sort of writing showing what you've agreed to do. You can do this informally. For example, you can simply send a confirming letter following a telephone call with the client summarizing the changes you both agreed to make. Be sure to keep a copy for your files.

However, if the amendment involves a contract provision that is very important—your payment, for example—insist on a written amendment and insist that you and the client sign it. The amendment should set forth all the changes and state that the amendment takes precedence over the original contract. (See Appendix 2 for an Amendment form.) ■

CHAPTER

# 19

# Drafting Your Own Client Agreement

This chapter guides you in creating a client agreement that you can use for almost any type of service you provide your clients. This is a full-blown formal contract. It covers nearly all concerns you would want covered in an agreement. (See Appendix 2 for a tear-out version.)

Some self-employed people prefer not to use formal contracts because they're afraid they will intimidate clients. This book provides a short, simple letter agreement to meet the needs of people who prefer them. (See Section D.)

However, it's a good idea to use a more comprehensive formal contract if:

- you are hired to do a project that is long and complex or involves a substantial amount of money, or
- you're dealing with a new client or a client who has been untrustworthy in the past.

## A. Essential Provisions

There are a number of provisions that should be included in most client agreements. These cover the important points your contract should address. They are the absolute minimum number of provisions that should be included in your agreement. All of these sample clauses are included in the Client Agreement. (See Section C for a sample of how an entire agreement might look when assembled.)

These provisions may be all you need for a basic agreement. Or you may need to combine them with some of your own clauses or one or more of the optional clauses discussed below. (See Section B.)

**Title of agreement.** You don't need a title for a client agreement, but if you want one, call it Independent Contractor Agreement, Client Agreement, Agreement for Professional Services or Consulting Agreement. Consulting Agreement may sound a little more high-toned than Independent Contractor Agreement and is often used by highly skilled professionals. Since you are not your client's employee, do not use Employment Agreement as a title.

> **SUGGESTED LANGUAGE:**
>
> INDEPENDENT CONTRACTOR AGREEMENT

**Names of parties.** Here at the beginning of your contract, it's best to refer to yourself by your full business name. Later on in the contract, you can use an obvious abbreviation.

If you're a sole proprietor, use the full name you use for your business. This can be your own name or a fictitious business name or assumed name you use to identify your business. (See Chapter 2.) For example, if consultant Al Brodsky calls his one-person marketing research business "ABC Marketing Research," he would use that name on the contract. Using a fictitious business name helps show you're a business, not an employee.

If your business is incorporated, use your corporate name, not your own name—for example: "John Smith, Incorporated" instead of "John Smith."

Similarly, if you've formed a limited liability company (see Chapter 2), use the name of the LLC, not your personal name—for example: "Jane Brown, a Limited Liabilty Corporation" instead of "Jane Brown."

Do not refer to yourself as an employee or to the client as an employer.

For the sake of brevity, it is usual to identify yourself and the client by shorter names in the rest of the agreement. You can use an abbreviated version of your full name—for example, ABC for ABC Marketing Research. Or you can refer to yourself simply as Contractor or Consultant.

Refer to the client initially by its company name and subsequently by a short version of the name or as Client or Firm.

Include the addresses of the principal place of businesses of both the client and yourself. If you or the client have more than one office or workplace, the principal place of business is the main office or workplace.

> **SUGGESTED LANGUAGE:**
>
> This Agreement is made between [Client's name] (Client) with a principal place of business at [Client's business address] and [Your name] (Contractor), with a principal place of business at [Your business address].

### 1. Services to Be Performed

The agreement should describe in as much detail as possible what you are expected to accomplish. Word the description carefully to emphasize the results you're expected to achieve. Don't describe the method by which you will achieve the results. As an independent contractor, it should be up to you to decide how to do the work. The client's control should be limited to accepting or rejecting your final results. The more control the client exercises over how you work, the more you'll look like an employee. (See Chapter 15.)

**EXAMPLE:** Jack hires Jill to prepare an index for his multi-volume history of ancient Sparta. Jill should describe the results she is expected to achieve like this: Contractor agrees to prepare an index of Client's History of Sparta of at least 100 single-spaced pages. Contractor will provide Client with a printout of the finished index and a 3.5 inch computer disk version in ASCII format.

The agreement should not tell Jill how to create the index: "Contractor will prepare an alphabetical three-level index of Client's History of Sparta. Contractor will first prepare 3 by 5 inch index cards listing every index entry beginning with Chapter One. After each chapter is completed, Contractor will deliver the index cards to Client for Client's approval. When index cards have been created for all 50 chapters, Contractor will create a computer version of the index using Complex Software Version 7.6. Contractor will then print out and edit the index and deliver it to Client for approval."

It's perfectly okay for the agreement to establish very detailed specifications for your finished work product. But the specs should only describe the end results you must achieve, not how to obtain those results. You can include the description in the main body of the Agreement. Or if it's a lengthy explanation, put it on a separate document and attach it to the agreement. (See Chapter 18, Section D3.)

> **SUGGESTED LANGUAGE—ALTERNATIVE A—IF SERVICES LISTED:**
>
> Contractor agrees to perform the following services: [Describe services you will perform.]

> **SUGGESTED LANGUAGE—ALTERNATIVE B—IF EXHIBIT ATTACHED:**
>
> Contractor agrees to perform the services described in Exhibit A, which is attached to this Agreement.

Choose Alternative B if the explanation of the services is attached to the main contract.

### 2. Payment

Self-employed people who provide services to others can be paid in many different ways. (See Chapter 7, Section A.) The two most common payment methods are:

- a fixed fee, or
- payment by unit of time.

### a. Fixed fee

In a fixed fee agreement, you charge an agreed amount for the entire project. (See Chapter 7, Section A1.)

> **SUGGESTED LANGUAGE—ALTERNATIVE A—FIXED FEE:**
>
> In consideration for the services to be performed by Contractor, Client agrees to pay Contractor $[State amount].

### b. Unit of time

Many self-employed people—for example, lawyers, accountants and plumbers—customarily charge by the hour, day or other unit of time. (See Chapter 7, Section A1.) Charging by the hour does not support your independent contractor status, but you can get away with it if it's a common practice in the field in which you work.

> **SUGGESTED LANGUAGE—ALTERNATIVE B—UNIT OF TIME:**
>
> In consideration for the services to be performed by Contractor, Client agrees to pay Contractor at the rate of $[State amount] per [hour, day, week or other unit of time].

### c. Capping your payment

Clients often wish to place a cap on the total amount they'll spend on the project when you're paid by the hour because they're afraid you might work slowly to earn a larger fee. If the client insists on a cap, make sure it allows you to work enough hours to get the project done.

> **OPTIONAL LANGUAGE—CAPPING PAYMENT**
>
> [OPTIONAL: Contractor's total compensation shall not exceed $____ without Client's written consent.]

## 3. Terms of Payment

Terms of payment means how you will bill the client and be paid. Before you can be paid for your work, submit an invoice to the client setting out the amount due. An invoice doesn't have to be fancy or filled with legalese. It should include: an invoice number, the dates covered by the invoice, the hours expended if you're being paid by the hour and a summary of the work performed. (See Chapter 7, Section B2, for a detailed discussion and sample invoice form.)

### a. Payment upon completing work

The following provision requires you to send an invoice after you complete work. The client is required to pay your fixed or hourly fee within a set number of days after you send the invoice. The time period of 30 days is a typical payment term, but it can be shorter or longer if you wish. Note that the time for payment starts to run as soon as you send your invoice, not when the client receives it. This will help you get paid more quickly.

> **SUGGESTED LANGUAGE—ALTERNATIVE A—PAYMENT ON COMPLETION:**
>
> Upon completing Contractor's services under this Agreement, Contractor shall submit an invoice. Client shall pay Contractor within ____ [10, 15, 30, 45, 60] days from the date of Contractor's invoice.

### b. Divided payments

You can also opt to be paid part of a fixed or hourly fee when the agreement is signed and the remainder when the work is finished. The amount of the upfront payment is subject to negotiation. Many self-employed people like to receive at least one-third to one-half of a fee before they start work. If the client is new or might have problems paying you, it's wise to get as much money in advance as you can.

The following provision requires that you be paid a specific amount when the client signs the agreement and then the rest when the work is finished.

> **SUGGESTED LANGUAGE—ALTERNATIVE B—DIVIDED PAYMENTS:**
>
> Contractor shall be paid $[State amount] upon signing this Agreement and the remaining amount due when Contractor completes the services and submits an invoice. Client shall pay Contractor within ____ [10, 15, 30, 45, 60] days from the date of Contractor's invoice.

### c. Fixed fee installment payments

If the project is long and complex, you may prefer to be paid in installments rather than waiting until the project is finished to receive the bulk of your payment. One way to do this is to break the job into phases or milestones and be paid a fixed fee when each phase is completed. Clients often like this pay-as-you-go arrangement, too.

To do this, draw up a schedule of installment payments tying each payment to your completion of specific services. It's usually easier to set forth the schedule in a separate document and attach it to the agreement as an exhibit. The main body of the agreement should simply refer to the attached payment schedule.

> **SUGGESTED LANGUAGE—ALTERNATIVE C—INSTALLMENT PAYMENTS:**
>
> Contractor shall be paid according to the Schedule of Payments set forth in Exhibit __ [A or B] attached to and made part of this agreement.

Following is a form of a schedule of payments you can complete and attach to your agreement. This schedule requires four payments: a down payment when the contract is signed and three installment payments. However, you and the client can have as many payments as you want.

> **SUGGESTED LANGUAGE—PAYMENT SCHEDULE EXHIBIT—SCHEDULE OF PAYMENTS**
>
> Client shall pay Contractor according to the following schedule of payments:
>
> 1) $[State sum] when this Agreement is signed.
>
> 2) $[State sum] when an invoice is submitted and the following services are completed:
>
> [Describe first stage of services]
>
> 3) $[State sum] when an invoice is submitted and the following services are completed:
>
> [Describe second stage of services]
>
> 4) $[State sum] when an invoice is submitted and the following services are completed:
>
> [Describe third stage of services]
>
> [ADD ANY ADDITIONAL PAYMENTS]
>
> All payments shall be due within ____ [10, 15, 30, 45, 60] days from the date of Contractor's invoice.

### d. Hourly payment for lengthy projects

Use the following clause if you're being paid by the hour or other unit of time and the project will last more than one month. Under this provision, you submit an invoice to the client each month setting forth how many hours you've worked and the client is required to pay you within a specific number of days from the date of each invoice.

> **SUGGESTED LANGUAGE—ALTERNATIVE D—PAYMENTS AFTER INVOICE:**
>
> Contractor shall send Client an invoice monthly. Client shall pay Contractor within [10, 15, 30, 45, 60] days from the date of each invoice.

## 4. Expenses

Expenses are the costs you incur that you can attribute directly to your work for a client. They include, for example, the cost of phone calls or traveling done on the client's behalf. Expenses do not include your normal fixed overhead costs such as your office rent or the cost of commuting to and from your office. They also do not include materials the client provides you to do your work. (See Section A5.)

In the recent past, the IRS viewed the payment of a worker's expenses by a client as a sign of employee status. However, the agency now downgrades the importance of this factor. IRS auditors will focus on whether a worker has any expenses that were not reimbursed, particularly fixed ongoing costs such as office rent or employee salaries. (See Chapter 15.)

Even though the IRS has changed its stance, other government agencies may consider payment of a worker's business or traveling expenses to be a strong indication of employment relationship. For this reason, it is usually best that your compensation be high enough to cover your expenses; you should not be separately reimbursed for them.

Setting your compensation at a level high enough to cover your expenses has another advantage: It frees you from having to keep records of your expenses. Keeping track of the cost of every phone call or photocopy you make for a client can be a real chore and may be more trouble than it's worth.

However, if a project will involve expensive traveling, you may wish to separately bill the client for the cost. The following provision contains an optional clause for this.

**SUGGESTED LANGUAGE—ALTERNATIVE A—IF IC RESPONSIBLE FOR EXPENSES:**

Contractor shall be responsible for all expenses incurred while performing services under this Agreement.

OPTIONAL: However, Client shall reimburse Contractor for all reasonable travel and living expenses necessarily incurred by Contractor while away from Contractor's regular place of business to perform services under this Agreement. Contractor shall submit an itemized statement of such expenses. Client shall pay Contractor within 30 days from the date of each statement.

However, some self-employed people customarily have expenses paid by their clients. For example, attorneys, accountants and many self-employed consultants typically charge their clients separately for photocopying charges, deposition fees and travel. Where there is an otherwise clear independent contractor relationship and payment of expenses is customary in your trade or business, you can probably get away with doing it.

**SUGGESTED LANGUAGE—ALTERNATIVE B—IF CLIENT RESPONSIBLE FOR EXPENSES:**

Client shall reimburse Contractor for the following expenses that are directly attributable to work performed under this Agreement:

- travel expenses other than normal commuting, including airfares, rental vehicles, and highway mileage in company or personal vehicles at __[state amount] cents per mile
- telephone, fax, online and telegraph charges
- postage and courier services
- printing and reproduction
- computer services, and
- other expenses resulting from the work performed under this Agreement.

Contractor shall submit an itemized statement of Contractor's expenses. Client shall pay Contractor within 30 days from the date of each statement.

## 5. Materials

Generally, you should provide all the materials and equipment necessary to complete a project. However, this might not always be possible. For example:

- a computer consultant may have to perform work on the client's computers
- a marketing consultant may need research materials from the client, or
- a freelance copywriter may need copies of the client's old sales literature.

Specify any materials you need from the client in your agreement.

### a. You provide all materials

If you furnish all the materials and equipment, use the following clause.

> **SUGGESTED LANGUAGE—ALTERNATIVE A—IF IC PROVIDES MATERIALS:**
>
> Contractor will furnish all materials and equipment used to provide the services required by this Agreement.

### b. Client provides materials or equipment

List the materials or equipment the client will provide. If you need these items by a specific date, specify the deadline as well.

> **SUGGESTED LANGUAGE—ALTERNATIVE B—IF CLIENT PROVIDES MATERIALS:**
>
> Client shall make available to Contractor, at Client's expense, the following materials, facilities and equipment:
> [List]________________________________________.
>
> These items will be provided to Contractor by [Date] ________.

## 6. Term of Agreement

The term of the agreement means when it begins and ends. Unless the agreement provides a specific starting date, it begins on the date it's signed. If you and the client sign on different dates, the agreement begins on the date the last person signed. You normally shouldn't begin work until the client signs the agreement, so it's usually best that the agreement not provide a specific start date that might be before the client signs.

The agreement should have a definite ending date. This ordinarily will mark the final deadline for you to complete your services. However, even if the project is lengthy, the end date should not be too far in the future. A good outside time limit is 12 months. A longer term makes the agreement look like an employment agreement, not an independent contractor agreement. If the work is not completed at the end of 12 months, you can negotiate and sign a new agreement.

> **SUGGESTED LANGUAGE:**
>
> This Agreement will become effective when signed by both parties and will end no later than
>
> _____, 19__.

## 7. Terminating the Agreement

Signing a contract doesn't make you bound by it forever no matter what happens. You and the client can agree to call off your agreement at any time. In addition, contracts typically contain provisions allowing either person to terminate the agreement under certain circumstances. Termination means either party can end the contract without the other side's agreement.

When a contract is terminated, both you and the client stop performing—that is, you discontinue your work and the client has no obligation to pay you for any work you may do after the effective date of termination. However, the client is legally obligated to pay you for any work you did prior to the termination date.

> **EXAMPLE:** Murray, a self-employed programmer, agrees to design an Internet site for Mary and signs a client agreement. About half-way through the project, Murray decides to terminate the agreement because Mary refuses to pay him the advance required by the contract. When the termination becomes effective, Murray has no obligation to do any further work for Mary, but he is still entitled to be compensated for the work he already did.

On the down side, however, you remain liable to the client for any damages it may have suffered due to your failure to perform as agreed before the termination date.

> **EXAMPLE:** Jill, a self-employed graphic artist, contracts with Bill to design the cover for a book Bill plans to publish. Bill terminates the agreement when Jill fails to deliver the cover by the contract deadline. Jill has no duty to create the cover. Nor is Bill required to pay Jill even if she does produce a cover. However, Jill is liable for any damages Bill suffered by her failure to live up to the agreement. It turned out that the delay in providing a cover cost Bill an extra $1,000 in printing bills. Jill is liable for this amount.

It's important to clearly define the circumstances under which you or the client may end the agreement.

In the past, the IRS viewed a termination provision giving either you, the client or both of you the right to terminate the agreement at any time to be strong evidence of an employment relationship. However, the agency no longer considers this to be such an important factor.

Even so, it's wise to place some limits on the client's right to terminate the contract. It's usually not in your best interest to give a client the right to terminate you for any reason or no reason at all, since the client may abuse that right.

Instead, both you and the client should be able to terminate the agreement without legal repercussions only if there is reasonable cause to do so; or, at most, only by giving written notice to the other.

### a. Termination with reasonable cause

Termination with reasonable cause means either you or the client have a good reason to end the agreement. A serious violation of the agreement is reasonable cause to terminate the agreement—but what is considered serious depends on the particular facts and circumstances.

A minor or technical contract violation is not serious enough to justify ending the contract for cause. For example, if a client promised to let you use office space a few hours a week, but failed to do so, the transgression would normally be viewed as minor and wouldn't justify terminating the agreement. However, if a self-employed programmer agreed to perform programming services for an especially low price because the client promised to let her use its mainframe computer, and the client then reneged and told the programmer to lease her own mainframe, the programmer likely would be justified in terminating the agreement.

Unless your contract provides otherwise, a client's failure to pay you on time may not necessarily constitute reasonable cause for you to terminate the agreement. You may add a clause to your contract providing that late payments are always reasonable cause for terminating the contract. The following clause provides that you may terminate the agreement if the client doesn't pay you what you're owed within 20 days after you make a written demand for payment. For example, if you send a client an invoice due within 30 days and the client fails to pay within that time, you may terminate the agreement 20 days after you send the client a written demand to be paid what you're owed. This may help give clients incentive to pay you.

The clause also makes clear that the client must pay you for the services you performed before the contract was terminated.

**SUGGESTED LANGUAGE—ALTERNATIVE A—REASONABLE CAUSE:**

With reasonable cause, either party may terminate this Agreement effective immediately by giving written notice of termination for cause. Reasonable cause includes:

- a material violation of this agreement, or
- nonpayment of Contractor's compensation after 20 days written demand for payment.

Contractor shall be entitled to full payment for services performed prior to the effective date of termination.

### b. Termination without cause

Sometimes you or the client just can't live with a limited termination right. Instead, you want to be able to get out of the agreement at any time without incurring liability. For example, a client's business plans may change and it may no longer need your services. Or you may have too much work and need to lighten your load.

If you want a broader right to end the work relationship, add a provision to the contract that gives either of you the right to terminate the agreement for any reason upon written notice. You need to provide at least a few days notice. Being able to terminate without notice tends to make you look like an employee. A period of 30 days is a common notice period, but shorter notice may be appropriate if the project is of short duration.

**SUGGESTED LANGUAGE—ALTERNATIVE B—WITHOUT CAUSE:**

Either party may terminate this Agreement at any time by giving ____ [5, 10, 15, 30, 45, 60] days written notice of termination. Contractor shall be entitled to full payment for services performed prior to the date of termination.

## 8. Independent Contractor Status

One of the most important functions of an independent contractor agreement is to help establish that you are an independent contractor, not your client's employee. The key to doing this is to make clear that you, not the client, have the right to control how the work will be performed.

You will need to emphasize the factors the IRS and other agencies consider in determining whether a client controls how the work is done. Of course, if you merely recite what you think the IRS wants to hear but fail to adhere to these understandings, agency auditors won't be fooled. Think of this clause as a reminder to you and your client about how to conduct your business relationship. (See Chapter 15.)

**SUGGESTED LANGUAGE:**

Contractor is an independent contractor, not Client's employee. Contractor's employees or subcontractors are not Client's employees. Contractor and Client agree to the following rights consistent with an independent contractor relationship.

- Contractor has the right to perform services for others during the term of this Agreement.
- Contractor has the sole right to control and direct the means, manner and method by which the services required by this Agreement will be performed.
- Contractor has the right to hire assistants as subcontractors, or to use employees to provide the services required by this Agreement.
- The Contractor or Contractor's employees or subcontractors shall perform the services required by this Agreement; Client shall not hire, supervise or pay any assistants to help Contractor.
- Neither Contractor nor Contractor's employees or subcontractors shall receive any training from Client in the skills necessary to perform the services required by this Agreement.
- Client shall not require Contractor or Contractor's employees or subcontractors to devote full time to performing the services required by this Agreement.
- Neither Contractor nor Contractor's employees or subcontractors are eligible to participate in any employee pension, health, vacation pay, sick pay or other fringe benefit plan of Client.

## 9. Local, State and Federal Taxes

The agreement should address federal and state income taxes, Social Security taxes and sales taxes.

### a. Income taxes

Your client should not pay or withhold any income or Social Security taxes on your behalf. Doing so is a very strong indicator that you are an employee, not an independent contractor. Indeed, some courts have classified workers as employees based upon this factor alone. Keep in mind that one of the best things about being self-employed is that you don't have taxes withheld from your paychecks. (See Chapter 8.)

Include a straightforward provision, such as the one suggested below, to help make sure the client understands that you'll pay all applicable taxes due on your compensation and the client should therefore not withhold taxes from your payments.

**SUGGESTED LANGUAGE:**

Contractor shall pay all income taxes, and FICA (Social Security and Medicare taxes) incurred while performing services under this Agreement. Client will not:

- withhold FICA from Contractor's payments or make FICA payments on Contractor's behalf
- make state or federal unemployment compensation contributions on Contractor's behalf, or
- withhold state or federal income tax from Contractor's payments.

### b. Sales taxes

A few states require self-employed people to pay sales taxes, even if they only provide their clients with services. These states include Hawaii, New Mexico and South Dakota. Many other states require sales taxes to be paid for certain specified services. (See Chapter 8, Section A2c.)

Whether or not you're required to collect sales taxes, include the following provision in your agreement making it clear that the client will have to pay these and similar taxes. States constantly change sales tax laws and more are beginning to look at services as a good source of sales tax revenue. So this provision could come in handy in the future even if you don't really need it now.

**SUGGESTED LANGUAGE:**

The charges included here do not include taxes. If Contractor is required to pay any federal, state or local sales, use, property or value added taxes based on the services provided under this Agreement, the taxes shall be separately billed to Client. Contractor shall not pay any interest or penalties incurred due to late payment or nonpayment of any taxes by Client.

## 10. Notices

When you want to do something important involving the agreement, you need to tell the client about it. This is called giving notice. For example, you need to give the client notice if you want to modify the agreement or terminate it.

The following provision gives you several options for providing the client with notice: by personal delivery, by mail, or by fax or telex followed by a confirming letter.

If you give notice by mail, it is not effective until three days after it's sent. For example, if you want to end the agreement on 30 days notice and mail your notice of termination to the client, the agreement will not end until 33 days after you mailed the notice.

**SUGGESTED LANGUAGE:**

All notices and other communications in connection with this Agreement shall be in writing and shall be considered given as follows:

- when delivered personally to the recipient's address as stated on this Agreement
- three days after being deposited in the United States mail, with postage prepaid to the recipient's address as stated on this Agreement, or
- when sent by fax or telex to the last fax or telex number of the recipient known to the person giving notice. Notice is effective upon receipt, provided that a duplicate copy of the notice is promptly given by first class mail, or the recipient delivers a written confirmation of receipt.

## 11. No Partnership

You want to make sure that you and the client are separate legal entities, not partners. If a client is viewed as your partner, you'll be liable for its debts and the client will have the power to make contracts that obligate you to others without your consent.

**SUGGESTED LANGUAGE:**

This Agreement does not create a partnership relationship. Neither party has authority to enter into contracts on the other's behalf.

## 12. Applicable Law

It's a good idea for your agreement to indicate the state law that will govern if you have a dispute with the client. This is particularly helpful if you and the client are in different states. There is some advantage to having the law of your own state control, since local attorneys will likely be more familiar with that law.

**SUGGESTED LANGUAGE:**

This Agreement will be governed by the laws of the state of [Indicate state in which you have your main office] __________.

## 13. Exclusive Agreement

When you put your agreement in writing, it is treated as the last word on the areas covered if you and the client intend it to be the final and complete expression of your agreement. The written agreement takes precedence over any written or oral agreements or promises previously made. This means neither you nor the client can bring up side letters, oral statements or other material not covered by the contract.

Business contracts normally contain a provision stating that the written agreement is the complete and exclusive agreement between those involved. This is to help make it clear to a court or a mediator or arbitrator that the parties intended the contract to be their final agreement. A clause such as this helps avoid claims that promises not contained in the written contract were made and broken.

Make sure that all documents containing any of the client's representations upon which you are relying are attached to the agreement as exhibits. (See Chapter 18, Section D3.) If they aren't attached, they likely won't be considered to be part of the agreement.

**SUGGESTED LANGUAGE:**

This is the entire Agreement between Contractor and Client.

## 14. Signatures

The end of the main body of the agreement should contain spaces for you to sign, write in your title and date. Make sure the person signing the agreement has the authority to do so. (See Chapter 17, Section D1.)

**SUGGESTED LANGUAGE:**

Client:

[NAME OF CLIENT]

By: ______________________________

(Signature)

______________________________

(Typed or Printed Name)

Title: ______________________________

Date: ______________________________

Contractor:

[NAME OF CONTRACTOR]

By: ______________________________

(Signature)

______________________________

(Typed or Printed Name)

Taxpayer ID Number: ______________

Date: ______________________________

**Signing by fax.** It is increasingly common to use faxed signatures to finalize contracts. If you use faxed signatures, include a specific provision authorizing them at the end of the agreement.

**OPTIONAL LANGUAGE—SIGNATURES BY FAX**

Contractor and Client agree that this Agreement will be considered signed when the signature of a party is delivered by facsimile transmission. Signatures transmitted by facsimile shall have the same effect as original signatures.

## B. Optional Provisions

There are several optional provisions you may wish to include in your agreement. They are not absolutely necessary for every client agreement, but they can be extremely helpful. You should carefully consider including them in your contracts. Pay especially close attention to the provisions regarding:

- resolving disputes
- contract changes, and
- attorney fees.

It's usually to your advantage to include all of these provisions in your agreement.

### 1. Resolving Disputes

As you probably know, court litigation can be very expensive. To avoid this cost, alternative forms of dispute resolution have been developed that don't involve going to court. These include mediation and arbitration.

The following clause requires you and the client to take advantage of these alternate forms of dispute resolution. You're first required to submit the dispute to mediation. You agree on a neutral third person to serve as mediator and try to help you settle your dispute. The mediator has no power to impose a decision, only to try to help you arrive at one. (See Chapter 21, Section A.)

If mediation doesn't work, you must submit the dispute to binding arbitration. Arbitration is usually like an informal court trial without a jury, but involves arbitrators instead of judges and is usually much faster and cheaper than courts. You may be represented by a lawyer, but it's not required. (See Chapter 21, Section B.)

You should indicate where the mediation or arbitration would occur. You'll usually want it in the city or county where your office is located. You don't want to have to travel a long distance to attend a mediation or arbitration.

However, every state has an alternative to mediation or arbitration that can be even cheaper and quicker than either of these approach: small claims court. Small claims courts are designed to help resolve disputes involving a relatively small amount of money. The amount ranges from $2,000 to $7,500, depending on the state in which you live. If your dispute involves more money than the small claims limit, you can waive the excess and still bring a small claims suit. You don't need a lawyer to sue in small claims court; indeed, lawyers are barred from small claims court in several states. Small claims court is particularly useful where a client owes you a relatively small amount of money. (See Chapter 7, Section B4.)

The following clause provides that you or the client can elect to skip mediation and arbitration—and take your dispute to small claims court.

> **SUGGESTED LANGUAGE:**
>
> If a dispute arises under this Agreement, the parties agree to first try to resolve the dispute with the help of a mutually agreed-upon mediator in __________ [State city or county where mediation will occur]. Any costs and fees other than attorney fees associated with the mediation shall be shared equally by the parties.
>
> If it proves impossible to arrive at a mutually satisfactory solution through mediation, the parties agree to submit the dispute to binding arbitration in __________ [Note the city or county where arbitration will occur] under the rules of the American Arbitration Association. Judgment upon the award rendered by the arbitrator may be entered in any court having jurisdiction to do so.
>
> However, the complaining party may refuse to submit the dispute to mediation or arbitration and instead bring an action in an appropriate Small Claims Court.

## 2. Modifying the Agreement

It's very common for both you and your client to want to change the terms of an agreement after work has started. For example, the client might want to make a change in the contract specifications which could require you to do more work. Or you might discover that you underestimated how much time the project will take and need to be paid more to complete it.

When you modify your agreement in this way, you should write down the changes on a separate document, have it signed and attach it to your original agreement. (See Chapter 13, Section D; and see the Appendix for a form for a contract amendment.

The following provision recognizes that the original agreement you enter with the client may have to be changed. Although oral changes to contracts are enforceable, it's a better idea to write them down. This provision states that you and the client must write down your changes and both sign the writing. Such a contract provision requiring modifications to be in writing is probably a bit of overkill—that is, both you and the client can still make changes without writing them down. However, making this requirement explicit does stress the importance of documenting changes in writing.

Neither you nor the client must accept a proposed change to a contract. But because you are obligated to deal with each other fairly and in good faith, you can't simply refuse all modifications without attempting to reach a resolution.

⚠ If you use this clause, check to be sure you have also included the optional provision on resolving disputes. (See Section B1.) That way, if you and the client can't agree on the changes, the agreement will require that you submit your dispute to mediation; and, if that doesn't work, to binding arbitration. This avoids expensive court litigation. (See Chapter 21. Section A.)

**SUGGESTED LANGUAGE:**

Client and Contractor recognize that:

- Contractor's original cost and time estimates may be too low due to unforeseen events, or to factors unknown to Contractor when this Agreement was made
- Client may desire a mid-project change in Contractor's services that would add time and cost to the project and possibly inconvenience Contractor, or
- Other provisions of this Agreement may be difficult to carry out due to unforeseen circumstances.

If any intended changes or any other events beyond the parties' control require adjustments to this Agreement, the parties shall make a good faith effort to agree on all necessary particulars. Such agreements shall be put in writing, signed by the parties and added to this Agreement.

## 3. Attorneys' Fees

If you have to sue the client in court to enforce the agreement and if you win, you normally will not be awarded the amount of your attorney fees unless your agreement requires it. Including an attorney fees provision in the agreement can be in your interest. It can help make filing a lawsuit economically feasible; you might even be able to convince a lawyer to file a case against your client without providing an upfront cash retainer. It will also give the client a strong incentive to negotiate with you if you have a good case.

Sometimes, however, an attorney fees provision can work against you. It may help your client find an attorney to sue you and make you more anxious to settle. If you think it's more likely you'll violate the agreement than the client will, an attorney fees provision is probably not a good idea.

Under the following provision, if either person has to sue the other in court to enforce the agreement and wins—that is, becomes the prevailing party—the loser is required to pay the other person's attorney fees and expenses.

**SUGGESTED LANGUAGE:**

If any legal action is necessary to enforce this Agreement, the prevailing party shall be entitled to reasonable attorney fees, costs and expenses in addition to any other relief to which he or she may be entitled.

## 4. Late Fees

Many self-employed people charge a late fee if the client doesn't pay within the time specified in the IC agreement or invoice. Charging late fees for overdue payments can get clients to pay on time. The late fee is normally expressed as a monthly interest charge. (See Chapter 7, Section B2.)

If you wish to charge a late fee, make sure it's mentioned in your agreement. You should also clearly state what your late fee is on all your invoices.

**State Restrictions on Late Fees**
Your state might have restrictions on how much you can charge as a late fee. You'll have to investigate your state laws to find out. Check the index to the annotated statutes for your state—sometimes called a code—available in any law library. (See Chapter 7, Section B2.) Look under the terms interest, usury or finance charges. Also, a professional or trade organization may have helpful information.

You can safely charge as a late fee at least as much as banks charge businesses to borrow money. Find out the current bank interest rate by calling your bank or looking in the business section of your local newspaper.

The math requires two steps. First, divide the annual interest rate by 12 to determine your monthly interest rate.

**EXAMPLE:** Sam, a self-employed consultant, decides to start charging clients a late fee for overdue payments. He knows banks are charging 12% interest per year on borrowed money and decides to charge the same. He divides this rate by 12 to determine his monthly interest rate: 1%.

Then, multiply the monthly rate by the amount due to determine the amount of the monthly late fee.

> **EXAMPLE:** Acme Corp. is 30 days late paying Sam a $10,000 fee. Sam multiples this amount by his 1% finance charge to determine his late fee: $100 (.01 x $10,000). He adds this amount to Acme's account balance. He does this every month for which the payment is late.

**SUGGESTED LANGUAGE:**

Late payments by Client shall be subject to late penalty fees of ____% per month from the due date until the amount is paid.

## 5. Liability to the Client

If something goes wrong with your work, you might end up getting sued by the client and having to pay damages.

> **EXAMPLE:** Julie, a self-employed computer programmer, designs an inventory accounting program for a cosmetics company. A bug in the program causes the program to crash and the company is unable to conduct normal business for several days, losing tens of thousands of dollars. The company sues Julie, claiming that her program design was deficient.

Such lawsuits could easily cost more than you were paid for your work and could even bankrupt you. To avoid this, many self-employed people include provisions in their agreements limiting their liability. This is particularly wise if problems with your work or services could cause the client substantial injuries or economic losses.

### a. Liability cap

The following optional clause limits your total liability for any damages to the client to a set dollar amount or to no more than you were paid, whichever is less. It also relieves you of liability for lost profits or other special damages to the client.

Such damages are also called incidental or consequential damages. These are damages that can far exceed the amount the client actually paid you for your work. They arise out of circumstances you knew about or should have foreseen when the contract was made. This type of damages often involves lost profits that logically result from failure to live up to your agreement. For example, if you knew that your failure to deliver your work on time could cost the client a valuable business opportunity, you could be required to make up the lost profits the client would have earned had you delivered the work on time.

**SUGGESTED LANGUAGE—ALTERNATIVE A—CAP ON LIABILITY:**

Contractor's total liability to Client under this Agreement for damages, costs and expenses, regardless of cause, shall not exceed $____ or the compensation received by Contractor under this Agreement, whichever is less. Contractor shall not be liable for Client's lost profits, or special, incidental or consequential damages.

### b. No liability to client

The following provision limits your liability to the client to the maximum extent possible. It provides that you are not liable to the client for any losses or damages arising from your work.

**SUGGESTED LANGUAGE—ALTERNATIVE B—NO LIABILITY:**

Contractor shall not be liable to Client for any loss, damages or expenses resulting from Contractor's services under this Agreement.

### 6. Liability to Others

The work that you do can affect people other than the client. Such people are called third parties.

Third parties—people you don't know and have never dealt with yourself—typically enter the picture when they are directly or indirectly injured by the work you've performed for a client. The injuries can be physical, economic or both. For example, if an elevator crashed due to faulty software a self-employed programmer designed for the elevator manufacturer, the injured elevator passengers would be third parties the programmer never met or contracted with, but the programmer might be liable to them.

You need to be particularly concerned about your liability to third parries if you're engaged in a hazardous or risky project that could result in injuries to others if something goes wrong.

If third parties are damaged as a result of the work you perform for a client, they'll likely sue everyone involved—including you. Both you and the client may be liable for the full amount of such claims. Indemnification provisions are used to require one person to pay the other's attorney fees and damages arising from such claims. Such provisions don't affect third parties who sue you or absolve you from liability to them for your actions. The clause can only make the client pay any amounts due as a result of such liability.

> **EXAMPLE:** Bart, a self-employed software engineer, creates an experimental software program designed to automate a chemical factory for BigCorp. The program fails and a chemical spill takes place, damaging nearby landowners.
>
> The property owners affected by the chemical spill would undoubtedly sue not only BigCorp, but Bart as well, claiming that he negligently designed the software. Bart will likely be forced to defend himself against lawsuits filed by total strangers.

You may wish to include the following provision in your agreement requiring the client to indemnify you against third party claims. This means the client will be responsible for defending any lawsuits and for all damages and injuries that third parties suffer if something goes wrong with your work.

To accomplish this, the provision uses a standard legal phrase that may be hard to understand. It states that the client shall "indemnify, defend, and hold harmless" the contractor against third party claims. This means the client is required to assume responsibility for dealing with third party claims and must repay you if they end up costing you anything.

Many clients will balk at including an indemnification provision in your agreement. Moreover, as a practical matter, such a provision is useless if the client doesn't have the money or insurance to pay the amount due. You may be better off charging the client enough to obtain your own liability insurance protecting you against third party claims. (See Chapter 6, Section D.)

**SUGGESTED LANGUAGE:**

Client will indemnify, defend and hold harmless Contractor against all liabilities, damages and expenses, including reasonable attorney fees, resulting from any third party claim or lawsuit arising from Contractor's performance under this Agreement.

### 7. Intellectual Property Ownership

If you're hired to create or contribute to the creation of intellectual property—for example, important business documents, marketing plans, software programs, graphics, designs, photos, music, inventions or trademarks—the agreement should specify who owns your work.

There are many options regarding ownership of intellectual property that self-employed people create. Typically, your client will want to own all the intellectual property rights in your work, but this doesn't have to be the case. For example, you could retain sole ownership and grant the client a license to use your work. The only limit on how you deal with ownership of your work is your own imagination. (See Chapter 17.)

### a. You retain ownership

Under the following clause, you keep ownership of your work and merely give the client a nonexclusive license to use it. This means that the client may use your work, but does not own it and may not sell it to others. The license is royalty-free—meaning that the sole payment you receive for it is the sum the client paid you for your services. The client will make no additional payments for the license. (See Chapter 17, Section B2.)

> **SUGGESTED LANGUAGE—ALTERNATIVE A—NONEXCLUSIVE TRANSFER:**
>
> Contractor grants to Client a royalty-free nonexclusive license to use anything created or developed by Contractor for Client under this Agreement (Contract Property). The license shall have a perpetual term and the Client may not transfer it. Contractor shall retain all copyrights, patent rights and other intellectual property rights to the Contract Property.

### b. You transfer ownership to client

Under the following clause you transfer all your ownership rights to the client. But you must first get all your compensation from the client.

You also agree to help prepare any documents necessary to help the client obtain any copyright, patent or other intellectual property rights at no charge to the client. This would probably amount to no more than signing a patent or copyright registration application. However, the client is required to reimburse you for your expenses.

> **SUGGESTED LANGUAGE—ALTERNATIVE B—TRANSFER ALL RIGHTS:**
>
> Contractor assigns to Client all patent, copyright and trade secret rights in anything created or developed by Contractor for Client under this Agreement. This assignment is conditioned upon full payment of the compensation due Contractor under this Agreement.
>
> Contractor shall help prepare any documents Client considers necessary to secure any copyright, patent or other intellectual property rights at no charge to Client. However, Client shall reimburse Contractor for reasonable out of pocket expenses.

### c. Reusable materials

Many self-employed people who create intellectual property for clients have certain materials they use over and over again for different clients. For example, computer programmers may have certain utilities or program tools they incorporate into the software they create for many different clients.

You may lose the legal right to reuse such materials if you transfer all your ownership rights in your work to the client. To avoid this, include the following provision in your agreement. It provides that you retain ownership of such materials and only gives the client a nonexclusive license to use them. The license is royalty-free—meaning that the sole payment you receive for it is the sum the client paid you for your services. The client will make no additional payments for the license. The license also has a perpetual term, meaning it will last as long as your copyright, patent or other intellectual property rights do.

If you know what such materials consist of in advance, it's a good idea to list them in an exhibit attached to the agreement; but this isn't absolutely necessary.

> **OPTIONAL LANGUAGE—RIGHT TO REUSE:**
>
> Contractor owns or holds a license to use and sublicense various materials in existence before the start date of this Agreement (Contractor's Materials).
>
> [OPTIONAL: Contractor's Materials include, but are not limited to, those items identified in Exhibit __, attached to and made part of this Agreement.]
>
> Contractor may, at its option, include Contractor's Materials in the work performed under this Agreement. Contractor retains all right, title and interest, including all copyrights, patent rights and trade secret rights in Contractor's Materials. Contractor grants Client a royalty-free nonexclusive license to use any Contractor's Materials incorporated into the work performed by Contractor under this Agreement. The license shall have a perpetual term and may not be transferred by Client.

## 8. Assignment and Delegation

An assignment is the process by which rights or benefits under a contract are transferred to someone else. For example, a client might assign the right to receive the benefit of your services to someone else. Such a person is called an assignee. When this occurs, the assignee steps into the original client's shoes. You must now work for the assignee, not the client with whom you contracted. If you fail to perform, the assignee may sue you for breach of contract.

> **EXAMPLE:** Terri, a self-employed designer, agrees to design a cover and chapter headings for several books published by Scrivener & Sons. Scrivener assigns this right to Pop's Books. This means that Terri must perform the work for Pop's instead of Scrivener. If Terri fails to do so, Pop's can sue her for breach of contract.

You may also assign the benefits you receive under an IC agreement to someone else.

> **EXAMPLE:** Jimmy agrees to provide Fastsoft 20 hours of computer programming services in exchange for being able to use its computer for 100 hours. Jimmy assigns, or transfers, the right to the computer time to his friend Kate. This means Fastsoft must let Kate use its computer.

Delegation is the flipside of assignment. Instead of transferring benefits under a contract, you transfer the duties. As long as the new person does the job correctly, all will be well. However, the person delegating duties under a contract usually remains responsible if the person to whom the delegation was made fails to perform.

> **EXAMPLE:** Jimmy delegates to Mindy his duty to perform 20 hours of programming services for Fastsoft. This means that Mindy, not Jimmy, will now do the work. But Jimmy remains liable if Mindy doesn't perform adequately.

### a. Legal restrictions

Unless a contract provides otherwise, you can ordinarily freely assign and delegate it subject to some important legal limitations. For example, a client can't assign the benefit of your services to someone else without your consent if it would increase the work you must do or otherwise magnify your burden under the contract. Similarly, you can't delegate your duties without the client's consent if it would decrease the benefits the client would receive.

One of the most important limitations for self-employed people is that contracts for personal services are ordinarily not assignable or delegable without the client's consent. This type of contract involves services that are personal in nature. Examples include contracts for the services of lawyers, physicians, architects, writers and artists. Courts consider it unfair for either a client or self-employed person to change horses in midstream in these cases.

> **EXAMPLE:** Arthur contracts with Betty, a freelance artist, to paint his portrait. He later attempts to assign his rights to his wife Carla—that is, require Betty to paint Carla's portrait instead of his. Neither Arthur nor Carla can force Betty to paint Carla's portrait because Betty's contract with Arthur was a personal services contract.

### b. Contract restrictions

Your contract may also place limits on assignment and delegation. Contractual limits on your right to delegate your duties to others are not supportive of your independent contractor status since they allow the client to control who will do the work. Moreover, it is often advantageous for you to have the right to delegate your contractual obligations to others. This gives you flexibility, for example, to hire someone else to do the work if you don't have time to do it.

However, some clients may balk at allowing you to delegate your contractual duties without the client's consent. This is usually where the client has hired you because of your special expertise, reputation for performance or financial stability and the client doesn't want some other person performing the services.

Also, there may be cases where you really don't want the client to have the right to assign the benefit of your services to someone else, who may turn out to be incompetent.

In this event, you may include the following provision in your agreement. It bars both you and the client from assigning your rights or delegating your duties without the other party's consent.

**SUGGESTED LANGUAGE:**

Neither party may assign any rights or delegate any duties under this Agreement without the other party's prior written approval.

## C. Sample Client Agreement

The following sample agreement is a fixed fee agreement calling for mediation and arbitration of disputes and payment of attorney fees. Note that the headings in the sample agreement include a citation in parentheses that coincides with the discussion earlier in the chapter.

**SAMPLE GENERAL IC AGREEMENT**

**Independent Contractor Agreement (See Section A)**

This Agreement is made between Acme Widget Co. (Client) with a principal place of business at 123 Main Street, Marred Vista, CA 90000 and ABC Consulting, Inc. (Contractor), with a principal place of business at 456 Grub Street, Santa Longo, CA 90001.

**Services to Be Performed (A1)**

Contractor agrees to perform the following services: Install and test Client's DX9-105 widget manufacturing press so that it performs according to the manufacturer's specifications.

**Payment (A2)**

In consideration for the services to be performed by Contractor, Client agrees to pay Contractor $20,000.

**Terms of Payment (A3)**

Upon completing Contractor's services under this Agreement, Contractor shall submit an invoice. Client shall pay Contractor within 30 days from the date of Contractor's invoice.

**Expenses (A4)**

Contractor shall be responsible for all expenses incurred while performing services under this Agreement.

**Materials (A5)**

Contractor will furnish all materials and equipment used to provide the services required by this Agreement.

**Term of Agreement (A6)**

This Agreement will become effective when signed by both parties and will end no later than May 1, 199_.

**Terminating the Agreement (A7)**

With reasonable cause, either party may terminate this Agreement effective immediately by giving written notice of termination for cause. Reasonable cause includes:

- a material violation of this agreement, or
- nonpayment of Contractor's compensation after 20 days written demand for payment.

Contractor shall be entitled to full payment for services performed prior to the effective date of termination.

**Independent Contractor Status (A8)**

Contractor is an independent contractor, not Client's employee. Contractor's employees or subcontractors are not Client's employees. Contractor and Client agree to the following rights consistent with an independent contractor relationship.

- Contractor has the right to perform services for others during the term of this Agreement.
- Contractor has the sole right to control and direct the means, manner and method by which the services required by this Agreement will be performed.
- Contractor has the right to hire assistants as subcontractors, or to use employees to provide the services required by this Agreement.
- The Contractor or Contractor's employees or subcontractors shall perform the services required by this Agreement; Client shall not hire, supervise or pay any assistants to help Contractor.
- Neither Contractor nor Contractor's employees or subcontractors shall receive any training from Client in the skills necessary to perform the services required by this Agreement.
- Client shall not require Contractor or Contractor's employees or subcontractors to devote full time to performing the services required by this Agreement.
- Neither Contractor nor Contractor's employees or subcontractors are eligible to participate in any employee pension, health, vacation pay, sick pay or other fringe benefit plan of Client.

**Local, State and Federal Taxes (A9a)**

Contractor shall pay all income taxes, and FICA (Social Security and Medicare taxes) incurred while performing services under this Agreement. Client will not:

- withhold FICA from Contractor's payments or make FICA payments on Contractor's behalf
- make state or federal unemployment compensation contributions on Contractor's behalf, or
- withhold state or federal income tax from Contractor's payments.

**Sales Taxes (A9b)**

The charges included here do not include taxes. If Contractor is required to pay any federal, state or local sales, use, property or value added taxes based on the services provided under this Agreement, the taxes shall be separately billed to Client. Contractor shall not pay any interest or penalties incurred due to late payment or nonpayment of any taxes by Client.

**Notices (A10)**

All notices and other communications in connection with this Agreement shall be in writing and shall be considered given as follows:

- when delivered personally to the recipient's address as stated on this Agreement
- three days after being deposited in the United States mail, with postage prepaid to the recipient's address as stated on this Agreement, or
- when sent by fax or telex to the last fax or telex number of the recipient known to the person giving notice. Notice is effective upon receipt provided that a duplicate copy of the notice is promptly given by first class mail, or the recipient delivers a written confirmation of receipt.

**No Partnership (A11)**
This Agreement does not create a partnership relationship. Neither party has authority to enter into contracts on the other's behalf.

**Applicable Law (A12)**
This Agreement will be governed by the laws of the state of California.

**Exclusive Agreement (A13)**
This is the entire Agreement between Contractor and Client.

**Resolving Disputes (B1)**
If a dispute arises under this Agreement, the parties agree to first try to resolve the dispute with the help of a mutually agreed-upon mediator in Mariposa County. Any costs and fees other than attorney fees associated with the mediation shall be shared equally by the parties.

If it proves impossible to arrive at a mutually satisfactory solution through mediation, the parties agree to submit the dispute to binding arbitration in Mariposa County under the rules of the American Arbitration Association. Judgment upon the award rendered by the arbitrator may be entered in any court having jurisdiction to do so.

However, the complaining party may refuse to submit the dispute to mediation or arbitration and instead bring an action in an appropriate Small Claims Court.

**Modifying the Agreement (B2)**
Client and Contractor recognize that:

- Contractor's original cost and time estimates may be too low due to unforeseen events, or to factors unknown to Contractor when this Agreement was made
- Client may desire a mid-project change in Contractor's services that would add time and cost to the project and possibly inconvenience Contractor, or
- Other provisions of this Agreement may be difficult to carry out due to unforeseen circumstances.

If any intended changes or any other events beyond the parties' control require adjustments to this Agreement, the parties shall make a good faith effort to agree on all necessary particulars. Such agreements shall be put in writing, signed by the parties and added to this Agreement.

**Attorneys' Fees (B3)**
If any legal action is necessary to enforce this Agreement, the prevailing party shall be entitled to reasonable attorney fees, costs and expenses in addition to any other relief to which he or she may be entitled.

**Signatures (A14)**

Client:
Acme Widget Co.
By: ______________________________
(Signature)
BASILIO CHEW
(Typed or Printed Name)
Title: President
Date: April 30, 199X
Contractor: ABC Consulting, Inc.
By: ______________________________
(Signature)
GEORGE BAILEY
(Typed or Printed Name)
Title: President
Taxpayer ID Number: 123-45-6789
Date: April 30, 199X

## D. Using Letter Agreements

Many self-employed people and their clients commonly use letter agreements instead of more formal standard contracts. A letter agreement is usually a short contract written in the form of a letter. Although letter agreements may lack the gravitas of standard agreements, they are perfectly valid, binding contracts.

In a typical arrangement where a letter agreement is used, you and a client will reach a tentative agreement on an assignment in a meeting, over the phone, by fax, electronic mail or a combination of all four. After the agreement is reached, one of you drafts a letter documenting the important terms, signs it and sends it to the other person to sign.

Some clients have their own form letter agreements they use with all self-employed people they hire and will insist on using them. Review the letter carefully to make sure it meshes with the client's and your own oral statements and does not contain unfair provisions. (See Chapter 20.)

However, many clients will be happy for you to take on the work of drafting the agreement. You'll almost always be better off if you draft the agreement yourself because you can:

- avoid including any terms that are unduly favorable to the client, and
- make sure the agreement is completed and sent out quickly.

Take the initiative and offer to draw up the letter agreement. Explain that this is part of your service and that using an agreement you've drafted helps establish that you're not the client's employee.

Use the information in this chapter to draft one or more appropriate form letters you can use over and over again with minor alterations. This will be particularly easy to do if you use a word processor and keep the forms on disk.

**Don't Begin Work Until the Client Signs on Bottom Line**

No matter who drafts a letter agreement, don't begin work until you have a copy signed by the client. You don't want to begin work only to discover that the client wants to cancel the project or make major changes in your agreement.

### 1. Pros and Cons of Letter Agreements

Letter agreements are usually shorter and easier to draft than regular contracts. They are also less formal looking. As a result, they often seem less intimidating to clients. Many clients who are fearful of signing a formal contract without having a lawyer review it will sign a letter agreement with no hesitation.

And in some fields, using letter agreements is the commonly accepted practice for doing business. For example, letter agreements are commonly used when freelance writers accept short assignments from magazines or other publications. If this is the case in your field of work, you may have to use letter agreements as a matter of course.

Because letter agreements are usually much shorter than standard client agreements, they are particularly useful for brief projects where relatively little money is involved. A potential client could well think you're crazy if you insist on a lengthy formal contract for a simple one-day project. Clients don't want to spend a lot of time on legal documentation for minor projects—and you probably don't want to, either.

However, in the interests of brevity, letter agreements typically make no mention of many provisions contained in longer standard agreements that could prove very useful if a problem arises—for example, provisions concerning dispute resolution, how the agreement may be terminated and provisions cementing your status as an independent contractor.

Both you and your client may be better off using a standard client agreement if:

- the project is a large and complex one that involves a substantial amount of money
- you're dealing with a new client you're not sure you can trust, or
- you are otherwise worried that problems or disputes may occur. (See Chapter 18, Section A.)

### 2. What to Include

A letter agreement can be as short as one-half page. At a minimum, however, it should contain:

- a description of the services you will perform
- the deadline by which you must complete your services
- the fees you will charge, and
- when you will be paid.

Your agreement doesn't have to end here, however. Depending on the nature of your services and the client, there may be several other provisions you'll want to include. (See Section B5.)

#### a. Services to be performed

The single most important part of the agreement is the description of the services you'll perform for the client. This description will set out the specifics of the work you're required to do and will serve as the yardstick to measure whether your performance was satisfactory.

Describe in as much detail as possible the work you're expected to accomplish. However, word the description carefully to emphasize the results you're expected to achieve. Don't describe the method by which you will achieve the results. It should be up to you to decide how to do the work. The client's control should be limited to accepting or rejecting your final results.

It's fine for the agreement to establish very detailed specifications for your finished work product. But the specs should only describe the end results you must achieve, not how to obtain those results.

You can include the description in the main body of the agreement. Or if it's a lengthy explanation, put it on a separate document and attach it to the agreement. (See Chapter 18, Section D3.)

**SUGGESTED LANGUAGE—ALTERNATIVE A—DESCRIPTION IN AGREEMENT:**

I will perform the following services on your behalf: [Describe services you will perform.]

**SUGGESTED LANGUAGE—ALTERNATIVE B—DESCRIPTION ATTACHED:**

I will perform the services described in the Exhibit attached to this Agreement.

Choose Alternative B if the explanation of the services is attached to the agreement.

### b. Deadlines

The agreement should also make clear when your work will be completed and delivered to the client. Make sure you give yourself enough time to complete the job. It's better to err on the side of caution and give yourself more time than you think you'll need.

**SUGGESTED LANGUAGE:**

I agree to complete these services on or before _______ [date].

### c. Payment

Self-employed people can be paid in many different ways. (See Chapter 7, Section A).

The two most common payment methods are:

- a fixed fee, and
- payment by unit of time.

#### Fixed fee

In a fixed fee agreement, you charge an agreed amount for the entire project. (See Chapter 7, Section A1a.)

**SUGGESTED LANGUAGE—ALTERNATIVE A—FIXED FEE:**

In consideration of my performance of these services, you agree to pay me $[State amount].

#### Unit of time

Many self-employed people charge by the hour, day or other unit of time—for example, lawyers, accountants and plumbers. (See Chapter 7, Section A1b.) Charging by the hour does not support the idea that you are an independent contractor, but you can get away with it if it's a common practice in your field.

**SUGGESTED LANGUAGE—ALTERNATIVE B—UNIT OF TIME:**

In consideration of my performance of these services, you agree to pay me at the rate of $[State amount] per [Hour, day, week or other unit of time].

Clients often wish to place a cap on the total amount they'll spend on the project when you're paid by the hour because they're afraid you might work slowly just to earn a larger fee. If the client insists on a cap, make sure it allows you to work enough hours to get the project done.

> **OPTIONAL LANGUAGE—CAP ON PAYMENT:**
>
> My total compensation shall not exceed $____[State amount] without your written consent.

## d. Terms of payment

Terms of payment means how you will bill the client and be paid. The client will not likely pay you for your work until you submit an invoice setting out the amount due. An invoice doesn't have to be fancy. It should include: an invoice number, the dates covered by the invoice, the hours expended if you're being paid by the hour and a summary of the work performed. (See Chapter 7, Section B, for a detailed discussion and sample invoice form. A blank invoice form is included in Appendix 1.

### Full payment upon completing work

The following provision requires you to send an invoice after you complete work. The client is required to pay your fixed or hourly fee within a set number of days after you send the invoice. Thirty days is a typical payment term, but it can be shorter or longer if you wish. Note that the time for payment starts to run as soon as you send your invoice, not when the client receives it. This will help you get paid more quickly.

> **SUGGESTED LANGUAGE—ALTERNATIVE A—PAYMENT ON COMPLETION:**
>
> I will submit an invoice after my services are completed. You shall pay me within ____ [10, 15, 30, 45, 60] days from the date of the invoice.

### Divided payments

You can also opt to be paid part of your fee when the agreement is signed and the remainder when the work is finished. When you're paid by the hour, such an upfront payment is often called a retainer. The amount of the upfront payment is subject to negotiation. Many self-employed people like to receive at least one-third to one-half of their fees before they start work. If the client is new or might have problems paying you, it's wise to get as much money in advance as you can.

The following provision requires that you be paid a specific amount when the client signs the agreement and then the rest when the work is finished.

> **SUGGESTED LANGUAGE—ALTERNATIVE B—DIVIDED PAYMENTS:**
>
> I will be paid in two installments. The first installment shall be $[state amount] and is payable by [due date]. The remaining $____ [state amount] will be due within [10, 15, 30, 45, 60] days after I complete my services and submit an invoice.

## c. Optional provisions

There are a number of other provisions you can add to a letter agreement. These aren't absolutely necessary, but they can benefit you. They are all discussed in detail in Chapter 19. They include provisions:

- requiring the client to reimburse you for expenses you incur in performing the work (see Chapter 19, Section A4)
- requiring the client to provide you with materials, equipment or facilities (see Chapter 19, Section A5)
- requiring mediation and arbitration of disputes (see Chapter 19, Section B1)
- allowing you to obtain attorney fees if you sue the client and win (see Chapter 19, Section B3)

- requiring the client to pay a late fee for late payments (see Chapter 19, Section B4)
- limiting your liability to the client if something goes wrong (see Chapter 19, Section B5), and
- restricting the client's ability to assign its benefits or delegate its duties under the agreement (see Chapter 19, Section B8).

In addition, if your work involves the creation of intellectual property—for example, any type of work of authorship, such as an article or other written work—you should include a provision in your agreement stating who will own your work. (See Chapter 19, Section B7.)

### 3. Putting the Agreement Together

There are two ways to handle a letter agreement. The old-fashioned way is to prepare and sign two copies and mail or deliver them to the client to sign. The client signs both copies, then returns one signed copy to you by mail or messenger and retains one copy for its records. Both copies are original, binding contracts.

Today, however, it's very common for an IC to draft a letter agreement, sign it and then fax a copy to the client. The client signs the letter and faxes a copy back to you. This has the advantage of speed, but you don't have the client's original signature on the letter, only a copy.

A faxed signature is probably legally sufficient if you and the client don't dispute that it is a fax of an original signature. However, if a client claims that a faxed signature was forged, it could be difficult or impossible to prove it's genuine since it very easy to forge a faxed signature with modern computer technology. Forgery claims are rare, however, so this is usually not a problem. Even so, it's a good practice for you and the client to follow up the fax with signed originals exchanged by mail or air express.

### 4. Sample Letter Agreement

The following letter agreement is between a self-employed public relations consultant and an oil company. The consultant agrees to create a marketing plan for the company's new oil additive called Zotz. The work will be performed for a fixed fee paid in two installments.

MALONEY & ASSOCIATES
1000 GRUB STREET
MARRED VISTA CA 90000

February 1, 199x

Jerry Wellhead
Vice President
Acme Oil Co.
1000 Greasy Way
Tulsa, OK 10000

Dear Jerry:

I am pleased to have the opportunity to provide my services. This letter will serve as our agreement.

I will perform the following services on your behalf: I will create a marketing plan for the rollout of Acme's new oil additive called Zotz. The plan will include guidelines for magazine, radio and television advertising.

I agree to complete these services on or before March 1, 199x.

In consideration of my performance of these services, you agree to pay me $5,000.

I will be paid in two installments. The first installment shall be $2,500 and is payable by February 5, 199x. The remaining $2,500 will be paid within 30 days after I complete my services and submit an invoice.

If this Agreement meets with your approval, please sign below to make this a binding contract between us. Please sign both copies and return one to me. The other signed copy is for your records.

Sincerely

Susan Maloney

Agreed to:

Acme Oil Co.

By: ______________________________

(Signature)

Jerry Wellhead

Title: Vice President

Date: __________

■

CHAPTER

# 20

# Reviewing a Client's Agreement

Many clients have their own agreements they will want to use. This is particularly likely if you work for firms that often use third parties to perform services or for large companies that have their own legal departments.

A client may present you with a lengthy, complex agreement, hand you a pen and tell you to sign. You may be told that the agreement is only a standard form that all non-employees who work for the client sign. However, signing a client agreement is never a mere technical formality. A client agreement is not simply a bunch of words on a piece of paper. It's a binding, legal document that will have important consequences for you in the real world.

Since client-drafted agreements are usually written with the client's best interests in mind, not yours, you'll almost always be better off if you use your own agreement. A client will be more willing to do this if you:

- provide a well-drafted agreement of your own (see Chapter 19), and
- point out that if the client is audited, an IRS or other government auditor will be far more impressed by an agreement you drafted than a standard form prepared by the client; using your agreement helps establish that you're an independent contractor, not the client's employee, and may help the client avoid assessments and penalties in the event of an audit.

If the client insists on using its own agreement, be sure to review the document before you accept the assignment. Read the agreement carefully and make sure you understand it and are comfortable with it before signing. If there are any provisions you don't understand, ask the client to explain them to you and rewrite them so that you do understand them.

No matter what the client may say, no agreement is engraved in stone, even if it's a "standard" agreement the client claims everybody signs. You can always request that an unfair or unduly burdensome provision should be deleted or changed. If the client refuses, you have the option of turning down the assignment or going ahead anyway, but you lose nothing by asking.

When you review a client's agreement, make sure it:

- jibes with the client's statement and your own oral statement (see Section A)
- contains all necessary provisions (see Section B), and
- does not contain unfair provisions (see Section C).

There may also be provisions you want to add. (See Section D.)

### Careless Doctor Done In by an Agreement

One emergency room physician learned the hard way that it's always necessary to carefully read and understand an agreement before signing it. The doctor, who worked on the east coast, received an offer to work as an independent contractor for a hospital in Hawaii. The doctor agreed to take the job and signed a lengthy independent contractor agreement prepared by the hospital without reading it carefully. The agreement provided a two-year term. She thought this meant she had guaranteed work for two years and this would justify the expense of moving to Hawaii.

The doctor moved to Hawaii and started work. But within three months, she had serious disagreements with hospital officials over various clinical and administrative issues. The hospital notified her that she was being terminated. She protested, pointing out that her contract was for two years. When she took a closer look, however, she discovered that the agreement included a provision allowing the hospital to terminate her if it concluded it was necessary for the emergency department to operate efficiently.

## A. Agreement Consistent With Promises

Unfortunately, some clients are in the habit of telling people they hire one thing to get them to accept an assignment, and then writing something very different in the agreements they prepare.

The language in your written agreement will be an important map in your relationship with the client. If a client says one thing to you in person and the agreement says something else, the agreement ordinarily will control. For example, if the client tells you that your work must be completed in two months, but the agreement imposes a one month deadline, the work will have to be done in one month.

For this reason, make absolutely sure the agreement meshes with what the client has told you and you have told the client. If there are differences, point them out. And if they're important differences, change the document to reflect your true agreement.

### How to Change an Agreement

If you want to delete all or part of a provision, you can simply cross it out. Minor wording changes can be written in by hand or typed. Both you and the client should sign your initials as near the deletions or additions as possible.

If you wish to add an extensive amount of new wording, it's best to retype the entire agreement to prevent it from becoming illegible or downright confusing. This is easy to do if the agreement is written on a word processor.

Another approach is to write the changes on a separate piece of paper, called an attachment, that you and the client sign. If you use an attachment, state that if there is a conflict between the attachment and the main contract, the attachment will prevail. (See Chapter 18, Section D3.)

## B. Reviewing the Contract

You should also make sure that the agreement contains all necessary provisions. At a bare minimum, it should include:

- your name and address
- the client's name and address
- the dates the contract begins and ends
- a description of the services you'll perform
- how much you'll be paid, and
- how you'll be paid.

These and other standard provisions are normally included in client agreements. If the client's agreement lacks any of these provisions, you should add them. (See Chapter 19, Section A.)

If you're creating or helping to create intellectual property—for example, writings, photos, graphics, music, software programs, designs or inventions—the agreement should also contain a clause making it clear who will own your work. (See Chapter 19, Section B7.)

## C. Provisions to Avoid

Clients' agreements sometimes contain provisions that are patently unfair to you. You should seek to delete these entirely or at least replace them with provisions that are more equitable. Examine the agreement carefully for provisions such as the following.

### 1. Indemnifying the Client

Indemnification is a fancy legal word that means a promise to repay someone for their losses or damages if a specified event occurs. Some contracts may contain indemnification provisions that require you to indemnify the client if various problems occur—for example, if a problem with your work injures a third party who sues the client. In effect, these provisions require you to act as the client's insurer.

When it comes to indemnifying the client, your rule should be to just say no. Examine the client's agreement carefully to see if it contains such a provision. If you find one, try to delete it. Indemnification clauses can be hard to spot and even harder to understand. They'll usually contain the words indemnification or hold harmless, but not always. Any provision that requires you to defend or repay the client is an indemnification provision.

#### a. Problems with the IRS and other agencies

If the IRS or another government agency determines that the client has misclassified you as an independent contractor, the client may have to pay back taxes, fines and penalties. (See Chapter 15.) Some hiring firms try to shift the risk of IRS or other penalties to the independent contractor's shoulders by including an indemnification clause in their agreements. Such provisions typically require you to repay the hiring firm for any losses suffered if you are reclassified as an employee. Here's an example of such a provision.

> If Contractor is determined to be Client's employee, Contractor shall indemnify and hold Client harmless from any and all liabilities, costs and expenses Client may incur, including attorney fees and penalties.

Do not sign an agreement that contains such a provision. The cost of fighting an IRS or other government audit and paying the possible penalties for worker misclassification can be enormous. This provision makes you responsible for paying all of these costs. If you are presented with an agreement containing such a provision, strike it out or refuse to sign the contract.

#### b. Injuries and damages arising from your services

Your work or services on the client's behalf may damage or injure third parties—that is, people other than you and the client. For example, a passerby might be injured by a dropped hammer while walking by a construction project undertaken by a self-employed building contractor. Or a software program written by an outside programmer designed to run an elevator might fail and cause injuries. It's likely that the people injured in these situations—the passerby and the elevator passengers—would sue the client to pay for the costs of their medical care and other expenses related to their injuries.

Clients often include indemnification provisions in their contracts making the people they hire responsible for all damages and injuries that other people suffer if something goes wrong with their work. Many of these provisions are so broadly written they require you to indemnify the client even if the claim is frivolous, mainly the client's fault or already covered by the client's insurance.

> **EXAMPLE:** Art, a self-employed engineer, designs and installs a new type of widget in BigCorp's factory. Art signed an agreement prepared by BigCorp's lawyers containing the following indemnification provision:
>
> Contractor shall indemnify and hold Client harmless from any and all claims, losses, actions, damages, interest, penalties, and reasonable attorney fees and costs arising by reason of Contractor's performance under this Agreement.
>
> One of BigCorp's employees alters the widget's setting without Art's permission. As a result, the widget explodes and injures a visitor at the factory who sues BigCorp. Art has to pay all of BigCorp's costs of defending the lawsuit and any damages the injured person recovers even though the explosion wasn't his fault. This is because the indemnification clause requires Art to repay BigCorp. for any claim brought by anyone that could be said to "arise" from Art's services for BigCorp.

Most indemnification clauses are even more convoluted and harder to read than the one in the example above. The wisest course is to strike out any such provision from the client's agreement.

If the client insists on keeping such a provision, you should seek to add a provision to the agreement limiting your total liability to the client to a specified dollar amount or no more than the client pays you. This will prevent you from losing everything you own or going bankrupt. (See Chapter 19, Section B5.)

Also, make sure you have enough liability insurance to cover any potential claims. Feel free to charge the client more to cover your increased insurance costs. (See Chapter 7.)

### c. Intellectual property infringement

If you create or help create intellectual property for a client, the client may seek to have you indemnify it for the costs involved if other people claim that your work infringes their copyright, patent, trade secret or other intellectual property rights.

> **EXAMPLE:** Jennifer, a self-employed computer programmer, creates a program for Acme Corp. A few months later, BigCorp., Jennifer's former employer, claims that she stole substantial portions of the program from software it owned and sues Jennifer and Acme for copyright infringement. Since Jennifer's contract with Acme contained an indemnification provision, she is legally obligated to pay Acme's attorney fees for defending the lawsuit and any money Acme may have to pay BigCorp. as damages or to settle the claim.

Intellectual property indemnity clauses are routinely included in publishing contracts, software consulting agreements and almost any other type of agreement involving the creation of intellectual property. You'd be better off without such a clause—that is, less of your own money will be at risk on the job—but it's often hard to get clients to remove them. After all, the clauses are mainly aimed at ensuring proper behavior. You shouldn't commit intellectual property infringement and clients do not to want to pay any damages if you do.

Instead of deleting the clause, your best approach may be to add a provision limiting your total liability to the client to a specified dollar amount or no more than the client pays you. (See Chapter 19, Section B5a.)

## 2. Insurance Requirements

Look carefully to see if the client's agreement contains a provision requiring you to maintain insurance coverage. Many clients want all self-employed people they hire to have extensive insurance coverage because it helps eliminate an injured person's motivation to attempt to recover from the client for fear you won't be able to pay. It's not unreasonable for a client to require you to have liability insurance. (See Chapter 6, Section B.)

However, some clients go overboard and require you to obtain an excessive amount of insurance or obtain unusual and expensive policies. For example, one self-employed courier recently contracted with a courier firm to make document deliveries using his own car. The contract included an insurance clause requiring him to obtain cargo insurance. His insurance agent told him this type of coverage was usually obtained only by trucking firms and would cost several thousand dollars per year. It was ridiculous for a document courier to be required to obtain such coverage, since it provided far more coverage than the client needed and cost far more than the courier could afford to pay. The courier simply ignored the contract and never obtained the cargo insurance.

The better practice is to delete provisions requiring excessive insurance from your contract or demand substantially more compensation to pay for the extra insurance. Make it clear to the client that you have to charge more than you usually do because it's requiring you to carry so much insurance coverage.

Following is a very reasonable provision requiring you to carry liability insurance you can add to a client's agreement in place of an unreasonable insurance clause.

**SUGGESTED LANGUAGE:**

Client shall not provide any insurance coverage for Contractor or Contractor's employees or contract personnel. Contractor agrees to maintain an insurance policy to cover any negligent acts committed by Contractor or Contractor's employees or agents while performing services under this Agreement.

## 3. Noncompetition Restrictions

Businesses that hire self-employed people sometimes want to restrict them from performing similar services for their competitors. To do this, they include a noncompetition clause in a client agreement barring them from working for competitors. Try to eliminate such provisions since they limit your ability to earn a living.

At most, you might agree to the following provision barring you from performing the same services for named competitors of the client while you're performing them for the client.

**SUGGESTED LANGUAGE:**

Contractor agrees that, while performing services required by this Agreement, Contractor will not perform the exact same services for the following competitors of Client: [LIST COMPETITORS].

## 4. Confidentiality Provisions

Many clients routinely include confidentiality provisions in their agreements. These provisions bar you from disclosing to others the client's trade secrets—for example, marketing plans, information on products under development, manufacturing techniques or customer lists. It's not unreasonable for a client to want you to keep its secrets away from the eyes and ears of competitors.

Unfortunately, however, many of these provisions are worded so broadly that they can make it difficult for you to work for other clients without fear of violating your duty of confidentiality. If, like most self-employed people, you make your living by performing similar services for many firms in the same industry, insist on a confidentiality provision that is reasonable in scope and defines precisely what information you must keep confidential. Such a provision should last for only a limited time—five years at the most.

### a. Unreasonable provision

A general provision barring you from making any unauthorized disclosure or using any technical, financial or business information you obtain directly or indirectly from the client is unreasonable. Such broad restrictions can make it very difficult for you to do similar work for other clients without violating the confidentiality clause. Here's an example of an overbroad provision.

> Contractor may be given access to Client's proprietary or confidential information while working for Client. Contractor agrees not to use or disclose such information except as directed by Client.

Such a provision doesn't make clear what information is and is not the client's confidential trade secrets, so you never know for sure what information you must keep confidential and what you can disclose when working for others.

Also, since this provision bars you from later using any of the client's confidential information to which you have access, it could prevent you from using information you already knew before working with the client. It could also bar you from using information that becomes available to the public. You would then be in the absurd position of not being allowed to use information that the whole world knows about. Always attempt to delete or rewrite such a provision.

Specifically, do not sign a contract requiring you to keep confidential any information:

- you knew about before working with the client
- you learn from a third person who has no duty to keep it confidential
- you develop independently even though the client later provides you with similar or identical information, or
- that becomes publicly known though no fault of your own—for example, you wouldn't have to keep a client's manufacturing technique confidential after it is disclosed to the public in an article in a trade journal written by someone other than you.

### b. Reasonable provision

A reasonable nondisclosure provision makes clear that, while you may not use confidential information the client provides, you have the right to freely use information you obtain from other sources or that the public learns later.

The following nondisclosure provision enables you to know for sure what material is, and is not, confidential by requiring the client to mark confidential any document you get in the course of that work. A client who tells you confidential information must later write it down and deliver it to you within 15 days.

> **SUGGESTED LANGUAGE:**
>
> During the term of this Agreement and for ____ [6 months to 5 years] afterward, Contractor will use reasonable care to prevent the unauthorized use or dissemination of Client's confidential information. Reasonable care means at least the same degree of care Contractor uses to protect its own confidential information from unauthorized disclosure.
>
> Confidential information is limited to information clearly marked as confidential, or disclosed orally and summarized and identified as confidential in a writing delivered to Contractor within 15 days of disclosure.
>
> Confidential information does not include information that:
>
> - the Contractor knew before Client disclosed it
> - is or becomes public knowledge through no fault of Contractor
> - Contractor obtains from sources other than Client who owe no duty of confidentiality to Client, or
> - Contractor independently develops.

### 5. Unfair Termination Provisions

Your agreement can always be terminated if you or the client breaches one of its major terms—for example, you seriously fail to satisfy the project specifications. Ordinarily, however, neither you nor the client can terminate the agreement just because you feel like it. Some clients add termination provisions to their contracts allowing them to terminate the agreements at will—that is, for any reason or no reason at all. For example, the agreement may provide that the client has the right to terminate the agreement on ten days written notice.

If you sign an agreement with such a provision, you lose the security of knowing the client must allow you to complete your assignment and pay you for it provided you live up to the terms of your agreement. Instead, you can be fired at any time, just like an employee.

If the client insists on such a provision, fairness dictates that it be mutual—that is, you should have the same termination rights as the client. The client should also be required to give you reasonable notice of the termination. How long the notice should be depends on the length of the project. For lengthy projects, 30 days notice may be appropriate. For short projects, it may make sense to require just a few days notice. Finally, the agreement should make clear that the client must pay you for all the work you performed prior to termination. (See Chapter 19, Section A7a.)

### 6. Time of the Essence

Examine the client's agreement carefully to see if it contains the phrase "time is of the essence." You'll often find such clauses in the portion of the contract dealing with the project deadlines.

These simple words can have a big legal impact. Ordinarily, a delay in performance of your contractual obligations is not considered important enough to constitute a material breach of the agreement. This means the client can sue you for any damages sustained due to your lateness, but is not entitled to terminate the contract.

**EXAMPLE:** Barney, a construction contractor, contracts to build a new wing on the AAA Motel. The contract provides that the wing is to be completed by April 1. Barney completes the new wing four weeks late. AAA may sue him for any damages caused by the delay. But since the contract lacks a time is of the essence clause, it may not terminate the contract and is legally obligated to pay Barney the contract price.

However, if the contract includes a time is of the essence provision, most courts hold that even a slight delay in performance will constitute a material breach. The client cannot only sue you for damages, but can terminate the contract. This means the client need not perform its contractual obligations—for example, the client need not pay you.

**EXAMPLE:** Assume that Barney's contract in the above example included a time is of the essence clause. This would mean that the AAA Motel was legally entitled to terminate the contract when Barney missed the completion deadline and sue Barney for breach of contract.

If you want to be able to have a little slippage in your deadlines, delete any time is of the essence clause from the client's agreement.

## D. Provisions You May Wish to Add

There are a number of provisions that benefit you that you may wish to add to the client's agreement. These provisions are all discussed in detail in Chapter 19 and include:

- requiring mediation and arbitration of disputes (see Chapter 19, Section B1)
- recognizing that the contract may have to be modified in the future and providing a mechanism to do so (see Chapter 19, Section B2)
- allowing you to obtain attorney fees if you sue the client and win (see Chapter 19, Section B3)
- requiring the client to pay a late fee for late payments (see Chapter 19, Section B4)
- limiting your liability to the client if something goes wrong (see Chapter 19, Section B5), and
- restricting the client's ability to assign its benefits or delegate its duties under the agreement (see Chapter 19, Section B8).

## E. Client Purchase Orders

Not all clients use standard contracts. Instead, you may be handed a pre-printed form that looks very different from the contracts in this book. Such a form is usually called a purchase order. Some clients use purchase orders instead of, or in addition to, standard contracts.

A purchase order is an internal form developed by a client authorizing you to perform work and bill for it. Typically, purchase orders are used by larger companies that have separate accounting departments. Accounting departments often don't want to have to deal with lengthy or confusing client agreements.

Purchase orders are designed to provide the minimum information a company needs to document the services you'll perform and how much you'll be paid. They typically contain much the same information as a letter agreement: a description of the services you'll perform, payment terms and deadlines. (See Chapter 18.) The order should be signed by the client. You should include the purchase order number on your invoices and all correspondence with the client.

Some companies use purchase orders in conjunction with standard contracts or letter agreements. That is, either you or the client will prepare a contract or letter agreement and the client will also prepare a purchase order. In this event, make sure the terms of the purchase order are consistent with your client agreement.

Other companies use purchase orders alone for small projects because they don't want to go to the trouble of drafting a client agreement. Make certain the purchase order is properly filled out. This should include an accurate description of the services you'll perform, the due date and the terms of payment.

Some companies' accounting departments will not pay you unless you have a signed purchase order. This is so even though you have a signed letter agreement or standard client agreement.

Before you start work for a client, find out if it uses purchase orders. If it does, insist on being provided a signed order before you start work.

Following is an example of a typical purchase order for services.

**PURCHASE ORDER**

Acme, Inc.

P.O. #: 123

Vendor: Gerard & Associates
123 Solano Avenue
Berkeley, CA 99999
510-555-5555

Date: 8/1/9_

Delivery Date: 9/1/9_

Terms: 18¢ per word translated. Total price not to exceed $6,084.

Description of Services: Contractor will translate Acme instruction manual from the English language into idiomatic Russian using the Cyrillic alphabet. The translated material will be provided in Word Perfect format. Contractor will provide one disk copy and one printed copy of all translated material.

Authorized by: ______________________________
Joe Jones

■

CHAPTER

21

# Help Beyond This Book

If a client claims that your work doesn't meet the contract specifications, insists that you've missed a deadline or fails to pay you for any reason, you've got a potential legal dispute on your hands. The way you handle such disputes will have a big impact on your self-employment success. The best way to handle them is usually through informal negotiations. Call or meet with the client and talk out your differences. If you reach a settlement, promptly write it down and have all involved sign it.

**EXAMPLE:** Gwen, a self-employed computer programmer, is hired by Acme Corp. to create a custom software program for $10,000. Gwen delivers the program on time. However, Acme refuses to pay Gwen because it claims the program doesn't fully live up to the contract specifications. Gwen meets with Acme's president and admits that the program doesn't do everything it's supposed to do.

However, she demonstrates that the program fulfills at least 80% of Acme's requirements. Gwen offers, therefore, to accept $8,000 as full payment instead of the $10,000 stated as full payment in the original contract. Acme's president agrees. Gwen drafts a short agreement stating that Acme will pay her $8,000 in full settlement of their dispute.

## A. Help Resolving Disputes

If informal negotiations don't work, you have a number of options:

- alternative dispute resolution, which includes mediation and arbitration, and
- filing a lawsuit in court.

This chapter discusses those options in detail. It also provides guidance on how to find and use more specific legal resources, including lawyers and other knowledgeable experts. And finally, it explains the basics of doing your own legal research.

### 1. Mediation and Arbitration

Mediation and arbitration—often lumped together under the term alternative dispute resolution, or ADR—are two methods for settling disputes without resorting to expensive court litigation. People often confuse the two, but they are in fact very different. Mediation is always nonbinding on the participants, whereas arbitration usually is binding and often takes the place of a court action.

#### a. Mediation

In mediation, a neutral third person called a mediator meets with the people involved in the dispute and makes suggestions as to how to resolve their controversy. Typically, the mediator either sits both sides down together and tries to provide an objective view of their dispute, or shuttles between them as a hopefully cool conduit for what may be hot opinions.

Where the underlying problem is actually a personality conflict or simple lack of communication, a good mediator can often help those involved in the dispute find their own compromise settlement. Where the argument is more serious, a mediator may at least be able to lead them to a mutually satisfactory ending of both the dispute and their relationship that will obviate time-consuming and expensive litigation.

If you've ever had a dispute with a friend or relative that another friend or relative helped resolve by meeting with you both and helping you talk things over, you've already been through a process very like mediation.

Mediation is nonbinding. That means that if either person involved in the dispute doesn't like the outcome of the mediation, he or she can ask for binding arbitration or go to court.

### b. Arbitration

If those involved in a dispute cannot resolve it by mediation, they often submit it to arbitration. The arbitrator—again, a neutral third person—is either selected directly by those involved in the dispute or is designated by an arbitration agency.

An arbitrator's role is very different from that of a mediator. Unlike a mediator who seeks to help the parties resolve their dispute themselves, an arbitrator personally imposes a solution.

The arbitrator normally hears both sides at an informal hearing. You can be represented by a lawyer at the hearing, but it's not required. The arbitrator acts as both judge and jury: after the hearing, he or she issues a decision called an award. The arbitrator follows the same legal rules a judge or jury would follow in deciding whether you or the other side has a valid legal claim and should be awarded money.

Arbitration can be either binding or nonbinding. If arbitration is nonbinding, either person named in the award can take the matter to court if he or she doesn't like the outcome. Binding arbitration is usually final. You can't go to court and try the dispute again if you don't like the arbitrator's decision—except in unusual cases where you can show the arbitrator was guilty of fraud, misconduct or bias. In effect, binding arbitration takes the place of a court trial.

If the losing party to a binding arbitration doesn't pay the money required by an arbitration award, the winner can easily convert the award into a court judgment which can be enforced just like any other court judgment—in other words, a binding arbitration award is just as good as a judgment you could get from a court.

#### FINDING A MEDIATOR OR ARBITRATOR

It is usually up to you and the client to decide who should serve as a mediator or arbitrator. You can normally choose anyone you want unless your contract restricts your choice.

It can be a professional mediator or arbitrator or just someone you both respect. A professional organization may be able to refer you to a good mediator or arbitrator. Businesses often use private dispute resolution services that maintain a roster of mediators and arbitrators—often retired judges, attorneys or business people with expertise in a particular field. Two of the best known of these services are:

- **The American Arbitration Association.** This is the oldest and largest private dispute resolution service, with offices in most major cities. It handles both mediations and arbitrations. The main office is in New York City: 212-484-4000. The American Arbitration Association also has a very informative site on the Internet at www.adr.org.
- **Judicate.** This Philadelphia-based firm emphasizes mediation and handles disputes in all 50 states. It can be reached at 800-631-9900.

You can obtain a directory of arbitration and mediation organizations called the *Dispute Resolution Directory* from the American Bar Association. The full directory costs $63, but the ABA will send you the pages for your city free of charge. Contact the American Bar Association Standing Committee on Dispute Resolution, 1800 M Street, NW, Suite 290N, Washington, DC 20036; 202-331-2258.

### c. Agreeing to mediation and arbitration

No one can be forced into arbitration or mediation; you must agree to it, either in the contract or later when a dispute arises. Business contracts today commonly include an arbitration provision and many also require mediation. This is primarily because of two truths: speed and cost.

Mediation and arbitration are usually much faster than court litigation. Most arbitrations and mediations are concluded in less than six months. Court litigation often takes years.

No one can ever tell how much a court case will cost, but it's usually a lot. Business lawyers typically charge from $150 to $250 per hour. Unless the amount of money involved is small and the case can be tried in small claims court (see Section A2c), arbitration and mediation are usually far cheaper than a lawsuit. A private dispute resolution company will typically charge about $500 to $1,000 for a half-day of arbitration or mediation.

For detailed guidance on mediation and arbitration, see *How to Mediate Your Dispute*, by Peter Lovenheim (Nolo Press).

## 2. Filing a Lawsuit

If your attempts to settle the dispute through informal negotiations or mediation fail and the client won't agree to binding arbitration, your remaining alternative is to sue the client in court. Or you or the client may choose to skip informal negotiations or arbitration altogether and immediately go to court.

Most legal disputes between self-employed people and clients involve a breach of contract. A person who fails to live up to the terms of a contract is said to have breached it. In a typical breach of contract case, the person who sues—called the plaintiff—asks the judge to issue a judgment against the person being sued—called the defendant. Usually, the plaintiff wants money, also known as damages.

### a. What you need to prove

Proving a breach of contract case is not complicated. You must first show that the contract existed. If it is written, the document itself should be presented to the court. If the contract is oral, you'll have to testify as to its terms. You must also show that you did everything you were required to do under the contract.

You must then make clear how the client breached the contract and clearly show the amount of damages you have suffered as a result. In many situations, this amounts to no more than showing that the client committed itself to buy certain services from you, that you provided those services, and that the client has not paid a legitimate bill for a stated number of dollars.

**EXAMPLE:** Ted, a self-employed graphic designer, contracted with the Acme Sandblasting Company to redesign its logo and newsletter. Ted completed the work, but Acme refused to pay him. After informal negotiations failed, Ted sued Acme in court. To win his case, Ted should produce the written contract with Acme, a decent-looking sample of the redesigned newsletter and a letter from someone with expertise in the field stating that the work met or exceeded industry standards. Ted would also be wise to try to rebut the likely points the client might make. For example, if the design work was a few weeks late, Ted would want to present a good excuse, such as the fact that Acme asked for time-consuming changes.

### b. What you can sue for

If you sue a client for breach of contract, forget about collecting the huge awards you hear about people getting when they're injured in accidents and sue for personal injuries. Damages for breach of contract are strictly limited by law. As a general rule, you'll get just enough to compensate you for your direct economic loss—that is, the amount of money you lost because the client failed to live up to its promises. For example, if a client promised to pay you $1,000 for your services, but failed to pay after you performed them, you'd be entitled to $1,000 in damages.

You can't get punitive damages—special damages designed to punish wrongdoers—or damages to compensate you for your emotional pain or suffering, even though clients who breach their contracts can really be a pain. On the bright side, however, if a client sues you for breach of contract, damages are limited in the same way.

### c. Small claims court

Most business contract suits are filed in state court. All states have a special court especially designed to handle disputes where only a small amount of money is involved. These are called small claims courts. The amount for which you can sue in small claims court varies from state to state—but usually ranges between $2,500 to $5,000.

Small claims court has the same advantages as arbitration: it's usually fast and inexpensive. You don't need a lawyer to go to small claims court. Indeed, some states don't allow lawyers to represent people in small claims court.

Small claims court is particularly well suited to help you collect against clients who fail to pay you, provided the amount is relatively small. (See Chapter 7, Section B4.)

For detailed guidance on how to represent yourself in small claims court, see *Everybody's Guide to Small Claims Court*, by Ralph Warner, National and California editions (Nolo Press).

### d. Suing in other courts

If your claim exceeds the small claims court limit for your state, you'll need to file your lawsuit in another court. Most business lawsuits are handled in state courts. Every state has its own trial court system with one or more courts that deal with legal disputes between people and businesses. These courts are more formal than small claims courts and the process usually takes longer. You may be represented by a lawyer, but you don't have to be. Many people have successfully handled their own cases in state trial courts.

For detailed guidance on how to represent yourself in court, see:

- *Represent Yourself in Court: How to Prepare and Try a Winning Case*, by Paul Bergman and Sara Berman-Barrett (Nolo Press). This book explains how to handle a civil case yourself, without a lawyer, from start to finish.
- *How to Sue for Up to $25,000 . . . and Win!*, by Judge Roderic Duncan (Nolo Press). This books shows you how to handle claims up to $25,000 in California courts.

## B. Finding and Using a Lawyer

An experienced attorney may help answer your questions and allay your fears about setting up and running your business. Many different areas of law are involved when you're an self-employed, including:

- federal tax law
- state tax law
- contract law, and
- general business law.

Fortunately, there are attorneys who specialize in advising small businesses. These lawyers are a bit like general practitioner doctors: they know a little about a lot of different areas of law. A lawyer with plenty of experience working with businesses like yours should be able to answer your questions.

Such a lawyer can help you:

- start your business—review incorporation documents, for example
- analyze zoning ordinances, land use regulations and private title documents that may restrict your ability to work at home
- review client agreements
- coach or represent you in lawsuits or arbitrations where the stakes are high or the legal issues complex
- deal with intellectual property issues—copyrights, trademarks, patents, trade secrets and business names, and
- look over a proposed office lease.

### 1. Finding a Lawyer

When you begin looking for a lawyer, try to find someone with experience representing businesses similar to yours. It's usually not wise to start your search by consulting phone books, legal directories or advertisements. Lawyer referral services operated by bar associations are usually equally unhelpful. Often, they simply supply the names of lawyers who have signed onto the service, accepting the lawyer's own word for what types of skills he or she has.

The best way to locate a lawyer is through referrals from other self-employed people in your community. Industry associations and trade groups are also excellent sources of referrals. If you already have or know a lawyer, he or she might also be able to refer you to an experienced person who has the qualifications you need. Other people, such as your banker, accountant or insurance agent, may know of good business lawyers.

### 2. Paying a Lawyer

Whenever you hire a lawyer, insist upon a written explanation of how the fees and costs will be paid.

Most business lawyers charge by the hour. Hourly rates vary, but in most parts of the United States, you can get competent services for your business for $150 to $250 an hour. Comparison shopping among lawyers will help you avoid overpaying. But the cheapest hourly rate isn't necessarily the best. A novice who charges only $80 an hour may take three hours to review a consulting contract. A more experienced lawyer who charges $200 an hour may do the same job in half an hour and make better suggestions. If a lawyer will be delegating some of the work on your case to a less experienced associate, paralegal or secretary, that work should be billed at a lower hourly rate. Be sure to get this information recorded in your initial written fee agreement.

Sometimes, a lawyer may quote you a flat fee for a specific job. For example, a lawyer may offer to incorporate your business for a flat fee of $2,000. You pay the same amount regardless of how much time the lawyer spends. This can be cheaper than paying an hourly fee, but not always.

Alternatively, some self-employed people hire lawyers on retainer—that is, they pay a flat annual fee in return for the lawyer handling all their routine legal business. However, few small businesses can afford to keep a lawyer on retainer.

**Using a Lawyer as a Legal Coach**

One way to keep your legal costs down is to do as much work as possible yourself and simply use the lawyer as your coach. For example, you can draft your own agreements, giving your lawyer the relatively quick and inexpensive task of reviewing them.

But get a clear understanding about who's going to do what. You don't want to do the work and get billed for it because the lawyer duplicated your efforts. And you certainly don't want any crucial elements to fall through cracks because you each thought the other person was attending to the work.

## C. Help From Other Experts

Lawyers aren't the only ones who can help you deal with the legal issues involved in being self-employed. Tax professionals, members of trade groups and the Small Business Administration can also be very helpful.

### 1. Tax Professionals

Tax professionals include tax attorneys, certified public accountants and enrolled agents. Tax pros can answer your tax questions and help you with tax planning, preparing your tax returns and dealing with IRS audits. (See Chapter 8, Section B2.)

### 2. Industry and Trade Associations

Business or industry trade associations or similar organizations can be useful sources of information and services. Many such groups track federal and state laws, lobby Congress and state legislatures and even help members deal with the IRS and other federal and state agencies. Many also offer their members insurance and other benefits and have useful publications.

There are hundreds of such organizations representing every conceivable occupation—for example, the American Society of Home Inspectors, the Association of Independent Video and Filmmakers and the Graphic Artists Guild. There are also national membership organizations that allow all types of self-employed people to join—for example, the National Association of the Self Employed and the Home Office Association of America.

If you don't know the name and address of an organization you may be eligible to join, ask other self-employed people. Or check out the *Encyclopedia of Associations* (Gale Research); it should be available in your public library. Also, many of these organizations have Web sites on the Internet, so you may be able to find the one you want by doing an Internet search. (See Section E.)

### 3. Small Business Administration

The U.S. Small Business Administration, or SBA, is an independent federal agency that helps small businesses. The SBA is best known for providing loan guaranties to bolster small businesses that want to start or expand, but it provides several other useful services for small business, including:

- **SBA Answer Desk.** The Answer Desk is a nationwide, toll-free information center that helps callers with questions and problems about starting and running businesses. Service is provided through a computerized telephone message system augmented by staff counselors. It is available 24 hours a day, seven days a week, with counselors available Monday through Friday, 9 am to 5 pm Eastern Time. The Answer Desk can be reached at 800-8-ASK-SBA.
- **Publications.** The SBA also produces and maintains a library of publications, videos and computer programs. These are available by mail to SBA customers for a nomi-

nal fee. A complete listing of these products is in the *Resource Directory for Small Business Management.* SBA field offices also offer free publications that describe SBA programs and services.

- **SBA Internet site.** You can download SBA publications from the SBA Internet site and obtain information about SBA programs and services, points of contact and calendars of local events. The Internet address is www.sbaonline.sba.gov.
- **SCORE program.** The Service Corps of Retired Executives, or SCORE, is a group of retired business people who volunteer to help others in business. To obtain more information, call the SBA Answer Desk at 800-8-ASK-SBA or call or visit your local SBA office.

The SBA has offices in all major cities. Look in the phone book under U.S. Government for the office nearest you.

## D. Doing Your Own Legal Research

If you decide to investigate the law on your own, your first step should be to obtain a good guide to help you understand legal citations, use the law library and understand what you find there. There are a number of sources that provide a good introduction to legal research, including:

- *Legal Research: How to Find and Understand the Law*, by Stephen Elias and Susan Levinkind (Nolo Press). This book simply explains how to use all major legal research tools and helps you frame your research questions
- *Legal Research Made Easy: A Roadmap Through the Law Library Maze*, by Robert C. Berring (Legal Star/Nolo Press). This is a videotape with a six-step strategy for legal research. It's available from many public and law library video collections—or directly through Nolo Press.

Next, you need to find a law library that's open to the public. Your county should have a public law library, often at the county courthouse. Public law schools often contain especially good collections and generally permit the public to use them. Some private law schools grant access to their libraries—sometimes for a modest fee. The reference department of a major public or university library may have a fairly decent legal research collection. Finally, don't overlook the law library in your own lawyer's office. Many lawyers will agree to share their books with their clients.

### 1. Researching Federal Tax Law

Many resources are available to augment and explain the tax information in this book. Some are free, and others are reasonably priced. Tax publications for professionals are expensive, but are often available at public libraries or law libraries.

#### a. IRS Web site

Somewhat surprisingly, the IRS has perhaps the most useful and colorful Internet site of any government agency. It contains virtually every IRS publication and tax form, IRS announcements and a copy of the new *IRS Audit Manual on Independent Contractors.* It's almost worth getting on the Internet just to use this site. The Internet address is www.irs.ustreas.gov.

#### b. IRS booklets

The IRS also publishes over 350 free booklets explaining the tax code, and many are clearly written and useful. These IRS publications range from several pages to several hundred pages in length. Many of the most useful IRS publications are cited in the tax chapters in this book.

The following IRS publications cover basic tax information that every IC should know about:

- Publication 334, *Tax Guide for Small Businesses*
- Publication 505, *Tax Withholding and Estimated Tax*
- Publication 937, *Employment Taxes and Information Returns,* and
- Publication 15, *Circular E, Employer's Tax Guide.*

IRS publications are available in IRS offices, by calling 800-TAX-FORM or by sending in an order form. They can also be downloaded from the IRS's Internet site at www.irs.ustreas.gov.

**Don't Rely Exclusively on the IRS**

IRS publications are useful to obtain information on IRS procedures and to get the agency's view of the tax law. But keep in mind that these publications only present the IRS's interpretation of the law, which may be very one-sided and even be contrary to court rulings. Don't rely exclusively on IRS publications for information.

## c. IRS telephone information

The IRS offers a series of prerecorded tapes of tax topics on a toll-free telephone service called TELETAX. Call 800-829-4477. See IRS Publication 910 for a list of topics.

You can talk to an IRS representative at 800-829-1040, but expect difficulty getting though from January through May. Doublecheck anything an IRS representative tells you over the phone because the IRS is notorious for giving misleading or outright wrong answers to taxpayers' questions. The IRS does not stand behind oral advice that turns out to be incorrect.

## d. Directories of IRS rulings

The IRS has issued thousands of rulings on how workers in every conceivable occupation should be classified for tax purposes. Tax experts have collected and categorized them by occupation. By using these publications, you can find citations to IRS rulings involving workers similar to you. Two such publications are available:

- *Employment Status—Employee v. Independent Contractor*, 391-2nd T.M., by Helen Marmoll, summarizes and provides citations to IRS rulings on classification of workers in 374 different occupations—everything from accountants to yacht sales agents. This guide is expensive, around $80. You may be able to find a copy in a good law library; if not, it may be a good investment for you. It's published by Tax Management Inc. Call 800-372-1033.
- Many state Chambers of Commerce publish guides for companies that hire ICs that list IRS rulings for various occupations. For example, the California Chamber of Commerce publishes a guide called *Independent Contractors: A Manager's Guide and Audit Reference.* It categorizes thousands of IRS rulings by occupation and also covers rulings by California state agencies. Call your state Chamber of Commerce to see if it publishes such a guide for your state; it probably has an office in your state capitol.

An industry trade group or association may also be aware of, or even have copies of, helpful IRS rulings and court decisions.

## e. Tax guides

Dozens of privately-published self-help tax guides are available. The most detailed and authoritative are:

- *Master Tax Guide* (Commerce Clearing House)
- *Master Federal Tax Manual* (Research Institute of America), and
- *Federal Tax Guide* (Prentice-Hall).

You can find these in many public libraries.

## 2. Researching Other Areas of Law

Many fields of law other than federal tax law are involved when you're self-employed. For example, your state laws may control how you form a sole proprietorship or corporation, protect trade names, form contracts and resolve disputes.

If you have questions about your state workers' compensation, tax law or employment laws, first contact the appropriate state agency for more information. Many of these agencies publish informative pamphlets. See the Appendix of this book for lists of state workers' compensation agencies and sales and income tax departments.

In-depth research into your state law will require that you review:

- legislation, also called statutes, passed by your state legislature
- administrative rules and regulations issued by state administrative agencies such as your state tax department and unemployment compensation agency, and
- published decisions of your state courts.

Many states, particularly larger ones, have legal encyclopedias or treatises that organize summaries of state case law and some statutes alphabetically by subject. Through citation footnotes, you can locate the full text of the cases and statutes. These works are a good starting point for in-depth state law research.

It's also helpful if you can find a treatise on the subject you're researching. A treatise is a book that covers a specific area of law. The West Publishing Company publishes a series of short paperback treatises called the Nutshell Series. If you are facing a possible contract dispute, you may want to look at *Contracts in a Nutshell*, by Gordon A. Schaber and Claude D. Rohwer, and *Corporations in a Nutshell*, by Robert Hamilton.

A relatively unknown resource for quickly locating state business laws is the United States Law Digest volume of the *Martindale-Hubbel Law Directory*. It contains a handy summary of laws for each state. Dozens of business law topics are covered, including Corporations, Insurance, Leases, Statute of Frauds and Trademarks, Trade Names and Service Marks. The *Martindale-Hubbel Law Directory* is in most public libraries.

# E. Online Resources

The online world includes the Internet, commercial online services such as America Online and CompuServe and specialized computer databases such as Westlaw and Lexis. All contain useful information for the computer-savvy self-employed.

## 1. Internet Resources

A vast array of information for small business owners is available on the Internet, a global network of computer networks. To get access to the Internet, you need a computer and modem, appropriate software and an account with an Internet access provider. You can get Internet access free in many university and some public libraries.

## How to Read a Case Citation

To locate a published court decision, you must understand how to read a case citation. A citation provides the names of the people or companies involved on each side of the case, the volume of the legal publication—called a reporter—in which the case can be found, the page number on which it begins and the year in which the case was decided. Here is an example of what a legal citation looks like: *Smith v. Jones Int'l,* 123 F.3d 456 (1995). Smith and Jones are the names of the people having the legal dispute. The case is reported in volume 123 of the Federal Reporter, Third Series, beginning on page 456; the court issued the decision in 1995.

**Federal court decisions.** There are several different federal courts and the decisions of each are published in a different reporter. Opinions by the federal district courts are in a series called the Federal Supplement, or F.Supp.

Any case decided by a federal court of appeals is found in a series of books called the Federal Reporter. Older cases are contained in the first series of the Federal Reporter, or F. More recent cases are contained in the second or third series of the Federal Reporter, F.2d or F.3d.

Cases decided by the U.S. Supreme Court are found in three publications: United States Reports (identified as U.S.), the Supreme Court Reporter identified as S.Ct.) and the Supreme Court Reports, Lawyer's Edition (identified as L.Ed.). Supreme Court case citations often refer to all three publications.

There are also federal courts that specialize in handling tax disputes, including the United States Tax Court and United States Claims Court—formerly Court of Claims. Published decisions of the United States Tax Court can be found in the Tax Court Reports, or TC, published by the U.S. Government Printing Office. Tax Court decisions can also be found in a reporter called Tax Court Memorandum Decisions, or TCM, published by Commerce Clearing House, Inc.

Decisions from all federal courts involving taxation can be found in a reporter called U.S. Tax Cases, or USTC, published by Commerce Clearing House, Inc.

**State court decisions.** Most states publish their own official state reports. All published state courts' decisions are also included in the West Reporter System. West has divided the country into seven regions—and publishes all the decisions of the supreme and appellate state courts in the region together. These reporters are:

A. and A.2d. Atlantic Reporter (First and Second Series), which includes decisions from Connecticut, Delaware, the District of Columbia, Maine, Maryland, New Hampshire, New Jersey, Pennsylvania, Rhode Island and Vermont.

N.E. and N.E.2d. Northeastern Reporter (First and Second Series), which includes decisions from New York, Illinois, Indiana, Massachusetts and Ohio.

N.W. and N.W.2d. Northwestern Reporter (First and Second Series), which includes decisions from Iowa, Michigan, Minnesota, Nebraska, North Dakota, South Dakota and Wisconsin.

P. and P.2d. Pacific Reporter (First and Second Series), which includes decisions from Alaska, Arizona, California, Colorado, Hawaii, Idaho, Kansas, Montana, Nevada, New Mexico, Oklahoma, Oregon, Utah, Washington and Wyoming.

S.E. and S.E.2d. Southeastern Reporter (First and Second Series), which includes decisions from Georgia, North Carolina, South Carolina, Virginia and West Virginia.

So. and So.2d. Southern Reporter (First and Second Series), which includes decisions from Alabama, Florida, Louisiana and Mississippi.

S.W. and S.W.2d. Southwestern Reporter (First and Second Series), which includes decisions from Arkansas, Kentucky, Missouri, Tennessee and Texas.

All California appellate decisions are published in a separate volume, the California Reporter (Cal. Rptr.) and all decisions from New York appellate courts are published in a separate volume, New York Supplement (N.Y.S.).

There are hundreds of Internet sites dealing with small business issues such as starting a small business, marketing and business opportunities. Beware, however, that no one checks these sites for accuracy. A good way to find these sites is through an Internet directory such as Yahoo. You can access Yahoo at www.yahoo.com. Click on the Business and Economy category and then on the Small Business Information listing.

A growing number of court decisions are also available on the Internet for free or at nominal cost. Finding these decisions can be difficult, however. For detailed guidance, see *Law on the Net,* by James Evans (Nolo Press).

**NOLO PRESS INTERNET SITE**

Nolo Press maintains an Internet site that is useful for the self-employed. The site contains helpful articles, information about new legislation, book excerpts and the Nolo Press catalog. The site also includes a legal encyclopedia with specific information for people who are self-employed. The Internet address is www.nolo.com.

Yet another part of the Internet are Usenet newsgroups. These are collections of electronic-mail messages, called postings, on specific topics that can be read by anybody with access to the Internet. Most newsgroups are completely open—meaning anybody can just jump into the discussion by posting anything they want—although users are usually encouraged to keep to the topic of the newsgroup. Other newsgroups are moderated, meaning that there is a moderator who reviews postings sent before allowing them to appear in that newsgroup. Moderated newsgroups almost always contain more focused discussion since the moderators want to keep the conversation on the track.

You can use newsgroups to network with other self-employed people, ask specific questions and even find work. Some of the many newsgroups of interest to the self-employed include:

- misc.taxes.moderated
- misc.jobs.contract
- misc.entrepreneurs
- misc.business.consulting
- alt.computer.consultants, and
- alt.computer.consultants.moderated.

If you don't know how to access Internet newsgroups, review the instructions for your Web browser for guidance.

## 2. Commercial Online Services

Some of the best known parts of the online world are commercial online services such as CompuServe, Prodigy and America Online. To access these systems, a person must become a subscriber and pay a monthly and hourly fee. These systems typically offer online chats with other users logged onto the system, posting of public messages on various topics and vast collections of electronic databases. All of these services have special areas devoted to small business people and consultants. For example, CompuServe has a Working at Home Forum, and America Online has a Small Business area. You can also obtain information on taxes and download copies of IRS tax forms.

### Westlaw and Lexis

One quick way to find IRS rulings and many court decisions is to use computer databases. There are two main commercially owned legal databases: Westlaw and Lexis. Unfortunately, if you are not connected with a large law office or law school, you will likely find it difficult to get access to either of these systems. A small but growing number of public law libraries offer these services. Those that do offer them usually require a sizable advance deposit or a credit card; you pay as you go.

If you wish to use one of these systems, either ask a law librarian where the nearest publicly accessible terminal is located, or write to the companies at the addresses below. If you do find an available service, be prepared to pay as much as $300 per hour.

For more information about Westlaw, write to: West Publishing Co., 50 West Kellogg Boulevard, P.O. Box 3526, St. Paul, MN 55165. For Lexis, write to: Lexis, Mead Data Central, 200 Park Avenue, New York, NY 10017.

■

APPENDIX

1

# Documents and Agencies

**Documents**

- Asset Log
- Expense Journal
- Income Journal
- Invoice Form
- Application for Employer Identification Number—Form SS4
- Request for Taxpayer Identification Number and Certification—Form W-9

**Agencies**

- State Offices Providing Small Business Help
- State Unemployment Tax Agencies
- State Sales Tax Agencies
- Patent and Trademark Depository Libraries
- State Trademark Agencies and Statutes

# ASSET LOG

| Description of Property | Date Placed in Service | Cost or Other Basis | Business/ Investment Use % | Section 179 Deduction | Depreciation Prior Years | Basis for Depreciation | Method/ Convention | Recovery Period | Rate or Table% | Depreciation on Deduction |
|---|---|---|---|---|---|---|---|---|---|---|
| | | | | | | | | | | |
| | | | | | | | | | | |
| | | | | | | | | | | |
| | | | | | | | | | | |
| | | | | | | | | | | |
| **TOTAL** | | | | | | | | | | |

# EXPENSE JOURNAL

EXPENSE JOURNAL

| | | | | 1 | 2 | 3 |
|---|---|---|---|---|---|---|
| Date | Check No. | Transaction | Amount | Advertising | Supplies, Postage, Etc. | Outside Contractors |
| | | | | | | |
| | | | | | | |
| | | | | | | |
| | | | | | | |
| | | | | | | |
| | | | | | | |
| | | | | | | |
| | | | | | | |
| | | | | | | |
| | | | | | | |
| | | | | | | |
| | | | | | | |
| | | | | | | |
| | | | | | | |
| | | | | | | |
| | | | | | | |
| | | | | | | |
| | | | | | | |
| | | | | | | |
| | | | | | | |
| | | | | | | |
| | | | | | | |
| | | | | | | |
| | | | | | | |
| | | | | | | |
| | | | | | | |
| | | | | | | |
| | | | | | | |
| | | | | | | |
| | | | | | | |
| **TOTAL THIS PAGE** | | | | | | |
| **TOTAL YEAR TO DATE** | | | | | | |

| 4 | 5 | 6 | 7 | 8 | 9 | 10 | 11 |
|---|---|---|---|---|---|---|---|
| **Travel** | **Equipment** | **Rent** | **Utilities** | **Meals and Entertainment** | | | **Misc.** |
| | | | | | | | |
| | | | | | | | |
| | | | | | | | |
| | | | | | | | |
| | | | | | | | |
| | | | | | | | |
| | | | | | | | |
| | | | | | | | |
| | | | | | | | |
| | | | | | | | |
| | | | | | | | |
| | | | | | | | |
| | | | | | | | |
| | | | | | | | |
| | | | | | | | |
| | | | | | | | |
| | | | | | | | |
| | | | | | | | |
| | | | | | | | |
| | | | | | | | |
| | | | | | | | |
| | | | | | | | |
| | | | | | | | |
| | | | | | | | |
| | | | | | | | |
| | | | | | | | |
| | | | | | | | |
| | | | | | | | |
| | | | | | | | |
| | | | | | | | |
| | | | | | | | |

# INCOME JOURNAL

| Source | Invoice | Date | Amount |
|---|---|---|---|
| | | | |
| | | | |
| | | | |
| | | | |
| | | | |
| | | | |
| | | | |
| | | | |
| | | | |
| | | | |
| | | | |
| | | | |
| | | | |
| | | | |
| | | | |
| | | | |
| | | | |
| | | | |
| | | | |
| | | | |
| | | **TOTAL** | |

# INVOICE

________________________________

________________________________

________________________________

________________________________

Date: ________________________________

Invoice Number: ________________________________

Your Order Number: ________________________________

Terms: ________________________________

Time period of: ________________________________

To: ________________________________

________________________________

________________________________

________________________________

Services: ________________________________

________________________________

________________________________

Material Costs: ________________________________

Expenses: ________________________________

**TOTAL AMOUNT OF THIS INVOICE**: ________________________________

Signed by: ________________________________

Form **SS-4**

(Rev. December 1995)

Department of the Treasury
Internal Revenue Service

# Application for Employer Identification Number

(For use by employers, corporations, partnerships, trusts, estates, churches, government agencies, certain individuals, and others. See instructions.)

► Keep a copy for your records.

EIN

OMB No. 1545-0003

Please type or print clearly.

1 Name of applicant (Legal name) (See instructions.)

2 Trade name of business (if different from name on line 1)

3 Executor, trustee, "care of" name

4a Mailing address (street address) (room, apt., or suite no.)

5a Business address (if different from address on lines 4a and 4b)

4b City, state, and ZIP code

5b City, state, and ZIP code

6 County and state where principal business is located

7 Name of principal officer, general partner, grantor, owner, or trustor—SSN required (See instructions.) ►

8a Type of entity (Check only one box.) (See instructions.)

- ☐ Sole proprietor (SSN)
- ☐ Partnership
- ☐ REMIC
- ☐ State/local government
- ☐ Other nonprofit organization (specify) ►
- ☐ Other (specify) ►
- ☐ Personal service corp.
- ☐ Limited liability co.
- ☐ National Guard
- ☐ Estate (SSN of decedent)
- ☐ Plan administrator-SSN
- ☐ Other corporation (specify) ►
- ☐ Trust
- ☐ Federal Government/military
- ☐ Farmers' cooperative
- ☐ Church or church-controlled organization

(enter GEN if applicable)

8b If a corporation, name the state or foreign country (if applicable) where incorporated | State | Foreign country

9 Reason for applying (Check only one box.)

- ☐ Started new business (specify) ►
- ☐ Hired employees
- ☐ Created a pension plan (specify type) ►
- ☐ Banking purpose (specify) ►
- ☐ Changed type of organization (specify) ►
- ☐ Purchased going business
- ☐ Created a trust (specify) ►
- ☐ Other (specify) ►

10 Date business started or acquired (Mo., day, year) (See instructions.)

11 Closing month of accounting year (See instructions.)

12 First date wages or annuities were paid or will be paid (Mo., day, year). Note: If applicant is a withholding agent, enter date income will first be paid to nonresident alien. (Mo., day, year) . . . . . . . . . . . . . . . . . ►

| 13 Highest number of employees expected in the next 12 months. Note: If the applicant does not expect to have any employees during the period, enter -0-. (See instructions.) . . . . ► | Nonagricultural | Agricultural | Household |
|---|---|---|---|
| | | | |

14 Principal activity (See instructions.) ►

15 Is the principal business activity manufacturing? . . . . . . . . . . . . . . . . . . . . . . . . ☐ Yes ☐ No
If "Yes," principal product and raw material used ►

16 To whom are most of the products or services sold? Please check the appropriate box. ☐ Business (wholesale)
☐ Public (retail) ☐ Other (specify) ► ☐ N/A

17a Has the applicant ever applied for an identification number for this or any other business? . . . . . . . . ☐ Yes ☐ No
Note: If "Yes," please complete lines 17b and 17c.

17b If you checked "Yes" on line 17a, give applicant's legal name and trade name shown on prior application, if different from line 1 or 2 above.
Legal name ► Trade name ►

17c Approximate date when and city and state where the application was filed. Enter previous employer identification number if known.

| Approximate date when filed (Mo., day, year) | City and state where filed | Previous EIN |
|---|---|---|
| | | |

Under penalties of perjury, I declare that I have examined this application, and to the best of my knowledge and belief, it is true, correct, and complete.

**Business telephone number (include area code)**

**Fax telephone number (include area code)**

Name and title (Please type or print clearly.) ►

Signature ► Date ►

Note: Do not write below this line. For official use only.

| Please leave blank ► | Geo. | Ind. | Class | Size | Reason for applying |
|---|---|---|---|---|---|

For Paperwork Reduction Act Notice, see page 4. Cat. No. 16055N Form SS-4 (Rev. 12-95)

# General Instructions

Section references are to the Internal Revenue Code unless otherwise noted.

## Purpose of Form

Use Form SS-4 to apply for an employer identification number (EIN). An EIN is a nine-digit number (for example, 12-3456789) assigned to sole proprietors, corporations, partnerships, estates, trusts, and other entities for filing and reporting purposes. The information you provide on this form will establish your filing and reporting requirements.

## Who Must File

You must file this form if you have not obtained an EIN before and:

- You pay wages to one or more employees including household employees.
- You are required to have an EIN to use on any return, statement, or other document, even if you are not an employer.
- You are a withholding agent required to withhold taxes on income, other than wages, paid to a nonresident alien (individual, corporation, partnership, etc.). A withholding agent may be an agent, broker, fiduciary, manager, tenant, or spouse, and is required to file Form 1042, Annual Withholding Tax Return for U.S. Source Income of Foreign Persons.
- You file Schedule C, Profit or Loss From Business, or Schedule F, Profit or Loss From Farming, of Form 1040, U.S. Individual Income Tax Return, and have a Keogh plan or are required to file excise, employment, information, or alcohol, tobacco, or firearms returns.

The following must use EINs even if they do not have any employees:

- State and local agencies who serve as tax reporting agents for public assistance recipients, under Rev. Proc. 80-4, 1980-1 C.B. 581, should obtain a separate EIN for this reporting. See Household employer on page 3.
- Trusts, except the following:

1. Certain grantor-owned revocable trusts. (See the Instructions for Form 1041.)

2. Individual Retirement Arrangement (IRA) trusts, unless the trust has to file Form 990-T, Exempt Organization Business Income Tax Return. (See the Instructions for Form 990-T.)

3. Certain trusts that are considered household employers can use the trust EIN to report and pay the social security and Medicare taxes, Federal unemployment tax (FUTA) and withheld Federal income tax. A separate EIN is not necessary.

- Estates
- Partnerships
- REMICs (real estate mortgage investment conduits) (See the Instructions for Form 1066, U.S. Real Estate Mortgage Investment Conduit Income Tax Return.)
- Corporations
- Nonprofit organizations (churches, clubs, etc.)
- Farmers' cooperatives
- Plan administrators (A plan administrator is the person or group of persons specified as the administrator by the instrument under which the plan is operated.)

## When To Apply for a New EIN

New Business.— If you become the new owner of an existing business, do not use the EIN of the former owner. IF YOU ALREADY HAVE AN EIN, USE THAT NUMBER. If you do not have an EIN, apply for one on this form. If you become the "owner" of a corporation by acquiring its stock, use the corporation's EIN.

Changes in Organization or Ownership.— If you already have an EIN, you may need to get a new one if either the organization or ownership of your business changes. If you incorporate a sole proprietorship or form a partnership, you must get a new EIN. However, do not apply for a new EIN if you change only the name of your business.

Note: If you are electing to be an "S corporation," be sure you file Form 2553, Election by a Small Business Corporation.

File Only One Form SS-4.— File only one Form SS-4, regardless of the number of businesses operated or trade names under which a business operates. However, each corporation in an affiliated group must file a separate application.

EIN Applied For, But Not Received.— If you do not have an EIN by the time a return is due, write "Applied for" and the date you applied in the space shown for the number. Do not show your social security number as an EIN on returns.

If you do not have an EIN by the time a tax deposit is due, send your payment to the Internal Revenue Service Center for your filing area. (See Where To Apply below.) Make your check or money order payable to Internal Revenue Service and show your name (as shown on Form SS-4), address, type of tax, period covered, and date you applied for an EIN. Send an explanation with the deposit.

For more information about EINs, see Pub. 583, Starting a Business and Keeping Records, and Pub. 1635, Understanding Your EIN.

## How To Apply

You can apply for an EIN either by mail or by telephone. You can get an EIN immediately by calling the Tele-TIN phone number for the service center for your state, or you can send the completed Form SS-4 directly to the service center to receive your EIN in the mail.

Application by Tele-TIN.— Under the Tele-TIN program, you can receive your EIN over the telephone and use it immediately to file a return or make a payment. To receive an EIN by phone, complete Form SS-4, then call the Tele-TIN phone number listed for your state under Where To Apply. The person making the call must be authorized to sign the form. (See Signature block on page 4.)

An IRS representative will use the information from the Form SS-4 to establish your account and assign you an EIN. Write the number you are given on the upper right-hand corner of the form, sign and date it.

Mail or FAX the signed SS-4 within 24 hours to the Tele-TIN Unit at the service center address for your state. The IRS representative will give you the FAX number. The FAX numbers are also listed in Pub. 1635.

Taxpayer representatives can receive their client's EIN by phone if they first send a facsimile (FAX) of a completed Form 2848, Power of Attorney and Declaration of Representative, or Form 8821, Tax Information Authorization, to the Tele-TIN unit. The Form 2848 or Form 8821 will be used solely to release the EIN to the representative authorized on the form.

Application by Mail.— Complete Form SS-4 at least 4 to 5 weeks before you will need an EIN. Sign and date the application and mail it to the service center address for your state. You will receive your EIN in the mail in approximately 4 weeks.

## Where To Apply

The Tele-TIN phone numbers listed below will involve a long-distance charge to callers outside of the local calling area and can be used only to apply for an EIN. THE NUMBERS MAY CHANGE WITHOUT NOTICE. Use 1-800-829-1040 to verify a number or to ask about an application by mail or other Federal tax matters.

| If your principal business, office or agency, or legal residence in the case of an individual, is located in: | Call the Tele-TIN phone number shown or file with the Internal Revenue Service Center at: |
|---|---|
| Florida, Georgia, South Carolina | Attn: Entity Control<br>Atlanta, GA 39901<br>(404) 455-2360 |
| New Jersey, New York City and counties of Nassau, Rockland, Suffolk, and Westchester | Attn: Entity Control<br>Holtsville, NY 00501<br>(516) 447-4955 |
| New York (all other counties), Connecticut, Maine, Massachusetts, New Hampshire, Rhode Island, Vermont | Attn: Entity Control<br>Andover, MA 05501<br>(508) 474-9717 |
| Illinois, Iowa, Minnesota, Missouri, Wisconsin | Attn: Entity Control<br>Stop 57A<br>2306 E. Bannister Rd.<br>Kansas City, MO 64131<br>(816) 926-5999 |
| Delaware, District of Columbia, Maryland, Pennsylvania, Virginia | Attn: Entity Control<br>Philadelphia, PA 19255<br>(215) 574-2400 |
| Indiana, Kentucky, Michigan, Ohio, West Virginia | Attn: Entity Control<br>Cincinnati, OH 45999<br>(606) 292-5467 |
| Kansas, New Mexico, Oklahoma, Texas | Attn: Entity Control<br>Austin, TX 73301<br>(512) 460-7843 |

| | |
|---|---|
| Alaska, Arizona, California (counties of Alpine, Amador, Butte, Calaveras, Colusa, Contra Costa, Del Norte, El Dorado, Glenn, Humboldt, Lake, Lassen, Marin, Mendocino, Modoc, Napa, Nevada, Placer, Plumas, Sacramento, San Joaquin, Shasta, Sierra, Siskiyou, Solano, Sonoma, Sutter, Tehama, Trinity, Yolo, and Yuba), Colorado, Idaho, Montana, Nebraska, Nevada, North Dakota, Oregon, South Dakota, Utah, Washington, Wyoming | Attn: Entity Control Mail Stop 6271-T P.O. Box 9950 Ogden, UT 84409 (801) 620-7645 |
| California (all other counties), Hawaii | Attn: Entity Control Fresno, CA 93888 (209) 452-4010 |
| Alabama, Arkansas, Louisiana, Mississippi, North Carolina, Tennessee | Attn: Entity Control Memphis, TN 37501 (901) 365-5970 |

If you have no legal residence, principal place of business, or principal office or agency in any state, file your form with the Internal Revenue Service Center, Philadelphia, PA 19255 or call 215-574-2400.

## Specific Instructions

The instructions that follow are for those items that are not self-explanatory. Enter N/A (nonapplicable) on the lines that do not apply.

Line 1.—Enter the legal name of the entity applying for the EIN exactly as it appears on the social security card, charter, or other applicable legal document.

Individuals.—Enter the first name, middle initial, and last name. If you are a sole proprietor, enter your individual name, not your business name. Do not use abbreviations or nicknames.

Trusts.—Enter the name of the trust.

Estate of a decedent.—Enter the name of the estate.

Partnerships.—Enter the legal name of the partnership as it appears in the partnership agreement. Do not list the names of the partners on line 1. See the specific instructions for line 7.

Corporations.—Enter the corporate name as it appears in the corporation charter or other legal document creating it.

Plan administrators.—Enter the name of the plan administrator. A plan administrator who already has an EIN should use that number.

Line 2.—Enter the trade name of the business if different from the legal name. The trade name is the "doing business as" name.

Note: Use the full legal name on line 1 on all tax returns filed for the entity. However, if you enter a trade name on line 2 and choose to use the trade name instead of the legal name, enter the trade name on all returns you file. To prevent processing delays and errors, always use either the legal name only or the trade name only on all tax returns.

Line 3.—Trusts enter the name of the trustee. Estates enter the name of the executor, administrator, or other fiduciary. If the entity applying has a designated person to receive tax information, enter that person's name as the "care of" person. Print or type the first name, middle initial, and last name.

Line 7.—Enter the first name, middle initial, last name, and social security number (SSN) of a principal officer if the business is a corporation; of a general partner if a partnership; or of a grantor, owner, or trustor if a trust.

Line 8a.—Check the box that best describes the type of entity applying for the EIN. If not specifically mentioned, check the "Other" box and enter the type of entity. Do not enter N/A.

Sole proprietor.—Check this box if you file Schedule C or F (Form 1040) and have a Keogh plan, or are required to file excise, employment, information, or alcohol, tobacco, or firearms returns. Enter your SSN in the space provided.

REMIC.—Check this box if the entity has elected to be treated as a real estate mortgage investment conduit (REMIC). See the Instructions for Form 1066 for more information.

Other nonprofit organization.—Check this box if the nonprofit organization is other than a church or church-controlled organization and specify the type of nonprofit organization (for example, an educational organization).

If the organization also seeks tax-exempt status, you must file either Package 1023 or Package 1024, Application for Recognition of Exemption. Get Pub. 557, Tax-Exempt Status for Your Organization, for more information.

Group exemption number (GEN).—If the organization is covered by a group exemption letter, enter the four-digit GEN. (Do not confuse the GEN with the nine-digit EIN.) If you do not know the GEN, contact the parent organization. Get Pub. 557 for more information about group exemption numbers.

Withholding agent.—If you are a withholding agent required to file Form 1042, check the "Other" box and enter "Withholding agent."

Personal service corporation.—Check this box if the entity is a personal service corporation. An entity is a personal service corporation for a tax year only if:

- The principal activity of the entity during the testing period (prior tax year) for the tax year is the performance of personal services substantially by employee-owners, and
- The employee-owners own 10% of the fair market value of the outstanding stock in the entity on the last day of the testing period.

Personal services include performance of services in such fields as health, law, accounting, or consulting. For more information about personal service corporations, see the Instructions for Form 1120, U.S. Corporation Income Tax Return, and Pub. 542, Tax Information on Corporations.

Limited liability co.—See the definition of limited liability company in the Instructions for Form 1065. If you are classified as a partnership for Federal income tax purposes, mark the "Limited liability co." checkbox. If you are classified as a corporation for Federal income tax purposes, mark the "Other corporation" checkbox and write "Limited liability co." in the space provided.

Plan administrator.—If the plan administrator is an individual, enter the plan administrator's SSN in the space provided.

Other corporation.—This box is for any corporation other than a personal service corporation. If you check this box, enter the type of corporation (such as insurance company) in the space provided.

Household employer.—If you are an individual, check the "Other" box and enter "Household employer" and your SSN. If you are a state or local agency serving as a tax reporting agent for public assistance recipients who become household employers, check the "Other" box and enter "Household employer agent." If you are a trust that qualifies as a household employer, you do not need a separate EIN for reporting tax information relating to household employees; use the EIN of the trust.

Line 9.—Check only one box. Do not enter N/A.

Started new business.—Check this box if you are starting a new business that requires an EIN. If you check this box, enter the type of business being started. Do not apply if you already have an EIN and are only adding another place of business.

Hired employees.—Check this box if the existing business is requesting an EIN because it has hired or is hiring employees and is therefore required to file employment tax returns. Do not apply if you already have an EIN and are only hiring employees. For information on the applicable employment taxes for family members, see Circular E, Employer's Tax Guide (Publication 15).

Created a pension plan.—Check this box if you have created a pension plan and need this number for reporting purposes. Also, enter the type of plan created.

Banking purpose.—Check this box if you are requesting an EIN for banking purposes only, and enter the banking purpose (for example, a bowling league for depositing dues or an investment club for dividend and interest reporting).

Changed type of organization.—Check this box if the business is changing its type of organization, for example, if the business was a sole proprietorship and has been incorporated or has become a partnership. If you check this box, specify in the space provided the type of change made, for example, "from sole proprietorship to partnership."

Purchased going business.—Check this box if you purchased an existing business. Do not use the former owner's EIN. Do not apply for a new EIN if you already have one. Use your own EIN.

Created a trust.—Check this box if you created a trust, and enter the type of trust created.

Note: Do not file this form if you are the grantor/owner of certain revocable trusts. You must use your SSN for the trust. See the Instructions for Form 1041.

Other (specify).—Check this box if you are requesting an EIN for any reason other than those for which there are checkboxes, and enter the reason.

Line 10.—If you are starting a new business, enter the starting date of the business. If the business you acquired is already operating, enter the date you acquired the business. Trusts should enter the date the trust was legally created. Estates should enter the date of death of the decedent whose name appears on line 1 or the date when the estate was legally funded.

Line 11.—Enter the last month of your accounting year or tax year. An accounting or tax year is usually 12 consecutive months, either a calendar year or a fiscal year (including a period of 52 or 53 weeks). A calendar year is 12 consecutive months ending on December 31. A fiscal year is either 12 consecutive months ending on the last day of any month other than December or a 52-53 week year. For more information on accounting periods, see Pub. 538, Accounting Periods and Methods.

Individuals.—Your tax year generally will be a calendar year.

Partnerships.—Partnerships generally must adopt the tax year of either (a) the majority partners; (b) the principal partners; (c) the tax year that results in the least aggregate (total) deferral of income; or (d) some other tax year. (See the Instructions for Form 1065, U.S. Partnership Return of Income, for more information.)

REMIC.—REMICs must have a calendar year as their tax year.

Personal service corporations.—A personal service corporation generally must adopt a calendar year unless:

- It can establish a business purpose for having a different tax year, or
- It elects under section 444 to have a tax year other than a calendar year.

Trusts.—Generally, a trust must adopt a calendar year except for the following:

- Tax-exempt trusts,
- Charitable trusts, and
- Grantor-owned trusts.

Line 12.—If the business has or will have employees, enter the date on which the business began or will begin to pay wages. If the business does not plan to have employees, enter N/A.

Withholding agent.—Enter the date you began or will begin to pay income to a nonresident alien. This also applies to individuals who are required to file Form 1042 to report alimony paid to a nonresident alien.

Line 13.—For a definition of agricultural labor (farmworker), see Circular A, Agricultural Employer's Tax Guide (Publication 51).

Line 14.—Generally, enter the exact type of business being operated (for example, advertising agency, farm, food or beverage establishment, labor union, real estate agency, steam laundry, rental of coin-operated vending machine, or investment club). Also state if the business will involve the sale or distribution of alcoholic beverages.

Governmental.—Enter the type of organization (state, county, school district, municipality, etc.).

Nonprofit organization (other than governmental).—Enter whether organized for religious, educational, or humane purposes, and the principal activity (for example, religious organization—hospital, charitable).

Mining and quarrying.—Specify the process and the principal product (for example, mining bituminous coal, contract drilling for oil, or quarrying dimension stone).

Contract construction.—Specify whether general contracting or special trade contracting. Also, show the type of work normally performed (for example, general contractor for residential buildings or electrical subcontractor).

Food or beverage establishments.—Specify the type of establishment and state whether you employ workers who receive tips (for example, lounge—yes).

Trade.—Specify the type of sales and the principal line of goods sold (for example, wholesale dairy products, manufacturer's representative for mining machinery, or retail hardware).

Manufacturing.—Specify the type of establishment operated (for example, sawmill or vegetable cannery).

Signature block.— The application must be signed by (a) the individual, if the applicant is an individual, (b) the president, vice president, or other principal officer, if the applicant is a corporation, (c) a responsible and duly authorized member or officer having knowledge of its affairs, if the applicant is a partnership or other unincorporated organization, or (d) the fiduciary, if the applicant is a trust or estate.

## Some Useful Publications

You may get the following publications for additional information on the subjects covered on this form. To get these and other free forms and publications, call 1-800-TAX-FORM (1-800-829-3676). You should receive your order or notification of its status within 7 to 15 workdays of your call.

Use your computer.— If you subscribe to an on-line service, ask if IRS information is available and, if so, how to access it. You can also get information through IRIS, the Internal Revenue Information Services, on FedWorld, a government bulletin board. Tax forms, instructions, publications, and other IRS information, are available through IRIS.

IRIS is accessible directly by calling 703-321-8020. On the Internet, you can telnet to fedworld.gov. or, for file transfer protocol services, connect to ftp.fedworld.gov. If you are using the WorldWide Web, connect to http://www.ustreas.gov

FedWorld's help desk offers technical assistance on accessing IRIS (not tax help) during regular business hours at 703-487-4608. The IRIS menus offer information on available file formats and software needed to read and print files. You must print the forms to use them; the forms are not designed to be filled out on-screen.

Tax forms, instructions, and publications are also available on CD-ROM, including prior-year forms starting with the 1991 tax year. For ordering information and software requirements, contact the Government Printing Office's Superintendent of Documents (202-512-1800) or Federal Bulletin Board (202-512-1387).

Pub. 1635, Understanding Your EIN

Pub. 15, Employer's Tax Guide

Pub. 15-A, Employer's Supplemental Tax Guide

Pub. 538, Accounting Periods and Methods

Pub. 541, Tax Information on Partnerships

Pub. 542, Tax Information on Corporations

Pub. 557, Tax-Exempt Status for Your Organization

Pub. 583, Starting a Business and Keeping Records

Package 1023, Application for Recognition of Exemption

Package 1024, Application for Recognition of Exemption Under Section 501(a) or for Determination Under Section 120

## Paperwork Reduction Act Notice

We ask for the information on this form to carry out the Internal Revenue laws of the United States. You are required to give us the information. We need it to ensure that you are complying with these laws and to allow us to figure and collect the right amount of tax.

The time needed to complete and file this form will vary depending on individual circumstances. The estimated average time is:

| | |
|---|---|
| Recordkeeping | 7 min. |
| Learning about the law or the form | 18 min. |
| Preparing the form | 45 min. |
| Copying, assembling, and sending the form to the IRS | 20 min. |

If you have comments concerning the accuracy of these time estimates or suggestions for making this form simpler, we would be happy to hear from you. You can write to the Tax Forms Committee, Western Area Distribution Center, Rancho Cordova, CA 95743-0001. Do not send this form to this address. Instead, see Where To Apply on page 2.

Form **W-9**
(Rev. December 1996)
Department of the Treasury
Internal Revenue Service

# Request for Taxpayer Identification Number and Certification

**Give form to the requester. Do NOT send to the IRS.**

Please print or type

Name (If a joint account or you changed your name, see **Specific Instructions** on page 2.)

Business name, if different from above. (See **Specific Instructions** on page 2.)

Check appropriate box: ☐ Individual/Sole proprietor ☐ Corporation ☐ Partnership ☐ Other ▶ ..................................

Address (number, street, and apt. or suite no.)

Requester's name and address (optional)

City, state, and ZIP code

## Part I Taxpayer Identification Number (TIN)

Enter your TIN in the appropriate box. For individuals, this is your social security number (SSN). However, if you are a resident alien OR a sole proprietor, see the instructions on page 2. For other entities, it is your employer identification number (EIN). If you do not have a number, see **How To Get a TIN** on page 2.

**Note:** *If the account is in more than one name, see the chart on page 2 for guidelines on whose number to enter.*

**Social security number**

OR

**Employer identification number**

List account number(s) here (optional)

## Part II For Payees Exempt From Backup Withholding (See the instructions on page 2.)

▶

## Part III Certification

Under penalties of perjury, I certify that:

1. The number shown on this form is my correct taxpayer identification number (or I am waiting for a number to be issued to me), **and**
2. I am not subject to backup withholding because: **(a)** I am exempt from backup withholding, or **(b)** I have not been notified by the Internal Revenue Service (IRS) that I am subject to backup withholding as a result of a failure to report all interest or dividends, or **(c)** the IRS has notified me that I am no longer subject to backup withholding.

**Certification Instructions.**—You must cross out item **2** above if you have been notified by the IRS that you are currently subject to backup withholding because you have failed to report all interest and dividends on your tax return. For real estate transactions, item **2** does not apply. For mortgage interest paid, acquisition or abandonment of secured property, cancellation of debt, contributions to an individual retirement arrangement (IRA), and generally, payments other than interest and dividends, you are not required to sign the Certification, but you must provide your correct TIN. (See the instructions on page 2.)

**Sign Here** **Signature ▶** **Date ▶**

**Purpose of Form.**—A person who is required to file an information return with the IRS must get your correct taxpayer identification number (TIN) to report, for example, income paid to you, real estate transactions, mortgage interest you paid, acquisition or abandonment of secured property, cancellation of debt, or contributions you made to an IRA.

Use Form W-9 to give your correct TIN to the person requesting it (the requester) and, when applicable, to:

**1.** Certify the TIN you are giving is correct (or you are waiting for a number to be issued),

**2.** Certify you are not subject to backup withholding, or

**3.** Claim exemption from backup withholding if you are an exempt payee.

**Note:** *If a requester gives you a form other than a W-9 to request your TIN, you must use the requester's form if it is substantially similar to this Form W-9.*

**What Is Backup Withholding?**—Persons making certain payments to you must withhold and pay to the IRS 31% of such payments under certain conditions. This is called "backup withholding." Payments that may be subject to backup withholding include interest, dividends, broker and barter exchange transactions, rents, royalties, nonemployee pay, and certain payments from fishing boat operators. Real estate transactions are not subject to backup withholding.

If you give the requester your correct TIN, make the proper certifications, and report all your taxable interest and dividends on your tax return, payments you receive will not be subject to backup withholding. Payments you receive **will** be subject to backup withholding if:

**1.** You do not furnish your TIN to the requester, or

**2.** The IRS tells the requester that you furnished an incorrect TIN, or

**3.** The IRS tells you that you are subject to backup withholding because you did not report all your interest and dividends on your tax return (for reportable interest and dividends only), or

**4.** You do not certify to the requester that you are not subject to backup withholding under 3 above (for reportable interest and dividend accounts opened after 1983 only), or

**5.** You do not certify your TIN when required. See the Part III instructions on page 2 for details.

Certain payees and payments are exempt from backup withholding. See the Part II instructions and the separate **Instructions for the Requester of Form W-9.**

## Penalties

**Failure To Furnish TIN.**—If you fail to furnish your correct TIN to a requester, you are subject to a penalty of $50 for each such failure unless your failure is due to reasonable cause and not to willful neglect.

**Civil Penalty for False Information With Respect to Withholding.**—If you make a false statement with no reasonable basis that results in no backup withholding, you are subject to a $500 penalty.

**Criminal Penalty for Falsifying Information.**— Willfully falsifying certifications or affirmations may subject you to criminal penalties including fines and/or imprisonment.

**Misuse of TINs.**—If the requester discloses or uses TINs in violation of Federal law, the requester may be subject to civil and criminal penalties.

## Specific Instructions

**Name.**—If you are an individual, you must generally enter the name shown on your social security card. However, if you have changed your last name, for instance, due to marriage, without informing the Social Security Administration of the name change, enter your first name, the last name shown on your social security card, and your new last name.

If the account is in joint names, list first and then circle the name of the person or entity whose number you enter in Part I of the form.

*Sole Proprietor.*—You must enter your **individual** name as shown on your social security card. You may enter your business, trade, or "doing business as" name on the **business name** line.

*Other Entities.*—Enter the business name as shown on required Federal tax documents. This name should match the name shown on the charter or other legal document creating the entity. You may enter any business, trade, or "doing business as" name on the business name line.

### Part I—Taxpayer Identification Number (TIN)

You must enter your TIN in the appropriate box. If you are a resident alien and you do not have and are not eligible to get an SSN, your TIN is your IRS individual taxpayer identification number (ITIN). Enter it in the social security number box. If you do not have an ITIN, see **How To Get a TIN** below.

If you are a sole proprietor and you have an EIN, you may enter either your SSN or EIN. However, using your EIN may result in unnecessary notices to the requester.

**Note:** *See the chart on this page for further clarification of name and TIN combinations.*

**How To Get a TIN.**—If you do not have a TIN, apply for one immediately. To apply for an SSN, get **Form SS-5** from your local Social Security Administration office. Get **Form W-7** to apply for an ITIN or **Form SS-4** to apply for an EIN. You can get Forms W-7 and SS-4 from the IRS by calling 1-800-TAX-FORM (1-800-829-3676).

If you do not have a TIN, write "Applied For" in the space for the TIN, sign and date the form, and give it to the requester. For interest and dividend payments, and certain payments made with respect to readily tradable instruments, you will generally have 60 days to get a TIN and give it to the requester. Other payments are subject to backup withholding.

**Note:** *Writing "Applied For" means that you have already applied for a TIN* ***OR*** *that you intend to apply for one soon.*

### Part II—For Payees Exempt From Backup Withholding

Individuals (including sole proprietors) are **not** exempt from backup withholding. Corporations are exempt from backup withholding for certain payments, such as interest and dividends. For more information on exempt payees, see the separate Instructions for the Requester of Form W-9.

If you are exempt from backup withholding, you should still complete this form to avoid possible erroneous backup withholding. Enter your correct TIN in Part I, write "Exempt" in Part II, and sign and date the form.

If you are a nonresident alien or a foreign entity not subject to backup withholding, give the requester a completed **Form W-8**, Certificate of Foreign Status.

### Part III—Certification

For a joint account, only the person whose TIN is shown in Part I should sign (when required).

**1. Interest, Dividend, and Barter Exchange Accounts Opened Before 1984 and Broker Accounts Considered Active During 1983.** You must give your correct TIN, but you do not have to sign the certification.

**2. Interest, Dividend, Broker, and Barter Exchange Accounts Opened After 1983 and Broker Accounts Considered Inactive During 1983.** You must sign the certification or backup withholding will apply. If you are subject to backup withholding and you are merely providing your correct TIN to the requester, you must cross out item **2** in the certification before signing the form.

**3. Real Estate Transactions.** You must sign the certification. You may cross out item **2** of the certification.

**4. Other Payments.** You must give your correct TIN, but you do not have to sign the certification unless you have been notified that you have previously given an incorrect TIN. "Other payments" include payments made in the course of the requester's trade or business for rents, royalties, goods (other than bills for merchandise), medical and health care services (including payments to corporations), payments to a nonemployee for services (including attorney and accounting fees), and payments to certain fishing boat crew members.

**5. Mortgage Interest Paid by You, Acquisition or Abandonment of Secured Property, Cancellation of Debt, or IRA Contributions.** You must give your correct TIN, but you do not have to sign the certification.

### Privacy Act Notice

Section 6109 of the Internal Revenue Code requires you to give your correct TIN to persons who must file information returns with the IRS to report interest, dividends, and certain other income paid to you, mortgage interest you paid, the acquisition or abandonment of secured property, cancellation of debt, or contributions you made to an IRA. The IRS uses the numbers for identification purposes and to help verify the accuracy of your tax return. The IRS may also provide this information to the Department of Justice for civil and criminal litigation and to cities, states, and the District of Columbia to carry out their tax laws.

You must provide your TIN whether or not you are required to file a tax return. Payers must generally withhold 31% of taxable interest, dividend, and certain other payments to a payee who does not give a TIN to a payer. Certain penalties may also apply.

## What Name and Number To Give the Requester

| For this type of account: | Give name and SSN of: |
|---|---|
| 1. Individual | The individual |
| 2. Two or more individuals (joint account) | The actual owner of the account or, if combined funds, the first individual on the account [1] |
| 3. Custodian account of a minor (Uniform Gift to Minors Act) | The minor [2] |
| 4. a. The usual revocable savings trust (grantor is also trustee) | The grantor-trustee [1] |
| b. So-called trust account that is not a legal or valid trust under state law | The actual owner [1] |
| 5. Sole proprietorship | The owner [3] |
| **For this type of account:** | **Give name and EIN of:** |
| 6. Sole proprietorship | The owner [3] |
| 7. A valid trust, estate, or pension trust | Legal entity [4] |
| 8. Corporate | The corporation |
| 9. Association, club, religious, charitable, educational, or other tax-exempt organization | The organization |
| 10. Partnership | The partnership |
| 11. A broker or registered nominee | The broker or nominee |
| 12. Account with the Department of Agriculture in the name of a public entity (such as a state or local government, school district, or prison) that receives agricultural program payments | The public entity |

[1] List first and circle the name of the person whose number you furnish. If only one person on a joint account has an SSN, that person's number must be furnished.

[2] Circle the minor's name and furnish the minor's SSN.

[3] You must show your individual name, but you may also enter your business or "doing business as" name. You may use either your SSN or EIN (if you have one).

[4] List first and circle the name of the legal trust, estate, or pension trust. (Do not furnish the TIN of the personal representative or trustee unless the legal entity itself is not designated in the account title.)

**Note:** *If no name is circled when more than one name is listed, the number will be considered to be that of the first name listed.*

# State Offices Providing Small Business Help

**Alabama**
Economic and Community Affairs
Existing Business and Industry
State Capitol
Montgomery, AL 36130
(800) 248-0033 (205) 242-0400

**Alaska**
Division of Economic Development
Department of Commerce and Economic Development
PO Box 110804
Juneau, AK 99811-0804
(907) 465-2017

**Arizona**
Office of Business Finance
Department of Commerce
3800 North Central Avenue
Suite 1500
Phoenix, AZ 85012
(602) 280-1341

**Arkansas**
Small Business Information Center
Industrial Development Commission
State Capitol Mall
Room 4C-300
Little Rock, AR 72201
(501) 682-5275

**California**
Office of Small Business
Department of Commerce
801 K Street, Suite 1700
Sacramento, CA 95814
(916) 327-4357 (916) 445-6545

**Colorado**
One-Stop Assistance Center
1560 Broadway, Suite 1530
Denver, CO 80202
(800) 333-7798 (303) 592-5920

**Connecticut**
Small Business Services
Department of Economic Development
865 Brook Street
Rocky Hill, CN 06067
(203) 258-4200

**Delaware**
Development Office
PO Box 1401
99 Kings Highway
Dover, DE 19903
(302) 739-4271

**District of Columbia**
Office of Business and Economic Development
Twelfth Floor
717 14th Street NW
Washington, DC 20005
(202) 727-6600

**Florida**
Bureau of Business Assistance
Department of Commerce
107 West Gaines Street, Room 443
Tallahassee, FL 32399-2000
(800) 342-0771 (904) 488-9357

**Georgia**
Department of Community Affairs
100 Peachtree Street, Suite 1200
Atlanta, GA 30303
(404) 656-6200

**Hawaii**
Small Business Information Service
2404 Maile Way
Room A 202
University of Hawaii
Honolulu, HI 996822
(808) 956-7363

**Idaho**
Economic Development Division
Department of Commerce
700 W. State Street
Boise, ID 83720-0093
(208) 334-2470

**Illinois**
Small Business Assistance Bureau
Department of Commerce and Community Affairs
620 East Adams Street
Springfield, IL 62701
(800) 252-2923 (217) 785-5947

**Indiana**
Ombudsman's Office
Community Development Division
Department of Commerce
One North Capitol, Suite 700
Indianapolis, IN 46204-2288
(800) 824-2476 (317) 232-7304

**Iowa**
Bureau of Small Business Development
Department of Economic Development
200 East Grand Avenue
Des Moines, IA 50309
(800) 532-1216 (515) 242-4899

**Kansas**
Division of Existing Industry Development
700 SW Harrison, Suite 1300
Topeka, KN 66603
(913) 296-3712

**Kentucky**
Division of Small Business
Capitol Plaza Tower
Frankfort, KY 40601
(800) 626-2250 (502) 564-4252

**Louisiana**
Development Division
Office of Commerce and Industry
PO Box 94185
Baton Rouge, LA 70804-9185
(504) 342-5365

**Maine**
Business Development Division
Department of Economic and Community Development
State House 59
187 State Street
Augusta, ME 04333
(800) 872-3838 (207) 287-3153

**Maryland**
Division of Business Development
Department of Economic and Employment Development
217 East Redwood Street
Baltimore, MD 21202
(800) 873-7232 (301) 333-6970

**Massachusetts**
Office of Business Development
1 Ashburton Place
Boston, MA 02202
(617) 727-3206

**Michigan**
Michigan Jobs Commission
Ombudsman's Office
201 N. Washington Square
Lansing, MI 48913
(517) 373-9808

**Minnesota**
Small Business Assistance Office
Department of Trade and Economic Development
500 Metro Square Building
121 E. 7th Place
St. Paul, MN 55101-2146
(800) 652-9747 (612) 296-3871

**Mississippi**
Small Business Bureau
Research and Development Center
PO Box 849
Jackson, MS 39205
(601) 359-3552

**Missouri**
Small Business Development Office
Department of Economic Development
PO Box 118
Jefferson City, MO 65102
(314) 751-4982 (314) 751-8411

**Montana**
Business Assistance Division
Department of Commerce
1424 Ninth Avenue
Helena, MT 59620
(800) 221-8015 (406) 444-3923

**Nebraska**
Existing Business Division
Department of Economic Development
PO Box 94666
301 Centennial Mall South
Lincoln, NE 68509-4666
(402) 471-3782

**Nevada**
Nevada Commission on Economic Development
Capitol Complex
Carson City, NV 89710
(702) 687-4325

**New Hampshire**
Small Business Development Center
108 McConnell Hall
University of New Hampshire
15 College Road
Durham, NH 03824
(603) 862-2200

**New Jersey**
Office of Small Business Assistance
Department of Commerce and Economic Development
20 West State Street, CN 835
Trenton, NJ 08625
(609) 984-4442

**New Mexico**
Economic Development Division
Department of Economic Development
1100 St. Francis Drive
Santa Fe, NM 87503
(505) 827-0300

**New York**
Division for Small Business
Department of Economic Development
1515 Broadway
51st Floor
New York, NY 10036
(212) 827-6150

**North Carolina**
Business and Industry Division
Department of Commerce
Dobbs Building, Room 2019
430 North Salisbury Street
Raleigh, NC 27611
(919) 733-4151

**North Dakota**
Small Business Coordinator
Economic Development Commission
1833 E. Bismark Expressway
Bismark, ND 58504
(701) 224-2810

**Ohio**
Small and Developing Business Division
Department of Development
77 S. High Street
Columbus, OH 43215
(800) 248-4040 (614) 466-4232

**Oklahoma**
Oklahoma Department of Commerce
PO Box 26980
6601 N. Broadway Extension
Oklahoma City, OK 73126-0980
(800) 477-6552 (405) 843-9770

**Oregon**
Economic Development Department
775 Summer Street NE
Salem, OR 97310
(800) 233-3306 (503) 373-1200

**Pennsylvania**
Bureau of Small Business and Appalachian Development
Department of Commerce
461 Forum Building
Harrisburg, PA 17120
(717) 783-5700

**Puerto Rico**
Commonwealth Department of Commerce, Box S
4275 Old San Juan Station
San Juan, PR 00905
(809) 721-3290

**Rhode Island**
Business Development Division
Department of Economic Development
Seven Jackson Walkway
Providence, RI 02903
(401) 277-2601

**South Carolina**
Enterprise Development
PO Box 1149
Columbia, SC 29202
(803) 737-0888

**South Dakota**
Governor's Office of Economic Development
Capital Lake Plaza
711 Wells Avenue
Pierre, SD 57501
(800) 872-6190 (605) 773-5032

**Tennessee**
Small Business Office
Department of Economic and Community Development
320 Sixth Avenue North
Seventh Floor
Rachel Jackson Building
Nashville, TN 37219
(800) 872-7201 (615) 741-2626

**Texas**
Small Business Division
Department of Commerce
Economic Development Commission
PO Box 12728
Capitol Station
410 East Fifth Street
Austin, TX 78711
(800) 888-0511 (512) 472-5059

**Utah**
Small Business Development Center
102 West 500 South, Suite 315
Salt Lake City, UT 84101
(801) 581-7905

**Vermont**
Agency of Development and Community Affairs
The Pavilion
109 State Street
Montpelier, VT 05609
(800) 622-4553 (802) 828-3221

**Virginia**
Small Business and Financial Services
Department of Economic Development
PO Box 798
1021 E. Cary Street
11th Floor
Richmond, VA 23206-0798
(804) 371-8252

**Washington**
Small Business Development Center
Krugel Hall
Room 135
Washington State University
Pullman, WA 99164-4727
(509) 335-1576

**West Virginia**
Small Business Development Center Division
1115 Virginia Street East
Charleston, WV 25301
(304) 558-2960

**Wisconsin**
Public Information Bureau
Department of Development
PO Box 7970
123 West Washington Avenue
Madison, WI 53707
(800) 435-7287 (608) 266-1018

**Wyoming**
Economic Development and Stabilization Board
Barrett Building
2301 Central Ave.
Cheyenne, WY 82002
(307) 777-7284

---

Source: National Association for the Self-Employed, USA TODAY research.

# State Unemployment Tax Agencies

**Alabama**
Department of Industrial Relations
649 Monroe Street
Montgomery, AL 36131
334-242-8371

**Alaska**
Employment Security Division
P.O. Box 25509
Juneau, AK 99802-5509
907-465-5937

**Arizona**
Department of Economic Security
2801 North 33rd Avenue
Phoenix, AZ 85009
602-255-4755

**Arkansas**
Employment Security Division
P.O. Box 2981
Little Rock, AR 72203
501-682-3253

**California**
Employment Development Department, MIC-90
P.O. Box 942880
Sacramento, CA 94280-0001
916-653-1528

**Colorado**
Department of Labor and Employment
1515 Arapahoe, Tower 3,
Suite 400
Denver, CO 80202-2117
303-620-4793

**Connecticut**
Employment Security Division
Labor Department
200 Folley Brook Blvd.
Wethersfield, CT 06109
203-566-2128

**Delaware**
Department of Labor
Division of Unemployment Insurance
P.O. Box 9029
Newark, DE 19714
302-368-6731

**District of Columbia**
Department of Employment Services
500 C Street, NW, Room 501,
Washington, DC 20001
202-724-7462

**Florida**
Department of Labor and Employment Security
102 Caldwell Building
Tallahassee, FL 32399-0211
904-921-3108

**Georgia**
Department of Labor
148 International Blvd.
Atlanta, GA 30303
404-656-6225

**Hawaii**
Department of Labor and Industrial Relations
800 Punchbowl Street
Honolulu, HI 96813
808-586-8927

**Idaho**
Department of Employment
317 Main Street
Boise, ID 83735
208-334-6240

**Illinois**
Bureau of Employment Security
401 South State Street
Chicago, IL 60605
312-793-1916

**Indiana**
Division of Workforce Development
10 North Senate Avenue
Indianapolis, IN 46204
317-232-7698

**Iowa**
Department of Job Services
1000 East Grand Avenue
Des Moines, IA 50319
515-281-8200

**Kansas**
Department of Human Resources
401 Topeka Avenue
Topeka, KS 66603
913-296-5026

**Kentucky**
Division of Unemployment Insurance
P.O. Box 948
Frankfort, KY 40602
502-564-6838

**Louisiana**
Office of Employment Security
P.O. Box 98146
Baton Rouge, LA 70804
504-342-2992

**Maine**
Maine Department of Labor
P.O. Box 309
Augusta, ME 04332-0309
207-287-1239

**Maryland**
Office of Unemployment Insurance
1100 North Eutaw Street
Baltimore, MD 21201
410-767-2488

**Massachusetts**
Department of Employment and Training
19 Staniford Street
Boston, MA 02114
617-727-5054

**Michigan**
Employment Security Division
7310 Woodward Avenue
Detroit, Ml 48202
313-876-5131

**Minnesota**
Department of Economic Security
390 North Robert Street
St. Paul, MN 55101
612-296-3736

**Mississippi**
Employment Security Commission
P.O. Box 22781
Jackson, MS 39225-2781
601-961-7755

**Missouri**
Division of Employment Security
Box 59
Jefferson City, MO 65104
314-751-3328

**Montana**
Unemployment Insurance Division
P.O. Box 1728
Helena, MT 59604
406-444-3686

**Nebraska**
Division of Employment
Box 94600 State House Station
Lincoln, NE 68509
402-471-9839

**Nevada**
Department of Employment, Training, and Rehabilitation
500 East Third Street
Carson City, NV 89713
702-687-4599

**New Hampshire**
Department of Employment Security
32 South Main Street
Concord, NH 03301
603-224-3311 (Ext. 270)

**New Jersey**
New Jersey Department of Labor
CN 947
Trenton, NJ 08625-0947
609-292-2810

**New Mexico**
Employment Security Department
P.O. Box 2281
Albuquerque, NM 87103
505-841-8568

**New York**
State Department of Labor
State Campus, Building 12
Albany, NY 12240
518-457-4120

**North Carolina**
Employment Security Commission
P.O. Box 26504
Raleigh, NC 27611
919-733-7395

**North Dakota**
Job Service of North Dakota
P.O. Box 5507
Bismarck, ND 58502
701-328-2814

**Ohio**
Bureau of Employment Services
P.O. Box 923
Columbus, OH 43216
614-466-2578

**Oklahoma**
Employment Security Commission
Will Rogers Memorial Office Building
Oklahoma City, OK 73105
405-557-7135

**Oregon**
Employment Department
875 Union Street, NE
Salem, OR 97311
503-378-3257

**Pennsylvania**
Department of Labor and Industry
Labor and Industry Building
7th and Forster Street
Harrisburg, PA 17121
717-787-2097

**Rhode Island**
Department of Employment & Training
1 01 Friendship Street
Providence, RI 02903
401-277-3688

**South Carolina**
Employment Security Commission
P.O. Box 995
Columbia, SC 29202
803-737-3070

**South Dakota**
Department of Employment Security
P.O. Box 4730
Aberdeen, SD 57401
605-626-2312

**Tennessee**
Department of Employment Security
500 James Robertson Parkway
8th Floor, Volunteer Plaza Building
Nashville, TN 37245-3500
615-741-2346

**Texas**
Employment Commission
TEC Building
Austin, TX 78778
512-463-2712

**Utah**
Department of Employment Security
P.O. Box 45288
Salt Lake City, UT 84145
801-536-7755

**Vermont**
Department of Employment Security
P.O. Box 488
Montpelier, VT 05602
802-828-4242

**Virginia**
Employment Commission
P.O. Box 1358
Richmond, VA 23211
804-786-1256

**Washington**
Employment Security Department
P.O. Box 9046
Olympia, WA 98507-9046
206-753-3822

**West Virginia**
Unemployment Compensation Division
112 California Avenue
Charleston, WV 25305-0112
304-558-2675

**Wisconsin**
Department of Industry, Labor, and Human Relations
P.O. Box 7942–GEF 1
Madison, WI 53702
608-266-3177

**Wyoming**
Employment Resources Division
P.O. Box 2760
Casper, WY 82606
307-235-3201

# State Sales Tax Agencies

Depending on the nature of your occupation and the state in which you live, you may have to pay state sales taxes. Check with your state sales tax office for the current requirements. If sales taxes are required, you'll have to fill out an application to obtain a state sales tax number, also called a seller's permit in many states. (See Chapter 5, Section A2c.)

**The following states have no sales tax: Alaska, Delaware, Montana, New Hampshire and Oregon.**

**Alabama**
Department of Revenue
Sales and Use Taxes
P.O. Box 327710
Montgomery, AL 36132-7710
Phone: 205-242-1490

**Arizona**
Department of Revenue
Licensing Section
1600 West Monroe
Phoenix, AZ 85007
Phone: 602-255-2060

**Arkansas**
Department of Finance and Administration
Sales and Use Tax Section
P.O. Box 1272
Little Rock, AK 72203
Phone: 501-682-7104

**California**
State Board of Equalization
P.O. Box 942879
Sacramento, CA 94279-0001
Phone: 916-322-2010

**Colorado**
Department of Revenue
Sales and Use Tax Division
1375 Sherman Street
Denver, CO 80261
Phone: 303-534-1208

**Connecticut**
Department of Revenue Services
92 Farmington Avenue
Hartford, CT 06105
Phone: 203-566-8520

**District of Columbia**
Department of Finance and Revenue
Sales and Audit Division
P.O. Box 556
Room 570 North
Washington, DC 20044
Phone: 202-727-6566

**Florida**
Department of Revenue
Sales and Use Tax Division
5050 West Tennessee Street
Building K
Tallahassee, FL 32399-0100
Phone: 904-488-9750

**Georgia**
Department of Revenue
Sales and Use Tax Division
P.O. Box 740390
Atlanta, GA 30374-0390
Phone: 404-656-4065

**Hawaii**
Department of Taxation
Taxpayers' Services Branch
P.O. Box 259
Honolulu, HI 96809
Phone: 808-587-1455

**Idaho**
State Tax Commission
800 Park Boulevard
Boise, ID 83722
Phone: 208-334-7660

**Illinois**
Department of Revenue
Retailers' Occupation Tax
101 W. Jefferson Street
Springfield, IL 62796
Phone: 217-782-7897

**Indiana**
Department of Revenue
Sales Tax Division
100 N. Zenith Street
Indianapolis, IN 46204
Phone: 317-233-4015

**Iowa**
Department of Revenue and Finance
Taxpayer Services
Hoover State Office Building
Des Moines, IA 50319
Phone: 515-281-3114

**Kansas**
Department of Revenue
Sales Tax
P.O. Box 2001
Topeka, KS 66625
Phone: 913-296-2461

**Kentucky**
Revenue Cabinet
Sales Tax Division
P.O. Box 1274
Station 53
Frankfurt, KY 40602
Phone: 502-564-5170

**Lousiana**
Department of Revenue and Taxation
Sales Tax Division
P.O. Box 201
Baton Rouge, LA 70821-0201
Phone: 504-925-7356

**Maine**
Bureau of Taxation
Sales Tax Section
P.O. Box 1065
Augusta, ME 04332-1065
Phone: 207-287-2336

**Maryland**
State Comptroller
Sales and Use Tax Division
301 West Preston
Baltimore, MD 21201-2383
Phone: 410-225-1300

**Massachusetts**
Department of Revenue
Sales Tax Division
Room 200
100 Cambridge Street
Boston, MA 00204
Phone: 617-727-4490

**Michigan**
Department of Treasury
Sales, Use, and Withholding Taxes Division
Treasury Building
Lansing, MI 48922
Phone: 517-373-3190

**Minnesota**
Department of Revenue
Sales Tax Division
Mail Station 1110
St. Paul, MN 55146-1110
Phone: 612-296-3781

**Mississippi**
State Tax Commission
Sales Tax Division
P.O. Box 1033
Jackson, MS 39215
Phone: 601-359-1133

**Missouri**
Department of Revenue
Sales and Use Tax Bureau
P.O. Box 840
Jefferson City, MO 65105-0840
Phone: 314-751-4450

**Nebraska**
Department of Revenue
Taxpayer Assistance
P.O. Box 94818
Lincoln, NE 68509-4818
Phone: 402-471-5729

**Nevada**
Department of Taxation
Revenue Office
Capitol Complex
Carson City, NV 89710-0003
Phone: 702-687-4820

**New Jersey**
Department of Treasury
Division of Taxation
Taxpayer Information Service
Office of Communications
CN 281
Trenton, NJ 08646-0281
Phone: 609-588-3800

**New Mexico**
Department of Taxation and Revenue
P.O. Box 630
Santa Fe, NM 87509-0630
Phone: 505-827-0834

**New York**
Department of Taxation and Finance
Technical Services Bureau
Building 9, Room 104
W.A. Harriman State Office Building
Campus
Albany, NY 12227
Phone: 518-457-1250

**North Carolina**
Department of Revenue
Sales and Use Tax Division
P.O. Box 25000
Raleigh, NC 27640
Phone: 919-733-3661

**North Dakota**
State Tax Commission
Sales and Special Taxes Division
600 East Boulevard Avenue
Bismarck, ND 58505
Phone: 701-224-3470

**Ohio**
Department of Taxation
Sales and Use Tax Division
30 East Broad Street, 20th Floor
Columbus, OH 43215
Phone: 614-466-4810

**Oklahoma**
Tax Commission
Business Tax Division
2501 N. Lincoln Boulevard
Oklahoma City, OK 73194
Phone: 405-521-3279

**Pennsylvania**
Department of Revenue
Bureau of Business Trust
Dept. 280901
Harrisburg, PA 17128-0901
Phone: 717-783-0312

**Rhode Island**
Department of Administration
Division of Taxation
One Capital Hill
Providence, RI 02908-5800
Phone: 401-277-2950

**South Carolina**
Tax Commission
Public Assistance
P.O. Box 125
Columbia, SC 29214
Phone: 803-737-4788

**South Dakota**
Department of Revenue
Sales Taxes
700 Governors Drive
Pierre, SD 57501
Phone: 605-773-3311

**Tennessee**
Department of Revenue
Sales Tax
Andrew Jackson State Office
Building
500 Deaderick Street
Nashville, TN 37242
Phone: 615-741-3581

**Texas**
State Comptroller
Taxpayer Assistance
P.O. Box 13528
Austin, TX 78711
Phone: 800-252-5555

**Utah**
Tax Commission
Taxpayer Assistance
160 E. 300 South
1st Floor, South Center
Salt Lake City, UT 84134
Phone: 801-530-4848

**Vermont**
Department of Taxes
Business Tax Division
P.O. Box 547
Montpelier, VT 05601-0547
Phone: 802-828-2551

**Virginia**
Department of Taxation
Sales Tax
P.O. Box 6-L
Richmond, VA 23282
Phone: 804-367-8037

**Washington**
Department of Revenue
Taxpayer Information and Education
Section
P.O. Box 47470
Olympia, WA 98504-7470
Phone: 206-753-5525

**West Virginia**
Department of Tax and Revenue
Sales and Use Tax
P.O. Box 1826
Charleston, WV 25327-1826
Phone: 304-558-3333

**Wisconsin**
Department of Revenue
Sales and Use Taxes
P.O. Box 8902
Madison, WI 53708
Phone: 608-266-2776

**Wyoming**
Department of Revenue and Taxation
Sales Taxes
122 W. 25th Street
Cheyenne, WY 82002
Phone: 307-777-7961

# Patent and Trademark Depository Libraries

|  |  |  |  |
|---|---|---|---|
| **Alabama** | Auburn University Libraries | 205-844-1747 | |
| | Birmingham Public Library | 205-226-3620 | |
| **Alaska** | Anchorage: Z.J. Loussac Public Library | 907-562-7323 | |
| **Arizona** | Tempe: Noble Library, Arizona State University | 602-965-7010 | |
| **Arkansas** | Little Rock: Arkansas State Library | 501-682-2053 | |
| **California** | Los Angeles Public Library | 213-228-7220 | |
| | Sacramento: California State Library | 916-654-0069 | |
| | San Diego Public Library | 619-236-5813 | |
| | San Francisco Public Library | 415-557-4500 | |
| | Santa Rosa: Bruce Sawyer Center (not a PTDL, but useful) | 707-524-1773 | |
| | Sunnyvale Center for Innovation | 408-730-7290 | |
| **Colorado** | Denver Public Library | 303-640-6220 | |
| **Delaware** | Newark: University of Delaware Library | 302-831-2965 | |
| **District of Columbia** | Washington: Howard University Libraries | 202-806-7252 | |
| **Florida** | Fort Lauderdale: Broward County Main Library | 954-357-7444 | |
| | Miami: Dade Public Library | 305-375-2665 | |
| | Orlando: Univ. of Central Florida Libraries | 407-823-2562 | |
| | Tampa: Tampa Campus Library, University of South Florida | 813-974-2726 | |
| **Georgia** | Atlanta: Price Gilbert Memorial Library, Georgia Institute of Technology | 404-894-4508 | |
| **Hawaii** | Honolulu: Hawaii State Public Library System | 808-586-3477 | |
| **Idaho** | Moscow: University of Idaho Library | 208-885-6235 | |
| **Illinois** | Chicago Public Library | 312-747-4450 | |
| | Springfield: Illinois State Library | 217-782-5659 | |
| **Indiana** | Indianapolis: Marion County Public Library | 317-269-1741 | |
| | West Lafayette: Purdue University Libraries | 317-494-2872 | |
| **Iowa** | Des Moines: State Library of Iowa | 515-281-4118 | |
| **Kansas** | Wichita: Ablah Library, Wichita State University | 316-978-3155 | |
| **Kentucky** | Louisville Free Public Library | 502-574-1611 | |
| **Louisiana** | Baton Rouge: Troy H. Middleton Library, Louisiana State University | 504-388-8875 | |
| **Maine** | Orono: Raymond H. Fogler Library, University of Maine | 207-581-1678 | |
| **Maryland** | College Park: Engineering and Physical Sciences Library, University of Maryland | 301-405-9157 | |
| **Massachusetts** | Amherst: Physical Sciences Library, University of Massachusetts | 413-545-1370 | |
| | Boston Public Library | 617-536-5400, | Ext. 265 |
| **Michigan** | Ann Arbor: Media Union Library, University of Michigan | 313-647-5735 | |
| | Big Rapids: Abigail S. Timme Library, Ferris State University | 616-592-3602 | |
| | Detroit Public Library | 313-833-3379 | |
| **Minnesota** | Minneapolis Public Library and Information Center | 612-372-6570 | |
| **Mississippi** | Jackson: Mississippi Library Commission | 601-359-1036 | |
| **Missouri** | Kansas City: Linda Hall Library | 816-363-4600 | |
| | St. Louis Public Library | 314-241-2288, | Ext. 390 |
| **Montana** | Butte: Montana College of Mineral Science & Technology Library | 406-496-4281 | |
| **Nebraska** | Lincoln: Engineering Library, University of Nebraska | 402-472-3411 | |
| **Nevada** | Reno: University of Nevada-Reno Library | 702-784-6500, | Ext. 257 |

| | | | |
|---|---|---|---|
| **New Hampshire** | Concord: New Hampshire State Library | 603-271-2239 | |
| **New Jersey** | Newark Public Library | 201-733-7782 | |
| | Piscataway: Library of Science & Medicine, Rutgers University | 908-445-2895 | |
| **New Mexico** | Albuquerque: University of New Mexico General Library | 505-277-4412 | |
| **New York** | Albany: New York State Library | 518-474-5355 | |
| | Buffalo and Erie County Public Library | 716-858-7101 | |
| | New York Science, Industry & Business Library | 212-592-7000 | |
| **North Carolina** | Raleigh: D.H. Hill Library, North Carolina State University | 919-515-3280 | |
| **North Dakota** | Grand Forks: Chester Fritz Library, University of North Dakota | 701-777-4888 | |
| **Ohio** | Akron: Summit County Public Library | 330-643-9075 | |
| | Cincinnati and Hamilton County, Public Library of | 513-369-6936 | |
| | Cleveland Public Library | 216-623-2870 | |
| | Columbus: Ohio State University Libraries | 614-292-6175 | |
| | Toledo/Lucas County Public Library | 419-259-5212 | |
| **Oklahoma** | Stillwater: Oklahoma State University Library | 405-744-7086 | |
| **Oregon** | Portland: Paul L. Boley Law Library, Lewis & Clark College | 503-768-6786 | |
| **Pennsylvania** | Philadelphia, The Free Library of | 215-686-5331 | |
| | Pittsburgh, Carnegie Library of | 412-622-3138 | |
| | University Park: Pattee Library, Pennsylvania State University | 814-865 4861 | |
| **Puerto Rico** | Mayaguez General Library, University of Puerto Rico | 787-832-4040, | Ext. 3459 |
| **Rhode Island** | Providence Public Library | 401-455-8027 | |
| **South Carolina** | R.M. Cooper Library, Clemson University | 864-656-3024 | |
| **South Dakota** | Rapid City: Devereaux Library, South Dakota School of Mines and Technology | 605-394-6822 | |
| **Tennessee** | Memphis & Shelby County Public Library and Information Center | 901-725-8877 | |
| | Nashville: Stevenson Science Library, Vanderbilt University | 615-322-2717 | |
| **Texas** | Austin: McKinney Engineering Library, University of Texas at Austin | 512-495-4500 | |
| | College Station: Sterling C. Evans Library, Texas A & M University | 409-845-3826 | |
| | Dallas Public Library | 214-670-1468 | |
| | Houston: The Fondren Library, Rice University | 713-527-8101, | Ext. 2587 |
| | Lubbock: Texas Tech University | not yet operational | |
| **Utah** | Salt Lake City: Marriott Library, University of Utah | 801-581-8394 | |
| **Vermont** | Burlington: Bailey/Howe Library, University of Vermont | not yet operational | |
| **Virginia** | Richmond: James Branch Cabell Library, Virginia Commonwealth University | 804-828-1104 | |
| **Washington** | Seattle: Engineering Library, University of Washington | 206-543-0740 | |
| **West Virginia** | Morgantown: Evansdale Library, West Virginia University | 304-293-2510 | Ext. 113 |
| **Wisconsin** | Madison: Kurt F. Wendt Library, University of Wisconsin | 608-262-6845 | |
| | Milwaukee Public Library | 414-286-3051 | |
| **Wyoming** | Casper: Natrona County Public Library | 307-237-4935 | |

# State Trademark Agencies and Statutes

**Alabama**
Ala. Code § 8-12-6 to 8-12-19 (1984)
Secretary of State
Lands & Trademark Division
Room 528, State Office Building
Montgomery, AL 36130-7701
205-242-5325

**Alaska**
Alaska Stat. 45.50.101 et seq.
Department of Commerce and
Economic Development
Corporations Section
P.O. Box D
Juneau, AK 99811
907-465-2530

**Arizona**
Ariz. Rev. Stat. 44-1441 et seq.
Office of Secretary of State
1700 W. Washington St.
Phoenix, AZ 85007
602-542-6187

**Arkansas**
Ark. Code A. §§ 4-71-101 thru 4-71-114
Secretary of State
State Capitol
Little Rock, AR 72201-1094
501-682-3405
FAX 501-682-3481

**California**
Cal. Bus. & Prof. Code § 14200 et seq.
Secretary of State
Attn: Trademark Unit
State of California
1230 "J" Street
Sacramento, CA 95814
916-445-9872

**Colorado**
Colo. Rev. Stat. §§ 7-70-102 to 7-70-113
Colorado Secretary of State
Corporations Office
1560 Broadway, Suite 200
Denver, CO 80202
303-894-2251

**Connecticut**
Conn. Gen. Stats, 621a §§ 35-11a et seq.
and 622a §§ 35-18a et seq.
Secretary of State
Division of Corporations, UCC
& Trademarks
Attn: Trademarks
State of Connecticut
30 Trinity St.
Hartford, CT 06106
203-566-1721

**Delaware**
6 Del. C. § 3301 et seq.
State of Delaware
Department of State
Division of Corporations
Attn: Trademark Filings
Townsend Building
P.O. Box 898
Dover, DE 19903
302-739-3073

**Florida**
Fla. Stat. ch. 495.011 et seq.
Corporation Records Bureau
Division of Corporations
Department of State
P.O. Box 6327
Tallahassee, FL 32301
904-487-6051

**Georgia**
O.C.G.A. §§ 10-1-440 et seq.
Office of Secretary of State
State of Georgia
306 W. Floyd Towers
2 MLK Drive
Atlanta, GA 30334
404-656-2861

**Hawaii**
Hawaii Revised Stats. § 482 et seq.
Department of Commerce and
Consumer Affairs
Business Registration Division
1010 Richards St.
Honolulu, HA 96813
808-586-2730

**Idaho**
Idaho Code §§ 48-501 et seq. (1979)
Secretary of State
Room 203
Statehouse
Boise, ID 83720
208-334-2300
FAX 208-334-2282

**Illinois**
Ill. Rev. Stat. 1987, ch. 140, §§ 8-22
Illinois Secretary of State
The Index Dept., Trademark Division
111 E. Monroe
Springfield, IL 62756
217-782-7017

**Indiana**
Indiana Code § 24-2-1-1 et seq.
Secretary of State of Indiana
Trademark Division
Rm 155, State House
Indianapolis, IN 46204
317-232-6540

**Iowa**
Iowa Code ch. 548
Secretary of State
Corporate Division
Hoover Bldg.
Des Moines, IA 50319
(515) 281-5204

**Kansas**
K.S.A. § 81-111 et seq.
Secretary of State
Statehouse Bldg., Room 235N
Topeka, KS 66612
913-296-2034

**Kentucky**
K.R.S. 365.560 to 365.625
Office of Kentucky Secretary of State
Frankfort, KY 40601
502-564-2848

**Louisiana**
La. Rev. Stat. Ann. 51:211 et seq.
Secretary of State
Corporation Division
P.O. Box 94125
Baton Rouge, LA 70804-9125
504-925-4704

**Maine**
10 M.R.S.A. § 1521-1532
State of Maine
Department of State
Division of Public Administration
State House Station 101
Augusta, ME 04333
207-287-4195

**Maryland**
Md. Ann. Code Art. 41, §§ 3-101 thru 3-114
Secretary of State
State House
Annapolis, MD 21404
301-974-5521

**Massachusetts**
Mass. Laws Ann., Ch. 110 B, § 1-16.
Office of Secretary of State
Trademark Division
Rm. 1711
One Ashburton Place
Boston, MA 02108
617-727-8329

**Michigan**
Mich. Compiled Laws §§ 429.31 et seq.
Michigan Department of Commerce
Corporations and Securities Bureau
Corporation Division
P.O. Box 30054
Lansing, MI 48909
517-334-6302

**Minnesota**
M.S.A. §§ 333.001-333.54
Secretary of State of Minnesota
Corporations Division
180 State Office Bldg.
St. Paul, MN 55155
612-296-3266

**Mississippi**
Miss. Code Ann., § 75-25-1 (1971)
Office of Secretary of State
P.O. Box 1350
Jackson, MS 39215
601-359-1350
FAX 601-359-6344

**Missouri**
Missouri Rev. Stat. 1978 §§ 417.005 et seq.
Office of Secretary of State
Attn: Trademark Division
P.O. Box 778
Jefferson City, MO 65101
314-751-4756

**Montana**
Mont. Code Ann., §§ 30-13-301 et seq. (1985)
Office of Secretary of State
Montana State Capitol
Helena, MT 59620
406-444-3665
FAX 406-444-3976

**Nebraska**
N.R.S. 1943, ch. 87 §§ 87.101 et seq.
Secretary of State
State Capitol Bldg.
Lincoln, NE 68509
402-471-4079

**Nevada**
Nev. Rev. Stat. 600.240 et seq.
Secretary of State of Nevada
Capitol Complex
Carson City, NV 89710
702-687-5203

**New Hampshire**
RSA 350-A
Corporation Division
Office of Secretary of State
State House Annex
Concord, NH 03301
603-271-3244

**New Jersey**
N. J. Stat. § 56:3-13.1 thru 56:3-13-15
Secretary of State
State House
CN-300
West State Street
Trenton, NJ 08625
609-984-1900

**New Mexico**
N.M.S.A. 57-3-1 thru 57-3-14
Secretary of State
Capitol Bldg. Rm. 400
Santa Fe, NM 87503
505-827-3600

**New York**
N. Y. Gen. Bus. Law § 360 et seq.
Secretary of State
Department of State
Miscellaneous Records
162 Washington Avenue
Albany, NY 12231
518-473-2492

**North Carolina**
N.C.G.S. § 80-1 et seq.
Trademark Division
Office of Secretary of State
300 N. Salisbury Street
Raleigh, NC 27611
919-733-4161

**North Dakota**
N.D.C.C., ch. 47-22
Secretary of State
State Capitol
Bismark, ND 58505
701-328-4284
FAX 701-328-2992

**Ohio**
ORC, ch. 1329.54 thru 1329.68
Secretary of State
Corporations Department
30 E. Broad St., 14th Floor
Columbus, OH 43215-0418
614-466-3910

**Oklahoma**
78 Okla. St. Ann. § 21 thru 34
Office of the Secretary of State
State of Oklahoma
101 State Capitol Bldg.
Oklahoma City, OK 73105
405-521-3911

**Oregon**
ORS 647.005 thru 647.105(i) and 647.115
Director, Corporation Division
Office of Secretary of State
158 12th Street N.E.
Salem, OR 97310-0210
503-986-2200

**Pennsylvania**
54 Pa. Cons. Stat. Ann. § 1101-1126
Department of State
(Purdon 1987 Supp.)
Corporation Bureau
308 North Office Bldg.
Harrisburg, PA 17120
717-787-1057

**Puerto Rico**
Title 10, Laws of P. R. Ann. § 191-195
Secretary of State of Puerto Rico
P.O. Box 3271
San Juan, PR 00904
809-722-2121, Ext. 337

**Rhode Island**
R. I. Gen. Laws §§ 6-2-1 thru 6-2-18
Secretary of State
The Trademarks Division
100 No. Main St.
Providence, RI 02903
401-277-2340

**South Carolina**
S. C. Code Ann. § 39-15-120 et seq.
Office of Secretary of State
P.O. Box 11350
Columbia, SC 29211
803-734-2158

**South Dakota**
SDCL ch 37-6
Secretary of State
State Capitol Bldg.
500 East Capitol
Pierre, SD 57501
605-773-3537

**Tennessee**
Tenn. Code Ann. § 47-25-501 et seq.
Secretary of State
Suite 500
James K. Polk Bldg.
Nashville, TN 37219
615-741-0531

**Texas**
Tex. Bus. & Com. Code § 16.01 thru 16.28
Secretary of State
Corporations Section, Trademark Office
Box 13697, Capitol Station
Austin, TX 78711-3697
512-463-5576

**Utah**
U.C.A. 70-3-1 et seq.
Division of Corporations
& Commercial Code
Heber M. Wells Bldg.
160 E. 300 South St
Salt Lake City, UT 84111
801-530-4849

**Virginia**
Va. Code § 59.1-77 et seq.
State Corp. Commission
Division of Securities and Retail Franchises
1220 Bank Street
Richmond, VA 23209
804-271-9051

**Vermont**
§ 9 V.S.A. §§ 2521 thru 2532
Vermont Secretary of State
Corporations Division
Redstone Bldg., 26 Terrace St.
Mail: State Office Bldg.
Montpelier, VT 05602-2199
802-828-2386

**Washington**
R.C.W. 19.77.010 et seq.
Corporations Division
Office of Secretary of State
Republic Building—2nd Floor
505 E. Union St
Olympia, WA 98504
206-753-7120

**West Virginia**
W. Va. Code § 47-2-1- et seq. and § 47-3-1 et seq.
Secretary of State
Corporations Division
State Capitol
Charleston, WV 25305
304-558-8000

**Wisconsin**
Wisconsin Stat. § 132.01 et seq.
Secretary of State
Trademark Records
P.O. Box 7848
Madison, WI 53707
608-266-5653

**Wyoming**
W.S. §§ 40-1-101 et seq.
Office of Secretary of State
Corporation Division
Capitol Bldg.
Cheyenne, WY 82002
307-777-7311 ■

APPENDIX

2

# Sample Independent Contractor Agreement and Amendment

# GENERAL INDEPENDENT CONTRACTOR AGREEMENT

This Agreement is made between ______________________________ (Client)
with a principal place of business at ______________________________
and ______________________________ (Contractor), with a
principal place of business at ______________________________.

**Services to Be Performed**

(Check and complete applicable provision.)

☐ Contractor agrees to perform the following services:

______________________________

______________________________

OR

☐ Contractor agrees to perform the services described in Exhibit A, which is attached to and made part of this Agreement.

______________________________

______________________________

**Payment**

(Check and complete applicable provision.)

☐ In consideration for the services to be performed by Contractor, Client agrees to pay Contractor
$______________

OR

☐ In consideration for the services to be performed by Contractor, Client agrees to pay Contractor at the rate of
$______________ per ______________________________.

(Check if applicable.)

☐ Contractor's total compensation shall not exceed $______________ without Client's written consent.

**Terms of Payment**

(Check and complete applicable provision.)

☐ Upon completing Contractor's services under this Agreement, Contractor shall submit an invoice. Client shall pay Contractor within ________ [10, 15, 30, 45, 60] days from the date of Contractor's invoice.

OR

☐ Contractor shall be paid $______________ [State amount] upon signing this Agreement and the remaining amount due when Contractor completes the services and submits an invoice. Client shall pay Contractor within ______________________________ [10, 15, 30, 45, 60] days from the date of Contractor's invoice.

OR

☐ Contractor shall be paid according to the Schedule of Payments set forth in Exhibit __________ [A or B] attached to and made part of this Agreement.

OR

☐ Contractor shall send Client an invoice monthly. Client shall pay Contractor within [10, 15, 30, 45, 60] days from the date of each invoice.

**Late Fees**

(Check and complete if applicable.)

☐ Late payments by Client shall be subject to late penalty fees of ______________% per month from the due date until the amount is paid.

**Expenses**

(Check applicable provision.)

☐ Contractor shall be responsible for all expenses incurred while performing services under this Agreement.

☐ However, Client shall reimburse Contractor for all reasonable travel and living expenses necessarily incurred by Contractor while away from Contractor's regular place of business to perform services under this Agreement. Contractor shall submit an itemized statement of such expenses. Client shall pay Contractor within 30 days from the date of each statement.

OR

☐ Client shall reimburse Contractor for the following expenses that are directly attributable to work performed under this Agreement:

- travel expenses other than normal commuting, including airfares, rental vehicles, and highway mileage in company or personal vehicles at ______ cents per mile
- telephone, facsimile (fax), online and telegraph charges
- postage and courier services
- printing and reproduction
- computer services, and
- other expenses resulting from the work performed under this Agreement.

Contractor shall submit an itemized statement of Contractor's expenses. Client shall pay Contractor within 30 days from the date of each statement.

**Materials**

(Check and complete if applicable.)

☐ Contractor will furnish all materials and equipment used to provide the services required by this Agreement.

☐ Client shall make available to Contractor, at Client's expense, the following materials, facilities and equipment: ______________________________________________ [List]. These items will be provided to Contractor by ______ [Date].

**Term of Agreement**

This Agreement will become effective when signed by both parties and will end no later than ______, 19___.

**Terminating the Agreement**

(Check applicable provision.)

☐ With reasonable cause, either party may terminate this Agreement effective immediately by giving written notice of termination for cause. Reasonable cause includes:

- a material violation of this agreement, or
- nonpayment of Contractor's compensation after 20 days written demand for payment.

Contractor shall be entitled to full payment for services performed prior to the effective date of termination

OR

☐ Either party may terminate this Agreement at any time by giving ______ [5, 10, 15, 30, 45, 60] days written notice of termination without cause. Contractor shall be entitled to full payment for services performed prior to the effective date of termination.

**Independent Contractor Status**

Contractor is an independent contractor, not Client's employee. Contractor's employees or subcontractors are not Client's employees. Contractor and Client agree to the following rights consistent with an independent contractor relationship.

- Contractor has the right to perform services for others during the term of this Agreement.
- Contractor has the sole right to control and direct the means, manner and method by which the services required by this Agreement will be performed.
- Contractor has the right to hire assistants as subcontractors, or to use employees to provide the services required by this Agreement.
- The Contractor or Contractor's employees or subcontractors shall perform the services required by this Agreement; Client shall not hire, supervise or pay any assistants to help Contractor.
- Neither Contractor nor Contractor's employees or subcontractors shall receive any training from Client in the skills necessary to perform the services required by this Agreement.
- Client shall not require Contractor or Contractor's employees or subcontractors to devote full time to performing the services required by this Agreement.
- Neither Contractor nor Contractor's employees or subcontractors are eligible to participate in any employee pension, health, vacation pay, sick pay or other fringe benefit plan of Client.

**Local, State and Federal Taxes**

Contractor shall pay all income taxes, and FICA (Social Security and Medicare taxes) incurred while performing services under this Agreement. Client will not:

- withhold FICA (Social Security and Medicare taxes) from Contractor's payments or make FICA payments on Contractor's behalf
- make state or federal unemployment compensation contributions on Contractor's behalf, or
- withhold state or federal income tax from Contractor's payments.

The charges included here do not include taxes. If Contractor is required to pay any federal, state or local sales, use, property or value added taxes based on the services provided under this Agreement, the taxes shall be separately billed to Client. Contractor shall not pay any interest or penalties incurred due to late payment or nonpayment of such taxes by Client.

**Notices**

All notices and other communications in connection with this Agreement shall be in writing and shall be considered given as follows:

- when delivered personally to the recipient's address as stated on this Agreement
- three days after being deposited in the United States mail, with postage prepaid to the recipient's address as stated on this Agreement, or
- when sent by fax or telex to the last fax or telex number of the recipient known to the person giving notice. Notice is effective upon receipt, provided that a duplicate copy of the notice is promptly given by first class mail, or the recipient delivers a written confirmation of receipt.

**No Partnership**

This Agreement does not create a partnership relationship. Neither party has authority to enter into contracts on the other's behalf.

**Applicable Law**

This Agreement will be governed by the laws of the state of ______________________________.

**Exclusive Agreement**

This is the entire Agreement between Contractor and Client.

**Dispute Resolution**

(Check if applicable)

☐ If a dispute arises under this Agreement, the parties agree to first try to resolve the dispute with the help of a mutually agreed-upon mediator in the following location ______________________________ [List city or county where mediation will occur]. Any costs and fees other than attorney fees associated with the mediation shall be shared equally by the parties.

If it proves impossible to arrive at a mutually satisfactory solution through mediation, the parties agree to submit the dispute to binding arbitration in the following location ______________________________ [List city or county where arbitration will occur] under the rules of the American Arbitration Association. Judgment upon the award rendered by the arbitrator may be entered in any court having jurisdiction to do so.

However, the complaining party may refuse to submit the dispute to mediation or arbitration and instead bring an action in an appropriate Small Claims Court.

**Contract Changes**

(Check if applicable)

☐ Client and Contractor recognize that:

- Contractor's original cost and time estimates may be too low due to unforeseen events, or to factors unknown to Contractor when this Agreement was made
- Client may desire a mid-project change in Contractor's services that would add time and cost to the project and possibly inconvenience Contractor, or
- Other provisions of this Agreement may be difficult to carry out due to unforeseen circumstances.

If any intended changes or any other events beyond the parties' control require adjustments to this Agreement, the parties shall make a good faith effort to agree on all necessary particulars. Such agreements shall be put in writing, signed by the parties and added to this Agreement.

**Attorneys' Fees**

If any legal action is necessary to enforce this Agreement, the prevailing party shall be entitled to reasonable attorney fees, costs and expenses in addition to any other relief to which he or she may be entitled.

**Signatures**

Client:

[NAME OF CLIENT]

________________________________________
Name of Client

By: ________________________________________
Signature

________________________________________
Typed or Printed Name

Title: ________________________________________

Date: ____________________

Contractor

[NAME OF CONTRACTOR]

________________________________________
Name of Contractor

By: ________________________________________
Signature

________________________________________
Typed or Printed Name

Title: ________________________________________

Taxpayer ID Number: ____________________________

Date: ____________________

**If Agreement Is Faxed:**

Contractor and Client agree that this Agreement will be considered signed when the signature of a party is delivered by facsimile transmission. Signatures transmitted by facsimile shall have the same effect as original signatures.

# CONTRACT AMENDMENT FORM

## AMENDMENT

This Amendment is made between ______________________________ and

______________________________ to amend the the Original Agreement titled

______________________________ signed by them on

______________________ .

The Original Agreement is amended as follows:

All provisions of the Original Agreement, except as modified by this Amendment, remain in full force and effect and are reaffirmed. If there is any conflict between this Amendment and any provision of the Original Agreement, the provisions of this Amendment shall control.

**Client:**

______________________________

By: ______________________________
(Signature)

______________________________
Typed or Printed Name

Title: ______________________________

Date: ______________________

**Contractor:**

______________________________

By: ______________________________
(Signature)

______________________________
Typed or Printed Name

______________________________

Title: ______________________________

Date: ______________________

# Index

## D



## E



## F

## G



## H



## I

## J



## K



## L

## M



## N



## O



## P

## Q



## R



## S

# T

## U



## W



## Z

# CATALOG

**...more from Nolo Press**

## BUSINESS

| | | PRICE | CODE |
|---|---|---|---|
| | The California Nonprofit Corporation Handbook | $29.95 | NON |
| | The California Professional Corporation Handbook | $34.95 | PROF |
| | The Employer's Legal Handbook | $29.95 | EMPL |
| | Form Your Own Limited Liability Company | $34.95 | LIAB |
| ▣ | Hiring Independent Contractors: The Employer's Legal Guide, (Book w/Disk—PC) | $29.95 | HICI |
| ▣ | How to Form a CA Nonprofit Corp.—w/Corp. Records Binder & PC Disk | $49.95 | CNP |
| ▣ | How to Form a Nonprofit Corp., Book w/Disk (PC)—National Edition | $39.95 | NNP |
| ▣ | How to Form Your Own Calif. Corp.—w/Corp. Records Binder & Disk—PC | $39.95 | CACI |
| | How to Form Your Own California Corporation | $29.95 | CCOR |
| ▣ | How to Form Your Own Florida Corporation, (Book w/Disk—PC) | $39.95 | FLCO |
| ▣ | How to Form Your Own New York Corporation, (Book w/Disk—PC) | $39.95 | NYCO |
| ▣ | How to Form Your Own Texas Corporation, (Book w/Disk—PC) | $39.95 | TCOR |
| | Take Charge of Your Workers' Compensation Claim (California Edition) | $29.95 | WORK |
| | How to Market a Product for Under $500 | $29.95 | UN500 |
| | How to Mediate Your Dispute | $18.95 | MEDI |
| | How to Write a Business Plan | $21.95 | SBS |
| | The Independent Paralegal's Handbook | $29.95 | PARA |
| | Legal Guide for Starting & Running a Small Business, Vol. 1 | $24.95 | RUNS |
| ▣ | Legal Guide for Starting & Running a Small Business, Vol. 2: Legal Forms | $29.95 | RUNS2 |
| | Marketing Without Advertising | $19.00 | MWAD |

▣ Book with disk
● Book with CD-ROM

| | | PRICE | CODE |
|---|---|---|---|
| ◘ | The Partnership Book: How to Write a Partnership Agreement, (Book w/Disk—PC) | $34.95 | PART |
| | Sexual Harassment on the Job | $18.95 | HARS |
| | Starting and Running a Successful Newsletter or Magazine | $24.95 | MAG |
| ◘ | Taking Care of Your Corporation, Vol. 1, (Book w/Disk—PC) | $26.95 | CORK |
| ◘ | Taking Care of Your Corporation, Vol. 2, (Book w/Disk—PC) | $39.95 | CORK2 |
| | Tax Savvy for Small Business | $28.95 | SAVVY |
| | Trademark: Legal Care for Your Business and Product Name | $29.95 | TRD |
| | Wage Slave No More: Law & Taxes for the Self-Employed | $24.95 | WAGE |
| | Your Rights in the Workplace | $19.95 | YRW |

## CONSUMER

| | | PRICE | CODE |
|---|---|---|---|
| | Fed Up With the Legal System: What's Wrong & How to Fix It | $9.95 | LEG |
| | How to Win Your Personal Injury Claim | $24.95 | PICL |
| | Nolo's Everyday Law Book | $21.95 | EVL |
| | Nolo's Pocket Guide to California Law | $11.95 | CLAW |
| | Trouble-Free Travel...And What to Do When Things Go Wrong | $14.95 | TRAV |

## ESTATE PLANNING & PROBATE

| | | PRICE | CODE |
|---|---|---|---|
| | 8 Ways to Avoid Probate (Quick & Legal Series) | $15.95 | PRO8 |
| | How to Probate an Estate (California Edition) | $34.95 | PAE |
| | Make Your Own Living Trust | $24.95 | LITR |
| ◘ | Nolo's Will Book, (Book w/Disk—PC) | $29.95 | SWIL |
| | Plan Your Estate | $24.95 | NEST |
| | The Quick and Legal Will Book | $15.95 | QUIC |
| | Nolo's Law Form Kit: Wills | $14.95 | KWL |

## FAMILY MATTERS

| | | PRICE | CODE |
|---|---|---|---|
| | A Legal Guide for Lesbian and Gay Couples | $24.95 | LG |
| | California Marriage Law | $19.95 | MARR |
| | Child Custody: Building Parenting Agreements that Work | $24.95 | CUST |
| | Divorce & Money: How to Make the Best Financial Decisions During Divorce | $26.95 | DIMO |
| | Get A Life: You Don't Need a Million to Retire Well | $18.95 | LIFE |
| | The Guardianship Book (California Edition) | $24.95 | GB |

◘ Book with disk

● Book with CD-ROM

| | PRICE | CODE |
|---|---|---|
| How to Adopt Your Stepchild in California | $22.95 | ADOP |
| How to Do Your Own Divorce in California | $24.95 | CDIV |
| How to Do Your Own Divorce in Texas | $19.95 | TDIV |
| How to Raise or Lower Child Support in California | $18.95 | CHLD |
| The Living Together Kit | $24.95 | LTK |
| Nolo's Law Form Kit: Hiring Childcare & Household Help | $14.95 | KCHLD |
| Nolo's Pocket Guide to Family Law | $14.95 | FLD |
| Practical Divorce Solutions | $14.95 | PDS |

## GOING TO COURT

| | | |
|---|---|---|
| Collect Your Court Judgment (California Edition) | $24.95 | JUDG |
| How to Seal Your Juvenile & Criminal Records (California Edition) | $24.95 | CRIM |
| How to Sue For Up to 25,000...and Win! | $29.95 | MUNI |
| Everybody's Guide to Small Claims Court in California | $18.95 | CSCC |
| Everybody's Guide to Small Claims Court (National Edition) | $18.95 | NSCC |
| Fight Your Ticket ... and Win! (California Edition) | $19.95 | FYT |
| How to Change Your Name (California Edition) | $29.95 | NAME |
| Mad at Your Lawyer | $21.95 | MAD |
| Represent Yourself in Court: How to Prepare & Try a Winning Case | $29.95 | RYC |
| The Criminal Law Handbook: Know Your Rights, Survive the System | $24.95 | KYR |

## HOMEOWNERS, LANDLORDS & TENANTS

| | | |
|---|---|---|
| The Deeds Book (California Edition) | $16.95 | DEED |
| Dog Law | $14.95 | DOG |
| ▣ Every Landlord's Legal Guide (National Edition) | $34.95 | ELLI |
| Every Tenant's Legal Guide | $24.95 | EVTEN |
| For Sale by Owner (California Edition) | $24.95 | FSBO |
| How to Buy a House in California | $24.95 | BHCA |
| The Landlord's Law Book, Vol. 1: Rights & Responsibilities (California Edition) | $34.95 | LBRT |
| The Landlord's Law Book, Vol. 2: Evictions (California Edition) | $34.95 | LBEV |
| Leases & Rental Agreements (Quick & Legal Series) | $18.95 | LEAR |
| Neighbor Law: Fences, Trees, Boundaries & Noise | $16.95 | NEI |
| Tenants' Rights (California Edition) | $19.95 | CTEN |
| Stop Foreclosure Now in California | $29.95 | CLOS |

▣ Book with disk
● Book with CD-ROM

| | | PRICE | CODE |
|---|---|---|---|

## HUMOR

| | | PRICE | CODE |
|---|---|---|---|
| | 29 Reasons Not to Go to Law School | $9.95 | 29R |
| | Poetic Justice | $9.95 | PJ |

## IMMIGRATION

| | | PRICE | CODE |
|---|---|---|---|
| | How to Get a Green Card: Legal Ways to Stay in the U.S.A. | $24.95 | GRN |
| | U.S. Immigration Made Easy | $39.95 | IMEZ |

## MONEY MATTERS

| | | PRICE | CODE |
|---|---|---|---|
| | 101 Law Forms for Personal Use: Quick and Legal Series (Book w/disk) | $24.95 | 101LAW |
| | Chapter 13 Bankruptcy: Repay Your Debts | $29.95 | CH13 |
| | Credit Repair (Quick & Legal Series) | $15.95 | CREP |
| | The Financial Power of Attorney Workbook | $24.95 | FINPOA |
| | How to File for Bankruptcy | $26.95 | HFB |
| | Money Troubles: Legal Strategies to Cope With Your Debts | $19.95 | MT |
| | Nolo's Law Form Kit: Personal Bankruptcy | $14.95 | KBNK |
| | Stand Up to the IRS | $24.95 | SIRS |

## PATENTS AND COPYRIGHTS

| | | PRICE | CODE |
|---|---|---|---|
| | The Copyright Handbook: How to Protect and Use Written Works | $29.95 | COHA |
| | Copyright Your Software | $39.95 | CYS |
| ▫ | License Your Invention (Book w/Disk) | $39.95 | LICE |
| | The Patent Drawing Book | $29.95 | DRAW |
| | Patent, Copyright & Trademark | $24.95 | PCTM |
| | Patent It Yourself | $44.95 | PAT |
| ● | Software Development: A Legal Guide (Book with CD-ROM) | $44.95 | SFT |
| | The Inventor's Notebook | $19.95 | INOT |

## RESEARCH & REFERENCE

| | | PRICE | CODE |
|---|---|---|---|
| ● | Government on the Net, (Book w/CD-ROM—Windows/Macintosh) | $39.95 | GONE |
| ● | Law on the Net, (Book w/CD-ROM—Windows/Macintosh) | $39.95 | LAWN |
| | Legal Research: How to Find & Understand the Law | $19.95 | LRES |
| | Legal Research Made Easy (Video) | $89.95 | LRME |

▫ Book with disk

● Book with CD-ROM

| | PRICE | CODE |
|---|---|---|

## SENIORS

| Title | PRICE | CODE |
|---|---|---|
| Beat the Nursing Home Trap | $18.95 | ELD |
| Social Security, Medicare & Pensions | $19.95 | SOA |
| The Conservatorship Book (California Edition) | $29.95 | CNSV |

## SOFTWARE

**Call or check our website for special discounts on Software!**

| Title | PRICE | CODE |
|---|---|---|
| California Incorporator 2.0—DOS | $79.95 | INCI |
| Living Trust Maker 2.0—Macintosh | $79.95 | LTM2 |
| Living Trust Maker 2.0—Windows | $79.95 | LTWI2 |
| ● Small Business Legal Pro Deluxe CD—Windows/Macintosh CD-ROM | $79.95 | SBCD3 |
| Nolo's Partnership Maker 1.0—DOS | $79.95 | PAGI1 |
| Personal RecordKeeper 4.0—Macintosh | $49.95 | RKM4 |
| Personal RecordKeeper 4.0—Windows | $49.95 | RKP4 |
| Patent It Yourself 1.0—Windows | $229.95 | PYP12 |
| ● WillMaker 6.0—Windows/Macintosh CD-ROM | $69.95 | WD6 |

- ▫ Book with disk
- ● Book with CD-ROM

## ORDER FORM

| Code | Quantity | Title | Unit price | Total |
|---|---|---|---|---|
| | | | | |
| | | | | |
| | | | | |
| | | | | |
| | | | | |
| | | | | |
| | | Subtotal | | |
| | | California residents add Sales Tax | | |
| | | Basic Shipping (*$6.00 for 1 item; $7.00 for 2 or more*) | | |
| | | UPS RUSH delivery $7.50–any size order* | | |
| | | TOTAL | | |

Name

Address

(UPS to street address, Priority Mail to P.O. boxes)

* Delivered in 3 business days from receipt of order. S.F. Bay Area use regular shipping.

## FOR FASTER SERVICE, USE YOUR CREDIT CARD AND OUR TOLL-FREE NUMBERS

| | |
|---|---|
| Order 24 hours a day | 1-800-992-6656 |
| Fax your order | 1-800-645-0895 |
| e-mail | cs@nolo.com |
| General Information | 1-510-549-1976 |
| Customer Service | 1-800-728-3555, Mon.-Fri. 9am-5pm, PST |

## METHOD OF PAYMENT

☐ Check enclosed

☐ VISA ☐ MasterCard ☐ Discover Card ☐ American Express

Account # Expiration Date

Authorizing Signature

Daytime Phone

PRICES SUBJECT TO CHANGE.

## VISIT OUR OUTLET STORES!

You'll find our complete line of books and software, all at a discount.

**BERKELEY**
950 Parker Street
Berkeley, CA 94710
1-510-704-2248

**SAN JOSE**
111 N. Market Street, #115
San Jose, CA 95113
1-408-271-7240

## VISIT US ONLINE!

on the Internet

**www.nolo.com**

**NOLO PRESS 950 PARKER ST., BERKELEY, CA 94710**

more from

# NOLO PRESS

## Small Business Legal Pro 3

 *Windows/Macintosh CD-ROM*

As a small business owner, you face complex legal questions every day. Now, for less than the cost of an hour with a lawyer, get clear and concise answers instantly with *Small Business Legal Pro*. Includes the complete, searchable text of six bestselling Nolo books: *Legal Guide for Starting and Running a Small Business, Volumes 1 & 2, Tax Savvy for Small Business, Everybody's Guide to Small Claims Court, Marketing Without Advertising*, and *The Employer's Legal Handbook*.

In addition to all this practical advice, you'll get more than 120 forms you'll use every day, with plain-English, step-by-step instructions to fill them out. The newly revised and updated Version 3.0 includes hundreds of pages of high-impact, low-cost marketing strategies to help your business grow and succeed.

**"With Small Business Legal Pro, *you'll have a virtual lawyer at your fingertips.*"**

--ENTREPRENEUR MAGAZINE

***"This encyclopedic package provides clear and concise information with a clean interface."***

--HOME OFFICE COMPUTING

**SBCD3/ ~~79.95~~ 47.97**

**CALL 800-992-6656 OR USE THE ORDER FORM IN THIS BOOK**

## Simplified Statement of Cash Flows
### (Exhibit 4.16, p. 184; Problem 4-5 for Self-Study)

| | |
|---|---|
| **Operations** | |
| Cash Receipts from Customers | (1) |
| Less: Cash Payments to Suppliers, Employees, and Others | −(2) |
| Cash Flow from Operations [= (1) − (2)] | S1 |
| Reconciliation of Net Income to Cash Flow from Operations | |
| Net Income | (3) |
| Additions to Net Income to Compute Cash Flow from Operations | +(4) |
| Subtractions from Net Income to Compute Cash Flow from Operations | −(5) |
| Cash Flow from Operations [= (3) + (4) − (5)] | S1 |
| **Investing** | |
| Proceeds from Dispositions of "Investing" Assets | +(6) |
| Cash Used to Acquire "Investing" Assets | −(7) |
| Cash Flow from Investing [= (6) − (7)] | S2 |
| **Financing** | |
| Cash Provided by Increases in Debt or Capital Stock | +(8) |
| Cash Used to Reduce Debt or Capital Stock | −(9) |
| Cash Used for Dividends | −(10) |
| Cash Flow from Financing [= (8) − (9) − (10)] | S3 |
| Net Change in Cash [= S1 + S2 + S3] | (11) |
| Cash, Beginning of the Period | S4 |
| Cash, End of the Period [= (11) + S4] | S5 |

## Reporting Comprehensive Income: One-Statement Approach
### (Exhibit 12.4 Excerpt, p. 596)

| **Statement of Net Income and Comprehensive Income** | | |
|---|---|---|
| Revenues | | $100,000 |
| Expenses | | (25,000) |
| Gain on Sale of Securities | | 2,000 |
| Other Gains and Losses | | 8,000 |
| Earnings from Continuing Operations before Income Tax | | $ 85,000 |
| Income Tax Expense | | (21,250) |
| Earnings before Discontinued Operations and Extraordinary Items | | $ 63,750 |
| Discontinued Operations, Net of Tax | | 30,000 |
| Extraordinary Items, Net of Tax | | (30,500) |
| **Net Income (or, as preferred by the FASB, Earnings)** | | **$ 63,250** |
| Other Comprehensive Income, Net of Tax: | | |
| Foreign Currency Translation Adjustments | | 7,000 |
| Unrealized Gains and Losses on Securities: | | |
| Unrealized Holding Gains Arising during Period | $13,000 | |
| Less: Reclassification Adjustment for Gain Included in Net Income (Earnings) | (1,500) | 11,500 |
| Minimum Pension Liability Adjustment | | (2,500) |
| Other Comprehensive Income (Loss) | | **$ 16,000** |
| **Comprehensive Income (Loss)** | | **$ 79,250** |

## *Terminology* Note

The FASB faced an issue in naming the amount that we call *net income* in the traditional income statement. Use of the term *net* implies that it is a bottom-line number. Yet firms that append other comprehensive income to the bottom of their income statements need a label to differentiate traditional net income from the combined amounts of traditional net income plus other comprehensive income. The FASB expressed a preference to eliminate the term *net income* altogether, to label traditional net income as *earnings*, and to label the combined amounts of earnings and other comprehensive income as *comprehensive income*. The FASB gave firms wide latitude, not only in their choice of reporting formats, but in their choice of terminology. Most firms choose to follow the third reporting format above, leaving the traditional income statement unchanged. Thus, we will continue to use *net income* to designate the bottom line of the traditional income statement.

# Financial Accounting

## An Introduction to Concepts, Methods, and Uses

12th Edition

Clyde P. Stickney
Dartmouth College

Roman L. Weil
University of Chicago

Australia · Canada · Mexico · Singapore · Spain · United Kingdom · United States

# Financial Accounting, 12e
Stickney / Weil

Executive Editors:
Michele Baird, Maureen Staudt &
Michael Stranz

Project Development Manager:
Linda deStefano

Sr. Marketing Coordinators:
Lindsay Annett and Sara Mercurio

Production/Manufacturing Manager:
Donna M. Brown

Production Editorial Manager:
Dan Plofchan

Pre-Media Services Supervisor:
Becki Walker

Rights and Permissions Specialists:
Kalina Hintz and Connee Draper

Cover Image
Getty Images*



Printed in the
United States of America
1 2 3 4 5 6 7 8 08 07

For more information, please contact Thomson Custom Solutions, 5191 Natorp Boulevard, Mason, OH 45040. Or you can visit our Internet site at http://www.thomsoncustom.com



The Adaptable Courseware Program consists of products and additions to existing Thomson products that are produced from camera-ready copy. Peer review, class testing, and accuracy are primarily the responsibility of the author(s).

ISBN 978-0-324-67956-4
(0-324-67956-4)

International Divisions List

Asia (Including India):
Thomson Learning
(a division of Thomson Asia Pte Ltd)
5 Shenton Way #01-01
UIC Building
Singapore 068808
Tel: (65) 6410-1200
Fax: (65) 6410-1208

Australia/New Zealand:
Thomson Learning Australia
102 Dodds Street
Southbank, Victoria 3006
Australia

Latin America:
Thomson Learning
Seneca 53
Colonia Polano
11560 Mexico, D.F., Mexico
Tel (525) 281-2906
Fax (525) 281-2656

Canada:
Thomson Nelson
1120 Birchmount Road
Toronto, Ontario
Canada M1K 5G4
Tel (416) 752-9100
Fax (416) 752-8102

UK/Europe/Middle East/Africa:
Thomson Learning
High Holborn House
50-51 Bedford Row
London, WC1R 4LS
United Kingdom
Tel 44 (020) 7067-2500
Fax 44 (020) 7067-2600

Spain (Includes Portugal):
Thomson Paraninfo
Calle Magallanes 25
28015 Madrid
España
Tel 34 (0)91 446-3350
Fax 34 (0)91 445-6218

*Unless otherwise noted, all cover images used by Thomson Custom Solutions have been supplied courtesy of Getty Images with the exception of the *Earthview* cover image, which has been supplied by the National Aeronautics and Space Administration (NASA).

## *Custom Table of Contents*

CHAPTER 1

# Introduction to Business Activities and Overview of Financial Statements and the Reporting Process

## Learning Objectives

1. Develop an understanding of four principal activities of business firms: (a) establishing goals and strategies, (b) obtaining financing, (c) making investments, and (d) conducting operations.

2. Develop an understanding of the purpose and content of the three principal financial statements that business firms prepare to measure and report the results of their business activities: (a) balance sheet, (b) income statement, and (c) statement of cash flows.

3. Develop a sensitivity to financial reporting issues, including the following: (a) the potential conflict of interest between management's self-interest for job security and career enhancement with its responsibility to shareholders, (b) the alternative approaches to establishing accounting measurement and reporting standards, and (c) the role of the independent audit of a firm's financial statements.

4. Develop a sensitivity to ethical issues in financial reporting using a framework for thinking about ethical questions.

*E*nron. Global Crossing. WorldCom. Qwest Communication. Adelphia. Freddie Mac. The list seems to go on and on of companies accused of issuing fraudulent or misleading financial statements in recent years. Yet, the proper allocation of resources in an economy depends on reliable and relevant information about the profitability and risk of firms. How does one know if a firm has "cooked its books" in an effort to mislead users of its financial statements? What safeguards exist to minimize the risk of such financial shenanigans?

You are about to embark on a study of financial accounting. You will learn the important concepts underlying financial statements, the accounting principles firms use to measure the results of their business activities, and some tools for analyzing financial statements. You will consider the judgments that firms make in measuring and reporting business activities and the opportunity such judgments provide for biasing the reported amounts. Our perspective throughout this study is that of a user of financial statements. We presume that you will not become an accountant, or even an expert at detecting misleading financial reports. Our aim is to equip you with sufficient understanding of the concepts, methods, and uses of financial statements so that you can use accounting data effectively. We also describe recent legislated efforts to enhance the quality of financial information and reduce the risk of financial reporting abuses.

This chapter overviews the whole book. We begin by studying how a typical firm carries out its business activities. We next see how the firm measures the results of these business activities and reports them in the financial statements. We then raise several issues about the financial reporting process. This chapter introduces material that later chapters will cover in greater depth. At this point in your study of financial accounting, you will not fully understand all of the concepts and terms discussed in this chapter. As the chapter title suggests, we introduce the big picture of the concepts, methods, and uses to be discussed later.

## Overview of Business Activities

Firms prepare financial statements for various external users: owners, lenders, regulators, and employees. The financial statements attempt to present meaningful information about a firm's business activities. Understanding these financial statements requires an understanding of the business activities that they attempt to portray.

**Example 1** Wal-Mart Stores, Inc. (Wal-Mart) is the largest retailing firm in the world. It emphasizes an everyday-low-price strategy and operates through three principal store concepts:

**Wal-Mart Stores:** discount department stores that offer clothing, housewares, electronic equipment, pharmaceuticals, health and beauty products, sporting goods, and similar items.

**Wal-Mart Supercenters:** full-line supermarkets offering grocery products combined with Wal-Mart's traditional discount stores.

**Sam's Clubs:** members-only warehouse stores that offer food, household, automotive, electronic, sporting, and other products in unusually large quantities (for example, toothpaste by the quart) or containers at wholesale prices.

The following sections describe some of the more important business activities that the management of firms such as Wal-Mart must understand.

### Establishing Corporate Goals and Strategies

A firm's **goals** state the targets, or end results, toward which the firm directs its energies. A firm's **strategies** state the means for achieving these goals. The firm sets goals and strategies in light of the economic, institutional, and cultural environment in which it intends to operate. For example, a firm might set goals to

- maximize the return to the owners of the firm;
- provide a stimulating and stable lifetime working environment for employees; and
- contribute to and integrate with governmental goals and policies.

Management sets strategies for a firm as a whole. For example, a firm might choose to operate in one industry or several industries. It might integrate backward into production of raw materials or forward into distribution of products to customers. It might operate in the United States only or in other countries too.

Management also sets strategies for each business unit or product. For example, a firm might attempt to find a niche for each of its products. Such a strategy might permit the firm to obtain favorable selling prices for its products relative to competitors so that it can pass along cost increases to its customers. Common business terminology refers to this as a *product differentiation strategy*. Alternatively, the firm might choose primarily to emphasize cost control and might strive to be the low-cost producer in its industry. Such a strategy might permit it to charge aggressively low prices and generate high volumes. Terminology refers to this as a *low-cost leadership strategy,* which Wal-Mart follows. Some products lend themselves more naturally to one strategy or the other, whereas the firm might pursue either or both strategies for other products.

When establishing goals and strategies, the firm must consider external environmental factors, such as the following:

1. Who are the firm's competitors, and what goals and strategies do they pursue?
2. What barriers, such as patents or large investments in buildings and equipment, might preclude new firms from entering the industry? Or, are entry barriers low?
3. Is the demand for products within the industry increasing rapidly, such as for Internet software and biotechnology products, or is the demand relatively stable, such as the demand for groceries?
4. Is the industry subject to government regulation, such as for food and pharmaceutical products, or is it unregulated, such as for clothes and paper towels?

Wal-Mart and other publicly traded firms have a common goal—to increase the value of the firm and the wealth of its owners. Wal-Mart likely pursues other goals, such as providing for employees' ongoing development and welfare as well as contributing to the needs in the community. Its annual report to shareholders suggests that Wal-Mart's strategies to accomplish these goals include the following:

1. Wal-Mart will continue to convert existing discount stores into supercenters and open new supercenters to capitalize on one-stop shopping by customers and to gain efficiencies in product distribution, stocking, and advertising.
2. Wal-Mart will expand into other countries primarily by acquiring existing retail chains instead of opening new stores on its own. Wal-Mart will then install its information, purchasing, and distribution systems into these retail chains in an effort to add value by reducing costs.
3. Wal-Mart will open smaller grocery stores in neighborhoods too thinly populated to justify a supercenter.

## Obtaining Financing

Before Wal-Mart can implement these growth and expansion strategies, it must obtain the necessary financing. **Financing activities** involve obtaining funds from two principal sources: owners and creditors.

**Owners** Owners provide funds to a firm and in return receive some evidence of their ownership. For a corporation, shares of common stock provide evidence of ownership, and the owners are called *shareholders* or *stockholders*.[1] The firm need not repay the owners at a particular future date. Instead, the owners receive distributions, called **dividends,** from the firm only when the firm decides to pay them. The owners also have a claim on all the firm's increases in value resulting from future profitable operations.

**Creditors** Creditors provide funds but, unlike owners, require that the firm repay the funds, usually with interest, in specific amounts at specific dates. The length of time that elapses until final repayment varies.

Long-term creditors may provide funds and not require repayment for 20 years or more. A bond usually evidences such borrowings. In a bond agreement, the borrowing company promises to make payments to the creditors at specific dates in the future, some of which represents interest on the amounts borrowed and some represents repayments of the amount borrowed.

Banks usually lend for periods between several months and several years. A note, in which the borrowing company promises to repay the amount borrowed plus interest at some future date, provides evidence of bank borrowings.

Suppliers of raw materials or merchandise do not always view themselves as supplying funds to a firm. Yet when they sell raw materials or merchandise but do not require payment for 30 days, they implicitly provide funds—the firm gets raw materials or merchandise now but need not pay cash until later. Likewise, employees paid weekly or monthly, and governmental units requiring only monthly or quarterly tax payments, provide funds by not demanding payment hourly or daily.

Each firm will make its own financing decisions, choosing the proportion of funds it will obtain from owners, long-term creditors, and short-term creditors. Corporate finance courses cover such financing decisions.

---

[1]When a firm operates as a partnership, its owners are partners. When a firm operates as a sole proprietorship, its owner is a sole proprietor. This book focuses on the corporate form of legal entity.

## Making Investments

Once a firm has obtained funds, it usually invests them to carry out its business activities. Such **investing activities** involve acquiring some of the following:

1. **Land, buildings, and equipment.** These investments, which provide a firm with a capacity to manufacture and sell its products, usually take years to provide all of the potential services for which the firm acquired them.
2. **Patents, licenses, and other contractual rights.** These investments provide a firm with the legal right to use certain property or processes in pursuing its business activities.
3. **Common shares or bonds of other firms.** A firm might purchase shares or bonds of another firm (thereby becoming one of its owners or creditors). A firm might acquire these holdings for a few months with temporarily excess cash or might invest for longer-term purposes, such as to secure a source for critical raw materials or gain entrance into an emerging technology.
4. **Inventories.** To satisfy the needs of customers as they arise, firms must maintain an inventory of products to sell. A firm does not usually invest in specific inventory items for long because it will soon sell the items to customers. Because firms must have at least small amounts of inventory on hand, however, they must always invest some of their funds in inventory items.
5. **Accounts receivable from customers.** When a firm sells its products but does not require customers to pay immediately, the firm provides funds to its customers in the sense that it saves its customers from having to raise funds immediately to pay for purchases. Carrying some amount of accounts receivable may be in the best interest of the firm if, by doing so, the firm increases its sales and earnings. In extending credit to customers, the firm forgoes collecting its cash right away, but if it did not extend the credit, it might not make the sale in the first place. Insofar as the firm delays the collection of needed cash from its customers, it must obtain funds elsewhere. Carrying accounts receivable, therefore, implies that as the firm needs cash, it must raise funds elsewhere. In this sense, carrying accounts receivable uses funds.
6. **Cash.** Most firms will leave a portion of their funds as cash in checking accounts so that they can pay their current bills. Sophisticated cash management systems can keep these amounts small.

Managerial accounting and corporate finance courses cover the techniques that firms use to make investment decisions.

## Carrying Out Operations

A firm obtains financing and invests the funds in various resources to generate earnings. The **operating activities** of the firm comprise the following:

1. **Purchasing.** The purchasing department of a merchandising firm acquires products needed by its retail stores. The purchasing department of a manufacturing firm acquires raw materials needed for production.
2. **Production.** The production department in a manufacturing firm combines raw materials, labor services, and other manufacturing inputs to produce the products, or outputs, of a firm. A firm that offers services combines labor services with other inputs in creating its product.
3. **Marketing.** The marketing department oversees selling and distributing a firm's products to customers.
4. **Administration.** The administrative activity of a firm supports purchasing, production, marketing, and other operating departments. Administrative activities include data processing, legal services, research and development, and other support services.

Managerial accounting, marketing, and production courses cover the appropriate bases for making operating decisions.

## Summary of Business Activities

**Figure 1.1** summarizes the principal business activities discussed in this section. These business activities appear as interconnected cogwheels that must move in concert with each other if a firm is to succeed. Figure 1.1 also shows the relation of these business activities to the three principal financial statements, a topic discussed in the next section.

FIGURE 1.1 Overview of Business Activities

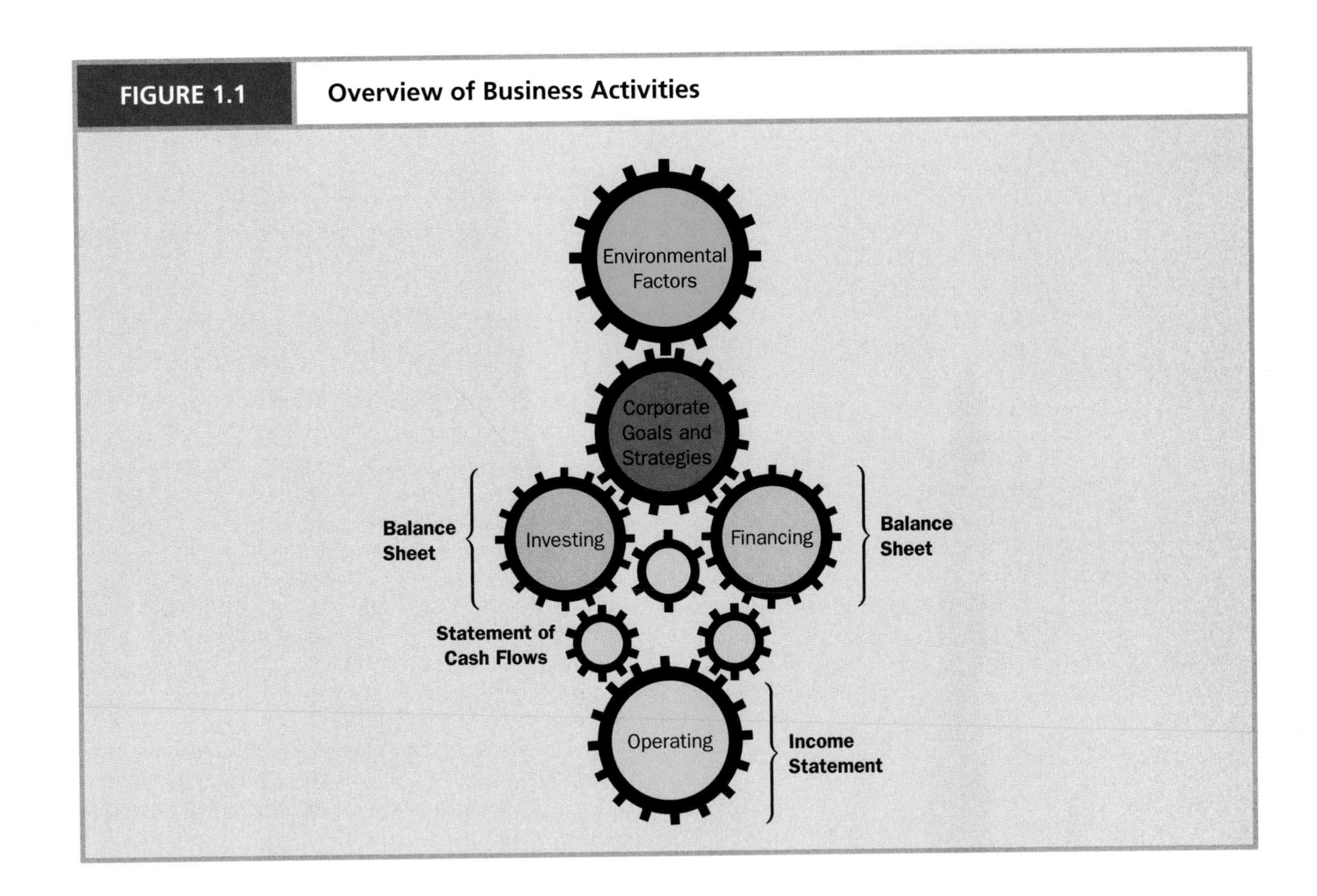


## Principal Financial Statements

Firms typically communicate the results of their business activities through the **annual report to shareholders**. (See the annual report of Wal-Mart at www.wal-mart.com). The annual report usually begins with a letter from the chairperson of the firm's board of directors and from its chief executive officer summarizing the activities of the past year and assessing the firm's prospects for the coming year. Descriptions and pictures of the firm's products, facilities, and employees usually follow. An important section of the annual report is the **Management Discussion and Analysis,** called the **MD&A,** in which a firm's management discusses reasons for changes in profitability and risk during the past year. The final section of the annual report includes the firm's financial statements and supplementary information, including the following:

1. Balance sheet;
2. Income statement;
3. Statement of cash flows;
4. Notes to the financial statements, including various supporting schedules; and
5. Opinion of the independent certified public accountant.

The following sections of this chapter briefly discuss each of these five items.

### Balance Sheet

One question that a user of financial statements might ask is: *"What is the financial position, or financial health, of a firm?"* The **balance sheet** attempts to answer this question by presenting a snapshot at a moment in time of the results of the firm's investing and financing activities. **Exhibit 1.1** presents a comparative balance sheet for Wal-Mart as of January 31, Year 13, and January 31, Year 14. Firms typically present balance sheets as of the beginning and end of their most recent year. Note several aspects of the balance sheet.

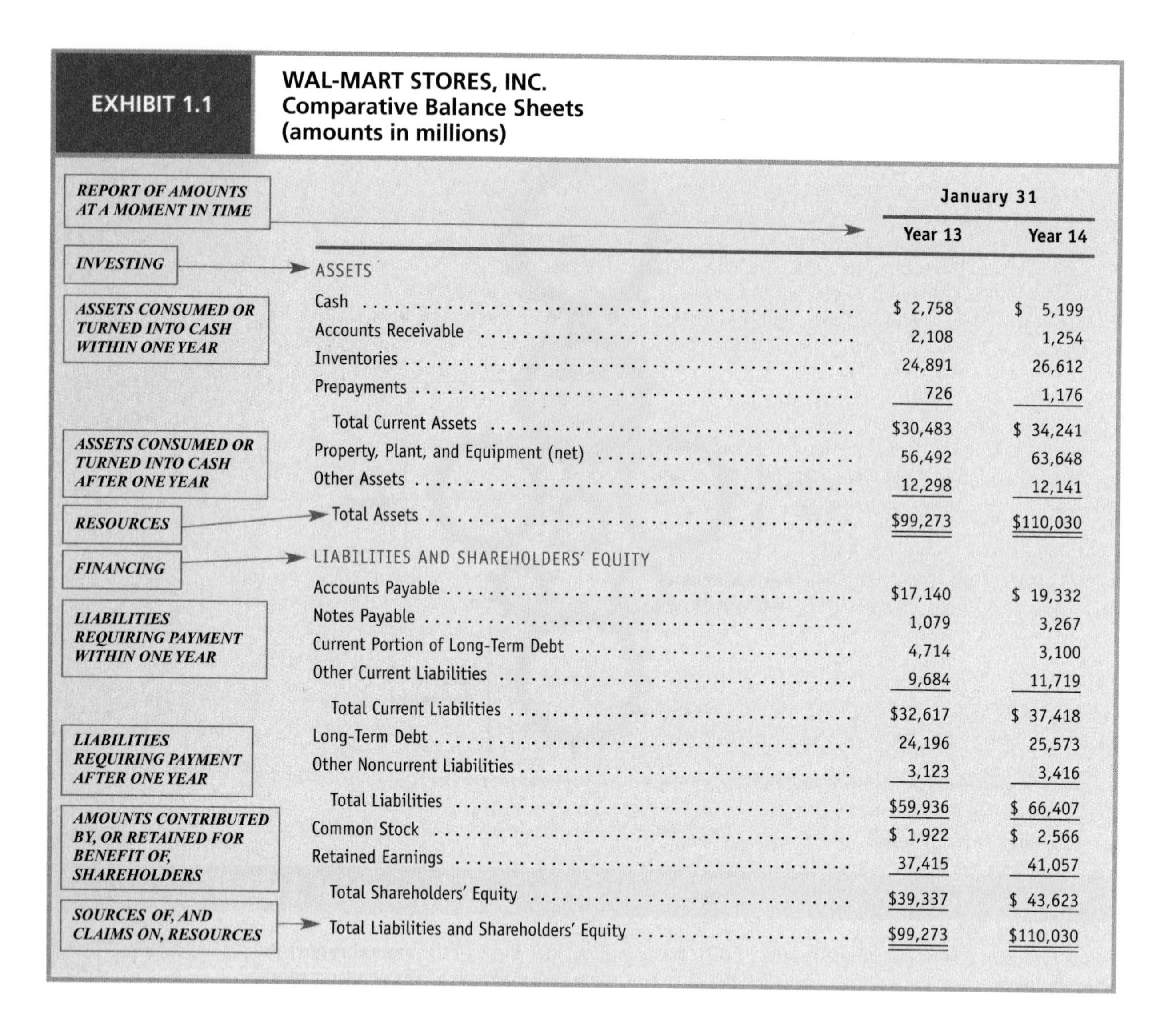

**EXHIBIT 1.1**

**WAL-MART STORES, INC.**
**Comparative Balance Sheets**
**(amounts in millions)**



| | January 31 | |
|---|---|---|
| | Year 13 | Year 14 |
| ASSETS | | |
| Cash | $ 2,758 | $ 5,199 |
| Accounts Receivable | 2,108 | 1,254 |
| Inventories | 24,891 | 26,612 |
| Prepayments | 726 | 1,176 |
| Total Current Assets | $30,483 | $ 34,241 |
| Property, Plant, and Equipment (net) | 56,492 | 63,648 |
| Other Assets | 12,298 | 12,141 |
| Total Assets | $99,273 | $110,030 |
| LIABILITIES AND SHAREHOLDERS' EQUITY | | |
| Accounts Payable | $17,140 | $ 19,332 |
| Notes Payable | 1,079 | 3,267 |
| Current Portion of Long-Term Debt | 4,714 | 3,100 |
| Other Current Liabilities | 9,684 | 11,719 |
| Total Current Liabilities | $32,617 | $ 37,418 |
| Long-Term Debt | 24,196 | 25,573 |
| Other Noncurrent Liabilities | 3,123 | 3,416 |
| Total Liabilities | $59,936 | $ 66,407 |
| Common Stock | $ 1,922 | $ 2,566 |
| Retained Earnings | 37,415 | 41,057 |
| Total Shareholders' Equity | $39,337 | $ 43,623 |
| Total Liabilities and Shareholders' Equity | $99,273 | $110,030 |

**Snapshot at a Moment in Time** The balance sheet presents a firm's investments and financing on a particular date (January 31, Year 13, and January 31, Year 14, for Wal-Mart). A snapshot permits the user to compare the types of investments made with the manner in which the firm financed those investments at that moment. The user must recognize, however, that the financial position of the firm at other times during the year can differ substantially from that depicted in this particular snapshot.

**Concepts of Assets, Liabilities, and Shareholders' Equity** The balance sheet presents a listing of a firm's assets, liabilities, and shareholders' equity.

**Assets** are economic resources with the ability or potential to provide future benefits to a firm. For example, Wal-Mart can use cash to purchase inventory, new stores, and equipment. The firm can sell inventory to customers for an amount it hopes will provide positive earnings. Wal-Mart can use equipment to transport inventory from its warehouses to its retail stores.

**Liabilities** are creditors' claims on the assets of a firm and show some of the sources of the funds the firm uses to acquire the assets.

- Wal-Mart has purchased merchandise inventories from its suppliers but has not paid for a portion of the purchases. As a result, these creditors have provided funds to the firm and have a claim on its assets. Wal-Mart includes its obligations to suppliers in Accounts Payable.
- Employees have provided labor services for which Wal-Mart has not made payment as of January 31, Year 13 and Year 14. These employees likewise have provided funds to the firm and have a claim on its assets. Wal-Mart includes its obligation to employees in Other Current Liabilities.

Creditors' claims, or liabilities, result from a firm's having previously received benefits (cash, inventories, labor services) for which the firm must pay a specified amount on a specified date.

**Shareholders' equity** shows the amounts of funds owners have provided and, in parallel, their claims on the assets of a firm. Unlike creditors, the owners have only a residual interest; that is, owners have a claim on all assets in excess of those required to meet creditors' claims. The shareholders' equity generally comprises two parts: contributed capital and retained earnings. **Contributed capital** reflects the funds invested by shareholders for an ownership interest. By January 31, Year 14, the owners have contributed $2,566 million for shares of Wal-Mart's common stock.

**Retained earnings** represent the source of funds a firm derives from its earnings that exceed the dividends it has distributed to shareholders since its formation. When a firm has earnings, it generates new assets. It can distribute those assets to owners as dividends. To the extent it keeps the new assets in the firm, it has retained earnings. The Retained Earnings designation in shareholders' equity shows the source of the funding for assets reinvested by management for the benefit of shareholders and the shareholders' claims on those assets. Management attempts to direct the use of a firm's assets so that over time it generates positive earnings, that is, it receives more assets than it consumes in operations. This increase in assets, after any claims of creditors, belongs to the firm's owners. As of January 31, Year 14, Wal-Mart's cumulative earnings exceed its cumulative dividends by $41,057 million. An amount of assets equal to retained earnings does not appear on any single line in Wal-Mart's balance sheet. Instead, Wal-Mart has used the cash generated by the retention of earnings to acquire inventories, new stores, equipment, and other assets. Almost all successful firms use a large percentage of the assets they generate by earnings to replace assets and to grow, rather than to pay dividends.

**Equality of Assets and Liabilities Plus Shareholders' Equity** As the balance sheet for Wal-Mart shows, assets equal liabilities plus shareholders' equity.

$$\text{Assets} = \text{Liabilities} + \text{Shareholders' Equity}$$

A firm invests every dollar of resources it obtains from financing. The balance sheet views the same resources from two angles: in a list of the assets the firm currently holds, having acquired them with its funds, and in a list of the parties (creditors and owners) who provided the funds and so have a claim on those assets. Thus,

$$\text{Assets} = \text{Liabilities} + \text{Shareholders' Equity,}$$

or

$$\begin{aligned} \text{Investing} &= \text{Financing} \\ \text{Resources} &= \text{Sources of Resources} \\ \text{Resources} &= \text{Claims on Resources} \end{aligned}$$

The *asset mix* (that is, the proportions of total assets represented by accounts receivable, inventories, equipment, and other assets, respectively) reflects a firm's investment decisions, and the mix of liabilities plus shareholders' equity reflects a firm's financing decisions, each measured at a moment in time.

**Balance Sheet Classification** The balance sheet classifies assets and liabilities as being either current or noncurrent.

Current assets include cash and assets that a firm expects to turn into cash, sell, or consume within approximately one year from the date of the balance sheet. Cash, temporary investments in securities, accounts receivable from customers, and inventories are the most common current assets. Current liabilities represent obligations a firm expects to pay within one year. Examples are notes payable to banks, accounts payable to suppliers, salaries payable to employees, and taxes payable to governments.

Noncurrent assets, typically held and used for several years, include land, buildings, equipment, patents, and long-term investments in securities. Noncurrent liabilities and shareholders' equity are a firm's longer-term sources of funds.

**Balance Sheet Valuation** Assets, liabilities, and shareholders' equity items appear on the balance sheet at monetary amounts. Accountants measure the amounts using one of two conceptual bases: (1) a **historical valuation,** which reflects the acquisition cost of assets or the amounts of funds originally obtained from creditors or owners, or (2) a **current valuation,** which reflects the current cost of acquiring assets or the current market value of creditors' and shareholders' claims on a firm. As Chapter 2 discusses, accountants can more easily verify historical valuations and will disagree less often as to their amounts, whereas financial statement users probably find current valuations more relevant for decision making. Ascertaining current valuations requires, however, greater subjectivity than does ascertaining historical costs.

The balance sheet reports cash as the amount of cash on hand or in the bank (a current valuation). Accounts receivable appear at the amount of cash the firm expects to collect from customers (often, approximating a current valuation). Liabilities generally appear at present value of the cash required to pay liabilities.

The remaining assets appear either at acquisition cost or at acquisition cost net of accumulated depreciation or amortization (a historical valuation). For example, inventories and land usually appear at the amount of cash or other resources that the firm originally sacrificed to acquire those assets. Buildings, equipment, and patents appear at acquisition cost, adjusted downward to reflect the portion of the assets' services that the firm has used since acquisition.[2]

Common stock appears at the amount invested by owners when the firm first issued common stock (a historical valuation). Retained earnings shows the sum of all prior years' earnings in excess of dividends (a combination of historical and current valuations). **Chapters 3** and **12** discuss further the valuation, or measurement, of retained earnings.

**Analysis of the Balance Sheet** Firms typically finance current assets with current liabilities and finance noncurrent assets with noncurrent liabilities and shareholders' equity. Current liabilities require payment within one year. Current assets generally convert into cash within one year. Firms can use this near-term cash flow to pay current liabilities. Long-term liabilities require payment over some number of future years. Noncurrent assets, such as buildings and equipment, generate cash flows over longer periods. Firms can use these more extended cash inflows to repay the noncurrent liabilities as they come due. Firms generally should not finance noncurrent assets with current liabilities. Noncurrent assets do not generate cash quickly enough to repay debt within one year. The balance sheet of Wal-Mart on January 31, Year 14 reveals the following:

| | | | |
|---|---|---|---|
| Current Assets | $ 34,241 | Current Liabilities | $ 37,418 |
| Noncurrent Assets | 75,789 | Noncurrent Liabilities and Shareholders' Equity | 72,612 |
| Total | $110,030 | Total | $110,030 |

Thus, Wal-Mart used short-term financing to raise funds approximately equal to its current assets and it uses long-term financing to raise an amount of funds approximately equal to the cost of its noncurrent assets. Current liabilities exceed current assets, an unusual relation. **Chapters 2** and **5** discuss more fully the analysis of a firm's financial health using the balance sheet.

## INCOME STATEMENT

Another question that a user of financial statements might ask is: *"How profitable is the firm?"* The second principal financial statement, the **income statement,** attempts to answer this question. **Exhibit 1.2** shows the income statement for Wal-Mart for the years ending January 31, Year 12, Year 13, and Year 14. The income statement indicates the **net income** or **earnings** for those time periods. Net income is revenues minus expenses. An analysis of revenues and expenses permits the user to understand why a firm's profitability changes over time and how it differs from that of other firms. Note several aspects of the income statement.

**Report of Amounts for a Period of Time** We noted earlier that the balance sheet reports amounts at a moment in time. The income statement, on the other hand, reports amounts for a period of time. Most firms use the calendar year as their reporting period. Wal-Mart uses an accounting period ending on January 31 each year, after the end of the busy holiday shopping season.

**Concepts of Net Income, Revenue, and Expense** The terms *net income* and *earnings* are synonyms used interchangeably in corporate annual reports and throughout this text. Most

[2] Sometimes, balance sheet amounts for inventories, buildings, land, equipment, and patents will reflect amounts lower than historical cost when these items decline in value after the firm acquires them. Later chapters discuss these exceptions, called *impairments*.

**EXHIBIT 1.2**

**WAL-MART STORES, INC.**
**Comparative Income Statements**
**(amounts in millions)**

*REPORT OF AMOUNTS FOR A PERIOD OF TIME*

*INFLOWS ON NET ASSETS FROM SALES OF GOODS AND SERVICES*

*OUTFLOWS OF NET ASSETS REQUIRED TO GENERATE REVENUES*

*NET INCOME: REVENUES – EXPENSES*

| | Year Ended January 31 | | |
|---|---|---|---|
| | **Year 12** | **Year 13** | **Year 14** |
| **Revenues** | | | |
| Sales Revenue | $217,799 | $244,524 | $256,329 |
| Other Revenues | 2,013 | 2,139 | 2,516 |
| Total Revenues | $219,812 | $246,663 | $258,845 |
| **Expenses** | | | |
| Cost of Goods Sold | $171,562 | $191,838 | $198,747 |
| Selling and Administrative | 36,356 | 41,236 | 44,909 |
| Interest | 1,326 | 1,063 | 996 |
| Income Taxes | 3,897 | 4,487 | 5,118 |
| Total Expenses | $213,141 | $238,624 | $249,770 |
| **Net Income** | $ 6,671 | $ 8,039 | $ 9,075 |

business firms aim to generate earnings from operating activities. The income statement reports a firm's success in achieving this goal for a given time span. The income statement reports the sources and amounts of a firm's revenues and the nature and amount of a firm's expenses. The excess of revenues over expenses equals the earnings for the period.

**Revenues** measure the inflows of assets (or reductions in liabilities) from selling goods and providing services to customers. During Years 12 to 14, Wal-Mart generated almost all of its revenues from the sale of merchandise. It generated a minor amount of revenue from other sources, primarily interest revenue on its bank accounts. From its customers, Wal-Mart received either cash or promises that it would receive cash in the future, called Accounts Receivable. Both are assets. Thus the firm generated revenues and increased assets. A firm cannot generate revenues without simultaneously generating net assets.

**Expenses** measure the outflow of assets (or increases in liabilities) used in generating revenues. Cost of Goods Sold (an expense) measures the cost of inventories sold to customers. Selling and administrative expenses measure the cash paid or the liabilities incurred to make future cash payments for selling and administrative services received during the period. For each dollar of expense, either an asset decreases or a liability increases.

A firm strives to generate an excess of net asset inflows from revenues over net asset outflows from expenses required to generate the revenues. Net income indicates a firm's accomplishments (revenues) relative to the efforts required (expenses) in pursuing its operating activities. When expenses for a period exceed revenues, a firm incurs a **net loss**.

**Classification of Revenues and Expenses** The income statement for Wal-Mart classifies some expenses by the department that carried out the firm's operating activities (retailing, selling, administration) and some expenses by their nature (interest, income taxes). Alternatively, the firm might classify all expenses by their nature (for example, salaries, depreciation, utilities). Firms classify revenues and expenses in their income statements in different ways.

**Relation to Balance Sheet** The income statement links the balance sheet at the beginning of the period with the balance sheet at the end of the period. The balance sheet amount for retained earnings represents the sum of all prior earnings of a firm in excess of dividends. The amount of net income for the current period helps explain the change in retained earnings between the beginning and the end of the period. During Year 14, Wal-Mart had net income of $9,075 million. It declared and paid dividends of $1,569 million. Wal-Mart also repurchased shares of its common stock and retired them, reducing retained earnings by $3,864 during Year

14 (**Chapter 12** discusses the accounting for common stock repurchases). Retained earnings during Year 14, therefore, increased by $3,642 as follows:

| | |
|---|---|
| Retained Earnings, January 31, Year 13 | $37,415 |
| Add Net Income for Year 14 | 9,075 |
| Subtract Dividends Declared and Paid during Year 14 | (1,569) |
| Subtract Amounts Related to Repurchase of Wal-Mart Common Stock | (3,864) |
| Retained Earnings, January 31, Year 14 | $41,057 |

**Analysis of the Income Statement** A percentage comparison of net income to revenues indicates how much of each dollar of revenues remains to benefit the common shareholders after the firm covers all expenses. The net income to revenues, or *profit margin,* percentages for Wal-Mart are as follows:

Year 12: $6,671/$219,812 = 3.0%

Year 13: $8,039/$246,663 = 3.3%

Year 14: $9,075/$258,845 = 3.5%

The profit margin percentage increased during the three-year period. By calculating expenses to revenues percentages for each expense item, we can identify the reason for any changes in the profit margin. **Chapter 5** further examines the reasons for the increasing profit margin for Wal-Mart.

## STATEMENT OF CASH FLOWS

Another question that users of financial statements might ask is: *"Is the firm generating sufficient cash flows from its customers to finance operations and to acquire buildings and equipment? Or, must it seek new funds from lenders or owners?"* The third principal financial statement, the **statement of cash flows,** attempts to answer this question. **Exhibit 1.3** presents the statement of cash flows for Wal-Mart for the years ending January 31, Year 12, Year 13, and Year 14. This statement reports the net cash flows derived from (or used by) operating, investing, and financing activities for those time periods. Of what significance is a statement explaining or analyzing the change in cash during a period of time? The following example illustrates the usefulness.

**Example 2** Diversified Technologies Corporation began business four years ago. In its first four years, it generated net income of $100,000, $300,000, $800,000, and $1,500,000, respectively. The company retained all of the assets generated by its earnings for growth (reflected on the balance sheet as increases in both net assets and retained earnings). Early in the fifth year, the company learned that, despite paying no dividends, it was running out of cash. Analysis revealed the company was expanding accounts receivable, inventories, buildings, and equipment so fast that operations and external financing were not generating cash quickly enough to keep pace with the growth.

This example illustrates a common phenomenon for business firms. A firm might not generate cash in sufficient amounts or at the proper times to finance all ongoing or growing operations. If a firm is to continue operating successfully, it must generate more cash than it spends. In some cases it can borrow from creditors to replenish its cash, but future operations must generate cash to repay these loans. Often a growing business finds that it can't collect cash from customers fast enough to pay its own bills even though the customers owe the firm more cash than the firm owes to its suppliers. A simple example of this occurs when a firm must pay its own suppliers faster, say, in 20 days, than it collects from its own customers, say, in 30 days.

**Report of Amounts for a Period of Time** The statement of cash flows reports cash inflows and outflows for a period of time. This financial statement, like the income statement and unlike the balance sheet, shows amounts over time.

**Classification of Items in the Statement of Cash Flows** **Exhibit 1.3,** a standard format for the statement of cash flows, classifies cash flows for Wal-Mart using the three-way classification of the principal business activities described earlier in the chapter: operations, investing, and financing. **Figure 1.2** depicts these various sources and uses of cash.

**EXHIBIT 1.3**

**WAL-MART STORES, INC.**
**Comparative Statements of Cash Flows**
**(amounts in millions)**

*REPORT AMOUNTS FOR A PERIOD OF TIME*

| | Year Ended January 31 | | |
|---|---|---|---|
| | Year 12 | Year 13 | Year 14 |
| **Operations** | | | |
| Net Income | $ 6,671 | $ 8,039 | $ 9,075 |
| Depreciation | 3,290 | 3,432 | 3,852 |
| (Increase) Decrease in Accounts Receivable | (210) | (101) | 373 |
| (Increase) in Inventories | (1,235) | (2,236) | (1,973) |
| (Increase) Decrease in Prepayments | (180) | 745 | (450) |
| Increase in Accounts Payable | 368 | 1,447 | 2,803 |
| Increase in Other Current Liabilities | 1,556 | 1,206 | 2,411 |
| Cash Flow from Operations | $10,260 | $12,532 | $ 16,091 |
| **Investing** | | | |
| Acquisition of Property, Plant, and Equipment | $ (8,383) | $ (9,355) | $ (10,308) |
| Other | 1,237 | (354) | 2,046 |
| Cash Flow from Investing | $ (7,146) | $ (9,709) | $ (8,262) |
| **Financing** | | | |
| Increase (Decrease) in Short-Term Borrowing | $ (1,533) | $ 1,836 | $ 688 |
| Increase in Long-Term Borrowing | 4,591 | 2,044 | 4,099 |
| Decrease in Long-Term Borrowing | (3,686) | (1,479) | (3,541) |
| Acquisition of Common Stock | (1,214) | (3,232) | (5,182) |
| Dividends | (1,249) | (1,328) | (1,569) |
| Other | 84 | (67) | 117 |
| Cash Flow from Financing | $ (3,007) | $ (2,226) | $ (5,388) |
| Change in Cash | $ 107 | $ 597 | $ 2,441 |
| Cash, Beginning of Year | 2,054 | 2,161 | 2,758 |
| Cash, End of Year | $ 2,161 | $ 2,758 | $ 5,199 |

*NET CASH INFLOW FROM SALES TO CUSTOMERS* → Cash Flow from Operations

*NET CASH OUTFLOW FROM INVESTMENT IN RETAILING CAPACITY* → Cash Flow from Investing

*NET CASH OUTFLOW FROM SHORT- AND LONG-TERM CREDITORS AND OWNERS* → Cash Flow from Financing

*CHANGE IN CASH = NET CASH FLOW FROM OPERATING, INVESTING, AND FINANCING* → Change in Cash

*AMOUNTS FOR CASH ON THE BALANCE SHEET* → Cash, Beginning of Year; Cash, End of Year

1. **Operations.** Most firms expect their primary source of cash to result from the excess of cash they receive from customers over the amount of cash they pay to suppliers, employees, and others in carrying out the firms' operating activities. Wal-Mart generated positive cash flow from operations in amounts larger than annual net income in each of the three years. Chapter 4 discusses the adjustments required to convert net income to cash flow from operations.
2. **Investing.** Firms that expect either to maintain current operating levels or to grow must continually acquire buildings, equipment, and other noncurrent assets. Just to keep productive capacity constant requires cash to replace assets used up. To grow requires even more. A firm can obtain some of the cash needed from selling existing land, buildings, and equipment. The firm's cash needs, however, usually exceed the cash proceeds of such disposals. Wal-Mart increased its cash outflows for the acquisition of property, plant, and equipment during the last three years as it built new superstores.
3. **Financing.** Firms obtain additional financing to support operating and investing activities by issuing bonds or common stock. The firm uses cash to pay dividends and to retire old financing, such as repaying long-term debt when it comes due. Wal-Mart generated more than sufficient

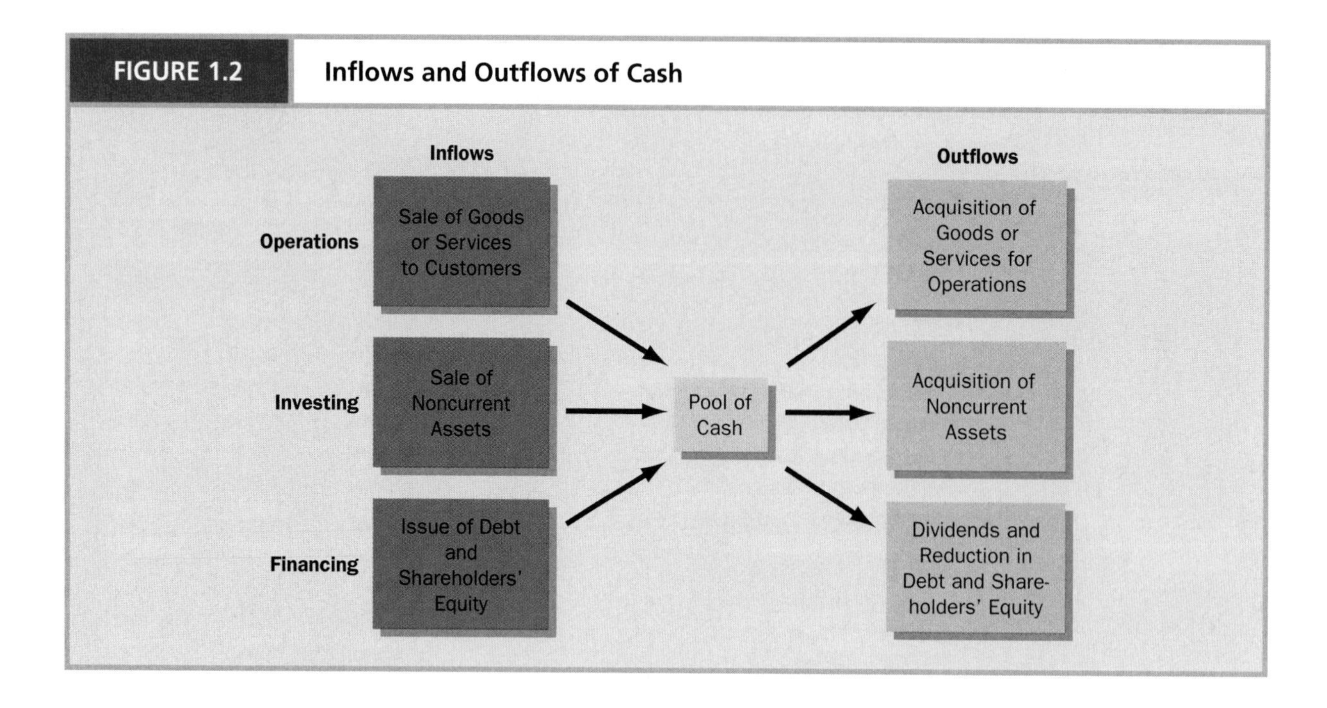


cash flow from operations during Year 12 to Year 14 to finance expenditures on property, plant, and equipment. It used this excess cash flow and the cash received from issuing long-term debt in excess of that needed to retire long-term liabilities to pay dividends to shareholders and acquire shares of its common stock.

**Relation to Balance Sheet and Income Statement** The statement of cash flows explains the change in cash between the beginning and the end of the period. Note on the bottom of the statement of cash flows for Wal-Mart that the amount of cash on the balance sheet was $2,758 million on January 31, Year 13, and $5,199 million on January 31, Year 14. The statement of cash flows shows the reasons for this $2,441 million increase in cash, as the following analysis shows:

| | |
|---|---|
| Cash on the Balance Sheet, January 31, Year 13 | $ 2,758 |
| Cash Flow from Operations during Year 14 | 16,091 |
| Cash Flow for Investing during Year 14 | (8,262) |
| Cash Flow for Financing during Year 14 | $(5,388) |
| Cash on the Balance Sheet, January 31, Year 14 | $ 5,199 |

The statement of cash flows also shows the relation between net income and cash flow from operations. Cash flow from operations exceeds net income each year for Wal-Mart. **Chapter 4** discusses the reasons for this excess.

## Summary of Principal Financial Statements

Refer again to **Figure 1.1** on page 7, which depicts the principal business activities as interconnected cogwheels. Note that the balance sheet reports the results of investing and financing activities at a moment in time. The income statement reports the success of a firm in using assets to generate earnings for a period of time. The statement of cash flows reports the net cash inflow or outflow from operating, investing, and financing activities for the same period of time. The financial statements seek to measure the success of a firm's business activities by providing information about a firm's:

Financial position (balance sheet)
Profitability (income statement)
Cash-generating ability (statement of cash flows)

FIGURE 1.3 Relations Among Principal Financial Statements

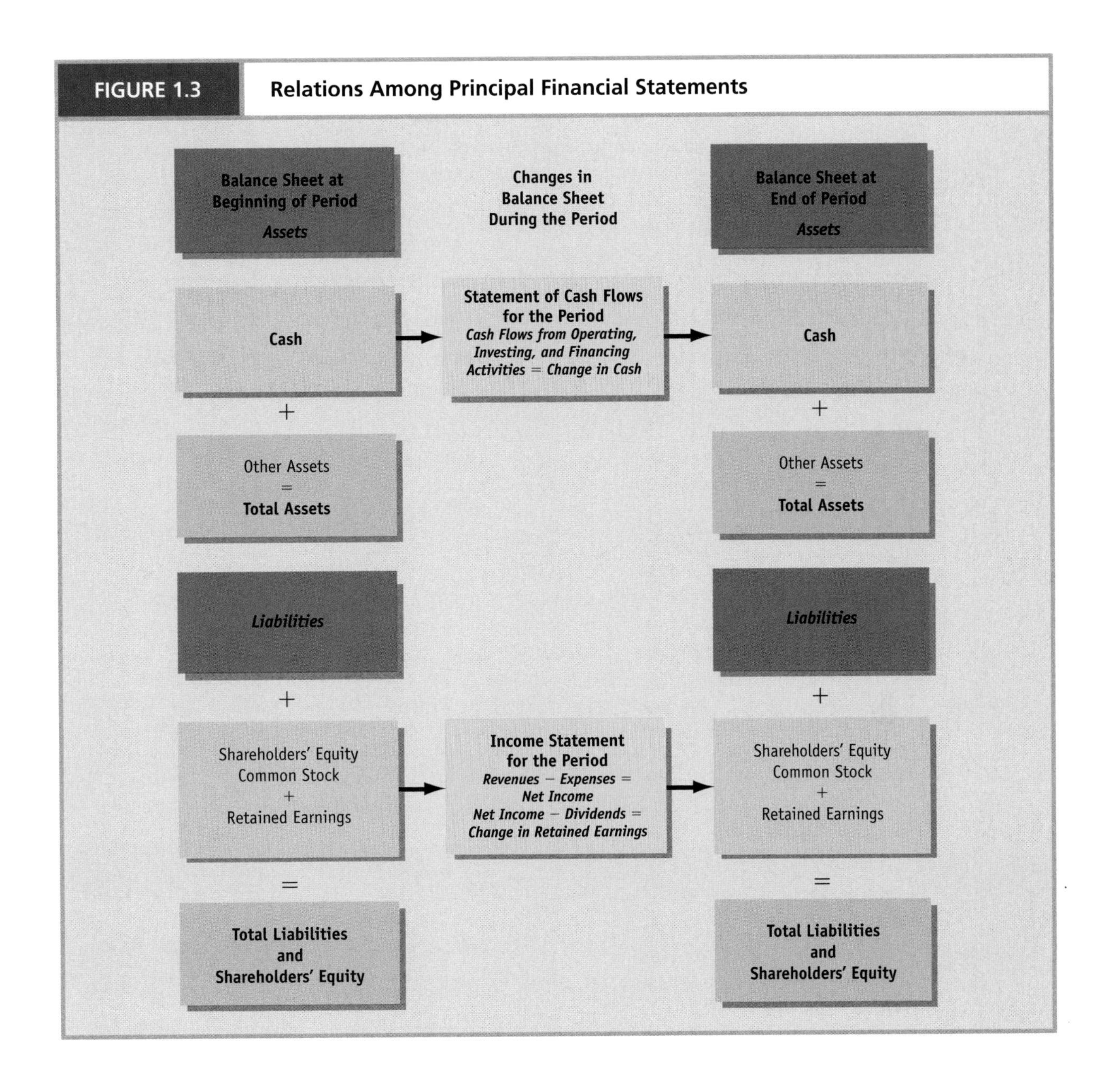


**Figure 1.3** depicts the relation between the three principal financial statements over time. The balance sheet reports the assets, liabilities, and shareholders' equity at a moment in time. The income statement reports net income for a period of time and helps explain the change in retained earnings on the balance sheet between the beginning and end of the period. The statement of cash flows explains the change in cash on the balance sheet between the beginning and end of the period.

## Other Items in Annual Reports

### Supporting Schedules and Notes

The balance sheet, income statement, and statement of cash flows present information condensed to ease comprehension by the typical reader. Some readers desire additional details omitted from these condensed versions. Firms therefore usually include with their financial statements additional schedules that provide more detail for some of the items reported in the three main statements. For example, most firms present separate schedules to explain the change in contributed capital and retained earnings.

## Problem 1.1 for Self-Study

**Preparing a balance sheet and an income statement.** The accounting records of Gateway, a manufacturer and direct marketer of personal computers, reveal the following (all dollar amounts in thousands):

| | December 31 | |
|---|---|---|
| | Year 3 | Year 4 |
| **Balance Sheet Items** | | |
| Accounts Payable | $411,788 | $ 488,717 |
| Accounts Receivable | 449,723 | 510,679 |
| Buildings and Equipment (net of accumulated depreciation) | 227,477 | 315,038 |
| Cash | 447,718 | 593,601 |
| Common Stock | 290,829 | 295,535 |
| Computer Software Development Cost (net of accumulated amortization) | 31,873 | 39,998 |
| Goodwill from Corporate Acquisitions | 35,631 | 118,121 |
| Inventories | 278,043 | 249,224 |
| Land | 14,888 | 21,431 |
| Long-Term Debt | 7,244 | 7,240 |
| Notes Payable | 15,041 | 13,969 |
| Other Current Assets | 74,216 | 152,531 |
| Other Current Liabilities | 207,336 | 315,292 |
| Other Noncurrent Liabilities | 50,857 | 98,081 |
| Retained Earnings | 410,870 | 634,509 |
| Royalties Payable | 125,270 | 159,418 |
| Short-Term Cash Investments | 0 | 38,648 |
| Taxes Payable | 40,334 | 26,510 |
| **Income Statement Items** | | |
| Cost of Goods Sold | | $5,217,239 |
| Income Tax Expense | | 93,823 |
| Interest Expense | | 1,740 |
| Interest Revenue | | 28,929 |
| Sales | | 6,293,680 |
| Selling and Administrative Expenses | | 786,168 |

**a.** Prepare a comparative balance sheet for Gateway as of December 31, Year 3 and Year 4. Classify the balance sheet items into the following categories: current assets, noncurrent assets, current liabilities, noncurrent liabilities, and shareholders' equity. Refer to the Glossary at the back of the book if you have difficulty with any of the accounts.

**b.** Prepare an income statement for Gateway for the year ending December 31, Year 4. Classify income statement items into revenues and expenses.

**c.** Calculate the amount of dividends declared and paid to common shareholders during the year ending December 31, Year 4.

**d.** Did the increase in cash between December 31, Year 3, and December 31, Year 4, primarily result from operating activities, investing activities, or financing activities? Explain.

Solutions to self-study problems appear at the end of the chapter.

Every set of published financial statements also contains explanatory notes as an integral part of the statements. As later chapters make clear, a firm must select the accounting methods followed in preparing its financial statements from a set of generally accepted methods. The notes indicate the actual accounting methods the firm uses and also disclose additional information that elaborates on items presented in the three principal statements. To understand fully a firm's balance sheet, income statement, and statement of cash flows requires understanding the notes. No such notes appear here for the financial statements of Wal-Mart because they would not mean much at this stage. Do not let this omission lead you to conclude that the notes are unimportant.

## AUDITOR'S OPINION

Regulatory bodies generally require firms whose shares of common stock publicly trade in the capital market to obtain an audit by an independent auditor. The audit involves two essential steps:

1. An examination to assess the effectiveness of a firm's internal control system for measuring and reporting business transactions.
2. An examination to assess whether the financial statements and notes present fairly a firm's financial position, results of operations, and cash flows in accordance with generally accepted accounting principles.

Responsibility for both the internal control system and the financial statements rests with the firm's management. The auditor conducts various tests of the accounting records in order to form opinions about the effectiveness of the internal control system and the fairness of the financial statements. The auditor's opinions for these two items generally describe the scope of the auditing performed, the standards used when conducting the audits, and the auditor's resulting judgment, or opinion. The auditor's opinion on the fairness of the financial statements of Wal-Mart reads as follows:

> We have audited the accompanying consolidated balance sheet of Wal-Mart Stores, Inc. as of January 31, Year 13 and Year 14, and the related consolidated statements of net income and cash flows for each of the three years ended January 31, Year 14. These financial statements are the responsibility of the Company's management. Our responsibility is to express an opinion on these financial statements based on our audits.
>
> We conducted our audit in accordance with the standards of the Public Company Accounting Oversight Board (United States). Those standards require that we plan and perform the audit to obtain reasonable assurance about whether the financial statements are free of material misstatement. An audit includes examining, on a test basis, evidence supporting the amounts and disclosures in the financial statements. An audit also includes assessing the accounting principles used and significant estimates made by management, as well as evaluating the overall financial statement presentation. We believe that our audit provides a reasonable basis for our opinion.
>
> In our opinion, the financial statements referred to above present fairly, in all material respects, the consolidated financial position of Wal-Mart Stores, Inc. at January 31, Year 13 and Year 14, and the consolidated results of their operations and their cash flows for each of the three years ended January 31, Year 14, in conformity with generally accepted accounting principles in the United States.
>
> Ernst & Young LLP Tulsa, Oklahoma
>
> March 19, Year 14.

The first paragraph indicates the financial presentations covered by the opinion and indicates that the responsibility for the financial statements rests with management. The second paragraph affirms that the auditor has followed auditing standards and practices generally accepted by the accounting profession unless otherwise noted and described. Exceptions to the statement that the auditor conducted the examination "in accordance with the standards of the Public Company Accounting Oversight Board" are rare. There are occasional references to the auditor having relied on financial statements examined by other auditors, particularly for subsidiaries or for data from prior periods.

The auditor's opinion expressed in the third paragraph is the heart of the report. It may be an **unqualified** or a **qualified opinion**. Most opinions are unqualified; that is, the auditor describes no exceptions or qualifications to its opinion that the statements "present fairly . . . the financial position . . . and the results of its operations and its cash flows . . . in conformity with generally accepted accounting principles." Qualifications to the opinion result primarily from material uncertainties regarding realization or valuation of assets, outstanding litigation or tax liabilities, or accounting inconsistencies between periods caused by changes in the application of accounting principles.

A qualification so material that the auditor cannot express an opinion as to the fairness of the financial statements as a whole must result in either a **disclaimer of opinion** or an

**adverse opinion**. Disclaimers of opinion and adverse opinions rarely appear in published annual reports.

The auditor's opinion on the effectiveness of a firm's internal control system follows a similar format, except that the wording reflects the differing scope and standards of the internal control assessment.

## Financial Reporting Issues

The financial reporting process involves activities of four principal participants:

1. Business firms and their managers.
2. Accounting standard setting and regulatory bodies.
3. Independent auditors.
4. Security analysts and other users of financial statements.

This section discusses the role of each of these participants and issues raised for the financial reporting process by the sometimes-conflicting goals or interests of each.

### Business Firms and Their Managers

Business firms receive funds from owners with the expectation that managers will use the funds to increase the value of the firm and the wealth of its owners. From a legal perspective, managers are agents of the shareholders, and are responsible for safeguarding and properly using the firm's resources. Managers establish internal control procedures to ensure the proper recording of business transactions and the appropriate measurement and reporting of the results of those transactions. Management thereby carries out its stewardship, or fiduciary, responsibility to owners.

Managers also have a self-interest in promoting job security, high salaries, and bonuses as well as professional career enhancement. The self-interest of management and management's responsibility to shareholders sometimes conflict. For example, a firm might experience a period of weak profitability that could negatively affect stock prices and the shareholders' assessment of management. In an effort to improve reported results in the short term, management might sell strategically important assets at a gain or accelerate the recording of sales of a subsequent period in order to turn a weak profitability record into a seemingly stronger one in an attempt to keep shareholders happy. Shareholders need to know quickly, however, if management does not perform well.

One approach to dealing with management's potential conflict of interest is to design compensation systems that encourage management to act in the best interests of shareholders. Providing bonuses or awarding stock options to management if the firm meets certain earnings or stock price targets are means of aligning management and stockholder interests. Ideally, these compensation plans should reward management for real economic value enhancement and not simply short-term gaming by managers to protect their jobs.

### Accounting Standard Setting and Regulatory Bodies

Firms use various accounting standards, or methods, in preparing their financial statements. One issue is who should decide which alternative accounting measurement and reporting standards firms can use and which they cannot use. For example, should a governmental body, practicing accountants, financial statement users, or the reporting firms make these decisions?

A governmental body might set such accounting standards and use the legislative power of the government to enforce them. With this approach to standard setting, one would worry that government employees might represent neither preparer nor user perspectives and thereby might specify accounting standards that were either impracticable to apply or irrelevant to user needs. One would also worry that a government body could encounter political pressures in **establishing acceptable accounting standards** that both meet the information needs of financial statement users and permit the government to collect tax payments from taxpayers.

A private-sector body lacks legal enforcement power for its accounting standards but can more likely incorporate viewpoints of various preparer and user groups. Placing the standard-setting process in the private sector does not, however, eliminate political pressures. Firms reporting about themselves sometimes prefer to report less accounting information when it is costly to prepare or when it provides competitors with otherwise secret information. Users desire more accounting information because they do not bear the cost of preparing the additional information and can ignore information that turns out to be unhelpful. The standard-setting body must deal with these conflicting viewpoints in gaining acceptance for its pronouncements.

A second issue concerns the desired degree of **uniformity in selecting accounting methods** across firms. Should standard-setters require all firms to follow the same accounting method for similar transactions? Such an approach might cause less confusion for financial statement users than if standard-setting bodies permitted firms a choice of several alternative methods. Requiring uniformity also reduces a firm's flexibility to manage reported earnings through its selection of accounting methods. An opposing argument is that because the economic characteristics of firms differ, management should be free, within certain prescribed limits, to select those accounting methods that best capture the firm's particular economics. This view suggests greater **flexibility in selecting accounting methods**.

A third issue concerns the type of accounting standards that standard-setting bodies should adopt. One approach is to specify **general principles** for a particular reporting topic and permit firms and their independent auditors to make judgments as to the application of those general principles to the particular circumstances. At the opposite extreme is a detailed **rules-based approach** in which standard-setting bodies attempt to anticipate the various games firms will play to work around the particular reporting requirements. The appeals of the general principles approach are the likely greater consistency of reporting standards with the broad concepts and principles of financial reporting and the greater opportunity for independent accountants to make professional judgments. The concern with the general principles approach is that it gives managers more "wiggle room" to avoid reporting in ways viewed as undesirable by management. The detailed rules approach attempts to constrain such opportunistic actions by management by closing perceived loopholes in reporting standards. This approach, however, can result in lengthy and complex reporting standards. It can also lead to a mentality of "show me where it says I can't report the way I want to," causing management to fine-tune transactions to skirt the rules put into effect to achieve uniformity.

## Independent Auditors

An audit of a firm's financial statements involves (1) an assessment of the capability of a firm's accounting system to accumulate, measure, and synthesize transactional data properly, and (2) an assessment of the operational effectiveness of this accounting system. The auditor obtains evidence for the first assessment by studying the procedures and internal controls built into the accounting system. The auditor obtains evidence for the second assessment by examining a sample of actual transactions.

Employees of a firm might conduct such audits (called **internal audits**). The employees' knowledge and familiarity with the activities of their firm probably enhance the quality of the audit work and increase the likelihood that the audit will generate suggestions for improving operations. Because internal auditors work for the firm, their audits do not add as much credibility to the firm's financial statements as an **external, independent audit** provides. Because the managers of a firm have incentives to report as favorable a picture as possible in the financial statements, an external audit by independent auditors acts to control, but not always successfully, management's optimistic inclinations.

To serve effectively as a control on management's actions, the auditors must maintain independence from the firm. Yet, business firms hire and compensate their auditors, thereby putting pressure on the client-auditor relationship and potentially compromising the auditor's independence. To maintain the integrity of the audit, auditors must have access to all information that might impact the financial statements. In cases where confrontations with management arise, auditors must have access to members of the firm's board of directors who are not managers or employees of the firm. The auditor works for the Audit Committee of the Board of Directors, whose members are themselves independent of management. Auditors need access to a higher-level body, such as an administrative agency of the government, to adjudicate conflicts that arise in the audit.

## Security Analysts and Other Users of Financial Statements

Financial statements should serve as reliable signals of value changes of firms so that investors, security analysts, and other users can make wise, economic decisions. Minimizing intentional management bias through appropriately applied accounting standards and effective, independent audits enhances the reliability of the financial statements.

Yet, security analysts may exhibit conflicts of interests when using accounting information. Security analysts often work for securities firms that invest in the business firms for which analysts make investment recommendations. Also, securities firms often compensate their analysts for short-term performance, providing security analysts with the same incentives as management to hide poor performance in a particular year.

## Ethical Issues in the Financial Reporting Process

Recent financial reporting abuses have raised ethical issues for the various participants in the financial reporting process. Consider the following examples of financial reporting choices made by three firms.

**Example 3.** Centennial Technologies, a designer, manufacturer, and marketer of PC cards for computers, printers, and other software-based equipment, engaged in various actions to inflate its earnings, including recording fictitious sales, accelerating revenues from a later year into an earlier one, and manipulating the count of inventories at year end.

**Example 4.** Enron Corporation set up various entities to which Enron sold some of its securities and derivatives at a gain, increasing earnings. If these entities were independent of Enron, then recognition of the gain was appropriate. If these entities were not independent of Enron, then Enron should have consolidated its financial statements with those of these other entities and delayed recognition of any gain until these other entities sold the securities and derivatives to outsiders. Enron did not disclose sufficient information in its financial statements for users to decide if the entities were independent.

**Example 5.** Delta Air Lines uses the straight-line method to depreciate its aircraft, even though an accelerated depreciation method better reflects the economic decline in the value of the aircraft. Generally accepted accounting principles (GAAP) allows firms to choose between these two depreciation methods.

Does an ethical issue arise in each of these situations? The response depends on an individual's value system. The actions of Centennial Technologies were clearly intended to mislead users of the financial statements and were therefore fraudulent. The proper functioning of the capital system requires reliable financial information. Society, through its elected representatives, has judged that fraudulent actions are not in the best interest of society and has ruled them illegal. Does the legality of an action make it ethical?

Consider next the actions of Enron. Whether one entity is independent of another is a matter of degree and requires judgment. Enron, desiring to recognize the gains on sales to these entities, sought to structure the entities so that they were sufficiently, but not totally, independent of Enron. Enron consulted with its auditor, Arthur Andersen, as it set up these entities to ensure that Arthur Anderson concurred with Enron's treatment of these entities as "independent." These consultations led to certain changes in the structure of these entities until Arthur Andersen concurred with Enron's treatment of the entities as independent. Is it ethical for supposedly independent auditors to be so closely involved in decisions and actions made by management? Assume now that the evidence provided by Enron to Arthur Andersen was sufficient to justify not consolidating these entities. Was it ethical not to disclose sufficient information about these entities for users of the financial statements to make their own judgments about independence? Is a legal action, accompanied by full disclosure, sufficient to constitute ethical behavior?

Finally, consider the choice of depreciation method made by Delta Air Lines. GAAP allows various accounting methods for certain items without prescribing the specific conditions firms should use to choose between them. Delta's choice of the straight-line depreciation method permits it to report larger cumulative earnings at any time prior to sale or disposal of the aircraft. Delta's depreciation method is also consistent with that used by its principal competitors. Is it ethical for Delta to use a depreciation method allowed by GAAP even though another depreciation method more closely reflects the decline in the services of aircraft? Does ethical behavior require legality, adequate disclosure, and selection of accounting methods that best reflect economic reality?

These three examples illustrate the difficulties managers and accountants often encounter in identifying ethical issues and ethical behavior. A person whose value system says that any action that is legal is also ethical would not view Delta's reporting choice as involving ethical concerns.

If GAAP allows a choice, then any choice is appropriate and therefore an ethical issue does not arise. A person whose value system places heavy weight on legality and adequate disclosure so that users can make their own judgments would judge Enron's actions as unethical because of its inadequate disclosure. However, if Enron had made fuller disclosure, such a person would view Enron as satisfying its ethical responsibility. A person whose value system requires that financial statements reflect economic reality in as reliable a manner as possible would argue that Delta has not satisfied its ethical responsibility.

Ethical issues arise for other participants in the financial reporting process. Consider briefly the following questions:

1. Is it ethical for a member of the family of the independent auditor to hold shares of common stock in a client of the independent auditing firm?
2. Is it ethical for the independent auditing firm of Delta Air Lines in **Example 5** to express an unqualified audit opinion even though it judges that an accelerated depreciation method better reflects the earnings and financial position of Delta?
3. Is it ethical for a standard-setting body to issue an accounting standard requiring an accounting method preferred by reporting firms or legislators even though it concludes that another accounting method better reflects the economics of the transaction?
4. Is it ethical for security analysts of an investment-banking firm to make investment recommendations on a particular firm's securities when the investment-banking firm is also selling those securities?

For some of the ethical issues raised in this section, either laws or pronouncements from professional bodies have defined ethical behavior. An individual, however, may choose a higher standard of ethical behavior than that legislated. In practice, employees who demand standards of behavior beyond those of a particular corporation's culture will often lose their jobs. External pressure to conform coupled with personal standards can place employees in an uncomfortable position. We do not attempt in this book to impose our value systems on you. Our purpose in raising ethical concerns is to increase your sensitivity to situations where ethical judgments likely arise and encourage you to apply your value system in a thoughtful and directed manner.

## The Financial Reporting Process in the United States

The **Securities and Exchange Commission (SEC),** an agency of the federal government, has the legal authority to set acceptable accounting methods, or standards, in the United States. The SEC looks to the **Financial Accounting Standards Board (FASB),** a private-sector body, for leadership in establishing such standards. The seven-member FASB contains various combinations of financial statement users, practicing accountants, and representatives from business firms, academia, and the government. Board members work full-time for the FASB and sever all relations with their previous employers. Thus, a private-sector body containing representatives from a broad set of constituencies establishes acceptable accounting standards.[3]

Common terminology includes the pronouncements of the FASB (and its predecessors) in the compilation of accounting rules, procedures, and practices known as **generally accepted accounting principles (GAAP)**. GAAP (a singular noun) includes, as well, writings of the SEC, the Emerging Issues Task Force (a committee jointly sponsored by the FASB and the SEC), the AICPA, and scholarly writings. The FASB issues its major pronouncements in the form of ***Statements of Financial Accounting Standards***. The FASB has issued more than 150 statements since it began operating in 1973. For some financial reporting topics, the FASB requires the use of a uniform accounting method by all firms. In other cases, firms enjoy some freedom to select from a limited set of alternative methods. Thus, the current standard-setting approach lies somewhere between uniformity and flexibility, tilting toward uniformity.

With the exception of one method of accounting for inventories and cost of goods sold (discussed in **Chapter 7**), the accounting methods that firms use for financial reporting can differ from the methods used in calculating taxable income. Permitting different methods recognizes that the goal of providing useful information to financial statement users may require accounting rules that differ from those that aid the taxation authorities in their attempt to efficiently raise revenues.

---

[3]You may want to visit the web site of the Financial Accounting Standards Board at http://www.fasb.org to learn more about the purpose, procedures, and pronouncements of the FASB.

As the FASB contemplates a reporting issue, its due-process procedure ensures that it receives input from various preparer and user groups. Involving these constituencies in the deliberations also identifies concerns and increases the likelihood of acceptance of the final reporting standards.

The FASB recognizes that it needs some theoretical structure to guide its deliberations. The Board has developed a conceptual framework to use, not as a rigorous deductive scheme but as a guiding mechanism for setting accounting standards. The conceptual framework contains guidance about the following:

1. Objectives of financial reporting.
2. Qualitative characteristics of accounting information.
3. Elements of financial statements.
4. Recognition and measurement principles.

The FASB reports the components of its conceptual framework in ***Statements of Financial Accounting Concepts***. **Figure 1.4** summarizes the **financial reporting objectives** established by the FASB and their relation to the principal financial statements.

1. **Financial reporting should provide information useful for making rational investment and credit decisions.** This general-purpose objective simply states that financial reporting should aim primarily at investors and creditors and should help these individuals in their decision making.
2. **Financial reporting should provide information to help investors and creditors assess the amount, timing, and uncertainty of future cash flows.** This objective flows from the first by defining "useful information" more fully. It states that investors and creditors primarily want to know about the cash they will receive from investing in a firm. The ability of the firm to generate cash flows affects the amount of cash that will likely flow to investors and creditors.
3. **Financial reporting should provide information about the economic resources of a firm and the claims on those resources.** The balance sheet attempts to accomplish this objective.
4. **Financial reporting should provide information about a firm's operating performance during a period.** The income statement attempts to accomplish this objective.
5. **Financial reporting should provide information about how an enterprise obtains and uses cash.** The statement of cash flows accomplishes this objective.
6. **Financial reporting should provide information about how management has discharged its stewardship responsibility to owners.** Stewardship refers to the prudent use of resources entrusted to a firm. No single statement helps in assessing stewardship. Rather, owners assess stewardship using information from all three financial statements and the notes.
7. **Financial reporting should include explanations and interpretations to help users understand the financial information provided.** Supporting schedules and notes to the financial statements attempt to satisfy this objective.

Concerns over the quality of financial reporting have led, and continue to lead, to government initiatives to deal with the problems. The Congress passed the **Sarbanes-Oxley Act** in 2002. The Act set up the **Public Company Accounting Oversight Board (PCAOB),** which has responsibility for monitoring the quality of audits and the financial reporting process. The Act requires the PCAOB to register firms conducting independent audits, establish or adopt acceptable auditing, quality control, and independence standards, and provide for periodic "audits" of the auditors. Initiatives to improve the quality of security analysts' services and the operations of mutual funds are also underway.

## An International Perspective

The processes for setting accounting principles in countries other than the United States vary. In some countries, the measurement rules followed in preparing the financial statements match those followed in computing taxable income. Thus, the legislative branch of the government sets acceptable accounting principles. In other countries, an agency of the government sets acceptable accounting

*(continued)*

FIGURE 1.4 Summary of Reporting Process and Principal Financial Statements

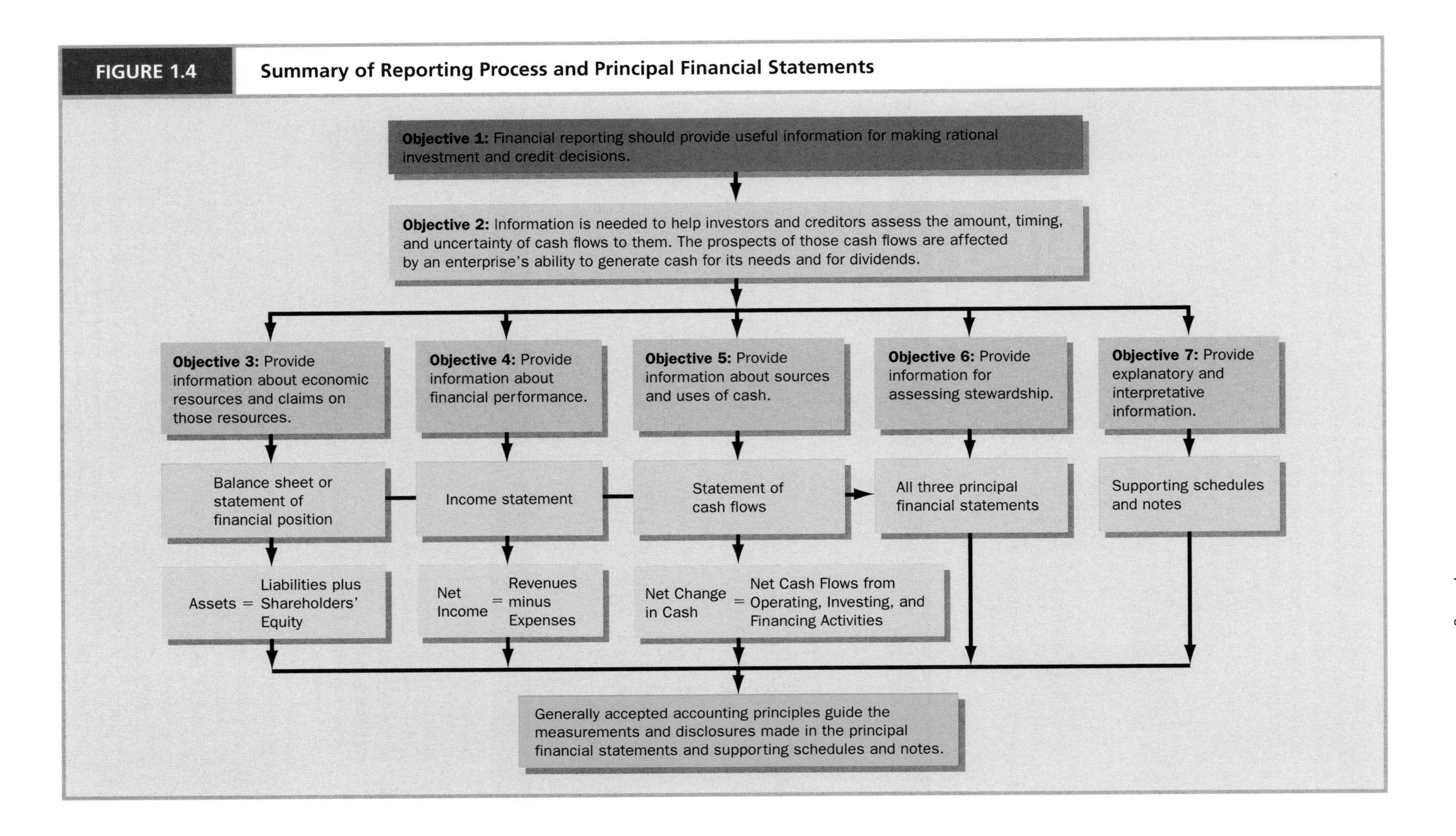

principles, but the measurement rules differ from those used in computing taxable income. In still other countries, the accounting profession, through its various boards and committees, plays a major role in setting accounting principles. Not surprisingly, the differing objectives of these standard-setting bodies (for example, taxation, accomplishment of government policy objectives, fair reporting to investors) result in diverse sets of accounting principles across countries.

The globalization of capital markets has increased the need for comparable and understandable financial statements across countries. To address this need, the **International Accounting Standards Board (IASB),** an independent entity comprised of 14 members and full-time professional staff, has played an increasingly important role in developing acceptable accounting principles worldwide. Although the IASB has no legal authority, it encourages its members to exert influence on the standard-setting process within their own countries to reduce diversity. Stock exchanges in many countries now allow foreign firms to list their securities on the exchanges as long as the financial statements of these firms reflect pronouncements of the IASB. Standard-setting boards in several countries now accept IASB pronouncements as allowable accounting principles within their countries. The European Commission has ruled that firms within member countries must conform their financial statements to the reporting standards of the IASB by 2005.[4]

## Summary

This chapter shows how a business's activities relate to the principal financial statements included in its annual reports to shareholders. The chapter raises questions that it does not fully answer. It provides you with a broad overview of the various financial statements before you examine the concepts and procedures underlying each statement.

**Chapters 2** through **4** discuss and illustrate the concepts and procedures of the balance sheet, income statement, and statement of cash flows, respectively. **Chapter 5** considers techniques for analyzing and interpreting these financial statements. **Chapters 6** through **14** explore more fully the principles of accounting for individual assets, liabilities, and shareholders' equities. The web site for this book contains additional information on particular topics not covered fully in this book (go to: www.thomsonedu.com/accounting/stickney).

Now we turn to the study of financial accounting. To comprehend the concepts and procedures in the book, you should study the numerical examples presented in each chapter and prepare solutions to several problems, including the self-study problems. You may find the glossary of terms at the back of the book useful.

## Solution to Self-Study Problem

### SUGGESTED SOLUTION TO PROBLEM 1.1 FOR SELF-STUDY

(Gateway: preparing a balance sheet and an income statement)

**a.** See Exhibit 1.4 on facing page.

**b.** See Exhibit 1.5 on facing page.

**c.** Gateway did not declare or pay a dividend to the common shareholders during Year 4 because the change in retained earnings equals net income, as the following analysis shows (amounts in thousands):

| | |
|---|---|
| Retained Earnings, December 31, Year 3 | $410,870 |
| Plus Net Income for Year 4 | 223,639 |
| Less Dividend Declared | 0 |
| Retained Earnings, December 31, Year 4 | $634,509 |

[4] You may want to visit the web site of the International Accounting Standards Board at http://www.iasb.org to learn more about the purpose, procedures, and pronouncements of the IASB. The Big 4 accounting firm, Deloitte, has a web site full of useful information about international accounting developments: http://www.iasplus.com.

EXHIBIT 1.4

**GATEWAY**
**Comparative Balance Sheet**
**December 31, Year 3 and Year 4**
**(Problem 1.1 for Self-Study)**

| | December 31 | |
|---|---|---|
| | **Year 3** | **Year 4** |
| ASSETS | | |
| **Current Assets** | | |
| Cash | $ 447,718 | $ 593,601 |
| Short-Term Cash Investments | 0 | 38,648 |
| Accounts Receivable | 449,723 | 510,679 |
| Inventories | 278,043 | 249,224 |
| Other Current Assets | 74,216 | 152,531 |
| Total Current Assets | $1,249,700 | $1,544,683 |
| **Noncurrent Assets** | | |
| Land | $ 14,888 | $ 21,431 |
| Buildings and Equipment (net of accumulated depreciation) | 227,477 | 315,038 |
| Computer Software Development Cost (net of accumulated amortization) | 31,873 | 39,998 |
| Goodwill from Corporate Acquisitions | 35,631 | 118,121 |
| Total Noncurrent Assets | $ 309,869 | $ 494,588 |
| Total Assets | $1,559,569 | $2,039,271 |
| LIABILITIES AND SHAREHOLDERS' EQUITY | | |
| **Current Liabilities** | | |
| Accounts Payable | $ 411,788 | $ 488,717 |
| Notes Payable | 15,041 | 13,969 |
| Royalties Payable | 125,270 | 159,418 |
| Taxes Payable | 40,334 | 26,510 |
| Other Current Liabilities | 207,336 | 315,292 |
| Total Current Liabilities | $ 799,769 | $1,003,906 |
| **Noncurrent Liabilities** | | |
| Long-Term Debt | $ 7,244 | $ 7,240 |
| Other Noncurrent Liabilities | 50,857 | 98,081 |
| Total Noncurrent Liabilities | $ 58,101 | $ 105,321 |
| Total Liabilities | $ 857,870 | $1,109,227 |
| **Shareholders' Equity** | | |
| Common Stock | $ 290,829 | $ 295,535 |
| Retained Earnings | 410,870 | 634,509 |
| Total Shareholders' Equity | $ 701,699 | $ 930,044 |
| Total Liabilities and Shareholders' Equity | $1,559,569 | $2,039,271 |

EXHIBIT 1.5

**GATEWAY**
**Income Statement**
**For the Year Ended December 31, Year 4**
**(Problem 1.1 for Self-Study)**

| | |
|---|---|
| **Revenues** | |
| Sales | $6,293,680 |
| Interest Revenue | 28,929 |
| Total Revenues | $6,322,609 |
| **Expenses** | |
| Cost of Goods Sold | $5,217,239 |
| Selling and Administrative Expenses | 786,168 |
| Interest Expense | 1,740 |
| Income Tax Expense | 93,823 |
| Total Expenses | $6,098,970 |
| **Net Income** | $ 223,639 |

**d.** Gateway experienced only minor changes in accounts related to financing activities (notes payable, long-term debt, common stock). The firm experienced increases in accounts related to investing activities (land, buildings and equipment, computer software development costs, goodwill from acquisitions), which would likely have reduced cash. Thus, the increase in cash primarily resulted from profitable operations. Note, though, that the increase in cash and short-term investments of cash of $184,531 [= ($593,601 + $38,648) − $447,718] is less than net income for the year of $223,639.

## Key Terms and Concepts

Goals contrasted with strategies
Financing activities
Dividends
Investing activities
Operating activities
Annual report to shareholders
Management Discussion and Analysis (MD&A)
Balance sheet
Assets
Liabilities
Shareholders' equity
Contributed capital
Retained earnings
Historical valuation
Current valuation
Income statement
Net income, earnings
Revenues
Expenses
Net loss
Statement of cash flows
Unqualified, qualified opinion
Disclaimer of opinion
Adverse opinion
Establishing acceptable accounting standards
Uniformity versus flexibility in selecting accounting methods
General principles versus rules-based approach to establishing accounting standards
Internal audits versus external audits
Securities and Exchange Commission (SEC)
Financial Accounting Standards Board (FASB)
Generally accepted accounting principles (GAAP)
*Statements of Financial Accounting Standards*
*Statements of Financial Accounting Concepts*
Financial reporting objectives
Sarbanes-Oxley Act
Public Company Accounting Oversight Board (PCAOB)
International Accounting Standards Board (IASB)

## Questions, Short Exercises, Exercises, Problems, and Cases

For additional student resources, content, and interactive quizzes for this chapter visit the FACMU website: **www.thomsonedu.com/accounting/stickney**

### QUESTIONS

1. Review the meaning of the terms and concepts listed in Key Terms and Concepts.
2. The chapter describes four activities common to all entities: setting goals and strategies, financing activities, investing activities, and operating activities. How would these four activities likely differ for a charitable organization versus a business firm?
3. "The photographic analogy for the balance sheet is a snapshot, and for the income statement and statement of cash flows, it is a motion picture." Explain.
4. "Asset valuation and income measurement are closely related." Explain.
5. A student states, "It is inconceivable to me that a firm could report increasing net income yet run out of cash." Clarify this seeming contradiction.
6. Does an unqualified, or "clean," opinion of an independent auditor indicate that the financial statements are free of errors and misrepresentations? Explain.
7. Suggest reasons why the format and content of financial accounting reports tend to be more standardized than the format and content of accounting reports that firms prepare for their internal decision-making, planning, and control purposes.
8. In some countries, governmental entities establish acceptable accounting standards and generally require conformity between financial reporting and tax reporting. Describe the advantages and disadvantages of this approach to standard setting.

**9.** "Prescribing a single method of accounting for a particular financial statement item will result in uniform reporting across firms." Do you agree? Why or why not?

**10.** "Politics is inherent to the accounting standard-setting process." Explain.

**11.** "Research on the efficiency of capital markets suggests that market prices adjust quickly and in an unbiased manner to new information about a firm. The efficiency of capital markets therefore eliminates the need for financial accounting reports." Do you agree? Why or why not?

## SHORT EXERCISES

**12. Balance sheet relations.** The balance sheet of Harley-Davidson Company, a manufacturer of motorcycles, showed current assets of $2,067 million, current liabilities of $990 million, shareholders' equity of $2,233 million, and noncurrent assets of $1,794 million at the end of Year 12. Net income for Year 12 was $580 million. Compute the amount of noncurrent liabilities on the balance sheet at the end of Year 12.

**13. Balance sheet relations.** The balance sheet of Mattel, a creator and manufacturer of toys and games, showed noncurrent assets of $2,071 million, noncurrent liabilities of $832 million, shareholders' equity of $1,979 million, and current liabilities of $1,649 million at the end of Year 12. Net loss for Year 12 was $22 million. Compute the amount of current assets on the balance sheet at the end of Year 12.

**14. Retained earnings relations.** The balance sheet of Reebok, a designer and manufacturer of athletic shoes, showed retained earnings of $411 million at the end of Year 11 and $446 million at the end of Year 12. The income statement for Year 12 showed net income of $52 million. Compute the amount of dividends declared during Year 12.

**15. Retained earnings relations.** The balance sheet of Ruby Tuesday, a restaurant chain, showed retained earnings of $326 million at the end of Year 12. Net income for Year 12 was $58 million and dividends declared and paid during Year 12 totaled $3 million. Compute the amount in retained earnings at the end of Year 11.

**16. Cash flow relations.** The statement of cash flows for Yahoo, Inc., an Internet service provider, showed a net cash inflow from operations of $303 million, a net cash outflow for investing of $346 million, and a net cash outflow for financing of $19 million for Year 12. The balance sheet at the end of Year 11 showed a balance in cash of $373 million. Compute the amount of cash on the balance sheet at the end of Year 12.

**17. Cash flow relations.** The statement of cash flows for Nike, a designer and manufacturer of athletic shoes, showed a net cash inflow from operations of $917.4 million and a net cash outflow for financing of $643.3 million for Year 12. The comparative balance sheets showed a balance in cash of $575.5 million at the end of Year 11 and $634.0 million at the end of Year 12. Net income for Year 12 was $474.0 million. Compute the net amount of cash provided or used for investing activities for Year 12.

## EXERCISES

**18. Preparing a personal balance sheet.** Prepare a balance sheet of your personal assets, liabilities, and owner's equity. How does the presentation of owner's equity on your balance sheet differ from that in **Exhibit 1.1**?

**19. Classifying financial statement accounts.** The balance sheet or income statement classifies various items in one of the following ways:

CA—Current assets
NA —Noncurrent assets
CL—Current liabilities
NL—Noncurrent liabilities
CC—Contributed capital
RE—Retained earnings
NI—Income statement item (revenue or expense)
X—Item generally not appearing on a balance sheet or an income statement

Using the letters, indicate the classification of each of the following items:

**a.** Factory
**b.** Interest revenue
**c.** Common stock issued by a corporation
**d.** Goodwill developed by a firm (see Glossary)

**e.** Automobiles used by sales staff
**f.** Cash on hand
**g.** Unsettled damage suit against a firm
**h.** Commissions earned by sales staff
**i.** Supplies inventory
**j.** Note payable, due in three months
**k.** Increase in market value of land held
**l.** Dividends declared
**m.** Employee payroll taxes payable
**n.** Note payable, due in six years

**20. Balance sheet relations.** Selected balance sheet amounts for Lowe's, a retailer of home construction and repair supplies, for four recent years appear below (amounts in millions):

| | Year 9 | Year 10 | Year 11 | Year 12 |
|---|---|---|---|---|
| Noncurrent Assets | $3,759 | $5,302 | $ 7,201 | ? |
| Shareholders' Equity | ? | 4,695 | 5,495 | $ 6,674 |
| Total Assets | ? | ? | 11,376 | ? |
| Current Liabilities | 1,765 | 2,386 | ?[a] | ?[b] |
| Current Assets | 2,586 | ? | ?[a] | ?[b] |
| Noncurrent Liabilities | 1,444 | ? | ? | 4,045 |
| Total Liabilities and Shareholders' Equity | ? | 9,012 | ? | 13,736 |

[a] Current Assets − Current Liabilities = $1,246.
[b] Current Assets − Current Liabilities = $1,903.

**a.** Compute the missing balance sheet amounts for each of the four years.
**b.** How did the mix of total assets (that is, the proportion of current versus noncurrent assets) change over the four-year period? What might account for such a change?
**c.** How did the mix of current liabilities, noncurrent liabilities, and shareholders' equity change over the four-year period? What might account for such a change?

**21. Balance sheet relations.** Selected balance sheet amounts for PepsiCo, a manufacturer of soft drinks and snacks, for four recent years appear below (amounts in millions):

| | Year 8 | Year 9 | Year 10 | Year 11 |
|---|---|---|---|---|
| Total Assets | $22,660 | ? | ? | $21,695 |
| Noncurrent Liabilities | 8,345 | $ 6,882 | ? | 8,023 |
| Noncurrent Assets | ? | 13,378 | $13,735 | ? |
| Total Liabilities and Shareholders' Equity | ? | 17,551 | ? | ? |
| Current Liabilities | 7,914 | ? | 3,935 | 4,998 |
| Shareholders' Equity | ? | 6,881 | 7,249 | ? |
| Current Assets | 4,362 | ? | 4,604 | 5,853 |

**a.** Compute the missing balance sheet amounts for each of the four years.
**b.** How did the mix of total assets (that is, the proportion of current versus noncurrent assets) change over the four-year period? What might account for such a change?
**c.** How did the mix of current liabilities, noncurrent liabilities, and shareholders' equity change over the four-year period? What might account for such a change?

**22. Balance sheet relations.** Selected balance sheet amounts for Johnson & Johnson, a consumer products company, for four recent years appear below (amounts in millions):

| | Year 8 | Year 9 | Year 10 | Year 11 |
|---|---|---|---|---|
| Current Assets | $11,132 | ?[a] | $15,450 | $18,473 |
| Noncurrent Assets | ? | ? | 15,871 | ? |
| Total Assets | ? | $29,163 | ? | ? |
| Current Liabilities | 8,162 | 7,454 | 7,140 | ?[c] |
| Noncurrent Liabilities | 4,459 | ? | 5,373 | ? |

*(continued)*

| | Year 8 | Year 9 | Year 10 | Year 11 |
|---|---|---|---|---|
| Contributed Capital . . . . . . . . . . . . . . . . . . . . | ? | $ 458 | ? | $ 1,727 |
| Retained Earnings . . . . . . . . . . . . . . . . . . . . . | $13,556 | 15,755 | ?[b] | ?[d] |
| Total Liabilities and Shareholders' Equity . . . . . | 26,211 | ? | ? | 34,488 |

[a] Current Assets − Current Liabilities = $5,746.
[b] Net income for Year 10 is $4,276 and dividends are $1,724.
[c] Current Assets − Current Liabilities = $10,429.
[d] Net income for Year 11 is $6,246 and dividends are $2,047.

**a.** Compute the missing balance sheet amounts for each of the four years.
**b.** How did the mix of total assets (that is, the proportion of current versus noncurrent assets) change over the four-year period? What might account for such a change?
**c.** How did the mix of current liabilities, noncurrent liabilities, and shareholders' equity change over the four-year period? What might account for such a change?

**23. Balance sheet relations.** Selected balance sheet amounts for Intel, a developer and manufacturer of semiconductors, for four recent years appear below (amounts in millions):

| | Year 9 | Year 10 | Year 11 | Year 12 |
|---|---|---|---|---|
| Current Assets . . . . . . . . . . . . . . . . . . . . . . . | ? | $21,150 | $17,633 | ? |
| Noncurrent Assets . . . . . . . . . . . . . . . . . . . . . | $26,030 | ? | 26,762 | ? |
| Total Assets . . . . . . . . . . . . . . . . . . . . . . . . . | ? | 47,945 | ? | ? |
| Current Liabilities . . . . . . . . . . . . . . . . . . . . . | 7,099 | ? | 6,570 | $6,595 |
| Noncurrent Liabilities . . . . . . . . . . . . . . . . . . | 4,095 | ? | 1,995 | 2,161 |
| Contributed Capital . . . . . . . . . . . . . . . . . . . . | ? | 8,389 | ? | 7,578 |
| Retained Earnings . . . . . . . . . . . . . . . . . . . . . | 25,349 | 28,933 | ?[a] | ?[b] |
| Total Liabilities and Shareholders' Equity . . . . . | 43,849 | ? | ? | ? |
| Current Assets/Current Liabilities . . . . . . . . . . | ? | 2.445 | ? | 2.870 |

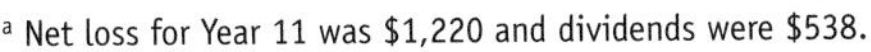

[a] Net loss for Year 11 was $1,220 and dividends were $538.
[b] Net income for Year 12 was $1,248 and dividends were $533.

**a.** Compute the missing amounts for each of the four years.
**b.** Identify changes in the mix of total assets (that is, the proportion of current versus noncurrent assets) and in the mix of total liabilities versus shareholders' equity over the four years above. Suggest the events or transactions that might explain these changes.

**24. Retained earnings relations.** Hewlett-Packard Corporation develops and manufactures a diversified line of computer products. Its sales tend to vary with changes in the business cycle. Selected data from its financial statements for four recent years appear below (amounts in millions):

| | Year 9 | Year 10 | Year 11 | Year 12 |
|---|---|---|---|---|
| Retained Earnings, Beginning of Year . . . . . . . . | $16,909 | ? | $14,097 | $13,693 |
| Net Income . . . . . . . . . . . . . . . . . . . . . . . . . | 2,026 | $ 3,697 | ? | ? |
| Dividends Declared and Paid . . . . . . . . . . . . . . | 650 | ? | 621 | 801 |
| Retained Earnings, End of Year . . . . . . . . . . . . | ? | 14,097 | 13,693 | 11,973 |

**a.** Compute the missing amounts for each year.
**b.** What is the likely reason for the variations in net income and net loss?
**c.** What might explain the unusual pattern of dividends over the four-year period?

**25. Retained earnings relations.** Selected data affecting retained earnings for Ford Motor Company, an automobile manufacturer, for four recent years appear below (amounts in millions):

| | Year 9 | Year 10 | Year 11 | Year 12 |
|---|---|---|---|---|
| Retained Earnings, January 1 . . . . . . . . . . . . . | $12,221 | ? | ? | $10,502 |
| Net Income . . . . . . . . . . . . . . . . . . . . . . . . . | 7,237 | $ 3,467 | ? | ? |
| Dividends Declared and Paid . . . . . . . . . . . . . . | 2,290 | ? | $ 1,929 | 743 |
| Retained Earnings, December 31 . . . . . . . . . . . | ? | 17,884 | 10,502 | 8,659 |

**a.** Compute the missing amounts for each of the four years.
**b.** What is the likely reason for the pattern of net income and net loss over the four-year period?
**c.** What is the likely reason for the pattern of dividends over the four-year period?

**26. Relating net income to balance sheet changes.** The comparative balance sheets for Target Corporation, a discount retailer, as of January 31, Year 11; January 31, Year 12; and January 31, Year 13, appear below (amounts in millions):

TARGET CORPORATION
Comparative Balance Sheet
January 31, Year 11, Year 12, and Year 13

| | January 31 | | |
|---|---|---|---|
| | Year 11 | Year 12 | Year 13 |
| Total Assets | $19,490 | $24,154 | $28,603 |
| Liabilities | $12,971 | $16,294 | $19,160 |
| Common Stock | 977 | 1,173 | 1,336 |
| Retained Earnings | 5,542 | 6,687 | 8,107 |
| Total Liabilities and Shareholders' Equity | $19,490 | $24,154 | $28,603 |

Target Corporation declared and paid dividends of $203 million during fiscal Year 12 and $218 million during fiscal Year 13.

**a.** Compute net income for fiscal Year 12 and Year 13 by analyzing the change in retained earnings.
**b.** Demonstrate that the following relation holds:

Net Income = Increase in Assets − Increase in Liabilities
− Increase in Contributed Capital + Dividends

**27. Income statement relations.** Selected income statement information for Dell Computer, a manufacturer of personal computers, for three recent years appears below (amounts in millions):

| | Year 10 | Year 11 | Year 12 |
|---|---|---|---|
| Sales | $31,888 | ? | $35,404 |
| Cost of Goods Sold | 25,445 | $25,661 | ? |
| Selling and Administrative Expenses | 3,193 | 2,784 | 3,050 |
| Research and Development Expense | 482 | 452 | 455 |
| Income Tax Expense | 830 | 636 | 846 |
| Net Income | ? | 1,635 | 1,998 |

**a.** Compute the missing amounts for each of the three years.
**b.** Prepare a common-size income statement for each year, in which sales is equal to 100 percent and each expense and net income are expressed as a percentage of sales. What factors appear to explain the change in the ratio of net income to sales?

**28. Income statement relations.** Selected income statement information for Pfizer, a pharmaceutical company, for three recent years appears below (amounts in millions):

| | Year 10 | Year 11 | Year 12 |
|---|---|---|---|
| Sales | $26,045 | ? | $32,373 |
| Cost of Goods Sold | 3,755 | $3,823 | ? |
| Selling and Administrative Expenses | 9,566 | 9,717 | 10,846 |
| Research and Development Expense | 4,374 | 4,776 | 5,176 |
| Income Tax Expense | 3,074 | 2,720 | 2,830 |
| Net Income | ? | 7,988 | 9,476 |

a. Compute the missing amounts for each of the three years.
b. Prepare a common-size income statement for each year, in which sales equals 100 percent and each expense and net income are expressed as a percentage of sales. What factors appear to explain the change in the ratio of net income to sales?

**29. Statement of cash flows relations.** Selected data from the statement of cash flows for Amazon.com, an on-line retailer, for Year 10, Year 11, and Year 12 appear below (amounts in thousands):

| | Year 10 | Year 11 | Year 12 |
|---|---|---|---|
| **Inflows of Cash** | | | |
| Increase in Short-term Debt | $ 0 | $ 0 | $ 0 |
| Issue of Long-term Debt | 681,499 | 10,000 | 0 |
| Issue of Common Stock | 44,697 | 116,456 | 121,689 |
| Revenues from Operations Increasing Cash | 2,753,398 | 3,143,165 | 3,892,988 |
| Sale of Marketable Securities (net) | 361,269 | 0 | 0 |
| Other Financing Activities | 0 | 0 | 38,471 |
| **Outflows of Cash** | | | |
| Acquisition of Property and Equipment | 134,758 | 50,321 | 39,163 |
| Purchase of Marketable Securities (net) | 0 | 196,775 | 82,521 |
| Decrease in Long-term Debt | 33,049 | 19,575 | 14,795 |
| Acquisition of Investments in Other Companies | 62,533 | 6,198 | 0 |
| Expenses for Operations Decreasing Cash | 2,883,840 | 3,262,947 | 3,718,697 |
| Other Financing Activities | 37,557 | 15,958 | 0 |
| Increase (Decrease) in Cash | 689,126 | (282,153) | 197,972 |

a. Prepare a statement of cash flows for Amazon.com for the three years using the format in **Exhibit 1.3**. Set cash flow from operations equal to revenues providing cash minus expenses using cash. The balance in cash at the beginning of Year 10 was $133,309. Changes in marketable securities are investing activities.
b. Comment on the relation among cash flows from operating, investing, and financing activities during the three years.

**30. Statement of cash flows relations.** Selected data from the statement of cash flows for AMR Corporation, the parent company of American Airlines, for three recent years appear below (amounts in millions):

| | Year 10 | Year 11 | Year 12 |
|---|---|---|---|
| **Inflows of Cash** | | | |
| Proceeds from Long-term Borrowings | $ 836 | $ 5,096 | $ 3,190 |
| Revenues from Operations Increasing Cash | 19,536 | 19,083 | 17,233 |
| Issue of Common Stock | 67 | 37 | 3 |
| Sale of Property, Plant, and Equipment | 332 | 401 | 220 |
| Other Investing Transactions | 71 | 0 | 268 |
| Total Inflows | $20,842 | $24,617 | $20,914 |
| **Outflows of Cash** | | | |
| Acquisition of Property, Plant, and Equipment | $ 3,678 | $ 3,640 | $ 1,881 |
| Expenses for Operations Decreasing Cash | 16,394 | 18,541 | 18,344 |
| Repayments of Long-term Debt | 766 | 922 | 687 |
| Other Investing Activities | 0 | 1,501 | 0 |
| Total Outflows | $20,838 | $24,604 | $20,912 |
| Increase (Decrease) in Cash | $ 4 | $ 13 | $ 2 |

a. Prepare a statement of cash flows for AMR for each of the three years using the format in **Exhibit 1.3**. Set cash flow from operations equal to revenues providing cash minus expenses using cash. The balance in cash at the beginning of Year 10 was $85 million.

**b.** Comment on the relation among cash flows from operating, investing, and financing activities during the three years.

**31. Relations between financial statements.** Compute the missing information in each of the following four independent cases. The letters in parentheses refer to the following:

BS—Balance sheet
IS—Income statement
SCF—Statement of cash flows

| | | |
|---|---|---|
| **a.** | Accounts Receivable, Jan. 1, Year 2 (BS) | $ 630 |
| | Sales on Account for Year 2 (IS) | 3,290 |
| | Collections from Customers on Account during Year 2 (SCF) | 2,780 |
| | Accounts Receivable, Dec. 31, Year 2 (BS) | ? |
| **b.** | Income Taxes Payable, Jan. 1, Year 2 (BS) | $ 1,240 |
| | Income Tax Expense for Year 2 (IS) | ? |
| | Payments to Governments during Year 2 (SCF) | 8,290 |
| | Income Taxes Payable, Dec. 31, Year 2 (BS) | 1,410 |
| **c.** | Building (net of depreciation), Jan. 1, Year 2 (BS) | $ 89,000 |
| | Depreciation Expense for Year 2 (IS) | ? |
| | Purchase of Building during Year 2 (SCF) | 17,600 |
| | Building (net of depreciation), Dec. 31, Year 2 (BS) | 102,150 |
| **d.** | Retained Earnings, Jan. 1, Year 2 (BS) | $ 76,200 |
| | Net Income for Year 2 (IS) | 14,200 |
| | Dividends Declared and Paid during Year 2 (SCF) | ? |
| | Retained Earnings, Dec. 31, Year 2 (BS) | 83,300 |

## PROBLEMS AND CASES

**32. Preparing a balance sheet and income statement.** The accounting records of eBay appear below (amounts in thousands):

| | December 31 | |
|---|---|---|
| **Balance Sheet Items** | **Year 11** | **Year 12** |
| Cash | $ 723,419 | $1,199,003 |
| Accounts Receivable | 101,703 | 131,453 |
| Other Current Assets | 58,683 | 138,002 |
| Property, Plant, and Equipment (net of depreciation) | 142,349 | 218,028 |
| Investment in Securities (noncurrent) | 416,612 | 604,871 |
| Goodwill and Other Intangibles | 198,639 | 1,735,489 |
| Other Noncurrent Assets | 37,124 | 97,598 |
| Accounts Payable to Suppliers | 33,235 | 47,424 |
| Other Current Liabilities | 146,904 | 338,800 |
| Long-term Debt | 12,008 | 13,798 |
| Other Noncurrent Liabilities | 57,244 | 167,949 |
| Common Stock | 1,275,517 | 3,108,754 |
| Retained Earnings | 153,621 | 447,719 |

*(continued)*

| Income Statement Items | For Year 12 |
|---|---|
| Net Operating Revenues | $1,214,100 |
| Interest and Other Revenues | 96,352 |
| Cost of Net Operating Revenues | 213,876 |
| Product Development Expenses | 104,636 |
| Selling Expenses | 349,650 |
| Administrative Expenses | 171,785 |
| Other Operating Expenses | 28,969 |
| Interest Expense | 1,492 |
| Income Taxes | 145,946 |
| **Dividend Information** | |
| Dividends Declared and Paid | $ 0 |

**a.** Prepare an income statement for eBay for the year ending December 31, Year 12. Refer to **Exhibit 1.2** for help in designing the format of the statement.

**b.** Prepare a comparative balance sheet for eBay on December 31, Year 11, and December 31, Year 12. Refer to **Exhibit 1.1** for help in designing the format of the statement.

**c.** Prepare an analysis of the change in retained earnings during the year ending December 31, Year 12.

**33. Preparing a balance sheet and income statement.** The accounting records of The GAP, a retail clothing store, reveal the following for a recent year (amounts in thousands):

| | January 31 | |
|---|---|---|
| Balance Sheet Items | Year 9 | Year 10 |
| Accounts Payable | $ 684,130 | $ 805,945 |
| Cash | 565,253 | 450,352 |
| Common Stock | 354,719 | 135,034 |
| Merchandise Inventory | 1,056,444 | 1,462,045 |
| Notes Payable to Banks (due within one year) | 90,690 | 168,961 |
| Long-term Debt | 496,455 | 784,925 |
| Other Current Assets | 250,127 | 285,393 |
| Other Current Liabilities | 778,283 | 777,973 |
| Other Noncurrent Assets | 215,725 | 275,651 |
| Other Noncurrent Liabilities | 340,682 | 417,907 |
| Property, Plant, and Equipment (net) | 1,876,370 | 2,715,315 |
| Retained Earnings | 1,218,960 | 2,098,011 |

| Income Statement Items | For the Year Ended January 31, Year 10 |
|---|---|
| Administrative Expenses | $ 803,995 |
| Cost of Goods Sold | 6,775,262 |
| Income Tax Expense | 657,884 |
| Interest Expense | 31,755 |
| Sales Revenue | 11,635,398 |
| Selling Expenses | 2,239,437 |

**a.** Prepare a comparative balance sheet for The GAP as of January 31, Year 9 and Year 10. Classify each balance sheet item into one of the following categories: current assets, noncurrent assets, current liabilities, noncurrent liabilities, and shareholders' equity.

**b.** Prepare an income statement for The GAP for the year ended January 31, Year 10. Separate income items into revenues and expenses.

**c.** Prepare a schedule explaining the change in retained earnings between January 31, Year 9 and January 31, Year 10.

**d.** Compare the amounts on The GAP's balance sheet on January 31, Year 9 and January 31, Year 10. Identify the major changes and suggest possible explanations for these changes.

**34. Preparing a balance sheet and income statement.** The accounting records of Southwest Airlines reveal the following for a recent year (amounts in thousands):

| | December 31 | |
|---|---|---|
| **Balance Sheet Items** | **Year 8** | **Year 9** |
| Accounts Payable | $ 157,415 | $ 156,755 |
| Accounts Receivable | 88,799 | 73,448 |
| Cash | 378,511 | 418,819 |
| Common Stock | 352,943 | 449,934 |
| Current Maturities of Long-term Debt | 11,996 | 7,873 |
| Inventories | 50,035 | 65,152 |
| Long-term Debt | 623,309 | 871,717 |
| Other Current Assets | 56,810 | 73,586 |
| Other Current Liabilities | 681,242 | 795,838 |
| Other Noncurrent Assets | 4,231 | 12,942 |
| Other Noncurrent Liabilities | 844,116 | 984,142 |
| Property, Plant, and Equipment (net) | 4,137,610 | 5,008,166 |
| Retained Earnings | 2,044,975 | 2,385,854 |

| **Income Statement Items** | **For the Year Ended December 31, Year 9** |
|---|---|
| Fuel Expense | $ 492,415 |
| Income Tax Expense | 299,233 |
| Interest Expense | 22,883 |
| Interest Revenue | 14,918 |
| Maintenance Expense | 367,606 |
| Other Operating Expenses | 1,638,753 |
| Sales Revenue | 4,735,587 |
| Salaries and Benefits Expense | 1,455,237 |

**a.** Prepare a comparative balance sheet for Southwest Airlines as of December 31, Year 8 and Year 9. Classify each balance sheet item into one of the following categories: current assets, noncurrent assets, current liabilities, noncurrent liabilities, and shareholders' equity.

**b.** Prepare an income statement for Southwest Airlines for the year ended December 31, Year 9. Separate income items into revenues and expenses.

**c.** Prepare a schedule explaining the change in retained earnings between December 31, Year 8 and December 31, Year 9.

**d.** Compare the amounts on Southwest Airlines' balance sheet on December 31, Year 8 and December 31, Year 9. Identify the major changes and suggest possible explanations for the changes.

**35. Relations between net income and cash flows.** The ABC Company starts the year in fine shape. The firm makes widgets—just what the customer wants. It makes them for $0.75 each and sells them for $1.00. The ABC Company keeps an inventory equal to shipments of the past 30 days, pays its bills promptly, and collects cash from customers within 30 days after the sale. The sales manager predicts a steady increase in sales of 500 widgets each month beginning in February. It looks like a great year, and it begins that way.

January 1 — Cash, $875; receivables, $1,000; inventory, $750

January — In January the firm sells, on account for $1,000, 1,000 widgets costing $750. Net income for the month is $250. The firm collects receivables outstanding at the beginning of the month. Production equals 1,000 units at a total cost of $750. The books at the end of January show the following:

February 1 — Cash, $1,125; receivables, $1,000; inventory, $750

February — This month's sales jump, as predicted, to 1,500 units. With a corresponding step-up in production to maintain the 30-day inventory, ABC Company makes 2,000 units at a cost of $1,500. All receivables from January sales are collected. Net income so far is $625. Now the books look like this:

| | |
|---|---|
| March 1 | Cash, $625; receivables, $1,500; inventory, $1,125 |
| March | March sales are even better, increasing to 2,000 units. Collections are on time. Production, to adhere to the inventory policy, is 2,500 units. Operating results for the month show net income of $500. Net income to date is $1,125. The books show the following: |
| April 1 | Cash, $250; receivables, $2,000; inventory, $1,500 |
| April | In April, sales jump another 500 units to 2,500, and the manager of ABC Company shakes the sales manager's hand. Customers are paying right on time. Production is pushed to 3,000 units, and the month's business nets $625 for a net income to date of $1,750. The manager of ABC Company takes off for Miami before the accountant's report is issued. Suddenly a phone call comes from the treasurer: "Come home! We need money!" |
| May 1 | Cash, $0; receivables, $2,500; inventory, $1,875 |

**a.** Prepare an analysis that explains what happened to ABC Company. (*Hint:* Compute the amount of cash receipts and cash disbursements for each month during the period January 1 to May 1.)
**b.** How can a firm show increasing net income but a decreasing amount of cash?
**c.** What insights are provided by the problem about the need for all three financial statements—balance sheet, income statement, and statement of cash flows?
**d.** What actions would you suggest that ABC Company take to deal with its cash flow problem?

**36. Balance sheet and income statement relations.** (Prepared by Professor Wesley T. Andrews Jr. and reproduced, with adaptation, by permission.)[5]

Once upon a time many, many years ago, a feudal landlord lived in a small province of central Europe. The landlord, called the Red-Bearded Baron, lived in a castle high on a hill. This benevolent fellow took responsibility for the well-being of many peasants who occupied the lands surrounding his castle. Each spring, as the snow began to melt, the Baron would decide how to provide for all his serf dependents during the coming year.

One spring, the Baron was thinking about the wheat crop of the coming growing season. "I believe that 30 acres of my land, being worth five bushels of wheat per acre, will produce enough wheat for next winter," he mused, "but who should do the farming? I believe I'll give Ivan the Indefatigable and Igor the Immutable the task of growing the wheat." Whereupon he summoned Ivan and Igor, two gentry noted for their hard work and not overly active minds, for an audience.

"Ivan, you will farm on the 20-acre plot of ground, and Igor will farm the 10-acre plot," the Baron began. "I will give Ivan 20 bushels of wheat for seed and 20 pounds of fertilizer. (Twenty pounds of fertilizer are worth two bushels of wheat.) Igor will get 10 bushels of wheat for seed and 10 pounds of fertilizer. I will give each of you an ox to pull a plow, but you will have to make arrangements with Feyador, the Plowmaker, for a plow. The oxen, incidentally, are only three years old and have never been used for farming, so they should have a good 10 years of farming ahead of them. Take good care of them, because an ox is worth 40 bushels of wheat. Come back next fall and return the oxen and the plows along with your harvest." Ivan and Igor bowed and withdrew from the Great Hall, taking with them the things provided by the Baron.

The summer came and went. After the harvest Ivan and Igor returned to the Great Hall to account to their master for the things given them in the spring. Ivan, pouring 223 bushels of wheat onto the floor, said, "My Lord, I present you with a slightly used ox, a plow broken beyond repair, and 223 bushels of wheat. I, unfortunately, owe Feyador, the Plowmaker, three bushels of wheat for the plow I got from him last spring. And, as you might expect, I used all the fertilizer and seed you gave me last spring. You will also remember, my Lord, that you took 20 bushels of my harvest for your own personal use."

Igor, who had been given 10 acres of land, 10 bushels of wheat, and 10 pounds of fertilizer, spoke next. "Here, my Lord, is a partially used-up ox, the plow for which I gave Feyador, the Plowmaker, three bushels of wheat from my harvest, and 105 bushels of wheat. I, too, used all my seed and fertilizer last spring. Also, my Lord, you took 30 bushels of wheat several days ago for your own table. I believe the plow is good for two more seasons."

[5]*Accounting Review* (April 1974). Reproduced by permission of the American Accounting Association.

"Knaves, you did well," said the Red-Bearded Baron. Blessed with this benediction, the two serfs departed. After the servants had taken their leave, the Red-Bearded Baron, watching the two hungry oxen slowly eating the wheat piled on the floor, began to contemplate what had happened. "Yes," he thought, "they did well, but I wonder which one did better?"

**a.** What measuring unit should the Red-Bearded Baron use to measure financial position and operating performance?

**b.** Prepare a balance sheet for Ivan and for Igor at both the beginning and the end of the period.

**c.** Prepare an income statement for Ivan and for Igor for the period.

**d.** Prepare a schedule reconciling the change in owner's equity between the beginning and the end of the period.

**e.** Did Ivan or Igor perform better during the period? Explain.

**37. Ethical issues in financial reporting.** A multinational computer equipment manufacturer reported the following amounts for two recent years (in millions):

| | Year 9 | Year 10 |
|---|---|---|
| Revenues | $87,548 | $88,396 |
| Expenses | (75,791) | (76,862) |
| Income Before Income Taxes | $11,757 | $11,534 |
| Income Tax Expense | (4,045) | (3,441) |
| Net Income | $ 7,712 | $ 8,093 |

**a.** Compute the ratio of net income divided by revenues for each year.

**b.** Compute the ratio of income before taxes divided by revenues for each year.

**c.** Compute the ratio of income tax expense divided by income before taxes, a ratio called the *effective tax rate*.

**d.** What do these ratios suggest as to the principal reason for the change in profitability between Year 9 and Year 10?

**e.** This firm reported increased revenues, net income, and profit margin between Year 9 and Year 10, despite weak conditions in technology industries. The explanation for the increased net income and profit margins was a reduction in the firm's effective tax rate.

To understand the reason for the reduced effective tax rate, the user of the financial statements would need to read the note to the financial statements on income taxes. Generally accepted accounting principles require firms to show the reasons why their effective tax rate differs from the statutory tax rate of 35 percent. The income tax note for this firm showed that the decreased effective tax rate was due to lower foreign taxes, the result of shifting certain foreign operations to countries with lower income tax rates. Understanding the income tax note requires knowledge of accounting and taxes usually attained by security analysts but not typically by the average investor.

Identify any ethical issues that this scenario raises and express your view as to whether this firm reported in an ethical manner.

2

*facmu* 12

# Accounting Concepts and Methods

CHAPTER 2

# Balance Sheet: Presenting the Investments and Financing of a Firm

## Learning Objectives

1. **Understand the accounting concepts of assets, liabilities, and shareholders' equity, including the conditions when firms recognize such items (recognition issues), their amounts (valuation issues), and where these items appear on the balance sheet (classification issues).**
2. **Understand the dual-entry recording framework and learn to use it to record a series of transactions, ending with the balance sheet.**
3. **Develop skills to analyze a balance sheet, focusing on the relations between assets, liabilities, and shareholders' equity that one would expect for financially healthy firms in different industries.**

The user of the financial statements might raise questions such as the following with respect to items on the balance sheet:

1. Why does the stock market value Wal-Mart's common stock at $210.1 billion when the balance sheet reports total shareholders' equity of only $43.6 billion?
2. Why do current liabilities comprise 60 percent of total financing (= total liabilities plus shareholders' equity) of Interpublic Group, a marketing services firm, whereas they comprise only 24 percent of total financing of American Airlines?
3. Why does long-term debt comprise 41 percent of total financing of American Airlines, whereas it comprises only 10 percent of total financing for Merck, a pharmaceutical company?
4. Why can savings and loan (S&L) associations face financial difficulties when their assets comprise loans receivable for residential housing and their financing includes savings accounts and certificates of deposit?

Understanding the balance sheet permits the user to answer these and other questions about the financial position of a firm. This chapter discusses concepts underlying the balance sheet, illustrates procedures for preparing the balance sheet, and demonstrates relations the user should look for when analyzing the balance sheets of healthy firms in different industries. This chapter and this book take the perspective of a financial statement user. An understanding of the principal concepts underlying the balance sheet aids the user in analyzing and interpreting published balance sheets. Mastery of the concepts requires some understanding of the accounting methods (or procedures) that accountants follow in preparing the balance sheet.

## Underlying Concepts

**Chapter 1** introduced the balance sheet, one of the three principal financial statements. Common terminology in some countries refers to this financial statement as a *statement of financial position*. The balance sheet presents a snapshot of the investments of a firm (assets) and the financing of those investments (liabilities and shareholders' equity) as of a specific time. The balance sheet shows the following balance, or equality:

Assets = Liabilities + Shareholders' Equity

This equation states that a firm's assets balance with the sources of funds (the financing) provided by creditors and owners that the firm used to acquire those assets. The balance sheet presents resources from two angles: a listing of the specific forms in which a firm holds them (for example, cash, inventory, equipment), and a listing of the people or entities (for example, suppliers, employees, governments, shareholders) who have a claim on the assets because they provided the financing—they are the sources of the financing. Accountants often refer to the sum of liabilities plus shareholders' equity as *total equities* or *total financing*. You can think of the sum as total *sources of financing,* or as *claims to assets*. The introduction to the balance sheet in **Chapter 1** left several questions unanswered:

1. Which resources does a firm recognize as assets?
2. What valuations does it place on these assets—that is, what number appears on the balance sheet for the asset?
3. How does it classify, or group, assets within the balance sheet?
4. Which claims against a firm's assets appear on the balance sheet as liabilities?
5. What valuations does a firm place on these liabilities?
6. How does a firm classify liabilities within the balance sheet?
7. What valuation does a firm place on shareholders' equity?
8. How does a firm disclose the shareholders' equity within the balance sheet, and what is the economic significance, if any, of the differences among the various components of shareholders' equity?

The following discussion not only provides a background for understanding the balance sheet as currently prepared but also permits the reader to assess alternative methods for measuring financial position.

### Asset Recognition

An asset is a resource that has the potential for providing a firm with a future economic benefit—the ability to generate future cash inflows or to reduce future cash outflows. A firm will recognize a resource as an asset only if (1) the firm has acquired rights to its use in the future as a result of a past transaction or exchange, and (2) the firm can measure or quantify the future benefits with a reasonable degree of precision.[1] All assets are future benefits; however, not all future benefits are assets.

**Example 1** Miller Corporation sold merchandise and received a note from the customer, who agreed to pay $2,000 within four months. This note receivable is an asset of Miller Corporation because Miller has a right to receive a definite amount of cash in the future as a result of the previous sale of merchandise.

**Example 2** Miller Corporation acquired manufacturing equipment costing $40,000 and agreed to pay the seller over three years. After the final payment, but not until then, the seller

[1]Financial Accounting Standards Board, *Statement of Financial Accounting Concepts No. 6,* "Elements of Financial Statements," 1985, par. 25. See the **Glossary** for the Board's actual definition of an asset.

will transfer legal title to the equipment to Miller Corporation. Even though Miller Corporation will not have legal title for three years, the equipment is Miller's asset because Miller has the rights, responsibilities, rewards, and risks of ownership and can maintain those rights as long as it makes payments on schedule. The seller exchanged the use of the equipment in return for cash and rights to receive cash in the future.

**Example 3** Miller Corporation has developed a well-known brand name for its products. Management expects this brand name to provide benefits to the firm in future sales of its products to customers. A brand name developed by a firm is generally, however, not an accounting asset. Although Miller Corporation has made various expenditures in the past to develop the brand name, U.S. accounting judges the future benefits too difficult to quantify with sufficient precision to allow Miller to recognize the brand name as an asset.

**Example 4** Miller Corporation makes research and development expenditures each year to develop new technologies and create new products. Accounting views the future benefits of this research to be too uncertain to justify recognition of an asset.

As **Examples 3** and **4** illustrate, accountants sometimes encounter difficulty in quantifying future benefits with sufficient precision to justify recognizing an asset. Let's extend **Examples 3** and **4** by assuming that Miller Corporation plans to acquire another firm with recognizable brand names and proven technologies. Miller Corporation hires a valuation appraisal firm to value these brand names and proven technologies and assist Miller Corporation in deciding how much to pay for the other firm. Accounting requires Miller to allocate portions of the purchase price to the brand names and proven technologies and report them as assets because the independent appraisal and the negotiation between an independent buyer and seller as to the purchase price establish the existence and value of the future benefits.

**Example 5** Miller Corporation plans to acquire a fleet of new trucks next year to replace those that are wearing out. These new trucks are not assets now because Miller Corporation has made no exchange with a supplier and, therefore, has not established a right to the future use of the trucks.

Another difficulty that accountants encounter in deciding which items to recognize as assets relates to mutually unexecuted contracts, sometimes called **executory contracts**. An executory contract involves a contractual exchange of promises to perform in the future but neither party has yet performed. In **Example 5,** suppose that Miller Corporation entered into a contract with a local truck dealer to acquire the trucks next year at a cash price of $80,000. Miller Corporation has acquired rights to future benefits, but the contract remains unexecuted by both the truck dealer (who must deliver the trucks) and Miller Corporation (who must pay the agreed cash price). Accounting does not generally recognize executory contracts, *mere exchanges of promises*, as assets. Miller Corporation will recognize an asset for the trucks when it receives them next year.

To take the illustration one step further, assume that Miller Corporation advances the truck dealer $15,000 of the purchase price at the time it signs the contract. Miller Corporation has acquired rights to future benefits and has exchanged cash. Thus, the contract is partially executed. Current accounting practice treats the $15,000 advance on the purchase of equipment as an asset reported under a title such as Advances to Suppliers. The trucks would not be assets at this time, however, because Miller Corporation has not yet received sufficient future rights to the services of the trucks to justify their inclusion in the balance sheet. Similar asset-recognition questions arise when a firm leases buildings and equipment for its own use under long-term leases or when a firm contracts with a transport company to deliver all of the firm's products to customers for some period of years. Later chapters discuss these issues more fully.

## Asset Valuation

Accounting must assign a monetary amount to each asset in the balance sheet. The accountant might use several methods of computing this amount.

**Acquisition or Historical Cost** The amount of cash paid (or the cash equivalent value of other forms of payment) in acquiring an asset is the **acquisition (historical) cost** of the asset. The accountant can typically ascertain this amount by referring to contracts, invoices, and canceled checks related to the acquisition of the asset. Nothing compels a firm to acquire a given asset, so accountants assume that the firm expects the future benefits from an acquired asset to be at least as large as the acquisition cost. At the time of acquisition, historical cost sets the lower limit on the value of the asset's future benefits to the firm.

**Current Replacement Cost** Each asset might appear on the balance sheet at the current cost of replacing it. Because the firm has to pay this amount to gain entry to use the asset and enjoy its future benefits, accountants sometimes refer to **current replacement cost** as an **entry value**.

For assets purchased frequently, such as merchandise inventory, the accountant can often calculate current replacement cost by consulting suppliers' catalogs or price lists. But accountants have difficulty in measuring the replacement costs of assets purchased less frequently—assets such as land, buildings, and equipment. A major obstacle to using current replacement cost as the valuation basis is the absence of well-organized secondhand markets for many used assets. When a firm cannot easily find similar used assets for sale, ascertaining current replacement cost requires finding the cost of a similar new asset and then adjusting that amount downward for the services of the asset already used. Difficulties can arise, however, in finding a similar asset. With technological improvements and other quality changes, equipment purchased currently will likely differ from equipment that a firm acquired 10 years earlier but still uses. Consider, for example, the difficulties in ascertaining the current replacement cost of a three-year-old computer, computer software package, or cellular phone—you just can't find a new item with the same (limited) features of the old one. An accountant who cannot find the replacement cost of the specific old one might substitute the current replacement cost of an asset capable of rendering equivalent services. This approach, however, requires subjectivity in identifying assets with equivalent service potential.

**Current Net Realizable Value** The net amount of cash (selling price less selling costs) that a firm would receive currently if it sold an asset is its **current net realizable value**. Because the firm would receive this amount if it exited ownership, accountants refer to net realizable value as an **exit value**. In measuring net realizable value, one generally assumes that the firm sells the asset as a willing seller, in an orderly fashion, rather than through a forced sale at some distress price.

Measuring net realizable value entails difficulties similar to those encountered in measuring current replacement cost. Without well-organized secondhand markets for used equipment, the accountant cannot readily measure net realizable value, particularly for equipment specially designed for a firm's needs. In this case, the future benefits to the firm from using the asset (value in use) will generally exceed the current selling price of the asset (value in exchange).

**Present Value of Future Net Cash Flows** An asset provides a future benefit. This future benefit results from the asset's ability either to generate future net cash receipts or to reduce future cash expenditures. Accounting could measure the net **present value** of the expected cash flows and use that amount for the valuation. For example, accounts receivable from customers will lead directly to future cash receipts. The firm can sell merchandise inventory for cash or promises to pay cash. The firm can use equipment to manufacture products that it can sell for cash. A building that the firm owns reduces future cash outflows for rental payments. Because these cash flows represent the future services, or benefits, of assets, the accountant might base asset valuations on them.

Because cash can earn interest over time, today's value of a stream of future cash flows, called its *present value,* is less than the sum of the future cash amounts that a firm will receive or save over time. The balance sheet shows asset valuations measured as of a current date. If valuation of an asset as of a current date reflects its future cash flows, then the accountant must discount the future net cash flows to find their present value as of the date of the balance sheet. **Chapters 9** and **10** and the **Appendix** discuss the discounting methodology. The following example presents the general approach.

**Example 6** Miller Corporation sold merchandise to a reliable customer, General Models Company, who promised to pay \$10,000 one year after the date of sale. General Models Company signed a promissory note to that effect and gave the note to Miller Corporation. Miller Corporation judges that the current borrowing rate of General Models Company is 10 percent per year. That is, if Miller Corporation voluntarily made a loan to General Models Company, the loan would carry an interest rate of 10 percent. Miller Corporation will receive \$10,000 one year from today. The \$10,000 includes both the amount lent initially plus interest on that amount for one year. Today's value of the $10,000 to be received in one year is not $10,000 but about $9,091; that is, $9,091 plus 10 percent interest on $9,091 equals $10,000 (= 1.10 × $9,091). Hence, the present value of $10,000 to be received one year from today is $9,091. (Miller Corporation is indifferent between receiving approximately $9,091 today and a promise from General Models Company for $10,000 due one year from today.) The asset represented by General Models Company's promissory note has a present value of $9,091. If the balance sheet states the note at the present value of the future cash flows, it would appear at approximately $9,091 on the date of sale.

Using discounted cash flows to value individual assets requires solving several problems. First, one must deal with the uncertainty of the amounts of future cash flows. The amounts a firm will receive can depend on whether competitors introduce new products, the rate of inflation, and other factors. Second, one must decide how to allocate the cash receipts from the sale of a single item of merchandise inventory to all of the assets involved in its production and distribution (for example, equipment, buildings, sales staff's automobiles). Third, one must decide how to select the appropriate rate to use to discount the future cash flows to the present. Should the interest rate be the firm's borrowing rate? Or, should the firm use the rate at which it could invest excess cash? Or, should the firm use its cost of capital (a concept introduced in managerial accounting and finance courses)? In **Example 6** above, the facts make General Models's borrowing rate the right one.

## SELECTING THE APPROPRIATE VALUATION BASIS

The valuation basis selected depends on the purpose of the financial report.

**Example 7** Miller Corporation prepares its income tax return for the current year. The Internal Revenue Code and Regulations specify that firms must use acquisition or adjusted acquisition cost valuations in most instances.

**Example 8** A fire recently destroyed the manufacturing plant, equipment, and inventory of Miller Corporation. The firm's fire insurance policy provides coverage in an amount equal to the cost of replacing the assets destroyed. Thus, current replacement cost at the time of the fire provides the appropriate support for the insurance claim.

**Example 9** Miller Corporation plans to dispose of a manufacturing division that has operated unprofitably. In deciding on the lowest price to accept for the division as a unit, the firm considers the net realizable value of each asset, which it will sum to estimate a lower bound on the acceptable selling price for the division.

**Example 10** Brown Corporation considers purchasing Miller Corporation. The highest price that Brown Corporation should pay is the present value of the future net cash flows it can realize from owning Miller Corporation.

**Example 11** Miller Corporation discovers that the demand for land it owns has declined so much that the original cost of the land exceeds the sum of all expected rental receipts for the indefinite future. Generally accepted accounting principles (GAAP) require Miller Corporation to show the land on the balance sheet at the net present value of the expected cash flows, discounted using a rate adjusted for the risk of the expected cash flows—the more certain the cash flows, the lower the discount rate.[2]

## GENERALLY ACCEPTED ACCOUNTING ASSET VALUATION BASES

The asset valuation basis appropriate for financial statements issued to shareholders and other investors is not so obvious. The financial statements currently prepared by publicly held firms use two valuation bases for assets that have not declined substantially in value since the firm acquired them—one basis for monetary assets and a different basis for nonmonetary assets.

**Monetary assets** include cash and claims to specified amounts of cash to be received in the future, such as accounts receivable. Monetary assets generally appear on the balance sheet at their net present value—their current cash, or cash equivalent, value. Cash appears at the amount

[2]Financial Accounting Standards Board, *Statement of Financial Accounting Standards No. 121,* "Accounting for the Impairment of Long-Lived Assets and for Long-Lived Assets to Be Disposed Of," 1995.

of cash on hand or in the bank. Accounts receivable from customers appear at the amount of cash the firm expects to collect in the future. If the time until a firm collects a receivable spans more than one year, the firm discounts the expected future cash inflow to a present value. Most firms, however, collect their accounts receivable within one to three months. The amount of future cash flows (undiscounted) approximately equals the present (discounted) value of these flows; thus accounting ignores the discounting process on the basis of a lack of materiality.

**Nonmonetary assets** will generate unknown, rather than specified, amounts of cash in the future and include merchandise inventory, land, buildings, and equipment. Nonmonetary assets generally appear at acquisition costs, in some cases adjusted downward to reflect that the firm has consumed some of the assets' services and, in others, to recognize some declines in market value. **Chapters 3, 7,** and **8** discuss these adjustments, called *depreciation* when the firm has used some of the services from the asset, and called *holding losses* or *impairments* when market value has declined even more than the amount of the depreciation. Some nonmonetary assets, such as holdings of marketable securities, appear on the balance sheet at current market value.

The acquisition cost of an asset includes more than its invoice price. Acquisition cost includes all expenditures made or obligations incurred to prepare an asset for its intended use by a firm. Transportation costs, installation costs, handling charges, and any other necessary and reasonable costs incurred until the firm puts the asset into service are part of the total cost assigned to the asset. For example, the accountant might calculate the cost of an item of equipment as follows:

| | |
|---|---|
| Invoice Price of Equipment | $120,000 |
| Less: 2 Percent Discount for Prompt Cash Payment | (2,400) |
| Net Invoice Price | $117,600 |
| Transportation Cost | 3,260 |
| Installation Cost | 7,350 |
| Total Cost | $128,210 |

The accountant records the acquisition cost of this equipment as $128,210.

Instead of disbursing cash or incurring a liability, the firm might give (or swap or barter) other forms of consideration (for example, common stock, merchandise inventory, or land) in acquiring an asset. In these cases, acquisition cost will reflect the market value of the consideration the firm gives or the market value of the asset it receives, whichever market value the firm can more reliably measure.

**Example 12** Miller Corporation issued 1,000 shares of its common stock to the former owner in acquiring a used machine. The common stock of Miller Corporation traded on a stock exchange for $15 per share on the day of the exchange. The firm records the machine on the books of Miller Corporation for $15,000.

**Foundations for Acquisition Cost** Accounting's use of acquisition cost valuations for nonmonetary assets rests on three important concepts or conventions.

First, accounting assumes that a firm is a **going concern**. In other words, accounting assumes a firm will remain in operation long enough to carry out all of its current plans. The firm will realize any increases in the market value of assets held in the normal course of business when the firm receives higher prices for its products. Imagine a furniture maker who, several years ago, purchased now-rare Honduran mahogany. The mahogany has increased in value, but the furniture maker will not recognize that increase until later when a customer purchases, say, a table made from that mahogany whose price reflects the current value of the mahogany, not its original cost. Accounting generally assumes that the current values of the individual assets have little relevance for users.

Second, acquisition cost valuations provide more reliable information than do the other valuation methods. **Reliability** in accounting refers to the ability of a measure, such as acquisition cost, to faithfully represent what it purports to measure (that is, the economic value sacrificed to acquire an asset). Reliability also encompasses the ability to verify, or audit, the measured amount. Different accountants will likely agree on the acquisition cost of an asset. Differences among observers can arise in ascertaining an asset's current replacement cost, current net realizable value, and present value of future cash flows.

Third, acquisition cost generally provides more conservative (that is, lower) valuations of assets (and measures of earnings) than do the other valuation methods. Many accountants believe that financial statements will less likely mislead users if balance sheets report assets at lower rather than higher amounts. Thus, **conservatism** has evolved as a convention to justify acquisition cost valuations (and subsequent downward, but generally not upward, adjustments to acquisition cost valuations).

The general acceptance of these valuation bases does not justify them. Research has not provided guidance as to the valuation basis—chosen from acquisition cost, current replacement cost, current net realizable value, and present value of future net cash flows—that has most relevance to financial statement users. Many preparers of financial statements prefer acquisition cost valuations to those based on market values because market valuations cause reported income to be more volatile than do acquisition cost valuations.

## Asset Classification

The classification of assets within the balance sheet varies in published annual reports. The following discussion gives the principal asset categories.

**Current Assets** Cash and other assets that a firm expects to turn into cash or to sell or consume during the normal operating cycle of the business are **current assets**. The operating cycle refers to the period of time during which a given firm converts cash into salable goods and services, sells those goods and services to customers, and receives cash from customers in payment for their purchases. The operating cycle for most manufacturing, retailing, and service firms spans one to three months, whereas for firms in some industries, such as building construction or liquor distilling, the operating cycle may span several years. Except for firms that have an operating cycle longer than one year, conventional accounting practice uses one year as the dividing line between current assets and noncurrent assets. Current assets include the following: cash; marketable securities held for the short term; accounts and notes receivable; inventories of merchandise, raw materials, supplies, work in process, and finished goods; and prepaid operating costs, such as insurance and rent paid for in advance. Prepaid costs, or prepayments, are current assets because if the firm did not pay for them in advance, it would use current assets within the next operating cycle to acquire those services.

**Investments** A second section of the balance sheet, labeled *Investments,* includes long-term (noncurrent) investments in securities of other firms. For example, a firm might purchase shares of common stock of a supplier to help ensure continued availability of raw materials. Or, it might acquire shares of common stock of a firm in another area of business activity to permit the acquiring firm to diversify its operations. When one corporation (the parent) owns more than 50 percent of the voting stock in another corporation (the subsidiary), it usually prepares a single set of consolidated financial statements; that is, the firm merges, or consolidates, the specific assets, liabilities, revenues, and expenses of the subsidiary with its own in the financial statements. The securities shown in the Investments section of the balance sheet therefore represent investments in firms whose assets and liabilities the parent or investor firm has not consolidated with its own. **Chapter 11** discusses consolidated financial statements and other methods of accounting for intercorporate investments.

**Property, Plant, and Equipment** The phrase *property, plant, and equipment* (sometimes called **plant,** or **fixed, assets**) designates the tangible, long-lived assets a firm uses in its operations over a period of years, having ordinarily not acquired them for resale. This category includes land, buildings, machinery, automobiles, furniture, fixtures, computers, and other equipment. The balance sheet shows these items (except land) at acquisition cost reduced by the cumulative (or *accumulated,* to use common accounting terminology) depreciation since the firm acquired the assets. (**Chapter 8** discusses additional downward adjustments if the assets have declined in value, beyond that indicated by depreciation calculations.) Frequently, only the net balance, or book value, appears on the balance sheet. Land usually appears at acquisition cost.

**Intangible Assets** Such items as patents, trademarks, franchises, and goodwill are **intangible assets**. Accountants generally do not recognize as assets the benefits resulting from expenditures that a firm makes in developing intangibles because of the difficulty of ascertaining the existence and value of future benefits. Consider, for example, the difficulty of identifying whether future benefits exist when a firm expends cash to research new technologies or advertise its products. Although a firm would not voluntarily engage in a portfolio of research projects unless the firm expected future benefits, the accountant cannot tell which expenditures have future benefits and which do not. Accountants do, however, recognize as assets those specifically identifiable intangibles that firms acquire in market exchanges with other entities—intangibles such as a patent or brand name acquired from its holder. Accounting's recognition of an asset in the latter case presumes that a firm would not purchase a patent, brand name, or other intangible from another entity unless the firm expected future benefits from the specific purchased intangible. The exchange between an independent purchaser and seller provides reliable evidence of the value of the future benefits. **Chapter 8** discusses more fully the accounting used for internally developed intangibles versus that used for externally purchased intangibles, a topic that remains controversial.

## Problem 2.1 for Self-Study

**Asset recognition and valuation.** The transactions listed below relate to Coca-Cola Company. Indicate whether or not each transaction immediately gives rise to an asset of the company under GAAP. If accounting recognizes an asset, state the account title and amount.

**a.** The company spends $10 million to develop a new soft drink. No commercially feasible product has yet evolved, but the company hopes that such a product will evolve in the near future.

**b.** The company signs a contract with United Can Corporation for the purchase of $4 million of soft-drink cans. It makes a deposit of $400,000 on signing the contract.

**c.** The company spends $2 million for advertisements that appeared during the past month: $500,000 to advertise the Coca-Cola name and $1,500,000 for specific brand advertisements, such as those for Diet Coke.

**d.** The company issues 50,000 shares of its common stock, valued on the market at $2.5 million, in the acquisition of all the outstanding stock of Corning Glass Company, a supplier of soft-drink bottles.

**e.** The company spends $800,000 on educational-assistance programs for its middle-level managers to obtain MBAs. Historically, 80 percent of the employees involved in the program receive their MBAs and remain with the company 10 years or more.

**f.** The company acquires and occupies land and a building near Atlanta by signing a mortgage payable for $75 million. Because the company has not yet paid the mortgage, the title document for the land and building remains in the vault of the bank, the holder of the mortgage note.

Solutions to self-study problems appear at the end of the chapter.

## Liability Recognition

A liability arises when a firm receives benefits or services and in exchange promises to pay the provider of those goods or services a reasonably definite amount at a reasonably definite future time. The firm usually pays cash but may give goods or services.[3] All liabilities are obligations; however, not all obligations are accounting liabilities.

**Example 13** Miller Corporation purchased merchandise inventory and agreed to pay the supplier $8,000 within 30 days. This obligation is a liability because Miller Corporation received the goods and must pay a definite amount at a reasonably definite future time.

**Example 14** Miller Corporation borrowed $4 million by issuing long-term bonds. On December 31 of each year it must make annual interest payments of 10 percent of the amount borrowed, and it must repay the $4 million principal in 20 years. This obligation is a liability because Miller Corporation received the cash and must repay the debt in definite amounts at definite future times.

**Example 15** Miller Corporation received an advance of $600 from a customer for products that Miller Corporation will manufacture next year. The cash advance creates a liability of $600. Miller

[3] *SFAC No. 6,* par. 35. See the **Glossary** for the Board's actual definition of a liability.

Corporation must manufacture and deliver the products next year or return the cash advance. (In some transactions of this sort, the manufacturer does not have the option to return the cash. If it does not provide the goods as promised, it may find itself liable for court-awarded damages based on the economic harm the customer suffered from not getting the promised items.)

**Example 16** Miller Corporation provides a three-year warranty on its products. The promise to maintain the products under warranty plans creates an obligation. The selling price for its products includes a charge for future warranty services even when the price tag or invoice does not explicitly show a separate charge for the promise. As customers pay the selling price, Miller Corporation receives a benefit (that is, the cash collected). Past experience provides a basis for estimating the amount of the liability. Miller Corporation can estimate the proportion of customers who will seek services under the warranty agreement and the expected cost of providing the warranty services. Thus, Miller Corporation can measure the amount of the obligation with a reasonable degree of accuracy and will show it as a liability.

**Example 17** A customer has sued Miller Corporation, claiming damages of $10 million from faulty products manufactured by Miller Corporation. The case has not yet gone to trial. Accounting typically does not recognize unsettled lawsuits as liabilities because of uncertainty regarding both the need to pay and the amount, as well as the timing, of any payment. Miller Corporation will disclose in notes to its financial statements the fact of the lawsuit and the potential for future payments.

**Example 18** Miller Corporation has guaranteed the payment of a personal loan from a local bank by its chief executive officer (CEO) if the CEO cannot make payment as the loan agreement requires. Accounting does not recognize loan guarantees as liabilities because of the uncertainty regarding the need to pay.

One of the troublesome questions of liability recognition relates to **contingent obligations**. Contingent obligations require some future event to occur before accounting can establish the existence or amount of an obligation. The unsettled lawsuit in **Example 17** and the loan guarantee in **Example 18** are contingent obligations. GAAP requires the recognition of a liability only when the payment is probable. As **Chapter 9** discusses, most accountants interpret probable to mean greater than 80 or 85 percent. The obligation under the warranty plan in **Example 16** meets this probable criterion because past experience shows that a predictable percentage of customers will make claims of an estimated amount.

**Example 19** Miller Corporation signed an agreement with its employees' labor union, promising to increase wages by 6 percent and to provide for medical and life insurance. Although this agreement creates an obligation, it does not immediately create an accounting liability. Employees have not yet provided labor services that would require the firm to pay wages and insurance. As employees work, a liability arises.

A second troublesome question of liability recognition relates to obligations under mutually unexecuted contracts. The labor union agreement in **Example 19** is a mutually unexecuted contract. Other examples include some leases, purchase order commitments, and employment contracts. Accounting does not usually recognize as liabilities the obligations created by mutually unexecuted contracts. As **Example 5** illustrates, accounting likewise does not recognize assets related to mutually unexecuted contracts. **Chapter 10** discusses the accounting treatment of these off-balance-sheet financing arrangements.

## Liability Valuation

Most liabilities are monetary, requiring payments of specific amounts of cash. Those due within one year or less appear at the amount of cash the firm expects to pay to discharge the obligation. If the payment dates extend more than one year into the future (for example, as in the case of the bonds in **Example 14**), the liability appears at the present value of the future cash outflows.

A liability that requires delivering goods or rendering services, rather than paying cash, is nonmonetary. The warranty liability in **Example 16** is nonmonetary and appears on the balance sheet at the estimated cost of providing the warranty services. The cash advance in **Example 15** is also nonmonetary but appears on the balance sheet at the amount of cash received. The seemingly inconsistent valuation of these two nonmonetary liabilities results from accounting's view that the warranty liability relates to products that the firm has already sold, whereas the cash advance relates to products that the firm will manufacture and deliver to customers next year. Other examples of nonmonetary liabilities arising from cash advances include amounts received by magazine publishers for future magazine subscriptions, by theatrical and sports teams for future performances or games, and by landlords for future rental services. Account titles frequently used for liabilities of this type are Advances from Customers, Advances from Subscribers, or Advances from Season Ticket Patrons.

## LIABILITY CLASSIFICATION

The balance sheet typically classifies liabilities in one of the following categories.

**Current Liabilities** Obligations that a firm expects to pay or discharge during the normal operating cycle of the firm, usually one year, are **current liabilities**. In general, the firm uses current assets to pay current liabilities. This category includes liabilities to merchandise suppliers, employees, and governmental units. It also includes notes and bonds payable to the extent that they will require the use of current assets within the next year.

**Long-Term Debt** Obligations arising from borrowings having due dates, or maturities, more than one year (or more than the normal operating cycle, if longer) after the balance sheet date are long-term debt. Long-term debt includes bonds, mortgages, and similar debts, as well as some obligations under long-term leases.

**Other Long-Term Liabilities** Obligations not properly considered as current liabilities or long-term debt appear as other long-term liabilities, which include items such as deferred income taxes and some retirement obligations.

## Problem 2.2 for Self-Study

**Liability recognition and valuation.** The transactions listed below relate to the New York Times Company. Indicate whether or not each transaction immediately gives rise to a liability of the company under GAAP. If the company recognizes a liability, state the account title and amount.

**a.** The company receives $10 million for newspaper subscriptions covering the one-year period beginning next month.
**b.** The company receives an invoice for $4 million from its advertising agency for television advertisements that appeared last month promoting the New York Times.
**c.** The company signs a one-year lease for rental of new delivery vehicles it will use in Brooklyn. It pays $50,000 of the annual rental of $80,000 at the signing.
**d.** Attorneys have notified the company that a New York City resident, seriously injured by one of the company's delivery vehicles in lower Manhattan, has sued the company for $10 million. Company lawyers have predicted that the court is likely to find the company liable in the lawsuit, but the company carries sufficient insurance to cover any losses.
**e.** Refer to part **d** above. Assume now that the company carries no insurance against such losses.
**f.** A two-week strike by employees has closed down newspaper publishing operations. As a result, the company could not deliver subscriptions totaling $2 million.

## SHAREHOLDERS' EQUITY VALUATION AND DISCLOSURE

The shareholders' equity in a firm is a residual interest or claim[4]—that is, the owners have a claim on all assets not required to meet the claims of creditors.[5] The valuation of the assets and liabilities included in the balance sheet therefore determines the valuation of total shareholders' equity.

[4]Although shareholders' equity is equal to assets minus liabilities, accounting provides an independent method for computing the amount. This and the next chapter present this method.
[5]*SFAC No. 6,* par. 49.

Accounting distinguishes between funds contributed by owners and assets generated by operations and retained by a firm. The balance sheet for a corporation generally separates the amount that shareholders contribute directly for an interest in the firm (that is, common stock) from earnings the firm subsequently realizes in excess of dividends declared (that is, retained earnings).

In addition, the balance sheet usually further disaggregates the amount received from shareholders into the **par** or **stated value** of the shares and the amounts contributed in excess of par value or stated value. The par or stated value of a share of stock is an amount assigned to comply with the corporation laws of each state and rarely equals the market price of the shares at the time the firm issues them.

The distinctions between common stock and retained earnings, and between the par or stated value and the amounts received in excess of par or stated value, sometimes have legal implications but not economic significance. Shareholders have a claim on all assets in excess of all liabilities, which equals total shareholders' equity. (**Chapter 12** discusses details of accounting for shareholders' equity.)

Clear thinking about the balance sheet requires clear understanding that shareholders' equity represents sources of funds and other assets, not funds or assets themselves. Shareholders contribute cash; the balance sheet shows the cash as an asset and indicates in shareholders' equity the source of the cash as Common Stock. A firm operates profitably and generates cash and other assets from its profits. That cash and other assets appear as assets. The shareholders' equity section shows the source of those new assets under the title Retained Earnings. Many students stumble over the concept of Retained Earnings, which are not assets but the name of the source of the assets. When you understand the concept that Retained Earnings are not assets, you will not likely make the common error of some business participants in saying, "The company should pay out some of its retained earnings." The speaker means that the company should pay out some cash or other assets, which the firm generated from profitable operations. A firm cannot pay out retained earnings. It can pay out assets, such as cash.

**Example 20** Stephens Corporation legally incorporated on January 1, Year 1. It issued 15,000 shares of $10 par value common stock for $10 cash per share. During Year 1, Stephens Corporation generated net income of $30,000 and paid dividends of $10,000 to shareholders. The shareholders' equity section of the balance sheet of Stephens Corporation on December 31, Year 1, is as follows:

| | |
|---|---|
| Common Stock (par value of $10 per share, 15,000 shares issued and outstanding) | $150,000 |
| Retained Earnings | 20,000 |
| Total Shareholders' Equity | $170,000 |

**Example 21** Instead of issuing $10 par value common stock as in **Example 20**, Stephens Corporation issued 15,000 shares of $1 par value common stock for $10 cash per share. (The market price of a share of common stock depends on the economic value of the firm, not on the par value of the shares.) The shareholders' equity section of the balance sheet of Stephens Corporation on December 31, Year 1, is as follows:

| | |
|---|---|
| Common Stock (par value of $1 per share, 15,000 shares issued and outstanding) | $ 15,000 |
| Additional Paid-in Capital (or Capital Contributed in Excess of Par Value) | 135,000 |
| Retained Earnings | 20,000 |
| Total Shareholders' Equity | $170,000 |

Firms legally organized as partnerships or sole proprietorships, instead of as corporations, do not make a distinction between contributed capital and retained earnings in their balance sheets. Rather, the owners' equity section of the balance sheet combines each owner's share of capital contributions and each owner's share of earnings in excess of distributions.

**Example 22** Refer to **Examples 20** and **21**. Assume that William Kinsey and Brenda Stephens organized this business firm as a partnership, with Kinsey and Stephens as equal partners. The owners' equity section of the firm's balance sheet on December 31, Year 1, is as follows:

| | |
|---|---|
| William Kinsey, Capital | $ 85,000 |
| Brenda Stephens, Capital | 85,000 |
| Total Owners' Equity | $170,000 |

A sole proprietorship would, by definition, have only one owner.

# Accounting Procedures for Preparing the Balance Sheet

Keeping in mind the concepts underlying assets, liabilities, and shareholders' equity, we can now consider how accounting applies these concepts in preparing the balance sheet. A user of the financial statements does not need to understand the record-keeping procedures that accountants follow to prepare a balance sheet. However, the user needs some type of analytical framework to understand how various transactions or events affect a firm's financial position, profitability, and cash flows. This chapter introduces an analytical framework based on the balance sheet equation and applies it to the preparation of the balance sheet. **Chapter 3** extends the analytical framework to include income transactions. **Chapter 4** extends the analytical framework to the statement of cash flows.

Our presentation of the analytical framework proceeds in three steps:

1. Demonstrate that all transactions and events have a dual effect on the balance sheet equation.
2. Illustrate the use of T-accounts for accumulating information about the dual effects of transactions and events in order to prepare a balance sheet.
3. Illustrate the use of journal entries for summarizing the dual effects of any particular transaction or event.

**Appendix 2.1** illustrates the use of a spreadsheet as an alternative to T-accounts for accumulating information about the dual effects of transactions and events to prepare a balance sheet.

## Dual Effects of Transactions on the Balance Sheet Equation

Firms engage in various transactions, or exchanges, with other entities or individuals during a period. For example, firms acquire merchandise from suppliers, pay employees for labor services, sell merchandise to customers, pay taxes to governments, and so on. Other events occur during the period that do not necessarily involve an exchange with another entity. For example, firms consume the services of buildings and equipment as they use them in operations and incur an obligation to pay interest on outstanding loans. These transactions and events are the initial building blocks for the financial statements. Accountants record the effects of *each* of these transactions and events as they occur during the period and then *accumulate* the effects of all transactions and events for presentation in the financial statements. Thus,

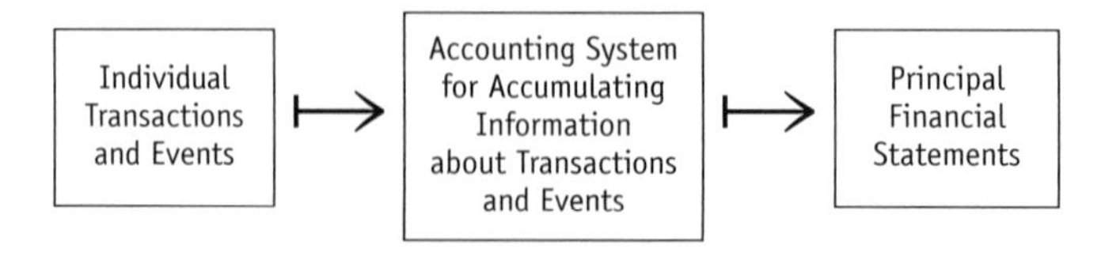


The analytical framework that we use throughout this book for understanding the effects of transactions and events on the financial statements relies on the balance sheet equation as its structural base. As this and subsequent chapters illustrate, the balance sheet equation provides a rich structural base for the analytical framework because it incorporates investing, financing, and operating transactions—three key activities of business firms. As we apply the analytical framework to successively more complex transactions in later chapters, you will want to recall that the basic balance sheet equation underlies the recording of all business transactions and events.

The balance sheet equation shows the equality of assets with liabilities plus shareholders' equity. The balance sheet equation maintains this equality by reporting the effects of each transaction in a dual manner. Any single transaction will have one of the following four effects or some combination of these effects:

1. It increases an asset and increases either a liability or shareholders' equity.
2. It decreases an asset and decreases either a liability or shareholders' equity.
3. It increases one asset and decreases another asset.
4. It increases one liability or shareholders' equity and decreases another liability or shareholders' equity.

To understand the dual effects of various transactions on the balance sheet equation, consider the following transactions for Miller Corporation during January as it prepares to open for business on February 1:

**(1)** On January 1, the firm issues 10,000 shares of $10 par value common stock for $100,000 cash.
**(2)** On January 5, it pays $60,000 cash to purchase equipment.
**(3)** On January 15, Miller Corporation purchases merchandise inventory costing $15,000 from a supplier, agreeing to pay later.
**(4)** On January 21, it pays the supplier in **(3)** $8,000 of the amount due.
**(5)** On January 25, the supplier in **(3)** accepts 700 shares of common stock at par value in settlement of the $7,000 amount still owed.
**(6)** On January 31, the firm pays $600 cash for a one-year insurance premium for coverage beginning February 1.
**(7)** On January 31, Miller Corporation receives $3,000 from a customer for merchandise to be delivered during February.

**Exhibit 2.1** illustrates the dual effects of these transactions on the balance sheet equation. Note that after each transaction, assets equal liabilities plus shareholders' equity.

Each transaction has two effects—an outflow and an inflow. For example, in transaction **(1)** Miller Corporation issues common stock to shareholders and receives cash. In transaction **(2)** the firm makes a cash expenditure and receives equipment. In transaction **(3)** the firm promises to make a future cash payment to a supplier and receives merchandise inventory. Most transactions and events recorded in the accounting system result from exchanges. The accounting records reflect the inflows and outflows arising from these exchanges.

The recording of each transaction maintains the balance sheet equality. Transactions **(1)**, **(3)**, and **(7)** increase assets and increase either a liability or shareholders' equity. Transaction **(4)** decreases assets and a liability. Transactions **(2)** and **(6)** result in an increase in one asset and a decrease in another asset. Transaction **(5)** results in an increase in shareholders' equity and a decrease in a liability.

*At this early stage in your study of accounting, we emphasize the importance of identifying the dual effects of every transaction on the balance sheet equation. You will not fully understand the business transaction or event unless you can identify these dual effects.*

## Problem 2.3 for Self-Study

**Dual effects of transactions on balance sheet equation.** Using the format in **Exhibit 2.1**, indicate the effects of each of the following transactions of Gaines Corporation on the balance sheet equation.

1. The firm issues 20,000 shares of $10 par value common stock for $12 cash per share.
2. The firm issues $100,000 principal amount of bonds for $100,000 cash.
3. The firm acquires, with $220,000 in cash, land costing $40,000 and a building costing $180,000.

*(continued)*

4. The firm acquires, on account, equipment costing $25,000 and merchandise inventory costing $12,000.
5. The firm signs an agreement to rent equipment from its owner and pays $1,500 rental in advance.
6. The firm pays $28,000 to the suppliers in **4**.

**EXHIBIT 2.1**

**MILLER CORPORATION**
**Illustration of Dual Effects of Transactions on Balance Sheet Equation**

| | Transaction | Assets | = | Liabilities | + | Shareholders' Equity |
|---|---|---|---|---|---|---|
| (1) | On January 1, Miller Corporation issues 10,000 shares of $10 par value common stock for $100,000 cash (an increase in an asset and a shareholders' equity). | +$100,000 | = | $ 0 | | +$100,000 |
| | Subtotal | $100,000 | = | $ 0 | + | $100,000 |
| (2) | On January 5, the firm pays $60,000 cash to purchase equipment (an increase in one asset and a decrease in another asset). | − 60,000<br>+ 60,000 | | | | |
| | Subtotal | $100,000 | = | $ 0 | + | $100,000 |
| (3) | On January 15, the firm purchases merchandise inventory costing $15,000 from a supplier on account (an increase in an asset and a liability). | + 15,000 | | + 15,000 | | |
| | Subtotal | $115,000 | = | $15,000 | + | $100,000 |
| (4) | On January 21, the firm pays the supplier in **(3)** $8,000 of the amount due (a decrease in an asset and a liability). | − 8,000 | | − 8,000 | | |
| | Subtotal | $107,000 | = | $ 7,000 | + | $100,000 |
| (5) | On January 25, the supplier in **(3)** accepts 700 shares of common stock in settlement of the $7,000 still owed (an increase in a shareholders' equity and a decrease in a liability). | | | − 7,000 | | + 7,000 |
| | Subtotal | $107,000 | = | $ 0 | + | $107,000 |
| (6) | On January 31, the firm pays $600 cash for a one-year insurance premium for coverage beginning February 1 (an increase in one asset and a decrease in another asset). | + 600<br>− 600 | | | | |
| | Subtotal | $107,000 | = | $ 0 | + | $107,000 |
| (7) | On January 31, the firm receives $3,000 from a customer for merchandise to be delivered during February (an increase in an asset and a liability). | + 3,000 | | + 3,000 | | |
| | Total—January 31 | $110,000 | = | $ 3,000 | + | $107,000 |

## PURPOSE AND USE OF ACCOUNTS

Although **Exhibit 2.1** illustrates the dual effects of individual transactions, it does not provide a useful framework for preparing a balance sheet at the end of a period. We cannot easily know how much of the total assets of $110,000 represents cash, how much represents inventory, how much represents equipment, and so on. We need to move to an additional level of disaggregation of assets, liabilities, and shareholders' equity to obtain the needed amounts for a balance sheet. To do this, we move next in our development of the analytical framework to the use of T-accounts for accumulating information about individual transactions and events during a period.

**Requirement for an Account** A balance sheet item can only increase or decrease or remain the same during a period of time. Thus, an account must provide for accumulating the increases and decreases (if any) that occur during the period for a single balance sheet item. By convention, the accounts separate the increases from the decreases. The total additions during the period increase the balance carried forward from the previous statement, the total subtractions decrease it, and the result is the new balance for the current balance sheet.

**Form of an Account** The account may take many possible forms, and accounting practice commonly uses several. Perhaps the most useful form of the account for textbooks, problems, and examinations is the **T-account**. Actual practice does not use this form of the account, except perhaps for memoranda or preliminary analyses. However, the T-account satisfies the requirement of an account and is easy to use. As the name indicates, the T-account looks like the letter T, with a horizontal line bisected by a vertical line. The name or title of the account appears on the horizontal line. One side of the space formed by the vertical line records increases in the item and the other side records decreases. Dates and other information can appear as well.

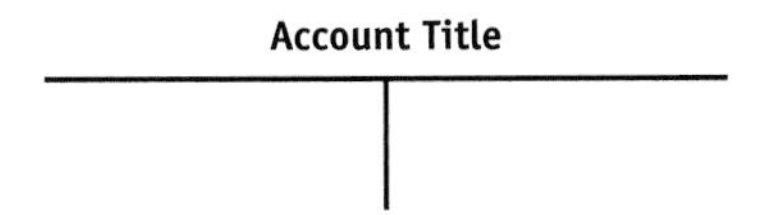


**Placement of Increases and Decreases in the Account** Given the two-sided account, we must choose the side used to record increases and the side for decreases. Long-standing custom follows three rules:

1. Increases in assets appear on the left side and decreases in assets appear on the right side.
2. Increases in liabilities appear on the right side and decreases in liabilities appear on the left side.
3. Increases in shareholders' equity appear on the right side and decreases in shareholders' equity appear on the left side.

This custom reflects the fact that in the balance sheet equation, assets appear to the left of the equal sign and liabilities and shareholders' equity appear to the right. Following this format, asset balances should appear on the left side of accounts; liability and shareholders' equity balances should appear on the right. Asset balances will appear on the left only if the left side of the account records asset increases. Similarly, liability and shareholders' equity balances appear on the right only if accounting records liability and shareholders' equity increases on the right side of accounts. When the accountant properly analyzes each transaction into its dual effects on the accounting equation and follows the three rules for recording the transaction, every transaction results in recording equal amounts in entries on the left-hand and the right-hand sides of the various accounts. This equality of the amounts recorded on the left and right for any single transaction provides a powerful check for the accuracy of record keeping. When you analyze a transaction and get unequal amounts for left and right amounts, you will know you have erred.

**Debit and Credit** Accountants use two convenient abbreviations: debit (Dr.) and credit (Cr.). **Debit,** used as a verb, means "record an entry on the left side of an account" and, used as a noun or an adjective, means "an entry on the left side of an account." **Credit,** used as a verb, means "record an entry on the right side of an account" and, used as a noun or an adjective, means "an entry on the right side of an account." Often, however, accountants use the word *charge* instead of debit, both as a noun and as a verb. In terms of balance sheet categories, a debit or charge indicates (1) an increase in an asset, (2) a decrease in a liability, or (3) a decrease in a shareholders' equity item. A credit indicates (1) a decrease in an asset, (2) an increase in a liability, or (3) an increase in a shareholders' equity item.

To maintain the equality of the balance sheet equation, the accountant must be sure that the amounts debited to various accounts for each transaction equal the amounts credited to various accounts. As a result, the sum of balances in accounts with debit balances at the end of each period must equal the sum of balances in accounts with credit balances. Debits equal credits: you've heard it before. It applies to each individual transaction and to their aggregation into the balance sheet as a whole.

**Summary of Account Terminology and Procedure** The following T-accounts summarize the conventional use of the account form and the terms debit and credit. Customarily, a checkmark in an account indicates a balance.

| Any Asset Account | |
|---|---|
| ✓ Beginning Balance<br>Increases<br>+<br>Dr. | Decreases<br>−<br>Cr. |
| ✓ Ending Balance | |

| Any Liability Account | |
|---|---|
| Decreases<br>−<br>Dr. | Beginning Balance ✓<br>Increases<br>+<br>Cr. |
| | Ending Balance ✓ |

| Any Shareholders' Equity Account | |
|---|---|
| Decreases<br>−<br>Dr. | Beginning Balance ✓<br>Increases<br>+<br>Cr. |
| | Ending Balance ✓ |

## Reflecting the Dual Effects of Transactions in the Accounts

We can now see how the dual effects of transactions change the accounts. We use three separate T-accounts: one for assets, one for liabilities, and one for shareholders' equity. The dual effects of the transactions of Miller Corporation for January appear in the T-accounts shown in **Exhibit 2.2**.

The amount entered on the left side of (or debited to) the accounts for each transaction equals the amount entered on the right side of (or credited to) the accounts. Recording equal amounts of debits and credits for each transaction ensures that the balance sheet equation will always balance. At the end of January, the assets account has a debit balance of $110,000. The balances in the liabilities and shareholders' equity accounts sum to a credit balance of $110,000.

To provide a direct computation of the amount of each asset, liability, and shareholders' equity item requires a separate account for each balance sheet item, rather than one for each of the three broad categories. The recording procedure is the same, except that it uses specific asset or equity accounts, not a single account.

**Exhibit 2.3** records the transactions of Miller Corporation for January using separate T-accounts for each balance sheet item. The numbers in parentheses refer to the seven transactions during January for Miller Corporation. Most accountants would use the checkmark to indicate a balance, as in **Exhibit 2.2,** rather than spell out the word *balance,* as we do in **Exhibit 2.3**.

The total assets of Miller Corporation of $110,000 as of January 31 comprise $34,400 in cash, $15,000 in merchandise inventory, $600 in prepaid insurance (advances *to* the insurance provider), and $60,000 in equipment. Total liabilities plus shareholders' equity of $110,000 comprise $3,000 of advances *from* customers and $107,000 of common stock.

One can prepare the balance sheet using the amounts shown as balances in the T-accounts. The balance sheet of Miller Corporation after the seven transactions of January appears in **Exhibit 2.4**.

**EXHIBIT 2.2**

**MILLER CORPORATION**
**Summary T-Accounts Showing Transactions during January**

| | Assets | | = Liabilities | | + Shareholders' Equity | |
|---|---|---|---|---|---|---|
| | Increases (Dr.) | Decreases (Cr.) | Decreases (Dr.) | Increases (Cr.) | Decreases (Dr.) | Increases (Cr.) |
| **(1)** Issue common stock for cash | 100,000 | | | | | 100,000 |
| **(2)** Purchase equipment with cash | 60,000 | 60,000 | | | | |
| **(3)** Purchase merchandise on account | 15,000 | | | 15,000 | | |
| **(4)** Pay cash to supplier in **(3)** | | 8,000 | 8,000 | | | |
| **(5)** Issue common stock to supplier in **(3)** | | | 7,000 | | | 7,000 |
| **(6)** Pay insurance premium in advance | 600 | 600 | | | | |
| **(7)** Receive cash from customer in advance | 3,000 | | | 3,000 | | |
| Balance | ✓ 110,000 | | | 3,000 ✓ | | 107,000 ✓ |

**EXHIBIT 2.3**

**MILLER CORPORATION**
**Individual T-Accounts Showing Transactions**

**Cash (Asset)**

| | Increases (Dr.) | Decreases (Cr.) | |
|---|---|---|---|
| (1) | 100,000 | 60,000 | (2) |
| (7) | 3,000 | 8,000 | (4) |
| | | 600 | (6) |
| Balance | 34,400 | | |

**Accounts Payable (Liability)**

| | Decreases (Dr.) | Increases (Cr.) | |
|---|---|---|---|
| (4) | 8,000 | 15,000 | (3) |
| (5) | 7,000 | | |
| | | 0 | Balance |

**Merchandise Inventory (Asset)**

| | Increases (Dr.) | Decreases (Cr.) |
|---|---|---|
| (3) | 15,000 | |
| Balance | 15,000 | |

**Advance from Customer (Liability)**

| Decreases (Dr.) | Increases (Cr.) | |
|---|---|---|
| | 3,000 | (7) |
| | 3,000 | Balance |

**Prepaid Insurance (Asset)**

| | Increases (Dr.) | Decreases (Cr.) |
|---|---|---|
| (6) | 600 | |
| Balance | 600 | |

**Common Stock (Shareholders' Equity)**

| Decreases (Dr.) | Increases (Cr.) | |
|---|---|---|
| | 100,000 | (1) |
| | 7,000 | (5) |
| | 107,000 | Balance |

**Equipment (Asset)**

| | Increases (Dr.) | Decreases (Cr.) |
|---|---|---|
| (2) | 60,000 | |
| Balance | 60,000 | |

**EXHIBIT 2.4**

**MILLER CORPORATION**
**Balance Sheet**
**January 31**

| | |
|---|---|
| ASSETS | |
| **Current Assets** | |
| Cash | $ 34,400 |
| Merchandise Inventory | 15,000 |
| Prepaid Insurance | 600 |
| Total Current Assets | $ 50,000 |
| **Property, Plant, and Equipment** | |
| Equipment | 60,000 |
| Total Assets | $110,000 |
| LIABILITIES AND SHAREHOLDERS' EQUITY | |
| **Current Liabilities** | |
| Advances from Customer | $ 3,000 |
| **Shareholders' Equity** | |
| Common Stock | 107,000 |
| Total Liabilities and Shareholders' Equity | $110,000 |

## Problem 2.4 for Self-Study

**T-accounts for various transactions.** Set up T-accounts for the following accounts:

| | |
|---|---|
| Cash | Bonds Payable |
| Merchandise Inventory | Land |
| Prepaid Rent | Buildings |
| Equipment | Common Stock—Par Value |
| Accounts Payable | Additional Paid-in Capital |

Indicate whether each account is an asset, a liability, or a shareholders' equity item. Enter in the T-accounts the transactions of Gaines Corporation in **Problem 2.3 for Self-Study**.

## Analysis of Business Transactions Using Journal Entries

The T-accounts illustrated in the preceding section work well when we want to aggregate the effects of many transactions during a period and then prepare a balance sheet (or income statement). We use T-accounts throughout **Chapters 2** and **3** for this purpose. In many settings, however, we want to observe the effect of a single transaction, or a small number of related transactions, on the financial statements. For example, consider the effect on the balance sheet of a firm issuing long-term debt and common stock in the acquisition of land, buildings, and equipment. Or, consider a more complicated transaction from **Chapter 10:** a lessee or tenant makes a payment to a lessor or landlord and must recognize the amount of the payment, the use of the asset, the amount of interest, and the amount of debt repaid. Setting up T-accounts for these single, although complex, transactions is unnecessarily time consuming and detracts from your thinking about the transaction itself and its effect on the financial statements. For this and similar purposes, we summarize the effects of the transaction on the financial statements using **journal entries**. Journal entries:

- Rely on the concept that every transaction or event has a dual effect on the balance sheet equation (as illustrated in **Exhibit 2.1**).
- Require knowledge of the accounts on the financial statements that a particular transaction affects and whether those accounts increase or decrease (as illustrated in **Exhibit 2.3**).

Journal entries add to these two building blocks already discussed a standard format for summarizing the effects of a particular business transaction. The standard format is as follows:

| | | |
|---|---|---|
| Account Title .................................. | Amount | |
| Account Title ............................ | | Amount |

The journal entry indicates the dual effects of a transaction as to both accounts and amounts. Accountants refer to the first line of the journal entry as the debit line and the second (indented) line of the journal entry as the credit line. Thus,

| | | |
|---|---|---|
| Account Debited ................................ | Amount Debited | |
| Account Credited ........................... | | Amount Credited |

The recording of journal entries follows the same rules for increases and decreases in asset, liability, and shareholders' equity accounts as we illustrated earlier for entries in T-accounts:

1. Debits (a) increase an asset account, or (b) decrease a liability or shareholders' equity account.
2. Credits (a) decrease an asset account, or (b) increase a liability or shareholders' equity account.

Journal entries may have multiple debit or multiple credit lines, or both. The dual effects rule requires that the sum of the amounts debited must equal the sum of the amounts credited. Journal entries often include the date of the transaction and an explanation for the transactions journalized. The journal entries in this book also show the effect of transactions on the balance sheet equation. This equation is not normally part of journal entries made in the accounting records, but we find it helps students to understand the transaction better. Thus, the standard format for journal entries in this book is:

---

Date

| | | |
|---|---|---|
| Account Debited . . . . . . . . . . . . . . . . . . . . . . . . . . . . . . . . . . . . | Amount | |
| Account Credited . . . . . . . . . . . . . . . . . . . . . . . . . . . . . . . . . | | Amount |

| Assets | = | Liabilities | + | Shareholders' Equity | (Class.) |
|---|---|---|---|---|---|
| | | | | | |

Journal entry explanation.

---

The column label (Class.), for Classification, refers to the type of shareholders' equity account affected by the transaction.

The journal entries for the seven transactions of Miller Corporation appear next.

---

**(1)** January 1

| | | |
|---|---|---|
| Cash . . . . . . . . . . . . . . . . . . . . . . . . . . . . . . . . . . . . . . . . . . . . . . | 100,000 | |
| Common Stock . . . . . . . . . . . . . . . . . . . . . . . . . . . . . . . . . . | | 100,000 |

| Assets | = | Liabilities | + | Shareholders' Equity | (Class.) |
|---|---|---|---|---|---|
| +100,000 | | | | +100,000 | ContriCap |

Issue 10,000 shares of $10 par value common stock for cash.

---

The issue of common stock increases the contributed capital (ContriCap) portion of shareholders' equity.

---

**(2)** January 5

| | | |
|---|---|---|
| Equipment . . . . . . . . . . . . . . . . . . . . . . . . . . . . . . . . . . . . . . . . . . | 60,000 | |
| Cash . . . . . . . . . . . . . . . . . . . . . . . . . . . . . . . . . . . . . . . . . . . | | 60,000 |

| Assets | = | Liabilities | + | Shareholders' Equity | (Class.) |
|---|---|---|---|---|---|
| +60,000 | | | | | |
| −60,000 | | | | | |

Purchase equipment costing $60,000 for cash.

---

**(3)** January 15

| | | |
|---|---|---|
| Merchandise Inventory . . . . . . . . . . . . . . . . . . . . . . . . . . . . . . . | 15,000 | |
| Accounts Payable . . . . . . . . . . . . . . . . . . . . . . . . . . . . . . . . . | | 15,000 |

| Assets | = | Liabilities | + | Shareholders' Equity | (Class.) |
|---|---|---|---|---|---|
| +15,000 | | +15,000 | | | |

Purchase merchandise inventory costing $15,000 on account.

---

(4) January 21

| | | |
|---|---|---|
| Accounts Payable | 8,000 | |
| Cash | | 8,000 |

| Assets | = | Liabilities | + | Shareholders' Equity | (Class.) |
|---|---|---|---|---|---|
| −8,000 | | −8,000 | | | |

Pay liabilities of $8,000 with cash.

(5) January 25

| | | |
|---|---|---|
| Accounts Payable | 7,000 | |
| Common Stock | | 7,000 |

| Assets | = | Liabilities | + | Shareholders' Equity | (Class.) |
|---|---|---|---|---|---|
| | | −7,000 | | +7,000 | ContriCap |

Issue 700 shares of $10 par value common stock in settlement of $7,000 accounts payable.

(6) January 31

| | | |
|---|---|---|
| Prepaid Insurance (or Advances to Insurance Company) | 600 | |
| Cash | | 600 |

| Assets | = | Liabilities | + | Shareholders' Equity | (Class.) |
|---|---|---|---|---|---|
| +600 | | | | | |
| −600 | | | | | |

Pay one-year insurance premium of $600 in advance.

(7) January 31

| | | |
|---|---|---|
| Cash | 3,000 | |
| Advances from Customers | | 3,000 |

| Assets | = | Liabilities | + | Shareholders' Equity | (Class.) |
|---|---|---|---|---|---|
| +3,000 | | +3,000 | | | |

Receive $3,000 from customer for merchandise to be delivered in February.

## Problem 2.5 for Self-Study

**Journal entries for various transactions.** Prepare journal entries for each of the six transactions of Gaines Corporation in **Problem 2.3 for Self-Study**.

# Balance Sheet Account Titles

This section describes the balance sheet account titles commonly used. The descriptions should help you understand the nature of various assets, liabilities, and shareholders' equities as well as select appropriate account names to use when you solve problems. You can use alternative account titles. The list does not show all the account titles we use in this book or that appear in the financial statements of publicly held firms. Many beginning students become overly concerned about precise wording for account titles. We require of our students only that the titles be descriptive and unambiguous and that students use the identical (or similar) account titles for identical (or similar) items. Later chapters discuss more fully the use of some of the account titles described below.

## ASSETS

*Cash:* coins and currency and items such as bank checks and money orders (the latter items are merely claims against individuals or institutions but by custom are called cash), bank deposits against which the firm can draw checks, and time deposits, usually savings accounts and certificates of deposit.

*Marketable Securities:* government bonds or corporate stocks and bonds that the firm plans to hold for a relatively short time. The word *marketable* implies that the firm can buy and sell them readily through a security exchange such as the New York Stock Exchange.

*Accounts Receivable:* amounts due from customers of a business from the sale of goods or services. The collection of cash occurs sometime after the sale. These accounts are also known as "charge accounts" or "open accounts." The general term Accounts Receivable used in the balance sheet describes the figure representing the total amount receivable from all customers. The firm, of course, keeps a separate record for each customer.

*Notes Receivable:* amounts due from customers or from others to whom a firm has made loans or extended credit. The customer or other borrower puts the claim into writing in the form of a formal note (which distinguishes the claim from an open account receivable).

*Interest Receivable:* interest—on assets such as promissory notes or bonds—that has accrued (or come into existence) through the passing of time but that the firm has not yet collected as of the date of the balance sheet.

*Merchandise Inventory:* goods on hand purchased for resale, such as canned goods on the shelves of a grocery store or suits on the racks of a clothing store.

*Raw Materials Inventory:* materials as yet unused for manufacturing products.

*Supplies Inventory:* lubricants, abrasives, and other incidental materials used in manufacturing operations; stationery, computer drives, flash memory cards, pens, and other office supplies; bags, twine, boxes, and other packaging supplies; gasoline, oil, spare parts, and other delivery supplies.

*Work-in-Process Inventory:* partially completed manufactured products.

*Finished Goods Inventory:* completed, but unsold, manufactured products.

*Advances to Suppliers:* payments made in advance for goods or services that a firm will receive at a later date. If the firm does not make a cash expenditure when it places an order, it does not recognize an asset until it receives the goods or services ordered.

*Prepaid Rent:* rent paid in advance for the future use of land, buildings, or equipment. Beginning accounting students sometimes get confused with the title Prepaid Rent because it could mean the asset on the balance sheet of the tenant or the liability on the balance sheet of the landlord. Using the title Advances to Landlord for the asset and Advances from Tenants for the liability avoids this confusion.

*Prepaid Insurance:* insurance premiums paid for future coverage. One could call this Advances to Insurance Company.

*Investment in Securities:* bonds or shares of common or preferred stock that the firm plans to hold for a relatively long time, typically longer than one year.

*Land:* land used in operations or occupied by buildings used in operations.

*Buildings:* factory buildings, store buildings, garages, warehouses, and so forth.

*Equipment:* lathes, ovens, machine tools, boilers, computers, bins, cranes, conveyors, automobiles, and so forth.

*Furniture and Fixtures:* desks, tables, chairs, counters, showcases, scales, and other selling and office equipment.

*Accumulated Depreciation:* the cumulative amount of the acquisition cost of long-term assets (such as buildings and equipment) allocated to the costs of production or to current and prior periods in measuring net income. The amount in this account reduces the acquisition cost of the long-term asset to which it relates when measuring the net book value of the asset shown in the balance sheet.

*Leasehold:* the right to use property owned by someone else.

*Organization Costs:* amounts paid for legal and incorporation fees, for printing the certificates for shares of stock, and for accounting and other costs incurred in organizing a business so that it can function. GAAP requires firms to expense organization costs in the year incurred.[6]

*Patents:* rights granted for up to 20 years by the federal government to exclude others from manufacturing, using, or selling certain processes or devices. Under current GAAP, the firm must expense research and development costs in the year incurred rather than recognize them as assets with future benefits.[7] As a result, a firm that develops a patent will not normally show it as an asset. On the other hand, a firm that purchases a patent from another firm or from an individual will recognize the patent as an asset. **Chapter 8** discusses this inconsistent treatment of internally developed and externally purchased patents.

*Goodwill:* When one firm acquires another firm, it allocates the purchase price to individual identifiable assets acquired and liabilities assumed, using their current market values. When the purchase price exceeds the amounts assigned to identifiable assets and identifiable liabilities, accounting assumes that the excess relates to intangibles that the acquiring firm cannot separately identify, such as customer goodwill or a well-trained labor force, and calls the excess *goodwill*. Accounting generally does not recognize as assets the good reputation and other desirable attributes that a firm creates or develops for itself. However, when one firm acquires another firm, accounting recognizes these desirable attributes as assets insofar as they cause the amount paid for the acquired firm to exceed the sum of the values assigned to all the other assets, less liabilities, identified in the acquisition.

## Liabilities

*Accounts Payable:* amounts owed for goods or services acquired under an informal credit agreement. The firm must pay these accounts fairly soon after the balance sheet date, usually within one or two months, although amounts due as much as 12 months later are current liabilities. The same items appear as Accounts Receivable on the creditor's books.

*Notes Payable:* the face amount of promissory notes given in connection with loans from a bank or with the purchase of goods or services. The same items appear as Notes Receivable on the creditor's (lender's) books. Most Accounts Payable, Accounts Receivable, Notes Payable, and Notes Receivable are current items, due within a year of the balance sheet date. Similar items due more than one year after the balance sheet date could have the same account titles, but would appear in the noncurrent section of the balance sheet.

*Interest Payable:* interest—on obligations—that has accrued or accumulated with the passage of time but that the firm has not yet paid as of the date of the balance sheet. The liability for interest customarily appears separately from the face amount of the obligation. The same items appear as Interest Receivable on the creditor's books.

*Income Taxes Payable:* the estimated liability for income taxes, accumulated and unpaid, based on the taxable income of the business from the beginning of the taxable year to the date of the balance sheet.

*Advances from Customers:* the general name used to indicate the obligation incurred when the firm receives payments in advance for goods or services it will furnish to customers in the future; a nonmonetary liability. The firm has an obligation to deliver goods or services, not return the cash. Even so, the firm records this liability at the amount of cash it receives. If the firm does not receive cash when a customer places an order, it does not record a liability; such a contract is mutually unexecuted and is called *executory*.

*Advances from Tenants, or Rent Received in Advance:* another example of a nonmonetary liability. For example, a firm owns a building that it rents to a tenant. The tenant prepays the rental charge for several months in advance. The firm cannot include the amount applicable to future months as a component of income until the firm renders a rental service as time passes. Meanwhile the advance payment results in a liability payable in services (that is, in the use of the building). On the records of the tenant, the same amount appears as an asset, Prepaid Rent (or Advances to Landlord).

*Mortgage Payable:* long-term promissory notes that the borrower has protected by pledging specific pieces of property as security for payment. If the borrower does not pay the loan or interest according to the agreement, the lender can require the sale of the property to generate funds to repay the loan.

---

[6]American Institute of Certified Public Accountants, *Statement of Position 98-5*, "Reporting the Cost of Start-up Activities," 1998.

[7]Financial Accounting Standards Board, *Statement of Financial Accounting Standards No. 2*, "Accounting for Research and Development Costs," 1974.

*Bonds Payable:* amounts borrowed by a business for a relatively long period under a formal written contract called an *indenture*. The borrower usually raises the funds from a number of lenders, all of whom receive written evidence of their share of the loan.

*Convertible Bonds:* bonds payable that the holder can convert into, or trade in for, shares of common stock. The bond indenture specifies the number of shares the lenders will receive when they convert their bonds into stock, the dates when conversion can occur, and other details.

*Capitalized Lease Obligations:* the present value of the commitment to make future cash payments in return for the right to use property owned by someone else. **Chapter 10** discusses the conditions under which a firm recognizes lease obligations as liabilities.

*Deferred Income Taxes:* particular income tax amounts that are delayed beyond the current accounting period. **Chapter 10** discusses this item, which appears on the balance sheet of most U.S. corporations.

## SHAREHOLDERS' EQUITY

*Common Stock:* a measure of the amount of cash or other assets received equal to the par or stated value of a firm's principal class of voting stock.

*Preferred Stock:* a measure of the amount of cash or other asserts received for the par value of a class of a firm's stock that has some preference relative to the common stock, usually in the area of dividends and assets in the event of corporate liquidation. Sometimes preferred shareholders may convert the stock into common stock.

*Additional Paid-in Capital:* a measure of the amount of cash or other assets received in the issuance of common or preferred stock in excess of par value or stated value. Some firms use for this account the alternative title "Capital Contributed in Excess of Par (or Stated) Value." Accountants and analysts often refer to the sum of the amounts in this account and the two accounts above as *contributed capital,* because the sum of these three items represents the sources of funds directly provided by shareholders to a firm.

*Retained Earnings:* since the time a business began operations, net assets (= all assets − all liabilities) increase as the firm generates earnings in excess of cash (or other net assets) distributed by dividend declarations. Retained Earnings is the account title for the account with a credit balance, which indicates that the owners provided (are the source of) those net assets and have claims on them. Retained earnings provide financing just as amounts explicitly contributed by owners do. When a firm declares dividends, net assets decrease (liability for dividends payable increases), and retained earnings decrease by an equal amount. As **Chapters 4** and **12** discuss, a firm does not generally hold net assets generated from retained earnings as cash. Beginning students commonly confuse Retained Earnings (an account with a credit balance) with Cash (an account with a debit balance).

*Accumulated Other Comprehensive Income:* an account summarizing gains and losses on some assets that the firm has not yet reported as part of income, so these gains and losses do not appear in Retained Earnings. We delay discussion of accumulated other comprehensive income until **Chapter 12**.

*Treasury Shares:* the cost of shares of stock that a firm originally issued but subsequently reacquires. Treasury shares do not receive dividends, and accountants do not identify them as outstanding shares. The cost of treasury shares almost always appears on the balance sheet as a deduction from the total of the other shareholders' equity accounts. **Chapter 12** discusses the accounting for treasury shares.

## Problem 2.6 for Self-Study

**Journal entries, T-accounts, and balance sheet preparation.** Electronics Appliance Corporation begins operations on September 1. The firm engages in the following transactions during the month of September:

**(1)** September 1: Issues 4,000 shares of $10 par value common stock for $12 cash per share.

**(2)** September 2: Gives 600 shares of $10 par value common stock to acquire a patent from another firm. The two entities agree on a price of $7,200 for the patent.

**(3)** September 5: Pays $10,000 as two months' rent in advance on a factory building that is leased for the three years beginning October 1. Monthly rental payments are $5,000.

*(continued)*

**(4)** September 12: Purchases raw materials on account for $6,100.
**(5)** September 15: Receives a check for $900 from a customer as a deposit on a special order for equipment that Electronics plans to manufacture. The contract price is $4,800.
**(6)** September 20: Acquires office equipment with a list price of $950. After deducting a discount of $25 in return for prompt payment, it issues a check in full payment.
**(7)** September 28: Issues a cash advance totaling $200 to three new employees who will begin work on October 1.
**(8)** September 30: Purchases factory equipment costing $27,500. It issues a check for $5,000 and assumes a long-term mortgage liability for the balance.
**(9)** September 30: Pays $450 for the labor costs of installing the new equipment in **(8)**.

**a.** Prepare journal entries for each of the nine transactions.
**b.** Set up T-accounts and enter each of the nine transactions. Note that the amounts for the balance sheet at the beginning of September are zero.
**c.** Prepare a balance sheet for Electronics Appliance Corporation as of September 30.

## Analysis of the Balance Sheet

Later chapters will caution you not to analyze the balance sheet in isolation from the income statement, the statement of cash flows, and other information about a firm. The balance sheet does, however, provide useful insights about the nature and mix of assets as well as the nature and mix of financing for those assets. A careful user should also be alert to asset and liability recognition and valuation questions that impact the interpretation of the balance sheet.

Analysts often use a **common-size balance sheet** to study the nature and mix of assets and their financing. In a common-size balance sheet, the analyst expresses each balance sheet item as a percentage of total assets or total liabilities plus shareholders' equity. **Exhibit 2.5** presents common-size balance sheets for Wal-Mart Stores (discount stores, superstores, and warehouse clubs), American Airlines (airline services), Merck (pharmaceuticals), Interpublic Group (advertising services), and Citigroup (financial institution). **Exhibit 2.5** also shows the market-to-book-value ratio for each firm at the end of a recent year. The market-to-book-value ratio compares the market value of each firm's common shares with the book value of shareholders' equity as reported on the balance sheet.

### Assessing the Nature and Mix of Financing

The balance sheet reflects the effects of a firm's investing and financing decisions. One question the user of the balance sheet should ask is: Does the firm have the right mix of financing for its assets? Two principles guide decisions about financing:

1. Firms generally use short-term financing for assets that a firm expects to turn into cash soon, such as accounts receivable and inventories, and long-term financing (debt or shareholders' equity) for assets that a firm expects to turn into cash over a longer period, such as property, plant, and equipment.
2. The mix of long-term financing depends on the nature of long-term assets and the amount of operating risk in the business. Firms with tangible long-term assets and predictable cash flows, such as electric utilities, tend to use a high proportion of long-term debt. The property, plant, and equipment serve as collateral for the borrowing (that is, the lender can repossess or confiscate the assets if the firm fails to make debt payments on time). The predictable cash flows reduce the risk that the firm will not have sufficient cash to make interest and principal payments when due. Firms with tangible long-term assets but less predictable cash flows, such as auto manufacturers and steel companies, whose sales vary with changes in economic conditions, tend to use a more balanced mix of long-term debt and shareholders' equity financing. The property, plant, and equipment serve as collateral for the borrowing, but the more uncertain cash flows suggest a lower proportion of long-term debt in the capital structure. Firms with high proportions of intangibles, whether recognized as assets on the balance sheet or not, tend to use low proportions of long-term debt. Lacking collateral, lenders must rely on cash flows from operations to service the debt. The less predictable these cash flows, the smaller is the proportion of long-term debt.

**EXHIBIT 2.5** **Common-Size Balance Sheets for Selected Companies**

| | Wal-Mart Stores | American Airlines | Merck | Interpublic Group | Citigroup |
|---|---|---|---|---|---|
| ASSETS | | | | | |
| Cash | 2.8% | 6.4% | 10.5% | 7.9% | 3.1% |
| Accounts Receivable | 2.1 | 4.9 | 11.4 | 38.3 | 82.2 |
| Inventories | 25.1 | 2.1 | 7.2 | — | — |
| Prepayments | 0.7 | 2.9 | 2.1 | 7.4 | — |
| Total Current Assets | 30.7% | 16.3% | 31.2% | 53.6% | [a] |
| Investments in Securities | — | — | 15.3 | 3.0 | — |
| Property, Plant, and Equipment | 56.9 | 62.5 | 29.8 | 7.0 | 2.1 |
| Intangible Assets | 8.9 | 4.3 | 15.2 | 29.3 | 3.2 |
| Other Assets | 3.5 | 16.9 | 8.5 | 7.1 | 9.4 |
| Total Assets | 100.0% | 100.0% | 100.0% | 100.0% | 100.0% |
| LIABILITIES AND SHAREHOLDERS' EQUITY | | | | | |
| Accounts Payable | 17.3% | 4.0% | 5.1% | 43.5% | 68.5% |
| Notes Payable | 1.1 | 2.9 | 7.7 | 1.8 | — |
| Other Current Liabilities | 14.5 | 17.0 | 13.2 | 14.8 | 21.3 |
| Total Current Liabilities | 32.9% | 23.9% | 26.0% | 60.1% | [a] |
| Long-term Debt | 24.4 | 40.7 | 10.3 | 15.4 | — |
| Other Noncurrent Liabilities | 3.1 | 32.2 | 15.1 | 6.1 | 2.3 |
| Total Liabilities | 60.4% | 96.8% | 51.4% | 81.6% | 92.1% |
| Shareholders' Equity | 39.6 | 3.2 | 48.6 | 18.4 | 7.9 |
| Total Liabilities and Shareholders' Equity | 100.0% | 100.0% | 100.0% | 100.0% | 100.0% |
| Market-value-to-book-value Ratio | 4.4 | 0.4 | 7.1 | 2.6 | 2.2 |

[a]Financial institutions typically do not distinguish current and noncurrent assets and liabilities.

The underlying rationale for these generalizations about the nature and mix of financing relates to lenders' and investors' assessments of the overall risk profile of a business. Firms incur risk both from the asset, or operating, side and from the financing side of their business. Risks on the operating side include variability in sales from changing economic conditions (cyclicality risk) or from short product life cycles (because of technological change or changes in consumer taste, for example) and variability of earnings from having a high proportion of fixed costs that do not change as sales change, such as depreciation expense on capital-intensive manufacturing facilities. When firms incur substantial risk on the operating side, they tend not to take on still higher risk on the financing side in the form of long-term debt. The firm must make principal and interest payments on the debt. The more variable the cash flows from operating activities, the more risk the firm will not have sufficient cash to service its debt. The less risk on the operating side, the more the firm will borrow for the long term.

Refer to the data in **Exhibit 2.5**. Wal-Mart Stores maintains a large percentage of its assets in merchandise inventories, which it expects to sell within a period of one to two months. It matches these investments in inventories with a similar proportion of short-term financing that appears in current liabilities. Wal-Mart also maintains a high percentage of its assets in property, plant, and equipment. It uses approximately 60 percent debt and 40 percent shareholders' equity financing for these long-term assets. Wal-Mart generates relatively predictable cash flows from its stores. With the stores serving as collateral, Wal-Mart appears able to manage a higher proportion of long-term debt than **Exhibit 2.5** shows. Although Wal-Mart uses long-term debt to finance much of its growth in new stores, the retention of earnings has caused shareholders' equity to increase to an unusually high percentage of long-term financing.

American Airlines invests a large percentage of its assets in property, plant, and equipment. It finances these assets with long-term sources of financing (long-term debt plus shareholders' equity). Airlines tend to use more long-term debt than shareholders' equity to finance the acquisition of equipment because (1) the equipment serves as collateral for the borrowing, and (2) long-term debt usually has a lower economic cost to the firm than do funds provided by shareholders. Economic conditions, however, impact the demand for air travel and cause sales to vary over time. With the high proportion of fixed costs from depreciation and, to a lesser extent, compensation, airlines tend not to want to add too much long-term debt with its fixed interest costs to their capital structures. Varying sales combined with high fixed costs increase the probability of bankruptcy. Thus, although airlines rely heavily on long-term debt, they include more shareholders' equity funds than if their sales were more predictable and less variable. The proportion of long-term financing provided by shareholders appears low in this case because American Airlines operated at a net loss for several years, lowering its assets and the account representing the source of those reduced assets, retained earnings.

Merck also invests a high proportion of its balance sheet assets in property, plant, and equipment. It has substantial investments in the drug patents it has developed within the firm, but these patents do not appear on the balance sheet. Pharmaceutical companies tend to maintain capital-intensive, automated manufacturing facilities to ensure quality control of their products. Unlike airlines, however, pharmaceutical companies tend not to carry much long-term debt. One reason for not using debt results from the nature of the resources of a pharmaceutical company. Key resources include its research scientists, who could leave the firm at any time, and its patents on pharmaceutical products, which competitors could render worthless at any moment by developing new, superior products. Given the risk inherent in these resources, which do not appear on the balance sheet, pharmaceutical firms tend not to add risk on the financing side of their balance sheets by taking on debt, which requires fixed interest and principal payments.

Interpublic Group provides advertising services for clients. It purchases time or space in various media (television, newspapers, magazines), for which it incurs an obligation (reported as Accounts Payable). It develops advertising copy for clients and sells them media time or space to promote their products, resulting in a receivable from the clients (Accounts Receivable). Thus, current receivables dominate the asset side of the balance sheet and current payables dominate the financing side of the balance sheet. Service firms such as Interpublic Group have few assets other than their employees, which accounting does not recognize as assets. Interpublic Group made several acquisitions of other marketing services firms and allocated a portion of the purchase price to goodwill and other intangibles. Thus, intangibles comprise a larger portion of Interpublic Group's total assets.

Citigroup, like most financial institutions, does not distinguish current from noncurrent assets and liabilities. Loans to customers for houses dominate the asset side of the balance sheet. These loans have maturities up to 25 years. Deposits by customers, denoting the source of funds that the customers may withdraw at any time, dominate the financing side. The use of short-term financing to support acquisition of long-term assets got S&Ls in trouble in the late 1980s; customers, fearing the bankruptcy of the S&Ls because many of the S&Ls' investments lost market value, began withdrawing their deposits. The S&Ls were not able to turn their long-term loans into cash quickly enough to satisfy deposit withdrawals. Although the balance sheets of S&Ls still contain their inherent imbalance, government guarantees of deposits up to certain limits and stronger oversight of the financial condition of S&Ls by regulators have reinstated confidence in them.

## Assessing the Impact of Asset and Liability Recognition and Valuation Issues

This chapter indicates that the balance sheet does not always provide data useful for analysts because, as a result of following GAAP,

- Not all resources of a firm appear on the balance sheet as assets,
- Assets on the balance sheet do not typically reflect current market valuations,
- Not all obligations of a firm appear on the balance sheet as liabilities,
- Liabilities on the balance sheet do not necessarily reflect current market valuations.

The analyst should recognize these weaknesses of the balance sheet when attempting to assess the financial condition of a firm. The market sets the price of a firm's common stock unconstrained by GAAP. The market price will likely increase when a pharmaceutical firm discovers a new drug, even though GAAP does not permit the firm to show it as an asset on the balance sheet. We can compare the market-to-book-value ratios for each of the firms in **Exhibit 2.5** to identify possible asset and liability recognition and valuation issues for these firms. Note that the competitive position of each of these firms and their growth potential will also impact the market-to-book-value ratio. We do not consider these factors in the discussion below.

Wal-Mart has a market-to-book-value ratio of 4.4. One reason for this ratio exceeding 1.0 is that the property, plant, and equipment of Wal-Mart appear on the balance sheet at acquisition cost adjusted downward for depreciation to date. The current market values of fixed assets likely exceed their book values, particularly for the land underlying stores. Wal-Mart also has several intangibles that do not appear on its balance sheet, including its brand name, its buying power with suppliers due to its size, its information systems that tie stores worldwide to each other and to suppliers, and its efficient distribution system from regional warehouses.

American Airlines has a market-to-book-value ratio less than 1.0. Its property, plant, and equipment appear at acquisition cost adjusted downward for depreciation. Airlines in recent years have increased the depreciable lives of their equipment, reducing the rate at which they write down equipment, perhaps resulting in an overvaluation of its fixed assets. American Airlines also has an obligation under its frequent flier program that appears on the balance sheet at the estimated incremental cost of providing free flights. Perhaps the market values this liability at an amount higher than that on the balance sheet. American Airlines also operated at a net loss for several years, increasing the proportion of liabilities on the balance sheet and increasing the probability of bankruptcy. The market recognized this risk and reduced its valuation of the common stock of American Airlines.

Merck has a market-to-book-value ratio of 7.1. This large ratio reflects GAAP's requirement that firms cannot recognize the value of internally developed patents and other technologies as assets on the balance sheet. The market incorporates estimates of the values of these technologies in setting Merck's market price. Merck recognizes the value of patents when it acquires other firms, which it reports in intangible assets. Unsettled lawsuits against the firm that Merck does not recognize as liabilities until the cash outflows to settle a claim are probable will likely reduce the market value of Merck's shares and depress the market-to-book-value ratio.

Interpublic Group has a market-to-book-value ratio of 2.6. The key resource of Interpublic Group is the creative talent of its employees. GAAP does not permit firms to recognize this resource as an asset on the balance sheet. The various marketing services agencies that comprise Interpublic Group also have recognized brand names, such as Foote, Cone, & Belding, and loyal corporate customers that enhance the market value of the firm even though they do not appear on the balance sheet as an asset. Interpublic Group has acquired several marketing services firms in recent years. A large portion of the price paid for these firms is for their brand names and creative talent, which Interpublic Group includes in Goodwill, an intangible asset.

Citigroup trades at a market-to-book-value ratio of 2.2. The assets of financial institutions include primarily investments in securities and loans receivable, which appear on the balance sheet at amounts close to market value. One would therefore expect financial institutions to have a market-to-book-value ratio close to 1.0. The larger ratio for Citigroup likely reflects the value of the firm's brand name, its dominating presence in most of the markets it serves, and the value of its skilled workforce.

These asset and liability recognition and valuation concerns might tempt a user of the financial statements to ignore the balance sheet entirely when evaluating a firm. Although the user should keep these limitations in mind, he or she should also keep in mind that GAAP governs the preparation of the balance sheet across firms and across time. Thus, firms prepare their balance sheets on a consistent basis. By incorporating the known biases injected by GAAP into their assessments, users can gain insight about the nature and mix of assets and their related financing.

## An International Perspective

The format of the balance sheet in some countries differs from that discussed in this chapter. In Germany, France, and some other European countries, property, plant, and equipment and other noncurrent assets appear first, followed by current assets. On the equities side, shareholders' equity appears first, followed by noncurrent liabilities, and then current liabilities. **Exhibit 2.6** presents the balance sheet of BMW, the German automobile manufacturer, for two recent years. Note that this balance sheet maintains the equality of investments and financing. Note also that some terms differ from those commonly used in the United States.

EXHIBIT 2.6

**BMW CORPORATION**
**Comparative Balance Sheet**
**(in millions of Euros)**

| | December 31 | |
|---|---|---|
| | **Year 11** | **Year 12** |
| **ASSETS** | | |
| Intangible Assets | € 2,419 | € 2,741 |
| Property, Plant, and Equipment | 7,355 | 8,578 |
| Financial Assets | 786 | 498 |
| Leased Products | 7,908 | 7,012 |
| Total Noncurrent Assets | €18,468 | €18,829 |
| Inventories | € 4,501 | € 5,197 |
| Trade Receivables | 2,135 | 1,818 |
| Receivables from Sales Financing | 17,398 | 19,493 |
| Other Receivables | 4,208 | 6,056 |
| Marketable Securities | 907 | 1,105 |
| Cash and Cash Equivalents | 2,437 | 2,333 |
| Total Current Assets | €31,586 | €36,002 |
| Prepayment and Other Assets | € 1,205 | € 680 |
| Total Assets | €51,259 | €55,511 |
| **SHAREHOLDERS' EQUITY AND LIABILITIES** | | |
| Subscribed Capital | € 673 | € 674 |
| Capital Reserve | 1,937 | 1,954 |
| Revenue Reserves | 9,405 | 11,075 |
| Accumulated Other Equity | (1,245) | 168 |
| Total Shareholders' Equity | €10,770 | €13,871 |
| Provisions | € 6,824 | € 7,666 |
| Debt | 25,665 | 26,262 |
| Trade Payables | 3,015 | 3,069 |
| Other Liabilities | 4,985 | 4,643 |
| Total Liabilities | €40,489 | €41,640 |
| Total Shareholders' Equity and Liabilities | €51,259 | €55,511 |

| Term Used in Exhibit 2.6 | Common Term Used in the United States |
|---|---|
| Financial Assets | Investment in Securities |
| Trade Receivables | Accounts Receivable |
| Subscribed Capital | Common Stock |
| Capital Reserve | Additional Paid-in Capital |
| Revenue Reserves | Retained Earnings |
| Accumulated Other Equity | Accumulated Other Comprehensive Income |
| Provisions | Other Liabilities |
| Debt | Bonds Payable |
| Trade Payables | Accounts Payable |

In the United Kingdom, the balance sheet presents the following form of the balance sheet equation.

$$\text{Noncurrent Assets} + \left(\text{Current Assets} - \text{Current Liabilities}\right) - \text{Noncurrent Liabilities} = \text{Shareholders' Equity}$$

**Exhibit 2.7** presents a balance sheet for Rolls-Royce, a British manufacturer of engines for aircraft, ships, and other large equipment, for two recent years. This form of balance sheet does not permit a direct comparison of investments with financing. The analyst must rearrange such balance sheets to obtain the desired information. A balance sheet for Rolls-Royce in the format discussed in this chapter and using terminology typically found in the United States appears in **Exhibit 2.8**. Terms used in the balance sheet in **Exhibit 2.7** that differ from those discussed in this chapter appear below.

| Term Used in Exhibit 2.7 | Common Term Used in the United States |
|---|---|
| Tangible Assets | Property, Plant, and Equipment |
| Stocks | Inventories |
| Debtors | Accounts Receivable |
| Borrowings | Notes Payable, Bonds Payable |
| Called-up Share Capital | Common Stock |
| Share Premium Account | Additional Paid-in Capital |
| Profit and Loss Account | Retained Earnings |

**EXHIBIT 2.7**

**ROLLS-ROYCE**
**Comparative Balance Sheet in U.K. Format and Terminology (in millions of pounds)**

| | August 31 | |
|---|---|---|
| | **Year 11** | **Year 12** |
| FIXED ASSETS | | |
| Intangible Assets | £ 823 | £ 868 |
| Tangible Assets | 1,732 | 1,876 |
| Investments | 234 | 266 |
| Total Fixed Assets | £ 2,789 | £ 3,010 |
| CURRENT ASSETS | | |
| Stocks | £ 1,222 | £ 1,158 |
| Debtors | 2,450 | 2,413 |
| Short-Term Deposits and Investments | 301 | 84 |
| Cash at Bank and In Hand | 578 | 634 |
| **Creditors Due within One Year** | | |
| Borrowings | (276) | (275) |
| Other Creditors | (2,720) | (2,727) |
| Net Current Assets | £ 1,555 | £ 1,287 |
| Total Assets Less Current Liabilities | £ 4,344 | £ 4,297 |
| **Creditors Due after More than One Year** | | |
| Borrowings | (1,104) | (1,038) |
| Other Creditors | (288) | (450) |
| Provisions for Liabilities | (882) | (772) |
| Total Assets Less Total Liabilities | £ 2,070 | £ 2,037 |
| **Capital and Reserves** | | |
| Called-up Share Capital | £ 320 | £ 323 |
| Share Premium Account | 636 | 634 |
| Revaluation Reserve | 103 | 100 |
| Other Reserves | 189 | 195 |
| Profit and Loss Account | 820 | 783 |
| Minority Interests | 2 | 2 |
| Total Shareholders' Equity | £ 2,070 | £ 2,037 |

**EXHIBIT 2.8**

**ROLLS-ROYCE**
**Comparative Balance Sheet in U.S. Format and Terminology**
**(in millions of pounds)**

| | August 31 | |
|---|---|---|
| | **Year 11** | **Year 12** |
| ASSETS | | |
| Cash | £ 578 | £ 634 |
| Marketable Securities | 301 | 84 |
| Accounts Receivable | 2,450 | 2,413 |
| Inventories | 1,222 | 1,158 |
| Total Current Assets | £4,551 | £4,289 |
| Investments | 234 | 266 |
| Property, Plant, and Equipment | 1,732 | 1,876 |
| Intangible Assets | 823 | 868 |
| Total Assets | £7,340 | £7,299 |
| LIABILITIES AND SHAREHOLDERS' EQUITY | | |
| Notes Payable | £ 276 | £ 275 |
| Other Current Liabilities | 2,720 | 2,727 |
| Total Current Liabilities | £2,996 | £3,002 |
| Long-Term Debt | 1,104 | 1,038 |
| Other Noncurrent Liabilities | 1,170 | 1,222 |
| Total Liabilities | £5,270 | £5,262 |
| SHAREHOLDERS' EQUITY | | |
| Common Stock | £ 320 | £ 323 |
| Additional Paid-in Capital | 636 | 634 |
| Retained Earnings | 820 | 783 |
| Revaluation Reserve | 103 | 100 |
| Other Reserves | 189 | 195 |
| Minority Interest | 2 | 2 |
| Total Shareholders' Equity | £2,070 | £2,037 |
| Total Liabilities and Shareholders' Equity | £7,340 | £7,299 |

One account reported in **Exhibit 2.7** seldom appears on balance sheets in the United States and most other countries: Revaluation Reserve. Common practice in most countries reports non-monetary assets (for example, inventories and property, plant, and equipment) at acquisition, or historical, cost. Accounting standards in the United Kingdom and in a few other countries permit the periodic revaluation of property, plant, and equipment to current market values. Firms obtain appraisals of the market values of their tangible fixed assets and intangibles such as brand names at periodic intervals (every three to five years).

At the end of Year 11, Rolls-Royce reports a balance in the Revaluation Reserve account of £103 million, suggesting that the current market values of its property, plant, and equipment and some intangibles such as brand names exceed acquisition cost by £103 million. The Revaluation Reserve decreased to £100 at the end of Year 12. An analyst wanting to convert the balance sheet of Rolls-Royce to U.S. accounting standards would decrease the reported amount of Tangible Fixed Assets and the Revaluation Reserve by £103 at the end of Year 11 and by £100 at the end of Year 12. The journal entry to record a decrease in market values subtracts £3 from Tangible Fixed Assets and from Revaluation Reserve during Year 12.

You will likely find the discussion of different balance sheet formats, terminology, and restatement entries in this section somewhat difficult to follow at this early stage in your study of financial accounting. The message here is that a solid grasp of important balance sheet concepts, as discussed throughout most of this chapter, will permit you to apply those concepts to balance sheets for firms around the world.

## Summary

The balance sheet comprises three major classes of items: assets, liabilities, and shareholders' equity.

Resources become accounting assets when a firm has acquired the rights to their future use as a result of a past transaction or exchange, and when it can measure the value of the future benefits with a reasonable degree of precision. Monetary assets appear, in general, at their current cash, or cash equivalent, values. Nonmonetary assets appear at acquisition cost, in some cases adjusted downward either for the cost of services that a firm has consumed or for the effects of technological or other obsolescence. Liabilities represent obligations of a firm to make payments of a reasonably definite amount at a reasonably definite future time for benefits already received. Shareholders' equity, the difference between total assets and total liabilities, for corporations typically comprises contributed capital and retained earnings.

Recording the effects of each transaction in a dual manner in the accounts maintains the equality of total assets and total liabilities plus shareholders' equity. The accountant initially records the dual effects of each transaction using the following format:

| | | |
|---|---|---|
| Debit Account (Asset Increases and Liability and Shareholders' Equity Decreases) . . . . . . . . . . . . . . . . . . . . . . . . . . . . . . . . . | Debit Amount | |
| Credit Account (Asset Decreases and Liability and Shareholders' Equity Increases) . . . . . . . . . . . . . . . . . | | Credit Amount |

When analyzing a balance sheet, one looks for a reasonable match between the nature and mix of assets and the nature and mix of liabilities plus shareholders' equity. The proportion of short-term versus long-term financing should match the proportion of current assets versus noncurrent assets. The mix of long-term debt versus shareholders' equity should reflect the degree of operating risk in the business. The larger the operating risk from cyclical sales, short product life cycles, high proportions of fixed costs, and similar factors, the more shareholders' equity capital as opposed to long-term debt one should look for on the balance sheet. The user, however, must recognize biases injected into the balance sheet as a result of applying GAAP in its preparation.

## Appendix 2.1: Using a Spreadsheet to Record Business Transactions

This chapter illustrated the use of T-accounts to accumulate information about the effects of business transactions on individual balance sheet accounts and to prepare the balance sheet. Some instructors and students might prefer to use a computer spreadsheet program to accomplish the same objectives. This appendix illustrates the use of a spreadsheet using the transactions of Miller Corporation discussed in the chapter and summarized in **Exhibit 2.1**.

**Exhibit 2.1** presents the balance sheet equation horizontally. That is,

Assets = Liabilities + Shareholders' Equity

Recall though that the balance sheet reports amounts at a moment in time. Picture now the balance sheet equation presented vertically at the beginning and end of a period. Thus,

This vertical arrangement of the balance sheet equation corresponds with the depiction in **Figure 1.3** in **Chapter 1** of the relation between the balance sheet, income statement, and statement

**EXHIBIT 2.9** **Transactions Spreadsheet Template**

| Balance Sheet Accounts | Balance: Beginning of Period | Transactions, By Number and Description: Describe Transaction Here | | | | | | | Balance: End of Period |
|---|---|---|---|---|---|---|---|---|---|
| | | 1 | 2 | 3 | 4 | 5 | 6 | 7 | |
| ASSETS | | | | | | | | | |
| **Current Assets:** | | | | | | | | | |
| | | | | | | | | | |
| | | | | | | | | | |
| Total Current Assets | | | | | | | | | |
| **Noncurrent Assets:** | | | | | | | | | |
| | | | | | | | | | |
| | | | | | | | | | |
| Total Noncurrent Assets | | | | | | | | | |
| Total Assets | | | | | | | | | |
| LIABILITIES AND SHAREHOLDERS' EQUITY | | | | | | | | | |
| **Current Liabilities:** | | | | | | | | | |
| | | | | | | | | | |
| | | | | | | | | | |
| Total Current Liabilities | | | | | | | | | |
| **Noncurrent Liabilities:** | | | | | | | | | |
| | | | | | | | | | |
| | | | | | | | | | |
| Total Noncurrent Liabilities | | | | | | | | | |
| Total Liabilities | | | | | | | | | |
| **Shareholders' Equity:** | | | | | | | | | |
| | | | | | | | | | |
| | | | | | | | | | |
| Total Shareholders' Equity | | | | | | | | | |
| Total Liabilities and Shareholders' Equity | | | | | | | | | |
| | | | | | | | | | |
| **Imbalance, *If Any*** | | | | | | | | | |
| | | | | | | | | | |
| **Income Statement Accounts** | | | | | | | | | |

of cash flows for a period of time. We use this vertical arrangement to design a spreadsheet for recording the dual effects of transactions on the balance sheet equation.

**Exhibit 2.9** presents the spreadsheet template, which you can download from the web site for this book at www.thomsonedu.com/accounting/stickney. The column Balance: Beginning of Period reports amounts on the balance sheet at the beginning of a period and the column Balance: End of Period reports amounts on the balance sheet at the end of the period. The rows of the spreadsheet show the various categories of assets (current and noncurrent), liabilities (current and noncurrent), and shareholders' equity. We leave space to add individual accounts, such as for cash, accounts payable, and common stock. The amounts showing dashes represent various subtotals and totals. The columns labeled Transactions by Number and Description refer to transactions during a period that transform the balance sheet at the beginning of the period to the balance sheet at the end of the period. We use this spreadsheet to show the dual effects of a firm's transactions, entering the appropriate account titles in the rows in the first column. The next-to-last row of the spreadsheet, labeled Imbalance, If Any, serves as an arithmetical check to ensure that the user records each transaction in a dual manner to maintain the balance sheet equation. You can ignore the row labeled Income Statement Accounts, which we discuss in **Appendix 3.1** in **Chapter 3**.[8]

We illustrate the use of this spreadsheet template using the seven transactions for Miller Corporation for the month of January as it prepares to open for business on February 1. We indicate whether each account is an asset (A), a liability (L), or a shareholders' equity (SE) account. **Exhibit 2.10** shows the completed spreadsheet incorporating the seven transactions.

Transaction 1: On January 1, Miller Corporation issues 10,000 shares of $10 par value common stock for $100,000 cash. This transaction increases cash by $100,000 and common stock, a shareholders' equity account, by an equal amount. Thus,

| | |
|---|---|
| (A) Cash .......................... | +100,000 |
| (SE) Common Stock .................. | +100,000 |

We enter the account title, Cash, on the first row under Current Assets and the account title, Common Stock, on the first row under Shareholders' Equity and place 100,000 in each of these accounts under the Transaction 1 column. A non-zero amount in the Imbalance, If Any row would indicate an incorrect recording of the transaction, which we must correct before moving to the next transaction.

Transaction 2: On January 5, Miller Corporation pays $60,000 cash to purchase equipment. This transaction decreases cash by $60,000 and increases equipment, a noncurrent asset, by $60,000. Thus,

| | |
|---|---|
| (A) Cash .......................... | − 60,000 |
| (A) Equipment ...................... | + 60,000 |

Transaction 3: On January 15, the firm purchases merchandise costing $15,000 from a supplier on account, agreeing to pay later. This transaction increases inventory by $15,000 and accounts payable, a current liability, by $15,000. Thus,

| | |
|---|---|
| (A) Inventory ...................... | + 15,000 |
| (L) Accounts Payable ................. | + 15,000 |

Transaction 4: On January 21, the firm pays the supplier in Transaction 3 above $8,000 of the amount due. This transaction reduces cash and accounts payable by $8,000. Thus,

| | |
|---|---|
| (A) Cash .......................... | − 8,000 |
| (L) Accounts Payable ................. | − 8,000 |

Transaction 5: On January 25, the supplier in Transaction 3 above accepts 700 shares of common stock in settlement of the $7,000 still owed. This transaction reduces accounts payable and increases common stock by $7,000. Thus,

| | |
|---|---|
| (L) Accounts Payable ................. | − 7,000 |
| (SE) Common Stock .................. | + 7,000 |

[8]The transactions template follows the usual procedures for the use of an Excel spreadsheet. If a particular problem has more than eight transactions, the Insert Column command allows for additional columns. The Insert Row command allows for additional rows for more accounts than the space provided. Insert columns and rows as indicated on the transactions template to ensure that summation formulas include the effects of additional rows and columns. The Edit Delete Row and Edit Delete Column commands allow for the deletion of rows and columns not needed for a particular problem.

**EXHIBIT 2.10** Transactions Spreadsheet for Miller Corporation

| Balance Sheet Accounts | Transactions, By Number and Description | | | | | | | | |
|---|---|---|---|---|---|---|---|---|---|
| | **Balance: Beginning of Period** | Issue Common Stock for Cash | Purchase Equip. for Cash | Purchase Inv. on Account | Pay Cash to Supplier | Give Common Stock to Supplier | Prepay Insurance | Receives Advance from Customer | **Balance: End of Period** |
| | | **1** | **2** | **3** | **4** | **5** | **6** | **7** | |
| ASSETS | | | | | | | | | |
| **Current Assets:** | | | | | | | | | |
| Cash | | 100,000 | −60,000 | | −8,000 | | −600 | 3,000 | 34,400 |
| Inventory | | | | 15,000 | | | | | 15,000 |
| Prepaid Insurance | | | | | | | 600 | | 600 |
| Total Current Assets | | | | | | | | | 50,000 |
| **Noncurrent Assets:** | | | | | | | | | |
| Equipment | | | 60,000 | | | | | | 60,000 |
| Total Noncurrent Assets | | | | | | | | | 60,000 |
| Total Assets | | | | | | | | | 110,000 |
| LIABILITIES AND SHAREHOLDERS' EQUITY | | | | | | | | | |
| **Current Liabilities:** | | | | | | | | | |
| Accounts Payable | | | | 15,000 | −8,000 | −7,000 | | | |
| Advance from Customer | | | | | | | | 3,000 | 3,000 |
| Total Current Liabilities | | | | | | | | | 3,000 |
| **Noncurrent Liabilities:** | | | | | | | | | |
| | | | | | | | | | |
| Total Noncurrent Liabilities | | | | | | | | | |
| Total Liabilities | | | | | | | | | 3,000 |
| **Shareholders' Equity:** | | | | | | | | | |
| Common Stock | | 100,000 | | | | 7,000 | | | 107,000 |
| | | | | | | | | | |
| Total Shareholders' Equity | | | | | | | | | 107,000 |
| Total Liabilities and Shareholders' Equity | | | | | | | | | 110,000 |
| | | | | | | | | | |
| **Imbalance, *If Any*** | | | | | | | | | |
| | | | | | | | | | |
| **Income Statement Accounts** | | | | | | | | | |

Transaction 6: On January 31, the firm pays $600 cash for a one-year insurance premium for coverage beginning February 1. This transaction reduces cash by $600 and increases Prepaid Insurance, a current asset, for the future benefits of insurance coverage. Thus,

| | | |
|---|---|---|
| (A) Cash . . . . . . . . . . . . . . . . . . . . . . . . . . . | − | 600 |
| (A) Prepaid Insurance . . . . . . . . . . . . . . . . . . | + | 600 |

Transaction 7: On January 31, the firm receives $3,000 from a customer for merchandise to be delivered during February. This transaction increases cash and increases Advances from Customers, a current liability, by $3,000. Thus,

| | | |
|---|---|---|
| (A) Cash . . . . . . . . . . . . . . . . . . . . . . . . . . . | + | 3,000 |
| (L) Advances from Customers . . . . . . . . . . . . . | + | 3,000 |

The last column on the right shows the amounts for the balance sheet on January 31. The amounts result from summing the amounts from the balance sheet at the beginning of the period (zero in this case) and the amounts for each transaction during the period. We can then use the amounts in the spreadsheet to prepare a balance sheet. Exhibit 2.4 in the chapter presents the balance sheet for Miller Corporation on January 31.

## Problem 2.7 for Self-Study

**Spreadsheet analysis of business transactions.** Design a spreadsheet similar to that in **Exhibit 2.9** or use the transactions spreadsheet template available with this text at www.thomsonedu.com/accounting/stickney. Enter each of the transactions of Gaines Corporation in **Problem 2.3 for Self-Study** and derive the amounts for the balance sheet after the six transactions.

# Suggested Solutions to Self-Study Problems

### SUGGESTED SOLUTION TO PROBLEM 2.1 FOR SELF-STUDY

(Coca-Cola Company; asset recognition and valuation)

**a.** GAAP does not recognize research and development expenditures as assets because of the uncertainty of future benefits that a firm can measure with reasonable precision.
**b.** Deposit on Containers, $400,000. This is a partially executed contract, which accountants recognize as an asset to the extent of the partial performance.
**c.** GAAP does not allow firms to capitalize advertising expenditures as assets because of the uncertainty of future benefits which firms can measure with reasonable precision.
**d.** Investment in Common Stock, $2.5 million.
**e.** GAAP does not recognize an asset for the same reasons as in part **a** above.
**f.** Land and Building, $75 million. The accountant must allocate the aggregate purchase price between the land and the building because the building is depreciable and the land is not. Legal passage of title is not necessary to justify recognition of an asset. Coca-Cola has acquired the rights to use the land and building and can sustain those rights as long as it makes the required payments on the mortgage obligation.

### SUGGESTED SOLUTION TO PROBLEM 2.2 FOR SELF-STUDY

(New York Times Company; liability recognition and valuation)

**a.** Subscription Fees Received in Advance, $10 million.
**b.** Accounts Payable, $4 million. Other account titles are also acceptable.
**c.** GAAP does not recognize a liability in this case because the life of the vehicles substantially exceeds the one-year rental period. **Chapter 10** discusses the criteria for recognition of leases as liabilities.
**d.** GAAP does not recognize a liability because of the low probability, given the insurance coverage, that the firm will make a future cash payment.

**e.** GAAP requires the recognition of a liability when a cash payment is "probable." GAAP provides no specific guidelines as to how high the probability needs to be to recognize a liability. Anecdotal evidence suggests that practicing accountants use 80 to 85 percent.

**f.** The firm likely previously recorded the $2 million in the account Subscription Fees Received in Advance. The strike will probably extend the subscription period by two weeks. Thus, the firm has already recognized a liability.

## SUGGESTED SOLUTION TO PROBLEM 2.3 FOR SELF-STUDY

(Gaines Corporation; dual effects of transactions on balance sheet equation)

| Transaction | Assets | = | Liabilities | + | Shareholders' Equity |
|---|---|---|---|---|---|
| **1.** The firm issues 20,000 shares of $10 par value common stock for $12 cash per share . . . . . . . . . . | + $240,000 | | 0 | | + $240,000 |
| **2.** The firm issues $100,000 principal amount of bonds for $100,000 cash . . . . . . . . . . . . . . . . . . . . . . . | + 100,000 | | + $100,000 | | 0 |
| **3.** The firm acquires, with $220,000 cash, land costing $40,000 and a building costing $180,000 . . . . . . . | + 220,000<br>− 220,000 | | 0 | | 0 |
| **4.** The firm acquires, on account, equipment costing $25,000 and merchandise inventory costing $12,000 | + 37,000 | | + 37,000 | | 0 |
| **5.** The firm signs an agreement to rent equipment from its owner and pays $1,500 rental in advance . . . . . | + 1,500<br>− 1,500 | | 0 | | 0 |
| **6.** The firm pays $28,000 to the suppliers in **4** . . . . . . | − 28,000 | | − 28,000 | | |
| Totals . . . . . . . . . . . . . . . . . . . . . . . . . . . . . . . | $349,000 | + | $109,000 | = | $240,000 |

## SUGGESTED SOLUTION TO PROBLEM 2.4 FOR SELF-STUDY

(Gaines Corporation; T-accounts for various transactions)

**Cash (Asset)**

| | Debit | Credit | |
|---|---|---|---|
| (1) | 240,000 | 220,000 | (3) |
| (2) | 100,000 | 1,500 | (5) |
| | | 28,000 | (6) |

**Merchandise Inventory (Asset)**

| | Debit | Credit | |
|---|---|---|---|
| (4) | 12,000 | | |

**Prepaid Rent (Asset)**

| | Debit | Credit | |
|---|---|---|---|
| (5) | 1,500 | | |

**Land (Asset)**

| | Debit | Credit | |
|---|---|---|---|
| (3) | 40,000 | | |

**Buildings (Asset)**

| | Debit | Credit | |
|---|---|---|---|
| (3) | 180,000 | | |

**Equipment (Asset)**

| | Debit | Credit | |
|---|---|---|---|
| (4) | 25,000 | | |

**Accounts Payable (Liability)**

| | Debit | Credit | |
|---|---|---|---|
| (6) | 28,000 | 37,000 | (4) |

**Bonds Payable (Liability)**

| | Debit | Credit | |
|---|---|---|---|
| | | 100,000 | (2) |

**Common Stock Par Value (Shareholders' Equity)**

| | Debit | Credit | |
|---|---|---|---|
| | | 200,000 | (1) |

**Additional Paid-in Capital (Shareholders' Equity)**

| | Debit | Credit | |
|---|---|---|---|
| | | 40,000 | (1) |

## SUGGESTED SOLUTION TO PROBLEM 2.5 FOR SELF-STUDY

(Gaines Corporation; journal entries for various transactions)

| | Debit | Credit |
|---|---|---|
| (1) Cash . . . . . . . . . . . . . . . . . . . . . . . . . . . . . . . . . . . . . . . . . . . . . | 240,000 | |
| Common Stock . . . . . . . . . . . . . . . . . . . . . . . . . . . . . . . . . . . . | | 200,000 |
| Additional Paid-in Capital . . . . . . . . . . . . . . . . . . . . . . . . . . . | | 40,000 |

| Assets | = | Liabilities | + | Shareholders' Equity | (Class.) |
|---|---|---|---|---|---|
| +240,000 | | | | +240,000 | ContriCap |

Issue 20,000 shares of $10 par value common stock for $12 cash per share.

| | | |
|---|---|---|
| **(2)** Cash | 100,000 | |
| Bonds Payable | | 100,000 |

| Assets | = | Liabilities | + | Shareholders' Equity | (Class.) |
|---|---|---|---|---|---|
| +100,000 | | +100,000 | | | |

Issue $100,000 principal amount of bonds for $100,000 cash.

| | | |
|---|---|---|
| **(3)** Land | 40,000 | |
| Building | 180,000 | |
| Cash | | 220,000 |

| Assets | = | Liabilities | + | Shareholders' Equity | (Class.) |
|---|---|---|---|---|---|
| +220,000 | | | | | |
| −220,000 | | | | | |

Acquires for $220,000 land costing $40,000 and a building costing $180,000.

| | | |
|---|---|---|
| **(4)** Equipment | 25,000 | |
| Merchandise Inventory | 12,000 | |
| Accounts Payable | | 37,000 |

| Assets | = | Liabilities | + | Shareholders' Equity | (Class.) |
|---|---|---|---|---|---|
| +37,000 | | +37,000 | | | |

Purchases equipment costing $25,000 and merchandise inventory costing $12,000 on account.

| | | |
|---|---|---|
| **(5)** Prepaid Rent | 1,500 | |
| Cash | | 1,500 |

| Assets | = | Liabilities | + | Shareholders' Equity | (Class.) |
|---|---|---|---|---|---|
| +1,500 | | | | | |
| −1,500 | | | | | |

Paid $1,500 as advance rental on equipment.

| | | |
|---|---|---|
| **(6)** Accounts Payable | 28,000 | |
| Cash | | 28,000 |

| Assets | = | Liabilities | + | Shareholders' Equity | (Class.) |
|---|---|---|---|---|---|
| −28,000 | | −28,000 | | | |

Pays $28,000 to the supplier in **(4)**.

## SUGGESTED SOLUTION TO PROBLEM 2.6 FOR SELF-STUDY

(Electronics Appliance Corporation; journal entries, T-accounts, and balance sheet preparation)

**a.** Journal entries for the nine transactions follow:

**(1)** Sept. 1

| | | |
|---|---|---|
| Cash | 48,000 | |
| Common Stock | | 40,000 |
| Additional Paid-in Capital | | 8,000 |

| Assets | = | Liabilities | + | Shareholders' Equity | (Class.) |
|---|---|---|---|---|---|
| +48,000 | | | | +40,000 | ContriCap |
| | | | | +8,000 | ContriCap |

Issue 4,000 shares of $10 par value common stock for $12 cash per share.

**(2)** Sept. 2

| | | |
|---|---|---|
| Patent | 7,200 | |
| Common Stock | | 6,000 |
| Additional Paid-in Capital | | 1,200 |

| Assets | = | Liabilities | + | Shareholders' Equity | (Class.) |
|---|---|---|---|---|---|
| +7,200 | | | | +6,000 | ContriCap |
| | | | | +1,200 | ContriCap |

Issue 600 shares of $10 par value common stock in the acquisition of a patent.

**(3)** Sept. 5

| | | |
|---|---|---|
| Prepaid Rent or Advances to Landlord | 10,000 | |
| Cash | | 10,000 |

| Assets | = | Liabilities | + | Shareholders' Equity | (Class.) |
|---|---|---|---|---|---|
| −10,000 | | | | | |
| +10,000 | | | | | |

Prepay rent for October and November on factory building.

**(4)** Sept. 12

| | | |
|---|---|---|
| Raw Materials Inventory | 6,100 | |
| Accounts Payable | | 6,100 |

| Assets | = | Liabilities | + | Shareholders' Equity | (Class.) |
|---|---|---|---|---|---|
| +6,100 | | +6,100 | | | |

Purchase raw materials costing $6,100 on account.

**(5)** Sept. 15

| | Dr. | Cr. |
|---|---|---|
| Cash | 900 | |
| Advances from Customers | | 900 |

| Assets | = | Liabilities | + | Shareholders' Equity | (Class.) |
|---|---|---|---|---|---|
| +900 | | +900 | | | |

Receive an advance of $900 from a customer as a deposit on equipment to be manufactured in the future.

**(6)** Sept. 20

| | Dr. | Cr. |
|---|---|---|
| Equipment | 925 | |
| Cash | | 925 |

| Assets | = | Liabilities | + | Shareholders' Equity | (Class.) |
|---|---|---|---|---|---|
| +925 | | | | | |
| −925 | | | | | |

Acquire equipment with a list price of $950 for $925 after taking a discount for prompt payment.

**(7)** Sept. 28

| | Dr. | Cr. |
|---|---|---|
| Advances to Employees | 200 | |
| Cash | | 200 |

| Assets | = | Liabilities | + | Shareholders' Equity | (Class.) |
|---|---|---|---|---|---|
| +200 | | | | | |
| −200 | | | | | |

Give cash advances of $200 to employees beginning work on Oct. 1.

**(8)** Sept. 30

| | Dr. | Cr. |
|---|---|---|
| Equipment | 27,500 | |
| Cash | | 5,000 |
| Mortgage Payable | | 22,500 |

| Assets | = | Liabilities | + | Shareholders' Equity | (Class.) |
|---|---|---|---|---|---|
| +27,500 | | +22,500 | | | |
| −5,000 | | | | | |

Acquire equipment for $5,000 cash and assume a $22,500 mortgage for the balance of the purchase price.

**(9)** Sept. 30

| | Dr. | Cr. |
|---|---|---|
| Equipment | 450 | |
| Cash | | 450 |

| Assets | = | Liabilities | + | Shareholders' Equity | (Class.) |
|---|---|---|---|---|---|
| +450 | | | | | |
| −450 | | | | | |

Pay installation cost of $450 on equipment acquired in **(8)**.

b. **Exhibit 2.11** presents T-accounts for Electronics Appliance Corporation and shows the recording of the nine entries in the accounts. The letters A, L, and SE after the account titles indicate the balance sheet category of the accounts.
c. **Exhibit 2.12** presents a balance sheet as of September 30.

**EXHIBIT 2.11**

**ELECTRONICS APPLIANCE CORPORATION**
**T-Accounts and Transactions during September**
**(Problem 2.6 for Self-Study)**

**Cash (A)**

| | Debit | Credit | |
|---|---|---|---|
| (1) | 48,000 | 10,000 | (3) |
| (5) | 900 | 925 | (6) |
| | | 200 | (7) |
| | | 5,000 | (8) |
| | | 450 | (9) |
| ✓ | 32,325 | | |

**Advances to Employees (A)**

| | Debit | Credit | |
|---|---|---|---|
| (7) | 200 | | |
| ✓ | 200 | | |

**Raw Materials Inventory (A)**

| | Debit | Credit | |
|---|---|---|---|
| (4) | 6,100 | | |
| ✓ | 6,100 | | |

**Prepaid Rent (A)**

| | Debit | Credit | |
|---|---|---|---|
| (3) | 10,000 | | |
| ✓ | 10,000 | | |

**Equipment (A)**

| | Debit | Credit | |
|---|---|---|---|
| (6) | 925 | | |
| (8) | 27,500 | | |
| (9) | 450 | | |
| ✓ | 28,875 | | |

**Patent (A)**

| | Debit | Credit | |
|---|---|---|---|
| (2) | 7,200 | | |
| ✓ | 7,200 | | |

**Accounts Payable (L)**

| | Debit | Credit | |
|---|---|---|---|
| | | 6,100 | (4) |
| | | 6,100 | ✓ |

**Advances from Customers (L)**

| | Debit | Credit | |
|---|---|---|---|
| | | 900 | (5) |
| | | 900 | ✓ |

**Mortgage Payable (L)**

| | Debit | Credit | |
|---|---|---|---|
| | | 22,500 | (8) |
| | | 22,500 | ✓ |

**Common Stock (SE)**

| | Debit | Credit | |
|---|---|---|---|
| | | 40,000 | (1) |
| | | 6,000 | (2) |
| | | 46,000 | ✓ |

**Additional Paid-in Capital (SE)**

| | Debit | Credit | |
|---|---|---|---|
| | | 8,000 | (1) |
| | | 1,200 | (2) |
| | | 9,200 | ✓ |

**EXHIBIT 2.12**

**ELECTRONICS APPLIANCE CORPORATION**
**Balance Sheet**
**September 30**
**(Problem 2.6 for Self-Study)**

| ASSETS | | |
|---|---|---|
| **Current Assets** | | |
| Cash | $32,325 | |
| Advances to Employees | 200 | |
| Raw Materials Inventory | 6,100 | |
| Prepaid Rent | 10,000 | |
| Total Current Assets | | $48,625 |
| **Property, Plant, and Equipment** | | |
| Equipment | | 28,875 |
| **Intangibles** | | |
| Patent | | 7,200 |
| Total Assets | | $84,700 |

| LIABILITIES AND SHAREHOLDERS' EQUITY | | |
|---|---|---|
| **Current Liabilities** | | |
| Accounts Payable | $ 6,100 | |
| Advances from Customers | 900 | |
| Total Current Liabilities | | $ 7,000 |
| **Long-Term Debt** | | |
| Mortgage Payable | | 22,500 |
| Total Liabilities | | $29,500 |
| **Shareholders' Equity** | | |
| Common Stock, $10 Par Value | $46,000 | |
| Additional Paid-in Capital | 9,200 | |
| Total Shareholders' Equity | | 55,200 |
| Total Liabilities and Shareholders' Equity | | $84,700 |

### SUGGESTED SOLUTION TO PROBLEM 2.7 FOR SELF-STUDY

(Gaines Corporation; spreadsheet analysis of business transactions) See **Exhibit 2.13** on page 80.

## Key Terms and Concepts

Executory contract
Acquisition (historical) cost
Current replacement cost (entry value)
Current net realizable value (exit value)
Present value
Monetary assets
Nonmonetary assets
Going concern
Reliability
Conservatism
Current assets
Plant, or fixed, assets
Intangible assets
Contingent obligations
Current liabilities
Par or stated value
T-account
Debit
Credit
Journal entry
Common-size balance sheet

## Questions, Short Exercises, Exercises, Problems, and Cases

### QUESTIONS

1. Review the meaning of the terms and concepts listed above in Key Terms and Concepts.
2. Conservatism is generally regarded as a convention in accounting. Indicate whom this might hurt.
3. One of the criteria for the recognition of an asset or a liability is that there be an exchange. What justification can you see for this requirement?
4. Accounting typically does not recognize either assets or liabilities for mutually unexecuted contracts. What justification can you see for this treatment?
5. Accounting treats cash discounts taken on the purchase of merchandise or equipment as a reduction in the amount recorded for the assets acquired. What justification can you see for this treatment?
6. A group of investors owns an office building that it rents unfurnished to tenants. It purchased the building five years previously from a construction company. At that time, it expected the building to have a useful life of 40 years. Indicate the procedures you might follow to ascertain the valuation amount for this building under each of the following valuation methods:
   **a.** Acquisition cost.
   **b.** Adjusted acquisition cost (reduced for services already consumed).
   **c.** Current replacement cost.
   **d.** Current net realizable value.
   **e.** Present value of future net cash flows.
7. Some of the assets of one firm correspond to the liabilities of another firm. For example, an account receivable on the seller's balance sheet is an account payable on the buyer's balance sheet. For each of the following items, indicate whether it is an asset or a liability and give the corresponding account title on the balance sheet of the other party to the transaction:
   **a.** Advances by Customers.
   **b.** Bonds Payable.
   **c.** Interest Receivable.
   **d.** Prepaid Insurance.
   **e.** Rental Fees Received in Advance.

For additional student resources, content, and interactive quizzes for this chapter visit the FACMU website: **www.thomsonedu.com/accounting/stickney**

### SHORT EXERCISES

8. **Asset and liability recognition and valuation.** The St. Louis Cardinals sign Albert Pujols to a five-year contract beginning next year. The contract calls for payments of $50 million each year for the next five years, beginning next year. Assume that these payments have a present value of $200 million. The contract also provides that the St. Louis Cardinals give Albert Pujols an Italian sports car valued at $250,000 at the time of signing the contract. Indicate the treatment of this contract at the time of signing.

**EXHIBIT 2.13**

**Transactions Spreadsheet for Gaines Corporation (Problem 2.7 for Self-Study)**

| Balance Sheet Accounts | Balance: Beginning of Period | Issue Common Stock for Cash | Issue Bond for Cash | Acquire Land and Building for Cash | Acquire Equip. and Inventory on Account | Pay Rental in Advance | Pay Supplier | Balance: End of Period |
|---|---|---|---|---|---|---|---|---|
| | | **1** | **2** | **3** | **4** | **5** | **6** | |
| ASSETS | | | | | | | | |
| **Current Assets:** | | | | | | | | |
| Cash | | 240,000 | 100,000 | −220,000 | | −1,500 | −28,000 | 90,500 |
| Merchandise Inventory | | | | | 12,000 | | | 12,000 |
| Prepaid Rent | | | | | | 1,500 | | 1,500 |
| Total Current Assets | | | | | | | | 104,000 |
| **Noncurrent Assets:** | | | | | | | | |
| Land | | | | 40,000 | | | | 40,000 |
| Building | | | | 180,000 | | | | 180,000 |
| Equipment | | | | | 25,000 | | | 25,000 |
| Total Noncurrent Assets | | | | | | | | 245,000 |
| Total Assets | | | | | | | | 349,000 |
| LIABILITIES AND SHAREHOLDERS' EQUITY | | | | | | | | |
| **Current Liabilities:** | | | | | | | | |
| Accounts Payable | | | | | 37,000 | | -28,000 | 9,000 |
| Total Current Liabilities | | | | | | | | 9,000 |
| **Noncurrent Liabilities:** | | | | | | | | |
| Bonds Payable | | | 100,000 | | | | | 100,000 |
| Total Noncurrent Liabilities | | | | | | | | 100,000 |
| Total Liabilities | | | | | | | | 109,000 |
| **Shareholders' Equity:** | | | | | | | | |
| Common Stock | | 200,000 | | | | | | 200,000 |
| Additional Paid-in Capital | | 40,000 | | | | | | 40,000 |
| Total Shareholders' Equity | | | | | | | | 240,000 |
| Total Liabilities and Shareholders' Equity | | | | | | | | 349,000 |
| **Imbalance, *If Any*** | | | | | | | | |
| **Income Statement Accounts** | | | | | | | | |

9. **Asset recognition and valuation.** Citigroup, a financial services firm, provides tuition support for employees who pursue an MBA degree in the evenings. Employees receiving the support must continue working for Citigroup for five years or repay a ratable portion of the benefit. During the current year, Citigroup paid $10 million in tuition support to various employees. Citigroup estimates, based on experience, that the MBA education enhances employees' skills by $15 million. Indicate Citigroup's treatment of the $10 million.
10. **Liability recognition and valuation.** Delta Air Lines offers frequent flyer benefits to customers who reach specified levels of miles flown. The present value of the future loss in revenues due to customers earning free flights during the current year is $200 million. The present value of the estimated cost of providing the free flights is $25 million. Indicate the amount, if any, that Delta Air Lines should recognize as a liability for its frequent flyer program for the current year.
11. **Asset valuation.** Pizza Hut purchased a used van from Domino's Pizza to deliver pizzas. Pizza Hut paid $14,500 for the van. It paid $1,200 to have the Domino's Pizza name removed from the van and replaced with the Pizza Hut colors and name. It paid $75 for the license fee and $680 for insurance for the first year of use, which begins next month. How much should Pizza Hut record as the acquisition cost of this van? Describe the treatment of any of the above amounts that you did not include in the acquisition cost of the van.
12. **Dual effects on balance sheet equation.** Target Stores engaged in the following three transactions: (1) purchased and received merchandise costing $25 million on account from various suppliers, (2) returned merchandise costing $1.5 million because of damage or incorrect shipments, (3) paid the suppliers the amount due from purchases on account. Indicate the effects of each of these three transactions on the balance sheet equation.
13. **Journal entry for acquisition.** Leonard Corporation gave $10 million cash and shares of its common stock valued at $60 million for all of the assets of Newsom Corporation. The assets received consisted of inventory valued at $20 million and land, buildings, and equipment valued at $40 million. Leonard Corporation also assumed responsibility for $25 million of liabilities of Newsom Corporation. Give the journal entry to record this transaction.

## EXERCISES

14. **Asset recognition and valuation.** The following transactions relate to IBM Corporation, a manufacturer of electronic equipment and provider of computer services. Indicate whether or not each transaction immediately gives rise to an asset of the company under GAAP. If accounting recognizes an asset, state the account title, the amount, and the classification of the asset on the balance sheet as either a current asset or a noncurrent asset.
    - **a.** The firm invests $8,000,000 in a government bond. The bond has a maturity value of $10,000,000 in three years, and IBM intends to hold the bond to maturity.
    - **b.** The firm sends a check for $600,000 to a landlord for two months' rent in advance on warehouse facilities. The rental period begins next month.
    - **c.** The firm writes a check for $1,000,000 to obtain an option to purchase a tract of land. The price of the land is $10,000,000.
    - **d.** The firm signs a four-year employment agreement with its president for $30,000,000 per year. The contract period begins next month.
    - **e.** The firm purchases a patent on a laser printer from its creator for $8,500,000.
    - **f.** The firm receives a patent on a new computer processor that it developed. The firm spent $3,200,000 to develop the patented invention.
    - **g.** The firm received notice that a supplier had shipped by freight memory chips billed at $12,000,000, with payment due in 30 days. The seller retains title to the memory chips until received by the buyer.
15. **Asset recognition and valuation.** The transactions listed below relate to Delta Air Lines, a provider of airline services. Indicate whether or not each transaction immediately gives rise to an asset of Delta Air Lines under GAAP. If accounting recognizes an asset, give the account title, the amount, and the classification of the asset on the balance sheet as either a current asset or a noncurrent asset.
    - **a.** The firm places an order with Boeing for 10 Boeing 767 airplanes, costing $80 million each.
    - **b.** Delta writes a check for a total of $5 million as a deposit on the aircraft it ordered in part **a**.
    - **c.** The firm spends $4 million to obtain landing rights for the next five years at Boston's Logan Airport.

**d.** The firm writes a check for $1.5 million and assumes a mortgage from its bank for $8.5 million to purchase new ground equipment costing $10 million.

**e.** The firm signs a four-year employment agreement with its president for $15 million per year. The contract period begins next month.

**f.** The firm issues common stock with a market value of $60 million to acquire used aircraft from a bankrupt regional airline. The equipment had a book value on the bankrupt airline's books of $75 million.

**g.** The firm spends $2 million for advertisements that appeared last month to promote flights to the Bahamas that it intends to add to its flight schedule next month.

**h.** The firm invests $4 million in a government bond. The bond has a maturity value of $5 million in three years, and Delta intends to hold the bond to maturity.

**16. Asset recognition and valuation.** The transactions listed below relate to McDonald's, a fast-food chain. Indicate whether or not each transaction immediately gives rise to an asset under GAAP. If accounting recognizes an asset, state the account title, the amount, and the classification of the asset on the balance sheet as a current asset or a noncurrent asset.

**a.** The firm spends $2,500,000 to develop and test-market a new dinner meal. It intends to launch the new product nationally next month.

**b.** The firm spends $3,900,000 to acquire rights to prepare and sell a new sugar-free dessert bar.

**c.** The firm spends $2,800,000 to obtain options to purchase land as future sites for its McDonald's restaurants.

**d.** The firm spends $920,000 for television advertisements that appeared last month.

**e.** The firm issues shares of its common stock currently selling on the market for $20,000,000 for 40 percent of the shares of Philadelphia Markets, another restaurant chain. Recent appraisals suggest that a 40 percent share of Philadelphia Markets is worth between $18,000,000 and $22,000,000. McDonald's intends to hold these shares as a long-term investment.

**f.** The firm acquires land and a building costing $5,000,000 by paying $1,200,000 in cash and signing a promissory note for the remaining $3,800,000 of the purchase price. McDonald's expends $80,000 for a title search and other legal fees, $6,000 in recording fees with the state of Illinois, and $180,000 to destroy the building. McDonald's intends to use the land for a parking lot.

**17. Asset recognition and valuation.** The transactions below relate to Office Depot, an office supply retailer. Indicate whether or not each transaction immediately gives rise to an asset under GAAP. If accounting recognizes an asset, state the account title, the amount, and the classification of the asset on the balance sheet as either a current asset or a noncurrent asset.

**a.** The firm rents retail space in a local shopping center for the five-year period beginning next month. Office Depot pays $255,000, which includes $125,000 as rent for the first year and $130,000 as a security deposit against future damages and unpaid rent. The firm commits to paying $500,000 later, for the last four years of the lease.

**b.** Refer to part a. The firm spends $10,000 to install partitions between administrative and retail space, $6,500 to paint the walls colors that are consistent with other Office Depot stores, and $20,000 to install carpeting.

**c.** The firm purchases display counters with a list price of $30,000. It pays for the display counters in time to take a 2 percent discount for prompt payment. Costs to transport the display counters to the new location total $1,200, and costs to install them total $800.

**d.** The firm hires a store manager at an annual salary of $60,000.

**e.** The firm spends $1,500 in newspaper and television advertisements that appeared this month.

**f.** The firm purchases inventory with an invoice price of $160,000. It pays for merchandise with an original invoice price of $120,000 in time to take a 2 percent discount for prompt payment. It has not yet paid for the remainder of the merchandise purchased. The firm treats cash discounts taken as a reduction in the acquisition cost of merchandise. An inspection of the merchandise reveals that merchandise with an original invoice price of $12,000 is defective. Office Depot returns this merchandise to the supplier, having not yet paid for it.

**18. Liability recognition and valuation.** Indicate whether or not each of the following events immediately gives rise to a liability under GAAP. If accounting recognizes a liability, state the account title, the amount, and the classification of the liability on the balance sheet as either a current liability or a noncurrent liability.

**a.** A landscaper agrees to improve land owned by a firm during the next year. The agreed price for the work is $7,500. Consider from the standpoint of the firm owning the land.
**b.** A magazine publisher receives a check for $72 for a one-year subscription to a magazine. The subscription period begins next month. Consider from the standpoint of the publisher.
**c.** A construction company agrees to build a bridge for $10 million. It receives a down payment of $2 million upon signing the contract; it is entitled to the remainder when it completes the bridge, which is expected to be in three years.
**d.** A firm issues additional common stock with a par value of $3,000,000 for $7,600,000.
**e.** A firm receives a 60-day, 6 percent loan of $100,000 from a local bank. Consider from the standpoint of the firm receiving the loan.
**f.** A firm signs a contract to purchase at least $60,000 of merchandise from a particular supplier during the next year. Consider from the standpoint of the purchaser of the merchandise.
**g.** Refer to part **f.** The firm places an order for $15,000 of merchandise with the supplier.

**19. Liability recognition and valuation.** The transactions below relate to Kansas City Royals, Inc., owner of a professional baseball team and Kauffman Stadium. Indicate whether or not each of the following transactions immediately gives rise to a liability of the firm under GAAP. If accounting recognizes a liability, state the account title, the amount, and the classification of the liability on the balance sheet as either a current liability or a noncurrent liability.
**a.** The firm signs a five-year contract with Joe Superstar for $7,400,000 per year. The contract period begins on February 1 of next year.
**b.** The firm receives $2,700,000 from sales of season tickets for the baseball season starting April 1 of next year.
**c.** The firm issues bonds in the principal amount of $8,000,000 for $8,400,000. The bonds mature in 20 years and bear interest at 8 percent per year. The firm intends to use the proceeds to expand Kauffman Stadium.
**d.** The firm receives a bill for utility services received last month totaling $3,400.
**e.** The firm receives notice that a former player has filed suit against Kansas City Royals, Inc., alleging nonperformance of contractual terms. The player claims $10,000,000 in damages.
**f.** The firm orders new uniforms for the team for the baseball season beginning next spring. The contract calls for a $10,000 deposit upon signing the contract and a $15,000 payment on delivery of the uniforms in February of next year. The firm signs the contract and sends a check for $10,000.

**20. Liability recognition and valuation.** The transactions below relate to the activities of a local college. Indicate whether or not each of the following transactions immediately gives rise to a liability under GAAP. If accounting recognizes a liability, state the account title, the amount, and the classification of the liability on the balance sheet as either a current liability or a noncurrent liability.
**a.** The college issues bonds with a principal amount and issue price of $10,000,000. The college must pay interest at a rate of 7 percent of the principal amount each year and repay the principal in 15 years. The college intends to use the proceeds to construct a new dormitory.
**b.** The college contracts with a construction company to build the dormitory in part **a** for a contract price of $12,000,000. The college pays $2,000,000 at the time of signing the contract and must pay the remainder as construction progresses. Construction begins next month and should take three years to complete.
**c.** The college institutes a guaranteed tuition plan for students entering next fall. The tuition for the academic year beginning next fall is $15,000 per year. Although the college expects tuition to increase each year, any entering student who pays $45,000 in advance will receive rights to a four-year undergraduate education without additional tuition payments. The college receives $1,800,000 from students who are entering next fall and who signed up for the guaranteed tuition plan.
**d.** The college bookstore receives textbooks for the coming academic year with an invoice price of $170,000.
**e.** The college owes employees $280,000 in compensation for the last pay period. The college is also responsible for payment of payroll taxes of 6 percent of compensation.
**f.** The college receives a grant from the Carnegie Foundation for $1,500,000 to enhance undergraduate teaching. The college intends to disburse the funds to faculty members to develop new teaching materials.

21. **Liability recognition and valuation.** The events below relate to the Chicago Symphony Orchestra. Indicate whether or not each of the following events immediately gives rise to a liability under GAAP. If accounting recognizes a liability, state the account title, the amount, and the classification of the liability on the balance sheet as either a current liability or a noncurrent liability.
   **a.** The firm receives $340,000 for season tickets sold for next season.
   **b.** The firm places an order with a printing company totaling $85,000 for symphony performance programs for next season.
   **c.** The firm receives the programs ordered in part **b,** along with an invoice for $85,000.
   **d.** The firm receives notice from its attorneys that a loyal customer attending a concert last season and sitting in the first row of the symphony hall has sued the Chicago Symphony Orchestra for $10 million claiming hearing loss. The customer normally sits further back but was asked to move forward for this particular concert because of damage to the regular seat.
   **e.** The firm signs a three-year contract with its first violinist at a salary of $140,000 per year.
   **f.** For $225,000, the firm issues long-term bonds with a par value of $200,000.
   **g.** The firm receives a 90-day, 8 percent loan of $40,000 from a local bank.
22. **Balance sheet classification.** GAAP classifies items on the balance sheet in one of the following ways:
   **(1)** Asset.
   **(2)** Liability.
   **(3)** Shareholders' equity.
   **(4)** Item that would not appear on the balance sheet as conventionally prepared under GAAP.

   Using these numbers, indicate the appropriate classification of each of the following items:
   **a.** Salaries payable.
   **b.** Retained earnings.
   **c.** Notes receivable.
   **d.** Unfilled customers' orders.
   **e.** Land.
   **f.** Interest payable.
   **g.** Work-in-process inventory.
   **h.** Mortgage payable.
   **i.** Organization costs.
   **j.** Advances by customers.
   **k.** Advances to employees.
   **l.** Patents.
   **m.** Good credit standing.
   **n.** Common stock.
23. **Balance sheet classification.** GAAP classifies items on the balance sheet in one of the following ways:
   **(1)** Asset.
   **(2)** Liability.
   **(3)** Shareholders' equity.
   **(4)** Item that would not appear on the balance sheet as conventionally prepared under GAAP.

   Using these numbers, indicate the appropriate classification of each of the following items:
   **a.** Preferred stock.
   **b.** Furniture and fixtures.
   **c.** Potential liability under lawsuit (case has not yet gone to trial).
   **d.** Prepaid rent.
   **e.** Capital contributed in excess of par value.
   **f.** Cash on hand.
   **g.** Goodwill.
   **h.** Estimated liability under warranty contract.
   **i.** Raw materials inventory.
   **j.** Rental fees received in advance.
   **k.** Bonds payable.
   **l.** Prepaid insurance.

**m.** Income taxes payable.
**n.** Treasury stock.

## PROBLEMS AND CASES

**24. Dual effects of transactions on balance sheet equation and journal entries.** A firm engages in the following six transactions.

**(1)** The firm issues 5,000 shares of $10 par value common stock for $60,000 cash.
**(2)** It purchases merchandise costing $29,200 on account.
**(3)** The firm acquires store equipment costing $32,700. It issues a check for $5,000, with the balance payable over three years under an installment contract.
**(4)** The firm issues a check for $4,500 covering two months' rent in advance.
**(5)** Refer to transaction **(3)**. The firm issues 2,000 shares of its $10 par value common stock with a market value of $27,700 in full settlement of the installment contract.
**(6)** The firm pays the merchandise supplier in transaction **(2)** the amount due.

**a.** Indicate the effects of these six transactions on the balance sheet equation using this format:

| Transaction Number | Assets | = | Liabilities | + | Shareholders' Equity |
|---|---|---|---|---|---|
| (1) | +$60,000 | | $0 | | +$60,000 |
| Subtotal | $60,000 | = | $0 | + | $60,000 |

**b.** Give the journal entry for each of the six transactions.

**25. Dual effects of transactions on balance sheet equation and journal entries.** A firm engages in the following six transactions.

**(1)** The firm issues 12,000 shares of $1 par value common stock for $12 cash per share.
**(2)** The firm acquires land costing $50,000 and a building costing $900,000. It gives $110,000 in cash and 70,000 shares of its $1 par value common stock valued at $840,000 to acquire the land and building.
**(3)** The firm pays $6,000 cash for a one-year insurance policy on the land and building. The policy period commences next month.
**(4)** The firm acquires merchandise inventory costing $150,000 on account from various suppliers.
**(5)** The firm pays $147,000 in cash to the suppliers in transaction **(4)** for its previous purchases on account. The $3,000 difference between the original inventory cost and the amount paid represents a discount for prompt payment, which the firm treats as a reduction in the cost of the merchandise.
**(6)** The firm received $1,300 from a customer as an advance on inventory that the firm will deliver to the customer next month.

**a.** Indicate the effects of these six transactions on the balance sheet equation using this format:

| Transaction Number | Assets | = | Liabilities | + | Shareholders' Equity |
|---|---|---|---|---|---|
| (1) | +$144,000 | | $0 | | +$144,000 |
| Subtotal | $144,000 | = | $0 | + | $144,000 |

**b.** Give the journal entry for each of the six transactions.

**26. Journal entries for various transactions.** Present journal entries for each of the following transactions of Kirkland Corporation. You may omit dates and explanations for the journal entries.

**(1)** Issues 10,000 shares of $10 par value common stock in the acquisition of the assets of Ragsdale Corporation. The common shares given in the exchange have a market value of $250,000 on the New York Stock Exchange. The identifiable assets received and their market values are: inventories, $30,000; land, $15,000; buildings, $125,000; equipment, $60,000.
**(2)** Orders merchandise costing $36,000, paying $5,000 in advance to the supplier.
**(3)** Receives the merchandise ordered in **(2)** and an invoice for the $31,000 due.
**(4)** Inspects the merchandise and discovers that merchandise costing $3,500 is damaged and returns it to the supplier. The account has not yet been paid.
**(5)** Pays the amount due to the supplier in transactions **(2), (3),** and **(4)** after subtracting a 2 percent discount for prompt payment. The discount applies to the full amount of purchases

($36,000) net of returns ($3,500). Kirkland Corporation treats cash discounts taken as a reduction in the acquisition cost of merchandise.

**(6)** Receives an order from a customer for $8,000 of merchandise to be delivered next month. The customer sends an $800 check with the order as a deposit on the purchase.

**(7)** Acquires an automobile costing $18,000 by paying $2,000 in cash and signing a note payable for the balance. The note bears interest at 8 percent per year and is due in 12 months.

**(8)** Pays $24,000 for property and liability insurance for the one-year period beginning next month.

**27. Journal entries for various transactions.** Express the following transactions of Winkle Grocery Store, Inc., in journal entry form. If an entry is not required, indicate the reason. You may omit explanations for the journal entries. The store:

**(1)** Receives $30,000 from John Winkle in return for 1,000 shares of the firm's $30 par value common stock.

**(2)** Gives a 60-day, 8 percent note to a bank and receives $5,000 cash from the bank.

**(3)** Rents a building and pays the annual rental of $12,000 in advance.

**(4)** Acquires display equipment costing $8,000 and issues a check in full payment.

**(5)** Acquires merchandise inventory costing $25,000. The firm issues a check for $12,000, with the remainder payable in 30 days.

**(6)** Signs a contract with a nearby restaurant under which the restaurant agrees to purchase $2,000 of groceries each week. The firm receives a check for the first two weeks' orders in advance.

**(7)** Obtains a fire insurance policy providing $50,000 coverage beginning next month. The firm pays the one-year premium of $1,200.

**(8)** Pays $600 for advertisements that will appear in newspapers next month.

**(9)** Places an order with suppliers for $35,000 of merchandise to be delivered next month.

**28. Journal entries for various transactions.** The transactions below relate to Wendy's International, Inc., a restaurant chain. Express each transaction in journal entry form. If an entry is not required, indicate the reason. You may omit explanations for the journal entries.

**(1)** The firm places an order for restaurant supplies totaling $1,600,000 to be delivered next month.

**(2)** The firm receives the supplies ordered in transaction **(1)** on open account.

**(3)** The firm discovers supplies received in transaction **(2)** costing $40,000 are defective and returns them to the supplier for full credit.

**(4)** The firm pays for some of the purchases in transactions **(1)** and **(2)** originally invoiced for $1,400,000 in time to take advantage of 2 percent discounts for prompt payment. The firm treats discounts taken as a reduction in the cost of supplies.

**(5)** The firm pays for the remaining purchases—see transactions **(1)** through **(4)**.

**(6)** The firm places an order, totaling $900,000, for refrigerated salad bars for its restaurants. It pays $200,000 at the time of signing the contract and agrees to pay the remainder at delivery.

**(7)** The firm receives the salad bars ordered in transaction **(6)** and pays the amount due.

**(8)** The firm pays $27,000 to install the salad bars in its restaurants.

**(9)** The firm discovers that salad bars costing $22,000 are defective and returns them to the supplier, expecting a cash refund.

**(10)** The firm receives a check for $22,000 from the supplier in transaction **(9)**.

**29. Recording transactions and preparing a balance sheet.** Moulton Corporation engaged in the following seven transactions during December, Year 12 in preparation for opening the business on January 1, Year 13.

**(1)** Issued for cash 80,000 shares of $10 par value common stock at par.

**(2)** Acquired for cash land costing $50,000 and a building costing $450,000. The building has an expected useful life of 25 years beginning on January 1, Year 13.

**(3)** Purchased merchandise inventory costing $280,000 on account from various suppliers.

**(4)** Paid for inventory purchased in **(3)** with an original invoice price of $250,000 in time to take advantage of a 2 percent discount for prompt payment. The firm treats discounts taken as a reduction in the cost of inventories. The firm has not yet paid for the remaining $30,000 of purchases on account.

**(5)** Paid $12,000 for a one-year insurance policy on the land and building. The insurance coverage begins January 1.

(6) Borrowed $300,000 from a bank on December 31, Year 12. The loan bears interest at an annual rate of 8 percent and is due in five years. The interest is payable on January 1 of each year, beginning January 1, Year 14, and the $300,000 amount borrowed is due on December 31, Year 17.
(7) Acquired equipment on December 31 costing $80,000 and signed a 6 percent note payable to the supplier. The note is due on June 30, Year 13. The equipment has an estimated useful life of five years.

**a.** Depending on the preference of your instructor, record these seven transactions either in T-accounts or a transactions spreadsheet.
**b.** Prepare a balance sheet for Moulton Corporation as of December 31, Year 12.
*Note:* ***Problem 3.33*** *extends the problem to income transactions for Year 13.*

**30. Recording transactions and preparing a balance sheet.** Veronica Regaldo creates a new business firm in Mexico on January 2, Year 8, to operate a retail store. Transactions of Regaldo Department Stores during January in preparation for opening its first retail store in February appear below. (Ps means Mexican pesos.)
(1) January 2: Receives Ps500,000 from Veronica Regaldo for all of the common stock of Regaldo Department Stores. The stock has no par or stated value.
(2) January 5: Pays another firm Ps20,000 for a patent and pays the Mexican government Ps4,000 to register the patent.
(3) January 10: Orders merchandise from various suppliers at a cost of Ps200,000. See transactions **(5)**, **(6)**, and **(7)** below for later information regarding these merchandise orders.
(4) January 15: Signs a lease to rent land and a building for Ps30,000 a month. The rental period begins February 1. Regaldo pays Ps60,000 for the first two months' rent in advance.
(5) January 20: Receives the merchandise ordered on January 10. Regaldo delays payment for the merchandise until it receives an invoice from the supplier—see transaction **(7)** below.
(6) January 21: Discovers that merchandise costing Ps8,000 is defective and returns the items to the supplier.
(7) January 25: Receives invoices for Ps160,000 of the merchandise received on January 20. After subtracting an allowed discount of 2 percent of the invoice for paying promptly, Regaldo pays the suppliers the amount due of Ps156,800 (0.98 × Ps160,000). The firm treats cash discounts taken as a reduction in the acquisition cost of the merchandise.
(8) January 30: Obtains fire and liability insurance coverage from Windwards Islands Insurance Company for the period beginning February 1, Year 8. It pays the one-year insurance premium of Ps12,000.

**a.** Depending on the preference of your instructor, record these eight transactions either in T-accounts or a transactions spreadsheet.
**b.** Prepare a balance sheet for Regaldo Department Stores on January 31, Year 8.
*Note:* ***Problem 3.34*** *extends this problem to income transactions for February.*

**31. Recording transactions and preparing a balance sheet.** Patterson Corporation begins operations on January 1, Year 13. See the assumptions given at the end of the list. The firm engages in the following transactions during January:
(1) Issues 15,000 shares of $10 par value common stock for $210,000 in cash.
(2) Issues 28,000 shares of common stock in exchange for land, building, and equipment. The land appears at $80,000, the building at $220,000, and the equipment at $92,000 on the balance sheet.
(3) Issues 2,000 shares of common stock to another firm to acquire a patent.
(4) Acquires merchandise inventories with a list price of $75,000 on account from suppliers.
(5) Acquires equipment with a list price of $6,000. It deducts a $600 discount and pays the net amount in cash. The firm treats cash discounts as a reduction in the acquisition cost of equipment.
(6) Pays freight charges of $350 for delivery of the equipment in **(5)**.
(7) Discovers that merchandise inventories with a list price of $800 are defective and returns them to the supplier for full credit. The merchandise inventories had been purchased on account—see **(4)**—and no payment had been made as of the time that the goods are returned.
(8) Signs a contract for the rental of a fleet of automobiles beginning February 1. Pays the $1,400 rental for February in advance.

(9) Pays invoices for merchandise inventories purchased in **(4)** with an original list price of $60,000, after deducting a discount of 3 percent for prompt payment. The firm treats cash discounts as a reduction in the acquisition cost of merchandise inventories.

(10) Obtains fire and liability insurance coverage from Southwest Insurance Company. The two-year policy, beginning February 1, carries a $2,400 premium that has not yet been paid.

(11) Signs a contract with a customer for $20,000 of merchandise that Patterson plans to deliver in the future. The customer advances $4,500 toward the contract price.

(12) Acquires a warehouse costing $60,000 on January 31. The firm makes a down payment of $7,000 and assumes a 20-year, 6 percent mortgage for the balance. Interest is payable on January 31 each year.

(13) Discovers that merchandise inventories with an original list price of $1,500 are defective and returns them to the supplier. This inventory was paid for in **(9)**. The returned merchandise inventories are the only items purchased from this particular supplier during January. A cash refund has not yet been received from the supplier.

(14) On January 31, the firm purchases 6,000 shares of $10 par value common stock of the General Cereal Corporation for $95,000. This purchase is a short-term use of excess cash. The shares of General Cereal Corporation trade on the New York Stock Exchange.

The following assumptions will help you resolve certain accounting uncertainties:

- Transactions **(2)** and **(3)** occur on the same day as transaction **(1)**.
- The invoices paid in **(9)** are the only purchases for which suppliers made discounts available to the purchaser.

**a.** Depending on the preference of your instructor, enter these 14 transactions either in T-accounts or a transactions spreadsheet.

**b.** Prepare a balance sheet as of January 31, Year 13.

*Note:* ***Problem 3.35*** *extends this problem to income transactions for February.*

**32. Recording transactions and preparing a balance sheet.** Whitley Products Corporation begins operations on April 1. The firm engages in the following transactions during April:

(1) Issues 25,000 shares of $10 par value common stock for $15 per share in cash.

(2) Acquires land costing $25,000 and a building costing $275,000 by paying $50,000 in cash and signing a note payable to a local bank for the remainder of the purchase price. The note bears interest at 8 percent per year and matures in three years.

(3) Acquires equipment costing $125,000 for cash.

(4) Pays $2,800 to transport the equipment to the office of Whitley Products Corporation.

(5) Pays $3,200 to install and test the equipment.

(6) Pays the one-year premium of $12,000 for property and liability insurance on the building and equipment for coverage beginning May 1.

(7) Agrees to manufacture custom-order merchandise for a particular customer beginning in May at a selling price of $15,000. The customer advances $1,500 of the selling price with the order.

(8) Orders raw materials costing $60,000 from various suppliers.

(9) Receives notification from the suppliers that the raw materials ordered in transaction **(8)** were shipped. The merchandise belongs to the suppliers until received by Whitley Products Corporation.

(10) Receives the raw materials shipped in transaction **(9)**.

(11) Discovers that raw materials costing $8,000 are damaged, and returns them to the supplier. The firm has not yet paid the supplier.

(12) Pays the raw materials suppliers in transactions **(8)**, **(9)**, **(10)**, and **(11)** the amounts due, after subtracting 2 percent for prompt payment. The firm treats cash discounts as a reduction in the acquisition cost of the raw materials.

**a.** Depending on the preference of your instructor, enter these 12 transactions either in T-accounts or a transactions spreadsheet.

**b.** Prepare a balance sheet for Whitley Products Corporation as of April 30.

**33. Effect of recording errors on the balance sheet equation.** Using the notation O/S (overstated), U/S (understated), or No (no effect), indicate the effects on assets, liabilities, and shareholders' equity of *failing to record or recording incorrectly* each of the following transactions or events. For example, a failure to record the issuance of common stock for $10,000 cash would be shown as follows:

- Assets—U/S $10,000.
- Liabilities—No.
- Shareholders' equity—U/S $10,000.

**(1)** A firm ordered $23,000 of merchandise from a supplier but did not record anything in its accounts.

**(2)** The firm received the merchandise in transaction **(1)** and recorded it by debiting Merchandise Inventory and crediting Accounts Payable for $32,000.

**(3)** The firm acquired an automobile costing $20,000 by paying $2,000 in cash and signing a note payable for the remainder of the purchase price. It recorded the acquisition by debiting Automobile for $20,000, crediting Cash for $18,000, and crediting Note Payable for $2,000.

**(4)** The firm paid the $1,800 annual insurance premium on the automobile in transaction **(3)** by debiting Automobile and crediting Cash for $1,800. The insurance period begins next month.

**(5)** The firm received an order from a customer for $5,500 of merchandise that the firm will deliver next month. The customer included a check for $1,500. The firm made no entry for this transaction.

**(6)** The firm issued 2,000 shares of its $10 par value common stock having a market value of $32,000 in exchange for land. It recorded the transaction by debiting Land and crediting Common Stock for $20,000.

**(7)** The firm signed a three-year employment agreement with its chief executive officer at an annual salary of $275,000. The employment period begins next month. The firm did not record anything in its accounts related to this agreement.

**34. Effect of recording errors on the balance sheet equation.** A firm recorded various transactions with the journal entries shown below. Using the notation O/S (overstated), U/S (understated), or No (no effect), indicate the effects on assets, liabilities, and shareholders' equity of any errors in recording each of these transactions. For example, if a firm recorded the issue of $10,000 of common stock by debiting Cash and crediting Bonds Payable, the effects of the error are shown as follows:

- Assets—No.
- Liabilities—O/S $10,000.
- Shareholders' equity—U/S $10,000.

| | | |
|---|---|---|
| **(1)** Equipment . . . . . . . . . . . . . . . . . . . . . . . . . . . . . . . . . . . . . . . . . | 10,000 | |
| Cash . . . . . . . . . . . . . . . . . . . . . . . . . . . . . . . . . . . . . . . . . . . | | 2,000 |
| Note Receivable . . . . . . . . . . . . . . . . . . . . . . . . . . . . . . . . . . | | 8,000 |

| Assets | = | Liabilities | + | Shareholders' Equity | (Class.) |
|---|---|---|---|---|---|
| +10,000 | | | | | |
| - 2,000 | | | | | |
| - 8,000 | | | | | |

To record acquisition of equipment using $2,000 cash and signing of an $8,000 promissory note for the balance.

| | | |
|---|---|---|
| **(2)** Equipment . . . . . . . . . . . . . . . . . . . . . . . . . . . . . . . . . . . . . . . . . | 4,000 | |
| Cash . . . . . . . . . . . . . . . . . . . . . . . . . . . . . . . . . . . . . . . . . . . | | 1,000 |
| Note Payable . . . . . . . . . . . . . . . . . . . . . . . . . . . . . . . . . . . . | | 3,000 |

| Assets | = | Liabilities | + | Shareholders' Equity | (Class.) |
|---|---|---|---|---|---|
| +4,000 | | +3,000 | | | |
| −1,000 | | | | | |

To record the placing of an order for equipment to be delivered next month. The firm made a $1,000 deposit with the order.

| | | |
|---|---|---|
| **(3)** Cash | 800 | |
| Accounts Receivable | | 800 |

| Assets | = | Liabilities | + | Shareholders' Equity | (Class.) |
|---|---|---|---|---|---|
| +800 | | | | | |
| −800 | | | | | |

To record an advance from a customer on merchandise to be shipped next month. The customer did not owe the firm any amounts at the time of this transaction.

| | | |
|---|---|---|
| **(4)** Prepaid Rent | 1,000 | |
| Rent Payable | | 1,000 |

| Assets | = | Liabilities | + | Shareholders' Equity | (Class.) |
|---|---|---|---|---|---|
| +1,000 | | +1,000 | | | |

To record the signing of a rental agreement for warehouse space for a one-year period beginning next month. The monthly rental fee of $1,000 is due on the first day of each month.

| | | |
|---|---|---|
| **(5)** Patent | 2,500 | |
| Cash | | 2,500 |

| Assets | = | Liabilities | + | Shareholders' Equity | (Class.) |
|---|---|---|---|---|---|
| +2,500 | | | | | |
| −2,500 | | | | | |

To record the issuance of common stock in the acquisition of a patent.

| | | |
|---|---|---|
| **(6)** Merchandise Inventories | 4,900 | |
| Cash | | 4,900 |

| Assets | = | Liabilities | + | Shareholders' Equity | (Class.) |
|---|---|---|---|---|---|
| +4,900 | | | | | |
| −4,900 | | | | | |

To record the acquisition of office equipment for cash.

35. **Balance sheet format, terminology, and accounting methods. Exhibit 2.14** presents a balance sheet for Marks and Spencer, PLC, a department store chain headquartered in the United Kingdom. This balance sheet appears in the format and uses terminology and accounting methods commonly found in the United Kingdom.
    **a.** Prepare a balance sheet for Marks and Spencer on March 30, Year 4, and March 30, Year 5, following the format, terminology, and accounting methods commonly found in the United States. Use the category Other Noncurrent Assets for assets that do not fit any typical U.S. grouping.
    **b.** Comment on the mix of Marks and Spencer's assets relative to its financing.
36. **Balance sheet format, terminology, and accounting methods. Exhibit 2.15** presents a balance sheet for United Breweries Group, a Danish brewing company. This balance sheet

**EXHIBIT 2.14**

**MARKS AND SPENCER**
**Balance Sheet**
**(in millions of pounds)**
**(Problem 35)**

| | March 30 | |
|---|---|---|
| | **Year 4** | **Year 5** |
| **Fixed Assets** | | |
| Land and Buildings (Note 3) | £2,094 | £2,193 |
| Fixtures, Fittings, and Equipment | 375 | 419 |
| Total Fixed Assets | £2,469 | £2,612 |
| **Current Assets** | | |
| Stocks | £ 374 | £ 351 |
| Debtors (Note 1) | 538 | 618 |
| Investments | 28 | 29 |
| Cash at Bank and in Hand | 266 | 293 |
| Total Current Assets | £1,206 | £1,291 |
| **Current Liabilities** | | |
| Creditor Amounts Falling Due within One Year (Note 2) | (925) | (897) |
| Net Current Assets | £ 281 | £ 394 |
| Total Assets Less Current Liabilities | £2,750 | £3,006 |
| Creditor Amounts Falling Due after One Year | (565) | (550) |
| Other Noncurrent Liabilities and Provisions | (4) | (19) |
| Net Assets | £2,181 | £2,437 |
| **Capital and Reserve** | | |
| Called-Up Share Capital | £ 675 | £ 680 |
| Share Premium Account | 50 | 69 |
| Revaluation Reserve (Note 3) | 458 | 460 |
| Profit and Loss Account | 998 | 1,228 |
| Total Capital Employed | £2,181 | £2,437 |

**Note 1:** Debtors include the following:

| | Year 4 | Year 5 |
|---|---|---|
| Amounts Falling Due within One Year: | | |
| Accounts Receivable | £ 192 | £ 212 |
| Prepayments | 134 | 142 |
| Amounts Falling Due after One Year: | | |
| Accounts Receivable | 174 | 217 |
| Prepayments | 38 | 47 |
| | £ 538 | £ 618 |

**Note 2:** Creditor amounts falling due within one year include the following:

| | Year 4 | Year 5 |
|---|---|---|
| Accounts Payable | £ 187 | £ 168 |
| Bank Loans | 100 | 107 |
| Other | 638 | 622 |
| | £ 925 | £ 897 |

**Note 3:** The revaluation reserve arises from the restatement of land and buildings to current market value. None of the restatement affected net income.

**EXHIBIT 2.15**

**UNITED BREWERIES**
**Balance Sheet**
**(in millions of kronor)**
**(Problem 36)**

| | December 31 | |
|---|---|---|
| | **Year 8** | **Year 9** |
| ASSETS | | |
| **Fixed Assets** | | |
| Tangible Fixed Assets | Kr 3,934 | Kr 4,106 |
| Investments in Securities | 422 | 573 |
| Total Fixed Assets | Kr 4,356 | Kr 4,679 |
| **Current Assets** | | |
| Stocks | Kr 1,290 | Kr 1,393 |
| Trade Debtors | 1,413 | 1,444 |
| Prepayments | 317 | 285 |
| Securities | 3,018 | 3,460 |
| Cash at Bank and in Hand | 810 | 1,224 |
| Total Current Assets | Kr 6,848 | Kr 7,806 |
| Total Assets | Kr 11,204 | Kr 12,485 |
| CAPITAL AND LIABILITIES | | |
| **Capital and Reserves** | | |
| Share Capital | Kr 976 | Kr 976 |
| Revaluation Reserves (Note 1) | 416 | 561 |
| Other Reserves (Note 2) | 3,169 | 3,672 |
| Total Capital and Reserves | Kr 4,561 | Kr 5,209 |
| Provisions (Note 3) | Kr 1,149 | Kr 1,425 |
| Long-Term Debt | Kr 1,805 | Kr 1,723 |
| Short-Term Debt | | |
| Bank Debt | Kr 619 | Kr 986 |
| Trade Creditors | 913 | 902 |
| Other | 2,157 | 2,240 |
| Total Short-Term Debt | Kr 3,689 | Kr 4,128 |
| Total Capital and Liabilities | Kr 11,204 | Kr 12,485 |

**Note 1:** The Revaluation Reserves result from restating tangible fixed assets to current market values.

**Note 2:** Other Reserves represent cumulative net income in excess of dividends.

**Note 3:** Provisions represent noncurrent liabilities for income taxes and miscellaneous obligations.

appears in the format and uses the terminology and accounting methods commonly found in Denmark.

**a.** Prepare a balance sheet for United Breweries Group on December 31, Year 8, and December 31, Year 9, following the format, terminology, and accounting methods commonly found in the United States.

**b.** Comment on the mix of United Breweries Group's assets relative to its financing.

**37. Interpreting balance sheet changes. Exhibit 2.16** presents a common-size balance sheet for Outback Steakhouse, a restaurant chain, for two recent years.

**a.** Identify the ways in which the structure of Outback Steakhouse's assets and the structure of its financing correspond to what one would expect of a restaurant chain. What aspects of the structure of its assets and the structure of its financing are not what one would expect?

**EXHIBIT 2.16** **OUTBACK STEAKHOUSE**
**Common-Size Balance Sheet**
**(Problem 37)**

| | December 31 | |
|---|---|---|
| | **Year 11** | **Year 12** |
| ASSETS | | |
| Cash | 7.6% | 12.3% |
| Accounts Receivable | 0.9 | 0.9 |
| Inventories | 3.3 | 2.6 |
| Other Current Assets | 1.7 | 1.5 |
| Total Current Assets | 13.5% | 17.3% |
| Property, Plant, and Equipment | 68.1 | 67.9 |
| Other Assets | 18.4 | 14.8 |
| Total Assets | 100.0% | 100.0% |
| LIABILITIES AND SHAREHOLDERS' EQUITY | | |
| **Current Liabilities** | | |
| Accounts Payable | 4.0% | 4.0% |
| Notes Payable | 1.1 | 1.3 |
| Other Current Liabilities | 10.8 | 12.4 |
| Total Current Liabilities | 15.9% | 17.7% |
| **Noncurrent Liabilities** | | |
| Bonds Payable | 1.2% | 1.1% |
| Other Noncurrent Liabilities | 3.9 | 3.1 |
| Total Noncurrent Liabilities | 5.1% | 4.2% |
| Total Liabilities | 21.0% | 21.9% |
| **Shareholders' Equity** | | |
| Common Stock | 15.1% | 11.1% |
| Retained Earnings | 63.9 | 67.0 |
| Total Shareholders' Equity | 79.0% | 78.1% |
| Total Liabilities and Shareholders' Equity | 100.0% | 100.0% |

**b.** Identify the major changes in the nature and mix of assets and the nature and mix of financing between the two year-ends and suggest possible reasons for the changes.

**c.** "An increase in the common-size balance sheet percentage between two year-ends for a particular balance sheet item (for example, cash) does not necessarily mean that its dollar amount increased." Explain.

**38. Interpreting balance sheet changes. Exhibit 2.17** presents a common-size balance sheet for Pacific Gas and Electric (PG&E), an electric and gas utility, for two recent years.

**a.** In what ways is the common-size balance sheet of PG&E on December 31, Year 9, typical of an electric and gas utility?

**b.** Identify the major changes in the nature and mix of assets and the nature and mix of financing between the two year-ends and suggest possible reasons for the changes.

**39. Relating market value to book value of shareholders' equity.** Firms prepare their balance sheets using GAAP for the recognition and valuation of assets and liabilities. Accountants refer to the total common shareholders' equity appearing on the balance sheet as the *book value of shareholders' equity*. The *market value of shareholders' equity* equals the number of shares of common stock outstanding times the market price per share. Financial analysts frequently examine the ratio of the market value of shareholders' equity to the book value of shareholders' equity, referred to as the *market-to-book-value ratio,* in assessing current market prices. Theoretical and empirical research suggests that the size of the market-to-book-value

**EXHIBIT 2.17**

**PACIFIC GAS AND ELECTRIC COMPANY**
**Common-Size Balance Sheet**
**(Problem 38)**

| | December 31 | |
|---|---|---|
| | **Year 9** | **Year 10** |
| ASSETS | | |
| **Current Assets** | | |
| Cash | 1.6% | 7.2% |
| Accounts Receivable | 5.0 | 9.6 |
| Inventories | 1.5 | 1.1 |
| Prepayments | 4.0 | 13.7 |
| Total Current Assets | 12.1% | 31.6% |
| Property, Plant, and Equipment | 67.0 | 57.1 |
| Other Assets | 20.9 | 11.3 |
| Total Assets | 100.0% | 100.0% |
| LIABILITIES AND SHAREHOLDERS' EQUITY | | |
| **Current Liabilities** | | |
| Accounts Payable | 2.4% | 10.7% |
| Note Payable | 8.0 | 20.4 |
| Other Current Liabilities | 10.2 | 17.9 |
| Total Current Liabilities | 20.6% | 49.0% |
| **Noncurrent Liabilities** | | |
| Bonds Payable | 49.6% | 38.3% |
| Other Noncurrent Liabilities | 3.8 | 1.5 |
| Total Noncurrent Liabilities | 53.4% | 39.8% |
| Total Liabilities | 74.0% | 88.8% |
| **Shareholders' Equity** | | |
| Preferred Stock | 1.6% | 1.4% |
| Common Stock | 17.7 | 15.0 |
| Retained Earnings | 6.7 | (5.2) |
| Total Shareholders' Equity | 26.0% | 11.2% |
| Total Liabilities and Shareholders' Equity | 100.0% | 100.0% |

ratio is related to (1) a firm's ability to generate higher rates of profitability than its competitors, (2) its rate of growth, and (3) its use of GAAP in measuring assets and liabilities, which net to the book value of shareholders' equity.

**Exhibit 2.18** presents balance sheet information for five firms at the end of a recent year. It also shows their market-to-book-value ratios. Additional information regarding the five companies follows:

**(1)** *Coca-Cola (Coke):* Coke markets soft drinks worldwide. It has grown primarily by internal expansion rather than by acquiring other soft-drink firms. Coke maintains less than 50 percent ownership in a large number of its bottlers.

**(2)** *Bristol-Myers Squibb (Bristol):* Bristol generates approximately 75 percent of its revenues from prescription drugs and medical devices and 25 percent from nonprescription health products, toiletries, and beauty aids.

**(3)** *Bankers Trust (Bankers):* Bankers obtains funds primarily from depositors (reported on the Accounts Payable line in **Exhibit 2.18**) and invests them in short-term liquid assets or lends them to businesses and consumers. It also engages in investment activities on its own account.

**(4)** *International Paper (IP):* IP has the largest holdings of forest lands of any nongovernmental entity in the United States. It processes timber into wood products for the con-

**EXHIBIT 2.18** **Balance Sheets for Selected Companies (all dollar amounts in millions) (Problem 39)**

| | Coca-Cola | Bristol-Myers Squibb | Bankers Trust | International Paper | Walt Disney |
|---|---|---|---|---|---|
| ASSETS | | | | | |
| Cash and Marketable Securities | $ 1,531 | $ 2,423 | $79,048 | $ 270 | $ 1,510 |
| Accounts and Notes Receivable | 1,525 | 2,043 | 11,249 | 2,241 | 1,671 |
| Inventories | 1,047 | 1,397 | — | 2,075 | 2,264 |
| Other Current Assets | 1,102 | 847 | — | 244 | — |
| Total Current Assets | $ 5,205 | $ 6,710 | $90,297 | $ 4,830 | $ 5,445 |
| Investments in Securities | 3,928 | — | — | 1,032 | 630 |
| Property, Plant, and Equipment | 4,080 | 3,666 | 915 | 9,941 | 5,814 |
| Other Noncurrent Assets | 660 | 2,534 | 5,804 | 2,033 | 937 |
| Total Assets | $13,873 | $12,910 | $97,016 | $17,836 | $12,826 |
| LIABILITIES AND SHAREHOLDERS' EQUITY | | | | | |
| Accounts Payable | $ 2,564 | $ 693 | $24,939 | $ 1,204 | $ 2,475 |
| Short-Term Borrowing | 2,083 | 725 | 55,166 | 2,083 | — |
| Other Current Liabilities | 1,530 | 2,856 | 5,502 | 747 | 967 |
| Total Current Liabilities | $ 6,177 | $ 4,274 | $85,607 | $ 4,034 | $ 3,442 |
| Long-Term Debt | 1,426 | 644 | 6,455 | 4,464 | 2,937 |
| Other Noncurrent Liabilities | 1,035 | 2,288 | — | 2,824 | 939 |
| Total Liabilities | $ 8,638 | $ 7,206 | $92,062 | $11,322 | $ 7,318 |
| Preferred Stock | — | — | $ 645 | — | — |
| Common Stock | $ 1,600 | $ 451 | 1,401 | $ 1,914 | $ 945 |
| Retained Earnings | 10,708 | 7,299 | 3,324 | 4,711 | 5,849 |
| Treasury Stock | (7,073) | (2,046) | (416) | (111) | (1,286) |
| Total Shareholders' Equity | $ 5,235 | $ 5,704 | $ 4,954 | $ 6,514 | $ 5,508 |
| Total Liabilities and Shareholders' Equity | $13,873 | $12,910 | $97,016 | $17,836 | $12,826 |
| Market Value/Book Value Ratio | 12.6 | 5.2 | 1.0 | 1.3 | 4.4 |

struction industry and processes pulp from the timber into various types of commodity and specialty papers.

(5) *Walt Disney (Disney):* Disney produces motion picture films and operates theme parks.

**a.** The market-to-book-value ratios differ from 1.0 in part because the rates of profitability and growth of these five firms differ from those of their competitors. This problem does not provide you with sufficient information to assess the impact of these two factors on the market-to-book-value ratio. The ratios also differ from 1.0 because of the use of GAAP for assets and liabilities, which this chapter discussed. Identify the GAAP that most likely explains the market-to-book-value ratios for each of the five firms (that is, identify which accounting principles cause the book values of assets and liabilities to differ from the market value of shareholders' equity).

**b.** Discuss the likely rationale for the nature and mix of assets and the nature and mix of financing for each of the five firms.

**40. Relating market value to book value of shareholders' equity.** Firms prepare their balance sheets using GAAP for the recognition and valuation of assets and liabilities. Accountants refer to the total common shareholders' equity appearing on the balance sheet as the *book value of shareholders' equity*. The *market value of shareholders' equity* equals the number of shares of common stock outstanding times the market price per share. Financial analysts frequently examine the ratio of the market value of shareholders' equity to the book value of

**EXHIBIT 2.19** **Common-Size Balance Sheets for Selected Companies (Problem 40)**

| | Pfizer | Nestlé | Promodes | Deutsche Bank | British Airways | New Oji Paper Co. |
|---|---|---|---|---|---|---|
| ASSETS | | | | | | |
| Cash and Marketable Securities | 20.5% | 11.5% | 12.9% | 44.2% | 6.5% | 2.1% |
| Accounts and Notes Receivable | 16.4 | 18.9 | 22.6 | 50.4 | 13.5 | 19.8 |
| Inventories | 8.1 | 13.4 | 18.4 | — | 0.7 | 8.5 |
| Other Current Assets | 6.3 | 1.4 | — | — | — | 1.5 |
| Total Current Assets | 51.3% | 45.2% | 53.9% | 94.6% | 20.7% | 31.9% |
| Investments in Securities | 7.5 | 7.1 | 8.7 | 2.7 | 6.6 | 24.5 |
| Property, Plant, and Equipment | 28.2 | 43.7 | 28.3 | 1.2 | 72.7 | 43.6 |
| Other Noncurrent Assets | 13.0 | 4.0 | 9.1 | 1.5 | — | — |
| Total Assets | 100.0% | 100.0% | 100.0% | 100.0% | 100.0% | 100.0% |
| LIABILITIES AND SHAREHOLDERS' EQUITY | | | | | | |
| Accounts Payable | 5.1% | 11.2% | 44.3% | 66.5% | 9.4% | 10.2% |
| Short-Term Borrowing | 12.8 | 18.1 | 3.3 | 16.0 | 5.3 | 16.2 |
| Other Current Liabilities | 17.8 | 9.6 | 12.6 | 4.9 | 15.6 | 10.7 |
| Total Current Liabilities | 35.7% | 38.9% | 60.2% | 87.4% | 30.3% | 37.1% |
| Long-Term Debt | 3.4 | 6.7 | 7.7 | 1.2 | 36.8 | 16.9 |
| Other Noncurrent Liabilities | 12.9 | 9.2 | 4.1 | 8.0 | 4.0 | 4.8 |
| Total Liabilities | 52.0% | 54.8% | 72.0% | 96.6% | 71.1% | 58.8% |
| Common Stock | 27.5% | 3.0% | 0.8% | 1.7% | 8.1% | 17.3% |
| Retained Earnings | 43.9 | 42.7 | 27.2 | 1.7 | 20.8 | 23.9 |
| Treasury Stock | (23.4) | (0.5) | — | — | — | — |
| Total Shareholders' Equity | 48.0% | 45.2% | 28.0% | 3.4% | 28.9% | 41.2% |
| Total Liabilities and Shareholders' Equity | 100.0% | 100.0% | 100.0% | 100.0% | 100.0% | 100.0% |
| Market Value/Book Value Ratio | 8.2 | 3.3 | 4.6 | 1.7 | 2.4 | 1.4 |

shareholders' equity, referred to as the *market-to-book-value ratio,* in assessing current market prices. Theoretical and empirical research suggests that the size of the *market-to-book-value ratio* is related to (1) a firm's ability to generate higher rates of profitability than its competitors, (2) its rate of growth, and (3) its use of GAAP in measuring assets and liabilities, which net to the book value of shareholders' equity.

**Exhibit 2.19** presents *common-size* balance sheet information for six firms at the end of a recent year. It also shows their market-to-book-value ratios. Additional information regarding the six companies follows:

**(1)** Pfizer is a pharmaceutical company headquartered in the United States.

**(2)** Nestlé is a consumer products company headquartered in Switzerland. In addition to its chocolate products, it manufactures and distributes beverages (Nestea, Poulin Springs mineral water), frozen foods (Stouffer's), milk products (infant formulas), and pet foods (Alpo).

**(3)** Promodes is a French company that operates chains of supermarkets (Champion), hypermarkets (Continent, Continente), convenience stores (Promocash, Punt&Cash), and restaurant supply stores (Prodirest).

**(4)** Deutsche Bank is a German commercial bank that provides both traditional commercial banking services (deposit taking, loan making) and investment banking services (investment management, financial consulting).

(5) British Airways is headquartered in the United Kingdom and provides air transportation services.
(6) New Oji Paper Co. is a Japanese forest-products company. It purchases wood pulp from Canada and the United States and processes it into various papers for sale in Japan.

a. The market-to-book-value ratios differ from 1.0 in part because the rates of profitability and growth of these six firms differ from those of their competitors. This problem does not provide you with sufficient information to assess the impact of these two factors on the market-to-book-value ratio. The ratios also differ from 1.0 because of the use of GAAP for assets and liabilities, which this chapter discussed. Identify the GAAP that most likely explains the market-to-book-value ratios for each of the six firms (that is, identify which accounting principles cause the book values of assets and liabilities to differ from the market value of shareholders' equity).

b. Discuss the likely rationale for the nature and mix of assets and the nature and mix of financing for each of the six firms.

41. **Identifying industries using common-size balance sheet percentages. Exhibit 2.20** presents common-size balance sheets for five firms. The firms and a description of their activities follow:
(1) Commonwealth Edison: Generates and sells electricity to businesses and households.
(2) Hewlett-Packard: Develops, assembles, and sells computer hardware and printers. The firm outsources many of its computer and printer components.
(3) Household International: Lends money to consumers for periods ranging from several months to several years.
(4) May Department Stores: Operates department store chains and offers its own credit card.
(5) Newmont Mining: Mines for gold and other minerals, utilizing heavy equipment.
Use whatever clues that you can to match the companies listed above with the firms listed in **Exhibit 2.20**. Describe your reasoning.

42. **Ethical issues in asset revaluations.** GAAP in the United Kingdom permits firms to report property, plant, and equipment on the balance sheet either at acquisition cost (net of depreciation) or current replacement cost. Firms choosing to revalue property, plant, and equipment must revalue all assets of the same class, defined as those with a similar nature, function, or use. Firms must revalue such assets periodically, but not necessarily every year.

Unilever PLC is a consumer foods company headquartered in the United Kingdom. Its balance sheet for Year 12 reports total assets of £43,318 million, total liabilities of £39,689 million, and total shareholders' equity of £3,629 million. Total assets on this date include property, plant, and equipment of £9,240 million, which Unilever reports at acquisition cost

**EXHIBIT 2.20** **Common-Size Balance Sheets for Five Companies (Problem 41)**

| | (1) | (2) | (3) | (4) | (5) |
|---|---|---|---|---|---|
| Cash and Marketable Securities | 8.5% | 0.1% | 9.5% | 12.1% | 1.1% |
| Receivables | 81.9 | 2.8 | 1.4 | 25.7 | 24.1 |
| Inventories | — | 2.0 | 9.0 | 23.2 | 23.6 |
| Property, Plant, and Equipment (net) | 1.2 | 74.3 | 62.6 | 20.0 | 41.4 |
| Other Assets | 8.4 | 20.8 | 17.5 | 19.0 | 9.8 |
| Total Assets | 100.0% | 100.0% | 100.0% | 100.0% | 100.0% |
| Current Liabilities | 38.8% | 8.4% | 10.8% | 38.3% | 29.1% |
| Long-Term Debt | 50.0 | 47.7 | 28.1 | 9.3 | 28.3 |
| Other Noncurrent Liabilities | — | 13.9 | 6.7 | 3.9 | 6.2 |
| Shareholders' Equity | 11.2 | 30.0 | 54.4 | 48.5 | 36.4 |
| Total Liabilities and Shareholders' Equity | 100.0% | 100.0% | 100.0% | 100.0% | 100.0% |

net of depreciation. The firm discloses in notes to the financial statements that the current replacement cost of this property, plant, and equipment is £10,529 million. Unilever has reported the acquisition cost and current replacement cost of its property, plant, and equipment in notes to its financial statements for many years.

**a.** Lenders to a firm often calculate the ratio of total liabilities to total assets to assess the risk of bankruptcy. Calculate this ratio for Unilever using the amounts reported on its balance sheet.

**b.** Compute this ratio assuming that Unilever chose to revalue its property, plant, and equipment to current replacement cost.

**c.** Assume that Unilever has a restriction in its borrowing agreements that stipulates that major borrowings become immediately due if the total liabilities to total assets ratio exceeds 90 percent at any time. Before issuing financial statements for Year 12 with the amounts shown above, Unilever contemplates changing the accounting for its property, plant, and equipment from acquisition cost (net of depreciation) to current replacement cost. Identify any ethical issues that you think Unilever should consider in this decision. What advice would you give to the firm?

CHAPTER 3

# Income Statement: Reporting the Results of Operating Activities

## Learning Objectives

1. **Understand the accrual basis of accounting and why accountants believe that the accrual basis provides measures of operating performance usually superior to those provided by the cash basis of accounting.**

2. **Deepen your understanding of the accrual basis of accounting by examining when firms recognize revenues and expenses and how they measure these two components of net income.**

3. **Build on your skills in recording transactions and other events in the accounts by extending the dual-entry recording framework studied in Chapter 2 to include income statement items.**

4. **Develop skills to analyze the income statement, focusing on the relations between revenues, expenses, and net income for various types of businesses.**

*U*sers of financial statements often look at the relation between net income and revenues, referred to as the **profit margin percentage,** when evaluating the profitability of a firm. Consider the following firms and their profit margin percentages for a recent year:

- AK Steel (steel manufacturing): 1.4 percent.
- Wal-Mart Stores (discount retailing): 3.3 percent.
- Kellogg (branded food processing): 8.7 percent.
- Omnicom Group (marketing services): 9.4 percent.
- Pfizer (pharmaceuticals): 27.5 percent.

Did AK Steel and Wal-Mart Stores perform more poorly than, and did Pfizer perform better than, Kellogg and Omnicom Group? Do these different profit margins signal differences in operating performance and do they have economic or strategic explanations? Answering these questions requires a deeper understanding of the income statement than the overview that **Chapter 1** provides. Recall that the income statement reports **net income,** or earnings, for a particular period of time. Net income equals revenues minus expenses. **Revenues** measure the net assets (assets less liabilities) that flow into a firm when it sells goods or renders services. **Expenses** measure the net assets that a firm consumes in the process of generating revenues. As a measure of operating performance, revenues reflect the services rendered by a firm, and expenses indicate the efforts required.

This chapter considers the measurement principles and accounting procedures that underlie the income statement and the insights it provides about the operating performance of a firm. We begin by considering the concept of an accounting period, the span of time over which accountants measure operating performance. Next, the chapter describes and illustrates two common approaches to measuring operating performance, the cash basis and the accrual basis. It then illustrates the accounting procedures used in applying the accrual basis of accounting for a merchandising firm. Finally, the chapter discusses relations between revenues, expenses, and net income for different types of businesses.

## The Accounting Period Convention

The income statement reports operating performance for a specific time period. Years ago, the length of this period varied among firms. Firms prepared income statements at the completion of some activity, such as after the completion of a construction project.

The operating activities of most modern firms do not divide so easily into distinguishable projects. Instead, the income-generating activity occurs continually. For example, a firm acquires a plant and uses it in manufacturing products for a period of forty years or more. A firm purchases delivery equipment and uses it in transporting merchandise to customers for four, or five, or more years. If preparing the income statement awaited the completion of all operating activities, the firm might prepare the report only when it ceased to exist and the report would be too late to help a reader appraise operating performance. An **accounting period** of uniform length facilitates timely comparisons and analyses among firms. One might think of an accounting period as the time elapsing between balance sheet dates. Balance sheets prepared at the end of the day on December 31 of one year and at the end of the day on December 31 of the next year bound a calendar-year accounting period. Balance sheets prepared at the end of the day on November 30 and at the end of the day on December 31 bound a one-month accounting period—the month of December.

An accounting period of one year underlies the principal financial statements distributed to shareholders and potential investors. Most firms prepare their annual reports using the calendar year as the accounting period. Many firms, however, use a **natural business year** or **fiscal period** to attempt to measure performance at a time when they have concluded most operating activities for the period. The ending date of a natural business year varies from one firm to another and usually occurs when inventories are at their lowest level during the year. For example, JCPenney, a retailer, uses a natural business year ending near the end of January, which comes after the Christmas shopping season and before the start of the Easter season. Winnebago Industries, a manufacturer of recreational vehicles, uses a year ending in late August, the end of its model year.

Firms frequently prepare reports of performance for periods shorter than a year to indicate progress during the year. These are known as *interim reports* or *reports for interim periods*. Preparing interim reports does not eliminate the need to prepare an annual report.

## Accounting Methods for Measuring Performance

**Figure 3.1** depicts the operating process for the acquisition and sale of merchandise. A retailing firm, such as Wal-Mart, purchases merchandise and then holds the merchandise in inventory until a customer purchases it. At the time of sale, the inventory changes hands. The firm need not, however, receive cash right away. Thus, a second period may commence before the firm collects cash from sales made during the first period.

Some operating activities both start and finish within a given accounting period. For example, operating process B in **Figure 3.1** occurs entirely in Year 2. These cases present few difficulties in measuring operating performance. The difference between the cash received from customers and the cash disbursed to acquire, hold, sell, and deliver the merchandise represents earnings from this series of transactions.

Many operating activities, however, start in one accounting period and finish in another. Operating process A in **Figure 3.1** begins during Year 1 and finishes in Year 2. The firm pays for the merchandise and other operating costs in Year 1 but receives cash from customers in Year 2. Operating process C begins in Year 2 and finishes in Year 3. The firm pays out cash in Year 2 to purchase, hold, and sell the merchandise but receives cash from customers in Year 3. To measure performance for a specific accounting period, such as Year 2 in **Figure 3.1,** requires measuring the amount of revenues and expenses from operating activities already begun by the beginning of the period and other activities not yet complete at the end of the period.

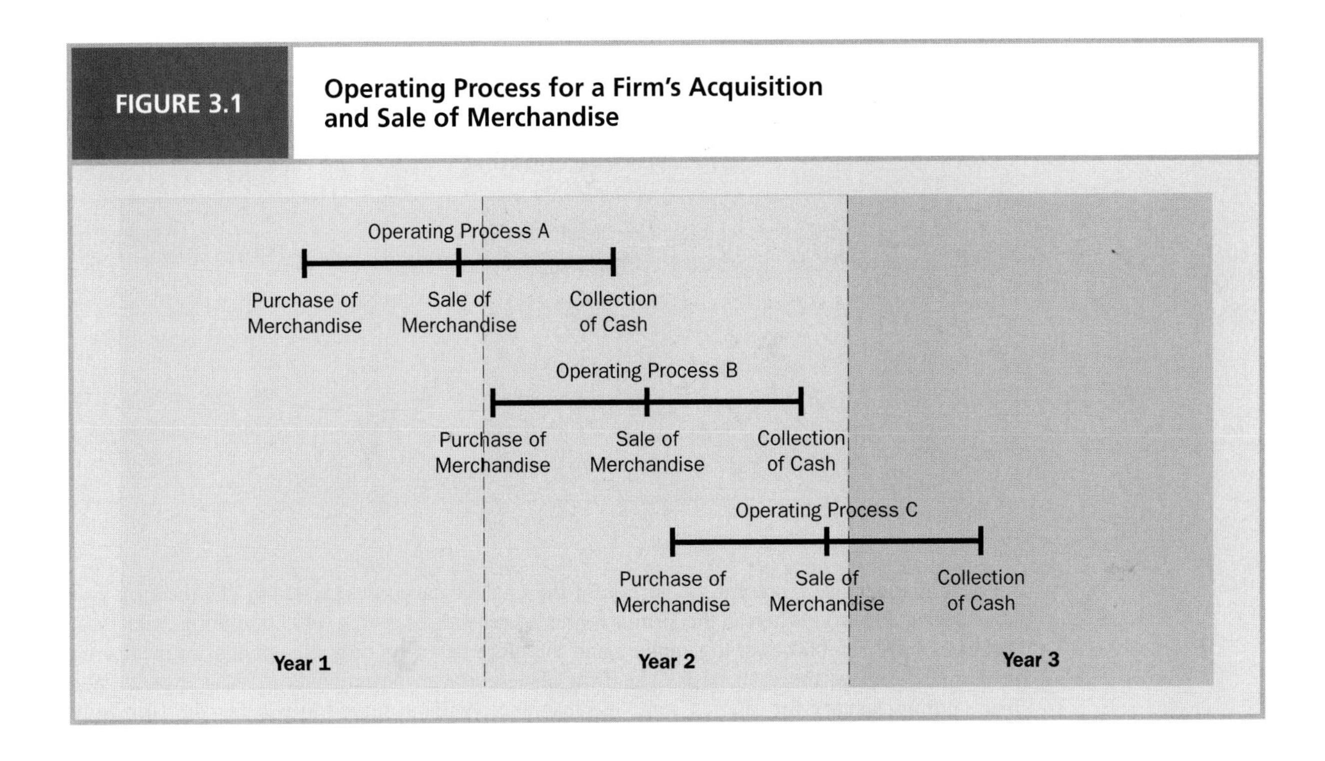


**Figure 3.1** oversimplifies the problem in measuring operating performance for discrete accounting periods. For example, a firm may acquire a building in one period and use it for thirty or more years. Thus, the length of the operating process line may extend over many periods. Firms may purchase merchandise in one accounting period, pay for it in a second, sell it during a third, and collect cash from customers in a fourth. Moreover, the events need not occur in that order. Cash payments can precede the acquisition of merchandise, such as for the acquisition of buildings to store and display the merchandise, or follow it, such as payments to sell and deliver the merchandise and provide warranty services. Cash collection can precede the sale of merchandise, as occurs when customers make advance payments, or follow it, as occurs with sales on account. To measure performance for a specific accounting period requires measuring the amount of revenues and expenses from operating activities that span more than that one accounting period. Two approaches to measuring operating performance are (1) the cash basis of accounting and (2) the accrual basis of accounting.

## Cash Basis of Accounting

Under the **cash basis of accounting,** a firm measures performance from selling goods and providing services as it receives cash from customers and makes cash expenditures to providers of goods and services. To understand the measurement of performance under the cash basis of accounting, consider the following information.

Donald and Joanne Allens open a hardware store on January 1, Year 1. The firm receives $20,000 in cash from the Allens in exchange for all of the common stock of the firm and borrows $12,000 from a local bank. It must repay the loan on June 30, Year 1, with interest charged at the rate of 8 percent per year. The firm rents a store building on January 1 and pays two months' rent of $4,000 in advance. On January 1, it also pays the premium of $2,400 for property and liability insurance coverage for the year ending December 31, Year 1. During January it acquires merchandise costing $40,000, of which it purchases $26,000 for cash and $14,000 on account for payment in February. Sales to customers during January total $50,000, of which $34,000 is for cash and $16,000 is on account for collection in February and March. The cost of the merchandise sold during January was $32,000. The firm paid $5,000 in salaries to various employees.

**Example 1** **Exhibit 3.1** presents an income statement for Allens' Hardware Store for the month of January, Year 1, using the cash basis of accounting. Cash receipts of $34,000 from

**EXHIBIT 3.1**

**ALLENS' HARDWARE**
**Income Statement**
**For the Month of January, Year 1**
**Cash Basis of Accounting**

| | | |
|---|---|---|
| Revenues from Cash Receipts for Sales of Merchandise | | $34,000 |
| Less Expenses from Cash Disbursements for Merchandise and Services: | | |
| Merchandise | $26,000 | |
| Salaries | 5,000 | |
| Rent | 4,000 | |
| Insurance | 2,400 | |
| Total Expenses | | (37,400) |
| Net Income on a Cash Basis | | $(3,400) |

sales of merchandise represent the portion of the total January sales, $50,000, that the firm collects during January. Whereas the firm acquires merchandise costing $40,000 during January, it disburses only $26,000 cash to suppliers and therefore subtracts only this amount in measuring performance under the cash basis. The firm also subtracts the amounts of cash expenditures made during January for salaries, rent, and insurance, without regard to whether the firm fully consumes the services by the end of the month. Cash expenditures for merchandise and services exceeded cash receipts from customers during January by $3,400. Note that under the cash basis, the income statement does not include cash received from owners or from borrowing, both of which are financing transactions.

As a basis for measuring performance for a particular accounting period (here, January, Year 1, for Allens' Hardware Store), the cash basis of accounting has three weaknesses.

**Inadequately Matches Inflows and Outflows** First, the cash basis does not adequately match the cost of the efforts required in generating inflows with the inflows themselves. The cash outflows of one period can relate to operating activities whose cash inflows occur in preceding or succeeding periods. The store rental payment of $4,000 provides rental services for both January and February, but the cash basis subtracts the full amount in measuring performance during January, but none for February. As a result, February's performance could look better than January's for no reason other than the happenstance of timing of cash payments for rent. Likewise, the annual insurance premium paid in January provides coverage for the full year, whereas the cash basis of accounting subtracts none of this insurance cost in measuring performance during February through December.

The longer the period over which a firm receives future benefits, the more deficient becomes the cash basis for reporting. Consider, for example, a capital-intensive firm's investments in buildings and equipment that it will use for twenty or more years. The length of time between the purchase of these assets and the collection of cash for goods produced and sold can span many years and distort the measurement of performance during both the period of the cash expenditure and subsequent periods. The longer the accounting period (for example, one year versus one month), the less the cash basis suffers from mismatching through time.

**Unnecessarily Delays the Recognition of Revenues** Second, the cash basis of accounting unnecessarily postpones the time when the firm recognizes revenue. A firm should recognize revenue when it has delivered goods or services and finished the parts of the operating process that are difficult to control and predict. To complete the operating process involves the following actions by the seller:

- Acquiring goods and labor.
- If a manufacturing firm, converting the goods and labor into sellable products.
- Persuading the customer to buy the goods or services.
- Delivering the goods or services to the customer.
- Collecting cash from the customer.

**EXHIBIT 3.2**

**ALLENS' HARDWARE STORE**
**Income Statement**
**For the Month of January, Year 1**
**Accrual Basis of Accounting**

| | | |
|---|---|---|
| Sales Revenue | | $50,000 |
| Less Expenses: | | |
| Cost of Goods Sold | $32,000 | |
| Salaries Expense | 5,000 | |
| Rent Expense | 2,000 | |
| Insurance Expense | 200 | |
| Interest Expense | 80 | |
| Total Expenses | | 39,280 |
| Net Income on an Accrual Basis | | $10,720 |

- Waiting, in the case of sales of goods, until the period of warranty (or the buyer's right to return) expires.

The most difficult parts of the operating process for manufacturing and retailing firms are persuading the customer to buy the goods and delivering the goods. The most difficult parts for service firms are persuading the customer to use the services and then rendering the services. Most firms have little difficulty in predicting the amounts and timing

- of cash collections or returns from customers, and
- of expenditures to fulfill warranty promises.

In these cases, awaiting the measurement of performance until the firm collects cash often results in reporting the effects of operating activities one or more periods after the critical revenue-generating activity—having the customer purchase the goods—has occurred. For example, sales to customers during January by Allens' Hardware Store totaled $50,000. Under the cash basis of accounting, the firm will not recognize $16,000 of this amount until it collects the cash during February or later. If the firm checks the creditworthiness of customers before making the sales on account, it will probably collect the cash, or at least a predictable fraction of the cash, and therefore need not postpone recognition of the revenue until the time it actually collects the cash.

## Provides Opportunities to Distort the Measurement of Operating Performance

A third criticism of the cash basis is it provides an opportunity for firms to distort the measurement of operating performance by timing their cash expenditures. Firms might delay cash expenditures near the end of the accounting period in an effort to increase earnings and appear more profitable. Firms might accelerate cash expenditures to appear less profitable and perhaps discourage competitors from entering the industry or labor unions from negotiating higher wages for employees.

Who uses the cash basis of accounting? Lawyers, accountants, and other professionals, who have relatively small investments in inventories and multi-period assets, such as buildings and equipment, and usually collect cash from clients soon after they render services, are the most frequent users of the cash basis. Most of these firms actually use a modified cash basis of accounting, under which they treat the costs of buildings, equipment, and similar items as assets when purchased. They then recognize a portion of the acquisition cost as an expense when they consume services of these assets. Except for the treatment of these long-lived assets, such firms measure performance at the times they receive and disburse cash.

Most individuals use the cash basis of accounting for the purpose of computing personal income and personal income taxes. When a firm, such as one in merchandising or manufacturing, uses inventories as an important factor in generating revenues, the Internal Revenue Code prohibits the firm from using the cash basis of accounting in its income tax returns.

## ACCRUAL BASIS OF ACCOUNTING

The **accrual basis of accounting** typically recognizes revenue when a firm sells goods (manufacturing and retailing firms) or renders services (service firms). At these points the firm has usually identified a customer, agreed upon a price, and delivered the good or service. At these points the firm also knows the costs of assets already used in producing revenues (such as the acquisition cost of merchandise sold) and can estimate any future costs it will incur (such as for warranties) in connection with the good or service sold. The firm recognizes these costs as expenses in the period when the firm recognizes the revenues that the costs helped produce. Thus accrual accounting attempts to match expenses with associated revenues. When the usage of an asset's future benefits does not match with particular revenues, the firm recognizes the costs of these assets as expenses of the period during which the firm uses the benefits provided by the assets.

**Example 2** **Exhibit 3.2** presents an income statement for Allens' Hardware Store for January of Year 1 using the accrual basis of accounting. The firm recognizes the entire $50,000 of sales during January as revenue, even though it has received only $34,000 in cash by the end of January. The firm will probably collect the remaining accounts receivable of $16,000 in February or later periods. Therefore, the sale of the goods, rather than the collection of cash from customers, triggers the recognition of revenue.[1] The merchandise sold during January cost $32,000. Recognizing this amount as an expense (cost of goods sold) matches the cost of the merchandise sold with revenue from sales of those goods. Of the advance rental payment of $4,000, only $2,000 applies to the cost of benefits consumed during January. The remaining rental of $2,000 purchases benefits for the month of February and will therefore appear on the balance sheet on January 31 as an asset. Likewise, only $200 of the $2,400 insurance premium represents the cost of coverage used up during January. The remaining $2,200 of the insurance premium provides coverage for February through December and will become an expense during those months. In the meantime, it appears as an asset on the balance sheet on January 31. The interest expense of $80 represents one month's interest on the $12,000 bank loan at an annual rate of 8 percent (= $12,000 × 0.08 × 1/12).[2] Although the firm will not pay this interest until the loan comes due (technically, *matures*) on June 30, Year 1, the firm benefited from using the funds during January; it should therefore recognize one month of the total interest cost on the loan as an expense of January. The firm reports its obligation to pay the $80 as a liability on the balance sheet on January 31. The salaries, rent, insurance, and interest expenses, unlike the cost of merchandise sold, do not directly match sales revenues recognized during the period. These costs therefore become expenses of January to the extent that the firm consumed services during the month.

An important concept for understanding accrual accounting is the *matching principle*. The data for Allens' Hardware Store helps in understanding the subtleties of the matching principle. In January, sales were $50,000, of which the store collected $34,000 in cash and expected to collect the rest later. The cost of the goods sold to customers was $32,000, of which the store paid $26,000 in cash and expected to pay the rest later. First, consider comparing the $50,000 they have collected or will collect for those goods delivered in January to the $32,000 cost of goods put into the hands of customers in January. Next, consider comparing the $34,000 cash the store received in January from customers to the $26,000 it paid in January to suppliers. Which comparison provides a more useful measure of January's operating activities? If you can see that the comparison of the expected total collections to the expected total costs of those goods provides more useful information about operating performance than does comparing the $34,000 cash collected during the month to the $26,000 cash paid to suppliers during the month, then you understand the matching principle. That principle suggests that the *eventual total* cash inflows and outflows for operations of a period provide information more useful for assessing performance of that period than does comparing the actual cash inflows and outflows that occurred during the period.

More advanced applications of the matching principle relate all the costs of carrying out a specific set of activities (for example, salaries, interest, taxes) to all the benefits from performing

---

[1]This example assumes that the firm collects all its receivables from sales. Later, we relax this assumption and assume merely that the firm collects a predictable proportion of its sales on account. When the firm collects a predictable proportion of its sales on account, we can still say that the sale triggers revenue recognition, but the amount of the revenue results from calculations about expected cash collections.

[2]By convention in business practice, interest rates on loans almost always refer to annual interest rates. Also, to simplify the calculation of interest, convention assumes a year equal to twelve months of thirty days each, or 360 days.

those activities, no matter when the firm paid cash or when it collects cash. Cash flows for a period provide useful information, as **Chapter 4** shows, but they do not measure operating performance for short time periods as well as do the results of applying the matching principle in accrual accounting.

Thus, the accrual basis of accounting provides a better measure of operating performance for Allens' Hardware Store for the month of January than does the cash basis for two reasons:

1. Revenues more accurately reflect the results of sales activity during January than does cash received from customers during that period.
2. Expenses more closely match reported revenues than expenditures match receipts.

Likewise, the accrual basis will provide a superior measure of performance for future periods because activities of those periods will bear their share of the costs of rental, insurance, and other services the firm will consume. Thus the accrual basis focuses on inflows of net assets from operations (revenues) and the use of net assets in operations (expenses), independent of whether the firm has collected cash for those inflows and spent cash for the outflows of net assets.

Many accountants argue that the accrual basis provides fewer opportunities than the cash basis to distort the measurement of earnings. The matching of expenses with associated revenues serves as a guiding principle that constrains a firm's efforts to time the recognition of expenses to suit its particular reporting objectives. As later chapters make clear, however, timing revenue recognition and measuring related expenses under the accrual basis both often involve judgment. Thus, even the accrual basis provides opportunities for earnings management.

Most business firms, particularly those involved in merchandising and manufacturing activities, use the accrual basis of accounting. From this point on in the book, all discussions assume use of the accrual basis of accounting. We introduced the cash basis of accounting in this section primarily to demonstrate the deficiencies of the cash basis and the related rationale for the accrual basis. The next section examines the measurement principles of accrual accounting.

## Problem 3.1 for Self-Study

**Cash versus accrual basis of accounting.** Thompson Hardware Store commences operations on January 1, Year 5. J. Thompson invests $10,000 for all of the common stock of the firm, and the firm borrows $8,000 from a local bank. The firm must repay the loan on June 30, Year 5, with interest at the rate of 6 percent per year.

The firm rents a building on January 1 and pays two months' rent in advance in the amount of $2,000. On January 1, it also pays the $1,200 premium for property and liability insurance coverage for the year ending December 31, Year 5.

The firm purchases $28,000 of merchandise inventory on account on January 2 and pays $10,000 of this amount on January 25. A physical inventory indicates that the cost of merchandise on hand on January 31 is $15,000.

During January, the firm makes cash sales to customers totaling $20,000 and sales on account totaling $9,000. The firm collects $2,000 from these credit sales by the end of January.

The firm pays other costs during January as follows: utilities, $400; salaries, $650; and taxes, $350.

**a.** Prepare an income statement for January, assuming that Thompson uses the accrual basis of accounting and recognizes revenue at the time goods are sold (delivered).
**b.** Prepare an income statement for January, assuming that Thompson uses the cash basis of accounting.
**c.** Which basis of accounting do you believe provides a better indication of the operating performance of the firm during January? Why?

Solutions to the self-study problems appear at the end of the chapter.

# Measurement Principles of Accrual Accounting

Under the accrual basis of accounting, one must consider when a firm recognizes revenues and expenses (timing questions) and how much it recognizes (measurement questions).

## TIMING OF REVENUE RECOGNITION

Refer again to **Figure 3.1**. A firm could conceivably recognize revenue, a measure of the increase in net assets from selling goods or providing services, at the time of purchase of merchandise, at the time of sale to customers, at the time of cash collection from customers, at some point(s) between these events, or even continually. Answering the timing question requires a set of criteria for revenue recognition.

**Criteria for Revenue Recognition** The accrual basis of accounting recognizes revenue when both of the following events have occurred:

1. A firm has performed all, or most of, the services it expects to provide.
2. The firm has received cash or some other asset such as a receivable, whose cash-equivalent value it can measure with reasonable precision.

Accrual accounting requires matching expenses with related revenues, which requires that the firm either know or can estimate the expenses it will incur in generating revenues. Delaying revenue recognition until the firm has substantially completed performance increases the likelihood that it can measure expenses, and thereby net income, accurately. Thus, the first criterion for revenue recognition provides that firms must meet a minimum threshold for measuring expenses before they can justify recognizing revenues. Most firms involved in selling goods and services recognize revenue at the time of sale (delivery). The firm has transferred the goods to a buyer or has performed the services. The firm can estimate with reasonable precision the cost of future services, such as for warranties and the amounts it will have to give back because customers return goods.

The second criterion for revenue recognition focuses on measuring accurately the amount of cash the firm will ultimately receive. The exchange between an independent buyer and seller provides a reliable initial measure of the amount of revenue. If the firm makes the sale on account, past experience and an assessment of credit standings of customers provide a basis for predicting the amount and timing of the initial selling price the firm will collect in cash.

Thus, the time of sale usually meets the criteria for revenue recognition. In general, the firm will recognize revenue when it has no further significant uncertainty about the amount and timing of cash inflows and outflows from the sales transaction.

Understand the importance of these criteria by considering an example of a firm that should not recognize revenue at the time of sale, such as a software developer shipping a new product, where the customers have a right of return. The developer can estimate neither how many of the customers will return the product nor the cost of debugging the software once customers start to use it. Hence, that developer cannot estimate with reasonable precision the eventual cash inflows (that is, revenues) and cash outflows (that is, expenses).

## MEASUREMENT OF REVENUE

A firm measures the amount of revenue by the cash or cash-equivalent value of other assets it receives from customers. As a starting point, this amount is the agreed-upon price between buyer and seller at the time of sale. If a firm recognizes revenue in a period before it collects the cash, however, it will likely need to make adjustments to the agreed-upon price in measuring revenue to recognize the effects of the time delay.

**Uncollectible Accounts** A firm that sells to hundreds of customers, or more, does not expect to collect from all of them. If a firm expects not to collect some portion of the sales for a period, it must adjust the amount of revenue recognized during that period for estimated uncollectible accounts arising from those sales. This adjustment of revenue occurs in the period when the firm recognizes revenue and not in the later period when it identifies specific customers' accounts as uncollectible. If the firm postpones the adjustment, earlier decisions to extend credit to customers will affect income of subsequent periods. A failure to recognize anticipated uncollectibles at the time of sale incorrectly measures the performance of the firm for both the period of sale and the period when the firm judges the account uncollectible. **Chapter 6** considers these problems further.

**Sales Discounts and Allowances** Customers may take advantage of discounts for prompt payment, and the seller may grant allowances for unsatisfactory merchandise. In these cases, the firm will eventually receive cash in a smaller amount than the stated selling price. The firm must estimate these amounts and make appropriate reductions, in the period of the sale, in measuring the amount of revenue it recognizes.

**Sales Returns** Customers may return goods. If so, the firm will eventually receive cash in an amount smaller than the original sales price. The firm must estimate the expected returns and make appropriate reductions in revenue measured at the time of sale.

## Timing of Expense Recognition

Assets provide future benefits to a firm. Expenses measure the assets consumed in generating revenue. Assets are **unexpired costs,** and expenses are **expired costs** or "gone assets." We focus on *when* the asset expiration takes place. The critical question is, "When do the benefits of an asset expire, leaving the balance sheet, and become expenses, entering the income statement as reductions in net income and retained earnings, a shareholders' equity account?"

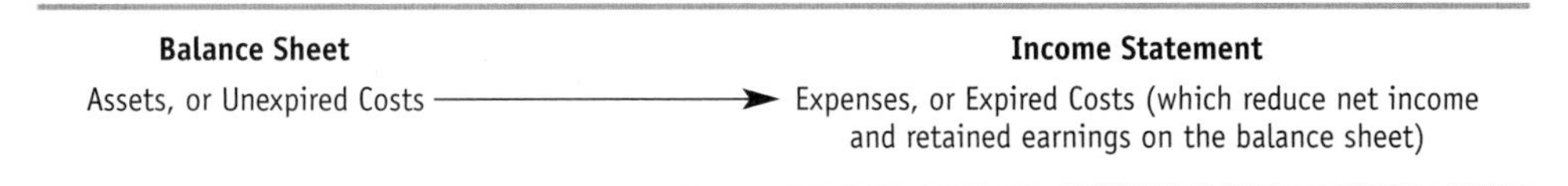


**Criteria for Expense Recognition** Accountants recognize expenses as follows:

1. If the event triggering an asset expiration relates to the recognition of a particular revenue, that expiration becomes an expense in the period when the firm recognizes the revenue. For example, the sale of merchandise usually results in the seller's recognizing revenue and the immediate disappearance, as cost of goods sold, of the firm's benefits of having held the merchandise inventory. This treatment—the **matching principle**—matches cost expirations with revenues.
2. If the event triggering an asset expiration does not relate to a particular revenue, that expiration becomes an expense of the period when the firm consumes the benefits of that asset in operations.

**Product Costs** The matching principle most clearly applies to the cost of goods sold because the firm can easily associate the asset expiration with the revenue from selling the goods. At the time of sale, the asset physically changes hands. The firm recognizes sales revenue, and the cost of the goods sold becomes an expense.

A merchandising firm purchases inventory and later sells it without changing its physical form. The inventory appears initially as an asset stated at acquisition cost on the balance sheet. Later, when the firm sells the inventory, the same amount of acquisition cost appears as an expense (cost of goods sold) on the income statement.

A manufacturing firm, on the other hand, incurs various costs as it changes the physical form of the goods it produces. These costs are not yet expenses—gone assets—but represent the transformation of one form of asset into another. A firm incurs three types of costs in manufacturing: (1) direct material costs, (2) direct labor costs, and (3) manufacturing overhead costs (sometimes called "indirect manufacturing costs"). A firm incurs direct material and direct labor costs because it manufactures particular products, so the firm can associate those costs with the particular products manufactured. Manufacturing overhead includes a mixture of costs that provides a firm with a capacity to produce. Examples of manufacturing overhead costs are expenditures for supervisors' salaries, utilities, property taxes, and insurance on the factory, as well as depreciation on manufacturing plant and equipment. The firm uses the services of each of these items during a period when it creates new assets—the inventory of goods it works on or holds for sale.

Accountants call these costs *indirect* because they jointly benefit all goods produced during the period, not any one particular item.

The manufacturing process transfers the benefits from direct material, direct labor, and manufacturing overhead to the asset represented by units of inventory. Because the firm shows the inventory items as assets until it sells them to customers, the various direct material, direct labor, and manufacturing overhead costs incurred in producing the goods remain as assets on the balance sheet in the manufacturing inventory under the titles Work-in-Process Inventory and Finished Goods Inventory. Such costs, called **product costs,** are assets transformed from one form to another. They remain on the balance sheet as assets until the firm sells the goods that they embody; then, they become expenses. **Chapter 7** discusses more fully the accounting for manufacturing costs.

**Marketing Costs** The costs a firm incurs in marketing or selling its products during a period relate primarily to the units it sells during the period. In generating current revenue, a firm incurs costs for salaries and commissions of the sales staff, for sales literature used, and for advertising. Because these marketing costs associate with the revenues of the period, accounting reports them as expenses in the period when the firm uses their services. One might argue that some marketing costs, such as for advertising and other sales promotion, provide future-period benefits for a firm and that therefore the firm should continue to treat them as assets. Accountants cannot readily distinguish the portion of the cost relating to the current period (an expense) from the portion relating to future periods (an asset). They therefore treat most marketing activity costs as expenses of the period when the firm uses the services. Even when such costs enhance the future marketability of a firm's products, accounting treats these marketing costs as **period expenses** rather than assets—an expedient practice, not one derived from fundamental concepts.

**Administrative Costs** The costs of administering the activities of a firm include the president's salary, accounting and information systems costs, and the costs of conducting various supportive activities such as legal services, employee training, and corporate planning. These costs do not relate directly to particular units produced or sold, so accounting treats administrative costs, like marketing costs, as period expenses.

## Measurement of Expenses

Expenses result from the firm's consuming assets during the period, so the amount of an expense equals the portion of the asset's cost consumed. The basis for expense measurement is the same as for asset measurement. Because accounting reports assets primarily at acquisition cost on the balance sheet, it measures expenses by the acquisition cost of the assets sold or used during the period.

## Classification of Expenses

Firms vary in the classification of expenses in the income statement. Some firms classify expenses by the function within the firm: cost of goods sold for the merchandise or manufacturing function, selling expenses for marketing activities, and administrative expenses for administrative activities. Other firms classify their expenses by the nature of the expense, such as salary expense, depreciation expense, and so on. Most firms use a mixture of classifications, but maintain those classifications over time. **Chapters 6** and **12** discuss income statement classification more fully.

## Summary

Over sufficiently long time periods, the amount of net income equals cash inflows minus cash outflows from operating, investing, and debt servicing activities; that is, the amount of net income equals the difference between the cash received from customers and the amount of cash paid to suppliers, employees, and other providers of goods and services.[3] Users of financial

[3]Sometimes we state this fundamental principle of accrual accounting more succinctly: Over long enough time periods, net income equals cash-in less cash-out, other than transactions with owners. (Consider, for example, shareholders paying for shares of stock or receiving dividends, which do not affect net income.)

statements, however, desire information about a firm's operating performance for short time periods, such as a year or less. Thus, accountants slice up this total amount of net income and allocate a portion to each period. Using the timing of cash flows to dictate the allocation of this total income usually provides a poor measure of operating performance. Cash receipts from customers do not always occur in the same accounting period as the related cash expenditures to the providers of goods and services. The accrual basis allocates net income so that outflows of net assets more closely match inflows of net assets of the period, independent of whether the net assets are currently in the form of cash.

The accrual basis determines the timing of income recognition. The accrual basis typically recognizes revenue at the time of sale (delivery). Costs that associate directly with particular revenues become expenses in the period when a firm recognizes the revenues. A firm treats the cost of acquiring or manufacturing inventory items in this manner. Costs that do not closely associate with particular revenue streams become expenses of the period when a firm consumes the goods or services in operations. Most marketing and administrative costs receive this treatment.

The next section considers the preparation of the income statement.

### Problem 3.2 for Self-Study

**Revenue and expense recognition.** A firm uses the accrual basis of accounting and recognizes revenues at the time it sells goods or renders services. Indicate the amount of revenue or expense that the firm recognizes during April in each of the following transactions:

**a.** Collects $4,970 cash from customers during April for merchandise sold and delivered in March.
**b.** Sells merchandise to customers during April for $14,980 cash.
**c.** Sells to customers during April merchandise totaling $5,820 that the firm expects to collect in cash during May.
**d.** Pays suppliers $2,610 during April for merchandise received by the firm and sold to customers during March.
**e.** Pays suppliers $5,440 during April for merchandise received and sold to customers during April.
**f.** Receives from suppliers and sells to customers during April merchandise that cost $2,010 and that the firm expects to pay for during May.
**g.** Receives from suppliers during April merchandise that cost $1,570 and that the firm expects to pay and sell to customers during May.

## Accounting Procedures for Preparing the Income Statement

As we discussed in connection with the balance sheet in **Chapter 2,** you need not be able to prepare an income statement in order to use it in assessing a firm's operating performance. To assess a firm's performance, you must understand how various business transactions affect a firm's net income for a period and its financial position at the end of the period. We extend the analytical framework discussed in **Chapter 2** to include income transactions. We use both the T-accounts and journal entries. **Appendix 3.1** illustrates the extension of the transactions spreadsheet discussed in **Appendix 2.1** to incorporate income items. Recall that this analytical framework relies on the balance sheet equation as its building block.

### RELATION BETWEEN BALANCE SHEET AND INCOME STATEMENT

**Chapter 2** discussed the balance sheet, which reports the assets of a firm and the financing of those assets by creditors and owners at a specific moment in time. This chapter discusses the income statement, which measures for a period of time the excess of revenues (net asset inflows) over expenses (net asset outflows) from selling goods and providing services. Firms may distribute net assets to shareholders each period as a dividend. Accountants view dividends as a distribution of net assets generated by earnings and not as a cost incurred in generating earnings.

Thus, dividends are not expenses on the income statement. The Retained Earnings account on the balance sheet measures the cumulative excess of net income over dividends since the firm began operations. The following disaggregation of the balance sheet equation shows the relation of revenues, expenses, and dividends to the components of the balance sheet.

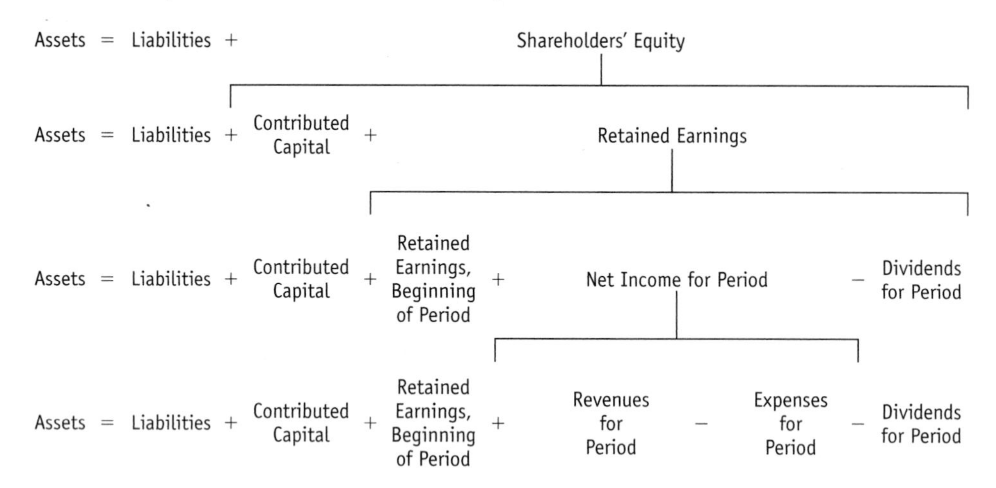


## Purpose and Use of Individual Revenue and Expense Accounts

The accountant could record revenue and expense amounts directly in the Retained Earnings account. For example, the sale of merchandise on account has at least two pairs of effects on the balance sheet:

1. An inflow of assets—accounts receivable, in this example, or cash, in other cases—and an increase in retained earnings (sales revenue).
2. A decrease in assets (merchandise inventory) and a decrease in retained earnings (cost of goods sold).

Recording revenues and expenses directly in the Retained Earnings account would make measuring the mere amount of net income easy. The following equation would derive net income:

$$\text{Net Income} = \text{Retained Earnings at End of Period} - \text{Retained Earnings at Beginning of Period} + \text{Dividends}$$

The income statement does not, however, report just the amount of net income. One can derive the amount of net income, as the equation above shows, using balance sheet amounts (and the amount of dividends, if any). More importantly, the income statement reports the sources and amounts of a firm's revenues and the nature and amounts of a firm's expenses that net to earnings for the period. Knowing the components of a firm's net income helps both in understanding the causes of past performance and in forecasting future performance. Knowing the purpose of the income statement—to help explain why income was what it was—will help you understand the need for some of the procedures for preparing it.

To help prepare the income statement, accountants maintain individual revenue and expense accounts during the accounting period. These accounts begin the accounting period with a zero balance. During the period, the accountant records revenues and expenses in the accounts. At the end of the period, the balances in revenue accounts represent the cumulative revenues for the period. Similarly, at the end of the period, the balances in expense accounts represent the cumulative expenses for the period. The accountant reports these revenues and expenses, which aggregate to the period's net income, in the income statement.

Revenues and expenses are components of retained earnings, temporarily labeled with descriptive account titles so that the accountant can prepare an income statement. After preparing the income statement at the end of the period, the accountant transfers the balance in each revenue and expense account to the Retained Earnings account. Common terminology describes this transfer as *closing* the revenue and expense accounts because after closing (or transfer), each revenue and expense account has a zero balance. Retained earnings will increase by the amount of net income (or decrease by the net loss) for the period.

Maintaining separate revenue and expense accounts during the period and transferring their balances to the Retained Earnings account at the end of the period has the same result as initially recording revenues and expenses directly in the Retained Earnings account. Using separate revenue and expense accounts enables the firm to show specific types of revenue and expenses in the income statement, which otherwise could show the total amount of income, but not its components. Once revenue and expense accounts serve their purpose, the need for separate accounts for a given accounting period ends. The accountant closes these accounts, so that they begin the following accounting period with a zero balance, ready for the revenue and expense entries of the new period.

The **closing process** transfers the balances in revenue and expense accounts to retained earnings. The term *closing* describes the process of reducing the balance to zero in each revenue and expense account. Because revenue and expense accounts accumulate amounts for only a single accounting period, accountants often refer to them as **temporary accounts**. In contrast, the accounts on the balance sheet reflect the cumulative changes in each account from the time the firm first began business. The balance sheet accounts remain open each period. Because the balances in these accounts at the end of one period carry over as the beginning balances for the following period, accountants often describe them as **permanent accounts**.

## Debit and Credit Procedures for Revenues, Expenses, and Dividends

Because revenues, expenses, and dividends increase or decrease retained earnings, the recording procedures for these items are the same as for any other transaction affecting shareholders' equity accounts.

| Shareholders' Equity | |
|---|---|
| Decreases (Debit) | Increases (Credit) |
| Expenses<br>Dividends | Revenues<br>Issues of Capital Stock |

A transaction generating revenue increases net assets (increase in assets or decrease in liabilities) and increases shareholders' equity. The usual journal entry to record a revenue transaction is therefore as follows:

| | | |
|---|---|---|
| Asset Increase or Liability Decrease (or both) | Amount | |
| Revenue | | Amount |

| Assets | = | Liabilities | + | Shareholders' Equity | (Class.) |
|---|---|---|---|---|---|
| + | or | − | | + | IncSt → RE |

Typical entry to recognize revenue.

The designation, IncSt → RE, indicates an income statement account that the accountant subsequently closes to retained earnings.

A transaction generating an expense decreases net assets (decrease in assets or increase in liabilities) and decreases shareholders' equity. The usual journal entry to record an expense transaction is therefore as follows:

| | | |
|---|---|---|
| Expense | Amount | |
| Asset Decrease or Liability Increase (or both) | | Amount |

| Assets | = | Liabilities | + | Shareholders' Equity | (Class.) |
|---|---|---|---|---|---|
| − | or | + | | − | IncSt → RE |

Typical entry to recognize expense.

Dividends result in a decrease in net assets and a decrease in shareholders' equity. As **Chapter 12** discusses, a firm may pay dividends either in cash or in other assets. Although the accounting procedures for dividends do not depend on the form of the distribution, we assume that the firm pays dividends in cash unless we have contrary information. The usual entry to record the declaration of a dividend by the board of directors of a corporation is as follows:

| | | |
|---|---|---|
| Retained Earnings . . . . . . . . . . . . . . . . . . . . . . . . . . . . . . . . . . . . . . . | Amount | |
|     Dividend Payable . . . . . . . . . . . . . . . . . . . . . . . . . . . . . . . . . . | | Amount |

| Assets | = | Liabilities | + | Shareholders' Equity | (Class.) |
|---|---|---|---|---|---|
| | | + | | − | RE |

Typical entry to record dividend declaration.

When the firm pays the dividend, the journal entry is as follows:

| | | |
|---|---|---|
| Dividend Payable . . . . . . . . . . . . . . . . . . . . . . . . . . . . . . . . . . . . . . . | Amount | |
|     Cash . . . . . . . . . . . . . . . . . . . . . . . . . . . . . . . . . . . . . . . . . . . . | | Amount |

| Assets | = | Liabilities | + | Shareholders' Equity | (Class.) |
|---|---|---|---|---|---|
| − | | − | | | |

Typical entry to record dividend payment.

Dividends are not expenses, even though the journal entries resemble those for expenses. They are not costs incurred in generating revenues. Rather, they represent distributions of assets, arising from current and previous years' operations, to the owners of the firm. Because dividends are not expenses, they do not affect the measurement of net income for the period.

Before we illustrate the recording procedures for revenues and expenses, we provide a brief overview of the steps in the accounting process.

## Overview of the Accounting Process

A firm's accounting system generally involves the following operations:

1. Entering the results of each transaction in a file or other record in the form of a journal entry, a process called *journalizing*. **Chapter 2** introduced journal entries, which we use again in this chapter.
2. Transferring the amount from the journal entries to individual balance sheet and income statement accounts, a process called *posting*. In computer systems, posting occurs instantly after journalizing. **Chapter 2** also introduced T-accounts, which serve the function of accumulating information about the effect of business transactions on each balance sheet and income statement account.

These first two steps occur each day throughout the accounting period.

3. Making adjusting and correcting journal entries to the accounts at the end of the accounting period to properly measure net income for the period and financial position at the end of the period. The recording of adjusting and correcting entries results in virtually simultaneous posting of these entries to individual balance sheet and income statement accounts.
4. Preparing the income statement for the period from amounts in revenue and expense accounts.
5. Closing revenue and expense accounts to retained earnings.
6. Preparing the balance sheet from amounts in balance sheet accounts.

These last four steps typically occur at the end of the accounting period. **Figure 3.2** shows these operations, which the next section illustrates using the transactions of Miller Corporation during

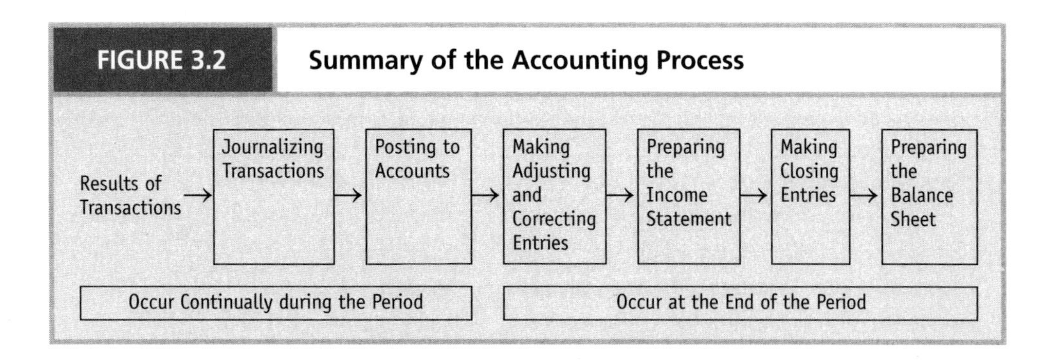


February. A seventh step in the accounting process involves the preparation of the statement of cash flows, which **Chapter 4** discusses.

## ILLUSTRATION OF THE ACCOUNTING PROCESS FOR INCOME TRANSACTIONS

**Chapter 2** illustrated the transactions of Miller Corporation for the month of January as it prepared to open for business on February 1. None of the transactions during January involves income statement accounts. Note in **Exhibits 2.3** and **2.4** that there is no amount in the Retained Earnings account at the end of January. The firm had not yet generated revenues, nor incurred expenses, nor declared a dividend. This section illustrates the accounting process for Miller Corporation during February.

**Journalizing** Miller Corporation engages in seven transactions during February.

*Transaction 1:* On February 5, Miller Corporation purchases an additional $25,000 of merchandise on account. **Chapter 2** illustrated the entry to record this type of transaction:

| | | |
|---|---|---|
| **(1)** Merchandise Inventory | 25,000 | |
| Accounts Payable | | 25,000 |

| Assets | = | Liabilities | + | Shareholders' Equity | (Class.) |
|---|---|---|---|---|---|
| +25,000 | | +25,000 | | | |

Purchase of inventory costing $25,000 on account.

*Transaction 2:* During the month of February, Miller Corporation sells merchandise to customers for $50,000. Of this amount, $3,000 represents sales to customers who paid $3,000 to Miller Corporation on January 31, which we recorded at the end of January as Advances from Customer. Miller Corporation makes the remaining $47,000 of sales on account. Retail firms typically recognize revenue at the time of delivery of merchandise to customers, regardless of whether or not they have yet received cash from customers. The journal entry to record these sales is:

| | | |
|---|---|---|
| **(2)** Advances from Customer | 3,000 | |
| Accounts Receivable | 47,000 | |
| Sales Revenue | | 50,000 |

| Assets | = | Liabilities | + | Shareholders' Equity | (Class.) |
|---|---|---|---|---|---|
| +47,000 | | −3,000 | | +50,000 | IncSt → RE |

Sales of merchandise for $50,000, of which $3,000 was received during January and $47,000 is on account.

Sales Revenue is a temporary income statement account that the firm will close to Retained Earnings at the end of February.

*Transaction 3:* The historical, or acquisition, cost of the merchandise sold to customers in Transaction 2 is $30,000. Because Miller Corporation no longer holds this merchandise in inventory, it is no longer an asset. We must reduce the inventory account and recognize the cost of the inventory as an expense to match against the sales revenue in Transaction 2. Thus,

| | | |
|---|---|---|
| **(3)** Cost of Goods Sold | 30,000 | |
| Merchandise Inventory | | 30,000 |

| Assets | = | Liabilities | + | Shareholders' Equity | (Class.) |
|---|---|---|---|---|---|
| −30,000 | | | | −30,000 | IncSt → RE |

The cost of merchandise sold to customers during February is $30,000.

Cost of Goods Sold is likewise a temporary income statement account that the firm will close to the Retained Earnings account at the end of February.

*Transaction 4:* Miller Corporation incurs and pays selling and administrative expenses in the aggregate of $14,500 during February. The journal entry to record this transaction is:

| | | |
|---|---|---|
| **(4)** Selling and Administrative Expenses | 14,500 | |
| Cash | | 14,500 |

| Assets | = | Liabilities | + | Shareholders' Equity | (Class.) |
|---|---|---|---|---|---|
| −14,500 | | | | −14,500 | IncSt → RE |

Selling and administrative expenses paid in cash during February total $14,500.

Miller Corporation presumably received all of the benefits of these selling and administrative services during February. Thus, the full amount is an expense for the month, which we match against revenues. None of the expenditure results in an asset that would appear on the balance sheet at the end of February.

*Transaction 5:* Miller Corporation collects $35,000 from customers for sales previously made on account. The firm recognized revenue from these sales at the time of sale (see Transaction 2 for February). The collection of the cash increases cash and decreases accounts receivable.

| | | |
|---|---|---|
| **(5)** Cash | 35,000 | |
| Accounts Receivable | | 35,000 |

| Assets | = | Liabilities | + | Shareholders' Equity | (Class.) |
|---|---|---|---|---|---|
| +35,000 | | | | | |
| −35,000 | | | | | |

Cash collections of $35,000 from sales previously made on account.

*Transaction 6:* The firm pays $20,000 to suppliers for merchandise previously purchased on account. The journal entry to record this payment is:

| | | |
|---|---|---|
| **(6)** Accounts Payable | 20,000 | |
| Cash | | 20,000 |

| Assets | = | Liabilities | + | Shareholders' Equity | (Class.) |
|---|---|---|---|---|---|
| −20,000 | | −20,000 | | | |

Cash payments of $20,000 for purchases previously made on account.

*Transaction 7:* Miller Corporation declares and pays a dividend to shareholders of $1,000. The entry to record the dividend is:

| | | |
|---|---|---|
| **(7)** Retained Earnings | 1,000 | |
| Cash | | 1,000 |

| Assets | = | Liabilities | + | Shareholders' Equity | (Class.) |
|---|---|---|---|---|---|
| −1,000 | | | | −1,000 | RE |

Dividends declared and paid during February total $1,000.

Note that the effect of a dividend on retained earnings is the same as an expense. Dividends, however, do not appear on the income statement as an expense in measuring net income.

These seven journal entries summarize the activities of Miller Corporation during the month of February. Each entry involves an exchange between Miller Corporation and its suppliers, its customers, or its shareholders. A typical firm would record entries such as the above, except for the dividend, continually during the month as they occur each day. We have simplified the journal entries by aggregating the daily transactions into summary entries for the month.

**Posting** The accounting system next posts these journal entries to the particular balance sheet and income statement accounts affected. We use T-accounts to show the balances in various accounts at the beginning of February and the posting of these seven transactions during the month of February for Miller Corporation. **Exhibit 3.3** shows the T-accounts, with the seven entries recorded in blue. We discuss the entries designated in red and gold next. Note that the income statement accounts have no balance at the beginning of February but reflect the effect of income transactions during February.

**Adjusting and Correcting Entries** Transactions 1 to 7 involve transactions, or exchanges, with other entities or individuals during the month of February. The transactions trigger the making of an entry in the accounts. The passage of time may also trigger the need to make entries. For example, interest expense on borrowing accrues as time passes. The costs of rent and insurance accrue as firms use these services over time. Most firms do not make entries related to the passage of time during the accounting period, but instead wait until the end of the period to make adjusting entries for them. Adjusting entries change the amounts in balance sheet and income statement accounts so that accountants can measure net income for the period and financial position at the end of the period. Adjusting entries usually involve at least one income statement account and one balance sheet account. Miller Corporation makes three adjusting entries at the end of February.

*Transaction 8:* The firm recognizes depreciation on the $60,000 of equipment purchased during January. Miller Corporation purchased the equipment during January and recorded it as an asset. It did not begin consuming the services of the equipment until it started using it on February 1. Miller Corporation estimates that the equipment has a five-year life and zero salvage at the end of five years. It will depreciate, or write off, the equipment straight line over the useful life. Thus, depreciation each month is $1,000 (= $60,000/60 months). Miller Corporation includes depreciation of equipment as part of selling and administrative expenses. The firm could record depreciation by reducing the equipment account by $1,000 and recognizing an expense of $1,000. Instead of reducing the equipment account directly, firms typically use an account called Accumulated Depreciation to record the write-down in the cost of the equipment for services used during the period. The Accumulated Depreciation account appears on the balance sheet as a subtraction from the acquisition cost of the equipment. Accountants refer to accounts such as Accumulated Depreciation as *contra accounts,* because they appear as subtractions from the amounts in other related accounts. Thus:

| | | |
|---|---|---|
| **(8)** Selling and Administrative Expenses | 1,000 | |
| Accumulated Depreciation | | 1,000 |

| Assets | = | Liabilities | + | Shareholders' Equity | (Class.) |
|---|---|---|---|---|---|
| −1,000 | | | | −1,000 | IncSt → RE |

Depreciation of equipment for February of $1,000.

**EXHIBIT 3.3**

**MILLER CORPORATION**
**T-Accounts Showing Transactions during February**

| | Cash (A) | | |
|---|---|---|---|
| ✓ | 34,400 | | |
| (5) | 35,000 | 14,500 | (4) |
| | | 20,000 | (6) |
| | | 1,000 | (7) |
| ✓ | 33,900 | | |

| | Accounts Receivable (A) | | |
|---|---|---|---|
| ✓ | 0 | | |
| (2) | 47,000 | 35,000 | (5) |
| ✓ | 12,000 | | |

| | Merchandise Inventory (A) | | |
|---|---|---|---|
| ✓ | 15,000 | | |
| (1) | 25,000 | 30,000 | (3) |
| ✓ | 10,000 | | |

| | Prepaid Insurance (A) | | |
|---|---|---|---|
| ✓ | 600 | | |
| | | 50 | (9) |
| ✓ | 550 | | |

| | Equipment (A) | | |
|---|---|---|---|
| ✓ | 60,000 | | |
| ✓ | 60,000 | | |

| | Accumulated Depreciation (XA) | | |
|---|---|---|---|
| | | 0 | ✓ |
| | | 1,000 | (8) |
| | | 1,000 | ✓ |

| | Accounts Payable (L) | | |
|---|---|---|---|
| | | 0 | ✓ |
| (6) | 20,000 | 25,000 | (1) |
| | | 5,000 | ✓ |

| | Advances from Customer (L) | | |
|---|---|---|---|
| | | 3,000 | ✓ |
| (2) | 3,000 | | |
| | | 0 | ✓ |

| | Income Tax Payable (L) | | |
|---|---|---|---|
| | | 0 | ✓ |
| | | 1,780 | (10) |
| | | 1,780 | ✓ |

| | Common Stock (SE) | | |
|---|---|---|---|
| | | 107,000 | ✓ |
| | | 107,000 | ✓ |

| | Retained Earnings (SE) | | |
|---|---|---|---|
| | | 0 | ✓ |
| (7) | 1,000 | 2,670 | (11) |
| | | 1,670 | ✓ |

| | Sales Revenue (SE) | | |
|---|---|---|---|
| | | 0 | ✓ |
| (11) | 50,000 | 50,000 | (2) |
| | | 0 | ✓ |

| | Cost of Goods Sold (SE) | | |
|---|---|---|---|
| ✓ | 0 | | |
| (3) | 30,000 | 30,000 | (11) |
| ✓ | 0 | | |

| | Selling and Administrative Expenses (SE) | | |
|---|---|---|---|
| ✓ | 0 | | |
| (4) | 14,500 | | |
| (8) | 1,000 | | |
| (9) | 50 | 15,550 | (11) |
| ✓ | 0 | | |

| | Income Tax Expense (SE) | | |
|---|---|---|---|
| ✓ | 0 | | |
| (10) | 1,780 | 1,780 | (11) |
| ✓ | 0 | | |

Black = Balances
Blue = Transaction Entries
Red = Adjusting Entries
Gold = Closing Entries

Note that the minus sign under Assets in the balance sheet equation on page 115 indicates the effect on assets, not on accumulated depreciation. Likewise, the minus sign under Shareholders' Equity indicates the effect on shareholders' equity, not on selling and administrative expense.

*Transaction 9:* Miller Corporation records the portion of the cost of the prepaid insurance of $600 attributable to insurance services received during February. The firm paid the one-year insurance premium on January 31 for insurance coverage from February 1 of this year through January 31 of next year. Assuming the allocation of an equal amount of this insurance premium to each month of the year, $50 (= $600/12) applies to February. Miller Corporation includes the cost of insurance in selling and administrative expenses. The entry is:

| | | |
|---|---|---|
| **(9)** Selling and Administrative Expenses | 50 | |
| Prepaid Insurance | | 50 |

| Assets | = | Liabilities | + | Shareholders' Equity | (Class.) |
|---|---|---|---|---|---|
| −50 | | | | −50 | IncSt → RE |

The cost of insurance services received during February is $50.

The remaining $550 of prepaid insurance becomes an expense during the next eleven months. Note that firms typically reduce the Prepaid Insurance account directly, instead of using an account similar to accumulated depreciation for equipment.

*Transaction 10:* Miller Corporation recognizes income tax expense on the net income before income taxes for February. Assume an income tax rate of 40 percent. Net income before income taxes for February equals $4,450 (= $50,000 − $30,000 − $14,500 − $1,000 − $50). Income tax expense is therefore $1,780 (= 0.40 × $4,450). Firms pay their income taxes quarterly. Thus, the income taxes remain unpaid at the end of February. Even though the income taxes are not yet payable, they represent an expense that Miller Corporation must match against the net income before taxes for February under the accrual basis of accounting. The entry is:

| | | |
|---|---|---|
| **(10)** Income Tax Expense | 1,780 | |
| Income Tax Payable | | 1,780 |

| Assets | = | Liabilities | + | Shareholders' Equity | (Class.) |
|---|---|---|---|---|---|
| | | +1,780 | | −1,780 | IncSt → RE |

Income tax expense on net income before taxes for February total $1,780.

As the accountant reviews the entries made during the period, recording errors may surface. For example, property taxes on the corporate headquarters buildings may appear as a debit to cost of goods sold instead of selling and administrative expenses. A payment to a supplier for purchases on account may appear as a debit to Accounts Receivable instead of Accounts Payable. If the accountant discovers recording or other errors, adjusting entries correct these errors. There were no such errors for Miller Corporation during February.

The accountant next posts adjusting entries to the particular accounts affected. Adjusting entries appear in red in **Exhibit 3.3**.

**Preparation of the Income Statement** The revenue and expense accounts show the effects of income transactions during February and adjusting entries at the end of February. We can now prepare the income statement for the month. **Exhibit 3.4** shows the income statement for Miller Corporation.

**EXHIBIT 3.4**

**MILLER CORPORATION**
**Income Statement**
**For the Month of February**

| | |
|---|---|
| Sales Revenue | $ 50,000 |
| Cost of Goods Sold | (30,000) |
| Selling and Administrative Expenses | (15,550) |
| Income before Income Taxes | $ 4,450 |
| Income Tax Expense | (1,780) |
| Net Income | $ 2,670 |

EXHIBIT 3.5

**MILLER CORPORATION**
**Comparative Balance Sheet**

| | January 31 | February 28 |
|---|---|---|
| ASSETS | | |
| Cash | $ 34,400 | $ 33,900 |
| Accounts Receivable | — | 12,000 |
| Merchandise Inventory | 15,000 | 10,000 |
| Prepaid Insurance | 600 | 550 |
| Total Current Assets | $ 50,000 | $ 56,450 |
| Equipment (at acquisition cost) | $ 60,000 | $ 60,000 |
| Less Accumulated Depreciation | — | (1,000) |
| Equipment-Net | $ 60,000 | $ 59,000 |
| Total Assets | $110,000 | $115,450 |
| LIABILITIES AND SHAREHOLDERS' EQUITY | | |
| Accounts Payable | $ — | $ 5,000 |
| Advances from Customer | 3,000 | — |
| Income Tax Payable | — | 1,780 |
| Total Current Liabilities | $ 3,000 | $ 6,780 |
| Noncurrent Liabilities | — | — |
| Total Liabilities | $ 3,000 | $ 6,780 |
| Common Stock | $107,000 | $107,000 |
| Retained Earnings | — | 1,670 |
| Total Shareholders' Equity | $107,000 | $108,670 |
| Total Liabilities and Shareholders' Equity | $110,000 | $115,450 |

**Closing Entries** The separate revenue and expense accounts have now served their purpose in providing the information to prepare the income statement. We next transfer the amounts in each revenue and expense account to the retained earnings—we close each revenue and expense account for the period. This step involves zeroing out the amount in each income statement account and transferring the differences between debits to revenue accounts and credits to expense accounts to retained earnings. The difference is the amount of net income for the period. **Exhibit 3.3** shows closing entries in gold.

**Preparation of the Balance Sheet** The accounts with balances are all balance sheet accounts and provide the information for preparing the balance sheet at the end of the period. **Exhibit 3.5** presents the comparative balance sheet for Miller Corporation on January 31 and February 28. Note that retained earnings increases for the month by $1,670, from zero at the end of January to $1,670 at the end of February. The change in retained earnings equals net income of $2,670 minus dividends of $1,000. Retained earnings begins the month of March with a balance of $1,670. Net income minus dividends for the month of March added to $1,670 will yield the balance in retained earnings at the end of March, assuming that Miller declares a dividend. Thus, the retained earnings account, as with all other balance sheet accounts, reflects the *cumulative* effect of transactions affecting that account.

## Problem 3.3 for Self-Study

**Journal Entries, T-accounts, Income Statement and Balance Sheet Preparation.** The balance sheet of Rybowiak's Building Supplies on June 30, Year 12, appears in **Exhibit 3.6**.

The following transactions occurred during the month of July.

**(1)** Sold merchandise on account for a total selling price of $85,000.
**(2)** Purchased merchandise inventory on account from various suppliers for $46,300.

**EXHIBIT 3.6**

**RYBOWIAK'S BUILDING SUPPLIES**
**Balance Sheet**
**June 30, Year 12**

| | |
|---|---|
| **Assets** | |
| Cash | $ 44,200 |
| Accounts Receivable | 27,250 |
| Merchandise Inventory | 68,150 |
| Prepaid Insurance | 400 |
| Total Current Assets | $140,000 |
| Equipment (at cost) | $210,000 |
| Less Accumulated Depreciation | (84,000) |
| Equipment-Net | $126,000 |
| Total Assets | $266,000 |
| **Liabilities and Shareholders' Equity** | |
| Accounts Payable | $ 33,100 |
| Note Payable | 5,000 |
| Salaries Payable | 1,250 |
| Total Current Liabilities | $ 39,350 |
| Common Stock | $150,000 |
| Retained Earnings | 76,650 |
| Total Shareholders' Equity | $226,650 |
| Total Liabilities and Shareholders' Equity | $266,000 |

**(3)** Paid rent for the month of July of $11,750.
**(4)** Paid salaries to employees during July of $20,600.
**(5)** Collected accounts receivable of $34,150.
**(6)** Paid accounts payable of $38,950.

Information affecting adjusting entries at the end of July is as follows:

**(7)** The firm paid the premium on a one-year insurance policy on March 1, Year 12, with coverage beginning on that date. This is the only insurance policy in force on June 30, Year 12.
**(8)** The firm depreciates its equipment over a ten-year life. Estimated salvage value of the equipment is negligible.
**(9)** Employees earned salaries of $1,600 during the last two days of July but were not paid. These are the only unpaid salaries at the end of July.
**(10)** The note payable is a ninety-day, 6 percent note issued on June 30, Year 12.
**(11)** Merchandise inventory on hand on July 31, Year 12, totals $77,950. The cost of goods sold for July equals merchandise inventory on June 30, Year 12, plus purchases of merchandise during July minus merchandise inventory on July 31, Year 12.

**a.** Prepare journal entries to reflect the transactions and other events during July. The firm classifies expenses by their nature (that is, insurance, depreciation). Revenues and expenses should appear with an indication of the specific revenue or expense account debited or credited. Be sure to indicate whether each entry increases or decreases assets, liabilities, or shareholders' equity.
**b.** Enter the amounts from the June 30, Year 12 balance sheet and the effects of the eleven items above in T-accounts.
**c.** Prepare an income statement for the month of July. Ignore income taxes.
**d.** Enter closing entries in the T-accounts from part **b**.
**e.** Prepare a comparative balance sheet as of June 30, and July 31, Year 12.

## Interpreting and Analyzing the Income Statement

The income statement provides information for assessing the operating profitability of a firm. One tool for analysis, a **common-size income statement,** expresses each expense and net income as a percentage of revenues. A common-size income statement permits an analysis of changes or differences in the relations between revenues, expenses, and net income and identifies relations that the analyst should explore further.

### TIME SERIES ANALYSIS

**Exhibit 3.7** presents a common-size income statement for Bed Bath & Beyond, a retailing firm specializing in home furnishings, for a recent five-year period. The changes in sales from the preceding year were as follows: Year 10, 34.4 percent; Year 11, 29.0 percent; Year 12, 22.2 percent; Year 13, 25.2 percent.

Bed Bath & Beyond experienced a steadily increasing profit margin percentage (= net income divided by sales) during this five-year period. The increasing profit margin results primarily from a decreasing selling-and-administrative-expense-to-sales percentage. The cost-of-goods-sold-to-sales percentage remained relatively stable, suggesting that the firm priced its products at a relatively constant markup on the cost of purchasing the merchandise. Sales increased significantly each year. Many selling and administrative expenses do not increase proportionally with sales. Individual stores need not add employees at the same rate as the increases in sales, for example, if employees are working at less than capacity. The cost of information systems and accounting and legal costs need not increase proportionally with sales. Thus, the rapid sales increases coupled with less-than-proportional increases in selling and administrative expenses led to a decline in this expense percentage.

### CROSS-SECTION ANALYSIS

A common-size income statement also provides information about differences in the economic characteristics and business strategies of firms. **Exhibit 3.8** presents common-size income statements for AK Steel (steel manufacturing), Wal-Mart (discount retailing), Kellogg (branded food processing), Omnicom Group (marketing services), and Pfizer (pharmaceuticals).

Pfizer has the highest profit margin of the five firms. Pharmaceutical products enjoy patent protection, which gives pharmaceutical companies a monopoly on a particular prescription drug until the patent period expires or another firm discovers a superior product. Customers needing the drug view it as a necessity. Medical insurance often pays for all or most of the cost of the drug. Thus, demand for prescription drugs is relatively price inelastic, so changes in selling prices do not have a significant effect on demand. Pfizer also incurs research and development costs up front before obtaining approval to sell the drug. A significant portion of research and

**EXHIBIT 3.7**

**BED BATH & BEYOND**
**Common-Size Income Statement**

| | Year 9 | Year 10 | Year 11 | Year 12 | Year 13 |
|---|---|---|---|---|---|
| Sales | 100.0% | 100.0% | 100.0% | 100.0% | 100.0% |
| Cost of Goods Sold | (58.3) | (58.7) | (58.8) | (58.8) | (58.6) |
| Selling and Administrative Expenses | (30.3) | (30.0) | (29.8) | (29.4) | (28.3) |
| Net Interest Revenue (Expense) | 0.3 | 0.3 | 0.4 | 0.4 | 0.3 |
| Income Tax Expense | (4.7) | (4.5) | (4.6) | (4.7) | (5.2) |
| Net Income | 7.0% | 7.1% | 7.2% | 7.5% | 8.2% |

**EXHIBIT 3.8** Common-Size Income Statements for Five Firms

| | AK Steel | Wal-Mart Stores | Kellogg | Omnicom Group | Pfizer |
|---|---|---|---|---|---|
| Sales | 100.0% | 100.0% | 100.0% | 100.0% | 100.0% |
| Other Revenues | — | 0.9 | 0.3 | 0.2 | 1.2 |
| Cost of Goods or Services Sold | (88.3) | (78.5) | (55.0) | (65.7) | (12.5) |
| Selling and Administrative Expenses | (6.3) | (16.9) | (26.8) | (19.6) | (33.5) |
| Research and Development Expense | — | — | — | — | (16.0) |
| Interest Expense | (3.0) | (0.4) | (4.7) | (0.4) | (0.8) |
| Income Tax Expense | (1.0) | (1.8) | (5.1) | (5.1) | (10.9) |
| Net Income | 1.4% | 3.3% | 8.7% | 9.4% | 27.5% |

development costs does not result in viable products. Pharmaceutical companies argue that they need the large profit margins to induce them to take the risks inherent in incurring high costs researching new drugs without promise of payback. Thus, the need to make up-front expenditures on research and development with an uncertain outcome requires a higher profit rate than do less risky businesses to induce investors to provide financing. Note that the cost of manufacturing pharmaceutical products is a relatively small percentage of the selling price—12.5 percent for Pfizer. The cost of marketing the product to physicians using a sales force and to consumers through advertising and administrative costs represents the largest single expense item. The cost of researching and developing new profitable drugs is larger in this year than the manufacturing costs of the drugs sold. Note that interest expense is a relatively small percentage of sales revenue, suggesting that Pfizer does not borrow extensively. Its most important assets are its employees, who can leave the firm at any time, and its patents, which can become obsolete quickly. Thus, Pfizer and most other pharmaceutical companies do not have significant assets that can serve as collateral for borrowing. In addition, profitable pharmaceutical firms do not likely need to engage in extensive borrowing.

Omnicom Group has the next-largest profit margin percentage. It differentiates its product (marketing services) by the quality of the creative talent it employs. The individual agencies that comprise Omnicom Group have maintained a reputation for developing innovative and effective advertising campaigns. Its customers return and willingly pay for its services, which provide a profit margin larger than that earned by companies without such reputations. Note that interest expense is a relatively small percentage of revenues, similar to Pfizer. The most important asset of marketing services firms is their employees, who cannot serve as collateral for borrowing.

Kellogg has the next-highest profit margin percentage, almost as large as that of Omnicom Group. Whereas a pharmaceutical company enjoys patent protection and a marketing services firm enjoys a reputation for creative employees, Kellogg enjoys trademark protection for its brand names, which provides a degree of monopoly power. Although close substitutes for most branded food products exist, advertising, experience in using the product, and other factors create the impression of a differentiated product and keep customers coming back. Those loyal customers provide Kellogg with a higher profit rate than that earned by similar companies with less well-known brands. Branded products provide predictable cash flows and permit consumer products companies to borrow extensively for new product development and corporate acquisitions.

Wal-Mart and AK Steel both have relatively small profit margin percentages. Neither company sells a differentiated product. Customers can obtain virtually identical products from competitors. Business terminology refers to their products as *commodities*. Demand is relatively price elastic. If either firm raised its prices, customers would likely switch to a competitor. The strategy of both firms is to keep their costs as low as possible and attempt to price their products lower than those of competitors. They strive to make up for the low profit margins by selling higher volumes of products. High profit margins would induce new competitors who, in the case of discount retailers, need only retail space and inventory to enter. Thus competition, both actual and potential, is intense. The manufacturing facilities for steel production are capital-intensive. The need to use the fixed manufacturing facilities as fully as possible induces steel firms to price low in an effort to gain market share. When excess manufacturing capacity exists in the

**EXHIBIT 3.9** **Economic Characteristics of Firms with Low and High Profit**

| Characteristic | Low Profit Margins | High Profit Margins |
|---|---|---|
| Nature of Product | Undifferentiated | Differentiated |
| Barriers to Competition | Low | High |
| Extent of Competition | Highly Competitive | More Monopolistic |
| Price Elasticity of Demand | Relatively Elastic | Relatively Inelastic |

industry, as was the case in this year, competition is intense and profit margins decline. Note the differences in the selling and administrative expense percentages for Wal-Mart and AK Steel. Wal-Mart spends a higher percentage of its sales on marketing and administration, primarily in promoting its stores to consumers. AK Steel spends less on attempting to create demand and relies on its sales force primarily to take customers' orders. The capital-intensive nature of steel manufacturing results in a high expense percentage for interest to finance the manufacturing facilities, which serve as collateral for the borrowing.

**Exhibit 3.9** summarizes the factors that tend to differentiate firms and industries with low profit margins from those with high profit margins. Cross-sectional comparisons of profit margins of various firms are meaningful only when the analysis considers the particular economic characteristics of the industries in which the firms compete.

We explore interpretation and analysis of the income statement more fully in **Chapters 5** and **6**.

## Summary

Net income for a period is the difference between revenues from selling goods and services and the expenses incurred to generate those revenues. In an effort to match expenses with associated revenues, firms typically measure net income using the accrual basis of accounting. Under the accrual basis, firms recognize revenue when they can predict with reasonable assurance the amount of cash they will ultimately receive from customers and the amount of cash they will ultimately expend to make and sell the good or service. They then attempt to match associated expenses with the revenues, to the extent possible, or with the period when they consume assets in operations.

Measurement of net income for a period directly affects financial position at the end of a period. Revenues result in increases in assets or decreases in liabilities. Expenses decrease assets or increase liabilities. Because revenues increase retained earnings, revenue transactions result in credits (increases) to shareholders' equity and in debits either to assets (an increase) or to liabilities (a decrease). Expenses decrease retained earnings and result in debits (decreases) to shareholders' equity and credits either to assets (decreases) or to liabilities (increases).

Some events will not enter the regular day-to-day recording process during the period because no explicit transaction between the firm and some external party (such as a customer, or creditor, or governmental unit) has taken place to require a journal entry. Such events require adjusting entries at the end of the period so that the firm's periodic income and its financial position properly appear in the financial statements reported on an accrual basis.

The accrual basis of accounting uses information about the amount of cash received and expended to measure revenues and expenses. However, the accrual basis does not use the receipt or expenditure of cash to trigger the recognition of revenue or expense. **Exhibit 3.10** summarizes the relation between cash inflows/outflows and the recognition of revenues/expenses.

Interpreting the income statement involves studying the relation between revenues, expenses, and net income both over time and across firms. Changes in economic conditions, technologies, government policies, competitive conditions, and similar factors can affect relations between these items over time. Differences in profit margins across firms and industries relate to the degree of product differentiation, barriers to competition, the extent of competition, and the price elasticity of demand.

**EXHIBIT 3.10**

**Relation of Cash Flows (Receipts and Expenditures) to Recognition of Revenues and Expenses**

| Transaction | Journal Entry Each Year: Year 1 | Year 2 | Year 3 |
|---|---|---|---|
| 1. Cash received from customer in Year 1 for services to be performed in Year 2 | (T)[a] Cash ....... X<br>Liability ..... X | (A)[b] Liability ...... X<br>Revenue ...... X | |
| 2. Cash received from customer in Year 2 for services performed in Year 2 | | (T) Cash ........ X<br>Revenue ...... X | |
| 3. Services performed in Year 2 for which cash will not be received until Year 3 | | (A) Accounts Receivable ... X<br>Revenue ..... X | (T) Cash ........ X<br>Accounts Receivable .... X |
| 4. Cash expended in Year 1 for services consumed in Year 2 | (T) Asset ....... X<br>Cash ........ X | (A) Expense ...... X<br>Asset ........ X | |
| 5. Cash expended in Year 2 for services consumed in Year 2 | | (T) Expense ...... X<br>Cash ........ X | |
| 6. Services consumed in Year 2 for which cash will not be expended until Year 3 | | (A) Expense ...... X<br>Liability ...... X | (T) Liability ...... X<br>Cash ......... X |

[a](T) is a transaction entry made during the period.
[b](A) is an adjusting entry made at the end of a period.

# Appendix 3.1: Using a Spreadsheet to Record Business Transactions

**Appendix 2.1** illustrated the use of a spreadsheet to reflect the effect of business transactions on the balance sheet. This appendix extends the spreadsheet to include income statement transactions and events. Refer to **Exhibit 1.3,** which shows the relation between the balance sheet at the beginning and end of a period and the income statement and statement of cash flows for the period. Note the following relations:

| **Balance Sheet at Beginning of Period** | **Changes in Balance Sheet during the Period** | **Balance Sheet at End of Period** |
|---|---|---|
| Assets | | Assets |
| = | | = |
| Liabilities | | Liabilities |
| + | | + |
| Common Stock | | Common Stock |
| + | | + |
| Retained Earnings ⟶ | Net Income Minus Dividends ⟶ | Retained Earnings |

Retained earnings change between the beginning and end of a period by the amount of net income for the period in excess of dividends declared to shareholders. The income statement shows the amount of net income for the period, but this is not its principal purpose. The purpose of the income statement is to help users understand *why* net income was what it was. A user desiring to assess a firm's past operating performance, project its likely future profitability, and make comparisons of performance across companies will need to know not just the amount of net income, but the revenues and expenses that net to this net income amount. Thus, preparing the income statement requires that we accumulate information about the various types of revenues (sales, interest) and the various types of expenses (cost of goods sold, selling and administrative

**EXHIBIT 3.11** **Transactions Spreadsheet for Miller Corporation for February**

| Balance Sheet Accounts | Balance: Beginning of Period | Transactions, By Number and Description | | | | | | | | | | Balance: End of Period |
|---|---|---|---|---|---|---|---|---|---|---|---|---|
| | | Purchase Inventory On Account | Sell Merchandise To Customers | Recognize Cost of Goods Sold | Pay S&A Expense | Collect Cash from Customer | Pay Cash to Suppliers | Declare and Pay Dividend | Recognize Depreciation Expense | Recognize Insurance Expense | Recognize Inc. Tax Expense | |
| | | 1 | 2 | 3 | 4 | 5 | 6 | 7 | 8 | 9 | 10 | |
| ASSETS | | | | | | | | | | | | |
| **Current Assets:** | | | | | | | | | | | | |
| Cash | 34,400 | | | | −14,500 | 35,000 | −20,000 | −1,000 | | | | 33,900 |
| Accounts Receivable | | | 47,000 | | | −35,000 | | | | | | 12,000 |
| Merchandise Inventory | 15,000 | 25,000 | | −30,000 | | | | | | | | 10,000 |
| Prepaid Insurance | 600 | | | | | | | | | −50 | | 550 |
| Total Current Assets | 50,000 | | | | | | | | | | | 56,450 |
| **Noncurrent Assets:** | | | | | | | | | | | | |
| Equipment | 60,000 | | | | | | | | | | | 60,000 |
| Accumulated Depreciation | | | | | | | | | −1,000 | | | (1,000) |
| Total Noncurrent Assets | 60,000 | | | | | | | | | | | 59,000 |
| Total Assets | 110,000 | | | | | | | | | | | 115,450 |
| LIABILITIES AND SHAREHOLDERS' EQUITY | | | | | | | | | | | | |
| **Current Liabilities:** | | | | | | | | | | | | |
| Accounts Payable | | 25,000 | | | | | −20,000 | | | | | 5,000 |
| Advances from Customer | 3,000 | | −3,000 | | | | | | | | | – |
| Income Tax Payable | | | | | | | | | | | 1,780 | 1,780 |
| Total Current Liabilities | 3,000 | | | | | | | | | | | 6,780 |
| Noncurrent Liabilities: | | | | | | | | | | | | |
| | | | | | | | | | | | | – |
| Total Noncurrent Liabilities | – | | | | | | | | | | | – |
| Total Liabilities | 3,000 | | | | | | | | | | | 6,780 |
| **Shareholders' Equity:** | | | | | | | | | | | | |
| Common Stock | 107,000 | | | | | | | | | | | 107,000 |
| Retained Earnings | | | 50,000 | −30,000 | −14,500 | | | −1,000 | −1,000 | −50 | −1,780 | 1,670 |
| Total Shareholders' Equity | 107,000 | | | | | | | | | | | 108,670 |
| Total Liabilities and Shareholders' Equity | 110,000 | | | | | | | | | | | 115,450 |
| **Imbalance, *If Any*** | – | – | – | – | – | – | – | – | – | – | – | – |
| **Income Statement Accounts** | | | Sales Rev | COGS | S&A Exp | | | | S&A Exp | S&A Exp | Inc Tax Exp | |

expenses, interest expense, income tax expense). We can therefore expand the Retained Earnings line above as follows:

| Retained Earnings at Beginning of Period | + | Revenues (various types) | − | Expenses (various types) | − | Dividends | = | Retained Earnings at End of Period |
|---|---|---|---|---|---|---|---|---|

Although we accumulate information about particular revenues and expenses during a period, revenues and expenses are elements of the Retained Earnings account. Dividends also reduce retained earnings. Their arithmetic effect on retained earnings is the same as expenses. However, accounting views dividends as a distribution of net assets generated by operating activities and not as a cost of generating revenues. Thus, dividends do not appear as expenses on the income statement.

Recall from **Chapter 2** that every transaction or event has a dual effect on the balance sheet equation. Revenues increase retained earnings and must also either increase an asset or reduce a liability. The sale of goods or services increases sales revenues and thereafter retained earnings and either increases Cash (an asset) or Accounts Receivable (an asset) or reduces Advances from Customers (a liability). Expenses decrease retained earnings and must also either decrease an asset or increase a liability. Compensation expense for employees reduces retained earnings and either reduces Cash (an asset) or increases Compensation Payable (a liability). We enter revenues, expenses, and dividend declarations on the retained earnings line of the spreadsheet instead of entering revenues and expenses on separate rows of the spreadsheet and later closing their amounts to the retained earnings line. To enhance preparation of the income statement, we indicate in the cell at the bottom of the spreadsheet the title of the specific revenue or expense account affected (for example, sales revenue, income tax expense). We illustrate the recording of income transactions in the next section.

## Illustration of the Recording of Income Transactions

**Exhibit 3.11** presents a transactions spreadsheet for Miller Corporation for the month of February. The first column of **Exhibit 3.11** shows the amounts from the balance sheet at the end of January. Note that retained earnings show a balance of zero on this date. The firm has not yet generated revenues, nor incurred expenses, nor declared a dividend. **Exhibit 3.11** also shows the effects of the transactions described earlier in this chapter for the month of February. We discuss next the effect of each of the ten transactions or events of Miller Corporation on the spreadsheet.

*Transaction 1:* On February 5, Miller Corporation purchases an additional $25,000 of merchandise on account. This transaction increases inventory and accounts payable. Thus,

**Spreadsheet Entry**

| | |
|---|---|
| (A) Merchandise Inventory . . . . . . . . . . . . . . . . . | +25,000 |
| (L) Accounts Payable . . . . . . . . . . . . . . . . . . . . . | +25,000 |

*Transaction 2:* During the month of February, Miller Corporation sells merchandise to customers for $50,000. Of this amount, $3,000 represents sales to customers who paid $3,000 to Miller Corporation on January 31, which we recorded at the end of January as Advances from Customers. Miller Corporation makes the remaining $47,000 of sales on account. Retail firms typically recognize revenue at the time of delivery of merchandise to customers, regardless of whether or not they have yet received cash from the customers. The spreadsheet entry to reflect these sales is:

**Spreadsheet Entry**

| | |
|---|---|
| (L) Advances from Customer . . . . . . . . . . . . . . . . | −3,000 |
| (A) Accounts Receivable . . . . . . . . . . . . . . . . . . | +47,000 |
| (SE) Retained Earnings (Sales Revenue) . . . . . . . . | +50,000 |

Assets minus liabilities, or net assets, increase by $50,000 [= $47,000 − (−$3,000)] and retained earnings, a shareholders' equity account, increases by $50,000. We enter Sales Rev in the Income Statement Accounts row of the spreadsheet to designate the name of the income account affected by this transaction.

*Transaction 3:* The historical, or acquisition, cost of the merchandise sold to customers in Transaction 2 is $30,000. Because Miller no longer holds this merchandise in inventory, it is no longer an asset. We must reduce the inventory account and recognize the cost of the inventory as an expense to match against the sales revenue in Transaction 2. Thus,

**Spreadsheet Entry**

| | |
|---|---|
| (A) Merchandise Inventory | −30,000 |
| (SE) Retained Earnings (Cost of Goods Sold) | −30,000 |

Firms generally use the account, Cost of Goods Sold, for the cost of items sold and do not include the word "Expense" in the account title, although the account is an expense. Note that expenses reduce retained earnings and shareholders' equity. Thus, the minus sign on the amount for Cost of Goods Sold does not mean that the expense decreases, but that retained earnings and shareholders' equity decrease. We enter COGS in the Income Statement Accounts row of the spreadsheet to remind ourselves of the account title when we are ready to prepare the income statement for February.

*Transaction 4:* Miller Corporation incurs and pays selling and administrative expenses in the aggregate of $14,500 during February. We record this transaction as follows:

**Spreadsheet Entry**

| | |
|---|---|
| (A) Cash | −14,500 |
| (SE) Retained Earnings (Selling and Administrative Expenses) | −14,500 |

Miller Corporation presumably received all of the benefits of these selling and administrative services during February. Thus, the full amount is an expense for the month, which we match against revenues. None of the expenditure results in an asset that would appear on the balance sheet at the end of February. We enter S&A Exp in the Income Statement Accounts row of the spreadsheet to facilitate preparation of the income statement.

*Transaction 5:* Miller Corporation collects $35,000 from customers for sales previously made on account. The firm recognized revenue from these sales at the time of sale (see Transaction 2 for February). The collection of the cash increases cash and decreases accounts receivable. Thus,

**Spreadsheet Entry**

| | |
|---|---|
| (A) Cash | +35,000 |
| (A) Accounts Receivable | −35,000 |

*Transaction 6:* The firm pays $20,000 to suppliers for merchandise previously purchased on account. The entry in the spreadsheet is as follows:

**Spreadsheet Entry**

| | |
|---|---|
| (A) Cash | −20,000 |
| (L) Accounts Payable | −20,000 |

*Transaction 7:* Miller Corporation declares and pays a dividend to shareholders of $1,000. The entries to record the dividend are:

**Spreadsheet Entry**

| | |
|---|---|
| (A) Cash | −1,000 |
| (SE) Retained Earnings (Dividend) | −1,000 |

Note that the effect of a dividend on retained earnings is the same as an expense. Dividends, however, do not appear on the income statement as an expense in measuring net income.

Transactions 1 to 7 involved transactions, or exchanges, with other entities during the month of February. Firms must also make entries, called *adjusting entries,* at the end of the period to measure net income for the period and financial position at the end of the period. Miller Corporation makes the next three entries at the end of February.

*Transaction 8:* The firm recognizes depreciation on the $60,000 of equipment purchased during January. Miller Corporation purchased the equipment during January and recorded it as an asset. It did not begin consuming the services of the equipment until it started using it on February 1. Miller Corporation estimates that the equipment has a five-year life and zero salvage at the end of five years. It will depreciate, or write off, the equipment straight line over the useful life. Thus, depreciation each month is $1,000 (= $60,000/60 months). Miller Corporation includes depreciation of equipment as part of selling and administrative expenses. The firm could record depreciation by reducing the equipment account by $1,000 and reducing retained earnings by $1,000. Instead of reducing the equipment account directly, firms typically use an account called Accumulated Depreciation to record the write-down in the cost of the equipment for services used during the period. The Accumulated Depreciation account appears on the balance sheet as a subtraction from the acquisition cost of the equipment. Thus:

**Spreadsheet Entry**

| | |
|---|---|
| (A) Accumulated Depreciation . . . . . . . . . . . . . . . | −1,000 |
| (SE) Retained Earnings (Selling and Administrative Expenses) . . . . . . | −1,000 |

Note that the minus sign on the Accumulated Depreciation account in the spreadsheet entry indicates the effect on assets, not on accumulated depreciation. Likewise, the minus sign on Selling and Administrative Expenses indicates the effect on retained earnings and shareholders' equity, not on depreciation expense.

*Transaction 9:* Miller Corporation records the portion of the cost of the prepaid insurance of $600 attributable to insurance services received during February. The firm paid the one-year insurance premium on January 31 for insurance coverage from February 1 of this year through January 31 of next year. Assuming the allocation of an equal amount of this insurance premium to each month of the year, $50 (= $600/12) applies to February. Miller Corporation includes the cost of insurance in selling and administrative expenses. The entries are:

**Spreadsheet Entry**

| | |
|---|---|
| (A) Prepaid Insurance . . . . . . . . . . . . . . . . . . . . | −50 |
| (SE) Retained Earnings (Selling and Administrative Expenses) . . . . . . | −50 |

The remaining $550 of prepaid insurance becomes an expense during the next eleven months. Note that firms typically reduce the Prepaid Insurance account directly, instead of using an account similar to accumulated depreciation for equipment.

*Transaction 10:* Miller Corporation recognizes income tax expense on the net income before income taxes for February. Assume an income tax rate of 40 percent. Net income before income taxes for February equals $4,450 (= $50,000 − $30,000 − $14,500 − $1,000 − $50). Income tax expense is therefore $1,780 (= 0.40 × $4,450). Firms pay their income taxes quarterly. Thus, the income taxes remain unpaid at the end of February. Even though the income taxes are not yet payable, they represent an expense that Miller Corporation must match against the net income before taxes for February under the accrual basis of accounting. The entries are:

**Spreadsheet Entry**

| | |
|---|---|
| (L) Income Tax Payable . . . . . . . . . . . . . . . . . . | +1,780 |
| (SE) Retained Earnings (Income Tax Expense) . . . . . . . . . . . . . . . . . | −1,780 |

This completes the spreadsheet for Miller Corporation for February. The amounts on the retained earnings row show the revenues, expenses, and dividends that net to the change in retained earnings for the month. Using the account titles on the Income Statement Accounts row and the corresponding amounts for revenues and expenses on the Retained Earnings row, we can prepare an income statement for the month of February. The final column on the right in **Exhibit 3.11** shows the amounts for the balance sheet at the end of February. We presented earlier in this chapter the income statement for February **(Exhibit 3.4)** and a comparative balance sheet for January 31 and February 28 **(Exhibit 3.5)**.

## Problem 3.4 for Self-Study

**Preparation of Transactions Spreadsheet.** Refer to **Problem 3.3 for Self-Study**. Prepare a transactions spreadsheet for Rybowiak's Building Supplies for the month of July, Year 12.

## Solutions to Self-Study Problems

### SUGGESTED SOLUTION TO PROBLEM 3.1 FOR SELF-STUDY

(J. Thompson; cash versus accrual basis of accounting.)

**a.** and **b.**

| | a. Accrual Basis | b. Cash Basis |
|---|---|---|
| Sales Revenue | $29,000 | $22,000 |
| Less Expenses: | | |
| Cost of Merchandise Sold | $13,000 | — |
| Payments on Merchandise Purchased | — | $10,000 |
| Rental Expense | 1,000 | 2,000 |
| Insurance Expense | 100 | 1,200 |
| Interest Expense | 40 | — |
| Utilities Expense | 400 | 400 |
| Salaries Expense | 650 | 650 |
| Taxes Expense | 350 | 350 |
| Total Expenses | $15,540 | $14,600 |
| Net Income | $13,460 | $ 7,400 |

**c.** The accrual basis gives a better measure of operating performance because it matches revenue generated during January with costs incurred in generating that revenue. The cash basis emphasizes cash inflows and outflows, regardless of whether inflows and outflows relate to the same goods or services sold. Note that the capital contributed by Thompson and the bank loan do not give rise to revenue under either basis of accounting because they are financing, not operating, activities.

### SUGGESTED SOLUTION TO PROBLEM 3.2 FOR SELF-STUDY

(Revenue and expense recognition.)

**a.** None.
**b.** $14,980.
**c.** $5,820.
**d.** None.
**e.** $5,440.
**f.** $2,010.
**g.** None.

### SUGGESTED SOLUTION TO PROBLEM 3.3 FOR SELF-STUDY

(Rybowiak Building Supplies; journal entries, T-accounts, income statement and balance sheet preparation.)

**a.**

| | | |
|---|---|---|
| **(1)** Accounts Receivable | 85,000 | |
| Sales Revenue | | 85,000 |

| Assets | = | Liabilities | + | Shareholders' Equity | (Class.) |
|---|---|---|---|---|---|
| +85,000 | | | | +85,000 | IncSt → RE |

Sales of merchandise on account.

(2) Merchandise Inventory . . . . . . . . . . 46,300
Accounts Payable . . . . . . . . . . 46,300

| Assets | = | Liabilities | + | Shareholders' Equity | (Class.) |
|---|---|---|---|---|---|
| +46,300 | | +46,300 | | | |

Purchase of merchandise inventory on account.

(3) Rent Expense . . . . . . . . . . 11,750
Cash . . . . . . . . . . 11,750

| Assets | = | Liabilities | + | Shareholders' Equity | (Class.) |
|---|---|---|---|---|---|
| −11,750 | | | | −11,750 | IncSt → RE |

Paid rent for July.

(4) Salaries Payable . . . . . . . . . . 1,250
Salary Expense . . . . . . . . . . 19,350
Cash . . . . . . . . . . 20,600

| Assets | = | Liabilities | + | Shareholders' Equity | (Class.) |
|---|---|---|---|---|---|
| −20,600 | | −1,250 | | −19,350 | IncSt → RE |

Paid salaries during July.

(5) Cash . . . . . . . . . . 34,150
Accounts Receivable . . . . . . . . . . 34,150

| Assets | = | Liabilities | + | Shareholders' Equity | (Class.) |
|---|---|---|---|---|---|
| +34,150 | | | | | |
| −34,150 | | | | | |

Collected accounts receivable.

(6) Accounts Payable . . . . . . . . . . 38,950
Cash . . . . . . . . . . 38,950

| Assets | = | Liabilities | + | Shareholders' Equity | (Class.) |
|---|---|---|---|---|---|
| −38,950 | | −38,950 | | | |

Paid accounts payable.

| | | |
|---|---|---|
| **(7)** Insurance Expense | 50 | |
| Prepaid Insurance | | 50 |

| Assets | = | Liabilities | + | Shareholders' Equity | (Class.) |
|---|---|---|---|---|---|
| −50 | | | | −50 | IncSt → RE |

Recognize insurance expense for July: $50 = $400/8. The Prepaid Insurance account has a balance of $400 on June 30, Year 12. Because the firm paid the one-year insurance premium on March 1, Year 12, we know that the policy has eight remaining months on June 30, Year 12.

| | | |
|---|---|---|
| **(8)** Depreciation Expense | 1,750 | |
| Accumulated Depreciation | | 1,750 |

| Assets | = | Liabilities | + | Shareholders' Equity | (Class.) |
|---|---|---|---|---|---|
| −1,750 | | | | −1,750 | IncSt → RE |

Recognize depreciation for July: $1,750 = $210,000/120.

| | | |
|---|---|---|
| **(9)** Salary Expense | 1,600 | |
| Salaries Payable | | 1,600 |

| Assets | = | Liabilities | + | Shareholders' Equity | (Class.) |
|---|---|---|---|---|---|
| | | +1,600 | | −1,600 | IncSt → RE |

Recognize unpaid salaries for July.

| | | |
|---|---|---|
| **(10)** Interest Expense | 25 | |
| Interest Payable | | 25 |

| Assets | = | Liabilities | + | Shareholders' Equity | (Class.) |
|---|---|---|---|---|---|
| | | +25 | | −25 | IncSt → RE |

Recognize unpaid interest for July: $25 = $5,000 × .06 × 30/360.

| | | |
|---|---|---|
| **(11)** Cost of Goods Sold | 36,500 | |
| Merchandise Inventory | | 36,500 |

| Assets | = | Liabilities | + | Shareholders' Equity | (Class.) |
|---|---|---|---|---|---|
| −36,500 | | | | −36,500 | IncSt → RE |

To recognize cost of goods sold for July: $36,500 = $68,150 + $46,300 − $77,950.

**b.** and **d. Exhibit 3.12** presents T-accounts for Rybowiak Building Supplies.

**EXHIBIT 3.12**

**RYBOWIAK BUILDING SUPPLIES**
**T-Accounts**
**(Problem 3.3 for Self-Study)**

**Cash**

| | Debit | Credit | |
|---|---|---|---|
| Bal. | 44,200 | 11,750 | (3) |
| (5) | 34,150 | 20,600 | (4) |
| | | 38,950 | (6) |
| Bal. | 7,050 | | |

**Accounts Receivable**

| | Debit | Credit | |
|---|---|---|---|
| Bal. | 27,250 | 34,150 | (5) |
| (1) | 85,000 | | |
| Bal. | 78,100 | | |

**Merchandise Inventory**

| | Debit | Credit | |
|---|---|---|---|
| Bal. | 68,150 | 36,500 | (11) |
| (2) | 46,300 | | |
| Bal. | 77,950 | | |

**Prepaid Insurance**

| | Debit | Credit | |
|---|---|---|---|
| Bal. | 400 | 50 | (7) |
| Bal. | 350 | | |

**Equipment**

| | Debit | Credit | |
|---|---|---|---|
| Bal. | 210,000 | | |
| Bal. | 210,000 | | |

**Accumulated Depreciation**

| | Debit | Credit | |
|---|---|---|---|
| | | 84,000 | Bal. |
| | | 1,750 | (8) |
| | | 85,750 | Bal. |

**Accounts Payable**

| | Debit | Credit | |
|---|---|---|---|
| | | 33,100 | Bal. |
| (6) | 38,950 | 46,300 | (2) |
| | | 40,450 | Bal. |

**Note Payable**

| | Debit | Credit | |
|---|---|---|---|
| | | 5,000 | Bal. |
| | | 5,000 | Bal. |

**Salaries Payable**

| | Debit | Credit | |
|---|---|---|---|
| (4) | 1,250 | 1,250 | Bal. |
| | | 1,600 | (9) |
| | | 1,600 | Bal. |

**Common Stock**

| | Debit | Credit | |
|---|---|---|---|
| | | 150,000 | Bal. |
| | | 150,000 | Bal. |

**Retained Earnings**

| | Debit | Credit | |
|---|---|---|---|
| | | 76,650 | Bal. |
| | | 13,975 | (12) |
| | | 90,625 | Bal. |

**Sales Revenue**

| | Debit | Credit | |
|---|---|---|---|
| (12) | 85,000 | 85,000 | (1) |

**Rent Expense**

| | Debit | Credit | |
|---|---|---|---|
| (3) | 11,750 | 11,750 | (12) |

**Salaries Expense**

| | Debit | Credit | |
|---|---|---|---|
| (4) | 19,350 | | |
| (9) | 1,600 | | |
| | 20,950 | 20,950 | (12) |

**Insurance Expense**

| | Debit | Credit | |
|---|---|---|---|
| (7) | 50 | 50 | (12) |

**Depreciation Expense**

| | Debit | Credit | |
|---|---|---|---|
| (8) | 1,750 | 1,750 | (12) |

**Interest Expense**

| | Debit | Credit | |
|---|---|---|---|
| (10) | 25 | 25 | (12) |

**Interest Payable**

| | Debit | Credit | |
|---|---|---|---|
| | | 25 | (10) |
| | | 25 | Bal. |

**Cost of Goods Sold**

| | Debit | Credit | |
|---|---|---|---|
| (11) | 36,500 | 36,500 | (12) |

c.

RYBOWIAK'S BUILDING SUPPLIES
Income Statement
For the Month of July, Year 12

| | | |
|---|---|---|
| Sales Revenue | | $85,000 |
| Less Expenses: | | |
| Cost of Goods Sold | $36,500 | |
| Salaries Expense ($19,350 + $1,600) | 20,950 | |
| Rent Expense | 11,750 | |
| Depreciation Expense | 1,750 | |
| Insurance Expense | 50 | |
| Interest Expense | 25 | (71,025) |
| Net Income | | $13,975 |

e.

RYBOWIAK'S BUILDING SUPPLIES
Balance Sheet
June 30 and July 31, Year 12

| | June 30 | July 31 |
|---|---|---|
| ASSETS | | |
| **Current Assets:** | | |
| Cash | $ 44,200 | $ 7,050 |
| Accounts Receivable | 27,250 | 78,100 |
| Merchandise Inventory | 68,150 | 77,950 |
| Prepaid Insurance | 400 | 350 |
| Total Current Assets | $140,000 | $163,450 |
| **Noncurrent Assets:** | | |
| Equipment—at Cost | $210,000 | $210,000 |
| Less Accumulated Depreciation | (84,000) | (85,750) |
| Total Noncurrent Assets | $126,000 | $124,250 |
| Total Assets | $266,000 | $287,700 |
| **LIABILITIES AND SHAREHOLDERS' EQUITY** | | |
| **Current Liabilities:** | | |
| Accounts Payable | $ 33,100 | $ 40,450 |
| Note Payable | 5,000 | 5,000 |
| Salaries Payable | 1,250 | 1,600 |
| Interest Payable | — | 25 |
| Total Current Liabilities | $ 39,350 | $ 47,075 |
| **Shareholders' Equity:** | | |
| Common Stock | $150,000 | $150,000 |
| Retained Earnings | 76,650 | 90,625 |
| Total Shareholders' Equity | $226,650 | $240,625 |
| Total Liabilities and Shareholders' Equity | $266,000 | $287,700 |

## SUGGESTED SOLUTION TO PROBLEM 3.4 FOR SELF-STUDY

(Rybowiak's Building Supplies; preparation of transactions spreadsheet.)

See **Exhibit 3.13.**

## EXHIBIT 3.13 Transactions Spreadsheet for Rybowiak Building Supplies for July (Problem 3.4 for Self-Study)

| Balance Sheet Accounts | | Transactions, by Number and Description | | | | | | | | | | | |
|---|---|---|---|---|---|---|---|---|---|---|---|---|---|
| | **Balance: Beginning of Period** | Sell Merchandise on Account | Purchase Merchandise on Account | Pay July Rent | Pay Salaries during July | Collect Accounts Receivable | Pay Accounts Payment | Recognize Insurance for July | Recognize Depreciation for July | Recognize Unpaid Sal. for July | Recognize Interest for July | Recognize COGS for July | **Balance: End of Period** |
| | | **1** | **2** | **3** | **4** | **5** | **6** | **7** | **8** | **9** | **9** | **10** | |
| ASSETS | | | | | | | | | | | | | |
| **Current Assets:** | | | | | | | | | | | | | |
| Cash | 44,200 | | | −11,750 | −20,600 | 34,150 | −38,950 | | | | | | 7,050 |
| Accounts Receivable | 27,250 | 85,000 | | | | −34,150 | | | | | | | 78,100 |
| Merchandise Inventory | 68,150 | | 46,300 | | | | | | | | | −36,500 | 77,950 |
| Prepaid Insurance | 400 | | | | | | | −50 | | | | | 350 |
| Total Current Assets | 140,000 | | | | | | | | | | | | 163,450 |
| **Noncurrent Assets:** | | | | | | | | | | | | | |
| Equipment | 210,000 | | | | | | | | | | | | 210,000 |
| Accumulated Depreciation | −84,000 | | | | | | | | −1,750 | | | | (85,750) |
| Total Noncurrent Assets | 126,000 | | | | | | | | | | | | 124,250 |
| Total Assets | 266,000 | | | | | | | | | | | | 287,700 |
| LIABILITIES AND SHAREHOLDERS' EQUITY | | | | | | | | | | | | | |
| **Current Liabilities:** | | | | | | | | | | | | | |
| Accounts Payable | 33,100 | | 46,300 | | | | −38,950 | | | | | | 40,450 |
| Notes Payable | 5,000 | | | | | | | | | | | | 5,000 |
| Salaries Payable | 1,250 | | | | −1,250 | | | | | 1,600 | | | 1,600 |
| Interest Payable | | | | | | | | | | | 25 | | 25 |
| Total Current Liabilities | 39,350 | | | | | | | | | | | | 47,075 |
| **Noncurrent Liabilities:** | | | | | | | | | | | | | |
| | | | | | | | | | | | | – | |
| Total Noncurrent Liabilities | – | | | | | | | | | | | | – |
| Total Liabilities | 39,350 | | | | | | | | | | | | 47,075 |
| **Shareholders' Equity:** | | | | | | | | | | | | | |
| Common Stock | 150,000 | | | | | | | | | | | | 150,000 |
| Retained Earnings | 76,650 | 85,000 | | −11,750 | −19,350 | | | −50 | −1,750 | −1,600 | −25 | −36,500 | 90,625 |
| Total Shareholders' Equity | 226,650 | | | | | | | | | | | | 240,625 |
| Total Liabilities and Shareholders' Equity | 266,000 | | | | | | | | | | | | 287,700 |
| | | | | | | | | | | | | | |
| ***Imbalance, If Any*** | – | – | – | – | – | – | – | – | – | – | – | – | |
| **Income Statement Accounts** | | Sales Rev | | Rent Exp | Salary Exp | | | Insurance Exp | Depre. Exp | Salary Exp | Interest Exp | COGS | |

## Key Terms and Concepts

Profit margin percentage
Net income or net loss
Revenues
Expenses
Accounting period
Natural business year or fiscal period
Cash basis of accounting
Accrual basis of accounting
Unexpired costs
Expired costs
Matching principle
Product costs
Period expenses
Closing process
Temporary accounts
Permanent accounts
Dividends
Common-size income statement
Times Series Analysis
Cross Section Analysis

## Questions, Short Exercises, Exercises, Problems, and Cases

For additional student resources, content, and interactive quizzes for this chapter visit the FACMU website: **www.thomsonedu.com/accounting/stickney**

### QUESTIONS

1. Review the meaning of the terms and concepts listed above in Key Terms and Concepts.
2. "If the accounting period were long enough, the distinction between the cash and accrual basis of accounting would disappear." Do you agree? Why or why not?
3. "Cash flows determine the amount of revenue and expense but not the timing of their recognition." Explain.
4. "The valuation of assets and liabilities relates closely to the measurement of revenues and expenses." Explain.
5. Distinguish between a cost and an expense.
6. "The key term for understanding the accrual basis of accounting is matching." Explain.
7. Both interest expense on borrowing and dividends on common stock reduce net assets and reduce shareholders' equity. Yet, accountants treat interest as an expense in measuring net income but do not treat dividends on common stock as an expense. Explain the rationale for this apparent inconsistency.
8. What is the purpose of using contra accounts? What is the alternative to using them?

### SHORT EXERCISES

9. **Analyzing changes in accounts receivable.** Microsoft Corporation reported a balance of $5,129 million in accounts receivable at the beginning of Year 13 and $5,196 million at the end of Year 13. Its income statement reported sales revenue of $32,120 million for Year 13. Assuming that Microsoft Corporation makes all sales on account, compute the amount of cash collected from customers during Year 13.
10. **Analyzing changes in inventory.** General Electric Company reported a balance of $8,565 million in inventory at the beginning of Year 12 and $9,247 million at the end of Year 12. Its income statement reported cost of goods sold of $63,072 million for Year 12. Compute the cost of inventory either purchased or manufactured during Year 12.
11. **Analyzing changes in inventory and accounts payable.** Ann Taylor Stores reported a balance in inventories of $180.1 million at the beginning of Year 13 and $185.5 million at the end of Year 13. It reported a balance in accounts payable for merchandise purchases of $52.0 million at the beginning of Year 13 and $57.1 million at the end of Year 13. Its income statement showed cost of goods sold of $633.5 million for Year 13. Compute the amount of cash paid to suppliers of inventory during Year 13 for purchases made on account.
12. **Analyzing changes in salaries payable.** AMR, the parent company of American Airlines, reported a balance in salaries payable of $721 million at the beginning of Year 12 and $705 million at the end of Year 12. Its income statement reported salary expense of $8,392 million for Year 12. Compute the amount of cash paid to employees for salaries during Year 12.
13. **Analyzing changes in retained earnings.** Johnson & Johnson, a pharmaceutical and medical products company, reported a balance in retained earnings of $23,066 million at the beginning of Year 12 and $26,571 million at the end of Year 12. It reported dividends declared and paid of $3,092 million for Year 12. Compute the amount of net income for Year 12.

14. **Computing interest expense.** Gillette borrowed $250 million on October 1, Year 12. The debt carries an annual interest rate of 6 percent and is payable in two installments, on April 1 and October 1 of each year. The debt matures on October 1, Year 32. Gillette's accounting period ends on December 31 of each year. Compute the amount of interest expense for Year 12 assuming that Gillette uses the accrual basis of accounting.
15. **Computing income tax expense and income taxes paid.** Radio Shack reported a balance in Income Taxes Payable of $78.1 million at the beginning of Year 13 and $60.1 million at the end of Year 13. Net income before income taxes for Year 13 totaled $424.9 million. The firm is subject to an income tax rate of 38 percent. Compute the amount of cash payments made for income taxes during Year 13.
16. **Journal entry for prepaid rent.** A firm reported a balance in its Prepaid Rent account of $1,200 on January 1, Year 12. On February 1, Year 12, the firm paid $18,000 as the annual rental for the period from February 1, Year 12 to January 31, Year 13. It recorded this rental payment by debiting Rent Expense and crediting Cash for $18,000. This is the only entry made in the accounting records during Year 12 related to rent. The firm uses the calendar year as its reporting period. Give the journal entry that this firm must make on December 31, Year 12, to correctly report Prepaid Rent and Rent Expense in its accounts.
17. **Journal entry to correct recording error.** A firm acquired equipment on January 1, Year 12 for $20,000. The equipment has an expected useful life of 5 years and zero salvage value. The firm recorded the acquisition by debiting Retained Earnings (Equipment Expense) and crediting Cash for $20,000. Give the journal entries that the firm must make on December 31, Year 12 to correct its initial recording error and any related effects (ignore income tax effects).

## EXERCISES

18. **Revenue recognition.** Neiman Marcus uses the accrual basis of accounting and recognizes revenue at the time it sells merchandise or renders services. Indicate the amount of revenue (if any) the firm recognizes during the months of February, March, and April in each of the following transactions. The firm does the following:
    - **a.** Collects $800 cash from a customer during March for a custom-made suit that the firm will make and deliver to the customer in April.
    - **b.** Collects $2,160 cash from customers for meals served in the firm's restaurant during March.
    - **c.** Collects $39,200 cash from customers during March for merchandise sold and delivered in February.
    - **d.** Sells merchandise to customers during March on account, for which the firm will collect $59,400 cash from customers during April.
    - **e.** Rents space in its store to a travel agency for $9,000 a month, effective March 1. Receives $18,000 cash on March 1 for two months' rent.
    - **f.** Same as part **e,** except that it receives the check for the March and April rent on April 1.
19. **Revenue recognition.** Indicate which of the following transactions or events immediately gives rise to the recognition of revenue under the accrual basis of accounting.
    - **a.** A custom T-shirt manufacturer completes production of an order for an upcoming rock concert. It has not yet delivered the T-shirts to the purchaser or received any cash.
    - **b.** The publisher of *Business Week* ships magazines that subscribers paid for in advance.
    - **c.** The Florida Marlins, a professional baseball team, sells season tickets for next year's games.
    - **d.** General Mills earns interest on a bond issued by Ford Motor Company. Ford will pay the interest next year.
    - **e.** Sears collects cash from customers for sales made on account last year.
    - **f.** Interpublic Group performs advertising services for clients on account. The advertisements will appear in three months.
20. **Expense recognition.** Sun Microsystems uses the accrual basis of accounting and recognizes revenue at the time it sells goods or renders services. Indicate the amount of expenses (if any) the firm recognizes during the months of June, July, and August in each of the following transactions. The firm does the following:
    - **a.** Pays rent of $180,000 on July 1 for one year's rent on a warehouse beginning on that date.

**b.** Receives a utility bill on July 2 totaling $4,560 for services received during June. It pays the utility bill during July.

**c.** Purchases office supplies on account costing $12,600 during July. It pays $5,500 for these purchases during July and the remainder during August. Office supplies on hand on July 1 cost $2,400, on July 31 cost $9,200, and on August 31 cost $2,900.

**d.** Pays $7,200 on July 15 for property taxes on office facilities for the current calendar year.

**e.** Pays $2,000 on July 15 as a deposit on a custom-made delivery van that the manufacturer will deliver in September.

**f.** Pays $4,500 on July 25 as an advance on the August salary of an employee.

**g.** Pays $6,600 on July 25 for advertisements that appeared in computer journals during June.

**21. Expense recognition.** Kroger Stores uses the accrual basis of accounting and recognizes revenue at the time it sells goods or renders services. Indicate the amount of expense recognized during October (if any) from each of the following transactions or events:

**a.** Pays $4,670 on October 5 for advertisements that appeared in local newspapers during September.

**b.** Pays $12,000 on October 6 for refrigerators delivered to its stores on September 30. The firm expects the refrigerators to last for five years and have no salvage value.

**c.** Pays $24,000 on October 10 for property taxes for the period from October 1 of this year to September 30 of next year.

**d.** Pays $12,900 on October 15 for cleaning supplies purchased on October 10. Cleaning supplies on hand on October 1 cost $2,900 and on October 31 cost $3,500.

**e.** Pays $1,200 on October 20 for repairs done to a forklift truck on October 1. The truck had a remaining useful life of five years on October 1.

**f.** Pays $10,000 on October 25 as a deposit on land Kroger plans to purchase for a new store.

**g.** Pays $20,000 on October 31 as rent on a warehouse for October and November.

**22. Identifying missing half of journal entries.** In the business world, many transactions are routine and repetitive. Because accounting records business transactions, many accounting entries are also routine and repetitive. Knowing one-half of an entry in the double-entry recording system often permits a reasoned guess about the other half. The items below give the account name for one-half of an entry. Indicate your best guess as to the nature of the transaction being recorded and the name of the account of the routine other half of the entry. Also indicate whether the transaction increases or decreases the other account.

**a.** Debit: Cost of Goods Sold.

**b.** Debit: Accounts Receivable.

**c.** Credit: Accounts Receivable.

**d.** Debit: Accounts Payable.

**e.** Credit: Accounts Payable.

**f.** Credit: Accumulated Depreciation.

**g.** Debit: Retained Earnings.

**h.** Credit: Prepaid Insurance.

**i.** Debit: Property Taxes Payable.

**j.** Debit: Merchandise Inventory.

**k.** Debit: Advances from Customers.

**l.** Credit: Advances to Suppliers.

**23. Journal entries for notes receivable and notes payable.** Corner Grocery Store borrows $100,000 from Citizen's Bank on December 1, Year 12, using a ninety-day note as evidence of the loan. The note bears interest at an annual rate of 6 percent and is due on March 1, Year 13. Both firms use the calendar year as their reporting period.

**a.** Present journal entries for the Corner Grocery Store from December 1, Year 12, through payment at maturity. Assume each calendar month contains thirty days.

**b.** Present journal entries for Citizen's Bank from December 1, Year 12, through repayment at maturity.

**24. Journal entries for inventories.** On January 1, Year 12, the Merchandise Inventories account of Dollar General had a balance of $1,131 million. During Year 12, Dollar General purchased inventories on account for $4,368 million. On December 31, Year 12, it finds that merchandise inventory still on hand amounts to $1,123 million. The Accounts Payable account had a balance of $322 million on January 1, Year 12, and $341 million on December

31, Year 12. Present journal entries to account for all changes in the Merchandise Inventories and Accounts Payable accounts during Year 12.

25. **Journal entries for insurance.** Eason Corporation commenced business on September 1, Year 12. It paid the one-year insurance premium of $3,600 for property and liability protection on this date and debited Prepaid Insurance. On September 1, Year 13, it renewed the insurance policy and paid the $4,800 one-year insurance premium, again debiting Prepaid Insurance. Eason Corporation uses the calendar year as its reporting period. Give the journal entries that Eason Corporation would make on September 1, Year 12, December 31, Year 12, September 1, Year 13, and December 31, Year 13.

26. **Effect of errors on financial statements.** Using the notations O/S (overstated), U/S (understated), and NO (no effect), indicate the effects (direction and amount) on assets, liabilities, and shareholders' equity as of December 31, Year 5, of the following independent errors or omissions. Ignore income tax implications.
    **a.** On December 1, Year 5, a firm paid $12,000 for rental of a building for December, Year 5 and January, Year 6. The firm debited Rent Expense and credited Cash on December 1 and made no further entries with respect to this rental during December or January.
    **b.** On December 15, Year 5, a firm received $1,200 from a customer as a deposit on merchandise the firm expects to deliver to the customer in January, Year 6. The firm debited Cash and credited Sales Revenue on December 15 and made no further entries with respect to this deposit during December or January.
    **c.** On December 1, a firm acquired a used forklift truck costing $4,800. The truck was expected to have a two-year life and zero salvage value. The firm recorded the transaction by debiting Repair Expense and crediting Cash for $4,800 and made no further entries during December with respect to the acquisition.
    **d.** On December 15, a firm purchased office supplies costing $8,600. It recorded the purchase by debiting Office Supplies Expense and crediting Cash. The Office Supplies Inventory account on December 1 had a balance of $2,700. The firm took a physical inventory of office supplies on December 31 and found that office supplies costing $2,450 were on hand. The firm made no entries in its accounts with respect to office supplies on December 31.
    **e.** A firm incurs expense for interest of $1,500 for the month of December on a forty-five-day loan obtained on December 1. The firm properly recorded the loan on its books on December 1 but made no entry to record interest on December 31. The loan is payable with interest on January 15, Year 6.
    **f.** A firm purchased merchandise on account costing $11,600 on December 23, Year 5, debiting Merchandise Inventory and crediting Accounts Payable. The firm paid for this purchase on December 28, Year 5, debiting Cost of Goods Sold and crediting Cash. The merchandise had not been sold as of December 31, Year 5.

27. **Effect of recording errors on financial statements.** Forgetful Corporation neglected to make various adjusting entries on December 31, Year 8, the end of its accounting period. Indicate the effects on assets, liabilities, and shareholders' equity on December 31, Year 8, of failing to adjust for the following independent items as appropriate, using the notations O/S (overstated), U/S (understated), and NO (no effect). Also, give the amount of the effect. Ignore income tax implications.
    **a.** On December 15, Year 8, Forgetful Corporation received a $1,400 advance from a customer for products to be manufactured and delivered in January, Year 9. The firm recorded the advance by debiting Cash and crediting Sales Revenue and has made no adjusting entry as of December 31, Year 8.
    **b.** On July 1, Year 8, Forgetful Corporation acquired a machine for $5,000 and recorded the acquisition by debiting Cost of Goods Sold and crediting Cash. The machine has a five-year useful life and zero estimated salvage value.
    **c.** On November 1, Year 8, Forgetful Corporation received a $2,000 note receivable from a customer in settlement of an account receivable. It debited Notes Receivable and credited Accounts Receivable on receipt of the note. The note is a six-month note due April 30, Year 9, and bears interest at an annual rate of 12 percent. Forgetful Corporation made no other entries related to this note during Year 8.
    **d.** Forgetful Corporation paid its annual insurance premium of $1,200 on October 1, Year 8, the first day of the year of coverage. It debited Prepaid Insurance $900, debited Insurance Expense $300, and credited Cash for $1,200. It made no other entries related to this insurance during Year 8.

**e.** The Board of Directors of Forgetful Corporation declared a dividend of $1,500 on December 31, Year 8. The dividend will be paid on January 15, Year 9. Forgetful Corporation neglected to record the dividend declaration.

**f.** On December 1, Year 8, Forgetful Corporation purchased a machine on account for $50,000, debiting Machinery and crediting Accounts Payable for $50,000. Ten days later, the account was paid, and the company took the allowed 2 percent discount. Cash was credited $49,000, Miscellaneous Revenue was credited $1,000, and Accounts Payable was debited $50,000. It is the policy of Forgetful Corporation to record cash discounts taken as a reduction in the cost of assets. On December 28, Year 8, the machine was installed for $4,000 in cash; Maintenance Expense was debited and Cash was credited for $4,000. The machine started operation on January 1, Year 9. Since the machine was not placed into operation until January 1, Year 9, as appropriate, no depreciation was recorded for Year 8.

## PROBLEMS AND CASES

**28. Cash versus accrual basis of accounting.** Ailawadi Corporation began operations on January 1, Year 6. The firm's cash account revealed the following transactions for the month of January.

| Date | Transaction | Amount |
|---|---|---|
| **Cash Receipts** | | |
| Jan. 1 | Investment by Praveen Ailawadi for 100 percent of Ailawadi Corporation's Common Stock | $ 75,000 |
| Jan. 1 | Loan from Upper Valley Bank, due June 30, Year 6, with interest at 6 percent per year | 40,000 |
| Jan. 15 | Advance from a customer for merchandise scheduled for delivery in February, Year 6 | 1,200 |
| Jan. 1–31 | Sales to customers | 60,000 |
| | **Cash Disbursements** | |
| Jan. 1 | Rental of retail space at a monthly rental of $3,000 | (6,000) |
| Jan. 1 | Purchase of display equipment (three-year life, zero salvage value) | (36,000) |
| Jan. 1 | Premium on property and liability insurance for coverage from January 1 to December 31, Year 6 | (1,200) |
| Jan. 15 | Payment of utility bills | (750) |
| Jan. 16 | Payment of salaries | (3,500) |
| Jan. 1–31 | Purchase of merchandise | (44,800) |
| | Balance, January 31, Year 6 | $ 83,950 |

The following information relates to Ailawadi Corporation as of January 31, Year 6:

**(1)** Customers owe the firm $9,500 from sales made during January.
**(2)** The firm owes suppliers $3,900 for merchandise purchased during January.
**(3)** Unpaid utility bills total $260, and unpaid salaries total $1,260.
**(4)** Merchandise inventory on hand totals $5,500.

**a.** Prepare an income statement for January, assuming that Ailawadi Corporation uses the accrual basis of accounting and recognizes revenue at the time it sells goods to customers.

**b.** Prepare an income statement for January, assuming that Ailawadi Corporation uses a cash basis of accounting.

**c.** Which basis of accounting do you believe provides a better indication of the operating performance of the firm during January? Why?

**29. Cash versus accrual basis of accounting.** McKindly Consultants, Inc., opens a consulting business on July 1, Year 2. Dick Gray and Paula Mosenthal each contribute $50,000 cash for shares of the firm's common stock. The corporation borrows $60,000 from a local bank on August 1, Year 2. The loan is repayable on July 31, Year 3, with interest at the rate of 8 percent per year. The firm rents office space on August 1, paying two months' rent in advance. It pays the remaining monthly rental fees of $1,500 per month on the first of each month, beginning October 1. The firm purchases office equipment with a five-year life for cash on August 1 for $24,000. The firm renders consulting services for clients between August 1 and December 31, Year 2, totaling $135,000. It collects $109,000 of this amount by year-end. It incurs and pays other costs by the end of the year as follows: utilities, $3,460; salaries,

$98,500; supplies, $2,790. It has unpaid bills at year-end as follows: utilities, $580; salaries, $11,300; supplies, $840. The firm used all the supplies it had acquired.

**a.** Prepare an income statement for the five months ended December 31, Year 2, assuming that the corporation uses the accrual basis of accounting and recognizes revenue at the time it renders services.

**b.** Prepare an income statement for the five months ended December 31, Year 2, assuming that the corporation uses the cash basis of accounting.

**c.** Which basis of accounting do you believe provides a better indication of the operating performance of the consulting firm for the period? Why?

**30. Preparing income statement and balance sheet using accrual basis.** Bob Hansen opens a retail store on January 1, Year 8. Hansen invests $50,000 for all of the common stock of the firm. The store borrows $40,000 from a local bank. The store must repay the loan with interest for both Year 8 and Year 9 on December 31, Year 9. The interest rate is 10 percent per year. The store purchases a building for $60,000 cash for use as a retail store. The building has a thirty-year life, zero estimated salvage value, and is to be depreciated using the straight-line method. The store purchases $125,000 of merchandise on account during Year 8 and pays $97,400 of the amount by the end of Year 8. A physical inventory taken on December 31, Year 8 indicates $15,400 of merchandise is still on hand.

During Year 8, the store makes cash sales to customers totaling $52,900 and sales on account totaling $116,100. Of the sales on account, the store collects $54,800 by December 31, Year 8. The store incurs and pays other costs as follows: salaries, $34,200; utilities, $2,600. It has unpaid bills at the end of Year 8 as follows: salaries, $2,400; utilities, $180. The firm is subject to an income tax rate of 40 percent. Income taxes for Year 8 are payable on March 15, Year 9.

**a.** Prepare an income statement for Hansen Retail Store for Year 8, assuming that the retail store uses the accrual basis of accounting and recognizes revenue at the time of sale. Show supporting computations for each revenue and expense.

**b.** Prepare a balance sheet for Hansen Retail Store as of December 31, Year 8. Show supporting computations for each balance sheet item.

**31. Miscellaneous transactions and adjusting entries.** Give the journal entry to record (1) each of the following transactions as well as (2) any necessary adjusting entries on December 31, Year 6, assuming that the firm uses a calendar-year accounting period. You may omit explanations for the journal entries.

**a.** Sung Corporation gives a sixty-day note to a supplier on December 2, Year 6. The note in the face amount of $6,000 replaces an open account payable of the same amount. The note is due on January 30, Year 7, with interest at 10 percent per year.

**b.** Allstate Insurance Company sells a two-year insurance policy on September 1, Year 6, receiving the two-year premium of $18,000 in advance. It credits a liability account.

**c.** Blaydon Company acquires a machine on October 1, Year 6, for $40,000 cash. It expects the machine to have a $4,000 salvage value and a four-year life.

**d.** Pyke Electronics Company acquires an automobile on July 1, Year 6, for $24,000 cash. It expects the automobile to have a $3,000 salvage value and a three-year life.

**e.** Devine Company rents needed office space for the three-month period beginning December 1, Year 6. It pays the three months' total rent of $12,000 on this date, debiting an asset account.

**f.** Hall Corporation begins business on November 1, Year 6. It acquires office supplies costing $7,000 on account, debiting an asset account. Of this amount, it pays $5,000 by year-end. A physical inventory indicates that office supplies costing $1,500 are on hand on December 31, Year 6.

**32. Miscellaneous transactions and adjusting entries.** Give the journal entry to record (1) each of the following transactions as well as (2) any necessary adjusting entries on December 31, Year 3, assuming that the firm uses a calendar-year accounting period. You may omit explanations for the journal entries.

**a.** Gale Company rents out excess office space on October 1, Year 3. It receives on that date the annual rental of $48,000 for the period from October 1, Year 3, to September 30, Year 4, and credits a liability account.

**b.** Whitley Company receives a $10,000, two-month, 6-percent note on December 1, Year 3, in full payment of an open account receivable.

**c.** The balance in the Prepaid Insurance account of Pierce Company on January 1, Year 3, was $500. On March 1, Year 3, the company renews its only insurance policy for another two years, beginning on that date, by paying the $6,600 two-year premium. It debits an asset account for the payment.

**d.** The Repair Parts Inventory account of Kelly Company showed a balance of $4,000 on January 1, Year 3. During Year 3, the firm purchases parts costing $14,900 and charges them to Repair Expense. An inventory of repair parts at the end of December reveals parts costing $3,800 on hand.

**e.** Roberts Company acquires with cash an office machine on July 1, Year 3, costing $200,000. It estimates that the machine will have a ten-year life and a $20,000 residual value. It uses the straight-line depreciation method.

**f.** Lovejoy Company pays its property taxes for the year ending December 31, Year 3, of $12,000 on September 1, Year 3. It debits Property Tax Expense at the time of the payment.

**33. Analysis of transactions and preparation of income statement and balance sheet.** Refer to the information for Moulton Corporation as of December 31, Year 12 in **Problem 29** of **Chapter 2**. Moulton Corporation opened for business on January 1, Year 13. It uses the accrual basis of accounting. Transactions and events during Year 13 were as follows:

**(1)** During Year 13: Purchased inventory on account costing $1,100,000 from various suppliers.

**(2)** During Year 13: Sold merchandise to customers for $2,000,000 on account.

**(3)** During Year 13: The cost of merchandise sold to customers totaled $1,200,000.

**(4)** During Year 13: Collected $1,400,000 from customers for sales made previously on account.

**(5)** During Year 13: Paid merchandise suppliers $950,000 for purchases made previously on account.

**(6)** During Year 13: Paid various suppliers of selling and administrative services $625,000. The firm consumed all of the benefits of these services during Year 13.

**(7)** June 30, Year 13: Repaid the note payable to a supplier with interest (see transaction **(7)** in **Problem 2.29**).

**(8)** December 31, Year 13: Recognized interest on the long-term bank loan (see transaction **(6)** in **Problem 2.29**).

**(9)** December 31, Year 13: Recognized insurance expense for Year 13 (see transaction **(5)** in **Problem 2.29**).

**(10)** December 31, Year 13: Recognized depreciation expense for Year 13 (see transactions **(2)** and **(7)** of **Problem 2.29**).

**(11)** December 31, Year 13: Recognized income tax expense and income tax payable for Year 13. The income tax rate is 40 percent. Assume that income taxes for Year 13 are payable by March 15, Year 14.

**a.** Using either T-accounts or a transactions spreadsheet, depending on the preference of your instructor, enter the balances in balance sheet accounts on January 1, Year 13 (see **Problem 29** in **Chapter 2**) and the effects of the eleven transactions above.

**b.** Prepare an income statement for Year 13.

**c.** Prepare a comparative balance sheet as of December 31, Year 12, and December 31, Year 13.

**34. Analysis of transactions and preparation of income statement and balance sheet.** Refer to the information for Regaldo Department Stores as of January 31, Year 8 in **Problem 30** of **Chapter 2**. Regaldo Department Stores opened for business during February, Year 8. It uses the accrual basis of accounting. Transactions and events during February were as follows.

**(1)** February 1: Purchased display counters and computer equipment for Ps90,000. The firm borrowed Ps90,000 from a local bank to finance the purchases. The bank loan bears interest at a rate of 12 percent each year and is repayable with interest on February 1, Year 9.

**(2)** During February: Purchased merchandise on account totaling Ps217,900.

**(3)** During February: Sold merchandise costing Ps162,400 to various customers for Ps62,900 cash and Ps194,600 on account. *Hint:* Enter the sales transaction and the recognition of the cost of goods sold in separate columns on the spreadsheet.

**(4)** During February: Paid to employees compensation totaling Ps32,400 for services rendered during the month.

**(5)** During February: Paid utility (electric, water, gas) bills totaling Ps2,700 for services received during February.

**(6)** During February: Collected Ps84,600 from customers for sales on account (see transaction **(3)** above).

(7) During February: Paid invoices from suppliers of merchandise (see transaction **(2)** above) with an original purchase price of Ps210,000 in time to receive a 2 percent discount for prompt payment and Ps29,000 to other suppliers after the discount period had elapsed. The firm treats discounts taken as a reduction in the acquisition cost of merchandise. *Hint:* Enter the payment of accounts payable within the discount period and after the discount period in separate columns on the spreadsheet.

(8) February 28: Compensation that employees earned during the last several days in February and that the firm will pay early in March totaled Ps6,700.

(9) February 28: Utility services that the firm used during February and that the firm will not pay until March totaled Ps800.

(10) February 28: The display counters and computer equipment purchased in transaction **(1)** have an expected useful life of five years and zero salvage value at the end of the five years. The firm depreciates such equipment on a straight-line basis over the expected life and uses an Accumulated Depreciation account.

(11) February 28: The firm recognizes an appropriate portion of the prepaid rent as of January 31 (see **Problem 30** in **Chapter 2**).

(12) February 28: The firm recognizes an appropriate portion of the prepaid insurance as of January 31 (see **Problem 30** in **Chapter 2**).

(13) February 28: The firm amortizes (that is, recognizes as an expense) the patent over sixty months (see **Problem 30** in **Chapter 2**). The firm does not use a separate Accumulated Amortization account for the patent.

(14) February 28: The firm recognizes an appropriate amount of interest expense on the loan in transaction **(1)** above.

(15) February 28: The firm is subject to an income tax rate of 30 percent of net income before income taxes. The income tax law requires firms to pay income taxes on the fifteenth day of the month after the end of each quarter (that is, April 15, July 15, October 15, and January 15).

**a.** Using either T-account or a transactions spreadsheet, depending on the preference of your instructor, enter the balances in balance sheet accounts on February 1, Year 8 (see **Problem 30** in **Chapter 2**) and the effects of the fifteen transactions above.

**b.** Prepare an income statement for the month of February, Year 8.

**c.** Prepare a comparative balance sheet as of January 31 and February 28, Year 8.

**35. Analysis of transactions and preparation of income statement and balance sheet.** Refer to the information for Patterson Corporation for January, Year 13, in **Problem 31** in **Chapter 2**. The following transactions occur during February.

(1) February 1: The firm pays the two-year insurance premium of $2,400 for fire and liability coverage beginning February 1.

(2) February 5: Acquires merchandise costing $1,050,000. Of this amount, $1,455 is from suppliers to whom Patterson returned defective merchandise during January but for which the firm had not yet received a refund for amounts paid. Patterson Corporation acquired the remaining purchases on account.

(3) During February: Sells merchandise to customers totaling $1,500,000. Of this amount, $4,500 was to customers who had advanced Patterson Corporation cash during January. Patterson Corporation makes the remaining sales on account.

(4) During February: The cost of the goods sold in transaction **(3)** was $950,000.

(5) During February: Pays in cash selling and administrative expenses of $235,000.

(6) During February: Collects $1,206,000 from customers for sales previously made on account.

(7) During February: Pays $710,000 to suppliers of merchandise for purchases previously made on account.

(8) February 28: Recognizes rent expense for February.

(9) February 28: Recognizes depreciation expense of $2,500 for February. Patterson Corporation uses an Accumulated Depreciation account.

(10) February 28: Recognizes amortization expense of $450 on the patent. Patterson Corporation does not use an Accumulated Amortization account for patents.

(11) February 28: Recognizes an appropriate amount of insurance expense for February.

(12) February 28: Recognizes interest expense on the mortgage payable (see **Problem 31** in **Chapter 2**).

(13) February 28: Recognizes income tax expense for February. The income tax rate is 40 percent. Income taxes for February are payable by April 15.

**a.** Using either T-accounts or a transactions spreadsheet, depending on the preference of your instructor, enter the balances in balance sheet accounts on February 1, Year 13 (see **Problem 31** in **Chapter 2**), and the effects of the thirteen transactions above.

**b.** Prepare an income statement for the month of February, Year 13.

**c.** Prepare a comparative balance sheet as of January 31 and February 28, Year 13.

**36. Analysis of transactions and preparation of income statement and balance sheet.** Zealock Bookstore opened a bookstore near a college campus on July 1, Year 10. Transactions and events of Zealock Bookstore during Year 10 appear below. The firm uses the calendar year as its reporting period.

**(1)** July 1: Receives $25,000 from Quinn Zealock for 25,000 shares of the bookstore's $1 par value common stock.

**(2)** July 1: Obtains a $30,000 loan from a local bank for working capital needs. The loan bears interest at 6 percent per year. The loan is repayable with interest on June 30, Year 11.

**(3)** July 1: Signs a rental agreement for three years at an annual rental of $20,000. Pays the first year's rent in advance.

**(4)** July 1: Acquires bookshelves for $4,000 cash. The bookshelves have an estimated useful life of five years and zero salvage value.

**(5)** July 1: Acquires computers for $10,000 cash. The computers have an estimated useful life of three years and $1,000 salvage value.

**(6)** July 1: Makes security deposits with various book distributors totaling $8,000. The deposits are refundable on June 30, Year 11 if the bookstore pays on time all amounts due for books purchased from the distributors between July 1, Year 10 and June 30, Year 11.

**(7)** During Year 10: Purchases books on account from various distributors costing $160,000.

**(8)** During Year 10: Sells books costing $140,000 for $172,800. Of the total sales, $24,600 is for cash and $148,200 is on account. *Hint:* Enter the sales transaction in a separate column in the spreadsheet than the recognition of the cost of goods sold.

**(9)** During Year 10: Returns unsold books and books ordered in error costing $14,600. The firm had not yet paid for these books.

**(10)** During Year 10: Collects $142,400 from sales on account.

**(11)** During Year 10: Pays employees compensation of $16,700.

**(12)** During Year 10: Pays book distributors $139,800 of the amounts due for purchases on account.

**(13)** December 28, Year 10: Receives advances from customers of $850 for special order books that the bookstore will order and expects to receive during Year 11.

**(14)** December 31, Year 10: Records an appropriate amount of interest expense on the loan in **(2)** for Year 10.

**(15)** December 31, Year 10: Records an appropriate amount of rent expense for Year 10.

**(16)** December 31, Year 10: Records an appropriate amount of depreciation expense on the bookshelves in **(4)**.

**(17)** December 31, Year 10: Records an appropriate amount of depreciation expense on the computers in **(5)**.

**(18)** December 31, Year 10: Records an appropriate amount of deposit expense from **(6)**.

**(19)** December 31, Year 10: Records an appropriate amount of income tax expense for Year 10. The income tax rate is 40 percent. The taxes are payable on March 15, Year 11.

**a.** Using either T-accounts or a transactions spreadsheet, depending on the preference of your instructor, enter the nineteen transactions and events above.

**b.** Prepare an income statement for Year 10.

**c.** Prepare a balance sheet on December 31, Year 10. Classify assets and liabilities as current and noncurrent.

**d.** Evaluate the operating performance for Year 10 and financial health of Zealock Bookstore at the end of Year 10.

**37. Analysis of transactions and preparation of comparative income statements and balance sheets.** Refer to the information for Zealock Bookstore in **Problem 36**. The following transactions relate to Year 11.

**(1)** March 15: Pays income taxes for Year 10.

**(2)** June 30: Repays the bank loan with interest.

**(3)** July 1: Obtains a new bank loan for $75,000. The loan is repayable on June 30, Year 12, with interest due at maturity of 8 percent.

**(4)** July 1: Receives the security deposit back from the book distributors.

**(5)** July 1: Pays the rent due for the period July 1, Year 11, to June 30, Year 12.

**(6)** During Year 11: Purchases books on account costing $310,000.
**(7)** During Year 11: Sells books costing $286,400 for $353,700. Of the total sales, $24,900 is for cash, $850 is from special orders received during December Year 10, and $327,950 is on account.
**(8)** During Year 11: Returns unsold books costing $22,700. The firm had not yet paid for these books.
**(9)** During Year 11: Collects $320,600 from sales on account.
**(10)** During Year 11: Pays employees compensation of $29,400.
**(11)** During Year 11: Pays book distributors $281,100 for purchases of books on account.
**(12)** December 31, Year 11: Declares and pays a dividend of $4,000.

**a.** Using either T-accounts or a transactions spreadsheet, depending on the preference of your instructor, enter the amounts for the balance sheet on December 31, Year 10, from **Problem 36**, the effects of the twelve transactions above, and any required entries on December 31, Year 11, to properly measure net income for Year 11 and financial position on December 31, Year 11.
**b.** Prepare a comparative income statement for Year 10 and Year 11.
**c.** Prepare a comparative balance sheet for December 31, Year 10, and December 31, Year 11.
**d.** Evaluate the operating performance and financial health of Zealock Bookstore.

**38. Working backward to the balance sheet at the beginning of the period. (Problems 38** through **40** derive from problems by George H. Sorter.) The following data relate to the Prima Company.
**(1) Exhibit 3.14**: Balance sheet at December 31, Year 2.
**(2) Exhibit 3.15**: Statement of net income and retained earnings for Year 2.
**(3) Exhibit 3.16**: Statement of cash receipts and disbursements for Year 2.

Purchases of merchandise during the period, all on account, were $127,000. All "Other Operating Expenses" were credited to Prepayments.

**EXHIBIT 3.14**

**PRIMA COMPANY**
**Balance Sheet**
**December 31, Year 2**
**(Problem 38)**

| | |
|---|---|
| **Assets** | |
| Cash | $ 10,000 |
| Marketable Securities | 20,000 |
| Accounts Receivable | 25,000 |
| Merchandise Inventory | 30,000 |
| Prepayments for Miscellaneous Services | 3,000 |
| Total Current Assets | $ 88,000 |
| Land, Buildings, and Equipment (at cost) | 40,000 |
| Less Accumulated Depreciation | (16,000) |
| Land, Buildings, and Equipment (net) | $ 24,000 |
| Total Assets | $112,000 |
| **Liabilities and Shareholders' Equity** | |
| Accounts Payable (for merchandise) | $ 25,000 |
| Interest Payable | 300 |
| Taxes Payable | 4,000 |
| Total Current Liabilities | $ 29,300 |
| Note Payable (6 percent, long-term) | 20,000 |
| Total Liabilities | $ 49,300 |
| Common Stock | $ 50,000 |
| Retained Earnings | 12,700 |
| Total Shareholders' Equity | $ 62,700 |
| Total Liabilities and Shareholders' Equity | $112,000 |

**EXHIBIT 3.15**

**PRIMA COMPANY**
**Statement of Net Income and Retained Earnings**
**For Year 2**
**(Problem 38)**

| | | |
|---|---|---|
| Sales | | $200,000 |
| **Less Expenses:** | | |
| Cost of Goods Sold | $130,000 | |
| Depreciation Expense | 4,000 | |
| Taxes Expense | 8,000 | |
| Other Operating Expenses | 47,700 | |
| Interest Expense | 1,200 | |
| Total Expenses | | 190,900 |
| Net Income | | $ 9,100 |
| Less Dividends | | 5,000 |
| Increase in Retained Earnings | | $ 4,100 |

**EXHIBIT 3.16**

**PRIMA COMPANY**
**Statement of Cash Receipts and Disbursements**
**For Year 2**
**(Problem 38)**

| | | |
|---|---|---|
| **Cash Receipts** | | |
| Cash Sales | $ 47,000 | |
| Collection from Credit Customers | 150,000 | |
| Total Receipts | | $197,000 |
| **Cash Disbursements** | | |
| Payment to Suppliers of Merchandise | $128,000 | |
| Payment to Suppliers of Miscellaneous Services | 49,000 | |
| Payment of Taxes | 7,500 | |
| Payment of Interest | 1,200 | |
| Payment of Dividends | 5,000 | |
| Purchase of Marketable Securities | 8,000 | |
| Total Disbursements | | 198,700 |
| Excess of Disbursements over Receipts | | $ 1,700 |

Prepare a balance sheet for January 1, Year 2. (*Hint:* Using either T-accounts or a transactions spreadsheet, depending on the preference of your instructor, enter the December 31, Year 2 amounts from the balance sheet. Using the information in the income statement and statement of cash receipts and disbursements, reconstruct the transactions that took place during the year and enter the amounts in the appropriate places in the T-accounts or transactions template. Finally, compute the amounts on the January 1, Year 2 balance sheet.)

39. **Working backward to cash receipts and disbursements. Exhibit 3.17** presents the comparative balance sheet of The Secunda Company as of the beginning and end of Year 2. **Exhibit 3.18** presents the income statement for Year 2. The company makes all sales on account and purchases all goods and services on account. The Other Operating Expenses account includes depreciation charges and expirations of prepayments. The company debits dividends declared during the year to Retained Earnings.

Prepare a schedule showing all cash transactions for Year 2. (*Hint:* Using either T-accounts or a transactions spreadsheet, depending on the preference of your instructor, enter

**EXHIBIT 3.17**

**THE SECUNDA COMPANY**
**Balance Sheet**
**January 1 and December 31, Year 2**
**(Problem 39)**

| | 1/1/Year 2 | 12/31/Year 2 |
|---|---|---|
| **Assets** | | |
| Cash | $ 20,000 | $ 9,000 |
| Accounts Receivable | 36,000 | 51,000 |
| Merchandise Inventory | 45,000 | 60,000 |
| Prepayments | 2,000 | 1,000 |
| Total Current Assets | $103,000 | $121,000 |
| Land, Buildings, and Equipment (at cost) | $ 40,000 | $ 40,000 |
| Less Accumulated Depreciation | (16,000) | (18,000) |
| Land, Buildings, and Equipment (net) | $ 24,000 | $ 22,000 |
| Total Assets | $127,000 | $143,000 |
| **Liabilities and Shareholders' Equity** | | |
| Interest Payable | $ 1,000 | $ 2,000 |
| Accounts Payable | 30,000 | 40,000 |
| Total Current Liabilities | $ 31,000 | $ 42,000 |
| Mortgage Payable | 20,000 | 17,000 |
| Total Liabilities | $ 51,000 | $ 59,000 |
| Common Stock | $ 50,000 | $ 50,000 |
| Retained Earnings | 26,000 | 34,000 |
| Total Shareholders' Equity | $ 76,000 | $ 84,000 |
| Total Liabilities and Shareholders' Equity | $127,000 | $143,000 |

**EXHIBIT 3.18**

**THE SECUNDA COMPANY**
**Income Statement**
**For Year 2**
**(Problem 39)**

| | | |
|---|---|---|
| Sales Revenue | | $100,000 |
| **Less Expenses:** | | |
| Cost of Goods Sold | $50,000 | |
| Interest Expense | 3,000 | |
| Other Operating Expenses | 29,000 | |
| Total Expenses | | 82,000 |
| Net Income | | $ 18,000 |

the amounts shown as of January 1, Year 2, and December 31, Year 2. Starting with the revenue and expense accounts, reconstruct the transactions that took place during the year, and enter the amounts in the places in the T-accounts or transactions spreadsheet. Note that the Retained Earnings account in the balance sheet on December 31, Year 2, reflects the effects of earnings and dividends activities during Year 2.)

**40. Working backward to the income statement.** Tertia Company presents balance sheets at the beginning and end of Year 2 (**Exhibit 3.19**), as well as a statement of cash receipts and disbursements (**Exhibit 3.20**). Prepare a combined statement of income and retained earnings for Year 2. (*Hint:* Using either T-accounts or a transactions spreadsheet, depending on

EXHIBIT 3.19

**TERTIA COMPANY**
**Balance Sheets**
**January 1 and December 31, Year 2**
**(Problem 40)**

| | 1/1/Year 2 | 12/31/Year 2 |
|---|---|---|
| **Assets** | | |
| Cash | $ 40,000 | $ 67,800 |
| Accounts and Notes Receivable | 36,000 | 41,000 |
| Merchandise Inventory | 55,000 | 49,500 |
| Interest Receivable | 1,000 | 700 |
| Prepaid Miscellaneous Services | 4,000 | 5,200 |
| Building, Machinery, and Equipment | 47,000 | 47,000 |
| Accumulated Depreciation | (10,000) | (12,000) |
| Total Assets | $173,000 | $199,200 |
| **Liabilities and Shareholders' Equity** | | |
| Accounts Payable (miscellaneous services) | $ 2,000 | $ 2,500 |
| Accounts Payable (merchandise purchases) | 34,000 | 41,000 |
| Property Taxes Payable | 1,000 | 1,500 |
| Mortgage Payable | 35,000 | 30,000 |
| Total Liabilities | $ 72,000 | $ 75,000 |
| Common Stock | $ 25,000 | $ 25,000 |
| Retained Earnings | 76,000 | 99,200 |
| Total Shareholders' Equity | $101,000 | $124,200 |
| Total Liabilities and Shareholders' Equity | $173,000 | $199,200 |

EXHIBIT 3.20

**TERTIA COMPANY**
**Statement of Cash Receipts and Disbursements**
**For Year 2**
**(Problem 40)**

| | Year 2 |
|---|---|
| **Cash Receipts** | |
| 1. Collection from Credit Customers | $144,000 |
| 2. Cash Sales | 63,000 |
| 3. Collection of Interest | 1,000 |
| Total Cash Receipts | $208,000 |
| **Less Cash Disbursements** | |
| 4. Payment to Suppliers of Merchandise | $114,000 |
| 5. Repayment on Mortgage | 5,000 |
| 6. Payment of Interest | 500 |
| 7. Prepayment to Suppliers of Miscellaneous Services | 57,500 |
| 8. Payment of Property Taxes | 1,200 |
| 9. Payment of Dividends | 2,000 |
| Total Cash Disbursements | $180,200 |
| Increase in Cash Balance for Year 2 | $ 27,800 |

the preference of your instructor, enter the amounts shown as of January 1, Year 2, and December 31, Year 2. Starting with the cash receipts and disbursements for the year, reconstruct the transactions that took place during the year, and enter them in the appropriate places in the T-accounts or transactions spreadsheet. The Retained Earnings account reflects the effect of earnings activities and dividends for Year 2.)

**41. Reconstructing the income statement and balance sheet.** (Adapted from a problem by Stephen A. Zeff.) Portobello Co., a firm that sells merchandise to retail customers, is in its tenth year of operation. On December 28, Year 10, three days before the close of its fiscal year, a flash flood devastated the company's administrative office and destroyed almost all of its accounting records. The company was able to save the balance sheet on December 31, Year 9 (see **Exhibit 3.21**), the checkbook, the bank statements, and some soggy remains of the specific accounts receivable and accounts payable balances. Based on a review of the surviving documents and a series of interviews with company employees, you obtain the following information.

**(1)** The company's insurance agency advises that a four-year insurance policy has six months to run as of December 31, Year 10. The policy cost $12,000 when the company paid the four-year premium during Year 7.

**(2)** During Year 10, the company's board of directors declared $6,000 of dividends, of which the firm paid $3,000 in cash to shareholders during Year 10 and will pay the remainder during Year 11. Early in Year 10, the company also paid dividends of $1,800 cash that the board of directors had declared during Year 9.

**EXHIBIT 3.21**

**PORTOBELLO CO.**
**Balance Sheet**
**December 31, Year 9**
**(Problem 41)**

| | |
|---|---|
| **Assets** | |
| Cash | $ 18,600 |
| Accounts Receivable | 33,000 |
| Notes Receivable | 10,000 |
| Interest Receivable | 600 |
| Merchandise Inventories | 22,000 |
| Prepaid Insurance | 4,500 |
| Total Current Assets | $ 88,700 |
| **Computer System:** | |
| At Cost | $ 78,000 |
| Less Accumulated Depreciation | (26,000) |
| Net | $ 52,000 |
| Total Assets | $140,700 |
| **Liabilities and Shareholders' Equity** | |
| Accounts Payable for Merchandise | $ 36,000 |
| Dividend Payable | 1,800 |
| Salaries Payable | 6,500 |
| Taxes Payable | 10,000 |
| Advances from Customers | 600 |
| Total Liabilities | $ 54,900 |
| Common Stock | $ 40,000 |
| Retained Earnings | 45,800 |
| Total Shareholders' Equity | $ 85,800 |
| Total Liabilities and Shareholders' Equity | $140,700 |

**(3)** On April 1, Year 10, the company received from Appleton Co. $10,900 cash, which included principal of $10,000 and interest, in full settlement of Appleton's nine-month note dated July 1, Year 9. According to the terms of the note, Appleton paid all interest at maturity on April 1, Year 10.

**(4)** The amount owed by the company to merchandise suppliers on December 31, Year 10, was $20,000 less than the amount owed on December 31, Year 9. During Year 10, the company paid $115,000 to merchandise suppliers. The cost of merchandise inventory on December 31, Year 10, based on a physical count, was $18,000 larger than the balance in the Merchandise Inventory account on the December 31, Year 9, balance sheet. On December 8, Year 10, the company exchanged shares of its common stock for merchandise inventory costing $11,000. The company's policy is to purchase all merchandise on account.

**(5)** The company purchased delivery trucks on March 1, Year 10, for $60,000. To finance the acquisition, it gave the seller a $60,000 four-year note that bears interest at 10 percent per year. The company must pay interest on the note each six months, beginning September 1, Year 10. The company made the required payment on this date. The delivery trucks have an expected useful life of ten years and an estimated salvage value of $6,000. The company uses the straight-line depreciation method.

**(6)** The company's computer system has a six-year total expected life and zero expected salvage value.

**(7)** The company makes all sales on account and recognizes revenue at the time of shipment to customers. During Year 10, the company received $210,000 cash from its customers. The company's accountant reconstructed the Accounts Receivable subsidiary ledger, the detailed record of the amount owed to the company by each customer. It showed that customers owed the company $51,000 on December 31, Year 10. A close examination revealed that $1,400 of the cash received from customers during Year 10 applies to merchandise that the company will not ship until Year 11. Also, $600 of the cash received from customers during Year 9 applies to merchandise not shipped to customers until Year 10.

**(8)** The company paid $85,000 in cash to employees during Year 10. Of this amount, $6,500 relates to services that employees performed during Year 9, and $4,000 relates to services that employees will perform during Year 11. Employees performed the remainder of the services during Year 10. On December 31, Year 10, the company owes employees $1,300 for services performed during the last several days of Year 10.

**(9)** The company paid $27,000 in cash for property and income taxes during Year 10. Of this amount, $10,000 relates to income taxes applicable to Year 9, and $3,000 relates to property taxes applicable to Year 11. The company owes $4,000 in income taxes on December 31, Year 10.

**(10)** The company entered into a contract with a management consulting firm for consulting services. The total contract price is $48,000. The contract requires the company to pay the first installment of $12,000 cash on January 1, Year 11, and the company intends to do so. The consulting firm had performed 10 percent of the estimated total consulting services under the contract by December 31, Year 10.

Prepare an income statement for Year 10 and a balance sheet on December 31, Year 10. (*Hint:* Using either T-accounts or a transactions spreadsheet, depending on the preference of your instructor, enter the amounts from the balance sheet on December 31, Year 9. Next, enter the information for each of the ten items listed above in either the T-accounts or transactions spreadsheet, adding income statement and additional balance sheet accounts as needed.)

**42. Reconstructing the income statement and balance sheet.** Computer Needs, Inc., operates a retail store that sells computer hardware and software. It began operations on January 2, Year 8, and operated successfully during its first year, generating net income of $8,712 and ending the year with $15,600 in its bank account. **Exhibit 3.22** presents an income statement for Year 8, and **Exhibit 3.23** presents a balance sheet as of the end of Year 8.

As Year 9 progressed, the owners and managers of Computer Needs, Inc., felt that they were doing even better. Sales seemed to be running ahead of Year 8, and customers were always in the store. Unfortunately, a freak lightning storm hit the store on December 31, Year 9, and completely destroyed the computer on which Computer Needs, Inc., kept its records. It now faces the dilemma of figuring how much income it generated during Year 9 in order to assess its operating performance and figure out how much income taxes it owes for the year.

**EXHIBIT 3.22**

**COMPUTER NEEDS, INC.**
**Income Statement**
**For the Year Ended December 31, Year 8**
**(Problem 42)**

| | |
|---|---|
| Sales | $152,700 |
| Cost of Goods Sold | (116,400) |
| Selling and Administrative Expenses | (17,400) |
| Depreciation | (2,800) |
| Interest | (4,000) |
| Income Taxes | (3,388) |
| Net Income | $ 8,712 |

**EXHIBIT 3.23**

**COMPUTER NEEDS, INC.**
**Balance Sheet**
**December 31, Year 8**
**(Problem 42)**

| | |
|---|---|
| **Assets** | |
| Cash | $ 15,600 |
| Accounts Receivable | 32,100 |
| Inventories | 46,700 |
| Prepayments | 1,500 |
| Total Current Assets | $ 95,900 |
| Property, Plant, and Equipment: | |
| At Cost | $ 59,700 |
| Less Accumulated Depreciation | (2,800) |
| Net | $ 56,900 |
| Total Assets | $152,800 |
| **Liabilities and Shareholders' Equity** | |
| Accounts Payable—Merchandise Suppliers | $ 37,800 |
| Income Tax Payable | 3,388 |
| Other Current Liabilities | 2,900 |
| Total Current Liabilities | $ 44,088 |
| Mortgage Payable | 50,000 |
| Total Liabilities | $ 94,088 |
| Common Stock | $ 50,000 |
| Retained Earnings | 8,712 |
| Total Shareholders' Equity | $ 58,712 |
| Total Liabilities and Shareholders' Equity | $152,800 |

You are asked to prepare an income statement for Year 9 and a balance sheet at the end of Year 9. To assist in this effort, you obtain the following information.

**(1)** The bank at which Computer Needs, Inc., maintains its account provided a summary of the transactions during Year 9, as shown in **Exhibit 3.24**.

**(2)** Collections received during January, Year 10, from third-party credit card companies and from customers for sales made during Year 9 totaled $40,300. This is your best estimate of accounts receivable outstanding on December 31, Year 9.

**(3)** A physical inventory of merchandise was taken on January 1, Year 10. Using current catalogs from suppliers, you estimate that the merchandise has an approximate cost of $60,700.

**EXHIBIT 3.24**

**COMPUTER NEEDS, INC.**
**Analysis of Changes in Bank Accounts**
**For the Year Ended December 31, Year 9**
**(Problem 42)**

| | |
|---|---|
| Balance, January 1, Year 9 | $ 15,600 |
| Receipts: | |
| Cash from Cash Sales | 37,500 |
| Checks Received from Third-Party Credit Cards and Customers | 151,500 |
| Disbursements: | |
| To Merchandise Supplier | (164,600) |
| To Employees and Other Providers of Selling and Administrative Activities | (21,000) |
| To U.S. Government for Income Taxes for Year 8 | (3,388) |
| To Bank for Interest ($4,000) and Principal on Mortgage ($800) | (4,800) |
| To Supplier of Equipment | (6,000) |
| Balance, December 31, Year 9 | $ 4,812 |

**(4)** Computer Needs, Inc., had paid its annual insurance premium on October 1, Year 9 (included in the amounts in **Exhibit 3.24**). You learn that $1,800 of the insurance premium applies to coverage during Year 10.

**(5)** Based on depreciation claimed during Year 8 and new equipment purchased during Year 9, you approximate that depreciation expense for Year 9 was $3,300.

**(6)** Bills received from merchandise suppliers during January, Year 10, totaled $45,300. This is your best estimate of accounts payable outstanding to these suppliers on December 31, Year 9.

**(7)** Other Current Liabilities represent amounts payable to employees and other providers of selling and administrative services. Other Current Liabilities as of December 31, Year 9, total $1,200.

**a.** Prepare an income statement for Computer Needs, Inc., for Year 9 and a balance sheet on December 31, Year 9. The income tax rate is 28 percent. (*Hint:* Using either T-accounts or a transactions spreadsheet, depending on the preference of your instructor, enter the amounts from the balance sheet on December 31, Year 8. Next, enter the information in items **(1)** to **(7)** above in either the T-accounts or transactions spreadsheet. Finally, derive the income statement amounts for Year 9.)

**b.** How well did Computer Needs, Inc., perform during Year 9?

**43. Interpreting common-size income statements. Exhibit 3.25** presents common-size income statements for The Gap and Limited Brands, two apparel retailing firms, for three recent years. In addition to the cost of merchandise sold, both firms include occupancy expense for their stores (rent, utilities, depreciation) in cost of goods sold. The Gap tends to rely on print advertising to create demand, whereas Limited Brands relies more heavily on in-store promotions (two-for-one discounts and special daily price reductions). The Gap recently expanded into other countries, whereas Limited Brands operates almost exclusively in the United States.

**a.** Suggest possible reasons for the decreasing cost of goods sold to sales percentages for the two firms during the three-year period.

**b.** Suggest possible reasons why the cost of goods sold to sales percentages for The Gap are less than those for Limited Brands.

**c.** Suggest possible reasons for the increasing selling and administrative expenses to sales percentages for the two firms for the three-year period.

**d.** Suggest possible reasons why the selling and administrative expenses to sales percentages for Limited Brands are less than those for The Gap.

**e.** Suggest possible reasons for the different pattern of changes in the interest expense to sales percentages for the two firms during the three-year period.

**f.** Suggest possible reasons for the increasing income tax expense to sales percentages for the two firms during the three-year period.

**EXHIBIT 3.25** Common-Size Income Statements for The GAP and Limited Brands (Problem 43)

| | The GAP | | | Limited Brands | | |
|---|---|---|---|---|---|---|
| | Year 8 | Year 9 | Year 10 | Year 8 | Year 9 | Year 10 |
| Sales | 100.0% | 100.0% | 100.0% | 100.0% | 100.0% | 100.0% |
| Other Revenues | 0.2 | 0.1 | 1.2 | 0.4 | 0.6 | 0.5 |
| Cost of Goods Sold or Services Sold | (59.0) | (56.8) | (56.0) | (66.2) | (64.1) | (62.8) |
| Selling and Administrative Expenses | (25.8) | (26.5) | (26.7) | (23.1) | (24.6) | (25.3) |
| Interest Expense | (3.0) | (2.2) | (2.2) | (5.3) | (5.4) | (5.5) |
| Income Tax Expense | (4.2) | (5.5) | (6.6) | (2.0) | (2.3) | (2.4) |
| Net Income/Sales | 8.2% | 9.1% | 9.7% | 3.8% | 4.2% | 4.5% |

g. Are the profit margin percentages of The Gap or those of Limited Brands closer to the level you would expect for a specialty apparel retailer? Explain your reasoning. You may wish to refer to the data for the five firms in **Exhibit 3.8** in responding to this question.

**44. Interpreting common-size income statements.** The Coca-Cola Company (Coke) and PepsiCo dominate the nonalcoholic beverage segment in the United States. Most beverage manufacturing involves adding water to a previously prepared syrup. This mixing process usually occurs at the stage just prior to bottling the beverage. Coke relies on independent companies to conduct the mixing, bottling, and distribution operations, with Coke selling the syrup to the bottlers. PepsiCo relies more heavily on owning its bottlers and thereby remains involved in both manufacturing and distribution. **Exhibit 3.26** present common-size income statements for Coke and PepsiCo for three recent years.

**a.** Suggest possible reasons why the cost of goods sold to sales percentages for Coke are significantly lower than those of PepsiCo.

**b.** Suggest reasons why the selling and administrative expense to sales percentages steadily increased for Coke but steadily decreased for PepsiCo.

**c.** Suggest possible reasons for the decreasing interest expense to sales percentages for both firms.

**EXHIBIT 3.26** Common-Size Income Statements for Coke and PepsiCo (Problem 44)

| | Coke | | | PepsiCo | | |
|---|---|---|---|---|---|---|
| | Year 10 | Year 11 | Year 12 | Year 10 | Year 11 | Year 12 |
| Sales | 100.0% | 100.0% | 100.0% | 100.0% | 100.0% | 100.0% |
| Other Revenues | 0.9 | 3.3 | 1.2 | 0.8 | 0.9 | 1.3 |
| Cost of Goods Sold | (35.7) | (34.4) | (36.3) | (45.8) | (45.7) | (45.8) |
| Selling and Administrative Expenses | (34.7) | (35.0) | (35.8) | (36.3) | (35.5) | (34.5) |
| Interest Expense | (2.6) | (1.6) | (1.0) | (1.2) | (0.9) | (0.7) |
| Income Tax Expense | (10.0) | (9.6) | (7.8) | (5.6) | (6.4) | (6.5) |
| Net Income | 17.9% | 22.7% | 20.3% | 11.9% | 12.4% | 13.8% |
| Sales Growth | 1.8% | 6.5% | 8.2% | 22.7% | 5.3% | 6.8% |

**EXHIBIT 3.27** **Common-Size Income Statements for Nokia (Problem 45)**

| | Year 7 | Year 8 | Year 9 |
|---|---|---|---|
| Sales | 100.0% | 100.0% | 100.0% |
| Other Revenues | 1.6 | 1.2 | 0.5 |
| Cost of Goods Sold | (64.6) | (62.3) | (61.8) |
| Selling and Administrative Expenses | (10.4) | (10.3) | (9.5) |
| Research and Development Expense | (9.0) | (8.8) | (9.1) |
| Interest Expense | (1.9) | (1.5) | (1.2) |
| Income Tax Expense | (4.0) | (5.7) | (6.1) |
| Net Income/Sales | 11.7% | 12.6% | 12.8% |

**d.** Coke's income tax expense to sales percentages are larger than those for PepsiCo, suggesting that Coke has a higher income tax burden. For a different perspective, compute the ratio of income tax expense to net income before income taxes for each firm. For example, this percentage for Coke for Year 10 is 35.8 percent [= 10.0%/(17.9% + 10.0%)]. What insight does this measure provide regarding the income tax burden of Coke versus PepsiCo?

**e.** Compare the net income to sales percentages for Coke and PepsiCo with those of the five firms in **Exhibit 3.8**. Suggest reasons for the levels of the profit margins of Coke and PepsiCo relative to these other five firms.

**45. Interpreting common-size income statements.** Nokia has a 28 percent worldwide market share in cellular phones, the largest share of any company. Its sales have grown approximately 50 percent annually in recent years. Nokia's cellular phones use digital technology, which permits them to interact with computers and the Internet. **Exhibit 3.27** presents common-size income statements for Nokia for three recent years. Discuss the likely reasons for the increasing profit margin for Nokia over the three-year period.

**46. Interpreting common-size income statements.** McDonald's is the largest fast-food restaurant chain in the United States. This segment of the restaurant market has experienced increased competition in recent years with aggressive actions from existing firms and the entry of new restaurant concepts. **Exhibit 3.28** presents common-size income statements for McDonald's for three recent years. Discuss the likely reasons for the decreasing profit margin for McDonald's over the three-year period.

**EXHIBIT 3.28** **Common-Size Income Statements for McDonald's (Problem 46)**

| | Year 10 | Year 11 | Year 12 |
|---|---|---|---|
| Sales | 100.0% | 100.0% | 100.0% |
| Other Revenues | 0.6 | 1.4 | 0.0 |
| Cost of Goods Sold | (64.6) | (66.9) | (69.0) |
| Selling and Administrative Expenses | (11.1) | (11.1) | (11.2) |
| Interest Expense | (3.1) | (3.1) | (3.1) |
| Income Tax Expense | (7.1) | (6.4) | (5.0) |
| Net Income/Sales | 14.7% | 13.9% | 11.7% |

**47. Identifying industries using common-size income statement percentages. Exhibit 3.29** presents common-size income statements for six firms. The six firms and a description of their operations follow.

**a.** Commonwealth Edison: generates and sells electricity to businesses and households in capital-intensive plants.

**b.** Delta Air Lines: provides airline transportation services.

**c.** Gillette: manufactures and sells a variety of branded consumer personal care and household products.

**d.** Hewlett-Packard: manufactures and sells computers, printers, and other hardware.

**e.** Kroger Stores: operates a chain of grocery stores nationwide.

**f.** Kelly Services: provides temporary office services to businesses and other firms. Sales revenue represents amounts billed to customers for temporary help services and cost of goods and services sold includes amounts paid to Kelly temporary help employees.

Use whatever clues you can to match the companies in **Exhibit 3.29** with the six firms listed above. Explain your reasoning.

**48. Preparing adjusting entries.** To achieve efficient recording of day-to-day cash receipts and disbursements relating to operations, a firm may credit all cash receipts to revenue accounts and debit all cash disbursements to expense accounts. The efficiency stems from treating all receipts in the same way and all disbursements in the same way. The firm can program its computer to automatically record operating cash receipts and disbursements in this way. In the day-to-day recording of transactions, the computer program need not be concerned with whether a specific cash transaction reflects settlement of a past accrual, a revenue or expense correctly assigned to the current period, or a prepayment relating to a future period. At the end of the period, accountants analyze the existing account balances and construct the adjusting entries required to correct them. This process results in temporarily incorrect balances in some balance sheet and income statement accounts during the accounting period.

Construct the adjusting entry required for each of the following scenarios.

**a.** On September 1, Year 2, a tenant paid $24,000 rent for the one-year period starting at that time. The tenant debited the entire amount to Rent Expense and credited Cash. The tenant made no adjusting entries for rent between September 1 and December 31. Construct the adjusting entry to be made on December 31, Year 2, to recognize the proper balances in the Prepaid Rent and Rent Expense accounts. What is the amount of Rent Expense for Year 2?

**b.** Refer to part **a**. The tenant's books for December 31, Year 2, after adjusting entries, show a balance in the Prepaid Rent account of $16,000. This amount represents rent for the period January 1 through August 31, Year 3. On September 1, Year 3, the tenant paid $30,000 for rent for the one-year period starting September 1, Year 3. The tenant debited this amount to Rent Expense and credited Cash but made no adjusting entries for rent

**EXHIBIT 3.29** **Common-Size Income Statements for Six Firms (Problem 47)**

| | (1) | (2) | (3) | (4) | (5) | (6) |
|---|---|---|---|---|---|---|
| Sales | 100.0% | 100.0% | 100.0% | 100.0% | 100.0% | 100.0% |
| Other Revenues | — | — | 0.5 | 0.9 | — | — |
| Cost of Goods and Services Sold | (73.6) | (81.4) | (60.6) | (63.0) | (61.3) | (34.4) |
| Selling and Administrative Expenses | (19.9) | (14.1) | (17.9) | (24.0) | (5.0) | (40.9) |
| Depreciation | (2.1) | (0.8) | (5.1) | (3.4) | (8.7) | (3.9) |
| Interest | (1.4) | — | (8.0) | (0.9) | (7.3) | (0.8) |
| Income Taxes | (1.4) | (1.5) | (3.3) | (2.9) | (6.7) | (7.4) |
| Net Income | 1.6% | 2.2% | 5.6% | 6.7% | 11.0% | 12.6% |

during Year 3. Construct the adjusting entry required on December 31, Year 3. What is Rent Expense for Year 3?

**c.** Refer to part **b**. The tenant's books for December 31, Year 3, after adjusting entries, show a balance in the Prepaid Rent account of $20,000. This amount represents rent for the period January 1 through August 31, Year 4. On September 1, Year 4, the tenant paid $18,000 for rent for the six-month period starting September 1, Year 4. The tenant debited this amount to Rent Expense and credited Cash but made no adjusting entries during Year 4. Construct the adjusting entry required on December 31, Year 4. What is Rent Expense for Year 4?

**d.** Whenever the firm makes payments for wages, it debits Wage Expense. At the start of April, the Wages Payable account had a balance of $5,000, representing wages earned but not paid during the last few days of March. During April, the firm paid $30,000 in wages, debiting the entire amount to Wage Expense. At the end of April, analysis of amounts earned since the last payday indicates that employees have earned wages of $4,000 that they have not received. These are the only unpaid wages at the end of April. Construct the required adjusting entry. What is Wage Expense for April?

**e.** A firm purchased an insurance policy providing one year's coverage from May 1, Year 1, and debited the entire amount to Insurance Expense. After the firm made adjusting entries, the balance sheet on December 31, Year 1, correctly showed Prepaid Insurance of $3,000. Construct the adjusting entry that the firm must make on January 31, Year 2, if the firm closes its books monthly and prepares a balance sheet for January 31, Year 2.

**f.** The record-keeping system for an apartment building instructs the bookkeeper always to credit rent revenue when the firm collects a payment from tenants. At the beginning of Year 3, the liability account Advances from Tenants had a credit balance of $25,000, representing collections from tenants for rental services to be rendered during Year 3. During Year 3, the firm collected $250,000 from tenants; it debited Cash and credited Rent Revenue. It made no adjusting entries during Year 3. At the end of Year 3, analysis of the individual accounts indicates that of the amounts already collected, $30,000 represents collections for rental services to be provided to tenants during Year 4. Present the required adjusting entry. What is Rent Revenue for Year 3?

**g.** When the firm acquired new equipment costing $10,000 on January 1, Year 1, the bookkeeper debited Depreciation Expense and credited Cash for $10,000 but made no further entries for this equipment during Year 1. The equipment has an expected service life of five years and an estimated salvage value of zero. Construct the adjusting entry required before the accountant can prepare a balance sheet for December 31, Year 1.

**49. Ethical issues in accounting choices.** Firms must make various choices in applying generally accepted accounting principles under the accrual basis of accounting (for example, depreciable lives for buildings and equipment, estimated uncollectibles for accounts receivable, estimated warranty costs). Selected data from the financial statements of AMR (parent company of American Airlines), Delta Air Lines, and UAL (parent company of United Airlines) appear below (in millions):

| | AMR | Delta | UAL |
|---|---|---|---|
| (1) Average Assets Subject to Depreciation and Amortization | $29,826.5 | $24,919.5 | $22,835.5 |
| (2) Depreciation and Amortization Expense | $1,336.0 | $1,181.0 | $970.0 |
| (3) Average Depreciation and Amortization Life: (3) = (1)/(2) | 22.3 years | 21.1 years | 23.5 years |
| (4) Estimated Uncollectible Accounts as a Percentage of Accounts Receivable | 7.1% | 10.2% | 3.5% |

These firms provide similar types of airline services with similar types of assets. They each received unqualified opinions from their independent auditors. Yet, UAL appears to apply its accounting principles more aggressively in income-enhancing ways relative to AMR and Delta. Discuss any ethical issues suggested by the above data.

CHAPTER 4

# Statement of Cash Flows: Reporting the Effects of Operating, Investing, and Financing Activities on Cash Flows

## Learning Objectives

1. Understand why using the accrual basis of accounting to prepare the balance sheet and income statement creates the need for a statement of cash flows.
2. Understand the types of transactions that result in cash flows from operating, investing, and financing activities.
3. Develop an ability to prepare a statement of cash flows from a comparative balance sheet and income statement.
4. Distinguish between the direct and indirect methods of reporting and analyzing cash flows from operations.
5. Develop an ability to analyze the statement of cash flows, including the relation among cash flows from operating, investing, and financing activities for businesses in various stages of their growth.

*W*hat do General Cinema, United Airlines, and MCI have in common? Each of these firms filed for bankruptcy during the late 1900s and early 2000s. Yet each of these firms operated profitably for most of the years preceding its bankruptcy filing. The bankruptcies occurred because these firms were unable to generate sufficient cash to cover operating costs, debt service costs, and capital expenditures. This chapter discusses the statement of cash flows, which reports the impact of a firm's operating, investing, and financing activities on cash flows during an accounting period.

## Need for a Statement of Cash Flows

How can a profitable firm run out of cash? Two explanations, at least, suggest answers.

1. **Net income for a particular period does not equal cash flow from operations. Chapter 3** points out that most firms use the accrual basis of accounting in measuring operating performance. Firms typically recognize revenue at the time of sale, even though they may receive cash from customers prior to, or coincident with, or after the time of sale. Thus, revenues on the income statement for any period will not likely equal cash received from customers. Likewise,

firms match expenses either with associated revenues or with the period they consume goods or services in operations. The cash outflow for the expense may occur prior to, or coincident with, or subsequent to the recognition of the expense. Thus, expenses on the income statement will not likely equal cash paid to suppliers of goods and services each period. Firms typically make most of the cash outflows for expenses prior to or coincident with the sale of the good or service, whereas they usually receive cash after the sale. This lag between cash outflows and cash inflows can lead to cash shortfalls, particularly for a growing firm. Consider the typical firm where cash disbursements to employees and suppliers precede cash collections from customers. The faster such a firm grows, the shorter of cash it will find itself, until it takes steps to provide the funds necessary to pay suppliers while it awaits cash collections from its own customers. The use of the accrual basis of accounting in measuring net income creates the need for a separate financial statement that reports the impact of operations on cash flows so the reader can judge a firm's cash flow needs and how it has dealt with them.

2. **Firms receive cash inflows and disburse outflows because of investing and financing activities, which the income statement does not report directly, although it provides hints of such activities.** Firms building their productive capacity generally use cash to acquire property, plant, and equipment. Short- and long-term borrowing comes due, requiring cash. Firms that pay dividends regularly to shareholders are reluctant to curtail the dividends when cash is tight. Thus, a profitable firm may generate positive cash flow from operations and still have cash shortages.

The balance sheet reports the balance in cash at the beginning and end of the year, but does not explain how it changed during the period. The income statement measures the value added from selling goods and services for more than their costs. Although a primary goal of business firms is to generate a profit, firms need adequate cash to carry out this goal. The statement of cash flows explains how a firm obtains and uses cash.

## Overview of the Statement of Cash Flows

**Exhibit 4.1**, which is similar to **Exhibit 1.3** discussed briefly in **Chapter 1**, presents a statement of cash flows for Wal-Mart Stores for three recent years. Note the following aspects of this statement.

### The Statement Explains the Reasons for the Change in Cash between Balance Sheet Dates

The last two lines of the statement of cash flows report the amount of cash on Wal-Mart's balance sheet at the beginning and the end of each year. The FASB requires that the statement of cash flows explain changes in cash and cash equivalents. Cash equivalents represent short-term, highly liquid investments in which a firm has temporarily placed excess cash. Throughout this text, we use the term *cash flows* to refer to flows of cash and cash equivalents.[1] The remaining lines show the inflows and outflows of cash during the year, which explain the net change between the two balance sheet dates. Thus, the statement of cash flows reports flows, or changes over time, whereas the balance sheet reports amounts at a moment in time.

### The Statement Classifies the Reasons for the Change in Cash as an Operating, or Investing, or Financing Activity

Various inflows and outflows of cash during the year appear in the statement of cash flows in one of three categories: operating, investing, and financing. **Figure 4.1** presents the three major types of cash flows, which the following sections describe.

**Operations** Selling goods and providing services are the most important ways for a financially healthy company to generate cash. Assessed over several years, the cash flow from operations indicates the extent to which operating activities generate more cash than they use. A firm can use the excess **cash flow from operations** to acquire buildings and equipment, pay dividends, retire long-term debt, and conduct other investing and financing activities.

[1]See Financial Accounting Standards Board (FASB), *Statement of Financial Accounting Standards No. 95*, "Statement of Cash Flows," 1987.

**EXHIBIT 4.1**

**WAL-MART STORES, INC.**
**Comparative Statement of Cash Flows with Reconciliation of Net Income to Cash Flow from Operations (all dollar amounts in millions)**

| | Year Ended January 31 | | | | |
|---|---|---|---|---|---|
| | Year 12 | Change | Year 13 | Change | Year 14 |
| **Cash Provided by Operating Activities** | | | | | |
| Sources of Cash | | | | | |
| Cash Received from Customers | $ 219,602 | $ 26,960 | $ 246,562 | $12,656 | $ 259,218 |
| Uses of Cash | | | | | |
| Cash Paid for Goods Available for Sale | (172,429) | (20,198) | (192,627) | (5,290) | (197,917) |
| Cash Paid for Selling and Administrative Items | (31,690) | (4,163) | (35,853) | (3,243) | (39,096) |
| Cash Paid for Interest | (1,326) | 263 | (1,063) | 67 | (996) |
| Cash Paid for Income Taxes | (3,897) | (590) | (4,487) | (631) | (5,118) |
| Net Cash Provided by Operating Activities | $ 10,260 | $ 2,272 | $ 12,532 | $ 3,559 | $ 16,091 |
| **Reconciliation of Net Income to Cash Provided by Operating Activities** | | | | | |
| Net Income | $ 6,671 | $ 1,368 | $ 8,039 | $ 1,036 | $ 9,075 |
| Adjustments to Reconcile Net Income to Net Cash Provided | | | | | |
| Depreciation | 3,290 | 142 | 3,432 | 420 | 3,852 |
| (Increase) Decrease in Accounts Receivable | (210) | 109 | (101) | 474 | 373 |
| (Increase) Decrease in Inventories | (1,235) | (1,001) | (2,236) | 263 | (1,973) |
| (Increase) Decrease in Prepayments | (180) | 925 | 745 | (1,195) | (450) |
| Increase in Accounts Payable | 368 | 1,079 | 1,447 | 1,356 | 2,803 |
| Increase in Other Current Liabilities | 1,556 | (350) | 1,206 | 1,205 | 2,411 |
| Net Cash Provided by Operating Activities | $ 10,260 | $ 2,272 | $ 12,532 | $ 3,559 | $ 16,091 |
| **Investing** | | | | | |
| Acquisition of Property, Plant, and Equipment | $ (8,383) | | $ (9,355) | | $ (10,308) |
| Other | 1,237 | | (354) | | 2,046 |
| Cash Flow from Investing | $ (7,146) | | $ (9,709) | | $ (8,262) |
| **Financing** | | | | | |
| Increase (Decrease) in Short-Term Borrowing | $ (1,533) | | $ 1,836 | | $ 688 |
| Increase in Long-Term Borrowing | 4,591 | | 2,044 | | 4,099 |
| Increase in Common Stock | 0 | | 0 | | 0 |
| Decrease in Long-Term Borrowing | (3,686) | | (1,479) | | (3,541) |
| Acquisition of Common Stock | (1,214) | | (3,232) | | (5,182) |
| Dividends | (1,249) | | (1,328) | | (1,569) |
| Other | 84 | | (67) | | 117 |
| Cash Flow from Financing | $ (3,007) | | $ (2,226) | | $ (5,388) |
| Change in Cash | $ 107 | | $ 597 | | $ 2,441 |
| Cash, Beginning of Year | 2,054 | | 2,161 | | 2,758 |
| Cash, End of Year | $ 2,161 | | $ 2,758 | | $ 5,199 |

**Investing** The second section of the statement of cash flows shows the amount of **cash flow from investing activities**. The acquisition of noncurrent assets, particularly property, plant, and equipment, usually represents a major ongoing use of cash. A firm must replace such assets as they wear out, and it must acquire additional noncurrent assets if it is to grow. A firm obtains part of the cash needed to acquire noncurrent assets from sales of existing noncurrent assets. Such cash inflows seldom, however, cover the entire cost of new acquisitions. Firms not experiencing rapid growth can usually finance capital expenditures with cash flow from operations. Firms growing rapidly must often borrow funds or issue common stock to finance their acquisitions of noncurrent assets.

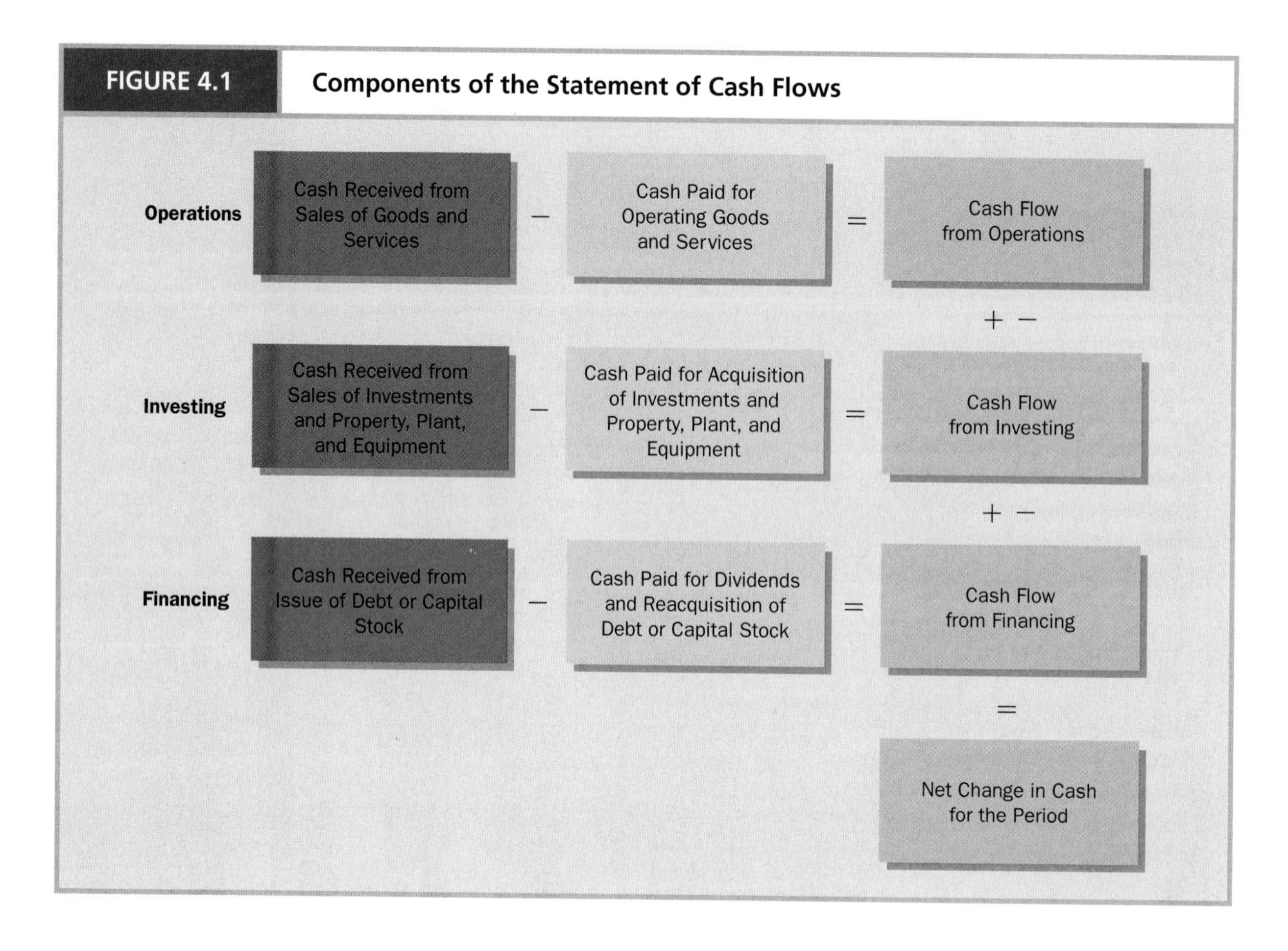


**Financing** Third, a firm obtains cash from short- and long-term borrowing and from issues of common or preferred stock. It uses cash to pay dividends to shareholders, to repay short- or long-term borrowing, and to reacquire shares of outstanding common or preferred stock. These amounts appear as **cash flow from financing activities** in the statement of cash flows.

Refer to **Exhibit 4.1**. Wal-Mart reported positive cash flow from operations in each of the three years. Cash flow from operations exceeded the cash outflow needed to acquire additional property, plant, and equipment each year. Business terminology refers to an excess of cash flow from operations over cash flow investing as *free cash flow*. Firms can use free cash flow to repay borrowing, pay a dividend, repurchase common stock, and add to cash on the balance sheet. Wal-Mart used the free cash flow, plus the cash received from additional borrowing in excess of the cash used to repay old borrowing, to pay dividends and reacquire shares of its common stock. **Chapter 5** explores more fully why Wal-Mart uses its excess cash, in particular, to reacquire its common stock.

**Ambiguities in Classifying Cash Flows** Cash flows do not always fit unambiguously into only one of these three categories of cash flows. For example, you might think of cash received from interest and dividend revenues generated by investments in securities as coming from operating activities. The logic for this treatment is that interest and dividends appear as revenues in the income statement, the financial statement that reports the results of a firm's operations. Alternatively, cash received from interest and dividends might appear as coming from investing activities. The logic for this treatment is that cash flows related to the purchase and sale of investments in securities, other than trading securities (a distinction discussed in **Chapter 11**) appear as investing activities. The Financial Accounting Standards Board's *Statement of Financial Accounting Standards No. 95* requires firms to classify the receipt of cash from interest and dividend revenues as an operating activity but the cash related to the purchase and sale of investments in securities as an investing activity.[2]

Similar ambiguities arise with interest expense on debt. Should the cash outflow for interest expense appear as an operating activity (to achieve consistency with its inclusion in the income

[2] Ibid., para 16, 17, 22.

statement as an expense) or as a financing activity (to achieve consistency with the classification of debt issues and retirements as a financing activity)? FASB *SFAS No. 95* requires that firms classify interest expense as an operating activity but the issue or redemption of debt as a financing activity.[3] Dividends that a firm pays to its shareholders appear, however, as a financing activity.[4] The classification of interest expense on debt as an operating activity and dividends paid on common or preferred stock as a financing activity appears inconsistent. The FASB's likely rationale is that accountants treat interest as an expense in computing net income, whereas dividends represent a distribution of assets generated by net income, not an expense reducing net income.

Similar ambiguities arise in classifying purchases and sales of marketable securities (treated as investing, not operating, activities) and increases and decreases in short-term bank borrowings (treated as financing, not operating, activities). Later chapters discuss these items more fully.

**Treatment of Noncash Transactions** Firms sometimes engage in investing and financing transactions that do not directly involve cash. For example, a firm may acquire a building by assuming a mortgage obligation or might exchange a tract of land for equipment. Holders of a firm's debt might convert the debt into common stock. These transactions do not appear in the statement of cash flows as either investing or financing activities because they are not factors in explaining the change in cash. Firms must disclose noncash investing and financing activities in a separate schedule or note.[5]

## The Statement Reconciles Net Income with Cash Flow from Operations

The first section of the statement of cash flows derives cash flow from operations. FASB *Statement No. 95* permits firms to report cash flow from operations in either of two ways:

1. **Direct Method** The direct method reports the amounts of cash received from customers less cash disbursed to various suppliers, employees, lenders for interest payments, and taxing authorities. The top panel of **Exhibit 4.1** illustrates the direct method. The FASB has said that it prefers this method, but most firms use the alternative, described next.
2. **Indirect Method** The indirect method begins with net income for a period and then shows adjustments to net income to convert revenues to cash received from customers and to convert expenses to cash disbursed to various suppliers of goods and services. The boxed section of **Exhibit 4.1** illustrates the indirect method of reporting cash flow from operations.

*Statement No. 95* permits firms to report cash flow from operations using either the direct or the indirect method. A firm that presents the preferred, but less often used, direct method must show a reconciliation (see the boxed section of **Exhibit 4.1**) between net income and cash flow from operations either at the bottom of the statement of cash flows or in a separate note.

The majority of firms report cash flow from operations using the indirect method because, before the FASB expressed a preference for the direct method in 1987, most firms used the indirect method, so both preparers and users had become familiar with it. Experienced financial analysts who have used the statement of cash flows generally understand the adjustments required to convert net income to cash flow from operations. Our experience in teaching the statement of cash flows indicates, however, that students have difficulty on their initial exposure to the statement of cash flows understanding these adjustments, whereas they find the direct method easier to comprehend. We therefore illustrate the computation of cash flow from operations using both the direct and indirect methods.

**Overview of Adjustments to Net Income to Compute Cash Flow from Operations under the Indirect Method** This section presents an overview of the types of required adjustments to net income to compute cash flow from operations. Later sections of the chapter illustrate these adjustments more fully. The reconciliation of net income to cash flow from operations that a firm using the direct method must provide is essentially the indirect method. In this chapter, you will usually see a boxed Reconciliation in the statement of cash flows, which is, even though not explicitly described as such, the indirect method.

---

[3] Ibid., para 19, 20, 23.
[4] Ibid., para 20.
[5] Ibid., para 32.

Refer to the reconciliation of net income to cash flow from operations in the boxed section of **Exhibit 4.1**. The first line in the reconciliation shows the amount of net income for each year from the income statement. Accountants use the accrual basis of accounting to compute net income. The reconciliation shows the adjustments required to convert net income, measured on an accrual basis, to the amount of net cash flow generated from operations during the period.

Wal-Mart shows an addition to net income each year for depreciation. To understand the reconciliation's adjustment, first examine the top section deriving net cash flow as the excess of receipts over disbursements. Depreciation has no place in that section because it uses no cash this period. Wal-Mart used cash in some earlier period to acquire a store building or equipment, and in that period reported the use of cash as an investing activity. Depreciation appears as an expense on the income statement because operating retail stores requires using a portion of the service potential of buildings and equipment. The events causing the firm to recognize depreciation expense, however, consume not cash but other assets (buildings, equipment). Each year, Wal-Mart recognizes Depreciation Expense, decreasing shareholders' equity, and Accumulated Depreciation, decreasing total assets. The reconciliation starts with net income, which had a subtraction for depreciation expense. Because depreciation reduces net income but does not use cash, the reconciliation adds back the amount to derive cash flow from operations. Subtracting depreciation expense in arriving at the amount of net income on the first line of the reconciliation and then adding the same amount on the second line results in a zero net effect on cash flow from operations. Thus, the reconciliation's adjustment for depreciation removes the amount of a noncash expense.

**Conceptual Note** Once you understand the preceding paragraph and its implications, you will have mastered most of the difficulty that financial statement readers have with the indirect method of computing cash flow from operations. Perhaps this will help your understanding: Imagine a firm whose only activity for a period was the recording of depreciation expense of $100—nothing else. That firm would show negative net income, a loss, of $100, but it would have had no change in cash. Cash flow from operations is zero. So, the reconciliation must start with the −$100 for net income, then add back the $100 depreciation expense, in order to reconcile net income of −$100 with the zero cash flow from operations.

The reconciliation for Wal-Mart shows a subtraction for the change (increase) in accounts receivable. How would you derive the amount of cash Wal-Mart collected from its customers during a particular period? Wal-Mart has made sales in periods before this one as well as in this one. (For now, assume Wal-Mart does not receive cash from customers in the current period as an advance on goods it will deliver in future periods.) Wal-Mart received cash this period from customers who purchased before this period where the revenue appeared in an earlier period's income statement. This resulted in an increase in Wal-Mart's accounts receivable in the earlier period, with a decrease this period when the firm collected cash. Wal-Mart received cash also from customers who both purchased and paid this period, resulting in no change in accounts receivable. Finally, Wal-Mart made sales this period, which it will collect in future periods, resulting in an increase in accounts receivable this period. So the amount of cash Wal-Mart received from customers in a period equals the amount of sales for that period, reduced by the increase in accounts receivable during the period or increased by the decrease in accounts receivable during the period.

**Exercises to Cement Understanding.** Look at the data in **Exhibits 1.2** and **1.3**. Copy down the amount of total revenues for Year 13 from **Exhibit 1.2**, $246,663. Note the amount by which Accounts Receivable increased between Year 12 and Year 13 in **Exhibit 1.3**, $101. Subtract that $101 from total revenues of $246,663. Note this difference, $246,562, equals the Cash Received from Customers in **Exhibit 4.1**. Now, repeat this exercise for Year 14: Copy down the total revenues for Year 14 from **Exhibit 1.2**. Note the amount by which Accounts Receivable *decreased* between Years 13 and 14.[6] *Add* the amount of the decrease to total revenues and see that the sum

---

[6]A more advanced issue: if you were to look at the Wal-Mart balance sheet in **Exhibit 1.1**, you would see that the amounts for Accounts Receivable reported there for Year 14 decreased by $854, from $2,108 to $1,254, while the Statement of Cash Flows in **Exhibit 1.3** shows a decrease of only $373. These numbers might differ because Wal-Mart sold some receivables during the year in financing transactions or as part of a disinvesting transaction. **Chapter 6** discusses the raising of cash by selling, or otherwise using, accounts receivable. Sometimes a comparison of the balance sheet and statement of cash flows amounts for changes in accounts receivable shows a larger balance sheet increase than does the statement of cash flows. This difference likely results from the company's buying an entire business during the year and acquiring all the acquired company's assets, including its accounts receivable.

equals Cash Received from Customers in **Exhibit 4.1**. When Accounts Receivable decrease, the firm has collected an amount of cash greater than its sales on account for the year.

Income includes all sales, even if cash collections for those sales did not occur this period. So, the reconciliation, which starts with net income, including all sales, must have a subtraction for the increase in accounts receivable, and cash will come in a later period. This chapter discusses these and other adjustments to reconcile net income with cash flow from operations.

### Problem 4.1 for Self-Study

**Classifying cash flows by type of activity.** Indicate whether each of the following transactions of the current period would appear as an operating, investing, or financing activity in the statement of cash flows. If any transaction would not appear in the statement of cash flows, suggest the reason.

**a.** Disbursement of $96,900 to merchandise suppliers.
**b.** Receipt of $200,000 from issuing common stock.
**c.** Receipt of $49,200 from customers for sales made this period.
**d.** Receipt of $22,700 from customers this period for sales made last period.
**e.** Receipt of $1,800 from a customer for goods the firm will deliver next period.
**f.** Disbursement of $16,000 for interest expense on debt.
**g.** Disbursement of $40,000 to acquire land.
**h.** Issue of common stock with market value of $60,000 to acquire land.
**i.** Disbursement of $25,300 as compensation to employees for services rendered this period.
**j.** Disbursement of $7,900 to employees for services rendered last period but not paid for last period.
**k.** Disbursement of $53,800 for a patent purchased from its inventor.
**l.** Acquisition of a building by issuing a note payable to a bank.
**m.** Disbursement of $19,300 as a dividend to shareholders.
**n.** Receipt of $12,000 from the sale of equipment that originally cost $20,000 and had $8,000 of accumulated depreciation at the time of sale.
**o.** Disbursement of $100,000 to redeem bonds at maturity.
**p.** Disbursement of $40,000 to acquire shares of IBM common stock.
**q.** Receipt of $200 in dividends from IBM relating to the shares of common stock acquired in transaction **p** above.

## Preparing the Statement of Cash Flows

Firms could prepare their statement of cash flows directly from entries in their cash account. To do so would require them to classify each transaction affecting cash as an operating, or investing, or financing activity. As the number of transactions affecting cash increases, however, this approach becomes cumbersome. Most firms design their accounting systems to accumulate the information needed to prepare income statements and balance sheets. They then use a work sheet at the end of the period to transform information from the income statement and balance sheet into a statement of cash flows.

We present a **T-account work sheet** for preparing the statement of cash flows. The work sheet computes cash flow from operations using both the direct and the indirect methods. Later, we show how to derive the direct method's presentation.

### THE CASH CHANGE EQUATION

Fundamental to the preparation of a statement of cash flows is an understanding of how changes in cash relate to changes in noncash accounts. The accounting equation states:

Assets = Liabilities + Shareholders' Equity

Cash + Noncash Assets = Liabilities + Shareholders' Equity

**Balance Sheet Equation (Eq. 1)**

This equation must be true for balance sheets constructed at both the start of the period and the end of the period. If the start-of-period and end-of-period balance sheets maintain the accounting equation, then the following equation must also be valid:

**Balance Sheet Change Equation (Eq. 2)**

$$\text{Change in Cash} + \text{Change in Noncash Assets} = \text{Change in Liabilities} + \text{Change in Shareholders' Equity}$$

Rearranging terms in this equation, we obtain the equation for changes in cash:

**Cash Change Equation (Eq. 3)**

$$\text{Change in Cash} = \text{Change in Liabilities} + \text{Change in Shareholders' Equity} - \text{Change in Noncash Assets}$$

The left-hand side of the **Cash Change Equation** (Eq. 3) represents the change in cash. The right-hand side of the equation, reflecting changes in all noncash accounts, must also net to the change in cash. The equation states that the changes in cash (left-hand side) equal the changes in liabilities plus the changes in shareholders' equity less the changes in noncash assets (right-hand side). For example, a loan from a bank or another lender increases cash and increases a liability. The issue of common stock increases cash and increases shareholders' equity. The purchase of equipment decreases cash and increases noncash assets. Increases in noncash assets carry a negative sign on the right-hand side of the equation. *Thus, we can identify the causes of the change in cash by studying the changes in noncash accounts and classifying those changes as operating, investing, and financing activities.* Focus on the preceding italicized sentence. It provides the overview, the big picture, of the procedures that underlie the T-account work sheet for the statement of cash flows.

The cash change equation results from rearranging the balance sheet equation. When we focus on the balance sheet effects of a transaction, we give the balance sheet equation's components. When we focus on the cash flow effects of a transaction, we give the cash flow equations components. For some complex transactions, we show both.

## Data for Illustrations

To illustrate the preparation of the statement of cash flows in this chapter, we use information for Solinger Electric Corporation for Year 4. **Exhibit 4.2** presents the balance sheet at the beginning and end of Year 4. **Exhibit 4.3** presents the income statement for Year 4. Solinger Electric Corporation declared and paid $7,000 in dividends during Year 4.

## T-Account Work Sheet

One can prepare the statement of cash flows by examining every transaction affecting the cash account, as in **Figure 4.1**, and classifying each one as an operating activity or investing activity or financing activity. If a firm's record-keeping system incorporates the appropriate classification codes into the initial recording of transactions in the Cash account, then preparing the statement of cash flows becomes straightforward.

Given the large number of transactions affecting the Cash account during a period, most firms prefer to prepare the statement of cash flows after they have prepared the income statement and the balance sheet. This section presents a step-by-step procedure for preparing the statement of cash flows.

The T-account work sheet for preparing the statement of cash flows, discussed next, provides built-in checks to ensure the full recognition of the effects of each transaction on various accounts. The T-account work sheet is also a direct extension of the T-accounts used in **Chapters 2** and **3**. The accountant prepares a T-account work sheet at the end of the period after preparing the balance sheet and income statement. The work sheet provides the information for preparing the statement of cash flows.

We begin by illustrating the preparation of a T-account work sheet using the data for Solinger Electric Corporation in **Exhibits 4.2** and **4.3**. We then add some complexities to illustrate the use of the T-account work sheet in a more realistic situation.

**Step 1** Obtain balance sheets for the beginning and end of the period covered by the statement of cash flows. **Exhibit 4.2** presents the comparative balance sheets of Solinger Electric Corporation for December 31, Year 3 and Year 4.

*Cash Change Equation (Eq. 3)*

| Change in Cash | = | Change in Liabilities | + | Change in Shareholders' Equity | – | Change in Noncash Assets |
|---|---|---|---|---|---|---|

**EXHIBIT 4.2**

**SOLINGER ELECTRIC CORPORATION**
**Comparative Balance Sheet**
**December 31, Years 3 and 4**

| | December 31 | |
|---|---|---|
| | Year 3 | Year 4 |
| ASSETS | | |
| **Current Assets** | | |
| Cash | $ 30,000 | $ 6,000 |
| Accounts Receivable | 20,000 | 55,000 |
| Merchandise Inventory | 40,000 | 50,000 |
| Total Current Assets | $ 90,000 | $111,000 |
| **Noncurrent Assets** | | |
| Buildings and Equipment (Cost) | $100,000 | $225,000 |
| Accumulated Depreciation | (30,000) | (40,000) |
| Total Noncurrent Assets | $ 70,000 | $185,000 |
| Total Assets | $160,000 | $296,000 |
| EQUITIES | | |
| **Current Liabilities** | | |
| Accounts Payable—Merchandise Suppliers | $ 30,000 | $ 50,000 |
| Accounts Payable—Other Suppliers | 10,000 | 12,000 |
| Salaries Payable | 5,000 | 6,000 |
| Total Current Liabilities | $ 45,000 | $ 68,000 |
| **Noncurrent Liabilities** | | |
| Bonds Payable | $ 0 | $100,000 |
| **Shareholders' Equity** | | |
| Common Stock | $100,000 | $100,000 |
| Retained Earnings | 15,000 | 28,000 |
| Total Shareholders' Equity | $115,000 | $128,000 |
| Total Equities | $160,000 | $296,000 |

**EXHIBIT 4.3**

**SOLINGER ELECTRIC CORPORATION**
**Income Statement**
**For Year 4**

| | |
|---|---|
| Sales Revenue | $125,000 |
| Cost of Goods Sold | (60,000) |
| Depreciation Expense | (10,000) |
| Salary Expense | (20,000) |
| Interest Expense | (4,000) |
| Other Expenses | (11,000) |
| Net Income | $ 20,000 |

**Step 2** Prepare a T-account work sheet. An example of such a T-account work sheet appears in **Exhibit 4.4**. The top of the work sheet shows a master T-account titled Cash. Note that this T-account has sections labeled Operations, Investing, and Financing, which we use to classify transactions that affect cash. Enter the beginning and ending amounts of cash in the master T-account. (The beginning and ending amounts of cash for Solinger Electric Corporation are $30,000 and $6,000, respectively.) The number at the top of the T-account is the beginning balance; the one at the bottom is the ending balance. The check marks indicate that the figures are balances. The master T-account, Cash, represents the left-hand side of **Equation (3)** for changes in cash.

After preparing the master T-account for Cash (as at the top of **Exhibit 4.4**), complete the work sheet by preparing T-accounts for each noncash asset, liability, and shareholders' equity account. The lower portion of **Exhibit 4.4** shows the T-accounts for each noncash account. Enter the beginning and ending balances in each account for the period given in the balance sheet (see **Exhibit 4.2**). The sum of the changes in these individual T-accounts expresses the right-hand side of **Equation (3)** for changes in cash.

**Step 3** Explain the change in the master cash account between the beginning and the end of the period by accounting for the change in each noncash account during the period. Accomplish this step by reconstructing the entries originally recorded in the accounts during the period and entering them into the same accounts on the T-account work sheet as the accounts originally used during the period. The only extension is that entries in the master account for cash require classi-

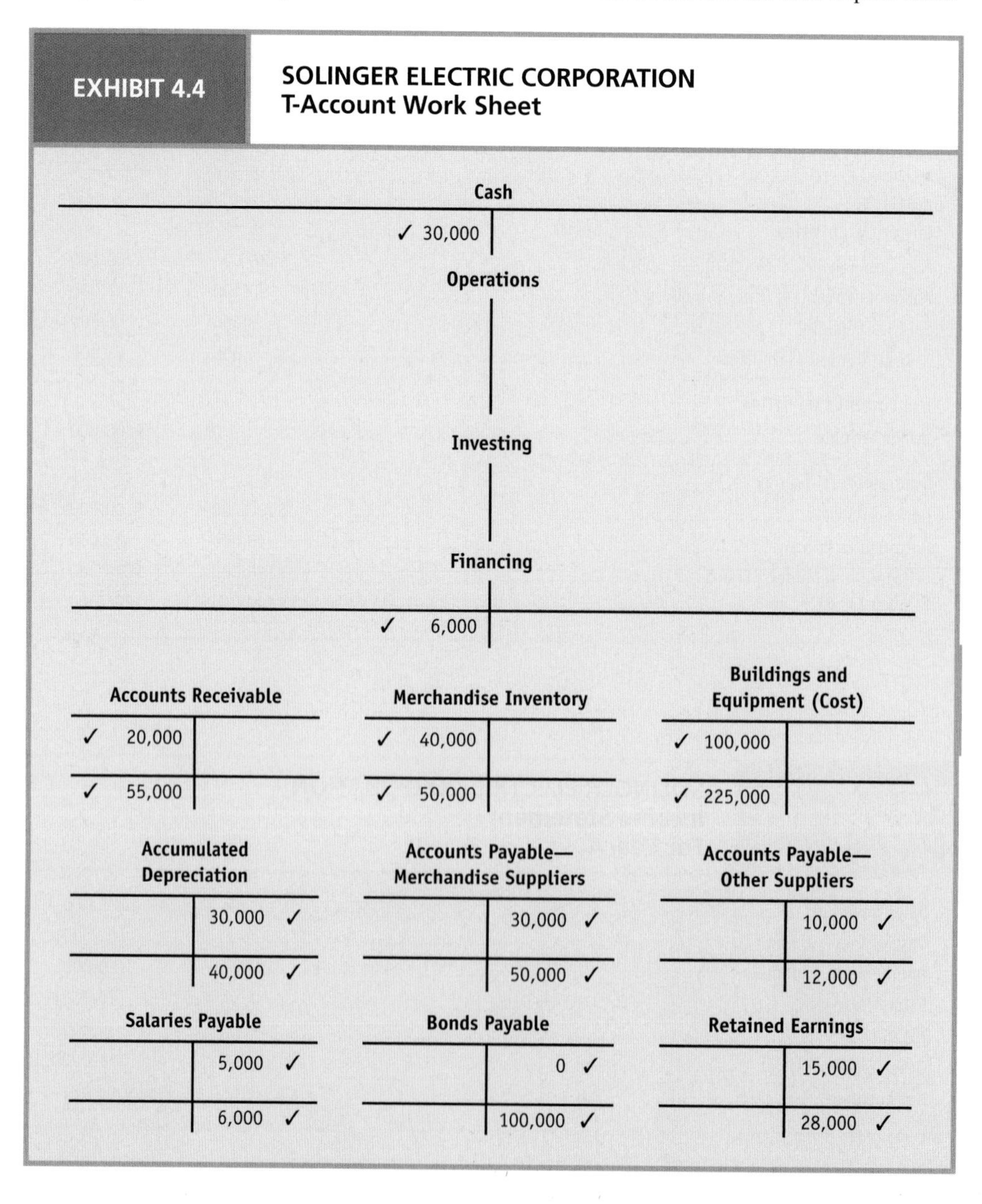

*Cash Change Equation (Eq. 3)*

| Change in Cash | = | Change in Liabilities | + | Change in Shareholders' Equity | − | Change in Noncash Assets |
|---|---|---|---|---|---|---|

fication as an operating, or investing, or financing activity. Once this procedure has accounted for the net change in each of the noncash accounts, it will have generated sufficient information to account for the net change in cash. In other words, if the reconstructed transactions explain the changes in the right-hand side of the Cash Change Equation **(Eq. 3)**, they will have explained the causes of the changes in cash itself on the left-hand side. In constructing the T-account work sheet, we make analytic entries similar in form to journal entries, but entries that we do not actually enter into a firm's record-keeping system. These analytic entries appear on the work sheet; once we have completed the T-account work sheet and the statement of cash flows, the analytic entries have served their purpose, and we discard them or, perhaps, save them in a file to remind us next period what we did this period.

The amounts debited to various accounts on the T-account work sheet must equal amounts credited to various accounts. A common source of error when preparing the T-account work sheet results from the partial recording of a transaction in which debits do not equal credits. The error becomes evident only on completion of the work sheet when the preparer discovers that the entries in one or more accounts on the T-account work sheet do not explain the change in the account during the period. The preparer must then retrace each of the entries to discover the source of the error. A careful initial recording of analytic entries in the T-account work sheet can save considerable time afterward.

Reconstructing the transactions during the year usually proceeds more easily by accounting first for supplementary information. The following information applies to Solinger Electric Corporation for Year 4:

1. Net income is $20,000.
2. Depreciation expense is $10,000.
3. Dividends declared and paid total $7,000.

The analytic entry to record the information concerning net income is as follows:

| | | |
|---|---|---|
| **(1)** Cash (Operations: Net Income) . . . . . . . . . . . . . . . . . . . . . . . . . . . . . . | 20,000 | |
| Retained Earnings . . . . . . . . . . . . . . . . . . . . . . . . . . . . . . . . . | | 20,000 |
| Analytic entry recorded in T-account work sheet. | | |

We do not know the effects of analytic entries on the balance sheet equation, because the firm does not record this and similar entries in the journal. Use these entries and the analysis in preparing the T-account work sheet for the statement of cash flows.

To understand this analytic entry, review the process of recording revenues and expenses and the closing entries for those temporary accounts from **Chapter 3**. All of the journal entries that together record the process of earning $20,000 in net income are equivalent to the following single journal entry:

| | | |
|---|---|---|
| Net Assets (= All Assets Minus All Liabilities) . . . . . . . . . . . . . . . . . . . . | 20,000 | |
| Retained Earnings . . . . . . . . . . . . . . . . . . . . . . . . . . . . . . . . . | | 20,000 |
| Summary entry equivalent to recording earnings of $20,000. | | |

| Change in Cash | = | Change in Liabilities | + | Change in Shareholders' Equity | − | Change in Noncash Assets |
|---|---|---|---|---|---|---|
| +20,000 (Opns.) | | 0 | | +20,000 | | 0 |

The T-account work sheet starts with the assumption that all earnings produce cash from operations. Subsequent additions and subtractions correct for transactions where the assumption is invalid.

We show the effect of the entry on the Cash Change Equation. Where Cash changes, we classify (in parentheses in the amount in the Change in Cash column) the change as caused by operating activities (Opns.), by investing activities (Invst.), or by financing activities (Finan.). Here, the earnings activities are operating.

The summary journal entry debits Net Assets. We assume at this stage of preparing the statement of cash flows that all of the net assets generated by the earnings process were cash. Thus, in analytic entry **(1)**, the debit shows a provisional increase in cash from operations in an amount equal to net income for the period. This is an analytic entry; we make entry **(1)** on the T-account work sheet to help reconstruct the transactions affecting cash.

Not all of the items recognized as expenses and deducted in calculating net income decrease cash. To calculate the net amount of cash from operations, we must add back to the provisional increase in cash any amounts of the expenses that do not use cash (but instead use other noncash net assets in this period). For example, one expense that does not require an operating cash outflow this period is depreciation, illustrated in entry **(2)**:

| | | |
|---|---|---|
| **(2)** Cash (Operations: Depreciation Expense Addback) . . . . . . . . . . . . . . . . . . | 10,000 | |
| Accumulated Depreciation . . . . . . . . . . . . . . . . . . . . . . . . . . . . . . | | 10,000 |
| Analytic entry recorded in T-account work sheet. | | |

Depreciation expense, deducted in calculating net income, did not reduce cash this period. Some time ago, the firm used cash to acquire the fixed assets it now depreciates. The use of cash appeared on the statement for the earlier period in the Investing section. This analytic entry adds back depreciation expense to net income in calculating the amount of cash flow from operations.

Next, record the supplementary information concerning dividends of $7,000 declared and paid:

| | | |
|---|---|---|
| **(3)** Retained Earnings . . . . . . . . . . . . . . . . . . . . . . . . . . . . . . . . . . . . . . | 7,000 | |
| Cash (Financing: Dividends) . . . . . . . . . . . . . . . . . . . . . . . . . . . . | | 7,000 |
| Analytic entry recorded in T-account work sheet. | | |

Dividends reduce retained earnings and cash. Paying dividends appears on the statement of cash flows as a financing activity.

Once the T-account work sheet reflects the supplementary information, one must make inferences about the reasons for the remaining changes in the noncash accounts on the balance sheet. (The preparer of a statement of cash flows for an actual firm will likely not need to make such inferences because sufficient information regarding the change in each account will probably appear in the firm's accounting records.) Explanations for the changes in noncash accounts appear below, in balance sheet order.

The Accounts Receivable account shows an increase of $35,000. The analytic entry to record this information in the work sheet is as follows:

| | | |
|---|---|---|
| **(4)** Accounts Receivable . . . . . . . . . . . . . . . . . . . . . . . . . . . . . . . . . . . . | 35,000 | |
| Cash (Operations: Subtractions) . . . . . . . . . . . . . . . . . . . . . . . . . | | 35,000 |
| Analytic entry recorded in T-account work sheet. | | |

The operations of the period generated sales, but not all of these sales resulted in an increase in cash. Some of the sales increased accounts receivable. Because we start the statement of cash flows with net income, provisionally assuming that all sales generated cash, we must subtract that portion of revenues that did not produce cash (that is, the excess of sales on account over cash collections from customers) in deriving the amount of actual cash from operations.

The next noncash account that changed, Merchandise Inventory, shows an increase during the year of $10,000. As the operations of the firm have expanded, so has the amount carried in inventory. The analytic entry in the work sheet to explain the change in Merchandise Inventory is as follows:

| | | |
|---|---|---|
| **(5)** Merchandise Inventory . . . . . . . . . . . . . . . . . . . . . . . . . . . . . . . . . . | 10,000 | |
| Cash (Operations: Subtractions) . . . . . . . . . . . . . . . . . . . . . . . . . | | 10,000 |
| Analytic entry recorded in T-account work sheet. | | |

*Cash Change Equation (Eq. 3)*

| Change in Cash | = | Change in Liabilities | + | Change in Shareholders' Equity | − | Change in Noncash Assets |
|---|---|---|---|---|---|---|

Solinger Electric Corporation found it necessary to increase the amount of inventory carried to make possible increased future sales. Increasing inventory ordinarily uses cash, and the preparation process provisionally assumes that the firm used cash to acquire all inventory. Later, the process will adjust for the amounts the firm will pay for later and for amounts where the firm has already used cash because it paid in advance. The cost of goods sold has reduced the amount of net income; because we start the statement of cash flows with net income, in deriving the amount of cash from operations we must subtract from net income the increase in inventories during the year (that is, the excess of purchases over the cost of goods sold).

The next noncash account, Buildings and Equipment (Cost), shows a net increase of \$125,000 (= \$225,000 − \$100,000). Because we have no other information, we must assume that the firm acquired buildings and equipment costing \$125,000 during the year. The analytic entry is as follows:

| | | |
|---|---|---|
| **(6)** Buildings and Equipment (Cost) | 125,000 | |
| Cash (Investing: Acquisitions of Buildings and Equipment) | | 125,000 |
| Analytic entry recorded in T-account work sheet. | | |

The next noncash account showing a change is Accounts Payable—Merchandise Suppliers. As the amounts carried in inventory increase, so do amounts owed to suppliers of inventory. The analytic entry to explain the increase in the amount of Accounts Payable—Merchandise Suppliers is as follows:

| | | |
|---|---|---|
| **(7)** Cash (Operations: Additions) | 20,000 | |
| Accounts Payable—Merchandise Suppliers | | 20,000 |
| Analytic entry recorded in T-account work sheet. | | |

Ordinarily, acquiring inventory requires cash, and in analytic entry **(5)** we assumed that the firm paid cash in the period of acquisition for all inventory it acquired. Suppliers who allow a firm to pay later for goods and services received now in effect supply the firm with cash. Thus, an increase in the amount of accounts payable for inventory results from a transaction in which inventory increases but cash will decrease later. This is equivalent to saying that an increase in payables provides cash, even if it is only temporary. Accounting classifies the increase in cash resulting from increased payables for inventory as an operating source of cash.

The next noncash account showing a change is Accounts Payable—Other Suppliers. As the scope of operations has increased, so has the amount owed to others. The analytic entry to explain the increase in the amount of Accounts Payable—Other Suppliers is as follows:

| | | |
|---|---|---|
| **(8)** Cash (Operations: Additions) | 2,000 | |
| Accounts Payable—Other Suppliers | | 2,000 |
| Analytic entry recorded in T-account work sheet. | | |

The reasoning behind this analytic entry is the same as for analytic entry **(7)**. Creditors who permit a firm to increase amounts owed temporarily provide cash.

The same reasoning applies to an increased amount of Salaries Payable, the next noncash account showing a change. The analytic entry to record the increase in Salaries Payable is as follows:

| | | |
|---|---|---|
| **(9)** Cash (Operations: Additions) | 1,000 | |
| Salaries Payable | | 1,000 |
| Analytic entry recorded in T-account work sheet. | | |

**EXHIBIT 4.5**

**SOLINGER ELECTRIC CORPORATION**
**T-Account Work Sheet**

**Cash**

| | | Debit | Credit | | |
|---|---|---|---|---|---|
| | ✓ | 30,000 | | | |
| | | **Operations** | | | |
| Net Income | (1) | 20,000 | 35,000 | (4) | Increased Accounts Receivable |
| Depreciation Expense Addback | (2) | 10,000 | 10,000 | (5) | Increased Merchandise Inventory |
| Increased Accounts Payable to Merchandise Suppliers | (7) | 20,000 | | | |
| Increased Accounts Payable to Other Suppliers | (8) | 2,000 | | | |
| Increased Salaries Payable | (9) | 1,000 | | | |
| | | **Investing** | | | |
| | | | 125,000 | (6) | Acquisition of Buildings and Equipment |
| | | **Financing** | | | |
| Long-Term Bond Issue | (10) | 100,000 | 7,000 | (3) | Dividends |
| | ✓ | 6,000 | | | |

**Accounts Receivable**

| | Debit | Credit | |
|---|---|---|---|
| ✓ | 20,000 | | |
| (4) | 35,000 | | |
| ✓ | 55,000 | | |

**Merchandise Inventory**

| | Debit | Credit | |
|---|---|---|---|
| ✓ | 40,000 | | |
| (5) | 10,000 | | |
| ✓ | 50,000 | | |

**Buildings and Equipment (Cost)**

| | Debit | Credit | |
|---|---|---|---|
| ✓ | 100,000 | | |
| (6) | 125,000 | | |
| ✓ | 225,000 | | |

**Accumulated Depreciation**

| | Debit | Credit | |
|---|---|---|---|
| | | 30,000 | ✓ |
| | | 10,000 | (2) |
| | | 40,000 | ✓ |

**Accounts Payable—Merchandise Suppliers**

| | Debit | Credit | |
|---|---|---|---|
| | | 30,000 | ✓ |
| | | 20,000 | (7) |
| | | 50,000 | ✓ |

**Accounts Payable—Other Suppliers**

| | Debit | Credit | |
|---|---|---|---|
| | | 10,000 | ✓ |
| | | 2,000 | (8) |
| | | 12,000 | ✓ |

**Salaries Payable**

| | Debit | Credit | |
|---|---|---|---|
| | | 5,000 | ✓ |
| | | 1,000 | (9) |
| | | 6,000 | ✓ |

**Bonds Payable**

| | Debit | Credit | |
|---|---|---|---|
| | | 0 | ✓ |
| | | 100,000 | (10) |
| | | 100,000 | ✓ |

**Retained Earnings**

| | Debit | Credit | |
|---|---|---|---|
| | | 15,000 | ✓ |
| (3) | 7,000 | 20,000 | (1) |
| | | 28,000 | ✓ |

Employees who do not demand immediate payment for earned salaries temporarily provide their employer with cash.

Bonds Payable, the final noncash account with a change not yet explained, shows a net increase of $100,000 for the year. Even without explicit information, one can deduce that the firm issued long-term bonds during the year. The analytic entry is as follows:

| | | |
|---|---|---|
| **(10)** Cash (Financing: Long-Term Bond Issue) . . . . . . . . . . . . . . . . . . . . . . | 100,000 | |
| Bonds Payable . . . . . . . . . . . . . . . . . . . . . . . . . . . . . . . . . . . . | | 100,000 |
| Analytic entry recorded in T-account work sheet. | | |

*Cash Change Equation (Eq. 3)*

| Change in Cash | = | Change in Liabilities | + | Change in Shareholders' Equity | – | Change in Noncash Assets |
|---|---|---|---|---|---|---|

**Exhibit 4.5** presents the completed T-account work sheet for Solinger Electric Corporation for Year 4. The 10 analytic entries explain all changes in the noncash T-accounts. If the work is correct, the causes of the change in the Cash account will appear in the entries in the master Cash account.

**Step 4** In the final step we use the information provided in the master T-account for Cash in the completed work sheet to prepare a formal statement of cash flows. **Exhibit 4.6** on page 170 presents the Solinger Electric Corporation statement, which we prepared earlier using the columnar work sheet approach.

## Depreciation Does Not Provide Cash

Because the indirect method adds depreciation expense to net income to calculate cash provided by operations, readers of financial statements might incorrectly conclude that depreciation charges provide cash. The recording of depreciation expense does not affect cash. A noncash asset decreases, and a shareholders' equity account decreases. Cash from operations results from selling goods and services to customers. If a firm makes no sales, there will be no cash provided by operations regardless of how large the depreciation charge.

To understand that *depreciation does not provide cash*, refer to the income statement of Solinger Electric Corporation (**Exhibit 4.3**) and the operations section of the statement of cash flows (**Exhibit 4.6**). **Exhibit 4.7** on page 170 reproduces them in condensed form. Ignore income taxes for a moment. Suppose that depreciation for Year 4 had been $25,000 rather than $10,000. The condensed income statement and cash flow from operations would then appear as in **Exhibit 4.8** on page 170. Note that the total cash flow provided by operations, which is receipts from customers minus all expenses that used cash, remains $8,000. Transactions involving long-term assets affect cash only when (1) a firm acquires a long-term asset for cash or (2) it sells the asset for cash.

At a more sophisticated level, when the firm considers income taxes, depreciation does affect cash flow. Depreciation affects the calculation of net income reported in the financial statements, and as a deduction, it also reduces taxable income on tax returns. The larger the amount of depreciation on tax returns, the smaller is the taxable income and the smaller the current payment for income taxes. **Chapters 8** and **10** discuss the effect of depreciation on income taxes.

### Problem 4.2 for Self-Study

**Preparing a T-account work sheet for a statement of cash flows. Exhibit 4.9** on page 171 presents a comparative balance sheet for Robbie Corporation as of December 31, Year 1 and Year 2. **Exhibit 4.10** on page 171 shows the income statement for Year 2. During Year 2 the firm sold no plant and equipment. It declared and paid dividends of $2,000. Prepare a T-account work sheet for the preparation of a statement of cash flows. Use the format shown in **Exhibit 4.5**.

## Direct Method

Although most firms present the operating section of their statement of cash flows using the indirect method, students often find the presentation confusing on initial exposure. To overcome interpretive problems of the indirect method, such as that arising from the addback of depreciation, the FASB prefers that firms use the direct method in reporting cash flow from operations. A note to the financial statements must, however, reconcile net income with cash flow from operations (that is, the indirect method). Next, we show how you can derive cash flow from

**EXHIBIT 4.6**

**SOLINGER ELECTRIC CORPORATION**
**Statement of Cash Flows**
**Year 4**
**Indirect Method**

| | | |
|---|---|---|
| **Operations** | | |
| Net Income | $ 20,000 | |
| Additions: | | |
| Depreciation Expense Not Using Cash | 10,000 | |
| Increased Accounts Payable: | | |
| To Suppliers of Merchandise | 20,000 | |
| To Other Suppliers | 2,000 | |
| Increased Salaries Payable | 1,000 | |
| Subtractions: | | |
| Increased Accounts Receivable | (35,000) | |
| Increased Merchandise Inventory | (10,000) | |
| Cash Flow from Operations | | $ 8,000 |
| **Investing** | | |
| Acquisition of Buildings and Equipment | | (125,000) |
| **Financing** | | |
| Dividends Paid | $ (7,000) | |
| Proceeds from Long-Term Bonds Issued | 100,000 | |
| Cash Flow from Financing | | 93,000 |
| Net Change in Cash for Year | | $ (24,000) |
| Cash, January 1, Year 4 | | 30,000 |
| Cash, December 31, Year 4 | | $ 6,000 |

**EXHIBIT 4.7**

**SOLINGER ELECTRIC CORPORATION**
**Year 4**
**Depreciation, $10,000**

| **Income Statement** | | **Cash Flow from Operations** | |
|---|---|---|---|
| Revenues | $125,000 | Net Income | $ 20,000 |
| Expenses Except Depreciation | (95,000) | Additions: | |
| | $ 30,000 | Depreciation | 10,000 |
| Depreciation Expense | (10,000) | Other Additions | 23,000 |
| | | Subtractions | (45,000) |
| Net Income | $ 20,000 | Cash Flow Provided by Operations | $ 8,000 |

**EXHIBIT 4.8**

**SOLINGER ELECTRIC CORPORATION**
**Year 4**
**Depreciation, $25,000**

| **Income Statement** | | **Cash Flow from Operations** | |
|---|---|---|---|
| Revenues | $125,000 | Net Income | $ 5,000 |
| Expenses Except Depreciation | (95,000) | Additions: | |
| | $ 30,000 | Depreciation | 25,000 |
| Depreciation Expense | (25,000) | Other Additions | 23,000 |
| | | Subtractions | (45,000) |
| Net Income | $ 5,000 | Cash Flow Provided by Operations | $ 8,000 |

*Cash Change Equation (Eq. 3)*

| Change in Cash | = | Change in Liabilities | + | Change in Shareholders' Equity | − | Change in Noncash Assets |
|---|---|---|---|---|---|---|

**EXHIBIT 4.9**

**ROBBIE CORPORATION**
**Comparative Balance Sheet**
**December 31, Year 1 and Year 2**
**(all dollar amounts in thousands)**
**(Problem 4.2 for Self-Study)**

| | December 31 Year 1 | December 31 Year 2 |
|---|---|---|
| ASSETS | | |
| **Current Assets** | | |
| Cash | $10 | $ 25 |
| Accounts Receivable | 15 | 22 |
| Merchandise Inventories | 20 | 18 |
| Total Current Assets | $45 | $ 65 |
| **Noncurrent Assets** | | |
| Property, Plant, and Equipment | $50 | $ 66 |
| Less Accumulated Depreciation | (25) | (31) |
| Total Property, Plant, and Equipment | $25 | $ 35 |
| Total Assets | $70 | $100 |

| | December 31 Year 1 | December 31 Year 2 |
|---|---|---|
| LIABILITIES AND SHAREHOLDERS' EQUITY | | |
| **Current Liabilities** | | |
| Accounts Payable for Merchandise | $30 | $ 37 |
| Total Current Liabilities | $30 | $ 37 |
| **Long-Term Debt** | | |
| Bonds Payable | 10 | 18 |
| Total Liabilities | $40 | $ 55 |
| **Shareholders' Equity** | | |
| Common Stock | $10 | $ 20 |
| Retained Earnings | 20 | 25 |
| Total Shareholders' Equity | $30 | $ 45 |
| Total Liabilities and Shareholders' Equity | $70 | $100 |

**EXHIBIT 4.10**

**ROBBIE CORPORATION**
**Income Statement**
**Year 2**
**(all dollar amounts in thousands)**
**(Problem 4.2 for Self-Study)**

| | |
|---|---|
| Sales Revenue | $180 |
| Cost of Goods Sold | (140) |
| Selling and Administrative Expenses | (25) |
| Depreciation Expense | (6) |
| Interest Expense | (2) |
| Net Income | $ 7 |

operations displayed with the direct method from the T-account work sheet you have constructed to derive cash flow from operations displayed with the indirect method.

Refer to the T-account work sheet in **Exhibit 4.5**. **Exhibit 4.11** on page 172 derives the display for the direct method. The derivation takes three steps, which we show in three separate panels in **Exhibit 4.11**. After this first illustration, we show all three steps in a single panel.

**Step 1** Copy the income statement into the left-hand box. Copy to the bottom left of the middle box the amount for Cash Flow from Operations derived from the indirect method. This number

**EXHIBIT 4.11**

## SOLINGER ELECTRIC CORPORATION
## Deriving Direct Method Cash Flow from Operations Using Data from T-Account Work Sheet

Each step is in a separate panel. Later versions condense all work into a single panel.

1. Copy Income Statement and Cash Flow from Operations

| Operations | (a) | (b) | Changes in Related Balance Sheet Accounts from T-Account Work Sheet (c) | (d) |
|---|---|---|---|---|
| Sales Revenues | $125,000 | | | |
| Cost of Goods Sold | (60,000) | | | |
| Depreciation Expense | (10,000) | | | |
| Salary Expense | (20,000) | | | |
| Interest Expense | (4,000) | | | |
| Other Expense | (11,000) | | | |
| Net Income | $ 20,000 | | Totals | $ — |
| | | $ 8,000 | = Cash Flow from Operations Derived via Indirect Method | |

2. Copy Information from T-Account Work Sheet Next to Related Income Statement Item

| Operations | (a) | Copy from T-Account Work Sheet (b) | Changes in Related Balance Sheet Accounts from T-Account Work Sheet (c) | (d) |
|---|---|---|---|---|
| Sales Revenues | $125,000 | $(35,000) | = Accounts Receivable Increase | |
| Cost of Goods Sold | (60,000) | (10,000) | = Inventory Increase | |
| | | 20,000 | = Accounts Payable to Merchandise Suppliers Increase | |
| Depreciation Expense | (10,000) | 10,000 | (Expense Not Using Cash) | |
| Salary Expense | (20,000) | 1,000 | = Salaries Payable Increase | |
| Interest Expense | (4,000) | — | Interest Payable (no change in balance sheet) | |
| Other Expenses | (11,000) | 2,000 | = Accounts Payable to Other Suppliers Increase | |
| Net Income | $ 20,000 | $ 20,000 | Totals | $ — |
| | | $ 8,000 | = Cash Flow from Operations Derived via Indirect Method | |

3. Sum Across Rows to Derive Direct Receipts and Expenditures

| Operations | (a) | Indirect Method (b) | Changes in Related Balance Sheet Accounts from T-Account Work Sheet (c) | Direct Method (d) | From Operations: Receipts Less Expenditures |
|---|---|---|---|---|---|
| Sales Revenues | $125,000 | $(35,000) | = Accounts Receivable Increase | $ 90,000 | Receipts from Customers |
| Cost of Goods Sold | (60,000) | (10,000) | = Inventory Increase | $(50,000) | Payments for Goods Sold |
| | | 20,000 | = Accounts Payable to Merchandise Suppliers Increase | | |
| Depreciation Expense | (10,000) | 10,000 | (Expense Not Using Cash) | — | |
| Salary Expense | (20,000) | 1,000 | = Salaries Payable Increase | (19,000) | Payments for Salaries |
| Interest Expense | (4,000) | — | Interest Payable (no change in balance sheet) | (4,000) | Payments for Interest |
| Other Expenses | (11,000) | 2,000 | = Accounts Payable to Other Suppliers Increase | (9,000) | Payments for Other Expenses |
| Net Income | $ 20,000 | $ 20,000 | Totals | $ 8,000 | = Cash Flow from Operations Derived via Direct Method |
| | | $ 8,000 | = Cash Flow from Operations Derived via Indirect Method | | |

Note that the information in column (b) is cash flow from operations derived with the indirect method, with items in a different order.

*Cash Change Equation (Eq. 3)*

| Change in Cash | = | Change in Liabilities | + | Change in Shareholders' Equity | − | Change in Noncash Assets |
|---|---|---|---|---|---|---|

serves as a check on the derivation of the direct method's answer. If the direct method gives a different answer, we have erred.

**Step 2** Copy into the middle box all the additions and subtractions in the *operations section* of the master Cash account in the T-account work sheet, including net income. Place the number next to the income statement item to which the addition or subtraction relates: Accounts Receivable subtraction next to Sales Revenue, Depreciation next to Depreciation Expense, and so on. This is the only part of the procedure that is not mechanical. You need to understand which income statement item a given addition or subtraction refers to. In placing the number in column (b), use a plus sign if it is on the debit side of the T-account work sheet, which represents a tentative increase in cash. Place the number in column (b) with a minus sign if it comes from the credit side of the Cash account, which represents a tentative decrease in cash. To repeat: copy the data from the operations section only.

Add the numbers in column (b) to check that you have, indeed, shown the indirect method's computation of cash flow from operations. Note that column (b) resembles the Operating section of the indirect method for the statement of cash flows, except the numbers have a different order—net income is at the bottom, not at the top—and the other numbers are in income statement order, but still the same numbers.

**Step 3** Add across the numbers in columns (a) and (b), writing the sums in column (d). Give the resulting sum a label—either Receipts or Payments (or Expenditures). The label should describe the nature of the receipt or payment, but you can use any descriptive label that seems appropriate to you. Add the numbers in column (d). If you have done the work correctly, the sum in column (d) will match the sum in column (b) and both will equal cash flow from operations.

To understand that the indirect and direct methods provide the same data, but in different subtotals, consider the following. The indirect method starts with net income, which equals revenues minus expenses. Then, it adds amounts to net income for revenues (or expenses) producing (or using) cash in amounts different from the revenue (or expense) item. The direct method takes each revenue (and expense) and adds to or subtracts from each separately to derive the related receipt or payment. The additions and subtractions for the direct method are the same as for the indirect method. Only the order of computations and presentation differs.

## Problem 4.3 for Self-Study

**Deriving a direct method presentation of cash flows from operations from T-account work sheet.** Refer to **Problem 4.2 for Self-Study** and its solution. Prepare a derivation of cash flows from operations for Robbie Corporation for Year 2 using the direct method and the T-account work sheet in **Exhibit 4.17** on page 185. Use the format of **Exhibit 4.11**, but condense your work into a single panel.

## Extension of the Illustration

The illustration for Solinger Electric Corporation considered so far in this chapter is simpler than the typical published statement of cash flows in at least four respects:

1. There are only a few balance sheet accounts requiring explanation.
2. Several types of more complex transactions that affect cash flow from operations do not arise.
3. Each transaction recorded in Step **3** involves only one debit and one credit.
4. Except for the Retained Earnings account, each explanation of a noncash account change involves only one analytic entry on the work sheet.

Most of the complications that arise in interpreting published statements of cash flows relate to accounting events that later chapters discuss. These chapters will illustrate the corresponding effects on the statement of cash flows. We can illustrate here one complication, arising from an asset disposition revealed in a supplementary disclosure. Assume such disclosure tells us the firm sold some of its buildings and equipment during the year at their book value. That is, when the firm disposed of existing buildings and equipment, the cash proceeds from disposition equaled acquisition cost less the accumulated depreciation of the assets. With this assumption, no gain or loss on the disposition arises.

Reconsider the Solinger Electric Corporation example with the following new information. Solinger Electric Corporation retired some equipment during Year 4. The firm originally paid $10,000 for this equipment and then disposed of it for $3,000 at a time when accumulated depreciation on the equipment was $7,000. The actual entry made during the year to record the disposal of the equipment was as follows:

| | | |
|---|---|---|
| Cash . . . . . . . . . . . . . . . . . . . . . . . . . . . . . . . . . . . . . . . . . . . . | 3,000 | |
| Accumulated Depreciation . . . . . . . . . . . . . . . . . . . . . . . . . . . . . . . . | 7,000 | |
| Buildings and Equipment (Cost) . . . . . . . . . . . . . . . . . . . . . . . . . . | | 10,000 |

Journal entry for disposal of equipment at book value.

| Change in Cash | = | Change in Liabilities | + | Change in Shareholders' Equity | – | Change in Noncash Assets |
|---|---|---|---|---|---|---|
| +3,000 (Invst.) | | 0 | | 0 | | +7,000<br>–10,000 |

Here, we focus on the Statement of Cash Flows, so we show the Cash Change Equation which we derived from the Balance Sheet Equation.

Assume that the comparative balance sheets as shown in **Exhibit 4.2** are correct and thus the net decrease in cash for Year 4 is still $24,000. The entries in the T-accounts must differ to reflect this new information. Had this transaction occurred in Year 4, the analytic entry (1), on page 165, would still be the same because net income would still be $20,000. The following analytic entry in the T-account work sheet recognizes the effect of the disposal of equipment:

| | | |
|---|---|---|
| **(1a)** Cash (Investing: Disposal of Equipment) . . . . . . . . . . . . . . . . . . . . . . | 3,000 | |
| Accumulated Depreciation . . . . . . . . . . . . . . . . . . . . . . . . . . . . . . . . | 7,000 | |
| Buildings and Equipment (Cost) . . . . . . . . . . . . . . . . . . . . . . . . | | 10,000 |

Analytic entry recorded in T-account work sheet.

The debit to Cash (Investing: Disposal of Equipment) shows the proceeds of the disposal. As a result of analytic entry **(1a)**, the T-accounts for Buildings and Equipment (Cost) and Accumulated Depreciation appear as follows:

**Buildings and Equipment (Cost)**

| | Debit | Credit | |
|---|---|---|---|
| ✓ | 100,000 | | |
| | | 10,000 | **(1a)** |
| ✓ | 225,000 | | |

**Accumulated Depreciation**

| | Debit | Credit | |
|---|---|---|---|
| | | 30,000 | ✓ |
| **(1a)** | 7,000 | | |
| | | 40,000 | |

When the time comes to explain the change in Buildings and Equipment (Cost) account, the T-account indicates both a total increase of $125,000 and a credit (decrease) entry **(1a)** of $10,000 to recognize the disposal of equipment. To explain the net increase in the Buildings and Equipment (Cost) account, given the decrease already entered, one must assume that the firm acquired new buildings and equipment for $135,000 (= $125,000 + $10,000) during the period.

*Cash Change Equation (Eq. 3)*

| Change in Cash | = | Change in Liabilities | + | Change in Shareholders' Equity | − | Change in Noncash Assets |
|---|---|---|---|---|---|---|

The reconstructed analytic entry, which replaces analytic entry (6) above on page 167, to complete the explanation of the change in this account is as follows:

| | | |
|---|---|---|
| **(6a)** Buildings and Equipment (Cost) . . . . . . . . . . . . . . . . . . . . . . . . . . . . | 135,000 | |
| Cash (Investing: Acquisition of Buildings and Equipment) . . . . . . . | | 135,000 |
| Analytic entry recorded in T-account work sheet. | | |

Likewise, when the time comes to explain the change in the T-account for Accumulated Depreciation, there is a net credit change of $10,000 and a debit entry **(1a)** of $7,000 to recognize disposal. Thus the depreciation charge for Year 4 must have been $17,000 (= $10,000 + $7,000). The reconstructed analytic entry to complete the explanation of the change in the Accumulated Depreciation account is as follows:

| | | |
|---|---|---|
| **(2a)** Cash (Operations: Depreciation Expense Addback) . . . . . . . . . . . . . . . . . | 17,000 | |
| Accumulated Depreciation . . . . . . . . . . . . . . . . . . . . . . . . . . . . | | 17,000 |
| Analytic entry recorded in T-account work sheet. | | |

**Exhibit 4.12** on page 176 presents a revised T-account work sheet for Solinger Electric Corporation incorporating the new information on the disposal of equipment.

Another complication arises when firms sell buildings and equipment for an amount different from their book values. Assume, for example, that Solinger Electric Corporation sold the equipment discussed above for $2,000 instead of $3,000. The entry made during the year to record the disposal of the equipment is as follows:

| | | |
|---|---|---|
| Cash . . . . . . . . . . . . . . . . . . . . . . . . . . . . . . . . . . . . . . . . . . . . . . | 2,000 | |
| Loss on Disposal of Equipment . . . . . . . . . . . . . . . . . . . . . . . . . . . . | 1,000 | |
| Accumulated Depreciation . . . . . . . . . . . . . . . . . . . . . . . . . . . . . . . . | 7,000 | |
| Buildings and Equipment (Cost) . . . . . . . . . . . . . . . . . . . . . . . . . | | 10,000 |
| Journal entry to record disposal of equipment at a loss. | | |

| Assets | = | Liabilities | + | Shareholders' Equity | (Class.) |
|---|---|---|---|---|---|
| +2,000 | | | | −1,000 | IncSt →RE |
| +7,000 | | | | | |
| -10,000 | | | | | |

| Change in Cash | = | Change in Liabilities | + | Change in Shareholders' Equity | − | Change in Noncash Assets |
|---|---|---|---|---|---|---|
| +2,000 (Invst.) | | 0 | | −1,000 | | −10,000<br>+7,000 |

Journal entry for disposal of equipment for cash at a loss. We show both the Balance Sheet Equation and the Cash Change Equation.

This entry removes from the accounting records all amounts related to the equipment sold, which includes its acquisition cost of $10,000 and the $7,000 of depreciation recognized while Solinger Electric Corporation used the equipment. Accountants refer to the net amount appearing on the books for such equipment as its *book value*. The entry also records the cash received

**EXHIBIT 4.12**

**SOLINGER ELECTRIC CORPORATION**
**Revised T-Account Work Sheet**

**Cash**

| | | | | | |
|---|---|---|---|---|---|
| | ✓ | 30,000 | | | |
| | | **Operations** | | | |
| Net Income | (1) | 20,000 | 35,000 | (4) | Increased Accounts Receivable |
| Depreciation Expense Addback | (2a) | 17,000 | 10,000 | (5) | Increased Merchandise Inventory |
| Increased Accounts Payable to Merchandise Suppliers | (7) | 20,000 | | | |
| Increased Accounts Payable to Other Suppliers | (8) | 2,000 | | | |
| Increased Salaries Payable | (9) | 1,000 | | | |
| | | **Investing** | | | |
| Disposal of Equipment | (1a) | 3,000 | 135,000 | (6a) | Acquisition of Buildings and Equipment |
| | | **Financing** | | | |
| Long-Term Bond Issue | (10) | 100,000 | 7,000 | (3) | Dividends |
| | ✓ | 6,000 | | | |

**Accounts Receivable**

| | | | |
|---|---|---|---|
| ✓ | 20,000 | | |
| (4) | 35,000 | | |
| ✓ | 55,000 | | |

**Merchandise Inventory**

| | | | |
|---|---|---|---|
| ✓ | 40,000 | | |
| (5) | 10,000 | | |
| ✓ | 50,000 | | |

**Buildings and Equipment (Cost)**

| | | | |
|---|---|---|---|
| ✓ | 100,000 | | |
| (6a) | 135,000 | 10,000 | (1a) |
| ✓ | 225,000 | | |

**Accumulated Depreciation**

| | | | |
|---|---|---|---|
| | | 30,000 | ✓ |
| (1a) | 7,000 | 17,000 | (2a) |
| | | 40,000 | ✓ |

**Accounts Payable—Merchandise Suppliers**

| | | | |
|---|---|---|---|
| | | 30,000 | ✓ |
| | | 20,000 | (7) |
| | | 50,000 | ✓ |

**Accounts Payable—Other Suppliers**

| | | | |
|---|---|---|---|
| | | 10,000 | ✓ |
| | | 2,000 | (8) |
| | | 12,000 | ✓ |

**Salaries Payable**

| | | | |
|---|---|---|---|
| | | 5,000 | ✓ |
| | | 1,000 | (9) |
| | | 6,000 | ✓ |

**Bonds Payable**

| | | | |
|---|---|---|---|
| | | 0 | ✓ |
| | | 100,000 | (10) |
| | | 100,000 | ✓ |

**Retained Earnings**

| | | | |
|---|---|---|---|
| | | 15,000 | ✓ |
| (3) | 7,000 | 20,000 | (1) |
| | | 28,000 | ✓ |

from disposal of the equipment. The difference between the cash proceeds and the book value of the equipment is a loss of $1,000 [= $2,000 − ($10,000 − $7,000)].

The following analytic entry on the T-account work sheet would recognize the effect of the disposal of equipment for $2,000:

| | | |
|---|---|---|
| **(1a)** Cash (Investing: Disposal of Equipment) | 2,000 | |
| Cash (Operations: Loss on Disposal of Equipment Addback) | 1,000 | |
| Accumulated Depreciation | 7,000 | |
| Buildings and Equipment (Cost) | | 10,000 |
| Entry recorded in T-account work sheet. | | |

*Cash Change Equation (Eq. 3)*

| Change in Cash | = | Change in Liabilities | + | Change in Shareholders' Equity | – | Change in Noncash Assets |
|---|---|---|---|---|---|---|

The debit to Cash (Investing: Disposal of Equipment) shows the $2,000 proceeds of disposal. The debit to Cash (Operations: Loss on Disposal of Equipment Addback) for $1,000 adds back to net income the loss on disposal of equipment that did not use cash. Like the depreciation expense addback, the debit to Cash (Operations: Loss on Disposal of Equipment Addback) does not represent a source of cash (ignoring income taxes). The addback merely offsets the subtraction for the loss in computing net income. The loss did not use cash this period. The firm spent cash sometime in the past and now finds the assets purchased with that cash have market value lower than book value.

Consider the impact of reporting the disposal of equipment at a loss on various lines of the statement of cash flows:

| | |
|---|---|
| **Operations** | |
| Net Income (Loss on Disposal of Equipment) | ($1,000) |
| Loss on Disposal of Equipment Addback | 1,000 |
| Cash Flow from Operations | $ 0 |
| **Investing** | |
| Proceeds from Disposal of Equipment | $2,000 |
| Net Change in Cash | $2,000 |

One might view the recognition of a loss on the disposal of equipment as indicating that the firm took insufficient depreciation during the accounting periods before the disposal. If the firm had known for certain that it would receive $2,000 for the equipment, it would have recognized another $1,000 of depreciation during the periods while it used the equipment. The disposal of the equipment would then have resulted in no gain or loss. The firm would have shown the $1,000 additional depreciation as an addback to net income in computing cash flow from operations during the periods while it used the equipment.

Extending this illustration still further, assume that Solinger Electric Corporation received $4,500 for the equipment. The entry made to record the disposal of equipment is as follows:

| | Debit | Credit |
|---|---|---|
| Cash | 4,500 | |
| Accumulated Depreciation | 7,000 | |
| Buildings and Equipment (Cost) | | 10,000 |
| Gain on Disposal of Equipment | | 1,500 |

Journal entry to record disposal of equipment.

| Assets | = | Liabilities | + | Shareholders' Equity | (Class.) |
|---|---|---|---|---|---|
| +4,500 | | | | +1,500 | IncSt → RE |
| +7,000 | | | | | |
| −10,000 | | | | | |

| Change in Cash | = | Change in Liabilities | + | Change in Shareholders' Equity | – | Change in Noncash Assets |
|---|---|---|---|---|---|---|
| +4,500 (Invst.) | | 0 | | +1,500 | | +7,000<br>−10,000 |

This entry, like that above for disposal at a loss, removes the amounts on the books for the equipment and records the cash proceeds. In this case the cash proceeds exceed the book value of the equipment, resulting in a gain on disposal.

The following analytic entry in the T-account work sheet would recognize the effect of the disposal of equipment for $4,500:

| | | |
|---|---|---|
| **(1a)** Cash (Investing: Disposal of Equipment) | 4,500 | |
| Accumulated Depreciation | 7,000 | |
| Buildings and Equipment (Cost) | | 10,000 |
| Cash (Operations: Gain on Disposal of Equipment Subtraction) | | 1,500 |
| Analytic entry recorded in T-account work sheet. | | |

The debit to Cash (Investing: Disposal of Equipment) shows the $4,500 proceeds of disposal. The credit to Cash (Operations: Gain on Disposal of Equipment Subtraction) reduces net income for the gain on disposals of equipment that did not provide an operating cash inflow. Unless we subtract the $1,500 gain in the operations section of the work sheet, we overstate the amount of cash inflow from this transaction, as the following analysis summarizes:

| | |
|---|---|
| **Operations** | |
| Net Income (Gain on Disposal of Equipment) | $ 1,500 |
| Subtraction for Gain on Disposal of Equipment Not Providing an Operating Cash Inflow | (1,500) |
| Cash Flow from Operations | $ 0 |
| **Investing** | |
| Proceeds from Disposal of Equipment | $ 4,500 |
| Net Change in Cash from Transaction | $ 4,500 |

One might interpret the recognition of a gain on the disposal of equipment as indicating that the firm took too much depreciation during the accounting periods before the disposal. If the firm had known for certain that it would receive $4,500 for the equipment, it would have recognized $1,500 less depreciation during the periods while it used the equipment. The disposal of the equipment would then have resulted in no gain or loss. The firm would have shown $1,500 less depreciation as an addback to net income in computing cash flow from operations. The $1,500 credit in the operating section of the T-account work sheet in effect adjusts for the excess depreciation.

## An International Perspective

*Standard No. 7* of the International Accounting Standards Committee recommends the preparation of a statement of cash flows that reports cash flows from operating, investing, and financing activities.[7] Standard-setting bodies in most countries have adopted this standard, so that firms in these countries prepare a statement of cash flows following the format illustrated in this chapter.

## Problem 4.4 for Self-Study

**Preparing a statement of cash flows. Exhibit 4.13** presents a comparative balance sheet for Gordon Corporation as of December 31, Year 1 and Year 2. **Exhibit 4.14** on page 180 presents the income statement for Year 2.

During Year 2, the company declared and paid dividends of $120,000.

During Year 2, the company disposed of buildings and equipment that originally cost $55,000 and had accumulated depreciation at the time of disposal of $30,000.

[7]International Accounting Standards Committee, *International Accounting Standard No. 7*, "Cash Flow Statements," 1994.

*Cash Change Equation (Eq. 3)*

| Change in Cash | = | Change in Liabilities | + | Change in Shareholders' Equity | − | Change in Noncash Assets |
|---|---|---|---|---|---|---|

**a.** Prepare a T-account work sheet for the preparation of the statement of cash flows for Year 2 using the indirect method for cash flows from operations.

**b.** Derive a presentation for cash flows from operations using the format of **Exhibit 4.11**, but using a single panel.

**c.** Present a formal statement of cash flows for Year 2 using the direct method for cash flows from operations and including a reconciliation of net income to cash flows from operations.

---

**EXHIBIT 4.13**

**GORDON CORPORATION**
**Comparative Balance Sheet**
**December 31, Year 1 and Year 2**
**(all dollar amounts in thousands)**
**(Problem 4.4 for Self-Study)**

| | December 31 | |
|---|---|---|
| | Year 1 | Year 2 |
| ASSETS | | |
| **Current Assets** | | |
| Cash | $ 70 | $ 40 |
| Accounts Receivable | 320 | 420 |
| Merchandise Inventories | 360 | 470 |
| Prepayments | 50 | 70 |
| Total Current Assets | $ 800 | $1,000 |
| **Property, Plant, and Equipment** | | |
| Land | $ 200 | $ 250 |
| Buildings and Equipment (net of accumulated depreciation of $800 and $840) | 1,000 | 1,150 |
| Total Property, Plant, and Equipment | $1,200 | $1,400 |
| Total Assets | $2,000 | $2,400 |
| LIABILITIES AND SHAREHOLDERS' EQUITY | | |
| **Current Liabilities** | | |
| Accounts Payable | $ 320 | $ 440 |
| Income Taxes Payable | 60 | 80 |
| Other Current Liabilities | 170 | 360 |
| Total Current Liabilities | $ 550 | $ 880 |
| **Noncurrent Liabilities** | | |
| Bonds Payable | 250 | 200 |
| Total Liabilities | $ 800 | $1,080 |
| **Shareholders' Equity** | | |
| Common Stock | $ 500 | $ 540 |
| Retained Earnings | 700 | 780 |
| Total Shareholders' Equity | $1,200 | $1,320 |
| Total Liabilities and Shareholders' Equity | $2,000 | $2,400 |

**EXHIBIT 4.14**

**GORDON CORPORATION**
**Income Statement**
**Year 2**
**(all dollar amounts in thousands)**
**(Problem 4.4 for Self-Study)**

| | |
|---|---|
| Revenues | $1,600 |
| Less: | |
| Cost of Goods Sold | (900) |
| Depreciation Expense | (70) |
| Selling and Administrative Expense | (255) |
| Interest Expense | (30) |
| Loss on Disposal of Buildings and Equipment | (15) |
| Income Tax Expense | (130) |
| Net Income | $ 200 |

## Using Information in the Statement of Cash Flows

The statement of cash flows provides information that helps the reader in (1) assessing the impact of operations on liquidity and (2) assessing the relations among cash flows from operating, investing, and financing activities.

### IMPACT OF OPERATIONS ON LIQUIDITY

Perhaps the most important factor not reported on either the balance sheet or the income statement is how the operations of a period affected cash flows. Increased earnings do not always generate increased cash flow from operations. When increased earnings result from expanding operations (that is, more *units* sold in contrast to mere increases in selling price or reductions in cost), the firm usually has decreased cash flow from operations. A growing, successful firm—such as Wal-Mart, discussed in connection with **Exhibit 4.1**—may have increasing amounts for accounts receivable and inventories, resulting in a lag between earnings and cash flows. The need to await the collection of accounts receivable but to acquire and pay for inventory in anticipation of larger disposals in the future can cause cash shortages from operations. Successful, growing businesses will need permanent financing from long-term debt issues or common stock issues to cover short-term cash needs. Some successful firms find themselves constantly short of cash because they have not adequately obtained long-term financing to cover perpetual short-term needs. Financial managers need to anticipate that the more successful the firm becomes, the more long-term sources of funds it will need to finance operations.

On the other hand, increased cash flow can accompany reduced earnings. Consider, for example, a firm that is experiencing operating problems and reduces the scope of its activities. Such a firm likely will report reduced net income or even losses. However, it might experience positive cash flow from operations because it collects accounts receivable from prior periods but does not replace inventories, thus saving cash.

### RELATIONS AMONG CASH FLOWS FROM OPERATING, INVESTING, AND FINANCING ACTIVITIES

The relations among the cash flows from each of the three principal business activities likely differ depending on the characteristics of the firm's products and the maturity of its industry. Consider each of the four following patterns of cash flows.

| Cash Flows from: | A | B | C | D |
|---|---|---|---|---|
| Operations | $ (3) | $ 7 | $15 | $ 8 |
| Investing | (15) | (12) | (8) | (2) |
| Financing | 18 | 5 | (7) | (6) |
| Net Cash Flow | $ 0 | $ 0 | $ 0 | $ 0 |

*Cash Change Equation (Eq. 3)*

| Change in Cash | = | Change in Liabilities | + | Change in Shareholders' Equity | – | Change in Noncash Assets |
|---|---|---|---|---|---|---|

**Case A** illustrates a typical new, rapidly growing firm. It does not yet operate profitably, and it experiences buildups of its accounts receivable and inventories. Thus, it has negative cash flow from operations. To sustain its rapid growth, the firm must invest heavily in plant and equipment. During this stage, the firm must rely on external sources of cash to finance both its operating and its investing activities.

**Case B** illustrates a somewhat more seasoned firm than the one in **Case A**, but one that is still growing. It operates profitably, but because its rapid growth has begun to slow, it generates positive cash flow from operations. However, this cash flow from operations falls short of the amount the firm needs to finance acquisitions of plant and equipment. The firm therefore requires external financing.

**Case C** illustrates the cash flow pattern of a mature, stable firm. It generates a healthy cash flow from operations—more than enough to acquire new plant and equipment. It uses the excess cash flow to repay financing from earlier periods and, perhaps, to pay dividends.

**Case D** illustrates a firm in the early stages of decline. Its cash from operations has begun to decrease but remains positive because of decreases in accounts receivable and inventories. In the later stages of decline, its cash flow from operations may even turn negative as it finds itself unable to sell its products at a positive net cash flow. It cuts back significantly on capital expenditures because it is in a declining industry. It uses some of its excess cash flow to repay outstanding financing, and it has the remainder available for investment in new products or other industries.

Refer to the statements of cash flows for the four firms for a recent year in **Exhibit 4.15**. You will find the statement of cash flows more informative when you study cash flows for several years. These single-year statements illustrate some important relations for firms in different stages of growth.

**EXHIBIT 4.15** Excerpts from Statements of Cash Flows for Four Firms (all dollar amounts in millions)

| | Amazon.com | Discount Auto Parts | Anheuser-Busch | Levitz Furniture |
|---|---|---|---|---|
| OPERATIONS | | | | |
| Cash Flow from Operations | $(130) | $ 28 | $ 2,258 | $ (14) |
| INVESTING | | | | |
| Disposal of Fixed Assets | $ — | $ — | $ — | $ 12 |
| Acquisition of Fixed Assets | (135) | (69) | (1,075) | (6) |
| Sale of Marketable Securities | 299 | — | — | — |
| Other Investing Transactions | — | 5 | (43) | — |
| Cash Flow from Investing | $ 164 | $(64) | $(1,118) | $ 6 |
| FINANCING | | | | |
| Increase in Short-Term Borrowing | $ — | $ — | $ — | $ 670 |
| Increase in Long-Term Borrowing | 648 | 40 | 804 | — |
| Increase in Capital Stock | 45 | — | 135 | — |
| Decrease in Short-Term Borrowing | — | — | — | (659) |
| Decrease in Long-Term Borrowing | — | — | (514) | (2) |
| Decrease in Capital Stock | — | — | (986) | — |
| Dividends | — | — | (571) | — |
| Other Financing Transactions | (38) | — | — | — |
| Cash Flow from Financing | $ 655 | $ 40 | $(1,132) | $ 9 |
| Change in Cash | $ 689 | $ 4 | $ 8 | $ 1 |
| Cash, Beginning of Year | 133 | 8 | 152 | 5 |
| Cash, End of Year | $ 822 | $ 12 | $ 160 | $ 6 |

Amazon.com depicts typical cash flows of a new, rapidly growing firm. It operated at a net loss for the year and its operations consumed cash—it generated negative cash flow from operations. The firm also made expenditures on additional property, plant, and equipment to maintain its growth. It financed the negative cash flow from operations and capital expenditures by selling marketable securities and issuing additional long-term debt. The firm had issued common stock in an earlier year and invested the cash temporarily in marketable securities. It needed the cash this year and therefore sold a portion of its marketable securities. The proceeds of the new borrowing increased its cash balance on the balance sheet.

Discount Auto Parts depicts typical cash flows for a firm in the rapid growth phase, but somewhat more seasoned than Amazon.com. It posted a positive net income for the year and generated positive cash flow from operations. This cash flow from operations still fell short of the amounts it needed to finance acquisitions of property, plant, and equipment. Discount Auto Parts financed the shortfall with additional long-term borrowing, appropriate financing for noncurrent assets.

Anheuser-Busch depicts a cash flow pattern typical of a mature firm. It reports positive net income and positive cash flow from operations. Cash flow from operations exceeds the amounts the firm needs to finance investments in new property, plant, and equipment. The firm used the excess net cash flow from operating and investing activities to pay dividends and repurchase shares of its common stock.

The cash flows for Levitz Furniture are for a period just prior to its filing for bankruptcy and therefore in the late stage of the decline phase. It operated at a net loss and generated negative cash flow from operations. The investing section indicates the cash inflows from selling property, plant, and equipment. Given its poor financial condition, Levitz Furniture had to rely on short-term financing. Lenders required the firm to repay financing within one year and replace it with new financing.

These four cases do not, of course, cover all of the patterns of cash flows found in corporate annual reports. They illustrate, however, how the characteristics of a firm's products and industry can affect the interpretation of information in the statement of cash flows.

## Ethical Issues and the Statement of Cash Flows

Many analysts focus attention on cash flow from operations, thinking it as important as, or more important than, net income. A common misconception is that the management has little opportunity to manipulate transactions affecting the statement of cash flows. After all, goes the logic, cash is cash; how can you fake a cash flow? The manipulation possibilities arise, not from the amounts of cash flows, which indeed management cannot easily fudge, but from either (1) their timing, or (2) their classification and disclosure in the statement and related notes.

### Timing of Operating Cash Flows

Firms have some choice as to when they disburse cash. Firms that delay making payments to suppliers, employees, and others during the last several days of an accounting period conserve cash and thereby increase cash flow from operations for that period. When these firms make the cash payments during the early part of the next period, cash flow from operations declines. Thus, the firm increases cash flow from operations during the first period but decreases cash flow from operations during the second period. The firm, however, can delay making payments at the end of the second period to offset the negative cash flow effects of the payments made early in that period. As long as the firm continues to grow, it reports larger cash flow from operations each period than if it had not delayed making the cash payments at the end of each period.

Unless contracts or other agreements preclude the firm from delaying payments each period, the firm does not appear to engage in illegal activity. The ethical issues concern whether the firm portrays a misleading picture of its operating cash flows and whether it discloses sufficient information for the user to assess how much manipulation has occurred.

### Classification and Disclosure of Transactions

Firms have also contrived to improve the appearance of operating cash flow by manipulating borrowing with an asset as collateral—a financing transaction—into a sale of an operating asset, which appears as an operating source of cash.

*Example.* Enron raised funds by borrowing from Merrill Lynch using barges anchored off the coast of Nigeria as collateral. Rather than execute a note and raise cash with an explicit borrow-

*Cash Change Equation (Eq. 3)*

| Change in Cash | = | Change in Liabilities | + | Change in Shareholders' Equity | − | Change in Noncash Assets |
|---|---|---|---|---|---|---|

ing, Enron arranged to sell the barges to Merrill Lynch for cash with an unwritten understanding that Enron would repurchase the barges later at a price equal to the original sale plus an amount equivalent to the interest on the borrowed funds for the term of the loan. Enron treated the barges as operating assets, so it showed the proceeds from Merrill Lynch as operating cash inflows, rather than as financing cash inflows.

If there had been no commitment by Enron to repurchase, then treatment of the sale of the barges as operating cash flow is consistent with GAAP: Enron sold some of its operating assets, just as another company might sell its inventory. [Even so, one might argue such a sale is a disposition of a long-term asset, hence a (dis-)investing source of cash.] When, as actually happened, Enron has a firm commitment to repurchase the barges, the economic substance is a borrowing and GAAP requires the firm to account for it as such. The ethical issue here is whether the managers of Enron willingly disclosed the repurchase commitment to its auditors or simply waited for the auditors to ask if such a repurchase agreement exists. A related ethical issue is whether Enron disclosed sufficient information about this transaction for the user of the financial statement to assess if the accounting was appropriate.

## Summary

The statement of cash flows reports the effects of a firm's operating, investing, and financing activities on cash flows. Information in the statement helps in understanding

1. how operations affect the liquidity of a firm,
2. the level of capital expenditures needed to support ongoing and growing levels of activity, and
3. the major changes in the financing of a firm.

To prepare the statement of cash flows requires analyzing changes in balance sheet accounts during the accounting period, as represented by the Cash Change Equation **(Eq. 3)**. As a byproduct of correct double-entry recording of all transactions, the net change in cash will equal the net change in all noncash accounts.

The statement of cash flows usually presents cash flow from operations in the indirect format, beginning with net income for the period. The statement then adjusts for revenues not providing cash, for expenses not using cash, and for changes in operating working capital accounts. The result is cash flow from operations. Some firms follow a more direct approach to calculating cash flow from operations by listing all revenues that provide cash and subtracting all expenses that use cash. Such firms present a reconciliation of net income to cash flow from operations. The cash flows from investing activities and financing activities appear after cash flow from operations.

Interpreting a statement of cash flows requires an understanding of the economic characteristics of the industries in which a firm conducts its activities, including capital intensity, growth characteristics, and similar factors.

## Problem 4.5 for Self-Study

**Effect of transactions on the statement of cash flows. Exhibit 4.16** on page 184 shows a simplified statement of cash flows for a period. Numbers appear on eleven of the lines in the statement. Other lines (indicated with an "S") are various subtotals and grand totals; ignore these in the remainder of the problem. Assume that the accounting cycle is complete for the period and that the firm has prepared all of the financial statements. It then discovers that it has overlooked a transaction. It records that transaction in the accounts and corrects all of the financial statements. For each of the following transactions, indicate which of the numbered lines of the statement of cash flows change, and state the amount and direction of the change. If net income, line **(1)**, changes, be sure to indicate whether it decreases or increases. Ignore income tax effects.

*(continued)*

**EXHIBIT 4.16** **Simplified Statement of Cash Flows (Problem 4.5 for Self-Study)**

| | |
|---|---|
| **Operations** | |
| Cash Receipts from Customers | (1) |
| Less: Cash Payments to Suppliers, Employees, and Others | −(2) |
| Cash Flow from Operations [= (1) − (2)] | S1 |
| Reconciliation of Net Income to Cash Flow from Operations | |
| Net Income | (3) |
| Additions to Net Income to Compute Cash Flow from Operations | +(4) |
| Subtractions from Net Income to Compute Cash Flow from Operations | −(5) |
| Cash Flow from Operations [= (3) + (4) − (5)] | S1 |
| **Investing** | |
| Proceeds from Dispositions of "Investing" Assets | +(6) |
| Cash Used to Acquire "Investing" Assets | −(7) |
| Cash Flow from Investing [= (6) − (7)] | S2 |
| **Financing** | |
| Cash Provided by Increases in Debt or Capital Stock | +(8) |
| Cash Used to Reduce Debt or Capital Stock | −(9) |
| Cash Used for Dividends | −(10) |
| Cash Flow from Financing [= (8) − (9) − (10)] | S3 |
| Net Change in Cash [= S1 + S2 + S3] | (11) |
| Cash, Beginning of the Period | S4 |
| Cash, End of the Period [= (11) + S4] | S5 |

(*Hint:* First, construct the entry the firm would make in its accounts to record the transaction. Then, for each line of the journal entry, identify the line of **Exhibit 4.16** affected.)

**a.** Depreciation expense of $2,000 on an office computer
**b.** Purchase of machinery for $10,000 cash
**c.** Declaration of a cash dividend of $6,500 on common stock; the firm paid the dividend by the end of the fiscal year
**d.** Issue of common stock for $12,000 cash
**e.** Proceeds of the sale of a common stock investment, a noncurrent asset, for $15,000 cash; the firm sold the investment for its book value of $15,000

## Solutions to Self-Study Problems

### SUGGESTED SOLUTION TO PROBLEM 4.1 FOR SELF-STUDY

(Classifying cash flows by type of activity.)

**a.** Operating
**b.** Financing
**c.** Operating
**d.** Operating
**e.** Operating
**f.** Operating
**g.** Investing
**h.** Item does not affect cash flows during the current period and would therefore not appear in the statement of cash flows. The firm must disclose this transaction in a separate schedule or a note to the financial statements.

**i.** Operating
**j.** Operating
**k.** Investing
**l.** Item does not affect cash flows during the current period and would therefore not appear in the statement of cash flows. The firm must disclose this transaction in a separate schedule or note to the financial statements.
**m.** Financing
**n.** Investing
**o.** Financing
**p.** Investing
**q.** Operating

SUGGESTED SOLUTION TO PROBLEM 4.2 FOR SELF-STUDY

(Robbie Corporation; preparing a T-account work sheet for a statement of cash flows.) **Exhibit 4.17** presents a completed T-account work sheet for Robbie Corporation.

EXHIBIT 4.17

**ROBBIE CORPORATION**
**T-Account Work Sheet**
**(all dollar amounts in thousands)**
**(Problem 4.2 for Self-Study)**

**Cash**

| | Debit | Credit | |
|---|---|---|---|
| ✓ | 10 | | |
| **Operations** | | | |
| (2) | 2 | 7 | (1) |
| (4) | 6 | | |
| (5) | 7 | | |
| (8) | 7 | | |
| **Investing** | | | |
| | | 16 | (3) |
| **Financing** | | | |
| (6) | 8 | 2 | (9) |
| (7) | 10 | | |
| ✓ | 25 | | |

**Accounts Receivable**

| | Debit | Credit | |
|---|---|---|---|
| ✓ | 15 | | |
| (1) | 7 | | |
| ✓ | 22 | | |

**Merchandise Inventories**

| | Debit | Credit | |
|---|---|---|---|
| ✓ | 20 | | |
| | | 2 | (2) |
| ✓ | 18 | | |

**Property, Plant, and Equipment**

| | Debit | Credit | |
|---|---|---|---|
| ✓ | 50 | | |
| (3) | 16 | | |
| ✓ | 66 | | |

**Accumulated Depreciation**

| | Debit | Credit | |
|---|---|---|---|
| | | 25 | ✓ |
| | | 6 | (4) |
| | | 31 | ✓ |

**Accounts Payable for Merchandise**

| | Debit | Credit | |
|---|---|---|---|
| | | 30 | ✓ |
| | | 7 | (5) |
| | | 37 | ✓ |

**Bonds Payable**

| | Debit | Credit | |
|---|---|---|---|
| | | 10 | ✓ |
| | | 8 | (6) |
| | | 18 | ✓ |

**Common Stock**

| | Debit | Credit | |
|---|---|---|---|
| | | 10 | ✓ |
| | | 10 | (7) |
| | | 20 | ✓ |

**Retained Earnings**

| | Debit | Credit | |
|---|---|---|---|
| | | 20 | ✓ |
| (9) | 2 | 7 | (8) |
| | | 25 | ✓ |

EXHIBIT 4.18

**ROBBIE CORPORATION**
**Deriving Direct Method Cash Flow from Operations Using Data from T-Account Work Sheet**
**(Problem 4.3 for Self-Study)**

1. Copy Income Statement and Cash Flow from Operations
2. Copy Information from T-Account Work Sheet Next to Related Income Statement Item
3. Sum Across Rows to Derive Direct Receipts and Expenditures

| Operations | (a) | Indirect Method (b) | Changes in Related Balance Sheet Accounts from T-Account Work Sheet (c) | Direct Method (d) | From Operations: Receipts Less Expenditures |
|---|---|---|---|---|---|
| Sales Revenues ........ | $180,000 | $ (7,000) | = Accounts Receivable Increase (1) | $ 173,000 | Receipts from Customers |
| Cost of Goods Sold ...... | (140,000) | 2,000 | = Inventory Decrease (2) | $(131,000) | Payments for Merchandise |
| | | 7,000 | = Accounts Payable for Merchandise Increase (5) | | |
| Selling and Administrative Expenses .......... | (25,000) | | (No balance sheet account changes.) | (25,000) | Payments for Selling and Administrative Services |
| Depreciation Expense .... | (6,000) | 6,000 | (Expense Not Using Cash) (4) | — | |
| Interest Expense ....... | (2,000) | — | Interest Payable (no change in balance sheet) | (2,000) | Payments for Interest |
| Net Income .......... | $ 7,000 | $ 7,000 | Totals ............ | $ 15,000 | = Cash Flow from Operations Derived via Direct Method |
| | | $15,000 | = Cash Flow from Operations Derived via Indirect Method | | |

Note that the information in column (b) is cash flow from operations derived with the indirect method, with items in a different order.

SUGGESTED SOLUTION TO PROBLEM 4.3 FOR SELF-STUDY

(Robbie Corporation; deriving direct method presentation for cash flows from operations from T-account work sheet.)

**Exhibit 4.18** derives the direct method presentation for cash flows from operations from the T-account work sheet in **Exhibit 4.17**.

SUGGESTED SOLUTION TO PROBLEM 4.4 FOR SELF-STUDY

(Gordon Corporation; preparing a statement of cash flows.)
**Exhibit 4.19** presents a completed T-account work sheet for Gordon Corporation. **Exhibit 4.20**, on page 188, derives a direct method presentation of cash flows for operations. **Exhibit 4.21**, on page 189, presents a formal statement of cash flows.

SUGGESTED SOLUTION TO PROBLEM 4.5 FOR SELF-STUDY

(Effect of transactions on the statement of cash flows.) Preparing the journal entry for each transaction aids in understanding the effect on the 11 numbered lines in **Exhibit 4.16**.

| | | | |
|---|---|---|---|
| **a.** | Depreciation Expense ..................................... | 2,000 | |
| | Accumulated Depreciation ............................ | | 2,000 |

| Change in Cash | = | Change in Liabilities | + | Change in Shareholders' Equity | − | Change in Noncash Assets |
|---|---|---|---|---|---|---|
| 0 | | 0 | | −2,000 | | −2,000 |

This entry has no impact on operating cash flows, so no effect on lines **(1)** and **(2)**. It involves a debit to an income statement account, so line **(3)** decreases by $2,000. Depreciation expense reduces net income but does not affect cash line **(11)**. Thus, line **(4)** must increase by $2,000 for the addback of depreciation expense to net income. This addback eliminates the effect of depreciation on both cash flow from operations and cash.

**EXHIBIT 4.19**

**GORDON CORPORATION**
**T-Account Work Sheet**
**(all dollar amounts in thousands)**
**(Problem 4.4 for Self-Study)**

**Cash**

| | Debit | Credit | |
|---|---|---|---|
| ✓ | 70 | | |
| **Operations** | | | |
| (1) | 200 | 100 | (5) |
| (4) | 70 | 110 | (6) |
| (3) | 15 | 20 | (7) |
| (10) | 120 | | |
| (11) | 20 | | |
| (12) | 190 | | |
| **Investing** | | | |
| (3) | 10 | 50 | (8) |
| | | 245 | (9) |
| **Financing** | | | |
| (14) | 40 | 120 | (2) |
| | | 50 | (13) |
| ✓ | 40 | | |

**Accounts Receivable**

| | Debit | Credit | |
|---|---|---|---|
| ✓ | 320 | | |
| (5) | 100 | | |
| ✓ | 420 | | |

**Merchandise Inventory**

| | Debit | Credit | |
|---|---|---|---|
| ✓ | 360 | | |
| (6) | 110 | | |
| ✓ | 470 | | |

**Prepayments**

| | Debit | Credit | |
|---|---|---|---|
| ✓ | 50 | | |
| (7) | 20 | | |
| ✓ | 70 | | |

**Land**

| | Debit | Credit | |
|---|---|---|---|
| ✓ | 200 | | |
| (8) | 50 | | |
| ✓ | 250 | | |

**Buildings and Equipment**

| | Debit | Credit | |
|---|---|---|---|
| ✓ | 1,800 | | |
| (9) | 245 | 55 | (3) |
| ✓ | 1,990 | | |

**Accumulated Depreciation**

| | Debit | Credit | |
|---|---|---|---|
| | | 800 | ✓ |
| (3) | 30 | 70 | (4) |
| | | 840 | ✓ |

**Accounts Payable**

| | Debit | Credit | |
|---|---|---|---|
| | | 320 | ✓ |
| | | 120 | (10) |
| | | 440 | ✓ |

**Income Taxes Payable**

| | Debit | Credit | |
|---|---|---|---|
| | | 60 | ✓ |
| | | 20 | (11) |
| | | 80 | ✓ |

**Other Current Liabilities**

| | Debit | Credit | |
|---|---|---|---|
| | | 170 | ✓ |
| | | 190 | (12) |
| | | 360 | ✓ |

**Bonds Payable**

| | Debit | Credit | |
|---|---|---|---|
| | | 250 | ✓ |
| (13) | 50 | | |
| | | 200 | ✓ |

**Common Stock**

| | Debit | Credit | |
|---|---|---|---|
| | | 500 | ✓ |
| | | 40 | (14) |
| | | 540 | ✓ |

**Retained Earnings**

| | Debit | Credit | |
|---|---|---|---|
| | | 700 | ✓ |
| (2) | 120 | 200 | (1) |
| | | 780 | ✓ |

EXHIBIT 4.20

**GORDON CORPORATION**
**Deriving Direct Method Cash Flow from Operations Using Data from T-Account Work Sheet**
**(all dollar amounts in thousands)**
**(Problem 4.4 for Self-Study)**

1. Copy Income Statement and Cash Flow from Operations
2. Copy Information from T-Account Work Sheet Next to Related Income Statement Item
3. Sum Across Rows to Derive Direct Receipts and Expenditures

| Operations | (a) | Indirect Method (b) | Changes in Related Balance Sheet Accounts from T-Account Work Sheet (c) | Direct Method (d) | From Operations: Receipts Less Expenditures |
|---|---|---|---|---|---|
| Revenues | $1,600 | $(100) | = Accounts Receivable Increase | $1,500 | Receipts from Customers |
| Cost of Goods Sold | (900) | 120 | = Accounts Payable Increase | (890) | Payments for Merchandise |
| | | (110) | = Merchandise Inventory Increase | | |
| Depreciation Expense | (70) | 70 | (Expense Not Using Cash) | — | |
| Selling and Administrative Expenses | (255) | 190 | = Other Current Liabilities Increase | (85) | Payments for Selling and Administrative Services |
| | | (20) | = Prepayment Increase | | |
| Interest Expense | (30) | — | = Interest Payable (no change in balance sheet) | (30) | Payments for Interest |
| Loss on Disposal of Buildings and Equipment | (15) | 15 | (Loss Not Using Cash) | — | |
| Income Tax Expense | (130) | 20 | = Income Taxes Payable Increase | (110) | Payments for Income Taxes |
| Net Income | $ 200 | $ 200 | Totals | $ 385 | = Cash Flow from Operations Derived via Direct Method |
| | | $ 385 | = Cash Flow from Operations Derived via Indirect Method | | |

Note that the information in column (b) is cash flow from operations derived with the indirect method, with items in a different order.

*continued* SUGGESTED SOLUTION TO PROBLEM 4.5 FOR SELF-STUDY

**b.** 

| | Dr. | Cr. |
|---|---|---|
| Machinery | 10,000 | |
| Cash | | 10,000 |

| Change in Cash | = | Change in Liabilities | + | Change in Shareholders' Equity | – | Change in Noncash Assets |
|---|---|---|---|---|---|---|
| –10,000 (Invst.) | | 0 | | 0 | | 10,000 |

This entry does not involve receipts from customers nor payments to suppliers, so does not affect lines **(1)** and **(2)**. It involves a credit to Cash, so line **(11)** decreases by $10,000. Because line **(11)** is the net change in cash for the period, some other line must change as well. Acquisitions of equipment represent Investing activities, so line **(7)** increases by $10,000. Note that line **(7)** has a negative sign, so this means an increase to an amount subtracted; increasing this line reduces cash.

**c.** 

| | Dr. | Cr. |
|---|---|---|
| Retained Earnings | 6,500 | |
| Cash | | 6,500 |

EXHIBIT 4.21

**GORDON CORPORATION**
**Statement of Cash Flows**
**Year 2**
**(all dollar amounts in thousands)**
**(Problem 4.4 for Self-Study)**

| | | |
|---|---|---|
| **Operations** | | |
| Receipts from Customers | | $1,500 |
| Payments to Suppliers of Merchandise | | (890) |
| Paid for Selling and Administrative Expenses | | (85) |
| Payments to Lenders for Interest | | (30) |
| Payments for Income Taxes | | (110) |
| Cash Flow from Operating Activities | | $ 385 |
| **Reconciliation of Net Income to Cash provided by Operations** | | |
| Net Income | $ 200 | |
| Additions: | | |
| Depreciation Expense | 70 | |
| Loss on Disposal of Equipment | 15 | |
| Increase in Accounts Payable | 120 | |
| Increase in Income Taxes Payable | 20 | |
| Increase in Other Current Liabilities | 190 | |
| Subtractions: | | |
| Increase in Accounts Receivable | (100) | |
| Increase in Merchandise Inventories | (110) | |
| Increase in Prepayments | (20) | |
| Cash Flow from Operations | $ 385 | |
| **Investing** | | |
| Acquisition of Land | $ (50) | |
| Disposal of Buildings and Equipment | 10 | |
| Acquisition of Buildings and Equipment | (245) | |
| Cash Used for Investing | | (285) |
| **Financing** | | |
| Common Stock Issued | $ 40 | |
| Dividends Paid | (120) | |
| Repayment of Bonds | (50) | |
| Cash Used for Financing | | (130) |
| Net Decrease in Cash | | $ (30) |
| Cash, Beginning of Year 2 | | 70 |
| Cash, End of Year 2 | | $ 40 |

*continued* (Problem 4.5)

| Change in Cash | = | Change in Liabilities | + | Change in Shareholders' Equity | – | Change in Noncash Assets |
|---|---|---|---|---|---|---|
| –6,500 (Finan.) | | 0 | | –6,500 | | 0 |

This entry involves a credit to Cash, so line **(11)** decreases by $6,500. Dividends are a financing activity, so line **(10)** increases by $6,500.

**d.** Cash ........ 12,000
Common Stock ........ 12,000

| Change in Cash | = | Change in Liabilities | + | Change in Shareholders' Equity | – | Change in Noncash Assets |
|---|---|---|---|---|---|---|
| 12,000 (Finan.) | | 0 | | 12,000 | | 0 |

*continued* (Problem 4.5)

The debit to Cash means that line **(11)** increases by $12,000. Issuing stock is a financing transaction, so line **(8)** increases by $12,000.

| | | | |
|---|---|---|---|
| **e.** | Cash ........................................ | 15,000 | |
| | Investment in Securities ........................ | | 15,000 |

| Change in Cash | = | Change in Liabilities | + | Change in Shareholders' Equity | – | Change in Noncash Assets |
|---|---|---|---|---|---|---|
| 15,000 (Invst.) | | 0 | | 0 | | –15,000 |

The debit to Cash means that line **(11)** increases by $15,000. Selling investments in securities is an investing activity, so line **(6)** increases by $15,000.

## Key Terms and Concepts

Cash flow from operations
Cash flow from investing activities
Cash flow from financing activities
Direct method
Indirect method
Cash Change Equation
T-account work sheet

## Questions, Short Exercises, Exercises, Problems, and Cases

### QUESTIONS

For additional student resources, content, and interactive quizzes for this chapter visit the FACMU website: **www.thomsonedu.com/accounting/stickney**

1. Review the meaning of the terms and concepts listed above in Key Terms and Concepts.
2. "One can most easily accomplish the reporting objective of the income statement under the accrual basis of accounting and the reporting objective of the statement of cash flows by issuing a single income statement using the cash basis of accounting." Evaluate this proposal.
3. "The accrual basis of accounting creates the need for a statement of cash flows." Explain.
4. "The statement of cash flows provides information about changes in the structure of a firm's assets and equities." Explain.
5. A student remarked: "The direct method of computing cash flow from operations is easier to understand than the indirect method. Why do the vast majority of firms follow the indirect method in preparing their statements of cash flows?" Respond to this student.
6. The statement of cash flows classifies cash expenditures for interest expense as an operating activity but classifies cash expenditures to redeem debt as a financing activity. Explain this apparent paradox.
7. The statement of cash flows classifies cash expenditures for interest on debt as an operating activity but classifies cash expenditures for dividends to shareholders as a financing activity. Explain this apparent paradox.
8. The statement of cash flows classifies changes in accounts payable as an operating activity but classifies changes in short-term bank borrowing as a financing activity. Explain this apparent paradox.
9. The acquisition of equipment by assuming a mortgage is a transaction that firms cannot report in their statement of cash flows but must report in a supplemental schedule or note. Of what value is information about this type of transaction? What is the reason for its exclusion from the statement of cash flows?
10. One writer stated, "Depreciation expense is a firm's chief source of cash for growth." A reader criticized this statement by replying, "The fact remains that if companies had elected, in any year, to charge off $10 million more depreciation than they did charge off, they would not thereby have added one dime to the total of their cash available for expanding plants or for increasing inventories or receivables. Therefore, to speak of depreciation expense as a source of cash is incorrect and misleading." Comment on these statements, taking into account income tax effects.
11. A firm generated net income for the current year, but cash flow from operations was negative. How can this happen?

12. A firm operated at a net loss for the current year, but cash flow from operations was positive. How can this happen?
13. The disposal of equipment for an amount of cash greater than the book value of the equipment results in a cash receipt equal to the book value of the equipment plus the gain on the disposal, which appears in income. How might the accountant treat this transaction in the statement of cash flows? Consider both the direct and indirect methods.

## SHORT EXERCISES

14. **Derive sales revenue from data in the statement of cash flows and balance sheet.** Microsoft Corporation reported a balance of $5,196 million in accounts receivable at the beginning of Year 14 and $5,334 million at the end of Year 14. Its statement of cash flows using the direct method reported cash collections from customers of $33,551 million for Year 14. Assuming that Microsoft Corporation makes all sales on account, compute the amount of sales during Year 14.
15. **Derive cost of goods sold from data in the statement of cash flows.** The section showing cash flow from operations, using the indirect method, for General Electric Company reported an increase in inventories during Year 13 of $1,753 million and no change in accounts payable for inventories. The direct method would show cash payments for inventory, purchased and manufactured, totaling $64,713 million. Compute the cost of goods sold for Year 13.
16. **Derive cost of goods sold from data in the statement of cash flows.** The section showing cash flow from operations, using the indirect method, for Ann Taylor Stores reported an increase in inventories of $5.7 million during Year 13. It reported also that the balance in accounts payable for inventories increased by $5.9 million. The direct method would show cash payments for merchandise inventory purchased of $646.9 million. Compute the cost of goods sold for Year 13.
17. **Derive wages and salaries expense from data in the statement of cash flows.** AMR, the parent company of American Airlines, reported in its reconciliation of net income to cash flow from operations a decrease in wages and salaries payable of $21 million during Year 13. It provided data showing that cash payments for wages and salaries to employees for Year 13 were $8,853 million. Compute the amount of wages and salaries expense for Year 13.
18. **Derive cash disbursements for dividends.** Johnson & Johnson, a pharmaceutical and medical products company, reported a balance in retained earnings of $26,571 million at the beginning of Year 13 and $28,132 million at the end of Year 13. Its dividends payable account increased by $233 million during Year 13. It reported net income for Year 13 of $5,030 million. How much cash did Johnson & Johnson disburse for dividends during Year 13? Indicate where this information would appear in the simplified statement of cash flows in **Exhibit 4.16**.
19. **Effect of borrowing and interest on statement of cash flows.** Gillette borrowed $250 million on October 1, Year 12 by issuing bonds. The debt carries an annual interest rate of 6 percent and is payable in two installments, on April 1 and October 1 of each year. The debt matures on October 1, Year 32. Gillette's accounting period ends on December 31 of each year. Using the format of **Exhibit 4.16**, indicate the effects of all these transactions on Gillette's statement of cash flows for Year 12.
20. **Effect of income taxes on statement of cash flows.** Radio Shack reported a balance in Income Taxes Payable of $78.1 million at the beginning of Year 13, $60.1 million at the end of Year 13, and income tax expense for the year of $161.5 million. Using the format of **Exhibit 4.16**, indicate the effects of all these transactions on Radio Shack's statement of cash flows for Year 13.
21. **Effect of rent transactions on statement of cash flows.** A firm reported a balance in its Prepaid Rent (Advances to Landlord) account of $1,200 on January 1, Year 12, for use of the building for the month of January, Year 12. On February 1, Year 12, the firm paid $18,000 as the annual rental for the period from February 1, Year 12 to January 31, Year 13. It recorded this rental payment by debiting Prepaid Rent (Advances to Landlord) and crediting Cash for $18,000. At the end of Year 12, the firm made all proper adjusting entries to correctly report balance sheet and income statement amounts. Using the format of **Exhibit 4.16**, indicate the effects of all these transactions on the firm's statement of cash flows for Year 12.
22. **Calculating components of cash flow from operations. Exhibit 4.22**, on page 192, provides items from the financial statements of Information Technologies, a systems engineering firm, for Year 2. How much cash did Information Technologies collect from its customers during Year 2?

**EXHIBIT 4.22**

**INFORMATION TECHNOLOGIES**
**Data from Income Statement**
**Year 2**
**(Problem 22)**
**(all dollar amounts in thousands)**

| | |
|---|---|
| Sales | $ 14,508 |
| Cost of Goods Sold | (11,596) |
| Depreciation Expense | (114) |
| Other Expenses, Including Salaries and Wages Expense | (2,276) |
| Income Taxes | (210) |
| Net Income | $ 312 |
| **Data from Beginning of Year 2 and End of Year 2 Balance Sheets** | |
| Accounts Receivable | $782 Decrease |
| Inventories | 66 Decrease |
| Prepayments for Other Costs | 102 Decrease |
| Accounts Payable for Inventories | 90 Increase |
| Current Liabilities for Wages and Salaries Payable | 240 Decrease |

23. **Calculating components of cash flow from operations.** Refer to **Exhibit 4.22**, which provides items from the financial statements of Information Technologies. How much cash did Information Technologies pay during Year 2 to its suppliers of goods?
24. **Calculating components of cash flow from operations.** Refer to **Exhibit 4.22**, which provides items from the financial statements of Information Technologies. How much cash did Information Technologies pay during Year 2 to its employees and suppliers of other services?
25. **Working backward from changes in the Buildings and Equipment account.** The comparative balance sheets of American Airlines show a balance in the Buildings and Equipment account at cost on December 31, Year 4, of $17,369 million; at December 31, Year 3, the balance was $16,825 million. The Accumulated Depreciation account shows a balance of $5,465 million at December 31, Year 4, and of $4,914 million at December 31, Year 3. The statement of cash flows reports that expenditures for buildings and equipment during Year 4 totaled $1,314 million. The income statement indicates a depreciation charge of $1,253 million during Year 4. The firm sold buildings and equipment during the year at their book value.

    Calculate the acquisition cost and accumulated depreciation of the buildings and equipment that were retired during the year and the proceeds from the disposition.
26. **Preparing a statement of cash flows from changes in balance sheet accounts.** The comparative balance sheets of Southwest Airlines show the following information for a recent year (amounts in thousands):

| Change | Amount | Direction |
|---|---|---|
| Cash | $ 40,308[a] | Increase |
| Accounts Receivable | 15,351 | Decrease |
| Inventories | 15,117 | Increase |
| Prepayments | 16,776 | Increase |
| Property, Plant, and Equipment (at cost) | 1,134,644[b] | Increase |
| Accumulated Depreciation | 264,088[b] | Increase |
| Other Nonoperating Assets | 8,711 | Increase |
| Accounts Payable | 660 | Decrease |
| Other Current Liabilities | 114,596 | Increase |
| Long-Term Debt | 244,285 | Increase |
| Other Nonoperating Liabilities | 140,026 | Increase |
| Common Stock | 96,991 | Increase |
| Retained Earnings | 340,879[c] | Increase |

[a]Cash was $378,511 thousand at the beginning of the year and $418,819 thousand at the end of the year.
[b]Southwest Airlines did not sell any property, plant, and equipment during the year.
[c]Net income was $474,378 thousand.

a. Prepare a statement of cash flows for Southwest Airlines for the year. Treat changes in nonoperating assets as investing transactions and changes in nonoperating liabilities as financing transactions.

b. Discuss briefly the pattern of cash flows from operating, investing, and financing activities for Southwest Airlines for the year.

27. **Calculating and interpreting cash flow from operations.** The following items appear in the financial statements of Bamberger Enterprises, a firm offering IT services for Sarbanes-Oxley compliance, for Year 2 (amounts in thousands):

| | |
|---|---|
| Sales | $14,600 |
| Depreciation Expense | (210) |
| Income Taxes | (200) |
| Other Expenses | (13,900) |
| Net Income | $ 290 |

The changes in the current asset and current liability accounts were as follows:

| | | |
|---|---|---|
| Accounts Receivable | $780 | Decrease |
| Inventories | 80 | Decrease |
| Prepayments | 100 | Decrease |
| Accounts Payable | 90 | Increase |
| Other Current Liabilities | 240 | Decrease |

a. Compute the amount of cash flow from operations.

b. Comment on the major reasons why cash flow from operations exceeds net income.

28. **Calculating and interpreting cash flow from operations.** Selected data for Nokia, a Finnish cellular phone manufacturer, appear below (amounts in millions of euros):

| | Year 8 | Year 9 | Year 10 | Year 11 |
|---|---|---|---|---|
| Net Income (Loss) | €1,032 | €1,689 | €2,542 | €3,847 |
| Depreciation Expense | 465 | 509 | 665 | 1,009 |
| Increase (Decrease) in: | | | | |
| Accounts Receivable | 272 | 1,573 | 982 | 2,304 |
| Inventories | 121 | 103 | 362 | 422 |
| Prepayments | (77) | 17 | 33 | (49) |
| Accounts Payable | 90 | 140 | 312 | 458 |
| Other Current Liabilities | 450 | 1,049 | 867 | 923 |

a. Compute the amount of cash flow from operations for each of the four years using the indirect method.

b. Discuss briefly the most important reasons why cash flow from operations differs from net income or net loss for each year.

29. **Calculating and interpreting cash flows.** Omnicom Group is a marketing services firm. Omnicom Group creates advertising copy for clients and places the advertising in television, magazines, and other media. Accounts receivable represent amounts owed by clients and accounts payable represent amounts payable to various media. Omnicom Group has purchased other marketing services firms in recent years. Selected data for Omnicom Group for three recent years appear below (amounts in millions):

| | Year 3 | Year 4 | Year 5 |
|---|---|---|---|
| Net Income | $279 | $363 | $499 |
| Depreciation and Amortization Expense | 164 | 196 | 226 |
| Increase (Decrease) in Accounts Receivable | 238 | 648 | 514 |
| Increase (Decrease) in Inventories | 35 | 13 | 98 |
| Increase (Decrease) in Prepayments | 64 | (10) | 125 |

*(continued)*

| | Year 3 | Year 4 | Year 5 |
|---|---|---|---|
| Increase (Decrease) in Accounts Payable ........ | 330 | 786 | 277 |
| Increase (Decrease) in Other Current Liabilities .... | 70 | 278 | 420 |
| Acquisition of Property, Plant, and Equipment .... | 115 | 130 | 150 |
| Acquisition of Investments in Securities (noncurrent) | 469 | 643 | 885 |
| Dividends Paid .......................... | 88 | 104 | 122 |
| Long-Term Debt Issued .................... | 208 | 83 | 599 |
| Common Stock Issued (Reacquired) ............ | 42 | (252) | (187) |

**a.** Prepare a comparative statement of cash flows for Omnicom Group for the three years. Use the indirect method of computing cash flow from operations.

**b.** Discuss the relation between net income and cash flow from operations and the pattern of cash flow from operating, investing, and financing activities during the three years.

**30. Effects of gains and losses from sales of equipment on cash flows.** The text does not present material that shows how to answer these questions. You will need to think the answers through for yourself. **Exhibit 4.23** presents an abbreviated statement of cash flows for Largay Corporation for the current year (amounts in thousands). After preparing this statement of cash flows for the current year, you discover that the firm sold an item of equipment on the last day of the year but failed to record it in the accounts or to deposit the check received from the purchaser. The equipment originally cost \$50,000 and had accumulated depreciation of \$40,000 at the time of sale. Recast the statement of cash flows in the exhibit, assuming that Largay Corporation sold the equipment for cash in the following amounts (ignore income taxes):

**a.** \$10,000

**b.** \$12,000

**c.** \$8,000

**EXHIBIT 4.23**

**LARGAY CORPORATION**
**Statement of Cash Flows**
**Current Year**
**(Exercise 30)**
**(all dollar amounts in thousands)**

| | |
|---|---|
| **Operations** | |
| Net Income ........................................ | \$100 |
| Depreciation Expense ................................ | 15 |
| Changes in Working Capital Accounts .................. | (40) |
| Cash Flow from Operations ........................... | \$ 75 |
| **Investing** | |
| Acquisition of Buildings and Equipment ................ | (30) |
| **Financing** | |
| Repayment of Long-Term Debt ......................... | (40) |
| Change in Cash ..................................... | \$ 5 |
| Cash, Beginning of Year .............................. | 27 |
| Cash, End of Year ................................... | \$ 32 |

**31. Effect of various transactions on the statement of cash flows. Exhibit 4.16** shows a simplified statement of cash flows for a period. Numbers appear on eleven of the lines in the statement. Other lines are various subtotals and grand totals; ignore these in the remainder of the problem. Assume that the accounting cycle is complete for the period and that the firm has prepared all of the financial statements. It then discovers that it has overlooked a transaction. It records that transaction in the accounts and corrects all of the financial statements. For each of the following transactions, indicate which of the numbered lines of the statement of cash flows change, and state the amount and direction of the change. If net income, line **(3)**, changes, be sure to indicate whether it decreases or increases. Ignore income tax effects. (*Hint:* First, construct the entry the firm would enter in the accounts to record the transaction in the accounts. Then, for each line of the journal entry, identify the line of **Exhibit 4.16** affected.)

**a.** Amortization of a patent, treated as an expense, $600

**b.** Acquisition of a factory site financed by issuing capital stock with a market value of $50,000 in exchange

**c.** Purchase of inventory on account for $7,500; assume inventory had increased for the year before the firm recorded this overlooked transaction

**d.** Purchase of inventory for cash of $6,000; assume inventory had increased for the year before the firm recorded this overlooked transaction

**e.** Uninsured fire loss of merchandise inventory totaling $1,500; assume inventory had increased for the year before the firm recorded this overlooked transaction

**f.** Collection of an account receivable totaling $1,450; assume accounts receivable had increased for the year before the firm recorded this overlooked transaction

**g.** Issue of bonds for $10,000 cash

**h.** Disposal of equipment for cash at its book value of $4,500

## PROBLEMS AND CASES

**32. Inferring cash flows from financial statement data. Exhibit 4.24** presents data from the financial statements for Heidi's Hide-Out, a bar and video-game club, with private rooms for rent for parties. Heidi's deals with

- many employees, to some of whom it has made advances on wages and to some of whom it owes wages for past work;
- many landlords, to some of whom it has made advance payments and to some of whom it owes rent for past months;

**EXHIBIT 4.24**

**HEIDI'S HIDE-OUT**
**Selected Detail from Financial Statements**
**Current Year**
**(Problem 32)**

| | Beginning of Year | End of Year |
|---|---|---|
| BALANCE SHEETS | | |
| Cash | $22,000 | $ 10,000 |
| Accounts Receivable from Retail Customers | 8,000 | 8,900 |
| Inventory of Retail Merchandise | 11,000 | 10,000 |
| Advances to Employees | 1,000 | 1,500 |
| Advances to Landlords (Prepaid Rent) | 5,000 | 5,600 |
| Advances to Suppliers of Retail Merchandise | 10,000 | 10,500 |
| Total Assets | $57,000 | $ 46,500 |
| Accounts Payable to Suppliers of Retail Merchandise | $ 8,000 | $ 7,700 |
| Advances from Retail Customers | 9,000 | 10,000 |
| Rent Payable to Landlords | 6,000 | 5,300 |
| Wages Payable to Employees | 2,000 | 1,800 |
| Owners' Equity | 32,000 | 21,700 |
| Total Liabilities and Owners' Equity | $57,000 | $ 46,500 |
| INCOME STATEMENT FOR THE YEAR | | |
| Sales Revenue from Retail Customers | | $ 120,000 |
| Cost of Retail Merchandise Sold | $90,000 | |
| Rent Expense | 33,000 | |
| Wage Expense | 20,000 | |
| Less: Total Expenses | | (143,000) |
| Net Income | | $ (23,000) |

- many customers, some of whom have paid for special parties not yet held and some of whom have not yet paid for parties they have held; and
- many suppliers of goods, including food and beverages, some of whom Heidi's has paid for orders not yet received and some of whom have delivered goods for which Heidi's has not yet paid.

Heidi's and its customers, suppliers, and employees settle all transactions with cash, never with noncash assets.

**a.** Calculate the amount of cash received from retail customers during the current year.

**b.** Calculate the amount of cash Heidi's paid landlords during the current year for the rental of space.

**c.** Calculate the amount of cash Heidi's paid employees during the current year.

**d.** Calculate the amount of cash Heidi's paid suppliers of retail merchandise, which includes food and beverages it sells to retail customers, during the current year.

**33. Inferring cash flows from balance sheet and income statement data.** (Based on a problem prepared by Stephen A. Zeff.) You work for the Plains State Bank in Miles City, Montana, as an analyst specializing in the financial statements of small businesses seeking loans from the bank. Digit Retail Enterprises, Inc., provides you with its balance sheet on December 31, Year 4 and Year 5 (**Exhibit 4.25**), and its income statement for Year 5 (**Exhibit 4.26**).

**EXHIBIT 4.25**

**DIGIT RETAIL ENTERPRISES, INC.**
**Balance Sheet**
**(Problem 33)**

| | December 31 Year 4 | December 31 Year 5 |
|---|---|---|
| ASSETS | | |
| **Current Assets** | | |
| Cash | $ 36,000 | $ 50,000 |
| Accounts Receivable | 23,000 | 38,000 |
| Notes Receivable | 7,500 | — |
| Interest Receivable | 100 | — |
| Merchandise Inventory | 48,000 | 65,000 |
| Prepaid Insurance | 9,000 | 12,000 |
| Prepaid Rent | 2,000 | — |
| Total Current Assets | $125,600 | $165,000 |
| **Property, Plant, and Equipment** | | |
| At Cost | $100,000 | $ 90,000 |
| Less Accumulated Depreciation | (20,000) | (35,000) |
| Net | $ 80,000 | $ 55,000 |
| Total Assets | $205,600 | $220,000 |
| LIABILITIES AND SHAREHOLDERS' EQUITY | | |
| **Current Liabilities** | | |
| Accounts Payable—Merchandise Suppliers | $ 18,000 | $ 20,000 |
| Salaries Payable | 2,100 | 2,800 |
| Rent Payable | — | 3,000 |
| Advances from Customers | 8,500 | 6,100 |
| Note Payable | — | 5,500 |
| Dividends Payable | 4,200 | 2,600 |
| Other Current Liabilities | 1,300 | 3,700 |
| Total Current Liabilities | $ 34,100 | $ 43,700 |
| **Shareholders' Equity** | | |
| Common Stock | $160,000 | $164,500 |
| Retained Earnings | 11,500 | 11,800 |
| Total Shareholders' Equity | $171,500 | $176,300 |
| Total Liabilities and Shareholders' Equity | $205,600 | $220,000 |

**EXHIBIT 4.26**

**DIGIT RETAIL ENTERPRISES, INC.**
**Income Statement**
**Year 5**
**(Problem 33)**

| | |
|---|---|
| Sales Revenue | $270,000 |
| Gain on Sale of Property, Plant, and Equipment | 3,200 |
| Interest Revenue | 200 |
| Total Revenues | $273,400 |
| **Less Expenses:** | |
| Cost of Goods Sold | $145,000 |
| Salaries Expense | 68,000 |
| Rent Expense | 12,000 |
| Insurance Expense | 5,000 |
| Depreciation Expense | 20,000 |
| Other Expenses | 13,800 |
| Total Expenses | $263,800 |
| Net Income | $ 9,600 |

The firm's independent accountant has audited these financial statements and found them to be in conformity with generally accepted accounting principles. Digit Retail Enterprises acquired no new property, plant, and equipment during the year.

**a.** Calculate the amount of cash received from customers during the year.
**b.** Calculate the acquisition cost of merchandise purchased during the year.
**c.** Calculate the amount of cash paid to suppliers of merchandise during the year.
**d.** Calculate the amount of cash paid to salaried employees during the year.
**e.** Calculate the amount of cash paid to insurance companies during the year.
**f.** Calculate the amount of cash paid to landlords for rental of space during the year.
**g.** Calculate the amount of dividends paid during the year.
**h.** Calculate the amount of cash received when property, plant, and equipment were sold during the year.

**34. Preparing and interpreting a statement of cash flows using a T-account work sheet.** Condensed financial statement data for Hale Company for the current year appear in **Exhibits 4.27** and **4.28**, on page 198. During the current year, the firm sold for $5,000 equipment costing $15,000 with $10,000 of accumulated depreciation.

**a.** Prepare a statement of cash flows for Hale Company for the year using the indirect method of computing cash flow from operations. Support the statement with a T-account work sheet.
**b.** Derive a presentation of cash flows from operations using the direct method.
**c.** Present a statement of cash flows for Hale Company using the direct method for cash flows from operations. Include reconciliation of net income to cash flow from operations.
**d.** Comment on the pattern of cash flows from operations, investing, and financing activities.

**35. Preparing and interpreting a statement of cash flows using a T-account work sheet.** Financial statement data for Dickerson Manufacturing Company for the current year appear in **Exhibit 4.29**, on page 199.

Additional information includes the following:

**(1)** Net income for the year was $568,000; dividends declared and paid were $60,000.
**(2)** Depreciation expense for the year was $510,000 on buildings and machinery.
**(3)** The firm sold for $25,000 machinery originally costing $150,000 with accumulated depreciation of $120,000.
**(4)** The firm retired bonds during the year at their book value.

**a.** Prepare a statement of cash flows for Dickerson Manufacturing Company for the year using the indirect method to compute cash flow from operations. Support the statement with a T-account work sheet.
**b.** Comment on the pattern of cash flows from operating, investing, and financing activities.

**EXHIBIT 4.27**

**HALE COMPANY**
**Comparative Balance Sheet**
**(Problem 34)**

| | January 1 | December 31 |
|---|---|---|
| ASSETS | | |
| Cash | $ 52,000 | $ 58,000 |
| Accounts Receivable | 93,000 | 106,000 |
| Inventory | 151,000 | 162,000 |
| Land | 30,000 | 30,000 |
| Buildings and Equipment (cost) | 790,000 | 830,000 |
| Less Accumulated Depreciation | (460,000) | (504,000) |
| Total Assets | $ 656,000 | $ 682,000 |
| LIABILITIES AND SHAREHOLDERS' EQUITY | | |
| Accounts Payable for Inventory | $ 136,000 | $ 141,000 |
| Interest Payable | 10,000 | 8,000 |
| Mortgage Payable | 120,000 | 109,000 |
| Common Stock | 250,000 | 250,000 |
| Retained Earnings | 140,000 | 174,000 |
| Total Liabilities and Shareholders' Equity | $ 656,000 | $ 682,000 |

**EXHIBIT 4.28**

**HALE COMPANY**
**Statement of Income and Retained Earnings**
**Current Year**
**(Problem 34)**

| | | |
|---|---|---|
| Sales Revenues | | $1,200,000 |
| Expenses | | |
| Cost of Goods Sold | $788,000 | |
| Wages and Salaries | 280,000 | |
| Depreciation | 54,000 | |
| Interest | 12,000 | |
| Income Taxes | 22,000 | |
| Total | | $1,156,000 |
| Net Income | | $ 44,000 |
| Dividends on Common Stock | | (10,000) |
| Addition to Retained Earnings for Year | | $ 34,000 |
| Retained Earnings, January 1 | | 140,000 |
| Retained Earnings, December 31 | | $ 174,000 |

36. **Preparing and interpreting a statement of cash flows using a T-account work sheet.** GTI, Inc. manufactures parts, components, and processing equipment for electronics and semiconductor applications in the communication, computer, automotive, and appliance industries. Its sales tend to vary with changes in the business cycle since the sales of most of its customers are cyclical. **Exhibit 4.30** on page 200 presents balance sheets for GTI as of December 31, Year 7 through Year 9, and **Exhibit 4.31** on page 200 presents income statements for Year 8 and Year 9.

Notes to the firm's financial statements reveal the following (amounts in thousands):

**(1)** Depreciation expense, included in Administration Expenses, was $641 in Year 8 and $625 in Year 9.

**EXHIBIT 4.29**

**DICKERSON MANUFACTURING COMPANY**
**Comparative Balance Sheet**
**(Problem 35)**

| | January 1 | December 31 |
|---|---|---|
| ASSETS | | |
| **Current Assets** | | |
| Cash | $ 358,000 | $ 324,000 |
| Accounts Receivable | 946,000 | 1,052,000 |
| Inventory | 1,004,000 | 1,208,000 |
| Total Current Assets | $2,308,000 | $2,584,000 |
| **Noncurrent Assets** | | |
| Land | $ 594,000 | $ 630,000 |
| Buildings and Machinery | 8,678,000 | 9,546,000 |
| Less Accumulated Depreciation | (3,974,000) | (4,364,000) |
| Total Noncurrent Assets | $5,298,000 | $5,812,000 |
| Total Assets | $7,606,000 | $8,396,000 |
| LIABILITIES AND SHAREHOLDERS' EQUITY | | |
| **Current Liabilities** | | |
| Accounts Payable | $ 412,000 | $ 558,000 |
| Taxes Payable | 274,000 | 290,000 |
| Other Short-Term Payables | 588,000 | 726,000 |
| Total Current Liabilities | $1,274,000 | $1,574,000 |
| **Noncurrent Liabilities** | | |
| Bonds Payable | 1,984,000 | 1,934,000 |
| Total Liabilities | $3,258,000 | $3,508,000 |
| **Shareholders' Equity** | | |
| Common Stock | $1,672,000 | $1,704,000 |
| Retained Earnings | 2,676,000 | 3,184,000 |
| Total Shareholders' Equity | $4,348,000 | $4,888,000 |
| Total Liabilities and Shareholders' Equity | $7,606,000 | $8,396,000 |

**(2)** Other Noncurrent Assets represent patents. Patent amortization, included in Administrative Expenses, was $25 in Year 8 and $40 in Year 9.

**(3)** Changes in Other Current Liabilities and Other Noncurrent Liabilities are both operating transactions relating to Administrative Expenses.

**a.** Prepare a T-account work sheet for the preparation of a statement of cash flows for GTI, Inc. for Year 8 and Year 9.

**b.** Prepare a statement of cash flows for GTI, Inc. for Year 8 and Year 9. Present cash flows from operations using the indirect method.

**c.** Present a derivation of cash flows from operations for Year 8 using the direct method.

**d.** Discuss the relation between net income and cash flow from operations and the pattern of cash flows from operating, investing, and financing activities.

**37. Preparing and interpreting a statement of cash flows using a T-account work sheet.** Flight Training Corporation, a private-sector firm, provides fighter pilot training under contracts with the U.S. Air Force and U.S. Navy. The firm owns approximately 100 Lear jets that it equips with radar jammers and other sophisticated electronic devices to mimic enemy aircraft. The company recently experienced cash shortages to pay its bills. The owner and manager of Flight Training Corporation stated: "I was just dumbfounded. I never had an inkling that there was a problem with cash."

**Exhibit 4.32** (page 201) presents comparative balance sheets for Flight Training Corporation on December 31, Year 1 through Year 4, and **Exhibit 4.33** (page 201) presents income statements for Year 2 through Year 4.

EXHIBIT 4.30

**GTI, INC.**
**Balance Sheets**
**(all dollar amounts in thousands)**
**(Problem 36)**

| | December 31 | | |
|---|---|---|---|
| | Year 7 | Year 8 | Year 9 |
| ASSETS | | | |
| Cash | $ 430 | $ 475 | $ 367 |
| Accounts Receivable | 3,768 | 3,936 | 2,545 |
| Inventories | 2,334 | 2,966 | 2,094 |
| Prepayments | 116 | 270 | 122 |
| Total Current Assets | $ 6,648 | $ 7,647 | $5,128 |
| Property, Plant, and Equipment (net) | 3,806 | 4,598 | 4,027 |
| Other Noncurrent Assets | 193 | 559 | 456 |
| Total Assets | $10,647 | $12,804 | $9,611 |
| LIABILITIES AND SHAREHOLDERS' EQUITY | | | |
| Accounts Payable (to Suppliers of Inventory) | $ 1,578 | $809 | $ 796 |
| Notes Payable to Banks | 11 | 231 | 2,413 |
| Other Current Liabilities | 1,076 | 777 | 695 |
| Total Current Liabilities | $ 2,665 | $ 1,817 | $3,904 |
| Long-Term Debt | 2,353 | 4,692 | 2,084 |
| Other Noncurrent Liabilities | 126 | 89 | 113 |
| Total Liabilities | $ 5,144 | $ 6,598 | $6,101 |
| Preferred Stock | $ — | $ 289 | $ 289 |
| Common Stock | 83 | 85 | 85 |
| Additional Paid-in Capital | 4,385 | 4,392 | 4,395 |
| Retained Earnings | 1,035 | 1,440 | (1,259) |
| Total Shareholders' Equity | $ 5,503 | $ 6,206 | $3,510 |
| Total Liabilities and Shareholders' Equity | $10,647 | $12,804 | $9,611 |

EXHIBIT 4.31

**GTI, INC.**
**Income Statements**
**(all dollar amounts in thousands)**
**(Problem 36)**

| | Year 8 | Year 9 |
|---|---|---|
| Sales | $22,833 | $11,960 |
| Cost of Goods Sold | (16,518) | (11,031) |
| Selling and Administrative Expenses | (4,849) | (3,496) |
| Interest Expense | (459) | (452) |
| Income Tax Expense | (590) | 328 |
| Net Income | $ 417 | $ (2,691) |
| Dividends on Preferred Stock | (12) | (8) |
| Net Income Available to Common | $ 405 | $ (2,699) |

**EXHIBIT 4.32**

**FLIGHT TRAINING CORPORATION**
**Balance Sheets**
**(all dollar amounts in thousands)**
**(Problem 37)**

| | December 31 | | | |
|---|---|---|---|---|
| | Year 1 | Year 2 | Year 3 | Year 4 |
| **Current Assets** | | | | |
| Cash | $ 142 | $ 313 | $ 583 | $ 159 |
| Accounts Receivable | 2,490 | 2,675 | 4,874 | 6,545 |
| Inventories | 602 | 1,552 | 2,514 | 5,106 |
| Prepayments | 57 | 469 | 829 | 665 |
| Total Current Assets | $ 3,291 | $ 5,009 | $ 8,800 | $ 12,475 |
| Property, Plant, and Equipment | $17,809 | $24,039 | $76,975 | $106,529 |
| Less Accumulated Depreciation | (4,288) | (5,713) | (8,843) | (17,231) |
| Net | $13,521 | $18,326 | $68,132 | $ 89,298 |
| Other Noncurrent Assets | $ 1,112 | $ 641 | $ 665 | $ 470 |
| Total Assets | $17,924 | $23,976 | $77,597 | $102,243 |
| **Current Liabilities** | | | | |
| Accounts Payable | $ 939 | $ 993 | $ 6,279 | $ 12,428 |
| Notes Payable | 1,021 | 140 | 945 | — |
| Current Portion of Long-Term Debt | 1,104 | 1,789 | 7,018 | 60,590 |
| Other Current Liabilities | 1,310 | 2,423 | 12,124 | 12,903 |
| Total Current Liabilities | $ 4,374 | $ 5,345 | $26,366 | $ 85,921 |
| **Noncurrent Liabilities** | | | | |
| Long-Term Debt | 6,738 | 9,804 | 41,021 | — |
| Other Noncurrent Liabilities | — | 1,029 | 900 | — |
| Total Liabilities | $11,112 | $16,178 | $68,287 | $ 85,921 |
| **Shareholders' Equity** | | | | |
| Common Stock | $ 20 | $ 21 | $ 22 | $ 34 |
| Additional Paid-in Capital | 4,323 | 4,569 | 5,486 | 16,317 |
| Retained Earnings | 2,469 | 3,208 | 3,802 | (29) |
| Total Shareholders' Equity | $ 6,812 | $ 7,798 | $ 9,310 | $ 16,322 |
| Total Liabilities and Shareholders' Equity | $17,924 | $23,976 | $77,597 | $102,243 |

**EXHIBIT 4.33**

**FLIGHT TRAINING CORPORATION**
**Comparative Income Statement**
**Year Ended December 31**
**(all dollar amounts in thousands)**
**(Problem 37)**

| | December 31 | | |
|---|---|---|---|
| | Year 2 | Year 3 | Year 4 |
| **Continuing Operations** | | | |
| Sales | $20,758 | $36,597 | $54,988 |
| **Expenses** | | | |
| Cost of Services | 14,247 | 29,594 | 47,997 |
| Selling and Administrative | 3,868 | 2,972 | 5,881 |
| Interest | 1,101 | 3,058 | 5,841 |
| Income Taxes | 803 | 379 | (900) |
| Total Expenses | $20,019 | $36,003 | $58,819 |
| Net Income | $ 739 | $ 594 | $ (3,831) |

Notes to the financial statements indicate the following:

**(1)** The firm did not sell any aircraft during the three-year period.

**(2)** Changes in Other Noncurrent Assets are investing transactions.

**(3)** Changes in Other Noncurrent Liabilities are operating transactions.

**(4)** The firm violated covenants in its borrowing agreements during Year 4. The lenders can therefore require Flight Training Corporation to repay its long-term debt immediately. Although the banks have not yet demanded payment, the firm reclassified its long-term debt as a current liability.

**a.** Prepare T-account work sheets for the preparation of a statement of cash flows for Flight Training Corporation for each of the years ending December 31, Year 2 through Year 4.

**b.** Prepare a comparative statement of cash flows for Flight Training Corporation for each of the years ending December 31, Year 2 to Year 4. Use the indirect method for presenting cash flows from operations.

**c.** Comment on the relation between net income and cash flow from operations and the pattern of cash flows from operating, investing, and financing activities for each of the three years.

**d.** Describe the likely reasons for Flight Training Corporation's cash flow problems.

**38. Working backward through the statement of cash flows.** Quinta Company presents the balance sheet shown in **Exhibit 4.34** and the statement of cash flows shown in **Exhibit 4.35** (page 203) for Year 5. The firm sold investments, equipment, and land for cash at their net book value. The accumulated depreciation of the equipment sold was $20,000.

Prepare a balance sheet for the beginning of the year, January 1, Year 5.

**39. Interpreting the statement of cash flows. Exhibit 4.36** (page 204) presents a statement of cash flows for Nike, Inc., maker of athletic shoes, for three recent years.

**a.** Why did Nike experience increasing net income but decreasing cash flow from operations during this three-year period?

**b.** What is the likely explanation for the changes in Nike's cash flow from investing during the three-year period?

**c.** How did Nike finance its investing activities during the three-year period?

**d.** Evaluate the appropriateness of Nike's use of short-term borrowing during Year 9.

**EXHIBIT 4.34**

**QUINTA COMPANY**
**All Balance Sheet Accounts**
**December 31, Year 5**
**(Problem 38)**

| | |
|---|---|
| ASSETS | |
| Cash | $ 25,000 |
| Accounts Receivable | 220,000 |
| Merchandise Inventories | 320,000 |
| Land | 40,000 |
| Buildings and Equipment (at cost) | 500,000 |
| Less Accumulated Depreciation | (200,000) |
| Investments (noncurrent) | 100,000 |
| Total Assets | $1,005,000 |
| LIABILITIES AND SHAREHOLDERS' EQUITY | |
| Accounts Payable | $ 280,000 |
| Other Current Liabilities | 85,000 |
| Bonds Payable | 100,000 |
| Common Stock | 200,000 |
| Retained Earnings | 340,000 |
| Total Liabilities and Shareholders' Equity | $1,005,000 |

**EXHIBIT 4.35**

**QUINTA COMPANY**
**Statement of Cash Flows**
**Year 5**
**(Problem 38)**

| | | |
|---|---|---|
| **Operations** | | |
| Net Income | $ 200,000 | |
| Additions: | | |
| Depreciation Expense | 60,000 | |
| Increase in Accounts Payable | 25,000 | |
| Subtractions: | | |
| Increase in Accounts Receivable | (30,000) | |
| Increase in Merchandise Inventories | (40,000) | |
| Decrease in Other Current Liabilities | (45,000) | |
| Cash Flow from Operations | | $ 170,000 |
| **Investing** | | |
| Sale of Investments | $ 40,000 | |
| Sale of Buildings and Equipment | 15,000 | |
| Sale of Land | 10,000 | |
| Acquisition of Buildings and Equipment | (130,000) | |
| Cash Flow from Investing | | (65,000) |
| **Financing** | | |
| Common Stock Issued | $ 60,000 | |
| Bonds Issued | 40,000 | |
| Dividends Paid | (200,000) | |
| Cash Flow from Financing | | (100,000) |
| Net Change in Cash | | $ 5,000 |

**40. Interpreting the statement of cash flows. Exhibit 4.37** (page 205) presents a statement of cash flows for Boise Cascade Corporation, a forest products company, for three recent years.
- **a.** Boise Cascade operated at a net loss each year but generated positive cash flow from operations. Explain.
- **b.** What is the likely explanation for the changes in Boise Cascade's cash flow from investing activities during the three-year period?
- **c.** What is the likely explanation for the changes in long-term financing during Year 5 and Year 6?

**41. Interpreting the statement of cash flow relations. Exhibit 4.38** (page 206) presents statements of cash flow for eight companies for a recent year:
- **a.** American Airlines (airline transportation)
- **b.** American Home Products (pharmaceuticals)
- **c.** Interpublic Group (advertising and other marketing services)
- **d.** Procter & Gamble (consumer products)
- **e.** Reebok (athletic shoes)
- **f.** Texas Instruments (electronics)
- **g.** Limited Brands (specialty retailing)
- **h.** Upjohn (pharmaceuticals)

Discuss the relation between net income and cash flow from operations, and the pattern of cash flows from operating, investing, and financing activities for each firm.

**42. Northrop Grumman Corporation; interpreting direct and indirect methods.**
- **a.** Refer to **Exhibit 4.39** (page 207) for Northrop Grumman, which shows excerpts from its Statements of Cash Flows, with cash flow from operations presented with the indirect method, for three recent years. Write a short (no more than 50 words) explanation of why

EXHIBIT 4.36

**NIKE, INC.**
**Statement of Cash Flows**
**(all dollar amounts in millions)**
**(Problem 39)**

| | Year 7 | Year 8 | Year 9 |
|---|---|---|---|
| **Operations** | | | |
| Net Income | $ 167 | $ 243 | $ 287 |
| Depreciation and Amortization | 15 | 17 | 34 |
| Other Addbacks and Subtractions | (5) | 5 | 3 |
| Working Capital Provided by Operations | $ 177 | $ 265 | $ 324 |
| (Increase) Decrease in Accounts Receivable | (38) | (105) | (120) |
| (Increase) Decrease in Inventories | (25) | (86) | (275) |
| (Increase) Decrease in Other Operating Current Assets | (2) | (5) | (6) |
| Increase (Decrease) in Accounts Payable | 21 | 36 | 59 |
| Increase (Decrease) in Other Current Operating Liabilities | 36 | 22 | 32 |
| Cash Flow from Operations | $ 169 | $ 127 | $ 14 |
| **Investing** | | | |
| Sale of Property, Plant, and Equipment | $ 3 | $ 1 | $ 2 |
| Acquisition of Property, Plant, and Equipment | (42) | (87) | (165) |
| Acquisition of Investment | (1) | (3) | (48) |
| Cash Flow from Investing | $ (40) | $ (89) | $(211) |
| **Financing** | | | |
| Increase in Short-Term Debt | — | — | $ 269 |
| Increase in Long-Term Debt | — | $ 1 | 5 |
| Issue of Common Stock | $ 3 | 2 | 3 |
| Decrease in Short-Term Debt | (96) | (8) | — |
| Decrease in Long-Term Debt | (4) | (2) | (10) |
| Dividends | (22) | (26) | (41) |
| Cash Flow from Financing | $(119) | $ (33) | $ 226 |
| Change in Cash | $ 10 | $ 5 | $ 29 |
| Cash, Beginning of Year | 74 | 84 | 89 |
| Cash, End of Year | $ 84 | $ 89 | $ 118 |

Northrop Grumman's cash flow from operations declined by about 20 percent per year between Years 7 and 8 and then again between Years 8 and 9. If you cannot explain, then explain why that might be.

**b.** Refer to **Exhibit 4.40** (page 208) for Northrop Grumman, which shows excerpts from its Statements of Cash Flows, with cash flow from operations presented with the direct method, for three recent years, the same years as in part **a**. Write a short (no more than 50 words) explanation of why Northrop Grumman's cash flow from operations declined by about 20 percent per year between Years 7 and 8 and then again between Years 8 and 9. If you cannot explain, then explain why that might be.

**c.** Which method of presenting cash flow from operations do you find easier to interpret?

**43. Ethical issues in manipulating cash flows from operations.** Top financial management wants to increase cash flow from operations. It asks you to implement the following strategies. Which of these, if implemented, will increase cash flow from operations contrasted to the amount if you do not implement the strategy for the firm? Comment on the wisdom and ethics of these strategies.

**a.** The firm delays maintaining equipment until after the start of the next period.

**b.** The firm delays purchasing new equipment until after the start of the next period.

**EXHIBIT 4.37**

**BOISE CASCADE CORPORATION**
**Statement of Cash Flows**
**(all dollar amounts in millions)**
**(Problem 40)**

| | Year 5 | Year 6 | Year 7 |
|---|---|---|---|
| **Operations** | | | |
| Net Income (Loss) | $(154) | $ (77) | $ (63) |
| Depreciation | 266 | 268 | 236 |
| Other Addbacks (Subtractions) | (56) | (43) | 41 |
| (Increase) Decrease in Accounts Receivable | (46) | — | (68) |
| (Increase) Decrease in Inventories | (3) | (31) | 6 |
| Increase (Decrease) in Accounts Payable | 9 | 15 | 55 |
| Increase (Decrease) in Other Current Liabilities | 50 | (1) | 9 |
| Cash Flow from Operations | $ 66 | $ 131 | $ 216 |
| **Investing** | | | |
| Sale of Property, Plant, and Equipment | $ 202 | $ 24 | $ 171 |
| Acquisition of Property, Plant, and Equipment | (283) | (222) | (271) |
| Other Investing Transactions | (31) | 9 | (75) |
| Cash Flow from Investing | $(112) | $(189) | $(175) |
| **Financing** | | | |
| Increase (Decrease) in Short-Term Borrowing | $ (54) | $ 27 | $ 25 |
| Increase in Long-Term Debt | 131 | 84 | 139 |
| Increase in Preferred Stock | 191 | 287 | — |
| Decrease in Long-Term Debt | (164) | (269) | (116) |
| Dividends | (55) | (67) | (84) |
| Other Financing Transactions | (5) | (2) | 2 |
| Cash Flow from Financing | $ 44 | $ 60 | $ (34) |
| Change in Cash | $ (2) | $ 2 | $ 7 |
| Cash, Beginning of Year | 22 | 20 | 22 |
| Cash, End of Year | $ 20 | $ 22 | $ 29 |

**c.** The firm sells $1 million of accounts receivable for $980,000 cash to a financial institution, but agrees to reimburse the purchaser for the amount by which uncollectible accounts exceed $20,000.

**d.** The firm delays paying for its employees' insurance premiums until after the start of the next period.

**e.** The firm delays paying some suppliers until after the due date, and until after the start of the next period.

**f.** The firm sells goods for cash but promises the customers that they can return the goods for full refund after the start of the next period.

**EXHIBIT 4.38** **Statements of Cash Flows for Selected Companies (all dollar amounts in millions) (Problem 41)**

| | American Airlines | American Home Products | Interpublic Group | Procter & Gamble | Reebok | Texas Instruments | Limited Brands | Upjohn |
|---|---|---|---|---|---|---|---|---|
| **Operations** | | | | | | | | |
| Net Income (Loss) | $ (110) | $ 1,528 | $ 125 | $ 2,211 | $ 254 | $ 691 | $ 455 | $ 491 |
| Depreciation | 1,223 | 306 | 61 | 1,134 | 37 | 665 | 247 | 175 |
| Other Addbacks (Subtractions) | 166 | 71 | 23 | 196 | (4) | (9) | — | 7 |
| (Increase) Decrease in Receivables | 37 | 14 | (66) | 40 | (65) | (197) | (102) | 6 |
| (Increase) Decrease in Inventories | (27) | (157) | 16 | 25 | (82) | (60) | (74) | (21) |
| Increase (Decrease) in Payables | 34 | 325 | 59 | 98 | 35 | 330 | 118 | 63 |
| Increase (Decrease) in Other Current Liabilities | 54 | (185) | (15) | (55) | (2) | 112 | 110 | (11) |
| Cash Flow from Operations | $ 1,377 | $ 1,902 | $ 203 | $ 3,649 | $ 173 | $ 1,532 | $ 754 | $ 710 |
| **Investing** | | | | | | | | |
| Capital Expenditures (net) | $(2,080) | $ (473) | $ (79) | $(1,841) | $ (62) | $(1,076) | $(430) | $(224) |
| Sale (Acquisition) of Marketable Securities | 290 | 24 | 3 | 23 | — | (47) | — | (287) |
| Sale (Acquisition) of Other Businesses | — | (9,161) | — | (295) | — | — | (60) | 308 |
| Other Investing | 36 | (5) | (85) | 105 | (4) | — | — | (1) |
| Cash Flow from Investing | $(1,754) | $(9,615) | $(161) | $(2,008) | $ (66) | $(1,123) | $(490) | $(204) |
| **Financing** | | | | | | | | |
| Increase (Decrease) in Short-Term Borrowing | $ (380) | $ 8,640 | $ 35 | $ (281) | $ 37 | $ (1) | $(322) | $ 5 |
| Increase in Long-Term Debt | 730 | — | 42 | 414 | — | 1 | 150 | 15 |
| Increase in Capital Stock | 1,081 | 38 | 19 | 36 | 13 | 110 | 17 | — |
| Decrease in Long-Term Debt | (1,069) | — | (15) | (797) | (3) | (88) | — | (46) |
| Acquisition of Treasury Stock | — | (314) | (37) | (14) | (112) | — | — | (32) |
| Dividends | (49) | (903) | (36) | (949) | (25) | (79) | (102) | (264) |
| Other Financing | 82 | 11 | (14) | 1 | (12) | 4 | — | 37 |
| Cash Flow from Financing | $ 395 | $ 7,472 | $ (6) | $(1,590) | $(102) | $ (53) | $(257) | $(285) |
| Change in Cash | $ 18 | $ (241) | $ 36 | $ 51 | $ 5 | $ 356 | $ 7 | $ 221 |
| Cash, Beginning of Year | 45 | 1,937 | 256 | 2,322 | 79 | 404 | 34 | 281 |
| Cash, End of Year | $ 63 | $ 1,696 | $ 292 | $ 2,373 | $ 84 | $ 760 | $ 41 | $ 502 |

**EXHIBIT 4.39**

**NORTHROP GRUMMAN CORPORATION**
**Data Taken from Consolidated Statements of Cash Flows**
**(Shaded Columns Showing Changes Do Not Appear in Original)**
**(all dollar amounts in millions)**
**(Problem 42)**

**Indirect Method**

| Years ended December 31 | Year 7 | Change | Year 8 | Change | Year 9 |
|---|---|---|---|---|---|
| **Cash Provided by Operating Activities** | | | | | |
| Net income | $ 467 | *141* | $ 608 | *(181)* | $ 427 |
| Adjustments to Reconcile Net Income to Net Cash Provided by Operations: | | | | | |
| Depreciation | 193 | *(18)* | 175 | *91* | 266 |
| Amortization of Intangible Assets | 196 | *10* | 206 | *173* | 379 |
| Common Stock Issued to Employees | 2 | *6* | 8 | *38* | 46 |
| Loss on Disposal of Discontinued Operations | — | *56* | 56 | *(56)* | — |
| Loss (Gain) on Disposals of Property, Plant, and Equipment | 21 | *(8)* | 13 | *(20)* | (7) |
| Retiree Benefits Income | (249) | *(243)* | (492) | *223* | (269) |
| Decrease (Increase) in | | | | | |
| Accounts Receivable | 170 | *(849)* | (679) | *1,952* | 1,273 |
| Inventoried Costs | 172 | *(95)* | 77 | *(105)* | (28) |
| Prepaid Expenses and Other Current Assets | 45 | *(73)* | (28) | *45* | 17 |
| Increase (Decrease) in | | | | | |
| Advances from Customers on Long-Term Contracts | 21 | *645* | 666 | *(1,314)* | (648) |
| Accounts Payable and Accruals | (2) | *89* | 87 | *(783)* | (696) |
| Provisions for Contract Losses | (8) | *28* | 20 | *(85)* | (65) |
| Deferred income Taxes | 230 | *115* | 345 | *(171)* | 174 |
| Income Taxes Payable | 58 | *(30)* | 28 | *(41)* | (13) |
| Retiree Benefits | (129) | *37* | (92) | *17* | (75) |
| Other Noncash Transaction | 20 | *(8)* | 12 | *24* | 36 |
| **Net cash provided by operating activities** | $1,207 | *(197)* | $1,010 | *(193)* | $ 817 |

EXHIBIT 4.40

**NORTHROP GRUMMAN CORPORATION**
**Data Taken from Consolidated Statements of Cash Flows**
**(Shaded Columns Showing Changes Do Not Appear in Original)**
**(all dollar amounts in millions)**
**(Problem 42)**

**Direct Method (without Reconciliation of Net Income to Cash Flow from Operations)**

| Years ended December 31 | Year 7 | Change | Year 8 | Change | Year 9 |
|---|---|---|---|---|---|
| **Cash Provided by Operating Activities** | | | | | |
| Sources of Cash | | | | | |
| Cash Received from Customers | | | | | |
| Collections from Customers on Long-Term Contracts .... | $ 1,691 | *(253)* | $ 1,438 | *1,664* | $ 3,102 |
| Other Collections .... | 7,450 | *(447)* | 7,003 | *4,145* | 11,148 |
| Less: Cash Paid to Suppliers and Employees .... | (7,715) | *465* | (7,250) | *(6,001)* | (13,251) |
| Net Cash Margin .... | $ 1,426 | *(235)* | $ 1,191 | *(192)* | $ 999 |
| *Cash Contribution Margin Percentage* .... | *15.6%* | | *14.1%* | | *7.0%* |
| Proceeds from Litigation Settlement .... | — | *—* | — | *220* | 220 |
| Interest Received .... | 18 | *(1)* | 17 | *—* | 17 |
| Income Tax Refunds Received .... | 75 | *(60)* | 15 | *8* | 23 |
| Other Cash Receipts .... | 7 | *3* | 10 | *14* | 24 |
| Cash Provided by Operating Activities .... | $ 1,526 | *(293)* | $ 1,233 | *50* | $ 1,283 |
| Other Operating Uses of Cash | | | | | |
| Interest Paid .... | $ 216 | *(51)* | $ 165 | *168* | $ 333 |
| Income Taxes Paid .... | 85 | *(28)* | 57 | *69* | 126 |
| Other Cash Payments .... | 18 | *(17)* | 1 | *6* | 7 |
| Cash Used in Operating Activities .... | $ 319 | *(96)* | $ 223 | *243* | $ 466 |
| **Net cash provided by operating activities** .... | $ 1,207 | *(197)* | $ 1,010 | *(193)* | $ 817 |

CHAPTER 6

# Receivables and Revenue Recognition

## Learning Objectives

1. Develop an understanding of the quality of earnings as a concept for evaluating generally accepted accounting principles discussed in Chapters 6 to 12.
2. Understand why the allowance method for uncollectible accounts matches bad debts with revenues better than the direct write-off method.
3. Apply the allowance method for uncollectible accounts.
4. Analyze information on accounts receivable to evaluate a firm's management of its credit operation.
5. Develop a sensitivity to issues in recognizing and measuring revenues and expenses for various types of businesses.
6. Cement an understanding of the concept that net income over sufficiently long periods equals cash inflows minus cash outflows other than transactions with owners (a measurement issue), regardless of when firms recognize the revenues and expenses (a timing issue).

Let's pause to review what you've learned thus far. **Chapter 1** provided an overview of the principal activities of business firms and showed their relation to the three principal financial statements: balance sheet, income statement, and statement of cash flows. **Chapters 2** to **4** then discussed the concepts underlying each financial statement, illustrated the accounting procedures to prepare them, and introduced tools you can use to analyze and interpret them. **Chapter 5** developed these analytical tools more fully, focusing on the analysis of profitability and risk. At this stage in your study, you probably have the sense that financial accounting is fairly precise. The illustrations in the chapter and the exercises and problems at the end of the chapter generally asked for specific numeric answers. We have purposefully presented the material thus far with this seemingly high degree of precision to ensure a solid grounding in the basic concepts and principles of financial accounting.

We are about to shift gears. Part Three **(Chapters 6 to 12)** examines **generally accepted accounting principles (GAAP)**. **Chapter 1** described GAAP as the methods that firms use to measure income for a period and financial position at the end of a period. Standard-setting bodies frequently permit more than one method of accounting for a particular transaction. Furthermore, GAAP is not cookbook rules that accountants can easily and unambiguously apply. Rather, application of GAAP requires professional judgment. Different accountants can reach different conclusions on the appropriate application of GAAP. Thus, you will encounter a higher degree of subjectivity in the remaining chapters of this book relative to the seeming precision of the first five chapters. One objective of this part of the book is to help you identify reporting areas requiring judgment and develop skills as a user to assess the appropriateness of the judgments made.

The discussion in Part Three proceeds in approximate balance sheet order:

Chapter 6: receivables and revenue recognition
Chapter 7: inventories and cost of goods sold
Chapter 8: noncurrent assets and amortization
Chapters 9 and 10: liabilities and interest expense
Chapter 11: marketable securities and long-term investments in securities of other companies
Chapter 12: shareholders' equity

## Financial Reporting Environment

**Chapter 1** briefly described various aspects of the financial reporting environment. Let's revisit some of the issues raised there as an introduction to our study of GAAP.

1. **What is GAAP?** GAAP is the methods of accounting that firms use to measure the results of their business transactions, with a principal aim of measuring net income for a period and financial position at the end of a period. GAAP prefers the accrual basis over the cash basis, a principle that provides a foundation for accounting. The related concept of matching expenses with associated revenues guides application of the accrual basis. Often, however, GAAP provides specific guidance for specific issues. For example, **Chapter 7** discusses various inventory cost-flow assumptions for matching expenses with revenues. The cost-flow assumptions discussed there (first-in, first-out, or FIFO; last-in, first-out, or LIFO; weighted average) are also part of GAAP. In sum, GAAP refers to both broad principles and more specific methods for applying those broad principles to particular business transactions.
2. **Who sets GAAP?** Standard-setting bodies within each country currently set GAAP. The International Accounting Standards Board (IASB) has assumed an increasing role in recent years in promoting the establishment of more uniform GAAP worldwide. The Financial Accounting Standards Board (FASB) sets acceptable accounting principles in the United States. Standard-setting bodies refer to their pronouncements as *standards*, which are a part of GAAP.
3. **Are there alternative GAAP for a particular business transaction?** Standard-setting bodies sometimes concur that a particular method of accounting is theoretically superior to alternative methods and require that method as GAAP. For example, the allowance method of accounting for uncollectible accounts, discussed later in this chapter, provides superior matching of losses from uncollectible accounts with revenues relative to the direct charge-off method. Thus, GAAP requires the allowance method for uncollectible accounts in most countries. In other cases, standard-setting bodies recognize that the economic effects of certain transactions may differ between firms and that a single method may not provide the best measures of earnings and financial position. In these cases, they allow more than one method of accounting for the transaction. For example, firms may choose from among FIFO, LIFO, and weighted average cost-flow assumptions in measuring cost of goods sold, depending on which flow assumption managers feel provides the best matching of expenses with revenues.
4. **How detailed are the rules for applying GAAP?** Standard-setting bodies generally provide broad guidelines rather than detailed rules for applying their pronouncements. They recognize that economic differences may exist between firms and that firms should have latitude to apply pronouncements to reflect these economic differences. For example, standard-setting bodies stipulate that firms should write off, or depreciate, the acquisition cost of buildings and equipment over the expected useful lives of these assets. Firms have latitude to select the depreciable life to reflect the expected intensity of use, maintenance policies, and similar factors. Firms also have latitude to select the depreciation method that sets the pattern of write-offs of the acquisition cost over an asset's depreciable life.
5. **Does the flexibility of alternative GAAP and the latitude provided firms in applying GAAP have economic consequences?** Some view the flexibility and latitude to choose and

apply GAAP as allowing firms to report in a manner that best portrays their earnings and financial position. Then, according to this view, investors and lenders who use the financial statements to make decisions will allocate their capital in an economically desirable way. Critics of this view note that the flexibility and latitude provide managers with an opportunity to report as favorable results as possible within the constraints of GAAP. Managers motivated with the dual economic incentives of job security and compensation would rather report higher than lower earnings. If users of the financial statements can decipher and adjust for any games managers might have played with reported amounts, then investors will allocate capital in an economically desirable way. In this scenario, however, users of the financial statements bear the economic cost of adjusting the financial statements. If users of the financial statements cannot see through the financial statements to the games that managers might have played, then investors may misallocate capital. Thus, the flexibility and latitude of GAAP do have **economic consequences** to managers, investors, lenders, and others.

6. **What controls exist to constrain the opportunistic actions by management?** The audit by a firm's independent accountant serves as one control on management. The independent accountant makes judgments about the suitability of the accounting principles a firm selects and the reasonableness of the way the firm applies the accounting principles. For example, does an airline use longer depreciable lives for its airplanes than do its competitors? Does the firm appear to increase (or decrease) its depreciable lives over time in an effort to increase (or decrease) reported earnings? Another control is the oversight provided by government regulators, such as the Securities and Exchange Commission (SEC) in the United States. The SEC reviews the financial statements of firms with publicly traded securities and takes actions against firms that appear to violate GAAP. Security analysts who analyze financial statements and make buy, sell, or hold recommendations on a firm's common stock can also impede opportunistic actions by management.

## Quality of Earnings

Investors typically use earnings to value a firm. Among the questions that investors and other users of the financial statements should ask are:

1. How well do the current period's earnings reflect the economic value added by the firm during the period and the likely amount of future economic value added from the firm's operating activities?
2. Do managers of the firm have much latitude in measuring earnings within GAAP in their particular industry and therefore have the potential for distorting measurements of economic value added?
3. Is there evidence that managers have taken advantage of this latitude and reported unusually favorable or unfavorable earnings?
4. Do earnings contain unusual or nonrecurring items of income that the investors should downplay or ignore when using earnings to value the firm?

Financial periodicals make frequent references to the **quality of earnings**. The writers of these articles appear to use the term to refer to one or some combination of the four concerns suggested by these questions. We use the term in this book in its broadest sense to include

- the representative faithfulness of earnings as a measure of value added,
- the ability managers have to use discretion in measuring and reporting earnings in their particular industry,
- the extent to which the managers have exercised this discretion, and
- the extent to which earnings include unusual or nonrecurring items.

Affirmative answers to the four questions above raise questions for a particular firm about the quality of its earnings and suggest the need for users of the financial statements to take care in using reported earnings in their decisions.

The discretion in measuring earnings can involve any of the following:

- Selecting accounting principles, or standards, when GAAP allows a choice.
- Making estimates in the application of accounting principles.
- Timing transactions to allow recognizing nonrecurring items in earnings.

**Example 1** Retailing and manufacturing firms that carry inventories can use a FIFO, LIFO, and weighted averaged cost flow assumption in measuring cost of goods sold. Depending on the extent that acquisition or manufacturing costs change over time and the rate at which inventories turn over, this choice can significantly impact earnings. As **Chapter 7** discusses, FIFO (first-in, first-out) matches older acquisition or manufacturing costs against revenues, whereas LIFO (last-in, first-out) matches more recent such costs against revenues. Analysts generally view earnings measured using LIFO as providing a better measure of economic value added than FIFO because of this more current matching. **Chapter 7** points out, however, that firms that sell more inventory than they purchase during a particular period calculate cost of goods sold under LIFO in part using older purchase prices of earlier years, distorting the measure of economic value added. Thus, assessments of the quality of earnings sometimes require tradeoffs between these four dimensions of earnings quality.

**Example 2** Fixed asset-intensive firms, such as airlines, railroads, electric utilities, and steel manufacturers, find that depreciation expense represents a large expense on their income statements. Such firms must estimate the expected useful lives and residual values of their buildings and equipment, and allocate a portion of the cost minus salvage value of these assets to each period of their use. Because many years will pass until the firm knows for certain what that life will be, some managers choose long depreciable lives that portray current management in a favorable light. The SEC, for example, accused the management of Waste Management of manipulating the lives of its garbage trucks during the 1990s to increase reported earnings. The more earnings measurement requires estimates, the greater the opportunity for management to distort earnings and the lower the quality of earnings.

**Example 3** An unexpected recession in East Asia resulted in decreased earnings for a manufacturer of semiconductors. To increase earnings to the level initially anticipated for the year, the firm sold a parcel of land that it did not currently need in operations. The land had a market value larger than its book value, so selling it generated a realized gain to be reported in net income for the year. This component of net income is of low quality because the firm will not likely report similar gains on sales of land on a recurring basis.

Throughout the chapters composing Part Three you will encounter managers' choices—of accounting principles, of estimates, and of the timing of asset sales—that will affect the year-to-year reported income. To the extent managers make choices enhancing currently reported income, analysts will say the firm has a lower earnings quality. The analyst can then either adjust the reported earnings to remove any perceived bias or compensate for the bias in some other way (for example, by applying a lower multiple to earnings in arriving at an acceptable stock price). We shall say time and again, however, that over long enough time spans, reported income will equal the sum of cash inflows less the sum of cash outflows, other than cash flows resulting from transactions with owners. All the management choices that affect the quality of earnings affect *when* the firm reports its income, not its *total amount*, summed over time.

## Quality of Earnings and Ethical Issues

**Chapter 1** suggested some ways to think about ethical issues that confront management when they make financial reporting decisions. Among the questions that one might raise are:

1. Does the action violate a known law or regulation?
2. Has the firm provided sufficient disclosure about the action for the users of financial statements to make their own judgments about the ethics of such actions?

Each chapter in this book contains one or more questions or problems that require consideration of ethical issues. The remaining chapters of this book discuss GAAP for various business transactions. Some accounting choices that managers make require judgments as to whether the accounting falls within GAAP. Even when GAAP allows the accounting choice, the question arises as to the adequacy of disclosures.

One question of particular importance to **Chapters 6** to **12** is: Do ethical issues arise when a firm makes accounting choices within GAAP and adequately discloses information about the choice but which, however, decreases the quality of earnings? The measurement and reporting of the gain on the sale of the land by the semiconductor firm in **Example 3** conform with GAAP and, we presume, the firm adequately disclosed it. Still, the nonrecurring nature of the gain lowers the quality of earnings.

Some people would argue that an ethical issue does not arise. The reporting conforms with GAAP. The firm provided sufficient disclosures for a user to decide whether to include the gain in net income when evaluating the current period's performance and projecting future earnings. Other people would argue that an ethical issue arises because management took an action for the primary purpose of increasing earnings. Presumably, if the firm had met its earnings target without needing to include the gain, it would not have sold the land. Management purposefully managed, some would say distorted, earnings to make it and the firm look better. As with other ethical issues raised throughout this book, we do not attempt to impose our own value systems. We encourage you to identify and think carefully about possible ethical issues relating to the accounting topics discussed in this book.

## Review of Income Recognition Principles

Let's review the principles of income recognition under the accrual basis of accounting from **Chapter 3**. We use the term *income* here to mean net income or earnings. **Figure 6.1** shows the flow of operating activities for a typical manufacturing firm. A manufacturing firm

**(1)** acquires production facilities to permit it to engage in manufacturing and acquires raw materials for use in producing products;
**(2)** uses labor and other services to transform the raw materials into a salable product;
**(3)** identifies a customer, agrees to a selling price, and delivers the product to customers;
**(4)** awaits collection of cash from customers, in the meantime holding an account receivable;
**(5)** collects cash from customers; and
**(6)** refunds cash for products returned by customers, or spends cash and other assets to fulfill promises under warranty.

**Amount of Income** The *amount* of income from these operating activities is the difference between the cash received at point **(5)** and the cash paid to suppliers of goods and services at times **(1)**, **(2)**, **(3)**, and **(4)**, as well as the cash or other assets spent during **(6)**. Income measures the economic value added from operating activities. The excess of cash inflows over cash outflows measure this economic value added.

**Measurement of Income for Discrete Accounting Periods** Firms measure and report their income using quarterly and annual time periods. Because the cash inflow from customers—point **(5)**—can occur in an accounting period different from the ones during which the firm pays cash out to various suppliers—times **(1)** through **(4)** and **(6)**—accrual accounting must allocate the income to one or more accounting periods.

**Matching Expenses with Revenues** The guideline for allocating expenses to accounting periods is to match expenses as closely as possible with revenues. In this way, income matches

FIGURE 6.1 Operating Process for a Manufacturing Firm

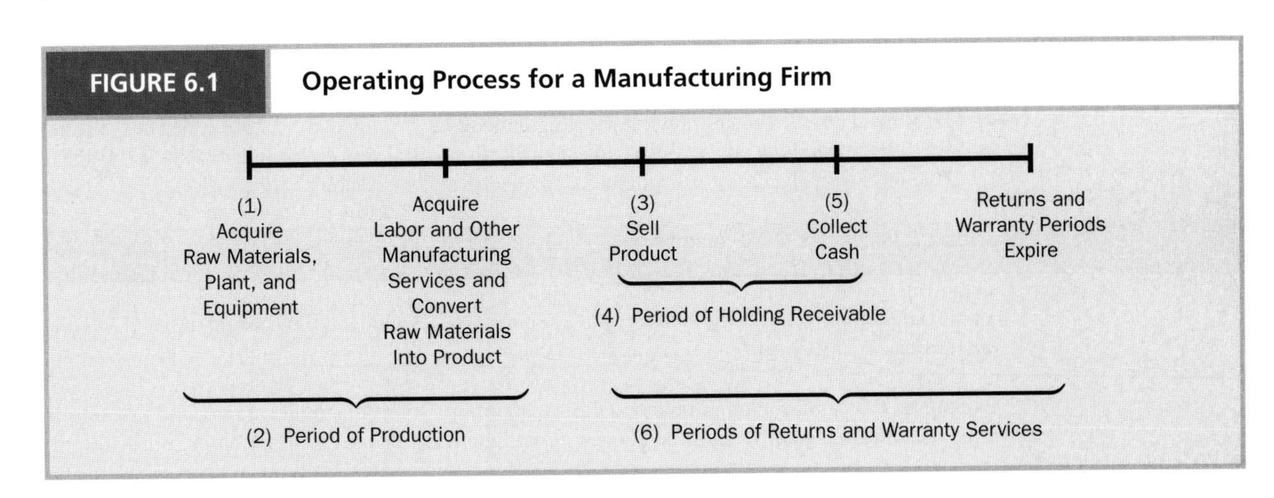

inputs with outputs, efforts with accomplishments. Most firms recognize revenue at the time of sale—time **(3)**. To properly measure income for the period of the sale, the firm must match against revenues the costs of producing the good or service incurred during times **(1)** and **(2)**, the cost of selling the product from time **(3)**, the cost of uncollectible accounts from time **(4)**, and the cost of returns and warranties from time **(6)**. A later section of this chapter discusses income recognition for a long-term construction contractor. We see there that construction contractors often recognize revenue during the period of construction—time **(2)**. These firms must then match appropriate portions of the total costs of construction as expenses against the revenue they recognize each period. We also see later that firms must sometimes wait until they collect cash from customers before recognizing revenue—at time **(5)**. These firms must likewise recognize all of the costs of creating and selling their good or service as expenses at time **(5)** to achieve matching.

**Timing of Revenue Recognition** The remaining question then is when to recognize revenue. Firms could conceivably recognize revenue at any one or more than one of the six times in **Figure 6.1**. To answer the revenue recognition timing question, GAAP sets two hurdles. As soon as a firm clears the following two hurdles for a particular operating activity, GAAP permits it to recognize revenue for that activity:

1. A firm has performed all, or a substantial portion of, the services it expects to provide or, in the case of product warranties, can forecast with reasonable precision the cost of providing the future services.
2. The firm has received cash, a receivable, or some other asset capable of reasonably precise measurement or, if the firm has offered to let the customer return the product for a refund, the firm can estimate the returns with reasonable precision.

Satisfying the first criterion for revenue recognition means that a firm can estimate the total expected cash outflows related to an operating activity. If a firm cannot estimate the total expected cash outflows, it will not know the amount of expenses to match against revenue, and therefore it will not know the amount of income. Satisfying the second criterion for revenue recognition means that a firm can estimate the amount of expected cash inflows from customers. If a firm cannot estimate expected cash inflows, it will not know the amount of revenues and therefore the amount of income. Stating these criteria more succinctly, revenue recognition requires that a firm have moved far enough down the time line in **Figure 6.1** so that it can measure all remaining uncertainty regarding cash inflows and cash outflows with reasonable precision.

The SEC in *Staff Accounting Bulletin No. 101* amplifies the criteria for revenue recognition as follows:

1. Persuasive evidence of an arrangement exists.
2. Delivery has occurred or services have been performed.
3. The seller's price to the buyer is fixed or determinable.
4. Collectibility is reasonably assured.

The second, third, and fourth criteria are similar to GAAP's two criteria stated above. The new dimension is the existence of an arrangement. The arrangement may take the form of a contract, prior business dealings with a particular customer, or customary business practices by a firm and its industry. The arrangement sets forth the responsibilities of the firm and its customers with respect to the nature and delivery of goods or services, the risks assumed by buyer and seller, the timing of cash payments, and similar factors. Having an arrangement in place permits a more informed judgment as to the second, third, and fourth criteria.

## Application of Income Recognition Principles

The examples below illustrate the application of the revenue recognition criteria to several businesses.

**Example 4** A magazine publisher sells subscriptions to its customers for periods of one to three years, receiving cash from customers at the beginning of the subscription period. Although the magazine publisher can measure the amount of revenue with little uncertainty, it must provide substantial future services to create, print, and deliver the magazine. Thus, it should delay the recognition of revenue and recognize proportionate parts of the subscription fee as it publishes each issue during the subscription period.

**Example 5** A winery harvests grapes each year and processes them into various wines. It then bottles the wine and places it in warehouses to age for several years. Although substantial performance has occurred by the time the aging process begins, uncertainties exist as to the amount of cash the firm will collect from customers when it sells the wine. The amount of cash received will depend on the quality of the aged wine, the quantity and quality of wines offered by competitors, and similar factors. The wine producer should therefore delay recognition of revenue not only until aging finishes, but also until it identifies a customer, agrees with the customer on a selling price, and delivers the wine.

**Example 6** Refer to **Example 5**. Suppose that the wine producer contracts to sell the wine to a customer at the beginning of the aging process. The wine producer agrees to age the wine in its warehouses and oversee the aging process. The selling price for the wine includes both the cost of initially producing the wine and the services involved in the aging process and subsequent bottling. The customer pays 30 percent of the selling price at the beginning of the aging process and the remainder when it takes delivery after aging. The wine producer should delay recognition of revenue until it completes the aging process and delivers the wine. The aging process is an important part of the services to be performed and the full cost of aging the wine remains uncertain. The customer might not take delivery and pay the remaining 70 percent of the selling price if the quality of the wine is poor.

**Example 7** A computer software company develops banking software and sells it to commercial banks. The company promises to do the necessary programming to adapt the software to a particular bank's needs and to train the bank's employees in its use. If the bank is not satisfied with the software, it can return it within one year and receive a full refund. The software company intends to continue developing the software and promises to provide customers with updates on an if-and-when-developed basis for a period of three years. Recognizing revenue at the time of initial delivery of the software appears premature. Installation and training comprise an important part of the services provided and the firm cannot estimate their costs with reasonable precision. Substantial uncertainty also likely exists regarding returns from customers. Assuming unpredictable returns, the earliest that the computer software company should recognize revenues is the end of the one-year return period. Another issue is whether customers normally expect and receive software updates. If so, then the software company should separate the revenue from software sales from those of after-market services.

A later section of this chapter discusses the application of the revenue recognition criteria to other types of businesses.

## Problem 6.1 for Self-Study

**Revenue recognition.** Indicate when each of the following firms should recognize revenue and briefly explain your reasoning.

**a.** Kodak enters into an agreement with Wal-Mart Stores to place its cameras in Wal-Mart Stores on a consignment basis. Wal-Mart, as consignee, promises to make its best efforts to sell the cameras. Kodak maintains title to the cameras until Wal-Mart sells them. When Wal-Mart sells a camera, it remits the selling price minus a fee to Kodak. Kodak takes back any damaged or defective cameras. When should Kodak recognize revenue?

**b.** Burlington Mills manufactures custom order athletic clothing for Nike. Near the end of the current year, Burlington Mills completed an order for athletic clothing. The order specified the goods Burlington Mills would deliver and the price Nike would pay. Nike could not take delivery at that time because it had sufficient inventory already in its distribution channel and did not have space to store the new items. Nike asked Burlington Mills to store the items in

*(continued)*

an independent warehouse for three months. Nike agreed to pay the storage costs. When should Burlington Mills recognize revenue from this transaction?

**c.** Nordstrom, a department store retailer, offers layaway plans to its customers. Under the plan, when the customer decides to buy the goods, Nordstrom sets them aside in its storeroom and collects a cash deposit from the customer. When the customer pays the remainder of the selling price, which they may do in several installments, Nordstrom releases the item. Nordstrom bears the risk of loss or damage to the stored items. When should Nordstrom recognize revenue?

**d.** Costco operates discount warehouses for sales of food and household products to customers. Purchasing from Costco requires payment of an annual membership fee, payable at the beginning of the year. Customers who cancel their memberships before year-end may receive a payment equal to a prorated portion of the membership fee. Costco can predict based on experience the likely amounts of refunds. When should Costco recognize revenue from membership fees?

**e.** Mall Development Corporation builds and operates shopping malls throughout the United States. It leases retail space in its shopping malls for an annual fixed rental, payable in equal monthly installments at the beginning of each month, plus an additional contingent rental, payable at the end of the year, equal to one percent of sales exceeding a stated amount. The stated amount varies by store depending on square footage leased. When should Mall Development Corporation recognize revenue from rentals?

## Revenue Recognition at Time of Sale

**Chapter 3** points out that most firms satisfy the criteria for revenue recognition at the time of sale or delivery of goods and services: point **(3)** in **Exhibit 6.1**. Firms have manufactured a product or created a service, identified a customer, delivered the product or service to the customer, and either received cash immediately or assessed the credit standing of customers and concluded that customers will likely pay the amount owed. The firm can estimate the amount of product returns, the cost of warranty repairs, and other after-sale costs.

When a firm satisfies the criteria for revenue recognition at the time of sale, it must estimate the effect of events that occur after the time of sale—that is, during times **(4)**, **(5)**, and **(6)** in **Figure 6.1**. Suppose, for example, that some customers do not pay the amounts owed to the firm. Or, suppose that some customers return unsatisfactory products. Recognizing revenue at the time of sale and properly matching expenses with revenues requires firms to estimate the cost of uncollectible accounts, returns, and similar items and recognize them as income reductions at the time of sale. This section discusses the accounting for such items.

### Uncollectible Accounts

Whenever a firm extends credit to customers, it will almost certainly never collect from some of its customers. Most businesses should prefer to have some customers who do not pay their bills; for most firms, the optimal amount of uncollectibles is not zero. If a firm is to have no uncollectible accounts, it must screen credit customers carefully, which is costly. Furthermore, the firm would deny credit to many customers who would pay their bills even though they could not pass a credit check designed to weed out all potential uncollectibles. When the firm denies credit to potential customers, some will take their business elsewhere, and the firm will lose sales. As long as the amount collected from credit sales to a given group of customers exceeds the cost of goods sold and the other costs of serving that group of customers, the firm will be better off selling to that group than losing the sales. The rational firm should prefer granting credit to a group of customers who have a high probability of paying their bills rather than losing their business, even though some of them will not pay and the firm will, as a result, have some uncollectible accounts.

For example, assume that gross margin—selling price less cost of goods sold—on credit sales to a new group of customers, such as college sophomores, is 40 percent of credit sales. A firm could then afford uncollectible accounts of up to 40 percent of the credit sales to the new customers and still show increased net income, as long as all other costs of serving customers remain constant.

This does not suggest, of course, that a firm should grant credit indiscriminately or ignore collection efforts aimed at those who have not paid their bills. A cost-benefit analysis of credit policy should dictate a strategy that results in uncollectible accounts of an amount that is reasonably predictable before the firm makes any sales.

The principal accounting issue related to uncollected accounts concerns when firms should recognize the loss from uncollectibles. The following example illustrates the nature of the issue.

**Example 8** Howarth's Home Center provides customer financing for sales of its appliances, furniture, and electronic equipment. Customers typically make monthly cash payments of a portion of the purchase price over 24 to 36 months. Despite efforts to collect amounts due and to repossess merchandise when customers fail to pay, Howarth's Home Center generally experiences losses from uncollectible accounts of 2 percent of total sales in the year of sale, 6 percent in the next year, and 4 percent in the third year. Thus the firm expects that it will ultimately not collect 12 (= 0.02 + 0.06 + 0.04) percent of sales made in any particular year. A summary of its recent experience appears below:

| | | Accounts Deemed Uncollectible in Year | | | | | |
|---|---|---|---|---|---|---|---|
| Year | Sales during Year | 2 | 3 | 4 | 5 | 6 | Total Uncollectibles |
| 2 | $ 800,000 | $16,000 | $48,000 | $ 32,000 | — | — | $ 96,000 |
| 3 | 1,200,000 | — | 24,000 | 72,000 | $ 48,000 | — | 144,000 |
| 4 | 1,500,000 | — | — | 30,000 | 90,000 | $60,000 | 180,000 |
| | | $16,000 | $72,000 | $134,000 | $138,000 | $60,000 | |

The accounting issue is,

- should the firm use the direct write-off method and recognize the *actual losses* from uncollectibles in the year it deems particular accounts to be uncollectible ($16,000 in Year 2, $72,000 in Year 3, $134,000 in Year 4, $138,000 in Year 5, and $60,000 in Year 6), or
- should the firm use the allowance method and recognize the total *expected losses* related to a particular year's sales in the year of the sale ($96,000 in Year 2, $144,000 in Year 3, and $180,000 in Year 4)?

**Direct Write-Off Method** The **direct write-off method** recognizes losses from uncollectible accounts in the period when a firm decides that specific customers' accounts are uncollectible. For example, Howarth's Home Center decides in Year 2 that specific customers' accounts totaling $16,000 have become uncollectible. It makes the following entry using the direct write-off method:

| | | |
|---|---|---|
| Bad Debt Expense | 16,000 | |
| Accounts Receivable | | 16,000 |

| Assets | = | Liabilities | + | Shareholders' Equity | (Class.) |
|---|---|---|---|---|---|
| −16,000 | | | | −16,000 | IncSt → RE |

To record losses from known uncollectible accounts.

The direct write-off method has three important shortcomings. First, it does not usually recognize the loss from uncollectible accounts in the period in which the sale occurs and the firm recognizes revenue. For example, Howarth recognizes bad debt expense of only $16,000 in Year 2 under the direct write-off method, even though it ultimately expects not to collect $96,000 of receivables from Year 2 sales. The remaining $80,000 (= $48,000 + $32,000) appears in bad debt expense in Year 3 and Year 4, clumsily matched against those periods' sales. Based on its own expectations of collectibility of accounts, the firm recognizes too little expense in the year of sale and too much in the period of write-off.

Second, the direct write-off method provides firms with an opportunity to manage earnings each period by deciding when particular customers' accounts become uncollectible. Establishing the uncollectible status of customers' accounts requires judgment of the customers' willingness and ability to pay and the intensity of the firm's collection efforts. Firms desiring to increase (decrease) earnings for a particular period can delay (accelerate) the write-off of specific customers' accounts.

Third, the amount of accounts receivable on the balance sheet under the direct write-off method does not reflect the amount a firm expects to collect in cash. Returning to **Example 8**, assume that Howarth's Home Center collects $200,000 of the $800,000 from sales on account during Year 2. Accounts receivable at the end of Year 2 under the direct write-off method are as follows:

| | |
|---|---|
| Total Sales on Account during Year 2 | $800,000 |
| Collections from Customers during Year 2 | (200,000) |
| Specific Customers' Accounts Written Off as Uncollectible | (16,000) |
| Accounts Receivable at End of Year 2 | $584,000 |

Howarth's Home Center expects to collect only $504,000 (= $584,000 − $48,000 − $32,000) of its accounts receivable outstanding at the end of Year 2. The $584,000 amount for accounts receivable on the balance sheet overstates the amount the firm expects to collect in cash.

Thus, the direct write-off method provides a poor matching of costs from uncollectible accounts with revenues and reports accounts receivable on the balance sheet at an amount exceeding the cash the firm expects to collect. (Double-entry recordkeeping causes these simultaneous income statement and balance sheet effects.) Consequently, GAAP does not allow firms to use the direct write-off method for financial reporting when they have significant amounts of expected uncollectible accounts and when the selling firm, such as a retail store, can reasonably predict them. Such firms must use the allowance method, explained below.

Firms must, however, use the direct write-off method for income tax reporting. Income tax laws permit firms to claim a deduction for bad debts only when a firm can demonstrate that particular customers will not pay. This is the first instance we have seen in this book of a firm's use of different accounting principles for financial reporting and for income tax reporting. This particular instance is mandatory—GAAP requires one treatment and the tax law another. **Chapter 10** discusses the accounting issues raised by such differences, called *temporary differences*.

**Allowance Method** When a firm can estimate with reasonable precision the amounts of uncollectibles, a necessary condition for recognizing revenue at the time of sale, GAAP requires an alternative procedure, the **allowance method**, for uncollectibles. The allowance method involves estimating the amount of uncollectible accounts that will occur over time in connection with the sales of each period. The firm recognizes this amount as an expense in the period of the sale, thereby matching expenses with associated revenues. The offsetting credit reduces accounts receivable to the amount of cash the firm expects to collect from customers.

The following discussion illustrates the allowance method. Refer to **Example 8** for Howarth's Home Center, which estimates, based on past experience, that it will not collect $96,000 of the $800,000 sales on accounts during Year 2. At the end of Year 2, the firm makes the following adjusting entry.

| | | |
|---|---|---|
| End of Year 2 | | |
| Bad Debt Expense | 96,000 | |
| Allowance for Uncollectible Accounts | | 96,000 |

| Assets | = | Liabilities | + | Shareholders' Equity | (Class.) |
|---|---|---|---|---|---|
| −96,000 | | | | −96,000 | IncSt → RE |

To provide for estimated uncollectible accounts relating to Year 2 sales.

Recognizing bad debt expense of $96,000 results in matching against Year 2 sales the amount the firm does not expect to collect in cash from Year 2 sales. The credit in this entry is not to Accounts Receivable because the firm is not writing off specific customers' accounts at this time. Instead, the credit account is Allowance for Uncollectible Accounts, a contra account to Accounts Receivable. Accounts Receivable is a control account—summing the amounts in individual customer's accounts. At the time the firm reduces the gross amount of receivables to its expected net amount, the firm still does not know which specific customers will not pay, so it cannot write off any particular customer's account. Thus, it cannot credit any of the individual accounts whose sums appear in Accounts Receivable. It must credit a separate contra account, which shows the subtraction from gross receivables to derive expected net receivables. A more descriptive title for this account might be Allowance for Uncollectible Accounts—We Know Not Whom.

## *Conceptual* Note

Views differ as to the type of account debited in the entry above. Some debit an expense account and others debit a revenue contra account when providing for estimated uncollectibles. Accounting for uncollectibles will reduce net income no matter which of the accounts the firm chooses to debit. Most firms debit Bad Debt Expense and include its amount among total expenses on the income statement. They reason that generating revenues implies a certain amount of bad debts. On the other hand, using a revenue contra account permits net sales to appear at the amount of cash the firm expects to collect. When a firm debits an expense account and includes it among total expenses on the income statement, net sales will overstate the amount of cash the firm expects to receive. Advocates of using a revenue contra account point out that uncollectible accounts cannot theoretically be an expense. An expense is a gone asset. Accounts that the firm did not expect to collect at the time of recording were never assets to begin with. Although we find the arguments for using a revenue contra account persuasive, we debit Bad Debt Expense in this book because most firms debit an expense, not a revenue contra, account.

Whenever a firm identifies particular customers' accounts that it judges to be uncollectible, it writes off the accounts by debiting Allowance for Uncollectible Accounts and crediting Accounts Receivable. For example, Howarth's makes the following entry at the end of Year 2 to write off the accounts of specific customers, the newly identified non-payers, totaling $16,000.

| | | |
|---|---|---|
| End of Year 2 | | |
| Allowance for Uncollectible Accounts . . . . . . . . . . . . . . . . . . . . . . . . . | 16,000 | |
| Accounts Receivable . . . . . . . . . . . . . . . . . . . . . . . . . . . . . . . | | 16,000 |

| Assets | = | Liabilities | + | Shareholders' Equity | (Class.) |
|---|---|---|---|---|---|
| +16,000 | | | | | |
| −16,000 | | | | | |

To write off specific customers' accounts.

If during Year 3 it identifies that specific customers' accounts of $72,000 are uncollectible, it makes the following entry:

| | | |
|---|---|---|
| End of Year 3 | | |
| Allowance for Uncollectible Accounts . . . . . . . . . . . . . . . . . . . . . . . . . | 72,000 | |
| Accounts Receivable . . . . . . . . . . . . . . . . . . . . . . . . . . . . . . . | | 72,000 |

| Assets | = | Liabilities | + | Shareholders' Equity | (Class.) |
|---|---|---|---|---|---|
| +72,000 | | | | | |
| −72,000 | | | | | |

Note that the write-off of specific customers' accounts using the allowance method does not affect income. The income effect occurs in the year of sale, when the firm provides for estimated uncollectible accounts. Note also that the write-off of specific customers' accounts has no effect on accounts receivable (net) of the allowance for uncollectible accounts. Accounts Receivable (gross) decreases but so does the amount in the contra account for estimated uncollectibles, which the accountant subtracts in reaching the net amount for receivables on the balance sheet.

The allowance method for uncollectible accounts overcomes the three shortcomings, discussed earlier, of the direct write-off method. First, during each period, the allowance method matches against sales revenue the amount the firm does not ultimately expect to collect in cash arising from that period's sales. The allowance method provides a better matching of revenues and expenses not only in the period of sale but also in subsequent periods. Losses from uncollectible amounts of sales from Year 2 of $48,000 in Year 3 and $32,000 in Year 4 are expenses of Year 2, not Year 3 or Year 4 as with the direct write-off method.

Second, the allowance method reduces a firm's opportunity to manage earnings through the timing of write-offs. The write-off of specific customers' accounts does not affect earnings or total assets.

Third, the allowance method results in reporting accounts receivable (net) of the allowance for uncollectible accounts on the balance sheet at the amount the firm expects to collect in cash in future periods. For example, Howarth's Home Center reports accounts receivable on the balance sheet at the end of Year 2 as follows:

| | |
|---|---|
| Accounts Receivable—Gross ($800,000 − $200,000 − $16,000) | $584,000 |
| Less Allowance for Uncollectible Accounts ($96,000 − $16,000) | (80,000) |
| Accounts Receivable—Net | $504,000 |

With respect to measuring economic value added each period, the allowance method, because of its better matching of bad debt expense and revenues, provides higher quality measures of earnings than the direct write-off method. The allowance method, however, requires firms to estimate the amount of uncollectibles before the time when it identifies the actual uncollectible accounts, which provides management with opportunities to manage earnings. Firms desiring to increase (decrease) earnings can underestimate (overestimate) bad debt expense. Such misestimates may not become evident for several accounting periods. Even then, management might attribute the misestimate to unexpected changes in economic conditions rather than earnings management. Thus, quality-of-earnings issues arise under both the direct write-off method and the allowance method.

**Procedural Note on the Allowance Method** Firms sometimes write off specific customers' accounts during an accounting period as the firm identifies the specific customers whose accounts have become uncollectible. Such firms generally wait until the end of the accounting period to recognize bad debt expense for the period. Thus, before making the provision for bad debts for the period, the firm may have a debit balance in the Allowance for Uncollectible Accounts account. This account always has a credit balance after recognizing the provision for estimated uncollectibles for the period. Because the firm will not prepare a balance sheet until after it makes its provision for the period, the debit balance in the Allowance account will never appear on the balance sheet. Still, the student may encounter a debit balance in the Allowance account before recognizing estimated uncollectibles for the period and should not let this debit balance cause confusion.

**Estimating the Amount of Uncollectible Accounts** Accountants use two procedures to calculate the amount of the adjustment for uncollectible accounts under the allowance method: the **percentage-of-sales procedure** and the **aging-of-accounts-receivable procedure**. The first requires fewer computations than the second, but the second provides a useful check on the first. Over time the two methods, correctly used, will give the same cumulative income and asset totals.

**Percentage-of-Sales Procedure** The percentage-of-sales procedure involves the following steps:

1. Estimate the amount of uncollectible accounts that will likely occur over time in connection with sales of each period.
2. Make an entry debiting Bad Debt Expense and crediting Allowance for Uncollectible Accounts for the amount in **step 1**.

Uncollectible account amounts will likely vary directly with the volume of credit business. The firm estimates the appropriate percentage by studying its own experience or by inquiring into the experience of similar firms. These rates generally fall within the range of ¼ percent to 2 percent of credit sales.

For example, sales on account at Howarth's Home Center during Year 2 totaled $800,000. Experience indicates that 12 percent of sales become uncollectible. The relatively high percentage reflects the age and financial condition of most of the firm's customers. The entry is as follows:

| | | |
|---|---|---|
| Bad Debt Expense | 96,000 | |
| Allowance for Uncollectible Accounts | | 96,000 |

| Assets | = | Liabilities | + | Shareholders' Equity | (Class.) |
|---|---|---|---|---|---|
| −96,000 | | | | −96,000 | IncSt → RE |

To provide for estimate of uncollectibles computed as a percentage of sales.

If cash sales occur in a relatively constant proportion to credit sales, the accountant can apply the estimated uncollectibles percentage, proportionately reduced, to the total sales for the period.

**Aging-of-Accounts-Receivable Procedure** The aging-of-accounts-receivable procedure involves the following steps:

1. Estimate the amount of outstanding accounts receivable that the firm does not expect to collect.
2. Adjust the balance in the Allowance for Uncollectible Accounts account so that, after the entry to recognize estimated uncollectibles, the balance in the account will equal the amount estimated in **step 1**.

Estimating the amount of outstanding accounts receivable that the firm does not expect to collect relies on information in an aging of accounts receivable. Aging accounts receivable involves classifying each customer's account as to the length of time the account has been uncollected. One set of intervals for classifying individual accounts receivable is as follows:

1. Not yet due.
2. Past due 30 days or fewer.
3. Past due 31 to 60 days.
4. Past due 61 to 180 days.
5. Past due more than 180 days.

The accountant presumes that the balance in Allowance for Uncollectible Accounts should be large enough to cover substantially all accounts receivable past due for more than, say, six months and smaller portions of the more recent accounts. A firm estimates the actual portions from experience.

As an example of the adjustments to be made, assume that the balance in accounts receivable of Howarth's Home Center at the end of Year 2, before providing for estimated uncollectible accounts, is $600,000 (= $800,000 from sales on account during Year 2 minus $200,000 of cash collections). **Exhibit 6.1** presents an aging of these accounts receivable. Estimated uncollectible accounts total $98,000. The entry at the end of Year 2, assuming that Howarth's Home Center uses aging of receivables instead of percentage of sales to estimate uncollectible accounts, is as follows:

| | | |
|---|---|---|
| End of Year 2 | | |
| Bad Debt Expense | 98,000 | |
| Allowance for Uncollectible Accounts | | 98,000 |

| Assets | = | Liabilities | + | Shareholders' Equity | (Class.) |
|---|---|---|---|---|---|
| −98,000 | | | | −98,000 | IncSt → RE |

To adjust the balance in the Allowance for Uncollectible Accounts to $98,000.

**EXHIBIT 6.1** **Illustration of Aging Accounts Receivable**

| | Amount | Estimated Uncollectible Percentage | Estimated Uncollectible Amounts |
|---|---|---|---|
| CLASSIFICATION OF ACCOUNTS | | | |
| Not yet due . . . . . . . . . . . . . . . . . . . . . . . . . | $444,000 | 8.1 | $36,000 |
| 1–30 days past due . . . . . . . . . . . . . . . . . . . | 75,000 | 20.0 | 15,000 |
| 31–60 days past due . . . . . . . . . . . . . . . . . . | 40,000 | 40.0 | 16,000 |
| 61–180 days past due . . . . . . . . . . . . . . . . . | 25,000 | 60.0 | 15,000 |
| More than 180 days past due . . . . . . . . . . . . | 16,000 | 100.0 | 16,000 |
| | $600,000 | | $98,000 |

Before this entry, the Allowance account had a zero balance, so the amount debited to expense and credited to the allowance is the full amount of uncollectibles estimated by the aging analysis. In reality, the Allowance account will have some balance just before this entry—perhaps a debit balance. Then, the amounts in the debit and credit of this entry will be the plug amounts necessary to bring the balance to the estimated amount—$98,000. If, for example, the Allowance account had a $10,000 debit balance just before the adjusting entry, the amounts would be $108,000 (= $10,000 + $98,000). If the Allowance account had a $23,000 credit balance just before the adjusting entry, the amounts would be $75,000 (= $98,000 − $23,000).

**Comparing Percentage-of-Sales and Aging Procedures** When the firm uses the percentage-of-sales procedure, the periodic provision for uncollectible accounts (for example, $96,000) increases (credits) the amounts provided in previous periods (zero in this case) in the account Allowance for Uncollectible Accounts. When the firm uses the aging procedure, it adjusts the balance in the account Allowance for Uncollectible Accounts (in this example, $98,000) to reflect the desired ending balance. If the percentage used under the percentage-of-sales method reasonably reflects collection experience, the balance in the allowance account should be approximately the same at the end of each period under both of these procedures of estimating uncollectible accounts.

**Exhibit 6.2** illustrates the operation of the allowance method for uncollectibles over two periods. In the first period the firm uses the percentage-of-sales procedure. In the second period it uses the aging procedure. Normally, a firm would use the same method in all periods.

**Dealing with Misestimates of Uncollectible Accounts** The allowance method requires estimates of uncollectible accounts. Changed economic conditions, more careful screening of credits, more aggressive collection efforts, and similar factors may cause the amount of actual uncollectible accounts to differ from the estimated amount. How should firms deal with such misestimates? Two possibilities are:

1. Make retroactive adjustments to net income of previous periods to correct for the misestimates.
2. Correct for the previous misestimates by adjusting the provision for bad debts during the current period.

GAAP requires firms to follow the second approach.[1] GAAP's reasoning is that the making of estimates is an integral part of measuring earnings under the accrual basis. Presuming that firms make conscientious estimates each year, adjustments for misestimates, although recurring, should be small. Following the first approach will likely confuse users of the financial statements because of the continual restatements that would likely take place. Firms could manage earnings upward in the current year and correct later with no negative effect on any year's income.

**Summary of Accounting for Uncollectible Accounts** The accounting for uncollectible accounts using the allowance method involves four transactions or events.

[1]Accounting Principles Board, *Opinion No. 20*, "Accounting Changes," 1971.

**(1)** Sale of Goods on Account

| | | |
|---|---|---|
| Accounts Receivable | Selling Price | |
| Sales Revenue | | Selling Price |

| Assets | = | Liabilities | + | Shareholders' Equity | (Class.) |
|---|---|---|---|---|---|
| +Selling Price | | | | +Selling Price | IncSt → RE |

**(2)** Collection of Cash from Customers

| | | |
|---|---|---|
| Cash | Amount Collected | |
| Accounts Receivable | | Amount Collected |

| Assets | = | Liabilities | + | Shareholders' Equity | (Class.) |
|---|---|---|---|---|---|
| +Amount Collected | | | | | |
| −Amount Collected | | | | | |

**(3)** Estimate of Expected Uncollectible Accounts

| | | |
|---|---|---|
| Bad Debt Expense | Estimated Uncollectible Amount | |
| Allowance for Uncollectible Accounts | | Estimated Uncollectible Amount |

| Assets | = | Liabilities | + | Shareholders' Equity | (Class.) |
|---|---|---|---|---|---|
| −Estimated Uncollectible Amount | | | | −Estimated Uncollectible Amount | |

The percentage-of-sales procedure computes directly the amount of the debit, with this same amount then increasing the amount in the Allowance for Uncollectible Accounts account as a credit. The aging-of-accounts-receivable procedure computes the credit balance needed in the Allowance for Uncollectible Accounts account at the end of the period. The amount credited to the Allowance for Uncollectible Accounts account is the amount needed to bring about this needed credit balance, with this same amount then debited to Bad Debt Expense.

**(4)** Write-off of Actual Uncollectible Accounts using the Allowance Method

| | | |
|---|---|---|
| Allowance for Uncollectible Accounts | Actual Uncollectible Amount | |
| Accounts Receivable | | Actual Uncollectible Amount |

| Assets | = | Liabilities | + | Shareholders' Equity | (Class.) |
|---|---|---|---|---|---|
| +Actual Uncollectible Amount | | | | | |
| −Actual Uncollectible Amount | | | | | |

**EXHIBIT 6.2** **Review of the Allowance Method of Accounting for Uncollectible Accounts**

**Transactions in the First Period**

(1) Sales are $1,000,000.

(2) The firm collects cash of $937,000 from customers in payment of their accounts.

(3) At the end of the first period, the firm estimates that uncollectibles will be 2 percent of sales; 0.02 × $1,000,000 = $20,000.

(4) The firm writes off specific accounts totaling $7,000 as uncollectible.

**Transactions in the Second Period**

(5) Sales are $1,200,000.

(6) The firm writes off specific accounts totaling $22,000 during the period as information on their uncollectibility becomes known. The debit balance of $9,000 will remain in the Allowance account until the firm makes an entry at the end of the period; see **(8)**.

(7) The firm collects cash of $1,100,000 from customers in payment of their accounts.

(8) An aging of the accounts receivable shows that the amount in the Allowance account should be $16,000. The amount of the adjustment is $25,000. It is the difference between the desired $16,000 credit balance and the current $9,000 debit balance in the Allowance account.

**(1) During First Period**

| | | |
|---|---|---|
| Accounts Receivable | 1,000,000 | |
| Sales Revenue | | 1,000,000 |

| Assets | = | Liabilities | + | Shareholders' Equity | (Class.) |
|---|---|---|---|---|---|
| +1,000,000 | | | | +1,000,000 | IncSt → RE |

To recognize sales on account.

**(2) During First Period**

| | | |
|---|---|---|
| Cash | 937,000 | |
| Accounts Receivable | | 937,000 |

| Assets | = | Liabilities | + | Shareholders' Equity | (Class.) |
|---|---|---|---|---|---|
| +937,000 | | | | | |
| −937,000 | | | | | |

To record cash collections from customers.

**(3) End of First Period**

| | | |
|---|---|---|
| Bad Debt Expense | 20,000 | |
| Allowance for Uncollectible Accounts | | 20,000 |

| Assets | = | Liabilities | + | Shareholders' Equity | (Class.) |
|---|---|---|---|---|---|
| −20,000 | | | | −20,000 | IncSt → RE |

To recognize estimated uncollectible accounts for first period using the percentage-of-sales procedure.

**(4) End of First Period**

| | | |
|---|---|---|
| Allowance for Uncollectible Accounts | 7,000 | |
| Accounts Receivable | | 7,000 |

| Assets | = | Liabilities | + | Shareholders' Equity | (Class.) |
|---|---|---|---|---|---|
| +7,000 | | | | | |
| −7,000 | | | | | |

To write off specific customers' accounts as uncollectible. The balance in Allowance for Uncollectible Accounts is $13,000 (= $20,000 − $7,000).

*(continued)*

**EXHIBIT 6.2** **Review of the Allowance Method of Accounting for Uncollectible Accounts (*Continued*)**

**(5) During Second Period**

| | | |
|---|---|---|
| Accounts Receivable | 1,200,000 | |
| Sales Revenue | | 1,200,000 |

| Assets | = | Liabilities | + | Shareholders' Equity | (Class.) |
|---|---|---|---|---|---|
| +1,200,000 | | | | +1,200,000 | IncSt → RE |

To recognize sales on account.

**(6) During Second Period**

| | | |
|---|---|---|
| Allowance for Uncollectible Accounts | 22,000 | |
| Accounts Receivable | | 22,000 |

| Assets | = | Liabilities | + | Shareholders' Equity | (Class.) |
|---|---|---|---|---|---|
| +22,000 | | | | | |
| −22,000 | | | | | |

To write off specific customers' accounts as uncollectible. The balance in Allowance for Uncollectible Accounts is now a debit of $9,000 (= $13,000 − $22,000).

**(7) During Second Period**

| | | |
|---|---|---|
| Cash | 1,100,000 | |
| Accounts Receivable | | 1,100,000 |

| Assets | = | Liabilities | + | Shareholders' Equity | (Class.) |
|---|---|---|---|---|---|
| +1,100,000 | | | | | |
| −1,100,000 | | | | | |

To record cash collections from customers.

**(8) End of Second Period**

| | | |
|---|---|---|
| Bad Debt Expense | 25,000 | |
| Allowance for Uncollectible Accounts | | 25,000 |

| Assets | = | Liabilities | + | Shareholders' Equity | (Class.) |
|---|---|---|---|---|---|
| −25,000 | | | | −25,000 | IncSt → RE |

To recognize estimated uncollectibles for second period using aging procedure. The entry converts a debit balance of $9,000 (see entry **(6)**) to the desired credit balance of $16,000.

Entries such as this can occur during the period, as the firm identifies specific uncollectible accounts, or at the end of the period, when the firm writes off a group of specific uncollectibles all at once, or both.

When a firm uses the direct write-off method, it makes the same entries for (1) and (2). It does not make entry (3). Entry (4) is as follows:

(4) Write-off of Actual Uncollectible Accounts using Direct Write-off Method

| | | |
|---|---|---|
| Bad Debt Expense | Actual Uncollectible Amount | |
| Accounts Receivable | | Actual Uncollectible Amount |

*(continued)*

| Assets | = | Liabilities | + | Shareholders' Equity | (Class.) |
|---|---|---|---|---|---|
| -Actual Uncollectible Amount | | | | -Actual Uncollectible Amount | IncST → RE |

## Problem 6.2 for Self-Study

**Accounting for uncollectible accounts.** Dee's Department Store opened for business on January 2, Year 4. Sales on account during Year 4 are $5,000,000. The firm estimates that uncollectible accounts will total 2 percent of sales. Dee's writes off $40,000 of accounts as uncollectible at the end of Year 4. Collections on account during Year 4 totaled $3,500,000.

**a.** Present the entry on December 31, Year 4, to apply the direct write-off method for uncollectible accounts.

**b.** Present the entries on December 31, Year 4, to apply the allowance method for uncollectible accounts.

**c.** Sales on account during Year 5 totaled $6,000,000. Dee's writes off $110,000 of accounts as uncollectible on December 31, Year 5. Prepare the entry on December 31, Year 5, to apply the direct write-off method for Year 5.

**d.** Refer to part **c**. Prepare the entries on December 31, Year 5, to apply the allowance method for Year 5.

**e.** Dee's ages its accounts receivable on December 31, Year 5, and discovers that it needs a balance in its Allowance for Uncollectible Accounts of $75,000 at the end of Year 5. Give the additional entries on December 31, Year 5, beyond that in part **d** above, to apply the allowance method for Year 5.

## THE ALLOWANCE METHOD APPLIES TO MANY TRANSACTIONS IN ACCOUNTING

We have just introduced the allowance method for uncollectibles. Accountants use the allowance method when the firm knows that at the time of sale, it will experience some reduction in future cash flows (in this case for uncollected accounts) but can estimate with reasonable precision the amount at the time of sale. The allowance method permits firms to reduce reported earnings in the period of sale to the amount of the expected net cash collections.

A similar phenomenon often occurs in business transactions:

- The customer has the right to return the product for a refund and the firm can estimate with reasonable precision the amount of returns at the time of sale, or
- the customer has the right to repairs or replacement under warranty if the purchased product is defective, and the firm can estimate with reasonable precision the amount of warranty costs at the time of sale.

In these cases, the firm will use a method analogous to the allowance method for uncollectibles. **Chapter 9** illustrates the allowance method for product warranties. You should master the concepts underlying the allowance method, which means mastering its procedures and journal entries to cement your understanding of the concepts, so that you can apply these concepts in other, similar transactions. The allowance method and its analogs underpin all revenue recognition where the firm recognizes revenue before it has resolved all uncertainty about future cash flows related to the sales transaction.

## SALES DISCOUNTS

Often the seller of merchandise offers a reduction from the invoice price for prompt payment. Such reductions, called **sales discounts** or cash discounts, indicate that goods may have two prices: a lower cash price and a higher price if the seller grants credit.[2] The seller offers a cash discount not only to provide an interest allowance on funds paid before the payment is due—the implied interest rate typically exceeds normal lending rates—but also to induce prompt payment so it can avoid additional bookkeeping and collection costs. The amount of sales discounts the seller expects the customers to take appears as an adjustment in measuring net sales revenue.

## Sales Returns and Allowances

When a customer returns merchandise, the return cancels the sale. The firm could make an entry to reverse the sale. In analyzing sales activities, however, management may be interested in the amount of goods returned. If so, it uses a sales revenue contra account, Sales Returns, to accumulate the amount of **sales returns** for a particular period. When the firm has significant amounts of returns and can reasonably estimate their amounts, it should use an allowance method for returns. The selling firm debits a revenue contra account during the period of sale for expected returns so as to report correctly the amount of cash it expects to collect from each period's sales. GAAP does not allow a firm to recognize revenue from sales when the customers have the right to return goods unless the firm can reasonably estimate the amount of returns and it uses an allowance method to do so.[3]

**Sales allowances** reduce the price charged to a customer, usually after the firm has delivered the goods and the customer has found them to be unsatisfactory or damaged. The sales allowance reduces sales revenue, but the firm may desire to accumulate the amount of such adjustments in a separate sales revenue contra account. It can use an account, Sales Allowances, for this purpose.

## Turning Receivables into Cash

A firm may find itself temporarily short of cash and unable to borrow from its usual sources. In such instances, it can convert accounts receivable into cash in various ways. A firm may **assign** its accounts receivable to a bank or a finance company to obtain a loan. The assigning (borrowing) company physically maintains control of the accounts receivable, collects amounts remitted by customers, and then forwards the proceeds to the lending institution. Alternatively, the firm may **pledge** its accounts receivable to the lending agency as collateral for a loan. If the borrowing firm cannot make loan repayments when due, the lending agency has the right to sell the accounts receivable to obtain payment. Finally, firms may **factor** (sell) the accounts receivable to a bank or a finance company to obtain cash. In this case, the firm sells accounts receivable to the lending institution, which physically controls the receivables and collects payments from customers. If the firm has pledged accounts receivable, a footnote to the financial statements should indicate this fact: the collection of such accounts receivable will not increase the liquid resources available to the firm to pay general trade creditors. Accounts receivable that the firm has factored or assigned do not appear on the balance sheet because the firm has sold them. **Chapter 10** discusses the difficulties in ascertaining whether a transfer of receivables is a sale (no liability appears on the balance sheet) or a collateralized loan (liability appears on the balance sheet). See *variable interest entity* in **Chapter 11** and the **Glossary**.

## Illustration of Balance Sheet Presentation

The balance sheet accounts presented so far in this chapter are Accounts Receivable and the Allowance for Uncollectible Accounts. **Exhibit 6.3** shows how transactions affecting these accounts appear in the income statement of Alexis Company. **Exhibit 6.4** illustrates the presentation of these items in the balance sheet, which includes all of the current assets for Alexis Company as of June 30, Year 1 and Year 2.

Accounts receivable appear on the balance sheet at the net amount the firm expects to collect. Firms use the allowance method to reduce the gross receivables to their expected collectible amount. The accounting involves a debit to Bad Debt Expense and a credit to the Allowance for

[2]See the **Glossary** at the back of the book for the definition of a discount and a summary of the various contexts in which accountants use this word.

[3]Financial Accounting Standards Board, *Statement of Financial Accounting Standards No. 48*, "Revenue Recognition When Right of Return Exists," 1981.

**EXHIBIT 6.3**

**ALEXIS COMPANY**
**Illustration of Sales and Sales Adjustments**
**Partial Income Statement**
**For the Year Ended June 30, Year 2**

| | | |
|---|---|---|
| REVENUES | | |
| Sales—Gross | | $515,200 |
| Less Sales Adjustments: | | |
| Discounts Taken | $23,600 | |
| Allowances | 11,000 | |
| Estimated Uncollectibles[a] | 10,300 | |
| Returns | 8,600 | |
| Total Sales Adjustments | | 53,500 |
| Net Sales | | $461,700 |

[a]Most companies call this account Bad Debt Expense and list it among the expenses.

**EXHIBIT 6.4**

**ALEXIS COMPANY**
**Illustration of Current Assets**
**Balance Sheet (Excerpts)**
**June 30, Year 1 and Year 2**

| | June 30, Year 1 | | June 30, Year 2 | |
|---|---|---|---|---|
| **Current Assets** | | | | |
| Cash and Certificates of Deposit | | $ 56,200 | | $ 63,000 |
| Accounts Receivable, Gross | $57,200 | | $58,100 | |
| Less Allowance for Uncollectible Accounts | (3,500) | | (3,600) | |
| Accounts Receivable, Net | | 53,700 | | 54,500 |
| Merchandise Inventory[a] | | 67,000 | | 72,000 |
| Prepayments | | 4,300 | | 4,800 |
| Total Current Assets | | $181,200 | | $194,300 |

[a]Additional required disclosures for this item do not appear here. See **Chapter 7**.

Uncollectible Accounts (asset contra) account in the period of sale. When the firm identifies a specific account receivable as uncollectible, it credits the specific account receivable to write it off and debits the Allowance for Uncollectible Accounts account. Periodically, the firm uses an aging-of-accounts-receivable balance to test the adequacy of the amounts debited to Bad Debt Expense.

## Analyzing Information on Accounts Receivable

The financial statements contain useful information for analyzing the collectibility of accounts receivable and the adequacy of a firm's provision for uncollectible accounts. **Exhibit 6.5** presents information taken from a recent annual report of Whirlpool Corporation, a manufacturer of household appliances. The analyst might address the following questions.

**EXHIBIT 6.5** **Selected Data for Whirlpool Corporation (amounts in millions)**

| | Year 4 | Year 5 | Year 6 | Year 7 |
|---|---|---|---|---|
| **Balance Sheet** | | | | |
| Accounts Receivable, net of allowance for uncollectible accounts of $2.4 at the end of Year 4, $2.6 at the end of Year 5, $3.3 at the end of Year 6, and $4.4 at the end of Year 7 . . . . . . . . . . . . . . . . . . . . . . . | $ 247.0 | $ 260.9 | $ 323.4 | $ 419.5 |
| **Income Statement** | | | | |
| Sales on Account . . . . . . . . . . . . . . . . . . . . | $3,987.6 | $4,008.7 | $4,179.0 | $4,314.5 |
| Bad Debt Expense . . . . . . . . . . . . . . . . . . . . | 5.4 | 6.0 | 17.7 | 11.5 |

1. **How quickly does the firm collect its accounts receivable?** The accounts receivable turnover ratio and the days accounts receivable outstanding, discussed in **Chapter 5**, provide the needed information.

| Year | Accounts Receivable Turnover Ratio | Days Accounts Receivable Outstanding |
|---|---|---|
| 5 . . . . . . . . . . | $4,008.7/0.5($247.0 + $260.9) = 15.8 | 365/15.8 = 23.1 days |
| 6 . . . . . . . . . . | $4,179.0/0.5($260.9 + $323.4) = 14.3 | 365/14.3 = 25.5 days |
| 7 . . . . . . . . . . | $4,314.5/0.5($323.4 + $419.5) = 11.6 | 365/11.6 = 31.5 days |

Thus, the collection period increased during the three-year period.

2. **What is the uncollectible accounts experience of the firm?** The analysts can compute the amount of accounts written off using information in the Allowance for Uncollectible Accounts account.

| Year | Balance at Beginning of Year | Increase for Bad Debt Expense | Accounts Written Off as Uncollectible | Balance at End of Year |
|---|---|---|---|---|
| 5 . . . . . . . . | $2.4 | $ 6.0 | $ 5.8 (plug) | $2.6 |
| 6 . . . . . . . . | $2.6 | $17.7 | $17.0 (plug) | $3.3 |
| 7 . . . . . . . . | $3.3 | $11.5 | $10.4 (plug) | $4.4 |

We can compare the amount of accounts written off to the average amount of gross accounts receivable each year.

| Year | Accounts Written Off as Uncollectible | Accounts Receivable Net | Allowance for Uncollectible Accounts | Accounts Receivable Gross | Write-offs/ Average Gross Accounts Receivable |
|---|---|---|---|---|---|
| 4 . . . . . . . . . | | $247.0 | $2.4 | $249.4 | |
| 5 . . . . . . . . . | $ 5.8 | $260.9 | $2.6 | $263.5 | 2.3% |
| 6 . . . . . . . . . | $17.0 | $323.4 | $3.3 | $326.7 | 5.8% |
| 7 . . . . . . . . . | $10.4 | $419.5 | $4.4 | $423.9 | 2.8% |

Thus, consistent with the slowing accounts receivable turnover ratio, the firm experienced increased write-offs of uncollectible accounts, particularly in Year 6.

3. **Has the firm adequately provided for estimated uncollectible accounts?** We can compare the amount of bad debt expense to sales and the balance in the Allowance for Uncollectible Accounts account to gross accounts receivable to help in responding to this question.

| Year | Bad Debt Expense (1) | Sales (2) | (1)/(2) = (3) | Balance in Allowance Accounts (4) | Gross Accounts Receivable (5) | (4)/(5) = (6) |
|---|---|---|---|---|---|---|
| 5 . . . . . . . . . . | $ 6.0 | $4,008.7 | 0.15% | $2.6 | $263.5 | 0.99% |
| 6 . . . . . . . . . . | $17.7 | $4,179.0 | 0.42% | $3.3 | $326.7 | 1.01% |
| 7 . . . . . . . . . . | $11.5 | $4.314.5 | 0.27% | $4.4 | $423.9 | 1.04% |

The ratio of bad debt expense to sales varies considerably from year to year but in a pattern similar to write-offs of uncollectible accounts computed in **question 2** above. The balance in the allowance accounts increased as a percentage of gross accounts receivable, which one would expect with a slowing of the accounts receivable turnover.

## Income Recognition at Times Different from Sale

Although most firms recognize revenue at the time of sale, many firms clear the hurdles mentioned earlier for revenue recognition at times other than the time of sale. Next, we provide examples of such firms and discuss some of the accounting issues they present.

### Income Recognition Before the Sale

Firms sometimes recognize revenues and expenses before the sale or delivery of a product—time **(2)** in **Figure 6.1**. In these cases, the firm has typically contracted with a particular customer, agreed on a selling price, and performed substantial work to create the product.

**Long-Term Contractors** The operating process for a long-term contractor (for example, building construction, shipbuilding) differs from that for a manufacturing firm (depicted in **Figure 6.1**) in three important respects:

1. The period of construction (production) may span several accounting periods.
2. The firm identifies a customer and agrees on a contract price in advance (or at least in the early stages of construction). The seller has little doubt about the ability of the customer to make the agreed-on payments.
3. The buyer often makes periodic payments of the contract price as work progresses.

The activities of contractors often meet the criteria for recognizing revenue from long-term contracts during the period of construction. The contract indicates that an arrangement exists (as required by the SEC), which shows that the contractor has identified a buyer and agreed on a price. Occasionally, the contractor collects cash in advance. More often, the contractor assesses the customer's credit and reasonably expects that the customer will pay the contract price in cash as construction progresses or afterward. Although at any given time these long-term construction contracts can require substantial future services, the contractor can estimate, with reasonable precision, the costs it will incur in providing these services. In agreeing to a contract price, the firm must have some confidence in its estimates of the total costs it will incur on the contract.

**Percentage-of-Completion Method** When a firm's construction activities meet the criteria for revenue recognition as construction progresses, the firm usually recognizes revenue during the construction period using the **percentage-of-completion method**. The percentage-of-completion method recognizes a portion of the contract price as revenue during each accounting period of construction. It bases the amount of revenue, expense, and income on the proportion of total work performed during the accounting period. It measures the proportion of total work carried out during the accounting period either from engineers' estimates of the degree of completion or from the ratio of costs incurred to date to the total costs expected for the entire contract.

The actual schedule of cash collections does not affect the revenue recognition process. Even if the contract specifies that the contractor will receive the entire contract price only on completing construction, the contractor may use the percentage-of-completion method so long as it can reasonably estimate the amount of cash it will receive and the remaining costs it expects to incur in completing the job.

As the contractor recognizes revenue for portions of the contract price, it recognizes equal portions of the total estimated contract costs as expenses. Thus, the percentage-of-completion method follows the accrual basis of accounting because the method matches expenses with related revenues.

To illustrate the percentage-of-completion method, we will assume that a firm agrees to construct a bridge for $5,000,000. The firm estimates the total cost at $4,000,000, to occur as follows: Year 1, $1,500,000; Year 2, $2,000,000; and Year 3, $500,000. Thus the firm expects a total profit of $1,000,000 (= $5,000,000 − $1,500,000 − $2,000,000 − $500,000).

Assume that the firm measures the percentage of completion by using the percentage of total costs incurred and that it incurs costs as anticipated. It recognizes revenue and expense from the contract as follows:

| Year | Degree of Completion | Revenue | Expense | Profit |
|---|---|---|---|---|
| 1 ......... | $1,500,000/$4,000,000 = 37.5% | $1,875,000 | $1,500,000 | $ 375,000 |
| 2 ......... | $2,000,000/$4,000,000 = 50.0% | 2,500,000 | 2,000,000 | 500,000 |
| 3 ......... | $ 500,000/$4,000,000 = 12.5% | 625,000 | 500,000 | 125,000 |
| | | $5,000,000 | $4,000,000 | $1,000,000 |

**Completed Contract Method** Some firms involved with construction contracts postpone revenue recognition until they complete the construction project and its sale. Construction contractors call this method of recognizing revenue the **completed contract method**. This is the same as the **completed sale method**, a term accountants use for the usual method of recognizing revenue when a retailing or manufacturing firm makes its sales. A firm using the completed contract method in the previous example recognizes no revenue or expense from the contract during Year 1 or Year 2. In Year 3, it recognizes contract revenue of $5,000,000 and contract expenses of $4,000,000 in measuring net income. Note that total income, or profit, is $1,000,000 under both the percentage-of-completion and the completed contract methods, equal to cash inflows of $5,000,000 less cash outflows of $4,000,000.

In some cases, firms use the completed contract method because the contracts take such a short time to complete (such as a few months) that earnings reported with the percentage-of-completion method and the completed contract method do not differ significantly. In these cases, firms use the completed contract method because they find it generally easier to implement. Firms must use the completed contract method in situations when they have not found a specific buyer while construction progresses, as sometimes happens in the construction of residential housing. These situations require future marketing effort. Moreover, substantial uncertainty may exist regarding the price that the contractor will ultimately ask and the amount of cash it will collect.

If uncertainty obscures the total costs the contractor will incur in carrying out the project, it will not use the percentage-of-completion method—even when it has a contract with a specified price. If a contractor cannot reasonably estimate total costs, it cannot estimate the total income for the contract, a GAAP requirement for recognizing revenue earlier than completion of the contract. Furthermore, a firm that uses the proportion of total costs incurred each period to measure the degree of completion will not have the necessary information to allocate this total income (revenues and expenses) to each period of construction activity.

**Quality-of-Earnings Issues for Long-Term Contractors** The percentage-of-completion method provides information on the profitability of a contractor as construction progresses. The completed contract method, in contrast, lumps all of the income from contracts into the period of completion. Because of the more timely reporting of operating performance in the percentage-of-completion method, GAAP requires the use of this method whenever firms can make reasonable estimates of revenues and expenses.[4]

[4]Financial Accounting Standards Board, *Statement of Financial Accounting Standards No. 111,* "Rescission of FASB Statement Number 32 and Technical Corrections," 1997.

The need to make estimates under the percentage-of-completion method provides management with opportunities to manage reported earnings. Managers can either underestimate total expenses or overestimate the degree of completion to accelerate the recognition of income. Recognizing too much income during the early years of a contract means recognizing too little in later years. Thus, although the percentage-of-completion method provides better measures of a firm's value added each period than does the completed contract method, enhancing earnings quality, the need to make estimates to apply the percentage-of-completion method lowers earnings quality.

## Income Recognition After the Sale

The operating process for some firms requires substantial performance after the time of sale. For other firms, it involves considerable uncertainty regarding the future amounts of cash inflows, cash outflows, or both. In these cases, recognizing income after the time of sale—at times **(4)**, **(5)**, or **(6)** in **Figure 6.1**—is appropriate.

**Insurance Companies** The operating process for an insurance company involves the following steps:

1. The insurance company sells insurance policies to customers, promising to provide insurance coverage (life, health, property, liability) in the future.
2. The insurance company collects cash (premiums) from policyholders during the periods of insurance coverage.
3. The insurance company invests the cash received from policyholders in stocks, bonds, real estate, and other investments, periodically receiving interest, dividends, and other revenues.
4. Policyholders make claims on their insurance policies, requiring the insurance company to disburse cash.

The insurance contract requires the insurance company to provide substantial future services after the sale of the contract. Also, the insurance company cannot estimate the amount of cash it will receive from investments. The uncertainty in these steps causes the accountant to delay the recognizing of income from the time of sale to some time after the company sells the insurance.

One alternative recognizes income at the end of the period of insurance coverage (for example, at the death of the insured in the case of life insurance). At that time, the insurance company can measure with certainty the cash it received from premiums and from investments and the amount of cash it disbursed for claims. Also, the insurance company has provided the services required by the insurance contract. Such an approach, however, may inappropriately delay the recognition of income. Most policyholders pay their insurance premiums as required by the insurance contract. Insurance companies generally receive cash each period for interest and dividends from their investments. Life insurance actuaries can usually estimate the amount and timing of insurance claims. Awaiting completion of the insurance coverage period provides insufficient signals regarding the operating performance of the insurance company while it is providing the insurance coverage.

Most insurance companies therefore recognize revenues and expenses during each period of insurance coverage: time **(6)** in **Figure 6.1**. The receipt of cash from policyholders and from investments creates a relatively high degree of certainty regarding the amount of revenue. Insurance companies must, however, match against this revenue each period a portion of the ultimate cash outflow to satisfy claims. Because this cash outflow may occur several or even many years later, measuring expenses requires estimates. Actuaries study demographic and other data to estimate the likely amount and timing of policyholder claims. Accountants use this information to measure expenses each period. The need to make such estimates, however, lowers the quality of earnings for insurance companies relative to companies whose cash inflows occur closer in time to its cash outflows.

**Franchisors** Companies (franchisors) such as McDonald's, Pizza Hut, and Taco Bell typically sell franchise rights to individuals (franchisees) in particular locales. These franchise rights permit the franchisee to use the franchisor's name, as well as advertising, consulting, and other services provided by the franchisor. The franchisee agrees to pay an initial franchise fee of $50,000 or even more. Most franchisees do not have the funds to pay this initial fee at the time they acquire the franchise rights. Instead, they sign a note agreeing to pay the fee over five or more years.

Franchisors usually provide most of the services required by the franchise arrangement by the time the franchisee opens for business (for example, selecting a location, constructing buildings, and acquiring equipment). Substantial uncertainty often exists, however, regarding the portion of the initial franchise fee that the franchisor will ultimately collect from its customer, the franchisee. New small businesses often fail. Franchisees can usually stop making payments and walk away from the franchise arrangement. A similar situation exists for other businesses, such as land-development companies, that sell products or services with delayed payment arrangements.

When substantial uncertainty exists at the time of sale regarding the amount (or timing or both) of cash that a firm will ultimately receive from customers, the firm delays the recognition of revenues and expenses until it receives cash: point **(5)** in **Figure 6.1**. Such sellers recognize revenue at the time of cash collection using either the installment method or the cost-recovery-first method. Unlike the cash method of accounting, however, these methods attempt to match expenses with revenues.

**Installment Method** The **installment method** recognizes revenue as the seller collects parts of the selling price in cash. In parallel, this method recognizes as expenses each period the same portion of the cost of the good or service sold as the portion of total revenue recognized.

**Example 9** Assume that a firm sells for \$100 merchandise costing \$60. Income from the transaction is \$40 (= \$100 − \$60), cash inflow less cash outflow. The buyer agrees to pay (ignoring interest) \$20, one-fifth of the total selling price, per month for five months. The installment method recognizes revenue of \$20 each month as the seller receives cash. The cost of goods sold is \$12 [= (\$20/\$100) × \$60] each month, with monthly expense measured as the same proportion of total expenses as the cash collection represents of its total, one-fifth in the example. By the end of five months, the income recognized totals \$40 [= 5 × (\$20 − \$12)].

**Cost-Recovery-First Method** When the seller has substantial uncertainty about the amount of cash it will collect, it can also use the **cost-recovery-first method** of income recognition. This method matches the costs of generating revenues dollar for dollar with cash receipts until the seller recovers all such costs. Expenses equal (match) revenues in each period until the seller recovers all costs. Only when cumulative cash receipts exceed total costs will income (that is, revenue without any matching expenses) appear in the income statement.

To illustrate the cost-recovery-first method, we use information from the previous example.

**Example 10** Assume the sale for \$100 of merchandise costing \$60, with the buyer promising to pay \$20 each month for five months. Whether the buyer will pay remains uncertain (or else we would not be considering this method). If the buyer makes all payments on schedule, during each of the first three months, the seller recognizes revenue of \$20 and expenses of \$20. By the end of the third month, cumulative cash receipts of \$60 exactly equal the cost of the merchandise sold. During the fourth and fifth months, the seller recognizes revenue of \$20 per month without an offsetting expense. For the five months as a whole, the seller recognizes total income of \$40, \$20 in each of the last two periods, so the pattern differs from that of the installment method.

**Example 11** Assume instead that the seller eventually collects only four, not five, of the expected monthly payments of \$20 each. Under the cost-recovery-first method, the seller will recognize no income in the first three months, \$20 in the fourth month, and \$20 in total. Under the installment method, the seller would recognize \$8 (= \$20 − \$12) each month for four months, \$32 in total. Then, when the seller learns in month five that it will collect no more cash, it will have to recognize a loss of \$12 (= \$32 − \$20) to report correctly the total income of only \$20.

**Use of Installment and Cost-Recovery-First Methods** GAAP permits the seller to use the installment and the cost-recovery-first methods only when the seller cannot make reasonably certain estimates of cash collection. For most sales of goods and services, past experience and an assessment of customers' credit standings provide a sufficient basis for estimating the amount of cash the seller will receive. If the seller can reasonably estimate the amount of cash it will receive, the seller may not use the installment method or the cost-recovery-first method for financial reporting and must recognize revenue no later than the time of sale.[5] When cash collection is sufficiently uncertain that GAAP allows either of these methods, then the cost-recovery-first method seems to reflect more accurately the substance of the uncertainty. The logic of the installment method assumes that the seller will eventually receive all of the promised cash or, if

[5]Accounting Principles Board, *Opinion No. 10*, "Omnibus Opinion—1966," par. 12, footnote 8.

not, that when payments cease, the seller can repossess the goods, and, at that time, the goods still retain sufficient market value to cover the seller's as-yet-uncollected costs. If those assumptions do not hold, GAAP forbids the installment method.

**Unifying Principle for Revenue Recognition** Although GAAP does not put the matter this way, we suggest you can understand revenue recognition with the following unifying principle. A firm can recognize revenue when it has delivered products or services to customers so long as the firm can estimate with reasonable statistical certainty the events remaining to complete the transaction (such as collecting cash, accepting returns, making warranty repairs and replacements). When significant uncertainty exists at the time of delivery about the events remaining to complete the transactions, firms must delay revenue recognition until the uncertainties resolve to the level of reasonable statistical certainty.

## Summary Illustration of Income Recognition Methods

**Exhibit 6.6** illustrates the various methods of income recognition discussed here and in **Chapter 3**. The illustration involves a contract for the construction of a bridge for $12 million. The expected and actual patterns of cash receipts and disbursements under the contract appear below.

| Period | Expected and Actual Cash Receipts | Expected and Actual Cash Expenditures |
|---|---|---|
| 1 | $ 1,000,000 | $1,600,000 |
| 2 | 1,000,000 | 4,000,000 |
| 3 | 2,000,000 | 4,000,000 |
| 4 | 4,000,000 | — |
| 5 | 4,000,000 | |
| Total | $12,000,000 | $9,600,000 |

The contractor completed the bridge in Period 3. **Exhibit 6.6** indicates the contractor's periodic revenues, expenses, and income recognized during each period under the contract using the following methods:

1. The cash basis of accounting.
2. The percentage-of-completion method.
3. The completed contract (completed sale) method.
4. The installment method.
5. The cost-recovery-first method.

No single firm could simultaneously justify all of the five methods of income recognition for financial reporting; we present them for illustrative purposes. Note that the total revenues, expenses, and income recognized for the five years are the same for all methods. Over sufficiently long time periods, total income equals cash inflows less cash outflows other than transactions with owners. The amount of income recognized each period differs, however, depending on the accounting method used.

## Problem 6.3 for Self-Study

**Income recognition for contractor.** Blount Construction Company contracted on May 15, Year 2, to build a bridge for a city for $4,500,000. Blount estimated that the cost of constructing the bridge would be $3,600,000. Blount incurred $1,200,000 in construction costs during Year 2, $2,000,000 during Year 3, and $400,000 during Year 4 in completing the bridge. The city paid $1,000,000 during Year 2, $1,500,000 during Year 3, and the remaining $2,000,000 of the contract price at the time the bridge was completed and approved in Year 4.

*(continued)*

**EXHIBIT 6.6**

## Comprehensive Illustration of Revenue and Expense Recognition (all dollar amounts in thousands)

| Period | Cash Basis of Accounting[a] Revenue | Expense | Income |
|---|---|---|---|
| 1 | $ 1,000 | $1,600 | $ (600) |
| 2 | 1,000 | 4,000 | (3,000) |
| 3 | 2,000 | 4,000 | (2,000) |
| 4 | 4,000 | — | 4,000 |
| 5 | 4,000 | — | 4,000 |
| Total | $12,000 | $9,600 | $2,400 |

| Period | Percentage-of-Completion Method | | | Completed Contract Method | | |
|---|---|---|---|---|---|---|
| | Revenue | Expense | Income | Revenue | Expense | Income |
| 1 | $ 2,000[d] | $1,600 | $ 400 | $ — | $ — | $ — |
| 2 | 5,000[e] | 4,000 | 1,000 | — | — | — |
| 3 | 5,000[e] | 4,000 | 1,000 | 12,000 | 9,600 | 2,400 |
| 4 | — | — | — | — | — | — |
| 5 | — | — | — | — | — | — |
| Total | $12,000 | $9,600 | $2,400 | $12,000 | $9,600 | $2,400 |

| Period | Installment Method[b] | | | Cost-Recovery-First Method[c] | | |
|---|---|---|---|---|---|---|
| | Revenue | Expense | Income | Revenue | Expense | Income |
| 1 | $ 1,000 | $ 800[f] | $ 200 | $ 1,000 | $1,000 | $ 0 |
| 2 | 1,000 | 800[f] | 200 | 1,000 | 1,000 | 0 |
| 3 | 2,000 | 1,600[g] | 400 | 2,000 | 2,000 | 0 |
| 4 | 4,000 | 3,200[h] | 800 | 4,000 | 4,000 | 0 |
| 5 | 4,000 | 3,200[h] | 800 | 4,000 | 1,600 | 2,400 |
| Total | $12,000 | $9,600 | $2,400 | $12,000 | $9,600 | $2,400 |

[a]The cash basis is not allowed for tax or financial reporting if inventories are a material factor in generating income.
[b]The installment method is allowed for financial reporting only if extreme uncertainty exists as to the amount of cash to be collected from customers. Its use for tax purposes is independent of the collectibility of cash.
[c]The cost-recovery-first method is allowed for financial reporting only if extreme uncertainty exists as to the amount of cash to be collected from customers. It is sometimes used for tax purposes.
[d]$1,600/$9,600 × $12,000.
[e]$4,000/$9,600 × $12,000.
[f]$1,000/$12,000 × $9,600.
[g]$2,000/$12,000 × $9,600.
[h]$4,000/$12,000 × $9,600.

**a.** Calculate Blount's net income (revenue less expenses) on the contract during Year 2, Year 3, and Year 4, assuming that it uses the percentage-of-completion method.
**b.** Repeat part **a**, assuming that it uses the completed contract method.
**c.** Repeat part **a**, assuming that it uses the installment method.
**d.** Repeat part **a**, assuming that it uses the cost-recovery-first method.

## Format and Classification within the Income Statement

Financial statement users study a firm's income statement to evaluate its past profitability and to project its likely future profitability. When assessing profitability, the analyst makes two types of distinctions regarding the nature of various income items:

1. Does the income item result from a firm's primary operating activity (creating and selling a good or service for customers) or from an activity incidental or peripheral to the primary operating activity (for example, periodic sales of equipment previously used by the firm in manufacturing)?
2. Does the income item result from an activity in which a firm will likely continue its involvement, or from an unusual transaction or event that is unlikely to recur regularly?

**Figure 6.2** depicts these distinctions, with examples of each.

A financial statement user who wants to evaluate a firm's ongoing operating profitability will likely focus on income items in the upper left cell. A financial statement user who wants to project net income of prior periods into the future would likely focus on the two "recurring income" cells. Income items in the nonrecurring cells should not affect ongoing assessments of profitability. This section considers the reporting of each type of income item.

### Distinguishing Revenue from Gains and Expenses from Losses

Accountants distinguish between revenues and expenses on the one hand and gains and losses on the other. Revenues and expenses result from the recurring, primary operating activities of a business (upper left cell in **Figure 6.2**). Income items discussed thus far in this book fall into this first category. Gains and losses result from either peripheral activities (lower left cell) or nonrecurring activities (upper and lower right cells). A second distinction is that accounting reports revenues and expenses in gross amounts, whereas it reports gains and losses as net amounts. The following examples illustrate this distinction.

**Example 12** IBM sells a computer to a customer for $400,000. The computer cost IBM $300,000 to manufacture. IBM records this sale as follows:

| | | |
|---|---|---|
| Cash . . . . . . . . . . . . . . . . . . . . . . . . . . . . . . . . . . . . . . . . . . . . . . . | 400,000 | |
| Sales Revenue . . . . . . . . . . . . . . . . . . . . . . . . . . . . . . . . . . . . . | | 400,000 |

| Assets | = | Liabilities | + | Shareholders' Equity | (Class.) |
|---|---|---|---|---|---|
| +400,000 | | | | +400,000 | IncSt → RE |

To record sale.

| | | |
|---|---|---|
| Cost of Goods Sold . . . . . . . . . . . . . . . . . . . . . . . . . . . . . . . . . . . . . | 300,000 | |
| Finished Goods Inventory . . . . . . . . . . . . . . . . . . . . . . . . . . . . | | 300,000 |

| Assets | = | Liabilities | + | Shareholders' Equity | (Class.) |
|---|---|---|---|---|---|
| -300,000 | | | | -300,000 | IncSt → RE |

To record the cost of goods sold.

This transaction fits into the upper left cell of **Figure 6.2** (primary/recurring). The income statement reports both gross sales revenue and gross cost of goods sold, providing information to the financial statement user regarding both the manufacturing cost of the computer and IBM's ability to mark up this cost in setting selling prices. The $100,000 net amount never appears as a separate line item in the body of the income statement. Pretax income includes the $100,000 as part of the difference between revenues and expenses.

**Example 13** The GAP, a retail clothing chain, sells computers previously used for processing data in its stores. This sale relates only peripherally to The GAP's primary operating activity,

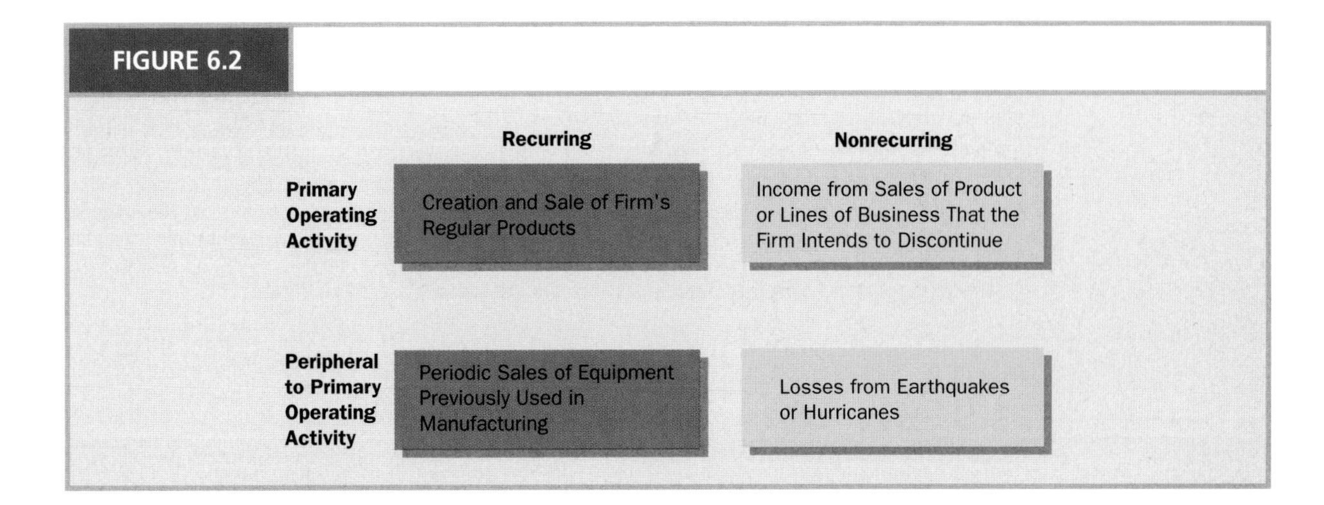


which is to sell casual clothes. Assume that the computers originally cost $500,000 and have $200,000 of accumulated depreciation at the time of sale. Thus, the computers have a net book value of $300,000. The disposal of these computers for $400,000 results in the following journal entry on The GAP's books:

| | | |
|---|---|---|
| Cash | 400,000 | |
| Accumulated Depreciation | 200,000 | |
| Equipment | | 500,000 |
| Gain on Sale of Equipment | | 100,000 |

| Assets | = | Liabilities | + | Shareholders' Equity | (Class.) |
|---|---|---|---|---|---|
| +400,000 | | | | +100,000 | IncSt → RE |
| +200,000 | | | | | |
| −500,000 | | | | | |

Disposal of computers for $100,000 more than book value.

This transaction fits into the lower left cell of **Figure 6.2** (peripheral/recurring). Illustrating **net versus gross reporting of income items**, the income statement reports only the $100,000 gain on the sale, not the selling price of $400,000 nor the book value of $300,000. The income statement reports gains and losses at net instead of gross amounts because, presumably, financial statement users do not need information on the components of either peripheral or nonrecurring income items.

Note that both revenues and gains appear as credits in journal entries and increase Retained Earnings. Both expenses and losses appear as debits in journal entries and reduce Retained Earnings. To repeat, report revenues and expenses at gross amounts but gains and losses at net amounts.

## An International Perspective

The IASB says that a firm should recognize revenue when

1. the seller has transferred significant risks and rewards of ownership to the buyer;
2. managerial involvement and control has passed from the seller to the buyer;
3. the seller can reliably measure the amount of revenue;
4. it is probable that the seller will receive economic benefits; and
5. the seller can reliably measure the costs (including future costs) of the transaction.[6]

*(continued)*

[6]International Accounting Standards Committee, *Statement No. 18*, "Revenue," 1984, revised and reissued 1995.

You can see that these criteria do not use the same words that we have used, but the concepts do not differ from those we discuss in this chapter. The IASB explicitly states that it prefers

- the allowance method for uncollectibles when the firm recognizes the revenue,
- the percentage-of-completion method when the seller provides services, and
- the matching principles for timing of expense recognition.[7]

For construction contracts, the IASB prefers the percentage-of-completion method when the firm can "measure or reliably estimate . . . total revenue, past and future costs, and the stage of completion." If the firm cannot measure the outcome reliably, then the firm should use the cost-recovery-first method.[8]

## Summary

**Chapter 5** points out that financial analysts care about the operating profitability of a firm. **Chapters 3** and **6** discuss the underlying concepts, accounting procedures, and uses of the income statement. The most important underlying concepts involve the measurement of revenues and expenses following the principles of the accrual basis of accounting. The accrual basis provides operating performance measures superior to those provided by the cash basis because the accrual basis matches revenues (the results of the firm's outputs) with expenses (its inputs). Imprecision in this matching occurs when the timing of cash receipts and disbursements deviates from the timing of revenue and expense recognition. Additional interpretative issues arise when a firm's income statement includes unusual or nonrecurring income items. The financial statement user must give adequate consideration to the recurring or nonrecurring nature of various income items in both assessing a firm's past operating performance and predicting its likely future performance.

## Appendix 6.1: Effects on the Statement of Cash Flows of Transactions Involving Accounts Receivable

From this point forward, most chapters contain an appendix describing the effects that the accounting treatments of the chapter have on the statement of cash flows. In describing effects in the derivation of cash flow from operations, we discuss both the direct and the indirect method. Recall that the direct method lists cash inflows from revenues and cash outflows for expenses. The indirect method begins the derivation of cash flow from operations with net income and then adjusts net income for noncash portions of revenues and expenses.

### Accounts Receivable

The question that an analyst would raise regarding cash flows and accounts receivable is: How much cash did a firm collect from customers during a period? To illustrate the adjustments to compute cash flow from operations, consider the data in **Exhibit 6.7**.

Cash flow from operations is $700. To compute this amount under the direct method, subtract from sales revenue the increase in accounts receivable (net) and the amount of bad debt expense, as the following analysis shows:

| | |
|---|---|
| Sales Revenue for Year 10 | $1,000 |
| Less Bad Debt Expense for Year 10 | (15) |
| Less Increase in Accounts Receivable, Net (= $365 – $80) during Year 10 | (285) |
| Cash Inflows from Sales during Year 10 | $ 700 |

[7] *Ibid.*

[8] International Accounting Standards Committee, *Statement No. 11*, "Construction Contracts," 1980.

**EXHIBIT 6.7** **Data Related to Accounts Receivable and Allowance for Uncollectible Accounts**

| | Accounts Receivable Gross | Allowance for Uncollectible Accounts | Accounts Receivable Net |
|---|---|---|---|
| Balance, January 1, Year 10 | $ 100 | $(20) | $ 80 |
| Sales on Accounts during Year 10 | 1,000 | — | 1,000 |
| Cash Collections during Year 10 | (700) | — | (700) |
| Bad Debt Expense for Year 10 | — | (15) | (15) |
| Write off of Uncollectible Accounts during Year 10 | (12) | 12 | — |
| Balance, December 31, Year 10 | $ 388 | $(23) | $ 365 |

Why subtract bad debt expense from gross sales in computing cash collected from customers? We subtract Bad Debt Expense in the income statement to reduce the gross amount of sales to the amount that the firm expects to collect in cash. In this case, the firm expects to collect only $985 in cash. Net income excludes the portion of gross sales that the firm never expects to collect in cash. Based on this rationale, some accountants treat the provision for uncollectible accounts as a revenue contra instead of as an expense. Thus, we might express the adjustment above as follows:

| | |
|---|---|
| Sales Revenue for Year 10 | $1,000 |
| Less Estimated Uncollectibles for Year 10 (revenue contra) | (15) |
| Net Sales | $ 985 |
| Less Increase in Accounts Receivable, Net (= $365 − $80) during Year 10 | (285) |
| Cash Inflows from Sales during Year 10 | $ 700 |

Once we compute net sales for the period, we can measure collections of that amount. If accounts receivable (net) do not change, we collect exactly that net revenue amount. If accounts receivable (net) increase, we collect the amount of net sales less the increase in accounts receivable (net). If accounts receivable (net) decrease, we collect the amount of net sales and, in addition, the amount by which net receivables decreased.

To compute cash flow from operations under the indirect method, subtract the increase in accounts receivable (net) from revenues to reach cash flow from operations, as the following analysis shows:

| | |
|---|---|
| Net Income for Year 10 includes Net Revenue of $1,000 − $15 | $ 985 |
| Less Increase in Accounts Receivable, Net (= $365 − $80) during Year 10 | (285) |
| Cash Flow from Operations during Year 10 | $ 700 |

Note that there is no need to subtract bad debt expense under the indirect method because net income already includes a subtraction for it. If the amount of accounts receivable (net) decreases during the period, add the amount of the decrease to net income in deriving cash flow from operations.

## Solutions to Self-Study Problems

**SUGGESTED SOLUTION TO PROBLEM 6.1 FOR SELF-STUDY**

**a.** Should Kodak recognize revenue (1) when it initially ships the cameras to Wal-Mart, or (2) wait until the consigned cameras are sold and Wal-Mart remits the selling price minus Wal-Mart's fee back to Kodak? At the time of initial shipment to Wal-Mart, Kodak has incurred most of its costs. Experience should provide Kodak with sufficient evidence to estimate the cost of lost and damaged cameras. Thus, the first criterion for revenue recognition of substantial performance appears to be met. The second criterion, however, has not been met. The amount that Kodak receives depends on the price at which Wal-Mart sells the cameras. This price depends on the demand for cameras in general, on new products offered by competitors, and similar factors. If Wal-Mart has to mark down the selling price in order to sell certain cameras, the amount that Kodak receives will decrease. The timing of this payment is also uncertain. Wal-Mart may fail to sell some cameras and then return them to Kodak. Thus, Kodak should recognize revenue when it receives the payment from Wal-Mart or, perhaps, earlier when Wal-Mart notifies Kodak of the amounts it will remit later for sales this period.

**b.** Should Burlington Mills recognize revenue (1) at the completion of manufacturing and delivery to the warehouse, or (2) upon final delivery to Nike? The criteria for revenue recognition appear to be met at the completion of manufacturing and delivery to the warehouse. Burlington Mills has performed most of the work it undertook to do. Nike bears the additional cost of storage. The custom nature of the products and storage in an independent warehouse preclude Burlington Mills from selling the products to others. Nike has promised to pay for these goods and Burlington Mills can estimate the amount and timing of cash it will receive from Nike. Thus, the rewards and risks of ownership appear to shift to Nike upon delivery to the warehouse. This sort of transaction, with important differences, served as the foundation for important frauds in the 1990s. The selling firm would bill the goods to the buyer to whom it promised an unconditional right of refusal or return. The amount of goods purchased exceeded the buyer's usual quantities. In this case, with substantial returns nearly certain and unpredictable, the firm should not recognize revenue.

**c.** Should Nordstrom recognize revenue (1) when it receives customers' initial deposits and places the products in a storeroom, or (2) when customers make the final payment and take the products? Because Nordstrom retains the rewards and risks of ownership until the customers make the final payments and accept the merchandise, Nordstrom should not recognize revenue at the time of initial deposit. An important part of substantial performance is providing storage services. The fact of initial deposit does not satisfy the second criterion for revenue recognition, the requirement that the seller receive cash or a receivable for a specific amount. Uncertainty exists as to if and when customers will make payments after the initial deposit. Nordstrom does not have an enforceable right to the subsequent payments.

**d.** Costco might recognize the membership fee as revenue (1) when it receives the fee at the beginning of the year, (2) ratably over the year, or (3) at the end of the year at completion of the refund period. Costco does not meet the first criterion for revenue recognition of substantial performance at the beginning of the year because it has not yet provided services. Costco provides services during the year and has a claim on proportionate parts of the membership fee as time passes. Thus, it should recognize revenue ratably over the year. Awaiting completion of the year before recognizing any revenue probably delays recognition of revenue past the time of satisfying the revenue recognition criteria. Even so, conservative accountants might delay revenue recognition until the end of the year.

**e.** Mall Development Corporation satisfies the criteria for revenue recognition with respect to the fixed monthly rental each month and should recognize that portion of the revenue monthly. The ultimate amount of the revenue it will recognize on the contingent portion of the rent depends on the sales of each lessee for the year. Although Mall Development Corporation likely estimates the probable amount of contingent rent when it sets the fixed monthly rent, too much uncertainty exists as to the amount of the contingent rent to justify recognizing that portion of revenue until it knows the amount. If lessees report sales monthly, then Mall Development Corporation can begin recognizing revenue from the contingent rents as soon as the lessee reaches the stated sales threshold. Periodic recognition of the contingent rent as revenue requires, however, that Mall Development Corporation estimate the amounts of sales returns and uncollectible accounts. If it cannot reasonably estimate these amounts before the end of the year, then it must delay recognition of the contingent rent until then.

### SUGGESTED SOLUTION TO PROBLEM 6.2 FOR SELF-STUDY

(Dee's Department Store; accounting for uncollectible accounts.)

**a.** Bad Debt Expense .......... 40,000
Accounts Receivable .......... 40,000

| Assets | = | Liabilities | + | Shareholders' Equity | (Class.) |
|---|---|---|---|---|---|
| −40,000 | | | | −40,000 | IncSt → RE |

To write off uncollectible accounts using the direct write-off method for Year 4.

**b.** Bad Debt Expense .......... 100,000
Allowance for Uncollectible Accounts .......... 100,000

| Assets | = | Liabilities | + | Shareholders' Equity | (Class.) |
|---|---|---|---|---|---|
| −100,000 | | | | −100,000 | IncSt → RE |

To provide for estimated uncollectible accounts using the allowance method for Year 4; $0.02 \times \$5{,}000{,}000 = \$100{,}000$.

Allowance for Uncollectible Accounts .......... 40,000
Accounts Receivable .......... 40,000

| Assets | = | Liabilities | + | Shareholders' Equity | (Class.) |
|---|---|---|---|---|---|
| +40,000 | | | | | |
| −40,000 | | | | | |

To write off uncollectible accounts using the allowance method for Year 4.

**c.** Bad Debt Expense .......... 110,000
Accounts Receivable .......... 110,000

| Assets | = | Liabilities | + | Shareholders' Equity | (Class.) |
|---|---|---|---|---|---|
| −110,000 | | | | −110,000 | IncSt → RE |

To write off uncollectible accounts using the direct write-off method for Year 5.

**d.** Bad Debt Expense .......... 120,000
Allowance for Uncollectible Accounts .......... 120,000

| Assets | = | Liabilities | + | Shareholders' Equity | (Class.) |
|---|---|---|---|---|---|
| −120,000 | | | | −120,000 | IncSt → RE |

To provide for estimated uncollectible accounts using the allowance method for Year 5; $0.02 \times \$6{,}000{,}000 = \$120{,}000$.

Allowance for Uncollectible Accounts .......... 110,000
Accounts Receivable .......... 110,000

| Assets | = | Liabilities | + | Shareholders' Equity | (Class.) |
|---|---|---|---|---|---|
| +110,000 | | | | | |
| −110,000 | | | | | |

To write off uncollectible accounts using the allowance method for Year 5.

*(continued)*

**e.** Bad Debt Expense .......... 5,000
Allowance for Uncollectible Accounts .......... 5,000

| Assets | = | Liabilities | + | Shareholders' Equity | (Class.) |
|---|---|---|---|---|---|
| -5,000 | | | | -5,000 | IncSt → RE |

To adjust the Allowance for Uncollectible Accounts to the desired ending balance of $75,000. The balance in the Allowance for Uncollectible Accounts after the entries in parts **b** and **d** is $70,000 (= $100,000 − $40,000 + $120,000 − $110,000).

Note: Most firms would likely combine the provision for bad debts in part **d** ($120,000) and part **e** ($5,000) into a single entry.

## SUGGESTED SOLUTION TO PROBLEM 6.3 FOR SELF-STUDY

(Blount Construction Company; income recognition for contractor.)

**a.** Percentage-of-completion method:

| Year | Incremental Percentage Complete | Revenue Recognized | Expenses Recognized | Net Income |
|---|---|---|---|---|
| 2 | 12/36 (0.333) | $1,500,000 | $1,200,000 | $300,000 |
| 3 | 20/36 (0.556) | 2,500,000 | 2,000,000 | 500,000 |
| 4 | 4/36 (0.111) | 500,000 | 400,000 | 100,000 |
| Total | 36/36 (1.000) | $4,500,000 | $3,600,000 | $900,000 |

**b.** Completed contract method:

| Year | Revenue Recognized | Expenses Recognized | Net Income |
|---|---|---|---|
| 2 | $ 0 | $ 0 | $ 0 |
| 3 | 0 | 0 | 0 |
| 4 | 4,500,000 | 3,600,000 | 900,000 |
| Total | $4,500,000 | $3,600,000 | $900,000 |

**c.** Installment method:

| Year | Cash Collected (= revenue) | Fraction of Cash Collected | Expenses (= fraction × total cost) | Net Income |
|---|---|---|---|---|
| 2 | $1,000,000 | 1.0/4.5 | $ 800,000 | $200,000 |
| 3 | 1,500,000 | 1.5/4.5 | 1,200,000 | 300,000 |
| 4 | 2,000,000 | 2.0/4.5 | 1,600,000 | 400,000 |
| Total | $4,500,000 | 1.0 | $3,600,000 | $900,000 |

**d.** Cost-recovery-first method:

| Year | Cash Collected (= revenue) | Expenses Recognized | Net Income |
|---|---|---|---|
| 2 | $1,000,000 | $1,000,000 | $ 0 |
| 3 | 1,500,000 | 1,500,000 | 0 |
| 4 | 2,000,000 | 1,100,000 | 900,000 |
| Total | $4,500,000 | $3,600,000 | $900,000 |

## Key Terms and Concepts

Generally accepted accounting principles (GAAP)
Economic consequences
Quality of earnings
Direct write-off method
Allowance method
Percentage-of-sales procedure for estimating uncollectibles
Aging-of-accounts-receivable procedure for estimating uncollectibles
Sales discounts
Sales returns
Sales allowances
Assigning accounts receivable
Pledging accounts receivable
Factoring accounts receivable
Percentage-of-completion method
Completed contract, or completed sale, method
Installment method
Cost-recovery-first method
Net versus gross reporting of income items

## Questions, Short Exercises, Exercises, Problems, and Cases

### QUESTIONS

For additional student resources, content, and interactive quizzes for this chapter visit the FACMU website: **www.thomsonedu.com/accounting/stickney**

1. Review the meaning of the terms and concepts listed above in Key Terms and Concepts.
2. Which of the two methods for treating uncollectible accounts (direct write-off or allowance) results in recognizing income reductions earlier rather than later? Explain.
3. The direct write-off and the allowance method for uncollectible accounts both involve matching. How does this matching differ between the two methods?
4. **a.** Under what conditions will the direct write-off method and the allowance method result in approximately the same bad debt expense each period?
   **b.** Under what conditions will the direct write-off method and the allowance method result in approximately the same amount for accounts receivable (net) on the balance sheet?
5. **a.** An old wisdom in tennis holds that if your first serves are always good, you are not hitting them hard enough. An analogous statement in business might be that if you have no uncollectible accounts, you probably are not selling enough on credit. Comment on the validity and parallelism of these statements.
   **b.** When are more uncollectible accounts better than less uncollectible accounts?
   **c.** When is a higher percentage of uncollectible accounts better than a lower percentage?
6. Under what circumstances will the Allowance for Uncollectible Accounts have a debit balance during the accounting period? The balance sheet figure for the Allowance for Uncollectible Accounts at the end of the period should never show a debit balance. Why?
7. Construction companies often use the percentage-of-completion method. Why doesn't a typical manufacturing firm use this method of income recognition?
8. Both the installment method and the cost-recovery-first method recognize revenue when a firm collects cash. Why, then, does the pattern of income (that is, revenues minus expenses) over time differ under these two methods?
9. **Chapter 3** said, "From this point on in the book, all discussions assume use of the accrual basis of accounting." This chapter introduces two methods of revenue recognition based on cash—the installment and cost recovery first methods. Explain why the statement quoted from **Chapter 3** remains true.
10. "When the total amount of cash that a firm expects to collect from a customer is highly uncertain, the cost-recovery-first method seems more appropriate than the installment method." Explain.
11. Economists typically define income as an increase in value, or wealth, while a firm holds assets. Accountants typically recognize income when the criteria for revenue recognition are satisfied. Why does the accountants' approach to income recognition differ from that of economists?
12. "The usefulness of the income statement declines as the quality of earnings declines." Do you agree? Why or why not?

SHORT EXERCISES

13. **Allowance method for uncollectible accounts.** Diversified Technologies opened for business on January 2, Year 1. Sales on accounts during Year 1 totaled $126,900. Collections from customers from sales on account during Year 1 totaled $94,300. Diversified Technologies estimates that it will ultimately not collect 4 percent of sales on account. During Year 1, the firm wrote off $2,200 of accounts receivable as uncollectible. The firm uses the allowance method for uncollectible accounts.
    **a.** Compute the amount of bad debt expense for Year 1.
    **b.** Compute the net amount at which accounts receivable will appear on the balance sheet on December 31, Year 1.
14. **Aging accounts receivable.** York Company's accounts receivable show the following balances:

| Age of Accounts | Balance Receivable |
|---|---|
| 0–30 Days | $1,200,000 |
| 31–60 Days | 255,000 |
| 61–120 Days | 75,000 |
| More than 120 Days | 30,000 |

York Company uses the aging-of-accounts-receivable procedure for uncollectible accounts. The credit balance in the Allowance for Uncollectible Accounts is now $16,000. Analysis of recent collection experience suggests that the following percentages be used to compute the estimates of amounts that will eventually prove uncollectible: 0–30 days, half of 1.0 percent; 31–60 days, 1.0 percent; 61–120 days, 10 percent; and more than 120 days, 30 percent. Prepare the indicated entry to provide for estimated uncollectible accounts.

15. **Reconstructing events when using the allowance method.** Selected data from the accounts of Seward Corporation appear below.

| | January 1 | December 31 |
|---|---|---|
| Accounts Receivable—Gross | $82,900 Dr. | $ 87,300 Dr. |
| Allowance for Uncollectible Accounts | 8,700 Cr. | 9,100 Cr. |
| Bad Debt Expense | — | 4,800 Dr. |
| Sales Revenue | — | 240,000 Cr. |

The firm makes all sales on account. There were no recoveries during the year of accounts written off in previous years.

Give the journal entries for the following transactions and events during the year:
**a.** Sales on account.
**b.** Provision for estimated uncollectible accounts.
**c.** Write-off of actual uncollectible accounts.
**d.** Collection of cash from customers from sales on account.

16. **Allowance method; reconstructing journal entry from events.** (From a problem by S. A. Zeff.) During Year 6, Pandora Company wrote off $2,200 of accounts receivable as uncollectible. Pandora Company collected no cash during Year 6 of amounts it had written off in previous years. The balance in the Allowance for Uncollectible Accounts account on the balance sheet was $3,500 at the beginning of Year 6 and $5,000 at the end of Year 6. Present the journal entry that the company made to provide for estimated uncollectibles during Year 6.
17. **Allowance method; reconstructing journal entries from events.** (From a problem by S. A. Zeff.) The balance sheets of Milton Corporation on December 31, Year 1 and Year 2, showed gross accounts receivable of $15,200,000 and $17,600,000, respectively. The balances in the Allowance for Uncollectible Accounts account at the beginning and end of Year 2 were credits of $1,400,000 and $1,550,000, respectively. The income statement for Year 2 shows that the expense for estimated uncollectible accounts was $750,000, which was 1 percent of sales. The firm makes all sales on account. There were no recoveries during Year 2 of accounts written off in previous years. Give all the journal entries made during Year 2 that have an effect on Accounts Receivable and Allowance for Uncollectible Accounts.

**18. Reconstructing events from journal entries.** Give the likely transaction or event that would result in making each of the independent journal entries that follow.

| | | |
|---|---|---|
| **a.** Bad Debt Expense | 2,300 | |
| Allowance for Uncollectible Accounts | | 2,300 |

| Assets | = | Liabilities | + | Shareholders' Equity | (Class.) |
|---|---|---|---|---|---|
| −2,300 | | | | −2,300 | IncSt → RE |

| | | |
|---|---|---|
| **b.** Allowance for Uncollectible Accounts | 450 | |
| Accounts Receivable | | 450 |

| Assets | = | Liabilities | + | Shareholders' Equity | (Class.) |
|---|---|---|---|---|---|
| +450 | | | | | |
| −450 | | | | | |

| | | |
|---|---|---|
| **c.** Bad Debt Expense | 495 | |
| Accounts Receivable | | 495 |

| Assets | = | Liabilities | + | Shareholders' Equity | (Class.) |
|---|---|---|---|---|---|
| −495 | | | | −495 | IncSt → RE |

**19. Percentage-of-completion and completed contract methods of income recognition.** The Shannon Construction Company agreed to build a warehouse for $6,000,000. Expected and actual costs to construct the warehouse were as follows: Year 1, $1,200,000; Year 2, $3,000,000; and Year 3, $600,000. The firm completed the warehouse in Year 3.

Compute revenue, expense, and income before income taxes for Year 1, Year 2, and Year 3 using the percentage-of-completion method and the completed contract method.

**20. Installment and cost-recovery-first methods of income recognition.** Cunningham Realty Partners sold a tract of land costing $80,000 to a manufacturing firm for $120,000. The manufacturing firm agreed to pay $30,000 per year for four years (plus interest).

Compute revenue, expense, and income before income taxes for each of the four years using the installment method and the cost-recovery-first method. Ignore interest.

## EXERCISES

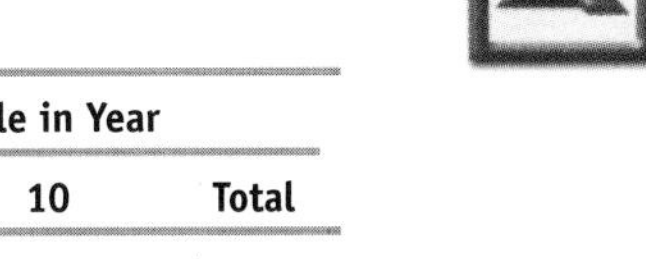

**21. Journal entries for the allowance method.** Data related to sales on account of Heath Company for its first three years of operations appear below:

| Year | Sales on Account | Accounts Written Off as Uncollectible in Year 6 | 7 | 8 | 9 | 10 | Total |
|---|---|---|---|---|---|---|---|
| 6 | $ 340,000 | $1,800 | $5,800 | $ 3,000 | — | — | $10,600 |
| 7 | 450,000 | — | 2,500 | 8,200 | $ 3,400 | — | 14,100 |
| 8 | 580,000 | — | — | 2,900 | 12,700 | $3,300 | 17,900 |
| | $1,370,000 | $1,800 | $8,300 | $14,100 | $16,100 | $3,300 | $42,600 |

Heath Company estimates that 3 percent of sales on account will ultimately become uncollectible. Uncollectible accounts generally occur within three years of the year of sale.

**a.** Prepare journal entries to recognize bad debt expense and to write off uncollectible accounts for Year 6, Year 7, and Year 8 using the allowance method.

**b.** Does 3 percent of sales on account appear to be a reasonable rate for estimating uncollectibles?

**22. Journal entries for the allowance method.** Data related to sales on account of Schneider Corporation appear below:

| Year | Sales on Account | Accounts Written Off as Uncollectible in Year | | | | | |
|---|---|---|---|---|---|---|---|
| | | 1 | 2 | 3 | 4 | 5 | Total |
| 1 .......... | $ 750,000 | $1,300 | $ 8,700 | $ 3,900 | — | — | $13,900 |
| 2 .......... | 1,200,000 | — | 2,500 | 16,600 | $ 3,800 | — | 22,900 |
| 3 .......... | 2,400,000 | — | — | 3,100 | 37,800 | $4,800 | 45,700 |
| | $4,350,000 | $1,300 | $11,200 | $23,600 | $41,600 | $4,800 | $82,500 |

Schneider Corporation estimates that 2 percent of sales on account will ultimately become uncollectible. Uncollectible accounts generally occur within three years of the year of sale.

**a.** Prepare journal entries to recognize bad debt expense and to write off uncollectible accounts for Year 1, Year 2, and Year 3 using the allowance method.

**b.** Does 2 percent of sales on account appear to be a reasonable rate for estimating uncollectibles?

**23. Reconstructing events when using the allowance method.** Selected data from the accounts of Logue Corporation before recognizing bad debt expense for the year appear below:

| | January 1 | December 31 |
|---|---|---|
| Accounts Receivable—Gross .............................. | $115,900 Dr. | $122,700 Dr. |
| Allowance for Uncollectible Accounts .......................... | 18,200 Cr. | 2,900 Dr. |
| Bad Debt Expense .................................... | — | — |
| Sales Revenue ........................................ | — | 450,000 Cr. |

Logue Corporation estimates that 6 percent of sales, which are all on account, will become uncollectible. There were no recoveries during the year of accounts written off in previous years.

Give journal entries for the following transactions or events of Logue Corporation that account for the changes in the accounts shown above:

**a.** Sales on account during the year.

**b.** Write-off of actual uncollectible accounts during the year.

**c.** Collection of cash from customers from sales on account during the year.

**d.** Provision for the year for estimated uncollectible accounts.

**24. Aging accounts receivable.** Dove Company's accounts receivable show the following balances by age:

| Age of Accounts | Balance Receivable |
|---|---|
| 0–30 Days .............................................. | $1,200,000 |
| 0–30 Days .............................................. | $ 400,000 |
| 31–60 Days .............................................. | 90,000 |
| 61–120 Days .............................................. | 40,000 |
| More than 120 Days .......................................... | 20,000 |

The credit balance in the Allowance for Uncollectible Accounts is now $17,200.

Dove Company's independent auditors suggest that the following percentages be used to compute the estimates of amounts that will eventually prove uncollectible: 0–30 days, half of 1.0 percent; 31–60 days, 1.0 percent; 61–120 days, 10 percent; and more than 120 days, 70 percent. Prepare a journal entry that will carry out the auditors' suggestion.

**25. Effects of transactions involving suppliers and customers on cash flows.** Refer to the excerpts from the accounting records of Home and Office Depot in **Exhibit 6.8**. These are the balances in various accounts at the beginning and the end of the year. The amounts shown for sales revenue, bad debt expense, and cost of merchandise sold are their income statement amounts for the year. Home and Office Depot deals with

- many retail customers, some of whom have not paid for goods they have purchased, and
- many suppliers of goods, some of whom the firm has paid for orders not yet received and some of whom have delivered goods for which the firm has not yet paid.

**EXHIBIT 6.8**

**HOME AND OFFICE DEPOT**
**Selected Details from Financial Statements**
**For the Current Year (Exercise 25)**

| | Beginning of Year | End of Year |
|---|---|---|
| PARTIAL BALANCE SHEETS | | |
| Accounts Receivable from Retail Customers | $ 8,000 | $ 8,600 |
| Less Allowance for Uncollectible Retail Receivables | (700) | (750) |
| Advances to Suppliers of Retail Merchandise | 10,000 | 10,400 |
| Inventory of Retail Merchandise | 11,000 | 11,200 |
| Total Assets | $28,300 | $ 29,450 |
| Accounts Payable to Suppliers of Retail Merchandise | 7,000 | 7,500 |
| All Other Accounts (net) | 21,300 | 21,950 |
| Totals | $28,300 | $ 29,450 |
| PARTIAL INCOME STATEMENT FOR THE CURRENT YEAR | | |
| Sales Revenue from Retail Customers | | 130,000 |
| Bad Debt Expense | 2,000 | |
| Cost of Retail Merchandise Sold | 85,000 | |
| Less: Total Expenses | | (87,000) |
| Income | | $ 63,000 |

Home and Office Depot settles all its accounts with customers and suppliers with cash, never with noncash assets.

**a.** Calculate the amount of cash the firm received from its customers during the year.

**b.** Calculate the amount of cash the firm paid to its suppliers of retail merchandise during the year.

**26. Revenue recognition for various types of businesses.** Discuss when each of the following types of businesses is likely to recognize revenue:

**a.** A shoe store.

**b.** A shipbuilding firm constructing an aircraft carrier under a government contract.

**c.** A real estate developer selling lots on long-term contracts with small down payments required.

**d.** A barbershop.

**e.** A citrus-growing firm.

**f.** A producer of television movies working under the condition that the rights to the movies are sold to a television network for the first three years and all rights thereafter revert to the producer.

**g.** A residential real estate developer who constructs only "speculative" houses and later sells the houses to buyers.

**h.** A producer of fine whiskey that is aged from 6 to 12 years before sale.

**i.** A savings and loan association lending money for home mortgages.

**j.** A travel agency that sells tickets in one period and has customers take trips or return tickets in the next period.

**k.** A printer who prints only custom-order stationery.

**l.** A seller to food stores of trading stamps redeemable by food store customers for various household products.

**m.** A wholesale food distributor.

**n.** A livestock rancher.

**o.** A shipping company that loads cargo in one accounting period, carries cargo across the ocean in a second accounting period, and unloads the cargo in a third period; the shipping is all done under contract, and cash collection of shipping charges is relatively certain.

**27. Income recognition for various businesses.** Discuss when each of the following businesses should recognize revenue and any income measurement issues that arise.

**a.** Company A develops software and sells it to customers for an up-front fee. Company A provides these customers with password-protected access to its web site for two years after delivery of the software. With this access, customers can download certain data and other software. Company A has an obligation to provide updates on its web site.

**b.** Company B develops software and sells it to newly formed storage application service providers (SAPs), who promise to pay for the software over the next two years. These SAPs in turn place the software on their web sites and sell rights to access the software to its customers.

**c.** Company C develops software that it places on its web site. It sells rights to access this software on-line to its customers for a period of two years. Customers pay an up-front fee for the right to access the software.

**d.** Company D maintains an auction site on the web. It charges customers an up-front fee to list products for sale and a transaction fee when a sale takes place. The transaction fee is refundable if the auction winner fails to honor its commitment to purchase the product.

**e.** Company E sells products of various companies on its web site. Company E transmits customers' purchase requests to the various supplier companies, who fill the orders. Customers charge for their purchases on third-party credit cards. Company E receives a fee from the supplier companies for each item sold.

**f.** Company F sells products of various companies on its web site. It promises to sell a minimum number of items each month and pays storage and insurance costs for that minimum number of units. Actual storage of these units takes place at the suppliers' warehouse. The suppliers also handle shipments to customers. Customers charge for their purchases on third-party credit cards.

**g.** Company G manufactures and sells personal computers (PCs). Customers receive a $400 rebate on the purchase of the computer if they will purchase Internet access services for three years after the purchase of the computer. Customers mail their rebate coupon to the Internet service provider (called an *ISP;* for example, AOL, AT&T). The ISP bears 90 percent on the initial cost of the rebate and the PC manufacturer bears the other 10 percent. If customers do not subscribe for the full three-year period, the parties reallocate the cost of the rebate, initially borne $360 (= 0.90 × $400) by the ISP and $40 (= 0.10 × $400) by the manufacturer, resulting in the PC manufacturer's paying the ISP to reduce the ISP's share of the rebate's cost from $360 to a smaller amount.

**h.** Company H sells advertising space on its web site to other companies. For an up-front fee, Company H guarantees a certain minimum number of hits, viewings, or click-throughs each month of the one-year contract period. It must return a pro rata portion of the fee if the hits and click-throughs fall short of the guarantee.

**i.** Company I sells advertising space on its web site to other companies. It recently received 10,000 shares of common stock of Upstart Company in payment for certain advertising space. Upstart Company intends to make an initial public offering of its common stock in six months. At the most recent financing round, venture capitalists paid $10 per share for the common stock.

**j.** Company J and Company K both maintain web sites. Each company "sells" advertising space to the other company for an agreed-upon period, with no money changing hands.

## PROBLEMS AND CASES

**28. Analyzing changes in accounts receivable.** Selected data from the financial statements of American Express appear below (amounts in millions).

| | Year 4 | Year 5 | Year 6 | Year 7 |
|---|---|---|---|---|
| **Balance Sheet** | | | | |
| Accounts and Notes Receivable, net of allowance for uncollectible accounts of $1,419 at the end of Year 4, $1,411 at the end of Year 5, $1,559 at the end of Year 6, and $1,728 at the end of Year 7 . . . . . . . . | $41,883 | $43,278 | $50,049 | $56,631 |
| **Income Statement** | | | | |
| Revenues on Account . . . . . . . . . . . . . . . . | | $19,132 | $21,278 | $23,675 |
| Bad Debt Expense . . . . . . . . . . . . . . . . . | | $ 2,187 | $ 2,212 | $ 2,439 |

**a.** Prepare journal entries for Year 5, Year 6, and Year 7 for the following events:

(1) Revenues on account.
(2) Provision for estimated uncollectibles.
(3) Write-off of actual uncollectible accounts.
(4) Collection of cash from customers.

**b.** Compute the amount of the following ratios:
(1) Accounts receivable turnover ratio for Year 5, Year 6, and Year 7.
(2) Bad debt expense divided by revenues on account for Year 5, Year 6, and Year 7.
(3) Allowance for uncollectible accounts divided by accounts receivable (gross) at the end of Year 5, Year 6, and Year 7.
(4) Write-offs of actual uncollectible accounts divided by average accounts receivable (gross) for Year 5, Year 6, and Year 7.

**c.** What do the ratios computed in part **b** suggest about the collection experience of American Express during the three-year period?

**29. Analyzing changes in accounts receivable.** The financial statements and notes for May Department Stores reveal the following for four recent years (amounts in millions):

| | Year 9 | Year 10 | Year 11 | Year 12 |
|---|---|---|---|---|
| Total Sales | $8,330 | $9,456 | $10,035 | $10,615 |
| Credit Sales/Total Sales | 62.2% | 64.9% | 66.9% | 68.7% |
| Bad Debt Expense | $ 57 | $ 64 | $ 82 | $ 96 |

| | End of Year | | | | |
|---|---|---|---|---|---|
| | Year 8 | Year 9 | Year 10 | Year 11 | Year 12 |
| Accounts Receivable, Gross | $1,592 | $2,099 | $2,223 | $2,456 | $2,607 |
| Less Allowance for Uncollectible Accounts | (47) | (61) | (66) | (84) | (99) |
| Accounts Receivable, Net | $1,545 | $2,038 | $2,157 | $2,372 | $2,508 |

**a.** Compute the amount of accounts written off as uncollectible during Year 9, Year 10, Year 11, and Year 12.

**b.** Compute the amount of cash collections from credit customers during Year 9, Year 10, Year 11, and Year 12.

**c.** Calculate the accounts receivable turnover ratio for Year 9, Year 10, Year 11, and Year 12 using total sales in the numerator and average accounts receivable (net) in the denominator.

**d.** Repeat part **c** but use credit sales in the numerator.

**e.** What are the likely reasons for the different trends in the two measures of the accounts receivable turnover ratio in parts **c** and **d**?

**30. Analyzing changes in accounts receivable.** Selected data from the financial statements of Sears, a retail department store chain, appear below (amounts in millions).

| | Year 6 | Year 7 | Year 8 | Year 9 |
|---|---|---|---|---|
| **Balance Sheet** | | | | |
| Accounts and Notes Receivable, net of allowance for uncollectible accounts of $1,113 at the end of Year 6, $974 at the end of Year 7, $760 at the end of Year 8, and $686 at the end of Year 9 | $19,843 | $17,972 | $18,033 | $17,317 |
| **Income Statement** | | | | |
| Sales on Account | | $39,953 | $39,484 | $40,937 |
| Bad Debt Expense | | $ 1,287 | $ 871 | $ 884 |

**a.** Prepare journal entries for Year 7, Year 8, and Year 9 for the following events:
(1) Sales on account.
(2) Bad debt expense.
(3) Write-off of actual uncollectible accounts.
(4) Collection of cash from customers.

**b.** Compute the amount of the following ratios:

**(1)** Accounts receivable turnover ratio for Year 7, Year 8, and Year 9.

**(2)** Bad debt expense divided by sales on account for Year 7, Year 8, and Year 9.

**(3)** Allowance for uncollectible accounts divided by accounts receivable (gross) at the end of Year 7, Year 8, and Year 9.

**(4)** Write-offs of actual uncollectible accounts divided by average accounts receivable (gross) for Year 7, Year 8, and Year 9.

**c.** What do the ratios computed in part **b** suggest about the collection experience of Sears during the three-year period?

**31. Reconstructing transactions affecting accounts receivable and uncollectible accounts.** The sales, all on account, of Pins Company in Year 10, its first year of operations, were $700,000. Collections totaled $500,000. On December 31, Year 10, Pins Company estimated that 2 percent of all sales would probably be uncollectible. On that date, Pins Company wrote off specific accounts in the amount of $8,000.

The balances in selected accounts on December 31, *Year 11*, are as follows:

| | | |
|---|---|---|
| Accounts Receivable (Dr.) . . . . . . . . . . . . . . . . . . . . . . . . . . . . . . . . . | 300,000 | |
| Allowance for Uncollectible Accounts (Dr.) . . . . . . . . . . . . . . . . . . . . . | 10,000 | |
| Bad Debt Expense . . . . . . . . . . . . . . . . . . . . . . . . . . . . . . . . . . . . . . | — | — |
| Sales Revenue (Cr.) . . . . . . . . . . . . . . . . . . . . . . . . . . . . . . . . . . . | | 800,000 |

The amounts shown for bad debt expense and sales revenue are the amount, if any, recorded in the accounts during Year 11. On December 31, Year 11, Pins Company carried out an aging of its accounts receivable balances and estimated that the Year 11 ending balance of accounts receivable contained $11,000 of probable uncollectibles. That is, the allowance account should have an $11,000 ending credit balance. It made adjusting entries appropriate for this estimate. Some of the $800,000 sales during Year 11 were for cash and some were on account; the problem purposefully does not give the amounts.

**a.** What was the balance in the Accounts Receivable (gross) account at the end of *Year 10*? Give the amount and whether it was a debit or a credit.

**b.** What was the balance in the Allowance for Uncollectible Accounts account at the end of *Year 10*? Give the amount and whether it was a debit or a credit.

**c.** What was bad debt expense for *Year 11*?

**d.** What was the amount of specific accounts receivable written off as being uncollectible during *Year 11*?

**e.** What were total cash collections in *Year 11* from customers (for cash sales and collections from customers who had purchased on account in either *Year 10* or *Year 11*)?

**f.** What was the net balance of accounts receivable included in the balance sheet asset total for December 31, *Year 11*?

**32. Effect of errors involving accounts receivable on financial statement ratios.** Indicate—using O/S (overstated), U/S (understated), or NO (no effect)—the pretax effect of each of the following errors on (1) the rate of return on assets, (2) the accounts receivable turnover, and (3) the debt equity ratio. Each of these ratios is less than 100 percent before discovering the error.

**a.** A firm using the allowance method neglected to provide for estimated uncollectible accounts at the end of the year.

**b.** A firm using the allowance method neglected to write off specific accounts as uncollectible at the end of the year.

**c.** A firm credited a check received from a customer to Advances from Customers even though the customer was paying for purchases previously made on account.

**d.** A firm recorded as a sale a customer's order received on the last day of the accounting period, even though the firm will not ship the product until the next accounting period.

**e.** A firm sold goods on account to a particular customer and properly recorded the transactions in the accounts. The customer returned the goods within a few days of the sale, before paying for them, but the firm neglected to record the return on the goods in its accounts. The firm normally treats sales returns as a reduction in Sales Revenue.

**33. Income recognition for a nuclear generator manufacturer.** General Electric Company agreed on June 15, Year 2, to construct a nuclear generator for Consolidated Edison Company. The contract price of $200 million is to be paid as follows: at the time of signing, $20 million;

on December 31, Year 3, $100 million; and at completion on June 30, Year 4, $80 million. General Electric Company incurred the following costs in constructing the generator: Year 2, $42 million; Year 3, $54 million; and Year 4, $24 million. These amounts conformed to original expectations.

**a.** Calculate the amount of revenue, expense, and income before income taxes for Year 2, Year 3, and Year 4 under each of the following revenue recognition methods:
  **(1)** Percentage-of-completion method.
  **(2)** Completed contract method.
  **(3)** Installment method.
  **(4)** Cost-recovery-first method.

**b.** Which method do you believe provides the best measure of General Electric Company's performance under the contract? Why?

**34. Income recognition for a contractor.** On October 15, Year 1, Flanikin Construction Company contracted to build a shopping center at a contract price of $180 million. The schedule of expected and actual cash collections and contract costs is as follows:

| Year | Cash Collections from Customers | Estimated and Actual Cost Incurred |
|---|---|---|
| 1 | $ 36,000,000 | $ 12,000,000 |
| 2 | 45,000,000 | 36,000,000 |
| 3 | 45,000,000 | 48,000,000 |
| 4 | 54,000,000 | 24,000,000 |
| | $180,000,000 | $120,000,000 |

**a.** Calculate the amount of revenue, expense, and net income for each of the four years under the following revenue recognition methods:
  **(1)** Percentage-of-completion method.
  **(2)** Completed contract method.
  **(3)** Installment method.
  **(4)** Cost-recovery-first method.

**b.** Which method do you believe provides the best measure of Flanikin Construction Company's performance under the contract? Why?

**35. Point-of-sale versus installment method of income recognition.** The J. C. Spangle catalog division began business on January 1, Year 8. Activities of the company for the first two years are as follows:

| | Year 8 | Year 9 |
|---|---|---|
| Sales, All on Account | $200,000 | $300,000 |
| Collections from Customers: | | |
| On Year 8 Sales | 90,000 | 110,000 |
| On Year 9 Sales | — | 120,000 |
| Purchase of Merchandise | 180,000 | 240,000 |
| Inventory of Merchandise at 12/31 | 60,000 | 114,000 |
| All Expenses Other Than Merchandise, Paid in Cash | 32,000 | 44,000 |

**a.** Prepare income statements for Year 8 and Year 9, assuming that the company uses the accrual basis of accounting and recognizes revenue at the time of sale.

**b.** Prepare income statements for Year 8 and Year 9, assuming that the company uses the accrual basis of accounting and recognizes revenue at the time of cash collection following the installment method of accounting. "All Expenses Other Than Merchandise, Paid in Cash" are period expenses.

**36. Revenue recognition for a franchise.** Pickin Chicken, Inc., and Country Delight, Inc., both sell franchises for their chicken restaurants. The franchisee receives the right to use the franchisor's products and to benefit from national training and advertising programs. The franchisee agrees to pay $50,000 for exclusive franchise rights in a particular city. Of this amount, the franchisee pays $20,000 on signing the franchise agreement and promises to pay

**EXHIBIT 6.9** Common-Size Income Statements for Selected Companies (Problem 37)

| | Amgen | Brown-Forman | Deere | Fluor | Golden West | Merrill Lynch | Rockwell |
|---|---|---|---|---|---|---|---|
| **Revenues** | | | | | | | |
| Sales of Goods | 98.7% | 99.9% | 83.6% | — | — | — | 99.3% |
| Sales of Services | — | — | — | 99.7% | 2.0% | 47.5% | — |
| Interest on Investments | 1.3 | 0.1 | 16.4 | 0.3 | 98.0 | 52.5 | 0.7 |
| Total Revenues | 100.0% | 100.0% | 100.0% | 100.0% | 100.0% | 100.0% | 100.0% |
| **Expenses** | | | | | | | |
| Cost of Goods or Services Sold | (14.3) | (35.5) | (69.4) | (95.6) | — | (43.3) | (77.4) |
| Selling and Administrative | (23.4) | (33.1) | (11.4) | (.6) | (15.9) | — | (12.6) |
| Other Operating[a] | (26.4) | (15.4) | (3.5) | — | (3.3) | — | — |
| Interest | (0.7) | (1.3) | (11.0) | (0.2) | (60.4) | (47.2) | (0.9) |
| Income before Income Taxes | 35.2% | 14.7% | 4.7% | 3.6% | 20.4% | 9.5% | 9.1% |
| Income Tax Expense | (16.1) | (5.9) | (1.6) | (1.3) | (8.4) | (3.9) | (3.5) |
| Net Income | 19.1% | 8.8% | 3.1% | 2.3% | 12.0% | 5.6% | 5.6% |
| **Revenues/Average Total Assets** | 0.9 | 1.3 | 0.7 | 3.1 | 0.1 | 0.1 | 1.2 |

[a]Represents research and development costs for Amgen and Deere, excise taxes for Brown-Forman, and a provision for loan losses by Golden West.

the remainder in five equal annual installments of $6,000 each starting one year after the initial signing payment.

Pickin Chicken, Inc., recognizes franchise revenue as it signs agreements, whereas Country Delight, Inc., recognizes franchise revenue on an installment basis. In Year 2, each company sold eight franchises. In Year 3, each sold five franchises. In Year 4, neither company sold a franchise.

**a.** Calculate the amount of revenue recognized by each company during Year 2, Year 3, Year 4, Year 5, Year 6, Year 7, and Year 8.

**b.** When do you think a franchisor should recognize franchise revenue? Why?

**37. Income recognition for various types of business.** Most business firms recognize revenues at the time of sale or delivery of goods and services and, following the principles of the accrual basis of accounting, match expenses either with associated revenues or with the period when they consume resources in operations. Users of financial statements should maintain a questioning attitude as to the appropriateness of recognizing revenues at the time of sale and as to the timing of expense recognition. **Exhibit 6.9** presents common-size income statements for seven firms for a recent year, with all amounts expressed as a percentage of total revenues. **Exhibit 6.9** also indicates the revenues generated by each firm for each dollar of assets in use on average during the year. A brief description of the activities of each firm follows.

**Amgen** engages in the development, manufacturing, and marketing of biotechnology products. Developing and obtaining approval of biotechnology products takes 10 or more years. Amgen has two principal products that it manufactures and markets and several more products in the development pipeline.

**Brown-Forman** is a distiller of hard liquors. After combining the ingredients, the company ages the liquors for five or more years before sale.

**Deere** manufactures farm equipment. It sells this equipment to a network of independent distributors, who in turn sell the equipment to final consumers. Deere provides financing and insurance services both to its distributors and to final consumers.

**Fluor** engages in construction services on multiyear construction projects. It subcontracts most of the actual construction work and receives a fee for its services.

**Golden West** is a savings and loan company. It takes deposits from customers and lends funds, primarily to individuals for home mortgages. Customers typically pay a fee (called "points") at the time of loan origination based on the amount borrowed. Their monthly mortgage payments include interest on the outstanding loan balance and a partial repayment of the principal of the loan.

**Merrill Lynch** engages in the securities business. It obtains funds primarily from short-term capital market sources and invests the funds primarily in short-term, readily marketable financial instruments. It attempts to generate an excess of investment returns over the cost of the funds invested. Merrill Lynch also offers fee-based services, such as financial consulting, buying and selling securities for customers, securities underwriting, and investment management.

**Rockwell** is a technology-based electronics and aerospace company. It engages in research and development on behalf of its customers, which include the U.S. government (space shuttle program, defense electronics) and private-sector entities. Its contracts tend to run for many years on a constantly renewed basis.

**a.** When should each of these companies recognize revenue? What unique issues does each company face in the recognition of expenses?

**b.** Suggest possible reasons for differences in the net income divided by revenues percentages for these companies.

**38. Meaning of Allowance for Uncollectible Accounts account.** Indicate whether each of the following accurately describes the meaning of the Allowance for Uncollectible Accounts account when properly used. If the description does not apply to this account, discuss why it does not.

**a.** Assets available in case customers don't later pay what they owe.

**b.** Cash available in case customers don't later pay what they owe.

**c.** Estimates of the amount that customers who purchased goods this period, but who have not yet paid, and won't later pay.

**d.** Estimates of the amount of goods purchased by customers, whether paid or not, for which we won't collect later.

**e.** Estimates of the amount that customers who purchased goods at any time, but who have not yet paid, and won't later pay.

**f.** Estimates of amount the firm will owe to others if its customers who purchased goods this period don't later pay what they owe.
**g.** Estimates of amount the firm will owe to others if its customers who ever purchased goods don't later pay what they owe.
**h.** Estimates of the amount of sales for the current period that will become bad debt expense for the current period.
**i.** An amount of deferred revenue.
**j.** A part of retained earnings.

**39. Understanding the purpose of the Allowance for Uncollectible Accounts account.** A member of the Audit Committee of a firm asks the Chief Financial Officer the following question: "How do you know the Allowance for Uncollectible Accounts is adequate?" Discuss the adequacy or inadequacy of each of the following independent responses.

**a.** "I think it much more likely than not that the amount of cash and marketable securities is adequate to cover any cash shortage caused by customers' not paying what they owe."
**b.** "I have checked the Bad Debt Expense for sales made this past period and found the amount reasonable."
**c.** "I have performed an aging of all accounts receivable for all sales this period and found the amount reasonable."
**d.** "I have performed an aging of all accounts receivable and found the amount reasonable."
**e.** "We performed a detailed confirmation of receivables from customers whose accounts the firm wrote off as uncollectible this period and found the decision to write them off suitable under the circumstances."
**f.** "We performed a detailed confirmation of receivables from customers whose accounts the firm has neither collected nor written off by the end of this period and found the decisions suitable under the circumstances."
**g.** "I know the Allowance account was correct at the end of last period and I checked the Bad Debt Expense for this period using a percentage of sales recommended for this class of customers by the top two credit reporting agencies."

**40. Ethical issues in accounting for uncollectible accounts.** Sage Department Stores operates a chain of department stores throughout the United States. It provides its customers with credit cards, both to stimulate sales and to save fees the firm would otherwise incur if customers used third-party credit cards. Sage Department Stores uses the percentage-of-sales procedure to estimate the amount of bad debt expense each period. That percentage has ranged from 2.5 to 3.2 percent in recent years. It also periodically ages its account receivable to assess the reasonableness of the balance in the Allowance for Uncollectibles Accounts account. Recession conditions in the economy during the current period caused Sage Department Stores and most of its competitors to operate at a net loss. Sage Department Stores increased its provision for estimated uncollectibles during the current year to 5 percent of sales. Although the increased provision resulted in a still larger net loss than it would have reported if it had used its usual percentage, security analysts did not seem to notice. The chief financial officer was prepared, however, to defend the higher percentage on the grounds that one might expect recession conditions to lead to more uncollectible accounts. Further, the chief financial officer reasoned that the firm could take a much smaller provision during the next few years if the higher level of uncollectibles did not materialize, which would increase future earnings and show even more earnings growth. The chief financial officer also reasoned that competitors were likely to play this earnings game. To remain competitive, Sage Department Stores must play it as well. Discuss any ethical issues that you see confronting the chief financial officer of Sage Department Stores.

**41. Ethical issues in accounting for long-contracts.** Halliburton Company provides engineering and construction services under multi-year contracts to energy, industrial, and governmental customers. It has used the percentage-of-completion method of accounting for its multi-year contracts for many years. In addition to the initial contract price, its contracts call for additional payments if the customer makes changes in the plans—so-called *change orders*. The firm has not, heretofore, recognized revenue from change orders until it completes the work and collects the cash. It now decides that it can predict with reasonable precision the amount it will collect under change orders and the costs of making the changes. So, it adds the expected collections from the change orders to cash collections expected from customers and the costs of the changes to the costs expected to complete various jobs. Thus, it now recognizes income from change orders as part of its income recognition for all multi-year contracts using the percentage-of-completion method. It calls reader's attention to this change in a hard-to-understand note to the financial statements. Discuss any ethical issues that you see confronting Halliburton Company in its decision to change its income recognition method for change orders.

42. **Ethical issues in income recognition.** Qwest Communications International (Qwest), a telecommunications company, signed a $100 million contract with the Arizona Schools Facilities Board to supply high-speed Internet access to public schools in Arizona. The contract required that Qwest supply and install the computer equipment and do the necessary steps to link the computers to the Internet, a process scheduled to take over 18 months. Qwest initially intended to recognize revenue from the contract on a percentage-of-completion basis as it completed installation at each school and received payment from the State. Soon after delivery but before installation of the equipment, Qwest decided to separate the recognition of revenue for sale of the equipment from the recognition of revenue for installation services. It recognized $33 million in revenue at the time of shipment and planned to recognize $67 million as installation occurred. This change accelerated the recognition relative to the pattern initially anticipated. To allow this change in revenue recognition, Qwest's auditor required that the customer, Arizona Schools Facilities Board, formally request a separation of the hardware sales contract from the installation contract. Personnel from Qwest drafted such a request, printed it on stationery of the Arizona Schools Facilities Board, and presented to the head of the Board to sign. The head of the Board hastily signed the letter at an airport after receiving assurance that the State wouldn't actually have to pay Qwest until Qwest installed the equipment. The head of the board stated, "There was a veiled threat that if we did not sign the letter, the equipment would not be available." Discuss any ethical issues that you see confronting managers at Qwest and its auditor.

CHAPTER 7

# Inventories: The Source of Operating Profits

## Learning Objectives

1. Apply the principles of cost inclusions for assets to both purchased and manufactured inventories.
2. Understand the effect of changes in the valuation of inventories subsequent to acquisition on both balance sheet and income statement amounts.
3. Understand why most firms either must make, or prefer to make, a cost flow assumption for inventories and cost of goods sold.
4. Compute cost of goods sold and ending inventory using a first-in, first-out (FIFO), a last-in, first-out (LIFO), and a weighted-average cost flow assumption.
5. Understand the effect of using FIFO, LIFO, and weighted-average cost flow assumptions on balance sheet and income statement amounts.
6. Analyze the effects of choices that firms make for inventories and cost of goods sold on assessments of profitability and risk.

*W*e continue our trek down the balance sheet and the income statement by considering the accounting for inventories, an asset, and cost of goods sold, the related expense. Inventories are a major asset for merchandising and manufacturing firms and cost of goods sold is typically their largest single expense. Such firms must make various choices in accounting for inventories. The choices made affect assessments of a firm's profitability and risk over time and can make two firms that are otherwise alike appear to differ.

**Example 1** **Exhibit 1.2** in **Chapter 1** indicates that Wal-Mart Stores reported cost of goods sold of $198,747 million and net income of $9,075 million in Year 14. The ratio of cost of goods sold to net income for Wal-Mart is 21.9 to 1 (= $198,747/$9,075). Assume that Wal-Mart must make inventory choices (discussed later in this chapter) that would reduce its cost of goods sold by, say, 2 percent. A 2 percent reduction in cost of goods sold results in a $3,975 (= 0.02 × $198,747) million decrease in cost of goods sold and a corresponding increase in income before taxes. Assuming an income tax rate of 35 percent, net income would increase by $2,584 [= (1.00 − 0.35) × $3,975] million to $11,659 (= $9,075 + $2,584) million. A 2 percent decrease in cost of goods sold increases net income by 28.5 percent (= $2,584/$9,075). Thus, a seemingly small change in cost of goods sold can significantly affect net income.

# A Review of Fundamentals

## Inventory Terminology

The term **inventory** means a stock of goods or other items that a firm owns and holds for sale or for further processing as part of its business operations. For example, lawnmowers are inventory of Toro, a lawnmower manufacturer, or of Lowe's, a home products store, but not of a firm that provides lawn-mowing services. Marketable securities are inventory of Merrill Lynch, a securities dealer, but not of Hewlett-Packard, a computer electronics company, which temporarily invests excess cash in such securities. *Merchandise inventory* denotes goods held for sale by a merchandising firm (retail or wholesale businesses). *Finished goods inventory* denotes goods held for sale by a manufacturing firm. The inventories of manufacturing firms also include *raw materials inventory* (materials being stored that will become part of goods to be produced) and *work-in-process inventory* (partially completed products in the factory). A later section discusses the accounting for inventories by manufacturing firms.

## Asset Valuation and Income Measurement

Assets are resources that provide future benefits. Inventories are assets because firms expect to sell the inventories for cash, or cash equivalents. At the time of sale, firms consume the benefits of inventories, receiving in exchange either cash or accounts receivable. Firms recognize revenue in an amount equal to the cash-equivalent value of the asset received and recognize an expense, cost of goods sold, for the current book value, usually acquisition cost, of the inventory sold. Firms usually recognize all of the difference between the selling price and the acquisition cost of the inventory as income in the period of sale.

**Example 2** Zale, Inc. operates a chain of retail jewelry stores. During the fall of Year 13, it purchased gold jewelry at an acquisition cost of $300,000. It initially marked this jewelry to sell for $750,000. A worldwide shortage of gold developed late in Year 13, leading to an increase in the amount that Zale, Inc. would have to pay to replace this jewelry to $500,000. Zale, Inc. consequently increased the selling price of the jewelry to $1,250,000. Zale, Inc. sold this jewelry in the spring and summer of Year 14 for $1,250,000. Zale, Inc. recognizes in earnings the difference between the selling price of $1,250,000 and the acquisition cost of $300,000 in Year 14 when it sells the jewelry, even though a portion of the increase in value of the jewelry occurred during Year 13. GAAP normally does not permit firms to recognize increases in the value of inventory until firms sell the inventory and a market transaction confirms the amount of the increase in value.

The market value of inventories sometimes declines below acquisition cost while firms hold the inventories awaiting sale to customers. GAAP requires firms in these situations to write down the inventories and recognize an expense in the amount of the writedown. The subsequent sale of the inventory results in income equal to the difference between the selling price and the new, lower book value.

**Example 3** Circuit City Stores sells computers and other electronic products. During the fall of Year 13, it purchased computer products at an acquisition cost of $300,000. It initially marked the computers to sell for $750,000. An unexpected introduction of a new computer chip late in Year 13 reduced the market value of these computers to $175,000. GAAP requires that Circuit City Stores write down these computers to $175,000 and recognize a loss of $125,000 (= $300,000 acquisition cost − $175,000 market value) in Year 13. The subsequent sale of these computers in Year 14 for $175,000 results in zero net income in that year from the computers sold.

Thus, the valuation of inventory on the balance sheet closely relates to the measurement of net income. A later section discusses the valuation of inventory subsequent to acquisition more fully.

## Inventory Equation

The **inventory equation** helps one understand accounting for inventory. The following equation measures all quantities in physical units:

$$\underbrace{\text{Beginning Inventory} + \text{Additions}}_{\text{Goods Available for Sale (or Use)}} - \text{Withdrawals} = \text{Ending Inventory}$$

If we begin a period with 2,000 pounds of sugar (beginning inventory) and purchase (add) 4,500 pounds during the period, then there are 6,500 (= 2,000 + 4,500) pounds available for sale or use. The term *goods available for sale* (or use) denotes the sum of Beginning Inventory plus Additions. If we use (withdraw) 5,300 pounds during the period, 1,200 pounds of sugar should remain at the end of the period (ending inventory).

The inventory equation can also appear as follows:

$$\underbrace{\text{Beginning Inventory} + \text{Additions}}_{\text{Goods Available for Sale (or Use)}} - \text{Ending Inventory} = \text{Withdrawals}$$

If we begin the period with 2,000 pounds of sugar, purchase 4,500 pounds during the period, and observe 1,200 pounds on hand at the end of the period, then we know the firm used 5,300 (= 2,000 + 4,500 − 1,200) pounds of sugar during the period.[1]

Financial statements report dollar amounts, not physical units such as pounds or cubic feet. The accountant must transform physical quantities for beginning inventory, additions, withdrawals, and ending inventory into dollar amounts in order to measure income for the period and financial position at the beginning and end of the period. When acquisition costs of inventory items remain constant, inventory accounting problems are minor because all items carry the same per-unit cost; physical quantities and dollar valuations change together. Any variation in the amounts recorded for inventories result only from changes in quantities. The major problems in inventory accounting arise because the per-unit acquisition costs of inventory items change over time.

**Example 4** Suppose that an appliance store has a beginning inventory of one television set, TV set 1, which costs $250. Suppose further that it purchases two TV sets during the period, TV set 2 for $290 and TV set 3 later in the period for $300, and that it sells one TV set for $550. The three TV sets are physically identical; the firm acquired them at different times as their acquisition costs changed, so only their costs differ.

We can write the inventory equation as follows, measuring all quantities in dollars of cost:

| Beginning Inventory | + | Purchases | − | Cost of Goods Sold | = | Ending Inventory |
|---|---|---|---|---|---|---|
| $250 | + | $590 | − | ? | = | ? |

If the firm using serial numbers or product bar codes identifies TV set 2 as the one sold, then cost of goods sold is $290 and ending inventory is $550 (= $250 + $300).

The physical similarity of some products, however, creates difficulties in identifying the products sold and the products in ending inventory. Even when technology, such as product bar codes, allows firms to keep track of the cost of each item in inventory, they may prefer not to incur the costs to do so. Firms may prefer instead to make a cost flow assumption. GAAP allows the appliance store to assume that it sold the first TV set purchased (referred to as a *first-in, first-out,* or FIFO cost flow assumption), or the last TV set purchased (referred to as a *last-in, first-out,* or LIFO cost flow assumption), or the average TV set purchased, that is, a TV carrying the average cost of all three TV sets (referred to as a *weighted-average* cost flow assumption). A later section discusses each of these cost flow assumptions more fully.

## Issues in Inventory Accounting

The remainder of this chapter discusses three issues in inventory accounting:

1. The costs included in the acquisition cost of inventory.
2. The treatment of changes in the market value of inventories subsequent to acquisition.
3. The cost flow assumption used to trace the movement of costs into and out of inventory.

---

[1]When a firm computes the cost of goods sold each time it sells an inventory item, it uses a *perpetual inventory system.* Bar codes and serial numbers on items sold permit such continual tracking of costs. When a firm computes the cost of goods sold at the end of each period by taking a physical inventory and presuming that any items not in ending inventory, withdrawals in the equation above, were sold, it uses a *periodic inventory system.* See the Glossary for a further description of these two inventory systems. We assume use of the periodic system throughout this book.

## Issue 1: Costs Included in Inventory

The guiding principle for cost inclusions is that the balance sheet amount for inventory should include all costs incurred to acquire goods and prepare them for sale.

**Merchandising Firms** Merchandising firms, such as Wal-Mart Stores, acquire inventory items in a physical condition ready for sale. Acquisition cost includes the invoice price less any cash discounts taken for prompt payment. Acquisition cost also includes the cost of transporting, receiving, unpacking, inspecting, and shelving, as well as any costs to record the **purchases** in the accounts. The example on **page 44** in **Chapter 2**, applying the cost-inclusion principle to equipment, applies to inventory as well.

**Manufacturing Firms** Manufacturing firms, such as Ford Motor Company, do not acquire inventory items in a physical form ready for sale. Instead, such firms transform raw materials into finished products in their factories. The acquisition cost of manufactured inventories includes three categories of costs:

- **Direct Materials**: The cost of materials that the manufacturing firm can trace directly to units of product it manufactures. Direct materials for Ford Motor Company include the engine, car body, tires, seat cushions, and other parts that physically become part of a manufactured automobile.
- **Direct Labor**: The cost of labor to transform raw materials into a finished product. This cost for Ford Motor Company includes the compensation of factory workers on the production line.
- **Manufacturing Overhead**: A variety of indirect costs (depreciation, insurance and taxes on manufacturing facilities, supervisory factory labor, and supplies for factory equipment) that firms cannot trace directly to products manufactured but which provide a firm with productive capacity. Manufacturing overhead for Ford Motor Company includes depreciation, insurance, and property taxes on the factory building and equipment; compensation of supervisory factory labor; electricity to light, heat, and cool the factory; and costs associated with setting up a manufacturing process to do a variety of jobs. In today's economy, these indirect costs constitute an ever-larger portion of total manufacturing costs and the treatment of these costs for product costing, performance evaluation, and other management purposes has become an art in itself. *Activity-based costing* is a modern name for the process of allocating indirect costs to processes and products.

Until the manufacturing firm sells the units and recognizes revenue, it treats all manufacturing costs as **product costs**, which are assets, and accumulates the costs in various inventory accounts.

A manufacturing firm, like a merchandising firm, also incurs various marketing costs (commissions for the sales staff, depreciation, insurance and taxes on the sales staff's automobiles) and administrative costs (salary of the chief executive officer, depreciation on computer facilities used in administration). Both merchandising and manufacturing firms treat selling and administrative costs as **period expenses**. **Figure 7.1** summarizes the nature and **flow of costs** for a manufacturing firm.

**Overview of the Accounting Process for a Manufacturing Firm** A manufacturing firm maintains separate inventory accounts to accumulate the product costs that it incurs at various stages of completion. The **Raw Materials Inventory** account includes (with debits) the cost of raw materials purchased but not yet transferred to the factory floor. The balance in the Raw Materials Inventory account indicates the cost of raw materials on hand in the raw materials storeroom or warehouse. When the manufacturer transfers raw materials from the raw material storeroom to the factory floor, it transfers the cost of the raw materials from the Raw Materials Inventory account (with credits) to the Work-in-Process Inventory account (with debits).

The **Work-in-Process Inventory** account accumulates (with debits) the cost of raw materials transferred from the raw materials storeroom, the cost of direct labor services used in production, and the manufacturing overhead costs incurred. When the firm completes manufacturing, it

FIGURE 7.1 Diagram of Cost Flows

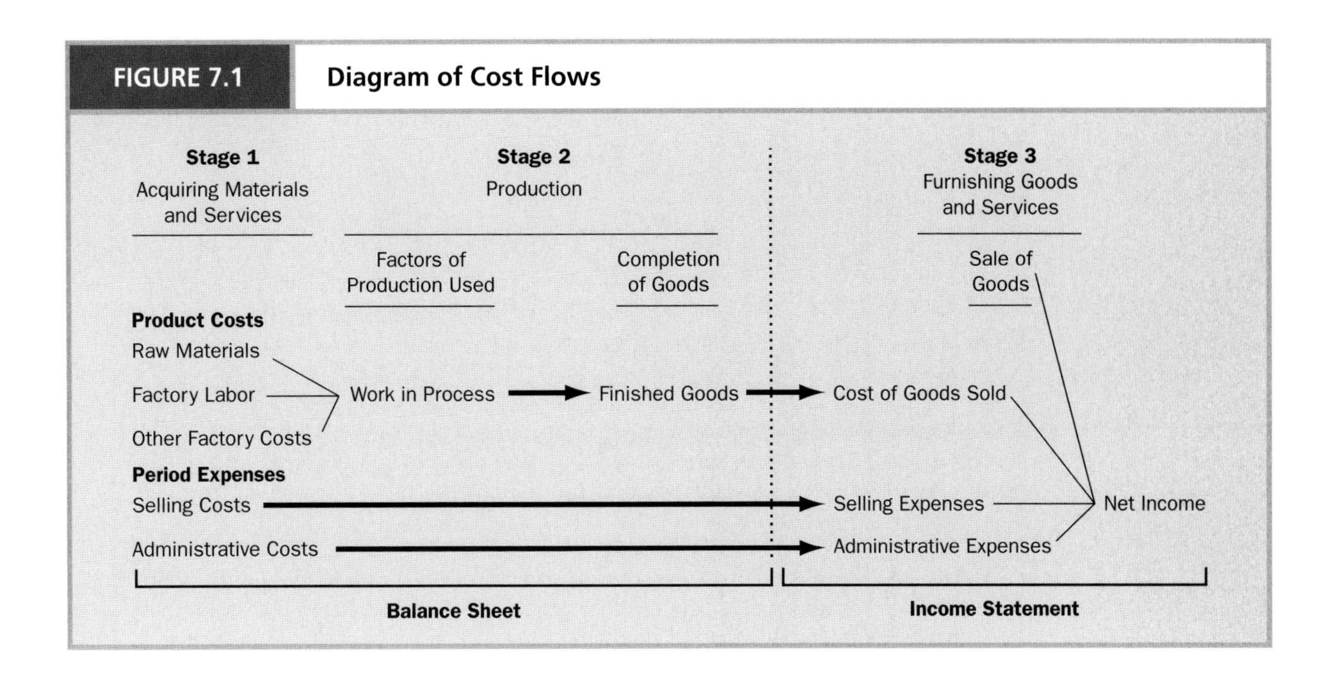


transfers the completed units from the factory floor to the finished goods storeroom. It reduces (with credits) the balance in the Work-in-Process Inventory account for the manufacturing costs assigned to the finished units and increases (with debits) the balance in the Finished Goods Inventory account. The balance in the Work-in-Process Inventory account indicates the product costs incurred thus far for units not yet finished as of the date of the balance sheet.

The **Finished Goods Inventory** account includes the total manufacturing cost of units completed but not yet sold. The sale of manufactured goods to customers results in a transfer, with credits, of their cost from the Finished Goods Inventory account to Cost of Goods Sold, an expense reducing Retained Earnings, with debits.

**Figure 7.2** shows the flow of manufacturing costs through the various inventory and other accounts. Compare the movement of amounts through the accounts in **Figure 7.2** with the parallel physical movement of inventory in **Figure 7.1**.

## Illustration of the Accounting Process for a Manufacturing Firm

This section illustrates the accounting process for a manufacturing firm using information about the operations of Moon Manufacturing Company. The company began operations on January 1 by issuing 10,000 shares of $10 par value common stock for $30 per share. Transactions during January and the appropriate journal entries follow:

**(1)** The firm acquires a building costing $200,000 and equipment costing $50,000 for cash.

| | Dr. | Cr. |
|---|---|---|
| **(1)** Building | 200,000 | |
| Equipment | 50,000 | |
| Cash | | 250,000 |

| Assets | = | Liabilities | + | Shareholders' Equity | (Class.) |
|---|---|---|---|---|---|
| +200,000 | | | | | |
| +50,000 | | | | | |
| −250,000 | | | | | |

**FIGURE 7.2** Flow of Manufacturing Costs through the Accounts

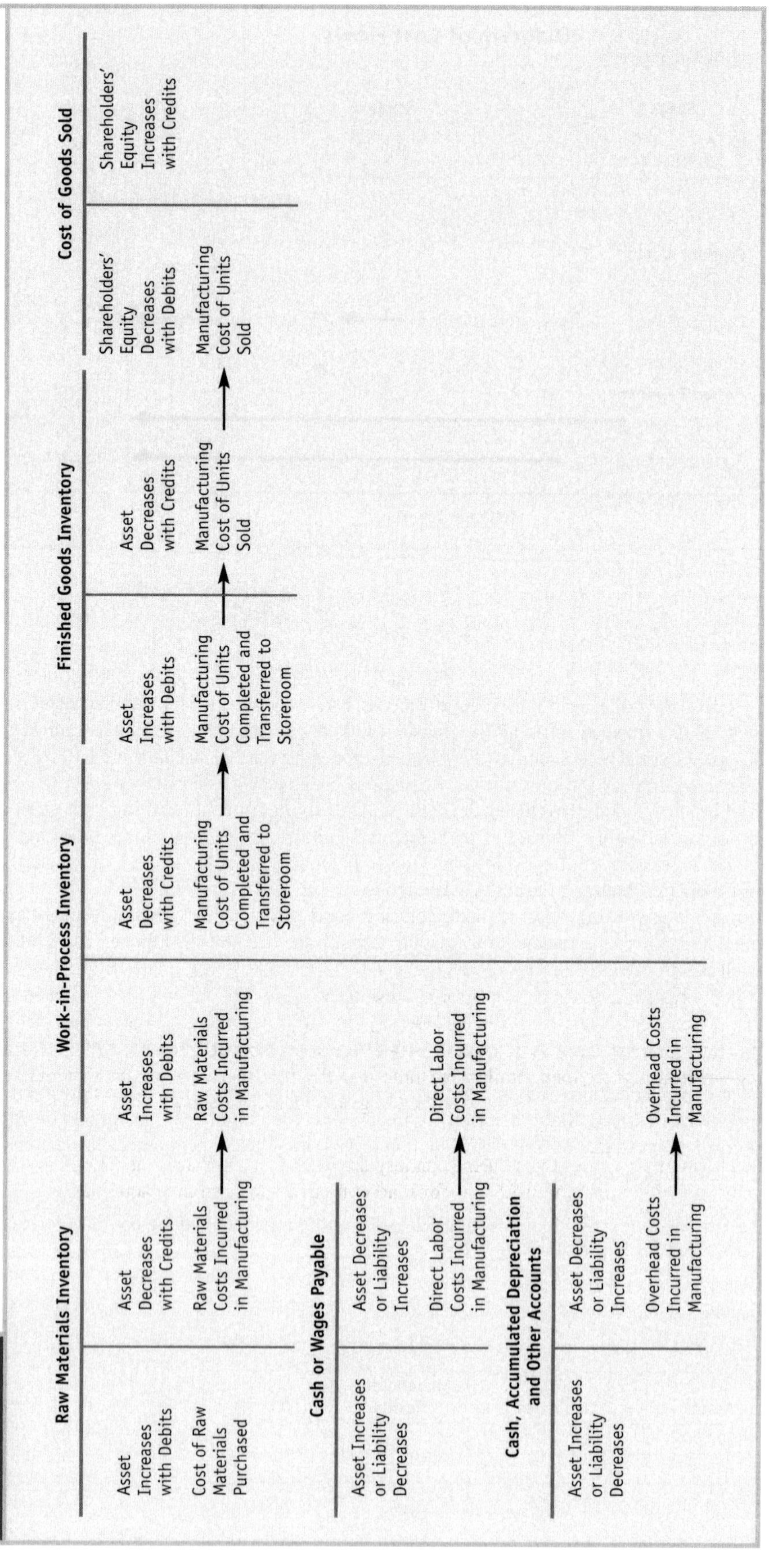

**Raw Materials Inventory**

| Debit | Credit |
| --- | --- |
| Asset Increases with Debits | Asset Decreases with Credits |
| Cost of Raw Materials Purchased | Raw Materials Costs Incurred in Manufacturing → |

**Cash or Wages Payable**

| Debit | Credit |
| --- | --- |
| Asset Increases or Liability Decreases | Asset Decreases or Liability Increases |
| | Direct Labor Costs Incurred in Manufacturing → |

**Cash, Accumulated Depreciation and Other Accounts**

| Debit | Credit |
| --- | --- |
| Asset Increases or Liability Decreases | Asset Decreases or Liability Increases |
| | Overhead Costs Incurred in Manufacturing → |

**Work-in-Process Inventory**

| Debit | Credit |
| --- | --- |
| Asset Increases with Debits | Asset Decreases with Credits |
| Raw Materials Costs Incurred in Manufacturing | Manufacturing Cost of Units Completed and Transferred to Storeroom → |
| Direct Labor Costs Incurred in Manufacturing | |
| Overhead Costs Incurred in Manufacturing | |

**Finished Goods Inventory**

| Debit | Credit |
| --- | --- |
| Asset Increases with Debits | Asset Decreases with Credits |
| Manufacturing Cost of Units Completed and Transferred to Storeroom | Manufacturing Cost of Units Sold → |

**Cost of Goods Sold**

| Debit | Credit |
| --- | --- |
| Shareholders' Equity Decreases with Debits | Shareholders' Equity Increases with Credits |
| Manufacturing Cost of Units Sold | |

**(2)** The firm purchases raw materials costing $25,000 on account.

| | Dr. | Cr. |
|---|---|---|
| **(2)** Raw Materials Inventory . . . . . . . . . . . . . . . . . . . . . . . . . . . . . . . . . | 25,000 | |
| Accounts Payable . . . . . . . . . . . . . . . . . . . . . . . . . . . . . . . . . | | 25,000 |

| Assets | = | Liabilities | + | Shareholders' Equity | (Class.) |
|---|---|---|---|---|---|
| +25,000 | | +25,000 | | | |

**(3)** It issues, to producing departments, raw materials costing $20,000.

| | Dr. | Cr. |
|---|---|---|
| **(3)** Work-in-Process Inventory . . . . . . . . . . . . . . . . . . . . . . . . . . . . . . . . . | 20,000 | |
| Raw Materials Inventory . . . . . . . . . . . . . . . . . . . . . . . . . . . . | | 20,000 |

| Assets | = | Liabilities | + | Shareholders' Equity | (Class.) |
|---|---|---|---|---|---|
| +20,000 | | | | | |
| −20,000 | | | | | |

**(4)** The total payroll for January is $60,000: $40,000 paid to factory workers and $20,000 paid to marketing and administrative personnel.

| | Dr. | Cr. |
|---|---|---|
| **(4)** Work-in-Process Inventory . . . . . . . . . . . . . . . . . . . . . . . . . . . . . . . . . | 40,000 | |
| Salaries Expense . . . . . . . . . . . . . . . . . . . . . . . . . . . . . . . . . . . . . . . . . | 20,000 | |
| Cash . . . . . . . . . . . . . . . . . . . . . . . . . . . . . . . . . . . . . . . . . | | 60,000 |

| Assets | = | Liabilities | + | Shareholders' Equity | (Class.) |
|---|---|---|---|---|---|
| +40,000 | | | | −20,000 | IncSt → RE |
| −60,000 | | | | | |

Recall that a manufacturing firm records nonmanufacturing costs as expenses of the period when the firm consumes the services because these costs rarely create assets with future benefits. Journal entry **(4)**, as well as entries **(5)** and **(6)**, which follow, illustrates the difference between the recording of a product cost and the recording of a period expense. The debits may, on first glance, look similar, but note that the first increases an asset account and the second increases an expense, thereby reducing shareholders' equity.

**(5)** The expenditures for utilities during January are $1,200. Of this amount, $1,000 is for manufacturing and the remaining $200 is for marketing and administrative activities.

| | Dr. | Cr. |
|---|---|---|
| **(5)** Work-in-Process Inventory . . . . . . . . . . . . . . . . . . . . . . . . . . . . . . . . . | 1,000 | |
| Utilities Expense . . . . . . . . . . . . . . . . . . . . . . . . . . . . . . . . . . . . . . . . | 200 | |
| Cash (Asset Decrease) . . . . . . . . . . . . . . . . . . . . . . . . . . . . . . . | | 1,200 |

| Assets | = | Liabilities | + | Shareholders' Equity | (Class.) |
|---|---|---|---|---|---|
| +1,000 | | | | −200 | IncSt → RE |
| −1,200 | | | | | |

These debits also split the expenditure between asset (product cost) and expense for the period.

**(6)** Depreciation on building and equipment during January is as follows: factory, $8,000; marketing and administrative, $2,000.

| | | |
|---|---|---|
| **(6)** Work-in-Process Inventory | 8,000 | |
| Depreciation Expense | 2,000 | |
| Accumulated Depreciation (Asset Decrease) | | 10,000 |

| Assets | = | Liabilities | + | Shareholders' Equity | (Class.) |
|---|---|---|---|---|---|
| +8,000 | | | | −2,000 | IncSt → RE |
| −10,000 | | | | | |

Note the split of the depreciation charge between product cost, Work-in-Process (asset), and period expense. The credit, to an asset contra account, enables the acquisition cost of the equipment to remain, unreduced, in the accounts.

**(7)** Units completed during January and transferred to the finished goods storeroom have a manufacturing cost of $48,500. The derivation of this amount requires the techniques of cost accounting.

| | | |
|---|---|---|
| **(7)** Finished Goods Inventory | 48,500 | |
| Work-in-Process Inventory | | 48,500 |

| Assets | = | Liabilities | + | Shareholders' Equity | (Class.) |
|---|---|---|---|---|---|
| +48,500 | | | | | |
| −48,500 | | | | | |

**(8)** Sales during January total $75,000, of which $25,000 is on account.

| | | |
|---|---|---|
| **(8)** Cash | 50,000 | |
| Accounts Receivable | 25,000 | |
| Sales Revenue | | 75,000 |

| Assets | = | Liabilities | + | Shareholders' Equity | (Class.) |
|---|---|---|---|---|---|
| +50,000 | | | | +75,000 | IncSt → RE |
| +25,000 | | | | | |

**(9)** The cost of the goods sold during January is $42,600.

| | | |
|---|---|---|
| **(9)** Cost of Goods Sold | 42,600 | |
| Finished Goods Inventory | | 42,600 |

| Assets | = | Liabilities | + | Shareholders' Equity | (Class.) |
|---|---|---|---|---|---|
| −42,600 | | | | −42,600 | IncSt → RE |

**Exhibit 7.1** presents a statement of income before taxes for the firm for January.

**EXHIBIT 7.1**

**MOON MANUFACTURING COMPANY**
**Income Statement**
**For the Month of January**

| | | |
|---|---|---|
| Sales Revenue | | $75,000 |
| Less Expenses: | | |
| Cost of Goods Sold | $42,600 | |
| Salaries Expense | 20,000 | |
| Utilities Expense | 200 | |
| Depreciation Expense | 2,000 | |
| Total Expenses | | 64,800 |
| Income Before Taxes | | $10,200 |

## Problem 7.1 for Self-Study

**Flow of manufacturing costs through the accounts.** The following data relate to the manufacturing activities of Haskell Corporation during March:

| | March 1 | March 31 |
|---|---|---|
| Raw Materials Inventory | $42,400 | $ 46,900 |
| Work-in-Process Inventory | 75,800 | 63,200 |
| Finished Goods Inventory | 44,200 | 46,300 |
| **Factory Costs Incurred during the Month** | | |
| Raw Materials Purchased | | $ 60,700 |
| Labor Services Received | | 137,900 |
| Heat, Light, and Power | | 1,260 |
| Rent | | 4,100 |
| **Expirations of Previous Factory Acquisitions and Prepayments** | | |
| Depreciation of Factory Equipment | | $ 1,800 |
| Prepaid Insurance Expired | | 1,440 |
| **Other Data Relating to the Month** | | |
| Sales Revenue | | $400,000 |
| Selling and Administrative Expenses | | 125,000 |

**a.** Calculate the cost of raw materials used during March.
**b.** Calculate the cost of units completed during March and transferred to the finished goods storeroom.
**c.** Calculate the cost of units sold during March.
**d.** Calculate income before taxes for March.

**Summary of the Accounting Process for Manufacturing Operations** The accounting procedures for the marketing and administrative costs of manufacturing firms resemble those for merchandising firms. The firm expenses these costs in the same period that it consumes the services. The accounting procedures for a manufacturing firm differ from those for a merchandising firm primarily in the treatment of inventories. A manufacturing firm incurs costs in transforming raw materials into finished products. Although a manufacturing firm consumes raw materials, labor services, and factory overhead services in the manufacturing process, the consumption results in the creation of an asset, partial or completed units of inventory. Until the manufacturing firm sells the units produced, it accumulates manufacturing costs in asset accounts—the Raw Materials Inventory account, the Work-in-Process Inventory account, or the Finished Goods Inventory account—depending on the stage of completion of each inventory item. The firm therefore debits product costs to inventory (asset) accounts until it sells the units produced.

## Issue 2: Valuation Subsequent to Acquisition

GAAP requires firms to record inventories initially at acquisition cost (**Issue 1** discussed above). The market value of inventories may change while the firm holds the inventory awaiting its sale to customers. This section discusses the treatment of increases and decreases in the market value of inventories subsequent to acquisition. Market value generally means **replacement cost**, the amount the firm would pay to acquire the inventory.

**Increases in Market Value** Inventories may increase in value subsequent to acquisition for a variety of reasons. A shortage of a key raw material may increase its costs and thereby increase the replacement cost of the inventory item of which the raw material is a component. A new labor agreement may increase labor cost and thereby increase the replacement cost of manufactured inventories. If a firm were to revalue its inventories to the higher replacement cost, or market value, it would make an entry such as the following:

| | | |
|---|---|---|
| Inventories . . . . . . . . . . . . . . . . . . . . . . . . . . . . . . . . . . . . . . . . . | X | |
| Unrealized Holding Gain on Inventories . . . . . . . . . . . . . . . . . . . | | X |

| Assets | = | Liabilities | + | Shareholders' Equity | (Class.) |
|---|---|---|---|---|---|
| +Increase in Value | | | | +Increase in Value | IncSt → RE* |

*This increase might be OCInc → AOInc, as described in Chapter 12.

GAAP, however, does not permit firms to revalue inventories above acquisition cost. GAAP's reasoning is that acquisition cost valuations provide more conservative measures of assets and net income during the periods prior to sale. The increase in the market value of the inventory likely permits the firm to raise its selling price. The benefit of the increase in market value affects net income in the period of sale when the firm realizes the benefit of a higher selling price instead of recognizing the benefit during the periods while market values increased.

**Decreases in Market Value** Inventories can decrease in value for a variety of reasons. A competitor introduces a technologically superior product that makes a particular firm's product offering obsolete. A firm's product includes materials found to contain a health hazard. The introduction of a lower-cost raw material lowers the manufacturing cost of a product using that raw material as a component.

In contrast to the treatment of increases in market value, GAAP requires firms to write down inventories when their replacement cost, or market value, declines below acquisition cost. GAAP refers to this valuation as the **lower of cost or market valuation basis.**[2] Once the firm writes down inventory, the new amount becomes the historical cost basis and the firm does not write up that amount for subsequent value increases, even if the new value remains less than the original cost.

A firm can recognize a decline of $5,000 in the market value of inventory with the following entry:

| | | |
|---|---|---|
| Unrealized Holding Loss on Inventory . . . . . . . . . . . . . . . . . . . . . . . . . | 5,000 | |
| Inventory . . . . . . . . . . . . . . . . . . . . . . . . . . . . . . . . . . . . . . . | | 5,000 |

| Assets | = | Liabilities | + | Shareholders' Equity | (Class.) |
|---|---|---|---|---|---|
| -5,000 | | | | -5,000 | IncSt → RE |

The account debited can appear on the income statement on a separate line. Some firms do not explicitly record the previous entry; instead, they include the unrealized holding loss as part of a

[2]AICPA, Committee on Accounting Procedures, *Accounting Research Bulletin No. 43*, "Inventory Pricing," 1953; Chapter 4, Statements 5 and 6.

**EXHIBIT 7.2** **Calculating Cost of Goods Sold Using Different Bases of Inventory Valuation**

| | Cost Basis | Lower-of-Cost-or Market Basis |
|---|---|---|
| Beginning Inventory | $ 19,000 | $ 19,000 |
| Purchases | 100,000 | 100,000 |
| Goods Available for Sale | $119,000 | $119,000 |
| Less Ending Inventory | (25,000) | (20,000) |
| Cost of Goods Sold | $ 94,000 | $ 99,000 |

higher cost of goods sold. Consider, for example, the calculations in **Exhibit 7.2** of cost of goods sold when beginning inventory is $19,000, purchases are $100,000, and ending inventory has a cost of $25,000 but has a market value of $20,000. Note that cost of goods sold is $5,000 larger when the firm values ending inventory at lower of cost or market than when it uses the acquisition cost basis for ending inventory. The loss of $5,000 does not appear separately, but income will be $5,000 smaller than when the firm uses the acquisition cost basis. To avoid misleading readers of the financial statements, the accountant should disclose in the notes the existence of large write-downs included in cost of goods sold.

The lower-of-cost-or-market basis for inventory valuation is a conservative accounting policy because (1) it recognizes losses from decreases in market value before the firm sells goods but it does not record gains from increases in market value before a sale takes place, and (2) it reports inventories on the balance sheet at amounts that are never greater, but may be less, than acquisition cost.[3] In other words, the lower-of-cost-or-market basis results in reporting unrealized holding losses on inventory items currently through lower net income amounts but delays reporting unrealized holding gains until the firm sells the goods.

**Summary of Effect of Inventory Revaluations** Over sufficiently long time spans, income equals cash inflows minus cash outflows (other than transactions with owners). For any one unit, there is only one total gain or loss figure: the difference between its selling price and its acquisition cost. The valuation rule determines when this gain or loss appears in the financial statements over the accounting periods between acquisition and final disposition. If a firm revalued inventory upward to current replacement cost (not allowed by GAAP in the United States), the net income of the period of the inventory write-up will be higher than if the firm had used the acquisition cost basis, but if so, the net income of the later period, when the firms sells the unit, will be lower. When a firm uses the lower-of-cost-or-market basis (generally required by GAAP in the United States), the net income for the period of an inventory writedown will be lower than if the firm had used the acquisition cost basis, but if so, the net income of a later period, when the firm sells the unit, will be higher. Thus, the amount of net income for only one period depends directly on the valuation of inventory on the balance sheet.

# Issue 3: Cost Flow Assumptions

## Specific Identification and the Need for a Cost Flow Assumption

The accounting records typically contain information on the cost of the beginning inventory for a period, which was the ending inventory of last period. The accounting records also contain information on purchases or production costs incurred during the current period. Thus, firms can easily

[3]Consult the **Glossary** for the definition of conservatism in accounting. Conservative accounting policies result in both lower asset totals and lower retained earnings totals, thus implying lower cumulative net income totals. Conservatism does not mean reporting lower income in every period. Over long-enough time spans, income will be cash-in less cash-out, other than transactions with owners. If a given accounting policy results in reporting lower income in earlier periods, it must report higher income in some subsequent periods. The conservative accounting policy results in lower income in the early periods. See **Exercise 27** at the end of this chapter.

measure the cost of goods available for sale or use. Firms can match units sold and units in ending inventory with specific purchases or production runs by using product bar codes. Using the code, the firm can trace the unit back to its purchase invoice or cost record. An automobile dealer or a dealer in diamonds or fur coats might compute the cost of ending inventory and cost of goods sold using **specific identification**. If so, then measuring the cost of goods sold and ending inventory presents no difficulties. Refer to **Example 4**. Bar codes on the TV sets permitted the firm to identify TV set 2 as the one sold and TV sets 1 and 3 as remaining in ending inventory.

The inventory items of some firms are sufficiently similar that the firms cannot feasibly use specific identification. Lumber in a building supply store, gasoline in a storage tank, and plants in a nursery are examples. An inventory costing problem arises because of two unknowns in the inventory equation:

| Beginning Inventory | + | Purchases | − | Cost of Goods Sold | = | Ending Inventory |
|---|---|---|---|---|---|---|
| (known) | | (known) | | (unknown) | | (unknown) |

Goods Available for Sale (or Use)

We know the costs of the beginning inventory and the purchases, but we do not know either the amount for cost of goods sold or the amount for ending inventory. Do we compute amounts for the units in ending inventory using the most recent costs, the oldest costs, the average cost, or some other choice? Or we could ask how to compute amounts for the cost of goods sold. Once we place an amount on one unknown quantity, the inventory equation automatically determines the amount for the other. The sum of the two unknowns, Cost of Goods Sold and Ending Inventory, must equal the cost of goods available for sale (= Beginning Inventory + Purchases). The higher the cost we assign to one unknown, the lower must be the cost we assign to the other.

If a firm finds specific identification not feasible, it must make some assumption about the flow of costs. Even when specific identification is feasible, GAAP does not require firms to use it but, instead, allow them to make a cost flow assumption. The accountant computes the acquisition cost applicable to the units sold and to the units remaining in inventory using one of the following **cost flow assumptions**:

1. First-in, first-out (FIFO).
2. Last-in, first-out (LIFO).
3. Weighted average.

The following demonstrations of each of these methods use the TV set example introduced earlier and repeated at the top of **Exhibit 7.3**. The example of the three TV sets illustrates most of the important points about the cost flow assumption required in accounting for inventories and cost of goods sold.

## First-In, First-Out

The **first-in, first-out (FIFO)** cost flow assumption assigns the costs of the earliest units acquired to the withdrawals and assigns the costs of the most recent acquisitions to the ending inventory. This cost flow assumes that the firm uses the oldest materials and goods first. This cost flow assumption conforms to good business practice in managing physical flows, especially in the case of items that deteriorate or become obsolete. Column (1) of **Exhibit 7.3** illustrates FIFO. FIFO assumes that the firm sells TV set 1 while TV sets 2 and 3 remain in inventory. The designation *FIFO* refers to the cost flow of the units sold. A parallel description for ending inventory is last-in, still-here, or LISH.

Think of FIFO as a conveyor belt. The first items put on the conveyor belt come off first for use or sale. The last items put on the conveyor belt remain there at the end of the period. Refer to the top panel of **Figure 7.3**. Cost of goods sold for Year 1 under FIFO uses purchases from the beginning of the year until the number of units purchased equal the number of units sold. The firm uses the cost of these purchases to measure cost of goods sold. The ending inventory for Year 1 uses the last purchases during Year 1. Cost of goods sold for Year 2 under FIFO uses the last purchases of Year 1 that comprised the ending inventory for Year 1 and the beginning inventory for Year 2, as well as sufficient purchases from Year 2 to cost all units sold. The ending inventory for Year 2 uses the last purchases during Year 2.

**EXHIBIT 7.3** **Comparison of Cost Flow Assumptions, Historical Cost Basis**

| **Assumed Data** | | | |
|---|---|---|---|
| Beginning Inventory: TV Set 1 Cost | | | $250 |
| Purchases: TV Set 2 Cost | | | 290 |
| TV Set 3 Cost | | | 300 |
| Cost of Goods Available for Sale | | | $840 |
| Sales: One TV set | | | $550 |
| | | **Cost Flow Assumption** | |
| | **FIFO** | **Weighted Average** | **LIFO** |
| **Financial Statements** | **(1)** | **(2)** | **(3)** |
| Sales | $550 | $550 | $550 |
| Cost of Goods Sold | 250[a] | 280[b] | 300[c] |
| Gross Margin on Sales | $300 | $270 | $250 |
| Ending Inventory | $590[d] | $560[e] | $540[f] |

[a]TV set 1 costs $250.
[b]Average TV sets costs $280 (= $840/3).
[c]TV set 3 costs $300.
[d]TV sets 2 and 3 cost $290 + $300 = $590.
[e]Two average TV sets cost 2 × $280 = $560.
[f]TV sets 1 and 2 cost $250 + $290 = $540.

## Last-In, First-Out

The **last-in, first out (LIFO)** cost flow assumption assigns the costs of the latest units acquired to the withdrawals and assigns the costs of the oldest units to the ending inventory. Some theorists argue that LIFO matches current costs to current revenues and, therefore, that LIFO better measures income. Column (3) of **Exhibit 7.3** illustrates LIFO. LIFO assumes that the firm sells TV set 3 costing $300 while the costs of TV sets 1 and 2 remain in inventory. The designation *LIFO* refers to the cost flow for the units sold. A parallel description for ending inventory is first-in, still-here, or FISH.

One might view LIFO as a coal pile. Newly purchased coal gets added to the top of the coal pile. Withdrawals likewise come off the top of the coal pile. Alternatively, think of LIFO as a stack of trays in a cafeteria. The last tray deposited on the stack is the first one next taken off. The lowest tray in the stack remains there as long as there are any trays left unused. It could, in principle, have been there for 30 years.

Refer to the lower panel of **Figure 7.3**. Cost of goods sold for Year 1 under LIFO uses purchases from the end of the year working backward through time until the number of units purchased equal the number of units sold. The firm uses the cost of these purchases to measure cost of goods sold. The ending inventory for Year 1 uses the first purchases during Year 1. Refer to the coal analogy: the coal on the bottom of the pile remains on hand at the end of Year 1. Cost of goods sold for Year 2 under LIFO likewise uses the last purchases from the end of the year backward until the number of units purchased equal the number of units sold. The ending inventory for Year 2 uses purchase prices from two time periods: (1) units equal to the number in ending inventory at the end of Year 1 use purchase prices at the beginning of Year 1, and (2) the increase in the number of units in ending inventory from Year 1 to Year 2 uses purchase prices at the beginning of Year 2. Refer to the coal analogy: a Year 2 layer of coal gets added to the layer left at the end of Year 1. The Year 1 layer carries purchase prices at the beginning of Year 1, and the Year 2 layer carries prices at the beginning of Year 2. Their sum is the total ending inventory for Year 2 under LIFO. So long as any lumps of coal remain in the pile, some will be from Year 1.

Firms have increasingly used LIFO since it first became acceptable for income tax reporting in 1939. In a period of consistently rising purchase prices and increasing inventory quantities, LIFO results in a higher cost of goods sold, a lower reported periodic income, and lower current income taxes than either the FIFO or the weighted-average cost flow assumption. You might be

FIGURE 7.3 Purchase Prices Used in Measuring Cost of Goods Sold and Ending Inventory Under FIFO and LIFO

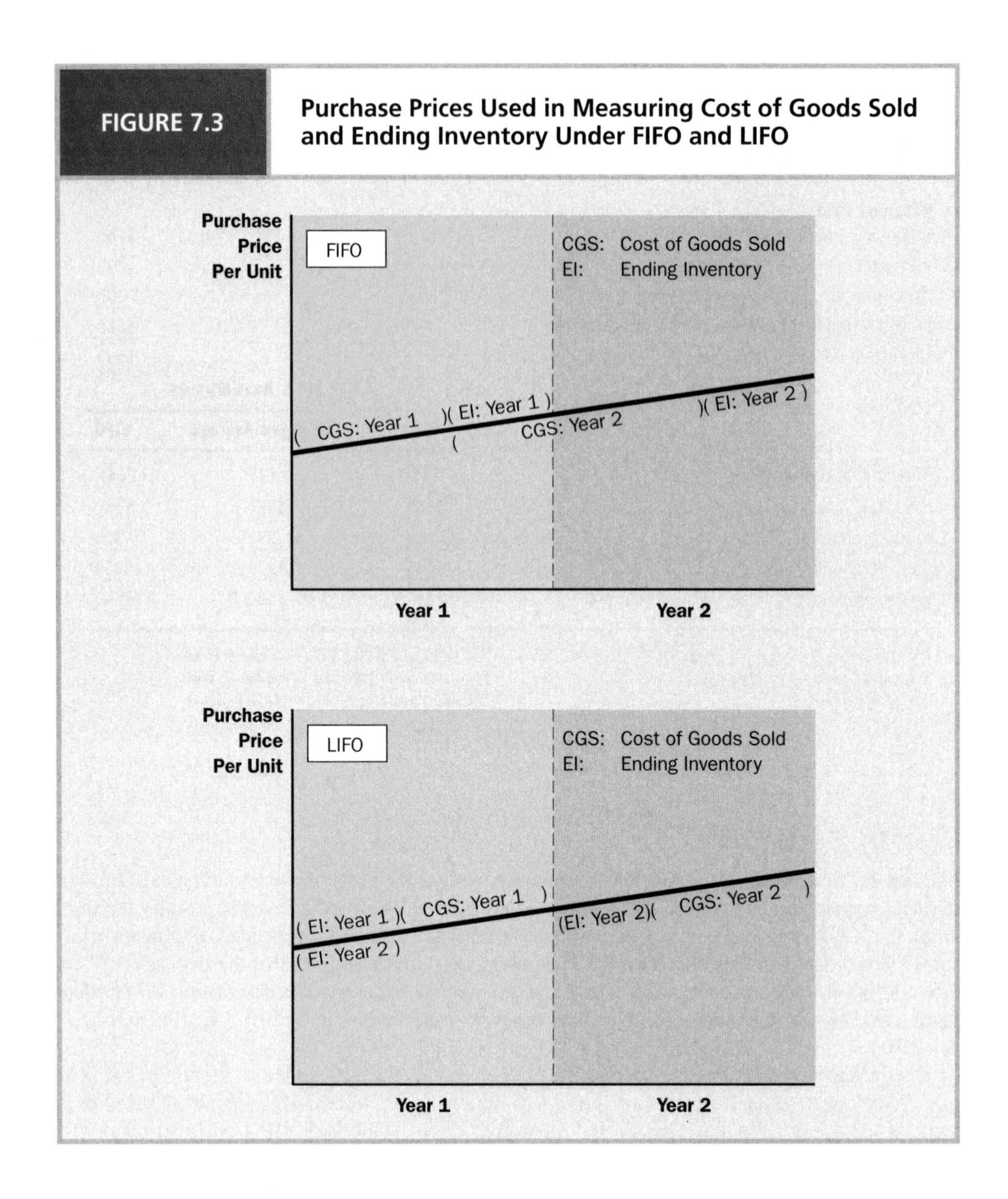


tempted to conclude, therefore, that LIFO always provides a lower amount of net income. Such a conclusion is incorrect, however, as we discuss later, in the context of "dips into LIFO layers."

LIFO usually does not reflect physical flows. Recall, however, that FIFO and LIFO are *cost* flow assumption, not *physical* flow assumption. Firms use LIFO not because it matches physical flows but because it produces a cost-of-goods-sold figure based on more up-to-date purchase prices and reduces income tax payments.

## Weighted Average

Under the **weighted-average** cost flow assumption, a firm calculates the average of the costs of all goods available for sale (or use) during the accounting period, including the cost applicable to the beginning inventory. The weighted-average cost applies to the units sold during the period and to units on hand at the end of the period. Column (2) in **Exhibit 7.3** illustrates the weighted-average cost flow assumption. The weighted-average cost of each TV set available for sale during the period is $280 [= ⅓ × ($250 + $290 + $300)]. Cost of Goods Sold is thus $280, and ending inventory is $560 (= 2 × $280).

## Comparison of Cost Flow Assumptions

Note from **Exhibit 7.3** that cost of goods sold plus ending inventory equals $840, the total cost of goods available for sale, in all three cases. When purchase prices change, no acquisition cost-based accounting method for costing ending inventory and cost of goods sold allows the accountant to show up-to-date costs on both the income statement and the balance sheet. For example, consider a period of rising prices. If measurements of cost of goods sold for the income statement use recent, higher acquisition prices, as occurs under LIFO, then older, lower acquisition prices must appear in costs of the ending inventory for the balance sheet. As long as accounting bases cost of goods sold and ending inventory on acquisition costs, financial statements can present current cost amounts in the income statement or the balance sheet but not in both.

Of the three cost flow assumptions, FIFO results in balance sheet figures that are the closest to current cost because the latest purchases dominate the ending inventory amounts. Remember LISH: last-in, still-here. The Cost of Goods Sold expense tends to be out-of-date, however, because FIFO assumes that the earlier purchase prices of the beginning inventory and the earliest purchases during the period become expenses. When purchase prices rise, FIFO usually leads to the highest reported net income of the three methods, and when purchase prices fall, it leads to the smallest.

LIFO ending inventory can contain costs of items acquired many years previously—think of those trays at the bottom of the stack. When purchase prices have been rising and inventory amounts increasing, LIFO produces balance sheet figures usually much lower than current costs. LIFO's cost-of-goods-sold figure closely approximates current costs. **Exhibit 7.4** summarizes the differences between FIFO and LIFO. Of the cost flow assumptions (LIFO, FIFO, weighted average), LIFO usually results in the smallest net income when purchase prices are rising (highest cost of goods sold) and the largest when purchase prices are falling (lowest cost of goods sold). Also, LIFO results in the least fluctuation in gross margins in businesses in which selling prices tend to change as purchase prices of inventory items change.

The weighted-average cost flow assumption falls between the other two in its effects, but it resembles FIFO more than LIFO in its effects on the financial statements. When inventory turns over rapidly, the weighted-average inventory cost flow provides amounts virtually identical to FIFO's amounts. You can see this for yourself if you work any of **Problems 29**, **39**, and **41** at the end of the chapter. The remaining discussion treats FIFO and weighted average similarly.

Differences in cost of goods sold and inventories under the cost flow assumptions relate in part to the rate of change in the acquisition costs of inventory items. Using older purchase prices for inventories under LIFO or using older purchase prices for cost of goods sold under FIFO has little impact if prices have been stable for several years. As the rate of price change increases, the effect of using older versus more recent prices increases, resulting in larger differences in cost of goods sold and inventories between FIFO and LIFO.

Differences in cost of goods sold also relate in part to the rate of inventory turnover. As the rate of inventory turnover increases, purchases during the period make up an increasing proportion of the cost of goods available for sale. Because purchases are the same regardless of the cost flow assumption, cost of goods sold amounts will not vary as much with the choice of cost flow assumption. Consider the impact of a choice of cost flow assumption for a florist or grocery store (small effect because of fast turnover with small ending inventory relative to annual sales) versus that for a distiller of liquor that requires aging (large effect because of slow turnover with large ending inventory relative to annual sales). Even with a rapid inventory turnover, inventory

**EXHIBIT 7.4** Age of Information about Inventory Items

| Cost Flow Assumption | Income Statement | Balance Sheet | Inventory-on-Hand Assumption |
|---|---|---|---|
| FIFO . . . . . . . . . . . . | Old Purchase Prices | Current Purchase Prices . . . . . . . . . . . | LISH |
| LIFO . . . . . . . . . . . . | Current Purchase Prices | Very[a] Old Purchase Prices . . . . . . . . . . . | FISH |

[a]The oldest purchase prices on the FIFO income statement are just over one year old in nearly all cases, and the average purchase price on the FIFO income statement (for a year) is slightly more than one-half year old. The larger the rate of inventory turnover, the closer the average age on the income statement is to one-half year. LIFO balance sheet items are generally much older than FIFO income statement items, with some costs from the 1940s in many cases.

amounts on the balance sheet can still differ significantly depending on the cost flow assumption. The longer a firm uses LIFO, likely the larger will be the difference between inventories based on LIFO and FIFO cost flow assumptions.

## Problem 7.2 for Self-Study

**Computing cost of goods sold and ending inventory under various cost flow assumptions. Exhibit 7.5** presents data on beginning inventory, purchases, withdrawals, and ending inventory for June and July.

**a.** Compute the cost of goods sold and the ending inventory for June using (1) FIFO, (2) LIFO, and (3) weighted-average cost flow assumptions.
**b.** Compute the cost of goods sold and the ending inventory for July using (1) FIFO, (2) LIFO, and (3) weighted-average cost flow assumptions.

## A Closer Look at LIFO's Effects on Financial Statements

LIFO usually presents a cost-of-goods-sold figure that reflects current costs. It also generally has the practical advantage of deferring income taxes. If a firm uses a LIFO assumption in its income tax return in the United States, it must also use LIFO in its financial reports to shareholders. This "LIFO conformity rule" results from Internal Revenue Service regulations for firms using LIFO for income tax reporting.

In the last 30 years, many firms, including DuPont, General Motors, Eastman Kodak, and Sears, have switched from FIFO to LIFO. Given the rapid rate of purchase price increases in the past, the switch from FIFO to LIFO has resulted in substantially lower cash payments for income taxes. For example, when DuPont and General Motors switched from FIFO to LIFO, they each lowered current income taxes by about $150 million. At the same time, these firms reported lower net income to shareholders than they would have reported if they had still used FIFO. The cash flow from the deferral of income taxes actually increased present value of the cash flows for these firms, despite reporting lower earnings to shareholders.

**EXHIBIT 7.5** Data for Inventory Calculations (Problem 7.2 for Self-Study)

| | Units | Unit Cost | Total Cost |
|---|---|---|---|
| ITEM X | | | |
| Beginning Inventory, June 1 | — | — | — |
| Purchases, June 1 | 100 | $10.00 | $1,000 |
| Purchases, June 7 | 400 | 11.00 | 4,400 |
| Purchases, June 18 | 100 | 12.50 | 1,250 |
| Total Goods Available for Sale | 600 | | $6,650 |
| Withdrawals during June | (495) | | ? |
| Ending Inventory (June 30) and Beginning Inventory (July 1) | 105 | | $ ? |
| Purchases, July 5 | 300 | 13.00 | 3,900 |
| Purchases, July 15 | 200 | 13.50 | 2,700 |
| Purchases, July 23 | 250 | 14.00 | 3,500 |
| Total Goods Available for Sale | 855 | | $ ? |
| Withdrawals during July | (620) | | ? |
| Ending Inventory (July 31) | 235 | | $ ? |

**EXHIBIT 7.6** **Data for Illustration of LIFO Layers (Inventory at January 1, Year 5)**

| LIFO Layers | | Cost | |
|---|---|---|---|
| Year Purchased | Number of Units | Per Unit | Total Cost |
| 1 . . . . . . . . . . . . . . . . . . . . . . . . . . . . . . . . . . . . | 100 | $ 50 | $ 5,000 |
| 2 . . . . . . . . . . . . . . . . . . . . . . . . . . . . . . . . . . . . | 110 | 60 | 6,600 |
| 3 . . . . . . . . . . . . . . . . . . . . . . . . . . . . . . . . . . . . | 120 | 80 | 9,600 |
| 4 . . . . . . . . . . . . . . . . . . . . . . . . . . . . . . . . . . . . | 130 | 100 | 13,000 |
| | 460 | | $34,200 |

**LIFO Layers** In any year when purchases exceed sales, the quantity of units in inventory increases. The amount added to inventory for that year is called a **LIFO inventory layer**. For example, assume that a firm acquires 100 TV sets each year and sells 98 TV sets each year for four years. Its inventory at the end of the fourth year contains 8 units. The cost of the 8 units under LIFO is the cost of sets numbered 1 and 2 (from the first year), 101 and 102 (from the second year), 201 and 202 (from the third year), and 301 and 302 (from the fourth year). Common terminology would say that this firm has four LIFO layers, each labeled with its year of acquisition. The physical units on hand would almost certainly be the units most recently acquired in Year 4, units numbered 393 through 400, but they would appear on the balance sheet at costs incurred for purchases during each of the four years.

To take another example, the data in **Exhibit 7.6** illustrate four LIFO inventory layers, one for each of the years shown.

**Dipping into LIFO Layers** A firm using LIFO must consider the implications of dipping into old LIFO layers. LIFO reduces current taxes in periods of rising purchase prices and rising inventory quantities. If inventory quantities decline, however, the opposite effect occurs in the year of the decline: older, lower costs per unit of prior years' LIFO layers leave the balance sheet and become expenses.

If a firm must for some reason reduce end-of-period physical inventory quantities below the beginning-of-period quantities, cost of goods sold will reflect the current period's purchases plus a portion of the older and lower costs in the beginning inventory. Such a firm will have lower cost of goods sold as well as larger reported income and larger income taxes in that period than if the firm had been able to maintain its ending inventory at beginning-of-period levels.

**Example 5** Assume that LIFO inventory at the beginning of Year 5 comprises 460 units with a total cost of $34,200, as in **Exhibit 7.6**. Assume that the cost at the end of Year 5 is $120 per unit and that the income tax rate is 40 percent. If, for some reason, Year 5 ending inventory drops to 100 units, all 360 units purchased in Year 2 through Year 4 will enter cost of goods sold. These 360 units cost $29,200 (= $6,600 + $9,600 + $13,000), but the current cost of comparable units is $43,200 (= 360 units × $120 per unit). Cost of goods sold will be $14,000 smaller (= $43,200 − $29,200) than if quantities had not declined because the firm dipped into old LIFO layers. Income subject to income taxes will be $14,000 larger and income after taxes will be $8,400 larger than if the firm had purchased inventory so that quantities had not declined from 460 to 100 units. LIFO results in firms deferring taxes as long as they do not dip into LIFO layers.

**Quality of Earnings and Ethical Issues from Dipping into LIFO Layers.** Because firms often control whether inventory quantities increase or decrease through their purchase or production decisions, LIFO affords firms an opportunity to manage their earnings in a particular year. Managements face the decision as to when to replenish inventories at year-end. Buy in December and these higher acquisition costs go into cost of goods sold. Wait until January of next year to purchase and the current year's cost of goods sold contains costs older, usually lower, than December's. Financially literate board members, who represent investors, and securities analysts should probe to understand how much of the current year's earnings results from managed timing of purchases or production. Analysts view firms that dip into LIFO layers to manage their earnings as having a lower quality of earnings than firms that use FIFO.

A related question is whether management faces an ethical issue when it decides to dip into LIFO layers. Some observers point out that some LIFO liquidations are either unavoidable, because of raw materials shortages for example, or good business practice, because of improved inventory control systems for example, and do not raise ethical issues. These same observers would also argue that firms that purposely dip into LIFO layers primarily to meet earnings targets do face ethical issues. Other observers point out that the timing of inventory purchases or production is so complex and subjective that disentangling these possible reasons for LIFO liquidations is not feasible.

**Annual Report Disclosure** Many firms using LIFO have inventory cost layers built up since the 1940s, and the costs of the early units are often as little as 10 percent of the current cost. For these firms, a dip into old layers will substantially increase income. A note from recent financial statements of The Timken Company illustrates typical disclosures when a firm dips into LIFO layers:

> During [the current year], inventory quantities were reduced as a result of ceasing manufacturing operations in Duston, England. This reduction resulted in a liquidation of LIFO inventory quantities carried at lower costs prevailing in prior years, compared to the cost of current purchases, the effect of which increased income . . . by approximately $5,700,000 or $0.09 per diluted share [approximately 10 percent].

**LIFO's Effects on Purchasing Behavior** Consider the quandary faced by a purchasing manager of a LIFO firm nearing the end of a year when the quantity sold has exceeded the quantity purchased so far during the year. If the year ends with units sold exceeding units purchased, the firm will dip into old LIFO layers and will have to pay increased taxes. Assume that the purchasing manager thinks current purchase prices for the goods are abnormally high and prefers to delay making purchases until prices drop, presumably in the next year. Waiting implies dipping into old LIFO layers during this period and paying higher taxes. Buying now may entail higher inventory and carrying costs.

**Example 6** Refer again to the data in **Exhibit 7.6**. Assume that the cost per unit at the end of Year 5 has risen to $120 per unit and that the inventory on hand at the end of Year 5 is 460 units, with costs as in **Exhibit 7.6**. The purchasing manager thinks the price will drop back to $105 per unit early in Year 6. An order of 50 new units arrives, which the firm will fill. The firm has two strategies:

1. The purchasing manager can refuse to acquire 50 more units for $120. Then, the 50 units sold will carry costs of $100 each (from the Year 4 layer). Taxable income will be $1,000 larger [= 50 units × ($120 − $100)] than if the firm acquires units at the end of Year 5 for $120 each.
2. The firm can acquire 50 more units for $120. Then, the firm will have paid $750 more [= 50 units × ($120 − $105)] for inventory than if it had waited until the beginning of Year 6 to acquire inventory. In this example, so long as income tax rates are less than 75 percent, the firm will probably prefer the first strategy. If the income tax rate is 35 percent, for example, the cost of the first strategy is $350 (= 0.35 × $1,000) in extra taxes, whereas the second strategy incurs extra costs of $750 to buy sooner rather than later. In reality, of course, the purchasing manager can seldom be sure about future prices and will have to make estimates for computations such as those above.

LIFO can induce firms to manage LIFO layers and cost of goods sold in a way that would be unwise in the absence of tax effects. LIFO also gives management the opportunity to manage income: under LIFO, end-of-year purchases, which the firm can manage, affect net income for the year.

**LIFO Balance Sheet** LIFO usually leads to a balance sheet amount for inventory so much less than current costs that it may mislead readers of financial statements. For example, in recent years various manufacturers of steel products have reported that the cost of their ending inventory based on LIFO is about 80 percent of what it would have been had FIFO been used. Inventory on the balance sheet, reported assuming LIFO cost flow, makes up about 10 percent of total assets.

The Securities and Exchange Commission (SEC) has worried that this out-of-date information might mislead the readers of financial statements. As a result, the SEC requires firms using LIFO to disclose, in notes to the financial statements, the amounts by which inventories based on FIFO or current cost exceed their amounts as reported on a LIFO basis.[4] A note to the financial statements of Nucor Corporation, a steel manufacturer, reports that if the FIFO method of inventory accounting had been used to value all inventories, they would have been $158 million higher than reported at the end of the year and $43 million higher at the beginning of the year. Nucor's beginning inventories under LIFO amounted to $589 million and its ending inventory amounted to $560 million. Its beginning inventory under FIFO would have been $632 (= $589 + $43) million, a 7.3 percent increase. Its ending inventory would have been $718 (= $560 + $158) million, a 28.2 percent increase.

## Problem 7.3 for Self-Study

**Assessing the impact of a LIFO layer liquidation.** Refer to the data in **Problem 7.2 for Self-Study**. During August, the firm purchased 600 units for $15 each and sold 725 units.

**a.** Compute the cost of goods sold and the ending inventory for August using (1) FIFO, (2) LIFO, and (3) weighted-average cost flow assumptions.

**b.** Calculate the effect of the LIFO liquidation on net income before income taxes for the year.

## Choosing the Cost Flow Assumption

Firms in the United States can choose a FIFO or LIFO or weighted-average cost flow assumption. The decision involves the following considerations:

1. The extent of changes in manufacturing or purchase costs: When such costs do not change significantly, then the three cost flow assumptions provide similar amounts for inventories and cost of goods sold.
2. The rate of inventory turnover: The faster the rate of inventory turnover, the smaller are the differences in inventories and cost of goods sold among the three cost flow assumptions.
3. The direction of expected changes in costs: FIFO results in higher net income and income taxes when costs increase and lower net income and income taxes when costs decrease.
4. The relative emphasis on reporting higher earnings to shareholders versus saving income taxes.
5. The opportunity that LIFO provides to manage earnings by delaying purchases toward the end of the accounting period.
6. The increased recordkeeping costs of LIFO (for example, keeping track of LIFO layers for all of a firm's products) and its inconsistency with the usual physical flow of inventories.
7. The requirement that firms must use LIFO for financial reporting if they use LIFO for income tax reporting.

An annual survey of the choices that firms make in their selection of accounting principles reported for a recent year that more than a third of the companies surveyed used a combination of cost flow assumptions. More than 60 percent use FIFO for a significant portion of their inventories, and about 40 percent use LIFO for a significant portion. Fewer than one-third use weighted average or specific identification.[5] The industries with the largest percentages of firms using LIFO included firms that rely on oil as a raw material, such as chemicals, petroleum, rubber and plastic products. Retailing firms also use LIFO extensively. The industries with the smallest proportion of firms using LIFO included technology-based firms, which experience decreasing production costs, such as computers and other electronic equipment.

---

[4]Some managers refer to the excess of FIFO or current cost over LIFO cost of inventories as the *LIFO reserve*. For reasons discussed in the Glossary at the back of the book, the term *reserve* is objectionable because it misleads some readers. Some firms nevertheless continue to use it. A term such as *inventory valuation allowance* is as descriptive as *LIFO reserve* and is less likely to mislead financial statement readers.

[5]American Institute of Certified Public Accountants, *Accounting Trends and Techniques*, 2005, page 170.

## Analyzing Inventory Disclosures

The financial statements and notes provide helpful information for analyzing the effects of inventories and cost of goods sold on assessments of profitability and risk. This section illustrates several such analyses.

### CONVERSION FROM LIFO TO FIFO

As indicated above, approximately one-third of firms in the United States use a LIFO cost flow assumption, with most of the remaining firms using FIFO or weighted average. Most countries other than the United States do not allow firms to use LIFO. With the disclosures of the excess of FIFO or current cost over LIFO inventories, the analyst can compute inventories and cost of goods sold on a FIFO basis and thereby make more valid comparisons between a LIFO firm and its FIFO competitor.

To illustrate the conversion, consider the information provided on **page 341** for the inventories of Nucor Corporation. **Exhibit 7.7** summarizes this information and shows the conversion of inventories and cost of goods sold from a LIFO to a FIFO basis. The inventory amounts in the first column under LIFO come from Nucor's balance sheet and cost of goods sold and sales come from its income statement. We compute purchases using the inventory equation. The amounts disclosed for the excess of FIFO over LIFO permit the computation of beginning and ending inventories under FIFO in the third column. Purchases remain the same under LIFO and FIFO. Using the inventory equation, cost of goods sold under FIFO is $5,882 million, compared to $5,997 million under LIFO. The gross margin percentage under FIFO is 6.1 percent (= $384/$6,266), as compared to 4.3 percent (= $269/$6,266) under LIFO. The larger gross margin under FIFO suggests that acquisition costs increased during the year.

We can also compute inventory turnover ratios under LIFO and FIFO. The computations are as follows:

LIFO: $5,997/0.5($589 + $560) = 10.4 times per year
FIFO: $5,882/0.5($632 + $718) = 8.7 times per year

The inventory turnover ratio under LIFO is misleading because it uses relatively current cost in the numerator and the older cost of LIFO layers in the denominator. The inventory turnover

**EXHIBIT 7.7**

**Derivation of FIFO Income Data for LIFO Company, Inventory Data from Financial Statements and Footnotes of Nucor Corporation (all dollar amounts in millions)**

(Amounts shown in **boldface** appear in Nucor's financial statements. We derive other amounts as indicated.)

| | LIFO Cost Flow Assumption (actually used) | + | Excess of FIFO over LIFO Amount | = | FIFO Cost Flow Assumption (hypothetical) |
|---|---|---|---|---|---|
| Beginning Inventory | **$ 589** | | **$ 43** | | $ 632 |
| Purchases | 5,968[a] | | 0 | | 5,968 |
| Cost of Goods Available for Sale | $ 6,557 | | **$ 43** | | $ 6,600 |
| Less Ending Inventory | **560** | | **158** | | 718 |
| Cost of Goods Sold | **$5,997** | | $(115) | | $ 5,882 |
| Sales | **$6,266** | | 0 | | **$6,266** |
| Less Cost of Goods Sold | **5,997** | | $(115) | | 5,882 |
| Gross Margin on Sales | $ 269 | | $ 115 | | $ 384 |

[a]Computation of Purchases not presented in financial statements:
Purchases = Cost of Goods Sold + Ending Inventory − Beginning Inventory
$5,968 = **$5,997** + **$560** − **$589**

under FIFO more accurately measures the actual turnover of inventory because it uses relatively current cost data in both the numerator and the denominator.

Consider also the current ratio (= current assets/current liabilities). Financial statement analysts use the current ratio to assess the short-term liquidity risk of a company. If a firm uses LIFO in periods of rising prices while inventory quantities increase, the amount of inventory included in the numerator of the current ratio will be smaller than it would be if the firm valued inventory at more current costs using FIFO. Hence, the unwary reader may underestimate the liquidity of a company that uses a LIFO cost flow assumption.

## Identifying Operating Margin and Holding Gains

In general, the reported net income under FIFO exceeds that under LIFO during periods of rising purchase prices. This higher reported net income results from including a larger realized holding gain in reported net income under FIFO than under LIFO. This section illustrates the significance of holding gains in the calculation of net income under FIFO and LIFO.

The conventionally reported gross margin (sales minus cost of goods sold) comprises (1) an operating margin and (2) a realized holding gain (or loss). The term **operating margin** denotes the difference between the selling price of an item and its replacement cost at the time of sale. This operating margin gives some indication of a particular firm's relative advantage, such as a reputation for quality or service, in the market for its goods. The term **realized holding gain** denotes the difference between the current replacement cost of an item at the time of sale and its acquisition cost. It reflects the change in cost of an item during the period the firm holds the inventory item. Holding gains indicate increasing purchase prices and the skill (or luck) of the purchasing department in timing acquisitions.

Holding inventory generates costs. The costs are both explicit, such as for storage, and implicit, such as for the earnings forgone because the firm invests cash to acquire the inventory. Even when a firm experiences holding gains, the amount of the gain need not be sufficient to compensate the firm for its costs to hold inventory. Management should want to know about both the costs of holding inventory and the benefits.

To understand the calculation of the operating margin and the holding gain, consider the TV set example discussed in this chapter. The acquisition cost of the three items available for sale during the period is $840. Assume that the firm sells one TV set for $550. The replacement cost of the TV set at the time of sale is $320. The current replacement cost at the end of the period for each item in ending inventory is $350, or $700 total for the two items remaining. The top portion of **Exhibit 7.8** separates the conventionally reported gross margin into the operating margin and the realized holding gain.

The operating margin is the difference between the $550 selling price and the $320 replacement cost at the time of sale. The operating margin of $230 is, by definition, the same under both the FIFO and the LIFO cost flow assumptions. The realized holding gain is the difference between cost of goods sold based on replacement cost and cost of goods sold based on acquisition cost. The realized holding gain under FIFO exceeds that under LIFO because FIFO assumes that the earlier purchases at lower costs enter cost of goods sold. The larger realized holding gain explains why net income under FIFO typically exceeds that under LIFO during periods of rising prices. The smaller realized holding gains and losses under LIFO explain why earnings under LIFO tend to fluctuate less over time than under FIFO (except when a firm dips into LIFO layers). In conventional financial statements, the realized holding gain does not appear separately, as in **Exhibit 7.8**. The term **inventory profit** sometimes denotes the realized holding gain on inventory. The amount of inventory profit varies from period to period as the rate of change in the purchase prices of inventories varies. The larger the inventory profit, the less sustainable are earnings and therefore the lower is the quality of earnings.

The calculation of an **unrealized holding gain** on units in ending inventory also appears in **Exhibit 7.8**. The unrealized holding gain shows the difference between the current replacement cost of the ending inventory and its acquisition cost.[6] This unrealized holding gain on ending inventory does not appear in the firm's income statement as presently prepared under GAAP. The

[6]The unrealized holding gain for a given year on items on hand both at the beginning and at the end of the year is the difference between year-end current cost and beginning-of-year current cost. The examples in this chapter do not illustrate this complication; all items on hand at the end of the year are purchased during the year. See the **Glossary** definition of *inventory profit* for an illustration of the computation of holding gain for a year in which the beginning inventory includes unrealized holding gains from preceding periods.

**EXHIBIT 7.8** **Reporting of Operating Margins and Holding Gains for TV Sets**

| | Cost Flow Assumption | | | |
|---|---|---|---|---|
| | FIFO | | LIFO | |
| Sales Revenue | $550 | | $550 | |
| Less Replacement Cost of Goods Sold | 320 | | 320 | |
| Operating Margin on Sales | | $230 | | $230 |
| **Realized Holding Gain on TV Sets** | | | | |
| Replacement Cost (at time of sale) of Goods Sold | $320 | | $320 | |
| Less Acquisition Cost of Goods Sold (FIFO—TV set 1; LIFO—TV set 3) | 250 | | 300 | |
| Realized Holding Gain on TV Sets (inventory profit) | | 70 | | 20 |
| Conventionally Reported Gross Margin[a] | | $300 | | $250 |
| **Unrealized Holding Gain** | | | | |
| Replacement Cost of Ending Inventory (2 × $350) | | $700 | | $700 |
| Less Acquisition Cost of Ending Inventory (FIFO—TV sets 2 and 3; LIFO—TV sets 1 and 2) | | 590 | | 540 |
| Unrealized Holding Gain on TV Sets | | 110 | | 160 |
| Economic Profit on Sales and Holding Inventory of TV Sets (not reported in financial statements) | | $410 | | $410 |

[a]Note that Exhibit 7.3 stops here.

unrealized holding gain under LIFO exceeds that under FIFO because earlier purchases with lower costs remain in ending inventory under LIFO. The sum of the operating margin plus all holding gains (both realized and unrealized), labeled "Economic Profit" in **Exhibit 7.8**, is the same under FIFO and LIFO. FIFO recognizes most of the holding gain (that is, the realized portion) in computing net income each period. LIFO does not recognize most of the holding gain (that is, the unrealized portion) in the income statement. Instead, under LIFO, the unrealized holding gain remains unreported as long as the older acquisition costs appear on the balance sheet as ending inventory.

The total increase in wealth for a period includes both realized and unrealized holding gains. That total increase, $410 in the example, does not depend on the cost flow assumption and does not appear in financial statements under currently accepted accounting principles.

## Problem 7.4 for Self-Study

**Identifying operating margins and holding gains.** Refer to the data in **Problem 7.2 for Self-Study**. Sales revenue during June was $7,500. The average replacement cost at the time of sale was $12 per unit. The replacement cost on June 30 is $12.60 per unit.

**a.** Calculate the operating margin, the realized holding gain, and the unrealized holding gain during June using (1) FIFO, (2) LIFO, and (3) weighted-average cost flow assumptions. Refer to **Exhibit 7.8** for the desired format for this analysis.

**b.** "A firm's cost flow assumption does not affect the operating margin." Explain.

**c.** "The FIFO cost flow assumption typically includes most of the holding gain in net income, whereas the LIFO cost flow assumption implicitly includes more of the holding gain in the balance sheet." Explain.

## Effect of Dipping into LIFO Layers

GAAP requires firms that dip into LIFO layers during the period to indicate the effect of the dip on cost of goods sold. Assume, for example, that Nucor Corporation disclosed the following in its notes for the year:

> Because of recession conditions in the economy, Nucor dipped into LIFO layers created in earlier years. The effect of these inventory reductions decreased cost of goods sold and increased pretax income by $15 million.

As an earlier section discussed, firms may delay purchases at the end of the year and dip into LIFO layers as a means of increasing reported earnings. Firms may also dip into LIFO layers because of improved inventory control systems that reduce the amount of inventory needed. The analyst can assess the effect of dipping into LIFO layers by recomputing cost of goods sold and earnings assuming no dipping had occurred.

Cost of goods sold for Nucor under LIFO was $5,997 million (see **Exhibit 7.7**). Without dipping into LIFO layers, cost of goods sold would have been $6,012 (= $5,997 + $15). The ratio of cost of goods sold to sales with dipping is 95.7 percent (= $5,997/$6,266). The ratio without dipping is 95.9 percent (= $6,012/$6,266).

## Relating Cost of Goods Sold Percentages and Inventory Turnover Ratios

Firms often experience changes in their cost of goods sold to sales percentages and their inventory turnover ratios simultaneously. Both ratios use cost of goods sold, so the analyst might expect a relation between them. Recall from **Chapter 5** that the rate of return on assets (ROA) is the product of the profit margin for ROA and the total assets turnover. Changes in the cost of goods sold to sales percentage affect the profit margin. Changes in inventory turnovers affect the total asset turnover. The sections below describe possible explanations for various combinations of changes in the cost of goods sold to sales percentage and the inventory turnover ratio.

**Increasing Cost of Goods Sold to Sales Percentage and Increasing Inventory Turnover Ratio** This combination results in a lower profit margin and an increased total assets turnover. The net effect on ROA depends on which of the two effects dominates. The firm may have shifted its sales mix toward faster moving, but more commodity-like, products that generate smaller gross margins. The firm may also have increased the proportion of manufacturing outsourced to other manufacturers. Outsourcing lowers inventories by reducing the need for raw materials and work-in-process inventories. However, the firm must share some of the profit margin with the supplier, the outsourcing firm.

**Decreasing Cost of Goods Sold to Sales Percentage and Increasing Inventory Turnover Ratio** This combination increases both the profit margin and the total assets turnover and thereby increases ROA. How might this combination occur? A firm institutes improved inventory control systems that increase its inventory turnover. The savings in storage costs and costs of obsolescence reduce the cost of goods sold to sales percentage. Another possibility is that a firm experiences an increase in demand for its products because of reductions in supply by its competitors. Inventory moves more quickly. Shortages in supply permit the firm to raise prices and thereby lower the cost of goods sold to sales percentage. Consider, also, a capital-intensive manufacturer (with variable costs lower than average costs because of significant fixed costs). The increase in demand permits the firm to lower its per-unit manufacturing costs, moving inventory more quickly and lowering its cost of goods sold to sales percentage.

**Increasing Cost of Goods Sold to Sales Percentage and Decreasing Inventory Turnover Ratio** This combination decreases both the profit margin and the total assets turnover and lowers ROA. How might this combination occur? The firm experiences a buildup of obsolete inventory because competitors introduce technologically superior products. The firm has to write down the cost of the inventory to lower-of-cost-or-market or reduce selling prices to sell the inventory. Consider, also, the capital-intensive manufacturing firm. Demand weakens, resulting in a curtailment of production. The weakening of demand slows the turnover of inventory.

**Decreasing Cost of Goods Sold to Sales Percentage and Decreasing Inventory Turnover Ratio** This combination increases the profit margin but decreases the total assets turnover. The net effect on ROA depends on which effect dominates. This pair could occur because the firm reduced the amount of outsourcing, permitting the firm to capture more of the gross margin. The need for more raw materials and work-in-process inventories slows the inventory turnover. Or, the firm shifts its sales mix toward higher margin differentiated products that turn over more slowly.

Other explanations could cause the various combinations discussed above. Additional combinations of changes in one of the two ratios and not in the other are also possible. Given the links between the two ratios, the analyst should remain alert to possible common explanations.

## Problem 7.5 for Self-Study

**Effect of cost flow assumptions on financial ratios.** Refer to the data in **Problems 7.2** and **7.3 for Self-Study**. Assume the following:

| | June 30 | July 31 | August 31 |
|---|---|---|---|
| Current Assets Excluding Inventories | $1,650 | $3,480 | $3,230 |
| Current Liabilities | 2,290 | 4,820 | 3,750 |

**a.** Compute the current ratio for June 30, July 31, and August 31 using (1) FIFO, (2) LIFO, and (3) weighted-average cost flow assumptions for inventories. Assume for this part that there are no differences in income taxes payable related to the three cost flow assumptions.

**b.** Compute the inventory turnover ratio (= cost of goods sold/average inventory) for July and August using (1) FIFO, (2) LIFO, and (3) weighted-average cost flow assumptions.

## An International Perspective

*Statement No. 2* of the International Accounting Standards Board supports the use of the lower-of-cost-or-market basis for the valuation of inventories, with market value based on net realizable values.[7] All major industrialized countries require the lower-of-cost-or-market method in the valuation of inventories, although the definition of market value varies across countries. *Statement No. 2* states a preference for the FIFO and weighted-average cost flow assumptions. LIFO is an acceptable alternative, but firms should disclose how LIFO inventories differ from those valued under the lower-of-cost-or-market method. Few countries, except the United States and Japan, allow LIFO. Even in Japan, few firms use LIFO. Firms in most countries use FIFO and weighted-average cost flow assumptions. The extensive use of LIFO in the United States results from income tax considerations.

## Summary

Inventory measurements affect both the cost of goods sold on the income statement for the period and the amount shown for inventory on the balance sheet at the end of the period. The sum of the two must equal the beginning inventory plus the cost of purchases, at least in accounting based on acquisition costs and market transactions. The allocation between

[7] International Accounting Standards Board, *Statement No. 2*: "Inventories," 1993.

expense and asset depends primarily on the cost flow assumption (FIFO, LIFO, or weighted average) used.

FIFO provides inventory amounts that closely approximate current replacement costs. Costs of goods sold amounts are somewhat out of date, reflecting the cost of purchases toward the end of the previous period and the beginning of the current period. The gross margin under FIFO includes a realized holding gain for the increase in replacement cost between the time of purchase and the time of sale. The realized holding gain does not represent sustainable economic value added if the firm must replace the inventory at its currently higher costs. The amount of the holding gain depends on the change in costs between the time of purchase and sale. When inventory turns over four times a year, the elapsed time between acquisition and sale is three months. When inventory turns over six times a year, the elapsed time is only two months.

LIFO generally provides amounts for cost of goods sold that approximate current replacement cost at the time of sale. An exception occurs, however, when a firm dips into LIFO layers acquired in earlier years. The cost of goods sold amount for that period will reflect a combination of current and older acquisition costs. Inventories under LIFO may reflect very old costs, depending on how long the firm has used LIFO. A substantial unrealized holding gain on LIFO inventories exists that GAAP does not report on the balance sheet.

The weighted-average cost flow assumption provides inventory amounts similar to FIFO and cost of goods sold amounts that generally fall between those of FIFO and LIFO. When deciding on a cost flow assumption, firms must consider the effect on earnings, on taxes paid, and on recordkeeping costs.

## Appendix 7.1: Effects on the Statement of Cash Flows of Transactions Involving Inventory

The question that an analyst would raise regarding cash flows and inventory is: How much cash did a firm use to purchase (merchandising firm) or produce (manufacturing firm) inventory during the year? **Exhibit 7.9** presents information to illustrate the adjustments to compute the cash paid for inventory purchases of a merchandising firm.

**EXHIBIT 7.9** Data to Illustrate Cash Used to Purchase Inventory

| | Inventory | Accounts Payable |
|---|---|---|
| Balance, Beginning of Year | $ 100 | $ 75 |
| Purchases of Inventory on Account | 1,000 | 1,000 |
| Cost of Goods Sold | (800) | — |
| Cash Payments for Purchases on Account | — | (900) |
| Writedown of Inventories to Lower of Cost or Market | (50) | — |
| Balance, End of Year | $ 250 | $ 175 |

The computation of cash payments for purchases of inventory under the direct method of computing cash flow from operations is as follows:

| | |
|---|---|
| Cost of Goods Sold | $(800) |
| Less Increase in Inventory ($250 − $100) | (150) |
| Plus Increase in Accounts Payable ($175 − $75) | 100 |
| Less Writedown to Lower of Cost or Market | (50) |
| Cash Payments for Purchases on Account | $(900) |

If firms include the loss from the writedown of inventories in cost of goods sold, as usually occurs, then the first line above becomes a negative $850. The analyst need not make a separate adjustment for the writedown of inventories in this case to compute cash payments for purchases.

The computation of cash flow from operations under the indirect method is as follows:

| | |
|---|---|
| Net Income (for Cost of Goods Sold and Loss from Writedown of Inventories) | $(850) |
| Less Increase in Inventory ($250 − $100) | (150) |
| Plus Increase in Accounts Payable ($175 − $75) | 100 |
| Cash Flow from Operations | $(900) |

The computation of cash paid for production of inventory follows these same procedures, but is more complex. The analyst must include changes in raw materials inventory, work-in-process inventory, and finished goods inventory in the change in inventory. In addition to changes in accounts payable from raw materials purchases, the analyst must include changes in other current liabilities, such as wages payable and taxes payable, to the extent that they relate to manufacturing operations. Firms seldom provide the necessary data to separate payables into the portions arising from manufacturing activities versus selling and administrative activities. Thus, computing cash paid for manufacturing costs is usually difficult.

## Solutions to Self-Study Problems

### SUGGESTED SOLUTION TO PROBLEM 7.1 FOR SELF-STUDY

(Haskell Corporation; flow of manufacturing costs through the accounts.)

| | | |
|---|---|---|
| **a.** | Beginning Raw Materials Inventory | $ 42,400 |
| | Raw Materials Purchased | 60,700 |
| | Raw Materials Available for Use | $103,100 |
| | Subtract Ending Raw Materials Inventory | 46,900 |
| | Cost of Raw Materials Used | $ 56,200 |
| **b.** | Beginning Work-in-Process Inventory | $ 75,800 |
| | Cost of Raw Materials Used (from part **a**) | 56,200 |
| | Direct Labor Costs Incurred | 137,900 |
| | Heat, Light, and Power Costs | 1,260 |
| | Rent Costs | 4,100 |
| | Depreciation of Factory Equipment | 1,800 |
| | Prepaid Insurance Costs Consumed | 1,440 |
| | Total Beginning Work-in-Process and Manufacturing Costs Incurred | $278,500 |
| | Subtract Ending Work-in-Process Inventory | (63,200) |
| | Cost of Units Completed and Transferred to Finished Goods Storeroom | $215,300 |
| **c.** | Beginning Finished Goods Inventory | $ 44,200 |
| | Cost of Units Completed and Transferred to Finished Goods Storeroom (from part **b**) | 215,300 |
| | Subtract Ending Finished Goods Inventory | (46,300) |
| | Cost of Goods Sold | $213,200 |
| **d.** | Income before taxes is $61,800 (= $400,000 − $213,200 − $125,000). | |

SUGGESTED SOLUTION TO PROBLEM 7.2 FOR SELF-STUDY

(Computing cost of goods sold and ending inventory under various cost flow assumptions.)

**a.** See **Exhibit 7.10**.

**EXHIBIT 7.10** (Suggested Solution to Problem 7.2 for Self-Study, part **a**)

| | Units | Unit Cost | Total Cost FIFO | LIFO | Weighted Average |
|---|---|---|---|---|---|
| Beginning Inventory | — | — | — | — | |
| Purchases, June 1 | 100 | $10.00 | $1,000 | $1,000 | $1,000 |
| Purchases, June 7 | 400 | 11.00 | 4,400 | 4,400 | 4,400 |
| Purchases, June 18 | 100 | 12.50 | 1,250 | 1,250 | 1,250 |
| Total Goods Available for Sale | 600 | | $6,650 | $6,650 | $6,650 |
| Withdrawals during June | (495) | | (5,345)[a] | (5,595)[c] | (5,486)[e] |
| Ending Inventory | 105 | | $1,305[b] | $1,055[d] | $1,164[f] |

[a](100 × $10.00) + (395 × $11.00) = $5,345.
[b](100 × $12.50) + (5 × $11.00) = $1,305.
[c](100 × $12.50) + (395 × $11.00) = $5,595.
[d](100 × $10.00) + (5 × $11.00) = $1,055.
[e]495 × ($6,650/600) = $5,486.
[f]105 × ($6,650/600) = $1,164.

**b.** See **Exhibit 7.11**.

**EXHIBIT 7.11** (Suggested Solution to Problem 7.2 for Self-Study, part **b**)

| | Units | Unit Cost | Total Cost FIFO | LIFO | Weighted Average |
|---|---|---|---|---|---|
| Beginning Inventory, July 1 | 105 | See **Exhibit 7.10** | $ 1,305 | $ 1,055 | $ 1,164 |
| Purchases, July 5 | 300 | $13.00 | 3,900 | 3,900 | 3,900 |
| Purchases, July 15 | 200 | 13.50 | 2,700 | 2,700 | 2,700 |
| Purchases, July 23 | 250 | 14.00 | 3,500 | 3,500 | 3,500 |
| Total Goods Available for Sale | 855 | | $11,405 | $11,155 | $11,264 |
| Withdrawals during July | (620) | | (8,115)[a] | (8,410)[c] | (8,168)[e] |
| Ending Inventory | 235 | | $ 3,290[b] | $ 2,745[d] | $ 3,096[f] |

[a]$1,305 + (300 × $13.00) + (200 × $13.50) + (15 × $14.00) = $8,115.
[b](235 × $14.00) = $3,290.
[c](250 × $14.00) + (200 × $13.50) + (170 × $13.00) = $8,410.
[d]$1,055 + (130 × $13.00) = $2,745.
[e]620 × ($11,264/855) = $8,168.
[f]235 × ($11,264/855) = $3,096.

### SUGGESTED SOLUTION TO PROBLEM 7.3 FOR SELF-STUDY

(Assessing the impact of a LIFO layer liquidation.)

**a.** **See Exhibit 7.12.**

**b.** 125 × ($15 − $13) = $250.

**EXHIBIT 7.12** **(Suggested Solution to Problem 7.3 for Self-Study, part a)**

| | Units | Unit Cost | Total Cost FIFO | Total Cost LIFO | Total Cost Weighted Average |
|---|---|---|---|---|---|
| Beginning Inventory | 235 | See Exhibit 7.11 | $ 3,290 | $ 2,745 | $ 3,096 |
| Purchases during August | 600 | $15 | 9,000 | 9,000 | 9,000 |
| Total Goods Available for Sale | 835 | | $ 12,290 | $ 11,745 | $ 12,096 |
| Withdrawals during August | (725) | | (10,640)[a] | (10,625)[c] | (10,503)[e] |
| Ending Inventory | 110 | | $ 1,650[b] | $ 1,120[d] | $ 1,593[f] |

[a]$3,290 + (490 × $15) = $10,640.
[b](110 × $15) = $1,650.
[c](600 × $15) + (125 × $13) = $10,625.
[d]$1,055 + (5 × $13) = $1,120.
[e]($12,096/835) × 725 = $10,503.
[f]($12,096/835) × 110 = $1,593.

### SUGGESTED SOLUTION TO PROBLEM 7.4 FOR SELF-STUDY

(Identifying operating margins and holding gains.)

**a.** See **Exhibit 7.13.**

**EXHIBIT 7.13** **(Suggested Solution to Problem 7.4 for Self-Study, part a)**

| | Cost Flow Assumption FIFO | Cost Flow Assumption LIFO | Cost Flow Assumption Weighted Average |
|---|---|---|---|
| OPERATING MARGIN | | | |
| Sales Revenue | $ 7,500 | $ 7,500 | $ 7,500 |
| Less Replacement Cost of Goods Sold (495 × $12) | (5,940) | (5,940) | (5,940) |
| Operating Margin | $ 1,560 | $ 1,560 | $ 1,560 |
| REALIZED HOLDING GAIN | | | |
| Replacement Cost of Goods Sold | $ 5,940 | $ 5,940 | $ 5,940 |
| Acquisition Cost of Goods Sold | (5,345) | (5,595) | (5,486) |
| Realized Holding Gain | $ 595 | $ 345 | $ 454 |
| Conventionally Reported Gross Margin | $ 2,155 | $ 1,905 | $ 2,014 |
| UNREALIZED HOLDING GAIN | | | |
| Replacement Cost of Ending Inventory (105 × $12.60) | $ 1,323 | $ 1,323 | $ 1,323 |
| Acquisition Cost of Ending Inventory | (1,305) | (1,055) | (1,164) |
| Unrealized Holding Gain | $ 18 | $ 268 | $ 159 |
| Economic Profit | $ 2,173 | $ 2,173 | $ 2,173 |

**b.** The operating margin relates sales to the replacement cost of goods sold. The replacement cost at the time of sale depends on prices for the items at that time (even though the firm purchased no items then) and not on the prices that the firm, earlier, actually paid. In contrast, in historical cost accounting, when the firm makes a cost flow assumption, it matches one of the earlier prices with the sales revenue at the time of sale.

**c.** The holding gain arises because the replacement costs of inventory items increase while firms hold them. The length of the assumed holding period for goods sold under a FIFO cost flow assumption exceeds the assumed holding period under LIFO (unless a firm dips into LIFO layers of earlier years). Thus, FIFO typically reports a larger realized holding gain as part of net income. The length of the assumed holding period for goods in ending inventory under a LIFO cost flow assumption exceeds the assumed holding period under FIFO. Thus, LIFO's inventory valuation on the balance sheet reflects a larger unrealized holding gain.

### SUGGESTED SOLUTION TO PROBLEM 7.5 FOR SELF-STUDY

(Effect of cost flow assumptions on financial ratios.)

**a. Current Ratio**

| June 30 | FIFO | LIFO | Weighted Average |
|---|---|---|---|
| ($1,650 + $1,305)/$2,290 | 1.29 | | |
| ($1,650 + $1,055)/$2,290 | | 1.18 | |
| ($1,650 + $1,164)/$2,290 | | | 1.23 |
| **Current Ratio** | | | |
| JULY 31 | | | |
| ($3,480 + $3,290)/$4,820 | 1.40 | | |
| ($3,480 + $2,745)/$4,820 | | 1.29 | |
| ($3,480 + $3,096)/$4,820 | | | 1.36 |
| AUGUST 31 | | | |
| ($3,230 + $1,650)/$3,750 | 1.30 | | |
| ($3,230 + $1,120)/$3,750 | | 1.16 | |
| ($3,230 + $1,593)/$3,750 | | | 1.29 |
| **b. Inventory Turnover Ratio** | | | |
| JULY | | | |
| $8,115/0.5($1,305 + $3,290) | 3.53 | | |
| $8,410/0.5($1,055 + $2,745) | | 4.43 | |
| $8,168/0.5($1,164 + $3,096) | | | 3.83 |
| AUGUST | | | |
| $10,640/0.5($3,290 + $1,650) | 4.31 | | |
| $10,625/0.5($2,745 + $1,120) | | 5.50 | |
| $10,503/0.5($3,096 + $1,593) | | | 4.48 |

## Key Terms and Concepts

Inventory
Inventory equation
Purchases
Direct material
Direct labor
Manufacturing overhead
Product costs
Period expenses
Flow of costs
Raw Materials Inventory
Work-in-Process Inventory
Finished Goods Inventory
Replacement cost
Lower-of-cost-or-market basis
Specific identification
Cost flow assumption

First-in, first-out (FIFO)
Last-in, first-out (LIFO)
Weighted average
LIFO inventory layer
Inventory valuation allowance
Operating margin
Realized holding gain
Inventory profit
Unrealized holding gain

## Questions, Short Exercises, Exercises, Problems, and Cases

For additional student resources, content, and interactive quizzes for this chapter visit the FACMU website: **www.thomsonedu.com/accounting/stickney**

### QUESTIONS

1. Review the meaning of the terms and concepts listed above in Key Terms and Concepts.
2. Identify the underlying accounting principle that guides the measurement of the acquisition cost of inventories, equipment, buildings, and other assets. What is the rationale for this accounting principle?
3. "Firms may treat depreciation on equipment either as a product cost or as a period expense, depending on the type of equipment." Explain.
4. Compare and contrast the Merchandise Inventory account of a merchandising firm and the Finished Goods Inventory account of a manufacturing firm.
5. "The total income from an inventory item is the difference between the cash received from selling the item and the cash paid to acquire it. The inventory valuation method (acquisition cost, current replacement cost, lower of cost or market) is therefore largely irrelevant." Do you agree? Why or why not?
6. Assume that a firm changes its cost basis for inventory from acquisition cost to current cost, that current costs exceed acquisition costs at the end of the year of change, and that the firm sells all year-end inventory during the following year. Ignore income tax effects. What will be the impact of the change on net income for the year of change? on income for the following year? on total income over the two years?
7. "Inventory computations require cost flow assumptions only because specific identification of items sold is costly. Specific identification is theoretically superior to any cost flow assumption and eliminates the possibility of income manipulation available with some cost flow assumptions." Comment.
8. Assume no changes in physical quantities during the period. During a period of rising purchase prices, will a FIFO or a LIFO cost flow assumption result in the higher ending inventory amount? the lower amount? Which cost flow assumption will result in the higher ending inventory amount during a period of declining purchase prices? the lower inventory amount?
9. **a.** During a period of rising purchase prices, will a FIFO or a LIFO cost flow assumption result in the higher cost of goods sold? the lower cost of goods sold? Assume no changes in physical quantities during the period.
   **b.** Which cost flow assumption, LIFO or FIFO, will result in the higher cost of goods sold during a period of declining purchase prices? the lower cost of goods sold?
10. "LIFO provides a more meaningful income statement than FIFO, even though it provides a less meaningful balance sheet." Does it ever? Does it always?
11. Explain each of the following statements:
    **a.** Differences between the effects of FIFO and LIFO on the financial statements increase as the rate of change in the cost of inventory items increases.
    **b.** Differences between the effects of FIFO and LIFO on cost of goods sold decrease as the rate of inventory turnover increases.
    **c.** Differences between FIFO and LIFO amounts for inventories on the balance sheet do not decrease as the rate of inventory turnover increases.
12. Identify the reasons a firm might dip into an old LIFO layer.
13. "A firm that changes its selling prices when the acquisition costs of its inventory items change will report smoother gross margins over time under LIFO than under FIFO." Explain this statement using the concepts of operating margins and realized holding gains.
14. Suggest reasons a firm would choose a FIFO instead of a LIFO cost flow assumption for financial reporting.
15. "A firm that dips into its LIFO layers and pays additional taxes in a particular year is still better off than if had used FIFO all along." Explain. Under what circumstances will the firm be worse off?

## SHORT EXERCISES

16. **Identifying inventory cost inclusions.** Intervest Corporation, a real estate developer, purchased a tract of land for $450,000. Intervest intends to develop and subdivide the land for construction of residential homes. Legal costs related to the acquisition totaled $12,000. Property taxes and insurance during the development of the land totaled $22,900. Development costs totaled $689,000. Advertising costs incurred to sell the land totaled $38,200. Identify the costs that Intervest should include in its Land Inventory account.
17. **Effect of inventory valuation on net income.** Selby Corporation, a commercial real estate developer, acquired a tract of land on September, Year 6 for $240,000. Forest fires on December, Year 6, reduced the value of this land to $160,000. In December, Year 7, a local hospital announced plans to build a new hospital near the land, increasing its market value to $420,000. Selby Corporation sold this land in February, Year 8 for $460,000. Compute the amount of income or loss related to this land for Year 6, Year 7, and Year 8 and for the three years as a whole, assuming that the firm reports the land inventory on its balance sheet using (1) acquisition cost, (2) lower of cost or market, and (3) market value.
18. **Income computation for a manufacturing firm.** Fun-in-the-Sun Tanning Lotion Company manufactures suntan lotion made from organic materials. During its first year of operations it purchased raw materials costing $78,200, of which it used $56,300 in manufacturing suntan lotions. It incurred manufacturing labor costs of $36,100 and manufacturing overhead costs of $26,800 during the year. Inventories taken at the end of the year revealed unfinished suntan lotions costing $12,700 and finished suntan lotions costing $28,500. Compute the amount of cost of goods sold for the year.
19. **Computations involving different cost flow assumptions.** Sun Health Food's purchases of vitamins during Year 10, its first year of operations, were as follows:

| | Quantity | Cost per Unit | Total Cost |
|---|---|---|---|
| January 5 Purchase | 460 | $4.30 | $ 1,978 |
| April 16 Purchase | 670 | 4.20 | 2,814 |
| August 26 Purchase | 500 | 4.16 | 2,080 |
| November 13 Purchase | 870 | 4.10 | 3,567 |
| Totals | 2,500 | | $10,439 |

The inventory on December 31 was 420 units. Compute the cost of the inventory on December 31 and the cost of goods sold for the year under each of the following cost flow assumptions:

**a.** FIFO.
**b.** Weighted average.
**c.** LIFO.

20. **Effect of LIFO on financial statements over several periods.** Harmon Corporation commenced operations on January 2, Year 8. It uses a LIFO cost flow assumption. Its purchases and sales for the first three years of operations appear below:

| | Purchases | | Sales | |
|---|---|---|---|---|
| | Units | Unit Cost | Units | Unit Price |
| Year 8 | 83,000 | $20.00 | 64,000 | $32.00 |
| Year 9 | 92,000 | $25.00 | 101,000 | $40.00 |
| Year 10 | 120,000 | $30.00 | 110,000 | $48.00 |

**a.** Compute the amount of ending inventory for each of the three years.
**b.** Compute the amount of income for each of the three years.

21. **Effect of dipping into LIFO inventories.** Robertson Corporation uses a LIFO cost flow assumption for inventories and cost of goods sold. Its beginning inventory for the current year totaled $48,900 and its ending inventory totaled $42,600. Cost of goods sold for the year totaled $286,700. A note to its financial statements discloses that income before taxes would have been $2,700 lower if it had not dipped into LIFO layers during the year. Compute the purchase price of merchandise that Robertson Corporation would have had to acquire to avoid dipping into its LIFO layers.

22. **Conversion from LIFO to FIFO.** United States Steel Corporation reported inventories using LIFO of $1,030 million at the beginning of Year 13, and $1,283 million at the end of Year 13. Its cost of goods sold for Year 13 totaled $8,469 million. Inventories on a FIFO basis would exceed its inventories on a LIFO basis by $310 million at the beginning of the year and $270 million at the end of the year. Compute the amount for cost of goods sold if the firm had used a FIFO instead of a LIFO cost flow assumption.
23. **Separating operating margins and holding gains.** During Year 8, its first year of operations, Miller Corporation purchased 10,000 units at $80 each and sold 9,000 units for $100 each. On December 31, Year 8, the units had a replacement cost of $84. The average replacement cost during Year 8 was $82. Compute the amount of operating margin, realized holding gain or loss, unrealized holding gain or loss, and economic profit for Year 8.

## EXERCISES

24. **Identifying inventory cost inclusions.** Trembly Department Store commenced operations on January 2. It engaged in the following transactions during January. Identify the amount that the firm should include in the valuation of merchandise inventory.
    **a.** Purchases of merchandise on account during January totaled $300,000.
    **b.** The freight cost to transport merchandise to Trembly's warehouse was $13,800.
    **c.** The salary of the purchasing manager was $3,000.
    **d.** Depreciation, taxes, insurance, and utilities for the warehouse totaled $27,300.
    **e.** The salary of the warehouse manager was $2,200.
    **f.** The cost of merchandise that Trembly purchased in part **a** and returned to the supplier was $18,500.
    **g.** Cash discounts taken by Trembly from purchases on account in part **a** totaled $4,900.
25. **Income computation for a manufacturing firm.** The following data relate to General Mills, a food processor, for a recent fiscal year ending May 31 (amounts in millions):

| | June 1, Year 10 | May 31, Year 11 |
|---|---|---|
| Raw Materials Inventory | $ 73.7 | $101.5 |
| Work-in-Process Inventory | 100.8 | 119.1 |
| Finished Goods Inventory | 286.2 | 322.3 |

The company incurred manufacturing costs (direct material, direct labor, manufacturing overhead) during the fiscal year totaling $2,752.0. Sales revenue was $6,700.2, selling and administrative expenses were $2,903.7, and interest expense was $151.9. The income tax rate is 35 percent. Compute net income for the fiscal year.

26. **Income computation for a manufacturing firm.** The following data relate to Dow Chemical Corporation for a single year (amounts in millions):

| | January 1 | December 31 |
|---|---|---|
| Raw Materials Inventory | $ 452 | $ 373 |
| Work-in-Process Inventory | 843 | 837 |
| Finished Goods Inventory | 2,523 | 2,396 |

The company incurred manufacturing costs (direct material used, direct labor, and manufacturing overhead) during the year totaling $28,044. Sales revenue was $32,632, marketing and administrative expenses were $2,436, and interest expenses were $828. The income tax rate is 35 percent. Compute net income for the year.

27. **Over sufficiently long time spans, income is cash-in less cash-out; cost basis for inventory.** Duggan Company began business on January 1, Year 1. Information concerning merchandise inventories, purchases, and sales for the first three years of operations follows:

| | Year 1 | Year 2 | Year 3 |
|---|---|---|---|
| Sales | $200,000 | $300,000 | $400,000 |
| Purchases | 210,000 | 271,000 | 352,000 |
| Inventories, December 31: | | | |
| At cost | 60,000 | 80,000 | 115,000 |
| At market | 50,000 | 65,000 | 120,000 |

**a.** Compute the gross margin on sales (sales minus cost of goods sold) for each year, using the lower-of-cost-or-market basis in valuing inventories.
**b.** Compute the gross margin on sales (sales minus cost of goods sold) for each year, using the acquisition cost basis in valuing inventories.
**c.** Indicate your conclusion about whether the lower-of-cost-or-market basis of valuing inventories is conservative.

**28. When goods available for sale exceed sales, firms can manipulate income even when they use specific identification.** Cypres, a discounter of consumer electronics, has 300 identical computers available for sale during December. It acquired these computers as follows: 100 in June for $300 each, 100 in August for $400 each, and 100 in November for $350 each. Assume that sales for December are 200 units at $600 each.
**a.** Compute gross margin for December assuming FIFO.
**b.** Compute gross margin for December assuming specific identification of computers sold to minimize reported income for tax purposes.
**c.** Compute gross margin for December assuming specific identification of computers sold to maximize reported income for the purpose of increasing the store manager's profit-sharing bonus for the year.

**29. Computations involving different cost flow assumptions.** Arnold Company's raw material purchases during January, its first month of operations, were as follows:

| | Quantity | Cost per Pound | Total Costs |
|---|---|---|---|
| 1/2 Purchased . . . . . . . . . . . . . . . . . . | 1,200 pounds | $2.20 | $ 2,640 |
| 1/8 Purchased . . . . . . . . . . . . . . . . . . | 2,200 pounds | 2.25 | 4,950 |
| 1/15 Purchased . . . . . . . . . . . . . . . . . | 2,800 pounds | 2.28 | 6,384 |
| 1/23 Purchased . . . . . . . . . . . . . . . . . | 1,500 pounds | 2.30 | 3,450 |
| 1/28 Purchased . . . . . . . . . . . . . . . . . | 3,000 pounds | 2.32 | 6,960 |
| Total Goods Available for Use . . . . . . . . . . . . . . . . . . . . . | 10,700 pounds | | $24,384 |

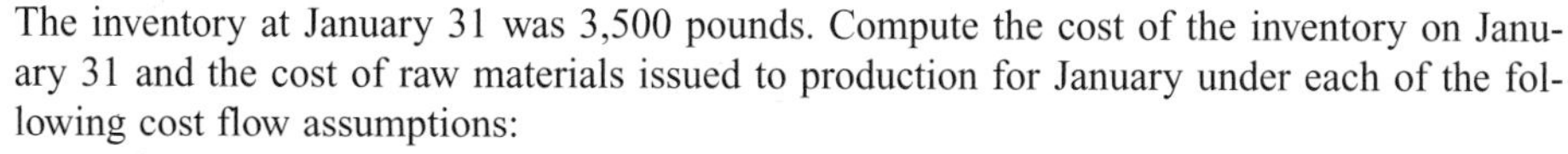

The inventory at January 31 was 3,500 pounds. Compute the cost of the inventory on January 31 and the cost of raw materials issued to production for January under each of the following cost flow assumptions:
**a.** FIFO.
**b.** Weighted average.
**c.** LIFO.

**30. Effect of LIFO on financial statements over several periods.** Howell Corporation commenced operations on January 2, Year 7. Its purchases and sales for the first four years of operations appear below:

| | Purchases | | Sales | |
|---|---|---|---|---|
| | Units | Unit Cost | Units | Unit Price |
| Year 7 . . . . . . . . . . . . . . . . . . . | 36,000 | $10.00 | 28,000 | $12.00 |
| Year 8 . . . . . . . . . . . . . . . . . . . | 40,000 | 10.50 | 38,000 | 12.60 |
| Year 9 . . . . . . . . . . . . . . . . . . . | 45,000 | 11.30 | 48,000 | 13.23 |
| Year 10 . . . . . . . . . . . . . . . . . . | 53,000 | 12.40 | 52,000 | 13.89 |

Howell Corporation uses a LIFO cost flow assumption. Ignore income taxes.
**a.** Compute the ending inventory for each of the four years.
**b.** Compute income for each of the four years.
**c.** Compute the ratio of income divided by sales for each of the four years.
**d.** Interpret the pattern of income-to-sales percentages computed in part **c**.

**31. Reconstructing financial statement data from information on effects of liquidations of LIFO layers.** The inventory footnote to the annual report of Chan Company reads in part as follows:

> Because of continuing high demand throughout the year, inventories were unavoidably reduced and could not be replaced. Under the LIFO system of accounting, used for many years by Chan Company, the net effect of all the inventory changes was to increase pretax income by $900,000 over what it would have been had inventories been maintained at their physical levels at the start of the year.

The acquisition cost of Chan Company's merchandise purchases was $26 per unit during the year, after having risen erratically over past years. Its inventory positions at the beginning and the end of the year follow.

| Date | Physical Count of Inventory | LIFO Cost of Inventory |
|---|---|---|
| January 1 . . . . . . . . . . . . . . . . . . . . . . . | 200,000 units | $ ? |
| December 31 . . . . . . . . . . . . . . . . . . . . . . | 150,000 units | $600,000 |

**a.** What was the average cost per unit of the 50,000 units removed from the January 1 LIFO inventory?
**b.** What was the January 1 LIFO cost of inventory?

32. **LIFO provides opportunity for income manipulation.** EKG Company, a manufacturer of medical supplies, began the year with 10,000 units of product that cost $8 each. During the year, it produced another 60,000 units at a cost of $15 each. Sales for the year were expected to total 70,000 units. During November, the company needs to plan production for the remainder of the year. The company might produce no additional units beyond the 60,000 units already produced. On the other hand, the company could produce up to 100,000 additional units; the cost would be $22 per unit regardless of the quantity produced. Assume that sales are 70,000 units for the year at an average price of $30 per unit.
**a.** What production level for the remainder of the year gives the largest cost of goods sold for the year? What is that cost of goods sold?
**b.** What production level for the remainder of the year gives the smallest cost of goods sold for the year? What is that cost of goods sold?
**c.** Compare the gross margins implied by the two production plans devised in the preceding parts.

33. **Analysis of annual report; usage of LIFO.** The notes to the financial statements in an annual report of Sears, Roebuck, and Company contained the following statement: "If the physical quantity of goods in inventory at [year-end and the beginning of the year] . . . were to be replaced, the estimated expenditures of funds required would exceed the amounts reported by approximately $670 million and $440 million, respectively." Sears uses LIFO.
**a.** How much higher or lower would Sears's pretax reported income have been if it had valued inventories at current costs rather than with a LIFO cost flow assumption?
**b.** Sears reported $606 million net income for the year. Assume tax expense equal to 35 percent of pretax income. By what percentage would Sears' net income have increased if a FIFO cost flow assumption had been used?

34. **Separating operating margin from holding gains.** On January 1, the merchandise inventory of Giles Computer Store comprised 200 units acquired for $300 each. During the year, the firm acquired 2,500 additional units at an average cost of $400 each and sold 2,300 units for $800 each. The replacement cost of these units at the time they were sold averaged $400 during the year. The replacement cost of units on December 31 was $500 per unit.
**a.** Calculate cost of goods sold under both FIFO and LIFO cost flow assumptions.
**b.** Prepare partial income statements showing gross margin on sales as revenues less cost of goods sold under both FIFO and LIFO cost flow assumptions.
**c.** Prepare partial income statements, separating the gross margin on sales into operating margins and realized holding gains, under both FIFO and LIFO.
**d.** Append to the bottom of the statements prepared in part **c** a statement showing the amount of unrealized holding gains and the total of realized income plus unrealized holding gains.
**e.** If you did the previous steps correctly, the totals in part **d** are the same for both FIFO and LIFO. Is this equality a coincidence? Why or why not?

35. **Effect of inventory errors.** On December 30, Year 1, Warren Company received merchandise costing $1,000 and counted it in the December 31 listing of all items on hand. The firm included the cost of the inventory in its ending inventory on the balance sheet on December 31, Year 1. The firm received an invoice on January 4, Year 2, when it recorded the acquisition as a Year 2 acquisition. It should have recorded the acquisition for Year 1. Assume that the firm never discovered the error. Indicate the effect (overstatement, understatement, none) on each of the following amounts (ignore income taxes):
**a.** Inventory, 12/31/Year 1.
**b.** Inventory, 12/31/Year 2.

**c.** Cost of goods sold, Year 1.
**d.** Cost of goods sold, Year 2.
**e.** Net income, Year 1.
**f.** Net income, Year 2.
**g.** Accounts payable, 12/31/Year 1.
**h.** Accounts payable, 12/31/Year 2.
**i.** Retained earnings, 12/31/Year 2.

## PROBLEMS AND CASES

**36. Preparation of journal entries and income statement for a manufacturing firm.** Best Furniture, Inc., a furniture manufacturer, showed the following amounts in its inventory accounts on January 1:

| | |
|---|---|
| Raw Materials Inventory | $226,800 |
| Work-in-Process Inventory | 427,900 |
| Finished Goods Inventory | 182,700 |

Best Furniture engaged in the following transactions during January:
**(1)** Acquired raw materials costing $667,200 on account.
**(2)** Issued, to producing departments, raw materials costing $689,100.
**(3)** Paid salaries and wages during January for services received during the month as follows:

| | |
|---|---|
| Factory Workers | $432,800 |
| Sales Personnel | 89,700 |
| Administrative Officers | 22,300 |

**(4)** Calculated depreciation on buildings and equipment during January as follows:

| | |
|---|---|
| Manufacturing Facilities | $182,900 |
| Selling Facilities | 87,400 |
| Administrative Facilities | 12,200 |

**(5)** Incurred and paid other operating costs in cash as follows:

| | |
|---|---|
| Manufacturing | $218,500 |
| Selling | 55,100 |
| Administrative | 34,700 |

**(6)** The cost of goods manufactured and transferred to the finished goods storeroom totaled $1,564,500.
**(7)** Sales on account during January totaled $2,400,000.
**(8)** A physical inventory taken on January 31 revealed a finished goods inventory of $210,600.

**a.** Present journal entries to record the transactions and events that occurred during January.
**b.** Prepare an income statement for Best Furniture, Inc. for January. Ignore income taxes.

**37. Flow of manufacturing costs through the accounts.** The following data related to the manufacturing activities of the Cornell Company during June:

| | June 1 | June 30 |
|---|---|---|
| Raw Materials Inventory | $ 46,900 | $ 43,600 |
| Factory Supplies Inventory | 7,600 | 7,700 |
| Work-in-Process Inventory | 110,900 | 115,200 |
| Finished Goods Inventory | 76,700 | 71,400 |

It incurred factory costs during the month as follows:

| | |
|---|---|
| Raw Materials Purchased | $429,000 |
| Supplies Purchased | 22,300 |
| Labor Services Received | 362,100 |
| Heat, Light, and Power | 10,300 |
| Insurance | 4,200 |

It experienced expirations of previous factory acquisitions and prepayments as follows:

| | |
|---|---|
| Depreciation on Factory Equipment | $36,900 |
| Prepaid Rent Expired | 3,600 |

Other information included the following:

| | |
|---|---|
| Sales | $1,350,000 |
| Selling and Administrative Expenses | 246,900 |
| Interest Expense | 47,100 |
| Income Tax Rate | 40% |

**a.** Calculate the cost of raw materials and factory supplies used during June.
**b.** Calculate the cost of units completed during June and transferred to the finished goods storeroom.
**c.** Calculate the cost of goods sold during June.
**d.** Calculate the amount of net income for the month of June. The income tax rate is 40 percent.

**38. Flow of manufacturing costs through the accounts.** The following data relate to the activities of Hickory Industries during April:

| | April 1 | April 30 |
|---|---|---|
| Raw Materials Inventory | $ 50,600 | $ 54,400 |
| Work-in-Process Inventory | 156,200 | 153,800 |
| Finished Goods Inventory | 76,800 | 79,400 |

It incurred factory costs during the month as follows:

| | |
|---|---|
| Raw Materials Purchased | $182,600 |
| Labor Services Received | 144,800 |
| Heat, Light, and Power | 6,200 |
| Depreciation | 35,600 |

Other data relating to the month included the following:

| | |
|---|---|
| Sales | $525,000 |
| Selling and Administrative Expenses | 82,600 |

**a.** Calculate the cost of raw materials used during April.
**b.** Calculate the cost of units completed during April and transferred to the finished goods storeroom.
**c.** Calculate net income for April. The income tax rate is 40 percent.

**39. Detailed comparison of various choices for inventory accounting.** Hartison Corporation commenced retailing operations on January 1, Year 1. It made purchases of merchandise inventory during Year 1 and Year 2 as follows:

| | Quantity Purchased | Unit Price | Acquisition Cost |
|---|---|---|---|
| 1/10/Year 1 | 1,200 | $10 | $12,000 |
| 6/30/Year 1 | 700 | 12 | 8,400 |
| 10/20/Year 1 | 400 | 13 | 5,200 |
| Total Year 1 | 2,300 | | $25,600 |

| | Quantity Purchased | Unit Price | Acquisition Cost |
|---|---|---|---|
| 2/18/Year 2 | 600 | $15 | $ 9,000 |
| 7/15/Year 2 | 1,200 | 16 | 19,200 |
| 12/15/Year 2 | 800 | 18 | 14,400 |
| Total Year 2 | 2,600 | | $42,600 |

Hartison Corporation sold 1,800 units during Year 1 and 2,800 units during Year 2.

**a.** Calculate the cost of goods sold for Year 1 using a FIFO cost flow assumption.
**b.** Calculate the cost of goods sold for Year 1 using a LIFO cost flow assumption.
**c.** Calculate the cost of goods sold for Year 1 using a weighted-average cost flow assumption.
**d.** Calculate the cost of goods sold for Year 2 using a FIFO cost flow assumption.
**e.** Calculate the cost of goods sold for Year 2 using a LIFO cost flow assumption.
**f.** Calculate the cost of goods sold for Year 2 using a weighted-average cost flow assumption.
**g.** For the two years taken as a whole, will FIFO or LIFO result in reporting the larger net income? What is the difference in net income for the two-year period under FIFO as compared with LIFO? Assume an income tax rate of 40 percent for both years.
**h.** Which method, LIFO or FIFO, should Hartison Corporation probably prefer and why?

**40. Continuation of preceding problem introducing current cost concepts.** (Do not attempt this problem until you have worked **Problem 39.**) Assume the same data for Hartison Corporation as given in the previous problem. In addition, assume the following:

| | |
|---|---|
| **Selling Price per Unit** | |
| Year 1 | $18 |
| Year 2 | 24 |
| **Average Current Replacement Cost** | |
| Year 1 | $12 |
| Year 2 | 16 |
| **Current Replacement Cost** | |
| December 31, Year 1 | $14 |
| December 31, Year 2 | 18 |

**a.** Prepare an analysis for Year 1 that identifies operating margins, realized holding gains and losses, and unrealized holding gains and losses for the FIFO, LIFO, and weighted-average cost flow assumptions.
**b.** Repeat part **a** for Year 2.
**c.** Demonstrate that over the two-year period, income plus holding gains before taxes of Hartison Corporation are independent of the cost flow assumption.

**41. Detailed comparison of various choices for inventory accounting.** Burton Corporation commenced retailing operations on January 1, Year 1. Purchases of merchandise inventory during Year 1 and Year 2 appear below:

| | Quantity Purchased | Unit Price | Acquisition Cost |
|---|---|---|---|
| 1/10/Year 1 | 600 | $10 | $ 6,000 |
| 6/30/Year 1 | 200 | 12 | 2,400 |
| 10/20/Year 1 | 400 | 15 | 6,000 |
| Total Year 1 | 1,200 | | $14,400 |

*(continued)*

| | Quantity Purchased | Unit Price | Acquisition Cost |
|---|---|---|---|
| 2/18/Year 2 | 500 | $14 | $ 7,000 |
| 7/15/Year 2 | 500 | 12 | 6,000 |
| 12/15/Year 2 | 800 | 10 | 8,000 |
| Total Year 2 | 1,800 | | $21,000 |

Burton Corporation sold 1,000 units during Year 1 and 1,500 units during Year 2.

**a.** Calculate the cost of goods sold for Year 1 using a FIFO cost flow assumption.
**b.** Calculate the cost of goods sold for Year 1 using a LIFO cost flow assumption.
**c.** Calculate the cost of goods sold for Year 1 using a weighted-average cost flow assumption.
**d.** Calculate the cost of goods sold for Year 2 using a FIFO cost flow assumption.
**e.** Calculate the cost of goods sold for Year 2 using a LIFO cost flow assumption.
**f.** Calculate the cost of goods sold for Year 2 using a weighted-average cost flow assumption.
**g.** Will FIFO or LIFO result in reporting the larger net income for Year 1? Explain.
**h.** Will FIFO or LIFO result in reporting the larger net income for Year 2? Explain.

**42. Continuation of preceding problem introducing current cost concepts.** (Do not attempt this problem until you have worked **Problem 41**.) Assume the same data for Burton Corporation as given in the previous problem. In addition, assume the following:

| | |
|---|---|
| **Selling Price per Unit** | |
| Year 1 | $25 |
| Year 2 | 22 |
| **Average Current Replacement Cost** | |
| Year 1 | $14 |
| Year 2 | 12 |
| **Current Replacement Cost** | |
| December 31, Year 1 | $16 |
| December 31, Year 2 | 10 |

**a.** Prepare an analysis for Year 1 that identifies operating margins, realized holding gains and losses, and unrealized holding gains and losses for the FIFO, LIFO, and weighted-average cost flow assumptions.
**b.** Repeat part **a** for Year 2.
**c.** Demonstrate that over the two-year period, income plus holding gains before taxes of Burton Corporation are independent of the cost flow assumption.

**43. Effect of FIFO and LIFO on income statement and balance sheet.** Hanover Oil Products (HOP) operates a gasoline outlet. It commenced operations on January 1. It prices its gasoline at 10 percent above its average purchase price for gasoline. Purchases of gasoline during January, February, and March appear below:

| | Gallons Purchased | Unit Price | Acquisition Cost |
|---|---|---|---|
| January 1 | 4,000 | $1.40 | $ 5,600 |
| January 13 | 6,000 | 1.46 | 8,760 |
| January 28 | 5,000 | 1.50 | 7,500 |
| Total | 15,000 | | $21,860 |

| | Gallons Purchased | Unit Price | Acquisition Cost |
|---|---|---|---|
| February 5 | 7,000 | $1.53 | $10,710 |
| February 14 | 6,000 | 1.47 | 8,820 |
| February 21 | 10,000 | 1.42 | 14,200 |
| Total | 23,000 | | $33,730 |

*(continued)*

| | Gallons Purchased | Unit Price | Acquisition Cost |
|---|---|---|---|
| March 2 | 6,000 | $1.48 | $ 8,880 |
| March 15 | 5,000 | 1.54 | 7,700 |
| March 26 | 4,000 | 1.60 | 6,400 |
| Total | 15,000 | | $22,980 |

Sales for each month were as follows:
January: $20,840 (13,000 gallons)
February: $35,490 (22,000 gallons)
March: $28,648 (17,000 gallons)

**a.** Compute the cost of goods sold for January using both a FIFO and a LIFO cost flow assumption.
**b.** Repeat part **a** for February.
**c.** Repeat part **a** for March.
**d.** Why does the cost flow assumption that provides the largest cost of goods sold amount change each month?
**e.** Compute the cost of goods sold to sales percentage for each month using both a FIFO and a LIFO cost flow assumption.
**f.** Which cost flow assumption provides the most stable cost of goods sold to sales percentage over the three months? Explain why this is the case.
**g.** HOP deliberately allowed its inventory to decline to 1,000 gallons at the end of March because of the high purchase cost. Assume for this part that HOP had purchased 6,000 gallons on March 26 instead of 4,000, thereby maintaining an ending inventory equal to the beginning inventory for the month of 3,000 gallons. Compute the amount of cost of goods sold for March using both a FIFO and a LIFO cost flow assumption. Why are your answers the same as, or different from, those in part **c** above? Explain.

**44. Dealing with LIFO inventory layers.** (Adapted from a problem by S. A. Zeff.) The Back Store has been using a LIFO cost flow assumption for several decades. On December 31, Year 30, the company's inventory comprised the following layers:

| | | |
|---|---|---|
| Year 10 Layer | 60,000 units at $3.00 = | $180,000 |
| Year 22 Layer | 20,000 units at $6.00 = | 120,000 |
| Year 30 Layer | 10,000 units at $9.00 = | 90,000 |
| | 90,000 units | $390,000 |

During Year 30, the company had bought 200,000 units at $9 each. At the end of the year, the replacement cost of units was $8. During Year 31, the company bought 250,000 units at $10 per unit, which was the average replacement cost of units for the entire year. The company sold 300,000 units at $15 each during Year 31. At the end of the year, the replacement cost of units was $11.

**a.** How many units did the company sell during Year 30?
**b.** What was the operating margin on sales, that is, the replacement cost gross margin, for Year 31?
**c.** What was the gross margin conventionally reported for Year 31 assuming LIFO?
**d.** What was the economic profit, that is, realized margin plus all holding gains, for Year 31?
**e.** If the company had used FIFO for both Year 30 and Year 31, what would have been the conventionally reported gross margin for Year 31?

**45. Reconstructing underlying events from ending inventory amounts.** (Adapted from CPA examination.) Burch Corporation began a merchandising business on January 1, Year 1. It acquired merchandise costing $100,000 in Year 1, $125,000 in Year 2, and $135,000 in Year 3. Information about Burch Corporation's inventory as it would appear on the balance sheet under different inventory methods follows:

| December 31 | LIFO Cost | FIFO Cost | Lower of FIFO Cost or Market |
|---|---|---|---|
| Year 1 | $40,200 | $40,000 | $37,000 |
| Year 2 | 36,400 | 36,000 | 34,000 |
| Year 3 | 41,800 | 44,000 | 44,000 |

In answering each of the following questions, indicate how you deduced the answer. You may assume that in any one year, prices moved only up or down but not both.

**a.** Did prices go up or down in Year 1?
**b.** Did prices go up or down in Year 3?
**c.** Which inventory method would show the highest income for Year 1?
**d.** Which inventory method would show the highest income for Year 2?
**e.** Which inventory method would show the highest income for Year 3?
**f.** Which inventory method would show the lowest income for all three years considered as a single period?
**g.** For Year 3, how much higher or lower would income be on the FIFO cost basis than on the lower-of-cost-or-market basis?

**46. LIFO layers influence purchasing behavior and provide opportunity for income manipulation.** Wilson Company sells chemical compounds made from expensium. The company has used a LIFO inventory flow assumption for many years. The inventory of expensium on December 31, Year 10, comprised 4,000 pounds from Year 1 through Year 10 at prices ranging from $30 to $52 per pound. **Exhibit 7.14** shows the layers of Year 10 ending inventory.

**EXHIBIT 7.14** WILSON COMPANY
Layers of Year 10 Year-End Inventory
(Problem 46)

| Year Acquired | Purchase Price per Pound | Year 10 Year-End Inventory Pounds | Year 10 Year-End Inventory Cost |
|---|---|---|---|
| Year 1 | $30 | 2,000 | $ 60,000 |
| Year 6 | 46 | 200 | 9,200 |
| Year 7 | 48 | 400 | 19,200 |
| Year 10 | 52 | 1,400 | 72,800 |
| | | 4,000 | $161,200 |

Expensium costs $62 per pound during Year 11, but the purchasing agent expects its price to fall back to $52 per pound in Year 12. Sales for Year 11 require 7,000 pounds of expensium. Wilson Company wants to carry a stock of 4,000 pounds of inventory. The purchasing agent suggests that the firm decrease the inventory of expensium from 4,000 to 600 pounds by the end of Year 11 and replenish it to the desired level of 4,000 pounds early in Year 12.

The controller argues that such a policy would be foolish. If the firm allows inventories to decrease to 600 pounds, the cost of goods sold will be extraordinarily low (because Wilson will consume older LIFO layers) and income taxes will be extraordinarily high. The controller suggests that the firm plan Year 11 purchases to maintain an end-of-year inventory of 4,000 pounds.

Assume that sales for Year 11 do require 7,000 pounds of expensium, that the prices for Year 11 and Year 12 are as forecast, and that the income tax rate for Wilson Company is 40 percent.

**a.** Calculate the cost of goods sold and the end-of-year LIFO inventory for Year 11, assuming that the firm follows the controller's advice and that inventory at the end of Year 11 is 4,000 pounds.
**b.** Calculate the cost of goods sold and the end-of-year LIFO inventory for Year 11, assuming that the firm follows the purchasing agent's advice and that inventory at the end of Year 11 is 600 pounds.

**c.** Assume the firm follows the advice of the controller, not the purchasing agent. Calculate the tax savings for Year 11 and the extra cash costs for inventory.

**d.** What should Wilson Company do? Consider quality of earnings and ethical issues in your response.

**e.** Management of Wilson Company wants to know what discretion it has to vary income for Year 11 by planning its purchases of expensium. If the firm follows the controller's policy, aftertax income for Year 11 will be $50,000. What is the range, after taxes, of income that the firm can achieve by the purposeful management of expensium purchases?

**47. Assessing the effect of LIFO versus FIFO on the financial statements. Exhibit 7.15** summarizes data taken from the financial statements and notes for Bethlehem Steel Company for three years (in millions):

**EXHIBIT 7.15**

**BETHLEHEM STEEL COMPANY**
**Financial Data**
**(Problem 47)**

| | | For the Year Ended December 31 | | |
|---|---|---|---|---|
| | | Year 9 | Year 10 | Year 11 |
| Sales | | $5,250.9 | $4,899.2 | $4,317.9 |
| Cost of Goods Sold (LIFO) | | $4,399.1 | $4,327.2 | $4,059.7 |
| Net Income (Loss) | | $ 245.7 | $ (463.5) | $ (767.0) |
| Income Tax Rate | | 34% | 34% | 34% |
| | December 31 | | | |
| | Year 8 | Year 9 | Year 10 | Year 11 |
| Inventories at FIFO Cost | $ 899.1 | $ 972.8 | $ 967.4 | $958.3 |
| Excess of FIFO Cost over LIFO Values | (530.1) | (562.5) | (499.1) | (504.9) |
| Inventories at LIFO Cost | $ 369.0 | $ 410.3 | $ 468.3 | $453.4 |
| Current Assets (LIFO) | $1,439.8 | $1,435.2 | $1,203.2 | $957.8 |
| Current Liabilities | $ 870.1 | $ 838.0 | $ 831.4 | $931.0 |

**a.** Compute the amount of cost of goods sold for Year 9, Year 10, and Year 11, assuming that Bethlehem Steel Company had used a FIFO instead of a LIFO cost flow assumption.

**b.** Compute the cost of goods sold divided by sales percentages for Year 9, Year 10, and Year 11, using both LIFO and FIFO cost flow assumptions.

**c.** Refer to part **b**. Explain the rank-ordering of the cost of goods sold percentages for each year—that is, why does the LIFO percentage exceed the FIFO percentage or vice versa? (*Hint:* attempt to ascertain the direction of the change in inventory quantities and manufacturing costs each year.)

**d.** Compute the inventory turnover ratios (cost of goods sold divided by average inventories) for Year 9, Year 10, and Year 11, using both a LIFO and a FIFO cost flow assumption.

**e.** Which inventory turnover ratio in part **d** more accurately measures the actual inventory turnover rate? Explain your reasoning.

**f.** Compute the current ratio (current assets divided by current liabilities) at the end of each year using both LIFO and FIFO cost flow assumptions. Assume that the extra income taxes paid if the firm had used FIFO result in a reduction in cash. (*Hint:* the adjustment for income taxes must reflect the cumulative income tax effect between the date that Bethlehem Steel Company adopted LIFO and each year-end, not just the additional taxes payable each year.)

**g.** Assess the short-term liquidity risk of Bethlehem Steel Company to the extent permitted by the information presented in this problem.

**48. Interpreting inventory disclosures.** DuPont is a diversified chemicals company. Its annual report discloses inventories under LIFO of $4,409 million at the beginning of the year and

$4,107 million at the end of the year and cost of goods sold of $19,476 million for the year. If the firm had used a FIFO cost flow assumption, its beginning inventory would have been $4,853 million and its ending inventory would have been $4,407 million. Sales for the year totaled $26,996 million.

**a.** Compute the amount for cost of goods sold for the year assuming DuPont had used a FIFO instead of a LIFO cost flow assumption.
**b.** Did the quantities of items in inventory increase or decrease during the year? Explain.
**c.** Did the manufacturing cost of items in inventory increase or decrease during the year? Explain.
**d.** Compute the inventory turnover ratio for DuPont under both LIFO and FIFO for the year.
**e.** Why does the inventory turnover ratio under LIFO exceed that under FIFO?
**f.** "The choice of inventory cost flow assumption affects the inventory turnover ratio but should not normally affect the accounts payable turnover ratio." Explain.

**49. Analyzing inventory disclosures.** Eli Lilly, a pharmaceutical company, uses a LIFO cost flow assumption. Amounts taken from a recent annual report appear in **Exhibit 7.16** (in millions):

**EXHIBIT 7.16** ELI LILLY Financial Data (Problem 49)

| | | For the Year Ended December 31 | | |
|---|---|---|---|---|
| | | Year 8 | Year 9 | Year 10 |
| Sales | | $9,236.8 | $10,002.6 | $10,862.2 |
| Cost of Goods Sold | | 2,015.1 | 2,098.0 | 2,055.7 |
| | | December 31 | | |
| | Year 7 | Year 8 | Year 9 | Year 10 |
| Raw Materials Inventory | $262.0 | $325.1 | $295.1 | $230.1 |
| Work-in-Process Inventory | 459.4 | 435.8 | 372.7 | 380.6 |
| Finished Goods Inventory | 191.0 | 236.3 | 224.7 | 284.3 |
| Total Inventories at FIFO Cost | $912.4 | $997.2 | $892.5 | $895.0 |
| Adjustment to LIFO Cost | (11.7) | 2.7 | 7.1 | (11.9) |
| Total Inventories at LIFO Cost | $900.7 | $999.9 | $899.6 | $883.1 |

**a.** Compute the cost of goods sold for Year 8, Year 9, and Year 10 based on a FIFO cost flow assumption.
**b.** Did Eli Lilly appear to dip into LIFO layers in any of the three years? Explain.
**c.** Why is the adjustment from FIFO to LIFO a positive amount in some years and a negative amount in other years?
**d.** Compute the ratio of cost of goods sold to sales under both LIFO and FIFO for Year 8, Year 9, and Year 10.
**e.** Compute the inventory turnover ratio defined as cost of goods sold divided by average total inventories under both LIFO and FIFO for Year 8, Year 9, and Year 10.
**f.** Compute the costs of goods manufactured and sent to the finished goods storeroom during Year 8, Year 9, and Year 10. Use the amounts based on FIFO.
**g.** Compute an inventory turnover ratio for finished goods inventory defined as cost of goods sold divided by average finished goods inventory. Use the amounts based on FIFO.
**h.** Compute an inventory turnover ratio for work-in-process inventory defined as costs of goods manufactured (from part **f**) divided by average work-in-process inventory. Use the amounts based on FIFO.
**i.** Suggest reasons for the change in the cost of goods sold percentage computed in part **d** and the change in the inventory turnover ratio computed in part **e**.

**50. Ethical issuing in accounting for inventory.** Managers of firms experience pressure from securities markets to achieve earnings targets each year. Discuss ethical issues in each of the following independent actions that increase earnings for the year. Assume that the effect is material.

**a.** On December 10 of the current year, Firm A receives an advance of $50,000 from a hockey team for 20,000 custom-made shirts with the team's logo, which the team intends to distribute to fans entering a hockey game during the first week in January. Firm A completes the manufacturing of the shirts on December 30, intending to ship them on December 31 before its accounting period ends. Unfortunately, a snowstorm on December 31 prevented their shipment. Firm A recorded this transaction as a sale for December, and reduced its inventory accordingly. It set the items aside in its shipping room on December 31 with a clear sign to its own personnel conducting a physical inventory on that date and to its auditors who were observing the count that the items were not to be counted as inventory.

**b.** Firm B uses large warehouses to store its finished goods ready for sale. After its personnel and auditors conducted a physical inventory of goods on one side of its warehouses, Firm B transported a portion of the inventory to another part of the warehouse, removing the inventory tags that indicated that the items had already been counted in inventory, and thereby included the items a second time in inventory. In this way, the firm overstated its ending inventory for the current year, understated its cost of goods sold, and overstated its earnings. This action resulted in an overstatement of the beginning inventory for the next year. Assuming a correct count of the ending inventory for the second year, the action has the result of overstating cost of goods sold for the second year and understating earnings. Net income for the two years combined, however, is correctly stated, the net result of an overstatement in the first year offset by an equal understatement in the second year.

**c.** Firm C manufactures high quality sunglasses that carry the endorsements of several sports personalities. In an effort to achieve sales targets for the fourth quarter of the year, Firm C pressured its independent distributors to make unusually large orders of the sunglasses. Low-priced imitations of these sunglasses hit the market soon thereafter, causing the distributors to accumulate large inventories. The distributors shipped these sunglasses back to Firm C. Firm C stored the returned sunglasses in a remote warehouse out of the view of its auditors and did not record them as returned goods.

CHAPTER 8

# Long-Lived Tangible and Intangible Assets: The Source of Operating Capacity

## Learning Objectives

1. Understand the concepts distinguishing expenditures that accountants capitalize as assets, and subsequently depreciate or amortize, from expenditures that accountants expense in the period incurred.

2. Understand the concepts underlying the measurement of acquisition cost for an asset.

3. Understand depreciation and amortization as processes of cost allocation, not of valuation.

4. Develop skills to compute depreciation and amortization under various commonly used methods.

5. Develop the skills to re-compute depreciation or amortization for changes in estimates of service lives and salvage values.

6. Develop the skills to compute an impairment loss on long-lived assets.

7. Develop the skills to record the retirement of assets at various selling prices.

Consider the following recent transactions of technology-based companies, which significantly affected their reported earnings and measures of their financial position.

**Example 1** WorldCom (now MCI) provides telecommunication services using, in part, the telecommunication lines of other carriers. It pays these other carriers fixed amounts plus amounts based on usage each year for access to the telecommunication lines. During 2000 and the first quarter of 2001, WorldCom capitalized these costs as part of its fixed assets. Capitalizing these costs and subsequently depreciating them had the effect of increasing earnings during 2000 by $53.1 billion and in 2001 by $17.1 billion, relative to a policy of expensing these costs in the period incurred. WorldCom subsequently concluded that, according to generally accepted accounting principles (GAAP), it should have treated these access costs as an expense each period. The restatement of earnings resulted in a restated net loss of $48.9 billion for 2000, compared to reported earnings of $4.2 billion. For 2001, the restated net loss was $15.6 billion, compared to reported earnings of $1.5 billion.

**Example 2** Hewlett-Packard Company (HP) acquired Compaq Computer (Compaq) in 2002 for $24.2 billion. HP allocated $793 million of the purchase price to in-process technologies, technologies not yet developed to the point of demonstrating technological feasibility. Because future benefits of expenditures on in-process technologies are uncertain, GAAP requires firms to expense these amounts in the year of the acquisition. Excluding this charge, HP generated a net loss of $135 million for the year. After the charge, which provided no income tax benefit, HP reported a net loss of $928 million. Most analysts probably viewed a charge of this magnitude as nonrecurring and eliminated it in their analysis, thereby resulting in a measure of earnings with higher quality because the components are more likely to recur.

**Example 3** Nortel Networks (Nortel) made 20 corporate acquisitions costing a total of $33.5 billion between 1997 and 2000 to build its Internet network. It allocated $14.5 billion of this aggregate purchase price to identifiable assets, such as accounts receivable, inventories, plant and equipment, and identifiable liabilities, such as accounts payable and long-term debt. Nortel allocated the remaining $19 billion to goodwill, an account used to represent the portion of the purchase price that Nortel could not allocate to identifiable assets and liabilities. Items in goodwill might include the skills of software engineers, customer contacts from previously installed telecommunications systems, loyal employees, and so on. Early in 2001, Nortel announced that it took a $12.3 billion charge against earnings to write down this goodwill because the acquired firms were not then worth the amount that Nortel paid for them. Combining the charge with the loss from operations resulted in a net loss of $19.2 billion for the quarter. Thus, ongoing operations produced a net loss of $6.9 billion. Analysts likely wondered whether Nortel timed the write-off of goodwill to occur in a period of otherwise poor performance, questioning the quality of earnings not only for the quarter but also for previous and succeeding periods. Analysts suspect managers of making loss periods worse by piling on losses whenever they must report any losses at all. Such piling on increases the likelihood of reporting profits in future periods.

These examples illustrate that the accounting for long-lived assets can significantly impact reported earnings and their quality. Long-lived assets include **tangible assets**, such as land, buildings, and equipment. Long-lived assets also include **intangible assets**, such as patents, brand names, customer lists, airport landing rights, franchise rights, and similar items. Long-lived assets also include investments in securities, a topic we discuss in **Chapter 11**. Questions such as the following arise in accounting for long-lived assets:

1. Should firms treat expenditures with potential long-term benefits as assets on the balance sheet or as expenses in the income statement in the year of the expenditure?
2. For expenditures treated as assets, should firms depreciate or amortize a portion of the assets' cost as expenses each period or wait until firms dispose of the assets to write off the cost?
3. For assets that firms depreciate or amortize, over what period and using what pattern of charges should firms write off the cost of the assets?
4. How should firms treat changes in expected useful lives and salvage values that become evident partway through assets' lives?
5. How should firms treat substantial declines in the market values of long-lived assets that occur because of technological change, introduction of new products by competitors, changing consumer tastes, changing government regulations, and similar factors?
6. What impact does the disposal of long-lived assets have on earnings?

The long lives and uncertain future benefits of long-lived assets create a challenge for accountants to answer these questions in a definitive, unambiguous way. Answering these questions for intangibles requires even more effort because of their lack of physical substance. The increased importance in recent years of technology and service businesses which rely more on intangibles than manufacturing businesses underscores the need to understand the accounting for long-lived assets, particularly intangibles. Users of the financial statements need to understand both the GAAP that guides firms' responses to the questions and the flexibility that firms have in applying these GAAP in assessing the quality of earnings.

## Capitalization versus Immediate Expensing

The first question addressed is: should firms treat expenditures with potential long-term benefits as (1) expenses in the period incurred, or (2) assets on the balance sheet (that is, "capitalize" their cost), which firms will subsequently write off as expenses over the estimated service life? Refer to **Figure 8.1**. Pattern **E** represents immediate write-off as an expense. Patterns **A, S, D,** and **N** represent various patterns for writing off the capitalized acquisition cost during its service life. **Figure 8.2** depicts these patterns in terms of the amount recognized as an expense each year. We discuss these patterns more fully later in the chapter.

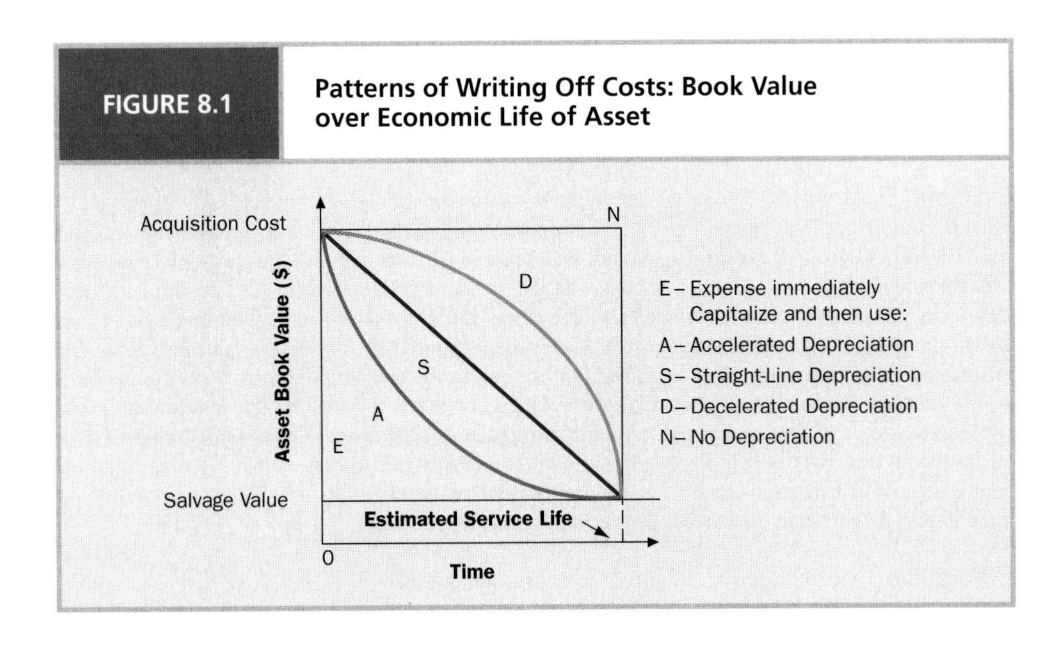


FIGURE 8.1 Patterns of Writing Off Costs: Book Value over Economic Life of Asset

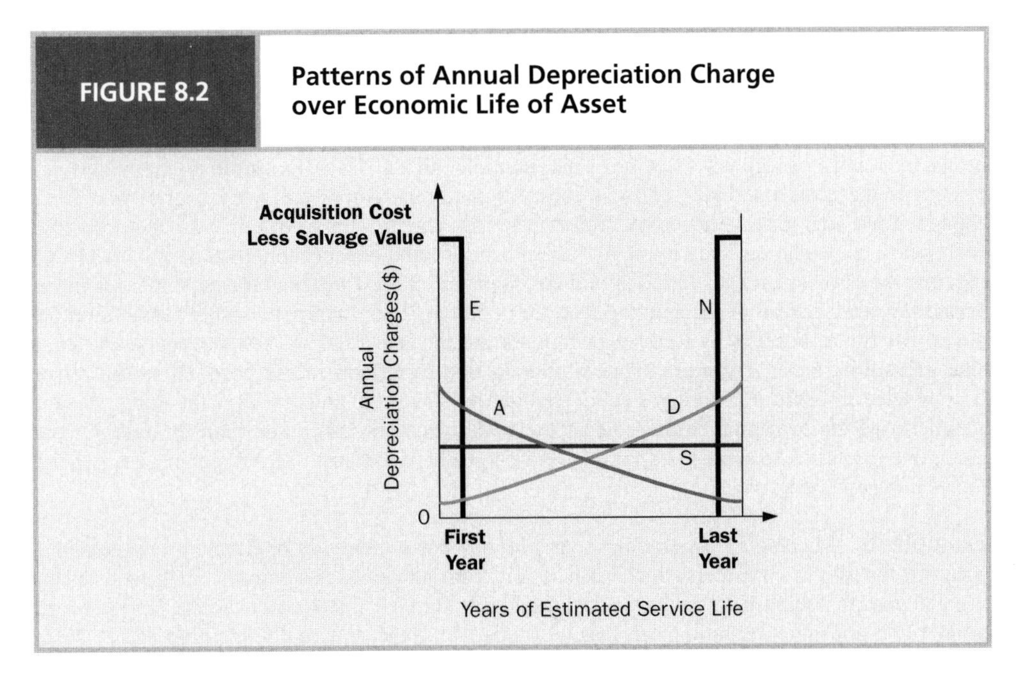


FIGURE 8.2 Patterns of Annual Depreciation Charge over Economic Life of Asset

Recall the criteria from **Chapter 2** for treating an expenditure as an asset. A firm

- has acquired rights to the future use of a resource as a result of a past transaction or exchange, and
- can measure or quantify the future benefits with a reasonable degree of precision.[1]

The exchange of cash for a resource with future service potential satisfies the first criterion. Satisfying the second criterion is more difficult because of the extended period of time that must pass before expected future benefits materialize. Also, satisfying the second criterion is even more difficult for intangible assets than for tangible assets because of the difficulty of observing the realization of benefits. The following examples illustrate the application of these two criteria for capitalization versus immediate expensing.

[1]This second criterion is not part of the FASB's formal definition of an asset (see the **Glossary**) but is a criterion for recognizing an asset in practice.

**Example 4** General Motors acquires land and a building for a factory. The land and building provide General Motors with future benefits. The external market exchange between an independent buyer and seller sets the minimum value of the expected future benefits. The land and building therefore appear as assets on General Motors' balance sheet.

**Example 5** Wal-Mart self-constructs new stores using individuals on its payroll as well as outside contractors. The new stores, once completed, provide Wal-Mart with future benefits. This example differs from **Example 4** in that Wal-Mart (1) incurs part of the cost internally, and (2) makes expenditures over time to construct the stores instead of acquiring a completed asset. Wal-Mart presumably decided to construct the stores itself instead of having an outside contractor do the work because of lower cost, or higher quality, or both. The accountant presumes that all expenditures incurred during construction are assets. If, at completion, the total cost of the stores exceeds their market value, Wal-Mart will write down the cost to the market value and recognize a loss. If the market value of the completed stores exceeds Wal-Mart's cost to construct them, Wal-Mart will not recognize a gain or construction profit. Rather, it will enjoy the future benefit of using the stores over time with depreciation charges less than they would have been had Wal-Mart paid an independent contractor to build it.

**Example 6** Merck, a pharmaceutical firm, acquires a patent on a new drug from its creator for $30 million. The patent gives Merck an exclusive legal right to manufacture and market the drug for 20 years, although the time-consuming drug approval process has reduced the economic life to a shorter time period. The external market transaction between an independent buyer and seller for the patent fixes both Merck's right to the patent and the minimum value to Merck. The patent therefore appears on Merck's balance sheet as an asset.

**Example 7** Merck spends $40 million each year on research to identify, develop, and test new drugs to combat infectious diseases. This example differs from **Example 6** in that Merck (1) incurs the costs internally, and (2) does not acquire a completed asset. Merck would not engage in research if it did not expect future benefits. The difficulty is identifying the portion of each year's expenditures that leads to future benefits and the portion that does not. GAAP requires the firm to treat all research and development (R&D) expenditures as expenses in the period incurred, instead of recognizing a portion of such expenditures as assets.[2] GAAP's rationale is that future benefits of R&D expenditures are too uncertain to justify recognizing as an asset. **Example 5** and **Example 7** are similar in that each firm incurs costs internally for an uncompleted resource. Accountants act as though they favor the physical over the intangible and are more willing to capitalize the cost of the building, a tangible asset, than the research, an intangible resource. Accountants do not have a choice here, because GAAP requires the expensing of the R&D costs.

**Example 8** Microsoft Corporation incurs various costs internally to develop new computer software for sale to customers. Such costs result from planning, designing, coding, and testing the software to establish its technological feasibility. GAAP treats costs incurred prior to the point of establishing technological feasibility like all other internally incurred R&D costs: Microsoft must expense the costs in the period incurred.[3] GAAP allows firms to capitalize as assets software costs incurred after the point of technological feasibility. Thus, a firm that takes existing software and adapts it to meet the needs of particular customers could capitalize the costs incurred between the technological feasibility stage of the software and the time it releases the adaptation to customers.

**Example 9** Merck acquires a biotechnology firm engaged in the early development of drugs using DNA technologies. Of the $25 million purchase price, it allocates $20 million to the value of potential drugs in the pipeline, called *in-process technologies.* The biotechnology firm has expensed its R&D costs each year as required by GAAP (see **Example 7**). Merck acquires the in-process technologies in an external market transaction with an independent seller, which establishes the existence and value of the technologies. The technologies, however, are not a completed resource. GAAP currently requires firms to expense the cost of in-process technologies in the year acquired, thereby not report-

[2]Financial Accounting Standards Board, *Statement of Financial Accounting Standards No. 2,* "Accounting for Research and Development Costs," 1974.

[3]Financial Accounting Standards Board, *Statement of Financial Accounting Standards No. 86,* "Accounting for the Costs of Computer Software to Be Sold, Leased or Otherwise Marketed," 1985.

ing any portion of the amount as an asset.[4] GAAP's reasoning is the same as for internally incurred R&D: future benefits are too uncertain to justify recognizing an asset. The FASB is considering requiring that firms capitalize, not expense, the cost of acquired in-process technologies, which have reached the stage of technological feasibility, similar to the treatment of software, above.

**Example 10** Refer to **Example 3**. Nortel negotiated with the selling firms to establish the purchase price of each acquired firm. Nortel used appraisals and other means to establish the market value of the identifiable assets and liabilities of each firm. GAAP requires the acquiring firm to allocate a portion of the purchase price to intangibles, even though they do not appear as assets on the books of the acquired company at the time of the acquisition. For the purchasing firm to recognize an intangible, the asset must (1) arise from contractual or other rights, or (2) be separable (that is, capable of being separated from the acquired enterprise and sold, transferred, licensed, rented, or exchanged).[5] Thus, Nortel might allocate a portion of the purchase price to contractual rights as a licensee to sell another firm's products in Japan. Nortel might also allocate a portion of the purchase price to specific proved technologies as long as the technologies meet the test of separability.

**Example 11** Refer to **Examples 3** and **10**. Nortel allocates to goodwill any excess of its purchase price for each entity over the amount assigned to identifiable net assets (= assets less liabilities). GAAP presumes that on the date of acquisition the negotiations between an independent buyer and seller establish the total value of the enterprise acquired. To the extent that Nortel can establish the existence and value of specific assets and liabilities, it allocates the purchase price to these items. Any remaining portion of the purchase price not so allocated must result from other valuable resources that Nortel cannot separately identify. Thus, goodwill appears on the balance sheet as an asset on the date of the acquisition to represent the totality of those assets it cannot identify.

**Figure 8.3** summarizes GAAP's treatment of expenditures on resources with potential long-term benefits. We can make the following generalizations:

1. Firms capitalize as assets expenditures made to acquire or self-construct tangible assets. The physical existence of tangible assets provides evidence of the existence of an asset. Unless information indicates otherwise, the presumption is that the amount the firm paid to acquire the tangible assets is the best evidence of its value on the date of acquisition.

**FIGURE 8.3** Treatment of Expenditures with Potential Long-Term Benefits

| | Nature of Resource | |
|---|---|---|
| | **Tangible** | **Intangible** |
| **Acquired Internally** | Self-constructed Buildings and Equipment (asset) | Research and Development (expense)<br>Advertising (expense)<br>Employee Training (expense)<br>Software Development Costs:<br>Pre-technological Feasibility (expense)<br>Post-technological Feasibility (asset) |
| **Acquired Externally** | Land, Buildings, and Equipment (asset) | Patent (asset)<br>Proved Technologies (asset)<br>In-Process Technologies (expense)<br>Goodwill (asset) |

[4]Financial Accounting Standards Board, *Interpretation No. 4*, "Applicability of FASB Statement No. 2 to Business Combinations Accounted for by the Purchase Method," 1975.

[5]Financial Accounting Standards Board, *Statement of Financial Accounting Standards No. 142*, "Goodwill and Other Intangible Assets," 2001.

2. Firms expense as incurred expenditures made internally to develop intangibles. The absence of an external market validation of the existence and value of a completed intangible leads GAAP to require the immediate expensing of these expenditures.
3. Firms can generally treat expenditures made to acquire completed intangibles externally as assets. A firm must expense in the period acquired the cost of in-process technologies. Although the firm acquired the in-process technologies voluntarily and in an arm's-length, external transaction, GAAP views these as unproved technologies. Goodwill is neither a contractual right nor a separable resource but appears as an asset on the date of a corporate acquisition.

The accounting for expenditures with potential long-term benefits has been, and continues to be, one of the most controversial topics faced by standard setters. Inconsistencies between the treatment of tangibles and intangibles and between costs incurred internally and externally permeate GAAP. The user of the financial statements should keep such inconsistencies in mind when comparing a manufacturing firm with significant tangible assets and a technology or service firm with significant intangibles, or when comparing a firm that develops brand names and other intangibles internally and one that acquires such intangibles by acquiring them from other firms.

## Measurement of Acquisition Cost

The acquisition cost of a long-lived asset, whether tangible or intangible, includes all costs incurred to prepare it for rendering services. The acquisition cost of a piece of equipment, for example, will be the sum of the invoice price (less any discounts), transportation costs, installation charges, and any other costs incurred before the equipment is ready for use. Also consider the following, more complex, example.

**Example 12** Refer to **Example 4**. General Motors incurs the following costs in searching for and acquiring the land and building:

1. Purchase price of land with an existing building, $1,000,000.
2. Fees paid to lawyer in handling purchase contracts, $10,000.
3. Transfer taxes paid to local real estate taxing authorities, $2,000.
4. Salaries earned by management personnel during the search for the site and the negotiation of its purchase, $8,000.
5. Operating expenditures for company automobiles used during the search, $75.
6. Depreciation charges for company automobiles used during the search, $65.
7. Fees paid to consulting engineer for a report on the structural soundness of the building, its current value, and the estimated cost of making needed repairs, $15,000.
8. Uninsured costs to repair automobiles damaged in a multi-vehicle accident during the search, $3,000.
9. Profits lost on sales the company failed to make because, during the search, management paid insufficient attention to a potential new customer, $20,000.

The first six cost items relate to the search for, and acquisition of, the land and building. General Motors should accumulate these items in a temporary Land and Building account. Some firms would treat items **5** and **6** as expenses of the period because of their immaterial amount, but strictly applying accounting theory will capitalize these costs as an asset. After completing the acquisition of the land and building, General Motors should allocate the accumulated costs of $1,020,140 (= $1,000,000 + $10,000 + $2,000 + $8,000 + $75 + $65) between these two assets. Accounting requires such division of costs, or allocation, because the firm will depreciate the building but not the land. The allocation uses the relative market values of the land and building. For example, if the engineer in item **7** estimates the current value of the existing building to be $250,000, then the firm might allocate 25 percent (= $250,000/$1,000,000) of the combined costs of $1,020,140 to the building and 75 percent to the land.

Item **7** relates to the building only, so the accountant includes the cost of the engineer's report in the cost of the building.

Accountants would not agree on the treatment of item **8**, repair costs for the accident. Some would think it is part of the cost of the site, to be allocated to the land and the building. Others would report it as an expense of the period. Probably all practicing accountants would classify it as an expense for financial reporting so that the firm could bolster its argument for calling it a current tax deduction in computing taxable income.

Item **9**, forgone profits, does not result from an expenditure. The firm did not incur this cost in an arm's-length transaction with outsiders. Historical cost accounting would not record this cost, called an *opportunity cost*, in any account.

**Example 13** Refer to **Example 6**. Merck's acquisition cost for the patent includes the $30 million paid to its creator. Merck also pays an attorney $800,000 in legal fees to evaluate its legal rights under the patent and $1,800 to register the patent. Merck's acquisition cost is therefore $30,801,800 (= $30,000,000 + $800,000 + $1,800).

**Self-Constructed Asset** When a firm, such as Wal-Mart Stores in **Example 5**, constructs its own buildings or equipment, it will record many entries in these asset accounts for the labor, material, and overhead costs incurred. GAAP requires the firm to include (capitalize) **interest paid during construction** as part of the cost of the asset it constructs.[6] Accounting's capitalizing interest results from reasoning that the firm must incur financing costs in self-constructing an asset just as it must incur labor and material costs. The accountant computes the amount of financing costs from the entity's actual borrowings and interest payments. The amount represents the interest cost that the firm incurred during periods of asset acquisition and that, in principle, the firm could have avoided if it had not acquired the assets. The inclusion of interest stops after the firm has finished constructing the asset.

If the firm borrows new funds in connection with the asset being constructed, it will use the interest rate on that borrowing. If the expenditures on construction exceed such new borrowings, then the interest rate applied to the excess is the weighted-average rate the firm pays for its other borrowings. The total amount of interest capitalized cannot exceed total interest costs for the period. The capitalization of interest in the acquisition cost of assets during construction reduces otherwise reportable interest expense and thereby increases income during periods of construction. In later periods, the asset will have higher depreciation charges, reducing income. Total expense over the life of the asset will equal the cash expenditure; capitalizing interest into the asset's cost delays expense recognition from the times of borrowing to the times of using the asset.

**Example 14** Refer to **Example 5**. Assume the following long-term debt structure for Wal-Mart Stores:

| | |
|---|---|
| Construction Loan at 15 Percent on Building under Construction | $1,000,000 |
| Other Borrowings at 12 Percent Average Rate | 3,600,000 |
| Total Long-Term Debt | $4,600,000 |

The account Building under Construction has an average balance during the year of $3,000,000. Wal-Mart Stores bases the amount of interest to be capitalized on all of the new construction-related borrowings ($1,000,000) and enough of the older borrowings ($2,000,000) to bring the total to $3,000,000. The firm computes interest capitalized as follows:

| | |
|---|---|
| $1,000,000 × 0.05 | $ 50,000 |
| $2,000,000 × 0.06 | 120,000 |
| $3,000,000 | $170,000 |

The entries to record interest and to capitalize the required amounts might be as follows:

| | | |
|---|---|---|
| Interest Expense | 266,000 | |
| Interest Payable | | 266,000 |

| Assets | = | Liabilities | + | Shareholders' Equity | (Class.) |
|---|---|---|---|---|---|
| | | +266,000 | | −266,000 | IncSt → RE |

To record all interest as expense: $266,000 [= (0.05 × $1,000,000) + (0.06 × $3,600,000) = $50,000 + $216,000].

*(continued)*

[6]Financial Accounting Standards Board, *Statement of Financial Accounting Standards No. 34*, "Capitalization of Interest Costs," 1979.

| | | |
|---|---|---|
| Building under Construction | 170,000 | |
| Interest Expense | | 170,000 |

| Assets | = | Liabilities | + | Shareholders' Equity | (Class.) |
|---|---|---|---|---|---|
| +170,000 | | | | +170,000 | IncSt → RE |

The amount capitalized reduces interest expense and increases the recorded cost of the building.

The firm might combine the preceding two entries into one as follows:

| | | |
|---|---|---|
| Interest Expense | 96,000 | |
| Building under Construction | 170,000 | |
| Interest Payable | | 266,000 |

| Assets | = | Liabilities | + | Shareholders' Equity | (Class.) |
|---|---|---|---|---|---|
| +170,000 | | +266,000 | | −96,000 | IncSt → RE |

The amount for Interest Expense is a plug.

The firm must disclose, in notes to the financial statements, both total interest for the year, $266,000, and the amount capitalized, $170,000. The income statement will report interest expense, $96,000 in this example. Interest capitalization in earlier years will cause the depreciation of the building in future years to exceed its amount without capitalization. Over the life of the asset, from construction through retirement, capitalizing interest does not change total income because larger depreciation charges later exactly offset the increased income that lower interest charges cause during the construction period. Over sufficiently long time spans, total expenses incurred for using assets must equal total cash expenditures made to acquire them.

## Problem 8.1 for Self-Study

**Calculating the acquisition cost of fixed assets.** Jensen Company purchased land with a building as the site for a new plant it planned to construct. The company received bids from several independent contractors for demolition of the old building and construction of the new one. It rejected all bids and undertook demolition and construction using company labor, facilities, and equipment.

It debited or credited amounts for all transactions relating to these properties to a single account, Construction in Process. Descriptions of various items in the Construction in Process account appear below. Jensen Company will close the Construction in Process account at the completion of construction. It will remove all amounts in it and reclassify them into the following accounts:

1. Land account.
2. Building account.
3. Revenue, gain, expense, or loss account.
4. Some balance sheet account other than Land or Building.

Reclassify the amounts of the following transactions into one or more of these accounts. If you use **4** (some other balance sheet account), indicate the nature of the account.

**a.** Cost of land, including old building.
**b.** Legal fees paid to bring about purchase of land and to transfer its title.
**c.** Invoice cost of materials and supplies used in construction.

*(continued)*

**d.** Direct labor and materials costs incurred in demolishing old building.
**e.** Direct costs of excavating raw land to prepare it for the foundation of the new building.
**f.** Discounts earned for prompt payment of item **c**.
**g.** Interest for the year on notes issued to finance construction.
**h.** Amounts equivalent to interest on Jensen Company's own funds that it used in construction but that it would have invested in marketable securities if it had used an independent contractor; it debited the amount to Construction in Process and credited Interest Revenue so that the cost of the real estate would be comparable to the cost if it had purchased the building from an independent contractor.
**i.** Depreciation during the construction period on trucks used both in construction and in regular company operations.
**j.** Proceeds of sale of materials salvaged from the old buildings; the firm had debited these to Cash and credited Construction in Process.
**k.** Cost of building permits.
**l.** Salaries of certain corporate engineering executives; these represent costs for both Salary Expense and Construction in Process, with the portion debited to Construction in Process representing an estimate of the portion of the time spent during the year on planning and construction activities for the new building.
**m.** Payments for property taxes on the plant site (its former owner owed these taxes but Jensen Company assumed responsibility to pay them).
**n.** Payments for property taxes on plant site during construction period.
**o.** Insurance premiums to cover workers engaged in demolition and construction activities; the insurance policy contains various deductible clauses, requiring the company to pay the first $5,000 of damages from any accident.
**p.** Cost of injury claims for $2,000 paid by the company because the amount was less than the deductible amount in the policy.
**q.** Costs of new machinery to be installed in the building.
**r.** Installation costs for the machinery in item **q**.
**s.** Profit on construction of the new building (computed as the difference between the lowest independent contractor's bid and the actual construction cost); the firm debited this to Construction in Process and credited Construction Revenue.

## Treatment of Acquisition Cost over Life of Asset

The second question we address in accounting for long-term assets is: should firms write off a portion of the acquisition costs of assets as expenses each period or wait until the firms dispose of the assets to write off the cost? Refer again to **Figures 8.1** and **8.2**. Writing off a portion of the acquisition cost each period corresponds to patterns **A**, **S**, and **D**, whereas writing off the acquisition at disposal corresponds to pattern **N**. The guiding principle to respond to this question is matching:

- If a firm consumes the services of long-lived assets over time in generating revenues, then the firm should write off a portion of the asset's cost each period to match against the revenues.
- If the firm does not consume the services of long-lived assets as time passes, then it should not write off a portion of its cost each period. Instead, the asset should remain on the balance sheet at acquisition cost. If the firm later sells or abandons the asset, it removes the acquisition cost from the asset account, offsetting that cost against the proceeds, if any, from selling the asset to measure the gain or loss on disposal.

Accountants refer to the periodic write-off of the acquisition cost of a tangible asset, such as buildings and equipment, as **depreciation**. Accountants use the term **amortization** to refer to the periodic write-off of intangible assets.

**Example 15** Refer to **Examples 5** and **14**. Wal-Mart Stores consumes the services of the store building over time in generating revenues. It should therefore write off, or depreciate, the acquisition cost minus estimated salvage value over the building's expected useful life. During the building's useful life, it would not write off the portion of the asset's acquisition cost equal to the estimated salvage value but instead match that cost against any proceeds from selling the building at the end of the asset's useful life.

**Example 16** Refer to **Examples 6** and **13** where Merck acquires a patent from its creator. Patents have a legal life of 20 years. However, new technologies may shorten the economic life of the patent before the 20-year legal life expires. Merck should write off, or amortize, the acquisition cost of the patent over its expected useful life, the shorter of the economic and the legal life.

**Example 17** Refer to **Examples 3** and **11** for Nortel. The portion of the purchase price of an acquired company that Nortel allocates to goodwill represents the excess cost of the acquired entity as a whole over the amounts that Nortel can attribute to identifiable assets and liabilities. As **Example 3** discusses, goodwill might include trained scientists, other knowledgeable employees, reputation for quality products, and loyal customers. GAAP views goodwill as having an *indefinite* life, which you should not confuse with an *infinite* life. Thus, GAAP does not require firms to amortize goodwill.[7] Firms must, however, test goodwill annually for a loss in value, a topic discussed later in the section on impairment losses.

## Depreciation and Amortization: Fundamental Concepts

The third question accounting addresses in dealing with long-term assets is: how should firms spread the acquisition cost of long-lived assets over their expected service lives? We address this question by discussing the fundamental concepts underlying depreciation and amortization and then describe and illustrate depreciation and amortization methods that firms use.

**Depreciation and Amortization: A Process of Cost Allocation** The acquisition cost of a long-lived asset is the cost of a series of future services. The asset is a prepayment, similar to prepaid rent or insurance, a payment in advance for services the firm will receive. As the firm uses the asset in each accounting period, it treats a portion of the cost of the asset as the cost of the service received.

Accountants cannot compute a uniquely correct amount for the periodic charge for depreciation or amortization. Long-lived assets benefit several accounting periods. Accountants and economists call such an asset a **joint cost** of several accounting periods because each of the periods of the asset's use benefits from its services. There is usually no single correct way to allocate a joint cost.

Firms select a depreciation or amortization method for a long-lived asset. That method allocates the cost minus salvage value to each period of expected useful life of the asset in a systematic, predetermined manner.

**Depreciation and Amortization: Return of Investment** A firm attempts to generate a return on investments the firm makes. Before a firm can earn a return, as measured by net income, it must generate larger revenues than all costs. One might view costs incurred as the investment required to generate revenues. Revenues cover the costs in that the net assets generated convert the costs of assets' services back into cash or other assets. Depreciation and amortization are the costs of the services of long-lived assets that revenue must cover before a firm earns a return on investment.

In historical cost accounting, the process of depreciation and amortization allows for a return of the acquisition cost of the asset, no more and no less. Because the costs of services provided by long-lived assets generally increase over time, accountants increasingly recognize that basing depreciation and amortization on acquisition costs will not, in most cases, expense amounts sufficiently large to maintain the productive capacity of the business. That is, if a firm sets aside an amount of cash each period equal to depreciation and amortization expense, it would not likely have enough cash at the end of the life of the long-lived asset to replace it. By now, you should be comfortable with the notion that recognizing depreciation expense has nothing to do with setting aside a fund of cash for acquiring replacement and new assets.

---

[7]Financial Accounting Standards Board, *Statement of Financial Accounting Standards No. 142*, "Goodwill and Other Intangible Assets," 2001.

**Depreciation and Amortization: Not a Measure of the Decline in Economic Value** Depreciation and amortization involve cost allocation, not valuation. In ordinary conversation and in economics, *depreciation* and *amortization* frequently mean a decline in **value**. Over the entire service life of a long-lived asset, the asset's value declines from acquisition until the firm retires it from service. The charge made to each accounting period does not measure the decline in economic value during the period but rather represents a process of cost allocation—a systematic process, but one in which the firm has some freedom to choose. If, in a given period, an asset increases in value, the firm still records depreciation and amortization during that period. There have been two partially offsetting processes: (1) a holding gain on the asset for the increase in market value, which historical cost-based accounting does not usually recognize, and (2) an allocation of the asset's acquisition cost to the period of benefit, which accounting does recognize.

## Depreciation and Amortization: Measurement

Measuring the amount of depreciation or amortization of long-lived assets requires firms to:

1. Measure the depreciable or amortizable basis of the asset.
2. Estimate its service life.
3. Decide the pattern of expiration of asset cost over its service life.

This section discusses each of the three items.

### Depreciable or Amortizable Basis of Long-Lived Assets: Cost Less Salvage Value

Historical cost accounting bases depreciation and amortization charges on the acquisition cost less the estimated salvage value of long-lived assets. The terms **salvage value** and **net residual value** refer to the estimated proceeds on the disposition of an asset less all removal and selling costs. At any time before a firm retires an asset, it can only estimate the salvage value. Hence, before retirement, the terms *salvage value* and *estimated salvage value* are synonyms. Because firms recover salvage value through the proceeds of sale, they need not depreciate or amortize that amount.

For buildings, common practice assumes a zero salvage value. This treatment rests on the assumption that the cost to be incurred in tearing down the buildings will approximate the sales value of the scrap materials recovered. Other tangible assets, however, may have substantial salvage value. For example, a car-rental firm will replace its automobiles at a time when other owners can use the cars for several years more. The rental firm will be able to recover a substantial part of acquisition cost from selling used cars. Intangible assets related to a contractual right, such as landing rights at an airport or franchise rights to sell a franchiser's products, generally expire at a specific time. The residual value of such intangibles is zero. Identifiable intangibles acquired in a corporate acquisition that embody the attribute of separability, such as customer lists or brand names, may have salvage values, depending on the length of time the firm expects to use the intangibles.

Salvage value can be negative. Consider, for example, the cost of dismantling a nuclear electricity-generating plant. At the time the firm acquires the plant, it estimates the fair value at that date of the dismantling cost. The fair value of dismantling costs can be either a current market price for the future cost or the present value of the expected cash outflows. A provider of nuclear plant dismantling services might be willing to assume all risks of dismantling the plant 25 or 30 years hence for an agreed-upon price at the time of acquisition of the plant, which provides evidence of a current market price for this future service. Alternatively, the owner of the power plant might estimate the expected cost of obtaining the dismantling services or engaging in the dismantling itself in 25 to 30 years and then discounting the expected cash outflows to a present value. It adds this fair value to the asset and recognizes a liability of equal amount. The firm computes depreciation based on both the cost of the plant assets and the fair value of the obligation to dismantle the nuclear power plant because the firm must recover this cost through depreciation during the asset's useful life.[8]

[8]Financial Accounting Standards Board, *Statement of Financial Accounting Standards No. 143*, "Accounting for Asset Retirement Obligations," 2001.

Estimates of salvage value are necessarily subjective. Disputes over salvage values have led in the past to many disagreements between the Internal Revenue Service and taxpayers. Partly to reduce such controversy, the Internal Revenue Code provides that firms may ignore salvage value entirely in calculating depreciation and amortization for tax reporting. When calculating depreciation and amortization in problems in this text, take the salvage value into account unless the problem gives explicit contrary instructions, such as for declining-balance depreciation methods.

## Estimating Service Life

The second factor in measuring depreciation and amortization is the expected **service life** of long-lived assets. In making the estimate, the accountant must consider both **physical and functional factors** that limit the life of the assets. Physical factors for tangible assets include such things as ordinary wear and tear from use, chemical action such as rust, and the effects of wind and rain. The most important functional factor for both tangible and intangible assets is obsolescence. Inventions, for example, may result in new processes that reduce the unit cost of production to the point where a firm finds continued operation of old equipment uneconomical, even though the equipment may be relatively unimpaired physically. Computers may work as well as ever, but firms replace them because new, smaller computers occupy less space and compute faster. Although display cases and storefronts may not have worn out, retail stores replace them to make the store look better. Technology-based intangibles can become obsolete overnight. Although the legal life of a patent on a drug is 20 years, the expected economic life of the drug may be as short as 8 or 10 years.

Despite abundant data from experience, estimating service lives for financial reporting presents the most difficult task in the depreciation and amortization calculation. Because obsolescence typically results from forces outside the firm, accountants have particular difficulty estimating its effect on the service life. This difficulty leads accountants to evaluate their estimates of service lives every few years and change the depreciation and amortization amounts going forward, a later topic in this chapter.

Estimates of service lives, like estimates of salvage value, have caused considerable controversy in the past between the Internal Revenue Service and taxpayers. The Congress created the **Modified Accelerated Cost Recovery System (MACRS)** to alleviate disagreements about service lives for depreciable, tangible assets. MACRS groups almost all depreciable assets into one of seven classes and specifies the depreciable lives and annual depreciation amounts for each class. The seven classes are as follows:

| Class | Examples |
| --- | --- |
| 3-year . . . . . . . . . . . . . . . . . . . . | Some racehorses; almost no others |
| 5-year . . . . . . . . . . . . . . . . . . . . | Cars, trucks, some manufacturing equipment, research and development property |
| 7-year . . . . . . . . . . . . . . . . . . . . | Office equipment, railroad cars, locomotives |
| 10-year . . . . . . . . . . . . . . . . . . . | Vessels, barges, and land improvements |
| 20-year . . . . . . . . . . . . . . . . . . . | Municipal sewers |
| 27.5-year . . . . . . . . . . . . . . . . . | Residential rental property |
| 39-year . . . . . . . . . . . . . . . . . . . | Nonresidential buildings acquired after 1993; otherwise 31.5 years |

The MACRS generally results in depreciating buildings, equipment, and other tangible assets faster than their economic service lives. Although most firms use MACRS for income tax reporting, they must use the estimated service life for financial reporting. The Internal Revenue Code specifies that firms must amortize intangibles acquired in a corporate asset acquisition over 15 years. Firms must amortize other intangibles acquired, such as a patent from its creator, over the expected useful life of the intangible.

## Earnings Quality Issue

Management estimates the depreciable lives and salvage values of depreciable assets. Management can shade those estimates to manage income. Management at Waste Management Inc. increased salvage value so much that over several years the company increased pretax income by about $600 million, when total pretax income for the period was on the order of $3.3 billion.

Thus, the fudged depreciation charges represented about 20 percent of pretax income for the period of the fraud. Whenever the amount of depreciation expenses will materially change income, as it does for a fixed-asset intensive business like Waste Management, the analyst will worry about earnings quality. In the case of Waste Management's fraud, the savvy analyst, at least with hindsight, can spot a large and increasing deviation between net income and cash flow from operations.

## PATTERN OF DEPRECIATION AND AMORTIZATION

After measuring the asset's acquisition cost and estimating both its salvage value and its service life, the firm has fixed the total of depreciation and amortization charges and the time span over which to charge those costs. If the firm estimates salvage value of zero, it will depreciate or amortize the entire cost. The firm must select the pattern for allocating those charges to the specific years of the service life.

Depreciation of tangible assets based on the passage of time follows one of three basic patterns. These appear as **A** (accelerated), **S** (straight-line), and **D** (decelerated) in **Figures 8.1** and **8.2**. Amortization of intangible assets follows pattern **S**, not always by GAAP rule but in usual practice. You can more easily understand the terms *accelerated* and *decelerated* if you compare in **Figure 8.2** the depreciation charges in the early years with the straight-line method. Accelerated depreciation methods write off the depreciable basis of tangible assets more quickly during the early years of the asset's life as compared with straight-line depreciation amounts. Accelerated depreciation methods recognize less depreciation during the later years relative to straight-line depreciation. Over the service life of depreciable assets, total depreciation charges using accelerated and straight-line methods are the same. Only their timing differs. GAAP does not allow decelerated depreciation for financial reporting. Managerial accounting texts discuss this method because of its useful managerial accounting applications.

The next section illustrates the calculation of depreciation using the following acceptable methods under GAAP.

1. Straight-line (time) method (pattern **S**).
2. Production or use (straight-line use) method.
3. Accelerated depreciation (pattern **A**):
   a. Declining-balance methods.
   b. Sum-of-the-years'-digits method.
   c. Modified Accelerated Cost Recovery System for income tax reporting.

When acquiring or retiring a long-lived asset during an accounting period, the firm should calculate depreciation and amortization only for that portion of the period during which it uses the asset.

**Straight-Line (Time) Method** Firms most commonly use the **straight-line (time) method** for financial reporting to depreciate tangible assets and amortize intangible assets. The straight-line method divides the acquisition cost of the assets, less an estimated salvage value, by the number of years of its expected service life, to calculate the annual depreciation or amortization.

$$\text{Annual Depreciation or Amortization} = \frac{\text{Cost Less Estimated Salvage Value}}{\text{Estimated Life in Years}}$$

For example, if a machine costs \$5,000, has an estimated salvage value of \$200, and has an expected service life of five years, the annual depreciation will be \$960 [= (\$5,000 − \$200)/5]. If a patent acquired from its creator for \$30,000 has an expected service life of five years and zero salvage value, the annual amortization will be \$6,000 (= \$30,000/5). Occasionally, instead of a positive salvage value, the cost or removal exceeds the gross proceeds on disposition. Add this excess of removal costs over the gross proceeds to the cost of the asset in making the calculation. Thus, if a firm constructs a building for \$3,500,000 and estimates a cost of \$500,000 to remove it at the end of 25 years, the annual depreciation will be \$160,000 [= (\$3,500,000 + \$500,000)/25]. By the end of the 22nd year, the accumulated depreciation will exceed the original cost of the building. Rather than show an Accumulated Depreciation account with a balance larger than the asset's cost, accounting shows the excess as a liability, such as Liability for Removal Costs.

**EXHIBIT 8.1** **200 Percent Declining-Balance Depreciation with Switch to Straight-Line as Appropriate**

**Asset Costing $5,000 with Five-Year Life and $200 Salvage Value**
**(Asset is retired at start of Year 6, when book value = $200.)**

| Start of Year | Acquisition Cost (1) | Accumulated Depreciation as of Jan. 1 (2) | Net Book Value as of Jan. 1 (3) | Double Declining-Balance Depreciation Rate (4) | Double Declining-Balance Depreciation Charge for the Year (5) | Net Book Value as of Jan. 1 Less Salvage Value of $200 (6) | Straight-Line Depreciation for Year (7) | Depreciation Charge for Year (8) |
|---|---|---|---|---|---|---|---|---|
| 1 | $5,000 | $ 0 | $5,000 | 40% | $2,000 | $4,800 | $960 | $2,000 |
| 2 | 5,000 | 2,000 | 3,000 | 40 | 1,200 | 2,800 | 700 | 1,200 |
| 3 | 5,000 | 3,200 | 1,800 | 40 | 720 | 1,600 | 533 | 720 |
| 4 | 5,000 | 3,920 | 1,080 | 40 | 432 | 880 | 440 | 440[a] |
| 5 | 5,000 | 4,360 | 640 | 40 | 256 | 440 | 440 | 440 |
| 6 | 5,000 | 4,800 | 200 | | | | | $4,800 |

Column (1), given.

Column (2) = column (2), previous period + column (8), previous period.

Column (3) = column (1) − column (2).

Column (4) = 2/life of asset in years = 2/5.

Column (5) = column (3) × column (4).

Column (6) = column (3) − salvage value, $200.

Column (7) = column (6)/remaining life in years: 5 for first row; 4 for second row; 3 for third row; 2 for fourth row; 1 for fifth row. This is the charge resulting from using straight-line for the first time in this year.

Column (8) = larger of column (5) and column (7).

[a]Firm switches to straight-line method in fourth year, when straight-line depreciation charges first exceed double declining-balance depreciation charges.

**Production or Use (Straight-Line Use) Method** Firms do not use all assets uniformly over time. Manufacturing plants often have seasonal variations in operations, so that they use certain machines 24 hours a day at one time of the year and 8 hours or less a day at another time of the year. Trucks do not receive the same amount of use in each year of their lives. The straight-line (time) method of depreciation may result in depreciation patterns unrelated to usage patterns.

When the rate of usage of a depreciable asset varies over time and the firm can estimate the total usage over its life, the firm can use a **straight-line (use) method** based on actual usage each period. For example, a firm could base depreciation of a truck for a period on the ratio of miles driven during the period to total miles the firm expects to drive over the truck's life. The depreciation cost per unit (mile) of use is as follows:

$$\text{Depreciation or Amortization Cost per Unit} = \frac{\text{Cost Less Estimated Salvage Value}}{\text{Estimated Number of Units}}$$

Assume that a truck cost $54,000, has an estimated salvage value of $4,000, and will provide 200,000 miles of use before retirement. The depreciation per mile is $0.25 [= ($54,000 − $4,000)/200,000]. If the truck operates 24,000 miles in a given year, the depreciation charge is $6,000 (= 24,000 × $0.25).

**Accelerated Depreciation** The earning power of some depreciable assets declines as they grow older. Cutting tools lose some of their precision; printing presses require more frequent shutdowns for repairs; rent receipts from an old office building fall below those from a new one. Some assets provide more and better services in the early years of their lives and require increasing amounts of maintenance as they grow older. These cases justify accelerated depreciation methods, which recognize larger depreciation charges in early years and smaller depreciation charges in later years. GAAP does not *require* firms to use an accelerated method when the firm expects asset usage to be accelerated. As a later section discusses, GAAP provides firms considerable flexibility in choosing their depreciation method.

Even when a firm computes depreciation charges for a year on an accelerated method, it allocates amounts within a year on a straight-line basis. Thus, depreciation charges for a month are 1/12 the annual amount.

**Declining-Balance Methods** Under **declining-balance methods**, the depreciation charge results from multiplying a fixed rate times the net book value of the asset (cost less accumulated depreciation without subtracting salvage value) at the start of each period. Because the net book value declines from period to period, the result is a declining periodic charge for depreciation throughout the life of the asset. Even though the method does not subtract salvage value from cost, the depreciation stops when net book value reaches salvage value. Because the declining-balance method applies a fixed rate to a declining, but always positive, book value, declining-balance methods never fully depreciate an asset's cost. A firm using a declining-balance method must therefore switch to straight-line at some point in the asset's life.

The fixed rate that firms use to apply declining-balance depreciation is usually some multiple of $1/n$, where $n$ is the expected service life of the asset. The double declining-balance method uses a multiple of 2. Thus, an asset with a five-year life will have a fixed rate of 40 percent (= 2 × 1/5). The depreciation charge is 40 percent times the book value of the asset at the start of each year. Refer to our earlier example illustrating the straight-line (time) method. The firm acquires a machine costing $5,000 with $200 salvage value and a five-year life. **Exhibit 8.1** illustrates the double declining-balance method. To depreciate the machine to its $200 salvage value at the end of five years, the firm switches to the straight-line method at the beginning of Year 4. Firms generally switch when depreciation under the straight-line method exceeds depreciation for that year under the double declining-balance method. If the firm continued to use the double declining-balance method for Year 4, depreciation would be $432 (= 0.4 × $1,080). Depreciation on the straight-line method is $440 [= ($1,080 − $200)/2].

Firms sometimes use fixed rates of 1.25 or 1.5 times $1/n$ when applying the declining-balance method.

**Sum-of-the-Years'-Digits Method** Under the **sum-of-the-years'-digits method**, the depreciation charge results from applying a fraction, which decreased from year to year, to the cost less salvage value of the asset. The fraction has a numerator and denominator as follows:

**EXHIBIT 8.2** **Sum-of-the-Years'-Digits Depreciation**

**Asset with Five-Year Life, $5,000 Cost, and $200 Estimated Salvage Value**

| Year | Acquisition Cost Less Salvage Value (1) | Remaining Life in Years (2) | Fraction =(2)/15 (3) | Depreciation Charge for the Year =(3)×(1) (4) |
|---|---|---|---|---|
| 1 ...................... | $4,800 | 5 | 5/15 | $1,600 |
| 2 ...................... | 4,800 | 4 | 4/15 | 1,280 |
| 3 ...................... | 4,800 | 3 | 3/15 | 960 |
| 4 ...................... | 4,800 | 2 | 2/15 | 640 |
| 5 ...................... | 4,800 | 1 | 1/15 | 320 |
| | | | | $4,800 |

- Numerator—the number of years of remaining life at the beginning of the year of the depreciation calculation.
- Denominator—the sum of all digits of the estimated service life.

If the service life is $n$ years, the denominator for the sum-of-the-years'-digits method is $1 + 2 + \ldots + n$. Consider the example above of a machine with a five-year life. The sum-of-the-years'-digits is 15 (= 1 + 2 + 3 + 4 + 5). **Exhibit 8.2** shows the depreciation charges under the sum-of-the-years'-digits method for the machine costing $5,000 and having a salvage value of $200.

**Modified Accelerated Cost Recovery System for Income Tax Reporting (MACRS)** Firms generally use MACRS to compute depreciation for income tax reporting. MACRS specifies the percentage of an asset's cost that firms may write off as depreciation each year for various defined classes of assets. MACRS derives these percentages using 1.5 and 2.0 declining-balance methods. MACRS assumes that firms acquire depreciable assets at the midpoint of the first year, regardless of the acquisition date. The MACRS percentages for five-year property, for example, are 0.20, 0.32, 0.192, 0.115, 0.115, and 0.058. Thus, the five years of depreciation run from the midpoint of the first year to the midpoint of the sixth year. Firms rarely judge MACRS appropriate for financial reporting.

## Problem 8.2 for Self-Study

**Calculating periodic depreciation.** Markam Corporation acquires a new machine costing $20,000 on January 2, Year 3. The firm expects

- To use the machine for five years.
- To operate it for 24,000 hours during that time.
- To recoup an estimated salvage value of $2,000 at the end of five years.

Calculate the depreciation charge for each of the five years using the following:

**a.** The straight-line (time) method.
**b.** The sum-of-the-years'-digits method.
**c.** The double declining-balance method, switching to the straight-line method in the fourth year.
**d.** The straight-line (use) method; expected operating times are 5,000 hours each year for four years and 4,000 hours in the fifth year.

## Factors in Choosing the Depreciation and Amortization Method

Depreciation and amortization affect both income reported in the financial statements and taxable income on tax returns. Firms generally use straight-line amortization for intangibles for both financial and tax reporting. GAAP allows firms more flexibility in choosing their depreciation method for financial reporting. Firms will likely choose different depreciation methods for financial and tax reporting purposes. If so, the difference between depreciation in the financial statements and on the tax return leads to an issue in accounting for income taxes, which **Chapter 10** discusses.

**Tax Reporting** In selecting depreciation methods for tax reporting, the firm should try to maximize the present value of the reductions in tax payments from claiming depreciation. When tax rates stay constant over time, earlier deductions have greater value than later ones because a dollar saved today has greater value than a dollar saved tomorrow. Congress has presented business firms with several permissible alternatives under MACRS for computing the amount of depreciation to deduct each year. A firm will usually choose the alternative that meets the general goal of allowing it to pay the least amount of tax, as late as possible, within the law. Accountants call this goal the *least and latest rule*.

**Financial Reporting** Financial reporting for long-lived assets seeks an income statement that realistically measures the expiration of the assets' benefits and provides a reasonable pattern of cost allocation. No one knows, however, just how much service potential of a long-lived asset expires in any one period. The cost of the long-lived asset jointly benefits all the periods of use, and accountants have found no single correct way to allocate such joint costs. Financial statements should report depreciation charges based on reasonable estimates of asset expirations. Most U.S. firms use the straight-line method for financial reporting. **Chapter 14** discusses more fully a firm's selection from alternative accounting methods, including the choice of depreciation methods. Analysts understand that accelerated methods provide higher-quality earnings measures than do straight-line methods, but the differences are minor. Analysts will not likely downgrade a company because it uses the straight-line depreciation method.

## Accounting for Periodic Depreciation and Amortization

In recording periodic depreciation and amortization, the accountant debits either an expense account or a product cost account. Depreciation of factory buildings and equipment used in manufacturing operations, a product cost, becomes part of the cost of work-in-process and finished goods inventories. The amortization of a patent on a semiconductor that a firm embeds in its product is likewise a product cost. The cost for amortization of a customer list becomes either amortization expense or selling expense, depending on whether the firm classifies expenses by their nature or by their function. The cost of depreciation of office equipment becomes either depreciation expense or administrative expense.

In recording amortization of intangibles, the accountant usually records the matching credit directly to the asset account, such as Patent or Customer List. In recording depreciation of tangible assets, the accountant could, in principle, likewise record the matching credit directly in the asset account, such as Buildings or Equipment. In practice, most firms credit a contra-asset account, Accumulated Depreciation. This leaves the acquisition cost of the asset undisturbed and permits the analyst to compute both the amount written off through depreciation and the undepreciated acquisition cost. When the firm credits the asset account directly, an analysis of the accounts would disclose only the second of these. A later section of this chapter shows how the analyst can use this information to make useful inferences. Whether the firm uses a contra-asset account does not affect net assets.

The entry to record periodic depreciation of office facilities, a period expense first illustrated in **Chapter 3**, is as follows:

| | | |
|---|---|---|
| Depreciation (or Administrative) Expense . . . . . . . . . . . . . . . . . . . . . . . . | 1,500 | |
| Accumulated Depreciation . . . . . . . . . . . . . . . . . . . . . . . . . . . . . | | 1,500 |

| Assets | = | Liabilities | + | Shareholders' Equity | (Class.) |
|---|---|---|---|---|---|
| −1,500 | | | | −1,500 | IncSt → RE |

The entry to record periodic depreciation of manufacturing facilities, a product cost first illustrated in **Chapter 7**, is as follows:

| | | |
|---|---|---|
| Work-in-Process Inventory . . . . . . . . . . | 1,500 | |
| Accumulated Depreciation . . . . . . . . . . | | 1,500 |

| Assets | = | Liabilities | + | Shareholders' Equity | (Class.) |
|---|---|---|---|---|---|
| +1,500 | | | | | |
| −1,500 | | | | | |

The entry to record a patent's amortization embedded in a product is as follows:

| | | |
|---|---|---|
| Work-in-Process Inventory . . . . . . . . . . | 1,500 | |
| Patent . . . . . . . . . . | | 1,500 |

| Assets | = | Liabilities | + | Shareholders' Equity | (Class.) |
|---|---|---|---|---|---|
| +1,500 | | | | | |
| −1,500 | | | | | |

The entry to record amortization of a customer list is as follows:

| | | |
|---|---|---|
| Amortization or Selling Expense . . . . . . . . . . | 1,500 | |
| Customer List . . . . . . . . . . | | 1,500 |

| Assets | = | Liabilities | + | Shareholders' Equity | (Class.) |
|---|---|---|---|---|---|
| −1,500 | | | | −1,500 | IncSt → RE |

The Work-in-Process Inventory account is an asset. Product costs, such as depreciation on manufacturing facilities, accumulate in the Work-in-Process Inventory account until the firm completes goods produced and transfers them to Finished Goods Inventory. The Accumulated Depreciation account remains open at the end of the period and appears on the balance sheet as a deduction from the asset account to which it is contra. The balance in the Accumulated Depreciation account usually represents the total charges in all accounting periods up through the balance sheet date for the depreciation on assets currently in use. The terms *book value* and *net book value* of an asset refer to the difference between the balance of an asset account and the balance of its accumulated depreciation account.

## Impact of New Information about Long-Lived Assets

This chapter has considered thus far the acquisition and depreciation or amortization of long-lived assets based on transactions and knowledge at the time of initial acquisition. New information often comes to light over the life of tangible or intangible assets that firms must incorporate into their accounting for these assets. This section discusses the following items:

1. Changes in expected service lives or salvage values.
2. Additional expenditures to maintain or improve the assets.
3. Changes in the market value of assets.

### Changes in Service Lives or Salvage Values

The original depreciation or amortization schedule for a particular asset sometimes requires changing. A firm periodically re-estimates the service life and salvage value. The accuracy of the estimates improves as retirement approaches. If changing from the old to the new estimates would have a material impact, the firm must change the depreciation or amortization schedule. The generally accepted procedure makes no adjustment for the past misestimate but spreads the remaining book value less the new estimate of salvage value over the new estimate of the remaining service life of the asset.

The rationale for adjusting current and future depreciation or amortization charges instead of correcting past charges rests on the nature and role of estimates in accounting. The accountant makes estimates for depreciable lives and salvage values, uncollectible accounts, warranty costs, and similar items based on the available information at the time of the estimate. Changes in estimates occur regularly and users should learn to expect such changes, most of which do not materially impact the financial statements. Requiring firms to restate previously issued financial statements each year for changes in estimates might confuse users and undermine the credibility of the statements. Retroactive restatement might also provide management with an opportunity to inflate earnings of earlier periods, knowing that any adjustment will not affect future earnings. **Chapter 12** discusses these issues more fully.

To understand the **treatment of changes in periodic depreciation or amortization**, assume the following facts, illustrated in **Figure 8.4**. A firm

- purchases an office machine on January 1, Year 1, for $9,200,
- estimates that it will use the machine for 15 years, and
- estimates a salvage value of $200.

**FIGURE 8.4** **Revised Depreciation Schedule**

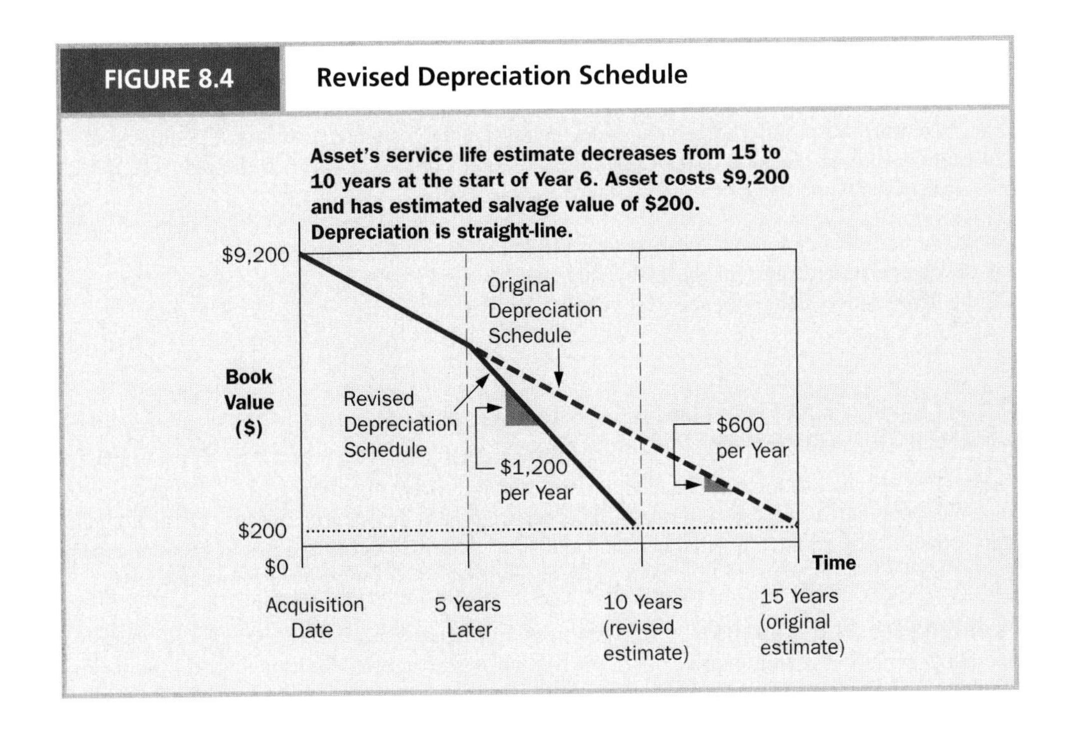


The depreciation charge recorded for each of the years from Year 1 through Year 5 under the straight-line method has been $600 [= ($9,200 − $200)/15]. On December 31, Year 6, before closing the books for the year, the firm decides that

- the machine will have a total useful life of only 10 years, and
- the salvage estimate of $200 remains reasonable.

The accepted procedure for recognizing this substantial decrease in service life revises the future depreciation so that the correct total will accumulate in the Accumulated Depreciation account at the end of the revised service life. The firm makes no adjustments of amounts previously recorded. In our example, the total amount of acquisition cost yet to be depreciated before the Year 6 change is $6,000 [= ($9,200 − $200) − (5 × $600)]. The new estimate of the remaining life is five years (the year just ended plus the next four), so the new annual depreciation charge is $1,200 (= $6,000/5). The accounting changes the amount recorded for annual depreciation in the current and future years, to $1,200 from $600. The depreciation entry on December 31, Year 6, and each year thereafter would be as follows:

| | | |
|---|---|---|
| Depreciation Expense | 1,200 | |
| Accumulated Depreciation | | 1,200 |

| Assets | = | Liabilities | + | Shareholders' Equity | (Class.) |
|---|---|---|---|---|---|
| −1,200 | | | | −1,200 | IncSt → RE |

To record depreciation for Year 6 based on revised estimates.

**Figure 8.4** illustrates the revised depreciation path.

## Problem 8.3 for Self-Study

**Adjustments for changes in estimates.** Central States Electric Company constructs a nuclear power-generating plant at a cost of $200 million. It expects the plant to last 50 years before retiring the plant from service. The company estimates that at the time it retires the plant from service, it will incur $20 million in "decommissioning" costs (costs to dismantle the plant and dispose of the radioactive materials). The firm computes and charges straight-line depreciation once per year, at year-end.

During the company's 11th year of operating the plant, Congress enacts new regulations governing nuclear waste disposal. The estimated decommissioning costs increase from $20 million to $24 million. During the 31st year of operation, the firm revises the life of the plant. It will last 60 years in total, not 50 years.

**a.** What is the depreciation charge for the first year?
**b.** What is the depreciation charge for the 11th year?
**c.** What is the depreciation charge for the 31st year?

## Additional Expenditures to Maintain or Improve Long-Lived Assets

Firms often spend cash for maintenance, repairs, and improvements during a tangible asset's life. Distinguishing between these purposes affects reported periodic income because firms immediately expense expenditures for maintenance and repairs but will debit to an asset account expenditures for improvements, and then depreciate or amortize those amounts in future periods.

**Maintenance and Repairs** A firm services and repairs an asset to maintain its expected operating condition. **Maintenance** includes routine costs such as for cleaning and adjusting. **Repairs** include the costs of restoring an asset's service potential after breakdowns or other damage. Expenditures for these items do not extend the estimated service life or otherwise increase productive capacity beyond the firm's original expectations for the asset. Because these expenditures restore future benefits to originally estimated levels, accounting treats such expenditures as expenses of the period when the firm makes the expenditure. In practice, to distinguish repairs from maintenance is difficult. Because expenditures for both become immediate expenses, accountants need not devote energy to distinguishing between them.

The repair policy a firm adopts will affect the depreciation rate for its assets. If, for example, a firm checks and services its assets often, the assets are likely to last longer than otherwise, and the depreciation rate will be lower.

**Improvements** Expenditures for **improvements**, sometimes called *betterments*, make an asset perform better than before. Such expenditures may increase the asset's life or reduce its operating cost or increase its rate of output. Because the expenditure improves the asset's service potential, accounting treats such an expenditure as an asset acquisition. The firm has acquired new future benefits. When the firm makes the expenditure, it can debit a new asset account or the existing asset account. Subsequent depreciation charges will increase.

**Example 18** Assume a firm suffers a loss to its building in a fire. The firm spends $200,000 to replace the loss. It judges that $160,000 of the expenditure replaces long-lived assets lost in the fire, and $40,000 represents improvements to the building. It could make the following single journal entry.

| | | |
|---|---|---|
| Building . . . . . . . . . . . . . . . . . . . . . . . . . . . . . . . . . . . . . . | 40,000 | |
| Loss from Fire . . . . . . . . . . . . . . . . . . . . . . . . . . . . . . . . . . . | 160,000 | |
| Cash . . . . . . . . . . . . . . . . . . . . . . . . . . . . . . . . . . . . . . . . | | 200,000 |

| Assets | = | Liabilities | + | Shareholders' Equity | (Class.) |
|---|---|---|---|---|---|
| +40,000 | | | | −160,000 | IncSt → RE |
| −200,000 | | | | | |

To record loss and subsequent expenditure.

The following two entries are equivalent and may be easier to understand:

| | | |
|---|---|---|
| Loss from Fire . . . . . . . . . . . . . . . . . . . . . . . . . . . . . . . . . . . | 160,000 | |
| Building . . . . . . . . . . . . . . . . . . . . . . . . . . . . . . . . . . . . . . | | 160,000 |
| To record loss from fire. | | |
| Building . . . . . . . . . . . . . . . . . . . . . . . . . . . . . . . . . . . . . . | 200,000 | |
| Cash . . . . . . . . . . . . . . . . . . . . . . . . . . . . . . . . . . . . . . . . | | 200,000 |

| Assets | = | Liabilities | + | Shareholders' Equity | (Class.) |
|---|---|---|---|---|---|
| −160,000 | | | | −160,000 | IncSt → RE |
| +200,000 | | | | | |
| −200,000 | | | | | |

To record expenditures on building.

**Distinguishing Maintenance and Repairs from Improvements** Accountants must distinguish maintenance and repairs from improvements because of the effect on periodic income. Some expenditures may both repair and improve. Consider expenditures to replace a roof damaged in a hurricane. The firm installs a new roof purposefully designed to be stronger than the old one so that it will support the air conditioning equipment the firm plans to install. Clearly, part of the expenditure represents repair and part, improvement.

The accountant must make judgments and allocate costs between maintenance and repairs and improvements. When in doubt, the accountant tends to call the expenditure a repair because of conservatism. The firm may also believe that expensing the item for financial reporting will support its immediate deduction for income tax reporting.

## Problem 8.4 for Self-Study

**Distinguishing repairs from improvements.** Purdy Company acquired two used trucks from Foster Company. Although the trucks were not identical, they both cost $15,000. Purdy knew when it negotiated the purchase price that the first truck required extensive engine repair, estimated to cost $4,000. The repair was made the week after acquisition and actually cost $4,200. Purdy Company thought the second truck was in normal operating condition when it negotiated the purchase price but discovered, after taking possession of the truck, that it required new bearings. The firm made this repair, costing $4,200, during the week after acquisition.

**a.** What costs should Purdy Company record in the accounts for the two trucks?
**b.** If the amounts recorded in part **a** are different, distinguish between the two repairs.

## CHANGES IN THE MARKET VALUE OF ASSETS

A firm acquires assets for their future benefits. The world changes, and expected future benefits change, sometimes increasing and sometimes decreasing in value. GAAP does not permit firms to write up the book value of their tangible and intangible long-lived assets when market values increase. This prohibition rests on conservatism. Accountants do not recognize increases in market value until the firm sells the assets, thereby realizing and validating the value increase.

Accounting treats value decreases differently. GAAP does not allow assets whose values have declined substantially to remain on the balance sheet at amortized acquisition cost. When a firm has information indicating that its assets have declined in market value or will provide a smaller future benefit than originally anticipated, GAAP may require the firm to recognize an impairment loss. GAAP distinguishes three categories of long-lived assets for purposes of measuring and recognizing impairment losses:

1. All long-lived assets except intangibles not subject to amortization. This category includes property, plant, equipment, patents, franchise rights, and similar assets. These assets provide a firm with benefits over a predictable, finite period of time.
2. Intangibles, other than goodwill, not subject to amortization. This category includes brand names and trademarks. Firms do not amortize these intangibles because they provide benefits to a firm over an *indefinite* period of time.
3. Goodwill.

**Asset Impairment for Long-Lived Assets Other Than Goodwill** GAAP requires a three-step procedure for measuring impairments for long-lived assets other than goodwill:

1. The test for an impairment loss on all long-lived assets except intangibles not subject to amortization (**category 1** above) compares the undiscounted cash flows from the assets with their book values. An asset **impairment loss** arises when the book values of the assets exceed the undiscounted cash flows. The amount of the impairment loss is the excess of the book values of the assets over their fair values (either market value or present value of expected cash flows). (If you are unfamiliar with the notion of discounting, see the **Appendix**.)
2. The test for an impairment loss on intangibles not subject to amortization, other than goodwill (**category 2** above), compares the book values of the assets to their market values. The amount of the impairment loss is the excess of the book values over the market values.
3. At the time the firm judges that an impairment loss has occurred, the firm writes down the book value of the asset to its current fair value. That fair value is (a) the market value of the asset or, (b) if the firm cannot ascertain the market value, the net present value of the expected future cash flows.[9]

In requiring that the firm use undiscounted cash flows to test for an asset impairment in **Step 1**, the FASB reasoned that a loss has not occurred if the firm can recover in future cash flows an amount at least equal to or larger than the current book value. The use of undiscounted, instead of discounted, cash flows seems theoretically unsound. The economic value of an asset may decline below its book value but the firm would recognize no impairment because the undiscounted future cash flows from the asset exceed its book value. We illustrate the application of the impairment test and the subsequent revaluations with examples.

**Basic Impairment Example** Miller Company owns an apartment building that originally cost \$20 million and by the end of the current period has accumulated depreciation of \$5 million, with net book value of \$15 (= \$20 − \$5) million. Miller Company had originally expected to collect rentals of \$1.67 million each year for 30 years before selling the building for \$8 million. Unanticipated placement of a new shopping center has caused Miller Company to reassess the future rentals. Miller Company expects the building to provide rentals for only 15 more years before Miller will sell it. Miller Company uses a discount rate of 12 percent per year in discounting expected rentals from the building.

**Example 19** Miller now expects to receive annual rentals of \$1.35 million per year for 15 years and to sell the building for \$5 million after 15 years; these payments, in total, have a present value of \$10.1 million when discounted at 12 percent per year. The building has a market value of \$10 million today.

---

[9]Financial Accounting Standards Board, *Statement of Financial Accounting Standards No. 121*, "Accounting for the Impairment of Long-Lived Assets and for Long-Lived Assets to Be Disposed Of," 1995.

GAAP deems that no impairment loss has occurred because the expected undiscounted future cash flows of \$25.25 [= (\$1.35 × 15) + \$5] million exceed the book value of \$15 million. The firm has suffered an economic loss but will not recognize any loss in its accounts.

**Example 20** Assume next that Miller expects to receive annual rentals of \$600,000 per year for 15 years and to sell the building for \$3 million after 15 years; these payments, in total, have a present value of \$4.6 million when discounted at 12 percent per year. The building has a market value of \$4 million today.

Miller has an asset impairment loss because the book value of \$15 million exceeds the expected undiscounted future cash flows of \$12.0 [= (\$0.6 × 15) + \$3] million. Miller measures the impairment loss as the excess of the book value of the building of \$15 million over its market value of \$4 million. Miller accounts for the impairment by removing the building's acquisition cost and the accumulated depreciation from the accounts and establishing a new asset cost equal to current fair value, which is market value because the firm can estimate that amount. The journal entries (with amounts in millions) would be as follows:

| | Dr. | Cr. |
|---|---|---|
| Accumulated Depreciation . . . . . . . . . . . . . . . . . . . . . . . . . . . . . . . . | 5.0 | |
| Apartment Building (New Valuation) . . . . . . . . . . . . . . . . . . . . . . . . . . | 4.0 | |
| Loss on Impairment . . . . . . . . . . . . . . . . . . . . . . . . . . . . . . . . . . . . | 11.0 | |
| Apartment Building (Acquisition Cost) . . . . . . . . . . . . . . . . . . . . | | 20.0 |

| Assets | = | Liabilities | + | Shareholders' Equity | (Class.) |
|---|---|---|---|---|---|
| +5.0 | | | | −11.0 | IncSt → RE |
| +4.0 | | | | | |
| −20.0 | | | | | |

The loss appears on the income statement. Note that the firm uses undiscounted amounts to decide there is an impairment loss but uses market values to measure the loss in this example.

**Example 21** As in **Example 20**, Miller expects to receive annual rentals of \$600,000 per year for 15 years and to sell the building for \$3 million after 15 years, which, discounted at 12 percent per year, has a present value of \$4.6 million. Because of new housing code regulations, Miller Company cannot readily find a buyer or anyone who will quote a market value. FASB *Statement of Financial Accounting Standards No. 121* says that in such a case, fair value is the present value of future cash flows, or \$4.6 million.

An asset impairment has occurred because the book value of \$15 million exceeds the expected undiscounted future cash flows of \$12.0 [= (\$0.6 × 15) + \$3] million. Miller Company measures the impairment loss as the excess of the book value of the building of \$15 million over the present value of the future cash flows of \$4.6 million. The firm accounts for the impairment loss by removing the building's acquisition cost and the accumulated depreciation from the accounts. It establishes a new asset cost equal to the current fair value, which is the present value of expected cash flows because the firm cannot estimate market value. The journal entries (with amounts in millions) could be as follows:

| | Dr. | Cr. |
|---|---|---|
| Accumulated Depreciation . . . . . . . . . . . . . . . . . . . . . . . . . . . . . . . . | 5.0 | |
| Apartment Building (New Valuation) . . . . . . . . . . . . . . . . . . . . . . . . . . | 4.6 | |
| Loss on Impairment . . . . . . . . . . . . . . . . . . . . . . . . . . . . . . . . . . . . | 10.4 | |
| Apartment Building (Acquisition Cost) . . . . . . . . . . . . . . . . . . . . | | 20.0 |

| Assets | = | Liabilities | + | Shareholders' Equity | (Class.) |
|---|---|---|---|---|---|
| +5.0 | | | | −10.4 | IncSt → RE |
| +4.6 | | | | | |
| −20.0 | | | | | |

The loss appears on the income statement.

**Asset Impairment for Goodwill** As an earlier section discussed, FASB *Statement of Financial Accounting Standards No. 142* does not allow firms to amortize goodwill in measuring net income each period. Firms must, however, test annually for impairment in the value of goodwill. If impairment occurs, firms must write down goodwill and recognize an impairment loss. Firms must first apply the procedures above for measuring and recognizing impairment losses on assets other than goodwill. Firms then follow these two steps:[10]

1. **Test for Impairment of Goodwill.** Ascertain the current fair market value of a reporting unit that includes goodwill. The reporting unit is generally a previously acquired company in which the acquiring firm allocated a portion of the purchase price to goodwill. If the current fair value of a reporting unit that includes goodwill on its balance sheet falls below the net book value of its assets (including goodwill) less liabilities, then an impairment loss on goodwill may have occurred. In usual business terminology, this loss occurs when the book value of a firm exceeds its market value. GAAP does not require that the reporting unit have shares that separately trade in the marketplace to apply the impairment test. In the absence of such shares, however, measuring the current fair value of the reporting unit presents a challenge.
2. **Measure the Amount of the Goodwill Impairment Loss.** Measuring the amount of the impairment loss of goodwill involves:
   a. allocating the current market value from **Step 1** to identifiable assets and liabilities of the reporting unit based on their current fair values (either market values or current values of expected cash flows),
   b. allocating any excess current market value to goodwill,
   c. comparing the amount allocated to goodwill in **Step 2b** with the book value of goodwill,
   d. recognizing an impairment loss on goodwill to write the book value of goodwill down to its current market value, computed in **Step 2b**.

This computation parallels the initial computation of goodwill when the reporting unit first recognized it at the time of a corporate acquisition. The accountant uses the allocations described in **Steps 2a** and **2b** solely for purposes of measuring the amount of the impairment loss to goodwill. The accountant does not record these allocated amounts in the accounts, except to the extent that they impact the amount of any impairment loss recognized before measuring an impairment loss on goodwill. The following example illustrates the application of this procedure.

**Example 22** Burns, Philp and Company Limited (B-P), a food ingredients company, spent $100 million to acquire the long-lived tangible assets and brand names of Tone's Spices. B-P intended to sell Tone's Spices in warehouse stores like Sam's Clubs and Costco warehouses. B-P allocated $15 million to long-lived tangible assets, $35 million to brand names, and the remaining $50 million to goodwill. Within days of the acquisition, market pressure from its competitor, McCormick Spices, forced B-P to offer higher-than-anticipated cash discounts and slotting fees (for the rights to put spices into the club stores and warehouses) to its large customers. B-P judged that the expected undiscounted future cash flows from the Tone's operations had declined to only $90 million and the market value of the Tone's operations had dropped to $60 million.

B-P first applies the provisions for identifying and measuring an impairment loss on the long-lived tangible assets and brand names. Assume that $25 million of the $90 million undiscounted cash flows relates to long-lived tangible assets, $50 million relates to brand names, and $15 million relates to goodwill. Assume also that the fair value of the long-lived tangible assets is $15 million and the fair value of the brand names is $25 million. No impairment loss arises on the long-lived tangible assets because the undiscounted cash flows of $25 million exceed the book value of $15 million. An impairment loss arises on the brand names because the book value of $35 million exceeds the fair value of $25 million. The firm recognizes an impairment loss on the brand names of $10 million (= $35 million − $25 million). The book value of the entity after recognizing the impairment loss is $90 million (= $15 million for long-lived tangible assets + $25 million for brand names + $50 million for goodwill).

B-P may have incurred an impairment loss on its goodwill because the market value of the entity of $60 million is less than the book value of $90 million (**Step 1** above).

B-P next allocates the $60 million current market value of Tone's to the long-lived tangible assets ($15 million—already measured), to the brand names ($25 million—already measured),

[10]Financial Accounting Standards Board, *Statement of Financial Accounting Standards No. 142*, "Goodwill and Other Intangible Assets," 2001.

and to goodwill ($20 million—plug) solely for the purpose of measuring the amount of the goodwill impairment loss. The book value of goodwill of $50 million exceeds its implied market value of $20 million by $30 million. B-P therefore recognizes an impairment loss on the goodwill. The journal entries would be as follows (with amounts in millions), with the first one made at the time of acquisition and the second on discovery of the impairment:

| | | |
|---|---|---|
| Land, Buildings, and Equipment . . . . . . . . . . | 15.0 | |
| Brand Names . . . . . . . . . . | 35.0 | |
| Goodwill on Acquisition of Tone's Spices . . . . . . . . . . | 50.0 | |
| Cash . . . . . . . . . . | | 100.0 |

| Assets | = | Liabilities | + | Shareholders' Equity | (Class.) |
|---|---|---|---|---|---|
| +15.0 | | | | | |
| +35.0 | | | | | |
| +50.0 | | | | | |
| −100.0 | | | | | |

To record the acquisition of Tone's Spices for $100 million cash.

| | | |
|---|---|---|
| Loss on Impairments . . . . . . . . . . | 40.0 | |
| Brand Names . . . . . . . . . . | | 10.0 |
| Goodwill on Acquisition of Tone's Spices . . . . . . . . . . | | 30.0 |

| Assets | = | Liabilities | + | Shareholders' Equity | (Class.) |
|---|---|---|---|---|---|
| −10.0 | | | | −40.0 | IncSt → RE |
| −30.0 | | | | | |

To record asset impairment losses.

The loss appears on the income statement. Recording this loss reduces the book value to $60 (= $15 + $25 + 20) million, its current fair value.

## Problem 8.5 for Self-Study

**Measuring impairment losses.** Real Estate Financing Corporation (REFC) acquired the assets of Key West Financing Corporation (KWFC) on June 1, Year 6, for $250 million. On the date of the acquisition, KWFC's assets consisted of loans receivable with a market value of $120 million and real estate leased to businesses and individuals with a market value of $60 million. Thus, REFC allocated $70 (= $250 − $120 − $60) million of the purchase price to goodwill. On October 15, Year 8, a hurricane hit Key West and severely damaged many homes and businesses. Information related to the assets of KWFC on October 15, Year 8, is as follows (amounts in millions):

| | Book Value | Undiscounted Cash Flows | Market Value |
|---|---|---|---|
| Loans Receivable . . . . . . . . . . | $140 | $160 | $125 |
| Real Estate . . . . . . . . . . | 80 | 65 | 50 |
| Goodwill . . . . . . . . . . | 70 | | |
| Total . . . . . . . . . . | $290 | | |

**a.** Assume that the market value of KWFC on October 15, Year 8, after the hurricane is $310 million. Compute the amount of any asset impairment losses.

**b.** Assume that the market value of KWFC on October 15, Year 8, after the hurricane is $220 million. Compute the amount of any asset impairment losses.

## Retirement of Assets

The final question asked regarding long-lived assets is: how does disposal of an asset affect earnings? Its owner can dispose of an asset through sale, or abandonment, or trade in on another asset.

**Sale of Asset** The firm records the consideration received from the sale (usually cash), eliminates any amounts, both debits and credits, in the accounts related to the asset sold, and recognizes a gain or loss for the difference. Because sales of long-lived assets are usually peripheral to a firm's principal business activities, accountants record the gain or loss net instead of gross. That is, the firm does not record the amount received as revenue and the book value of the asset sold as an expense. Rather, the firm nets the two amounts and reports only the gain or loss.

Before making the entry to record the sale, the firm records an entry to bring depreciation and amortization up to date. That is, the firm records the depreciation and amortization that has occurred between the start of the current accounting period and the date of disposition. When a firm retires an asset from service, it removes the cost of the asset and, in the case of tangible, depreciable assets, the related amount of accumulated depreciation from the books. As part of this entry, the firm records the amount received from the sale, a debit, and the amount of net book value removed from the books, a net credit (that is, a credit to the asset account and a smaller debit to the accumulated depreciation account). Typically, these two—debit for retirement proceeds and net credit to remove the asset from the accounts—differ from each other. The excess of the proceeds received on retirement over the book value is a gain (if positive) or a loss (if negative, that is, if net book value exceeds proceeds).

To understand the retirement of an asset, assume sales equipment costs $5,000, has an expected life of four years and a salvage value of $200. The firm has depreciated this asset on a straight-line basis at $1,200 [= ($5,000 − $200)/4] per year. The firm has recorded depreciation for two years and sells the equipment at midyear in the third year. The firm records the depreciation from the start of the accounting period to the date of sale: $600 [= 1/2 × ($5,000 − $200)/4].

| | Debit | Credit |
|---|---|---|
| Depreciation Expense . . . . . . . . . . . . . . . . . . . . . . . . . . . . . . . . . . | 600 | |
| Accumulated Depreciation . . . . . . . . . . . . . . . . . . . . . . . . . . . . | | 600 |

| Assets | = | Liabilities | + | Shareholders' Equity | (Class.) |
|---|---|---|---|---|---|
| −600 | | | | −600 | IncSt → RE |

To record depreciation charges up to the date of sale.

The book value of the asset is now its cost less 2 1/2 years of straight-line depreciation of $1,200 per year or $2,000 (= $5,000 − 2 1/2 × $1,200 = $5,000 − $3,000). The entry to record the retirement of the asset depends on the amount of the selling price.

1. Suppose the firm sells the equipment for cash at its book value of $2,000. The entry to record the sale would be as follows:

| | Debit | Credit |
|---|---|---|
| Cash . . . . . . . . . . . . . . . . . . . . . . . . . . . . . . . . . . . . . . . . . . . . . | 2,000 | |
| Accumulated Depreciation . . . . . . . . . . . . . . . . . . . . . . . . . . . . . . . | 3,000 | |
| Equipment . . . . . . . . . . . . . . . . . . . . . . . . . . . . . . . . . . . . . . . | | 5,000 |

| Assets | = | Liabilities | + | Shareholders' Equity | (Class.) |
|---|---|---|---|---|---|
| +2,000 | | | | | |
| +3,000 | | | | | |
| −5,000 | | | | | |

When firms sell a long-lived asset during its life at its book value or at the end of its service life for an amount equal to the expected salvage used in computing depreciation or amortization, then no gain or loss arises.

2. Suppose the firm sells the equipment for $2,300 cash, more than its book value. The entry to record the sale would be as follows:

| | | |
|---|---|---|
| Cash | 2,300 | |
| Accumulated Depreciation | 3,000 | |
| Equipment | | 5,000 |
| Gain on Retirement of Equipment | | 300 |

| Assets | = | Liabilities | + | Shareholders' Equity | (Class.) |
|---|---|---|---|---|---|
| +2,300 | | | | +300 | IncSt → RE |
| +3,000 | | | | | |
| −5,000 | | | | | |

The gain on retirement recognizes that, given the information now available, the firm charged too much depreciation. The gain appears on the income statement and, after closing entries, increases Retained Earnings. The gain restores to Retained Earnings the excess past depreciation charges.

3. Suppose the firm sells the equipment for $1,500 cash, less than its book value. The entry to record the sale would be as follows:

| | | |
|---|---|---|
| Cash | 1,500 | |
| Accumulated Depreciation | 3,000 | |
| Loss on Retirement of Equipment | 500 | |
| Equipment | | 5,000 |

| Assets | = | Liabilities | + | Shareholders' Equity | (Class.) |
|---|---|---|---|---|---|
| +1,500 | | | | −500 | IncSt → RE |
| +3,000 | | | | | |
| −5,000 | | | | | |

The loss on retirement recognizes that, given the information now available, the firm charged too little depreciation in the past. The loss appears on the income statement. The loss reduces Retained Earnings for the shortfall in past depreciation charged, in effect increasing depreciation charges all at once.

**Abandonment of Asset** Firms will sometimes abandon assets if there is no market for the asset. Examples include an automobile severely damaged in an accident or a machine requiring an overhaul that is not cost-effective. The firm eliminates the book value of the asset and recognizes a loss in an amount equal to the book value.

**Trading in an Asset** Instead of selling an asset to retire it from service, the firm may trade it in on a new asset. This is common practice for automobiles. Think of a trade-in transaction as a sale of the old asset followed by a purchase of the new asset. The accounting for trade-in transactions determines simultaneously the gain or loss on disposal of the old asset and the acquisition cost recorded for the new asset. The procedures depend on the similarity of the old and new assets, data available about the market value of the asset traded in, and the cash equivalent cost of the new asset. Intermediate accounting texts explain the various procedures. See also the **Glossary** entry for *trade-in transaction*.

## Financial Statement Presentation

This section discusses the presentation of long-lived assets in the balance sheet and income statement. **Appendix 8.1** discusses their presentation in the statement of cash flows.

## Balance Sheet

The balance sheet separates noncurrent from current assets. Tangible long-lived assets typically appear under the title Property, Plant, and Equipment, among the noncurrent assets. Firms generally disclose the assets' acquisition cost and accumulated depreciation in one of the three ways illustrated with the following data from a recent annual report of the General Electric Company (with dollar amounts in millions):

1. All information is listed in the balance sheet.

| | |
|---|---|
| Property, Plant, and Equipment: | |
| Acquisition Cost | $25,168 |
| Less Accumulated Depreciation | (11,557) |
| Property, Plant, and Equipment—Net | $13,611 |

2. Acquisition cost is omitted from the balance sheet.

| | |
|---|---|
| Property, Plant, and Equipment, less Accumulated Depreciation of $11,557 | $13,611 |

3. Acquisition cost and accumulated depreciation are omitted from the balance sheet but are detailed in the notes.

| | |
|---|---|
| Property, Plant, and Equipment—Net (Note 10) | $13,611 |

**Note 10 to the Financial Statement**
Property, plant, and equipment have an acquisition cost of $25,168 and accumulated depreciation of $11,557.

Firms typically disclose the amounts for land, buildings, and equipment separately in notes to the financial statements.

Because firms generally do not use an accumulated amortization account for intangibles, the balance sheet presents intangibles in a format similar to the last one above.

## Income Statement

Depreciation and amortization expenses appear in the income statement, sometimes disclosed separately and sometimes, particularly for manufacturing firms, as part of cost of goods sold expense. United States Steel Corporation lists depreciation expense separately from other costs of goods sold, explicitly noting that cost of goods sold excludes depreciation. General Electric includes depreciation expense in cost of goods sold and separately discloses the amount of depreciation in the notes. Gain or loss on retirement of plant assets appears on the income statement, but often the amounts are so small that some "Other" caption includes them.

# Analyzing Financial Statement Disclosures of Long-Lived Assets

## Property, Plant, and Equipment

The financial statements and notes provide useful information for analyzing changes in property, plant, and equipment. **Exhibit 8.3** presents information for Wal-Mart Stores for three recent years. This section illustrates insights provided by these disclosures.

**Fixed Asset Turnover** **Chapter 5** discussed the fixed asset turnover, a measure of the amount of sales generated from property, plant, and equipment. The computation of the fixed asset turnover for Wal-Mart Stores is as follows:

**EXHIBIT 8.3** **Selected Data for Wal-Mart Stores (all dollar amounts in millions)**

| | Year Ended January 31 | | | |
|---|---|---|---|---|
| | Year 11 | Year 12 | Year 13 | Year 14 |
| **Balance Sheet** | | | | |
| Land | $10,466 | $ 11,311 | $ 12,318 | $ 13,809 |
| Buildings and Improvements | 32,607 | 36,862 | 42,161 | 48,066 |
| Fixtures and Transportation Equipment | 13,843 | 15,224 | 16,739 | 19,130 |
| Total | $56,916 | $ 63,397 | $ 71,218 | $ 81,005 |
| Less Accumulated Depreciation | (11,499) | (12,868) | (14,726) | (17,357) |
| Total | $45,417 | $ 50,529 | $ 56,492 | $ 63,648 |

| | For the Year | | |
|---|---|---|---|
| | Year 12 | Year 13 | Year 14 |
| **Income Statement** | | | |
| Sales | $217,799 | $244,524 | $256,329 |
| Depreciation Expense | 3,290 | 3,432 | 3,852 |

| | |
|---|---|
| Year 12: $217,799/0.5($45,417 + $50,529) | 4.5 |
| Year 13: $244,524/0.5($50,529 + $56,492) | 4.6 |
| Year 14: $256,329/0.5($56,492 + $63,648) | 4.3 |

Sales increased 12.2 percent between Year 12 and Year 13, but only 4.8 percent between Year 13 and Year 14. **Chapter 5** suggested that Wal-Mart likely continued adding new stores in Year 14, but the decreased growth in sales caused its fixed asset turnover to decline.

**Proportion of Depreciable Assets Written Off** Because Wal-Mart Stores uses the straight-line depreciation method, we can compare the amount in accumulated depreciation with the acquisition cost of depreciable assets to approximate the fraction of the depreciable life of its assets Wal-Mart Stores has used.

| | |
|---|---|
| Year 11: $11,499/($32,607 + $13,843) | 24.8% |
| Year 12: $12,868/($36,862 + $15,224) | 24.7% |
| Year 13: $14,726/($42,161 + $16,739) | 25.0% |
| Year 14: $17,357/($48,066 + $19,130) | 25.8% |

One possible explanation for the increasing percentage is that Wal-Mart decreased the rate of growth in new buildings and equipment, allowing accumulated depreciation to catch up with new investments in depreciable assets. In fact, the growth in depreciable assets actually increased during the three-year period. Another possible explanation is that Wal-Mart increased the proportion of the shorter-lived fixtures and transportation relative to the longer-lived buildings. In fact, the proportion of short-lived assets decreased relative to buildings during the three-year period. One explanation consistent with the increasing percentage is that Wal-Mart had an increasing percentage of fully depreciated assets on its books. To make these calculations, you need to know both acquisition cost and accumulated depreciation.

**Average Life of Depreciable Assets** We can calculate the average total life of depreciable assets by dividing the average acquisition cost of depreciable assets by depreciation expense each year. The computations are as follows:

| | |
|---|---|
| Year 12: 0.5($32,607 + $13,843 + $36,862 + $15,224)/$3,290 .................... | 15.0 years |
| Year 13: 0.5($36,862 + $15,224 + $42,161 + $16,739)/$3,432 .................... | 16.2 years |
| Year 14: 0.5($42,161 + $16,739 + $48,066 + $19,130)/$3,852 .................... | 16.4 years |

Because depreciation expense is an amount for the year, we use the average amount of depreciable assets at acquisition cost in use during the year. We exclude land because it is not depreciated. The average life increased, reflecting a higher proportion of store buildings and warehouses with 20- to 25-year lives and a smaller proportion of fixtures and transportation equipment with 5- to 10-year lives.

**Average Age of Depreciable Assets** We can calculate the average age of depreciable assets by dividing the average amount of accumulated depreciation by depreciation expense each year. The computations are as follows:

| | |
|---|---|
| Year 12: 0.5($11,499 + $12,868)/$3,290 .................................. | 3.7 |
| Year 13: 0.5($12,868 + $14,726)/$3,432 .................................. | 4.0 |
| Year 14: 0.5($14,726 + $17,357)/$3,852 .................................. | 4.2 |

The increase in the average age reflects the increasing proportion of buildings and the decreasing proportion of fixtures and transportation equipment among depreciable assets.

## INTANGIBLES

Analysis of the disclosures regarding intangibles provides less insight for the following reasons:

1. Internally developed intangibles generally do not appear on the balance sheet.
2. Firms that grow through corporate acquisitions usually will report intangibles on the balance sheet, whereas companies that grow internally usually will not report intangibles.
3. Intangibles that do appear on the balance sheet usually appear net of amortization, so the analysts cannot perform the analyses illustrated above for property, plant, and equipment.

The absence of intangibles on the balance sheet, particularly those related to technologies and brand names, has caused many financial statement users to criticize GAAP in recent years. An investor in a firm's common stock may want to estimate the value of such intangibles when deciding to purchase or sell the firm's shares. The investor can use information in the financial statements to help in these valuations. One approach to valuation computes the present value of the cash flows derived from a brand name, technology, or other intangible.

**Exhibit 8.4** illustrates how one might value the Tootsie Roll family of brands (Tootsie Roll, Charms, and Junior Mints, among others) at more than $200 million based on the present value of the cash flows it generates. The method involves four steps, each labeled in the exhibit; all but one typically involves subjective judgment. The interaction of these judgments produces a range of final estimates, which shows the imprecision of the process. This imprecision makes U.S. accountants reluctant to audit and attest to such valuations.

1. The first step attempts to measure the cash flow that the brand, in this example Tootsie Roll, generates. The computation shows brand revenues less brand operating expenses, all of which ultimately result in cash flows. A more precise computation adjusts revenues and expenses for changes in receivables and payables, but **Exhibit 8.4** does not illustrate this sub-step. Brand revenues in excess of operating expenses for the year are $64.1 million.
2. The next step computes the cost of the physical capital—such as land, factories, inventories, and other components of working capital, assets for distributing the branded product—used to produce the cash flow. Tootsie Roll devotes about $188 million of assets to the production and distribution of Tootsie Roll products. (Computing this amount may require estimates and approximations.) **Exhibit 8.4** assumes that Tootsie Roll expects to earn 12 percent per year, before taxes, on its assets. Thus, Tootsie Roll must earn $22.6 (= 0.12 × $188) million from the Tootsie Roll family of products just to pay the cost of the physical capital devoted to the

**EXHIBIT 8.4**

**Illustration of Steps to Estimate the Value of a Brand Name Using Present Value of Discounted Cash Flows Caused by the Tootsie Roll Family of Brands (all dollar amounts in millions)**

| | | | |
|---|---|---|---|
| | Worldwide Sales | | $ 312.7 |
| | Cost of Goods Sold and Operating Expenses | | (248.6) |
| [1] | Operating Margin (20 percent) | | $ 64.1 |
| | Employed Physical Capital | $188.1 | |
| | Subtract Pretax Profit on Physical Capital Required at 12 Percent | | (22.6) |
| [2] | Profit Generated by Brand | | $ 41.5 |
| | Subtract Income Taxes at 37 Percent | | (15.3) |
| [3] | Net Brand Profits | | $ 26.2 |
| | Multiply by After-Tax Capitalization Factor | | × 8.0 |
| [4] | Estimated Brand Value | | $ 209.6 |

production of products carrying the Tootsie Roll brand. That leaves $41.5 (= $64.1 − $22.6) million of pretax profits attributable to the nonphysical assets, in this case the brands themselves. We base this calculation on the historical cost of the physical assets, but one might prefer to use some measure of their current cost.

3. **Exhibit 8.4** assumes that Tootsie Roll must pay income taxes at the rate of 37 percent of taxable income, so that after taxes, the brands generate $26.2 [= (1.00 − 0.37) × $41.5] million of after-tax cash flows for the year.
4. Now comes the toughest part of the estimation: how long does the analyst expect the cash flows from the brand to last, and how risky are they? If the analyst expected the brand-generated cash flow, $26.2 million per year, to persist for just one year, the analyst would value the brand at the amount of those cash flows—$26.2 million. If the analyst expected the brand values to last indefinitely and used an after-tax discount rate of 10 percent per year, then the analysis would result in a brand valuation of $262 million (= $26.2 million × 1/0.10). (See the discussion of the valuation of perpetuities in the **Appendix** at the back of the book, at **Examples 15** and following.) **Exhibit 8.4** shows a multiplier (like a price/earnings ratio) of 8, which suggests that the brand is worth about $209 million (= 8 × $26.2 million).[11]

## Problem 8.6 for Self-Study

**Sensitivity of brand valuation to estimates.** Refer to the computations of brand value for Tootsie Roll in **Exhibit 8.4** and observe the sensitivity of the final valuation to these estimates.

a. Recompute the brand value assuming a pretax charge for physical capital of 10 percent per year and a multiplier for brand cash flows of 10.
b. Recompute the brand value assuming a pretax charge for physical capital of 20 percent per year and a multiplier for brand cash flows of 6.

[11]Examine **Table 4** at the back of the book. Notice that a multiplier of 8 implies a discount rate of 8 percent for cash flows lasting 13 years or 10 percent for cash flows lasting 17 years or 12.5 percent for cash flows lasting forever (in perpetuity). The **Appendix** at the back of the book explains **Table 4** and how to use it. A comprehensive discussion of brand valuation from an accounting perspective appears in Patrick Barwise, Christopher Higson, Andrew Likierman, and Paul Marsh, *Accounting for Brands* (London: London Business School and The Institute of Chartered Accountants of England and Wales, 1989). See Carol J. Simon and Mary W. Sullivan, "The Measurement and Determinants of Brand Equity: A Financial Approach," *Marketing Science* 12, no. 1 (Winter 1993): 28–52, for the classic study of the economies of brand valuation. Simon and Sullivan find that of all the food-product brand names they examined, Tootsie Roll had the highest ratio of brand value to other asset value.

## An International Perspective

The accounting for long-lived assets in most developed countries closely parallels that in the United States. Firms commonly value long-lived assets at acquisition, or historical, cost. France permits periodic revaluations of buildings and equipment to current market values. Few firms perform these revaluations, however, because the French immediately tax the unrealized gains. As a result of the book-tax conformity requirement in France, few firms choose to revalue their buildings and equipment for financial reporting. Firms in Great Britain also periodically revalue their buildings and equipment to current market value. Unlike in France, in Great Britain the revaluations do not result in immediate taxation. Refer to **Exhibit 2.7**, a balance sheet for Rolls Royce. This firm revalued its tangible assets (it also purchased additional tangible assets). The offsetting credit for the revaluation is to the account Revaluation Reserve, shown among Capital and Reserves on the balance sheet.

Depreciation accounting in developed countries also closely parallels that in the United States. Firms use both straight-line and accelerated depreciation methods. In countries where financial and tax reporting closely conform (such as Japan and Germany), firms tend to use accelerated depreciation methods for both financial and tax reporting. When firms in Great Britain recognize depreciation on revalued plant assets, they charge the depreciation on the revalued amount above historical cost to the Revaluation Reserve account. Thus, neither the initial unrealized holding gain nor depreciation of this unrealized gain flows through the income statement. The income statement includes depreciation based on historical cost amounts only.

The International Accounting Standards Board (IASB) has specified accounting procedures for impairments somewhat different from the procedures used in the United States.[12] Firms compare the recoverable amount of an asset (its net selling price or present value of expected cash flows) to the book value of the asset to determine if an asset impairment has occurred. In the United States, recall, the test compares undiscounted cash flows to the book value of the asset.

The IASB has also specified accounting principles for intangibles somewhat different from those used in the United States.[13] For expenditures on development—the "D" of R&D, the stage following research where the firm has found something with commercial value—the IASB requires that the firm capitalize those costs into an asset account if the future economic benefits will flow from the expenditures and the firm can reliably measure the costs. This reporting standard precludes the recognition of intangibles, whether developed internally or acquired in external market transactions, if the intangible does not meet the definition of an asset. It specifically excludes from recognition as assets the results of expenditures on research, advertising, and employee training. *IRS No. 38* provides for the amortization of intangibles over their useful lives. It sets forth 20 years as the maximum amortization period unless firms can justify a longer life.

Both the FASB and the IASB have said they want to bring U.S. and international accounting standards closer to each other. The issues in the preceding paragraphs are likely areas for convergence. We shall not be surprised to see the FASB requiring a different treatment for development from that for research. The FASB has so much on its agenda that it is not likely to change its impairment tests in the direction of the IASB's, but we suspect that if the FASB were addressing this issue anew today, it would use the IASB's discounted tests, rather than the current undiscounted ones.

## Summary

This chapter started by setting forth a series of questions regarding the accounting for long-lived assets. We repeat those questions here and summarize GAAP's response to each.

1. Should firms treat expenditures with potential long-term benefits as assets on the balance sheet or as expenses in the income statement in the year of the expenditure? The criteria applied in practice to treat an expenditure as an asset requires that the firm:

[12]International Accounting Standards Board, *International Reporting Standards 36*, "Impairment of Assets," 1998.
[13]International Accounting Standards Board, *International Reporting Standards 38*, "Recognizing Intangibles," 1998.

- have acquired rights to the future use of a resource as a result of a past transaction or exchange, and
- be able to measure or quantify the future benefits with a reasonable degree of precision.

Firms easily apply these criteria to tangible assets, such as buildings and equipment, because of their physical attributes. Considerable controversy surrounds the recognition of many intangibles, particularly for research and development, brand names, and software development costs.

2. For expenditures treated as assets, should firms depreciate or amortize a portion of the assets as expenses each period or wait until firms dispose of the assets to write off the cost? The response to this question depends on when the firm expects to consume the service benefits of the assets. Firms depreciate or amortize the cost of long-lived assets whose service benefits diminish over time. Examples are buildings, equipment, and patents. Firms do not depreciate or amortize the cost of long-lived assets when service benefits do not diminish with time, such as for land and goodwill. Firms must test these, and all other, assets for impairment in value.
3. For assets that firms depreciate or amortize, over what period and using what pattern of charges should firms write off the cost of the assets? Firms write off the cost of assets over their expected service life. Most firms choose to write off an equal amount each year using the straight-line method. Some firms use accelerated methods for tangible assets.
4. How should firms treat new information about expected useful lives or salvage value? GAAP requires firms to depreciate or amortize the book value at the time of the change in estimate over the revised remaining useful life, thereby spreading the effect of the misestimate of the past over the current and future years.
5. How should firms treat substantial declines in the market values of long-lived assets that occur because of technological change, introduction of new products by competitors, changing consumer tastes, and similar factors? Firms must test assets for impairment in value. GAAP specifies different tests for goodwill and assets other than goodwill.
6. What impact does the disposal of long-lived assets have on earnings? Firms record any proceeds of disposal, eliminate the book value of the asset, and record a gain or loss for the difference (except for some trade-ins).

GAAP's responses to each of these questions provide guidance to firms in accounting for their long-lived assets. Firms have considerable latitude within GAAP, however, to apply the general principles. Firms must make estimates of service lives and salvage values for depreciation and amortization, choose their depreciation method, make estimates of market values in allocating the aggregate purchase price in corporate acquisitions and measuring asset impairments, and time asset sales for gain or loss. The opportunities open to firms to manage earnings with respect to long-lived assets cause analysts to consider the impact on the quality of earnings.

## Appendix 8.1: Effects on the Statement of Cash Flows of Transactions Involving Long-Lived Tangible and Intangible Assets

The statement of cash flows reports as investing activities both the cash used for acquisition of tangible or intangible assets and the cash provided by their retirement. It usually shows adjustments for depreciation, amortization, asset impairments, and gains and losses on disposal in deriving cash from operations when the firm uses the indirect method.

### Retirement Entries in the Statement of Cash Flows

When the firm sells a tangible or intangible asset for cash, it reports that cash in the Investing section as a nonoperating source, Proceeds from Disposition of Noncurrent Assets. Usually, the firm sells a tangible or intangible asset for an amount different from its book value at the time of sale. Thus, the retirement generates a loss (proceeds less than book value) or a gain (proceeds greater than book value).

### Loss on Retirement

The Investing section of the statement shows all the proceeds from selling the tangible or intangible asset. None of that cash results from operations. When firms report cash flow from operations using the direct method, they need not make adjustments for the loss on retirement. In the indirect format for operating cash flow, the statement of cash flows begins with net income, which includes the loss on sale. Because the sale is a (dis)investing—not an operating—activity, the presentation must add back the amount of the loss in deriving operating cash flows. The addback presentation resembles the addback for depreciation expense.

### Gain on Retirement

The gain on retirement increases net income, but the cash from retirement produces (dis)investing, not operating, cash flow. When the statement of cash flows presents cash from operations using the indirect format, it must subtract the amount of the gain from net income so that operating cash flow will exclude the cash from retirement, properly classified among investing activities.

## Depreciation and Amortization Charges

Whether depreciation and amortization charges appear in the statement of cash flows depends on whether the firm uses the direct method or the indirect method for reporting cash generated by operations. A firm using the direct method reports only receipts minus expenditures for operating activities. A firm using the indirect method reports net income—revenues minus expenses—along with various addbacks and subtractions to convert the amount into receipts minus expenditures. Because depreciation and amortization expenses reduce income but use no cash currently, an addback for depreciation and amortization expenses appears in the derivation of cash from operations. Most firms use the indirect method and show such an addback.

**Depreciation Charges for a Manufacturing Firm** **Chapter 7** introduced the notion, repeated in this chapter, that manufacturing firms typically do not debit all depreciation charges to depreciation expense. Rather, such firms debit depreciation on manufacturing facilities to the Work-in-Process Inventory account. The firm transfers the depreciation costs from Work-in-Process to Finished Goods Inventory at the same time that it transfers the goods from the production line to the finished goods warehouse. Then, the depreciation cost remains in Finished Goods Inventory until the firm sells the goods embodying the depreciation. Only then do the depreciation costs become expenses—part of Cost of Goods Sold. Conventionally, however, a manufacturing firm adds back all depreciation charges for a given period in its derivation of cash from operations for the same period in which the firm incurred those charges.

- If the firm has sold the goods embodying the depreciation, the addback cancels the depreciation included in Cost of Goods Sold.
- If the firm has not yet sold the inventory embodying the depreciation, then the current asset inventories will increase by the amount of any depreciation embodied in the Work-in-Process accounts. Recall that in the indirect method, the firm subtracts the increase in inventory accounts in deriving cash flows from operations. (Refer to **Exhibit 4.16**. Note that the addback on line 4 and the subtraction on line 5 increase by the same amount.)

Thus, whether the firm sells or keeps in inventory the manufactured goods embodying the depreciation charges, it will add back all depreciation charges, manufacturing and other, to net income in deriving operating cash flows. If the firm has not sold the goods, the subtraction for the increase in inventory will cancel just the right part—the depreciation charges in ending inventory—of the addback.

## Asset Impairment Losses

The recognition of an impairment loss involves a debit for the impairment loss and a credit to the asset account. The impairment loss reduces net income but does not use cash. Thus, like depreciation and amortization charges, the accountant adds back the impairment loss to net income in deriving cash flow from operations using the direct method.

## Noncash Acquisitions of Plant and Intangible Assets

Firms often borrow funds from a lender to acquire plant and intangible assets from a third party. A firm, for example, might borrow $5 million from an insurance company and use the cash to acquire a building from a developer. The loan appears on the statement of cash flows as a financing activity, and the acquisition of the building appears as an investing activity.

Firms sometimes give consideration other than cash to the seller (the developer in the above example) in acquiring plant and intangible assets. The consideration might be a mortgage note or the firm's shares. Because these investing and financing transactions involve a bartering that does not use cash, they do not appear on the statement of cash flows. The firm must report such transactions in a supplementary schedule or note.

## Solutions to Self-Study Problems

### SUGGESTED SOLUTION TO PROBLEM 8.1 FOR SELF-STUDY

(Jensen Company; calculating the acquisition cost of fixed assets.)

**(1)** a, b, d, j, l, m, o, p.
**(2)** c, e, f, g, i, k, l, n, o, p.
**(3)** h, i, p, s.
**(4)** i, q, r.

### COMMENTS AND EXPLANATIONS

**d.** Removing the old building makes the land ready to accept the new one. These costs apply to the land, not to the new building. For tax purposes, the firm might prefer to allocate this cost to the building because it can depreciate the building, but not the land.

**f.** The reduction in cost of materials and supplies will reduce the cost of the building. The actual accounting entries depend on the method used to record the potential discount. This book does not discuss these issues.

**h.** Although one capitalizes explicit interest, one may not capitalize opportunity-cost interest or interest imputed on one's own funds used. The adjusting entry credits Construction in Process and debits Interest Revenue. The debit reduces income, removing the revenue that had been recognized by the company.

**i.** Computation of the amounts to be allocated requires an estimate. Once the firm estimates amounts, it debits them to Building or to Depreciation Expense and Work-in-Process Inventory, as appropriate, for the regular company operations.

**j.** Credit to Land account, reducing its cost.

**l.** Allocate to Land and Building, based on an estimate of how time was spent. Given the description, most of these costs are probably for the building.

**m.** Include as part of the cost of the land.

**n.** Capitalize as part of the Building account for the same reasons that a firm capitalizes interest during construction. Some accountants would treat this item as an expense. In any case, the firm can treat this item as an expense for tax reporting.

**o.** Allocate the costs for insuring workers to the same accounts as the wages for those workers.

**p.** Most accountants would treat this as an expense or a loss for the period. Others would capitalize it as part of the cost of the building for the same reason that they capitalized the explicit insurance cost. If, however, the company was irrational in acquiring insurance policies with deductible clauses, this item would be an expense or loss. Accounting usually assumes that most managements make rational decisions most of the time. In any case, one can treat this item as an expense or a loss for tax reporting.

**q.** Debit to Machinery and Equipment account, an asset account separate from Building.

**r.** Treat the same as the preceding item; installation costs are part of the cost of the asset; see **Chapter 2**.

**s.** Recognizing revenue is incorrect. Credit the Construction in Process account and debit the Construction Revenue account.

### SUGGESTED SOLUTION TO PROBLEM 8.2 FOR SELF-STUDY

(Markam Corporation; calculating periodic depreciation.)

**a.** Straight-Line Method:

Years 3–7: ($20,000 − $2,000)/5 = $3,600 each year
Total: $3,600 × 5 = $18,000

**b.** Sum-of-the-Years'-Digits Method:

| | |
|---|---|
| Year 3: 5/15 × ($20,000 − $2,000) = | $ 6,000 |
| Year 4: 4/15 × ($20,000 − $2,000) = | 4,800 |
| Year 5: 3/15 × ($20,000 − $2,000) = | 3,600 |
| Year 6: 2/15 × ($20,000 − $2,000) = | 2,400 |
| Year 7: 1/15 × ($20,000 − $2,000) = | 1,200 |
| Total ..................... | $18,000 |

**c.** Double Declining-Balance Method:

| | |
|---|---|
| Year 3: 2/5 × $20,000 = | $ 8,000 |
| Year 4: 2/5 × ($20,000 − $8,000) = | 4,800 |
| Year 5: 2/5 × ($20,000 − $8,000 − $4,800) = | 2,880 |
| Year 6: 1/2 × ($4,320[a] − $2,000) = | 1,160 |
| Year 7: 1/2 × ($4,320 − $2,000) = | 1,160 |
| Total .......................... | $18,000 |

[a]$20,000 − $8,000 − $4,800 − $2,880 = $4,320

**d.** Units-of-Production Method:

Years 3–6: 5,000 × $0.75[a] = $3,750 per year
Year 7: 4,000 × $0.75[a] = $3,000
Total [($3,750 × 4) + $3,000] = $18,000

[a]($20,000 − $2,000)/24,000 = $0.75 per hour

## SUGGESTED SOLUTION TO PROBLEM 8.3 FOR SELF-STUDY

(Central States Electric Company; adjustments for changes in estimates.)
(All dollar amounts in millions.)

**a.** $4.4 per year = ($200 + $20)/50 years

**b.** $4.5 per year = [$200 + $20 + $4 − ($4.4 per year × 10 years)]/40 years remaining life
= ($224 − $44)/40
= $180/40

**c.** $3.0 per year = [$180 − ($4.5 × 20 years)]/30 years remaining life
= ($180 − $90)/30
= $90/30

## SUGGESTED SOLUTION TO PROBLEM 8.4 FOR SELF-STUDY

(Purdy Company; distinguishing repairs from improvements.)

**a.** Record the first truck at $19,200. Record the second truck at $15,000; debit $4,200 to expense or loss.

**b.** When Purdy Company acquired the first truck, it knew it would have to make the "repair," which is an improvement. The purchase price was lower because of the known cost to be incurred. At the time of acquisition, the firm anticipated the cost as required to produce the expected service potential of the asset. The fact that the cost was $4,200, rather than "about $4,000," does not seem to violate Purdy Company's expectations at the time it acquired the truck. If the repair had cost significantly more than $4,000—say, $7,000—then the excess could be a loss or an expense.

Purdy Company believed that the second truck was operable when it agreed on the purchase price. Purdy Company incurred the cost of the repair to achieve the level of service potential it thought it had acquired. There are no more future benefits after the repair than it had anticipated at the time of acquisition. Therefore, the $4,200 is an expense or a loss.

### SUGGESTED SOLUTION TO PROBLEM 8.5 FOR SELF-STUDY

(Real Estate Financing Corporation; measuring impairment losses.)

**a.** The undiscounted cash flows related to the loans receivable of $160 million exceed their book value of $140 million, so no impairment loss arises for the receivables. For real estate, the book value of $80 million exceeds their undiscounted cash flows of $65 million, so Real Estate Financing Corporation recognizes an impairment loss of $30 (= $50 − $80) million. The book value of the firm after recognizing the impairment loss is $260 million (= $140 for loans receivable + $50 million for real estate + $70 million for goodwill). The market value of the firm of $310 million exceeds the book value of $260 million, so no impairment loss on goodwill arises.

**b.** The answers for the loans receivable and the real estate in part **a** apply here as well. In this case, however, the book value of the firm of $260 million exceeds the market value of $220 million, so an impairment loss of goodwill arises. To measure the impairment loss, the accountant attributes $125 million of the market value of $220 million to loans receivable, $50 million to real estate, and $45 million to goodwill. Comparing the market value of goodwill of $45 million to the book value of goodwill of $70 million yields an impairment loss of $25 million.

### SUGGESTED SOLUTION TO PROBLEM 8.6 FOR SELF-STUDY

(Tootsie Roll; sensitivity of brand valuations to estimates)

TOOTSIE ROLL
Illustration of Steps to Estimate the Value of a Brand Name Using Present Value of Discounted Cash Flows Caused by the Tootsie Roll Family of Brands with Variations in Assumptions
(all dollar amounts in millions)

| | | Part a | Part b |
|---|---|---|---|
| Worldwide Sales | | $ 312.7 | $ 312.7 |
| Operating Expenses | | (248.6) | (248.6) |
| [1] Operating Margin (20 percent) | | $ 64.1 | $ 64.1 |
| Employed Physical Capital | $188.1 | | |
| Subtract Pretax Profit on Physical Capital Required at 10 (part **a**) or 20 (part **b**) Percent | | (18.8) | (37.6) |
| [2] Profit Generated by Brand | | $ 45.3 | $ 26.5 |
| Subtract Income Taxes at 37 Percent | | (16.8) | (9.8) |
| [3] Net Brand Profits | | $ 28.5 | $ 16.7 |
| Multiply by After-tax Capitalization Factor | | 10.0 | 6.0 |
| [4] Estimated Brand Value | | $ 285.0 | $ 100.2 |

## Key Terms and Concepts

Tangible assets
Intangible assets
Interest paid during construction
Depreciation
Amortization
Joint cost
Value
Salvage value
Net residual value
Service life
Physical and functional factors
Modified Accelerated Cost Recovery System (MACRS)
Straight-line (time and use) methods
Declining-balance methods
Sum-of-the-years'-digits method
Treatment of changes in periodic depreciation or amortization (estimates of useful lives and residual values of long-lived assets)
Maintenance
Repairs
Improvements
Impairment loss

## Questions, Short Exercises, Exercises, Problems, and Cases

### QUESTIONS

For additional student resources, content, and interactive quizzes for this chapter visit the FACMU website: **www.thomsonedu.com/accounting/stickney**

1. Review the meaning of the terms and concepts listed above in **Key Terms and Concepts**.
2. A particular firm's property department maintains buildings and equipment used in manufacturing, selling, and administrative activities. The salaries of personnel in the property department during the current period may appear as an expense during the current period, as an expense next period, or as an expense during several future periods. Explain.
3. A manufacturing firm receives a bid from a local construction company to build a warehouse for $300,000. The firm decides to build the warehouse itself and does so at a cost of $250,000. Why is it that GAAP does not permit the firm to record the warehouse at $300,000 and recognize revenue of $50,000? How will net income differ if the firm follows the generally accepted procedure versus recording the warehouse at $300,000?
4. **a.** What is the effect of capitalizing interest on reported net income summed over all the periods of the life of a given self-constructed asset, from building through use until eventual retirement? Contrast with a policy of expensing interest as incurred.
   **b.** Consider a company engaging in increasing dollar amounts of self-construction activity each period during periods when interest rates do not decline. What is the effect on reported income each year of capitalizing interest in contrast to expensing interest as incurred?
5. **a.** "Accounting for depreciating assets would be greatly simplified if accounting periods were only long enough or assets' lives short enough." What is the point of the quotation?
   **b.** "The major purpose of depreciation accounting is to provide funds for the replacement of assets as they wear out." Do you agree? Explain.
6. "Showing both acquisition cost and accumulated depreciation amounts separately provides a rough indication of the relative age of the firm's long-lived assets."
   **a.** Assume that the Dickens Company acquired an asset several years ago with a depreciable cost of $100,000 and no salvage value. Accumulated depreciation as of December 31, recorded on a straight-line basis, is $60,000. The depreciation charge for the year is $10,000. What is the asset's total depreciable life? How old is the asset?
   **b.** Assume straight-line depreciation. Devise a formula that, given the depreciation charge for the year and the asset's accumulated depreciation, you can use to estimate the age of the asset.
7. Some critics of the required accounting for changes in estimates of service lives of depreciable assets characterize it as the "always wrong" method. Why do you think these critics use this characterization?
8. "Applying the accounting for repairs versus betterments often comes down to a matter of materiality." Explain.
9. A firm expects to use a delivery truck for five years. At the end of three years, the transmission wears out and requires replacement at a cost of $4,000. The firm argues that it should capitalize this expenditure because without it the useful life is zero and with it the useful life will be another three years. Comment on the firm's reasoning relative to GAAP.
10. A firm sold for $10,000 a machine that originally cost $30,000 and had accumulated depreciation of $24,000 (book value = $6,000). Why does accounting include a gain on the sale of the machine of $4,000 in the income statement instead of showing sales revenue of $10,000 and cost of machine sold of $6,000?
11. Relate the concept of return of capital to the criterion under generally accepted accounting principles for deciding whether an impairment loss on assets other than intangibles not requiring amortization and goodwill has occurred.
12. Suggest reasons why generally accepted accounting procedures permit firms engaged in mineral exploration, such as for oil or gold, to capitalize and subsequently amortize exploration costs but require firms engaged in research and development, such as pharmaceutical companies, to expense the costs in the year incurred.
13. A firm that incurs research and development costs to develop a patented product must expense these costs as incurred. A firm that purchases that same patent from its creator, however, capitalizes the expenditure as an asset and amortizes it. What is the rationale for this different treatment of the patent?

## SHORT EXERCISES

14. **Calculating acquisition costs of long-lived assets.** Outback Steakhouse opened a new restaurant on the site of an existing building. It paid the owner $260,000 for the land and building, of which it attributes $52,000 to the land and $208,000 to the building. Outback incurred legal costs of $12,600 to conduct a title search and prepare the necessary legal documents for the purchase. It then paid $35,900 to renovate the building to make it suitable for Outback's use. Property and liability insurance on the land and building for the first year was $12,000, of which $4,000 applied to the period during renovation and $8,000 applied to the period after opening. Property taxes on the land and building for the first year totaled $15,000, of which $5,000 applied to the period during renovation and $10,000 applied to the period after opening. Calculate the amounts that Outback Steakhouse should include in the Land account and in the Building account.
15. **Calculating interest capitalized during construction.** Target Stores constructed new stores during Year 14. The average balance in the Construction-in-Process account excluding Year 14's capitalized interest costs was $3,400,000 during the year. Target Stores engaged in borrowing directly related to these stores in the amount of $2,000,000, which carries an interest rate of 6 percent. Target Stores has other borrowing outstanding totaling $8,000,000 at an average interest rate of 7 percent. Compute the amount of interest capitalized in the Construction-in-Process account during Year 14.
16. **Computing depreciation expense.** Office Depot acquired a forklift truck on January 1 of the current year for $40,000. The vehicle has an estimated useful life of six years and $4,000 estimated salvage value. Compute the amount of depreciation expense for the current year under the straight-line, double declining-balance, and sum-of-the-years'-digits depreciation methods. Round calculations to the nearest dollar.
17. **Change in depreciable life and salvage value.** Thomson Financial acquired a computer on January 1, Year 12, for $10,000,000. The computer had an estimated useful life of six years and $1,000,000 estimated salvage value. The firm uses the straight-line depreciation method. On January 1, Year 14, Thomson Financial discovers that new technologies make it likely that the computer will last only four years in total and that the estimated salvage value will be only $600,000. Compute the amount of depreciation expense for Year 14 for this change in depreciable life and salvage value. Assume that the change does not represent an impairment loss.
18. **Distinguishing repairs versus betterments.** Disney World experienced damage from a tornado at Space Mountain, one of its most popular attractions. It paid $30,200 to replace steel reinforcements to the structure damaged by the tornado, $86,100 for a new roof torn off by the tornado, $26,900 for a new air conditioning system that was housed on the roof, and $12,600 to replace carpeting damaged by water. Disney World estimates that higher quality steel used as replacements added 20 percent more structural support in terms of weight-bearing capacity. The new air conditioning system provides 25 percent more cooling power than the unit previously installed in the attraction. Compute the amount of these expenses that Disney World should treat as a repair and the amount it should treat as a betterment.
19. **Computing the amount of an impairment loss on tangible long-lived assets.** Wildwood Properties owns an apartment building that has a book value of $15,000,000 on January 1, Year 14. The highway department has decided to construct a new highway near the building, which substantially decreases its attractiveness to tenants. Wildwood Properties estimates that it will now collect rentals from the building of $1,400,000 a year for the next six years and that it will sell the building at the end of that time for $4,000,000. An appropriate interest rate to discount cash flows for this building is 10 percent. Assume that all cash flows occur at the end of the year. Compute the amount of any impairment loss that Wildwood Properties should recognize.
20. **Computing the gain or loss on sale of equipment.** Federal Express acquired a delivery truck on January 1, Year 10, for $48,000. It estimated that the truck would have a six-year useful life and $6,000 salvage value. Federal Express uses the straight-line depreciation method. On July 1, Year 14, Federal Express sells the truck for $14,000. Give the journal entries that Federal Express makes on July 1, Year 14, to recognize depreciation for Year 14 and the sale of the truck.

## EXERCISES

**21. Classifying expenditure as asset or expense.** For each of the following expenditures or acquisitions, indicate the type of account debited. Classify the account as (1) asset other than product cost, (2) product cost (Work-in-Process Inventory), or (3) expense. If the account debited is an asset account, specify whether it is current or noncurrent.

**a.** $150 for repairs of office machines.
**b.** $1,500 for emergency repairs to an office machine.
**c.** $250 for maintenance of delivery trucks.
**d.** $5,000 for a machine acquired in return for a three-year note.
**e.** $4,200 for research and development staff salaries.
**f.** $3,100 for newspaper ads.
**g.** $6,400 for wages of factory workers engaged in production.
**h.** $3,500 for wages of factory workers engaged in installing equipment the firm has purchased.
**i.** $2,500 for salaries of the office workforce.
**j.** $1,000 for legal fees incurred in acquiring an ore deposit.
**k.** $1,200 for a one-year insurance policy beginning next month.
**l.** $1,800 for U.S. Treasury notes, to be sold to pay the next installment due on income taxes.
**m.** $4,000 for royalty payment for the right to use a patent used in manufacturing.
**n.** $10,000 for purchase of a trademark.
**o.** $100 filing fee for copyright registration application.
**p.** $1,850 to purchase computer software used in recordkeeping.
**q.** $8,600 to purchase from its creator initial research on a possible drug to treat hypertension.

**22. Cost of self-constructed assets.** Assume that Bolton Company purchased a plot of land for $90,000 as a factory site. A small office building sits on the plot, conservatively appraised at $20,000. The company plans to use the office building after making some modifications and renovations (item **(4)** below). The company had plans drawn for a factory and received bids for its construction. It rejected all bids and decided to construct the factory itself. Management believes that plant asset accounts should include the following additional items:

|  |  |  |
|---|---|---|
| **(1)** | Materials and Supplies for Factory Building | $200,000 |
| **(2)** | Excavation of Land | 12,000 |
| **(3)** | Labor on Construction of Factory Building | 140,000 |
| **(4)** | Cost of Remodeling Old Building into Office Building | 13,000 |
| **(5)** | Interest Paid on Cash Borrowed by Bolton to Construct Factory[a] | 6,000 |
| **(6)** | Interest Forgone on Bolton's Own Cash Used | 9,000 |
| **(7)** | Cash Discounts on Materials Purchased for Factory Building | 7,000 |
| **(8)** | Supervision by Management on Factory Building | 10,000 |
| **(9)** | Workers' Compensation Insurance Premiums on Labor in **(3)** | 8,000 |
| **(10)** | Payment of Claims for Injuries during Construction of Factory Building Not Covered by Insurance | 3,000 |
| **(11)** | Clerical and Other Expenses on Construction of Factory Building | 8,000 |
| **(12)** | Paving of Streets and Sidewalks | 5,000 |
| **(13)** | Architect's Plans and Specifications of Factory Building | 4,000 |
| **(14)** | Legal Costs of Conveying Land | 2,000 |
| **(15)** | Legal Costs of Injury Claim during Construction of Factory Building | 1,000 |
| **(16)** | Income Credited to Retained Earnings Account (the difference between the forgone cost and the lowest contractor's bid) | 11,000 |

[a]This interest is the entire amount of interest paid during the construction period.

Show in detail the items Bolton should include in the following accounts: Land, Factory Building, Office Building, and Site Improvements. Explain the reason for excluding any of these items from the four accounts.

**23. Cost of self-developed product.** Duck Vehicle Manufacturing Company incurs various costs in developing a new, amphibious vehicle for use in providing tours on land and water. Indicate the accounting treatment for each of the following expenditures.

| | | |
|---|---|---|
| (1) | Salaries of company engineers to design the new vehicle | $325,000 |
| (2) | Cost of prototype of new vehicle built by external contractor | 278,200 |
| (3) | Cost of supplies and salaries of personnel to test prototype | 68,900 |
| (4) | Fees paid to Environmental Protection Agency to test emissions of new vehicle | 15,200 |
| (5) | Legal fees incurred to register and establish a patent on the new vehicle | 12,500 |
| (6) | Cost of castings, or molds, for metal parts of new vehicle | 46,000 |
| (7) | Cost of local permits to commence manufacturing the new vehicle | 5,000 |
| (8) | Cost of manufacturing the first vehicle for a customer | 167,600 |

24. **Amount of interest capitalized during construction.** Nucor, a steel manufacturer, self-constructs a new manufacturing facility in Vermont. At the start of Year 14, the Construction-in-Process account had a balance of $30 million. Construction activity occurred uniformly throughout the year. At the end of Year 14, the balance was $60 million before capitalization of interest for the year. The outstanding borrowings of the company during the year were as follows:

| | |
|---|---|
| New Construction Loans at 8 Percent per Year | $ 25,000,000 |
| Old Bond Issues Averaging 6 Percent Rate | 100,000,000 |
| Total Interest-Bearing Debt | $125,000,000 |

**a.** Compute the amount of interest capitalized in the Construction-in-Process account for Year 14.

**b.** Present journal entries for interest for Year 14.

**c.** On December 31, Year 15, Nucor completed the manufacturing facility and put it to work. Average Construction-in-Process for Year 15 was $110 million. The debt listed above remained outstanding throughout the construction project and the firm did not issue any additional interest-bearing debt during this time. Present journal entries for Year 15 related to interest expense and interest capitalization.

25. **Capitalizing interest during construction.** Nebok Company recently constructed a new headquarters building. Its Construction-in-Process account showed a balance of $23,186,000 on May 31, Year 9, and $68,797,000 on May 31, Year 10, before capitalization of interest for the year. Construction activity occurred evenly throughout the year. Nebok Company completed construction during Year 11. Nebok Company's note on long-term debt revealed the following (amounts in thousands):

| | May 31 | |
|---|---|---|
| | Year 9 | Year 10 |
| 8.5% unsecured term loan | $25,000 | $25,000 |
| 12.5% loan secured by real estate | 15,600 | 15,100 |
| 14.0% loan secured by real estate | 10,900 | 9,600 |
| Total | $51,500 | $49,700 |

None of this debt relates specifically to construction of the headquarters building. Interest expense before capitalization of interest for the year ending May 31, Year 10, is $5,480,000.

**a.** Compute the amount of interest capitalized in the Construction-in-Process account for the year.

**b.** Present journal entries for interest for the year.

**c.** Nebok Company's net income before interest expense and income taxes for the year is $16,300,000. Nebok Company is subject to a 35 percent income tax rate. Compute net income for the year ending May 31, Year 10.

**d.** Compute the interest coverage ratio (equals net income before interest expense and income taxes divided by interest expense) using the amounts computed in part **c**.

e. Repeat part **d** using the interest expense before capitalization of interest of $5,480,000 in the denominator.

f. Which measure of the interest coverage ratio provides a more appropriate measure for assessing risk?

26. **Calculations for various depreciation methods.** Alcoa acquires a machine for $88,800. It expects the machine to last six years and to operate for 30,000 hours during that time. Estimated salvage value is $4,800 at the end of the machine's useful life. Calculate the depreciation charge for each of the first three years using each of the following methods:
    a. The straight-line (time) method.
    b. The sum-of-the-years'-digits method.
    c. The declining-balance method using a 33 percent rate.
    d. The straight-line (use) method, with the following operating times: first year, 4,500 hours; second year, 5,000 hours; third year, 5,500 hours.

27. **Calculations for various depreciation methods.** On January 1, Year 8, assume that Luck Delivery Company acquired a new truck for $30,000. It estimated the truck to have a useful life of five years and no salvage value. The company closes its books annually on December 31. Indicate the amount of the depreciation charge for each year of the asset's life under the following methods:
    a. The straight-line method.
    b. The declining-balance method at twice the straight-line rate, with a switch to straight-line in Year 11.
    c. The sum-of-the-years'-digits method.
    d. MACRS depreciation, assuming that the truck belongs to the five-year property class. The proportions of the asset's cost written off each year are 0.20, 0.32, 0.192, 0.115, 0.115, 0.058. (*Hint:* charge depreciation for each of six calendar years.)
    e. Assume now that the firm acquired the truck on April 1, Year 8. Indicate the amount of the depreciation charge for each of the years from Year 8 to Year 13, using the sum-of-the-years'-digits method. (*Hint:* the firm allocates depreciation charges for a year, regardless of the depreciation method, on a straight-line basis to periods within a year.)

28. **Revision of estimated service life changes depreciation schedule.** Slow Poke Delivery Company buys a new machine for $80,000 on January 1, Year 12. It estimates that the machine will last 10 years and have a salvage value of $8,000. Early in Year 14, it discovers that the machine will last only an additional six years, or eight years total (with no change in estimated salvage value). Present a table showing the depreciation charges for each year from Year 12 to Year 15, using each of the following methods (round calculations to the nearest dollar):
    a. The straight-line method.
    b. The sum-of-the-years'-digits method.

29. **Journal entries for revising estimate of life.** Give the journal entries for the following selected transactions of Florida Manufacturing Corporation. The company uses the straight-line method of calculating depreciation and reports on a December 31 year-end.
    a. The firm purchases a cutting machine on November 1, Year 14, for $180,000. It estimates that the machine will have a useful life of 12 years and a salvage value of $7,200 at the end of that time. Give the journal entry for the depreciation at December 31, Year 14.
    b. Record the depreciation for the year ending December 31, Year 15.
    c. In August, Year 20, the firm estimates that the machine will probably have a total useful life of 14 years and a $3,840 salvage value. Record the depreciation for the year ending December 31, Year 20.
    d. The firm sells the machine for $40,000 on March 31, Year 25. Record the entries of that date, assuming that the firm records depreciation as indicated in part **c**.

30. **Working backward to derive proceeds from disposition of plant assets.** The balance sheets of Wilcox Corporation at the beginning and end of the year contained the following data:

| | Beginning of Year | End of Year |
|---|---|---|
| Property, Plant, and Equipment (at cost) | $400,000 | $550,000 |
| Accumulated Depreciation | 180,000 | 160,000 |
| Net Book Value | $220,000 | $390,000 |

During the year, Wilcox Corporation sold machinery and equipment at a gain of $4,000. It purchased new machinery and equipment at a cost of $230,000. Depreciation charges on machinery and equipment for the year amounted to $50,000. Calculate the proceeds Wilcox Corporation received from the sale of the machinery and equipment.

**31. Computing the amount of impairment loss.** Tillis Corporation acquired the assets of Kieran Corporation (Kieran) on January 1, Year 6, for $2,400,000. On this date, the market values of the assets of Kieran were as follows: land, $400,000; building, $600,000; equipment, $900,000. On June 15, Year 8, a competitor introduced a new product that will likely significantly affect future sales of Kieran's products. It will also affect the value of Kieran's property, plant, and equipment because of their specialized nature in producing Kieran's existing products. The following information relates to the property, plant, and equipment of Kieran on June 15, Year 8:

| | Book Value | Undiscounted Cash Flows | Market Value |
|---|---|---|---|
| Land | $ 550,000 | $575,000 | $550,000 |
| Building | 580,000 | 600,000 | 580,000 |
| Equipment | 1,200,000 | 950,000 | 800,000 |

The market value of Kieran as an entity on June 15, Year 8, is $2,200,000.

Compute the amount of impairment loss recognized on each of Kieran's property, plant, and equipment on June 15, Year 8.

**32. Journal entries to correct accounting errors.** Give correcting entries for the following situations. In each case, the firm uses the straight-line method of depreciation and closes its books annually on December 31. Recognize all gains and losses currently.

**a.** A firm purchased a computer for $3,000 on January 1, Year 3. It depreciated the computer at a rate of 25 percent of acquisition cost per year. On June 30, Year 5, it sold the computer for $800 and acquired a new computer for $4,000. The bookkeeper made the following entry to record the transaction:

| | Dr. | Cr. |
|---|---|---|
| Equipment | 3,200 | |
| Cash | | 3,200 |

| Assets | = | Liabilities | + | Shareholders' Equity | (Class.) |
|---|---|---|---|---|---|
| +3,200 | | | | | |
| −3,200 | | | | | |

**b.** A firm purchased a used truck for $7,000. Its cost, when new, was $12,000. The bookkeeper made the following entry to record the purchase:

| | Dr. | Cr. |
|---|---|---|
| Truck | 12,000 | |
| Accumulated Depreciation | | 5,000 |
| Cash | | 7,000 |

| Assets | = | Liabilities | + | Shareholders' Equity | (Class.) |
|---|---|---|---|---|---|
| +12,000 | | | | | |
| −5,000 | | | | | |
| −7,000 | | | | | |

**c.** A firm purchased a testing mechanism on April 1, Year 6 for $1,200. It depreciated the testing mechanism at a 10 percent annual rate. A burglar stole the testing mechanism on June 30, Year 8. The firm had not insured against this theft. The bookkeeper made the following entry:

| | Dr. | Cr. |
|---|---|---|
| Theft Loss | 1,200 | |
| Testing Mechanism | | 1,200 |

| Assets | = | Liabilities | + | Shareholders' Equity | (Class.) |
|---|---|---|---|---|---|
| −1,200 | | | | −1,200 | IncSt → RE |

**33. Effect of transaction on statement of cash flows.** (Requires coverage of **Appendix 8.1**.) Refer to the simplified statement of cash flows in **Exhibit 4.16**. Numbers appear on 11 of the lines in the statement. Ignore the unnumbered lines in responding to the questions that follow.

Assume that the accounting cycle is complete for the period and that the firm has prepared all of the financial statements. It then discovers that it has overlooked a transaction. It records that transaction in the accounts and corrects all of the financial statements. For each of the following transactions, indicate which of the numbered lines of the statement of cash flows change and the amount and direction of the change. Ignore income tax effects.

**a.** A firm sells for $3,000 cash a machine that cost $10,000 and that has $6,000 of accumulated depreciation.

**b.** A firm sells for $5,000 cash a machine that cost $10,000 and that has $6,000 of accumulated depreciation.

**c.** A firm trades in an old machine for a new machine. The old machine cost $10,000 and has $6,000 of accumulated depreciation. The new machine has a cash price of $9,000. The firm receives a trade-in allowance for the old machine of $5,000 and pays $4,000. Following GAAP, the firm records no gain or loss on trade-in but records the new machine at $8,000.

**d.** A fire destroys a warehouse. The loss is uninsured. The warehouse cost $90,000 and at the time of the fire had accumulated depreciation of $40,000.

**e.** Refer to the facts of part **d**. The fire also destroyed inventory costing $60,000. The loss is uninsured. Record the effects of only these new facts.

## PROBLEMS AND CASES

**34. Improvements versus repairs or maintenance.** The balance sheet of May Department Stores includes a building with an acquisition cost of $800,000 and accumulated depreciation of $660,000. The firm depreciates the building on a straight-line basis over 40 years. The remaining service life of the building and its depreciable life are both seven years. On January 2 of the current year, the firm makes an expenditure of $28,000 on the entrance ramps and stairs of the store. Indicate the accounting for the current year if May Department Stores made the expenditure of $28,000 under each of the following circumstances. Consider each of these cases independently, except where noted. Ignore income tax effects.

**a.** Management decided that improved entrances would make the store more attractive and that new customers would come. The ramps and stairs are a worthwhile investment.

**b.** A flood on New Year's Day ruined the entrance ramps and stairs previously installed. The firm carried no insurance coverage for this sort of destruction. The new ramps and stairs are physically identical to the old ones. The old ones had a book value of $28,000 at the time of the flood.

**c.** Vandals destroyed the ramps and stairs on New Year's Day. May carried no insurance for this sort of destruction. The ramps and stairs installed are physically identical to the old ones, which had a book value of $28,000 at the time of the destruction.

**d.** The old entrances were not handicapped-accessible. Management had previously considered replacing its old entrances with new, handicapped-accessible ones but had decided that there was zero benefit to the firm in doing so. It installed new entrance ramps and stairs because a new law required that all stores have such ramps and stairs on the street level. The alternative to installing the new ramps was to shut down the store. In responding to this part, assume zero benefits result from the new ramps. Part **e** considers the more realistic case of some benefits.

**e.** Management had previously considered replacing its old entrances with new ones but had decided that the handicapped-accessible ramps and stairs were worth only $7,000 (that is, would produce future benefits of only $7,000) and so were not a worthwhile investment.

The new law (see part **d**), however, now requires the store to do so (or else shut down), and management installs the new entrance facilities.

**35. Capitalizing versus expensing; if capitalized, what amortization period?** In each of the following situations, compute the amounts of revenue, gain, expense, and loss appearing on the income statement for the year and the amount of asset appearing on the balance sheet as of the end of the year. Show the journal entry or entries required, and provide reasons for your decisions. The firm uses straight-line amortization. The reporting period is the calendar year. Consider each of the situations independently, except where noted.

**a.** Because of a new fire code, MCB Upholstery Shop must install an additional fire escape on its building. Management previously considered installing the additional fire escape. It rejected the idea because it had already installed a modern sprinkler system, which was even more cost-effective. The new code gives management no alternative except to close the store. MCB acquires the fire escape for $28,000 cash on January 1. It expects to demolish the building seven years from the date it installs the fire escape.

**b.** Many years ago, a firm acquired shares of stock in General Electric Company for $100,000. On December 31, the firm acquired a building with an appraised value of $1 million. The company paid for the building by giving up its shares in General Electric Company at a time when equivalent shares traded on the New York Stock Exchange for $1,050,000.

**c.** Same data as in part **b**, except that the shares of stock represent ownership in Small Timers, Inc., whose shares trade on a regional stock exchange. The last transaction in shares of Small Timers, Inc., occurred on December 27. Using the prices of the most recent trades, the shares of stock of Small Timers, Inc., given in exchange for the building have a market value of $1,050,000.

**d.** Oberweis Dairy decides that it can save $3,500 a year for 10 years by switching from small panel trucks to larger delivery vans. To do so requires remodeling costs of $18,000 for various garages. The first fleet of delivery vans will last for five years, and the garages will last for 20 years. Oberweis remodeled the garages on January 1.

**e.** A company manufactures aircraft. During the current year, it made all sales to the government under defense contracts. The company spent $400,000 on institutional advertising to keep its name before the business community. It expects to resume sales of small jet planes to corporate buyers in two years.

**f.** AT&T runs a large laboratory that has, over the years, found marketable ideas and products worth tens of millions of dollars. On average, the successful products have a life of 10 years. Assume that expenditures for the laboratory this year were $1,500,000.

**g.** A textile manufacturer gives $250,000 to the Textile Engineering Department at Georgia Tech for basic research in fibers. The results of the research, if any, will belong to the general public.

**h.** On January 1, assume that Mazda incurs costs of $6 million for specialized machine tools necessary to produce a new-model automobile. Such tools last for six years, on average, but Mazda expects to produce the new model automobile for only three years.

**i.** On January 1, assume that United Airlines purchased a fleet of airplanes for $100 million cash. United Airlines expects the airplanes to have a useful life of 10 years and zero salvage value. At the same time, the airline purchased for cash $20 million of spare parts for use with those airplanes. The spare parts have no use, now or in the future, other than replacing broken or worn-out airplane parts. During the first year of operation, United Airlines used no spare parts.

**j.** Refer to the data in **i**. In the second year of operation, United Airlines used $1 million of spare parts.

**36. Accounting for intangibles.** In Year 1, Epstein Company acquired the assets of Falk Company, which included various intangibles. Discuss the accounting for the acquisition in Year 1, and in later years, for each of the following items.

**a.** Registration of the trademark Thyrom® for thyristors expires in three years. Epstein Company believes that the trademark has a fair market value of $100,000. It expects to continue making and selling Thyrom thyristors indefinitely.

**b.** The design patent covering the ornamentation of the containers for displaying Thyrom thyristors expires in five years. Epstein Company thinks that the design patent has a fair market value of $30,000 and expects to continue making the containers indefinitely.

**c.** Epstein Company views an unpatented trade secret on a special material used in manufacturing thyristors as having a fair market value of $200,000.

**d.** Refer to the trade secret in part **c**. Suppose that in Year 2 a competitor discovers the trade secret but does not disclose the secret to other competitors. How should Epstein Company change its accounting policies?

**e.** During Year 1, Epstein Company produced a sales promotion film, Using Thyristors for Fun and Profit, at a cost of $45,000. It licensed the film to purchasers of thyristors for use in training their employees and customers. Epstein has copyrighted the film.

**37. Interpreting disclosures regarding property, plant, and equipment.** Mead Corporation reports the following information in its financial statements and notes for a recent year (amounts in millions):

| | December 31 | |
|---|---|---|
| | Year 7 | Year 8 |
| Property, Plant, and Equipment (at cost) | $ 3,824.0 | $ 3,938.5 |
| Accumulated Depreciation | (1,803.7) | (1,849.3) |
| Property, Plant, and Equipment (net) | $ 2,020.3 | $ 2,089.2 |
| | For Year 8 | |
| Depreciation Expense | $ 188.1 | |
| Expenditures on Property, Plant, and Equipment | 315.6 | |
| Proceeds from Sales of Property, Plant, and Equipment | 38.7 | |
| Sales | 4,557.5 | |

**a.** Give the journal entries for the transactions and events that account for the changes in the Property, Plant, and Equipment account and the Accumulated Depreciation account for Year 8.

**b.** Mead Corporation uses the straight-line depreciation method for financial reporting. Estimate the average *total* life of property, plant, and equipment in use during Year 8.

**c.** Refer to part **b**. Estimate the average age to date of property, plant, and equipment in use during Year 8.

**d.** Compute the fixed asset turnover for Year 8.

**e.** Mead Corporation uses accelerated depreciation for income tax purposes. From information in the financial statement note on income taxes (discussed in **Chapter 10**), the balance in the accumulated depreciation account using accelerated depreciation would have been $2,810.2 million on December 31, Year 7, and $2,830.5 million on December 31, Year 8. Compute the fixed asset turnover ratio using accelerated depreciation.

**38. Interpreting disclosures regarding property, plant, and equipment.** PepsiCo reports the following information in its financial statements and notes in a recent year (amounts in millions):

| | December 31 | |
|---|---|---|
| | Year 6 | Year 7 |
| Property, Plant, and Equipment (at cost) | $14,250.0 | $16,130.1 |
| Accumulated Depreciation | (5,394.4) | (6,247.3) |
| Property, Plant, and Equipment (net) | $ 8,855.6 | $ 9,882.8 |
| | For Year 7 | |
| Sales | $28,472.4 | |
| Depreciation Expense | 1,200.0 | |
| Expenditures on Property, Plant, and Equipment | 2,253.2 | |
| Proceeds from Sales of Property, Plant, and Equipment | 55.3 | |

**a.** Give the journal entries for the transactions and events that account for the changes in the Property, Plant, and Equipment account and the Accumulated Depreciation account during Year 7.

**b.** PepsiCo uses the straight-line depreciation method for financial reporting. Estimate the average *total* life of property, plant, and equipment in use during Year 7.

**c.** Refer to part **b**. Estimate the average age to date of property, plant, and equipment in use during Year 7.

**d.** Compute the fixed asset turnover for Year 7.

**e.** PepsiCo uses accelerated depreciation for income tax purposes. From information in the financial statement note on income taxes (discussed in **Chapter 10**), the balance in the accumulated depreciation account using accelerated depreciation would have been \$7,018.8 on December 31, Year 6, and \$7,736.7 on December 31, Year 7. Compute the fixed asset turnover ratio using accelerated depreciation.

**39. Effect on net income of changes in estimates for depreciable assets.** American Airlines has \$3 billion of assets, including airplanes costing \$2.5 billion with net book value of \$1.6 billion. It earns net income equal to approximately 6 percent of total assets. American Airlines depreciates its airplanes for financial reporting purposes on a straight-line basis over 10-year lives to a salvage value equal to 10 percent of acquisition cost. American announces a change in depreciation policy; it will use 14-year lives and salvage values equal to 12 percent of original cost. The airplanes are all four years old. Assume an income tax rate of 35 percent.

Calculate the approximate impact on net income of the change in depreciation policy. Compute both dollar and percentage effects. What conjectures about the earnings quality of an airline can you derive from these computations?

**40. Recording transactions involving tangible and intangible assets.** Present journal entries for each of the following transactions of Moon Macrosystems:

**a.** Acquired computers costing \$400,000 and computer software costing \$40,000 on January 1, Year 6. Moon expects the computers to have a service life of 10 years and \$40,000 salvage value. It expects the computer software to have a service life of four years and zero salvage value.

**b.** Paid \$20,000 to install the computers in the office. Paid \$10,000 to install and test the computer software.

**c.** Recorded depreciation and amortization using the straight-line method for Year 6 and Year 7. Moon records a full year of depreciation in the year of acquisition. Treat depreciation and amortization as a period expense.

**d.** On January 1, Year 8, new software offered on the market made the software acquired above completely obsolete. Give any required journal entry.

**e.** On January 2, Year 8, Moon revised the depreciable life of the computers to a total of 14 years and the salvage value to \$56,000. Give the entry to record depreciation for Year 8.

**f.** On December 31, Year 9, Moon sold the computers for \$260,000. Give the required journal entries for Year 9.

**41. Recognizing and measuring impairment losses.** Give the journal entry to recognize an impairment loss, if appropriate, in each of the following cases. If a loss does not qualify as an impairment loss, explain the reason, and indicate the appropriate accounting.

**a.** Commercial Realty Corporation leases office space to tenants in Boston. One of its office buildings originally cost \$80 million and has accumulated depreciation of \$20 million. The city of Boston has announced its intention to construct an exit ramp from a nearby expressway on one side of the office building. Rental rates in the building will likely decrease as a result. The expected future undiscounted cash flows from rentals and from disposal of the building decreased from \$120 million before the announcement to \$50 million afterward. The market value of the building decreased from \$85 million before the announcement to \$32 million afterward.

**b.** Refer to part **a**. Assume that the undiscounted cash flows totaled \$70 million and that the market value totaled \$44 million after the announcement.

**c.** Medical Services Corporation plans, and then builds, its own office building and clinic. It originally anticipated that the building would cost \$15 million. The physicians in charge of overseeing construction had medical practices so busy that they did not closely track costs, which ultimately reached \$25 million. The expected future cash flows from using the building total \$22 million, and the market value of the building totals \$16 million.

**d.** Medco Pharmaceuticals acquired New Start Biotechnology two years ago for \$40 million. Medco allocated \$25 million to a patent held by New Start and \$15 million to goodwill. By the end of the current period, Medco has written down the book value of the patent to \$20 million. A competitor recently received approval for a biotechnology drug that will reduce the value of the patent that Medco acquired from New Start. The expected future undiscounted cash flows from sales of the patented drug total $18 million, and the market value of the patent is $12 million. The market value of the former New Start Biotechnology operation owned by Medco is now $25 million.

**e.** Chicken Franchisees, Inc., acquires franchise rights in the Atlanta area for Chicken Delight Restaurants, a national restaurant chain. The franchise rights originally cost $15 million; since acquisition, Chicken Franchisees has written down the book value to $10 million. Chicken Delight Restaurants recently received negative publicity because the chickens it delivered to its franchisees contained potentially harmful pesticides. As a result, business has declined. Chicken Franchisees estimates that the future undiscounted cash flows associated with the Chicken Delight name total $6 million and that the franchise rights have market value of $3 million.

**42. Expensing versus capitalizing research and development costs.** Pfizer, a pharmaceutical company, plans to spend $90 million on research and development (R&D) at the beginning of each of the next several years to develop new drugs. As a result of the R&D expenditure for a given year, it expects pre-tax income (not counting R&D expense) to increase by $36 million a year for three years, including the year of the expenditure itself. Pfizer has other pretax income of $30 million per year. The controller of Pfizer is curious about the effect on the financial statements of following one of two accounting policies with respect to R&D expenditures:

**(1)** Expensing the R&D costs in the year of expenditure (the policy required in the United States).

**(2)** Capitalizing the R&D costs and amortizing them over three years, including the year of the expenditure itself (the policy allowed in certain other countries).

Assume that the company does spend $90 million at the beginning of each of four years and that the planned increase in income occurs. Ignore income tax effects.

**a.** Prepare a four-year condensed summary of income before income taxes, assuming that Pfizer follows policy **(1)** and expenses R&D costs as incurred.

**b.** Prepare a four-year condensed summary of income before income taxes, assuming that Pfizer follows policy **(2)** and capitalizes R&D costs, then amortizes them over three years. Also compute the amount of Deferred R&D Costs (asset) appearing on the balance sheet at the end of each of the four years.

**c.** In what sense is policy **(1)** a conservative policy?

**d.** Ascertain the effect on income before income taxes and on the balance sheet if Pfizer continues to spend $90 million each year and the pretax income effects continue as in the first four years.

**43. Valuation of the brand name.** Ross Laboratories sells various formulations of infant baby food around the world under the brand name Similac. In a recent year, worldwide sales less operating expenses were $600 million. Ross Laboratories employed $500 million of physical capital to produce and sell this food. Estimate the value of the Similac brand assuming the following:

**a.** Ross Laboratories charges 10 percent, before taxes, for physical capital, pays income taxes at the rate of 40 percent of pretax income, and uses a price/earnings ratio (capitalization factor) of 17.

**b.** Ross Laboratories charges 20 percent, before taxes, for physical capital, pays income taxes at the rate of 40 percent of pretax income, and uses a price/earnings ratio (capitalization factor) of 8.

**44. Ethical issues in capitalization versus expensing policy.** Refer to **Example 1** involving the treatment of access costs by WorldCom.

**a.** Assume that you were the chief financial officer of WorldCom. What case might you make to your external auditors to justify the capitalization of the access costs in fixed assets?

**b.** What is GAAP's rationale for the expensing of such costs in the period incurred?

**c.** Do WorldCom's actions to capitalize these costs pose ethical issues? Explain.

CHAPTER 9

# Liabilities: Introduction

## Learning Objectives

1. **Understand and apply the concept of an accounting liability to obligations with uncertain payment dates and amounts.**
2. **Develop the skills to compute the issue price, book value, and current market value of various debt obligations in an amount equal to the present value of the future cash flows.**
3. **Understand the effective interest method and apply it to debt amortization for various long-term debt obligations.**
4. **Understand the accounting procedures for debt retirements, whether at or before maturity.**

**Chapters 9**, **10**, and **12** examine the accounting concepts and procedures for the right-hand side of the balance sheet, which shows the sources of a firm's financing. The funds used to acquire assets come from two sources: owners and nonowners. **Chapter 12** discusses shareholders' equity. This chapter and **Chapter 10** discuss obligations incurred by a business as a result of raising funds from nonowners. Banks and creditors providing debt on a long-term basis understand their role as providers of funds. Suppliers and employees who do not require immediate cash payment for goods provided or services rendered usually do not think of themselves as contributing to a firm's funds, but they do contribute. Likewise, customers who advance cash to a firm before delivery of a good or service provide funds to the firm. The obligations of a business to these nonowners who contribute funds are liabilities.

A thorough understanding of liabilities requires knowledge of compound interest and present value computations. These computations allow valid comparisons between payments made at different times because they take into account the interest that cash can earn over time. The **Appendix** at the back of the book describes and illustrates compound interest and present value. Understanding all the material in this chapter requires that you master the material in the **Appendix**. Without such mastery, you will have to accept on faith some of the computations in this chapter.

Lenders often, as part of the loan agreement, require the borrower to manage its operations and its financing so as to keep defined financial ratios above or below specified levels. If the ratios violate the constraints, the lender will have additional rights, such as to require early repayment of the loan. These lending constraints provide protection for the lender. For example, the lending agreement might require a debt-equity ratio no larger than 60 percent or an interest coverage ratio no less than 3.0. To set the limits for the ratios, the lender must understand GAAP's definition of a liability, so that it can write the lending agreement to achieve the goal the lender has in mind with respect to the borrower's acceptable and unacceptable levels of total debt. To illustrate, consider the data in **Exhibit 9.1** for Wal-Mart. The first column of amounts shows the reported liabilities on January 31, Year 14. Wal-Mart's debt-equity ratio, defined as total liabilities divided by total sources of financing, is 60.4 percent. The second column of

**EXHIBIT 9.1** **Selected Balance Sheet Data for Wal-Mart Stores January 31, Year 14 (amounts in millions)**

| | Reported | Restated |
|---|---|---|
| Current Liabilities | $ 37,418 | $ 37,418 |
| Long-Term Debt | 25,573 | 27,290 |
| Other Noncurrent Liabilities | 3,416 | 3,416 |
| Obligations under Operating Leases | — | 5,487 |
| Guarantees of Debt of Others | — | 2,110 |
| Total | $ 66,407 | $ 75,721 |
| Shareholders' Equity | 43,623 | 39,796 |
| Total | $110,030 | $115,517 |
| Debt-Equity Ratio (Total Liabilities/Total Sources of Financing) | 60.4% | 65.5% |

amounts shows additional information about Wal-Mart's obligations, some of which GAAP does not require Wal-Mart to report as a liability.

The restated amounts differ from the reported amounts for the following reasons:

1. Long-Term Debt: Interest rates have declined since Wal-Mart issued its long-term debt and the market value of this debt has consequently increased. (A later section of this chapter explains why this occurs.)
2. Obligations under Operating Leases: Wal-Mart has signed lease agreements for land, buildings, and equipment that do not appear on its balance sheet as liabilities. **Chapter 10** explains why such leases, called *operating leases,* do not appear on the balance sheet under GAAP. Yet, Wal-Mart has obligated itself to make these payments, which it will likely pay according to the schedule already known by the balance sheet date.
3. Guarantees of Debt of Others: Wal-Mart has guaranteed the debt of some of its suppliers. In the event the suppliers do not pay, Wal-Mart must pay the debt. A later section of this chapter explains why GAAP does not normally require firms to recognize obligations for guarantees of others' debt as a liability.

In computing the restated debt-equity ratio, we assume that the excess of market value over the book value of the long-term debt and the potential losses from the debt guarantees reduce shareholders' equity.

This example illustrates why you should understand the obligations that GAAP requires firms to recognize as liabilities and the potential obligations that GAAP allows firms to leave off of the balance sheet.

## Basic Concepts of Liabilities

### Liability Recognition

Nearly all liabilities are obligations, but not all obligations are liabilities. According to the Financial Accounting Standards Board (FASB), accounting generally recognizes an entity's obligation as a **liability** if the obligation meets three criteria.[1]

1. The obligation involves a probable future sacrifice of resources—a future transfer of cash, goods, or services or the forgoing of a future cash receipt—at a specified or ascertainable date. The firm can measure, with reasonable precision, the cash equivalent value of resources needed to satisfy the obligation.
2. The firm has little or no discretion to avoid the transfer.

[1] Financial Accounting Standards Board, *Statement of Financial Accounting Concepts No. 6,* "Elements of Financial Statements," 1985, par. 36.

3. The transaction or event giving rise to the entity's obligation has already occurred. (We suggest to our students that they understand this third FASB criterion by saying that for an obligation to be a liability, it must arise from something other than an executory contract, that is, the obligation arises from something other than a mere exchange of promises.)

**Example 1** General Electric has borrowed $75 million by issuing bonds that require it to pay interest of 4 percent every six months for 20 years and to repay the $75 million at the end of 20 years. This borrowing arrangement obligates General Electric to make definite cash payments in the future at specified times and is a liability, titled Bonds Payable.

**Example 2** Avon Products' employees have earned wages and salaries that Avon will not pay until the next payday, two weeks after the end of the current accounting period. Avon owes suppliers substantial amounts for materials they sold to Avon, but Avon need not pay these debts for 10 to 30 days after the end of the period. Avon owes the federal and state governments for taxes, but need not make payments until next month. Although the payment dates are not necessarily fixed, Avon can estimate those dates with reasonable precision. Each of these items therefore meets the three criteria for a liability. The items appear as liabilities under titles such as Wages Payable, Salaries Payable, Accounts Payable, and Taxes Payable.

**Example 3** When Sony USA sells television sets, it gives a warranty to repair or replace any faulty parts or faulty sets within one year after sale. This obligation meets the three criteria of a liability. Because some television sets will surely need repair, the future sacrifice of resources is probable. Sony has the obligation to make repairs. The transaction giving rise to the obligation, the sale of the television set, has already occurred. Sony does not know the amount with certainty, but Sony's experience with its television sets permits a reasonably precise estimate of the expected costs of repairs or replacements. The repairs or replacements will occur within a time span that Sony estimates with reasonable precision. Sony USA will thus show Estimated Warranty Liability on its balance sheet.

**Example 4** The Boston Red Sox receives $2,400,000 from the sale of tickets for the upcoming baseball season. *Sports Illustrated* receives $1,750,000 for the sale of magazine subscriptions for the next three years. The American Automobile Association receives $840,000 from the sale of memberships for the next two years. These advances from customers obligate the firms to provide services in the future. The amount of the cash advance serves as a measure of the liability until the firms provide the services.

**Example 5** Columbia Pictures has signed a binding contract to supply films to Cineplex Odeon, a chain of movie theaters, within the next six months. In this case, there is an obligation, definite time, and definite amount (of goods, if not cash), but no past or current transaction has occurred. **Chapter 2** pointed out that accounting does not recognize assets or liabilities for **executory contracts**—the mere exchange of promises with no mutual performance. Without some mutual performance, there is no current or past benefit. Thus no accounting liability arises in this case.

**Example 6** Online Corporation has signed a contract promising to employ its president for the next five years and to pay the president a salary of $750,000 per year. The salary will increase in future years at the same rate of increase as the Consumer Price Index, published by the U.S. government. Online Corporation has an obligation to make payments (although the president may quit at any time without penalty). The payments are for reasonably certain amounts and will occur at definite times. At the time Online signs the contract, no mutual performance has occurred; the contract is purely executory. Because no exchange has occurred, no liability appears on the balance sheet. A liability will, of course, arise as the president performs services over time and earns the salary.

**Example 7** ChevronTexaco Corporation has signed a contract with Tenneco Oil Pipeline Company to ship at least 10,000 barrels of crude oil per month for the next three years. ChevronTexaco must pay for the shipping services, even if it does not ship the oil. An arrangement such as this, called a *throughput contract,* does not qualify as an accounting liability because accounting views actual shipment, which has not yet occurred, as the event causing a liability. ChevronTexaco will disclose the obligation in notes. Similarly, obligations for take-or-pay contracts, in which the purchaser must pay for certain quantities of goods even if the purchaser does not take delivery of the goods, are not formal liabilities. Firms obligating themselves with take-or-pay contracts must disclose them.[2] Accounting views both throughput and take-or-pay contracts as executory.

**Example 8** Smokers have sued Philip Morris, a subsidiary of Altria, in a lawsuit alleging damages of $6 billion. The smokers claim that the company knowingly withheld information about the harmful effects of smoking and that misleading advertising about Marlboro cigarettes injured them. Shook, Hardy & Bacon, lawyers from Kansas City retained by the corporation, think that Philip Morris can mount an adequate defense to the charges. Because Philip Morris has no obligation to make a payment at a reasonably definite time, it has no liability. The notes to the financial statements will disclose the existence of the lawsuit, but normally no liability will appear on the balance sheet.

**Example 9** Citibank extends $800 million in lines of credit to its customers, agreeing to make loans up to this amount as customers need funds. Because the line of credit is an executory contract (a mere exchange of promises), accounting does not recognize it as a liability. Citibank discloses the amount of such lending commitments in the notes to its financial statements.

**Summary of Liability Recognition** **Figure 9.1** classifies obligations into six groups based on the degree to which they satisfy the criteria for liability recognition. Accounting recognizes obligations in the first four groups as liabilities. As the examples above illustrate, accounting typically does not recognize as liabilities obligations under mutually unexecuted contracts. The exclusion of executory contracts from liabilities has created controversy in accounting. **Chapter 10** discusses commitments arising from a long-term lease, an executory contract that accounting recognizes as a liability if the lease agreement satisfies certain conditions.

## CONTINGENCIES: POTENTIAL OBLIGATIONS

The question of whether to recognize contingent, or potential, obligations as liabilities causes controversy. One criterion for deciding whether an obligation qualifies as a liability is that the

**FIGURE 9.1** Classifications of Accounting Liabilities by Degree of Certainty

| Obligations with Fixed Payment Dates and Amounts | Obligations with Fixed Payment Amounts but Estimated Payment Dates | Obligations for Which the Firm Must Estimate Both Timing and Amount of Payment | Obligations Arising from Advances from Customers on Unexecuted Contracts and Agreements | Obligations under Mutually Unexecuted Contracts | Contingent Obligations |
|---|---|---|---|---|---|
| Notes Payable<br>Interest Payable<br>Bonds Payable | Accounts Payable<br>Taxes Payable | Warranties Payable[a] | Rental Fees Received in Advance<br>Subscription Fees Received in Advance | Purchase Commitments[b]<br>Employment Commitments | Unsettled Lawsuits[c]<br>Financial Instruments with Off-Balance-Sheet Risk[c] |

Most Certain ←——→ Least Certain

←— Recognized as Accounting Liabilities —→ ←— Not Generally Recognized as Accounting Liabilities —→

[a]If the firm cannot estimate the future warranty cost with reasonable precision, it will not recognize revenue at the time of sale, so there is no liability.
[b]If the market price of a purchase commitment for inventory is less than the contract price, the firm recognizes a loss and a liability in the amount of the loss.
[c]If an obligation meets certain criteria for a loss contingency, the firm recognizes this obligation as a liability. See the discussion in this chapter.

[2]Financial Accounting Standards Board, *Statement of Financial Accounting Standards No. 47*, "Disclosure of Long-Term Obligations," 1981.

firm will make a probable future sacrifice of resources. The world of business and law contains many uncertainties. At any given time a firm may find itself potentially liable for events that have occurred in the past. Accounting does not recognize contingencies of this nature as liabilities. They are potential future obligations rather than current obligations. The potential obligations arise from events that occurred in the past but whose outcome remains uncertain. Whether an item becomes a liability—and, if it does, the size of the liability—depends on a future event, such as the outcome of a lawsuit.

Suppose that an employee sues Mattel, the toy manufacturer, in a formal court proceeding for damages from an accident. The court has scheduled a trial sometime after the end of this accounting period. If Mattel's lawyers and auditors agree that the court will likely rule in favor of Mattel (or that any adverse settlement will be small), then Mattel will not recognize a liability on its balance sheet. The notes to the financial statements, however, must disclose significant contingencies.

Both the FASB and the International Accounting Standards Board (IASB) have said that a firm should recognize an estimated loss from a contingency in the accounts only if the contingency meets both of the following conditions: (a) information available before issuance of the financial statements indicates that it is probable that an event has impaired an asset or has caused the firm to incur a liability, and (b) the firm can reasonably estimate the amount of the loss.[3]

When information indicates that events have probably impaired an asset or led to a liability and when the firm has estimated the amount but the amount is a range, then the firm should use the lower end of the range. For example, if Mattel estimates the loss in a range from $1 million to $5 million, it must record the liability at $1 million.

## *Terminology* Note

In stating the conditions, quoted above, for the firm to record a liability, the FASB has said that the loss must be *probable*, but the FASB has not defined what it means by *probable*. In recent years, the FASB has discussed whether probable means "more likely than not"—any probability above 50 percent—or something larger, such as 80 to 85 percent. Currently, most accountants and auditors appear to use *probable* to mean 80 to 85 percent or larger. The FASB has used the phrase *more likely than not* when it means a probability greater than 50 percent. In contrast, the IASB has defined *probable* to mean "more likely than not."[4] The difference in meanings of the word *probable* can matter in applying accounting principles, and we expect the two standard-setting bodies to reconcile this difference within the next few years.

**Example 10** A toy manufacturer, such as Mattel, has sold products for which the Consumer Protection Agency has discovered a safety hazard. The toy manufacturer thinks it probable that it has incurred a liability. This potential obligation meets condition **b** if the firm's experience or other information enables it to make a reasonable estimate of the loss. The journal entry would be as follows:

| | | |
|---|---|---|
| Loss from Damage Claim | 1,000,000 | |
| Estimated Liability for Damages | | 1,000,000 |

| Assets | = | Liabilities | + | Shareholders' Equity | (Class.) |
|---|---|---|---|---|---|
| | | +1,000,000 | | −1,000,000 | IncSt → RE |

To recognize estimated liability for expected damage arising from safety hazard on toys sold.

[3]Financial Accounting Standards Board, *Statement of Financial Accounting Standards No. 5,* "Accounting for Contingencies," 1975; International Accounting Standards Board, *Statement No. 37,* "Provisions, Contingent Liabilities, and Contingent Assets," 1998.

[4]International Accounting Standards Board, *Statement No. 37.*

This entry debits a loss account (presented among Other Expenses on the income statement) and credits an **estimated liability**, which should appear on the balance sheet as a current liability, similar to the Estimated Warranty Liability account. In practice, an account with the title Estimated Liability for Damages would seldom, if ever, appear in published financial statements because firms fear that a court or jury might perceive it as an admission of guilt. Such an admission would likely affect the outcome of the lawsuit. Most firms would combine the liability account with others for financial statement presentation and GAAP approves this aggregated disclosure unless the amount is materially large.

**Contingency Disclosure** Accounting uses the term **contingency**, or sometimes *contingent liability,* only when it recognizes the item not in the accounts but rather in the notes. A recent annual report of Kodak illustrates the disclosure of contingencies. Its note on contingencies starts with 12 paragraphs over 3 pages giving moderate detail about environmental remediation issues, which end as follows:

> Estimates of the amount and timing of future costs of environmental remediation requirements are necessarily imprecise because of the continuing evolution of environmental laws and regulatory requirements, the availability and application of technology, the identification of presently unknown remediation sites and the allocation of costs among the potentially responsible parties. Based upon information presently available, such future costs are not expected to have a material effect on the Company's competitive or financial position. However, such costs could be material to results of operations in a particular future quarter or year.

Then, Kodak's note on contingencies presents 20 paragraphs covering another 3 pages on other commitments and contingencies, ending with:

> The Company and its subsidiary companies are involved in lawsuits, claims, investigations and proceedings, including product liability, commercial, environmental, and health and safety matters, which are being handled and defended in the ordinary course of business. There are no such matters pending that the Company and its General Counsel expect to be material in relation to the Company's business, financial position or results of operations.

## CONSTRUCTIVE LIABILITIES

As this book goes to press, the FASB is struggling with the concept of *constructive liability*, and whether firms should record them. You can understand the definition of constructive liability here, but you need to master some of the material in **Chapter 12** on recurring and nonrecurring income before you can understand why constructive liabilities have become an issue. Firms have discovered that by recording constructive liabilities, they give themselves more wiggle room, more flexibility, in reporting income and, more important, in reporting patterns of income over time.

A **constructive liability** arises, not from an obligation, but from management intent. Assume management has decided it will close a plant and knows that it will reduce the labor force, laying off current employees and making severance payments in amounts yet to be determined. It can recognize an expense, often euphemistically called *Restructuring Charges*, with a journal entry such as:

| | | |
|---|---|---|
| Restructuring Charges .................................... | 1,000 | |
| Liability for Severance Pay to Employees .................. | | 1,000 |

| Assets | = | Liabilities | + | Shareholders' Equity | (Class.) |
|---|---|---|---|---|---|
| | | +1,000 | | −1,000 | IncSt → RE |

Now the firm has established a liability. Later, when it discharges workers and makes a $200 cash payment for severance benefits, it can record the following entry:

| | | |
|---|---|---|
| Liability for Severance Pay to Employees | 200 | |
| Cash | | 200 |

| Assets | = | Liabilities | + | Shareholders' Equity | (Class.) |
|---|---|---|---|---|---|
| −200 | | −200 | | | |

No controversy arises from the above, conservative accounting. Issues arise from different circumstances. Assume management finds that the initial $1,000 charge for estimated severance payments was $300 too large, that the charge should have been only $700. At the later time when it finds the initial charge was too large, it will make the following entry to reverse part of the initial charge:

| | | |
|---|---|---|
| Liability for Severance Pay to Employees | 300 | |
| Reversal of Restructuring Charges | | 300 |

| Assets | = | Liabilities | + | Shareholders' Equity | (Class.) |
|---|---|---|---|---|---|
| | | −300 | | +300 | IncSt → RE |

Reversal of prior charge; reversal appears in income for the current year.

Management reduced income in the year of initial charge by $1,000 and then increased income in the reversal year by $300. The FASB suspects that firms are using this ability to record initial intent with later reversals as a way to manage income and requires firms to present the increase in income in the current year in the same place in the income statement as the decrease in income had appeared in the earlier year.[5] Analysts say that the mere ability of firms to reduce income and then later increase income based on its own intent, without obligation, reduces the quality of earnings.

## Liability Valuation

In historical cost accounting, liabilities appear on the balance sheet at the present value of payments that a firm expects to make in the future. The interest rate a firm uses in computing the present value amount throughout the life of a liability is the interest rate appropriate for the specific borrower at the time it initially incurred the liability. That is, the firm uses the historical interest rate.

As **Chapter 2** mentions, most current liabilities appear at the undiscounted amount payable because of the immaterially small difference between the amount ultimately payable and its present value.

## Liability Classification

The balance sheet generally classifies liabilities as current or noncurrent. Firms separate current from noncurrent liabilities based on the length of time that will elapse before the borrower must make payment. Current liabilities fall due within the operating cycle, usually one year, and noncurrent liabilities fall due later.

# Current Liabilities

Current liabilities, those due within the current operating cycle, normally one year, include accounts payable to creditors, short-term notes payable, payroll accruals, taxes payable, and current portion of long-term debt. A firm continually discharges current liabilities and replaces them with new ones in the course of business operations.

[5]See Paragraph 19 of *Statement of Financial Accounting Standards No. 146,* "Accounting for Costs Associated with Exit or Disposal Activities," 2002.

The firm will not pay these obligations for several weeks or months after the current balance sheet date. They have present value less than the amount to be paid. Even so, accountants show these items at the full amount owed because, considering the small difference between the amount owed and its present value, a separate accounting for this difference as interest expense would not be cost-effective.

## Accounts Payable to Creditors

A firm seldom pays for goods and services as received. It defers payment until it receives a bill from the supplier. Then it may not pay the bill immediately but instead accumulate it with other bills until a specified time of the month when it pays all bills. Because these accounts do not accrue explicit interest, management tries to obtain as much funding as possible from its creditors by delaying payment as long as possible. Failure to keep scheduled promises to creditors can lead to poor credit ratings and to restrictions on future credit. The Marmon Group, a Chicago-based conglomerate, has instructed its purchasing agents to negotiate for as long a grace period as they can before payments become due, but instructs the controllers always to pay bills by the due date.

## Short-Term Notes and Interest Payable

Firms obtain financing for less than a year from banks or other creditors in return for a short-term note called a *note payable.* The borrower records a liability. As time passes, the borrower makes entries, usually adjusting entries, debiting Interest Expense and crediting Interest Payable. When the borrower makes payments, the entry credits Cash and debits Interest Payable and, perhaps, Notes Payable. Payment amounts go first to reduce the liability for Interest Payable and then, if the amount exceeds the balance in the Interest Payable account, to reduce the principal for Notes Payable.

## Wages, Salaries, and Other Payroll Items

When employees earn wages, they owe part of their earnings to governments for income and other taxes. They may also owe other amounts for union dues and insurance plans. Although these amounts form part of the employer's wage expense, the employer does not pay them directly to employees but pays them, on employees' behalf, to the governments, unions, and insurance companies.

In addition, the employer must pay various payroll taxes and may have agreed to pay for other fringe benefits because employees earned wages.

Employers often provide paid vacations to employees who have worked for a specified period, such as six months or one year. The employer must accrue the costs of the earned but unused vacations (including the payroll taxes and fringe benefits on them) at the time the employees earn them, not at the later time when employees take vacations and receive their wages. This treatment results in charging each quarter of the year with a portion of the cost of vacations rather than allocating all costs to the summer, when most employees take the majority of their vacation days.

**Example 11** Assume that the Sacramento Radio Shack's employees earn $100,000 and that the employees owe taxes averaging 40 percent of these amounts for federal income taxes, state income taxes, Social Security taxes, and Medicare taxes. In addition, employees in aggregate owe $500 for union dues to be withheld by the employer and $3,000 for health insurance plans.

Radio Shack must pay various payroll taxes averaging 18 percent of gross wages. In addition, the employer owes $4,500 to Fireman's Fund for payments to provide life and health insurance coverage. Employees have earned vacation pay estimated to be $4,000; Radio Shack estimates employer payroll taxes and fringe benefits to be 18 percent of the gross amount.

The journal entries that follow record these wages. If production workers had earned some of the wages, the firm would debit some of the amounts to Work-in-Process Inventory rather than to Wages and Salaries Expense.

| Account | Debit | Credit |
|---|---|---|
| Wages and Salaries Expense | 100,000 | |
| Taxes Payable to Various Governments | | 40,000 |
| Withheld Dues Payable to Union | | 500 |
| Insurance Premiums Payable | | 3,000 |
| Wages and Salaries Payable | | 56,500 |

*(continued)*

| Assets | = | Liabilities | + | Shareholders' Equity | (Class.) |
|---|---|---|---|---|---|
| | | +40,000 | | −100,000 | IncSt → RE |
| | | +500 | | | |
| | | +3,000 | | | |
| | | +56,500 | | | |

To record wage expense; plug for $56,500 actually payable to employees.

| | | |
|---|---|---|
| Wages and Salaries Expense . . . . . . . . . . . . . . . . . . . . . . . . . . . . . . . | 22,500 | |
| Payroll Taxes Payable . . . . . . . . . . . . . . . . . . . . . . . . . . . . . . | | 18,000 |
| Insurance Premiums Payable . . . . . . . . . . . . . . . . . . . . . . . . . | | 4,500 |

| Assets | = | Liabilities | + | Shareholders' Equity | (Class.) |
|---|---|---|---|---|---|
| | | +18,000 | | −22,500 | IncSt → RE |
| | | +4,500 | | | |

To record employer's expense for amounts not payable directly to employees. The debit is a plug.

| | | |
|---|---|---|
| Wages and Salaries Expense . . . . . . . . . . . . . . . . . . . . . . . . . . . . . . . | 4,720 | |
| Estimated Vacation Wage and Fringes Payable . . . . . . . . . . . . . . | | 4,720 |

| Assets | = | Liabilities | + | Shareholders' Equity | (Class.) |
|---|---|---|---|---|---|
| | | +4,720 | | −4,720 | IncSt → RE |

To record estimate of vacation pay and fringes thereon earned during the current period: 0.18 × $4,000 = $720.

## Income Taxes Payable

Businesses organized as corporations must pay federal income taxes based on their taxable income from business activities. In contrast, businesses organized as partnerships or sole proprietorships do not pay income taxes. Instead the Internal Revenue Service taxes the income of the business entity as income of the individual partners or the sole proprietor. Each partner or sole proprietor adds his or her share of business income to income from all other (non-business) sources in preparing an individual income tax return. **Chapter 10** discusses more fully the accounting for income taxes.

## Deferred Performance Liabilities: Advances from Customers

Another current liability arises when a firm receives cash from customers for goods or services it will deliver in the future. It discharges this liability, unlike the preceding ones, by delivering goods or services rather than by paying cash. When the firm discharges the liability it will recognize revenue.

An example of this type of liability is the Houston Rockets' advance sale of basketball game tickets for ten games for, say, $2,000,000. At the time of the sale, the Rockets record the following entry:

| | | |
|---|---|---|
| Cash . . . . . . . . . . . . . . . . . . . . . . . . . . . . . . . . . . . . . . . . . . . . . . . | 2,000,000 | |
| Advances from Customers . . . . . . . . . . . . . . . . . . . . . . . . . . . . | | 2,000,000 |

| Assets | = | Liabilities | + | Shareholders' Equity | (Class.) |
|---|---|---|---|---|---|
| +2,000,000 | | +2,000,000 | | | |

The Rockets sell tickets for cash but earn no revenue until the organization renders the service when the team plays. These deferred performance obligations qualify as liabilities. The firm credits an account such as Advances from Customers or Liability for Advance Sales. After a game, the firm has discharged its obligation so it recognizes (credits) revenue and reduces (debits) the liability.

| | | |
|---|---|---|
| Advances from Customers . . . . . . . . . . | 200,000 | |
| Game Ticket Sales Revenue . . . . . . . . . . | | 200,000 |

| Assets | = | Liabilities | + | Shareholders' Equity | (Class.) |
|---|---|---|---|---|---|
| | | −200,000 | | +200,000 | IncSt → RE |

After the first game and the entry above, the Advances from Customers (Liability) account has a balance of $1,800,000.

Deferred performance liabilities arise also in connection with the sale of magazine subscriptions, airline tickets, and service contracts.

## Deferred Performance Liabilities: Product Warranties

A similar deferred performance liability arises when a firm provides a warranty for service or repairs for some period after a sale. At the time of the sale, the firm can only estimate the likely amount of warranty liability.

**Example 12** J&R Music World makes sales of $280,000 during the accounting period and estimates that it will eventually use an amount equal to 4 percent of the sales revenue to satisfy warranty claims. The expense for warranties is $11,200 (= 0.04 × $280,000). It will make the following entry:

| | | |
|---|---|---|
| Accounts Receivable . . . . . . . . . . | 280,000 | |
| Warranty Expense . . . . . . . . . . | 11,200 | |
| Sales . . . . . . . . . . | | 280,000 |
| Estimated Warranty Liability . . . . . . . . . . | | 11,200 |

| Assets | = | Liabilities | + | Shareholders' Equity | (Class.) |
|---|---|---|---|---|---|
| +280,000 | | +11,200 | | −11,200 | IncSt → RE |
| | | | | +280,000 | IncSt → RE |

To record sales and estimated liability for warranties on items sold.

Note that this entry recognizes the warranty expense in the period when the firm recognizes revenue, even though it will make the repairs in a later period. The accounting matches warranty expense with associated revenue. Because it recognizes the expense, it creates the liability. In this case, the firm knows neither the amount nor the due date of the liability for sure, but it can estimate both with reasonable precision. FASB *Statement of Financial Accounting Standards No. 5*, "Accounting for Contingencies," requires the accrual of the expense and the related warranty liability when the firm can "reasonably estimate" the amount.

As J&R makes expenditures of $1,750 in the next period for repairs under the warranty, the entry is as follows:

| | | |
|---|---|---|
| Estimated Warranty Liability . . . . . . . . . . | 1,750 | |
| Cash [or other assets consumed for repairs] . . . . . . . . . . | | 1,750 |

| Assets | = | Liabilities | + | Shareholders' Equity | (Class.) |
|---|---|---|---|---|---|
| −1,750 | | −1,750 | | | |

To record repairs made. Recognize no expense now; the firm recognized all expense in the period of sale.

Note that *actual* expenditures to satisfy warranty claims do not affect net income. The net income effect occurs in the year of sale when the firm estimates the *expected* amount of expenditures that it will make in all future periods to satisfy warranty claims arising from products sold in a particular year. With experience, the firm will adjust the percentage of sales that it charges to Warranty Expense. It attempts to maintain the Estimated Warranty Liability account at each balance sheet date with a credit balance that reasonably estimates the actual cost of repairs to be made under warranties outstanding at that time. The accounting for warranties resembles the allowance method for uncollectibles discussed in **Chapter 6**.

A firm may not use the allowance method of accounting for warranties (or for bad debts) in tax reporting. In tax reporting, taxpayers recognize expense in the period when they use cash or other assets to discharge the obligation. This creates a temporary difference between financial statement pretax income and taxable income, discussed in the next chapter.

### Problem 9.1 for Self-Study

**Journal entries for transactions involving current liabilities.** Prepare journal entries for each of the following transactions of Ashton Corporation during January, the first month of operations. The firm closes its books monthly.

**a.** January 2: The firm borrows $10,000 on a 9-percent, 90-day note from First National Bank.
**b.** January 3: The firm acquires merchandise costing $8,000 from suppliers on account.
**c.** January 10: The firm receives $1,500 from a customer as a deposit on merchandise that Ashton Corporation expects to deliver in February.
**d.** Month of January: The firm sells merchandise costing $6,000 to customers on account for $12,000.
**e.** Month of January: The firm pays suppliers $8,000 of the amount owed for purchases of merchandise on account and collects $7,000 of amounts owed by customers.
**f.** January 31: Products sold during January include a two-year warranty. Ashton Corporation estimates that warranty claims will equal 8 percent of sales. No customer made warranty claims during January.
**g.** January 31: Employees earned wages of $4,000 for the month of January. The firm must withhold from the amounts it pays to employees, as follows: 20 percent for income taxes, 10 percent for Social Security taxes, and $200 for union dues. In addition, the employer must pay Social Security taxes of 10 percent and unemployment taxes of 3.5 percent. These wages and taxes remain unpaid at the end of January.
**h.** January 31: The firm accrues interest expense on the bank loan (see transaction **a**).
**i.** January 31: The firm accrues income tax expense, but makes no payment, at a rate of 40 percent on net income during January.
**j.** February 1: Ashton Corporation pays employees their January wages net of withholdings.
**k.** February 10: The firm delivers merchandise costing $800 to the customer in transaction **c** in full satisfaction of its order.
**l.** February 15: The firm remits payroll taxes and union dues to government and union authorities.
**m.** February 20: A customer who purchased merchandise during January returns goods for warranty repairs. The repairs cost the firm $220, which it paid in cash.

## Long-Term Liabilities

The principal long-term liabilities are mortgages, notes, bonds, and leases. Borrowers usually pay interest on long-term liabilities at regular intervals during the life of a long-term obligation, whereas they pay interest on short-term debt in a lump sum at maturity. Borrowers often repay the principal of long-term obligations in installments or accumulate special funds for retiring long-term liabilities.

Accounting for all long-term liabilities generally rests on a single, unified set of concepts. We outline these concepts, and the related procedures, next and illustrate them throughout the rest of this chapter and **Chapter 10**.

## Concepts and Methods for Recording Long-Term Liabilities

In return for promising to make future payments, the borrower receives cash (or other assets with a measurable cash-equivalent value). The borrower initially records a long-term liability for that amount. By definition, the book value of the borrowing at any time equals the present value of all the then-remaining promised payments using the historical market interest rate applicable at the time the firm originally incurred the liability.

Sometimes a borrower knows the amount of cash received from a loan, the market interest rate at that time, and the various future times when it must pay. In this case, it uses the interest rate at the time of the loan and the elapsed time between payments to compute the amounts it must pay at the various future dates.

More often, however, a borrower knows the amount of cash received as well as the amounts and due dates of cash repayments, but the loan does not explicitly state the market interest rate or, perhaps, states it incorrectly. Finding the market interest rate implied by the receipt of a given amount of cash now in return for a series of promised future repayments requires a process called "finding the internal rate of return." The *internal rate of return* is the interest rate that discounts a series of future cash flows to its present value. Some refer to this rate as the *implicit rate of return*. The **Appendix** at the end of the book illustrates the process of finding the internal rate of return.

The borrower uses the original, or historical, market interest rate, either specified or computed at the time the borrower receives the loan, throughout the life of the loan to compute interest expense. When the borrower makes a cash payment, a portion (perhaps all) of the payment represents interest. Any excess of cash payment over interest expense reduces the borrower's liability for the principal amount. If a given payment is less than the interest expense accrued since the last payment date, then the liability principal increases by the excess of interest expense over cash payment.

Borrowers can retire long-term liabilities in several ways, but the accounting process is the same for all. The borrower debits the liability account for its current book value, credits cash (or other net assets given to satisfy the debt), and recognizes any difference as a gain (when book value exceeds cash disbursement, a credit) or a loss (when cash disbursement exceeds book value, a debit) on retirement of debt.

## Mortgages and Notes

In a **mortgage** contract, the lender (such as a bank) takes legal title to certain property of the borrower, with the provision that the title reverts to the borrower when the borrower repays the loan in full. (In a few states, the lender merely acquires a lien on the borrower's property rather than legal title to it.) The mortgaged property is **collateral** for the loan. The customary terminology designates the lender as the **mortgagee** and the borrower as the **mortgagor**.

As long as the mortgagor (borrower) makes the payments required by the mortgage agreement, the mortgagee (lender) does not have the ordinary rights of an owner to possess and use the property. If the mortgagor defaults by failing to make the periodic payments, the mortgagee can usually arrange to sell the property for its benefit through a process called *foreclosure*. The mortgagee has first rights to the proceeds from the foreclosure sale for satisfying

### *Terminology* Note

Accountants generally use the word *collateral*, rather than *security*, in this context. Whether the loan is secured requires legal judgment; moreover, accountants do not want to imply that the value of the collateral will be sufficient to satisfy the debt, as the word *secured* might imply. When you borrow money to finance a home purchase, you give the bank a mortgage, a form of IOU, and the bank accepts the mortgage and lends you cash. Common usage often says "the borrower gets a mortgage from the bank" whereas the correct statement is "the borrower gets a mortgage loan from the bank" because the borrower gives the mortgage to the bank, which provides the funds in return for the promise to pay built into the mortgage.

any unpaid claim. If funds remain, the mortgagee pays them to the mortgagor. If the proceeds are too small to pay the remaining loan, the lender becomes an unsecured creditor of the borrower for the unpaid balance.

A note payable resembles a mortgage payable except that the borrower does not pledge property as collateral.

## ILLUSTRATION OF ACCOUNTING FOR A MORTGAGE

Some of the more common problems in accounting for mortgages appear in the following illustration. On October 1, Year 1, Western Company borrows $125,000 for five years from the Home Savings and Finance Company to obtain funds for additional working capital. As collateral, Western Company pledges to Home Savings and Finance Company several parcels of land that it owns and that are on its books at a cost of $50,000. The mortgage specifies that interest will accrue on the unpaid balance of the amounts owed at a rate of 12 percent per year compounded semiannually (that is, 6 percent interest will accrue every six months). The borrower must make payments on April 1 and October 1 of each year. Western agrees to make 10 payments over the five years of the mortgage so that when it makes the last payment on October 1, Year 6, it will have paid the loan and all interest. The contract calls for the first nine payments to be $17,000 each. The tenth payment will discharge the balance of the loan. Western Company's accounting periods end on December 31 each year. (The derivation of the semiannual payment of $17,000 appears in **Example 11** in the **Appendix** at the end of the book.)

The entries from the time Western issues the mortgage through December 31, Year 2, follow:

10/1/Year 1

| | Dr. | Cr. |
|---|---|---|
| Cash | 125,000 | |
| Mortgage Payable | | 125,000 |

| Assets | = | Liabilities | + | Shareholders' Equity | (Class.) |
|---|---|---|---|---|---|
| +125,000 | | +125,000 | | | |

To record loan obtained from Home Savings and Finance Company for five years at 12 percent compounded semiannually.

As **Example 11** in the **Appendix** shows, $125,000 approximately equals the present value of 10 semiannual cash payments of $17,000, each discounted at 12 percent compounded semiannually.

12/31/Year 1

| | Dr. | Cr. |
|---|---|---|
| Interest Expense | 3,750 | |
| Interest Payable | | 3,750 |

| Assets | = | Liabilities | + | Shareholders' Equity | (Class.) |
|---|---|---|---|---|---|
| | | +3,750 | | −3,750 | IncSt → RE |

To record adjusting entry: interest expense on mortgage from 10/1/Year 1 to 12/31/Year 1.

Interest expense on the loan for the first six months is $7,500 (= 0.12 × $125,000 × 6/12). To simplify the calculations, accounting typically assumes that the interest accrues evenly over the six-month period. Thus interest expense for the three-month period (October, November, and December) equals one-half of the $7,500, or $3,750.

4/1/Year 2

| | Dr. | Cr. |
|---|---|---|
| Interest Expense | 3,750 | |
| Interest Payable | 3,750 | |
| Mortgage Payable | 9,500 | |
| Cash | | 17,000 |

(continued)

| Assets | = | Liabilities | + | Shareholders' Equity | (Class.) |
|---|---|---|---|---|---|
| -17,000 | | -3,750 | | -3,750 | IncSt → RE |
| | | -9,500 | | | |

To record cash payment made: interest expense on mortgage from 1/1/Year 2 to 4/1/Year 2, payment of six months' interest, and reduction of loan by the difference: $9,500 = $17,000 − $7,500.

After the cash payment on April 1, Year 2, the unpaid amount of the loan is $115,500 (= $125,000 − $9,500). Interest expense during the second six-month period accrues on this unpaid amount.

10/1/Year 2

| | | |
|---|---|---|
| Interest Expense | 6,930 | |
| Mortgage Payable | 10,070 | |
| Cash | | 17,000 |

| Assets | = | Liabilities | + | Shareholders' Equity | (Class.) |
|---|---|---|---|---|---|
| -17,000 | | -10,070 | | -6,930 | IncSt → RE |

To record cash payment made: interest expense for the period 4/1/Year 2 to 10/1/Year 2 is $6,930 [= 0.12 × ($125,000 − $9,500) × 6/12]; the loan balance declines by the difference: $10,070 = $17,000 − $6,930.

12/31/Year 2

| | | |
|---|---|---|
| Interest Expense | 3,163 | |
| Interest Payable | | 3,163 |

| Assets | = | Liabilities | + | Shareholders' Equity | (Class.) |
|---|---|---|---|---|---|
| | | +3,163 | | -3,163 | IncSt → RE |

To record adjusting entry: interest expense from 10/1/Year 2 to 12/31/Year 2 = [0.12 × ($125,000 − $9,500 − $10,070) × 3/12].

**Amortization Schedule** **Exhibit 9.2** presents an amortization schedule for this mortgage. For each period it shows the balance at the beginning of the period, the interest expense for the period, the cash payment for the period, the reduction in principal for the period, and the balance at the end of the period. (The last payment, $16,781 in this case, often differs slightly from the others because of the cumulative effect of rounding payments to the nearest dollar or hundred dollars.) All long-term liabilities have analogous amortization schedules, which aid in understanding the timing of payments to discharge the liability. Amortization schedules for various long-term liabilities appear throughout this chapter and the next.

## Problem 9.2 for Self-Study

**Accounting for an interest-bearing note.** Sapra Company receives cash of $112,434 in return for a three-year, $100,000 note, promising to pay $15,000 at the end of one year, $15,000 at the end of two years, and $115,000 at the end of three years. The market interest rate on the original issue date of the note is 10 percent.

**a.** Prepare an amortization schedule similar to **Exhibit 9.2** for the life of the note.

**b.** Prepare journal entries that Sapra Company would make on three dates: the date of issue, six months after the date of issue (assuming Sapra closes the books then), and one year after the date of issue (assuming Sapra makes a cash payment then).

**EXHIBIT 9.2** **Amortization Schedule for $125,000 Mortgage (or Note), Repaid in 10 Semiannual Installments of $17,000, Interest Rate of 12 Percent, Compounded Semiannually (6 percent compounded each six months)**

**Semiannual Journal Entry**

| | | |
|---|---|---|
| Interest Expense . . . . . . . . . . . . . . . . . . . . . . . . . . . . . . . . . . . . | Amount in Column **(3)** | |
| Mortgage (or Note) Payable . . . . . . . . . . . . . . . . . . . . . . . . . . . . . . . . | Amount in Column **(5)** | |
| Cash . . . . . . . . . . . . . . . . . . . . . . . . . . . . . . . . . . . . . . . . . | | Amount in Column **(4)** |

| Assets | = | Liabilities | + | Shareholders' Equity | (Class.) |
|---|---|---|---|---|---|
| −col(4) | | −col(5) | | −col(3) | IncSt → RE |

| 6-Month Period (1) | Loan Balance Start of Period (2) | Interest Expense for Period (3) | Payment (4) | Portion of Payment Reducing Principal (5) | Loan Balance End of Period (6) |
|---|---|---|---|---|---|
| 0 . . . . . . . . . . . . . . . . . . . . . . | | | | | $125,000 |
| 1 . . . . . . . . . . . . . . . . . . . . . . | $125,000 | $7,500 | $17,000 | $ 9,500 | 115,500 |
| 2 . . . . . . . . . . . . . . . . . . . . . . | 115,500 | 6,930 | 17,000 | 10,070 | 105,430 |
| 3 . . . . . . . . . . . . . . . . . . . . . . | 105,430 | 6,326 | 17,000 | 10,674 | 94,756 |
| 4 . . . . . . . . . . . . . . . . . . . . . . | 94,756 | 5,685 | 17,000 | 11,315 | 83,441 |
| 5 . . . . . . . . . . . . . . . . . . . . . . | 83,441 | 5,006 | 17,000 | 11,994 | 71,447 |
| 6 . . . . . . . . . . . . . . . . . . . . . . | 71,447 | 4,287 | 17,000 | 12,713 | 58,734 |
| 7 . . . . . . . . . . . . . . . . . . . . . . | 58,734 | 3,524 | 17,000 | 13,476 | 45,258 |
| 8 . . . . . . . . . . . . . . . . . . . . . . | 45,258 | 2,715 | 17,000 | 14,285 | 30,973 |
| 9 . . . . . . . . . . . . . . . . . . . . . . | 30,973 | 1,858 | 17,000 | 15,142 | 15,831 |
| 10 . . . . . . . . . . . . . . . . . . . . . . | 15,831 | 950 | 16,781 | 15,831 | 0 |

Note: In preparing this table, we rounded calculations to the nearest dollar.
Column (2) = column (6) from previous period.
Column (3) = 0.06 × column (2).
Column (4) is given, except row 10, where it is the amount such that column (4) = column (2) + column (3).
Column (5) = column (4) − column (3).
Column (6) = column (2) − column (5).

## CONTRACTS AND LONG-TERM NOTES: INTEREST IMPUTATION

Firms often finance the acquisition of buildings, equipment, and other fixed assets using interest-bearing notes, which the previous section discussed. Some borrowing arrangements do not state an explicit interest rate. Instead, the purchase price includes carrying charges, and the buyer pays the total over a specified number of months. The contract does not mention the charge for interest. The principal on the note in this case includes **implicit interest**.

Generally accepted accounting principles (GAAP) require that all long-term monetary liabilities, including those carrying no explicit interest, appear on the balance sheet at the present value of the future cash payments. To compute a present value requires an interest rate, sometimes in this context called a *discount rate*. In historical cost accounting, the discounting process uses the historical (or original) interest rate appropriate for the particular borrower at the time it incurred the obligation. That rate will have depended on the amount and terms of the borrowing arrangement as well as on the risk that the borrower will default on its obligations. The rate is known as the **implicit (imputed) interest rate**. The excess of the par, or face, value of the liability over its present value represents interest to be recognized over the period of the loan. The next two sections discuss two acceptable ways to compute the present value of the liability and the amount of imputed interest.

The first approach uses the market value of the asset acquired as a basis for computing the present value of the liability.

**Example 13** Hixon Saab can purchase a piece of equipment for $105,000 cash. The firm purchases the equipment in return for a single-payment note with a face amount of $160,000 payable in three years. The implied interest rate is about 15 percent per year; that is, $105,000 grows at 15 percent in one year to $120,750 (= $105,000 × 1.15), in two years to about $139,000 (= $120,750 × 1.15), and in three years to about $160,000 (= $139,000 × 1.15). The journal entry using this approach would be as follows:

| | | |
|---|---|---|
| Equipment | 105,000 | |
| Note Payable | | 105,000 |

| Assets | = | Liabilities | + | Shareholders' Equity | (Class.) |
|---|---|---|---|---|---|
| +105,000 | | +105,000 | | | |

To record purchase of equipment using the known cash price. The firm infers the interest rate for the note from the known cash price of the equipment and the face amount of the note.

At the end of each accounting period that intervenes between the acquisition of the equipment and repayment of the note, Hixon Saab makes journal entries for depreciation of the equipment. We do not show these here. The firm must also make journal entries to recognize interest expense. Assume that Hixon Saab issued the note at the beginning of a year. The entries for the three years would be as follows:

| | | |
|---|---|---|
| **(1)** Interest Expense | 15,750 | |
| Note Payable | | 15,750 |

| Assets | = | Liabilities | + | Shareholders' Equity | (Class.) |
|---|---|---|---|---|---|
| | | +15,750 | | −15,750 | IncSt → RE |

Entry made one year after issuance of note.

Interest is 0.15 × $105,000. The firm does not pay $15,750 in cash but adds it to the principal amount of the liability.

| | | |
|---|---|---|
| **(2)** Interest Expense | 18,113 | |
| Note Payable | | 18,113 |

| Assets | = | Liabilities | + | Shareholders' Equity | (Class.) |
|---|---|---|---|---|---|
| | | +18,113 | | −18,113 | IncSt → RE |

Entry made one year after entry (1), two years after issuance of note.

Interest is 0.15 × ($105,000 + $15,750).

| | | |
|---|---|---|
| **(3)** Interest Expense | 21,137 | |
| Note Payable | | 21,137 |

| Assets | = | Liabilities | + | Shareholders' Equity | (Class.) |
|---|---|---|---|---|---|
| | | −21,137 | | −21,137 | IncSt → RE |

Entry made one year after entry (2), at maturity of note to increase liability to its maturity amount, $160,000.

Interest = \$160,000 − (\$105,000 + \$15,750 + \$18,113), which is approximately equal to 0.15 × (\$105,000 + \$15,750 + \$18,113); the difference represents the accumulated rounding error caused by using 15 percent as the implicit interest rate rather than the exact rate, which is 15.074 percent.

| | Dr. | Cr. |
|---|---|---|
| **(4)** Note Payable | 160,000 | |
| Cash | | 160,000 |

| Assets | = | Liabilities | + | Shareholders' Equity | (Class.) |
|---|---|---|---|---|---|
| −160,000 | | −160,000 | | | |

To repay note at maturity.

Of the \$160,000 paid at maturity, \$55,000 represents interest accumulated on the note since its issue.

**Use of Market Interest Rate to Establish Market Value of Asset and Present Value of Note** If the firm purchased used equipment with the same three-year note, it might be unable to establish a reliable estimate of the current market value of the asset acquired. To find the present value of the note, the firm would discount the payments using the interest rate it would have to pay for a similar loan in the open market at the time that it acquired the equipment. The firm would continue to use this rate throughout the life of the note. GAAP allows this second method for quantifying the amount of the liability and computing the imputed interest.

**Example 14** Refer to the data in **Example 13**. When Hixon Saab borrows using single-payment notes with maturity of three years, it pays interest of 12 percent compounded annually, rather than 15 percent. The present value at 12 percent per year of the \$160,000 note due in three years is \$113,885 (= \$160,000 × 0.71178; see **Appendix Table 2** at the back of the book, 3-period row, 12-percent column). The entry to record the purchase of used equipment paid for with the note would be as follows:

| | Dr. | Cr. |
|---|---|---|
| Equipment | 113,885 | |
| Note Payable | | 113,885 |

| Assets | = | Liabilities | + | Shareholders' Equity | (Class.) |
|---|---|---|---|---|---|
| +113,885 | | +113,885 | | | |

To record purchase of equipment. The firm infers the cost of equipment from known interest rate.

The firm would make entries at the end of each period to recognize interest expense (using the 12 percent borrowing rate) and to increase the principal amount of the liability. After the third period, the principal amount of the liability would be \$160,000.

**Total Expense Is Independent of Interest Rate** In **Example 13**, the buyer records the equipment of \$105,000 and recognizes \$55,000 of imputed interest. In **Example 14**, the buyer records the equipment at \$113,885 and recognizes \$46,115 of imputed interest. The expense over the combined lives of the note and the equipment—interest plus depreciation—will total the same amount, \$160,000, no matter which interest rate the buyer uses. Over sufficiently long time periods, total expense equals total cash expenditure other than transactions with owners; accrual accounting affects only the timing of the expense recognition.

**Long-Term Notes Held as Receivables** A borrower's long-term note payable is the lender's long-term asset, a note receivable. GAAP requires the lender to show the asset in the Long-Term Note Receivable account at its present value. The rate at which the lender discounts the note should, in theory, equal the rate the borrower uses. In practice, the two rates sometimes differ because the lender and the borrower reach different conclusions regarding the borrower's default risk, sometimes called *credit risk*. The lender's accounting mirrors the borrower's: the lender has interest revenue, whereas the borrower has interest expense.

## Problem 9.3 for Self-Study

**Accounting for note with imputed interest rate.** Bens Company purchased a fork-lift truck from Roulstone's Auto Agency. The truck had a list price of $25,000, but sellers often offer discounts of 8 to 20 percent from list prices for purchases of this sort. Bens Company paid for the truck by giving a noninterest-bearing note due two years from the date of purchase. The note had a maturity value of $28,730. The rate of interest that Bens Company paid to borrow on secured two-year loans ranged from 10 percent to 15 percent during the period when the purchase occurred.

**a.** Record the acquisition of the truck on Bens Company's books, computing the fair market value of the truck using a 10 percent discount from list price.
**b.** What imputed interest rate will Bens use for computing interest expense throughout the loan if it records the acquisition as in part **a**?
**c.** Record the acquisition of the truck on Bens Company's books, assuming that the estimated interest rate Bens Company must pay to borrow is reliable and is 12 percent per year.
**d.** Record the acquisition of the truck on Bens Company's books, assuming that the interest rate Bens Company must pay to borrow is 1 percent per month.
**e.** Prepare journal entries to record the loan and to record interest over two years, assuming that Bens records the truck at $23,744 and that the interest rate implicit in the loan is 10 percent per year.
**f.** This book has stressed that over sufficiently long time periods, total expense is equal to total cash outflow. In what sense is the total expense for this transaction the same, independent of the interest rate (and, therefore, the interest expense)?

## BONDS

When a firm can borrow funds from one lender, the firm usually issues mortgages or notes. When it needs larger amounts, the firm may have to borrow from the general investing public with a bond issue. Corporations and governmental units are the usual issuers. Bond issues have the following features:

1. A firm or its investment banker writes a **bond indenture**, or agreement, which gives the terms of the loan and the rights and duties of the borrower and other parties to the contract. To provide some protection to bondholders, bond indentures typically limit the borrower's right to declare dividends, to make other distributions to owners, and to acquire other businesses.
2. The borrower can prepare engraved bond certificates, although this practice is disappearing. Each one represents a portion of the total loan. The usual minimum denomination is $1,000. The federal government has issued some bonds in denominations as small as $50.
3. If the borrower pledges property as collateral for the loan (such as in a mortgage bond), the indenture names a trustee to hold title to the property serving as collateral. The trustee, usually a bank, acts as the representative of the bondholders.
4. The borrower or trustee appoints an agent, usually a bank, to act as *registrar* and *disbursing agent*. The borrower deposits interest and principal payments with the disbursing agent, who distributes the funds to the bondholders.
5. Some bonds are **coupon bonds**. Coupons attached to the bond certificate represent promises to make periodic payments throughout the life of the bond. When a coupon comes due, the bondholder cuts it off and deposits it with a bank. Then the bank sends the coupon, as it would with a check it received, through the bank clearing system to the disbursing agent, who deposits the payment in the bondholder's account at the bank. In recent years, most bonds are registered, which means that the borrower records the name and address of the bondholder and sends the periodic payments to the registered owner, who need not clip coupons and redeem them.

6. The borrower usually issues the entire loan to an investment banking firm or to a group of investment bankers known as a *syndicate*, which takes over the responsibility for selling the bonds to the investing public. Members of the syndicate usually bear the risks and rewards of interest rate fluctuations during the period while they are selling the bonds to the public.

**Types of Bonds** Mortgage bonds carry a mortgage on real estate as collateral for the repayment of the loan. Issuers of *collateral trust bonds* usually hand over stocks and bonds of other corporations to the trustee. The most common type of corporate bond, except in the railroad and public utility industries, is the **debenture bond**. This type carries no special collateral; instead, the borrower issues it on the general credit of the business. For added protection to the bondholders, the bond indenture usually includes provisions that limit the dividends that the borrower can declare or the amount of subsequent long-term debt that it can incur. Common terminology refers to such provisions as *covenants* or *debt covenants*. **Convertible bonds** are debentures that the holder (lender) can exchange, possibly after some specific period of time has elapsed, for a specific number of common or, perhaps, preferred shares, of the borrower.

From the viewpoint of the issuer (borrower), all bonds require future cash outflows in return for a current or previous cash inflow. The pattern of future cash outflows varies across bonds depending on the provisions in the bond indenture.

**Example 15** Ford Motor Company issues $250 million of 10 percent, semiannual, 20-year bonds. The bond indenture requires Ford to pay interest of $12.5 million each six months for 20 years and to repay the $250 million principal at the end of 20 years.

**Example 16** DaimlerChrysler Corporation issues $180 million of 15-year bonds. The bond indenture requires DaimlerChrysler to pay $13.1 million each six months for 15 years. Each periodic payment includes interest plus repayment of a portion of the principal. (Common terminology refers to such bonds as **serial bonds**.)

**Example 17** General Motors Corporation (GM) issues bonds that do not require a periodic cash payment (common terminology refers to such bonds as **zero coupon bonds**), but it promises a single payment of $300 million at the end of 10 years. The $300 million paid after 10 years includes both interest and principal.

## *Computational* Note

By the time you have completed the Appendix to this book, you should be able to work out the implicit interest rate in this GM bond issue, given the amount of initial issue proceeds. To take an example, assume GM receives $140 million for this issue. Then the implicit interest rate is just under 8 percent per year. See Table 2 at the back of the book, 10-period column and 8-percent row; the factor is 0.46319. Multiply that factor by $300 million, and notice that you get slightly less than $140 million, which indicates that 8 percent slightly overstates the implicit rate, which is 7.92 percent. Later in this section, we show the derivation of proceeds for this note assuming a 10-percent rate. Over the life of the bond issue, GM will record $160 million of interest expense, all of which it will pay at maturity along with the $140 million originally borrowed.

The bond indenture indicates the par value or face value of a bond. The par value is usually an even amount (for example, $250 million in **Example 15**, $180 million in **Example 16**, and $300 million in **Example 17**). Bonds requiring periodic payments (called *coupon* payments if the bonds have coupons) use the par value as the base for computing the amount of the cash payment. For example, the Ford Motor Company bonds in **Example 15** require the payment of 10 percent of the par amount each year, payable in two installments. The semiannual payment is $12.5 million (= 0.10 × $250 million × 0.5).

## *Terminology* Note

Almost everyone in business refers to the periodic payments as *interest payments*. This term causes confusion because, as you will soon see, the amount of interest expense for a period almost never equals the amount of these payments for that same period. The periodic payment will always include some amount to pay interest to the lender, but not necessarily all interest accrued since the last payment. If the payment exceeds all accrued interest, then the payment will discharge some of the principal amount. Both payment of interest and payment of principal serve to reduce the debt, so one all-purpose term used for the payments is *debt service payments*. We urge you not to call them, or to think of them as, interest payments until you have understood why they do not equal interest expense. You will never be wrong to call them *debt service payments*.

### Proceeds of a Bond Issue

The price at which a firm issues bonds on the market depends on two factors: the future cash payments that the bond indenture requires the firm to make; and the discount rate that the market deems appropriate given the risk of the borrower, the general level of interest rates in the economy, and other factors. The issuing price equals the present value of the required cash flows discounted at the appropriate market interest rate.

**Example 18** Refer to **Example 15**. The time line for this semiannual bond covers 40 six-month periods and is as follows:

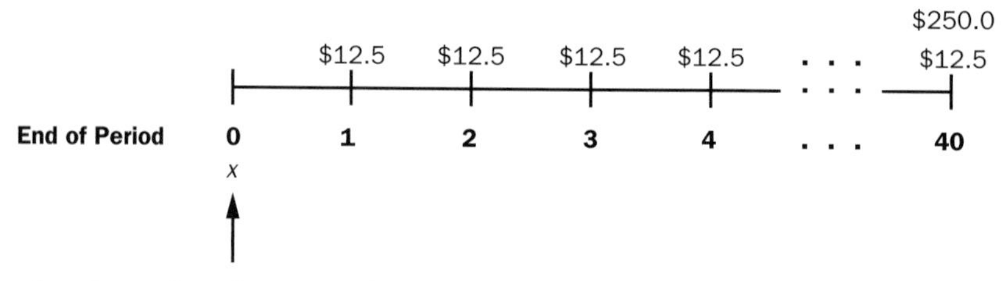


Assume that the market charges Ford an interest rate of 10 percent, compounded semiannually. Ford will receive $250 million for its bond issue, computed as follows:

| Required Cash Flows | Present Value Factor for 10 Percent Interest Rate Compounded Semiannually for 20 Years | Present Value of Required Cash Flows |
|---|---|---|
| $250.0 million at end of 20 years . . . . . . . . . . . . . . . . . | 0.14205[a] | $ 35.5 |
| $12.5 million every six months for 20 Years . . . . . . . . . . | 17.15909[b] | 214.5 |
| Issue Price . . . . . . . . . . . . . . . . . . . . . . . . . . . . . | | $250.0 |

[a]Table 2, 5-percent column and 40-period row.
[b]Table 4, 5-percent column and 40-period row.

**Example 19** Refer to **Example 16**. The time line for this serial bond covers 30 six-month periods and is as follows:

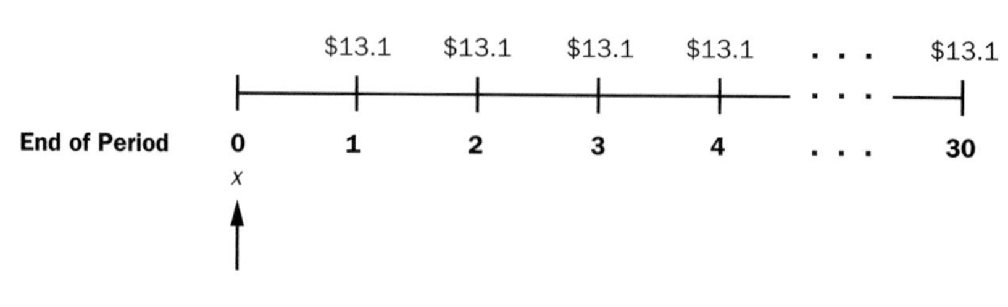

Assume that the market deems that an appropriate interest rate for DaimlerChrysler's bonds on the date of issue is also 10 percent, compounded semiannually. The issue price of Daimler-Chrysler's bonds is $201.4 million, computed as follows:

| Required Cash Flows | Present Value Factor for 10 Percent Interest Rate Compounded Semiannually for 15 Years | Present Value of Required Cash Flows |
|---|---|---|
| $13.1 million every six months for 15 years .......... | 15.37245[a] | $201.4 |
| Issue Price ............................. | | $201.4 |

[a]Table 4, 5-percent column and 30-period row.

**Example 20** Refer to **Example 17**. The time line for this zero coupon bond covers 10 one-year periods and is as follows:

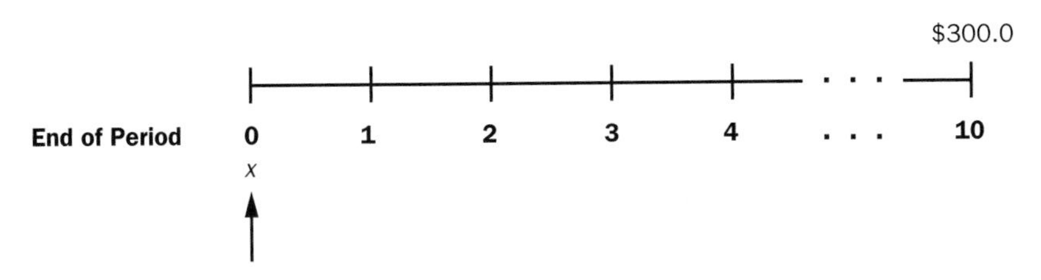


Assume that the market deems that 10 percent, compounded annually, is an appropriate interest rate on the date of issue for GM bonds. The issue price is $115.7 million, computed as follows:

| Required Cash Flows | Present Value Factor for 10 Percent Interest Rate Compounded Annually for 10 Years | Present Value of Required Cash Flows |
|---|---|---|
| $300 million at end of 10 years .................. | 0.38554[a] | $115.7 |
| Issue Price ............................. | | $115.7 |

[a]Table 2, 10-percent column and 10-period row.

The issue price will usually differ from the par value of the bonds. (The Ford example above, where the price and the value equal each other, is rare, but we include it for teaching purposes.) The difference arises because the coupon interest rate stated in the bond indenture differs from the interest rate that the market charges the issuer to borrow.

If the market required rate exceeds the coupon rate, the bonds will sell for less than par value. The difference between the par value and the selling price is known as the *discount* on the bonds. The zero coupon bonds of GM in **Example 20** have a par value of $300 million, an issue price of $115.7 million and a discount of $184.3 (= $300.0 − $115.7) million. The 10-percent rate required by the lenders before they will purchase these bonds exceeds the bonds' promised periodic payment rate, zero in this case. Purchasing the GM bonds for $115.7 million and receiving $300 million 10 years later yields the purchasers a return of 10 percent compounded annually. That is, $300 million is the future value of $115.7 million invested to yield 10 percent compounded annually for 10 years. The $184.3 million will be the amount of interest GM will accrue over the 10-year life of the bonds.

If the coupon rate exceeds the market-required rate, the bonds will sell for more than par value. The difference between the selling price and the par value is known as the *premium* on the bond. The DaimlerChrysler bonds in **Example 19** have an issue price of $201.4 million, a par value of $180 million, and a premium of $21.4 (= $201.4 − $180.0) million. Purchasing the DaimlerChrysler bonds for $201.4 million and receiving $13.1 million every six months for 15 years yields the lenders a return of 10 percent compounded semiannually.

If the coupon rate equals the market-required rate, a rare occurrence, the bonds will sell for par value. The Ford Motor Company bonds in **Example 18** have a coupon rate and a market required rate of 10 percent compounded semiannually. The issue price of $250 million equals the par value of the bonds. Purchasing these bonds for $250 million and receiving $12.5 million every six months for 20 years and $250 million at the end of 20 years yields the purchaser 10 percent compounded semiannually.

The presence of a discount or premium by itself indicates nothing about the credit risk of the borrower. A solid firm, such as General Electric, with a small credit risk that would enable it to borrow funds at 8 percent, might issue 7-percent bonds that would sell at a discount. In contrast, an untested firm with a lower credit standing that requires it to pay 15 percent on loans might issue 18-percent bonds that would sell at a premium.

**Issued at Par** The Macaulay Corporation issues $100,000 face value of 12-percent semiannual coupon debenture bonds on July 1, Year 1. Macaulay must repay the principal amount five years later, on July 1, Year 6. Macaulay owes periodic payments (coupons) on January 1 and July 1 of each year. The coupon payments promised at each payment date total $6,000. **Figure 9.2** presents a time line for the two sets of cash flows associated with this bond. Assume that the market rate of interest for Macaulay on July 1, Year 1, exactly equals 12 percent compounded semiannually. Then, the calculation of the loan proceeds would be as follows (the **Appendix** at the back of the book explains the present value calculations):

| | |
|---|---|
| **(a)** Present Value of $100,000 to Be Paid at the End of Five Years . . . . . . . . . . . . . . . . . . . . | $ 55,839 |
| (Table 2 at the back of the book shows the present value of $1 to be paid in 10 periods at 6 percent per period to be $0.55839; $100,000 × 0.55839 = $55,839.) | |
| **(b)** Present Value of $6,000 to Be Paid Each Six Months for Five Years . . . . . . . . . . . . . . . . . | 44,161 |
| (Table 4 shows the present value of an ordinary annuity of $1 per period for 10 periods discounted at 6 percent to be $7.36009; $6,000 × 7.36009 = $44,161.) | |
| Total Proceeds . . . . . . . . . . . . . . . . . . . . . . . . . . . . . . . . . . . . . . . . . . . . . . . . . | $100,000 |

FIGURE 9.2 **Time Line for Five-Year Semiannual Coupon Bonds with 12 Percent Annual Coupons (100,000 par value, issued at par)**

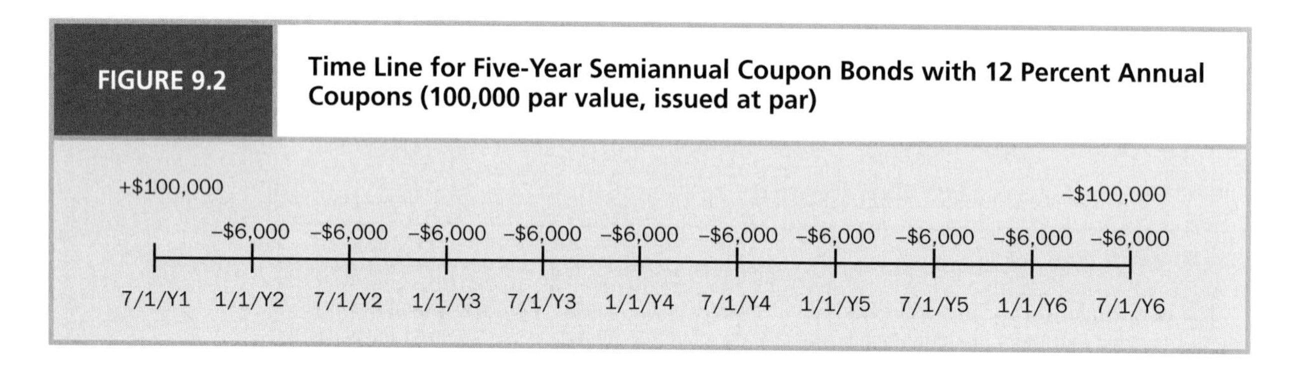


Common terminology quotes bond prices as a percentage of par value. The issue price of these bonds is 100.0 (that is, 100 percent of par), which implies that the market interest rate is 12 percent, compounded semiannually, the same as the coupon rate.

**Issued at Less Than Par** Assume that Macaulay Corporation issued these same bonds at a price to yield 14 percent compounded semiannually. The cash flows promised after July 1, Year 1, associated with these bonds (periodic payments plus repayment of principal) equal those in the time line in **Figure 9.2**. The market discounts these future cash flows to their present value using a 14 percent discount rate compounded semiannually. The calculation of the issue proceeds (that is, initial market price) follows:

| | |
|---|---|
| **(a)** Present Value of $100,000 to Be Paid at the End of Five Years . . . . . . . . . . . . . . . . . . . . | $50,835 |
| (Present value of $1 to be paid in 10 periods at 7 percent per period is $0.50835; $100,000 × 0.50835 = $50,835.) | |
| **(b)** Present Value of $6,000 to Be Paid Each Six Months for Five Years . . . . . . . . . . . . . . . . . | 42,141 |
| (Present value of an ordinary annuity of $1 per period for 10 periods discounted at 7 percent per period = $7.02358; $6,000 × 7.02358 = $42,141.) | |
| Total Proceeds . . . . . . . . . . . . . . . . . . . . . . . . . . . . . . . . . . . . . . . . . . . . . . . . . | $92,976 |

Assume Macaulay issues these bonds at 92.98 for $92,980. This implies a market yield of slightly less than 14 percent compounded semiannually.

**Issued at More than Par** Assume that Macaulay Corporation issued the bonds at a price to yield 10 percent compounded semiannually. The cash flows promised after July 1, Year 1, again equal those in **Figure 9.2**. The market discounts these future cash flows at 10 percent compounded semiannually. The calculation of the proceeds follows:

| | |
|---|---|
| **(a)** Present Value of $100,000 to Be Paid at the End of Five Years . . . . . . . . . . . . . . . . . . . . | $ 61,391 |
| (Present value of $1 to be paid in 10 periods at 5 percent per period is $0.61391; $100,000 × 0.61391 = $61,391.) | |
| **(b)** Present Value of $6,000 to Be Paid Each Six Months for Five Years . . . . . . . . . . . . . . . . . | 46,330 |
| (Present value of an ordinary annuity of $1 per period for 10 periods, discounted at 5 percent per period = $7.72173; $6,000 × 7.72173 = $46,330.) | |
| Total Proceeds . . . . . . . . . . . . . . . . . . . . . . . . . . . . . . . . . . . . . . . . . . . . . . . . | $107,721 |

Assume Macaulay issues these bonds at 107.72 (that is, 107.72 percent of par) for $107,720. The price would imply a market yield of slightly more than 10 percent compounded semiannually.

**Bond Tables** One need not make these tedious calculations every time one analyzes a bond issue. **Bond tables** show the results of calculations like those just described. **Tables 5** and **6** at the back of the book illustrate such tables. **Table 5** shows the price for 10-percent semiannual coupon bonds as a percent of par for various market interest rates (yields) and years to maturity. **Table 6** shows market rates and implied prices for 12-percent semiannual coupon bonds. (Most calculators make the calculations represented by these tables in a few seconds. Spreadsheet program financial functions make the calculations even faster. We urge you to learn how to use and interpret the tables because you will need the resulting understanding to know which financial functions to use in your spreadsheet and which data to input to the functions.)

The percentages of par shown in these tables represent the present value of the bond indicated. Because the tables give factors as a percent of par, the user must multiply them by 10 to find the price of a $1,000 bond. If you have never used bond tables before now, turn to **Table 6** after the Appendix at the back of the book, and find in the 5-year row the three different prices for the three different market yields used in the preceding examples. Notice further that a bond will sell at par if and only if it has a market yield equal to its coupon rate.

These tables apply both when a corporation issues a bond and when an investor later resells it. Computing the market price will be the same in either case. The following generalizations describe bond prices:

1. When the market interest rate equals the coupon rate, the market price will equal par.
2. When the market interest rate exceeds the coupon rate, the market price will be less than par.
3. When the market interest rate is less than the coupon rate, the market price will exceed par.

## Problem 9.4 for Self-Study

**Calculating the issue price of bonds.** Engel Corporation issues $1,000,000 face-value bonds on January 1, Year 1. The bonds carry a 10-percent coupon rate, payable in two installments on June 30 and December 31 each year. The bonds mature on December 31, Year 10.

Compute the issue price of these bonds assuming that the market requires the following yields:

**a.** 8 percent compounded semiannually
**b.** 10 percent compounded semiannually
**c.** 12 percent compounded semiannually

**Accounting for Bonds Issued at Par** The following illustrates bonds issued at par. We use the data presented in the previous sections for the Macaulay Corporation bonds issued at par, and we assume that the firm closes its books semiannually on June 30 and December 31. The entry at the time of issue follows:

7/1/Year 1

| | | |
|---|---|---|
| Cash | 100,000 | |
| Debenture Bonds Payable | | 100,000 |

| Assets | = | Liabilities | + | Shareholders' Equity | (Class.) |
|---|---|---|---|---|---|
| +100,000 | | +100,000 | | | |

$100,000 of 12-percent, five-year, semi-annual coupon bonds issued at par.

The borrower would make entries for interest at the end of the accounting period and on the payment dates. Entries through January 1, Year 2, would be as follows:

12/31/Year 1

| | | |
|---|---|---|
| Interest Expense | 6,000 | |
| Interest Payable | | 6,000 |

| Assets | = | Liabilities | + | Shareholders' Equity | (Class.) |
|---|---|---|---|---|---|
| | | +6,000 | | −6,000 | IncSt → RE |

To record interest expense for six months.

1/1/Year 2

| | | |
|---|---|---|
| Interest Payable | 6,000 | |
| Cash | | 6,000 |

| Assets | = | Liabilities | + | Shareholders' Equity | (Class.) |
|---|---|---|---|---|---|
| −6,000 | | −6,000 | | | |

To record payment of six months' interest.

**Bond Issued between Payment Dates** The actual date when a firm issues a bond seldom coincides with one of the periodic payment dates. Assuming that Macaulay brought these same bonds to market on September 1, rather than July 1, at par, the borrower would expect the purchasers of the bonds to pay for two months of periodic payments in advance. On the first coupon, Macaulay Corporation promises a full $6,000 for six months' debt service, but it would have had the use of the borrowed funds for only four months. The purchasers of the bonds would pay $100,000 plus two months' interest of $2,000 ($= 0.12 \times \$100{,}000 \times 2/12$) and would get the $2,000 back when Macaulay redeemed the first coupons. The journal entry made by Macaulay Corporation, the issuer, would be as follows:

9/1/Year 1

| | | |
|---|---|---|
| Cash | 102,000 | |
| Bonds Payable | | 100,000 |
| Interest Payable | | 2,000 |

| Assets | = | Liabilities | + | Shareholders' Equity | (Class.) |
|---|---|---|---|---|---|
| +102,000 | | +100,000 | | | |
| | | +2,000 | | | |

To record issue of bonds at par between payment dates.

The purchasers pay an amount equal to interest for the first two months but will get it back when the borrower pays for the first coupon.

**Accounting for Bonds Issued at Less than Par** The following illustrates bonds issued at less than par. Assume the data presented above where the Macaulay Corporation issued 12-percent, $100,000-par value, five-year bonds to yield 14 percent compounded semiannually. Earlier, we derived the issue price to be $92,976. The journal entry at issue would be as follows:

| | | |
|---|---|---|
| 7/1/Year 1 | | |
| Cash . . . . . . . . . . . . . . . . . . . . . . . . . . . . . . . . . . . . . . . . . . . . . | 92,976 | |
| Debenture Bonds Payable . . . . . . . . . . . . . . . . . . . . . . . . . . . . | | 92,976 |

| Assets | = | Liabilities | + | Shareholders' Equity | (Class.) |
|---|---|---|---|---|---|
| +92,976 | | +92,976 | | | |

The issuance of these bonds for $92,976, instead of for the $100,000 par value, indicates that lenders demand more than 12 percent from Macaulay Corporation, the borrower. If purchasers of these bonds paid the $100,000 par value, they would earn only 12 percent compounded semiannually, the coupon rate. Purchasers desiring a rate of return of 14 percent will not buy until the market price drops to $92,976. At this price, purchasers of the bonds will earn the 14 percent return they require. The return will comprise 10 coupon payments of $6,000 each over the next five years plus $7,024 (= $100,000 − $92,976) as part of the payment at maturity.

For Macaulay Corporation, the total interest expense over the life of the bonds will equal $67,024 (= periodic payments totaling $60,000 plus $7,024 paid at maturity). The accounting for these payments allocates the total interest expense of $67,024 to the periods of the loan using the effective interest method, explained next. Following the next discussion will be easier if you refer to **Exhibit 9.3**.

**Interest Expense under the Effective Interest Method** Under the **effective interest method**, interest expense each period equals the market interest rate at the time the firm initially issued the bonds (14 percent compounded semiannually, which is 7 percent per six months, in this example) multiplied by the book value of the liability at the beginning of the interest period. For example, interest expense for the period from July 1, Year 1, to December 31, Year 1, the first six-month period, is $6,508 (= 0.07 × $92,976). The bond indenture provides that the borrower repay only $6,000 (= 0.06 × $100,000) on January 1, Year 2. This amount equals the coupon rate times the par value of the bonds. The difference between the interest expense of $6,508 and the interest then payable of $6,000 increases the book value of the bond. The borrower will pay this amount as part of the principal payment at maturity. The journal entry made on December 31, Year 1, to recognize interest for the last six months of Year 1 follows:

| | | |
|---|---|---|
| 12/31/Year 1 | | |
| Interest Expense . . . . . . . . . . . . . . . . . . . . . . . . . . . . . . . . . . . . | 6,508 | |
| Interest Payable . . . . . . . . . . . . . . . . . . . . . . . . . . . . . . . . | | 6,000 |
| Debenture Bonds Payable . . . . . . . . . . . . . . . . . . . . . . . . . . | | 508 |

| Assets | = | Liabilities | + | Shareholders' Equity | (Class.) |
|---|---|---|---|---|---|
| | | +6,000 | | −6,508 | IncSt → RE |
| | | +508 | | | |

To recognize interest expense for six months.

The Interest Payable account will appear as a current liability on the balance sheet at the end of Year 1. Debenture Bonds Payable of $93,484 (= $92,976 + $508) will appear on the balance sheet as a noncurrent liability.

**EXHIBIT 9.3**

**Effective Interest Amortization Schedule for $100,000 of 12-Percent, Semiannual Coupon, Five-Year Bonds Issued for 92.976 Percent of Par to Yield 14 Percent, Interest Compounded Semiannually**

**Semiannual Journal Entry**

| | | |
|---|---|---|
| Interest Expense .......... | Amount in Column **(3)** | |
| Cash .......... | | Amount in Column **(4)** |
| Debenture Bonds Payable .......... | | Amount in Column **(5)** |

| Assets | = | Liabilities | + | Shareholders' Equity | (Class.) |
|---|---|---|---|---|---|
| −col(4) | | +col(5) | | −col(3) | IncSt → RE |

| Period (6-Month Intervals) (1) | Liability at Start of Period (2) | Effective Interest: 7 Percent per Period (3) | Coupon Rate: 6 Percent of Par (4) | Increase in Recorded Book Value of Liability (5) | Liability at End of Period (6) |
|---|---|---|---|---|---|
| 0 .......... | | | | | $ 92,976 |
| 1 .......... | $92,976 | $ 6,508 | $ 6,000 | $ 508 | 93,484 |
| 2 .......... | 93,484 | 6,544 | 6,000 | 544 | 94,028 |
| 3 .......... | 94,028 | 6,582 | 6,000 | 582 | 94,610 |
| 4 .......... | 94,610 | 6,623 | 6,000 | 623 | 95,233 |
| 5 .......... | 95,233 | 6,666 | 6,000 | 666 | 95,899 |
| 6 .......... | 95,899 | 6,713 | 6,000 | 713 | 96,612 |
| 7 .......... | 96,612 | 6,763 | 6,000 | 763 | 97,375 |
| 8 .......... | 97,375 | 6,816 | 6,000 | 816 | 98,191 |
| 9 .......... | 98,191 | 6,873 | 6,000 | 873 | 99,064 |
| 10 .......... | 99,064 | 6,936 | 6,000 | 936 | 100,000 |
| Total .......... | | $67,024 | $60,000 | $7,024 | |

Note: In preparing this table, we rounded calculations to the nearest dollar.
Column (2) = column (6) from previous period.
Column (3) = 0.07 × column (2), except for period 10, where it is a plug.
Column (4) is given.
Column (5) = column (3) − column (4), except for period 10, where it is a plug.
Column (6) = column (2) + column (5).

On January 1, Year 2, the borrower makes the first periodic cash payment.

| | | |
|---|---|---|
| 1/1/Year 2 | | |
| Interest Payable .......... | 6,000 | |
| Cash .......... | | 6,000 |

| Assets | = | Liabilities | + | Shareholders' Equity | (Class.) |
|---|---|---|---|---|---|
| −6,000 | | −6,000 | | | |

To record payment of interest for six months.

Interest expense for the second six months (from January 1, Year 2, through June 30, Year 2) is $6,544 (= 0.07 × $93,484). It exceeds the $6,508 for the first six months because the recorded book value of the liability at the beginning of the second six months has grown. The journal entry on June 30, Year 2, to record interest expense follows:

6/30/Year 2

| | | |
|---|---|---|
| Interest Expense | 6,544 | |
| Interest Payable | | 6,000 |
| Debenture Bonds Payable | | 544 |

| Assets | = | Liabilities | + | Shareholders' Equity | (Class.) |
|---|---|---|---|---|---|
| | | +6,000 | | −6,544 | IncSt → RE |
| | | +544 | | | |

To recognize interest expense for six months.

An amortization schedule for these bonds over their five-year life appears in **Exhibit 9.3**. Column **(3)** shows the periodic interest expense, and column **(6)** shows the book value that appears on the balance sheet at the end of each period.

The effective interest method of recognizing interest expense on a bond has the following financial statement effects:

1. Interest expense on the income statement will equal a constant percentage of the recorded liability at the beginning of each interest period. This percentage will equal the market interest rate for these bonds when the borrower issued them. When the borrower issues bonds for less than par value, the dollar amount of interest expense will increase each period as the recorded book value amount increases.
2. On the balance sheet at the end of each period, the bonds will appear at the present value of the remaining cash outflows discounted at the market interest rate measured when the borrower initially issued the bonds. For example, on July 1, Year 2, just after the borrower has made a coupon payment, the remaining cash payments have present value computed as follows:

| | |
|---|---|
| **(a)** Present Value of \$100,000 to Be Paid at the End of Four Years | \$58,201 |
| (Table 2 shows the present value of \$1 to be paid at the end of eight periods discounted at 7 percent to be \$0.58201; \$100,000 × 0.58201 = \$58,201.) | |
| **(b)** Present Value of Eight Remaining Semiannual Interest Payments Discounted at 14 Percent Compounded Semiannually | 35,827 |
| (Table 4 shows the present value of an ordinary annuity of \$1 per period for eight periods discounted at 7 percent to be \$5.97130; \$6,000 × 5.97130 = \$35,827.) | |
| Total Present Value | \$94,028 |

The amount \$94,028 appears in column **(6)** of **Exhibit 9.3** for the liability at the end of the second six-month period.

**Accounting for Bonds Issued at More than Par** The following discussion illustrates bonds issued at more than par. Assume the data presented where the Macaulay Corporation issued 12-percent, \$100,000-par value, five-year bonds to yield approximately 10-percent compounded semiannually. The issue price, derived previously, was \$107,721. The journal entry at the time of issue follows:

7/1/Year 1

| | | |
|---|---|---|
| Cash | 107,721 | |
| Debenture Bonds Payable | | 107,721 |

| Assets | = | Liabilities | + | Shareholders' Equity | (Class.) |
|---|---|---|---|---|---|
| +107,721 | | +107,721 | | | |

The firm borrows \$107,721. The issuance of these bonds for \$107,721, instead of the \$100,000 par value, indicates that 12 percent exceeds the interest rate the purchasers (lenders) demand. If

EXHIBIT 9.4

**Effective Interest Amortization Schedule for $100,000 of 12-Percent, Semiannual Coupon, Five-Year Bonds Issued for 107.721 Percent of Par to Yield 10 Percent, Compounded Semiannually**

**Semiannual Journal Entry**

| | | |
|---|---|---|
| Interest Expense | Amount in Column (3) | |
| Debenture Bonds Payable | Amount in Column (5) | |
| Cash | | Amount in Column (4) |

| Assets | = | Liabilities | + | Shareholders' Equity | (Class.) |
|---|---|---|---|---|---|
| −col(4) | | −col(5) | | −col(3) | IncSt → RE |

| Period (6-Month Intervals) (1) | Liability at Start of Period (2) | Effective Interest: 5 Percent per Period (3) | Coupon Rate: 6 Percent of Par (4) | Decrease in Recorded Book Value of Liability (5) | Liability at End of Period (6) |
|---|---|---|---|---|---|
| 0 | | | | | $107,721 |
| 1 | $107,721 | $ 5,386 | $ 6,000 | $ 614 | 107,107 |
| 2 | 107,107 | 5,355 | 6,000 | 645 | 106,462 |
| 3 | 106,462 | 5,323 | 6,000 | 677 | 105,785 |
| 4 | 105,785 | 5,289 | 6,000 | 711 | 105,074 |
| 5 | 105,074 | 5,245 | 6,000 | 746 | 104,328 |
| 6 | 104,328 | 5,216 | 6,000 | 784 | 103,544 |
| 7 | 103,544 | 5,177 | 6,000 | 823 | 102,721 |
| 8 | 102,721 | 5,136 | 6,000 | 864 | 101,857 |
| 9 | 101,857 | 5,093 | 6,000 | 907 | 100,950 |
| 10 | 100,950 | 5,050 | 6,000 | 950 | 100,000 |
| Total | | $52,279 | $60,000 | $7,721 | |

Note: In preparing this table, we rounded calculations to the nearest dollar.
Column (2) = column (6) from previous period.
Column (3) = 0.05 × column (2), except for period 10, where it is a plug.
Column (4) is given.
Column (5) = column (4) − column (3), except for period 10, where it is a plug.
Column (6) = column (2) − column (5).

purchasers of these bonds paid the $100,000 par value, they would earn 12 percent compounded semiannually, the coupon rate. Purchasers requiring a rate of return of only 10 percent will bid up the market price to $107,721. At this price, purchasers of the bonds will earn only the 10 percent return they demand. Their return comprises 10 coupon payments of $6,000 each over the next five years reduced by the $7,721 (= $107,721 − $100,000) paid as part of initial amount transferred to the borrower but not repaid at maturity.

For Macaulay Corporation, the total interest expense over the life of the bonds equals $52,279 (= periodic payments totaling $60,000 less $7,721 received at the time of original issue but not repaid at maturity). Following the next discussion will be easier if you refer to **Exhibit 9.4**.

**Interest Expense under the Effective Interest Method** Under the effective interest method, interest expense each period equals the market interest rate at the time the firm initially issued the bonds (10 percent compounded semiannually, equals 5 percent per six months, in this example), multiplied by the recorded book value of the liability at the beginning of the interest period. For example, interest expense for the period from July 1, Year 1, to December 31, Year 1, the first six-month period, is $5,386 (= 0.05 × $107,721). The bond indenture requires the borrower to pay $6,000 (= 0.06 × $100,000) on January 1, Year 2. This amount equals the coupon rate times the face value of the bonds. The difference between the payment of $6,000 and the interest expense of $5,386 reduces the amount of the liability. The journal entry made on December 31, Year 1, to recognize interest for the last six months of Year 1 follows:

| 12/31/Year 1 | | |
|---|---|---|
| Interest Expense | 5,386 | |
| Debenture Bonds Payable | 614 | |
| Interest Payable | | 6,000 |

| Assets | = | Liabilities | + | Shareholders' Equity | (Class.) |
|---|---|---|---|---|---|
| | | −614 | | −5,386 | IncSt → RE |
| | | +6,000 | | | |

To recognize interest expense for six months.

The Interest Payable account appears as a current liability on the balance sheet at the end of Year 1. Debenture Bonds Payable has a new balance of \$107,107 (= \$107,721 − \$614), which appears as a noncurrent liability.

On January 1, Year 2, the borrower makes the first periodic cash payment and the following entry:

| 1/1/Year 2 | | |
|---|---|---|
| Interest Payable | 6,000 | |
| Cash | | 6,000 |

| Assets | = | Liabilities | + | Shareholders' Equity | (Class.) |
|---|---|---|---|---|---|
| −6,000 | | −6,000 | | | |

To record payment of interest for six months.

Interest expense for the second six months (from January 1, Year 2, through June 30, Year 2) equals \$5,355 (= 0.05 × \$107,107). Because the amount of the liability has declined from the beginning of the preceding period to the beginning of the current period, the amount of interest expense for the period declines from \$5,386 for the first six months, and the borrower records it as follows:

| 6/30/Year 2 | | |
|---|---|---|
| Interest Expense | 5,355 | |
| Debenture Bonds Payable | 645 | |
| Interest Payable | | 6,000 |

| Assets | = | Liabilities | + | Shareholders' Equity | (Class.) |
|---|---|---|---|---|---|
| | | −645 | | −5,355 | IncSt → RE |
| | | +6,000 | | | |

To recognize interest expense for six months.

An amortization schedule for these bonds over their five-year life appears in **Exhibit 9.4**. Column **(3)** shows the periodic interest expense, and column **(6)** shows the book value that appears on the balance sheet at the end of the period.

The effective interest method of recognizing interest expense on a bond has the following financial statement effects:

1. Interest expense on the income statement equals a constant percentage of the recorded liability at the beginning of each interest period. This percentage equals the market interest rate when the borrower first issued the bonds. When the borrower issues bonds for more than par

value, the dollar amount of interest expense will decrease each period as the unpaid liability decreases to the amount to be paid at maturity.

2. On the balance sheet at the end of each period, the bonds will appear at the present value of the remaining cash flows discounted at the market interest rate when the borrower initially issued the bonds. For example, on July 1, Year 2, just after the borrower has made the coupon payment, the remaining cash payments have present value computed as follows:

| | |
|---|---|
| **(a)** Present Value of $100,000 to Be Paid at the End of Four Years ........ | $ 67,684 |
| (Table 2 shows the present value of $1 to be paid at the end of eight periods discounted at 5 percent to be $0.67684; $100,000 × 0.67684 = $67,684.) | |
| **(b)** Present Value of Eight Remaining Semiannual Interest Payments Discounted at 10 Percent, Compounded Semiannually ........ | 38,779 |
| (Table 4 shows the present value of an ordinary annuity of $1 per period for eight periods discounted at 5 percent to be $6.46321; $6,000 × 6.46321 = $38,779.) | |
| Total Present Value ........ | $106,463 |

The amount $106,462, different because of rounding effects, appears in column **(6)** of **Exhibit 9.4** for the liability at the end of the second six-month period.

## Problem 9.5 for Self-Study

**Preparing journal entries to account for bonds.** Refer to **Problem 9.4 for Self-Study.** Prepare the journal entries on January 1, June 30, and December 31, Year 1, to account for the bonds, assuming each of the following market-required interest rates at the time the firm issued the bonds:

**a.** 8 percent compounded semiannually
**b.** 10 percent compounded semiannually
**c.** 12 percent compounded semiannually

**Bond Retirement** Many bonds remain outstanding until the stated maturity date. Refer to **Exhibit 9.3**, where Macaulay Corporation issued the 12-percent coupon bonds to yield 14 percent. The company pays the final coupon, $6,000, and the face amount, $100,000, on the stated maturity date. The entries are as follows:

| | | |
|---|---|---|
| 7/1/Year 6 | | |
| Interest Expense ........ | 6,936 | |
| Cash ........ | | 6,000 |
| Debenture Bonds Payable ........ | | 936 |

| Assets | = | Liabilities | + | Shareholders' Equity | (Class.) |
|---|---|---|---|---|---|
| −6,000 | | +936 | | −6,936 | IncSt → RE |

See row for Period 10 of **Exhibit 9.3**.

| | | |
|---|---|---|
| Debenture Bonds Payable ........ | 100,000 | |
| Cash ........ | | 100,000 |

| Assets | = | Liabilities | + | Shareholders' Equity | (Class.) |
|---|---|---|---|---|---|
| −100,000 | | −100,000 | | | |

To record retirement at maturity of bonds.

**Retirement before Maturity** A firm sometimes purchases its own bonds on the open market before maturity. Because market interest rates constantly change, the purchase price will seldom equal the recorded book value of the bonds. Assume that Macaulay Corporation originally issued its 12-percent coupon bonds to yield 14 percent compounded semiannually. Assume that three years later, on June 30, Year 4, interest rates in the marketplace have increased so that the market currently requires Macaulay Corporation to pay a 15-percent interest rate. Refer to **Table 6** in the appendix at the back of the book, 2-year row, 15-percent column; there you see that 12-percent bonds with two years until maturity sell in the marketplace for 94.9760 percent of par when the current interest rate is 15 percent compounded semiannually.

The principles of historical cost accounting do not constrain the pricing of bonds in the marketplace. Even though Macaulay Corporation shows Debenture Bonds Payable on the balance sheet at $96,612 (see row for Period 6 of **Exhibit 9.3**), the marketplace puts a price of only $94,976 on the entire bond issue. From the point of view of the marketplace, these bonds are the same as two-year bonds issued on June 30, Year 4, at an effective yield of 15 percent, so they carry a discount of $5,024 (= $100,000 − $94,976).

If on June 30, Year 4, Macaulay Corporation purchased $10,000 of par value of its own bonds, it would have to pay only $9,498 (= 0.94976 × $10,000) for those bonds, which have a book value of $9,661. Macaulay would make the following journal entries at the time of purchase:

| 6/30/Year 4 | | |
|---|---|---|
| Interest Expense | 6,713 | |
| Interest Payable | | 6,000 |
| Debenture Bonds Payable | | 713 |

| Assets | = | Liabilities | + | Shareholders' Equity | (Class.) |
|---|---|---|---|---|---|
| | | +6,000 | | −6,713 | IncSt → RE |
| | | +713 | | | |

See row for Period 6 of **Exhibit 9.3**.

| | | |
|---|---|---|
| Interest Payable | 6,000 | |
| Cash | | 6,000 |

| Assets | = | Liabilities | + | Shareholders' Equity | (Class.) |
|---|---|---|---|---|---|
| −6,000 | | −6,000 | | | |

To record payment of coupons, as usual.

| | | |
|---|---|---|
| Debenture Bonds Payable | 9,661 | |
| Cash | | 9,498 |
| Gain on Retirement of Bonds | | 163 |

| Assets | = | Liabilities | + | Shareholders' Equity | (Class.) |
|---|---|---|---|---|---|
| −9,498 | | −9,661 | | +163 | IncSt → RE |

To record purchase of bonds for less than the current amount shown in the accounting records. Derive gain amount as a plug.

Historical cost accounting calls the entry to make debits equal credits in recording the retirement a *gain*. This gain arises because the firm can retire a liability recorded at one amount, $9,661, for a smaller cash payment, $9,498. The borrower earned this gain as interest rates increased between Year 1 and Year 4. Historical cost accounting reports the gain in the period when the borrower realizes it—that is, in the period when the borrower retires the bonds. This phenomenon parallels the economic events that occur when a firm invests in land, holds the land as its value increases, sells the land in a subsequent year, and reports all the gain in the year of sale. The phenomenon results from the historical cost accounting convention of recording amounts at historical cost and not recording increases in wealth until the firm realizes those increases in arm's-length transactions with outsiders.

**A Quality-of-Earnings Issue** During recent years, interest rates have increased from their levels a few years earlier. Many companies had issued bonds at prices near par, with coupon rates of only 3 or 4 percent per year. When interest rates increased, these bonds traded in the market for substantial discounts from face value. Companies can repurchase their own bonds, recording substantial gains in the process. (In one year, an airline reported profits after seven consecutive years of losses. The airline, now bankrupt, had gains on bond retirement that year in excess of the entire amount of reported net income.) Such gains do not form part of high-quality earnings because of their nonrecurring nature.

**Serial Bonds** The bond indenture may require the issuing firm to make a provision for partial early retirement of the bond issue. Such requirements generally take one of two forms. One, for serial bonds, provides that stated portions of the principal amount will come due on a series of maturity dates. The DaimlerChrysler Corporation bonds in **Example 16** are serial bonds.

**Sinking Fund Bonds** The other major type of partial early retirement provision in bond indentures requires the firm to accumulate a fund of cash or other assets to use to pay the bonds when the maturity date arrives or to reacquire and retire portions of the bond issue. Financial terminology refers to these funds as **sinking funds**, although *bond retirement funds* would be more descriptive. The trustee of the bond issue usually holds the assets of the sinking fund. The fund appears on the balance sheet of the borrower as a noncurrent asset in the Investments section.

**Right of Offset** Why do we show the sinking fund of the preceding paragraph as an asset separate from the liability that the firm will pay off with the fund of cash? To take a concrete example, assume a firm with outstanding bonds of $100 million, appearing as a liability, and a $60 million fund of cash, appearing as an asset in a separate fund. The firm plans to use the cash to pay off part of the bond debt. One might think that a superior financial disclosure on the balance sheet would be to show only the net liability of $40 (= $100 − $60) million. Managers prefer to offset liabilities with cash funds, such as this. To understand why, think about the effect on the debt-equity ratio of reducing liabilities and assets by the same amount: the debt-equity ratio will decline and the firm will appear less risky.[6] Standard setters are aware of managers' wish to offset liabilities with assets. GAAP allows it only in the rare case when the asset has a contractual **right of offset**, that is the borrower owes the lender and the lender owes the borrower and each agrees the other can offset the two items to discharge the debts. That is, the asset must be a receivable from the lender to whom the firm owes an offsetting liability and the asset specifically gives its holder the right to use the asset to discharge the liability.

**Refunded Bonds** Most bond indentures make no provision for **bond refunding**, that is, for making serial repayments or for accumulating sinking funds for bond retirement. The market judges these bonds to be protected with property held by trustees as collateral or with the high credit standing of the issuer. In such cases, the borrower may pay the entire bond liability at maturity out of cash in the bank at that time. Quite commonly, however, the firm refunds the bond issue. That is, it issues a new set of bonds to obtain the funds to retire the old ones when they come due. Accounting records the issue of the new bonds and the retirement of the old bonds as separate transactions.

**Callable Bonds** A common provision gives the issuing company the option to retire portions of the bond issue before maturity if it so desires, but the provision does not require the company to do so. To facilitate such reacquisition and retirement of the bonds, the bond indenture provides that the bonds be *callable*. That is, the issuing company has the right, but not the obligation, to reacquire its bonds before stated maturity at prices specified in the bond indenture. When the borrower calls the bonds, the trustee will not immediately pay a subsequent coupon presented for redemption. Rather, the trustee will notify the holder to present all remaining coupons and the bond principal for payments equal to the call price plus accrued interest.

The indenture usually sets the **call price** initially higher than the par value and provides a declining schedule for the call price as the maturity date approaches. Because the borrower may exercise the call provision at a time when the coupon rate exceeds the market interest rate, the market usually prices callable bonds at something less than the price of otherwise similar but noncallable bonds.

Assume, for example, that a firm issued 12-percent semiannual coupon bonds at a discount, implying a borrowing rate higher than 12 percent. Later, market interest rates and the firm's

---

[6]Firms with assets greater than liabilities, that is, all solvent firms, have debt-equity ratios less than 1.0. When we subtract equal amounts from the numerator and denominator of a fraction less than 1.0, the fraction gets smaller; refer to footnote **7** in **Chapter 5**, on page 230.

credit standing allow it to borrow at 10 percent. Assume the borrower has recorded all accrued interest expense and the bonds have a current book value of $98,000, including accrued interest. If the borrower calls $100,000 of par value of these bonds at 105 percent of par, the entry would be as follows:

| | | |
|---|---|---|
| Debenture Bonds Payable | 98,000 | |
| Loss on Retirement of Bonds | 7,000 | |
| Cash | | 105,000 |

| Assets | = | Liabilities | + | Shareholders' Equity | (Class.) |
|---|---|---|---|---|---|
| -105,000 | | -98,000 | | -7,000 | IncSt → RE |

To record bonds called and retired. Derive loss amount as a plug.

The market rate of interest that a firm must pay to borrow depends on two factors: the general level of interest rates and its own creditworthiness. If the market rate of interest has risen since the borrower issued the bonds (or if the firm's credit rating has declined), the bonds will sell in the market at less than issue price. A firm that wanted to retire such bonds would not call them because the call price would typically exceed the face value. Instead the firm would probably purchase its bonds on the open market and realize a gain on the bonds' retirement.

## Unifying Principles of Accounting for Long-Term Liabilities

Long-term liabilities obligate the borrowing firm to pay specified amounts at definite times more than one year in the future. We now state a single set of principles to describe the balance sheet presentation of all long-term liabilities and a single procedure for computing that amount and the amount of interest expense. First, we describe the amounts on the balance sheet, a "state description" (like a blueprint). Then, we describe the process for computing the amounts, a "process description" (like a recipe).

**State Description** All long-term liabilities appear on the balance sheet at the present value of the remaining future payments. The present value computations use the historical rate of interest—the market interest rate on the date the borrower incurred the obligation.

**Process Description** The methods of computing balance sheet amounts for all long-term liabilities and their related expenses embody identical concepts and follow the same procedures:

1. Record the liability at the cash (or cash equivalent) value received. This amount equals the present value of the future contractual payments discounted using the market interest rate for the borrower on the date the loan begins. (Sometimes the borrower must compute the original issue market interest rate by finding the internal rate of return. See the **Appendix** at the back of the book.)
2. At any subsequent time when the firm makes a cash payment or an adjusting entry for interest, it computes interest expense as the book value of the liability at the beginning of the period (which includes any interest added in prior periods) multiplied by the historical interest rate. The accountant debits this amount to Interest Expense and credits it to the liability account. If the firm makes a cash payment, the accountant debits the liability accounts and credits Cash.

The effect will change the book value of the liabilities to a number closer to par value (or leave the amount at par if book value already equals par value) for the start of the next period. If the accountant follows these procedures, the liabilities on the balance sheet will always satisfy the state description above. The liabilities will have book value equal to the present value of the remaining future payments discounted at the historical market interest rate.

Amortization schedules, such as in **Exhibits 9.2, 9.3, 9.4, 9.5**, and **10.1** (in **Chapter 10**), illustrate this unchanging procedure for a variety of long-term liabilities. The next problem for self-study focuses on the procedure.

## Problem 9.6 for Self-Study

**Unifying principles of accounting for long-term liabilities.** This problem illustrates the unifying principles of accounting for long-term liabilities described just above. Assume that a firm closes its books once per year, making adjusting entries once per year. On the date the firm borrows, the market requires it to pay interest at the rate of 10 percent per year, compounded annually for all loans spanning a two-year period. Note the following steps:

1. Compute the initial issue proceeds received by the firm issuing the obligation (that is, borrowing the cash) on the date of issue.
2. Give the journal entry for issue of the liability and receipt of cash.
3. Show the journal entry or entries for interest accrual and cash payment, if any, at the end of the first year, and re-compute the book value of all liabilities related to the borrowing at the end of the first year. Combine the liability accounts for the main borrowing and accrued interest into a single account called Monetary Liability.
4. Show the journal entry or entries for interest accrual and cash payment at the end of the second year, and re-compute the book value of all liabilities related to the borrowing at the end of the second year.

Perform the above steps for each of the following borrowings:

**a.** The firm issues a single-payment note on the first day of the first year, promising to pay $1,000 on the last day of the second year.
**b.** The firm issues a 10-percent annual coupon bond, promising to pay $100 on the last day of the first year and $1,100 (= $1,000 + $100) on the last day of the second year.
**c.** The firm issues an 8-percent annual coupon bond, promising to pay $80 on the last day of the first year and $1,080 (= $1,000 + $80) on the last day of the second year.
**d.** The firm issues a 12-percent annual coupon bond, promising to pay $120 on the last day of the first year and $1,120 (= $1,000 + $120) on the last day of the second year.
**e.** The firm issues a level-payment note (like a mortgage or installment note), promising to pay $576.19 on the last day of the first year and another $576.19 on the last day of the second year.

## Summary

A liability obligates a firm to make a (probable) future sacrifice of resources. The firm can estimate the amount of that obligation and its timing with reasonable certainty. The transaction causing the obligation has already occurred; in other words, the obligation arises from other than an executory contract.

Accounting initially records long-term liabilities at the cash-equivalent value the borrower receives (which equals the present value of the future cash flows the borrower promises, discounted at the market interest rate for that borrower at the time of issue). As the maturity date of the liability draws near, the book value of the liability approaches the amount of the final cash payment. To compute interest expense throughout the accounting life of the liability, historical cost accounting uses the borrower's market interest rate on the date the borrower incurred the liability. Interest expense for any period is the book value of the liability at the start of the period multiplied by the historical interest rate.

Borrowers retire liabilities in different ways. In each case, the borrower debits the liability retired for its book value, credits the asset given up, and recognizes gain or loss on retirement.

## Appendix 9.1: Effects on the Statement of Cash Flows of Transactions Involving Long-Term Liabilities

Because expenditures for interest do not usually equal interest expense, and debt service payments sometimes discharge principal as well as interest, you need to think about how the transactions for long-term liabilities affect the statement of cash flows.

## Interest Expense Exceeds Debt Service Payments—Example is Bonds Issued for Less than Par

Here, interest expense exceeds debt service payments, that is, coupon payments for a bond issue. Refer to **Exhibit 9.3** for the first period. Interest expense reported for the first six months is \$6,508 (= \$6,000 in coupon payments and \$508 in increased principal amount). Notice that the borrower used only \$6,000 of cash for the expense. Liabilities increased by \$6,508. A cash payment of \$6,000 followed. The remainder of the interest expense, \$508, increased the noncurrent liability Debenture Bonds Payable.

**Direct Method of Computing Cash Flows from Operations** The entire coupon payment of \$6,000, paid in cash, discharged part of interest expense of \$6,508, which exceeds the amount of the coupon payments. Hence, the expenditure for interest expense equals the coupon payments of \$6,000.

**Indirect Method of Computing Cash Flows from Operations** When the firm uses the indirect method, interest expense has already reduced the beginning number, net income, in the example, by \$6,508. So, in deriving cash flows from operations, we add back the part of the expense, \$508, which did not use cash.

The principles here apply to all long-term debt where interest expense exceeds debt service payments.

## Debt Service Payments Exceed Interest Expense—Example is Bonds Issued for More than Par

Here, expenditures for debt service, the coupons of a bond issue, exceed interest expense. Refer to **Exhibit 9.4** for the first period of a bond issued at a premium. Interest expense reported for the first six months is \$5,386 (= \$6,000 in coupon payments reduced by \$614 which pays back part of the borrowed principal). The borrower used only \$5,386 for interest expense, although it paid \$6,000 in cash. Cash flow from operations goes down by \$5,386 and Financing uses of cash to discharge debt increases by \$614. (In practice, some accountants ignore this distinction and show all the cash expenditure as operating. We do not illustrate this practice in this text as it is conceptually wrong, and therefore harder to understand.[7])

**Direct Method of Computing Cash Flows from Operations** The operating expenditure is the amount of interest expense, \$5,386. The remainder of the expenditure of cash, \$614, appears as a Financing use of cash, to reduce the principal amount of the debt.

**Indirect Method of Computing Cash Flows from Operations** Income has already been reduced by interest expense of \$5,386. The rest of the expenditure, \$614, appears, just as above, in the Financing use of cash, to reduce the principal amount of the debt.

The principles here apply to all long-term debt where debt service payments exceed interest expense.

## Gain or Loss on Debt Retirement

When a firm retires debt for cash, it reports the cash expenditure in the Financing section as a nonoperating use, Cash Used to Reduce Debt. In most cases, the amount of cash the firm uses to retire a liability differs from the book value of the liability at the time of retirement. Thus, the retirement generates a gain (when book value exceeds cash used) or a loss (when cash used exceeds book value).

The Financing section of the statement shows all the cash used to retire the liability; that use of cash had nothing to do with operations. When the statement of cash flows presents cash from operations with the indirect format, the computation starts with net income. That figure includes the gain or loss on the repurchase of the liability. Because the repurchase is a financing, not an operating, use of cash, the accountant subtracts the amount of the gain or adds back the amount of the loss in deriving operating cash flows. See line **(4)**, for addition of the loss, and line **(5)**, for the subtraction of the gain, of **Exhibit 4.16**.

[7]We think this practice stems from accountants' not being clear in their distinctions between interest payments, debt service payments, and interest expense.

## Solutions to Self-Study Problems

### SUGGESTED SOLUTION TO PROBLEM 9.1 FOR SELF-STUDY

(Ashton Corporation; journal entries for transactions involving current liabilities.)

**a. January 2**

| | Dr. | Cr. |
|---|---|---|
| Cash | 10,000 | |
| Note Payable | | 10,000 |

| Assets | = | Liabilities | + | Shareholders' Equity | (Class.) |
|---|---|---|---|---|---|
| +10,000 | | +10,000 | | | |

To record 90-day, 9-percent bank loan.

**b. January 3**

| | Dr. | Cr. |
|---|---|---|
| Merchandise Inventory | 8,000 | |
| Accounts Payable | | 8,000 |

| Assets | = | Liabilities | + | Shareholders' Equity | (Class.) |
|---|---|---|---|---|---|
| +8,000 | | +8,000 | | | |

To record purchases of merchandise on account.

**c. January 10**

| | Dr. | Cr. |
|---|---|---|
| Cash | 1,500 | |
| Advances from Customer | | 1,500 |

| Assets | = | Liabilities | + | Shareholders' Equity | (Class.) |
|---|---|---|---|---|---|
| +1,500 | | +1,500 | | | |

To record advance from customer on merchandise scheduled for delivery in February.

**d. Month of January**

| | Dr. | Cr. |
|---|---|---|
| Accounts Receivable | 12,000 | |
| Sales Revenue | | 12,000 |

| Assets | = | Liabilities | + | Shareholders' Equity | (Class.) |
|---|---|---|---|---|---|
| +12,000 | | | | +12,000 | IncSt → RE |

To record sales on account during January.

| | Dr. | Cr. |
|---|---|---|
| Cost of Goods Sold | 6,000 | |
| Merchandise Inventory | | 6,000 |

| Assets | = | Liabilities | + | Shareholders' Equity | (Class.) |
|---|---|---|---|---|---|
| −6,000 | | | | −6,000 | IncSt → RE |

To record the cost of merchandise sold.

**e. Month of January**

| | Dr. | Cr. |
|---|---|---|
| Accounts Payable | 8,000 | |
| Cash | | 8,000 |

| Assets | = | Liabilities | + | Shareholders' Equity | (Class.) |
|---|---|---|---|---|---|
| −8,000 | | −8,000 | | | |

To record payments to suppliers for purchases on account.

| | Dr. | Cr. |
|---|---|---|
| Cash | 7,000 | |
| Accounts Receivable | | 7,000 |

| Assets | = | Liabilities | + | Shareholders' Equity | (Class.) |
|---|---|---|---|---|---|
| +7,000 | | | | | |
| −7,000 | | | | | |

To record collections from customers for sales on account.

**f. January 31**

| | | |
|---|---|---|
| Warranty Expense . . . . . . . . . . . . . . . . . . . . . . . . . . . . . . . . . . . | 960 | |
| Estimated Warranty Liability . . . . . . . . . . . . . . . . . . . . . . . . | | 960 |

| Assets | = | Liabilities | + | Shareholders' Equity | (Class.) |
|---|---|---|---|---|---|
| | | +960 | | −960 | IncSt → RE |

To record estimated warranty cost for goods sold during January; 0.08 × $12,000 = $960.

**g. January 31**

| | | |
|---|---|---|
| Wages Expense . . . . . . . . . . . . . . . . . . . . . . . . . . . . . . . . . . . | 4,000 | |
| U.S. Withholding Taxes Payable . . . . . . . . . . . . . . . . . . . . . . | | 800 |
| Social Security Taxes Payable . . . . . . . . . . . . . . . . . . . . . . . | | 400 |
| Withheld Union Dues Payable . . . . . . . . . . . . . . . . . . . . . . . | | 200 |
| Wages Payable . . . . . . . . . . . . . . . . . . . . . . . . . . . . . . . . . | | 2,600 |

| Assets | = | Liabilities | + | Shareholders' Equity | (Class.) |
|---|---|---|---|---|---|
| | | +800 | | −4,000 | IncSt → RE |
| | | +400 | | | |
| | | +200 | | | |
| | | +2,600 | | | |

To record January wages net of taxes and union dues withheld.

| | | |
|---|---|---|
| Wages Expense . . . . . . . . . . . . . . . . . . . . . . . . . . . . . . . . . . . | 540 | |
| Social Security Taxes Payable . . . . . . . . . . . . . . . . . . . . . . . | | 400 |
| Unemployment Taxes Payable . . . . . . . . . . . . . . . . . . . . . . . | | 140 |

| Assets | = | Liabilities | + | Shareholders' Equity | (Class.) |
|---|---|---|---|---|---|
| | | +400 | | −540 | IncSt → RE |
| | | +140 | | | |

To record employer's share of payroll taxes.

**h. January 31**

| | | |
|---|---|---|
| Interest Expense . . . . . . . . . . . . . . . . . . . . . . . . . . . . . . . . . | 75 | |
| Interest Payable . . . . . . . . . . . . . . . . . . . . . . . . . . . . . . . . | | 75 |

| Assets | = | Liabilities | + | Shareholders' Equity | (Class.) |
|---|---|---|---|---|---|
| | | +75 | | −75 | IncSt → RE |

To record interest expense on notes payable for January; $10,000 × 0.09 × 30/360 = $75.

**i. January 31**

| | | |
|---|---|---|
| Income Tax Expense . . . . . . . . . . . . . . . . . . . . . . . . . . . . . . . | 170 | |
| Income Tax Payable . . . . . . . . . . . . . . . . . . . . . . . . . . . . . | | 170 |

| Assets | = | Liabilities | + | Shareholders' Equity | (Class.) |
|---|---|---|---|---|---|
| | | +170 | | −170 | IncSt → RE |

To accrue income taxes payable for January: 0.40 × ($12,000 − $6,000 − $960 − $4,000 − $540 − $75) = $170.

j. **February 1**

| | | |
|---|---|---|
| Wages Payable . . . . . . . . . . . . | 2,600 | |
| Cash . . . . . . . . . . . . | | 2,600 |

| Assets | = | Liabilities | + | Shareholders' Equity | (Class.) |
|---|---|---|---|---|---|
| −2,600 | | −2,600 | | | |

To pay employees their January wages net of withholdings.

k. **February 10**

| | | |
|---|---|---|
| Advances from Customers . . . . . . . . . . . . | 1,500 | |
| Sales Revenue . . . . . . . . . . . . | | 1,500 |

| Assets | = | Liabilities | + | Shareholders' Equity | (Class.) |
|---|---|---|---|---|---|
| | | −1,500 | | +1,500 | IncSt → RE |

To record the sale of merchandise to customer.

| | | |
|---|---|---|
| Cost of Goods Sold . . . . . . . . . . . . | 800 | |
| Merchandise Inventory . . . . . . . . . . . . | | 800 |

| Assets | = | Liabilities | + | Shareholders' Equity | (Class.) |
|---|---|---|---|---|---|
| −800 | | | | −800 | IncSt → RE |

To record the cost of merchandise sold.

l. **February 15**

| | | |
|---|---|---|
| U.S. Withholding Taxes Payable . . . . . . . . . . . . | 800 | |
| Social Security Taxes Payable . . . . . . . . . . . . | 800 | |
| Unemployment Taxes Payable . . . . . . . . . . . . | 140 | |
| Withheld Union Dues Payable . . . . . . . . . . . . | 200 | |
| Cash . . . . . . . . . . . . | | 1,940 |

| Assets | = | Liabilities | + | Shareholders' Equity | (Class.) |
|---|---|---|---|---|---|
| −1,940 | | −800 | | | |
| | | −800 | | | |
| | | −140 | | | |
| | | −200 | | | |

To record payment of payroll taxes and union dues.

m. **February 20**

| | | |
|---|---|---|
| Estimated Warranty Liability . . . . . . . . . . . . | 220 | |
| Cash . . . . . . . . . . . . | | 220 |

| Assets | = | Liabilities | + | Shareholders' Equity | (Class.) |
|---|---|---|---|---|---|
| −220 | | −220 | | | |

To record the cost of warranty repairs on products sold during January.

---

## SUGGESTED SOLUTION TO PROBLEM 9.2 FOR SELF-STUDY

(Sapra Company; accounting for an interest-bearing note.)

**a.** See **Exhibit 9.5**.

**EXHIBIT 9.5**

**Amortization Schedule for Note with Face Value of $100,000 Issued for $112,434, Bearing Interest at the Rate of 15 Percent of Face Value per Year, Issued to Yield 10 Percent (Problem 9.2 for Self-Study)**

| Yearly Periods (1) | Loan Balance Start of Period (2) | Interest Expense for Period (3) | Payment (4) | Portion of Payment Reducing Book Value of Liability (5) | Loan Balance End of Period (6) |
|---|---|---|---|---|---|
| 0 .................... | | | | | $112,434 |
| 1 .................... | $112,434 | $11,243 | $ 15,000 | $ 3,757 | 108,677 |
| 2 .................... | 108,677 | 10,868 | 15,000 | 4,132 | 104,545 |
| 3 .................... | 104,545 | 10,455 | 115,000 | 104,545 | 0 |

Column (2) = column (6) from previous period.
Column (3) = 0.10 × column (2).
Column (4) is given.
Column (5) = column (4) − column (3).
Column (6) = column (2) − column (5).

**b.** **Date of Issue**

| | | |
|---|---|---|
| Cash ............................................. | 112,434 | |
| Note Payable .................................... | | 112,434 |

| Assets | = | Liabilities | + | Shareholders' Equity | (Class.) |
|---|---|---|---|---|---|
| +112,434 | | +112,434 | | | |

Proceeds of issue of note.

**Six Months after Issue**

| | | |
|---|---|---|
| Interest Expense ..................................... | 5,622 | |
| Interest Payable .................................. | | 5,622 |

| Assets | = | Liabilities | + | Shareholders' Equity | (Class.) |
|---|---|---|---|---|---|
| | | +5,622 | | −5,622 | IncSt → RE |

See **Exhibit 9.5**; accrual of six months' interest = $11,243/2.

**One Year after Issue**

| | | |
|---|---|---|
| Interest Expense ..................................... | 5,621 | |
| Interest Payable ..................................... | 5,622 | |
| Note Payable ......................................... | 3,757 | |
| Cash ............................................. | | 15,000 |

| Assets | = | Liabilities | + | Shareholders' Equity | (Class.) |
|---|---|---|---|---|---|
| −15,000 | | −5,622 | | −5,621 | IncSt → RE |
| | | −3,757 | | | |

Interest expense for the remainder of the first year and cash payment made. Excess of cash payment over interest expense reduces note principal.

## SUGGESTED SOLUTION TO PROBLEM 9.3 FOR SELF-STUDY

(Bens Company; accounting for note with imputed interest rate.)

**a.** Truck ........ 22,500
Note Payable ........ 22,500

| Assets | = | Liabilities | + | Shareholders' Equity | (Class.) |
|---|---|---|---|---|---|
| +22,500 | | +22,500 | | | |

0.90 × $25,000 = $22,500.

**b.** $28,730/$22,500 = 1.27689. The truck has fair market value of $22,500 (= 0.90 × $25,000); $(1 + r)^2 = 1.27689$, which implies that $r = \sqrt{1.27689} - 1.00 = 0.13$, or 13 percent per year. In other words, $22,500 grows to $28,730 in two years when the interest rate is 13 percent per period.

**c.** Truck ........ 22,903
Note Payable ........ 22,903

| Assets | = | Liabilities | + | Shareholders' Equity | (Class.) |
|---|---|---|---|---|---|
| +22,903 | | +22,903 | | | |

$(1.12)^{-2} = 0.79719$; 0.79719 × 28,730 = $22,903.

**d.** $(1.01)^{-24} = 0.78757$. (See **Table 2** at the end of the book, 24-period row, 1-percent column.) 0.78757 × $28,730 = $22,627.

Truck ........ 22,627
Note Payable ........ 22,627

| Assets | = | Liabilities | + | Shareholders' Equity | (Class.) |
|---|---|---|---|---|---|
| +22,627 | | +22,627 | | | |

**e.** Truck ........ 23,744
Note Payable ........ 23,744

| Assets | = | Liabilities | + | Shareholders' Equity | (Class.) |
|---|---|---|---|---|---|
| +23,744 | | +23,744 | | | |

Year 1:

Interest Expense (= 0.10 × $23,744) ........ 2,374
Note Payable ........ 2,374

| Assets | = | Liabilities | + | Shareholders' Equity | (Class.) |
|---|---|---|---|---|---|
| | | +2,374 | | −2,374 | IncSt → RE |

Year 2:

Interest Expense [= 0.10 × ($23,744 + $2,374)] ........ 2,612
Note Payable ........ 2,612

| Assets | = | Liabilities | + | Shareholders' Equity | (Class.) |
|---|---|---|---|---|---|
| | | +2,612 | | −2,612 | IncSt → RE |

Balance in Note Payable Account at End of Year 2 ........ $28,730

**f.** Total expense equals interest expense on the note plus depreciation on the truck. Interest expense is $28,730 less the amount at which the firm records the truck. Depreciation expense is equal to the amount at which Bens records the truck less estimated salvage value. Thus, over the life of the truck or two years, whichever is longer, the total expense equals $28,730 less the salvage value of the truck.

## SUGGESTED SOLUTION TO PROBLEM 9.4 FOR SELF-STUDY

(Engel Corporation; calculating the issue price of bonds.)

**a.**

| Required Cash Flows | Present Value Factor for 8 Percent Interest Rate Compounded Semiannually for 10 Years | Present Value of Required Cash Flows |
|---|---|---|
| $1,000,000 at end of 10 years . . . . . . . . . . . . . . . | 0.45639[a] | 456,390 |
| $50,000 every six months for 10 years . . . . . . . . . | 13.59033[b] | 679,516 |
| Issue Price . . . . . . . . . . . . . . . . . . . . . . . . . . |  | $1,135,906 |

[a]Table 2, 4-percent column and 20-period row.
[b]Table 4, 4-percent column and 20-period row.

**b.**

| Required Cash Flows | Present Value Factor for 10 Percent Interest Rate Compounded Semiannually for 10 Years | Present Value of Required Cash Flows |
|---|---|---|
| $1,000,000 at end of 10 years . . . . . . . . . . . . . . . | 0.37689[a] | $376,890 |
| $50,000 every six months for 10 years . . . . . . . . . . | 12.46221[b] | 623,110 |
| Issue Price . . . . . . . . . . . . . . . . . . . . . . . . . . |  | $1,000,000 |

[a]Table 2, 5-percent column and 20-period row.
[b]Table 4, 5-percent column and 20-period row.

**c.**

| Required Cash Flows | Present Value Factor for 12 Percent Interest Rate Compounded Semiannually for 10 Years | Present Value of Required Cash Flows |
|---|---|---|
| $1,000,000 at end of 10 years . . . . . . . . . . . . . . . | 0.31180[a] | $311,800 |
| $50,000 every six months for 10 years . . . . . . . . . . | 11.46992[b] | 573,496 |
| Issue Price . . . . . . . . . . . . . . . . . . . . . . . . . . |  | $885,296 |

[a]Table 2, 6-percent column and 20-period row.
[b]Table 4, 6-percent column and 20-period row.

## SUGGESTED SOLUTION TO PROBLEM 9.5 FOR SELF-STUDY

(Engel Corporation; preparing journal entries to account for bonds.)

**a. January 1**

| | | |
|---|---|---|
| Cash . . . . . . . . . . . . . . . . . . . . . . . . . . . . . . . . . . . . . . . . . . . . | 1,135,906 | |
| Bonds Payable . . . . . . . . . . . . . . . . . . . . . . . . . . . . . . . . | | 1,135,906 |

| Assets | = | Liabilities | + | Shareholders' Equity | (Class.) |
|---|---|---|---|---|---|
| +1,135,906 | | +1,135,906 | | | |

To record issue of $1,000,000-par value, 10-percent semiannual coupon bonds priced to yield 8 percent compounded semiannually.

**June 30**

| | | |
|---|---|---|
| Interest Expense (= 0.04 × 1,135,906) . . . . . . . . . . . . . . . . . . . . . | 45,436 | |
| Bonds Payable . . . . . . . . . . . . . . . . . . . . . . . . . . . . . . . . . . . . . . . | 4,564 | |
| Cash (= 0.05 × $1,000,000) . . . . . . . . . . . . . . . . . . . . . . . . | | 50,000 |

(continued)

| Assets | = | Liabilities | + | Shareholders' Equity | (Class.) |
|---|---|---|---|---|---|
| -50,000 | | -4,564 | | -45,436 | IncSt → RE |

To record interest expense and amount paid for first six months.

**December 31**

| | | |
|---|---|---|
| Interest Expense [= 0.04 × ($1,135,906 − $4,564)] | 45,254 | |
| Bonds Payable | 4,746 | |
| Cash (= 0.05 × $1,000,000) | | 50,000 |

| Assets | = | Liabilities | + | Shareholders' Equity | (Class.) |
|---|---|---|---|---|---|
| -50,000 | | -4,746 | | -45,254 | IncSt → RE |

To record interest expense and amount payable for second six months.

**b. January 1**

| | | |
|---|---|---|
| Cash | 1,000,000 | |
| Bonds Payable | | 1,000,000 |

| Assets | = | Liabilities | + | Shareholders' Equity | (Class.) |
|---|---|---|---|---|---|
| +1,000,000 | | +1,000,000 | | | |

To record issue of $1,000,000-par value, 10-percent semiannual coupon bonds priced to yield 10 percent compounded semiannually.

**June 30**

| | | |
|---|---|---|
| Interest Expense (= 0.05 × $1,000,000) | 50,000 | |
| Cash (= 0.05 × $1,000,000) | | 50,000 |

| Assets | = | Liabilities | + | Shareholders' Equity | (Class.) |
|---|---|---|---|---|---|
| -50,000 | | | | -50,000 | IncSt → RE |

To record interest expense and amount paid for first six months.

**December 31**

| | | |
|---|---|---|
| Interest Expense (= 0.05 × $1,000,000) | 50,000 | |
| Cash (= 0.05 × $1,000,000) | | 50,000 |

| Assets | = | Liabilities | + | Shareholders' Equity | (Class.) |
|---|---|---|---|---|---|
| -50,000 | | | | -50,000 | IncSt → RE |

To record interest expense and amount paid for second six months.

**c. January 1**

| | | |
|---|---|---|
| Cash | 885,296 | |
| Bonds Payable | | 885,296 |

| Assets | = | Liabilities | + | Shareholders' Equity | (Class.) |
|---|---|---|---|---|---|
| +885,296 | | +885,296 | | | |

To record issue of $1,000,000-par value, 10-percent semiannual coupon bonds priced to yield 12 percent compounded semiannually.

**June 30**

| | | |
|---|---|---|
| Interest Expense (= 0.06 × $885,296) | 53,118 | |
| Cash (= 0.05 × $1,000,000) | | 50,000 |
| Bonds Payable | | 3,118 |

*(continued)*

| Assets | = | Liabilities | + | Shareholders' Equity | (Class.) |
|---|---|---|---|---|---|
| −50,000 | | +3,118 | | −53,118 | IncSt → RE |

To record interest expense and amount paid for first six months.

**December 30**

| | | |
|---|---|---|
| Interest Expense [= 0.06 × ($885,296 + $3,118)] . . . . . . . . . . . . . . | 53,305 | |
| Cash (= 0.05 × $1,000,000) . . . . . . . . . . . . . . . . . . . . . . . | | 50,000 |
| Bonds Payable . . . . . . . . . . . . . . . . . . . . . . . . . . . . . . . | | 3,305 |

| Assets | = | Liabilities | + | Shareholders' Equity | (Class.) |
|---|---|---|---|---|---|
| −50,000 | | +3,305 | | −53,305 | IncSt → RE |

To record interest expense and amount payable for second six months.

## SUGGESTED SOLUTION TO PROBLEM 9.6 FOR SELF-STUDY

(Unifying principles of accounting for long-term liabilities.)

**Exhibit 9.6** shows the accounting for five types of long-term monetary liabilities stated at the present value of future cash flows in columns labeled **a.** through **e.** The accounting for each of these monetary liabilities follows a common procedure.

1. Compute the initial amount of cash received by the borrower as well as the historical interest rate. Sometimes you will know both of these. Sometimes you will know the cash received and you must calculate the interest rate. Sometimes, as **Exhibit 9.6** illustrates in all five cases, you will know the interest rate and must compute the initial cash received.
   - **a.** To compute the initial amount of cash received, given the contractual payments and the market interest rate, multiply each of the contractual payments by the present value factor (as from **Table 2** at the back of the book) for a single payment of $1 to be received in the future. **Exhibit 9.6** shows the present value factors at 10-percent interest for payments to be received in 1 year (0.90909) and in 2 years (0.82645).
   - **b.** Computing the market interest rate, given the initial cash proceeds and the series of contractual payments, requires finding the *internal rate of return* of the series of cash flows. The Appendix at the back of the book illustrates this process. **Exhibit 9.6** shows that only the 10-percent coupon bond and the level-payment note have initial cash proceeds equal to $1,000. The difference in amounts arises because each of the items has a different present value, in spite of the fact that some people might, loosely speaking, call each a "$1,000 liability."
2. Record a journal entry debiting cash and crediting the monetary liability with the amount of cash received. This presentation showing the common theme uses the generic account title Monetary Liability, although in practice a firm would use more descriptive titles.
3. At every contractual payment date and at the end of an accounting period, compute interest expense as the book value of the liability at the beginning of the period (which includes the principal liability account and the Interest Payable account if the firm keeps these amounts in separate accounts) multiplied by the historical interest rate. Debit the computed amount to interest expense and credit the liability account.

   If the borrower makes a cash payment, credit cash and debit the liability. The book value of the liability is now equal to the beginning balance plus interest expense recorded less cash payments made, if any.

   **Exhibit 9.6** does not illustrate this fact directly, but if you were to return to **Step 1** at this point and compute the present value of the remaining contractual payments using the historical interest rate (10 percent in the examples), that present value would equal the book value computed after **Step 3**.
4. At each payment date, or at each period-end closing date, repeat **Step 3**. Eventually, when the borrower makes the final payment (as illustrated at the bottom of **Exhibit 9.6**), it will have discharged the entire amount of the liability plus interest. The remaining liability is zero. The accounting has amortized the liability to zero at the same time that the firm has extinguished its obligation.

## EXHIBIT 9.6 Accounting for Long-Term Monetary Liabilities Based on the Present Value of Future Cash Flows (Problem 9.6 for Self-Study)

| | a. Single-Payment Note of $1,000 Maturing in Two Years | | | b. Two-Year Annual Coupon Bond—10 Percent ($100) Coupons | | |
|---|---|---|---|---|---|---|
| | Amount | Dr. | Cr. | Amount | Dr. | Cr. |
| **(1)** Compute Present Value of Future Contractual Payments Using Historical Interest Rate on Day Monetary Liability is First Recorded. Rate is 10.0 Percent. | | | | | | |
| **(a)** 1 Year Hence | $ 0 | | | $ 100.00 | | |
| **(b)** 2 Years Hence | $1,000.00 | | | $1,100.00 | | |
| Multiply Payment by Present Value Factors (Table 2). | | | | | | |
| 0.90909 × (a) | $ 0 | | | $ 90.91 | | |
| 0.82645 × (b) | 826.45 | | | 909.09 | | |
| **(c)** Total Present Value | $ 826.45 | | | $1,000.00 | | |
| **(2)** Record Initial Liability and Cash or Other Assets Received from Step 1. | | | | | | |
| Cash or Other Assets | | 826.45 | | | 1,000.00 | |
| Monetary Liability | | | 826.45 | | | 1,000.00 |
| **(3)** First Recording (payment date or end of period): End of First Year. | | | | | | |
| **(a)** Compute Interest Expense as Monetary Liability × Historical Interest Rate. | | | | | | |
| Amount on Line 1(c) × 0.10 | $ 82.64 | | | $ 100.00 | | |
| **(b)** Record Interest Expense. | | | | | | |
| Interest Expense | | 82.64 | | | 100.00 | |
| Monetary Liability | | | 82.64 | | | 100.00 |
| **(c)** Record Cash Payment (if any). | | | | | | |
| Monetary Liability | | — | | | 100.00 | |
| Cash | | | — | | | 100.00 |
| **(d)** Compute Book Value of Monetary Liability. | | | | | | |
| Beginning Balance | $ 826.45 | | | $1,000.00 | | |
| Add Interest Expense | 82.64 | | | 100.00 | | |
| Subtotal | $ 909.09 | | | $1,100.00 | | |
| Subtract Cash Payment (if any) | — | | | (100.00) | | |
| = Ending Balance | $ 909.09 | | | $1,000.00 | | |
| **(4)** Second Recording: End of Second Year. | | | | | | |
| **(a)** Compute Interest Expense as Monetary Liability × Historical Interest Rate. | | | | | | |
| Amount on Line 3(d) × 0.10 | $ 90.91 | | | $100.00 | | |
| **(b)** Record Interest Expense. | | | | | | |
| Interest Expense | | 90.91 | | | 100.00 | |
| Monetary Liability | | | 90.91 | | | 100.00 |
| **(c)** Record Cash Payment (if any). | | | | | | |
| Monetary Liability | | 1,000.00 | | | 1,100.00 | |
| Cash | | | 1,000.00 | | | 1,100.00 |
| **(d)** Compute Book Value of Monetary Liability. | | | | | | |
| Beginning Balance | $ 909.09 | | | $1,000.00 | | |
| Add Interest Expense | 90.91 | | | 100.00 | | |
| Subtotal | $1,000.00 | | | $1,100.00 | | |
| Subtract Cash Payment (if any) | (1,000.00) | | | (1,100.00) | | |
| = Ending Balance | $ 0 | | | $ 0 | | |

| c. Two-Year Annual Coupon Bond—8 Percent ($80) Coupons | | | d. Two-Year Annual Coupon Bond—12 Percent ($120) Coupons | | | e. Two-Year Level-Payment Note—Annual Payments of $576.19 | | |
|---|---|---|---|---|---|---|---|---|
| **Amount** | **Dr.** | **Cr.** | **Amount** | **Dr.** | **Cr.** | **Amount** | **Dr.** | **Cr.** |
| $ 80.00 | | | $ 120.00 | | | $ 576.19 | | |
| $1,080.00 | | | $1,120.00 | | | $ 576.19 | | |
| $ 72.73 | | | $ 109.09 | | | $ 523.81 | | |
| 892.57 | | | 925.62 | | | 476.19 | | |
| $ 965.30 | | | $1,034.71 | | | $1,000.00 | | |
| | 965.30 | | | 1,034.71 | | | 1,000.00 | |
| | | 965.30 | | | 1,034.71 | | | 1,000.00 |
| $ 96.53 | | | $ 103.47 | | | $ 100.00 | | |
| | 96.53 | | | 103.47 | | | 100.00 | |
| | | 96.53 | | | 103.47 | | | 100.00 |
| | 80.00 | | | 120.00 | | | 576.19 | |
| | | 80.00 | | | 120.00 | | | 576.19 |
| $ 965.30 | | | $1,034.71 | | | $1,000.00 | | |
| 96.53 | | | 103.47 | | | 100.00 | | |
| $1,061.83 | | | $1,138.18 | | | $1,100.00 | | |
| (80.00) | | | (120.00) | | | (576.19) | | |
| $ 981.83 | | | $1,018.18 | | | $ 523.81 | | |
| $ 98.18 | | | $ 101.82 | | | $ 52.38 | | |
| | 98.18 | | | 101.82 | | | 52.38 | |
| | | 98.18 | | | 101.82 | | | 52.38 |
| | 1,080.00 | | | 1,120.00 | | | 576.19 | |
| | | 1,080.00 | | | 1,120.00 | | 576.19 | |
| $ 981.83 | | | $1,018.18 | | | $ 523.81 | | |
| 98.18 | | | 101.82 | | | 52.38 | | |
| $1,080.01 | | | 1,120.00 | | | $ 576.19 | | |
| (1,080.00) | | | (1,120.00) | | | (576.19) | | |
| $ 0[a] | | | $ 0 | | | $ 0 | | |

[a]Rounding error of $0.01.

## Key Terms and Concepts

- Liability
- Executory contract
- Estimated liability
- Contingency
- Constructive liability
- Mortgage
- Collateral
- Mortgagee, mortgagor
- Implicit interest
- Implicit (imputed) interest rate
- Bond indenture
- Coupon bonds
- Debenture bonds
- Convertible bonds
- Serial bonds
- Zero coupon bonds
- Bond tables
- Effective interest method
- Sinking fund
- Bond refunding
- Right of offset
- Call price

## Questions, Short Exercises, Exercises, Problems, and Cases

For additional student resources, content, and interactive quizzes for this chapter visit the FACMU website: **www.thomsonedu.com/accounting/stickney**

### QUESTIONS

1. Review the meaning of the terms and concepts listed above in **Key Terms and Concepts**.
2. For each of the following items, indicate whether the item meets all of the criteria of a liability. If so, how does the firm value it?
   a. Interest accrued but not paid on a note
   b. Advances from customers for goods and services to be delivered later
   c. Confirmed orders from customers for goods and services to be delivered later
   d. Bonds payable
   e. Product warranties
   f. Damages the company must pay if it loses a pending lawsuit
   g. Future costs of restoring strip-mining sites after completing mining operations
   h. Contractual promises to purchase natural gas for each of the next 10 years
   i. Promises by an airline to provide free flights in the future if customers accumulate a certain number of miles at regular fares
3. What is the amount of the liability recognized in each of the independent cases below?
   a. A plaintiff files a lawsuit against a company. The probability is 90 percent that the company will lose. If it loses, the amount of the loss will most likely be $100,000.
   b. A cereal company issues redeemable coupons for "free" boxes of cereal. Now, it issues one million coupons that promise the retailer who redeems the coupons $1 per coupon. The probability of redemption of any one coupon is 9 percent.
4. "Firms should obtain as much financing as possible from suppliers through accounts payable because it is a free source of capital." Do you agree? Why or why not?
5. The Francis W. Parker School, a private lower school, has a reporting year ending June 30. It hires teachers for a 10-month period: September of one year through June of the following year. It contracts to pay teachers in 12 monthly installments over the period September of one year through August of the next year. For the current academic year, suppose that the total contractual salaries to be paid to teachers is $3,600,000. How should the school account for this amount in the financial statements issued June 30, at the end of the academic year?
6. A magazine publisher offers a reduced annual subscription fee if customers pay for three years in advance. Under this subscription program, the magazine publisher receives from customers $45,000, which it credits to the account Advances from Customers. The estimated cost of publishing and distributing magazines for these customers is $32,000. Why does accounting report a liability of $45,000 instead of $32,000?
7. A noted accountant once remarked that the optimal number of faulty TV sets for Sony to sell is "not zero," even if Sony promises to repair all faulty Sony sets that break down, for whatever reason, within two years of purchase. Why could the optimal number be "not zero"?
8. Describe the similarities and differences between the allowance method for uncollectibles (see **Chapter 6**) and the allowance method for estimated warranty costs.
9. What factors determine the amount a firm actually receives when it offers a bond issue to the market?

**10.** Distinguish between the following sets of terms for bonds. Under what circumstances will they be the same? Under what circumstances will they differ?
   **a.** Par value and face value
   **b.** Par value and book value
   **c.** Book value and current market value

**11.** "Using the market interest rate at the time of issue to account for bonds in subsequent periods provides a book value for bonds that is consistent with using historical, or acquisition, cost valuations for assets." Explain.

**12.** If the Deer Valley Ski Association borrows $1,000,000 by issuing, at par, 20-year, 10-percent bonds with semiannual coupons, the total interest expense over the life of the issue is $2,000,000 (= 20 × 0.10 × $1,000,000). If Deer Valley undertakes a 20-year mortgage or note with an implicit borrowing rate of 10 percent, the annual payments are $1,000,000/8.51356 = $117,460. (See **Table 4** at the end of the book, 20-period row, 10-percent column.) The total mortgage payments are $2,349,200 (= 20 × $117,460), and the total interest expense over the life of the note or mortgage is $1,349,200 (= $2,349,200 − $1,000,000).

Why are the amounts of interest expense different for these two types of borrowing for the same length of time at identical interest rates?

**13.** Define *zero coupon bonds*. What are the advantages of a zero coupon bond to the issuer and to the investor?

**14.** A call premium is the difference between the call price of a bond and its par (face) value. What is the purpose of such a premium?

**15.** Critics of historical cost accounting for long-term debt argue that the procedures give management unreasonable opportunity to "manage" income with the timing of bond retirements. What phenomenon do these critics have in mind?

## SHORT EXERCISES

**16. Recognition of a loss contingency.** While shopping in a store on July 5, Year 6, a customer slips on the floor and sustains back injuries. On January 15, Year 7, the customer sues the store for $1 million. The case comes to trial on April 30, Year 7. The jury renders its verdict on June 15, Year 7, and finds the store liable for negligence. The jury grants a damage award of $400,000 to the customer. The store, on June 25, Year 7, appeals the decision to a higher court on the grounds that the lower court failed to admit certain evidence. The higher court rules on November 1, Year 7, that the trial court should have admitted the evidence. The trial court rehears the case beginning on March 21, Year 8. Another jury, on April 20, Year 8, again finds the store liable for negligence and awards $500,000. On May 15, Year 8, the store pays the $500,000 judgment. When should the store recognize a loss from these events? Explain your reasoning.

**17. Journal entries for payroll.** During the current period, suppose that McGee Associates' office employees earned wages of $700,000. McGee withheld 30 percent of this amount for payments for various income and payroll taxes. In addition, McGee must pay 10 percent of gross wages for the employer's share of various taxes. McGee has promised to contribute 4 percent of gross wages to a profit-sharing fund, which workers will share as they retire. Employees earned vacation pay estimated to be $14,000; estimated fringe benefits are 20 percent of that amount.
   **a.** Prepare journal entries for these wage-related items.
   **b.** What is total wage and salary expense?

**18. Allowance method for uncollectibles and warranties.** A firm estimates both uncollectible accounts and warranty repair costs using a percentage of sales. Assume that the combined Bad Debt Expense and Warranty Expense computed with the percentage-of-sales method totals $20,000. At the end of the period, the firm also performs aging analyses of both receivables and outstanding warranty obligations and finds that the implied total expense should be $19,000. What will be the firm's combined bad debt and warranty expense for the year? Explain.

**19. Amortization schedule for note where explicit interest differs from market rate of interest.** Blaydon Company acquires a computer from Orange Computer Company. The cash price (fair market value) of the computer is $36,157. Blaydon Company gives a three-year, interest-bearing note with maturity value of $40,000. The note requires annual interest payments of 8 percent of face value, or $3,200 per year. The interest rate implicit in the note is 12 percent per year.

a. Prepare an amortization schedule for the note.
b. Prepare journal entries for Blaydon Company over the life of the note.

20. **Computing the issue price of bonds.** Compute the issue price of each of the following bonds.
a. $10,000,000-face value, zero coupon bonds due in 20 years, priced on the market to yield 12 percent compounded semiannually.
b. $10,000,000-face value, serial bonds repayable in equal semiannual installments of $500,000, which includes coupon payments and repayment of principal, for 20 years, priced on the market to yield 6 percent.

21. **Using bond tables; computing interest expense.** Refer to **Table 5** at the back of the book for 10-percent semiannual coupon bonds. On January 1, Year 1, assume that Florida Edison Company issued $1 million face value, 10 percent semiannual coupon bonds maturing in 10 years (on December 31, Year 10) at a price to yield 12 percent per year, compounded semiannually. Use the effective interest method of computing interest expense.
a. What were the proceeds of the original issue?
b. What was the interest expense for the first half of Year 1?
c. What was the interest expense for the second half of Year 1?
d. What was the book value of the bonds on January 1, Year 6 (when the bonds have five years until maturity)?
e. What was the interest expense for the first half of Year 6?

22. **Amortization schedule for bonds.** Womack Company issues 10-percent semiannual coupon bonds maturing five years from the date of issue. The coupons, each dated for January 1 and July 1 of each year, promise 5 percent of the face value, 10 percent total for a year. The firm issues the bonds to yield 8 percent, compounded semiannually. The bonds have face value of $100,000.
a. What are the initial issue proceeds received by Womack Company?
b. Construct an amortization schedule, similar to that in **Exhibit 9.4**, for this bond issue.
c. Assume that at the end of the third year of the bond's life, Womack Company reacquires $10,000 face value of bonds for 103 percent of par and retires them. Give the journal entry to record the retirement.

## EXERCISES

23. **Accounting for uncollectible accounts and warranties.** Hurley Corporation sells household appliances (for example, refrigerators, dishwashers) to customers on account. The firm also provides warranty services on products sold. Hurley *estimates* that 2 percent of sales will ultimately become uncollectible and that warranty costs will equal 6 percent of sales. *Actual* uncollectible accounts and warranty expenditures generally occur within three years of the time of sale. Amounts in selected accounts appear below:

| December 31: | Year 8 | Year 9 | Year 10 |
|---|---|---|---|
| Accounts Receivable, net of Allowance for Uncollectible Accounts of $355 on December 31, Year 8, $405 on December 31, Year 9, and $245 on December 31, Year 10 | $7,000 | $7,750 | $6,470 |
| Estimated Warranty Liability | 1,325 | 1,535 | 1,720 |

| For the Year: | Year 9 | Year 10 |
|---|---|---|
| Sales Revenue | $18,000 | $16,000 |

a. Prepare an analysis that explains the change in the Allowance for Uncollectible Accounts account during Year 9 and Year 10.
b. Prepare an analysis that explains the change in the Estimated Warranty Liability account during Year 9 and Year 10.

24. **Journal entries for coupons.** Morrison's Cafeteria sells coupons that customers may use later to purchase meals. Each coupon book sells for $25 and has a face value of $30; that is, the customer can use the book to purchase meals with menu prices of $30. On January 1, redeemable unused coupons that Morrison's had sold for $4,000 were outstanding. Cash inflows during the next three months appear below:

| | January | February | March |
|---|---|---|---|
| Cash-Paying Customers | $48,000 | $48,500 | $50,000 |
| Sale of Coupon Books | 2,100 | 2,200 | 2,400 |
| Total Cash Receipts | $50,100 | $50,700 | $52,400 |

Customers returned coupons with a discounted face value for meals as follows: January, $1,600; February, $2,300; March $2,100.

**a.** Prepare journal entries for January, February, and March to reflect the above information.

**b.** What effect, if any, do the coupon sales and redemptions have on the liabilities on the March 31 balance sheet?

**25. Journal entries for service contracts.** Abson Corporation began business on January 1. Abson sells copiers to business firms. It also sells service contracts to maintain and repair copiers for $600 per year. When a customer signs a contract, Abson collects the $600 fee and credits Service Contract Fees Received in Advance. Abson recognizes revenues on a quarterly basis during the year in which the coverage is in effect. For purposes of computing revenue, Abson assumes that all sales of service contracts occur midway through each quarter. Sales of contracts and service expenses for its first year of operations appear below:

| | Sales of Contracts | Service Expenses |
|---|---|---|
| First Quarter | $180,000 (300 contracts) | $ 32,000 |
| Second Quarter | 300,000 (500 contracts) | 71,000 |
| Third Quarter | 240,000 (400 contracts) | 105,000 |
| Fourth Quarter | 120,000 (200 contracts) | 130,000 |

**a.** Prepare journal entries for the first three quarters of the year for Abson Corporation. Assume that the firm prepares quarterly reports on March 31, June 30, and September 30.

**b.** What is the balance in the Service Contract Fees Received in Advance account on December 31?

**26. Journal entries for estimated warranty liabilities and subsequent expenditures.** A new appliance introduced by Maypool Corporation carries a two-year warranty against defects. The firm *estimates* that the total cost of warranty claims over the two-year period on appliances sold in a particular year (for example, Year 1) will equal 4 percent of sales revenue in the year of sale (that is, Year 1). The firm will incur *actual* warranty costs over the two-year period following the time of sale. Sales (all on account) and actual warranty expenditures (all paid in cash) for the first two years of the appliance's life were as follows:

| | Sales | Actual Warranty Expenditures |
|---|---|---|
| Year 1 | $1,200,000 | $12,000 |
| Year 2 | 1,500,000 | 50,000 |

**a.** Prepare journal entries for the events of Year 1 and Year 2. Closing entries are not required.

**b.** What is the balance in the Estimated Warranty Liability account at the end of Year 2?

**27. Journal entries for estimated warranty liabilities and subsequent expenditures.** Global Motors Corporation offers three-year warranties against defects on the sales of its automobiles. The firm *estimates* that the total cost of warranty claims over the three-year warranty period on automobiles sold in a particular year (for example, Year 1) will equal 6 percent of sales revenue in the year of sale (that is, Year 1). The firm will incur *actual* warranty costs over the three-year period following the time of sale. Sales (all for cash) and actual warranty costs incurred on automobiles under warranty (60 percent in cash and 40 percent in parts) appear below:

| | Sales | Actual Warranty Costs Incurred during Year on Automobiles under Warranty |
|---|---|---|
| Year 1 | $ 800,000 | $22,000 |
| Year 2 | 1,200,000 | 55,000 |
| Year 3 | 900,000 | 52,000 |

a. Prepare journal entries for the events of Year 1, Year 2, and Year 3. Closing entries are not required.
b. What is the balance in the Estimated Warranty Liability account at the end of Year 3?

28. **Journal entry for short-term note payable.** On December 1, Sung Company obtained a 60-day loan for $50,000 from the City State Bank at an annual interest rate of 6 percent. On the maturity date, the bank renewed the note for another 30 days, and Sung Company issued a check to the bank for the accrued interest. Sung Company closes its books annually at December 31.
   **a.** Present entries on the books of Sung Company to record the issue of the note, the year-end adjustment, the renewal of the note, and the payment of cash at maturity of the renewed note.
   **b.** Present entries at the maturity date of Sung Company's original note for the following variations in the settlement of the note.
      **(1)** Sung pays the original note at maturity.
      **(2)** Sung Company renews the note for 30 days; the new note bears interest at 9 percent per annum. Sung did not pay interest on the old note at maturity.
29. **Computing the issue price of bonds.** Compute the issue price of each of the following bonds.
   **a.** $1,000,000-face value, zero coupon bonds due in 20 years, priced on the market to yield 12 percent compounded semiannually.
   **b.** $1,000,000-face value, serial bonds repayable in equal semiannual installments of $50,000 for 20 years, priced on the market to yield 6 percent.
   **c.** $1,000,000-face value, 10-percent semiannual coupon bonds, with interest payable each six months and the principal due in 20 years, priced on the market to yield 8 percent.
   **d.** $1,000,000-face value semiannual coupon bonds, with an annual coupon rate of 6 percent for the first ten years and 14 percent for the second ten years and the principal due in 20 years, priced on the market to yield 10 percent.
30. **Accounting for bonds.** Several years ago, Huergo Dooley Corporation (HDC) issued $2,000,000-face value, 8-percent semiannual coupon bonds on the market priced to yield 10 percent, compounded semiannually. The bonds require HDC to make semiannual payments of 4 percent of face value, on June 30 and December 31 of each year. The bonds mature on December 31, Year 5.
   **a.** Compute the book value of these bonds on January 1, Year 1.
   **b.** Give HDC's journal entry to recognize interest expense and cash payments on June 30, Year 1.
   **c.** Give HDC's journal entry to recognize interest expense and cash payments on December 31, Year 1.
   **d.** On January 1, Year 2, these bonds traded in the market at a price to yield 6 percent, compounded semiannually. On this date, HDC repurchased 20 percent of these bonds on the open market and retired them. Give its journal entry to record the repurchase.
31. **Computing the issue price of bonds and interest expense.** O'Brien Corporation issues $8,000,000-par value, 8-percent semiannual coupon bonds maturing in 20 years. The market initially prices these bonds to yield 6 percent compounded semiannually.
   **a.** Compute the issue price of these bonds.
   **b.** Compute interest expense for the first six-month period.
   **c.** Compute interest expense for the second six-month period.
   **d.** Compute the book value of the bonds after the second six-month period.
   **e.** Use present value computations to verify the book value of the bonds after the second six-month period as computed in part **d** above.
32. **Computing the issue price of bonds and interest expense.** Robinson Company issues $5,000,000-par value, 8-percent semiannual coupon bonds maturing in 10 years. The market initially prices these bonds to yield 10 percent compounded semiannually.
   **a.** Compute the issue price of these bonds.
   **b.** Compute interest expense for the first six-month period.
   **c.** Compute interest expense for the second six-month period.
   **d.** Compute the book value of the bonds after the second six-month period.
   **e.** Use present value computations to verify the book value of the bonds after the second six-month period as computed in part **d** above.
33. **Using bond tables.** Refer to **Table 6** for 12-percent semiannual coupon bonds issued to yield 11 percent per year compounded semiannually. All of the questions that follow refer to the $1 million face value of such bonds issued by Centrix Company.

**a.** What are the initial issue proceeds for bonds issued to mature in 25 years?
**b.** What is the book value of those bonds after five years?
**c.** What is the book value of the bonds when they have 15 years until maturity?
**d.** What are the initial issue proceeds for bonds issued to mature in 15 years? (Compare your answer to part **c**.)
**e.** Write an expression for interest expense for the last six months before maturity.
**f.** If the market rate of interest on the bonds is 13 percent compounded semiannually, what is the market value of the bonds when they have 15 years to maturity?
**g.** When the bonds have 10 years until maturity, they trade in the market for 112.46 percent of par. What is the effective market rate at that time?

**34. Journal entries for bond coupon payments and retirement.** (Adapted from a problem by S. Zeff.) On December 31, Year 7, at the close of Mendoza Corporation's fiscal year, the company has outstanding $1 million face value of 12-percent semiannual coupon bonds, with payments due on July 1 and December 31 each year through the bonds' maturity date of December 31, Year 16. The company issued the bonds at a market yield (interest rate) of 10 percent, compounded semiannually. On December 31, Year 7, the market interest rate for similar bonds is 14 percent, compounded semiannually. The company uses the effective interest method of accounting for these bonds, rounds computations to the nearest dollar, and closes its books once per year, on December 31.

**a.** Give the journal entries to record the company's interest expense for both the first and the second payments during Year 7.
**b.** Suppose the firm repurchases one-half of the bonds for cash in the open market on December 31, Year 7, at the price implied by the market interest rate of 14 percent compounded semiannually on that date. Give the journal entry, ignoring income taxes.
**c.** How would the gain or loss in **b** appear on the income statement?

**35. Amortization schedule for bonds.** Seward Corporation issues on January 2, Year 1, 8-percent semiannual coupon bonds maturing three years from the date of issue. The coupons, dated for January 1 and July 1 of each year, each promise 4 percent of the face value, 8 percent total for a year. The firm issues the bonds to yield 10 percent, compounded semiannually.

**a.** What are the issue proceeds received by Seward Corporation?
**b.** Construct an amortization schedule, similar to that in **Exhibit 9.3**, for this bond issue.
**c.** Give the journal entries relating to these bonds for the first year. Seward Corporation uses a calendar year for its accounting period.
**d.** Assume that on January 2, Year 3, Seward Corporation reacquires $20,000 face value of these bonds for 102 percent of par and retires them. Give the journal entry to record the retirement.

**36. Journal entries to account for bonds.** Brooks Corporation issues $100,000 face value, 10-year bonds on January 2, Year 2. The bonds' coupons, dated June 30 and December 31 of each year, each promise 4 percent of face value, 8 percent total for a year. The market initially prices the bonds to yield 6 percent compounded semiannually.

**a.** Compute the issue price of the bonds.
**b.** Give the journal entries to account for these bonds during Year 2.
**c.** Brooks Corporation reacquires these bonds on the open market on January 2, Year 3, at a time when the market prices the bonds to yield 10 percent compounded semiannually. Give the journal entry to record the reacquisition.

## PROBLEMS AND CASES

**37. Allowance method for warranties; reconstructing transactions.** Assume that Central Appliance sells appliances, all for cash. It debits all acquisitions of appliances during a year to the Merchandise Inventory account. The company provides warranties on all its products, guaranteeing to make required repairs, within one year of the date of sale, for any of its appliances that break down. The company has many years of experience with its products and warranties.

**Exhibit 9.7** contains summary data and financial statement excerpts for Central Appliance for the end of Year 1 and for some of the events during Year 2. The firm made entries to the Estimated Liability for Warranty Repairs account during Year 2 as it made repairs, which converted the credit balance at the end of Year 1 into a debit balance of $15,000 at the end of Year 2. That is, before the firm makes its entry to recognize warranty expense for the entire year, the Estimated Liability for Warranty Repairs account has a *debit* balance of $15,000.

**EXHIBIT 9.7** **CENTRAL APPLIANCE (Problem 37)**

| Balance Sheet Excerpts | End of Year 1 | |
|---|---|---|
| Merchandise Inventory | $ 100,000 | |
| All Other Accounts | 110,000 | |
| Total Assets | $ 210,000 | |
| Estimated Liability for Warranty Repairs | $ 6,000 | |
| All Other Accounts | 204,000 | |
| Total Liabilities and Owners' Equity | $ 210,000 | |
| **Income Statements Excerpts** | **Year 1** | **Year 2** |
| Sales Revenue | $ 800,000 Cr. | $1,000,000 Cr. |
| Warranty Expense | (18,000) Dr. | ? |

Also, the Merchandise Inventory account, to which the firm has debited all purchases of inventory, has a balance of $820,000 before the adjusting entry for Cost of Goods Sold, so that Goods Available for Sale totaled $820,000. Central Appliance makes its adjusting entries and closes its books only once each year, at the end of the year.

At the end of Year 2, the management of Central Appliance analyzes the appliances sold within the preceding 12 months. It classifies all appliances still covered by warranty as follows: those sold on or before June 30 (more than six months old), those sold after June 30 but on or before November 30 (more than one month but less than six months old), and those sold on or after December 1. Assume that it estimates that one-half of 1 percent of the appliances sold more than six months ago will require repair, 5 percent of the appliances sold one to six months before the end of the year will require repair, and 8 percent of the appliances sold within the last month will require repair. From this analysis, management estimates that $5,000 of repairs will still have to be made in Year 3 on the appliances sold in Year 2. Items remaining in ending inventory on December 31, Year 2, had cost $120,000.

**a.** What were the total acquisitions of merchandise inventory during Year 2?
**b.** What was the cost of goods sold for Year 2?
**c.** What was the dollar amount of repairs made during Year 2?
**d.** What was the warranty expense for Year 2?
**e.** Give journal entries for repairs made during Year 2, for the warranty expense for Year 2, and for cost of goods sold for Year 2.

**38. Accounting for zero-coupon debt.** When Time Warner, Inc., announced its intention to borrow about $500 million by issuing 20-year zero coupon (single-payment) notes, *The Wall Street Journal* reported the following:

> NEW YORK—Time Warner Inc. announced an offering of debt that could yield the company as much as $500 million . . . . The media and entertainment giant said that it would offer $1.55 billion principal amount of zero-coupon . . . notes due [in 20 years] . . . through Merrill Lynch. . . . Zero-coupon debt is priced at a steep discount to principal, [which] is fully paid at maturity . . . . A preliminary prospectus . . . didn't include the issue price and yield of the notes.[8]

Assume Time Warner borrows funds at the beginning of Year 1 and pays $1.55 billion in a single payment at the end of Year 20.

**a.** Assume the yield of the notes is 6 percent per year, compounded annually. What initial issue proceeds will Time Warner, Inc., realize from issuing these notes? That is, how much cash will Time Warner receive on issuing the notes?

[8]*The Wall Street Journal,* December 8, 1992, p. A6.

**b.** Assume the initial issue proceeds from these notes is $500 million. What is the annual yield on these notes?

**c.** Assume the initial issue proceeds from these notes is $400 million and their annual yield is 7 percent compounded annually. What interest expense will Time Warner record for Year 1, the first year the notes are outstanding?

**d.** Assume the initial issue proceeds from these notes is $400 million and their annual yield is 7 percent compounded annually. What interest expense will Time Warner record for Year 20, the last year the notes are outstanding?

**39. Accounting for long-term bonds.** The notes to the financial statements of Aggarwal Corporation for Year 4 reveal the following information with respect to long-term debt. *All interest rates in this problem assume semiannual compounding and the effective interest method of amortization.*

| | December 31 | |
|---|---|---|
| | Year 3 | Year 4 |
| $800,000 zero coupon notes due December 31, Year 13, initially priced to yield 10 percent . . . . . . . . . . . . . . . . . . . . . . . . . . . . . . | $ 301,512 | ? |
| $1,000,000, 7 percent bonds due December 31, Year 8. Interest is payable on June 30 and December 31. The bonds were initially priced to yield 8 percent . . . . . . . . . . . . . . . . . . . . . . . . . . . . . | ? | $966,336 |
| $1,000,000, 9 percent bonds due December 31, Year 19. Interest is payable on June 30 and December 31. The bonds were initially priced to yield 6 percent . . . . . . . . . . . . . . . . . . . . . . . . . . . . . | $1,305,832 | ? |

**a.** Compute the book value of the zero coupon notes on December 31, Year 4. A zero coupon note requires no periodic cash payments; only the face value is payable at maturity. Do not overlook the italicized sentence above.

**b.** Compute the amount of interest expense for Year 4 on the 7 percent bonds.

**c.** On July 1, Year 4, Aggarwal Corporation acquires half of the 9-percent bonds ($500,000 face value) in the market for $526,720 and retires them. Give the journal entry to record this retirement.

**d.** Compute the amount of interest expense on the 9-percent bonds for the second half of Year 4.

**40. Accounting for long-term bonds.** The notes to the financial statements of Wal-Mart Stores reveal the following information with respect to long-term debt. *All interest rates in this problem assume semiannual compounding and the effective interest method of amortization.*

| | January 31 | |
|---|---|---|
| | Year 11 | Year 12 |
| $100,000,000-par value of 9-percent debenture bonds due January 31, Year 20, initially priced on the market to yield 12 percent . . . . . . . . | $ 83,758,595 | ? |
| ? par value zero coupon bonds due July 31, Year 16, initially priced on the market to yield 8 percent . . . . . . . . . . . . . . . . . . . . | $162,395,233 | $175,646,684 |
| $400,000,000 face value 9.25-percent notes due January 31, Year 20, initially priced on the market to yield ? percent . . . . . . . . . . . . . . . | $400,000,000 | $400,000,000 |

**a.** Compute the amount of interest expense on the 9-percent coupon bonds for fiscal year Year 12. Do not overlook the italicized sentence above.

**b.** Compute the book value of the 9-percent bonds on January 31, Year 12.

**c.** Compute the par value of the zero coupon bonds.

**d.** Compute the initial market yield on the $400,000,000 notes.

**41. Comparison of straight-line and effective interest methods of amortizing debt discount.** IBM established IBM Credit Corporation (IBMCC) on May 1, Year 1. On July 1, Year 1, IBMCC issued $150 million of zero coupon notes due July 1, Year 8. These notes promise to pay a single amount of $150 million at maturity, seven years after issue date. IBMCC marketed these notes to yield 14 percent compounded semiannually. IBMCC may at any time redeem these notes, in part or in whole, for 100 percent of the principal amount.

IBMCC computes interest expense for financial reporting using the effective interest method but amortizes "original-issue discount" on the notes in equal semiannual amounts for tax reporting. That is, if IBMCC issued the $150 million face value of notes for $66 million, the *original-issue discount* is $84 (= $150 − $66) million and the semiannual interest expense each period is $6 [= $84/(7 years × 2 periods per year)] million per year. Assume an income tax rate of 40 percent. The financial statements of IBMCC present the following data about long-term debt on December 31, Year 1:

| (all dollar amounts in thousands) | |
|---|---|
| 14 3/8 Percent Notes Due July, Year 6 | $100,000 |
| Zero Coupon Notes Due July, Year 8 (Face Value of $150,000) | 62,238 |
| | $162,238 |

**a.** Calculate the proceeds to IBMCC from issuing the zero coupon notes.

**b.** Compute the amount of interest expense reported on the income statement for these notes for the six-month period ending December 31, Year 1.

**c.** Compute the amount of the interest deduction on the tax return for Year 1 resulting from these notes. Assuming IBMCC can fully deduct this amount on its income tax return, how much does this reduce the income taxes payable that IBMCC would otherwise compute?

**d.** What was the amount of "Unamortized Discount" that relates only to the zero coupon notes on the balance sheet at the end of Year 1?

**e.** A news story in Year 2 reported the following:

[T]he Treasury wants to plug a tax loophole that has enabled companies to borrow billions of dollars cheaply. . . . At issue is a tax break granted to companies issuing . . . "zero-coupon" bonds . . . and other deeply discounted debt instruments, which pay very low interest rates.

Describe the advantages to issuers and purchasers of zero coupon bonds. Give your interpretation of the tax loophole alleged by the U.S. Treasury. Be specific in your response by indicating the dollar amount of the "loophole" that IBMCC used in Year 1. What tax policy with respect to interest on zero coupon notes would you recommend?

**42. Ethical issues of managing income and the debt-equity ratio through bond retirement.** Suppose that Quaker Oats Company issued $40 million of 5-percent semiannual coupon bonds many years ago at par. The bonds now have 20 years until scheduled maturity. Because market interest rates have risen to 9 percent, the market value of the bonds has dropped to 63 percent of par. Quaker Oats has $5 million of current liabilities and $35 million of shareholders' equity in addition to the $40 million of long-term debt in its financial structure. The debt-equity ratio (= total liabilities/total liabilities plus shareholders' equity) is 56 percent [= ($40 + $5)/($5 + $40 + $35)]. (Shareholders' equity includes an estimate of the current year's income.) The president of Quaker Oats is concerned about boosting reported income for the year, which is about $8 million in the absence of any other actions. Also, the debt-equity ratio appears to be larger than that of other firms in the industry. The president wonders what the impact on net income and the debt-equity ratio would be if the company issues at par new 9-percent semiannual coupon bonds to mature in 20 years and uses the proceeds to retire the outstanding bond issue. Assume that such action is taken and that any gain on bond retirement is taxable immediately, at the rate of 40 percent.

**a.** Prepare the journal entries for the issue of new bonds in the amount required to raise funds to retire the old bonds, for the retirement of the old bonds, and for the income tax effects.

**b.** What is the effect on income for the year? Give both dollar and percentage amounts.

**c.** What is the debt-equity ratio after the transaction?

**d.** Discuss any ethical concerns raised by these considerations.

**43. Accounting for bonds in a troubled debt restructuring.** On January 1, 1985, First National Bank (FNB) acquired $10 million of face value bonds issued on that date by Occidental Oceanic Power Systems (OOPS). The bonds carried 12-percent semiannual coupons and were to mature 20 years from the issue date. OOPS issued the bonds at par.

By 2005, OOPS was in severe financial difficulty and threatened to default on the bonds. After much negotiation with FNB (and other creditors), it agreed to repay the bond issue but only on less burdensome terms. OOPS agreed to pay 5 percent per year, semiannually, for 25 years and to repay the principal on January 1, 2030, or 25 years after the negotiation. FNB will receive $250,000 every six months starting July 1, 2005, and $10 million on January 1, 2030. By January 1, 2005, OOPS was being charged 20 percent per year, compounded semiannually, for its new long-term borrowings.

**a.** What is the value of the bonds that FNB holds? In other words, what is the present value of the newly promised cash payments when discounted at OOPS's current borrowing rate of 20 percent per year, compounded semiannually?

**b.** Repeat part **a**, but use the market interest rate at the time of initial issue, 12 percent, compounded semiannually, to calculate the present value of the newly promised cash payments.

**c.** Consider three accounting treatments for this negotiation (called a "troubled debt restructuring" by the FASB in its *Statement of Financial Accounting Standards No. 114*).

**(1)** Write down the bonds to the value computed in part **a**, and base future interest revenue computations on that new book value and the new historical interest rate of 20 percent per year, compounded semiannually.

**(2)** Write down the bonds to the value computed in part **b**, and base future interest revenue computations on that new book value and the old historical interest rate of 12 percent per year, compounded semiannually.

**(3)** Make no entry to record the negotiation, and record interest revenue as the amount of cash, $250,000, that FNB receives semiannually.

Over the new life of the bond issue, how will total income vary as a function of the method chosen?

**d.** Which of the three methods listed in **c** would you recommend? Why?

**44. Discounting warranty obligations.** GAAP requires long-term monetary liabilities to appear at the present value of the future cash flows discounted at the market rate of interest appropriate to the monetary items at the time the firm initially recorded them. The Accounting Principles Board *Opinion No. 21* specifically excludes from present-value valuation those obligations that arise from warranties. The *Opinion* requires that warranties, being nonmonetary liabilities, be stated at the estimated cost of providing warranty goods and services in the future.

Assume that the estimated future costs of a three-year warranty plan on products sold during Year 1 are as follows:

| Year of Expected Expenditure | Expected Cost |
|---|---|
| 2 | $ 500,000 |
| 3 | 600,000 |
| 4 | 900,000 |
| Total | $2,000,000 |

Actual costs coincided with expectations as to both timing and amount.

**a.** Prepare the journal entries for each of the years 1 through 4 for this warranty plan following current GAAP.

**b.** Now assume that GAAP allows these liabilities to appear at their present value. Prepare the journal entries for each of the years 1 through 4 for this warranty plan, assuming that the firm states warranty liability at the present value of the future costs discounted at 10 percent. To simplify the calculations, assume that the firm incurs all warranty costs on December 31 of each year.

**c.** What theoretical arguments can you offer for the valuation basis in part **b**?

**45. Effects on statement of cash flows. Exhibit 4.16** in **Chapter 4** provides a simplified statement of cash flows with numbered lines. For each of the transactions that follow, indicate the number(s) of the line(s) affected by the transaction and state the amount and direction (increase or decrease) of the effect. If the transaction affects net income on line **(1)** or cash on line **(11)**, be sure to indicate if it increases or decreases the line. Ignore income tax effects.

**a.** The firm issues bonds for $100,000 cash.

**b.** The firm issues a note with a fair market value of $100,000 for a building.

**c.** The firm retires, for $90,000 cash, bonds with a book value of $100,000.

**d.** The firm calls for $105,000 and retires bonds with a book value of $100,000.

**e.** The firm records interest expense and expenditures for the first half of a year on bonds. The bonds have a face value of $100,000 and a book value at the beginning of the year of

$90,000. The coupon rate is 10 percent, paid semiannually in arrears, and the bonds were originally issued to yield 12 percent, compounded semiannually.

**f.** The firm records interest expense and expenditures for the first half of a year on bonds. The bonds have a face value of $100,000 and a book value at the beginning of the year of $105,000. The coupon rate is 12 percent, paid semiannually in arrears, and the bonds were originally issued to yield 10 percent, compounded semiannually.

**46. Ethical issues.** Several years ago, the firm issued bonds with an annual coupon rate of 5 percent. In today's market, those bonds yield 11 percent. The market value of the bonds had declined below book value by $50 million. Management considers issuing at par new bonds with coupon rate of 11 percent and using the proceeds to retire the outstanding issue, recognizing a gain of $50 million.

Comment on any ethical issues pertinent for management's consideration.

CHAPTER 10

# Liabilities: Off-Balance-Sheet Financing, Leases, Deferred Income Taxes, and Retirement Benefits

## Learning Objectives

1. Understand (a) why firms attempt to structure debt financing to keep debt off the balance sheet and (b) how standard-setters have refined the concept of an accounting liability to reduce off-balance-sheet financing abuses.

2. Distinguish between operating leases and capital leases on the bases of their economic characteristics, accounting criteria, and financial statement effects.

3. Understand why firms may recognize revenues and expenses for financial reporting in a period different from that used for tax reporting and the effect of such differences on the measurement of income tax amounts on the income statement and the balance sheet.

4. Understand the accounting issues related to retiree benefit plans (such as pensions and health care benefits).

*C*hapter 9 discussed the concept of an accounting liability and illustrated its application to current liabilities and long-term debt. This chapter explores liabilities further by considering four controversial accounting topics of the past decade: off-balance-sheet financing, leases, deferred taxes, and retirement benefits.

## Off-Balance-Sheet Financing

In 1980, *Forbes* magazine noted, "The basic needs of humans are simple: to get enough food, to find shelter, and to keep debt off the balance sheet." Accounting standard setters and financial engineers continually battle: the standard setters to devise rules to get debt onto the balance sheet and the financial engineers to devise transactions that both comply with the rules and keep debt off the balance sheet. Breaking the rules but making it appear otherwise is fraud; complying with the rules but keeping debt off the balance sheet is acceptable and, in some circles, highly regarded, even rewarded. This section discusses the rationale for off-balance-sheet financing and illustrates several financing arrangements to keep debt off the balance sheet.

## Rationale for Off-Balance-Sheet Financing

Firms often attempt to structure their financing to keep debt off the balance sheet (that is, to obtain funds without increasing liabilities on the balance sheet). They hope to show lower liabilities on the balance sheet and thereby to improve the debt ratios that analysts use to assess the financial risk of a firm. Reasons frequently cited for **off-balance-sheet financing** include the following:

1. It lowers the cost of borrowing. Unwary lenders may ignore off-balance-sheet financing and set lower interest rates for loans than the underlying risk levels warrant.
2. It avoids violating debt covenants. Covenants in existing debt contracts may restrict a firm's increasing debt ratios above defined levels. Structuring an off-balance-sheet financing arrangement permits the firm to obtain needed funds without affecting the debt ratios. It also provides a cushion in the event that the firm must engage in on-balance-sheet financing in the future.

The first rationale for off-balance-sheet financing assumes that some lenders, credit-rating agencies, and others who assess financial risks do not possess the knowledge, skills, and information needed to identify and deal appropriately with such financing arrangements. Even though firms have little evidence that financial statement users systematically ignore these obligations, firms often structure their financings as though they believe that assumption. Standard-setting bodies have required increased disclosures of off-balance-sheet financings in recent years to alert financial statement users.

## Structuring Off-Balance-Sheet Financing

Off-balance-sheet financings generally fall into one of the two categories of obligations that accounting does not recognize as liabilities: executory contracts and contingent obligations (see **Figure 9.1**).

**Executory Contracts** Firms frequently sign contracts promising to pay defined amounts in the future in return for future benefits. For a firm's obligation to qualify as an accounting liability, the firm must have received a past or current benefit in return for the obligation—the event or transaction must already have happened. If the firm will receive the benefit in the future, accounting treats the obligation as an executory contract and typically does not recognize a liability.

**Example 1** Delta Airlines needs additional aircraft to expand internationally. Delta could borrow the needed funds and purchase the aircraft. This arrangement places additional debt on the balance sheet. Instead, Delta signs a lease agreeing to pay the owner of the aircraft certain amounts each year for 12 years. Delta paints its name on the aircraft, uses them in operations, and makes the required lease payments. The usual accounting assumes that Delta receives benefits when it uses the aircraft, not when it initially signs the lease. Thus, Delta obtains financing for its flight equipment without showing a liability on the balance sheet.

**Example 2** Boise Cascade and Georgia-Pacific Corporation (forest products companies) need additional pulp-processing capacity. Each firm could borrow the needed funds and build its own manufacturing plant. Instead, they form a joint venture to build a pulp-processing plant suitable for their joint needs. Each firm agrees to use one-half of the new plant's capacity each year for 20 years and to pay half of all operating and debt service costs. The joint venture uses the purchase commitments of Boise Cascade and Georgia-Pacific to obtain a loan to build the facility. Accounting views the purchase commitments as executory contracts—all benefits occur in the future—and therefore neither firm will recognize a liability. The loan appears as a liability of the balance sheet of the joint venture. Thus, each firm obtains financing for the plant without showing a liability on its balance sheet.

**Contingent Obligations** As an alternative to borrowing funds and using a particular asset as collateral, a firm might obtain funds by "selling" the asset with the understanding that the firm will give cash back to the "purchaser" under stated conditions. Such sales usually provide that the selling firm must repay cash to the purchaser in the future if, for example, the asset sold generates less cash for the purchaser than anticipated at the time of "sale." If such payments do not meet the criteria for a loss contingency, as discussed in **Chapter 9**, then the firm treats the transaction as a sale, keeping debt off the balance sheet, instead of as a loan collateralized by the asset.

**Example 3** Sears extends credit to its customers to purchase appliances, furniture, and other goods. Sears could borrow from banks using its accounts receivable as collateral, thereby placing debt on the balance sheet. Sears would use collections from customers to repay the bank loans with interest. Instead, Sears "sells" the accounts receivable to the banks. The amount Sears expects to collect from receivables sold exceeds the amount of cash Sears receives from the banks. This excess provides the banks with their expected return, or interest income on their "loan" to Sears. Sears agrees to collect the receivables from customers and remit the cash to the banks. Sears will probably treat this transaction as a sale and thus avoid placing debt on the balance sheet.

**Example 4** Seagram Company, a distiller of liquors, ages its whiskeys for approximately ten years. The firm must pay the costs to produce the whiskey and to store it during aging. Using the whiskey as collateral, Seagram could borrow the necessary funds; however, this would lead to increased liabilities. Instead, it "sells" the whiskey to the banks. Seagram oversees the aging process on behalf of the banks. At the completion of the aging, Seagram assists banks in finding a buyer. The banks bear the risk of changes in selling prices for the whiskey. Seagram will probably treat this transaction as a sale and thereby avoid placing debt on the balance sheet.

**Example 5** **Chapter 4** described the Enron/Merrill Lynch Nigerian barge transactions. In these, Enron sold, to Merrill Lynch, barges anchored in Nigeria for a fair value greater than book value. Later, Enron repurchased the barges for a price roughly equal to the initial transaction price plus an amount equivalent to interest at a market rate. **Chapter 4** emphasized the effect of the transaction on income and cash flow from operations. Here, consider the transaction as a device of Enron's for excluding the debt owed to Merrill Lynch, the lender, from the balance sheet. Some would say that Enron borrowed funds from Merrill Lynch, using barges as collateral, and later repaid the financing.

The Financial Accounting Standards Board (FASB) has dealt with various off-balance-sheet financing arrangements on a case-by-case basis. A unifying theme appears in the FASB rules:

- First, the accountant identifies the party that enjoys the economic benefits of the item in each transaction and that bears the economic risks of holding it.
- Then, the accountant identifies the party that *needs* financing.

When the party *needing* financing controls the benefits and risks, the transaction usually leads to recognizing a liability on the balance sheet of the controlling party. When the party *providing* the financing controls the benefits and risks, the debt does not appear as a liability on the balance sheet of the firm needing financing.

**Example 6** Refer to **Example 1**, but assume that Delta signs a lease for 20 years. Also assume that each aircraft Delta leases has a useful life of approximately 20 years and that the required lease payments compensate the owner of the aircraft (that is, the lessor) for the cost of the aircraft plus a reasonable return for the level of risk incurred. Then Delta has rights to use nearly all the economic benefits of the aircraft and bears the risks of technological obsolescence, overcapacity during economic downturns, and similar factors. Because Delta controls the enjoyment of economic benefits and bears the economic risks, it will, following FASB rules discussed later, likely report its lease commitment as a liability. If, as in **Example 1**, the lease spans a period substantially shorter than the expected useful life of the aircraft and if the economics of the transaction virtually require that the lessor either sell the aircraft at the end of the lease period or re-lease them to another airline, then the lessor enjoys the benefits and bears the risks of ownership. In this case, the lease commitment will not appear as a liability on Delta's balance sheet. The next section of this chapter more fully discusses the accounting for leases.

**Example 7** Refer to **Example 2**. Suppose that the lender requires Boise Cascade to guarantee payment of the loan in case the joint venture defaults. Suppose also that Georgia-Pacific Corporation experiences severe financial difficulties and fails to pay its share of operating and debt service costs. Boise Cascade bears most of the economic risk of this joint venture and should probably recognize a liability. If the lender relies solely on the joint venture to repay the loan and does not require either Boise Cascade or Georgia-Pacific Corporation to guarantee the loan, then the debt will probably not appear on the balance sheet of either company. The loan would appear as a liability on the books of the joint venture only.

**Example 8** Refer to **Example 3**. Assume that Sears must transfer additional uncollected receivables to the lender/purchaser banks for any receivables judged uncollectible. Assume further that Sears must transfer additional uncollected receivables to the banks if interest rates rise above a specified level. In this case, Sears bears credit and interest rate risks and treats the transfer of receivables as a loan and not as a sale, with debt appearing on its balance sheet. If, instead, the banks bear credit and interest rate risks, Sears records the transfer of receivables as a sale; no debt related to the transaction appears on its balance sheet.

**Example 9** Refer to **Example 4**. Assume that Seagram guarantees an ultimate selling price that provides the lender with both a return of the original "purchase price" and coverage of storage and interest costs. Seagram bears the economic risks and must show a liability on its balance sheet. Assume, in contrast, that the lender does not require Seagram to guarantee a minimum selling price for the whiskey (for example, the lender might conclude that the quality of Seagram whiskey and a favorable market outlook for whiskey make such a guarantee unnecessary). Then Seagram will likely record the transaction as a sale and not a loan.

**Example 10** Refer to **Example 5**. Enron appears not to have given Merrill Lynch a written commitment to repay the loan, but a gentleman's agreement, a handshake. Even an unwritten agreement that Enron would repay would make the transaction a loan under GAAP. Without a written agreement, however, an accountant or auditor cannot easily discover the economic substance of the transaction.

Firms continue to create innovative financing schemes to keep debt off the balance sheet. The FASB continues to consider the appropriate accounting for each transaction as it arises. Some accountants think that there will never be a satisfactory solution until accounting requires the recording of a liability whenever a firm has an obligation to pay a reasonably definite amount at a reasonably definite time, independent of the executory nature of the contract.

## Balance Sheet Quality Issues

A parallel concept to quality of earnings is quality of the balance sheet. To what extent does the balance sheet portray the economic resources of a firm and the economic claims on those resources? Previous chapters discussed quality issues on the asset side of the balance sheet with respect to receivables arising from revenue recognition, LIFO inventories, valuation of long-lived assets at acquisition cost, and exclusion from the balance sheet of intangibles developed internally. Firms' attempts to keep debt off the balance sheet affect balance sheet quality as well. Understating the liabilities of a firm can mislead statement users when they assess the risk of the firm. GAAP attempts to define when an obligation is and is not a liability. However, obligations sometimes have some, but not all, attributes of a liability. GAAP requires firms to disclose information about such obligations in notes to the financial statements. Statement users should study these notes when assessing the risk of a firm and make their own judgments about whether to exclude such obligations.

## Ethical Issues in Off-Balance-Sheet Financing

Do firms confront ethical issues when engaging in off-balance-sheet financing? One view argues that ethical issues do not arise so long as firms structure their off-balance-sheet financing arrangements to comply with GAAP. GAAP sets the rules and firms do their best to comply. Another view argues that ethical issues do arise. The continuing efforts of standard-setting bodies to change and refine their standards to react to firms' efforts to work around the spirit, if not the letter, of the standards evidences that firms do not strive to reflect economic reality, but merely to keep debt off the balance sheet.

## Problem 10.1 for Self-Study

**Attempting off-balance-sheet financing.** Assume that International Paper Company (IP) needs $75 million of additional financing but, because of restrictions in existing debt covenants, cannot, except at prohibitive cost, put more debt on its balance sheet. Instead of borrowing funds, it creates a trust to which, on January 1, it transfers cutting rights to a mature timber tract. The trust will pay for these rights by borrowing $75 million for five years from a bank, with interest at 14 percent. The trust promises to make five equal installment payments, one on December 31 of each year.

*(continued)*

The trust will harvest and sell timber each year to obtain funds to make the loan payments and to pay operating costs. At current prices, the value of the standing wood exceeds by ten percent the amounts the trust will need to service the loan and to pay the ongoing operating costs (including wind, fire, and erosion insurance). The future selling price of timber will determine the trust's actions, as follows:

- If the selling price of timber declines in the future, the trust will harvest more timber and sell it to service the debt and to pay other operating costs.
- If the selling price of timber increases in the future, the trust will harvest timber at the level originally planned but will invest any cash left over after paying for debt service and operating costs, to provide a cushion for possible future price decreases. At the end of five years, the trust will distribute the value of any cash or uncut timber to IP.

IP will not guarantee the debt. The bank, however, has the right to inspect the tract at any time and to replace IP's forest-management personnel with managers of its own choosing if it feels that IP is mismanaging the tract.

**a.** Identify IP's economic returns and risks in this arrangement.
**b.** Identify the economic returns and risks in this arrangement for the bank lending the funds.
**c.** Should IP treat this transaction as a loan (a liability will appear on IP's balance sheet) or as a sale (no liability will appear on IP's balance sheet)? Explain your reasoning.

## Leases

Many firms acquire rights to use assets through long-term leases. **Examples 1** and **6** above, in which Delta Airlines signed leases to acquire the use of airplanes, illustrate such financing arrangements. A company seeking office space might agree to lease a floor of a building for 5 years or an entire building for 40 years, promising to pay a fixed periodic fee for the duration of the lease. Promising to make a series of lease payments commits the firm just as surely as a bond indenture or a mortgage and, often, results in similar accounting. This section examines two methods of accounting for long-term leases: the operating lease method, which keeps debt off the balance sheet, and the capital lease method, which puts it on.

To understand these two methods, suppose that Food Barn wants to acquire a computer that has a three-year life and costs \$45,000. Assume that Food Barn must pay 15 percent per year to borrow funds for three years. The computer manufacturer will sell the equipment for \$45,000 or lease it for three years. Food Barn must pay for property taxes, maintenance, and repairs of the computer whether it purchases or leases. Food Barn signs the lease on January 1, Year 1, and promises to make payments on the lease on December 31, Year 1, Year 2, and Year 3. In practice, a lessee (the "buyer" or "tenant") usually makes payments in advance, but assuming the payments occur at year-end makes the computations simpler in these illustrations. Compound interest computations show that each lease payment must be \$19,709.[1]

### Operating Lease Method

In an **operating lease**, the owner, or lessor (the "landlord"), enjoys the rewards and bears most of the risks of ownership. For example, the lease may require the lessee (the user or "tenant") to make fixed periodic payments. The lessor in this case benefits from decreases in interest rates (the lessor receives the fixed periodic amount) but bears the risk of interest rate increases (the lessor cannot increase the fixed periodic payment). The lease may specify that the lessee return the leased asset to the lessor at the end of the lease term. The lessor must then re-lease the asset to some other firm to obtain a portion of its return. The lessor bears the risk of technological change and other factors that would affect its ability to lease the asset to others. If the computer

[1]The present value of \$1 paid at the end of this year and each of the next two years equals \$2.28323 when discounted at an annual rate of 15 percent. See **Table 4** at the end of the book, 15-percent column, 3-period row. Because the lease payments must have a present value of \$45,000, each payment must equal \$45,000/2.28323 = \$19,709.

manufacturer, and not Food Barn, bears most of the risks of ownership, accounting considers the lease to be an executory contract and treats it as an operating lease. Food Barn would make no entry on January 1, Year 1, when it signs the lease. It makes the following entry on December 31, Year 1, Year 2, and Year 3:

*December 31 of each year*

| | | |
|---|---|---|
| Rent Expense . . . . . . . . . . . . . . . . . . . . . . . . . . . . . . . . . . . . . . . . | 19,709 | |
| Cash . . . . . . . . . . . . . . . . . . . . . . . . . . . . . . . . . . . . . . . . | | 19,709 |

| Assets | = | Liabilities | + | Shareholders' Equity | (Class.) |
|---|---|---|---|---|---|
| −19,709 | | | | −19,709 | IncSt → RE |

To recognize annual expense of leasing computer.

## Capital Lease Method

In a **capital lease**, the lessee enjoys the rewards and bears most of the risks of ownership. If the periodic rental payments vary with changes in interest rates, then Food Barn, not the computer manufacturer, bears interest rate risk. If the lease period approximately equals the useful life of the leased asset, then Food Barn bears the risk of technological changes and other factors that affect the market value of the asset. If Food Barn, not the computer manufacturer, bears most of the risks of ownership, accounting views the arrangement as an executed contract—a form of borrowing to purchase the computer. Food Barn must account for it as a capital lease.[2] This treatment recognizes the signing of the lease as the simultaneous acquisition of a long-term asset and the incurring of a long-term liability for lease payments. At the time Food Barn signs the lease, it records both the leased asset and the lease liability at the present value of the required cash payments, $45,000 in the example. The entry at the time Food Barn signed its three-year lease would be as follows:

*January 1, Year 1*

| | | |
|---|---|---|
| Leased Asset—Computer . . . . . . . . . . . . . . . . . . . . . . . . . . . . . . . . . | 45,000 | |
| Lease Liability . . . . . . . . . . . . . . . . . . . . . . . . . . . . . . . . . . . | | 45,000 |

| Assets | = | Liabilities | + | Shareholders' Equity | (Class.) |
|---|---|---|---|---|---|
| +45,000 | | +45,000 | | | |

To recognize acquisition of the asset and the related liability.

At the end of the year, Food Barn must make two separate entries—one related to the asset and one to the liability. First, it must depreciate the leased asset over its useful life. The first entry made at the end of each year records this depreciation. Assuming that Food Barn uses the straight-line depreciation method, it makes the following entries at the end of Year 1, Year 2, and Year 3:

*December 31 of each year*

| | | |
|---|---|---|
| Depreciation Expense (on Computer) . . . . . . . . . . . . . . . . . . . . . . . . . | 15,000 | |
| Accumulated Depreciation—Computer . . . . . . . . . . . . . . . . . . . . | | 15,000 |

| Assets | = | Liabilities | + | Shareholders' Equity | (Class.) |
|---|---|---|---|---|---|
| −15,000 | | | | −15,000 | IncSt → RE |

The second entry made at the end of each year recognizes that the debt service payment—the lease payment—pays interest and, in part, reduces the liability itself. The entries made at

[2]Financial Accounting Standards Board, *Statement of Financial Accounting Standards No. 13*, "Accounting for Leases," 1976, reissued and reinterpreted 1980.

the end of each of the three years, based on the amortization schedule in **Exhibit 10.1**, would be as follows:

*December 31, Year 1*

| | | |
|---|---|---|
| Interest Expense . . . . . . . . . . . . . . . . . . . . . . . . . . . . . . . . . . . . . . . . | 6,750 | |
| Lease Liability . . . . . . . . . . . . . . . . . . . . . . . . . . . . . . . . . . . . . . . . | 12,959 | |
| Cash . . . . . . . . . . . . . . . . . . . . . . . . . . . . . . . . . . . . . . . . | | 19,709 |

| Assets | = | Liabilities | + | Shareholders' Equity | (Class.) |
|---|---|---|---|---|---|
| -19,709 | | -12,959 | | -6,750 | IncSt → RE |

To recognize lease payment, interest on liability for the year (0.15 × $45,000 = $6,750), and the plug for reduction in the liability. The present value of the liability after this entry is $32,041 = $45,000 − $12,959.

*December 31, Year 2*

| | | |
|---|---|---|
| Interest Expense . . . . . . . . . . . . . . . . . . . . . . . . . . . . . . . . . . . . . . . . | 4,806 | |
| Lease Liability . . . . . . . . . . . . . . . . . . . . . . . . . . . . . . . . . . . . . . . . | 14,903 | |
| Cash . . . . . . . . . . . . . . . . . . . . . . . . . . . . . . . . . . . . . . . . | | 19,709 |

| Assets | = | Liabilities | + | Shareholders' Equity | (Class.) |
|---|---|---|---|---|---|
| -19,709 | | -14,903 | | -4,806 | IncSt → RE |

To recognize lease payment, interest on liability for the year (0.15 × $32,041 = $4,806), and the plug for reduction in the liability. The present value of the liability after this entry is $17,138 = $32,041 − $14,903.

*December 31, Year 3*

| | | |
|---|---|---|
| Interest Expense . . . . . . . . . . . . . . . . . . . . . . . . . . . . . . . . . . . . . . . . | 2,571 | |
| Lease Liability . . . . . . . . . . . . . . . . . . . . . . . . . . . . . . . . . . . . . . . . | 17,138 | |
| Cash . . . . . . . . . . . . . . . . . . . . . . . . . . . . . . . . . . . . . . . . | | 19,709 |

| Assets | = | Liabilities | + | Shareholders' Equity | (Class.) |
|---|---|---|---|---|---|
| -19,709 | | -17,138 | | -2,571 | IncSt → RE |

To recognize lease payment, interest on liability for the year (0.15 × $17,138 = $2,571), and the plug for reduction in the liability. The present value of the liability after this entry is zero (= $17,138 − $17,138).

**Exhibit 10.1** shows the amortization schedule for this lease. Note that it has the same form as the mortgage amortization schedule in **Exhibit 9.2**.

## Accounting Method Determines Timing, but Not Amount, of Total Expense

In the capital lease method, the expense over the three years totals $59,127, comprising $45,000 (= $15,000 + $15,000 + $15,000) for depreciation expense and $14,127 (= $6,750 + $4,806 + $2,571) for interest expense. This exactly equals the total rent expense of $59,127 recognized under the operating lease method described previously ($19,709 × 3 = $59,127). The capital lease method recognizes expense sooner than does the operating lease method, as **Exhibit 10.2** summarizes. But over sufficiently long time periods, expense equals cash expenditure. The operating lease method and the capital lease method differ in the timing, but not in the total amount, of expense. The capital lease method recognizes both the leased asset and the lease liability on the balance sheet, whereas the operating lease recognizes neither.

## Choosing the Accounting Method

When a transaction increases both an asset and a liability of a solvent firm, the debt-equity ratio (= total liabilities divided by total equities) increases, making the company appear more risky. Thus, given a choice, some managers prefer not to show an asset and a related liability on the

**EXHIBIT 10.1**

**Amortization Schedule for $45,000 Lease Liability, Accounted for as a Capital Lease, Repaid in Three Annual Installments of $19,709 Each, Interest Rate 15 percent, Compounded Annually**

**Annual Journal Entry**

| | | |
|---|---|---|
| Interest Expense .................................... | Amount in Column **(3)** | |
| Liability—Present Value of Lease Obligation .................... | Amount in Column **(5)** | |
| Cash .......................................... | | Amount in Column **(4)** |

| Year (1) | Lease Liability Start of Year (2) | Interest Expense for Year (3) | Payment (4) | Portion of Payment Reducing Lease Liability (5) | Lease Liability End of Year (6) |
|---|---|---|---|---|---|
| 0 ...................... | | | | | $45,000 |
| 1 ...................... | $45,000 | $6,750 | $19,709 | $12,959 | 32,041 |
| 2 ...................... | 32,041 | 4,806 | 19,709 | 14,903 | 17,138 |
| 3 ...................... | 17,138 | 2,571 | 19,709 | 17,138 | 0 |

Column (2) = column (6), previous period.
Column (3) = 0.15 × column (2).
Column (4) is given.
Column (5) = column (4) − column (3).
Column (6) = column (2) − column (5).

**EXHIBIT 10.2**

**Comparison of Expense Recognized under Operating and Capital Lease Methods**

| | Expense Recognized Each Year Under | |
|---|---|---|
| Year | Operating Lease Method | Capital Lease Method |
| 1 .............. | $19,709 | $21,750 (= $15,000 + $ 6,750) |
| 2 .............. | 19,709 | 19,806 (= 15,000 + 4,806) |
| 3 .............. | 19,709 | 17,571 (= 15,000 + 2,571) |
| Total ........... | $59,127[a] | $59,127 (= $45,000[b] + $14,127[c]) |

[a]Rent expense
[b]Depreciation expense
[c]Interest expense

balance sheet. Either borrowing funds to purchase an asset or obtaining an asset's service through a capital lease records new assets and liabilities on the balance sheet, increasing debt ratios. So, some managers prefer operating leases for acquiring asset services. Many managers would also prefer to recognize expenses later rather than sooner for financial reporting. Their preferences have led managers to structure asset acquisitions so that the financing takes the form of an operating lease. Meanwhile, the FASB has tried to specify rules for curtailing the use of the operating lease accounting treatment for leases that transfer risks and rewards of ownership from the lessor to the lessee. Analysts' reports suggest that the capital lease method provides higher quality measures of financial position.

**Conditions Requiring Capital Lease Accounting** The FASB has provided rules for classifying long-term leases.[3] A firm must account for a lease as a capital lease if the lease

[3]*Ibid.*, par. 7.

meets any one of four conditions. If the lease meets none of the four conditions, the firm treats the lease as an operating lease.

1. It transfers ownership to the lessee at the end of the lease term.
2. Transfer of ownership at the end of the lease term seems likely because the lessee (user) has a "bargain purchase" option. (A "bargain purchase" option gives the lessee the right to purchase the asset at a specified future time for a price less than the currently predicted fair market value of the asset at the future time.)
3. The lease extends for at least 75 percent of the asset's life.
4. The present value of the minimum contractual lease payments equals or exceeds 90 percent of the fair market value of the asset at the time the lessee signs the lease. The present value computation uses a discount rate appropriate for the creditworthiness of the lessee.

These criteria attempt to identify who enjoys the economic benefits and bears the economic risks of the leased asset. The FASB's rules reflect the concepts that when the lessor bears the risks of a lease, it is an operating lease, but that when the lessee bears the risks, it is a capital lease. If the leased asset, either automatically or for a bargain price, becomes the property of the lessee at the end of the lease period, then the lessee enjoys all of the economic benefits of the asset and incurs all risks of ownership. If the life of the lease extends for most of the expected useful life of the asset (the FASB specifies 75 percent or more), then the lessee enjoys most of the economic benefits of the asset, particularly when we measure them in present values, and incurs most of the risk of technological obsolescence.

Lessees who want to treat a lease as an operating lease, rather than a capital lease, can usually structure the leasing contract to avoid the first three conditions. Avoiding the fourth condition is more difficult because it requires lessors to bear more risk than they might desire. The fourth condition compares the present value of the lessee's contractual minimum lease payments with the fair market value of the asset at the time the lessee signs the lease. The lessor presumably could either sell the asset for its fair market value or lease it to the lessee for a set of lease payments set forth in the lease contract. The present value of the minimum lease payments has the economic character of a purchase price in that the lessee has committed to make payments just as it would commit to make payments in an installment purchase. The FASB says, in effect, that if this present value (lessee's purchase price) equals or exceeds 90 percent of the asset's fair market value (lessor's alternative selling price) at the inception of the lease, then the lessee has effectively purchased the asset and must account for the asset as though it had.

**Note on the Risks of the Leasing Business** What is the major long-term asset of Hertz or of Avis? Used cars: Hertz and Avis must deal on a daily basis with an inventory of used cars and will bear the loss should government regulations make obsolete the current stock of used cars because, for example, they consume too much gasoline or do not have passenger airbags. Thus, Hertz and Avis bear the risk of obsolescence and treat their leases of automobiles as operating leases. Hertz and Avis, of course, price their car rentals to compensate for the obsolescence risks they bear. A manufacturer of airplanes such as Boeing may find the new airplane business sufficiently risky that it does not want to bear the additional risks of being in the used airplane business, so Boeing might require its lease customers to sign long-term leases in which the lessee guarantees a certain minimum residual value for the leased asset. In this case, the lessee bears the risk of obsolescence; Boeing likely treats the leases as capital leases.

**Consequence of Lessors' Desire to Avoid Used Asset Risk** Lessors of assets, such as airplanes and computers, do not want to lease an asset under conditions in which they have more than 10 percent of the asset's original market value at risk. When the lessee returns the asset to the lessor, the lessor must renew the lease or sell it in order to capture the benefits originally expected from the asset and to realize all the profit inherent in its original manufacture. Many lessors will not accept such risks without additional payment from the lessee. A lease that meets the fourth condition above has transferred the risks and rewards of ownership from the lessor to the lessee. In economic substance, the lessee has acquired an asset and has agreed to pay for it under a long-term installment payment contract. Accounting recognizes that contract as the lessee's long-term liability.

**Effects on Lessor** Lessors, such as Boeing for aircraft and IBM for computers, manufacture and then lease assets to lessees. Other lessors, such as financial institutions, purchase assets from the manufacturer and then lease the asset to lessees.

The lessor (landlord) generally uses the same criteria as does the lessee (tenant) for classifying a lease as a capital lease or an operating lease. When the lessor signs an operating lease, the lessor recognizes rent revenue each period in amounts that mirror rent expense for the lessee. The lessor also depreciates the leased asset over time. When the lessor is the manufacturer and signs a capital lease, the lessor recognizes revenue in an amount equal to the present value of all future lease payments and recognizes expense (analogous to cost of goods sold) in an amount equal to the book value of the leased asset at the time of signing the capital lease. The difference between the revenue and the expense is the lessor's income on the "sale" of the asset. Only manufacturers of leased assets recognize this type of income. When the lessor is either the manufacturer or a financial institution, the capital lease is a means of financing for the lessee. The lessor records the lease receivable like any other long-term receivable—at the present value of the future cash flows. It recognizes interest revenue over the collection period of the payments with entries that are mirror images of the lessee's entries for interest expense. The entries made by the lessor of the computer to Food Barn appear below. These entries assume that the lessor manufactured the computer at a cost of $39,000.

**Operating Lease Method by Lessor**

*January 1, Year 1*

| | | |
|---|---|---|
| Equipment (Computer Leased to Customers) | 39,000 | |
| Inventory | | 39,000 |

| Assets | = | Liabilities | + | Shareholders' Equity | (Class.) |
|---|---|---|---|---|---|
| +39,000 | | | | | |
| −39,000 | | | | | |

To record the transfer of product from inventory to equipment (in hands of lessee).

*December 31 of each year*

| | | |
|---|---|---|
| Cash | 19,709 | |
| Rent Revenue | | 19,709 |

| Assets | = | Liabilities | + | Shareholders' Equity | (Class.) |
|---|---|---|---|---|---|
| +19,709 | | | | +19,709 | IncSt → RE |

To recognize annual revenue from renting computer.

*December 31 of each year*

| | | |
|---|---|---|
| Depreciation Expense | 13,000 | |
| Accumulated Depreciation—Computer | | 13,000 |

| Assets | = | Liabilities | + | Shareholders' Equity | (Class.) |
|---|---|---|---|---|---|
| −13,000 | | | | −13,000 | IncSt → RE |

To recognize depreciation on rented computer ($13,000 = $39,000/3).

**Capital Lease Method by Lessor**

*January 1, Year 1*

| | | |
|---|---|---|
| Lease Receivable | 45,000 | |
| Sales Revenue | | 45,000 |

| Assets | = | Liabilities | + | Shareholders' Equity | (Class.) |
|---|---|---|---|---|---|
| +45,000 | | | | +45,000 | IncSt → RE |

To recognize the "sale" of a computer for a series of future cash flows with a present value of $45,000.

| | | |
|---|---|---|
| Cost of Goods Sold | 39,000 | |
| Inventory | | 39,000 |

(continued)

| Assets | = | Liabilities | + | Shareholders' Equity | (Class.) |
|---|---|---|---|---|---|
| −39,000 | | | | −39,000 | IncSt → RE |

To record the cost of the computer "sold" as an expense.

*December 31, Year 1*

| | | |
|---|---|---|
| Cash | 19,709 | |
| Interest Revenue | | 6,750 |
| Lease Receivable | | 12,959 |

| Assets | = | Liabilities | + | Shareholders' Equity | (Class.) |
|---|---|---|---|---|---|
| +19,709 | | | | +6,750 | IncSt → RE |
| −12,959 | | | | | |

To recognize lease receipt, interest on receivable, and reduction in receivable for Year 1. See supporting calculations in the lessee's journal entries and in **Exhibit 10.1**.

*December 31, Year 2*

| | | |
|---|---|---|
| Cash | 19,709 | |
| Interest Revenue | | 4,806 |
| Lease Receivable | | 14,903 |

| Assets | = | Liabilities | + | Shareholders' Equity | (Class.) |
|---|---|---|---|---|---|
| +19,709 | | | | +4,806 | IncSt → RE |
| −14,903 | | | | | |

To recognize lease amounts for Year 2.

*December 31, Year 3*

| | | |
|---|---|---|
| Cash | 19,709 | |
| Interest Revenue | | 2,571 |
| Lease Receivable | | 17,138 |

| Assets | = | Liabilities | + | Shareholders' Equity | (Class.) |
|---|---|---|---|---|---|
| +19,709 | | | | +2,571 | IncSt → RE |
| −17,138 | | | | | |

To recognize lease amounts for Year 3.

Lessors tend to prefer capital lease accounting for financial reporting because it enables them to recognize all income from the sale of the asset on the date the parties sign the lease while spreading the interest revenue over the life of the lease. The operating lease method recognizes all lease revenue—which implicitly includes manufacturing profit and interest—gradually over time as the lessor receives lease payments. Lessors, however, recognize the preference of lessees to structure leases as operating leases. Because the lessor and lessee apply the same criteria to classify a lease as either an operating lease or a capital lease, lessors tend to accommodate the preferences of lessees, their customers, but set rental payments to compensate for any additional risk the lessor bears.

**Income Tax Consideration in Lease Arrangements** Leasing has become an industry separate from manufacturing. That is, companies such as General Electric Capital Services buy computers from, say, IBM and lease them to end-users. The leasing industry developed in part because the users of computers, airplanes, and other depreciable assets did not have sufficient taxable income to take advantage of accelerated depreciation deductions. A tax deduction has value only to the extent the taxpayer has taxable income against which to offset the deduction, reducing taxes otherwise payable. Any organization without taxable income, whether a not-for-profit, or a municipality, or a taxpayer with no earnings, cannot take advantage of a deduction, such as for depreciation. Other entities, such as financial institutions or manufacturers of leased equipment, have sufficient taxable income to benefit from these deductions. These parties structure leases as operating leases for income tax purposes so that the lessors claim depreciation deductions and save taxes. Lessees attempt to negotiate with the lessors to obtain lower lease payments, which in effect transfer some of the benefits of reduced taxes to the lessee.

The rules for classifying a lease as operating or capital for income tax purposes differ from the FASB rules discussed here for financial reporting. Thus, leases sometimes appear as operating leases for tax purposes and capital leases for financial reporting, or vice versa.

**Bundling in Lease Arrangements** In the last 50 years, leases have become a common method for users to finance acquisition of long-term assets. Think about the copying machine down the hall from where you work; likely your school or business chooses to lease it from the manufacturer, such as Canon or Xerox, rather than buy it. Increasingly, customers, such as your school, want to acquire a full package of services from the supplier. In the case of a copier, this would mean buying toner and other supplies, as well as a contract for ordinary maintenance and emergency service. As a result, companies such as Xerox quote a single price to the user to cover a complete package of goods and services: for example, the copier, toner, and combined maintenance/service contract. The customer makes a single monthly payment and Xerox confronts the accounting problem of separating the single payment into components—part for copier rental, part for toner, and part for the service/maintenance contract. GAAP provides little guidance other than require that the revenues from each of the components bear a reasonable relation to the fair market price of the item sold separately. In the case of Xerox, it has much data on separate sales of toner and maintenance/service contracts, but few separate transactions where it has sold a stand-alone copier, with no extras. Hence, the lessor must use some methods to disaggregate the total rental payment into the parts for toner and service, which is easy, and the rest into the part that is for interest and the part for equipment, which is hard. Xerox found itself in trouble with the SEC in how it split the total lease payment into a part for the sales price of the copier and part for the interest revenue implicit in the financing. One of the cases at the end of this chapter explores the issue. The FASB, with the EITF and the SEC, is attempting to provide more guidance to lessors who deal in bundled contracts.[4]

## Problem 10.2 for Self-Study

**Operating and capital lease methods for lessee and lessor**. On January 2, Year 1, Holt Book Store will acquire a delivery van that a local automobile dealer sells for $25,000. The dealer offers Holt Book Store the option of leasing the van for four years, with rentals of $8,231 due on December 31 of each year. Holt Book Store must return the van at the end of four years, although the automobile dealer anticipates that the resale value of the van after four years will be negligible. The automobile dealer acquired the van from the manufacturer for $23,500. The automobile dealer considers 12 percent an appropriate interest rate to charge Holt Book Store to finance the acquisition.

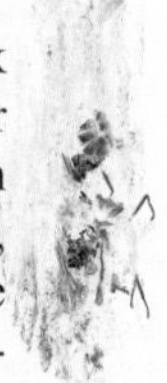

**a.** Does this lease qualify as an operating lease or as a capital lease for financial reporting according to the four criteria specified by the FASB? Explain.

**b.** Assume for this part that the lease qualifies as an operating lease. Give the journal entries made by Holt Book Store over the first two years of the life of the lease.

**c.** Repeat part **b** for the automobile dealership. Use straight-line depreciation and zero estimated salvage value.

**d.** Assume for this part that the lease qualifies as a capital lease. Give the journal entries made by Holt Book Store over the first two years of the life of the lease.

(continued)

[4]See *EITF 00–21: Revenue Arrangements with Multiple Deliverables*, FASB, January 2003.

CHAPTER 11

# Marketable Securities, Derivatives, and Investments

## Learning Objectives

1. Understand why firms acquire securities of other firms and how the purpose of the investment governs the method of accounting for that investment.
2. Develop skills to apply the market value method to minority, passive investments, including financial derivatives.
3. Develop skills to apply the equity method to minority, active investments, contrasting its financial statement effects with those of the market value method.
4. Understand the concepts underlying consolidated financial statements for majority, active investments, contrasting the financial statement effects of consolidation with those of the equity method.
5. Understand when a firm must consolidate a variable interest entity, for which it does not have conventional control, but does have parent-like characteristics.

*F*or a variety of reasons, corporations often acquire the securities (bonds, preferred stock, common stock) of other entities.

**Example 1** Southwest Airlines sells airline tickets and receives cash prior to providing transportation services. Rather than let the cash remain idle in its bank account, Southwest Airlines acquires U.S. Treasury Notes. The firm earns interest while it holds the notes and will sell the notes when it needs the cash for operations.

**Example 2** Merck, a pharmaceutical company, acquires shares of common stock of several firms engaged in biotechnology research. Merck will benefit from increases in the market prices of these shares if the research efforts are successful.

**Example 3** The Coca-Cola Company owns 40 percent of the common stock of Coca-Cola Enterprises, a bottler of its soft drinks. This ownership percentage permits The Coca-Cola Company to exert significant influence over the operations of Coca-Cola Enterprises and keep the affiliate's debt off the parent's balance sheet.

**Example 4** Walt Disney owns all of the common stock of ABC/Capital Cities. Walt Disney can control both the broad policy and the day-to-day business decisions of ABC/Capital Cities.

## Overview of the Accounting and Reporting of Investments in Securities

The accounting for investments in securities depends on (1) the expected holding period, and (2) the purpose of the investment.

### Expected Holding Period

The expected holding period determines where investments in securities appear in the balance sheet. Securities that firms expect to sell within the next year appear as **Marketable Securities** in the Current Assets section of the balance sheet. In **Example 1**, Southwest Airlines would likely include the U.S. Treasury Notes in Marketable Securities. Securities that firms expect to hold for more than one year from the date of the balance sheet appear in **Investments in Securities**, which firms include in a separate section of the balance sheet between Current Assets and Property, Plant, and Equipment. The investments in biotechnology companies by Merck in **Example 2** and the investment in Coca-Cola Enterprises by The Coca-Cola Company in **Example 3** appear in Investments in Securities. A later section explains that Walt Disney in **Example 4** would prepare consolidated financial statements with ABC/Capital Cities. The consolidation procedure requires Walt Disney to eliminate its Investment in ABC/Capital Cities account and replace it with that firm's individual assets and liabilities. Thus, the Investment in ABC/Capital Cities account will not appear on the balance sheet. Although we and others use the phrase *investment in securities* to refer broadly to all investments in the bonds, capital stock, and other financial instruments of other entities, the preferred usage reserves the word *investment* for holdings with a long-term purpose.

### Purpose of the Investments

The purpose of the investment in securities and the percentage of voting stock that one corporation owns of another determine the accounting for the investment. Refer to **Figure 11.1**, which identifies three types of investments.

**FIGURE 11.1** Types of Intercorporate Investments in Capital Stock

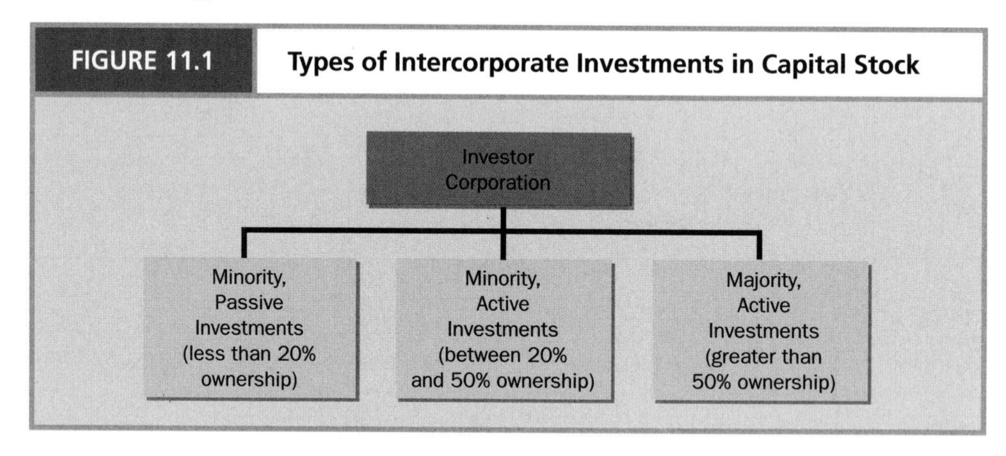


1. In **minority, passive investments**, an investor views acquiring bonds or shares of capital stock of another corporation as a worthwhile expenditure and acquires them for the interest, dividends and capital gains (increases in the market prices) anticipated from owning them. The acquiring company owns such a small percentage of the other corporation's shares that it cannot control or exert **significant influence** over the other company. Southwest Airlines' investment in U.S. Treasury notes in **Example 1** and Merck's investments in the common stock of biotechnology companies in **Example 2** are minority, passive investments. Occasionally, an owner of a small percentage of the shares has the contractual right to elect one or more members of the board of directors. If so, the acquiring company, even though a small percentage holder, could exercise significant influence. Generally accepted accounting principles (GAAP) view investments of less than 20 percent of the voting stock of another company as minority, passive investments in most cases.[1] The owner may intend to hold these

[1] Accounting Principles Board, *Opinion No. 18*, "The Equity Method of Accounting for Investments in Common Stock," 1971.

shares for relatively short time spans and classify them as current assets or for a longer time and classify them as investments.

2. In **minority, active investments**, an investor acquires shares of another corporation so that it can exert significant influence over the other company's activities. The investor might seek broad policy-making influence through representation on the other corporation's board of directors. Many different entities own the shares of most publicly held corporations, and most of these have not collaborated in voting their shares, although this is changing. Therefore, an owner can exert significant influence over another corporation with ownership of less than a majority of the voting stock. The investment by The Coca-Cola Company in the shares of Coca-Cola Enterprises in **Example 3** is a minority, active investment. GAAP views investments of between 20 and 50 percent of the voting stock of another company as minority, active investments "unless evidence indicates that significant influence cannot be exercised."[2] Minority, active investments appear under investments on the balance sheet.
3. In **majority, active investments**, an investor acquires shares of another corporation so that it can control the other company both at the broad policy-making level and at the day-to-day operational level. Refer to **Example 4**. Walt Disney acquired ABC/Capital Cities to add television broadcasting to its entertainment capabilities. GAAP views ownership of more than 50 percent of the voting stock of another company as implying an ability to control, unless there is evidence to the contrary. An investor cannot exercise control over a corporation, despite owning a majority of the voting stock, if a court effectively controls the corporation in bankruptcy proceedings or if the investor owns shares in a foreign company whose government restricts the withdrawal of assets from the country.

This chapter describes and illustrates the accounting for each of these three types of investments. Throughout our discussion, we will designate the acquiring corporation as P, for purchaser or for parent, depending on the context, and the acquired corporation as S, for seller or for subsidiary.

## Minority, Passive Investments

### Valuation of Securities at Acquisition

A firm initially records the acquisition of securities at acquisition cost, which includes the purchase price plus any commissions, taxes, and other costs incurred. For example, if a firm acquires securities classified as Marketable Securities for $10,000 plus $300 for commissions and taxes, the entry is as follows:

| | | |
|---|---|---|
| Marketable Securities | 10,300 | |
| Cash | | 10,300 |

| Assets | = | Liabilities | + | Shareholders' Equity | (Class.) |
|---|---|---|---|---|---|
| +10,300 | | | | | |
| −10,300 | | | | | |

Dividends on equity securities become revenue when declared. Interest on debt securities becomes revenue when earned. Assume that a firm holds equity securities earning $250 through dividend declarations and debt securities earning $300 from interest earned and that it has not yet received these amounts in cash. The entry is as follows:

| | | |
|---|---|---|
| Dividends and Interest Receivable | 550 | |
| Dividend Revenue | | 250 |
| Interest Revenue | | 300 |

| Assets | = | Liabilities | + | Shareholders' Equity | (Class.) |
|---|---|---|---|---|---|
| +550 | | | | +250 | IncSt → RE |
| | | | | +300 | IncSt → RE |

[2]Ibid.; Financial Accounting Standards Board, *Interpretation No. 35*, "Criteria for Applying the Equity Method of Accounting for Investments in Common Stock," 1981.

The valuation of minority, passive investments at the date of acquisition or the recording of dividends and interest presents no new issues. Valuing securities after acquisition, however, departs from acquisition cost accounting.

## VALUATION OF SECURITIES AFTER ACQUISITION

The Financial Accounting Standards Board (FASB) *Statement of Financial Accounting Standards No. 115*[3] requires firms to classify minority, passive investments in securities into three categories:

1. Debt securities for which a firm has both the positive intent and the ability to hold to maturity—shown on the balance sheet at an amount based on acquisition cost
2. Debt and equity securities, as well as derivatives, held as trading securities—shown on the balance sheet at market value, with changes in market value of securities held at the end of the accounting period reported each period in income
3. Debt and equity securities, as well as derivatives, held as securities available for sale—shown on the balance sheet at market value, with changes in market value of securities held at the end of the accounting period not affecting reported income until the firm sells, or otherwise disposes of, the securities

This three-way classification resulted from one of the most contentious battles ever to involve the attention of accounting standard-setters, including the FASB and several of its predecessor organizations. We discuss the controversy at the end of this section.

**Debt Securities Held to Maturity** Firms sometimes acquire debt securities with the intention of holding these securities until maturity, as in the next example.

**Example 5** Consolidated Edison, an electric utility, has $100 million of bonds payable outstanding that mature in five years. The utility acquires U.S. government securities whose periodic interest payments and maturity value exactly equal those on the utility's outstanding bonds. The firm intends to use the cash received from the government bonds to make required interest and principal payments on its own bonds.

**A Practice of Corporate Finance** Some students ask why a business would purchase bonds and then use the cash throwoff from those bonds to make debt service payments, rather than use cash to retire the bonds immediately. After firms issue bonds, the broad investing public holds these securities. The subsequent transaction costs the firm would incur if it were to attempt to contact all of the holders and then persuade them to turn in their bonds for cash make the effort too expensive to be cost-effective. Instead the firm uses the cash to acquire other bonds whose cash throwoffs approximately match the cash needs for debt service. Corporate finance courses treat transactions such as these. Here, the presentation emphasizes the accounting for the transaction, not why the firm enters into it.

**Accounting for Debt Held to Maturity** Debt securities for which a firm has a positive intent and ability to hold to maturity appear in the Investments section of the balance sheet at amortized acquisition cost. A firm initially records these debt securities at acquisition cost. This acquisition cost will differ from the maturity value of the debt if the coupon rate on the bonds differs from the required market yield on the bonds at the time the firm acquired them. The firm must amortize any difference between acquisition cost and maturity value over the life of the debt as an adjustment to interest revenue. The amortization procedure involves the same compound interest computation that **Chapter 9** introduced for the issuer of the bonds.

The holder of the debt records interest revenue each period at an amount equal to the book value of the investment in debt at the start of the period, multiplied by the market rate of interest applicable to that debt on the day the firm acquired it. It debits the Investment account and credits Interest Revenue, which increases Retained Earnings. Then, if it receives cash, it debits

---

[3]Financial Accounting Standards Board, *Statement of Financial Accounting Standards No. 115,* "Accounting for Certain Investments in Debt and Equity Securities," 1993.

Cash and credits the Investment account. The result of this process is a new book value (called the *amortized cost*) for use in the computations during the next period. See **Exhibit 9.6**. The holder of the debt securities in those illustrations makes the same entries, with reversed debits and credits.

## Problem 11.1 for Self-Study

**Accounting for an interest-bearing note receivable.** (Compare with **Problem 9.2** for Self-Study.) General Electric Capital Services (GECS) lends $112,434 to Sapra Company. GECS lends the cash of $112,434 in return for a three-year, $100,000 note, in which Sapra Company promises to pay $15,000 after one year, $15,000 after two years, and $115,000 after three years. The market interest rate on the original issue date of the note is 10 percent.

**a.** Prepare an amortization schedule, similar to that in **Exhibit 9.2**, for the life of the note. Change the column headings as follows: Column (2) is Note Receivable Balance Start of Period; Column (3) is Interest Revenue for Period; Column (6) is Note Receivable Balance End of Period.

**b.** Prepare journal entries that GECS would make on three dates: the date of issue, six months after the date of issue (assuming GECS closes the books then), and one year after the date of issue (assuming GECS receives a cash payment from Sapra then).

**Trading Securities** Firms sometimes purchase and sell debt and equity securities for the short-term profit potential—some would say for *speculation*. The term *trading* implies active and frequent buying and selling with the objective of generating profits from short-term differences in market prices. Acquisition and disposition of trading securities are operating activities. Commercial banks, for example, often trade securities in different capital markets worldwide to take advantage of temporary differences in market prices. Other financial firms, such as thrift institutions, insurance companies, and brokerage firms, also trade securities. Manufacturers, retailers, and other nonfinancial firms occasionally invest funds for trading purposes, but such situations are unusual. Firms include trading securities in Marketable Securities in the Current Assets section of the balance sheet.

Firms initially record trading securities at acquisition cost. FASB *Statement No. 115* requires firms to report trading securities at *market value* on the balance sheet. The FASB justifies this departure from acquisition cost accounting on two factors: (1) active securities markets provide objective measures of market values, and (2) market values provide financial statement users with the most relevant information for assessing the success of a firm's trading activities over time.

The income statement reports the debit (loss) for decreases in the market value and the credit (gain) for increases in the market value of trading securities in an account with a title such as *Unrealized Holding Loss (or Gain or Gains and Losses, net) on Valuation of Trading Securities.*

**Example 6** First Insurance acquired shares of Sun Microsystems' common stock on December 28, Year 3, for $400,000. The market value of these securities on December 31, Year 3, was $435,000. First Insurance sold these shares on January 3, Year 4, for $480,000. The journal entries to record these transactions appear below.

*December 28, Year 3*

| | Dr. | Cr. |
|---|---|---|
| Marketable Securities . . . . . . . . . . | 400,000 | |
| Cash . . . . . . . . . . | | 400,000 |

| Assets | = | Liabilities | + | Shareholders' Equity | (Class.) |
|---|---|---|---|---|---|
| +400,000 | | | | | |
| −400,000 | | | | | |

To record acquisition of trading securities.

(continued)

*December 31, Year 3*

| | | |
|---|---|---|
| Marketable Securities | 35,000 | |
| Unrealized Holding Gain on Trading Securities | | 35,000 |

| Assets | = | Liabilities | + | Shareholders' Equity | (Class.) |
|---|---|---|---|---|---|
| +35,000 | | | | +35,000 | IncSt → RE |

To revalue trading securities to market value and recognize an unrealized holding gain in income.

*January 3, Year 4*

| | | |
|---|---|---|
| Cash | 480,000 | |
| Marketable Securities | | 435,000 |
| Realized Gain on Sale of Trading Securities | | 45,000 |

| Assets | = | Liabilities | + | Shareholders' Equity | (Class.) |
|---|---|---|---|---|---|
| +480,000 | | | | +45,000 | IncSt → RE |
| −435,000 | | | | | |

To record the sale of trading securities at a gain.

The total income from the purchase and sale of these securities is $80,000 (equals cash inflows of $480,000 minus cash outflows of $400,000). The required accounting allocates $35,000 to Year 3, the change in market value during that year, and $45,000 to Year 4, the change in market value during that year.

**Securities Available for Sale** FASB *Statement No. 115* classifies securities that are neither debt securities held to maturity nor trading securities as *securities available for sale.* Securities available for sale that a firm intends to sell within one year appear in Marketable Securities in the Current Assets section of the balance sheet. All others appear in Investments in Securities. Acquisition and disposition of securities available for sale, whether current or noncurrent, are investing activities. The FASB requires firms to report these securities at market value on the balance sheet. Securities in this category often trade in active securities markets and therefore have easily measurable market values. Firms typically acquire these securities for an operating purpose (for example, investment of temporarily excess cash or investment in a supplier) rather than for their short-term profit potential, the typical reason for investments in trading securities.

**Accounting for Securities Available for Sale** Firms initially record investments in securities available for sale at acquisition cost. FASB *Statement No. 115* requires firms to report securities available for sale at market value on the balance sheet. *The unrealized holding gain or holding loss each period does not affect income immediately, as is the case with trading securities, but instead increases or decreases Accumulated Other Comprehensive Income, a separate shareholders' equity account.* Accumulated Other Comprehensive Income is a shareholders' equity account, which shows the sum of all increases and decreases in shareholders' equity that have not yet appeared in net income. **Chapter 12** discusses this account and the related accounts for its periodic components, called Other Comprehensive Income. Holding gains and losses on securities available for sale affect net income only when the firm sells these securities. The FASB's reasoning for delaying the recognition of holding gains and losses in earnings is that firms do not acquire securities classified as available for sale for the short-term returns, as with trading securities, but for support of an operating activity. Also, including unrealized gains and losses in earnings might result in significant earnings changes that could easily reverse during future periods.

**Example 7** Nike has temporarily excess cash and acquires common stock of Merck for $400,000 on November 1, Year 3. The market value of these shares is $435,000 on December 31, Year 3. Nike sells these shares on August 15, Year 4, for $480,000. The journal entries to record these transactions are as follows:

*November 1, Year 3*

| | | |
|---|---|---|
| Marketable Securities | 400,000 | |
| Cash | | 400,000 |

| Assets | = | Liabilities | + | Shareholders' Equity | (Class.) |
|---|---|---|---|---|---|
| +400,000 | | | | | |
| −400,000 | | | | | |

To record acquisition of securities available for sale.

*December 31, Year 3*

| | | |
|---|---|---|
| Marketable Securities | 35,000 | |
| Unrealized Holding Gain on Securities Available for Sale | | 35,000 |

| Assets | = | Liabilities | + | Shareholders' Equity | (Class.) |
|---|---|---|---|---|---|
| +35,000 | | | | +35,000 | OCInc → AOCInc |

To revalue securities available for sale to market value and record an unrealized gain. Don't overlook the fact that the increase in shareholders' equity appears directly in the Shareholders' Equity section and in other comprehensive income, but not in net income.

*August 15, Year 4*

| | | |
|---|---|---|
| Cash | 480,000 | |
| Marketable Securities | | 400,000 |
| Realized Gain on Sale of Securities Available for Sale | | 80,000 |

| Assets | = | Liabilities | + | Shareholders' Equity | (Class.) |
|---|---|---|---|---|---|
| +480,000 | | | | +80,000 | IncSt → RE |
| −400,000 | | | | | |

To record the sale of securities available for sale at a gain based on acquisition cost.

If Nike intended to hold the shares of Merck for more than one year, then it would debit or credit Investment in Securities instead of Marketable Securities in the three entries above.

The total income from the purchase and sale of these securities is $80,000 (equals cash inflow of $480,000 minus cash outflow of $400,000). The required accounting allocates the full gain to the year of sale, even though the balance sheet reports changes in market value of the assets as they occur. At the time of sale, or later in the period when the firm makes adjusting entries, it must make the following entry:

*August 15, Year 4 (or later, at time of adjusting entries)*

| | | |
|---|---|---|
| Unrealized Holding Gain on Securities Available for Sale | 35,000 | |
| Marketable Securities | | 35,000 |

| Assets | = | Liabilities | + | Shareholders' Equity | (Class.) |
|---|---|---|---|---|---|
| −35,000 | | | | −35,000 | OCInc → AOCInc |

To eliminate the previously recorded effects of changes in market values of securities available for sale.

The credit in this entry reduces the asset valuation: the firm no longer holds the security, so the accounting must remove from the balance sheet not only the asset's acquisition cost—as in the first August 15 entry above—but also the increases in market value included in the asset's valuation. The debit removes the unrealized gain from Accumulated Other Comprehensive

Income in shareholders' equity. Total shareholders' equity has increased by $80,000 while this firm held the securities, and the Gain on Sale account shows all that gain, which increases Retained Earnings.

We can summarize the effects as follows:

- In the year of purchase, asset valuation and shareholders' equity both increased by $35,000 from the increase in market value, but the increase has no effect on net income or Retained Earnings. The shareholders' equity account showing the increase is part of Accumulated Other Comprehensive Income.
- In the next year, asset valuation and shareholders' equity increased another $45,000. Retained Earnings increases by $80,000—the full gain on holding the securities. The shareholders' equity account for Unrealized Gain must decrease by $35,000 (= $80,000 − $45,000), just equal to the amounts of the market gain increase that the firm recorded in periods before the firm sold the security and realized the gain.

**Summary of Accounting for Securities Available for Sale** The accounting for securities available for sale involves three transactions or events.

**1.** Acquisition of Securities Available for Sale

| | | |
|---|---|---|
| Marketable Securities | Acquisition Cost | |
| Cash | | Acquisition Cost |

**2.** Revaluation to Market Value at End of Each Accounting Period

| | | |
|---|---|---|
| Marketable Securities | Increase in Market Value above Book Value | |
| Unrealized Holding Gain on Securities Available for Sale (Other Comprehensive Income) | | Increase in Market Value above Book Value |
| Unrealized Holding Loss on Securities Available for Sale (Other Comprehensive Income) | Decrease in Market Value below Book Value | |
| Marketable Securities | | Decrease in Market Value below Book Value |

The unrealized holding gain and loss accounts are Accumulated Other Comprehensive Income accounts, that is, shareholders' equity accounts, not income statement accounts. The accountant can make separate entries for each security held or a single, combined entry for the portfolio of securities.

**3.** Sale of Securities Available for Sale

| | | |
|---|---|---|
| Cash | Proceeds of Sale | |
| Marketable Securities | | Acquisition Cost |
| Realized Gain on Sale of Securities Available for Sale | | Plug Amount |
| Securities sold for amount larger than acquisition cost. | | |
| Cash | Proceeds of Sale | |
| Realized Loss on Sale of Securities Available for Sale | Plug Amount | |
| Marketable Securities | | Acquisition Cost |
| Securities sold for amount smaller than acquisition cost. | | |

The realized gain or loss accounts appear in the income statement as increases or decreases to Retained Earnings. Once the firm has sold the securities, it no longer has an unrealized gain or loss from holding these securities. The final step in the accounting eliminates from the balance sheet the recording of the unrealized holding gain or loss. The next entry eliminates any excess of the book value of the securities sold over their original acquisition cost.

| | | |
|---|---|---|
| Unrealized Holding Gain on Securities Available for Sale (Other Comprehensive Income) . . . . . . . . . . . . . . . . | Excess of Book Value over Acquisition Cost | |
| Marketable Securities . . . . . . . . . . . . . . . . . . . . . . . | | Excess of Book Value over Acquisition Cost |

These securities had risen in value between the time of acquisition and the time of sale. The firm had recorded an unrealized holding gain and an increase in the value of the securities. Now it eliminates both the unrealized holding gain and the securities themselves.

When the firm sells securities, it records the realized gain (or loss) as the difference between sales price and acquisition cost. It reduces the carrying amount of securities by the amount of acquisition cost in the entry recognizing the gain (or loss). The above entry reduces the carrying amount of the securities for the increases recorded since acquisition. After this entry, the balance in the securities account for the securities sold is zero because the firm no longer holds these specific securities. The firm can make this entry at the time of sale of an individual security or at the end of the period as part of the revaluation of the portfolio of securities.

If the firm had recorded unrealized holding losses on the securities before it sold them, the final entry, made at time of sale or the end of the period, would be:

| | | |
|---|---|---|
| Marketable Securities . . . . . . . . . . . . . . . . . . . . . . . . . . . . | Excess of Acquisition Cost over Book Value | |
| Unrealized Holding Loss on Securities Available for Sale (Other Comprehensive Income) . . . . . . . . . . | | Excess of Acquisition Cost over Book Value |

When the firm sells securities, it records the loss as the difference between sales price and acquisition cost. It reduces the carrying amount of securities by the amount of acquisition cost in the entry recognizing the loss, which creates a credit balance in the securities account for the security sold. This entry restores the balance in that account to zero. The firm can make this entry at the time of sale of an individual security or at the end of the period as part of the revaluation of the portfolio of securities.

**An Earnings Quality Issue** The unrealized holding gain on securities available for sale appears in other comprehensive income period by period, and its cumulative amount resides in the Accumulated Other Comprehensive Income account on the balance sheet. Whenever management wishes, it can sell securities with unrealized holding gains (or losses) and transfer through net income to retained earnings the entire unrealized holding gain (or loss). To the extent a firm has unrealized holding gains and losses in Accumulated Other Comprehensive Income, the firm can decide the amounts to report in any period as realized gains and losses, affecting the quality of earnings. Analysts do not have difficulty in understanding the potential for management to use the unrealized gains and losses to manage earnings and report possibly misleading financial statements.

## Reclassification of Securities

The firm's purpose for holding certain securities may change, requiring it to transfer securities from one of the three categories to another one. The firm transfers the security at its market value at the time of the transfer. FASB *Statement No. 115* prescribes the accounting for any unrealized gain or loss at the time of the transfer, a topic discussed in intermediate accounting principles textbooks.

## Problem 11.2 for Self-Study

**Accounting for securities available for sale.** Transactions involving Conlin Corporation's marketable securities available for sale appear in **Exhibit 11.1**.

**a.** Give the journal entries to account for these securities during Year 2 and Year 3.
**b.** How would the journal entries in part **a** differ if Conlin Corporation acquired these securities available for sale as long-term investments?
**c.** How would the journal entries in part **a** differ if Conlin Corporation classified these securities as trading securities?

EXHIBIT 11.1

**CONLIN CORPORATION**
**(Problem 11.2 for Self-Study)**

| | | | | | Market Value | |
|---|---|---|---|---|---|---|
| Security | Date Acquired | Acquisition Cost | Date Sold | Selling Price | Dec. 31, Year 2 | Dec. 31, Year 3 |
| A .................... | 2/3/Year 2 | $ 40,000 | — | — | $ 38,000 | $ 33,000 |
| B .................... | 7/15/Year 2 | 75,000 | 9/6/Year 3 | $78,000 | 79,000 | — |
| C .................... | 11/27/Year 2 | 90,000 | — | — | 93,000 | 94,000 |
| | | $205,000 | | | $210,000 | $127,000 |

## Disclosures about Securities

FASB *Statement No. 115* requires the following disclosures each period:

1. The aggregate market value, gross unrealized holding gains, gross unrealized holding losses, and amortized cost for debt securities held to maturity and debt and equity securities available for sale
2. The proceeds from sales of securities available for sale, and the gross realized gains and gross realized losses on those sales
3. The change during the period in the net unrealized holding gain or loss on securities available for sale included in a separate shareholders' equity account
4. The change during the period in the net unrealized holding gain or loss on trading securities included in earnings

In addition, firms often show the gross gains and gross losses included in earnings from transfers of securities from the available-for-sale category. **Exhibit 11.2**, expanded from **Exhibit 6.4**, illustrates these disclosures.

## Controversy Surrounding the Accounting for Marketable Securities

The accounting for minority, passive investments in securities has been controversial. We try to give the flavor of the controversy without the details. The accounting issues are as follows:

- Whether to report these securities at acquisition cost (or some method based on acquisition cost) or at market value on the balance sheet date
- If reported at market value, whether to report the changes in market value from period to period as part of that period's income, or await the period when the firm sells or otherwise disposes of the security to record the gain or loss in income

**EXHIBIT 11.2**

**ALEXIS COMPANY**
**Detailed Illustration of Current Assets**
**Balance Sheet (Excerpts)**
**June 30, Year 1 and Year 2**

| | | June 30, Year 1 | | June 30, Year 2 |
|---|---|---|---|---|
| **Current Assets** | | | | |
| Cash in Change and Petty Cash Funds | | $ 800 | | $ 1,000 |
| Cash in Bank | | 11,000 | | 13,000 |
| Cash Held as Compensating Balances | | 1,500 | | 1,500 |
| Certificates of Deposit | | 7,500 | | 8,000 |
| Marketable Securities at Market Value (see Note A) | | 25,000 | | 27,000 |
| Notes Receivable (see Note B) | | 10,000 | | 12,000 |
| Interest and Dividends Receivable | | 400 | | 500 |
| Accounts Receivable, Gross | $57,200 | | $58,100 | |
| Less Allowance for Uncollectible Accounts | (3,500) | | (3,600) | |
| Accounts Receivable, Net | | 53,700 | | 54,500 |
| Merchandise Inventory[a] | | 67,000 | | 72,000 |
| Prepayments | | 4,300 | | 4,800 |
| Total Current Assets | | $181,200 | | $194,300 |

**Note A.** Gross unrealized holding gains and gross unrealized holding losses on Marketable Securities are as follows:

| | June 30, Year 1 | June 30, Year 2 |
|---|---|---|
| Gross Unrealized Holding Gains | $ 7,000 | $ 7,000 |
| Gross Unrealized Holding Losses | (1,000) | (10,000) |
| Net Unrealized Holding Gain (Loss) | $ 6,000 | $ (3,000) |

The net unrealized holding gain (loss) changed during the year as follows:

| | |
|---|---|
| Net Unrealized Holding Gain, June 30, Year 1 | $ 6,000 |
| Unrealized Holding Gain on Securities Sold | 1,000 |
| Unrealized Holding Loss for Year Ending June 30, Year 2 | (10,000) |
| Net Unrealized Holding Loss, June 30, Year 2 | $ (3,000) |

**Note B.** The amount shown for Notes Receivable does not include notes with a face amount of $2,000 that have been discounted with recourse at The First National Bank. The company is contingently liable for these notes, should the makers not honor them at maturity. The estimated amount of the company's liability is zero.

[a]Additional required disclosures for this item do not appear here. See **Chapter 7**.

**Nonfinancial Firms** For nonfinancial firms, the primary controversy surrounds the treatment of financial derivatives that these firms use in reducing, that is, hedging, their risks of exposure to fluctuations in interest rates or in prices of raw materials that the firm will need to acquire later. The next section of this chapter discusses the accounting for such hedges. Before the FASB *Statement of Financial Accounting Standards No. 133,* "Accounting for Derivative Instruments and Hedging Activities," in 1998, nonfinancial firms engaging in hedges of forecasted transactions feared that the FASB would require them to show the gains and losses in income each period. The FASB had proposed such treatment; however, under pressure from nonfinancial firms, it ruled that such gains and losses need not appear in net income but only in comprehensive income, until subsequent events unfolded.

**Financial Firms** Financial firms, such as commercial and investment banks, generally hold substantial positions in marketable securities among their assets. The accounting for marketable securities impacts not only the amounts these firms reports as assets, but also the amounts they report as earnings and shareholders' equity. Financial firms typically carry higher debt-equity ratios than nonfinancial firms. The recording of marketable securities at market value results in an increase in the debt-equity ratio when market values decrease (shareholders' equity declines with no change in liabilities). Increased debt-equity ratios usually get the attention of bank regulators. To understand how the accounting interacts with the regulations to make financial firms, particularly banks, sensitive to market value accounting, refer to **Problem 52** at the end of this chapter.

## Derivative Instruments

Firms incur various risks in carrying out their business operations. A fire may destroy a warehouse of a retail chain and disrupt the flow of merchandise to stores. An automobile accident involving a member of the sales staff may injure the employee or others and damage the firm's automobile. A firm's products may injure customers and subject the firm to lawsuits. Most firms purchase property, medical, and liability insurance against such risks. The insurance shifts the risk of the loss, at least up to the limits in the insurance policy, to the insurance company. The firm pays insurance premiums for the right to shift the risk of these losses.

Firms engage in other transactions that subject them to risks. Consider the following scenarios.

**Example 8** Firm A, a U.S. firm, orders a machine on June 30, Year 1, for delivery on June 30, Year 2, from a British supplier for £10,000 (currency in Great Britain pounds sterling). The exchange rate between the U.S. dollar and the British pound is currently $1.60 per £1, indicating a purchase price of $16,000. Firm A worries that the value of the U.S. dollar will decline between June 30, Year 1, and June 30, Year 2, when it must convert U.S. dollars into British pounds, requiring it to pay more than $16,000 to purchase the machine.

**Example 9** Firm B holds a note receivable from a customer dated January 1, Year 1, for the customer's purchase of manufacturing equipment. The note has a face value of $100,000 and bears interest at the rate of 8 percent each year. Interest is receivable annually on December 31 and the note matures on December 31, Year 3. The customer has the option of repaying the note for its face value of $100,000, plus accrued interest, prior to maturity. Firm B knows that the customer will pay off early only if interest rates drop. In that event, the market value of the note would exceed $100,000, but Firm B would receive only $100,000 from the customer, not its higher market value. Firm B worries that the value of the note will increase if interest rates decrease but that if the customer pays off early, it will not capture the benefits of the higher interest rate it negotiated with the customer.

**Example 10** Firm C gives a note payable to a supplier on January 1, Year 1, to acquire manufacturing equipment. The note has a face value of $100,000 and bears interest at the prime lending rate. The prime lending rate is 8 percent on January 1, Year 1. The supplier resets the interest rate each December 31 to establish the interest charge for the next calendar year. Interest is payable on December 31 of each year and the note matures on December 31, Year 3. Firm C worries that interest rates will increase above 8 percent during the term of the note and negatively affect its cash flows.

**Example 11** Firm D holds 10,000 gallons of whiskey in inventory on October 31, Year 1. Firm D expects to complete aging this whiskey by March 31, Year 2, at which time it intends to sell the whiskey. Uncertainties about the quality of the aged whiskey and economic conditions at the time, however, make predicting the selling price on March 31, Year 2, difficult.

Most firms face risks of economic losses from changes in interest rate, foreign exchange rates, and commodity prices. Firms can purchase financial instruments to lessen these business risks. Typically, these are marketable securities and their accounting follows the principles for marketable securities, with some exceptions. The general term used for the financial instrument is a **derivative**. This section discusses the nature, use, accounting, and reporting of derivative instruments. Financial Accounting Standards Board *Statement of Financial Accounting Standards No. 133*[4] and Statement of *Financial Accounting Standards No. 138*[5] set forth the required accounting for derivative instruments, a form of marketable security.

---

[4]Financial Accounting Standards Board, *Statement of Financial Accounting Standards No. 133,* "Accounting for Derivative Instruments and Hedging Activities," 1998.

[5]Financial Accounting Standards Board, *Statement of Financial Accounting Standards No. 138,* "Accounting for Certain Derivative Instruments and Certain Hedging Activities, an Amendment to FASB Statement No. 133," 2000.

## NATURE AND USE OF DERIVATIVE INSTRUMENTS

A derivative is a financial instrument that obtains its value from some other financial item. An option to purchase a share of stock derives its value from the market price of the stock. A commitment to purchase a certain amount of foreign currency in the future derives its value from changes in the exchange rate for that currency. Firms use the derivative instrument to hedge the risk of losses from changes in interest rates, foreign exchange rates, and commodity prices. The general idea is that changes in the value of the derivative instrument offset changes in the value of an asset or liability or changes in future cash flows, thereby neutralizing, or at least reducing, the economic loss. Let's reconsider the four examples above.

**Example 12** Refer to **Example 8**. Firm A desires to incur a cost now to eliminate the effect of changes in the exchange rate between the U.S. dollar and British pound while it awaits delivery of the equipment. It purchases a **forward foreign exchange contract** from a bank on June 30, Year 1, in which it promises to pay a fixed U.S. dollar amount on June 30, Year 2, in exchange for £10,000 received then. The forward foreign exchange rate between U.S. dollars and British pounds on June 30, Year 1, for settlement on June 30, Year 2, establishes the number of U.S. dollars it must deliver. Assume that the forward rate on June 30, Year 1, for settlement of the forward contract on June 30, Year 2, is $1.64 per £1. By purchasing the forward contract, Firm A locks in the cost of the equipment at $16,400 (= £10,000 pounds × $1.64 per £1).

**Example 13** Refer to **Example 9**. Firm B wants to neutralize the effect of changes in the market value of the note receivable caused by changes in market interest rates. It engages in a **swap contract** with its bank. The swap has the effect of allowing Firm B to exchange its fixed interest rate obligation for a variable interest rate obligation. The market value of the note remains at $100,000 as long as the variable interest rate in the swap is the same as the variable rate used by the Firm B to revalue the note while it is outstanding.

**Example 14** Refer to **Example 10**. Firm C wants to protect itself against increases in the variable interest rate above the initial 8 percent rate. It, too, engages in a swap contract with its bank. The swap has the effect of allowing Firm C to exchange its variable interest rate obligation for a fixed interest rate obligation. The swap fixes its annual interest expense and cash expenditure to 8 percent of the $100,000 note. By engaging in the swap, Firm C cannot take advantage of decreases in interest rates below 8 percent, which it could have done with its variable rate note, but it no longer bears risk, with attendant costs, that the rate will rise above 8 percent.

**Example 15** Refer to **Example 11**. Firm D would like to fix the price at which it can sell the whiskey in its inventory on March 31, Year 2. It acquires a **forward commodity contract** in which it promises to sell 10,000 gallons of whiskey on March 31, Year 2, at a fixed price. The forward price of whiskey on October 31, Year 1, for delivery on March 31, Year 2, is $320 per gallon. Thus, it locks in a total cash inflow from selling the whiskey of $3,200,000.

Forward contracts and swap contracts illustrate two types of derivative instruments. Banks and other financial intermediaries structure derivatives to suit the particular needs of their customers. Thus, the nature and complexity of derivatives vary widely. We confine our discussion to swap contracts to illustrate the accounting and reporting of derivatives.

With these examples of derivatives in mind, consider the following elements of a derivative:

1. A derivative has one or more **underlyings**. An underlying is a variable such as a specified interest rate, or commodity price, or foreign exchange rate. The underlying in **Example 12** is the foreign exchange rate, in **Examples 13** and **14** is an interest rate, and in **Example 15** is the price of whiskey.
2. A derivative has one or more **notional amounts**. A notional amount is a number of currency units, bushels, shares, or other units specified in the contract. The notional amount in **Example 12** is £10,000, in **Examples 13** and **14** is the $100,000 face value of the note, and in **Example 15** is 10,000 gallons of whiskey.
3. A derivative often requires no initial investment, although it might. The firm usually acquires a derivative by exchanging promises with a **counterparty**, such as a commercial or investment bank. The exchange of promises is a mutually unexecuted contract.
4. Derivatives typically require, or permit, **net settlement**. Firm A in **Example 12** will not deliver $16,400 and receive in exchange £10,000. Firm A will actually purchase £10,000 in the market on June 20, Year 2, at the exchange rate on that date, when it needs the British pounds to purchase the equipment. Then, Firm A will receive cash from the counterparty to

the extent that the exchange rate on June 30, Year 2 exceeds \$1.64 per £1 and must pay the counterparty on this date to the extent that the exchange rate is less than \$1.64 per £1. Firm B in **Example 13** will receive from its customer the 8 percent interest established in the fixed rate note. If the variable interest rate used in the swap contract increases to 10 percent, the counterparty will pay Firm B an amount equal to 2 percent (= 0.10 − 0.08) of the notional amount of the note, \$100,000. Receiving interest of 8 percent from the customer and receiving cash of 2 percent from the counterparty results in net interest receipts of 10 percent. If the variable interest rate decreases to 5 percent, Firm B still receives the customer's interest of 8 percent as specified in the original note. It would then pay the counterparty 3 percent (= 0.08 − 0.05), resulting in total interest collection equal to the variable rate of 5 percent.

Because many derivatives require no initial investment, that is, no initial cash payment to the counterparty, historical cost accounting makes little sense for these instruments, which have zero initial cost, but potentially large positive or negative values later. This has led both the FASB and the IASB to require that firms record derivatives at their fair value on the balance sheet date.

## Accounting for Derivatives

A firm must recognize derivatives in its balance sheet as assets or liabilities, depending on the rights and obligations under the contract. The forward contract in **Example 12** is an asset and the commitment to purchase the equipment is a liability. The swap contracts in **Examples 13** and **14** may be assets or liabilities, depending on the level of interest rates—that is, whether at the balance sheet date the holder of the derivative would be entitled to receive assets or have to pay the counterparty. The forward contract in **Example 15** may be an asset or liability, depending on the price of whiskey.

Firms must revalue the derivatives to market value each period. (Common terminology refers to this as *marking to market*.) The revaluation amount, in addition to increasing or decreasing the derivative asset or liability, also affects either (1) net income immediately (like a trading security), or (2) other comprehensive income immediately and net income later (like securities available for sale).

The income effect of a change in the market value of a derivative depends on the nature of the hedge for which a firm acquires the derivative and whether the firm has chosen to use **hedge accounting**.

Generally accepted accounting principles (GAAP) classify derivatives as (1) speculative investments, or (2) fair value hedges, or (3) cash flow hedges. Firms usually acquire derivatives to hedge particular risks. Thus, firms typically classify derivatives as either fair value hedges or cash flow hedges. Firms must choose to designate each derivative as one or the other, depending on their general hedging strategy and purpose in acquiring the particular derivative instrument. If a firm chooses not to designate a particular derivative as either a fair value hedge or a cash flow hedge, GAAP requires that the firm account for the derivative as a speculative investment.

**Speculative Investment** Firms must revalue derivatives held as speculative investments to market value each period and, like trading securities, recognize the resulting gain or loss in earnings.

**Fair Value Hedges** Derivative instruments acquired to hedge exposure to changes in the fair value of an asset or liability are **fair value hedges**. Fair value hedges are of two general types: (1) hedges of a *recognized* asset or liability, and (2) hedges of an *unrecognized* firm commitment. Firm B in **Examples 9** and **13** entered into the interest swap agreement to neutralize the effect of changes in interest rates on the market value of its notes receivable, a recorded asset. Firm A in **Examples 8** and **12** acquired the forward foreign exchange contract to neutralize the effect of changes in exchange rates on its commitment to purchase the equipment. The FASB defines these derivative instruments as fair value hedges.

**Cash Flow Hedges** Derivative instruments acquired to hedge exposure to variability in expected future cash flows are **cash flow hedges**. Cash flow hedges are of two general types: (1) hedges of cash flows of an *existing* asset or liability, and (2) hedges of cash flows of *forecasted* transactions. Firm C in **Examples 10** and **14** entered into the interest swap agreement to neutralize changes in cash flows for interest payments on its variable rate notes payable. Firm D in **Examples 11** and **15** acquired the forward contract on whiskey to protect itself from changes in the selling price of whiskey on March 31, Year 2. These derivative instruments are therefore cash flow hedges.

A particular derivative could be either a fair value hedge or a cash flow hedge, depending on the firm's reason for engaging in the hedge. Both the forward foreign exchange contract in **Example 12** and the forward whiskey price contract in **Example 15** protect the firms' cash flows. The firms could conceivably classify both derivative instruments as cash flow hedges. Firm A in **Example 12** acquires the derivative to protect the value of equipment acquired and therefore classifies it as a fair value hedge. Firm D in **Example 15** acquires the derivative to protect its cash flows from changes in the price of whiskey and therefore classifies it as a cash flow hedge. One suspects that the FASB views the firm commitment to purchase the equipment by Firm A as having the economic substance of an asset and a liability, even though accounting treats it as a mutually unexecuted contract, whereas it views the forecasted transaction by Firm D to sell whiskey in the future as too uncertain to have the economic attributes of an asset at this point.

**Treatment of Hedging Gains and Losses** GAAP allows firms to choose whether to designate a particular derivative as hedging the risk of a change in market value (fair value hedges) or a change in cash flows (cash flow hedges). If firms choose not to designate the derivative as either a fair value hedge or a cash flow hedge, then firms must account for the derivative as a speculative investment. The derivative appears at market value, with fluctuations recognized in net income. The hedged item remains on the balance sheet at historical cost because, although the firm has acquired an instrument that economically hedges the item, the firm has chosen not to account for the pair as a hedge.

If firms designate a derivative as a fair value hedge, GAAP requires firms to revalue both the hedged item and the related derivative instrument to market value each period and to recognize gains and losses from changes in the market value in *net income* each period while the firm holds them. If the hedge is fully effective, the gain (loss) on the derivative will precisely offset the loss (gain) on the asset or liability hedged. The net effect on earnings is zero. If the hedge is not fully effective, the net gain or loss increases or decreases earnings to the degree the offset is incomplete.

If firms designate a derivative as a cash flow hedge, GAAP requires firms to revalue the derivative instrument to market value each period but to include gains and losses from changes in the market values of such derivatives in *other comprehensive income* each period to the extent the financial instrument is "highly effective" in neutralizing the risk. Firms must include the ineffective portion in net income currently. FASB *Statement No. 133* gives general guidelines but leaves to professional judgment the meaning of "highly effective." At the end of the period, the firm closes the Other Comprehensive Income account to the balance sheet account for Accumulated Other Comprehensive Income. The firm removes the accumulated amount in other comprehensive income related to a particular derivative instrument and transfers it to net income either periodically during the life of the derivative instrument or at the time of settlement, depending on the type of derivative instrument used.

The logic for the FASB's different treatment of gains and losses from changes in fair value of financial instruments results from applying the matching principle. In a fair value hedge of a recognized asset or liability, both the hedged asset (or liability) and its related derivative generally appear on the balance sheet. The firm revalues both the hedged asset (or liability) and its derivative to fair value each period and reports the gain or loss on the hedged asset (or liability) and the loss or gain on the derivative in net income. The net gain or loss indicates the effectiveness of the hedge in neutralizing the risk. In a cash flow hedge of an anticipated transaction, the hedged cash flow commitment does not appear on the balance sheet but the derivative instrument does appear. Recognizing a gain or loss on the derivative instrument in net income each period but recognizing the loss or gain on the anticipated transaction at the time an actual transaction occurs results in poor matching. Thus, the firm classifies the gain or loss on the derivative instrument in other comprehensive income and then, later, reclassifies it to net income when it records the actual transaction.

The logic for the FASB's different treatment would seem to break down for fair value hedges related to a firm commitment. GAAP usually does not recognize the firm commitment as an asset or a liability.

## Summary of Accounting for Derivatives as Marketable Securities

Derivatives appear on the balance sheet at fair market value.

- **Trading Security Treatment—Gains and losses on speculative securities, fair value hedges reported in current earnings, and the ineffective portion of cash flow hedges.**

The holder will include in current earnings the gains and losses on all speculative derivatives, fair value hedges, and the ineffective portion of cash flow hedges.

- **Securities Available for Sale Treatment—Gains and losses on effective cash flow hedges reported initially in other comprehensive income.** The holder will exclude from current earnings, but include in Accumulated Other Comprehensive Income, the gains and losses on derivatives used effectively to hedge cash flows.
- **Disclosures Related to Derivative Instruments** Several FASB pronouncements affect disclosures about derivative instruments. FASB *Statement No. 107*[6] requires firms to disclose the book, or carrying, value and the fair value of financial instruments. Financial instruments impose on one entity a right to receive cash and an obligation on another entity to pay cash. Financial instruments include accounts receivable, notes receivable, notes payable, bonds payable, forward contracts, swap contracts, and most derivatives. Fair value is the current amount of cash for which two willing parties would agree to exchange the instrument for cash.

FASB *Statement No. 133* requires the following disclosures (among others) with respect to derivatives.

1. A description of the firm's risk management strategy and how particular derivatives help accomplish the firm's hedging objectives. The description should distinguish between derivative instruments designated as fair value hedges, cash flow hedges, and all other derivatives.
2. For fair value and cash flow hedges, firms must disclose the net gain or loss recognized in earnings resulting from the hedge's ineffectiveness (that is, not offsetting the risk hedged) and the line on the income statement that includes this net gain or loss.
3. For cash flow hedges, firms must describe the transactions or events that will result in reclassifying gains and losses from other comprehensive income to net income and the estimated amount of such reclassifications during the next 12 months.
4. The net amount of gains and losses recognized in earnings because a hedged firm's commitment no longer qualifies as a fair value hedge or a hedged forecasted transaction no longer qualifies as a cash flow hedge.

FASB *Statement of Financial Accounting Standards No. 105*[7] requires a firm to disclose the nature and extent of risk of loss if the counterparty to a derivative instrument failed to perform according to the contractual terms. For example, a firm involved in an interest swap with a bank experiencing financial difficulty would need to disclose this credit risk and the amount of the loss that would result if the bank failed to perform. A firm contracting with a single bank for a large portion of its derivatives would need to disclose this concentration of credit risk.

## Dealing with Uncertainty by Using Measurement, Not Recognition

The business world is uncertain. For example, will the firm lose a lawsuit? Until recently, GAAP dealt with uncertainty through its rules for recognizing a liability, not measuring its amount. For example *SFAS No. 5*[8] says the firm will record a liability for the lawsuit only if losing is probable, which has come to mean a likelihood of 80 percent or greater. If the firm recognizes liability for such a probable loss, the amount will be the most likely amount for the loss, or the minimum amount if the most likely amount falls in a range with no number more likely than any other. Assume the mostly likely loss is $10 million. Then, the liability for the loss in a lawsuit will appear on the books either at zero or at $10 million. Assume, further, that the firm has a 60 percent chance of losing and if it loses the amount will be $10 million. Its expected loss is $6 (= 0.60 × $10) million, but, even so, the balance sheet number will be either zero or $10 million.

*Example.* Assume the firm faces 100 such lawsuits, each presenting a 60 percent chance of losing $10 million and a 40 percent chance of winning. Then, with near statistical certainty, the

[6]Financial Accounting Standards Board, *Statement of Financial Accounting Standards No. 107*, "Disclosures about Fair Value of Financial Instruments," 1991.

[7]Financial Accounting Standards Board, *Statement of Financial Accounting Standards No. 105*, "Disclosures of Information about Financial Instruments with Off-Balance-Sheet Risk and Financial Instruments with Concentrations of Credit Risk," 1990.

[8]Financial Accounting Standards Board, *Statement of Financial Accounting Standards No. 5*, "Accounting for Contingencies," 1975.

firm will have an obligation to pay about $600 million (= 100 × 0.60 × $10 million), but current GAAP will show no liability as the events individually do not rise to the level of certainty to warrant recognition.

If we were to record the liability for any one lawsuit at its expected value, $6 million, then we could say that the GAAP deals with uncertainty not by recognition—a binary choice of yes or no—but with measurement. The FASB *is* beginning to deal with uncertainty with measurement, not recognition. An example is the accounting for derivatives, which are neither assets nor liabilities until prices of the underlyings change. GAAP deals with the uncertainty of future price change, not with recognition criteria, but measurement rules: if the item results in an expected payment (or receipt), measure the liability (asset) with the market value of the amount.

## Minority, Active Investments

When an investor owns less than a majority of the voting stock of another corporation, the accountant must judge when the investor can exert significant influence. For the sake of uniformity, GAAP presumes that one company can significantly influence another company when the investor company owns 20 percent or more of the voting stock of the other company. An investor can exert significant influence even when owning less than 20 percent, but then its management must convince the independent accountants that the company can do so. Analysts become suspicious when they see an investor changing its ownership from just below 20 percent to just above, or vice versa. They suspect that the investor is trying to manage earnings, in ways discussed below, by changing its accounting techniques to suit the circumstances.

Recent trends in corporate governance are changing the way minority shareholders should view the voting power inherent in their shares. Consider a large, public company like IBM or General Electric whose shares are so widely owned that the largest single voting block of the shares might be 4 percent of the total outstanding. Many pension funds, mutual funds, and state retirement trusts (such as CALPERS, the California Public Employees Retirement System) own substantial investments in virtually all public companies. These fiduciaries collectively own enough shares that if they were to act in concert, they could control the boards and, hence, the actions of the companies whose shares they hold. These fiduciaries have increasingly acted in concert when deciding how to vote on pivotal corporate governance issues. When they act in concert, they can overwhelm the voting power of minority share holdings otherwise large enough to exert significant influence.

Minority, active investments, generally those where the investor owns between 20 and 50 percent, require the **equity method** of accounting. Under the equity method, the investor firm recognizes as revenue (expense) each period its share of the net income (loss) of the other firm. The investor recognizes dividends received from S as a return of investment, not as income. In the discussion that follows, we designate the investor firm as P and the firm whose shares it owns as S.

### Equity Method: Rationale

To understand why GAAP requires the equity method for minority, active investments, review the financial statement effects of using the **market value method** for such investments. Under the market value method for securities available for sale, P recognizes net income effects only when it receives a dividend (revenue) or sells some of the investment (gain or loss). (P does recognize amounts in comprehensive income for unrealized holding gains and losses, but the amount of comprehensive income for a period does not typically appear on the income statement.) Suppose, as often happens, that S follows a policy of financing its own growing operations through retaining earnings, consistently declaring dividends less than its net income. The market price of S's shares will probably increase to reflect this retention of assets. Under the market value method, P reports net income only from the dividends it receives. Because P, by assumption, exerts significant influence over S, it can affect S's dividend policy, which in turn affects P's net income. When P can so easily manage its own net income (via dividends from S), the market value method will not reasonably reflect P's earnings from investing in S. The equity method better measures a firm's earnings from its investment when, because of its ownership interest, it can exert significant influence over the operations and dividend policy of the investee firm. Most analysts agree that the equity method provides higher quality earnings than does the market value method when the owner can exert significant influence over the operations and financial policies of the affiliated company.

## EQUITY METHOD: PROCEDURES

The equity method records the initial purchase of an investment at acquisition cost, just as for minority, passive investments. Each period, Company P treats as revenue its proportionate share of the periodic earnings, not the dividends, of Company S. Company P treats dividends declared by S as a reduction in P's investment in S account.

Suppose that P acquires 30 percent of the outstanding shares of S for $600,000. The entry to record the acquisition is as follows:

| | Dr. | Cr. |
|---|---|---|
| **(1)** Investment in Stock of S | 600,000 | |
| Cash | | 600,000 |

| Assets | = | Liabilities | + | Shareholders' Equity | (Class.) |
|---|---|---|---|---|---|
| +600,000 | | | | | |
| −600,000 | | | | | |

Investment made in 30 percent of Company S.

Between the time of the acquisition and the end of P's next accounting period, S reports income of $80,000. P, using the equity method, records the following:

| | Dr. | Cr. |
|---|---|---|
| **(2)** Investment in Stock of S | 24,000 | |
| Equity in Earnings of Affiliate | | 24,000 |

| Assets | = | Liabilities | + | Shareholders' Equity | (Class.) |
|---|---|---|---|---|---|
| +24,000 | | | | +24,000 | IncSt → RE |

To record 30 percent of income earned by investee, accounted for using the equity method.

The account Equity in Earnings of Affiliate is a revenue account. If S declares and pays a dividend of $30,000 to holders of common stock, P receives $9,000 (= 0.30 × $30,000) and records the following:

| | Dr. | Cr. |
|---|---|---|
| **(3)** Cash | 9,000 | |
| Investment in Stock of S | | 9,000 |

| Assets | = | Liabilities | + | Shareholders' Equity | (Class.) |
|---|---|---|---|---|---|
| +9,000 | | | | | |
| −9,000 | | | | | |

To record dividends received from investee, accounted for using the equity method, and the resulting reduction in the Investment account.

Notice the credit to the Investment in S account. P records income earned by S as an increase in investment. The dividend returns part of the investment and decreases the Investment account.

*Hint:* Students often have difficulty understanding journal entry **(3)**, particularly the credit by the investor when the investee company pays a dividend. The transactions and entries resemble those for an individual's ordinary savings account at a local bank. Assume that you put $600,000 in a savings account, that later the bank adds interest of 4 percent (or $24,000) to the account, and that still later you withdraw $9,000 from the savings account. You can record journal entries **(1)** through **(3)** for these three events, with slight changes in the account titles: Investment in S changes to Savings Account, and Equity in Earnings of Affiliate changes to Interest Revenue. The cash withdrawal reduces the amount invested in the savings account. Similarly, the payment of a cash dividend by an investee company accounted for with the equity method reduces the investor's investment in the company. The investor, Company P, owns a sufficiently large percentage of the voting shares that it can effectively require Company S to pay a dividend, just as you can require the savings bank to remit cash to you almost whenever you choose.

Suppose that S subsequently reports earnings of $100,000 and also pays dividends of $40,000. P's entries are as follows:

| | | |
|---|---|---|
| **(4)** Investment in Stock of S . . . . . . . . . . | 30,000 | |
| Equity in Earnings of Affiliate . . . . . . . . . . | | 30,000 |

| Assets | = | Liabilities | + | Shareholders' Equity | (Class.) |
|---|---|---|---|---|---|
| +30,000 | | | | +30,000 | |

| | | |
|---|---|---|
| **(5)** Cash . . . . . . . . . . | 12,000 | |
| Investment in Stock of S . . . . . . . . . . | | 12,000 |

| Assets | = | Liabilities | + | Shareholders' Equity | (Class.) |
|---|---|---|---|---|---|
| +12,000 | | | | | |
| −12,000 | | | | | |

To record revenue and dividends from investee, accounted for using the equity method.

P's Investment in S account now has a balance of $633,000 as follows:

**Investment in Stock of S**

| | |
|---|---|
| **(1)** 600,000 | 9,000 **(3)** |
| **(2)** 24,000 | 12,000 **(5)** |
| **(4)** 30,000 | |
| Bal. 633,000 | |

Assume now that P sells one-fourth of its investment in S for $165,000. The entry is:

| | | |
|---|---|---|
| **(6)** Cash . . . . . . . . . . | 165,000 | |
| Investment in Stock of S . . . . . . . . . . | | 158,250 |
| Gain on Sale of Investment in S . . . . . . . . . . | | 6,750 |

| Assets | = | Liabilities | + | Shareholders' Equity | (Class.) |
|---|---|---|---|---|---|
| +165,000 | | | | +6,750 | IncSt → RE |
| −158,200 | | | | | |

Cost of investments sold is $158,250 = 1/4 × $633,000.

After the sale, the balance in the investment account is $474,750, as follows:

**Investment in Stock of S**

| | |
|---|---|
| **(1)** 600,000 | 9,000 **(3)** |
| **(2)** 24,000 | 12,000 **(5)** |
| **(4)** 30,000 | |
| Bal. 633,000 | |
| | 158,250 **(6)** |
| Bal. 474,750 | |

**Excess Purchase Price on Acquisition of Equity Method Investment** P may pay more for its investment than its proportion of the book value of the net assets (= assets − liabilities), or shareholders' equity, of S at the date of acquisition. For example, assume that P acquires 25 percent of the stock of S for $400,000 when S has total shareholders' equity of $1 million. P's cost exceeds book value acquired by $150,000 [= $400,000 − (0.25 × $1,000,000)]. P may pay an amount that differs from the book value of S's recorded net assets because the market values of the net assets differ from their book values or because of unrecorded assets (for example, trade secrets) or unrecorded liabilities (for example, an unsettled lawsuit). GAAP previously required firms to write off this excess acquisition cost over a period not to exceed 40 years.[9] Beginning with investments made after June 30, 2001, firms do not have the option to amortize this excess to the extent it relates to goodwill. Firms may also discontinue amortizing goodwill from investments made prior to July 1, 2001, as soon as they adopt *Statement No. 142*.[10] Firms must still amortize the excess to the extent it relates to other assets or liabilities. GAAP's rationale for the different treatment of goodwill is that it has an indefinite life, whereas other assets and liabilities typically have a definite life. **Chapter 8** discusses the accounting for goodwill and other long-lived assets, including their impairment.

**Summary of Equity Method** On the balance sheet, an investment accounted for with the equity method appears in the Investments section. The amount shown generally equals the acquisition cost of the shares plus P's share of S's undistributed earnings since the date P acquired the shares. On the income statement, P generally shows its share of S's income as revenue each period. The exceptions in the preceding sentence occur when the original purchase price exceeds the proportionate share of the book value acquired. The accounting method used by the investor, P, does not affect the financial statements of the investee, S.

## Problem 11.3 for Self-Study

**Journal entries to apply the equity method for long-term investments in securities.** **Exhibit 11.3** summarizes data about the minority, active investments of Equity Investing Group. Assume that any excess of acquisition cost over the book value of the net assets acquired relates to equipment with a 10-year remaining life on January 1, Year 1. Prepare the journal entries to do the following:

**a.** Record the acquisition of these securities on January 1, Year 1.
**b.** Apply the equity method for Year 1.
**c.** Apply the equity method for Year 2.
**d.** Record the sale of Security E on January 2, Year 3, for $190,000.

**EXHIBIT 11.3** EQUITY INVESTING GROUP (Problem 11.3 for Self-Study)

| Security | Date Acquired | Acquisition Cost | Ownership Percentage | Book Value of Net Assets on January 1, Year 1 | Earnings (Loss) | | Dividends | |
|---|---|---|---|---|---|---|---|---|
| | | | | | Year 1 | Year 2 | Year 1 | Year 2 |
| D ........ | 1/1/Year 1 | $ 80,000 | 40% | $200,000 | $ 40,000 | $50,000 | $10,000 | $12,000 |
| E ........ | 1/1/Year 1 | 190,000 | 30 | 500,000 | 120,000 | (40,000) | 30,000 | — |
| F ........ | 1/1/Year 1 | 200,000 | 20 | 800,000 | 200,000 | 50,000 | 60,000 | 60,000 |

## Majority, Active Investments

When one firm, P, owns more than 50 percent of the voting stock of another company such as S, P can control the activities of S. P can control both broad policy-making and day-to-day operations. Common usage refers to the majority investor as the **parent** and to the majority-owned

[9]Accounting Principles Board, *Opinion No. 17*, "Intangible Assets," 1970.

[10]Financial Accounting Standards Board, *Statement of Financial Accounting Standards No. 142*, "Goodwill and Other Intangible Assets," 2001.

company as the **subsidiary**. GAAP requires the parent to combine the financial statements of majority-owned companies with those of the parent, in **consolidated financial statements.**[11]

## Reasons for Legally Separate Corporations

Business firms have several reasons for preferring to operate as a group of legally separate corporations rather than as a single entity. From the standpoint of the parent company, the more important reasons for maintaining legally separate subsidiary companies include the following:

1. To reduce the parent's risk. Separate corporations may mine raw materials, transport them to a manufacturing plant, produce the product, and sell the finished product to the public. If any one part of the total process proves to be unprofitable or inefficient, losses from insolvency will fall only on the owners and creditors of the one subsidiary corporation. (Some drug firms acquire subsidiaries to make and market medical products in the hope that if something goes wrong in future years, the parent firm will not be liable for losses of the customers of the subsidiary. Individuals who believe the subsidiary's products have harmed them often sue the parent and sometimes succeed in "piercing the corporate veil.")
2. To meet more effectively the requirements of state corporation laws and tax legislation. If an organization does business in a number of states, it often faces overlapping and inconsistent taxation, regulations, and requirements. Organizing separate corporations to conduct operations in the various states can sometimes reduce administrative costs.
3. To expand or diversify with a minimum of capital investment. A firm may absorb another company by acquiring a controlling interest in its voting stock. The firm may accomplish this result with a substantially smaller capital investment, as well as with less difficulty, inconvenience, and risk, than if it constructs a new plant or starts a new line of business.
4. To sell an unwanted operation with a minimum of administrative, legal, and other costs. Firms generally save costs if they sell the common stock of a subsidiary rather than trying to sell each of its assets and transfer all known and, perhaps, unknown liabilities to a buyer.

## Purpose of Consolidated Statements

For a variety of reasons, then, a single economic entity may exist in the form of a parent and several legally separate subsidiaries. (General Electric Company, for example, comprises more than 50 large legally separate companies.) A consolidation of the financial statements of the parent and each of its subsidiaries presents the results of operations, financial position, and cash flows of an affiliated group of companies under the control of a parent as if the group of companies composed a single entity. The parent and each subsidiary are legally separate entities, but they operate as one centrally controlled **economic entity**. Consolidated financial statements generally provide more useful information to the shareholders of the parent corporation than do separate financial statements for the parent and each subsidiary.

Consolidated financial statements also generally provide more helpful information than does the equity method. The parent, because of its voting interest, can effectively control the use of all of the subsidiary's assets. Consolidation of the individual assets, liabilities, revenues, and expenses of both the parent and the subsidiary provides a more realistic picture of the operations and financial position of the single economic entity.

In a legal sense, consolidated statements merely supplement, and do not replace, the separate statements of the individual corporations, although common practice presents only the consolidated statements in published annual reports.

**Example 16** General Motors, General Electric, and IBM, among others, have wholly owned finance subsidiaries. These subsidiaries make a portion of their loans to customers who want to purchase the products of the parent company. The parent company consolidates the financial statements of these subsidiaries. These subsidiaries have billions of dollars of assets, mostly receivables. Statement readers may misunderstand the relative liquidity of firms preparing consolidated statements, which combine the parent's assets—largely noncurrent manufacturing plant and equipment—with the more liquid assets of the finance subsidiary. This suggests preparing separate statements for the manufacturing parent and for the financial subsidiary. The counterargument suggests that these entities operate as a single integrated unit, so that consolidated financial statements more accurately

[11]Financial Accounting Standards Board, *Statement of Financial Accounting Standards No. 94*, "Consolidation of All Majority-Owned Subsidiaries," 1987.

depict the nature of their operating relations. For many years, General Electric has shown the separate financial statements of its finance subsidiaries in addition to the consolidated statements.

**Example 17** A major mining corporation owns a mining subsidiary in South America. The government of the country enforces stringent control over cash payments outside the country. The company cannot control the use of all the assets, despite owning a majority of the voting shares. Therefore, it does not prepare consolidated statements with the subsidiary. The parent company will probably carry the investment on its books at cost.

## THE PURCHASE TRANSACTION

In a corporate acquisition, one corporation acquires all, or substantially all, of another corporation's common shares. The **purchase method** views a corporate acquisition as conceptually identical to the purchase of any single asset (for example, inventory or a machine). The purchaser records the intercorporate investment in the common stock of the acquired company at the *market value of the consideration given*. The purchaser allocates the purchase price to the identifiable assets and liabilities to record them at market values on the date of acquisition. The purchaser will record any assets acquired independent of whether the purchased company showed those assets on its books. For example, the purchased company owns patents and trademarks not recorded on its balance sheet. The purchasing company will show those assets on its balance sheet at their market value. The purchaser allocates any remaining excess to goodwill. That is, **goodwill** is the excess of purchase price of another company, or division, over the fair value of the net assets acquired, independent of whether those assets appeared on the books of the acquired company.

### Problem 11.4 for Self-Study

**Financial statement effects of the purchase method.** **Exhibit 11.4** presents balance sheet data for Powell Corporation and Steele Corporation as of January 1, Year 8. On this date, Powell Corporation exchanges 2,700 shares of its common stock, selling for $20 per share, for all of the common stock of Steele Corporation.

Prepare a consolidated balance sheet for Powell Corporation and Steele Corporation on January 1, Year 8, using the purchase method. Powell Corporation allocates any excess cost to property, plant, and equipment and to goodwill.

**EXHIBIT 11.4** **POWELL CORPORATION AND STEELE CORPORATION**
**Financial Statement Data for January 1, Year 8**
**(Problem 11.4 for Self-Study)**

| | Historical Cost | | Current Market Value |
|---|---|---|---|
| | Powell Corp. | Steele Corp. | Steele Corporation |
| ASSETS | | | |
| Current Assets | $10,000 | $ 7,000 | $ 7,000 |
| Property, Plant, and Equipment (net) | 30,000 | 18,000 | 23,000 |
| Goodwill | — | — | 40,000 |
| Total Assets | $40,000 | $25,000 | $70,000 |
| EQUITIES | | | |
| Liabilities | $25,000 | $16,000 | $16,000 |
| Common Stock ($1 par value) | 1,000 | 1,000 | 1,000 |
| Additional Paid-in Capital | 9,000 | 5,000 | 5,000 |
| Retained Earnings | 5,000 | 3,000 | 3,000 |
| Unrecorded Excess of Market Value over Historical Costs | — | — | 45,000 |
| Total Equities | $40,000 | $25,000 | $70,000 |

## THE BOUNDARIES OF THE CONSOLIDATED ENTITY

The usual criterion for preparing consolidated financial statements is that one entity has voting control, more than 50 percent, of another company's common stock. For more complicated business arrangements, voting control may not adequately measure control. Consider the following business arrangement of Company P, which wants to borrow to finance the acquisition of an airplane, but does not want the debt to appear on its balance sheet.

1. Company P gives $100,000 to an unrelated party; think of the local Red Cross. This is a one-time payment to induce the Red Cross to own 100 percent of the shares of Company S. The Red Cross buys these shares for $60,000 cash, contributing the cash to Company S in return for the shares.
2. Company S borrows $50 million from a group of Lenders; think of a group of banks, which it uses to buy the airplane that Company P wants to use. It agrees to repay the loan with monthly installments for 10 years, at the rate of $717,500 per month, which implies a loan at about 1 percent per month fully paid off with 120 (= 10 years × 12) payments.
3. Company P provides a written guarantee to the Lenders that it will make any debt service payment that Company S fails to make on time. Without the guarantee, the Lenders would not lend to Company S, as the airplane itself does not provide sufficient collateral to satisfy cautious lenders worried about possible loss in market value of an asset in such a volatile industry.
4. Company P makes a business arrangement with Company S; in this example, Company P agrees to use the airplane and pay $717,500 per month to rent the airplane. Company P might agree to pay Company S a bit more to cover Company S's transactions costs.

**Figure 11.2** depicts this structure.

In this arrangement, the Red Cross owns all Company S and controls it in the sense that voting shares convey control. Company P, however, enjoys the risks and rewards of using the airplane. If the airplane becomes obsolete, Company P bears the cost. If the rental arrangement turns out to be a bargain, Company P will enjoy the gains. Firms have organized entities such as Company S to turn accounts receivable into cash without explicit borrowing, and to finance research and development efforts. Ford, Kimberly-Clark, and Boeing have used them.

FIGURE 11.2 To Illustrate Why Voting Control Provides Inadequate Test for Consolidation

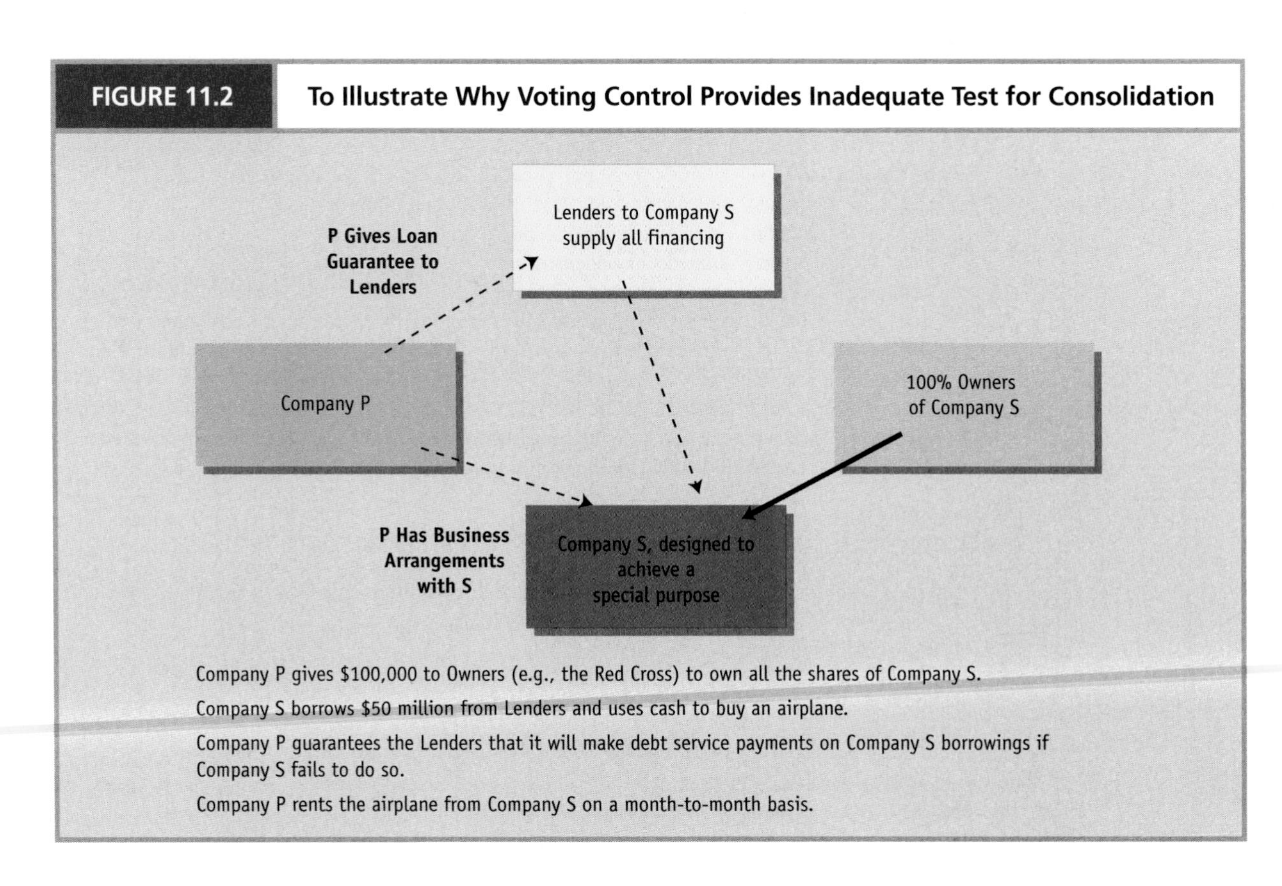


Company P gives $100,000 to Owners (e.g., the Red Cross) to own all the shares of Company S.

Company S borrows $50 million from Lenders and uses cash to buy an airplane.

Company P guarantees the Lenders that it will make debt service payments on Company S borrowings if Company S fails to do so.

Company P rents the airplane from Company S on a month-to-month basis.

Arrangements such as those described for leases in the illustration became popular in the late 1980s. Then, in 1990, the FASB and the SEC, through the Emerging Issues Task Force (EITF), formally recognized this sort of arrangement. It coined the phrases, which it never defined, *special purpose entity* and *special purpose vehicle* to describe the entity, such as Company S in this example. It specifically allowed Company P to engage in this sort of leasing arrangement, and keep the transaction off its books, other than the cash payments to Company S. Although that 1990 ruling, *EITF Consensus 90-15*,[12] applied only to leases, financial engineers soon turned its principles to other arrangements, such as securitizing accounts receivable. Eventually, after the collapse of Enron, which had built itself, in part, on shaky special purpose entities, the FASB tightened the rules for keeping such entities off the balance sheet.[13]

The FASB defined the new term ***variable interest entity (VIE)*** to refer to entities such as Company S in our illustration. A variable interest entity will exist whenever the owner holds equity so small that the entity requires other financial support to sustain its activities. The FASB means that the owner must have contributed assets larger than the expected losses of the entity. The FASB's rule suggests that unless the owners have contributed 10 percent or more of the total financing, then the entity is a VIE. (In the illustration, assuming a total financing of $50 million, the owner must have invested at least $5 million for Company S not to be a VIE.) The FASB requires that the ***primary beneficiary*** of the VIE consolidate the VIE. A company, such as Company P in the illustration, will be a primary beneficiary if it absorbs (or receives) the majority of the variability of the outcomes of the VIE. In the case where a company, like Company P, guarantees the debt of the VIE, then the guarantor will be the primary beneficiary.

## Disclosure of Consolidation Policy

The summary of significant accounting principles, a required part of the financial statement notes, must include a statement about the **consolidation policy** of the parent. If an investor does not consolidate a significant majority-owned subsidiary, the notes will disclose that fact. A recent annual report of American Home Products Corporation contained the following (bracketed remarks do not appear in the original):

> Notes to Consolidated Financial Statements
>
> **1.** *Summary of Significant Accounting Policies*: . . . Principles of Consolidation: The accompanying consolidated financial statements include the accounts of the Company and its subsidiaries with the exception of those subsidiaries described in **Note 3** which are accounted for on a cash basis . . . .
>
> **3.** *Provision for Impairment of Investment in Certain Foreign Locations*: [During the preceding year], the Corporation recorded a charge of $50,000,000 recognizing the impairment of its investment in its subsidiaries in South America, except for its investment in Brazil. The provision was made after determining that the continued imposition of constraints such as dividend restrictions, exchange controls, price controls, and import restrictions in these countries so severely impede management's control of the economic performance of the businesses that continued inclusion of these subsidiaries in the consolidated financial statements is inappropriate. The Company is continuing to operate these businesses, which for the most part are self-sufficient; however, the investments have been deconsolidated and earnings are recorded on a cash [similar to the market value method] basis. Net sales from these operations aggregated $97,790,000, $95,084,000, and $100,045,000 [in the last three years] . . . . Net income included in the consolidated statements of income was approximately $2,200,000 . . . and $2,000,000 [for the last two years].

## Understanding Consolidated Statements

This section discusses three concepts essential for understanding consolidated financial statements:

1. The need for intercompany eliminations
2. The meaning of consolidated net income
3. The nature of the external minority interest

---

[12]Emerging Issue Task Force, *EITF Issue 90-15*, "Impact of Nonsubstantive Lessors, Residual Value Guarantees, and Other Provisions in Leasing Transactions," July 1991.

[13]*FASB Interpretation No. 46*, "Consolidation of Variable Interest Entities," 2003.

**EXHIBIT 11.5** **Illustrative Data for Preparation of Consolidated Financial Statements**

| | Single-Company Statements | | | |
|---|---|---|---|---|
| | Company P (1) | Company S (2) | Combined (3) = (1) + (2) | Consolidated (4) |
| CONDENSED BALANCE SHEETS ON DECEMBER 31, YEAR 4 | | | | |
| **Assets** | | | | |
| Accounts Receivable | $ 200,000 | $ 25,000 | $ 225,000 | $ 213,000 |
| Investment in Stock of Company S (using equity method) | 705,000 | — | 705,000 | — |
| Other Assets | 2,150,000 | 975,000 | 3,125,000 | 3,125,000 |
| Total Assets | $3,055,000 | $1,000,000 | $4,055,000 | $3,338,000 |
| **Equities** | | | | |
| Accounts Payable | $ 75,000 | $ 15,000 | $ 90,000 | $ 78,000 |
| Other Liabilities | 70,000 | 280,000 | 350,000 | 350,000 |
| Common Stock | 2,500,000 | 500,000 | 3,000,000 | 2,500,000 |
| Retained Earnings | 410,000 | 205,000 | 615,000 | 410,000 |
| Total Equities | $3,055,000 | $1,000,000 | $4,055,000 | $3,338,000 |
| CONDENSED INCOME STATEMENTS FOR YEAR 4 | | | | |
| **Revenues** | | | | |
| Sales | $ 900,000 | $ 250,000 | $1,150,000 | $1,110,000 |
| Equity in Earnings of Company S | 48,000 | — | 48,000 | — |
| Total Revenues | $ 948,000 | $ 250,000 | $1,198,000 | $1,110,000 |
| **Expenses** | | | | |
| Cost of Goods Sold (excluding depreciation) | $ 440,000 | $ 80,000 | $ 520,000 | $ 480,000 |
| Depreciation Expense | 120,000 | 50,000 | 170,000 | 170,000 |
| Administrative Expense | 80,000 | 40,000 | 120,000 | 120,000 |
| Income Tax Expense | 104,000 | 32,000 | 136,000 | 136,000 |
| Total Expenses | $ 744,000 | $ 202,000 | $ 946,000 | $ 906,000 |
| Net Income | $ 204,000 | $ 48,000 | $ 252,000 | $ 204,000 |
| Dividends Declared | (50,000) | (13,000) | (63,000) | (50,000) |
| Increase in Retained Earnings for the Year | $ 154,000 | $ 35,000 | $ 189,000 | $ 154,000 |

We illustrate the concepts for understanding consolidated statements with the data in **Exhibit 11.5** for Company P and Company S.

- **Column (1)** shows the balance sheet and income statement for Company P.
- **Column (2)** shows the balance sheet and income statement for Company S.
- **Column (3)** sums the amounts from **columns (1)** and **(2)**, but these amounts do not represent the correct amounts for the consolidated statements.
- **Column (4)** presents the correct consolidated financial statements.

The following discussion explains why the correct amounts in **column (4)** differ from the sums in **column (3)**.

**Need for Intercompany Eliminations** State corporation laws typically require each legally separate corporation to maintain its own accounting records. Thus, during the accounting period, each corporation will record transactions of that entity with all other entities (both affiliated and nonaffiliated). At the end of the period, each corporation will prepare its own financial statements. The consolidation, or combining, of these financial statements basically involves summing the amounts for various financial statement items across the separate company statements. The accountant must adjust the amounts resulting from the summation, however, to eliminate double-counting resulting from **intercompany transactions**.

The guiding principle is that consolidated financial statements will reflect the results that the affiliated group would report if it were a single company. Consolidated financial statements reflect the transactions between the consolidated group of entities and others outside the entity. Thus, to take a simple example, if one affiliate company sells goods to another, the consolidation must remove from the financial statements the effects of this intercompany transaction.

**Eliminating Double-Counting of Intercompany Payables** Separate company records indicate that $12,000 of Company S's accounts receivable represent amounts receivable from Company P. **Column (3)** counts the current assets underlying this transaction twice: once as part of Accounts Receivable on Company S's books and a second time as Cash (Other Assets) on Company P's books. Also, the liability shown on Company P's books appears in the combined amount for Accounts Payable in **column (3)**. The consolidated group does not owe this $12,000 to an outsider. To eliminate double-counting on the asset side and to report Accounts Payable at the amount payable to outsiders, the consolidation process must eliminate $12,000 from the amounts for Accounts Receivable and Accounts Payable in **column (3)**. In **column (4)**, the consolidated Accounts Receivable and Accounts Payable both total $12,000 less than their sum in **column (3)**.

If either company holds bonds or long-term notes of the other, the consolidation process will eliminate the investment and related liability in the consolidated balance sheet. It will also eliminate the "borrower's" interest expense and the "lender's" interest revenue from the consolidated income statement.

**Eliminating Double-Counting of Investment** Company P's balance sheet shows an asset, Investment in Stock of Company S, which represents P's investment in S's net assets. The subsidiary's balance sheet shows its individual assets. When **column (3)** adds the two balance sheets, the sum shows both Company P's investment in Company S's assets and S's actual assets. The consolidation process must eliminate Company P's account, Investment in Stock of Company S, $705,000, from the sum of the balance sheets. Because the consolidated balance sheet must maintain the accounting equation, the process must make corresponding eliminations of $705,000 from the equities.

To understand the eliminations from the equities in the balance sheet, recall that the right-hand side shows the *sources* of the firm's financing. Financing for Company S comes from creditors (liabilities of $295,000) and from owners (shareholders' equity of $705,000). Company P owns 100 percent of Company S's voting shares. Thus the financing for the assets of the consolidated entity comes from the creditors of both companies and from Company P's shareholders. In other words, the equities of the consolidated entity are the liabilities of both companies plus the shareholders' equity of Company P alone. **Column (3)** adds the shareholders' equity accounts of Company S to those of Company P. It counts the financing from Company P's shareholders twice (once on the parent's books and once on the subsidiary's books). Hence, when the consolidation process eliminates Company P's investment account ($705,000) from the sum of the two companies' assets, it eliminates the shareholders' equity accounts of Company S ($500,000 of common stock and $205,000 of retained earnings). See **column (4)**.

**Learning Aid** When we prepare a consolidated balance sheet for two entities, we must deal with two sets of assets, two sets of liabilities, and two sets of shareholders' equity accounts. Which assets, liabilities, and shareholders' equity accounts provide a better understanding of the entity operating as one economic whole? The parent's balance sheet shows on one line, Investment in Subsidiary, the net assets of the subsidiary, whose individual amounts appear on the subsidiary's balance sheet. The individual components of net assets of the subsidiary provide more helpful information than does the single asset, Investment in Subsidiary. Thus, we eliminate the Investment in Subsidiary account and replace it with the individual assets and liabilities of the subsidiary. The shareholders' equity of the parent financed the subsidiary's shareholders' equity. Because the sources of the assets already appear as the parent's shareholders' equity, we eliminate the subsidiary's shareholders' equity when we eliminate the parent's investment account. As a result, a consolidated balance sheet reports the individual assets of the parent and the subsidiary, except the Investment in Subsidiary account, the individual liabilities of both entities, and the shareholders' equity of the parent.

**Back to Basics** By now, you should understand that Retained Earnings represents a source of financing for assets, not a pool of assets. Owners of P provide the financing to the consolidated group, so the consolidation process eliminates the retained earnings of S. Some students cannot understand what happens to S's retained earnings, which apparently disappear in consolidation. The difficulty stems from confusing Retained Earnings with assets, rather than understanding it as one of the sources of financing provided by shareholders, a component of shareholders' equity.

**Eliminating Intercompany Sales** The consolidation process must eliminate intercompany transactions from the sum of the income statements so that the consolidated income statement will present only the consolidated entity's transactions with outsiders. Consider intercompany sales. Separate company records indicate that Company S sold merchandise at its cost of $40,000 to Company P for $40,000 during the year. None of this inventory remains in Company P's inventory on December 31. Therefore, the merchandise inventory items sold appear in Sales Revenue both on Company S's books (sale to Company P) and on Company P's books (sale to outsiders). Thus, **column (3)** overstates sales of the consolidated entity to outsiders. Likewise, Cost of Goods Sold of both companies in **column (3)** counts twice the cost of the goods sold, first by Company S to Company P and then by Company P to outsiders. The consolidation process eliminates the effects of the intercompany sale from Company S to Company P. See **column (4)** for Sales and Cost of Goods Sold.

**A Realistic Complication** We simplified the above example by having Company S sell the goods to Company P at cost, $40,000, so that you can easily see that the consolidation process reduces both sales and cost of goods sold by $40,000. Often S sells the goods to P for a price larger than S's cost. Assume, now, that S sold goods costing $30,000 to P for $40,000, goods that P later sold to outsiders for $45,000. The consolidation process still eliminates $40,000 from both combined sales and combined cost of goods sold. The transactions remaining in the consolidated income statement will be P's sales to outsiders for $45,000 and S's cost of $30,000, just as if they were a single company selling, for $45,000, goods that cost $30,000. This realistic complication becomes more complex when Company P has not yet, by the end of the accounting period, sold all the goods it bought from Company S. In that case, Company S has recorded a profit on the goods remaining in inventory, which the consolidation must eliminate. We do not illustrate this complication.

**Eliminating Double-Counting of Income** Company P's accounts show Equity in Earnings of Company S of $48,000. Company S's records show individual revenues and expenses that net to $48,000. When **column (3)** sums the revenues and expenses of the two companies, it counts that income twice. The consolidation process must eliminate the account Equity in Earnings of Company S. See **column (4)** for Equity in Earnings.

**Consolidated Income** The amount of consolidated net income for a period exactly equals the amount that the parent would show on its separate company books if it used the equity method for all its subsidiaries. That is, consolidated net income is as follows:

| Parent Company's Net Income from Its Own Activities | + | Parent Company's Share of Subsidiaries' Net Income | − | Profit (or + Loss) on Intercompany Transactions |
|---|---|---|---|---|

A consolidated income statement differs from a parent's income statement using the equity method only in the components presented. The equity method for an unconsolidated subsidiary shows the parent's share of the subsidiary's net income minus gain (or plus loss) on intercompany transactions on a single line, Equity in Earnings of Unconsolidated Subsidiary.

## *Terminology* Note

Accountants sometimes refer to the equity method as a *one-line consolidation* because the revenues less the expenses of the subsidiary appear in the one account Equity in Earnings of Unconsolidated Subsidiary. Applying the equity method in realistic situations sometimes presents ambiguities. The guiding principle: treat the items in such a way that the parent's net income equals the same amount that it would show if it consolidated the investment rather than used the equity method for it.

A consolidated income statement combines the individual revenues and expenses of the subsidiary (less intercompany adjustments) with those of the parent. The consolidation process eliminates the account Equity in Earnings of Unconsolidated Subsidiary.

**External Minority Interest in Consolidated Subsidiary** Often, the parent does not own 100 percent of the voting stock of a consolidated subsidiary. The parent refers to the owners of the remaining shares of voting stock as the external minority shareholders or the **minority interest**. These shareholders have provided a portion or fraction of the subsidiary's financing and have a claim to this same fraction of its net assets (= total assets − total liabilities) shown on the subsidiary's separate corporate records. They also have a claim to the same fraction of the earnings of the subsidiary. Do not confuse this minority interest in a consolidated subsidiary with a firm's own minority investments, discussed earlier. The minority interest represents ownership by others outside the parent and its economic entity. The parent's minority investments represent its ownership of shares of other companies in which the parent owns less than 50 percent of the shares.

**The Minority Interest in Net Assets in the Consolidated Balance Sheet** Assume now that the parent owns less than 100 percent of the subsidiary's shares—80 percent to be concrete. The minority shareholders own 20 percent of the subsidiary. Should the parent's consolidated statements show (1) only its 80 percent fraction of each of the assets and liabilities of the subsidiary or (2) all of the subsidiary's assets and liabilities?

The parent, with its controlling voting interest, can direct the use of *all* the subsidiary's assets and liabilities, not merely the 80 percent it owns claim to. The generally accepted accounting principle, therefore, shows all of the assets and liabilities of the subsidiary. The consolidated balance sheet and income statement will disclose the interest of the minority shareholders in the consolidated, but less-than-wholly-owned subsidiary.

The amount of the **minority interest** appearing in the balance sheet generally results from multiplying the common shareholders' equity of the subsidiary by the minority's percentage of ownership. For example, if the common shareholders' equity (= assets − liabilities) of a consolidated subsidiary totals $500,000, and the minority owns 20 percent of the common stock, the minority interest will appear on the consolidated balance sheet as $100,000 (= 0.20 × $500,000). The minority interest appears in the shareholders' section of the consolidated balance sheet, clearly labeled to distinguish it from the sources of financing provided by the parent's shareholders.

The parent's consolidated income statement shows all the subsidiary's revenues less all the subsidiary's expenses and so includes all of the subsidiary's income even though that amount does not appear with a separate caption. The minority shareholders claim part of this income, typically an amount equal to the subsidiary's net income multiplied by the minority's percentage of ownership. The consolidated income statement subtracts the minority's share of the subsidiary's income in calculating consolidated net income, which shows the parent's shareholders' claim on the income of the entire group of companies.[14]

## Limitations of Consolidated Statements

The consolidated statements, provided for the parent's shareholders, do not replace the statements of individual corporations.

- Creditors must rely on the resources of the one corporation to which they loaned funds. They may misunderstand the protection for their loans if they use only a consolidated statement combining the data of a company that is sound with the data of one that is verging on insolvency.
- A corporation can declare dividends against only its own retained earnings.
- When the parent company does not own all of the shares of the subsidiary, the minority shareholders can judge the dividend constraints, both legal and financial, only by inspecting the subsidiary's statements.

---

[14]We discuss and illustrate more of the accounting for minority interest, as well as special purpose entities and proportional consolidation, topics too advanced to discuss in the text, on the web site that the publisher has made available for this book (see the **Preface**).

## Problem 11.5 for Self-Study

**Understanding consolidation concepts.** **Exhibit 11.6** presents income statement data and **Exhibit 11.7** presents balance sheet data for Parent and its 80-percent-owned Sub. The first two columns in each exhibit show amounts taken from the separate-company accounting records of each firm. The third column sums the amounts in the first two columns. The fourth column shows consolidated amounts for Parent and Sub after making intercompany eliminations.

**a.** Does the account Investment in Sub (equity method) on the Parent's books include any excess acquisition cost relative to the book value of the shareholders' equity of Sub?
**b.** Suggest four ways in which the data in **Exhibits 11.6** and **11.7** confirm that Parent owns 80 percent of Sub.
**c.** Why does the amount for accounts receivable in **column (3)** of **Exhibit 11.7** differ from the amount in **column (4)**?
**d.** Explain why the account Investment in Sub does not appear on the consolidated balance sheet.
**e.** Why does the total shareholders' equity of $692 on the consolidated balance sheet equal the shareholders' equity on Parent's separate-company books?
**f.** Compute the amount of intercompany sales during Year 1.
**g.** Explain why the $80 for Equity in Earnings of Sub does not appear on the consolidated income statement.
**h.** Why does the account Minority Interest in Earnings of Sub appear on the consolidated income statement but not on the income statements of either Parent or Sub?

## An International Perspective

Most countries outside of the United States require that the accounting for minority, passive investments use the lower-of-cost-or-market method, not the market value method. The accounting for minority and majority, active investments closely parallels practices in the United States: the equity method for minority, active investments and consolidation for majority investments. Practices for these active investments have become similar, however, only in recent years. Countries such as Germany and Japan have historically followed a strict legal definition of the corporate entity, with all intercorporate investments reported at acquisition cost and dividends received reported as the only revenues from these investments. The movement toward the use of the equity method and consolidation reflects increasing recognition that the economic entity likely differs from the legal entity and that financial statements based on the economic entity provide more useful information to users.

The International Accounting Standards Board (IASB) requires the equity method for investments where the holder exerts significant influence and has no plans to sell its investment in the near future.[15] The IASB expresses a preference for the consolidation of controlled entities, even if the investor owns less than 50 percent of the shares, and allows the cost method, the lower-of-cost-or-market method, the market value method, and the equity method for noncontrolling investments.[16]

As this book goes to press, the IASB is struggling with a treatment for derivative financial instruments—under what conditions to require these appear at market value. The IASB does not have an equivalent of a Statement of Other Comprehensive Income or a balance sheet account Accumulated Other Comprehensive Income. Instead, for items such as unrealized holding gains or losses on effective cash flow hedges, the IASB uses a separate account in shareholders' equity.

[15]International Accounting Standards Board, *International Accounting Standard No. 28*, "Investments in Associates," 1989, revised in 1998.
[16]International Accounting Standards Board, *International Accounting Standard No. 27*, "Consolidated Financial Statements," 1989.

**EXHIBIT 11.6**

**PARENT AND SUB**
**Income Statement Data for Year 1**
**(Problem 11.5 for Self-Study)**

| | Separate Company Books | | | |
|---|---|---|---|---|
| | Parent (1) | Sub (2) | Combined (3) = (1) + (2) | Consolidated (4) |
| Sales | $ 4,000 | $ 2,000 | $ 6,000 | $ 5,500 |
| Equity in Earnings of Sub | 80 | — | 80 | — |
| Cost of Goods Sold | (2,690) | (1,350) | (4,040) | (3,540) |
| Selling and Administrative Expenses | (1,080) | (480) | (1,560) | (1,560) |
| Interest Expense | (30) | (20) | (50) | (50) |
| Income Tax Expense | (70) | (50) | (120) | (120) |
| Minority Interest in Earnings of Sub | — | — | — | (20) |
| Net Income | $ 210 | $ 100 | $ 310 | $ 210 |

**EXHIBIT 11.7**

**PARENT AND SUB**
**Balance Sheet Data, December 31, Year 1**
**(Problem 11.5 for Self-Study)**

| | Separate Company Books | | | |
|---|---|---|---|---|
| | Parent (1) | Sub (2) | Combined (3) = (1) + (2) | Consolidated (4) |
| ASSETS | | | | |
| Cash | $ 125 | $ 60 | $ 185 | $ 185 |
| Accounts Receivable | 550 | 270 | 820 | 795 |
| Inventories | 460 | 210 | 670 | 670 |
| Investment in Sub (equity method) | 192 | — | 192 | — |
| Property, Plant, and Equipment (net) | 680 | 380 | 1,060 | 1,060 |
| Total Assets | $2,007 | $920 | $2,927 | $2,710 |
| LIABILITIES AND SHAREHOLDERS' EQUITY | | | | |
| Accounts Payable | $ 370 | $170 | $ 540 | $ 515 |
| Notes Payable | 400 | 250 | 650 | 650 |
| Other Current Liabilities | 245 | 60 | 305 | 305 |
| Long-Term Debt | 300 | 200 | 500 | 500 |
| Total Liabilities | $1,315 | $680 | $1,995 | $1,970 |
| Minority Interest in Net Assets of Sub | — | — | — | $ 48 |
| Common Stock | 200 | $ 50 | $ 250 | 200 |
| Retained Earnings | 492 | 190 | 682 | 492 |
| Total Shareholders' Equity | $ 692 | $240 | $ 932 | $ 740 |
| Total Liabilities and Shareholders' Equity | $2,007 | $920 | $2,927 | $2,710 |

## Summary

Businesses acquire the bonds or capital stock in other entities for a variety of reasons and in a variety of ways. The investor records the acquisition of the bonds or capital stock of another entity at the cash given or the market value of other consideration exchanged. The account debited for equity securities, either Marketable Securities or Investment in Securities, depends on the expected holding period.

| | | |
|---|---|---|
| Investment in S . . . . . . . . . . . . . . . . . . . . . . . . . . . . . . . . . . . . . . . | X | |
| Cash or Other Consideration Given . . . . . . . . . . . . . . . . . . . . . . . | | X |

The accounting for investments in equity securities subsequent to acquisition depends on the ownership percentage:

- The market value method generally applies when the parent owns less than 20 percent.
- The equity method generally applies when the parent owns at least 20 percent but not more than 50 percent of the stock of another company.
- The equity method applies also when the investor company can exercise significant influence even though it owns less than 20 percent.
- The investor generally prepares consolidated statements when it owns more than 50 percent of the voting shares of another company.

The accounting for investments in debt securities subsequent to acquisition depends on the ability and intent of the investor to hold the security until it matures:

- The market value method generally applies when the investor does not have both the ability and intent to hold to maturity, with treatment as security available for sale or as trading security determined by the holder's intent.
- The amortized cost method generally applies when the investor has both the ability and intent to hold the debt security to its maturity.

**Figure 11.3** summarizes the structure of accounting for holdings of securities, and **Exhibit 11.8** summarizes the accounting for investments subsequent to acquisition.

The market value method for securities available for sale recognizes net income only when the investor becomes entitled to receive interest or dividends, or when it sells the securities. Changes in the market value of securities available for sale appear in other comprehensive income, not net income. The market value method for trading securities recognizes in net income unrealized holding gains and losses from changes in the market value each period and any additional gain or loss realized at the time of sale.

Consolidated statements and the equity method have the same effect on net income. The parent shows as income its proportional share of the acquired firm's periodic income after acquisition, after eliminating the effects of intercompany transactions. In the equity method, this share appears on a single line of the income statement. Income statement amounts of revenues and expenses are larger in consolidation because the consolidated income statement combines the revenues and expenses of the acquired company with those of the parent. Balance sheet components under the consolidation method will exceed those under the equity method; the consolidated balance sheet substitutes the individual assets and liabilities of the acquired company for the parent company's Investment in the Subsidiary account.

## Appendix 11.1: Effects on the Statement of Cash Flows of Investments in Securities

### Investments in Debt with Both Ability and Intent to Hold to Maturity

**Amortized Cost Method** Acquisitions of debt securities are investing transactions. If the firm uses the direct method for cash flow from operations, it will show a cash inflow for the

FIGURE 11.3 Roadmap to Investments

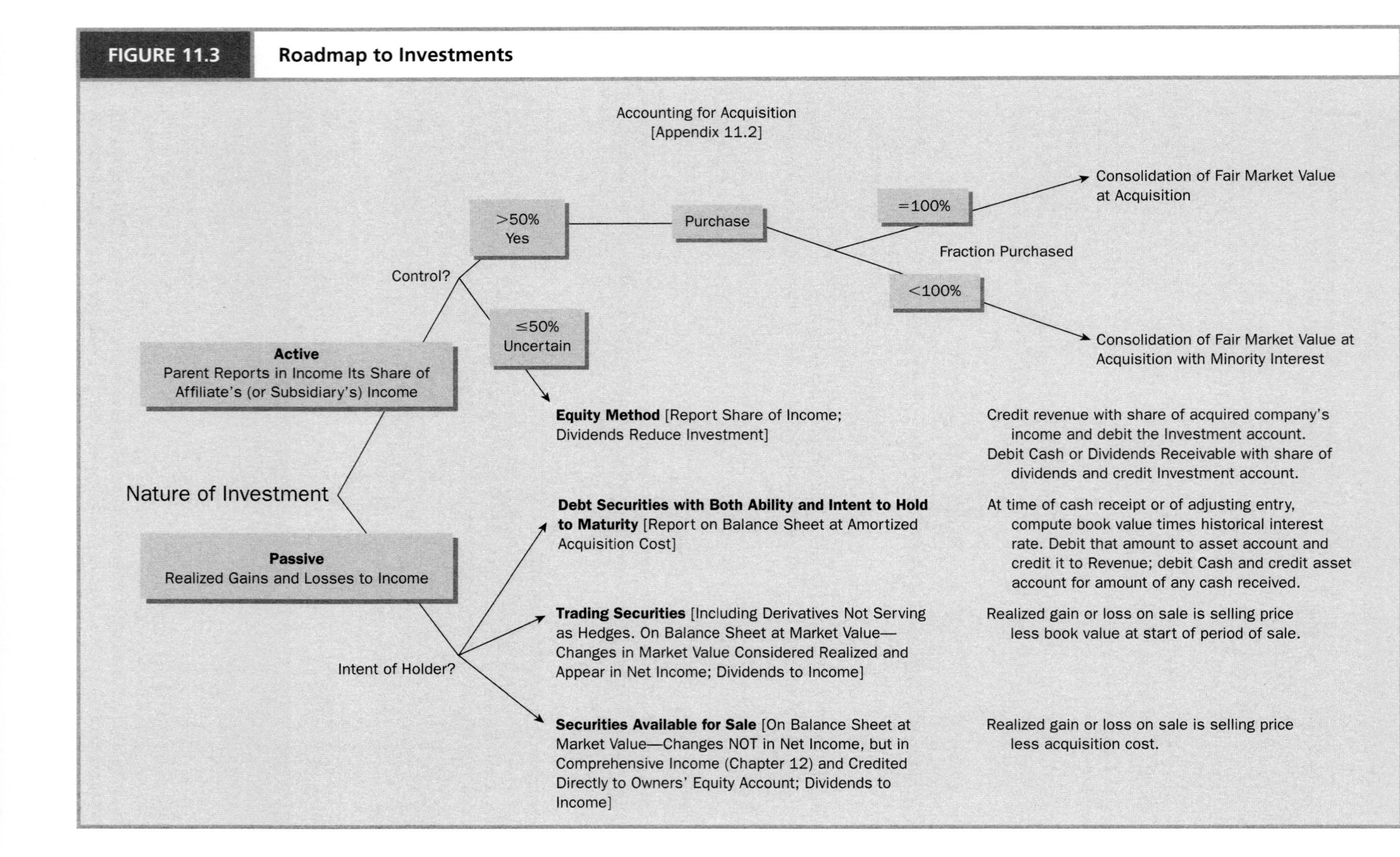

**EXHIBIT 11.8** **Effects of Various Methods of Accounting for Long-Term Investments in Corporate Securities**

| Method of Accounting | Balance Sheet | Income Statement | Statement of Cash Flows—Cash Flows from Operations |
|---|---|---|---|
| Market value method for securities available for sale (generally used when ownership percentage is less than 20 percent) | Investment account appears at current market value as a noncurrent asset. Unrealized holding gains and losses appear in a separate shareholders' equity account. | Dividends declared by investee appear as revenue of investor. Gains and losses (from acquisition cost) appear in income only when realized in arm's-length transactions with outsiders. | Dividends received from investee are included in cash provided by operations of investors. |
| Market value method for trading securities (generally used when ownership percentage is less than 20 percent) | Investment account appears at current market value as a noncurrent asset. Unrealized holding gains and losses appear in income statement and effects increase or decrease the Retained Earnings account. | Dividends declared by investee appear as revenue of investor. Gains and losses (from then-current book value) appear in income when realized in arm's-length transactions with outsiders. | Dividends received from investee are included in cash provided by operations of investors. Under indirect method, add back unrealized holding losses and subtract unrealized holding gains included in income of the period. |
| Equity method (generally used when ownership percentage is at least 20 percent but not more than 50 percent) | Investment account appears at acquisition cost plus share of investee's net income less share of investee's dividends since acquisition. More advanced: also, less amortization of excess of purchase price over book value acquired. | Equity in investee's net income is part of revenue in the period that investee earns income. Reduce income by the amount, if any, of amortization of excess purchase price over book value acquired. | Cash from operations increases only by the amount of dividend received. In indirect method, deduct equity in investee's undistributed earnings from net income to derive cash provided by operations of investor. More advanced: if amortization of excess purchase price, then add back the amount amortized during current period. |
| Consolidation (generally used when ownership percentage exceeds 50 percent) | Eliminate investment account and replace it with individual assets and liabilities of subsidiary. Show minority interest in subsidiary's net assets among equities. Eliminate intercompany assets and liabilities. | Combine individual revenues and expenses of subsidiary with those of parent. Subtract minority interest in subsidiary's net income. | Combine individual sources and uses of cash of subsidiary with those of parent. In indirect method, add minority interest in net income to obtain cash provided by operations. |
| Amortized Cost Method (used only for debt where holder has both intent and ability to hold to maturity) | Original cost plus accrued interest not yet received in cash. | Book value at start of period multiplied by historical interest rate—market interest rate on date of acquisition of debt. | If holder acquired debt at a price below par, then cash flow from operations includes only the interest payment received. If the holder acquired the debt at a price above par, then cash flow from operations will include the amount of interest for the period, with the remainder of the cash received appearing as a financing receipt for redemption of debt. |

debt service collections it receives so long as that amount does not exceed interest revenue for the period. Any excess reduces the amount of the asset for the investment in debt and appears as a cash inflow from an investing activity, as though the holder sold some of the asset. If the firm uses the indirect method, it will subtract from net income any amounts by which interest revenue exceeds debt service collections. If the debt service collection exceeds interest revenue, it will report the excess as a cash inflow from an investing activity.

## INVESTMENTS IN EQUITY SECURITIES AND DEBT WITHOUT ABILITY AND INTENT TO HOLD TO MATURITY

**Market Value Method** Acquisitions of trading securities are operating activities; acquisitions of securities available for sale are investing transactions.[17] When Company P uses the market value method to account for its investment in Company S, all dividend revenues recognized in computing net income also generally produce cash. Therefore, calculating cash flow from operations normally requires no adjustment to revenues (direct method) or net income (indirect method). When the investor does not receive cash equal to the amount of dividend revenue, and it uses the indirect method, it will subtract from (add to) net income the increase (decrease) in the account for Dividends Receivable. The investor using the direct method will show the cash receipts in deriving cash flows from operations.

Changes in the Marketable Securities account and the Investment in Securities account and in the account for Unrealized Holding Gains or Losses, which arise in the market value method, do not appear in the statement of cash flows for Securities Available for Sale. These changes in market value do not affect cash, so require no adjustment when computing cash flow from operations under either the direct or the indirect method. In contrast, holding gains and losses on trading securities do appear in income but do not affect cash flow, so they do require an adjustment to net income in deriving cash flow from operations if the firm uses the indirect method. For holding gains (losses) on trading securities, income has increased (decreased), but not cash from operations, so the indirect method shows a subtraction (addition) for those holding gains (losses).

**Equity Method** Under the direct method, the firm shows the dividends received as an operating cash flow. Accounting for investments using the equity method requires an adjustment to net income to compute cash flow from operations if the firm uses the indirect method in its statement of cash flows. Suppose that Company P prepares its financial statements at the end of a year during which occurred transactions **(1)** through **(5)** described in the section "Equity Method: Procedures" on page 531. Company P reported revenue from its investment in S of $54,000. This amount results from summing the revenue recognized in transactions **(2)** and **(4)**. P's income (ignoring income tax effects) increased $54,000 because of its investment. However, P's cash increased by only $21,000—transactions **(3)** and **(5)**—as a result of S's dividends. Consequently, in computing cash flow from operations, P must subtract from net income an amount equal to the excess of revenue over cash flow, or $33,000 (= $54,000 − $21,000), to show that cash did not increase by as much as the amount of revenue recognized under the equity method.

If a firm acquires an investment for an amount that exceeds the book value of the underlying net assets, the firm must write off the excess purchase price over the expected useful life of the item (other than goodwill). The expense does not use cash and requires an addback to net income when the firm computes cash flow from operations using the indirect method.

**Consolidation** The accountant prepares a consolidated statement of cash flows from the consolidated balance sheet and income statement. Advanced accounting texts discuss the adjustments unique to a consolidated statement of cash flows, such as for a minority interest in a consolidated subsidiary. You should be able to recognize, however, that the minority interest in income subtracted (or, in losses added) on the consolidated income statement does not affect cash flow from operations, so the firm using the indirect method will show an addback (subtraction) for the minority interest in income (loss).

# Solutions to Self-Study Problems

### SUGGESTED SOLUTION TO PROBLEM 11.1 FOR SELF-STUDY

(General Electric Capital Services and Sapra Company; accounting for an interest-bearing note receivable.)

**a.** See **Exhibit 9.5**; the numbers remain the same; only the column captions change.

[17]Financial Accounting Standards Board, *Statement of Financial Accounting Standards No. 102*, "Statement of Cash Flows—Exemption of Certain Enterprises and Classification of Cash Flows from Certain Securities Acquired for Resale," paragraph 26, 1989.

**b.** ***Date of Issue***

| | Dr. | Cr. |
|---|---|---|
| Notes Receivable | 112,434 | |
| Cash | | 112,434 |

| Assets | = | Liabilities | + | Shareholders' Equity | (Class.) |
|---|---|---|---|---|---|
| +112,434 | | | | | |
| −112,434 | | | | | |

GECS lends cash to Sapra and receives note.

***Six Months after Issue***

| | Dr. | Cr. |
|---|---|---|
| Interest Receivable | 5,622 | |
| Interest Revenue | | 5,622 |

| Assets | = | Liabilities | + | Shareholders' Equity | (Class.) |
|---|---|---|---|---|---|
| +5,622 | | | | +5,622 | IncSt → RE |

See **Exhibit 9.5**; accrual of six months' interest = $11,243/2.

***One Year after Issue***

| | Dr. | Cr. |
|---|---|---|
| Cash | 15,000 | |
| Interest Receivable | | 5,622 |
| Interest Revenue | | 5,621 |
| Notes Receivable | | 3,757 |

| Assets | = | Liabilities | + | Shareholders' Equity | (Class.) |
|---|---|---|---|---|---|
| +15,000 | | | | +5,621 | IncSt → RE |
| −5,622 | | | | | |
| −3,757 | | | | | |

Interest revenue for the remainder of the first year and cash received. Excess of cash payment over interest expense reduces note principal.

## SUGGESTED SOLUTION TO PROBLEM 11.2 FOR SELF-STUDY

(Conlin Corporation; accounting for securities available for sale.)

**a.** **(1)** *February 3, Year 2*

| | Dr. | Cr. |
|---|---|---|
| Marketable Securities [A] | 40,000 | |
| Cash | | 40,000 |

| Assets | = | Liabilities | + | Shareholders' Equity | (Class.) |
|---|---|---|---|---|---|
| +40,000 | | | | | |
| −40,000 | | | | | |

To record acquisition of Security A.

**(2)** *July 15, Year 2*

| | Dr. | Cr. |
|---|---|---|
| Marketable Securities [B] | 75,000 | |
| Cash | | 75,000 |

| Assets | = | Liabilities | + | Shareholders' Equity | (Class.) |
|---|---|---|---|---|---|
| +75,000 | | | | | |
| −75,000 | | | | | |

To record acquisition of Security B.

*(continued)*

**(3)** *November 27, Year 2*

Marketable Securities [C] . . . . . . . . . . 90,000

Cash . . . . . . . . . . 90,000

| Assets | = | Liabilities | + | Shareholders' Equity | (Class.) |
|---|---|---|---|---|---|
| +90,000 | | | | | |
| −90,000 | | | | | |

To record acquisition of Security C.

**(4)** *December 31, Year 2*

Unrealized Holding Loss on Security A Available for Sale . . . . . . . . . . 2,000

Marketable Securities [A] . . . . . . . . . . 2,000

| Assets | = | Liabilities | + | Shareholders' Equity | (Class.) |
|---|---|---|---|---|---|
| −2,000 | | | | −2,000 | OCInc → AOCInc |

To revalue Security A to market value.

**(5)** *December 31, Year 2*

Marketable Securities [B] . . . . . . . . . . 4,000

Unrealized Holding Gain on Security B Available for Sale . . . . . . . . . . 4,000

| Assets | = | Liabilities | + | Shareholders' Equity | (Class.) |
|---|---|---|---|---|---|
| +4,000 | | | | +4,000 | OCInc → AOCInc |

To revalue Security B to market value.

**(6)** *December 31, Year 2*

Marketable Securities [C] . . . . . . . . . . 3,000

Unrealized Holding Gain on Security C Available for Sale (Other Comprehensive Income) . . . . . . . . . . 3,000

| Assets | = | Liabilities | + | Shareholders' Equity | (Class.) |
|---|---|---|---|---|---|
| +3,000 | | | | +3,000 | OCInc → AOCInc |

To revalue Security C to market value.

---

*Note:* The accountant could combine entries **(4)**, **(5)**, and **(6)** above as follows:

---

*December 31, Year 2*

Marketable Securities . . . . . . . . . . 5,000

Unrealized Holding Gain (net) on Securities Available for Sale . . . . . . . . . . 5,000

| Assets | = | Liabilities | + | Shareholders' Equity | (Class.) |
|---|---|---|---|---|---|
| +5,000 | | | | +5,000 | OCInc → AOCInc |

To revalue the portfolio of marketable securities available for sale to market value.

**(7)** *September 6, Year 3*

Cash . . . . . . . . . . 78,000

Marketable Securities [B] . . . . . . . . . . 75,000

Realized Gain on Sale of Security B Available for Sale . . . . . . . . . . 3,000

| Assets | = | Liabilities | + | Shareholders' Equity | (Class.) |
|---|---|---|---|---|---|
| +78,000 | | | | +3,000 | IncSt → RE |
| −75,000 | | | | | |

To record sale of Security B at a gain equal to the difference between selling price and acquisition cost and include the realized gain in net income.

**(8)** *September 6, Year 3, or December 31, Year 3*

| | Dr. | Cr. |
|---|---|---|
| Unrealized Holding Gain on Security B Available for Sale | 4,000 | |
| Marketable Securities (B) | | 4,000 |

| Assets | = | Liabilities | + | Shareholders' Equity | (Class.) |
|---|---|---|---|---|---|
| −4,000 | | | | −4,000 | OCInc → AOCInc |

To eliminate the effects of changes previously recorded in the market values of securities available for sale.

**(9)** *December 31, Year 3*

| | Dr. | Cr. |
|---|---|---|
| Unrealized Holding Loss on Security A Available for Sale | 5,000 | |
| Marketable Securities [A] | | 5,000 |

| Assets | = | Liabilities | + | Shareholders' Equity | (Class.) |
|---|---|---|---|---|---|
| −5,000 | | | | −5,000 | OCInc → AOCInc |

To revalue Security A to market value.

**(10)** *December 31, Year 3*

| | Dr. | Cr. |
|---|---|---|
| Marketable Securities [C] | 1,000 | |
| Unrealized Holding Gain on Security C Available for Sale | | 1,000 |

| Assets | = | Liabilities | + | Shareholders' Equity | (Class.) |
|---|---|---|---|---|---|
| +1,000 | | | | +1,000 | OCInc → AOCInc |

To revalue Security C to market value.

*Note:* The accountant could combine entries **(8)**, **(9)**, and **(10)** above as follows:

*December 31, Year 3*

| | Dr. | Cr. |
|---|---|---|
| Unrealized Holding Gain (net) on Securities Available for Sale | 8,000 | |
| Marketable Securities | | 8,000 |

| Assets | = | Liabilities | + | Shareholders' Equity | (Class.) |
|---|---|---|---|---|---|
| −8,000 | | | | −8,000 | OCInc → AOCInc |

To revalue the portfolio of marketable securities available for sale to market value.

**b.** The journal entries are identical except that the account titled Marketable Securities becomes Investment in Securities.

**c.** The first three journal entries are identical. The unrealized holding gain or loss accounts in entries **(4)**, **(5)**, **(6)**, **(9)**, and **(10)** are income statement accounts when the firm classifies the securities as trading securities. Entry **(7)** is as follows:

*September 6, Year 3*

| | Dr. | Cr. |
|---|---|---|
| Cash | 78,000 | |
| Realized Loss from Sale of Trading Security B | 1,000 | |
| Marketable Securities [B] | | 79,000 |

| Assets | = | Liabilities | + | Shareholders' Equity | (Class.) |
|---|---|---|---|---|---|
| +78,000 | | | | −1,000 | IncSt → RE |
| −79,000 | | | | | |

To record sale of trading security for less than its book value at the time of sale.

If the security were a trading security, the firm would not make Entry **(8)**.

## SUGGESTED SOLUTION TO PROBLEM 11.3 FOR SELF-STUDY

(Equity Investing Group; journal entries to apply the equity method for long-term investments in securities.)

**a.**

| | Debit | Credit |
|---|---|---|
| Investment in Securities [D] | 80,000 | |
| Investment in Securities [E] | 190,000 | |
| Investment in Securities [F] | 200,000 | |
| Cash | | 470,000 |

| Assets | = | Liabilities | + | Shareholders' Equity | (Class.) |
|---|---|---|---|---|---|
| +80,000 | | | | | |
| +190,000 | | | | | |
| +200,000 | | | | | |
| −470,000 | | | | | |

**b.**

| | Debit | Credit |
|---|---|---|
| Investment in Securities [D] (0.40 × $40,000) | 16,000 | |
| Investment in Securities [E] (0.30 × $120,000) | 36,000 | |
| Investment in Securities [F] (0.20 × $200,000) | 40,000 | |
| Equity in Earnings of Affiliates | | 92,000 |

| Assets | = | Liabilities | + | Shareholders' Equity | (Class.) |
|---|---|---|---|---|---|
| +16,000 | | | | +92,000 | IncSt → RE |
| +36,000 | | | | | |
| +40,000 | | | | | |

| | Debit | Credit |
|---|---|---|
| Cash | 25,000 | |
| Investment in Securities [D] (0.40 × $10,000) | | 4,000 |
| Investment in Securities [E] (0.30 × $30,000) | | 9,000 |
| Investment in Securities [F] (0.20 × $60,000) | | 12,000 |

| Assets | = | Liabilities | + | Shareholders' Equity | (Class.) |
|---|---|---|---|---|---|
| +25,000 | | | | | |
| −4,000 | | | | | |
| −9,000 | | | | | |
| −12,000 | | | | | |

| | Debit | Credit |
|---|---|---|
| Depreciation Expense | 8,000 | |
| Investment in Securities [E] | | 4,000 |
| Investment in Securities [F] | | 4,000 |

| Assets | = | Liabilities | + | Shareholders' Equity | (Class.) |
|---|---|---|---|---|---|
| −4,000 | | | | −8,000 | IncSt → RE |
| −4,000 | | | | | |

| Security | Book Value of Investee on January 1, Year 1 | Ownership Percentage | Share of Book Value Acquired | Acquisition Cost of Investment | Excess Acquisition Cost | Annual Amortization for 10 Years |
|---|---|---|---|---|---|---|
| D ..... | $200,000 | 40% | $ 80,000 | $ 80,000 | — | — |
| E ..... | 500,000 | 30 | 150,000 | 190,000 | $40,000 | $4,000 |
| F ..... | 800,000 | 20 | 160,000 | 200,000 | 40,000 | 4,000 |

**c.** Investment in Securities [D] (0.40 × $50,000) . . . . . . . . . . . . . . . . . . 20,000
Investment in Securities [F] (0.20 × $50,000) . . . . . . . . . . . . . . . . . . 10,000
Investment in Securities [E] (0.30 × $40,000) . . . . . . . . . . . . . 12,000
Equity in Net Earnings of Affiliates . . . . . . . . . . . . . . . . . . . . . 18,000

| Assets | = | Liabilities | + | Shareholders' Equity | (Class.) |
|---|---|---|---|---|---|
| +20,000 | | | | +18,000 | IncSt → RE |
| +10,000 | | | | | |
| −12,000 | | | | | |

Cash . . . . . . . . . . . . . . . . . . . . . . . . . . . . . . . . . . . . . . . . . 16,800
Investment in Securities [D] (0.40 × $12,000) . . . . . . . . . . . . . 4,800
Investment in Securities [F] (0.20 × $60,000) . . . . . . . . . . . . . 12,000

| Assets | = | Liabilities | + | Shareholders' Equity | (Class.) |
|---|---|---|---|---|---|
| +16,800 | | | | | |
| −4,800 | | | | | |
| −12,000 | | | | | |

Depreciation Expense . . . . . . . . . . . . . . . . . . . . . . . . . . . . . . . 8,000
Investment in Securities [E] . . . . . . . . . . . . . . . . . . . . . . . . . 4,000
Investment in Securities [F] . . . . . . . . . . . . . . . . . . . . . . . . . 4,000

| Assets | = | Liabilities | + | Shareholders' Equity | (Class.) |
|---|---|---|---|---|---|
| −4,000 | | | | −8,000 | IncSt → RE |
| −4,000 | | | | | |

**d.** Cash . . . . . . . . . . . . . . . . . . . . . . . . . . . . . . . . . . . . . . . . . 190,000
Loss on Sale of Investment in Securities . . . . . . . . . . . . . . . . . . . . . 7,000
Investment in Securities [E] . . . . . . . . . . . . . . . . . . . . . . . . . 197,000

$197,000 = $190,000 + $36,000 − $9,000 − $4,000 − $12,000 − $4,000.

| Assets | = | Liabilities | + | Shareholders' Equity | (Class.) |
|---|---|---|---|---|---|
| +190,000 | | | | −7,000 | IncSt → RE |
| −197,000 | | | | | |

## SUGGESTED SOLUTION TO PROBLEM 11.4 FOR SELF-STUDY

(Powell Corporation and Steele Corporation; financial statement effects of the purchase method.)

**Exhibit 11.9** presents the consolidated balance sheet of January 1, Year 8, using the purchase method.

## SUGGESTED SOLUTION TO PROBLEM 11.5 FOR SELF-STUDY

(Parent and Sub; understanding consolidation concepts.)

**a.** No. The investment account shows a balance of $192, which equals 80 percent of Sub's shareholders' equity ($192 = 0.80 × $240).

**b.** **(1)** The account Investment in Sub has a balance of $192, which equals 80 percent of the shareholders' equity of Sub. This clue supports the 80 percent ownership only because no unamortized excess acquisition cost exists (see the response to question **a**).

**(2)** The minority interest in the net assets of Sub is $48, which equals 20 percent of the shareholders' equity of Sub ($48 = 0.20 × $240).

**(3)** The account Equity in Earnings of Sub on Parent's books has a balance of $80 for Year 1, which equals 80 percent of the net income of Sub for Year 1 ($80 = 0.80 × $100).

**(4)** The account Minority Interest in Earnings of Sub has a balance of $20 for Year 1 ($20 = 0.20 × $100).

**EXHIBIT 11.9**

**POWELL CORPORATION AND STEELE CORPORATION**
**Consolidated Balance Sheet, January 1, Year 8**
**(Problem 11.4 for Self-Study)**

| | Purchase Method |
|---|---|
| ASSETS | |
| Current Assets | $ 17,000 |
| Property, Plant, and Equipment (net) | 53,000 |
| Goodwill | 40,000 |
| Total Assets | $110,000 |
| EQUITIES | |
| Liabilities | $ 41,000 |
| Common Stock | 3,700[a] |
| Additional Paid-in Capital | 60,300[b] |
| Retained Earnings | 5,000 |
| Total Equities | $110,000 |

[a]$3,700 = $1,000 + (2,700 × $1).
[b]$60,300 = $9,000 + (2,700 × 19).

**c.** Parent and Sub have intercompany accounts receivable and accounts payable. Combined accounts receivable exceed consolidated accounts receivable by $25 (= $820 − $795), the same as the excess of combined accounts payable over consolidated accounts payable ($25 = $540 − $515).

**d.** Double-counting results if both the investment account and the individual assets and liabilities of Sub appear on the consolidated balance sheet.

**e.** The $240 of Sub's shareholders' equity disappears in consolidation. As a result the only remaining shareholders' equity amount is the $740 of Parent. The elimination of the $240 of Sub's shareholders' equity proceeds as follows:

| | |
|---|---|
| Elimination of Investment in Sub | $192 |
| Recognition of Minority Interest in Net Assets of Sub | 48 |
| Total | $240 |

**f.** $500 (= $6,000 − $5,500 or $4,040 − $3,540).

**g.** Consolidated amounts include individual revenues, expenses, and minority interest in earnings, which net to $80. Double-counting this earnings results if the accountant does not eliminate the equity in earnings account.

**h.** The separate-company income statements report the total revenues and expenses of each entity without regard to who owns the common stock of each company. The consolidated income statement shows the earnings allocable to the shareholders of the parent company. These shareholders have a claim on all of the earnings of Parent but on only 80 percent of the earnings of Sub. Consolidated revenues and expenses included the combined amounts for both companies, adjusted for intercompany transactions. The minority interest in net income of Sub shows the portion of Sub's net income not subject to a claim by Parent's shareholders.

## Key Terms and Concepts

Marketable securities
Investments in securities
Minority, passive investments
Significant influence
Minority, active investments
Majority, active investments
Derivative
Forward foreign exchange contract
Swap contract
Forward commodity contract

Underlyings
Notional amounts
Counterparty
Net settlement
Hedge accounting
Fair value hedge
Cash flow hedge
Equity method
Market value method
Parent
Subsidiary
Consolidated financial statements
Economic entity
Purchase method
Goodwill
Variable interest entity (VIE)
Primary beneficiary
Consolidation policy
Intercompany transactions
Minority interest (income statement)
Minority interest (balance sheet)

## Questions, Short Exercises, Exercises, Problems, and Cases

### QUESTIONS

For additional student resources, content, and interactive quizzes for this chapter visit the FACMU website: **www.thomsonedu.com/accounting/stickney**

1. Review the meaning of the terms and concepts listed above in Key Terms and Concepts.
2. "The classification of securities on the balance sheet as a current asset (Marketable Securities) or as a noncurrent asset (Investment in Securities) depends on a firm's intent." Explain.
3. Distinguish between the following pairs of terms:
   **a.** Debt securities classified as "held to maturity" versus "available for sale"
   **b.** Equity securities classified as "trading securities" versus "available for sale"
   **c.** Amortized acquisition cost versus market value of debt securities
   **d.** Unrealized holding gain or loss on trading securities versus on securities available for sale
   **e.** Realized gain or loss on trading securities versus on securities available for sale
4. What is the GAAP justification for including unrealized holding gains and losses on trading securities in income but reporting unrealized holding gains and losses on securities available for sale in Accumulated Other Comprehensive Income, a separate shareholders' equity account?
5. "The realized gain or loss from the sale of a particular security classified as available for sale will likely differ in amount from the realized gain or loss if the firm had classified that same security as a trading security." Explain.
6. "Reporting marketable securities available for sale at market values on the balance sheet but not including the unrealized holding gains and losses in income is inconsistent and provides opportunity for earnings management." Do you agree? Why or why not?
7. Compare and contrast each of the following pairs of accounts:
   **a.** For securities available for sale: Unrealized Holding Gain (or Loss) on Marketable Equity Securities and Unrealized Holding Gain (or Loss) on Investments in Securities
   **b.** Dividend Revenue, and Equity in Earnings of Unconsolidated Affiliates
   **c.** Equity in Earnings of Unconsolidated Affiliate, and Minority Interest in Earnings of Consolidated Subsidiary
   **d.** Minority Interest in Earnings of Consolidated Subsidiary, and Minority Interest in Net Assets of Consolidated Subsidiary
8. "Dividends received or receivable from another company are either a revenue in calculating net income or a return of investment, depending on the method of accounting the investor uses." Explain.
9. Why is the equity method sometimes described as a *one-line consolidation*? Consider both the balance sheet and the income statement in your response.
10. When is a derivative also a hedge? When is it not also a hedge? Can management manipulate earnings by its choice of whether to use hedge accounting?
11. Distinguish between a fair value hedge and a cash flow hedge.
12. When will a firm not show in income the periodic gain or loss caused by the change in fair value of a derivative?
13. Distinguish between minority investments in other companies and the minority interest in a consolidated subsidiary.
14. "Accounting for an investment in a subsidiary using the equity method and not consolidating it yields the same net income as consolidating the subsidiary. Total assets will differ, however, depending on whether or not the investor consolidates the subsidiary." Explain.
15. Why does GAAP require firms to write off over time any portion of an excess purchase price allocated to buildings and equipment but not goodwill?

## SHORT EXERCISES

**16. Accounting principles for marketable securities.** For each of the following items, describe its accounting using one of the following four systems:

**(1)** Measured at fair value with changes appearing in net income
**(2)** Measured at amortized cost
**(3)** Measured at fair value with changes appearing in other comprehensive income equity (FASB) or directly in shareholders' equity (IASB)
**(4)** Measurement depends on whether firm uses hedge accounting

**a.** A derivative judged to be effective used to hedge forecast sales
**b.** Derivatives appearing as liabilities; these derivatives do not hedge assets or liabilities or anticipated transactions.
**c.** Traded debt issued by others that the firm has purchased with ability to hold to maturity, but intent after the current year is uncertain. The firm frequently buys and sells debt of this sort.
**d.** Marketable equity securities held for an indefinite period as securities available for sale.

**17. Working backward from data on marketable securities transaction.** (Adapted from a problem by S. A. Zeff.) During Year 3, Fischer/Black Co. purchased equity securities classified as securities available for sale. On May 22, Year 4, the company recorded the following correct journal entry to record the sale of the equity securities:

| | | |
|---|---|---|
| Cash | 16,000 | |
| Realized Loss | 5,000 | |
| Unrealized Holding Loss | | 3,000 |
| Marketable Securities | | 18,000 |

| Assets | = | Liabilities | + | Shareholders' Equity | (Class.) |
|---|---|---|---|---|---|
| +16,000 | | | | −5,000 | IncSt → RE |
| −18,000 | | | | +3,000 | AOCInc |

**a.** What was the acquisition cost of these securities in Year 3?
**b.** What was the market value of these securities at the end of Year 3?
**c.** What is the total amount of securities gain or loss that Fischer/Black reports on the income statement for Year 4?

**18. Working backward from data on marketable securities transaction.** (Adapted from a problem by S. A. Zeff.) On December 12, Year 2, Canning had purchased 2,000 shares of Werther. By December 31, the market price of these shares had dropped by $1,000. On March 2, Year 3, Canning sold the 2,000 shares for $18,000 and reported a realized gain on the transaction of $4,000.

**a.** What was the acquisition cost of these securities if Canning had accounted for them as trading securities?
**b.** What was the acquisition cost of these securities if Canning had accounted for them as securities available for sale?

**19. Reconstructing events from journal entries.** Give the likely transaction or event that would result in making each of the independent journal entries that follow:

**a.**

| | | |
|---|---|---|
| Unrealized Loss on Securities Available for Sale | 4,000 | |
| Marketable Securities | | 4,000 |

| Assets | = | Liabilities | + | Shareholders' Equity | (Class.) |
|---|---|---|---|---|---|
| −4,000 | | | | −4,000 | OCInc → AOCInc |

**b.**

| | | |
|---|---|---|
| Cash | 1,100 | |
| Realized Loss on Sale of Securities Available for Sale | 200 | |
| Marketable Securities | | 1,300 |

*(continued)*

| Assets | = | Liabilities | + | Shareholders' Equity | (Class.) |
|---|---|---|---|---|---|
| +1,100 | | | | -200 | IncSt → RE |
| -1,300 | | | | | |

c.

| | | |
|---|---|---|
| Marketable Securities | 750 | |
| Unrealized Holding Gain on Securities Available for Sale | | 750 |

| Assets | = | Liabilities | + | Shareholders' Equity | (Class.) |
|---|---|---|---|---|---|
| +750 | | | | +750 | OCInc → AOCInc |

d.

| | | |
|---|---|---|
| Cash | 1,800 | |
| Marketable Securities | | 1,700 |
| Realized Gain on Sale of Securities Available for Sale | | 100 |

| Assets | = | Liabilities | + | Shareholders' Equity | (Class.) |
|---|---|---|---|---|---|
| +1,800 | | | | +100 | IncSt → RE |
| -1,700 | | | | | |

20. **Equity method entries.** Hanna Company purchased 100 percent of the common stock of Denver Company on January 2 for $550,000. The common stock of Denver Company at this date was $200,000, and the retained earnings balance was $350,000. During the year, net income of Denver Company was $120,000; dividends declared were $30,000. Hanna Company uses the equity method to account for the investment. Give the journal entries that Hanna Company made during the year to account for its investment in Denver Company.

21. **Working backward to consolidation relations.** Laesch Company, as parent, owns shares in Lily Company. Laesch Company has owned them since it formed Lily Company. Lily Company has never declared a dividend. Laesch Company has retained earnings from its own operations independent of intercorporate investments of $100,000. The consolidated balance sheet shows no goodwill and shows retained earnings of $156,000. Consider each of the following questions independently of the others:
    **a.** If the parent owns 80 percent of its consolidated subsidiary, what are the retained earnings of the subsidiary?
    **b.** If the subsidiary has retained earnings of $77,000, what fraction of the subsidiary does the parent own?
    **c.** If the parent had not consolidated the subsidiary but instead had accounted for it using the equity method, how much revenue would the parent company have recognized from the investment?

22. **Working backward from consolidated income statements.** Dealco Corporation published a consolidated income statement for the year, shown in **Exhibit 11.10**. The unconsolidated affiliate retained 25 percent of its earnings of $140 million during the year, having paid out the rest as dividends. The consolidated subsidiary earned $280 million during the year and declared no dividends.
    **a.** What percentage of the unconsolidated affiliate does Dealco Corporation own?
    **b.** What dividends did Dealco Corporation receive from the unconsolidated affiliate during the year?
    **c.** What percentage of the consolidated subsidiary does Dealco Corporation own?

## EXERCISES

23. **Classifying securities.** Firms must classify minority, passive investments in securities along two dimensions:

**EXHIBIT 11.10**

**DEALCO CORPORATION**
**Consolidated Income Statement**
**(Exercise 22)**

| | | |
|---|---|---|
| REVENUES | | |
| Sales | | $1,400,000 |
| Equity in Earnings of Unconsolidated Affiliate | | 56,000 |
| Total Revenues | | $1,456,000 |
| EXPENSES | | |
| Cost of Goods Sold (excluding depreciation) | | $ 910,000 |
| Administrative Expense | | 140,000 |
| Depreciation Expense | | 161,000 |
| Amortization of Goodwill | | 7,000 |
| Income Tax Expenses: | | |
| Currently Payable | $58,800 | |
| Deferred | 14,000 | 72,800 |
| Total Expenses | | $1,290,800 |
| Income of the Consolidated Group | | $ 165,200 |
| Less Minority Interest in Earnings of Consolidated Subsidiary | | (42,000) |
| Net Income to Shareholders | | $ 123,200 |

- Purpose of investment: debt securities held to maturity, trading securities, or securities available for sale.
- Length of expected holding period: current asset (Marketable Securities) or noncurrent asset (Investment in Securities).

Classify each of the securities below along each of these two dimensions.

**a.** A forest products company plans to construct a pulp-processing plant beginning in April of next year. It issues common stock for $200 million on December 10 of this year to help finance construction. The company invests this $200 million in U.S. government debt securities to generate income until it needs the cash for construction.

**b.** An electric utility has bonds payable outstanding for $100 million that mature in five years. The electric utility acquires U.S. government bonds that have a maturity value of $100 million in five years. The firm plans to use the proceeds from the government bonds to repay its own outstanding bonds.

**c.** A commercial bank acquires bonds of the state of New York to generate tax-exempt interest revenue. The bank plans to sell the bonds when it needs cash for loans and other ongoing operating needs.

**d.** A pharmaceutical company acquires common stock of a biogenetic engineering company that conducts research in human growth hormones. The pharmaceutical company hopes the investment will lead to strategic alliances or joint ventures in the future.

**e.** A commercial bank maintains a department that regularly purchases and sells securities on stock exchanges around the world. This department acquires common stock of Toyota on the New York Stock Exchange because it thinks the market price does not fully reflect favorable news about Toyota.

**f.** A U.S. computer company has bonds outstanding that are payable in French francs. The bonds mature in installments during the next five years. The computer company purchases a French winery's bonds, denominated in French francs, that mature in seven years. The computer company will sell a portion of the bonds of the French winery each year to obtain the French francs needed to repay its franc-denominated bonds.

**24. Journal entries to apply the market value method to short-term investments in securities.** Events related to Vermont Company's investment in the common stock of Texas Instruments appear below. Vermont Company closes its books on December 31 of each year.

8/21 Vermont Company purchases 1,000 shares of Texas Instruments' common stock for $45 per share as an investment of temporarily excess cash (classified as Securities Available for Sale).

9/13 The stockbroker for Vermont Company calls to report that the shares of Texas Instruments closed on the preceding day at $49 per share.
9/30 Texas Instruments declares a dividend of $0.50 per share.
10/25 Vermont Company receives a dividend check from Texas Instruments for $500.
12/31 The stockbroker calls Vermont Company to report that Texas Instruments' shares closed the year at $51 per share. Vermont Company closes its books for the year.
1/20 Vermont Company sells 600 shares for $55 per share, the closing price for the day.

Prepare dated journal entries as required by these events. Ignore income taxes.

25. **Journal entries to apply the market value method to short-term investments in securities.** Events related to Elston Corporation's investments of temporarily excess cash appear below. The firm classifies these investments as Securities Available for Sale.

| | | | Market Value on December 31 | | | |
|---|---|---|---|---|---|---|
| Security | Date Acquired | Acquisition Cost | Year 4 | Year 5 | Date Sold | Selling Price |
| A ...... | 10/15/Year 4 | $28,000 | $25,000 | — | 2/10/Year 5 | $24,000 |
| B ...... | 11/2/Year 4 | 49,000 | 55,000 | $53,000 | 7/15/Year 6 | 57,000 |

Elston received no dividends on Security A. It received dividends from Security B of $1,000 on December 31, Year 4, and $1,200 on December 31, Year 5. Prepare dated journal entries for the events related to these investments, assuming that the accounting period is the calendar year.

**a.** Acquisition of securities
**b.** Receipt of dividends
**c.** Revaluation on December 31
**d.** Sale of securities

26. **Journal entries to apply the market value method for short-term investments in securities.** Events related to Simmons Corporation's investments of temporarily excess cash appear below. The firm classifies these investments as Securities Available for Sale.

| | | | Market Value on December 31 | | | |
|---|---|---|---|---|---|---|
| Security | Date Acquired | Acquisition Cost | Year 6 | Year 7 | Date Sold | Selling Price |
| S ...... | 6/13/Year 6 | $12,000 | $13,500 | $15,200 | 2/15/Year 8 | $14,900 |
| T ...... | 6/13/Year 6 | 29,000 | 26,200 | 31,700 | 8/22/Year 8 | 28,500 |
| U ...... | 6/13/Year 6 | 43,000 | — | — | 10/11/Year 6 | 39,000 |

None of these three securities paid dividends. Prepare dated journal entries for the events related to these investments, assuming that the accounting period is the calendar year.

**a.** Acquisition of securities
**b.** Revaluation on December 31
**c.** Sale of securities

27. **Amount of income recognized under various methods of accounting for investments.** On January 1, Apollo Corporation acquired common stock of Venus Corporation. At the time of acquisition, the book value and the fair market value of Venus Corporation's net assets were $500 million. During the year, Venus Corporation earned $80 million and declared dividends of $20 million. The market value of shares increased by 15 percent during the year. How much income would Apollo Corporation report for the year related to its investment under the assumption that it took the following actions?

**a.** Paid $75 million for 15 percent of the common stock and uses the market value method to account for its investment in Venus Corporation as securities available for sale.
**b.** Paid $120 million for 15 percent of the common stock and uses the market value method to account for its investment in Venus Corporation as securities available for sale.
**c.** Paid $150 million for 30 percent of the common stock and uses the equity method to account for its investment in Venus Corporation.

**28. Balance sheet and income effects of alternative methods of accounting for investments.** On January 1, Trusco acquired common stock of USP Company. At the time of acquisition, the book value and the market value of USP's net assets were $400 million. During the current year, USP earned $50 million and declared dividends of $30 million. Indicate the amount shown for Investment in USP on the balance sheet on December 31 and the amount of income Trusco would report for the year related to its investment under the assumption that Trusco did the following:

**a.** Paid $40 million for a 10 percent interest in USP and uses the market value method for securities available for sale. The market value of USP on December 31 was $400 million.

**b.** Same as part **a** except that the market value of USP on December 31 was $390 million.

**c.** Paid $45 million for a 10 percent interest in USP and uses the market value method for securities available for sale. The market value of USP on December 31 was $450 million.

**d.** Paid $120 million for a 30 percent interest in USP and uses the equity method.

**e.** Paid $160 million for a 30 percent interest in USP and uses the equity method. The firm neither amortizes any goodwill nor finds it impaired.

**29. Journal entries to apply the market value method for long-term investments in securities.** The following information summarizes data about the long-term minority, passive investments in securities of Randle Company for its first two years of operations. Randle Company classifies these investments as Securities Available for Sale.

| | | | | | Market Value | |
|---|---|---|---|---|---|---|
| Security | Date Acquired | Acquisition Cost | Date Sold | Selling Price | Dec. 31, Year 1 | Dec. 31, Year 2 |
| M ......... | 4/10/Year 1 | $ 37,000 | 10/15/Year 2 | $43,000 | $ 35,000 | — |
| N ......... | 7/11/Year 1 | 31,000 | | | 38,000 | 45,000 |
| O ......... | 9/29/Year 1 | 94,000 | | | 87,000 | 89,000 |
| | | $162,000 | | | $160,000 | $134,000 |

Randle Company received dividends on December 31 of each year as follows:

| Security | Year 1 | Year 2 |
|---|---|---|
| M ......... | $1,500 | — |
| N ......... | 1,400 | $1,600 |
| O ......... | 5,000 | 4,000 |

Prepare journal entries to account for these investments in securities during Year 1 and Year 2 using the market value method.

**30. Journal entries to apply the market value method for long-term investments in securities.** The following information summarizes data about the long-term minority, passive investments in securities of Blake Company. Blake Company classifies these investments as Securities Available for Sale.

| | | | | | Market Value | |
|---|---|---|---|---|---|---|
| Security | Date Acquired | Acquisition Cost | Date Sold | Selling Price | Dec. 31, Year 4 | Dec. 31, Year 5 |
| F ......... | 7/9/Year 3 | $93,700 | 10/29/Year 4 | $89,700 | — | — |
| G ......... | 7/2/Year 4 | 42,800 | | | $38,300 | $36,900 |
| H ......... | 10/19/Year 4 | 29,600 | 9/17/Year 5 | 32,300 | 31,600 | — |
| I ......... | 2/9/Year 5 | 18,100 | | | — | 20,700 |

The market value of the investment in Security F totaled $91,200 on December 31, Year 3. Assume that none of the investees paid dividends during Year 4 or Year 5. Prepare journal entries to account for these investments in securities during Year 4 and Year 5 using the market value method.

**31. Journal entries to apply the equity method of accounting for investments in securities.** Wood Corporation made three long-term intercorporate investments on January 2. Data relating to these investments for the year appear below.

| Company | Percentage Acquired | Book Value and Market Value of Net Assets on January 2 | Acquisition Cost | Net Income (Loss) for the Year | Dividends Declared during the Year |
|---|---|---|---|---|---|
| Knox Corporation . . . . . . . . . . . . . . . . . . | 50% | $700,000 | $350,000 | $70,000 | $30,000 |
| Vachi Corporation . . . . . . . . . . . . . . . . . | 30 | 520,000 | 196,000 | 40,000 | 15,000 |
| Snow Corporation . . . . . . . . . . . . . . . . . | 20 | 400,000 | 100,000 | (24,000) | — |

Give the journal entries to record the acquisition of these investments and to apply the equity method during the year. The firm neither amortizes any goodwill nor finds it impaired.

**32. Journal entries to apply the equity method of accounting for investments in securities.** The following information summarizes data about the minority, active investments of Stebbins Corporation.

| Security | Date Acquired | Acquisition Cost | Ownership Percentage | Book Value of Net Assets on January 1, Year 1 | Earnings (Loss) | | Dividends | |
|---|---|---|---|---|---|---|---|---|
| | | | | | Year 1 | Year 2 | Year 1 | Year 2 |
| R . . . . . . . . | 1/1/Year 1 | $250,000 | 25% | $800,000 | $200,000 | $225,000 | $125,000 | $130,000 |
| S . . . . . . . . | 1/1/Year 1 | 325,000 | 40 | 750,000 | 120,000 | 75,000 | 80,000 | 80,000 |
| T . . . . . . . | 1/1/Year 1 | 475,000 | 50 | 950,000 | (150,000) | 50,000 | — | — |

Company R owns a building with 10 years of remaining life and with a market value exceeding its book value by $160,000. $40,000 of this amount applies to the shares Stebbins Corporation owns. Stebbins Corporation attributes the rest of any excess of purchase price over book value acquired to goodwill. The building has a 10-year remaining life. The market values of the recorded net assets of Company S and Company T equal their book values. The firm neither amortizes any goodwill nor finds it impaired.

**a.** Give the journal entries to record the acquisition of these investments and to apply the equity method during Year 1 and Year 2.

**b.** Stebbins Corporation sells Security R on January 1, Year 3, for $275,000. Give the journal entry to record the sale.

**33. Journal entries under various methods of accounting for investments.** Mulherin Corporation made three long-term investments on January 2. Data relating to these investments appear below.

| Company | Percentage Acquired | Book Value and Market Value of Net Assets on January 2 | Acquisition Cost | Net Income (Loss) for the Year | Dividends Declared during the Year | Market Value on Shares Held on December 31 |
|---|---|---|---|---|---|---|
| Hanson . . . . . . . . . . . . . . . | 15% | $2,000,000 | $ 320,000 | $200,000 | $ 40,000 | $ 305,000 |
| Maloney . . . . . . . . . . . . . . . . | 30 | 2,000,000 | 680,000 | 500,000 | 180,000 | 700,000 |
| Quinn . . . . . . . . . . . . . . . . | 100 | 2,000,000 | 2,800,000 | 600,000 | 310,000 | 1,950,000 |

Assume that these were the only intercorporate investments of Mulherin Corporation. Maloney Company has a patent developed from its internal research efforts that has a market value exceeding book value of zero on January 2. Mulherin's 30 percent share of that excess is $80,000. The patent has a 10-year remaining life on this date. Any excess purchase price from the acquisition of Quinn Company relates to goodwill. The firm neither amortizes any goodwill nor finds it impaired.

Give the journal entries on Mulherin Corporation's books to record these acquisitions of common stock and to account for the intercorporate investments under GAAP. Mulherin Corporation accounts for its investment in Quinn Corporation using the equity method on its separate-company books.

**34. Consolidation policy and principal consolidation concepts.** CAR Corporation manufactures computers in the United States. It owns 75 percent of the voting stock of Charles Electronics, 80 percent of the voting stock of Alexandre du France Software Systems (in France), and 90 percent of the voting stock of R Credit Corporation (a finance company). CAR Corporation prepares consolidated financial statements consolidating Charles Electronics, uses the equity method for R Credit Corporation, and treats its investment in Alexandre du France Software Systems as securities available for sale. Data from the annual reports of these companies appear below. There are no intercompany transactions.

| | Percentage Owned | Net Income | Dividends | Accounting Method |
|---|---|---|---|---|
| CAR Corporation Consolidated | — | $1,200,000 | $ 84,000 | — |
| Charles Electronics | 75% | 120,000 | 48,000 | Consolidated |
| Alexandre du France Software Systems[a] | 80 | 96,000 | 60,000 | Market Value (Securities Available for Sale) |
| R Credit Corporation | 90 | 144,000 | 120,000 | Equity |

[a]Market value of shares exceeds cost.

**a.** Which, if any, of the companies does CAR incorrectly account for according to GAAP?

Assuming the accounting methods and accounting itself for the three subsidiaries shown above are correct, answer the following questions:

**b.** How much of the net income reported by CAR Corporation Consolidated results from the operations of the three subsidiaries?

**c.** What is the amount of the minority interest now shown on the income statement, and how does it affect net income of CAR Corporation Consolidated?

**d.** If CAR had consolidated all three subsidiaries, what would have been the net income of CAR Corporation Consolidated?

**e.** If CAR had consolidated all three subsidiaries, what minority interest would appear on the income statement?

**35. Equity method entries, earnings quality, and ethics.** Bush Corporation acquired significant influence over Cheney Computer Company on January 2 by purchasing 20 percent of its outstanding stock for $100 million. Bush attributes the entire excess of cost over book value acquired to a patent, which it amortizes over 10 years. The shareholders' equity accounts of Cheney Computer Company appeared as follows on January 2 and December 31 of the current year (amounts in millions):

| | Jan. 2 | Dec. 31 |
|---|---|---|
| Common Stock | $300 | $300 |
| Retained Earnings | 120 | 190 |

Cheney Computer had earnings of $100 million and declared dividends of $30 million during the year. The accounts receivable of Bush Corporation at December 31 included $600,000 due from Cheney Computer. Bush Corporation accounts for its investment in Cheney Computer using the equity method.

**a.** Give the journal entries to record the acquisition of the shares of Cheney Computer and to apply the equity method during the year on the books of Bush Corporation.

**b.** Bush Corporation considers reducing its ownership from 20 percent to 19.5 percent so that it no longer has to use the equity method. Comment on this possibility and its implications for earnings quality and ethics.

**36. Equity method entries.** Vogel Company is a subsidiary of Joyce Company. Joyce Company accounts for its investment in Vogel Company using the equity method on its single-company books. Present journal entries for the following selected transactions. Record the set of entries on the books of Vogel Company separately from the set of entries on the books of Joyce Company.

**(1)** On January 2, Joyce Company acquired on the market, for cash, 100 percent of the common stock of Vogel Company. The outlay was $420,000. The total contributed capital of Vogel Company's stock outstanding was $300,000; the retained earnings balance was $80,000. Joyce attributes the excess of cost over book value acquired to an internally developed patent that has a ten-year remaining useful life on January 2.

**(2)** Vogel Company purchased materials for $29,000 from Joyce Company on account at the latter's cost.

**(3)** Vogel Company obtained an advance of $6,000 from Joyce Company. Vogel Company deposited the funds in the bank.
**(4)** Vogel Company paid $16,000 on the purchases in **(2)**.
**(5)** Vogel Company repaid $4,000 of the loan received from Joyce Company in **(3)**.
**(6)** Vogel Company declared and paid a dividend of $20,000 during the year.

**37. Working backwards from data which has eliminated intercompany transactions.** (Adapted from a problem by S. A. Zeff.) Alpha owns 100 percent of Omega and consolidates Omega in an entity called Alpha/Omega. Beginning in Year 2, Alpha sold merchandise to Omega at a price 50 percent larger than Alpha's costs. Omega sold some, but not all, of these goods to customers at a further markup. Excerpts from the single-company statements of Alpha and Omega and from the consolidated financial statements of Alpha/Omega appear below.

| | Single-Company Statements | | Consolidated Financial Statements |
|---|---|---|---|
| | Alpha | Omega | |
| Sales Revenue | $450,000 | $250,000 | $620,000 |
| Cost of Goods Sold | 300,000 | 210,000 | 430,000 |
| Merchandise Inventory | 60,000 | 50,000 | 100,000 |

**a.** What was the total sales price at which Alpha sold goods to Omega during Year 2?
**b.** What was Omega's cost of the goods it had purchased from Alpha but has not yet sold by the end of Year 2? What was Alpha's cost of those goods? Which of those two numbers appears in the total Merchandise Inventory on the consolidated balance sheet?

**38. Working backward from purchase data.** (Adapted from a problem by S. A. Zeff.) On May 1, Year 1, Homer acquired the assets and agreed to take on and pay off the liabilities of Tonga in exchange for 10,000 of Homer's common shares. Homer accounted for the acquisition of the net assets of Tonga using the purchase method. On the date of acquisition, Tonga's book value of depreciable assets exceeded Homer's estimate of their market value, but Homer judged all other items on Tonga's books to reflect market value on that date, so that the purchase price exceeded the fair value of the identifiable assets, generating goodwill. On the date of the acquisition, Tonga's shareholders' equity was $980,000 and its liabilities totaled $80,000. Tonga reported no goodwill on its balance sheet.

Homer made the following journal entry to record the acquisition:

| | | |
|---|---|---|
| Current Assets | 210,000 | |
| Depreciable Assets (net) | 700,000 | |
| Goodwill | 120,000 | |
| Liabilities | | 80,000 |
| Common Stock—Par | | 150,000 |
| Additional Paid-in Capital | | 800,000 |

| Assets | = | Liabilities | + | Shareholders' Equity | (Class.) |
|---|---|---|---|---|---|
| +210,000 | | +80,000 | | +150,000 | ContriCap |
| +700,000 | | | | +800,000 | ContriCap |
| +120,000 | | | | | |

**a.** What was the book value on Tonga's books of its total assets just before the acquisition?
**b.** What was the book value of depreciable assets of Tonga just before the acquisition?

**39. Financial statement effects of revaluations required by the purchase method.** Bristol-Myers Corporation and Squibb, both pharmaceutical companies, agreed to merge as of January 2, Year 10. Bristol-Myers exchanged 234 million shares of its common stock for the outstanding shares of Squibb. The shares of Bristol-Myers sold for $55 per share on the merger date, resulting in a transaction with a market value of $12.87 billion. Assume that the market value of buildings and equipment of Squibb exceeds their book value by $2,500 million, the market value of patents exceeds their book value by $6,400 million, and that the market values of all other recorded assets and liabilities equal their book values. The buildings and equipment have a ten-year remaining useful life and the patents have a five-year remaining useful life as of January 2, Year 10. Condensed balance sheet data on January 2, Year 10, appear below (amounts in millions).

| | Bristol-Myers | Squibb |
|---|---|---|
| Assets | $5,190 | $3,083 |
| Liabilities | 1,643 | 1,682 |
| Shareholders' Equity | 3,547 | 1,401 |
| | $5,190 | $3,083 |

**a.** Prepare a condensed consolidated balance sheet on the date of the merger assuming that the firms accounted for the merger using the purchase method. Show any amount for goodwill separately.

**b.** Projected net income for Year 10 before considering the effects of the merger are $1,225 for Bristol-Myers and $523 for Squibb. Compute the amount of consolidated net income projected for Year 10 for Bristol-Myers and Squibb using the purchase method.

**40. Effect of transactions involving the market value methods on the statement of cash flows.** Refer to the simplified statement of cash flows in **Exhibit 4.16**. Numbers appear on 11 of the lines in the statement. Ignore the unnumbered lines in considering the transactions below. Assume that the accounting cycle is complete for the period and that the firm has prepared all of its financial statements. It then discovers that it has overlooked a transaction. It records the transaction in the accounts and corrects all of the financial statements. For each of the following transactions or events, indicate which of the numbered lines of the statement of cash flows change and the amounts and directions of the changes (increase or decrease). Ignore income tax effects.

**a.** A firm purchased equity securities costing $59,800 during the period. The firm classifies these as short-term Securities Available for Sale.

**b.** A firm sold for $47,900 equity securities classified as short-term Securities Available for Sale. The securities originally cost $42,200 and had a book value of $44,000 at the time of sale.

**c.** A firm sold for $18,700 equity securities classified as short-term Securities Available for Sale. The securities originally cost $25,100 and had a book value of $19,600 at the time of sale.

**d.** A particular equity security purchased during the period for $220,500 had a market value of $201,500 at the end of the accounting period. The firm classifies the security as a short-term Security Available for Sale. The firm has already recorded the purchase.

**e.** Assume the same information as in part **d** except that the market value of the security at the end of the accounting period is $227,900.

**f.** A firm receives a dividend of $8,000 on shares held as a long-term Security Available for Sale and accounted for using the market value method.

**g.** A firm writes down, from $10,000 to $8,000, Securities Available for Sale accounted for with the market value method.

**41. Effect of transactions involving the equity method on the statement of cash flows.** Refer to the preceding exercise and use those instructions for the following items:

**a.** A 40-percent-owned affiliate accounted for using the equity method earns $25,000 and pays dividends of $10,000.

**b.** A 40-percent-owned affiliate accounted for using the equity method reports a loss for the year of $12,500.

**c.** A firm amortizes $3,000 of the excess of the purchase price over the book value of the underlying net assets in a 40-percent-owned affiliate. The excess related to a patent.

**42. Effect of errors involving securities available for sale on financial statement ratios.** Indicate using O/S (overstated), U/S (understated), or NO (no effect) the pre-tax effect of each of the following errors on (1) the rate of return on assets, and (2) the debt-equity ratio. Each of these ratios is between zero and 100 percent before management discovered the error.

**a.** A firm holding equity securities classified as short-term Securities Available for Sale neglected to write down the securities to market value at the end of the year.

**b.** A firm holding equity securities classified as long-term Securities Available for Sale neglected to write up the securities to market value at the end of the year.
**c.** A firm holding equity securities classified as long-term Securities Available for Sale recorded dividends received by debiting Cash and crediting the Investment account.
**d.** A firm holding equity securities accounted for using the equity method credited a check received for dividends from the investment to Dividend Revenue.
**e.** A firm holding equity securities accounted for using the equity method neglected to amortize excess purchase price related to undervalued depreciable assets.

**43. Effect of errors on financial statements.** Using the notation O/S (overstated), U/S (understated), or NO (no effect), indicate the effects on assets, liabilities, shareholders' equity, and net income of each of the independent errors that follow. Ignore income tax effects.
**a.** In applying the market value method to minority, passive investments in securities, a firm incorrectly credits dividends received to the investment account.
**b.** The market value of minority, passive investments in securities at the end of a firm's first year of operations was $5,000 less than cost. The firm neglected to make the required journal entry in applying the market value method.
**c.** In applying the equity method, P correctly accrues its share of S's net income for the year. However, when receiving a dividend, P credits Dividend Revenue.
**d.** P acquired 30 percent of S on January 1 of the current year for an amount in excess of the book value of S's net assets. The excess relates to patents. P correctly accounted for its share of S's net income and dividends for the year but neglected to amortize any of the excess purchase price.
**e.** During the current year, P sold inventory items to S, its wholly owned subsidiary, at a profit. S sold these inventory items, and S paid P for them before the end of the year. The firms made no elimination entry for this intercompany sale on the consolidation work sheet.
**f.** Refer to part **e**. Assume that S owes P $10,000 for intercompany purchases at year-end. The firm made no elimination entry for this intercompany debt.
**g.** P owns 90 percent of S. P treats the minority interest in consolidated subsidiaries as a liability. In preparing a consolidated work sheet, the firms made no entry to accrue the minority interest's share of S's net income or of S's net assets.

## PROBLEMS AND CASES

**44. Journal entries and financial statement presentation of short-term securities available for sale.** The following information summarizes data about Dostal Corporation's marketable securities held as current assets and as Securities Available for Sale:

| | | | | | Market Value | |
|---|---|---|---|---|---|---|
| Security | Date Acquired | Acquisition Cost | Date Sold | Selling Price | Dec. 31, Year 1 | Dec. 31, Year 2 |
| A | 2/5/Year 1 | $60,000 | 6/5/Year 2 | $72,000 | $66,000 | — |
| B | 8/12/Year 1 | 25,000 | — | — | 20,000 | $23,000 |
| C | 1/22/Year 2 | 82,000 | — | — | — | 79,000 |
| D | 2/25/Year 2 | 42,000 | 6/5/Year 2 | 39,000 | — | — |
| E | 3/25/Year 2 | 75,000 | — | — | — | 80,000 |

**a.** Give all journal entries relating to these marketable equity securities during Year 1 and Year 2, assuming that the accounting period is the calendar year.
**b.** Provide a suitable presentation of marketable securities in the balance sheet and related notes on December 31, Year 1.
**c.** Provide a suitable presentation of marketable securities in the balance sheet and related notes on December 31, Year 2.

**45. Journal entries and financial statement presentation of long-term securities available for sale.** The following information summarizes data about Rice Corporation's investments in equity securities held as noncurrent assets and as Securities Available for Sale:

| | | | | | Market Value | |
|---|---|---|---|---|---|---|
| Security | Date Acquired | Acquisition Cost | Date Sold | Selling Price | Dec. 31, Year 1 | Dec. 31, Year 2 |
| A ......... | 3/5/Year 1 | $40,000 | 10/5/Year 2 | $52,000 | $45,000 | — |
| B ......... | 5/12/Year 1 | 80,000 | — | — | 70,000 | $83,000 |
| C ......... | 3/22/Year 2 | 32,000 | — | — | — | 27,000 |
| D ......... | 5/25/Year 2 | 17,000 | 10/5/Year 2 | 16,000 | — | — |
| E ......... | 5/25/Year 2 | 63,000 | — | — | — | 67,000 |

**a.** Give all journal entries relating to these equity securities during Year 1 and Year 2, assuming that the accounting period is the calendar year.

**b.** Provide a suitable presentation of investments in securities in the balance sheet and related notes on December 31, Year 1.

**c.** Provide a suitable presentation of investments in securities in the balance sheet and related notes on December 31, Year 2.

**46. Reconstructing transactions involving short-term securities available for sale.** During Year 2, Zeff Corporation sold marketable securities for $14,000 that had a book value of $13,000 at the time of sale. The financial statements of Zeff Corporation reveal the following information with respect to Securities Available for Sale:

| | December 31 | |
|---|---|---|
| | Year 1 | Year 2 |
| **Balance Sheet** | | |
| Marketable Securities at Market Value .......................... | $187,000 | $195,000 |
| Net Unrealized Holding Gain on Securities Available for Sale .......... | 12,000 | 10,000 |
| **Income Statement** | Year 2 | |
| Realized Gain on Sale of Securities Available for Sale ............... | $4,000 | |

**a.** What was the cost of the marketable securities sold?

**b.** What was the unrealized holding gain on the securities sold at the time of sale?

**c.** What was the unrealized holding gain during Year 2 on securities still held by the end of Year 2?

**d.** What was the cost of marketable securities purchased during Year 2?

**47. Analysis of financial statement disclosures for securities available for sale. Exhibit 11.11** reproduces data about the marketable equity securities held as securities available for sale for Sunshine Mining Company for a recent year, with dates changed for convenience. Assume that Sunshine held no current marketable securities at the end of Year 1, sold no current marketable securities during Year 2, purchased no noncurrent marketable securities during Year 2, and transferred no noncurrent marketable securities to the current portfolio during Year 2. The income statement for Year 2 shows a realized loss on sale of noncurrent marketable securities of $3,068,000.

**a.** What amount of net unrealized holding gain or loss on noncurrent marketable securities appears on the balance sheet for the end of Year 1?

**b.** What amount of net unrealized gain or loss on noncurrent securities appears on the balance sheet for the end of Year 2?

**c.** What were the proceeds of the sale of noncurrent marketable securities sold during Year 2?

**d.** What amount of unrealized holding gain or loss on marketable securities appears on the income statement for Year 2?

**48. Effect of various methods of accounting for marketable equity securities.** Information related to marketable equity securities of Callahan Corporation appears below.

**EXHIBIT 11.11**

**SUNSHINE MINING COMPANY**
**Data on Marketable Equity Securities**
**(all dollar amounts in thousands)**
**(Problem 47)**

| Marketable Equity Securities | Acquisition Cost | Market Value |
|---|---|---|
| **At December 31, Year 2:** | | |
| Current Marketable Securities . . . . . . . . . . . . . . . . | $ 7,067 | $ 4,601 |
| Noncurrent Marketable Securities . . . . . . . . . . . . . . | $ 6,158 | $ 8,807 |
| **At December 31, Year 1:** | | |
| Noncurrent Marketable Securities . . . . . . . . . . . . . . | $21,685 | $11,418 |

| Security | Acquisition Cost in Year 1 | Dividends Received during Year 1 | Market Value on Dec. 31, Year 1 | Selling Price in Year 2 | Dividends Received during Year 2 | Market Value on Dec. 31, Year 2 |
|---|---|---|---|---|---|---|
| G . . . . . . . . . . . . . . | $18,000 | $ 800 | $16,000 | $14,500 | $ 200 | — |
| H . . . . . . . . . . . . . . | 25,000 | 1,500 | 24,000 | 26,000 | 500 | — |
| I . . . . . . . . . . . . . . | 12,000 | 1,000 | 14,000 | — | 1,500 | $17,000 |
| | $55,000 | $3,300 | $54,000 | $40,500 | $2,200 | $17,000 |

**a.** Assume that these securities represent trading securities. Indicate the nature and amount of income recognized during Year 1 and Year 2 and the presentation of information about these securities on the balance sheet on December 31, Year 1 and Year 2.

**b.** Repeat part **a** assuming that these securities are securities available for sale held as temporary investments of excess cash by Callahan Corporation.

**c.** Repeat part **a** assuming that these securities represent long-term investments by Callahan Corporation held as securities available for sale.

**d.** Compute the combined income for Year 1 and Year 2 under each of the three treatments of these securities in parts **a**, **b**, and **c**. Why do the combined income amounts differ? Will total shareholders' equity differ? Why or why not?

**49. Analysis of financial statement disclosures related to marketable securities and quality of earnings.** Citibank reports the following information relating to its marketable securities classified as Securities Available for Sale for a recent year (amounts in millions):

| | December 31 | |
|---|---|---|
| | Year 10 | Year 11 |
| Marketable Securities at Acquisition Cost . . . . . . . . . . . . . . . . . . . . . . | $14,075 | $13,968 |
| Gross Unrealized Holding Gains . . . . . . . . . . . . . . . . . . . . . . . . . . . | 957 | 1,445 |
| Gross Unrealized Holding Losses . . . . . . . . . . . . . . . . . . . . . . . . . . | (510) | (218) |
| Marketable Securities at Market Value . . . . . . . . . . . . . . . . . . . . . . . | $14,522 | $15,195 |

Cash proceeds from sales and maturities of marketable securities totaled $37,600 million in Year 11. Gross realized gains totaled $443 million and gross realized losses totaled $113 million during Year 11. The book value of marketable securities sold or matured totaled $37,008 million. Interest and dividend revenue during Year 11 totaled $1,081 million. Purchases of marketable securities totaled $37,163 million during Year 11.

**a.** Give the journal entries to record the sale of marketable securities during Year 11.

**b.** Analyze the change in the net unrealized holding gain from $447 million on December 31, Year 10, to $1,227 million on December 31, Year 11.

**c.** Compute the total income (both realized and unrealized) *occurring during Year 11* on Citibank's investments in securities.

**d.** How might the judicious selection of marketable securities sold during Year 11 permit Citibank to report an even larger net realized gain? Comment on the quality of earnings and ethics issues this option raises.

**50. Journal entries for various methods of accounting for intercorporate investments.** Rockwell Corporation acquired, as long-term investments, shares of common stock of Company R, Company S, and Company T on January 2. These are the only long-term investments in securities that Rockwell Corporation holds. Data relating to the acquisitions follow:

| Company | Percentage Acquired | Book Value and Market Value of Total Net Assets on January 2 | Acquisition Cost | Net Income for Year | Dividends Declared for Year | Market Value of Shares Owned on December 31 |
|---|---|---|---|---|---|---|
| R | 10% | $6,000,000 | $ 648,000 | $1,200,000 | $480,000 | $ 624,000 |
| S | 30 | 6,000,000 | 2,040,000 | 1,200,000 | 480,000 | 2,052,000 |
| T | 100 | 6,000,000 | 6,000,000 | 1,200,000 | 480,000 | 6,300,000 |

**a.** Give the journal entries made to acquire the shares of Company R and to account for the investment during the year, using the market value method.

**b.** Give the journal entries made to acquire the shares of Company S and to account for the investment during the year, using the equity method. Any excess cost relates to goodwill. The firm neither amortizes any goodwill nor finds it impaired.

**c.** Give the journal entries made to acquire the shares of Company T and to account for the investment during the year, using the equity method.

**51. Effect of intercorporate investment policies on financial statements.** Coca-Cola Company (Coke) and PepsiCo (Pepsi) dominate the soft drink industry in the United States and maintain leading market positions in many other countries. Coke has followed a policy of holding a 49 percent ownership interest in its bottlers, whereas Pepsi wholly owns its bottling operations. **Exhibit 11.12** presents selected balance sheet data for Coke and Pepsi for Year 11.

**EXHIBIT 11.12**

**COKE and PEPSI**
**Financial Statement Data for Year 11**
**(all dollar amounts in millions)**
**(Problem 51)**

| | Coke as Reported (1) | Coke's Bottlers (2) | Coke as Consolidated with Bottlers (3) | Pepsi as Reported (4) |
|---|---|---|---|---|
| ASSETS | | | | |
| Current Assets | $4,143 | $ 2,153 | $ 6,296 | $ 4,081 |
| Investment in Bottlers | 2,025 | — | — | — |
| Property, Plant, and Equipment | 3,112 | 8,957 | 12,069 | 5,711 |
| Goodwill | — | — | 310 | 5,845 |
| Total Assets | $9,280 | $11,110 | $18,675 | $15,637 |
| LIABILITIES AND SHAREHOLDERS' EQUITY | | | | |
| Current Liability | $4,296 | $ 2,752 | $ 7,048 | $ 3,264 |
| Long-Term Liabilities | 1,133 | 4,858 | 5,991 | 7,469 |
| External Interest in Bottlers | — | — | 1,785 | — |
| Shareholders' Equity | 3,851 | 3,500 | 3,851 | 4,904 |
| Total Liabilities and Shareholders' Equity | $9,280 | $11,110 | $18,675 | $15,637 |
| Net Income for Year 11 | $1,364 | $ 290 | $ 1,364 | $ 1,091 |
| Interest Expense for Year 11 | $ 231 | $ 452 | $ 683 | $ 689 |

The first column shows amounts for Coke as reported, with Coke using the equity method to account for investments in its bottlers. The second column shows amounts for Coke's bottlers as reflected in a note to Coke's financial statements. The third column shows consolidated amounts for Coke and its bottlers. The 51 percent external interest in these bottlers appears on a line between liabilities and shareholders' equity. The fourth column shows amounts for Pepsi as reported, with its bottlers consolidated.

**a.** Compute the rate of return on assets for Year 11 under each of the four treatments in **Exhibit 11.12**. Assume an income tax rate of 34 percent. Also assume that the amounts shown for total assets in **Exhibit 11.12** approximate the average assets for Year 11.

**b.** Compute the ratio of total liabilities to total assets at the end of Year 11 under each of the four treatments in **Exhibit 11.12**.

**c.** Evaluate the operating profitability and risk of Coke versus Pepsi using the ratios computed in questions **a** and **b**.

**d.** Suggest reasons why Coke might choose to own 49 percent of its bottlers whereas Pepsi holds 100 percent of its bottlers.

**52. Interaction of regulation and accounting rules for financial institutions, particularly banks. Exhibit 11.13** illustrates the key features of a banking or thrift institution. Because regulators and investors view the banking business—borrowing funds and lending them to others at higher rates—as less risky than most other businesses, they allow banks and thrifts to carry smaller shareholders' equity as a percentage of total equities than do most nonfinancial firms. That is, banks have, with the blessing of regulators, higher financial leverage than nonfinancial firms.

**Exhibit 11.13** illustrates a thrift institution that regulators allow to have shareholders' equity equal to only 8 percent of assets. (The effects illustrated here would be even more dramatic with a capital ratio of 5 percent or less, which many banks use.) Other realistic assumptions, used for illustration, are as follows:

- Liabilities can be 11.5 times as large as owners' equity; this follows from the assumption that owners' equity is 8 percent of total equities, so that liabilities can be 92 percent of total equities.
- Liabilities cost 5.5 percent (interest expense) each year.
- Assets earn 7.0 percent (interest revenues) each year.
- Variable operating expenses are 0.4 percent of total assets.
- Fixed operating expenses are 0.1 percent of the bank's starting asset position of $1,000.
- There are no income taxes.

Key features of a bank are that (1) they tend to borrow funds for short terms, such as through accepting cash into checking and savings accounts or by issuing short-term certificates of deposits, while (2) they lend funds for longer terms, such as for home mortgages or customers' auto purchases.

The top panel of **Exhibit 11.13** shows the balance sheet for the bank as it starts in business. The firm starts with contributed capital of $80; it borrows $920, giving it $1,000 of equities and funds available for lending. It lends $1,000, earning 7 percent on those assets. It pays interest on its $920 of borrowings, pays its operating expenses, and reports income of $14.4.

Now, let the market value of the bank's financial assets—loans—increase by 5 percent, or $50, as shown in the second panel. Show this increase on the balance sheet. The transactions are as follows:

**(1)** Assets increase in value by $50, so Retained Earnings increase by $50.

**(2)** The bank uses the extra owners' equity of $50 to borrow an additional $575 (= 11.5 × $50), which it lends to new customers, increasing total assets by $575.

The income statement shows more revenues and more expenses, but rates of return change little from those of the original bank. Note, however, the effect on the bank's size: a 5 percent increase in market value allows the bank to increase its assets and size by 62.5 percent, from $1,000 in assets to $1,625 in assets. This creates no problem for the bank. Bank managers are happy because their compensation often depends on bank size, not just bank profitability.

Now, go back to the original bank and let the market value of the bank's financial assets decrease by 5 percent, or $50, as shown in the third panel. Observe this decrease on the balance sheet. The transactions are as follows:

**(3)** Assets decrease in value by $50, so Retained Earnings decrease by $50.

**EXHIBIT 11.13**

### Effects of Changing Market Value of Assets on a Bank's Activities (Problem 52)
### Bank has Capital Ratio of 8 percent

Step [1]: Market Value of Assets Increases, Also Increasing Owners' Equity.
Step [2]: Bank Increases Lending to Maintain Capital (Leverage) Ratio at 8 Percent.
Step [3]: Market Value of Original Bank Decreases, Decreasing Owners' Equity.
Step [4]: Bank Decreases Lending to Maintain Capital (Leverage) Ratio at 8 Percent.
Operating Income Excludes Gains and Losses in Market Value of Assets Held.

*Original Bank, before Market Value Changes*

| Balance Sheet: Assets | Balance Sheet: Equities | | Partial Income Statement | | | Rate of Return On: | |
|---|---|---|---|---|---|---|---|
| | Liabilities: | | Revenues as % of Assets | | | | |
| $1,000 Original | Borrowings | $920 | | 7.0% | $70.0 | | |
| | | | Interest Expense as % of Borrowing | 5.5% | (50.6) | | |
| | Owners' Equity | | Operating Expense % of Assets | | | | |
| | Contributed Capital | 80 | | | | | |
| | Retained Earnings | 0 | | 0.4% | (4.0) | Assets[a] | 1.4% |
| | Total Owners' Equity | $80 | Fixed costs | | (1.0) | | |
| $1,000 | Totals | $1,000 | Operating Income | | $14.4 | Owners' Equity | 18.0% |

*Market Value of Assets Increases* 5.0%

| Balance Sheet: Assets | Balance Sheet: Equities | | Partial Income Statement | | | Rate of Return On: | |
|---|---|---|---|---|---|---|---|
| | Liabilities: | | Revenues as % of Assets | | | | |
| $1,000 Original | Original Borrowings | $920 | | 7.0% | $113.8 | | |
| | | | Interest Expense as % of Borrowing | 5.5% | (82.2) | | |
| [1] 50 Market Value Increase | | | Operating Expense % of Assets | | | | |
| [2] 575 New Lending | [2] New Borrowings | 575 | | 0.4% | (6.5) | | |
| | Owners' Equity | | Fixed costs | | (1.0) | | |
| | Contributed Capital | 80 | | | | | |
| | [1] Retained Earnings | 50 | | | | | |
| | Total Owners' Equity | $130 | | | | Assets[a] | 1.5% |
| $1,625 | Totals | $1,625 | Operating Income | | $24.1 | Owners' Equity | 18.5% |

*Market Value of Assets Decreases* −5.0%

| Balance Sheet: Assets | Balance Sheet: Equities | | Partial Income Statement | | | Rate of Return On: | |
|---|---|---|---|---|---|---|---|
| | Liabilities: | | Revenues as % of Assets | | | | |
| $1,000 Original | Original Borrowings | $920 | | 7.0% | $26.3 | | |
| | | | Interest Expense as % of Borrowing | | | | |
| [3] (50) Market Value Decline | | | | 5.5% | (19.0) | | |
| [4] (575) Reduce Lending | [4] Reduce Borrowings | (575) | Operating Expense % of Assets | | | | |
| | Owners' Equity | | | 0.4% | (1.5) | | |
| | Contributed Capital | 80 | Fixed costs | | (1.0) | | |
| | [3] Retained Earnings | (50) | | | | | |
| | Total Owners' Equity | $30 | | | | Assets[a] | 1.3% |
| $375 | Totals | $375 | Operating Income | | $4.8 | Owners' Equity | 16.0% |

[a]Banks do not add back interest expense when computing the rate of return on assets.

**(4)** The bank's decrease in owners' equity requires it to reduce borrowings by $575 (= 11.5 × $50). In order to pay back its creditors and to reduce assets, it must reduce its lendings by $575.

The income statement shows smaller revenues and smaller expenses, but rates of return change little from those of the original bank. Note, however, the effect on the bank's size: a 5 percent decrease in market value has forced the bank to reduce its size by 62.5 percent, from $1,000 in assets to $375 in assets. This creates a major problem for the bank. The problem arises because the bank must shrink in size, by a lot and quickly. The bank's assets comprise loans it has made to customers, typically long-term loans. The bank cannot easily recall those loans without incurring significant cost. (The typical loan gives the bank's customer the right to prepay the loan whenever the customer wants but does not allow the bank to call in the loan at the bank's whim.) Perhaps the bank will have to sell its assets to others, at a loss, to raise cash to reduce its own borrowings and its size.

Whatever it does, the bank has a problem. This problem arises because the bank shows the losses on its balance sheet and must reduce owners' equity because of this loss. Bankers understand this feature of market value accounting for assets, and they have, historically, been the most vocal opponents of such accounting for financial assets. For many years, bankers have prevailed in keeping the changes in market values of their assets off the balance sheet.

Many banks' assets are not marketable securities but are loans, which do not trade in the marketplace. The FASB does not require the marking to market value of nonmarketable financial assets. For such assets, banks use the amortized cost method, illustrated in **Problem 11.1** for Self-Study. For marketable debt securities, the bank will use market value accounting for trading securities or the amortized cost method (debt held to maturity), depending on its holding intention and ability. So long as the bank can show that it has both the ability and the intent to hold a loan until it matures, the bank need not show changes in that loan's market value on the balance sheet.

Repeat the analysis in **Exhibit 11.13**, using a capital ratio of 5 percent instead of 8 percent, and set the changes in market value of assets first to an increase of 4 percent and then to a decrease of 4 percent.

CHAPTER 12

# Shareholders' Equity: Capital Contributions, Distributions, and Earnings

## Learning Objectives

1. Understand the different priority claims of common and preferred shareholders on the assets of a firm and the disclosure of those claims in the shareholders' equity section of the balance sheet.
2. Understand the concepts underlying, and apply the accounting procedures for, the issuance of capital stock, particularly with respect to capital stock issued under various option arrangements.
3. Understand the concepts underlying, and apply the accounting procedures for, cash, property, and stock dividends.
4. Understand the concepts underlying, and apply the accounting procedures for, the acquisition and reissue of treasury stock.
5. Understand why the format for reporting income matters and master the concept that different kinds of income require different formats.
6. Understand the distinction between *earnings* and *comprehensive income.*
7. Develop the skills to interpret disclosures about changes in shareholders' equity accounts.

**Chapters 6** to **11** discussed generally accepted accounting principles for various assets and liabilities. Changes in assets and liabilities often cause shareholders' equity to change. Changes in shareholders' equity result from three types of transactions:

1. **Capital Contributions:** Firms issue common or preferred stock to obtain funds to finance various operating and investing activities.
2. **Distributions:** Firms distribute assets to shareholders either in the form of a dividend or the repurchase of common or preferred stock.
3. **Earnings Transactions:** Firms use assets financed by creditors and owners to generate earnings.

**Exhibit 12.1** presents the shareholders' equity section of the balance sheet of The Coca-Cola Company (Coke). Previous chapters discussed the items marked in boldface. Let's review the important concepts:

1. Shareholders' equity is a residual interest. It represents the shareholders' claim on the assets of a firm insofar as the assets exceed the amounts claimed by the higher-priority, often called

**EXHIBIT 12.1**

**THE COCA-COLA COMPANY**
**Disclosure of Shareholders' Equity**
**(amounts in millions)**

| | December 31 | |
|---|---|---|
| | **Year 11** | **Year 12** |
| Preferred Stock | none | none |
| **Common Stock: $0.25-par Value, 5,600 Shares Authorized, 3,491.8 Shares Issued** | $ 873 | $ 873 |
| **Additional Paid-in Capital** | 3,520 | 3,857 |
| **Retained Earnings** | 23,443 | 24,506 |
| Accumulated Other Comprehensive Income (Loss) | (2,788) | (3,047) |
| Total | $ 25,048 | $ 26,189 |
| Less Cost of Treasury Stock | (13,682) | (14,389) |
| Total Shareholders' Equity | $ 11,366 | $ 11,800 |

*senior*, creditors. Coke reports total assets of $24,501 million, liabilities of $12,701 million, and shareholders' equity of $11,800 million at the end of Year 12. Thus, lenders finance approximately half of Coke's assets and shareholders finance the other half.

2. All firms issue common stock. Firms may also issue preferred stock, which has a claim on the assets of a firm senior to that of the common shareholders. Coke does not have preferred stock outstanding in Year 11 and Year 12.
3. Common and preferred stock usually have a par or stated value. Coke's common stock has a $0.25 par value per share. Firms report amounts received from issuing common and preferred stock in excess of the par or stated value as Additional Paid-in Capital. Note that the amounts in Additional Paid-in Capital for Coke exceed the amounts in Common Stock, indicating that Coke issued shares for substantially more than $0.25 par value, a common practice among publicly traded firms.
4. Firms accumulate information about revenues and expenses during a reporting period to enable the preparation of the income statement. The amount of net income for a period increases net assets and retained earnings. Coke reported net income for Year 12 of $3,050 million.
5. Firms may periodically distribute net assets generated by earnings to shareholders as a dividend. Firms reduce net assets and retained earnings for the distribution. Coke declared dividends of $1,987 million during Year 12, representing a distribution equal to approximately two-thirds of net income for the year. Retained earnings increased $1,063 million duing Year 12 (= $3,050 − $1,987 = $24,506 − $23,443).
6. Retained earnings on the balance sheet provides a measure of the cumulative net assets generated by earnings in excess of dividends declared. Note that retained earnings for Coke substantially exceeds the capital provided by issuing common stock. The retention of net assets from operations represents a substantial source of funds for most businesses.

This chapter explores these concepts in greater depth and expands the discussion of shareholders' equity by considering:

1. The reasons for issuing preferred stock and the different rights which firms might grant to various preferred stockholders.
2. The use of various options arrangements as a means of enhancing the issuance of additional common stock.
3. The reasons for repurchasing shares of common or preferred stock and the accounting for such share repurchases. **Exhibit 12.1** indicates that Coke had repurchased $13,682 million of its own stock as of December 31, Year 11 and $14,389 million as of December 31, Year 12. These repurchases substantially reduced total shareholders' equity.
4. The reporting of earnings transactions, particularly the distinction between net income and other comprehensive income. **Exhibit 12.1** indicates that Coke reports a reduction in shareholders' equity for negative amounts of Accumulated Other Comprehensive Income (that is, losses) at the end of each year. What is the purpose of classifying certain earnings transac-

tions in Other Comprehensive Income instead of net income? What types of transactions or events give rise to Other Comprehensive Income?

We begin with capital contributions. We then move to distributions and finish the chapter with the reporting of earnings transactions.

## *Terminology* Note

Of all the words used in business, accounting, and finance, the word *capital* is one of the most ambiguous. It can mean *cash* ("The firm raised capital with a stock issue"), or *long-term assets* ("The firm's capital assets have depreciable lives between 7 and 10 years"), or all sources of funding, that is, all items on the right side of the balance sheet ("The firm's weighted average cost of capital is 11 percent"), or shareholders' equity or, as in this chapter, that part of shareholders' equity arising from owner's contributions of cash and other assets. We recommend that you use this word to mean only contributed capital, but the rest of the world will continue to use the word to mean whatever is convenient for the speaker at the time. You should be attuned to the various meanings and understand what a particular user means at a particular time by the word *capital*. See the discussion in the Glossary for *capital*.

## Capital Contributions

Most large, publicly traded firms operate as **corporations**. The corporate form has at least three advantages:

1. The corporate form provides the owner (shareholder) with limited liability; that is, should the corporation become insolvent, creditors can claim the assets of the corporate entity only. The corporation's creditors cannot claim the assets of the individual owners. On the other hand, to settle debts of partnerships and sole proprietorships, creditors have a claim on the owners' business and personal assets.
2. The corporate form allows the firm to raise funds by issuing shares. The general public can acquire the shares in varying amounts. Individual investments can range from a few dollars to billions of dollars.
3. The corporate form makes transfer of ownership interests relatively easy because owners can sell their shares to others without interfering with the ongoing operations of the firm. The transfer is a transaction between shareholders and does not directly involve the firm whose shares are exchanged.

The corporation has legal status separate from its owners. Individuals or other entities make capital contributions under a contract between themselves and the corporation. Because those who contribute funds receive certificates for shares of stock, business usage calls them *stockholders* or *shareholders*.

Various laws and contracts govern the rights and obligations of a shareholder:

1. The corporation laws of the state in which incorporation takes place.
2. The articles of incorporation or the **corporate charter.** (This contract sets out the agreement between the firm and the state in which the business incorporates. The state grants to the firm the privileges of operating as a corporation for certain stated purposes and of obtaining its capital through the issue of shares of stock.)
3. The **corporate bylaws.** (The board of directors adopts bylaws, which are the rules and regulations governing the internal affairs of the corporation.)
4. The **capital stock contract.** (Each type of **capital stock** has its own provisions on matters such as voting, sharing in earnings, distributing assets generated by earnings, and sharing in assets in case of dissolution.)

## Preferred Shareholders' Equity

Owners of **preferred stock** generally have a claim on the assets of a firm in the event of bankruptcy that is senior to the claim of common shareholders. Preferred shares also carry special rights. The senior status and special rights may induce certain investors to purchase preferred stock of a firm, even though they would be unwilling to purchase common stock of the same firm. The senior status and special rights reduce the risks of preferred shareholders relative to common shareholders. Preferred shareholders should therefore expect a lower return than common shareholders.

Although the rights of preferred shares vary from issue to issue, a preferred share usually entitles its holder to dividends at a certain rate, which the firm must pay before it can pay dividends to common shareholders. Firms may sometimes postpone or omit preferred dividends. Most preferred shares, however, have **cumulative dividend rights**, which means that a firm must pay all current and previously postponed preferred dividends before it can pay any dividends on common shares.

In recent years, corporations have made many of their preferred stock callable. The corporation can reacquire **callable preferred shares** at a specified price, which may vary according to a preset time schedule. If financing becomes available at a cost lower than the rate fixed for the preferred shares, the issuing firm may want to reduce its financing costs by issuing some new securities and then exercising its option to reacquire the callable shares at a fixed price. Some naïve critics think callability benefits only the firm issuing callable shares. The callability option provides a valuable alternative to the issuing firm but makes the shares less attractive to potential owners of the shares. Other things being equal, a firm will issue callable shares at a lower price than that for noncallable shares. The purchaser of the callable shares benefits from acquiring the callable shares at the lower price.

Firms sometimes issue preferred shares with a conversion feature. **Convertible preferred shares** give their holder the option to convert them into a specified number of common shares at a specified time. The conversion option may appear advantageous to both the individual shareholder and the corporation. The preferred shareholders enjoy the security of a relatively assured dividend as long as they hold the shares and senior status to common shareholders in the event of bankruptcy. The shareholders also have the option to realize capital appreciation by converting the shares into common stock if the market price of the common shares rises sufficiently. Because of this option, changes in the market price of the convertible preferred shares will often parallel changes in the market price of the common shares. The firm may also benefit from the conversion option. By including it in the issue, the firm is usually able to specify a lower dividend rate on the preferred stock than otherwise would have been required to issue the shares for a given price.

**Example 1** Global Crossing Ltd. is a telecommunications firm in bankruptcy. It reports several issues of convertible preferred stock outstanding at the end of Year 12. One issue carries a 6¾ percent cumulative dividend rate and is convertible into 6.31 shares of Global Crossing common stock. The preferred shares have senior rights as to dividends and in liquidation to all classes of common stock but are junior to all indebtedness of the firm.

Some preferred share issues carry mandatory redemption features; that is, the issuing firm must repurchase the shares from their holders, paying a specified dollar amount at a specified future time. Because such **redeemable preferred shares** have some of the characteristics of debt, firms must disclose them as liabilities on the balance sheet.[1]

**Example 2** Global Crossing Ltd. also reports mandatory redeemable preferred stock outstanding. This preferred stock carries a 10⅛ percent cumulative dividend based on a $100 liquidation value. The preferred stock outstanding at the end of Year 12 is subject to mandatory redemption on December 1, Year 18 at the $100 liquidation value plus any dividends in arrears. This preferred stock is senior to all other classes of preferred and common stock but junior to all indebtedness of the firm in the case of dividends and in liquidation.

## Common Shareholders' Equity

Firms need not issue preferred stock. All firms, however, must issue **common stock**. Common shareholders have a claim on the assets of a firm after creditors and preferred shareholders have received amounts promised to them. Frequently, corporations grant voting rights only to com-

[1]Financial Accounting Standards Board, *Statement of Financial Accounting Standards No. 150*, "Accounting for Certain Financial Instruments with Characteristics of both Liabilities and Equity," 2003.

mon shares, giving their holders the right to elect members of the board of directors and decide certain broad corporate policies (spelled out in the stock contract).

## Issuing Capital Stock

Firms may issue capital stock (preferred or common) for cash or for noncash assets or under various option arrangements.

### Issue for Cash

Firms usually issue shares for cash at the time of their initial incorporation and at periodic intervals as they need additional shareholder funds. Firms often issue shares to employees instead of paying them directly in cash for a portion of their compensation. The issue price for preferred stock usually approximates its par value. Firms generally issue common shares for amounts greater than **par (or stated) value**. Typically, individuals who purchase newly issued shares from a seasoned corporation pay a price larger than that paid by the original owners. The larger price compensates current shareholders for the additional assets accumulated through the retention of earnings, as well as for increases in the market values of assets over their book values. The firm credits the excess of issue proceeds over par (or stated) value to the **Additional Paid-in Capital account**.

If a firm issues par-value shares, the credit to the Additional Paid-in Capital account always equals the difference between the amount received and the par value of the shares issued. Thus the entry to record the issue of 1,000 common shares with a par value of $10 per share for $100,000 is as follows:

| | Debit | Credit |
|---|---|---|
| Cash | 100,000 | |
| Common Stock—$10 Par Value | | 10,000 |
| Additional Paid-in Capital | | 90,000 |

| Assets | = | Liabilities | + | Shareholders' Equity | (Class.) |
|---|---|---|---|---|---|
| +100,000 | | | | +10,000 | ContriCap |
| | | | | +90,000 | ContriCap |

To record the issue of common shares for cash in an amount greater than par value.

### Issue for Noncash Assets

Firms occasionally issue common stock for assets other than cash. For example, a firm might want to acquire the assets of another firm in an effort to expand or diversify its operations. The firm records the shares issued at an amount equal to the market value of the assets received or, if the firm cannot make a reasonable estimate, at the market value of the shares issued.[2]

Assume that a firm issues 1,000 shares of its common stock with a par value of $10 per share in the acquisition of another firm's assets having the following market values: accounts receivable, $6,000; inventories, $12,000; land, $10,000; building, $62,000; and equipment, $10,000. The journal entry to record the exchange is as follows:

| | Debit | Credit |
|---|---|---|
| Accounts Receivable | 6,000 | |
| Inventories | 12,000 | |
| Land | 10,000 | |
| Building | 62,000 | |
| Equipment | 10,000 | |
| Common Stock—$10 Par Value | | 10,000 |
| Additional Paid-in Capital | | 90,000 |

*(continued)*

[2]Accounting Principles Board, *Opinion No. 29*, "Accounting for Nonmonetary Transactions," 1973.

| Assets | = | Liabilities | + | Shareholders' Equity | (Class.) |
|---|---|---|---|---|---|
| +6,000 | | | | +10,000 | ContriCap |
| +12,000 | | | | +90,000 | ContriCap |
| +10,000 | | | | | |
| +62,000 | | | | | |
| +10,000 | | | | | |

To record the issue of common shares for noncash assets in amount greater than par value. The amount for Additional Paid-in Capital is a plug.

If the firm issues 100 shares with $10 par value to employees for $10,000 of compensation, instead of paying cash, the entry would be as follows:

| | | |
|---|---|---|
| Compensation Expense .................................... | 10,000 | |
| Common Stock—$10 Par Value .......................... | | 1,000 |
| Additional Paid-in Capital ............................ | | 9,000 |

| Assets | = | Liabilities | + | Shareholders' Equity | (Class.) |
|---|---|---|---|---|---|
| | | | | −10,000 | IncSt → RE |
| | | | | +1,000 | ContriCap |
| | | | | +9,000 | ContriCap |

To record the issue of common shares instead of cash for wages of $10,000 paid to employees.

## ISSUE UNDER OPTION ARRANGEMENTS

Corporations often give various individuals or entities the right, or option, to acquire shares of common stock at a price that is less than the market price of the shares at the time they exercise the option.

**Example 3** Ford, IBM, and other corporations with publicly traded common shares give **stock options** to their top managers each period as an element of their compensation. The stock options permit the employees to purchase shares of common stock at a price usually set equal to the market price of the stock at the time the firm grants the stock option. Employees exercise these stock options at a later time, after the stock price has increased. Firms adopt stock option plans to motivate employees to take actions that will increase the market value of a firm's common shares. Some firms issue employee stock options as a way to conserve cash. Likewise, firms issue **stock rights** to current shareholders, which give the shareholders the right to purchase shares of common or preferred stock at a specified price.

**Example 4** Firms sometimes issue bonds with **stock warrants** attached. The bond contract gives the holder the right to receive periodic interest payments and the principal amount at maturity. The stock warrant permits the holder to exchange the warrant and a specified amount of cash for shares of the firm's common stock. Attaching a stock warrant permits the firm to issue bonds at a lower interest cost than would be required for bonds without such a warrant attached.

**Example 5** Firms sometimes issue bonds or shares of preferred stock (see **Example 1** for Global Crossing, Ltd.) that are convertible into shares of common stock. The holders receive periodic interest or preferred dividend payments before conversion and usually maintain a senior claim on the firm's assets relative to the common shareholders. The conversion into common stock usually does not require any additional cash investment at the time of conversion. Including the conversion option in a debt or preferred stock issue permits the firm to pay a lower interest or preferred dividend rate than it would have to pay without the conversion option.

Each of these options has economic value. Accounting historically has not recognized the value of the options at the time firms grant them. Some accountants have argued that measuring the value of the option requires too much subjectivity to justify recognition in the accounts. Research in finance demonstrates valuation approaches that seem to meet accountants' standards for reliable measurement. The FASB now requires firms to recognize the cost of employee stock options in the accounting records, as discussed below.

**Employee Stock Option Plans** An understanding of the accounting for stock options requires several definitions. Refer to **Figure 12.1**. The *grant date* is the date a firm gives a stock option to employees. The *vesting date* is the first date employees can exercise their stock options. The *exercise date* is the date employees exchange the option and cash for shares of common stock. The *exercise price* is the price specified in the stock option contract for purchasing the common stock. The *market price* is the price of the stock as it trades in the market. Firms usually structure stock option plans so that a period of time elapses between the grant date and the vesting date. Firms may either preclude employees from exercising the option for one or more years (a service condition), require that the firm achieve specified levels of profitability (a performance condition), or achieve particular market prices for its common stock (a market condition) before employees can exercise their options. Delaying the vesting date increases the likelihood that the employee continues to work for the firm.

FIGURE 12.1 Graphic Depiction of Stock Option Arrangement

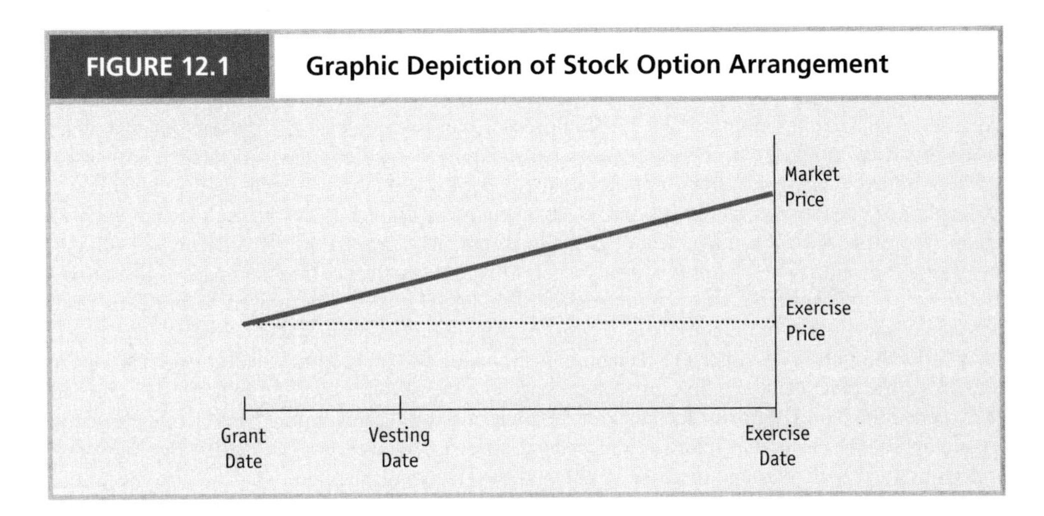


The value of a stock option results from two elements:

1. the benefit realized on the exercise date because the market price of the stock exceeds the exercise price (the **benefit element**), and
2. the length of the period during which the holder can exercise the option (the **time value element**).

One cannot know the amount of the benefit element before the exercise date. In general, stock options with exercise prices less than the current market price of the stock (described as *in the money*) have a higher value than stock options with exercise prices exceeding the current market price of the stock (described as *out of the money*). The time value element of an option's value results from the potential benefit it provides its holder for increases in the market price of the stock during the exercise period. An option provides the holder the right to enjoy the benefits of market price increases for the stock. This second element of an option will have more value the longer the exercise period and the more volatile the market price of the stock. Note that a stock option may have an exercise price that exceeds the current market price (zero value for the first element) but still have a value because of the possibility that the market price will exceed the exercise price on the exercise date (positive value for the second element). As the expiration date of the option approaches, the value of the second element approaches zero.[3]

**Example 6** In a recent year, Microsoft Corporation granted its employees 1,604 million options to purchase shares of its common stock. The weighted average option price for options granted during the year was $26.88. This weighted average option price compares to a low and high market price for the stock for the year of $21.42 and $29.12 respectively, suggesting that Microsoft likely set the option price approximately equal to the market price of the stock at the time of the grant. Microsoft applied an option-pricing model and derived a value of $12.08 per

[3]For an elaboration on the theory of option pricing, see Fischer Black and Myron Scholes, "The Pricing of Options and Corporate Liabilities," *Journal of Political Economy* (May–June 1973), pp. 637–654.

option, or \$19,376 million (= 1,604 × \$12.08) for all options granted during the year. Microsoft reported net income of \$9,993 million for the year.

**Political Pressures on Standard Setters** Employees accept stock options instead of cash compensation because of the potential economic benefit they can realize if the market price of the shares increases above the option price (the stock options ultimately become worthless if the market price of the stock never exceeds the option price). Thus, stock options have a potential economic opportunity cost to firms issuing them and a potential economic benefit to employees. The concepts of accounting studied throughout this book suggest that firms should recognize this economic cost as an expense during the period(s) that firms realize the benefits of employees' services related to the stock options.

Despite the conceptual rationale for expense recognition, firms have historically not recognized any expense relating to stock options. The justification generally offered for not recognizing an expense is that employees receive no benefit when the option price equals or exceeds the market price. This view, however, ignores the fair value of the options, which results from the potential future benefit because the market price might exceed the option price. Proponents acknowledge that some benefit might result, but firms cannot reliably measure that benefit at the time of the grant.

In the mid-1990s, the FASB proposed that firms recognize the cost of stock options as an expense in measuring earnings. The FASB proposal met with extensive resistance at that time. Firms found the proposed accounting would substantially reduce earnings. Although most large, publicly traded firms grant stock options to their employees, firms in technology industries in particular rely heavily on stock options as an element of employee compensation and lobbied against the proposed accounting. Although their arguments centered on the difficulty of reliably measuring the value of stock options, one suspects that the negative earnings impact of this attractive form of compensation was the real driver. The FASB at the time bowed to pressure from companies and from some members of Congress and altered its proposed accounting. The FASB gave firms the option of either (1) recognizing the value of stock options as an expense, or (2) continuing the practice at the time of recognizing no expense whenever the option price equaled or exceeded the market price of the stock at the time of granting the option. Most firms chose the second approach. Firms that followed this second approach, however, would still value the stock options granted each year and disclose in notes to the financial statement what net income would have been if they had reported under the first method. Microsoft in **Example 6** disclosed that its net income—if it had amortized the cost of options granted over the period of benefit—would have been \$7,531 million, instead of the \$9,993 million it reported on the income statement.

Reported earnings abuses stemming from Enron, Global Crossing, and other firms beginning around 2000 raised the public's interest in corporate reporting and particularly on the compensation of top management. Frequent articles appeared in the financial press indicating that most firms did not recognize the cost of stock options when measuring net income. Armed with disclosures similar to those of Microsoft above, the public became aware of the extensive use and cost of stock options and put pressure on the Congress to do something about it. Some firms, particularly those in non-technology industries for which the cost of stock options was relatively small, voluntarily switched their accounting to begin recognizing the cost of stock options as an expense. Technology firms continued to resist.

Public pressure and support from the Congress led the FASB and the IASB to reconsider the accounting for stock options. Both the FASB and the IASB have issued statements that require the firm to recognize, and report in measuring income, an expense for the issue of stock options.[4] The section below illustrates the accounting from these new reporting standards.

The accounting for stock options involves the following:

1. Measuring the fair value of stock options on the date of the grant using an option-pricing model. Option valuation models incorporate information about the current market price, the exercise price, the expected time between grant and exercise, the expected market price volatility of the stock, the expected dividends, and the risk-free interest rate.
2. Amortizing the fair value of the stock options over the expected period of benefit, which is usually the period from the date of the grant to the date of vesting. The firm debits Compensation Expense and credits Additional Paid-in Capital (Stock Options) for the amount amortized.

[4]International Accounting Standards Board, *International Financial Reporting Standard 2*, "Share-based Payment," 2004; Financial Accounting Standards Board, *Statement of Financial Accounting Standards Statement No. 123 (revised 2004)*, "Share-based Payment," 2004.

3. When employees exercise their options, the firm debits Cash for the proceeds, debits Additional Paid-in Capital (Stock Options) for any amounts credited to that account in step **2** above, credits Common Stock for the par value of the shares issued, and credits Additional Paid-in Capital for any excess of the value of the options plus the cash proceeds over the par value of the shares issued.

**Example 7** Fiala Corporation awards options to employees on January 1, Year 1, to acquire 1,000 shares of $5 par value common stock at an exercise price of $35 per share. The firm uses an option-pricing model to value the options at $8,000. It expects to receive benefits of enhanced employee services for the two-year period between the grant date and the vesting date. Employees exercise the options on December 31, Year 4, when the market price of the stock is $50 per share. Fiala Corporation makes the following entries:

*January 1, Year 1*

No entry.

*December 31, Year 1*

| | | |
|---|---|---|
| Compensation Expense | 4,000 | |
| Additional Paid-in Capital (Stock Options) | | 4,000 |

| Assets | = | Liabilities | + | Shareholders' Equity | (Class.) |
|---|---|---|---|---|---|
| | | | | −4,000 | IncSt → RE |
| | | | | +4,000 | ContriCap |

*December 31, Year 2*

| | | |
|---|---|---|
| Compensation Expense | 4,000 | |
| Additional Paid-in Capital (Stock Options) | | 4,000 |

| Assets | = | Liabilities | + | Shareholders' Equity | (Class.) |
|---|---|---|---|---|---|
| | | | | −4,000 | IncSt → RE |
| | | | | +4,000 | ContriCap |

*December 31, Year 4*

| | | |
|---|---|---|
| Cash (1,000 × $35) | 35,000 | |
| Additional Paid-in Capital (Stock Options) | 8,000 | |
| Common Stock—Par Value (1,000 × $5) | | 5,000 |
| Additional Paid-in Capital {$8,000 + [1,000 × ($35 − $5)]} | | 38,000 |

| Assets | = | Liabilities | + | Shareholders' Equity | (Class.) |
|---|---|---|---|---|---|
| +35,000 | | | | −8,000 | ContriCap |
| | | | | +5,000 | ContriCap |
| | | | | +38,000 | ContriCap |

This accounting recognizes an expense as the firm receives the benefits ($4,000 each in both Year 1 and Year 2), and increases contributed capital for both the cash equivalent value of employees' services rendered to obtain the common stock ($8,000) and the cash received when employees exercise their options ($35,000).

**Stock Rights** Like stock options, stock rights give their holder the right to acquire shares of stock at a specified price. The major differences between stock options and stock rights are as follows:

- Firms grant stock options to employees. Employees receive them as a form of compensation and may not transfer or sell them to others.
- Firms grant stock rights to current shareholders. The rights usually trade in public markets.

Firms issue stock rights to raise new capital from current shareholders. The granting of stock rights to current shareholders requires no accounting entries. GAAP ignores any value inherent in the stock right on the date of the grant. When holders exercise the rights, the firm records the issue of shares at the price paid just as it records the issue of new shares for cash.

**Stock Warrants** Firms issue stock warrants to the general investing public for cash. Assume that a corporation issues warrants for $15,000 cash. The warrants allow holders to purchase 10,000 shares for $20 each. The entry is as follows:

| | | |
|---|---|---|
| Cash | 15,000 | |
| Common Stock Warrants | | 15,000 |

| Assets | = | Liabilities | + | Shareholders' Equity | (Class.) |
|---|---|---|---|---|---|
| +15,000 | | | | +15,000 | |

To record the issue of warrants to the public. The accountant normally includes the Common Stock Warrants account with Additional Paid-in Capital for balance sheet presentation.

When warrant holders exercise their rights, the firm issues 10,000 shares of $5 par value common stock in exchange for the warrants plus $200,000 and records the following entry:

| | | |
|---|---|---|
| Cash | 200,000 | |
| Common Stock Warrants | 15,000 | |
| Common Stock—$5 Par Value | | 50,000 |
| Additional Paid-in Capital | | 165,000 |

| Assets | = | Liabilities | + | Shareholders' Equity | (Class.) |
|---|---|---|---|---|---|
| +200,000 | | | | −15,000 | ContriCap |
| | | | | +50,000 | ContriCap |
| | | | | +165,000 | ContriCap |

To record the issue of 10,000 shares for $200,000 cash and the redemption of warrants. (The amount originally received for the warrants transfers to Additional Paid-in Capital.)

If the warrants expire before the holders exercise them, the firm records the following entry:

| | | |
|---|---|---|
| Common Stock Warrants | 15,000 | |
| Additional Paid-in Capital | | 15,000 |

| Assets | = | Liabilities | + | Shareholders' Equity | (Class.) |
|---|---|---|---|---|---|
| | | | | −15,000 | ContriCap |
| | | | | +15,000 | ContriCap |

To record the expiration of common stock warrants and the transfer to permanent contributed capital.

Firms sometimes attach common stock warrants to a bond or preferred stock, which the holder can detach and redeem separately from the bond or preferred stock. The holder receives periodic interest or preferred dividends and holds an option to purchase common shares. When the accountant can objectively measure the value of the stock warrants separately from the value of the associated bond or preferred stock, the accounting allocates the issue price between the two securities.

To illustrate, assume that a firm issues 20-year, $1,000,000 bonds with 7-percent semiannual coupons. The bonds contain stock warrants, which their holders can either sell on the open market or exercise to acquire 10,000 shares of common stock for $200,000. The issue price for the bonds and warrants is $1,050,000. Immediately after issue, the bonds sell on the market for $1,035,000 and the warrants sell for $15,000. The accountant records the issue of the bonds as follows:

| | | |
|---|---|---|
| Cash | 1,050,000 | |
| Bonds Payable | | 1,035,000 |
| Common Stock Warrants | | 15,000 |

| Assets | = | Liabilities | + | Shareholders' Equity | (Class.) |
|---|---|---|---|---|---|
| +1,050,000 | | +1,035,000 | | +15,000 | ContriCap |

The subsequent accounting for the bonds follows the procedures discussed in **Chapter 9** for bonds issued above or below par value. The accounting for the warrants follows the procedures illustrated above for warrants issued for cash.

If the accountant cannot objectively measure the value of the stock warrants separately from the value of the bond or preferred stock, the accounting credits the full issue price to the bond or preferred stock and none of the price to the common stock warrants.

**Convertible Bonds or Preferred Stock** **Convertible bonds** and convertible preferred stock permit their owner either to hold the security as a bond or preferred stock or to convert the security into shares of common stock. The market value of the conversion option does not trade independently of the bond or preferred stock, as was the case above for bonds or preferred stock issued with a separable stock warrant. Thus, the accountant has greater difficulty allocating the issue price of a convertible bond or convertible preferred stock between its debt or preferred stock attributes and its conversion feature.

The required accounting allocates the full issue price to the bonds or preferred stock and none of the price to the conversion feature. Suppose, for example, that Johnson Company's credit rating would allow it to issue $100,000 of ordinary 10-year, 14-percent semiannual coupon bonds at par. The firm prefers to issue convertible bonds with a lower coupon rate. Assume that Johnson Company issues at par $100,000 of 10-year, 10-percent semiannual coupon bonds, but the holder of each $1,000 bond can convert it into 50 shares of Johnson Company $5 par value common stock. (Holders in aggregate can convert the entire issue into 5,000 shares.) Accounting requires the following entry:

| | | |
|---|---|---|
| Cash | 100,000 | |
| Convertible Bonds Payable | | 100,000 |

| Assets | = | Liabilities | + | Shareholders' Equity | (Class.) |
|---|---|---|---|---|---|
| +100,000 | | +100,000 | | | |

To record the issue of convertible bonds at par.

This entry effectively treats convertible bonds just like ordinary, nonconvertible bonds, and it records the value of the conversion feature at zero.

An alternative treatment, but one not currently permitted, allocates a portion of the issue price to the conversion feature. **Appendix Table 5** (for 10-percent coupon bonds) indicates that 10-percent, 10-year semiannual (nonconvertible) coupon bonds sell for about 79 percent of par for an issuer with a market borrowing rate of 14 percent. Thus, if the firm can issue 10-percent convertible bonds at par, the conversion feature must be worth about 21 (= 100 − 79) percent of par. This indicates that the bond buyers have paid 21 percent of the proceeds from the bond issue as a capital contribution for the right to acquire common stock later. If GAAP allowed the accountant to record the substance of the issue of these 10-percent convertible bonds at par, the entry would be as follows:

| | | |
|---|---|---|
| Cash | 100,000 | |
| Convertible Bonds Payable | | 79,000 |
| Additional Paid-in Capital | | 21,000 |

| Assets | = | Liabilities | + | Shareholders' Equity | (Class.) |
|---|---|---|---|---|---|
| +100,000 | | +79,000 | | +21,000 | ContriCap |

To record the issue of 10-percent semiannual coupon convertible bonds at a time when ordinary 10-percent bonds could be issued for 79 percent of par. GAAP does not allow this entry.

Notice that the calculation of the amounts for this entry requires knowing the proceeds of an issue of nonconvertible bonds with other features similar to the convertible bonds. Because auditors often believe that they are unable to reliably estimate this information, GAAP does not currently allow the previous journal entry.[5] The entry under GAAP simply debits Cash and credits Convertible Bonds Payable for $100,000.

The usual entry to record the conversion of bonds into shares ignores current market prices in the interest of simplicity and merely shows the swap of shares for bonds at their book value. For example, assume that the common stock of Johnson Company increases in the market to $30 a share, so that the holder of one $1,000 bond, convertible into 50 shares, can convert it into shares with a market value of $1,500. If holders convert all the convertible issue into common shares at this time, the firm would issue 5,000 shares of $5 par value stock on conversion and make the following journal entry:

| | | |
|---|---|---|
| Convertible Bonds Payable | 100,000 | |
| Common Shares—$5 Par | | 25,000 |
| Additional Paid-in Capital | | 75,000 |

| Assets | = | Liabilities | + | Shareholders' Equity | (Class.) |
|---|---|---|---|---|---|
| | | -100,000 | | +25,000 | ContriCap |
| | | | | +75,000 | ContriCap |

To record the conversion of 100 convertible bonds with book value of $100,000 into 5,000 shares of $5 par value stock.

An allowable alternative treatment recognizes that market prices provide information useful in quantifying the market value of the shares issued. Under the alternative treatment, with $30 market price per share and $150,000 fair market value of the 5,000 shares issued on conversion, the journal entry would be as follows:

| | | |
|---|---|---|
| Convertible Bonds Payable | 100,000 | |
| Loss on Conversion of Bonds | 50,000 | |
| Common Shares—$5 Par | | 25,000 |
| Additional Paid-in Capital | | 125,000 |

| Assets | = | Liabilities | + | Shareholders' Equity | (Class.) |
|---|---|---|---|---|---|
| | | -100,000 | | -50,000 | IncSt → RE |
| | | | | +25,000 | ContriCap |
| | | | | +125,000 | ContriCap |

To record the conversion of 100 convertible bonds into 5,000 shares of $5 par value stock at a time when the market price is $30 per share.

[5]As this book goes to press, the FASB is studying the accounting for convertible bonds. Given the FASB's recent moves to recognize the value of various options, the required accounting for convertible bonds in the future may follow this alternative treatment.

The alternative entry results in the same total shareholders' equity, with smaller retained earnings but larger contributed capital. It results from treating the conversion as the following two separate transactions:

| | | |
|---|---|---|
| Cash | 150,000 | |
|     Common Shares—$5 Par | | 25,000 |
|     Additional Paid-in Capital | | 125,000 |

| Assets | = | Liabilities | + | Shareholders' Equity | (Class.) |
|---|---|---|---|---|---|
| +150,000 | | | | +25,000 | ContriCap |
| | | | | +125,000 | ContriCap |

To record issue of 5,000 shares of $5 par value stock at $30 per share.

| | | |
|---|---|---|
| Convertible Bonds Payable | 100,000 | |
| Loss on Retirement of Bonds | 50,000 | |
|     Cash | | 150,000 |

| Assets | = | Liabilities | + | Shareholders' Equity | (Class.) |
|---|---|---|---|---|---|
| -150,000 | | -100,000 | | -50,000 | IncSt → RE |

To record the retirement by purchase for $150,000 of 100 convertible bonds carried on the books at $100,000.

## Problem 12.1 for Self-Study

**Journal entries for capital contributions.** Prepare the journal entries to record the following transactions for Healy Corporation during Year 1.

**a.** Issued 100,000 of $10 par value common stock for $14 per share on January 2.
**b.** Issued 10,000 shares of common stock on January 2 in the acquisition of a patent. The firm has no separate information about the fair value of the patent.
**c.** Issued 2,000 shares of $100 convertible preferred stock on March 1 for $100 per share. Holders may convert each share of preferred stock into four shares of common stock.
**d.** Sold 10,000 common warrants on the open market on June 1 for $5 per warrant. Holders can exchange each warrant and $24 in cash for a share of common stock.
**e.** Holders of 600 shares of convertible preferred stock (see **c**) exchanged their shares for common stock on September 15. The market price of the common stock on this date was $26 per share. Record the conversion using the book values.
**f.** Holders of 4,000 common stock warrants exchanged their warrants (see **d**) and $96,000 in cash for common stock on November 20. The market price of the common stock on this date was $32 per share.
**g.** Granted options to employees to purchase 5,000 shares of common stock for $35 per share on December 31. The value of these options is $25,000 and the firm expects to realize all benefits of these options in future years.

## Corporate Distributions

Firms use net assets (= assets – liabilities) to generate more net assets through the earnings process. Some firms retain some or all of the net assets generated by earnings, causing net assets to increase, along with retained earnings, the component of shareholders' equity showing the cause of that increase in net assets. The retention of assets generated by earnings generally results in an increase in the market price of the firm's common shares. Most firms do not retain all of the net assets generated by earnings but, instead, pay a dividend to the common shareholders. Each common shareholder receives the same dividend per share as all other common shareholders. This section discusses corporate dividend policies and the accounting for dividends.

Firms may also choose to use the net assets generated by earnings to repurchase shares of its common stock. Cash flows out of the firm to shareholders, as it does with a cash dividend. In the case of share repurchases, only shareholders selling their shares to the firm receive cash. This section discusses business reasons for stock repurchases and the accounting for such repurchases.

## DIVIDENDS

The board of directors has the legal authority to declare dividends. The directors, in considering whether to declare dividends, must conclude that declaring a dividend is both legal (under law and contract) and financially expedient.

**Legal Limits on Dividends—Statutory (by Law)** State corporation laws limit directors' freedom to declare dividends. These limitations attempt to protect a firm's creditors. Without these limits, directors might dissipate the firm's assets for the benefit of shareholders, harming the creditors. Neither directors nor shareholders are liable for the corporation's debts.

Generally, the laws provide that the board may not declare dividends "out of capital," that is, debited against the contributed capital accounts, which result from fund-raising transactions with owners, but must declare them "out of earnings" by debiting them against the retained earnings account, which results from operating transactions. The wording and the interpretation of this rule vary among states. "Capital" usually means the total amount paid in by shareholders. Some states allow corporations to declare dividends out of the earnings of the current period even though the Retained Earnings account has a debit (negative) balance because of accumulated losses from previous periods.

For most companies, statutory limits do not influence the accounting for shareholders' equity and dividends. A balance sheet does not provide all the legal details of amounts available for dividends, but it should disclose information necessary for the user to apply the legal rules of the corporation's state of incorporation. For example, state statutes can provide that the corporation "may acquire its own shares only in amounts less than retained earnings." In other words, dividends cannot exceed the amount of retained earnings reduced by the cost of currently held repurchased shares. If a firm acquires its own shares under these constraints, the amount of future dividends it might otherwise declare declines. The firm must disclose this fact in a note to the financial statements.[6]

The firm can meet the statutory requirements for declaring dividends by building up a balance in retained earnings. Such a balance does not mean that the firm has a fund of cash available for the dividends. Some readers of financial statements mistakenly believe that retained earnings represent cash available for dividends. This error often results from the confusion of assets (such as cash) and sources of assets (such as retained earnings). Retained earnings represent the source of the increased net assets, which do not necessarily, or even usually, take the form of cash.

Managing cash requires the techniques of corporate finance; a firm must anticipate cash needs for dividends just as it anticipates cash needs for the purchase of equipment, the retirement of debts, and so on. A prudent firm might borrow from the bank to pay the regular dividend if its financial condition justifies the resulting increase in liabilities.

**Legal Limits on Dividends—Contractual** Contracts with bondholders, other lenders, and preferred shareholders often limit dividend payments and thereby compel the retention of earnings. A bond contract may require that total liabilities not exceed the total amount of shareholders' equity or that the retirement of the debt be made "out of earnings." Such a provision involves curtailing dividends so that the necessary debt service payments, plus any dividends, do not exceed the amount of earnings for the period. This provision forces the shareholders to increase their investment in the business by limiting the amount of dividends that the board might otherwise declare for them. Financial statement notes must disclose significant limitations on dividend declarations.

The notes to a financial statement of Sears contain the following disclosure:

---

[6]Accounting Principles Board, *Opinion No. 6*, "Status of Accounting Research Bulletins," 1965.

> Dividend payments are restricted by several statutory and contractual factors, including: Certain indentures relating to the long-term debt of Sears, which represent the most restrictive contractual limitation on the payment of dividends, provide that the company cannot take specified actions, including the declaration of cash dividends, which would cause its consolidated unencumbered assets, as defined [in the indentures, not in the annual report], to fall below 150 percent of its consolidated liabilities, as defined. At . . . [year-end], $11.2 billion in retained income [of a total of $12.7 billion of retained earnings] could be paid in dividends to shareholders under the most restrictive indentures. [Sears declared dividends for the year of approximately $700 million.]

**Dividends and Corporate Financial Policy** Directors usually declare dividends less than the legal maximum. The directors may allow retained earnings to increase as a matter of corporate financial policy for several reasons:

1. Available cash did not increase by as much as the amount of earnings, so maximum dividends would require raising more cash.
2. Restricting dividends in prosperous years may permit continued level or steadily growing dividend payments in poor years.
3. The firm may need funds for expansion of working capital or plant and equipment.
4. Reducing the amount of borrowings, rather than paying dividends, may seem prudent.
5. The firm can distribute the funds to shareholders with lower tax burdens for them by using the cash to repurchase shares.

## ACCOUNTING FOR DIVIDENDS

A firm may pay dividends in cash, other assets, or shares of its common stock.

**Cash Dividends** When the board of directors declares a **cash dividend**, the entry is as follows:

| | | |
|---|---|---|
| Retained Earnings (Dividends Declared) . . . . . . . . . . . . . . . . . . . . . . . . | 150,000 | |
| Dividends Payable . . . . . . . . . . . . . . . . . . . . . . . . . . . . . . . . . . | | 150,000 |

| Assets | = | Liabilities | + | Shareholders' Equity | (Class.) |
|---|---|---|---|---|---|
| | | +150,000 | | −150,000 | RE |

To record declaration of dividends.

Once the board of directors declares a dividend, the dividend becomes a legal liability of the corporation. Dividends Payable appears as a current liability on the balance sheet if the firm has not yet paid the dividends at the end of the accounting period. When the firm pays the dividends, the entry is as follows:

| | | |
|---|---|---|
| Dividends Payable . . . . . . . . . . . . . . . . . . . . . . . . . . . . . . . . . . . . | 150,000 | |
| Cash . . . . . . . . . . . . . . . . . . . . . . . . . . . . . . . . . . . . . . . . . . . . | | 150,000 |

| Assets | = | Liabilities | + | Shareholders' Equity | (Class.) |
|---|---|---|---|---|---|
| −150,000 | | −150,000 | | | |

**Property Dividends** Corporations sometimes distribute assets other than cash when paying a dividend; such a dividend is known as a **dividend in kind** or a **property dividend**. The accounting for such dividends resembles that for cash dividends, except that when the firm pays the dividend, it credits the asset given up, rather than cash. The amount debited to Retained Earnings equals the fair market value of the assets distributed. When this market value differs from the book value of the asset distributed, a gain or loss arises. The firm reports any gain or loss as part of earnings for the period.

**Stock Dividends** The retention of earnings may lead to a substantial increase in shareholders' equity, as the firm accumulates assets which it keeps invested in the business. Many of these investments commit assets for the long term; they represent relatively permanent commitments by shareholders to the business. The permanency results from the firm's having invested the net assets generated by the earnings process in operating assets such as inventories and fixed assets. To indicate such a permanent commitment of assets generated by reinvested earnings, the board of directors may declare a **stock dividend**. The accounting involves a debit to the Retained Earnings account and a credit to the contributed capital accounts. The stock dividend does not affect total shareholders' equity. It reallocates amounts from Retained Earnings to the contributed capital accounts. When the firm declares a stock dividend, shareholders receive additional shares of stock in proportion to their existing holdings. If, for example, the firm issues a 5-percent stock dividend, each shareholder receives one additional share for every 20 shares held before the dividend.

GAAP requires firms to record the value of the newly issued shares based on the market value of the shares issued. For example, the directors of a corporation may decide to declare a stock dividend of 10,000 additional shares of common stock with a par value of $10 per share at a time when the market price of a share is $40. The entry would be as follows:

| | | |
|---|---|---|
| Retained Earnings (Dividend Declared) | 400,000 | |
| Common Stock—$10 Par | | 100,000 |
| Additional Paid-in Capital | | 300,000 |

| Assets | = | Liabilities | + | Shareholders' Equity | (Class.) |
|---|---|---|---|---|---|
| | | | | −400,000 | RE |
| | | | | +100,000 | ContriCap |
| | | | | +300,000 | ContriCap |

To declare a stock dividend, recorded using market price of shares to quantify the amounts: $40 × 10,000 shares = $400,000.

The stock dividend relabels a portion of the retained earnings that had been legally available for dividend declarations as a more permanent form of shareholders' equity. A stock dividend formalizes the fact that the firm has used some funds represented by past earnings to expand plant facilities, or to replace assets at increased prices, or to retire bonds. The firm does not have this cash available for cash dividends. The stock dividend does not affect the availability of cash that the firm has already invested; the stock dividend signals the commitment to investment, perhaps more clearly than before, to readers of the balance sheet.

Stock dividends have little economic substance for shareholders. More shares of common stock represent the same ownership percentage. If the shareholders each receive the same type of shares as they already own, each shareholder's proportionate interest in the capital of the corporation and proportionate voting power do not change. Although the book value per common share (total common shareholders' equity divided by number of common shares outstanding) decreases, the shareholder has a proportionately larger number of shares, so the total book value of each shareholder's interest remains unchanged. The market value per share should decline, but all else being equal, the total market value of an individual's shares will not change. To describe such a distribution of shares as a "dividend"—meaning a distribution of assets generated by earnings—may mislead some readers, but the terminology is generally accepted.

## Stock Splits

**Stock splits** (or, more technically, split-ups) resemble stock dividends. The corporation issues additional shares of stock to shareholders in proportion to their existing holdings. The firm receives no additional assets. A stock split usually results in a reduction in the par value of all the stock in the issued class. A corporation may, for example, have 1,000 shares of $10 par value stock outstanding and, by a stock split, exchange those shares for 2,000 shares of $5 par value stock (a two-for-one split) or for 4,000 shares of $2.50 par value stock (a four-for-one split) or for any number of shares of no-par stock. If the shares outstanding have no par value, the shareholders keep the existing certificates and receive the new ones. Firms sometimes engage in **reverse stock splits**, in which they increase the par value of stock and reduce the number of shares outstanding.

A stock split or reverse stock split does not require a journal entry if, as in the examples above, the par value changes in proportion to the new number of shares. If the change in par value is not in proportion to the new number of shares, the firm increases or decreases Additional Paid-in Capital. The amount of retained earnings usually does not change. The amount shown in the Common Stock account represents a different number of shares. Of course the firm must record the new number of shares held by each shareholder in the subsidiary capital stock records.

Distinguishing a stock dividend from a stock split can sometimes cause difficulties. For example, is a 50-percent increase in the number of shares accounted for as a stock dividend, using the market value of the stock, or as a 1.5-for-one stock split, using the par value of the stock? Usually firms treat small-percentage distributions, say less than a 25-percent increase in the number of shares, as stock dividends and larger ones as stock splits.

A stock split (or a stock dividend) usually reduces the market value per share, all else equal, in inverse proportion to the split (or dividend). Thus a two-for-one split could be expected to result in a 50-percent reduction in the market price per share. Likewise, a reverse stock split usually increases the market value per share in inverse proportion to the reverse split. Therefore, management has used stock splits and reverse stock splits to keep the market price per share in what management believes is an acceptable trading range. For example, the board of directors might think that a market price of $60 to $80 is an effective trading range for its stock. If the share price has risen to $150 in the market, the board of directors may declare a two-for-one split. (This subjective judgment almost never has supporting evidence. Warren Buffett, chairman of Berkshire-Hathaway, refused to split the common shares until they sold in the marketplace for over $33,000 per share. When he did make lower cost shares available, the new shares had restrictions that required them to trade in blocks costing over $10,000 per trade. Buffett appears not to believe that his company's share price suffers as a result.) For certain, stock splits and dividends result in increased record-keeping costs.

## Problem 12.2 for Self-Study

**Journal entries for dividends and stock splits.** The shareholders' equity section of the balance sheet of Baker Corporation on January 1, Year 5, appears below:

| Shareholders' Equity | |
|---|---|
| Common Stock, $10 par value, 25,000 shares issued and outstanding | $250,000 |
| Additional Paid-in Capital | 50,000 |
| Retained Earnings | 150,000 |
| Total | $450,000 |

Prepare journal entries for each of the following transactions of Baker Corporation for Year 5. Ignore income taxes.

**a.** March 31: The board of directors declares a cash dividend of $0.50 per share. The firm will pay the dividend on April 15.
**b.** April 15: The firm pays the dividend declared on March 31.
**c.** June 30: The board of directors declares and distributes a 10-percent stock dividend. The market price per share on this date is $15.
**d.** December 31: The board of directors declares a 2-for-1 stock split and changes the par value of the common shares from $10 to $5.

## Stock Repurchases

When a corporation reacquires its own previously issued common shares, accounting calls these **treasury stock** or **treasury shares**. Refer to the disclosure of shareholders' equity for Coke in **Exhibit 12.1**. Coke has repurchased shares of its common stock in an amount equal to just over one-half of its total shareholders' equity before treasury stock transactions. Reasons for reacquiring outstanding common stock include the following:

1. To use in various option arrangements. When holders of stock options, stock rights, stock warrants, and convertible securities exercise their options, firms usually receive less cash (or market value of other consideration) than the market value of the common stock at the time. Reacquiring treasury shares and then issuing an equal number of new shares under various option arrangements may keep the total number of shares outstanding the same, thereby avoiding dilution of the voting interest of ongoing shareholders and perhaps maintaining earnings per share.
2. To serve as a worthwhile investment of excess cash. Some firms believe that their own shares provide a good investment. Evidence supports the notion that share prices increase after news becomes public that a firm intends to or has reacquired its own shares. Such shares do not receive dividends, nor do they have voting rights, because corporation laws do not consider them to be outstanding shares for these purposes. Firms can reissue the treasury shares when they need cash.
3. To defend against an unfriendly takeover bid. Many firms received unfriendly takeover bids in recent years. To defend against the takeover attempt, firms used available cash to repurchase shares of common stock on the market. Two different motives appear to be at work here:

   - This action reduces the amount of common shareholders' equity and increased the proportion of debt in the capital structure, making the firm more risky and therefore less attractive to an unfriendly bidder. Some firms even borrow cash to repurchase shares, which affects the debt ratio even more than using already available cash to reacquire shares.
   - The acquisition of shares uses available cash and thereby reduces the attractiveness of the company to outsiders who have believed that the available cash makes the company an attractive target.

4. To distribute cash to shareholders in a tax-advantaged way. Rather than pay dividends to all shareholders, many of whom will owe personal income taxes on the entire dividend amount, the firm can buy back shares from those who wish to raise cash. Many shareholders will have lower tax rates on receipts from sales of shares than on dividend receipts.

## Accounting for Treasury Shares

Accounting for treasury shares—share repurchases—follows from the fundamental principle that a corporation does not report a gain or loss on transactions involving its own shares. Even though the firm may sell (technically, reissue) the shares for more than their acquisition cost, the accounting does not report the economic gain as accounting income. Similarly, the firm may subsequently reissue shares for less than their acquisition cost, but even so, the economic loss will not reduce net income. The required accounting views treasury stock purchases and sales as capital, not operating, transactions and therefore debits (for economic losses) or credits (for economic gains) the contributed capital accounts for the adjustments for reissue of treasury shares. The amounts bypass the income statement, comprehensive income, and, generally, the Retained Earnings account.

When a firm reacquires common shares, it debits a Treasury Shares—Common account (a shareholders' equity contra account) with the total amount paid to reacquire the shares.

| | | |
|---|---|---|
| Treasury Shares—Common | 11,000 | |
| Cash | | 11,000 |

| Assets | = | Liabilities | + | Shareholders' Equity | (Class.) |
|---|---|---|---|---|---|
| -11,000 | | | | -11,000 | ContriCap |

To record $11,000 paid to reacquire 1,000 common shares. Treasury Shares—Common is a contra account to shareholders' equity.

If the firm later reissues treasury shares for cash, it debits Cash with the amount received and credits the Treasury Shares—Common account with the cost of the shares. If it reissues treasury shares at the conversion of bonds or preferred stock into common stock, it debits the convertible bonds or preferred stock instead of debiting cash. The reissue price will usually differ from the amount paid to acquire the treasury shares. If the reissue price exceeds the acquisition price, the credit to make the entry balance is to the Additional Paid-in Capital account. Assuming that the above firm reissued for $14,000 the 1,000 shares reacquired previously, the entry would be as follows:

| | | |
|---|---|---|
| Cash . . . . . . . . . . . . . . . . . . . . . . . . . . . . . . . . . . . . . . . . . . . . . | 14,000 | |
| Treasury Shares—Common . . . . . . . . . . . . . . . . . . . . . . . . . . . | | 11,000 |
| Additional Paid-in Capital . . . . . . . . . . . . . . . . . . . . . . . . . . . | | 3,000 |

| Assets | = | Liabilities | + | Shareholders' Equity | (Class.) |
|---|---|---|---|---|---|
| +14,000 | | | | +11,000 | ContriCap |
| | | | | +3,000 | ContriCap |

To reissue 1,000 shares of treasury stock at a price greater than acquisition cost. The $3,000 economic "gain" does not appear on the income statement as an accounting gain.

If the amount paid for the treasury shares exceeds the reissue price, the firm debits the balance to Additional Paid-in Capital, so long as that account has a sufficiently large credit balance. To the extent the required debit exceeds the credit balance in the Additional Paid-in Capital account, the firm reduces that account to zero and debits the excess to retained earnings. This debit to the Retained Earnings account resembles a dividend; it does not appear as an expense or a loss reported on the income statement.

The Treasury Shares account appears as a subtraction from total shareholders' equity on the balance sheet; see **Exhibit 12.1** for Coke.

## Problem 12.3 for Self-Study

**Journal entries for treasury stock transactions.** Prepare journal entries for the following transactions of Crissie Corporation:

**a.** Reacquired 2,000 shares of $10 par value common stock on January 15 for $45 per share.
**b.** Issued 1,200 shares of treasury stock to employees upon the exercise of stock options at a price of $48 per share on April 26.
**c.** Reacquired 3,000 shares of $10 par value common stock for $52 per share on August 15.
**d.** Issued 1,600 shares of treasury stock to holders of 800 shares of convertible preferred stock, which had a book value of $80,000 on November 24. Crissie Corporation uses a first-in, first-out assumption on reissues of treasury stock and uses book values to record conversions of preferred stock.
**e.** Sold 1,500 shares of treasury stock on the open market for $47 per share on December 20.

## Reporting Earnings Transactions

What purpose does the income statement serve? We argue that it is *not* "to show earnings for the period." The reader of the financial statements can generally ascertain earnings by subtracting the beginning balance of the Retained Earnings account from its ending balance and adding dividends. Accountants provide an income statement so managers and investors can see the *causes* of earnings. Then a reader can compare a company's performance with other companies (cross-section analysis) or with the company itself over time (time-series analysis), to make more informed projections about the future.

Previous chapters concentrated on *measuring* the results of earnings transactions. Next, we focus on *reporting*, or *disclosing*, earnings transactions in the financial statements. To motivate an understanding of the issues involved in reporting earnings transactions, consider the earnings data for Bernard Corporation in **Exhibit 12.2**. To simplify the illustration, we assume that revenues result in immediate cash receipts and expenses require immediate cash expenditures. Thus, income flows equal cash flows.

**EXHIBIT 12.2**

**BERNARD COMPANY**
**Measurement of Firm's Market Value from Cash Flow Data**

| Activities of the Firm | Cash Flows Occur at the End of Each Period — Period Number 1 | 2 | 3 | 4 | 5 | 6 | 7 | 8 | | Present Value of Activity Using Discount Rate of 10% |
|---|---|---|---|---|---|---|---|---|---|---|
| 1. Recurring .......... | $100 | $100 | $100 | $100 | $100 | $100 | $100 | $100 | ...... | $1,000.00 |
| 2. Recurring, but Growing at 6% per Year ....... | 30 | 32 | 34 | 36 | 38 | 40 | 43 | 45 | ...... | 750.00 |
| 3. Cyclic .............. | 115 | 0 | 115 | 0 | 115 | 0 | 115 | 0 | ...... | 602.38 |
| 4. Nonrecurring ........ | 120 | 0 | 0 | 0 | 0 | 0 | 0 | 0 | ...... | 109.09 |
| 5. Recurring .......... | (40) | (40) | (40) | (40) | (40) | (40) | (40) | (40) | ...... | (400.00) |
| 6. Nonrecurring ........ | (70) | 0 | 0 | 0 | 0 | 0 | 0 | 0 | ...... | (63.64) |
| Income for Year 1 ...... | $255 | | | | | | | | | |
| | Present Value [= Fair Market Value] of Entire Firm .................. | | | | | | | | | $1,997.83 |

Suppose that an analyst wished to value Bernard Corporation using the present value of the cash flows of its individual activities, some recurring, some not. Assume that the cash flows in **Exhibit 12.2** are after tax. Also assume the discount rate appropriate for finding the present value of Bernard Company's cash flows is 10 percent per year. Corporate finance classes discuss the issues related to choosing the discount rate and the **Appendix** to this book introduces the techniques of present value analysis. You need not have studied the **Appendix** if you will take on faith the derivation of the numbers in the right-hand column of **Exhibit 12.2**.

Bernard Company engages in six activities, numbered 1 through 6, shown in **Exhibit 12.2**.

*Activity 1.* The first activity generates $100 per year, with the cash flow at the end of each year, indefinitely. The present value of this activity is $1,000 (= $100/0.10). See **Example 18** in the **Appendix**. Investors sometimes call this process of deriving market value from a series of future cash flows *capitalizing earnings*. The analyst might say that the earnings of $100 have a *price/earnings ratio* of 10 or that the earnings "deserve [or carry] a multiple of 10."

*Activity 2.* The second activity generates $30 at the end of the first year, and a cash flow that grows by 6 percent per year thereafter. The present value of this activity is $750 [= $30/(0.10 − 0.06)]. See **Example 19** in the **Appendix**. The price/earnings ratio (or multiple) for these cash flows is 25, because of their growth. Management that is interested in higher market prices for its firm's shares wants investors to think of the company as a growth stock, because investors put higher values on growing-earnings companies than on stable-earnings companies.

*Activity 3.* The third activity is cyclic, generating $115 per year at the end of each odd-numbered year. The present value of this activity is $602.38. See **Example 20** in the **Appendix**.

*Activity 4.* The fourth activity is nonrecurring, generating a single cash flow of $120 at the end of the first year, with present value of $109.09 (= $120/1.10) at the start of the first year.

*Activity 5.* The fifth activity, an expenditure (outflow), uses $40 of cash each year, at the end of each year. The present value of this activity is −$400 (= −$40/0.10).

*Activity 6.* The sixth activity, a single expenditure (outflow), uses $70 cash at the end of the first year and has present value of −$63.64 (= −$70/1.10).

The value of the firm is the sum of the present values of its individual activities, $1,998 in **Exhibit 12.2**. In this example, most of the value of this firm comes from the recurring activities. In deriving firm values, investors generally care about recurring activities more than nonrecurring ones, because recurring activities add value each year, while nonrecurring activities, by definition, happen once or infrequently.

**Exhibit 12.2** shows the net earnings for Year 1 of $255. How can analysts and investors deduce the value of the company from this one year's earnings statement? They can't. This firm is too complex for even a sophisticated user to derive the value from a single column of data, without further information. Investors and analysts need to know about the components of a firm's earnings and their recurring versus nonrecurring nature so that they can make estimates of the market value of the firm. Analysts feel more confident estimating the value of firms with recurring activities than with nonrecurring activities. Analysts may also want to include changes in the market value of a firm's assets and liabilities in their valuation of a company even though GAAP may not include these market value changes in earnings.

## Overview of GAAP Reporting of Earnings Transactions

The sections that follow discuss the reporting of various earnings items. We begin by providing an overview of this reporting.

1. GAAP requires that firms initially report the results of most earnings transactions in the income statement instead of bypassing the income statement and reporting the amounts in some other shareholders' equity account. This reporting reflects the emphasis most analysts and investors place on the income statement when evaluating a firm's earnings performance and the concern that statement users may overlook earnings transactions reported elsewhere.
2. GAAP recognizes that some earnings transactions are central to a firm's principal business activities and recur, while others are peripheral or nonrecurring. GAAP requires firms to report items in their income statements in various categories to inform statement users about the nature of income items.
3. Changes in the market value of assets and liabilities affect the value of a firm as they occur. GAAP recognizes some of these value changes in net income as they occur even though the firm has not yet sold the asset for cash or settled the liability, events that confirm the amount of the value change. GAAP delays reporting value changes of other assets and liabilities in net income until confirming events occur. In the meantime, GAAP includes such value changes in Other Comprehensive Income, a component of shareholders' equity.
4. Firms sometimes discover errors in amounts previously reported, change their accounting principles, or change estimates made in applying their accounting principles. GAAP requires firms to retroactively restate previously reported amounts for corrections of errors and changes in accounting principles but adjust current and future amounts for changes in accounting estimates.

We examine the reporting of four types of earnings transactions:

1. Recurring versus nonrecurring.
2. Central versus peripheral.
3. Unrealized versus realized gains and losses from changes in the market values of assets and liabilities.
4. Adjustments for errors and changes in accounting principles and accounting estimates.

The issues discussed affect the quality of earnings. Firms have incentives to report good earnings news in ways that make it appear recurring and central, but to report bad earnings news as non-recurring and peripheral. Analysts must sort through the various components of earnings, recognizing possible biases injected by management, and apply appropriate multiples to the earnings components in valuing firms. GAAP attempts to aid this process by requiring firms to classify earnings transactions in particular ways in the financial statements. In recent years firms often reported so-called *pro forma earnings*, which each firm defines in its own way. Typically pro forma earnings remove the effects of nonrecurring events which have reduced net income, so that pro forma earnings exceed GAAP earnings. We rarely see pro forma earnings lower than GAAP earnings. As this book goes to press, regulators express concern about the lack of reporting discipline in pro forma earnings and firms remain free to say what they please. Analysts learn to suspect the earnings of firms that report nonrecurring items year after year.

## Reporting Recurring/Nonrecurring and Central/Peripheral Activities

An analyst likely asks two questions when using a firm's past profitability to project its likely future profitability:

1. Does the earnings item result from an activity in which a firm will likely continue its involvement, or does the earnings item result from an unusual transaction or event that is unlikely to recur regularly?

2. Does the earnings item result from a firm's primary operating activity (creating and selling a good or service for customers) or from an activity incidental or peripheral to the primary operating activity (for example, periodic sales of equipment previously used by the firm in manufacturing)?

**Figure 12.2** depicts these distinctions, with examples of each.

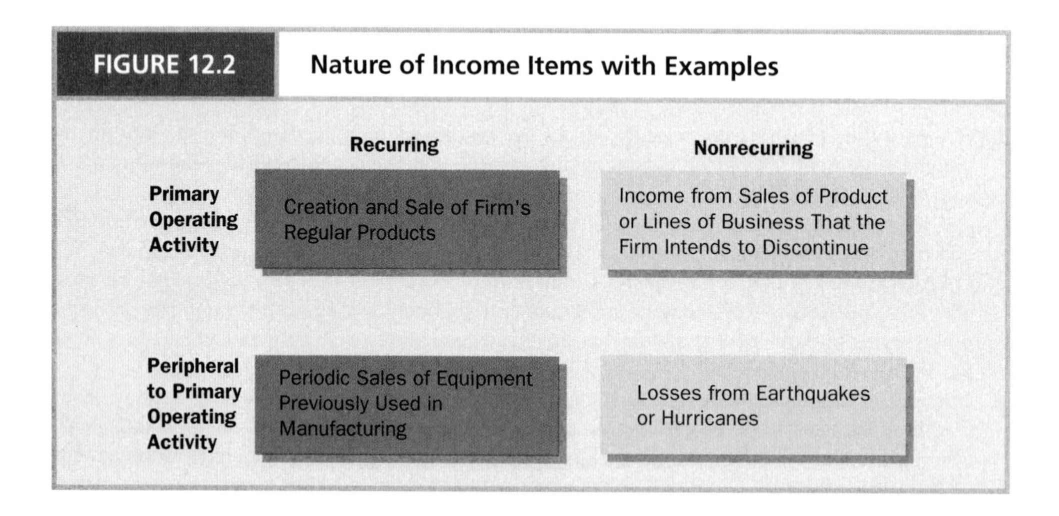


A financial statement user who wants to evaluate a firm's ongoing operating profitability will likely focus on earnings items in the upper left cell. A financial statement user who wants to project earnings of prior periods into the future would likely focus on the two *recurring earnings* cells. Earnings items in the nonrecurring cells should not affect ongoing assessments of profitability. This section considers the reporting of each type of earnings item.

## Measurement of Earnings Effect

Accountants distinguish between revenues and expenses on the one hand and gains and losses on the other. Revenues and expenses result from the recurring, primary operating activities of a business (upper left cell in **Figure 12.2**). Earnings items in this first category are the ordinary, recurring operating activities of the firm. Gains and losses result from either peripheral activities (lower left cell) or nonrecurring activities (upper and lower right cells). A second distinction is the reporting of revenues and expenses at gross amounts, whereas accounting reports gains and losses at net amounts. The following examples, which we introduced in Chapter 6, illustrate this distinction.

**Example 8** IBM sells a computer to a customer for $400,000. The computer cost IBM $300,000 to manufacture. IBM records this sale as follows:

| | | |
|---|---|---|
| Cash | 400,000 | |
| Sales Revenue | | 400,000 |

| Assets | = | Liabilities | + | Shareholders' Equity | (Class.) |
|---|---|---|---|---|---|
| +400,000 | | | | +400,000 | IncSt → RE |

To record sale.

| | | |
|---|---|---|
| Cost of Goods Sold | 300,000 | |
| Finished Goods Inventory | | 300,000 |

| Assets | = | Liabilities | + | Shareholders' Equity | (Class.) |
|---|---|---|---|---|---|
| −300,000 | | | | −300,000 | IncSt → RE |

To record the cost of goods sold.

This transaction fits into the upper left cell of **Figure 12.2** (primary/recurring). The income statement reports both sales revenue and cost of goods sold, providing information to the financial statement user regarding both the manufacturing cost of the computer and IBM's ability to mark up this cost in setting selling prices. Retained earnings increase by $100,000 as a result of this transaction. Notice that the $100,000 amount does not itself appear in the income statement. Rather, income increases by $100,000 as a result of sales of $400,000 offset with cost of goods sold of $300,000.

**Example 9** The GAP, a retail clothing chain, sells computers previously used for processing data in its stores. This sale relates only peripherally to The GAP's primary operating activity, which is to sell casual clothes. Assume that the computers originally cost $500,000 and have $200,000 of accumulated depreciation at the time of sale. Thus, the computers have a net book value of $300,000. The sale of these computers for $400,000 results in the following journal entry on The GAP's books:

| | | |
|---|---|---|
| Cash . . . . . . . . . . . . . . . . . . . . . . . . . . . . . . . . . . . . . . . . . . . . . | 400,000 | |
| Accumulated Depreciation . . . . . . . . . . . . . . . . . . . . . . . . . . . . . . . . | 200,000 | |
| Equipment . . . . . . . . . . . . . . . . . . . . . . . . . . . . . . . . . . . . . . | | 500,000 |
| Gain on Sale of Equipment . . . . . . . . . . . . . . . . . . . . . . . . . . . | | 100,000 |

| Assets | = | Liabilities | + | Shareholders' Equity | (Class.) |
|---|---|---|---|---|---|
| +400,000 | | | | +100,000 | IncSt → RE |
| +200,000 | | | | | |
| −500,000 | | | | | |

Disposal of computers for $100,000 more than book value.

This transaction fits into the lower left cell of **Figure 12.2** (peripheral/recurring). The income statement reports only the $100,000 gain on the sale, not the selling price of $400,000 and the book value of $300,000. The income statement reports gains and losses at net, instead of gross, amounts because, presumably, financial statement users do not care about information on the individual components comprising peripheral or nonrecurring earnings items. The gain on the sale increases retained earnings by $100,000.

Note that both revenues and gains appear as credits in journal entries and increase Retained Earnings. Both expenses and losses appear as debits in journal entries and reduce Retained Earnings. To repeat, revenues and expenses report operating or central items as gross amounts; gains and losses report nonrecurring or peripheral items as net amounts.

## Classifications in the Income Statement

Income statements contain some or all of the following sections or categories, depending on the nature of a firm's earnings for the period:

1. Earnings from continuing operations.
2. Earnings, gains, and losses from discontinued operations.
3. Extraordinary gains and losses.

The majority of income statements include only the first section. **Exhibit 12.3** presents an income statement for Hypothetical Company that includes all three sections.

**Earnings from Continuing Operations** Revenues, gains, expenses, and losses from the continuing areas of business activity of a firm appear in the first section of the income statement, **Earnings from Continuing Operations**. This section includes earnings derived from a firm's primary business activities as well as from activities peripherally related to operations. The key distinction is that the firm expects these sources of earnings to continue. Firms without the nonrecurring categories of earnings for a particular year (discussed next) need not use the title *Earnings from Continuing Operations* in their income statements. In this case, absence of nonrecurring types of earnings implies that all reported revenues, gains, expenses, and losses relate to continuing operations. Firms often report a subtotal within the continuing operations section of the income statement, labeled *operating income*, to report the revenues and expenses

**EXHIBIT 12.3**

**HYPOTHETICAL COMPANY**
**Income Statement**
**(all dollar amounts in millions)**

| | Year 6 | Year 7 | Year 8 |
|---|---|---|---|
| EARNINGS FROM CONTINUING OPERATIONS | | | |
| Sales | $ 240 | $ 265 | $ 295 |
| Cost of Goods Sold | (144) | (154) | (165) |
| Selling and Administrative Expenses | (50) | (58) | (67) |
| Operating Income | $ 46 | $ 53 | $ 63 |
| Interest Revenue | 4 | 5 | 7 |
| Interest Expense | (15) | (19) | (22) |
| Gain on Sale of Equipment | 4 | 9 | 3 |
| Income from Continuing Operations before Taxes | $ 39 | $ 48 | $ 51 |
| Income Taxes | (13) | (16) | (17) |
| Income from Continuing Operations | $ 26 | $ 32 | $ 34 |
| EARNINGS, GAINS, AND LOSSES FROM DISCONTINUED OPERATIONS | | | |
| Income (Loss) from Operations of Division Sold in Year 8 (net of income taxes) | $ 16 | $ (4) | $ 2 |
| Gain on Sale of Division (net of income taxes) | — | — | 40 |
| Income from Discontinued Operations | $ 16 | $ (4) | $ 42 |
| **Extraordinary Gains and Losses** | | | |
| Loss from Hurricane (net of income taxes) | — | $ (12) | — |
| Net Income | $ 42 | $ 16 | $ 76 |
| **Earnings per Common Share** | | | |
| Continuing Operations | $2.60 | $3.04 | $3.09 |
| Discontinued Operations | 1.60 | (0.38) | 3.82 |
| Extraordinary Items | — | (1.14) | — |
| Net Income | $4.20 | $1.52 | $6.91 |

from the firm's primary business activity of creating and selling goods or services. Revenues from marketable securities and investments in securities, interest expense on borrowings, and gains and losses from peripheral activities appear separately as *nonoperating income* in the continuing operations section of the income statement.

**Earnings, Gains, and Losses from Discontinued Operations** Sometimes a firm sells a major division or segment of its business during the year or contemplates its sale within a foreseeable time after the end of the accounting period. If so, it must disclose separately any earnings, gains, and losses related to that segment. The separate disclosure appears in the next section of the income statement, **Earnings, Gains, and Losses from Discontinued Operations**, alerting the financial statement reader that the firm does not expect this source of earnings to continue.[7] Firms report the earnings, gain, or loss net of income tax effects. This section follows the section presenting Earnings from Continuing Operations.

**Extraordinary Gains and Losses** A separate section of the income statement presents **Extraordinary Gains and Losses**. For an item to be extraordinary, it must generally meet both of the following:

1. Unusual in nature, and
2. Infrequent in occurrence.[8]

[7] Accounting Principles Board, *Opinion No. 30,* "Reporting the Results of Operations," 1973.
[8] *Ibid.*

An example of an item likely to be extraordinary for most firms is a loss from an earthquake or confiscation of assets by a foreign government. Firms report extraordinary items net of their tax effects.

## UNREALIZED GAINS AND LOSSES FROM CHANGES IN MARKET VALUES OF ASSETS AND LIABILITIES

The FASB has increasingly required firms to report certain assets and liabilities at their current market values at the end of each period instead of their historical, or acquisition, costs. Examples discussed in previous chapters include:

1. Valuation of inventories at lower of cost or market (**Chapter 7**).
2. Valuation of fixed assets and intangibles at current market value when evidence indicates that an asset impairment has occurred (**Chapter 8**).
3. Valuation of financial instruments, including derivatives used to hedge risk, at market value (**Chapter 11**).
4. Valuation of marketable equity securities at market value (**Chapter 11**).

When a firm writes assets and liabilities up or down to market value, the question arises as to how it should treat the offsetting credit (gain) or debit (loss). At the time of the revaluation, the firm has not yet **realized** the gains or losses. That is, the firm has not yet sold the asset or settled the liability with a transaction involving cash. In some cases, GAAP requires firms to **recognize** the gains and losses, that is, to include them, even though not yet realized, in earnings in the period of the revaluation. Firms include losses from the writedown of inventories, fixed assets, and intangibles in computing earnings in the period of the revaluation. Firms include gains and losses from the revaluation of financial instruments classified as fair value hedges in earnings each period as market values of the hedged financial instrument and its derivative change.

The FASB has been reluctant, however, to require firms to include in earnings all unrealized gains and losses from the revaluation of assets and liabilities. We suspect the reason has to do with the volatility that some unrealized gains and losses inject into the reported income of companies with a large fraction of their assets in the form of marketable securities and other financial instruments. The reluctance stems from the preferences of the firms preparing the statements, not from the preferences of the rule makers themselves. Including all unrealized gains and losses on marketable equity securities in earnings each period, for example, will cause reported earnings to fluctuate (in response to fluctuations in market prices) more than it would otherwise.

Many, probably most, managers prefer to report stable earnings in contrast to fluctuating earnings. All else equal, the less risky the earnings stream—that is, the less volatile are reported earnings—the higher will be the market price of the firm's shares. Financial institutions in particular do not like fluctuations in net income; see **Problem 52** at the end of **Chapter 11** for a discussion of why. Managers have succeeded over the years in lobbying the FASB to keep many fluctuating components of income out of reported earnings. Unrealized gains and losses have typically appeared on a separate line in the shareholders' equity section of the balance sheet.

The FASB has become increasingly concerned that users of financial statements might overlook these value changes if they appear on the comparative balance sheet only and not in the income statement. As a consequence, the FASB requires firms to disclose unrealized gains and losses that historically have bypassed the income statement in a category called **other comprehensive income**.[9]

Other comprehensive income *for a reporting period* includes changes in the market value of marketable equity securities available for sale and changes in the market value of derivatives used as cash flow hedges. Other comprehensive income also includes unrealized gains and losses from translating the financial statements of foreign units into U.S. dollars and certain changes in pension liabilities, topics covered in advanced financial accounting courses. **Accumulated other comprehensive income**, a shareholders' equity account on the balance sheet, reports the *cumulative* amounts of other comprehensive income as of the date of the balance sheet. **Comprehensive income** equals net income on the traditional income statement plus other comprehensive income for the period.

Firms have considerable flexibility as to how they report other comprehensive income each period. They can:

---

[9]Financial Accounting Standards Board, *Statement of Financial Accounting Standards No. 130,* "Reporting Comprehensive Income," 1997.

1. Include it in a combined statement of net income and other comprehensive income, in effect appending it to the bottom of the traditional income statement.
2. Include it in a separate statement of other comprehensive income.
3. Include it in a statement of changes in shareholders' equity.

**Exhibit 12.4** shows the required disclosure of comprehensive income in the first two of the allowed formats—as part of a combined earnings and comprehensive income statement, and as part of a separate statement of comprehensive income. We illustrate the third format, as part of a

**EXHIBIT 12.4 Reporting Comprehensive Income, Two Allowed Formats**

ONE-STATEMENT APPROACH

**Statement of Net Income and Comprehensive Income**

| | | |
|---|---|---|
| Revenues | | $100,000 |
| Expenses | | (25,000) |
| Gain on Sale of Securities | | 2,000 |
| Other Gains and Losses | | 8,000 |
| Earnings from Continuing Operations before Income Tax | | $ 85,000 |
| Income Tax Expense | | (21,250) |
| Earnings before Discontinued Operations and Extraordinary Items | | $ 63,750 |
| Discontinued Operations, Net of Tax | | 30,000 |
| Extraordinary Items, Net of Tax | | (30,500) |
| **Net Income (or, as preferred by the FASB, Earnings)** | | **$ 63,250** |
| Other Comprehensive Income, Net of Tax: | | |
| Foreign Currency Translation Adjustments | | 7,000 |
| Unrealized Gains and Losses on Securities: | | |
| Unrealized Holding Gains Arising during Period | $13,000 | |
| Less: Reclassification Adjustment for Gain Included in Net Income (Earnings) | (1,500) | 11,500 |
| Minimum Pension Liability Adjustment | | (2,500) |
| Other Comprehensive Income (Loss) | | **$ 16,000** |
| **Comprehensive Income (Loss)** | | **$ 79,250** |

TWO-STATEMENT APPROACH

**Statement of Net Income**

| | |
|---|---|
| Revenues | $100,000 |
| Expenses | (25,000) |
| Gain on Sale of Securities | 2,000 |
| Other Gains and Losses | 8,000 |
| Earnings from Continuing Operations before Income Tax | $ 85,000 |
| Income Tax Expense | (21,250) |
| Earnings before Discontinued Operations and Extraordinary Items | $ 63,750 |
| Discontinued Operations, Net of Tax | 30,000 |
| Extraordinary Items, Net of Tax | (30,500) |
| **Net Income (or, as preferred by the FASB, Earnings)** | **$ 63,250** |

**Statement of Comprehensive Income**

| | | |
|---|---|---|
| **Net Income (or, as preferred by the FASB, Earnings)** | | **$ 63,250** |
| Other Comprehensive Income, Net of Tax: | | |
| Foreign Currency Translation Adjustments | | 7,000 |
| Unrealized Gains and Losses on Securities: | | |
| Unrealized Holding Gains Arising during Period | $13,000 | |
| Less: Reclassification Adjustment for Gain Included in Net Income (Earnings) | (1,500) | 11,500 |
| Minimum Pension Liability Adjustment | | (2,500) |
| Other Comprehensive Income (Loss) | | **$ 16,000** |
| **Comprehensive Income (Loss)** | | **$ 79,250** |

statement explaining the reasons for changes in shareholders' equity accounts, later in this chapter in **Exhibit 12.6**.

## *Terminology* Note

The FASB faced an issue in naming the amount that we call *net income* in the traditional income statement. Use of the term *net* implies that it is a bottom-line number. Yet firms that append other comprehensive income to the bottom of their income statements need a label to differentiate traditional net income from the combined amounts of traditional net income plus other comprehensive income. The FASB expressed a preference to eliminate the term *net income* altogether, to label traditional net income as *earnings*, and to label the combined amounts of earnings and other comprehensive income as *comprehensive income*. The FASB gave firms wide latitude, not only in their choice of reporting formats, but in their choice of terminology. Most firms choose to follow the third reporting format above, leaving the traditional income statement unchanged. Thus, we will continue to use *net income* to designate the bottom line of the traditional income statement.

## ADJUSTMENTS FOR ERRORS AND ACCOUNTING CHANGES

Firms occasionally obtain new information about amounts included in net income of prior periods or change either their accounting principles or estimates used in applying their accounting principles. Consider the following examples:

1. Dell Computer Corporation discovers that it neglected to count $250,000 of inventory items stored in a warehouse in Texas at the end of last year. In consequence, Dell understated its earnings last period.
2. JC Penney has used a LIFO cost flow assumption for inventories and cost of goods sold for many years. It decides during the current year to change to a FIFO cost flow assumption. Earnings of prior years would differ if JC Penney had used a FIFO, instead of a LIFO, cost flow assumption.
3. American Airlines depreciates its aircraft over a 20-year life. More fuel-efficient aircraft now available on the market leads American Airlines to begin slowly replacing its existing aircraft with the new aircraft. American Airlines reduces the depreciable life of its existing aircraft, which increases current and future depreciation charges relative to the recent past.

Each of these three examples requires the accountant to decide whether to (1) retroactively restate prior years' earnings, (2) include an adjustment for the correction or accounting change in the current year's earnings, or (3) correct or adjust earnings of the current and future periods (so-called prospective adjustment).

Advocates of retroactive restatement view past earnings as useful to the extent they permit predictions of future earnings. Retroactive restatement results in computing earnings for prior years on the same basis as earnings of the current and future years, thereby enhancing earnings predictions.

Advocates of including adjustments for these items in the income statement of the current year argue that all earnings items should initially appear in the income statement of some period. In this way, the cumulative series of income statements includes all earnings items. The financial statement reader will less likely overlook the items if they appear in the income statement than if firms record the amounts as an adjustment of prior years' earnings. Adequate disclosure of the nature of each item in the income statement will permit the financial statement user to assess its importance when evaluating the firm's profitability. Probably with good reason, some on this side of the argument suggest that if managers have the discretion to leave such items out of the current year's income statement, many of them will find ways to justify restating prior years' earnings for income-reducing adjustments and including in earnings of the current period only income-enhancing adjustments. Virtually all theorists agree that giving managers the discretion to report good news and bad news as they see fit will lead to less useful financial statements unless auditors become more critical of their clients' accounting than they have been.

Advocates of adjusting earnings prospectively argue that restating previously reported amounts reduces the credibility of the financing reporting process. Advocates also argue that such adjustments are a normal, recurring part of the accounting process. Prospective adjustment avoids the connotation that prior years' earnings were computed incorrectly.

GAAP distinguishes the accounting for (1) corrections of errors, (2) adjustments for changes in accounting principles, and (3) adjustments for changes in accounting estimates.

**Reporting Correction of Errors** Errors result from such actions as miscounting inventories, miscalculations, and misapplying accounting principles. The example above for Dell Computer illustrates an accounting error. The accountant makes a **correction of errors**, if material, by retroactively restating earnings of prior periods and adjusting the beginning balance in Retained Earnings for the current period.[10] Dell Computer's understatement of inventory at the end of last period resulted in an overstatement of cost of goods sold last period, as the following analysis shows:

| Cost of Goods Sold | = | Beginning Inventory | + | Purchases | − | Ending Inventory |
|---|---|---|---|---|---|---|
| Overstated $250,000 | = | 0 | + | 0 | − | Understated $250,000 |

The firm retroactively restates net income of last period and makes the following entry (ignoring income tax effects) this period:

| | | |
|---|---|---|
| Merchandise Inventory . . . . . . . . . . . . . . . . . . . . . . . . . . . . . . . . . . | 250,000 | |
| Retained Earnings . . . . . . . . . . . . . . . . . . . . . . . . . . . . . . . . . | | 250,000 |

| Assets | = | Liabilities | + | Shareholders' Equity | (Class.) |
|---|---|---|---|---|---|
| +250,000 | | | | +250,000 | RE |

To correct inventory error. Firm overstated last period's cost of goods sold and understated earnings.

**Reporting Changes in Accounting Principles** JC Penney's change from a LIFO to a FIFO cost flow assumption is a **change in accounting principle**. Assuming that firms can compute the amount of earnings for prior periods under the new accounting principle, GAAP requires firms to retroactively restate prior years' earnings to reflect the new accounting principle.[11] Assume that inventory at the end the last year for JC Penney was $450 million under LIFO and $525 million under FIFO. The entry to record the change in accounting principle is as follows (in millions):

| | | |
|---|---|---|
| Merchandise Inventory . . . . . . . . . . . . . . . . . . . . . . . . . . . . . . . . . . | 75 | |
| Retained Earnings . . . . . . . . . . . . . . . . . . . . . . . . . . . . . . . . . | | 75 |

| Assets | = | Liabilities | + | Shareholders' Equity | (Class.) |
|---|---|---|---|---|---|
| +75 | | | | +75 | RE |

To restate inventory and retained earnings retroactively for a change from LIFO to FIFO.

JC Penney would also restate net income for each prior year reported in its financial statements using a FIFO cost flow assumption.

---

[10]Financial Accounting Standards Board, *Statement of Financial Accounting Standards No. 154,* "Accounting Changes and Error Corrections," 2005.

[11]Ibid.

**Reporting Changes in Accounting Estimates** As time passes, new information becomes available that sharpens estimates required to apply accounting principles. Examples include the amount of uncollectible accounts and the useful lives of depreciable assets. The example above for American Airlines is a change in accounting estimate. Earlier chapters have pointed out that accountants do not correct revenues and expenses of previous periods to incorporate new information. Instead, the accountant allows the effect of the change in estimate to affect current and future periods' earnings.[12] Refer, for example, to **Figure 8.4**, which illustrates the effect of a change in depreciable life on depreciation expense. Rather than correct Retained Earnings directly, the firm adjusts current and future depreciation charges, but not past ones, to take into account the book value at the time the new information arrives and the new information itself.

Accrual accounting requires frequent, ongoing changes in estimates. Restating the financial statements of previous years, which would require changing the opening balance of retained earnings for the current period, each time a firm changes an estimate will add complexity to the task of understanding financial statements and might reduce their credibility. This approach would also provide management with methods for boosting earnings of the current period by lowering the already-reported earnings of previous periods in order to raise current and future periods' earnings.

**Changes in estimates** do not always relate to recurring accrual accounting measurements, such as depreciable lives. Some changes in estimates concern unusual or nonrecurring events. Consider, for example, the following:

- A court this period finds a firm responsible for an act that occurred several years previously and caused injury. The damage award differs from the amount that the firm previously recognized with a debit to a loss and a credit to a liability.
- The Internal Revenue Service assesses additional income taxes on a firm's taxable income of previous years. The amount previously recognized as income tax expense for those years understates the final amount payable.

These events provide new information regarding measurements made in previous periods. Even though the events do not recur, GAAP treats them similarly to changes in estimates for recurring items. The earnings effect of these items appears in the income statement of the current period, appropriately disclosed, not in retained earnings as a direct adjustment.

### Problem 12.4 for Self-Study

**Journal entries for earnings and retained earnings transactions.** Prepare journal entries for each of the following transactions of Able Corporation for Year 5. Ignore income taxes.

**a.** January 15: As a result of a computer software error the preceding December, the firm failed to record depreciation on office facilities totaling $35,000.

**b.** March 20: An earthquake in California causes an uninsured loss of $70,000 to a warehouse.

**c.** December 31: The firm acquired its office building six years before December 31, Year 5. The building cost $400,000, had zero estimated salvage value, and had a 40-year life. The firm uses the straight-line depreciation method. Able Corporation now estimates that the building will have a total useful life of 30 years instead of 40 years. Record depreciation expense on the building for Year 5 and any required adjustment to depreciation of previous years.

## Earnings and Book Value per Share

Publicly held firms must show **earnings per common share** data in the body of the income statement.[13] Earnings per common share result from dividing net earnings minus preferred stock dividends by the average number of outstanding common shares during the accounting period. Firms reporting more than one of the three categories of earnings items must disclose earnings

[12]Ibid.

[13]Financial Accounting Standards Board, *Statement of Financial Accounting Standards No. 128,* "Earnings per Share," 1997.

per common share for each reported category. See **Exhibit 12.3** for categories and the disclosure. Issues in calculating earnings per common share go beyond the scope of this book. **Problem 40** at the end of this chapter illustrates some of the calculations.

Most firms also report **book value per common share** in their annual reports. Book value per common share equals total common shareholders' equity divided by the number of shares outstanding on the date of the balance sheet.

**Appendix 5.1** indicates that investors often apply multiples to earnings per common share and book value per common share in deciding on a reasonable market price for a firm's shares.

## An International Perspective

Firms in all industrialized countries and in most developing countries around the world prepare financial statements based on the accrual, rather than the cash, basis of accounting. The particular measurement rules used in applying the accrual basis differ among countries.

The format and classification of earnings items within the income statement vary across countries. The general measurement and disclosure principles discussed in this book should permit the reader of the financial statements to understand and interpret various income statement formats. Refer to the income statement for Wellcome PLC, a British pharmaceutical company, in **Exhibit 12.5**. Although this financial statement uses terminology different from that used in this book, the reader should be able to infer the nature of various items.

| Title Used in Exhibit 12.5 | Title Used in This Book |
|---|---|
| Group Profit and Loss Account | Income (or, sometimes, Earnings) Statement |
| Turnover | Sales Revenue |
| Stocks | Inventories |
| Trading Profit | Operating Earnings |
| Profit on Ordinary Activities | Earnings from Continuing Operations |

Wellcome's statement in **Exhibit 12.5** classifies operating costs according to their nature (for example, raw materials, labor, depreciation) instead of by the functional activity (for example, cost of goods sold, marketing expenses, administrative expenses). European companies often classify expenses by nature. One item seldom seen in annual reports of

*(continued)*

**EXHIBIT 12.5**

**WELLCOME PLC**
**Group Profit and Loss Account**
**(all currency amounts in millions)**

| | Year 7 | Year 8 |
|---|---|---|
| Turnover | £1,147.4 | £1,272.2 |
| Operating Costs: | | |
| Raw Materials and Consumables | (246.9) | (245.4) |
| Other External Charges | (401.5) | (424.4) |
| Staff Costs | (309.1) | (337.4) |
| Depreciation | (38.8) | (44.1) |
| Change in Stocks of Finished Goods and Work in Progress | 36.5 | 18.7 |
| Other Operating Charges | (10.3) | (6.7) |
| Trading Profit | £ 177.3 | £ 232.9 |
| Net Interest Payable | (8.2) | (11.7) |
| Profit on Ordinary Activities before Taxation | £ 169.1 | £ 221.2 |
| Tax on Profit from Ordinary Activities | (71.4) | (89.4) |
| Profit on Ordinary Activities Attributable to Shareholders | £ 97.7 | £ 131.8 |

U.S. companies is "Change in Stocks of Finished Goods and Work in Process." Wellcome reports among its operating costs on the income statement actual raw materials, labor, and overhead costs incurred during the year. Recall the inventory equation:

$$
\begin{aligned}
\text{Cost of Goods Sold} &= \text{Beginning Inventory} + \text{Purchases} - \text{Ending Inventory} \\
&= \text{Purchases} - (\text{Ending Inventory} - \text{Beginning Inventory}) \\
&= \text{Purchases} - (\text{Change in Inventory})
\end{aligned}
$$

Wellcome shows all purchases in its income statement, so it must subtract the increase in inventory to compute Cost of Goods Sold. For example, at the end of Year 8, £18.7 million of costs included in various operating cost items in the income statement actually apply to units in process or in finished goods at the end of Year 8 and not to units sold. Netting this £18.7 million against the actual total cost results in the proper matching of costs against revenues.

## Disclosure of Changes in Shareholders' Equity

The annual reports to shareholders must explain the changes in all shareholders' equity accounts.[14] The reconciliation of retained earnings may appear in the balance sheet, in a statement of earnings and retained earnings, or in a separate statement. The reconciliation of Other Comprehensive Income may also appear in a separate statement.

**Exhibit 12.6** shows the consolidated statement of shareholders' equity for Michigan Company (information adapted from the annual report of Ford Motor Company). The two-year comparative statement shows separate columns for common stock at par, additional paid-in capital, retained earnings, accumulated other comprehensive income, treasury shares, and total shareholders' equity. The statement shows opening balances, net income, cash dividends, unrealized gains and losses on securities available for sale, stock issued under employee option plans, common stock retired, and common stock issued on conversion of convertible bonds.

The FASB suggested that firms use a reporting format reconciling the beginning balance of Other Comprehensive Income, its components for a year, and the ending balance in a format such as the one below, which uses data for the General Electric Company (GE) for a recent year (dollar amounts in millions):

| | |
|---|---|
| Accumulated Other Comprehensive Income, Beginning of Year | $1,340 |
| Components of Other Comprehensive Income for Year (= $264 + $60; see below) | 324 |
| Accumulated Other Comprehensive Income, End of Year | $1,664 |

Some firms find this reporting distasteful because, we suspect, they do not want to point to any total other than net income as deserving the title *income*. Hence, they find various ways to re-label both the balance sheet beginning and ending balances as well as the components of other comprehensive income for the year. GE, for example, shows just under the income statement a separate Statement of Changes in Share Owners' Equity, which has the following components (dollar amounts in millions). Note: The information in brackets *does not appear* in the GE report; we add this information here for clarity.

| **Changes in Share Owners' Equity** | |
|---|---|
| Balance [of Share Owners' Equity], January 1 | $34,438 |
| Dividends and other transactions [treasury share acquisitions and dispositions] with share owners | (5,178) |

(*continued on page 603*)

[14]Accounting Principles Board, *Opinion No. 12*, "Omnibus Opinion—1967," 1967.

**EXHIBIT 12.6**

**MICHIGAN COMPANY (adapted from Ford Motor Company)**
**Consolidated Statement of Shareholders' Equity**
**(all dollar amounts in millions)**

| Line Number[a] | | Shares | Amount | Additional Paid-in Capital | Retained Earnings | Accumulated Other Comprehensive Income | Treasury Shares | Total Shareholders' Equity |
|---|---|---|---|---|---|---|---|---|
| | **Balance, January 1, Year 1** | 101.5 | $253.7 | $379.5 | $5,328.1 | $137.7 | $(56.5) | $6,042.5 |
| (1) | Net Income | | | | 906.5 | | | 906.5 |
| (2) | Unrealized Gain (Loss) on Securities Available for Sale | | | | | 46.2 | | 46.2 |
| | Comprehensive Income | | | | | | | 952.2 |
| (3) | Cash Dividends | | | | (317.1) | | | (317.1) |
| (4) | Common Stock Issued under Certain Employee Stock Plans | 0.2 | 0.5 | 9.9 | | | | 10.4 |
| (5) | Conversion of Debentures | | | 0.6 | | | 10.2 | 10.8 |
| (6) | Common Stock Retired | (2.5) | (6.2) | (9.5) | (140.9) | — | — | (156.6) |
| | **Balance, December 31, Year 1** | 99.2 | $248.0 | $380.5 | $5,776.6 | $183.9 | $(46.3) | $6,542.7 |
| (1) | Net Income | | | | 360.9 | | | 360.9 |
| (2) | Unrealized Gain (Loss) on Securities Available for Sale | | | | | (53.7) | | (53.7) |
| | Comprehensive Income | | | | | | | 307.2 |
| (3) | Cash Dividends | | | | (298.1) | | | (298.1) |
| (4) | Common Stock Issued under Certain Employee Stock Plans | 0.1 | 0.2 | 1.8 | | | | 2.0 |
| (5) | Conversion of Debentures | | | 1.4 | | | 30.6 | 32.0 |
| (6) | Common Stock Retired | (5.7) | (14.2) | (21.8) | (194.0) | — | — | (230.0) |
| | **Balance, December 31, Year 2** | 93.6 | $234.0 | $361.9 | $5,645.4 | $130.2 | $(15.7) | $6,355.8 |

[a]This caption and the line numbers do not appear on the original statement. The line numbers correspond to the journal entries in **Exhibit 12.7.**

| | | |
|---|---|---|
| **Changes other than transactions with share owners** | | |
| Increases attributable to net earnings [net income for year] .......... | $9,296 | |
| Unrealized gains (losses) on investment securities .................. | 264 | |
| Currency translation adjustments [another component of other comprehensive income] .................................. | 60 | |
| Total changes other than transactions with share owners ............. | | 9,620 |
| Balance at December 31 .................................... | | $38,880 |

What is other comprehensive income for the year? It is $324 (= $264 + $60), but unsophisticated readers of financial statements will likely not know this because of the method of disclosures. GE shows the beginning and ending balances of Other Comprehensive Income, which it calls *Nonowner changes other than earnings*, in the note reconciling all of the changes in the components of shareholders' equity.

## Journal Entries for Changes in Shareholders' Equity

**Exhibit 12.7** reconstructs the journal entries of Michigan Company for Year 1 and Year 2. The amounts in the entries in **Exhibit 12.7** represent millions of dollars, and the numbers of the journal entries correspond to the lines in **Exhibit 12.6** to which they correspond.

## An International Perspective on Shareholders' Equity

The accounting for shareholders' equity in most developed countries closely parallels that in the United States.

- Contributed capital accounts increase when firms issue shares of common or preferred stock.
- Revenues, gains, expenses, and losses affect the measurement of periodic earnings and retained earnings.
- Dividends reduce retained earnings and either reduce corporate assets (cash dividends or dividends in kind) or increase other shareholders' equity accounts (stock dividends).

Accounting in many other countries, unlike in the United States, uses **shareholders' equity reserve accounts**. Reserve accounts have credit balances and generally appear in the shareholders' equity section of the balance sheet.

Foreign firms often use a reserve account to disclose to financial statement readers that the firm will not declare dividends against a portion of retained earnings. Firms in the United States can achieve the same result by declaring a stock dividend.

## *Terminology* Note

GAAP in the United States discourages the use of the word *reserve*. Here's why. Think about the word *reserve* as used in ordinary English. Do you think it refers to something in accounting with a debit balance or a credit balance? When we ask this question, almost all students respond, "Debit," thinking of things like oil reserves and cash set aside for a rainy day. In accounting, the word *reserve* always means an account with a credit balance: Reserve for Depreciation means Accumulated Depreciation, not cash set aside to acquire new equipment; Reserve for Bad Debts means the Allowance for Uncollectible Accounts, not cash set aside in case customers do not pay their bills; Reserve for Warranties means the Estimated Liability for Warranties, not cash set aside to pay for warranty repairs. Since the word *reserve* causes so much confusion, most accountants in the United States do not use it. (We still encounter accountants, even young ones, who use the term, but we do not understand why anyone interested in clarity would use this word in accounting.)

**EXHIBIT 12.7**

**MICHIGAN COMPANY (adapted from Ford Motor Company)**
**Journal Entries Illustrating Transactions Involving Shareholders' Equity**
**Years 1 and 2**
**(all dollar amounts in millions)**

| Entry and Explanation | Year 1 | | Year 2 | |
|---|---|---|---|---|
| **(1)** No journal entry required because revenues and expenses for the year that net to net income are already recorded in retained earnings. | | | | |
| **(2)** Marketable Security Investments Available for Sale | 46.2 | | | |
| Unrealized Holding Gain (Loss) on Securities Available for Sale (Other Comprehensive Income) | | 46.2 | | |
| Unrealized Holding Gain (Loss) on Securities Available for Sale (Other Comprehensive Income) | | | 53.7 | |
| Marketable Security Investments Available for Sale | | | | 53.7 |
| To record increase in market value in Year 1 (decrease in market value in Year 2) with increase in asset account in Year 1 (decrease in asset account in Year 2). Matching credit in Year 1 (debit in Year 2) is to Other Comprehensive Income, which appears in the Statement of Other Comprehensive Income, bypassing the income statement, before closing to the balance sheet account Accumulated Other Comprehensive Income. | | | | |
| **(3)** Retained Earnings (Dividend Declared) | 317.1 | | 298.1 | |
| Cash (or Dividends Payable) | | 317.1 | | 298.1 |
| Cash dividends declared. | | | | |
| **(4)** Cash | 10.4 | | 2.0 | |
| Common Stock | | 0.5 | | 0.2 |
| Additional Paid-in Capital | | 9.9 | | 1.8 |
| Common stock issued under certain employee stock plans. | | | | |
| **(5)** Convertible Debentures (Bonds) | 10.8 | | 32.0 | |
| Treasury Stock | | 10.2 | | 30.6 |
| Additional Paid-in Capital | | 0.6 | | 1.4 |
| Common stock issued on conversion of convertible debentures (bonds). The shares "issued" on conversion were shares reissued from a block of treasury shares that had cost $10.2 (for Year 1) and $30.6 (for Year 2). | | | | |
| **(6)** Common Stock | 6.2 | | 14.2 | |
| Additional Paid-in Capital | 9.5 | | 21.8 | |
| Retained Earnings | 140.9 | | 194.0 | |
| Cash | | 156.6 | | 230.0 |
| Retirement of common stock acquired for cash. When a firm acquires shares for the treasury, it usually debits a Treasury Stock account, shown as a contra to all of shareholders' equity. In this case, the firm "retires" the shares, so the journal entry identifies the specific amounts for Common Stock, Additional Paid-in Capital, and Retained Earnings corresponding to these shares. The debits are to these accounts instead of to a single contra account. | | | | |

**Example 10** Japanese accounting standards require that firms declaring dividends transfer from retained earnings to a permanent legal capital account a specified percentage of the dividend each period. Japanese firms make this entry in addition to the usual entry for dividends. To illustrate, Oji Paper Company declared and paid a ¥10 million dividend during a recent year and made the following entry (amounts in millions):

| | | |
|---|---|---|
| Retained Earnings (Dividends Declared and Paid) | 10 | |
| Cash | | 10 |

| Assets | = | Liabilities | + | Shareholders' Equity | (Class.) |
|---|---|---|---|---|---|
| −10 | | | | −10 | RE |

To record declaration and payment of cash dividend.

The firm also made the following entry:

| | | |
|---|---|---|
| Retained Earnings . . . . . . . . . . . . . . . . . . . . . . . . . . . . . . . . . . . . . | 3 | |
| Permanent Legal Capital . . . . . . . . . . . . . . . . . . . . . . . . . . . . . | | 3 |

| Assets | = | Liabilities | + | Shareholders' Equity | (Class.) |
|---|---|---|---|---|---|
| | | | | −3 | RE |
| | | | | +3 | ContriCap |

To reclassify 30 percent of dividends out of retained earnings and into a permanent capital account.

The second entry effectively marks a portion of retained earnings as unavailable to support dividend declarations. Including the ¥3 million in a capital account indicates to the reader of the financial statements that the firm cannot declare dividends up to the full amount of retained earnings.

**Example 11** French accounting standards permit firms to reclassify a portion of retained earnings to a capital account to indicate that the firm cannot declare dividends up to the full amount of retained earnings, at least temporarily. An annual report of Société National Elf Acquitaine revealed the following:

| | |
|---|---|
| Reserves: | |
| Legal . . . . . . . . . . . . . . . . . . . . . . . . . . . . . . . . . . . . . . . . . . . . . . . . . . . . | € 100 |
| Long-Term Capital Gains . . . . . . . . . . . . . . . . . . . . . . . . . . . . . . . . . . . . . . . . | 3,885 |
| Operating Costs . . . . . . . . . . . . . . . . . . . . . . . . . . . . . . . . . . . . . . . . . . . . . | 7,300 |
| Unappropriated . . . . . . . . . . . . . . . . . . . . . . . . . . . . . . . . . . . . . . . . . . . . . . | 5,227 |

This disclosure indicates that the firm may declare dividends against €5,227 of retained earnings but not against retained earnings of €11,285 (= €100 + €3,885 + €7,300), which represent net assets permanently or temporarily not available for payment as dividends. For example, suppose a customer is suing the firm for €20,000. Although the firm feels that it has an adequate defense, it might lose. The firm could make the following entry when the customer files the suit:

| | | |
|---|---|---|
| Retained Earnings (unappropriated) . . . . . . . . . . . . . . . . . . . . . . . . . . | 20,000 | |
| Operating Risk Reserve . . . . . . . . . . . . . . . . . . . . . . . . . . . . . . . | | 20,000 |

| Assets | = | Liabilities | + | Shareholders' Equity | (Class.) |
|---|---|---|---|---|---|
| | | | | −20,000 | RE |
| | | | | +20,000 | RE |

To reclassify a portion of retained earnings into an operating risk reserve.

Assume now that the firm loses the suit and incurs an uninsured liability of €12,000. The entry, assuming immediate cash payment, is as follows:

| | | |
|---|---|---|
| Loss from Damage Suit . . . . . . . . . . . . . . . . . . . . . . . . . . . . . . . . . . . | 12,000 | |
| Cash . . . . . . . . . . . . . . . . . . . . . . . . . . . . . . . . . . . . . . . . . . . . . . | | 12,000 |

| Assets | = | Liabilities | + | Shareholders' Equity | (Class.) |
|---|---|---|---|---|---|
| −12,000 | | | | −12,000 | IncSt → RE |

Because the firm has now settled the lawsuit, it reclassifies the operating risk reserve back to retained earnings with the following entry:

| | | |
|---|---|---|
| Operating Risk Reserve | 20,000 | |
| Retained Earnings | | 20,000 |

| Assets | = | Liabilities | + | Shareholders' Equity | (Class.) |
|---|---|---|---|---|---|
| | | | | −20,000 | RE |
| | | | | +20,000 | RE |

A second use of reserve accounts relates to revaluations of assets. Accounting regulations in Great Britain and France permit firms to revalue periodically their fixed and other assets.

**Example 12** Refer to **Exhibit 2.7**, which shows the balance sheet of Rolls-Royce, a British company. During the current year, this firm revalued its fixed assets, with debits. It credited a revaluation reserve account. The firm will charge (debit) future depreciation and amortization of these revalued amounts to the revaluation reserve account, not to the earnings statement. Neither the initial revaluation nor the subsequent depreciation or amortization enters the earnings statement.

The reader of financial statements issued by firms outside of the United States will frequently encounter various accounts containing the word *reserve* appearing in the shareholders' equity section of the balance sheet. Reserve accounts, even outside the United States, *always* have a credit balance. The reader must be wary of these accounts for the following reasons:

1. Firms rarely set aside assets in the amount of the reserve for the purpose indicated by the title of the reserve (for example, "reserve for plant replacement," "reserve for business risks"). Because of the balance sheet equality of assets and equities, one can conclude only that there are assets in some form (for example, equipment or goodwill) equal to the amount of the reserve. But the assets would be there even if the amounts remained in the Retained Earnings account, without the firm's re-labeling them as reserves.
2. Firms increase and decrease reserves for a variety of purposes (such as reclassifying retained earnings and revaluing assets). The reader should understand the nature of each reserve and the events that cause its amounts to change.

## Comprehensive Presentation of Cash Flows, Income, and Changes in Shareholder Wealth

The book has discussed the purpose, underlying concepts, preparation, and interpretation of the three principal financial statements that firms prepare each period:

- Balance Sheet: a report on the financial position of a firm at a point in time.
- Income Statement: a report on the results of a firm's operating activities for a period of time.
- Statement of Cash Flows: a report on the net cash flows from a firm's operating, investing, and financing activities for a period of time.

**Exhibit 12.8** presents a summary statement for Wal-Mart Stores that relates various key measures from these three principal financial statements. **Exhibit 12.8** is not a required financial statement according to GAAP. We use it here primarily as a summary device. We discuss each of the elements in **Exhibit 12.8** next.

### Cash Flows from Operations

The top panel (shaded in green) shows the computation of cash flow from operations using the direct method (see the discussion in **Chapter 4**). The amounts in this panel result from cash transactions between Wal-Mart and its customers and suppliers. The existence of actual cash flows permits the auditor to trace and verify the amounts in this top panel with a high degree of reliability. Although firms have some flexibility to manage the timing of their operating cash flows (for exam-

**EXHIBIT 12.8**

**WAL-MART STORES**
**Consolidated Statement of Net Income, Comprehensive Income, Shareholders' Wealth (also showing Cash Flow from Operations) (amounts in millions)**

| Fiscal Year Ended January 31 | Year 11 | Year 12 | Year 13 |
|---|---|---|---|
| CASH PROVIDED BY OPERATING ACTIVITIES | | | |
| **Sources of Cash** | | | |
| Cash Received from Customer | $ 190,907 | $ 217,589 | $ 244,423 |
| Cash Received from Other Revenues | 1,966 | 2,013 | 2,139 |
| **Uses of Cash** | | | |
| Payments to Merchandise Suppliers | $(147,121) | $(169,139) | $(189,195) |
| Payments to Other Suppliers of Goods and Services | (31,082) | (34,980) | (39,285) |
| Payment of Interest | (1,374) | (1,326) | (1,063) |
| Payment of Income Taxes | (3,692) | (3,897) | (4,487) |
| Net Cash Provided by Operations | $ 9,604 | $ 10,260 | $ 12,532 |
| EXPENSES, GAINS, AND LOSSES TO CONVERT CASH FLOW FROM OPERATIONS TO NET INCOME | | | |
| Depreciation | (2,868) | (3,290) | (3,432) |
| CHANGES IN OPERATING ACCOUNTS TO CONVERT CASH FLOW FROM OPERATIONS TO NET INCOME | | | |
| Increase (Decrease) in: | | | |
| Accounts Receivable | 422 | 210 | 101 |
| Inventories | 1,795 | 1,235 | 2,236 |
| Prepayments | (75) | 180 | (745) |
| (Increase) Decrease in: | | | |
| Accounts Payable | (2,061) | (368) | (1,447) |
| Other Current Liabilities | (522) | (1,556) | (1,206) |
| Net Income | $ 6,295 | $ 6,671 | $ 8,039 |
| CHANGES IN OTHER COMPREHENSIVE INCOME | | | |
| Foreign Currency Translation | (1,126) | (472) | 1,113 |
| Cash Flow Hedges | 897 | (112) | (148) |
| Minimum Pension Liability | — | — | (206) |
| Comprehensive Income | $ 6,066 | $ 6,087 | $ 8,798 |
| CHANGES IN FAIR VALUES OF ASSETS AND LIABILITIES | | | |
| Land | ? | ? | ? |
| Buildings and Equipment | ? | ? | ? |
| Brand Names and Other Intangibles | ? | ? | ? |
| Long-Term Liabilities | ? | ? | ? |
| Total Change in Shareholder Wealth | ? | ? | ? |

ple, by delaying the payments of suppliers), the amounts in this panel are not as subject to management as amounts in other sections of this statement.

**Adjustment to Convert Cash Flow from Operations to Net Income** The middle section (shaded in blue) down to the line for Net Income shows the conversion of cash flow from operations to net income. Recall from **Chapter 4** that firms calculating cash flow from operations using the indirect method begin with net income and then convert it to cash flow from operations. This panel moves in the opposite direction. Revenue and expense that do not involve an operating cash flow in the period when firms recognize the revenue or expense appear in this section as additions to or subtractions from cash flow from operations. Depreciation, for example, is an expense in measuring net income but does not affect cash flow from operations. We therefore subtract depreciation from cash flow from operations to compute net income. Note that the indirect method of computing cash flow from operations adds depreciation to net income to zero out its effect when computing cash flow from operations. Because we are moving in the opposite direction in this case, we must now subtract depreciation. Other items that might appear in this section are deferred income taxes, gains and losses from sales of property, plant, and equipment, and equity in undistributed earnings of affiliates.

The middle section also includes changes in operating working capital accounts other than cash. These are the same items that appear in the conversion of net income to cash flow from operations using the indirect method but, because **Exhibit 12.8** moves in the opposite direction, the adjustments carry the opposite sign. For example, accounts receivable increase each year for Wal-Mart. The top panel of **Exhibit 12.8** shows the amount of cash received each year by Wal-Mart from sales to customers. An increase in accounts receivable for a year means that Wal-Mart sold more goods on account than it collected from customers. The conversion of cash flow from operations to net income requires that we convert cash collections from customers to sales revenue, resulting in an addition for the increase in accounts receivable.

The amounts shaded in blue in this section represent allocations of amounts from actual transactions. For example, depreciation allocates the cost of depreciable assets over their expected periods of benefit. Auditors can verify the acquisition cost of the depreciable assets by examining purchase invoices, cancelled checks, and other documents. However, firms must estimate the expected period of benefit of the depreciable assets and the depreciation method that will allocate this cost over those future periods. Take another example. The change in accounts receivable results from the net effects of sales on account, cash collections from customers, and estimates of uncollectible accounts. The sales transaction and the collection of cash are easily verifiable. Firms must estimate the amount of uncollectible accounts. To the extent that actual uncollectible accounts differ from estimates amounts, income gets shifted between accounting periods. The amounts in this section result from applying GAAP to compute net income on an accrual basis. Recall that accrual accounting attempts to match expenses with revenues to obtain a measure of operation performance that relates inputs to outputs. The estimates required to apply GAAP to achieve this matching, however, inject a degree of subjectivity into the measurement process and provide management with a greater opportunity to manage net income than to manage cash flow from operations.

**Adjustments to Convert Net Income to Comprehensive Income** The next section, also shaded in blue, shows changes in the values of assets and liabilities for which GAAP requires firms to revalue the asset or liability but does not permit inclusion of the resulting unrealized gain or loss in net income. Instead, firms include these unrealized gains and losses in other comprehensive income until the firm realizes the gains or losses in market transactions. Wal-Mart includes amounts for foreign currency translation, cash flow hedges, and minimum pension liability. This book discussed these items only briefly. A fuller study of these items would show that their computation requires information about exchange rates, interest rates, discount rates, and rates of return on investments, all of which can differ in the future from the current levels used to measure their amounts. The amounts included in other comprehensive income therefore represent only an estimate of the ultimate amount. Because they represent only estimates of amounts firms will realize later, the amounts are less reliable measures of wealth changes than cash flow from operations. Nonetheless, they do provide information to permit the user to track estimated wealth changes over time and, in that sense, provide more useful information than if firms made no such disclosures.

**Changes in Fair Values of Assets and Liabilities** The final section (shaded in pink) shows changes in the fair value of assets and liabilities for which GAAP generally requires valuation at acquisition cost or its equivalent (for example, land, buildings, equipment, long-term liabilities) or does not recognize the items as an asset or liability (for example, internally-developed intangibles or unsettled lawsuits). Changes in the value of these items affect the wealth of shareholders, even though GAAP does not permit the firm to revalue the asset (or liability) nor to include the unrealized gain (or loss) either in net income or in other comprehensive income. **Exhibit 12.8** shows no amounts for these items because GAAP generally does not require firms to report such fair values in notes to the financial statements. Users must obtain information about such value changes from other sources. For items with active markets, such as land and buildings, users can obtain more reliable measures of fair value changes than for items with less active markets, such as specially designed equipment or internally developed intangibles. We include this last section to emphasize the need to consider such value changes in measuring the change in wealth of shareholders.

**Summary of Comprehensive Presentation of Cash Flows, Income, and Changes in Shareholders' Wealth** **Exhibit 12.8** illustrates four categories of items that affect the value of a firm. GAAP treats each of these categories differently:

**Category 1:** Cash receipts and disbursements related to operations, which appear in the statement of cash flows.

**Category 2:** Accrual accounting adjustments to operating cash flows to derive net income. Firms report the net income amount in the income statement, showing the revenues and expenses for the period that net to this net income amount.

**Category 3:** Unrealized changes in the value of certain assets and liabilities that accountants can measure with sufficient reliability to justify revaluation of the asset or liability but whose ultimate amount when realized may differ. GAAP requires firms to include unrealized changes in value in Other Comprehensive Income.

**Category 4:** Unrealized changes in the value of certain assets and liabilities that accountants cannot measure with sufficient reliability to justify recognition of the value change either in the valuation of assets and liabilities or in shareholders' equity.

As one moves from category **1** to category **4**, the opportunity for management to manage earnings to its advantage increases. GAAP draws a line after category **2** as to amounts firms can include in reported net income, recognizing that even at this level firms must make choices in applying GAAP that provide opportunities for managing earnings. GAAP recognizes the potential information value of items in category **3** for tracking changes in the value of a firm but views the amounts as sufficiently tentative to justify excluding them from net income. GAAP views the value changes in category **4** as sufficiently unreliable to justify including them in the financial statements, even though these value changes may play an important role in the value of a firm.

## Summary

The shareholders' equity section of the balance sheet reports the sources of financing provided by preferred and common shareholders and their claims on the net assets of the firm. The equity of the preferred shareholders usually approximates the par value of the preferred shares. The remaining shareholders' equity accounts relate to the equity of the common shareholders. The equity of the common shareholders equals the sum of the amounts appearing in the Common Stock, Additional Paid-in Capital, Retained Earnings, Accumulated Other Comprehensive Income, Treasury Stock, and other common-share equity accounts. The user of the financial statements gains insight into capital contributions, earnings, other comprehensive income, dividends, and treasury stock transactions only by studying changes in the individual accounts.

## Appendix 12.1: Effects on the Statement of Cash Flows of Transactions Involving Shareholders' Equity

With the exception of earnings, most transactions affecting shareholders' equity accounts appear in the statement of cash flows as financing transactions.

### Capital Contributions

Firms that issue preferred or common stock for cash report the transaction as a financing activity. If the firm receives an asset other than cash (for example, building, equipment, or patent), it does not report the transaction as investing and financing activities that involve cash. The firm discloses the transaction in a supplementary schedule or note to the financial statements.

### Stock Options, Rights, and Warrants

The firm reports the cash received from the issuance of stock under options, rights, or warrants arrangements as a financing activity. When a firm recognizes a compensation expense related to stock options, it must add back the expense to earnings when computing cash flow from operations under the indirect method.

## Conversions of Bonds or Preferred Stock

The firm does not report the conversion of convertible bonds or convertible preferred stock into common stock as a financing activity that involves cash. The firm discloses this transaction in a supplementary schedule or note to the financial statements.

## Treasury Stock Transactions

Reacquisitions and reissuances of a firm's capital stock (that is, treasury stock) appear as financing activities in the statement of cash flows.

## Dividends

Dividends that a firm declares and pays in cash during a period appear as financing activities. If a portion of the dividend paid during the current period relates to dividends declared during the previous period or if the firm does not pay all dividends declared during the current period by the end of the period, then the amount in the Dividends Payable account will change. The change in the Dividends Payable account, a current liability, appears in the Financing section of the statement of cash flows as an adjustment to the amount of dividends declared. The adjustment converts dividends declared during the period to the amount of dividends paid in cash. Until this point of the book, all changes in current assets and liabilities, other than cash, marketable securities, and short-term borrowings, appeared as adjustments to net earnings in deriving cash provided by operations when the firm uses the indirect method in the statement of cash flows. Because all dividend activities are financing, the change in the Dividends Payable account must appear in the Financing section.

Firms sometimes issue property dividends (that is, dividends payable in inventory, equipment, or some other asset). Such dividends do not use cash and therefore appear in a supplementary schedule or note to the financial statements. If the firm recognized a gain or loss at the time that it distributed the property dividend, it subtracts the gain or adds back the loss to earnings when computing cash flow from operations under the indirect method.

Stock dividends, likewise, do not involve cash and therefore do not appear in the statement of cash flows. Nor do such dividends appear in a supplementary schedule or note because they do not usually change the equity ownership of the firm.

# Solutions to Self-Study Problems

### SUGGESTED SOLUTION TO PROBLEM 12.1 FOR SELF-STUDY

(Healy Corporation; journal entries for capital contributions.)

**a.** *January 2, Year 1*

| | | |
|---|---|---|
| Cash . . . . . . . . . . . . . . . . . . . . . . . . . . . . . . . . . . . . . . . . . . . . . | 1,400,000 | |
| Common Stock—$10 Par . . . . . . . . . . . . . . . . . . . . . . . . . . . . | | 1,000,000 |
| Additional Paid-in Capital . . . . . . . . . . . . . . . . . . . . . . . . . . . | | 400,000 |

| Assets | = | Liabilities | + | Shareholders' Equity | (Class.) |
|---|---|---|---|---|---|
| +1,400,000 | | | | +1,000,000 | ContriCap |
| | | | | +400,000 | ContriCap |

Issue of 100,000 shares of $10-par value common stock for $14 per share.

**b.** *January 2, Year 1*

| | | |
|---|---|---|
| Patent . . . . . . . . . . . . . . . . . . . . . . . . . . . . . . . . . . . . . . . . . . . . . | 140,000 | |
| Common Stock—$10 Par . . . . . . . . . . . . . . . . . . . . . . . . . . . . | | 100,000 |
| Additional Paid-in Capital . . . . . . . . . . . . . . . . . . . . . . . . . . . | | 40,000 |

*(continued)*

| Assets | = | Liabilities | + | Shareholders' Equity | (Class.) |
|---|---|---|---|---|---|
| +140,000 | | | | +100,000 | ContriCap |
| | | | | +40,000 | ContriCap |

Issue of 10,000 shares of $10-par value common stock in exchange for a patent. The value of the patent is not easily determinable, so use the issue price of $14 per share from part **a**.

**c.** *March 1, Year 1*

| | | |
|---|---|---|
| Cash | 200,000 | |
| Preferred Stock | | 200,000 |

| Assets | = | Liabilities | + | Shareholders' Equity | (Class.) |
|---|---|---|---|---|---|
| +200,000 | | | | +200,000 | ContriCap |

Issue of 2,000 shares of convertible preferred stock at par value.

**d.** *June 1, Year 1*

| | | |
|---|---|---|
| Cash | 50,000 | |
| Common Stock Warrants | | 50,000 |

| Assets | = | Liabilities | + | Shareholders' Equity | (Class.) |
|---|---|---|---|---|---|
| +50,000 | | | | +50,000 | ContriCap |

Issue of 10,000 common stock warrants for $5 per warrant.

**e.** *September 15, Year 1*

| | | |
|---|---|---|
| Preferred Stock | 60,000 | |
| Common Stock—$10 Par | | 24,000 |
| Additional Paid-in Capital | | 36,000 |

| Assets | = | Liabilities | + | Shareholders' Equity | (Class.) |
|---|---|---|---|---|---|
| | | | | −60,000 | ContriCap |
| | | | | +24,000 | ContriCap |
| | | | | +36,000 | ContriCap |

To record the conversion of 600 preferred shares into 2,400 common shares at book value.

**f.** *November 20, Year 1*

| | | |
|---|---|---|
| Cash | 96,000 | |
| Common Stock Warrants | 20,000 | |
| Common Stock—$10 Par | | 40,000 |
| Additional Paid-in Capital | | 76,000 |

| Assets | = | Liabilities | + | Shareholders' Equity | (Class.) |
|---|---|---|---|---|---|
| +96,000 | | | | −20,000 | ContriCap |
| | | | | +40,000 | ContriCap |
| | | | | +76,000 | ContriCap |

Issue of 4,000 shares of common stock in exchange for 4,000 stock warrants and $96,000 cash.

**g.** No entry in Year 1. The firm must amortize the $25,000 over the expected period of benefit in future years.

## SUGGESTED SOLUTION TO PROBLEM 12.2 FOR SELF-STUDY

(Baker Corporation; journal entries for dividends and stock splits.)

**a.** *March 31*

| | Dr. | Cr. |
|---|---|---|
| Retained Earnings | 12,500 | |
| Dividends Payable | | 12,500 |

| Assets | = | Liabilities | + | Shareholders' Equity | (Class.) |
|---|---|---|---|---|---|
| | | +12,500 | | −12,500 | RE |

To record declaration of cash dividend of $0.50 per share on 25,000 shares.

**b.** *April 15*

| | Dr. | Cr. |
|---|---|---|
| Dividends Payable | 12,500 | |
| Cash | | 12,500 |

| Assets | = | Liabilities | + | Shareholders' Equity | (Class.) |
|---|---|---|---|---|---|
| −12,500 | | −12,500 | | | |

To pay cash dividend declared on March 31.

**c.** *June 30*

| | Dr. | Cr. |
|---|---|---|
| Retained Earnings | 37,500 | |
| Common Stock | | 25,000 |
| Additional Paid-in Capital | | 12,500 |

| Assets | = | Liabilities | + | Shareholders' Equity | (Class.) |
|---|---|---|---|---|---|
| | | | | −37,500 | RE |
| | | | | +25,000 | ContriCap |
| | | | | +12,500 | ContriCap |

To record issuance of 10 percent stock dividend: $0.10 \times 25{,}000 = 2{,}500$ shares; $2{,}500 \times \$15 = \$37{,}500$.

**d.**

| | Dr. | Cr. |
|---|---|---|
| Common Stock, $10 par value | 275,000 | |
| Common Stock, $5 par value | | 275,000 |

| Assets | = | Liabilities | + | Shareholders' Equity | (Class.) |
|---|---|---|---|---|---|
| | | | | −275,000 | ContriCap |
| | | | | +275,000 | ContriCap |

To record 2-for-1 stock split. Alternatively, the firm might make no formal entry in the accounting records.

## SUGGESTED SOLUTION TO PROBLEM 12.3 FOR SELF-STUDY

(Crissie Corporation; journal entries for treasury stock transactions.)

**a.** *January 15*

| | | |
|---|---|---|
| Treasury Shares—Common | 90,000 | |
| Cash | | 90,000 |

| Assets | = | Liabilities | + | Shareholders' Equity | (Class.) |
|---|---|---|---|---|---|
| −90,000 | | | | −90,000 | ContriCap |

Paid $90,000 to reacquire 2,000 common shares at $45 per share.

**b.** *April 26*

| | | |
|---|---|---|
| Cash | 57,600 | |
| Treasury Shares—Common | | 54,000 |
| Additional Paid-in Capital | | 3,600 |

| Assets | = | Liabilities | + | Shareholders' Equity | (Class.) |
|---|---|---|---|---|---|
| +57,600 | | | | +54,000 | ContriCap |
| | | | | +3,600 | ContriCap |

Reissue of 1,200 shares of treasury stock costing $45 per share; treasury shares reissued to employees under stock option plan with an exercise price of $48 per share.

**c.** *August 15*

| | | |
|---|---|---|
| Treasury Shares—Common | 156,000 | |
| Cash | | 156,000 |

| Assets | = | Liabilities | + | Shareholders' Equity | (Class.) |
|---|---|---|---|---|---|
| −156,000 | | | | −156,000 | ContriCap |

Paid $156,000 to reacquire 3,000 common shares at $52 per share.

**d.** *November 24*

| | | |
|---|---|---|
| Preferred Stock | 80,000 | |
| Treasury Stock | | 77,600 |
| Additional Paid-in Capital | | 2,400 |

| Assets | = | Liabilities | + | Shareholders' Equity | (Class.) |
|---|---|---|---|---|---|
| | | | | −80,000 | ContriCap |
| | | | | +77,600 | ContriCap |
| | | | | +2,400 | ContriCap |

Reissue of 1,600 shares of treasury stock with a cost of $77,600 [= (800 × $45) + (800 × $52)] in exchange for convertible preferred stock with a book value of $80,000.

**e.** *December 20*

| | | |
|---|---|---|
| Cash | 70,500 | |
| Additional Paid-in Capital | 7,500 | |
| Treasury Stock | | 78,000 |

| Assets | = | Liabilities | + | Shareholders' Equity | (Class.) |
|---|---|---|---|---|---|
| +70,500 | | | | −7,500 | ContriCap |
| | | | | +78,000 | ContriCap |

Reissue of 1,500 shares of treasury stock costing $52 per share; treasury shares sold on the open market for $47 per share.

## SUGGESTED SOLUTION TO PROBLEM 12.4 FOR SELF-STUDY

(Able Corporation; journal entries for earnings and retained earnings transactions.)

**a.** *January 15*

| | | |
|---|---|---|
| Retained Earnings | 35,000 | |
| Accumulated Depreciation | | 35,000 |

| Assets | = | Liabilities | + | Shareholders' Equity | (Class.) |
|---|---|---|---|---|---|
| −35,000 | | | | −35,000 | RE |

To correct error in prior year's depreciation, increasing accumulated depreciation, with a direct charge to, reducing, retained earnings. (Recall that *charge* means *debit*.)

**b.** *March 20*

| | | |
|---|---|---|
| Loss from Earthquake | 70,000 | |
| Building | | 70,000 |

| Assets | = | Liabilities | + | Shareholders' Equity | (Class.) |
|---|---|---|---|---|---|
| −70,000 | | | | −70,000 | IncSt → RE |

To record loss from earthquake in an income statement account.

**c.** *December 31*

| | | |
|---|---|---|
| Depreciation Expense | 14,000 | |
| Accumulated Depreciation | | 14,000 |

| Assets | = | Liabilities | + | Shareholders' Equity | (Class.) |
|---|---|---|---|---|---|
| −14,000 | | | | −14,000 | IncSt → RE |

Original depreciation: $400,000/40 = $10,000 per year. Book value on January 1, Year 5 is $350,000 [= $400,000 − ($10,000 × 5)]. Depreciation for Year 5 is $14,000 (= $350,000/25).

## Key Terms and Concepts

- Corporation
- Corporate charter
- Corporate bylaws
- Capital stock contract
- Capital stock
- Preferred stock
- Cumulative dividend rights
- Callable preferred shares
- Convertible preferred shares
- Redeemable preferred shares
- Common stock
- Par or stated value
- Additional Paid-in Capital account
- Stock options
- Stock rights
- Stock warrants
- Benefit element in stock option
- Time value element in stock option
- Convertible bonds
- Cash dividend
- Dividend in kind or property dividend
- Stock dividend
- Stock split
- Reverse stock split
- Treasury shares, treasury stock
- Earnings from Continuing Operations
- Earnings, Gains, and Losses from Discontinued Operations
- Extraordinary Gains and Losses
- Realize v. recognize
- Other comprehensive income
- Accumulated other comprehensive income
- Comprehensive income
- Correction of errors
- Change in accounting principle
- Changes in estimates
- Earnings per common share
- Book value per common share
- Shareholders' equity reserve accounts

## Questions, Short Exercises, Exercises, Problems, and Cases

### QUESTIONS

For additional student resources, content, and interactive quizzes for this chapter visit the FACMU website: **www.thomsonedu.com/accounting/stickney**

1. Review the meaning of the terms and concepts listed in Key Terms and Concepts.
2. The chapter states that the amount in various common shareholders' equity accounts represents the equity of the common shareholders in the net assets of a firm. If a bankrupt firm sold its assets and used the proceeds to pay creditors, preferred shareholders, and common shareholders, would these common shareholders likely receive an amount equal to the amount in the common shareholders' equity accounts on the balance sheet? Explain.
3. "The accounting for stock options, stock dividends, and treasury stock cloud the distinction between capital transactions and operating transactions." Explain.
4. A firm contemplates issuing 10,000 shares of $100 par value preferred stock. The preferred stock promises a $4 per share annual dividend. The firm considers making this preferred stock callable, or convertible, or subject to mandatory redemption. Will the issue price be the same in each of these three cases? Explain.
5. Compare and contrast a stock option, a stock right, and a stock warrant. How does the accounting for these three differ?
6. Stock option valuation models indicate that the value of a stock option increases with the volatility of the stock, increases with the time between the grant date and the expected exercise date, and decreases with increases in the discount rate. Explain.
7. GAAP requires firms to amortize the value of stock options as an expense over the periods the firm expects to receive employee services as a result of granting the option. What is the theoretical rationale for this amortization?
8. Compare the position of a shareholder who receives a cash dividend with that of one who receives a stock dividend.
9. A firm that sells inventory for more than its acquisition cost realizes an economic gain that accountants include in earnings, but a firm that sells treasury stock for more than its acquisition cost realizes an economic gain that accountants exclude from earnings. What is the rationale for the difference in treatment of these economic gains?
10. A security analyst states: "Accountants could increase the usefulness of income statements if they included only recurring income items in the income statement and reported nonrecurring items directly in retained earnings, bypassing the income statement." Do you agree?

**11.** Why should GAAP exclude from earnings such items of other comprehensive income as holding gains and losses on securities held available for sale?

**12.** Distinguish between the nature of, and accounting for (1) a correction of an error in previously issued financial statements, (2) the adjustment for a change in accounting principle, and (3) the adjustment for a change in an accounting estimate made in preparing previously issued financial statements.

## SHORT EXERCISES

**13. Journal entry to issue common stock.** Office Depot issues 10 million shares of $2 par value common stock for $12.50 per share. Give the journal entry to record this transaction.

**14. Accounting for stock options.** Intel grants 100,000 stock options to employees on December 31, Year 13, permitting them to purchase 100,000 shares of Intel $1 par value common stock for $30 per share. An option-pricing model indicates that the value of each option on this date is $12. Intel expects to receive the benefit of enhanced employee services for the next three years. On December 31, Year 18, employees exercise these options when the market price of the stock is $40 per share. Compute the pretax effect of this option plan on the earnings of Intel for Year 13 through Year 18.

**15. Accounting for conversion of bonds.** Symantec has convertible bonds outstanding with a face value of $10,000,000 and a book value of $10,255,000. Holders of the bonds convert them into 100,000 shares of $10 par value common stock. The common stock sells for $105 per share on the market. Give the journal entries to record the conversion of the bonds using (1) the book value method, and (2) the market value method.

**16. Accounting for declaration and payment of a dividend.** Wyeth Corporation has 100,000 shares of $100 par value, 4-percent preferred stock outstanding. On December 10, Year 13, Wyeth Corporation declares the annual dividend on these preferred shares for Year 13. The dividend is paid on January 10, Year 14. Give the journal entries for these transactions.

**17. Accounting for treasury stock transactions.** On July 15, Year 13, PepsiCo purchased 75,000 shares of its common stock for $55 per share. On February 10, Year 14, PepsiCo reissued 50,000 shares of this common stock on the market for $62 per share. Give the journal entries to record these transactions.

**18. Computation of net income and comprehensive income.** Champion Enterprises reports the following items for Year 13 (amounts in millions): income from continuing operations, $126.7; extraordinary loss from hurricane, $10.3; other comprehensive income, $17.9; loss from discontinued operations, $26.9. Compute the amount of net income and the amount of comprehensive income for Year 13.

**19. Treatment of accounting errors, changes in accounting principles, and changes in accounting estimates.** A firm computes net income for Year 12 of $1,500 and for Year 13 of $1,800, its first two years of operations. Before issuing its financial statements for Year 13, the firm discovers that an item requires an income-reducing adjustment of $400 after taxes. Indicate the amount of net income for Year 12 and Year 13 assuming (1) the item is an error in the computation of depreciation expense for Year 12 (Year 13 depreciation expense is correct as computed), (2) the item is the change in net income for Year 12 as a result of adopting a new method of accounting for stock options in Year 13 (Year 13 stock option expense reflects the new accounting principle), and (3) the item is the change in estimated uncollectible accounts for Year 12 as a result of worsened credit losses experienced in Year 13; the firm included the adjustment amount in bad debt expense for Year 13.

## EXERCISES

**20. Journal entries to record the issuance of capital stock.** Prepare journal entries to record the issuance of capital stock in each of the independent cases below. You may omit explanations for the journal entries. A firm does the following:

**a.** Issues 50,000 shares of $5 par value common stock for $30 per share.

**b.** Issues 20,000 shares of $100 par value convertible preferred stock at par.

**c.** Issues 16,000 shares of $10 par value common stock in the acquisition of a patent. The shares of the firm traded on a stock exchange for $15 per share on the day of the transaction. The seller listed the patent for sale at $250,000.

**d.** Issues 25,000 shares of $1 par value common stock in exchange for convertible preferred stock with a par and book value of $400,000. The common shares traded on the market for $18 per share on the date of the transaction. Use the book value method to record the conversion.

**e.** Issues 5,000 shares of $10 par value common stock to employees as a bonus for reaching sales goals for the year. The shares traded for $12 per share on the day of the transaction.

**21. Journal entries for the issuance of capital stock.** Prepare journal entries to record the issuance of capital stock in each of the independent cases below. You may omit explanations for the journal entries. A firm does the following:

**a.** Issues 20,000 shares of $10 par value common stock in the acquisition of inventory with a market value of $175,000, land valued at $220,000, a building valued at $1,400,000, and equipment valued at $405,000.

**b.** Issues 10,000 shares of $100 par value preferred stock at par. The preferred stock is subject to mandatory redemption in five years.

**c.** Issues 5,000 shares of $1 par value common stock upon the exercise of stock warrants. The firm had issued the stock warrants several years previously for $8 per warrant and properly recorded the sale of the warrants in the accounts. The exercise price is $24 plus one warrant for each share of common stock.

**d.** Issues 20,000 shares of $10 par value common stock upon the conversion of 10,000 shares of $50 par value convertible preferred stock originally issued for par. Record the conversion using book values.

**22. Journal entries for employee stock options.** Morrissey Corporation grants 50,000 stock options to its managerial employees on December 31, Year 13, to purchase 50,000 shares of its $1 par value common stock for $60 per share. The market price of a share of common stock on this date is $50 per share. Employees must wait two years before the options vest and they can exercise the options and this two-year period is the expected period of benefit from the stock options. A financial consulting firm estimates that the market value of these options on the grant date is $400,000. On June 30, Year 15, holders of 30,000 options exercise their options at a time when the market price of the stock is $65 per share. On November 15, Year 16, holders of the remaining options exercise them at a time when the market price of the stock is $72 per share.

Present journal entries to record the effects of the transactions related to stock options during Year 13, Year 14, Year 15, and Year 16 following the market value method. The firm reports on a calendar-year basis. Ignore income tax effects.

**23. Journal entries for employee stock options.** Watson Corporation grants 20,000 stock options to its managerial employees on December 31, Year 6, to purchase 20,000 shares of its $10 par value common stock for $25 per share. The market price of a share of common stock on this date is $18 per share. Employees must work for another two years before they can exercise the options. A financial consulting firm estimates that the market value of these options on the grant date is $75,000. On April 30, Year 9, holders of 15,000 options exercise their options at a time when the market price of the stock is $30 per share. On September 15, Year 10, holders of the remaining options exercise them at a time when the market price of the stock is $38 per share.

Present journal entries to record these transactions on December 31, Year 6, Year 7, and Year 8; on April 30, Year 9; and on September 15, Year 10 following the market value method. Assume that the firm receives any benefits of the stock option plan during Year 6, Year 7, and Year 8 and that the firm reports on a calendar-year basis. Ignore income tax effects.

**24. Journal entries for stock warrants.** Kiersten Corporation sells 60,000 common stock warrants for $4 each on February 26, Year 12. Each warrant permits its holder to purchase a share of the firm's $10 par value common stock for $30 per share at any time during the next two years. The market price of the common shares was $20 per share on February 26, Year 12. Holders of 40,000 warrants exercised their warrants on June 6, Year 14, at a time when the market price of the stock was $38 per share. Kiersten Corporation experienced a major uninsured loss from a fire late in Year 14 and its market price fell immediately to $22 per share. The market price remained around $22 until the stock warrants expired on February 26, Year 16. Present journal entries on February 26, Year 12; June 6, Year 14; and February 26, Year 16, relating to these stock warrants.

**25. Journal entries for convertible bonds.** Higgins Corporation issues $1 million of 20-year, $1,000 face value, 10-percent semiannual coupon bonds at par on January 2, Year 1. Each $1,000 bond is convertible into 40 shares of $1 par value common stock. Assume that Higgins

Corporation's credit rating is such that it could issue 15-percent semiannual, nonconvertible bonds at par. On January 2, Year 5, holders convert their bonds into common stock. The common stock has a market price of $45 per share on January 2, Year 5.

**a.** Present the journal entries made under GAAP on January 2, Year 1, and January 2, Year 5, to record the issue and conversion of these bonds. Use the book value method to record the conversion.

**b.** Assume for this part that the firm allocates a portion of the issue price on January 2, Year 1, to the conversion option (GAAP does not allow this treatment). Give the journal entry on January 2, Year 1.

**26. Journal entries to correct errors and adjust for changes in estimates.** Prepare journal entries to record each of the following items for Uncertainty Corporation for Year 13. Ignore income tax effects.

**a.** Discovers on January 15, Year 13, that it neglected to amortize a patent during Year 12 in the amount of $12,000.

**b.** Discovers on January 20, Year 13, that it recorded the sale of a machine on December 30, Year 12, for $6,000 with the following journal entry:

| | | |
|---|---|---|
| Cash . . . . . . . . . . . . . . . . . . . . | 6,000 | |
| Loss on Sale of Machine . . . . . . . . . . . . | 4,000 | |
| Machine (acquisition cost) . . . . . . . . . . . | | 10,000 |

| Assets | = | Liabilities | + | Shareholders' Equity | (Class.) |
|---|---|---|---|---|---|
| +6,000 | | | | −4,000 | IncSt → RE |
| −10,000 | | | | | |

The machine had accumulated depreciation of $7,000 on the date of the sale.

**c.** Changes the depreciable life of a building as of December 31, Year 13, from a total useful life of 30 years to a total of 42 years. The building has an acquisition cost of $2,400,000 and is 11 years old as of December 31, Year 13. The firm has not recorded depreciation for Year 13. It uses the straight-line method and zero estimated salvage value.

**d.** The firm has used 2 percent of sales as its estimate of uncollectible accounts for several years. Its actual losses have averaged only 1.50 percent of sales. As a consequence, the Allowance for Estimated Uncollectibles account has a credit balance of $25,000 at the end of Year 13 before making the provision for Year 13. An aging of customers' accounts suggests that the firm needs $35,000 in the allowance account at the end of Year 13 to cover estimated uncollectibles. Sales for Year 13 are $1,000,000.

**27. Journal entries for dividends.** Give journal entries, if required, for the following transactions, which are unrelated unless otherwise specified:

**a.** A firm declares the regular quarterly dividend on its 6 percent, $100 par value preferred stock. There are 30,000 shares authorized and 15,000 shares issued, of which the firm has previously reacquired 2,000 shares and holds them in the treasury.

**b.** The firm pays the dividend on the preferred stock (see part **a**).

**c.** A company declares and issues a stock dividend of $300,000 of no-par common stock to its common shareholders.

**d.** The shares of no-par stock of the corporation sell on the market for $200 a share. To bring the market value down to a more popular price and thereby broaden the distribution of its stockholdings, the board of directors votes to issue four extra shares to shareholders for each share they already hold. The corporation issues the shares.

**28. Journal entries for dividends.** Prepare journal entries for the following transactions of Watt Corporation. The firm has 20,000 shares of $15 par value common stock outstanding on January 1, Year 6. The balance in the Additional Paid-in Capital account on this date is $200,000.

**a.** Declares a dividend of $0.50 per share on March 31, Year 6.

**b.** Pays the dividend in part **a** on April 15, Year 6.

**c.** Declares and distributes a 10 percent stock dividend on June 30, Year 6. The market price of the stock is $20 on this date.

**d.** Declares a $0.50 per share dividend on September 30, Year 6.

**e.** Pays the dividend in part **d** on October 15, Year 6.
**f.** Declares a 3-for-2 stock split on December 31, Year 6, but does not alter the par value.

**29. Journal entries for treasury stock transactions.** Prepare journal entries to record the following treasury stock transactions of Danos Corporation.
**a.** Purchases 10,000 shares of $10 par value common stock for $30 per share.
**b.** Issues 6,000 treasury shares to employees under stock option plans. The exercise price is $32 per share. Assume that the market price of the common stock on the exercise date is $35 per share. The stock options had a value of $6 a share when issued, which the firm has already amortized to expense.
**c.** Purchases 7,000 shares of common stock for $38 per share.
**d.** Issues 8,000 treasury shares in the acquisition of land valued at $300,000. Danos Corporation uses a FIFO assumption for reissues of treasury stock.
**e.** Sells the 3,000 remaining shares of treasury stock for $36 per share.

**30. Journal entries for treasury stock transactions.** Prepare journal entries to record the following treasury stock transactions of Melissa Corporation.
**a.** Purchases 10,000 shares of $5 par value common stock for $12 per share.
**b.** Issues 6,000 treasury shares upon the conversion of bonds with a book value of $72,000. Melissa Corporation records bond conversions using the book value method.
**c.** Purchases 20,000 shares of common stock for $15 per share.
**d.** Issues 24,000 treasury shares and 6,000 newly issued shares of common stock in the acquisition of land with a market value of $540,000.

**31. Effects on statement of cash flows.** Refer to the simplified statement of cash flows in **Exhibit 4.16**. Numbers appear on 11 of the lines in the statement. Ignore the unnumbered lines in considering the following transactions.

Assume that a firm has completed the accounting cycle for the period and prepared all of the financial statements. It then discovers that it has overlooked a transaction. It records the transaction in the accounts and corrects all of the financial statements. For each of the following transactions, indicate which of the numbered lines of the statement of cash flows changes and by how much. Ignore income tax effects.
**a.** The firm issues common shares for $200,000.
**b.** The firm repurchases for $75,000 common shares originally issued at par for $50,000 and retires them.
**c.** Holders of convertible bonds with a book value of $100,000 and a market value of $240,000 convert them into common shares with a par value of $10,000 and a market value of $240,000. Respond assuming that the firm uses (1) the book value method to record bond conversions, and (2) the market value method to record bond conversions.
**d.** The firm issues for $15,000 treasury shares that have been previously acquired for $20,000.
**e.** The directors declare a stock dividend. The par value of the shares issued is $1,000, and their market value is $300,000.
**f.** The directors declare a cash dividend of $70,000, but the firm has not yet paid it.
**g.** The firm pays a previously declared cash dividend of $70,000.
**h.** Holders of stock rights exercise them. The shares issued have a par value of $1,000 and a market value of $35,000 on the date of exercise. The firm receives the exercise price of $20,000 in cash.

## PROBLEMS AND CASES

**32. Transactions to incorporate and run a business.** The following events relate to shareholders' equity transactions of Wilson Supply Company during the first year of its existence. Present journal entries for each of the transactions.
**a.** January 2. The firm files articles of incorporation with the State Corporation Commission. The authorized capital stock consists of 5,000 shares of $100 par value preferred stock that offers an 8-percent annual dividend, and 50,000 shares of no-par common stock. The original incorporators acquire 300 shares of common stock at $30 per share; the firm collects cash for the shares. It assigns a stated value of $30 per share to the common stock.
**b.** January 6. The firm issues 2,000 shares of common stock for cash at $30 per share.
**c.** January 8. It issues 4,000 shares of preferred stock at par.
**d.** January 9. The firm issues certificates for the shares of preferred stock.

**e.** January 12. The firm acquires the tangible assets and goodwill of Richardson Supply, a partnership, in exchange for 1,000 shares of preferred stock and 12,000 shares of common stock. It values the tangible assets acquired as follows: inventories, $50,000; land, $80,000; buildings, $210,000; and equipment, $120,000.

**f.** July 3. The directors declare the semiannual dividend on preferred stock outstanding, payable July 25, to shareholders of record on July 12.

**g.** July 5. The firm operated profitably for the first six months, and it decides to expand. The company issues 25,000 shares of common stock for cash at $33 per share.

**h.** July 25. It pays the dividend on preferred stock declared on July 3.

**i.** October 2. The directors declare a dividend of $1 per share on the common stock, payable October 25, to shareholders of record on October 12.

**j.** October 25. The firm pays the dividend on common stock declared on October 2.

**33. Reconstructing transactions involving shareholders' equity.** Fisher Company began business on January 1. Its balance sheet on December 31 contained the shareholders' equity section in **Exhibit 12.9**.

**EXHIBIT 12.9**

**FISHER COMPANY**
**Shareholders' Equity as of December 31**
**(Problem 33)**

| | |
|---|---|
| Common Stock ($10 par value) | $60,000 |
| Additional Paid-in Capital | 31,440 |
| Retained Earnings | 12,000 |
| Less Unrealized Holding Loss on Securities Available for Sale | (2,000) |
| Less 360 Shares Held in Treasury | (7,200) |
| Total Shareholders' Equity | $94,240 |

During the year, Fisher Company engaged in the following transactions:

**(1)** Issued shares for $15 each.

**(2)** Acquired a block of 600 shares for the treasury in a single transaction.

**(3)** Reissued some of the treasury shares.

**(4)** Sold for $10,000 securities available for sale with original cost of $6,000. At the end of the year, securities available for sale had original cost of $12,000 and market value of $14,000.

Assuming that these were all of the common stock transactions during the year, answer the following questions:

**a.** How many shares did Fisher Company issue for $30?

**b.** What was the price at which it acquired the treasury shares?

**c.** How many shares did it reissue from the block of treasury shares?

**d.** What was the price at which it reissued the treasury shares?

**e.** What journal entries did it make during the year?

**f.** In which statement or statements will Fisher Company report the various gains and losses on its holdings of securities available for sale?

**34. Reconstructing transactions involving shareholders' equity.** Shea Company began business on January 1. Its balance sheet on December 31 contained the shareholders' equity section shown in **Exhibit 12.10**.

During the year, Shea Company engaged in the following transactions:

**(1)** Issued shares for $30 each.

**(2)** Acquired a block of 2,000 shares for the treasury in a single transaction.

**(3)** Reissued some of the treasury shares.

**(4)** Sold for $12,000 securities available for sale that had originally cost $14,000. At the end of the year, securities available for sale, still on hand, that had originally cost $25,000 had market value of $18,000.

Assuming that these were the only common stock transactions during the year, answer the following questions:

**a.** How many shares did Shea Company issue for $30 each?

**b.** What was the price at which it acquired the treasury shares?

**EXHIBIT 12.10**

**SHEA COMPANY**
**Shareholders' Equity as of December 31**
**(Problem 34)**

| | |
|---|---|
| Common Stock ($5 par value) | $100,000 |
| Additional Paid-in Capital | 509,600 |
| Retained Earnings | 50,000 |
| Less Unrealized Holding Loss on Securities Available for Sale | (7,000) |
| Less 1,200 Shares Held in Treasury | (33,600) |
| Total Shareholders' Equity | $619,000 |

**c.** How many shares did it reissue from the block of treasury shares?
**d.** What was the price at which it reissued the treasury shares?
**e.** What journal entries did it make during the year?
**f.** In which statement or statements will Shea Company report the various gains and losses on its holdings of securities available for sale?

**35. Accounting for stock options.** Lowe Corporation grants stock options to its managerial employees on December 31 of each year. Employees may acquire one share of common stock with each stock option. Lowe Corporation sets the exercise price equal to the market price of its common stock on the date of the grant. Employees must continue working for two years after the date of the grant before the options vest and employees can exercise them. This two-year period is the period of benefit. **Exhibit 12.11** presents information for the stock options granted by Lowe Corporation on December 31 of each year.

**EXHIBIT 12.11**

**Stock Option Data for Lowe Corporation**
**(Problem 35)**

| Year | Options Granted during Year | Option Price per Share | Option Value per Share |
|---|---|---|---|
| 1 | 5,000 | $18 | $2.40 |
| 2 | 6,000 | $22 | $3.00 |
| 3 | 7,000 | $25 | $3.14 |
| 4 | 8,000 | $30 | $3.25 |
| 5 | 9,000 | $38 | $5.33 |

Assume that Lowe Corporation accounts for its stock options using the market value method. Calculate the effect of the stock options on net income before income taxes for Years 1 to 5.

**36. Comprehensive review of accounting for shareholders' equity.** The shareholders' equity section of the balance sheet of Alex Corporation at December 31 is as follows:

| | |
|---|---|
| Shareholders' Equity | |
| Common Stock—$10 Par Value, 250,000 Shares Authorized and 50,000 Shares Outstanding | $ 500,000 |
| Additional Paid-in Capital | 250,000 |
| Retained Earnings | 1,500,000 |
| Total Shareholders' Equity | $2,250,000 |

**a.** Calculate the total book value and the book value per common share as of December 31.
**b.** For each of the following transactions or events, give the appropriate journal entry and compute the total book value and the book value per common share of Alex Corporation after the transaction. The transactions and events are independent of one another, except where noted.

**EXHIBIT 12.12**

**NESLIN COMPANY**
**Statement of Changes in Shareholders' Equity Accounts for Year 2**
**(Problem 37)**

| | Market Value per Share[a] | Number of Shares | Par Value | Additional Paid-in Capital | Retained Earnings | Accumulated Other Comprehensive Income | Treasury Stock | Total Shareholders' Equity |
|---|---|---|---|---|---|---|---|---|
| Balances, Dec. 31, Year 1 . . . . . . . . | $50 | 100,000 | $1,000,000 | $5,400,000 | $ 9,600,000 | $1,200,000 | | $17,200,000 |
| Events Causing Changes | | | | | | | | |
| (1) . . . . . . . . . . . . . . . . . . . . . | 52 | 20,000 | 200,000 | 840,000 | — | — | — | 1,040,000 |
| (2) . . . . . . . . . . . . . . . . . . . . . | 55 | (4,000) | — | — | — | — | $(220,000) | (220,000) |
| (3) . . . . . . . . . . . . . . . . . . . . . | 48 | 3,000 | — | (21,000) | — | — | 165,000 | 144,000 |
| (4) . . . . . . . . . . . . . . . . . . . . . | 60 | 1,000 | — | 5,000 | — | — | 55,000 | 60,000 |
| (5) . . . . . . . . . . . . . . . . . . . . . | 62 | — | — | — | 2,400,000 | — | — | 2,400,000 |
| (6) . . . . . . . . . . . . . . . . . . . . . | 63 | — | — | — | — | (150,000) | — | (150,000) |
| (7) . . . . . . . . . . . . . . . . . . . . . | 63 | — | — | — | (1,200,000) | — | — | (1,200,000) |
| Balances, Dec. 31, Year 2 . . . . . . . . | | 120,000 | $1,200,000 | $6,224,000 | $10,800,000 | $1,050,000 | — | $19,274,000 |

[a]Before event.

**(1)** Declares a 10-percent stock dividend when the market price of Alex Corporation's common stock is $30 per share.
**(2)** Declares a 2-for-1 stock split and reduces the par value of the common stock from $10 to $5 per share. The firm issues the new shares immediately.
**(3)** Purchases 5,000 shares of Alex Corporation's common stock on the open market for $25 per share and holds the shares as treasury stock.
**(4)** Purchases 5,000 shares of Alex Corporation's common stock on the open market for $15 per share and holds the shares as treasury stock.
**(5)** Sells the shares acquired in **(3)** for $35 per share.
**(6)** Sells the shares acquired in **(3)** for $20 per share.
**(7)** Sells the shares acquired in **(3)** for $15 per share.
**(8)** Officers exercise options to acquire 5,000 shares of Alex Corporation stock for $15 per share.
**(9)** Same as **(8)**, except that the exercise price is $50 per share.
**(10)** Holders of convertible bonds with a book value of $150,000 and a market value of $170,000 exchanged them for 10,000 shares of common stock having a market value of $17 per share. The firm recognizes no gain or loss on the conversion of bonds.
**(11)** Same as **(10)**, except that the firm recognizes gain or loss on the conversion of bonds into stock. Ignore income tax effects.

**c.** Using the results from part **b**, summarize the transactions and events that result in a reduction in
**(1)** total book value, and
**(2)** book value per share.

**37. Reconstructing events affecting shareholders' equity. Exhibit 12.12** reproduces the statement of changes in shareholders' equity accounts for Neslin Company for Year 2.

**a.** Identify the most likely events or transactions for each of the events numbered **(1)** to **(7)** in the exhibit. The events are not independent of one another.
**b.** Prepare journal entries for each of these events or transactions.

**38. Journal entries for changes in shareholders' equity. Exhibit 12.13** (on page 624) presents a statement of changes in shareholders' equity for Wal-Mart Stores for Year 11. Prepare journal entries for each of these transactions or events. The corporate acquisition involved an investment in the common stock of another entity.

**39. Treasury shares and their effects on performance ratios. Exhibit 12.14** (on page 625) presents the changes in common shareholders' equity of Merck for Year 3 through Year 5. Merck regularly purchases shares of its common stock and reissues them in connection with stock option plans. It will usually issue a small number of new common shares when it requires fractional shares to complete a stock option transaction. Earnings per common share were $2.70 in Year 3, $3.20 in Year 4, and $3.83 in Year 5.

**a.** Give the journal entries for Year 5 to record (1) the issue of common shares in connection with stock option plans, and (2) the purchase of treasury stock.
**b.** Compute the percentage change in net income and in earnings per share between Year 3 and Year 4, and between Year 4 and Year 5. Why do the percentage changes in earnings per share exceed the percentage changes in net income in both Year 4 and Year 5?
**c.** Compute the book value per outstanding common share at the end of Year 3, Year 4, and Year 5, and the percentage change in book value per share between Year 3 and Year 4, and between Year 4 and Year 5. Why are the percentage changes in book value per common share less than the percentage changes in both net income and earnings per share?
**d.** Compute the rate of return on common shareholders' equity for Year 3, Year 4, and Year 5.
**e.** Do the treasury stock purchases appear to be motivated primarily by the need to satisfy commitments under stock option plans? Explain.

**40. Case introducing earnings-per-share calculations for a complex capital structure.** The Layton Ball Corporation has a relatively complicated capital structure—that is, it raises funds using a variety of financing devices. In addition to common shares, it has issued stock options, warrants, and convertible bonds. **Exhibit 12.15** (on page 625) summarizes some pertinent information about these items. Net income for the year is $9,500, and the income tax rate used in computing income tax expense is 40 percent of pretax income.

**a.** First, ignore all items of capital except for the common shares. Calculate earnings per common share.
**b.** In past years, employees have been issued options to purchase shares of stock. **Exhibit 12.15** indicates that the price of the common stock throughout the current year has remained steady at $25 but that holders of the stock options could exercise them at any

**EXHIBIT 12.13**

**WAL-MART STORES**
**Statement of Changes in Shareholders' Equity Accounts for Year 11**
**(all dollar amounts in thousands)**
**(Problem 38)**

| | Number of Shares | Common Stock | Additional Paid-in Capital | Retained Earnings | Accumulated Other Comprehensive Income | Treasury Shares | Total Shareholders' Equity |
|---|---|---|---|---|---|---|---|
| Balance, January 31, Year 10 | 566,135 | $ 56,613 | $180,465 | $3,728,482 | $763,488 | — | $4,729,048 |
| (1) Net Income | — | — | — | 1,291,024 | — | — | 1,291,024 |
| (2) Cash Dividends | — | — | — | (158,889) | — | — | (158,889) |
| (3) Exercise of Stock Options | 662 | 66 | 3,754 | — | — | — | 3,820 |
| (4) Two-for-One Stock Split | 568,797 | 56,680 | (56,680) | — | — | — | — |
| (5) Shares Issued in Corporate Acquisition | 10,366 | 1,037 | 273,659 | — | — | — | 274,696 |
| (6) Purchase of Shares | — | — | — | — | — | $(25,826) | (25,826) |
| (7) Mark Securities to Market | — | — | — | — | (57,086) | — | (57,086) |
| Balance, January 31, Year 11 | 1,145,960 | $114,396 | $401,198 | $4,860,617 | $706,402 | $(25,826) | $6,056,787 |

EXHIBIT 12.14

**MERCK & CO.**
**Analysis of Changes in Common Shareholders' Equity**
**(all dollar amounts in millions)**
**(Problem 39)**

| | Common Stock[a] | | | Treasury Stock | | |
|---|---|---|---|---|---|---|
| | Shares | Amount | Retained Earnings | Shares | Amount | Total |
| December 31, Year 2 | 1,483.168 | $4,667.8 | $10,942.0 | (235.342) | $(4,470.8) | $11,139.0 |
| Net Income | — | — | 3,376.6 | — | — | 3,376.6 |
| Dividends | — | — | (1,578.0) | — | — | (1,578.0) |
| Stock Options Exercised | 0.295 | 74.7 | — | 14.104 | 294.3 | 369.0 |
| Treasury Stock Purchased | — | — | — | (33.377) | (1,570.9) | (1,570.9) |
| December 31, Year 3 | 1,483.463 | $4,742.5 | $12,740.6 | (254.615) | $(5,747.4) | $11,735.7 |
| Net Income | — | — | 3,870.5 | — | — | 3,870.5 |
| Dividends | — | — | (1,793.4) | — | — | (1,793.4) |
| Stock Options Exercised | 0.156 | 225.0 | — | 15.982 | 426.0 | 651.0 |
| Treasury Stock Purchased | — | — | — | (38.384) | (2,493.3) | (2,493.3) |
| December 31, Year 4 | 1,483.619 | $4,967.5 | $14,817.7 | (277.017) | $(7,814.7) | $11,970.5 |
| Net Income | — | — | 4,596.5 | — | — | 4,596.5 |
| Dividends | — | — | (2,094.8) | — | — | (2,094.8) |
| Stock Options Exercised | 0.307 | 286.5 | — | 14.183 | 427.6 | 714.1 |
| Treasury Stock Purchased | — | — | — | (27.444) | (2,572.8) | (2,572.8) |
| December 31, Year 5 | 1,483.926 | $5,254.0 | $17,319.4 | (290.278) | $(9,959.9) | $12,613.5 |

[a]Includes Additional Paid-in Capital.

EXHIBIT 12.15

**LAYTON BALL CORPORATION**
**Information on Capital Structure**
**for Earnings-per-Share Calculations**
**(Problem 40)**

Assume the following data about the capital structure and earnings for the Layton Ball Corporation for the year:

| | |
|---|---|
| Number of Common Shares Outstanding throughout the Year | 2,500 shares |
| Market Price per Common Share throughout the Year | $25 |
| Options Outstanding during the Year: | |
| Number of Shares Issuable on Exercise of Options | 1,000 shares |
| Exercise Price per Share | $15 |
| Warrants Outstanding during the Year: | |
| Number of Shares Issuable on Exercise of Warrants | 2,000 shares |
| Exercise Price per Share | $30 |
| Convertible Bonds Outstanding: | |
| Number (issued 15 years ago) | 100 bonds |
| Proceeds per Bond at Time of Issue (= par value) | $1,000 |
| Shares of Common Issuable on Conversion (per bond) | 10 shares |
| Coupon Rate (per year) | 4 percent |

time for $15 for each share. That is, the option allows the holder to surrender it along with $15 cash and receive one share in return. Thus the number of shares would increase, which would decrease the earnings-per-share figure. The company would, however, have more cash. Assume that the holders of options tender them, along with $15 each, to purchase shares. Assume that the company uses the cash to purchase shares for its own treasury at a price of $25 each. Compute a new earnings-per-share figure. The firm does not count shares in its own treasury in the denominator of the earnings-per-share calculation.

**c.** **Exhibit 12.15** indicates that there were also warrants outstanding in the hands of the public. The warrant allows the holder to turn in that warrant, along with $30 cash, to purchase one share of stock. If holders exercised the warrants, the number of outstanding shares would increase, which would reduce earnings per share. However, the company would have more cash, which it could use to purchase shares for the treasury, reducing the number of shares outstanding. Assume that all holders of warrants exercise them. Assume that the company uses the cash to purchase outstanding shares for the treasury. Compute a new earnings-per-share figure. Ignore the information about options and the calculations in part **b** at this point. Note that a rational warrants holder would not exercise the warrants for $30 when they can purchase a share for $25.

**d.** The firm also has convertible bonds outstanding. Each convertible bond entitles the holder to trade in that bond for 10 shares. If holders convert the bonds, the number of shares would increase, which would tend to reduce earnings per share. On the other hand, the company would not have to pay interest and thus would have no interest expense on the bond because it would no longer be outstanding. This would tend to increase income and earnings per share. Assume that all holders of convertible bonds convert their bonds into shares. Compute a new net income figure (do not forget income tax effects on income of the interest saved) and a new earnings-per-share figure. Ignore the information about options and warrants and the calculations in parts **b** and **c** at this point.

**e.** Now consider all the previous calculations. Which sets of assumptions from parts **b**, **c**, and **d** lead to the lowest possible earnings per share when they are all made simultaneously? Compute a new earnings per share under the most restrictive set of assumptions about reductions in earnings per share.

**f.** Accountants publish several earnings-per-share figures for companies with complicated capital structures and complicated events during the year. *The Wall Street Journal*, however, publishes only one figure in its daily columns (where it reports the price-earnings ratio—the price of a share of stock divided by its earnings per share). Which of the figures computed previously for earnings per share do you think *The Wall Street Journal* should report as the earnings-per-share figure? Why?

**41. Case for discussion: Value of stock options.** (The text does not give an explicit answer to this question but provides a sufficient basis to enable students to discuss the question.) Below is an excerpt from an article from the *San Francisco Examiner*, a leading Silicon Valley newspaper, which appeared at the height of the controversy over the accounting for employee stock options.

> For example, if StartUp Inc. recruits the brilliant software designer Joe Bithead . . . by offering him the option to buy 10,000 shares of StartUp's stock at its current price of a penny a share, what's the value of Joe's grant? If StartUp goes belly up, as 80 percent of new high-tech firms do, the grant is worthless. . . . If, on the other hand, after five years of struggle, StartUp manages to create a successful product and outperforms its competitors, the company's stock might sell for $10 a share on the public market. For a penny each, Joe can buy the 10,000 shares. . . . He unloads them in the market for a $100,000 profit.

The accounting question is, What cost, if any, does StartUp incur on Day 1 of the grant to Bithead of an option to acquire 10,000 shares five years hence for $0.01 per share? StartUp shares trade in public stock markets on the date of grant at $0.01 per share. Because the word *cost* has so many meanings (see *cost terminology* in the **Glossary**), make the question operational and specific by considering the following.

Imagine that you are the financial executive of StartUp and that Goldman Sachs offers to relieve you of the obligation to deliver the shares to Bithead. That is, Goldman will take a payment from you today and will deliver the shares to Bithead if he exercises the options but will do nothing otherwise, except keep your cash. How much are you willing to pay Goldman *today* to relieve you of your obligation to Bithead? That is, you pay Goldman now and they later deliver shares to Bithead if he exercises his options. No one can be sure of the exact answer, given the sketchy data, but which of the following ranges do you think most likely?

**a.** $0 to $10.
**b.** $10 to $100.
**c.** $100 to $1,000.
**d.** $1,000 to $10,000.
**e.** $10,000 to $100,000.
**f.** Some other answer (indicate answer).

**42. Ethical issues in accounting for stock options.** Throughout the deliberations on the accounting for stock options during the last decade, managers of business firms, particularly in technology industries, have argued against the recognition of an expense for stock options. These managers emphasize the difficulty in measuring reliably the cost of stock options until employees actually exercise them. Yet the managers of these firms receive a substantial proportion of their compensation in the form of stock options. Discuss any ethical issues raised by the position of these managers.

**43. Ethical issues in issuing stock options.** As this book goes to press, the business world is agog with scandal involving the backdating of stock options. Those who advocate the use of options as a part of an employee's compensation note that the employee gets nothing from the option unless the price increases, and if it does, then the shareholders, who hold more shares than do the employees, benefit even more than do the employees.

In granting options to employees, the Board can set whatever exercise or strike price it chooses. If that price is not the closing price on the grant date, however, the firm will suffer adverse income tax consequences.

We cannot be sure what the various SEC and internal investigations will uncover, but the following independent questions introduce the issues. Comment on any ethical matters raised in these scenarios.

**a.** During the first two weeks of March, members of the Board of Directors of Lucky Strike Company agreed, through a series of conference calls, that they would authorize a grant of options for 100,000 shares to the CEO at the March 18 Board meeting. On March 18, the current market price was $20 per share. During the meeting, the CEO asked a member of the compensation committee, "Our conference call on this was on the 5th, right? That's when we actually agreed to authorize the grant." On March 5, the market price of the shares was $12 per share and was the lowest closing price during March for the shares. The Board formally authorizes the options grant on March 18 so it must choose the exercise price for the options it grants on March 18.

**b.** During the day, the shares of Lucky Strike Company subject to option grants fluctuate in market price from $19 to $21 per share. The Board authorizes a grant of options to the CEO to acquire 100,000 at today's closing price, which turns out to be $20 per share. At the same time, during the day before the market closes for the day, management authorizes delay until tomorrow morning the release of good news about the FDA's granting approval of a new drug application, which it learned about only this morning. When the news becomes public on the next day, the share price increases, as management anticipated, from $20 per share to $30 per share. Assume that management, but not the Board, knows about the good news.

**c.** During the day, the shares of Lucky Strike Company subject to option grants have fluctuated in market price from $19 to $21 per share. Late in the day before the market closes for the day, it authorizes immediate release of bad news about the FDA's denying approval of a new drug application. When the news becomes public that afternoon, the share price drops, as anticipated, to $12 per share. The Board authorizes a grant of options to the CEO to acquire 100,000 at tomorrow's closing price, which turns out to be $11 per share. Assume that both management and the Board know about the bad news.

**d.** Management of Micrel, Inc. historically offered stock options to new employees with exercise price set equal to the closing price on the date of hire. The market price of its shares is so volatile, however, that employees hired within in a few days of each other received options of significantly differing values, because the closing prices on their days of hire vary so much. To deal with the morale-depressing inequity, management of Micrel, with the blessing of its Big 4 audit firm, changes its practice to grant the new employee options with an exercise price of equal to the lowest closing price during the 30 days following the date. Micrel, Inc. discloses this practice in its financial statements. (Microsoft had for many years a similar policy, which it stopped in 1999, for some of its option grants.) The accounting treated the grant date as the same as the date when the shares traded at their exercise price.

**e.** Refer to the data in the preceding item about Micrel, Inc. Several years after commencing this new policy, the Big 4 accounting firm rescinded its approval of the 30-day window for finding the low price to set as the exercise price. It required Micrel to restate its financial statements for the years when it had followed the policy.

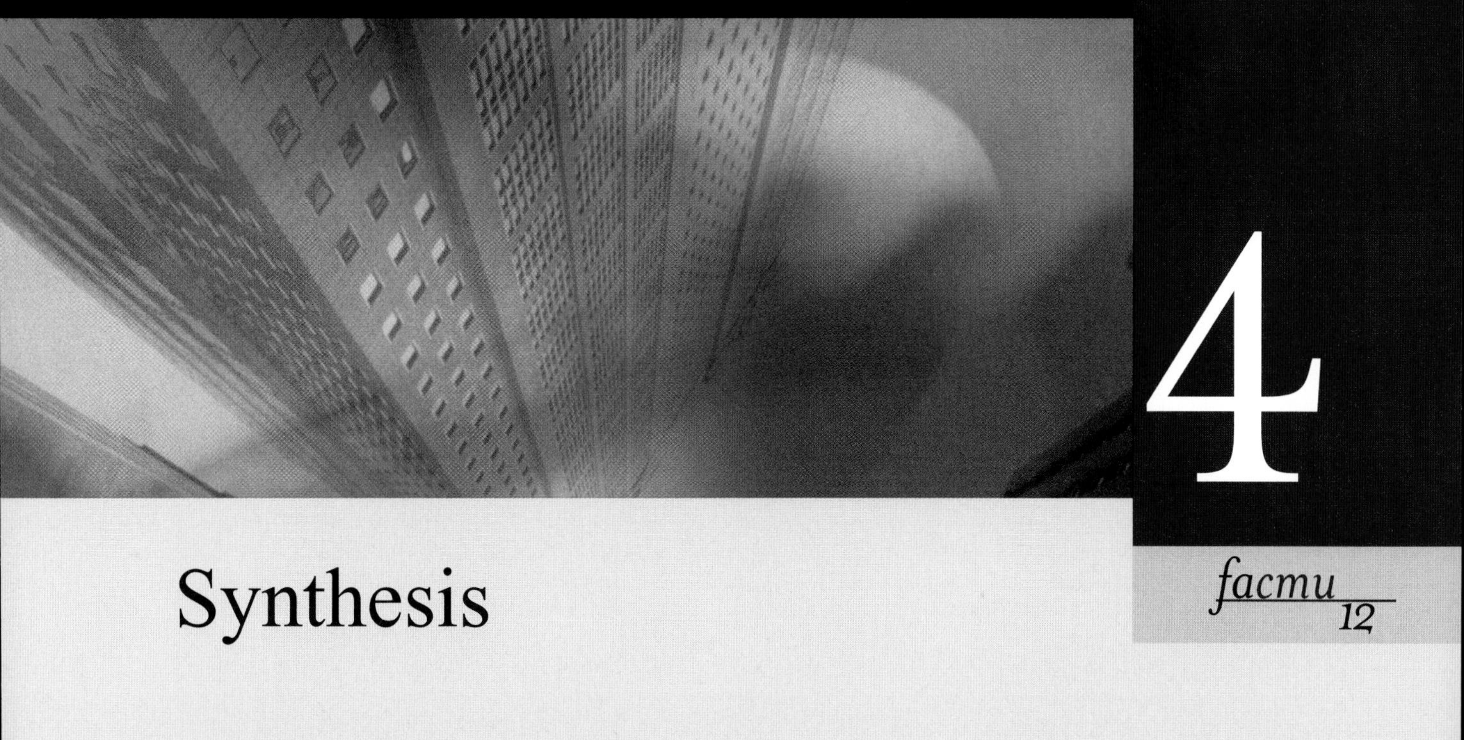

# 4

facmu 12

# Synthesis

CHAPTER 13

# Statement of Cash Flows: Another Look

## Learning Objectives

1. Review the rationale for the statement of cash flows, particularly regarding why net income differs from cash flows.
2. Review the T-account procedure for preparing the statement of cash flows introduced in Chapter 4.
3. Cement an understanding of the effect on the statement of cash flows of various transactions studied in Chapters 6 through 12.
4. Develop more effective skills in analyzing and interpreting the statement of cash flows.

**Chapter 4** introduced the statement of cash flows, discussing its rationale and illustrating a T-account approach for preparing this financial statement. Subsequent chapters briefly described the impact that various transactions have on the statement of cash flows. This chapter synthesizes these chapter-by-chapter discussions by providing a comprehensive example of a statement of cash flows.

## Review of Concepts Underlying the Statement of Cash Flows

**Chapter 4** discussed the following concepts underlying the statement of cash flows:

1. The statement of cash flows explains the reasons for the changes in cash and cash equivalents during a period. This statement classifies the reasons as relating to operating, investing, and financing activities.
2. Revenues from sales of goods or services to customers do not necessarily equal cash received from customers during a particular period. The receipt of cash may precede, coincide with, or follow the recognition of revenue. Expenses incurred to generate revenues do not necessarily equal cash expended for the goods and services consumed in operations during a particular period. The expenditure of cash may precede, coincide with, or follow the recognition of expenses. Thus, net income for a particular period will likely differ from cash flow from operations.
3. Firms typically report cash flows from operations using the indirect method. The indirect method starts with net income, then adds expenses that do not use cash in the amount of the expense, and subtracts revenues that do not provide cash in the amount of the revenue. The

adjustments to convert net income to cash flow from operations generally involve (1) adding the amount by which an expense exceeds the related cash expenditure for the period (for depreciation, the entire amount), (2) subtracting the amount by which a revenue exceeds the related cash receipt for the period (such as equity method earnings exceeding dividends), (3) adding credit changes in operating working capital accounts, such as accounts receivable, inventories, or accounts payable, and (4) subtracting debit changes in operating working capital accounts.

4. Cash flow from investing activities includes purchases and sales of marketable securities, investments in securities, property, plant, and equipment, and intangibles.
5. Cash flow from financing activities includes increases and decreases in short-term and long-term borrowing, increases and decreases in common and preferred stock, and dividends.

## Review of T-Account Procedure for Preparing the Statement of Cash Flows

The accountant prepares the statement of cash flows after completing the balance sheet and the income statement. **Chapter 4** describes and illustrates a procedure for preparing the statement of cash flows using a T-account work sheet. A summary of the procedure follows:

**Step 1** Obtain a balance sheet for the beginning and the end of the period spanned by the statement of cash flows.

**Step 2** Prepare a T-account work sheet. A master T-account for cash appears at the top of the work sheet. This master T-account has three sections, labeled Operations, Investing, and Financing. Enter the beginning and the ending balances in cash and cash equivalents in the master T-account. *Cash equivalents* represent short-term, highly liquid investments in which a firm has temporarily placed excess cash. Generally only investments with maturities of three months or less qualify as cash equivalents. We use the term *cash flows* to refer to changes in cash and cash equivalents. Complete the T-account work sheet by preparing a T-account for each balance sheet account other than cash and cash equivalents and enter the beginning and the ending balances.

**Step 3** Explain the change in the master cash account between the beginning and the end of the period by explaining, or accounting for, the changes in the other balance sheet accounts. Do this by reconstructing the entries originally made in the accounts during the period and entering them in appropriate T-accounts on the work sheet. By explaining the changes in balance sheet accounts other than cash, this process also explains the change in cash and cash equivalents. We make such extensive use of the Cash Change Equation in this chapter that we abbreviate words into symbols, as follows:

**Cash Change Equation (Eq. 3, Ch. 4)**

$$\text{Change in Cash} = \text{Change in Liabilities} + \text{Change in Shareholders' Equity} - \text{Change in Noncash Assets}$$

$$\Delta Cash = \Delta L + \Delta SE - \Delta N\$A$$

**Step 4** Prepare a statement of cash flows using information in the T-account work sheet.

## Comprehensive Illustration of the Statement of Cash Flows

The comprehensive illustration that follows uses data for Ellwood Corporation for Year 2. **Exhibit 13.1** presents an income statement for Year 2; **Exhibit 13.2** presents a comparative balance sheet for December 31, Year 1 and Year 2; and **Exhibit 13.3** presents a statement of cash flows. The calculation of cash flow from operations first presents the indirect method. The sections that follow explain each of the line items in **Exhibit 13.3**. **Exhibit 13.4** shows the T-account work sheet.

**EXHIBIT 13.1**

**ELLWOOD CORPORATION**
**Consolidated Income Statement**
**Year 2**

| | |
|---|---|
| REVENUES | |
| Sales | $10,500 |
| Interest and Dividends | 320 |
| Equity in Earnings of Affiliate | 480 |
| Gain on Disposal of Equipment | 40 |
| Total Revenues | $11,340 |
| EXPENSES | |
| Cost of Goods Sold | $ 6,000 |
| Selling and Administrative | 3,550 |
| Loss on Sale of Marketable Equity Securities | 30 |
| Interest Expense | 450 |
| Income Tax Expense | 300 |
| Total Expenses | $10,330 |
| Net Income | $ 1,010 |

## Line 1: Net Income

The income statement indicates net income for the period of $1,010. The work sheet entry presumes that cash provisionally increases by the amount of net income.

| | | |
|---|---|---|
| **(1a)** Cash (Operations—Net Income) | 1,010 | |
| Retained Earnings | | 1,010 |

The effect of net income on the Cash Change Equation is as follows:

| ΔCash | = | ΔL | + | ΔSE | − | ΔN$A |
|---|---|---|---|---|---|---|
| Operations + $1010 (1a) | = | $0 | + | $1,010 (1a) | − | $0 |

## Line 2: Depreciation of Buildings and Equipment

Internal records indicate that depreciation on manufacturing facilities totaled $450 and on selling and administrative facilities totaled $250 during the year. The firm included these amounts in cost of goods sold and selling and administrative expenses, respectively, in the income statement in **Exhibit 13.1**. None of this $700 of depreciation required an operating cash flow during Year 2. The firm reported cash expenditures for these assets as investing activities in the earlier periods when it acquired them. Thus the work sheet entry to explain the change in the Accumulated Depreciation account adds back depreciation to net income in deriving cash flow from operations.

| | | |
|---|---|---|
| **(2a)** Cash (Operations—Depreciation Expense Addback) | 700 | |
| Accumulated Depreciation | | 700 |

**Addback for Depreciation as a Product Cost** The addback for the $450 of depreciation on manufacturing facilities requires elaboration. **Chapter 7** explains that accountants count such depreciation charges as a product cost, not a period expense. The accountant debits Work-in-Process Inventory for this $450. If, during the period, the firm sells all the goods it produces, cost of goods sold includes this $450. Because cost of goods sold includes an amount that does not use cash, the addback to net income cancels the depreciation included in cost of goods sold.

**EXHIBIT 13.2**

**ELLWOOD CORPORATION**
**Consolidated Balance Sheet**

| | December 31 | |
|---|---|---|
| | **Year 1** | **Year 2** |
| ASSETS | | |
| **Current Assets** | | |
| Cash | $ 1,150 | $ 1,050 |
| Certificate of Deposit | 1,520 | 790 |
| Marketable Equity Securities Available for Sale | 280 | 190 |
| Accounts Receivable (net) | 3,400 | 4,300 |
| Inventories | 1,500 | 2,350 |
| Prepayments | 800 | 600 |
| Total Current Assets | $ 8,650 | $ 9,280 |
| **Investments** | | |
| Investment in Company A (15%—Available for Sale) | $ 1,250 | $ 1,280 |
| Investment in Company B (40%) | 2,100 | 2,420 |
| Total Investments | $ 3,350 | $ 3,700 |
| **Property, Plant, and Equipment** | | |
| Land | $ 1,000 | $ 1,000 |
| Buildings | 8,600 | 8,900 |
| Equipment | 10,840 | 11,540 |
| Less Accumulated Depreciation | (6,240) | (6,480) |
| Total Property, Plant, and Equipment | $14,200 | $14,960 |
| **Intangible Assets** | | |
| Patent | $ 2,550 | $ 2,550 |
| Less Accumulated Amortization | (600) | (750) |
| Total Intangible Assets | $ 1,950 | $ 1,800 |
| Total Assets | $28,150 | $29,740 |

Suppose, however, that the firm does not sell all the goods it produces during the period. The ending inventory of Work-in-Process Inventory or Finished Goods Inventory includes a portion of the $450 depreciation charge. Assume, for example, that the firm sold 80 percent of the units produced during the period. Cost of goods sold includes $360 of the depreciation, and inventory accounts include the remaining $90. The statement of cash flows adds back to net income the entire $450 of depreciation on manufacturing facilities for the period. The $90 of depreciation included in the cost of units not sold caused the inventory accounts to increase. Under the indirect method of computing cash flow from operations, the accountant subtracts this increase in inventories in computing cash flow from operations. The $450 addition for depreciation less the $90 subtraction for the increase in inventories nets to a $360 addition to income. Because cost of goods sold includes only $360 of depreciation, the addition required to cancel the depreciation included in cost of goods sold equals $360. Thus, the work sheet entry **2a** shows an addback for the full amount of depreciation for the period (both as a product cost and as a period expense), not just the amount included in cost of goods sold; then, line **13** of the statement of cash flows includes a subtraction for the $90 increase in inventories caused by adding depreciation to work in process.

## Line 3: Amortization of Patent

The effect of patent amortization on cash flow is conceptually identical to that of depreciation. Company records indicate that cost of goods sold for Year 2 includes patent amortization of $150. The work sheet entry to explain the change in the Accumulated Amortization account is as follows:

**EXHIBIT 13.2**

**ELLWOOD CORPORATION**
**Consolidated Balance Sheet (*continued*)**

| | December 31 | |
|---|---|---|
| | **Year 1** | **Year 2** |
| LIABILITIES AND SHAREHOLDERS' EQUITY | | |
| **Current Liabilities** | | |
| Bank Notes Payable | $ 2,000 | $ 2,750 |
| Accounts Payable (for inventory) | 2,450 | 3,230 |
| Warranties Payable | 1,200 | 900 |
| Advances from Customers | 600 | 1,000 |
| Total Current Liabilities | $ 6,250 | $ 7,880 |
| **Noncurrent Liabilities** | | |
| Bonds Payable | $ 2,820 | $ 1,370 |
| Capitalized Lease Obligation | 1,800 | 2,100 |
| Deferred Income Taxes | 550 | 650 |
| Total Noncurrent Liabilities | $ 5,170 | $ 4,120 |
| **Shareholders' Equity** | | |
| Preferred Stock | $ 1,000 | $ 1,200 |
| Common Stock | 2,000 | 2,100 |
| Additional Paid-in Capital | 4,000 | 4,200 |
| **Accumulated Other Comprehensive Income** | | |
| Unrealized Holding Loss on Marketable Securities | (30) | (40) |
| Unrealized Holding Gain on Investments in Securities | 50 | 80 |
| Retained Earnings | 9,960 | 10,580 |
| Total | $16,980 | $18,120 |
| Less Cost of Treasury Stock | (250) | (380) |
| Total Shareholders' Equity | $16,730 | $17,740 |
| Total Liabilities and Shareholders' Equity | $28,150 | $29,740 |

| | | |
|---|---|---|
| **(3a)** Cash (Operations—Amortization Expense Addback) | 150 | |
| Accumulated Amortization | | 150 |

## Line 4: Loss on Sale of Marketable Equity Securities

The accounting records indicate that Ellwood Corporation sold marketable equity securities held as securities available for sale during Year 2. Ellwood Corporation acquired these securities for $80 during Year 1, wrote them down to their market value of $70 at the end of Year 1, and sold them during Year 2 for $50. The firm made the following entries in the accounting records to record this sale:

| | | |
|---|---|---|
| **(4)** Cash | 50 | |
| Realized Loss on Sale of Marketable Equity Securities (IncSt) | 30 | |
| Marketable Equity Securities | | 80 |

| Assets | = | Liabilities | + | Shareholders' Equity | (Class.) |
|---|---|---|---|---|---|
| +50 | | | | −30 | IncSt → RE |
| −80 | | | | | |

(*continued*)

| Δ Cash | = | Δ L | + | Δ SE | − | Δ N$A |
|---|---|---|---|---|---|---|
| +50 (Opns.) | = | 0 | + | −30 | − | −80 |

In this chapter, journal entry numbers without letters, such as this one numbered (4), refer to actual entries Ellwood recorded in its books. For such entries, we give the effect on the Cash Change Equation, derived from the balance sheet equation. For some complex transactions, such as this one, we give the Balance Sheet Equation as well.

| | | |
|---|---|---|
| Marketable Equity Securities . . . . . . . . . . . . . . . . . . . . . . . . . . . . . . . . | 10 | |
| Unrealized Holding Loss on Marketable Equity Securities (Other Comprehensive Income) . . . . . . . . . . . . . . . . . . . . . . . . | | 10 |

| Assets | = | Liabilities | + | Shareholders' Equity | (Class.) |
|---|---|---|---|---|---|
| +10 | | | | +10 | OCInc → AOCInc |

| Δ Cash | = | Δ L | + | Δ SE | − | Δ N$A |
|---|---|---|---|---|---|---|
| 0 | = | 0 | + | +10 | − | +10 |

Unrealized holding loss on securities available for sale is part of Other Comprehensive Income, closed to Accumulated Other Comprehensive Income on the balance sheet.

The work sheet entries to reflect this transaction are the following:

| | | |
|---|---|---|
| **(4a)** Cash (Investing—Sale of Marketable Equity Securities) . . . . . . . . . . . . . | 50 | |
| Cash (Operations—Loss on Sale of Marketable Equity Securities Addback) . . . . . . . . . . . . . . . . . . . . . . . . . . . . . . . . . . . . | 30 | |
| Marketable Equity Securities . . . . . . . . . . . . . . . . . . . . . . . . . . . | | 80 |
| **(4b)** Marketable Equity Securities . . . . . . . . . . . . . . . . . . . . . . . . . . . . . . . . | 10 | |
| Unrealized Holding Loss on Marketable Equity Securities (Other Comprehensive Income) . . . . . . . . . . . . . . . . . . . . . . . . | | 10 |

The statement of cash flows classifies all $50 cash proceeds as an investing activity on line **15** and none as an operating activity. Net income on line **1** in **Exhibit 13.3** includes a subtraction for the loss on the sale of marketable equity securities. To avoid understating the amount of cash flow from operations, the accountant adds back the loss to net income. This addback offsets the loss included in the calculation of net income and eliminates its effect on cash flow from operations. Line **15** shows the cash proceeds from the sale as an investing activity. The analyst might reasonably view purchases and sales of marketable equity securities as operating activities because these transactions involve the use of temporarily excess cash. Most, but not all, firms consider these transactions sufficiently peripheral to the firms' principal operating activity—selling goods and services to customers—that they classify such purchases and sales as investing activities.

## Line 5: Deferred Income Taxes

Notes to the financial statements of Ellwood Corporation indicate that income tax expense of $300 comprises $200 currently payable taxes and $100 deferred to future periods. Ellwood Corporation made the following entry during the year to recognize income tax expense.

**EXHIBIT 13.3**

**ELLWOOD CORPORATION**
**Consolidated Statement of Cash Flows**
**Year 2**

| | | |
|---|---|---|
| OPERATIONS | | |
| (1) Net Income | $ 1,010 | |
| Noncash Revenues, Expenses, Gains, and Losses Included in Income: | | |
| (2) Depreciation of Buildings and Equipment | 700 | |
| (3) Amortization of Patent | 150 | |
| (4) Loss on Sale of Marketable Equity Securities | 30 | |
| (5) Deferred Income Taxes | 100 | |
| (6) Excess of Coupon Payments over Interest Expense | (50) | |
| (7) Gain on Disposal of Equipment | (40) | |
| (8) Equity in Undistributed Earnings of Affiliate | (320) | |
| (9) Decrease in Prepayments | 200 | |
| (10) Increase in Accounts Payable (for inventory) | 780 | |
| (11) Increase in Advances from Customers | 400 | |
| (12) Increase in Accounts Receivable (net) | (900) | |
| (13) Increase in Inventories | (850) | |
| (14) Decrease in Warranties Payable | (300) | |
| Cash Flow from Operations | | $ 910 |
| INVESTING | | |
| (15) Sale of Marketable Equity Securities | $ 50 | |
| (16) Sale of Equipment | 180 | |
| (17) Acquisition of Equipment | (1,300) | |
| Cash Flow from Investing | | (1,070) |
| FINANCING | | |
| (18) Short-Term Bank Borrowing | $ 750 | |
| (19) Long-Term Bonds Issued | 400 | |
| (20) Preferred Stock Issued | 200 | |
| (21) Retirement of Long-Term Debt at Maturity | (1,500) | |
| (22) Acquisition of Common Stock | (130) | |
| (23) Dividends | (390) | |
| Cash Flow from Financing | | (670) |
| Net Change in Cash | | $ (830) |
| Cash, Beginning of Year 2 | | 2,670 |
| Cash, End of Year 2 | | $ 1,840 |

| | | |
|---|---|---|
| **(5)** Income Tax Expense | 300 | |
| Cash | | 200 |
| Deferred Income Taxes | | 100 |

| Assets | = | Liabilities | + | Shareholders' Equity | (Class.) |
|---|---|---|---|---|---|
| −200 | | +100 | | −300 | IncSt → RE |

| Δ Cash | = | Δ L | + | Δ SE | − | Δ N$A |
|---|---|---|---|---|---|---|
| −200 (Opns.) | = | +100 | + | −300 | − | 0 |

**EXHIBIT 13.4**

## ELLWOOD CORPORATION
## T-Account Work Sheet

**Cash**

| | | | | | |
|---|---|---|---|---|---|
| | ✓ | 2,670 | | | |
| **Operations** | | | | | |
| Net Income | (1a) | 1,010 | 50 | (6a) | Excess Coupon Payments |
| Depreciation Expense | (2a) | 700 | 40 | (7a) | Gain on Sale of Equipment |
| Amortization Expense | (3a) | 150 | 320 | (8a) | Equity in Undistributed Earnings |
| Loss on Sale of Marketable Securities | (4a) | 30 | 900 | (12a) | Increase in Accounts Receivable (net) |
| Deferred Income Taxes | (5a) | 100 | 850 | (13a) | Increase in Inventories |
| Decrease in Prepayments | (9a) | 200 | 300 | (14a) | Decrease in Warranties Payable |
| Increase in Accounts Payable | (10a) | 780 | | | |
| Increase in Advances from Customers | (11a) | 400 | | | |
| **Investing** | | | | | |
| Sale of Marketable Securities | (4a) | 50 | 1,300 | (17a) | Acquisition of Equipment |
| Sale of Equipment | (7a) | 180 | | | |
| **Financing** | | | | | |
| Short-Term Borrowing | (18a) | 750 | 1,500 | (21a) | Retirement of Long-Term Debt |
| Long-Term Bonds Issued | (19a) | 400 | 130 | (22a) | Acquisition of Common Stock |
| Preferred Stock Issued | (20a) | 200 | 390 | (23a) | Dividends |
| | ✓ | 1,840 | | | |

**Marketable Equity Securities Available for Sale**

| | | | |
|---|---|---|---|
| ✓ | 280 | | |
| (4b) | 10 | 80 | (4a) |
| | | 20 | (24a) |
| ✓ | 190 | | |

**Accounts Receivable (net)**

| | | | |
|---|---|---|---|
| ✓ | 3,400 | | |
| (12a) | 900 | | |
| ✓ | 4,300 | | |

**Inventories**

| | | | |
|---|---|---|---|
| ✓ | 1,500 | | |
| (13a) | 850 | | |
| ✓ | 2,350 | | |

**Prepayments**

| | | | |
|---|---|---|---|
| ✓ | 800 | | |
| | | 200 | (9a) |
| ✓ | 600 | | |

**Investment in Company A Available for Sale**

| | | | |
|---|---|---|---|
| ✓ | 1,250 | | |
| (25a) | 30 | | |
| ✓ | 1,280 | | |

**Investment in Company B**

| | | | |
|---|---|---|---|
| ✓ | 2,100 | | |
| (8a) | 320 | | |
| ✓ | 2,420 | | |

**Land**

| | | | |
|---|---|---|---|
| ✓ | 1,000 | | |
| ✓ | 1,000 | | |

**Buildings**

| | | | |
|---|---|---|---|
| ✓ | 8,600 | | |
| (26a) | 300 | | |
| ✓ | 8,900 | | |

**Equipment**

| | | | |
|---|---|---|---|
| ✓ | 10,840 | | |
| (17a) | 1,300 | 600 | (7a) |
| ✓ | 11,540 | | |

*(continued)*

EXHIBIT 13.4

## ELLWOOD CORPORATION
## T-Account Work Sheet (*continued*)

**Accumulated Depreciation**

| | | | |
|---|---|---|---|
| | | 6,240 | ✓ |
| (7a) | 460 | 700 | (2a) |
| | | 6,480 | ✓ |

**Patent**

| | | | |
|---|---|---|---|
| ✓ | 2,550 | | |
| ✓ | 2,550 | | |

**Accumulated Amortization**

| | | | |
|---|---|---|---|
| | | 600 | ✓ |
| | | 150 | (3a) |
| | | 750 | ✓ |

**Bank Notes Payable**

| | | | |
|---|---|---|---|
| | | 2,000 | ✓ |
| | | 750 | (18a) |
| | | 2,750 | ✓ |

**Accounts Payable (for inventory)**

| | | | |
|---|---|---|---|
| | | 2,450 | ✓ |
| | | 780 | (10a) |
| | | 3,230 | ✓ |

**Warranties Payable**

| | | | |
|---|---|---|---|
| | | 1,200 | ✓ |
| (14a) | 300 | | |
| | | 900 | ✓ |

**Advances from Customers**

| | | | |
|---|---|---|---|
| | | 600 | ✓ |
| | | 400 | (11a) |
| | | 1,000 | ✓ |

**Bonds Payable**

| | | | |
|---|---|---|---|
| | | 2,820 | ✓ |
| (6a) | 50 | 400 | (19a) |
| (21a) | 1,500 | | |
| (27a) | 300 | | |
| | | 1,370 | ✓ |

**Capitalized Lease Obligation**

| | | | |
|---|---|---|---|
| | | 1,800 | ✓ |
| | | 300 | (26a) |
| | | 2,100 | ✓ |

**Deferred Income Taxes**

| | | | |
|---|---|---|---|
| | | 550 | ✓ |
| | | 100 | (5a) |
| | | 650 | ✓ |

**Preferred Stock**

| | | | |
|---|---|---|---|
| | | 1,000 | ✓ |
| | | 200 | (20a) |
| | | 1,200 | ✓ |

**Common Stock**

| | | | |
|---|---|---|---|
| | | 2,000 | ✓ |
| | | 100 | (27a) |
| | | 2,100 | ✓ |

**Additional Paid-in Capital**

| | | | |
|---|---|---|---|
| | | 4,000 | ✓ |
| | | 200 | (27a) |
| | | 4,200 | ✓ |

**Unrealized Holding Loss on Marketable Securities**

| | | | |
|---|---|---|---|
| ✓ | 30 | | |
| (24a) | 20 | 10 | (4b) |
| ✓ | 40 | | |

**Unrealized Holding Gain on Investments in Securities**

| | | | |
|---|---|---|---|
| | | 50 | ✓ |
| | | 30 | (25a) |
| | | 80 | ✓ |

**Retained Earnings**

| | | | |
|---|---|---|---|
| | | 9,960 | ✓ |
| (23a) | 390 | 1,010 | (1a) |
| | | 10,580 | ✓ |

**Treasury Stock**

| | | | |
|---|---|---|---|
| ✓ | 250 | | |
| (22a) | 130 | | |
| ✓ | 380 | | |

The $100 of deferred income taxes reduced net income but did not require a cash outflow during Year 2. To explain the change in the Deferred Income Taxes account, the work sheet must add back deferred income taxes to net income to derive cash flow from operations.

| | | |
|---|---|---|
| **(5a)** Cash (Operations—Deferred Tax Addback) . . . . . . . . . . . . . . . . . . . . . . . | 100 | |
| Deferred Income Taxes . . . . . . . . . . . . . . . . . . . . . . . . . . . . . . . | | 100 |

## LINE 6: EXCESS OF COUPON PAYMENTS OVER INTEREST EXPENSE

Bonds Payable on the balance sheet includes one series of bonds initially issued at a premium (that is, the coupon rate exceeded the required market rate of interest when Ellwood Corporation issued the bonds, so that initial issue proceeds exceeded face value). The amortization of bond premium makes interest expense over the life of the bonds less than the periodic debt service payments for coupons. The entry made in the accounting records for interest expense during the period was as follows:

| | | |
|---|---|---|
| **(6)** Interest Expense | 450 | |
| Bonds Payable | 50 | |
| Cash | | 500 |

| Δ Cash | = | Δ L | + | Δ SE | – | Δ N$A |
|---|---|---|---|---|---|---|
| –500 (Opns.) | | –50 | | –450 | | 0 |

The firm spent $500 of cash even though it subtracted only $450 of interest expense in computing net income. To explain the change in the Bonds Payable account, the work sheet subtracts an additional $50 from net income to derive cash flow from operations.

| | | |
|---|---|---|
| **(6a)** Bonds Payable | 50 | |
| Cash (Operations—Excess Coupon Payments Subtraction) | | 50 |

The statement of cash flows classifies cash used for interest expense as an operating activity because it views interest as a cost of carrying out operations. Some security analysts suggest that this $50 use of cash for principal repayment is a financing activity for debt service, not an operating activity, and would place it in the Financing section. The Financial Accounting Standards Board *Statement of Financial Accounting Standards No. 95*, however, classifies the $50 cash outflow as an operating activity.

## LINE 7: GAIN ON DISPOSAL OF EQUIPMENT

The accounting records indicate that the firm disposed of for $180 during Year 2 a machine originally costing $600, with accumulated depreciation of $460. The journal entry made to record this disposal was as follows:

| | | |
|---|---|---|
| **(7)** Cash | 180 | |
| Accumulated Depreciation | 460 | |
| Equipment | | 600 |
| Gain on Disposal of Equipment | | 40 |

| Assets | = | Liabilities | + | Shareholders' Equity | (Class.) |
|---|---|---|---|---|---|
| +180 | | | | +40 | IncSt → RE |
| +460 | | | | | |
| –600 | | | | | |

| Δ Cash | = | Δ L | + | Δ SE | – | Δ N$A |
|---|---|---|---|---|---|---|
| +180 (Invst.) | | 0 | | +40 | | +460<br>–600 |

Line **16** shows all the cash proceeds of $180 as an increase in cash from an investing activity. Line **1** includes the $40 gain on sale. To avoid overstating the amount of cash derived from this sale, the accountant subtracts the $40 gain from net income in computing cash flow from operations.

| | | |
|---|---|---|
| **(7a)** Cash (Investing—Sale of Equipment) . . . . . . . . . . . . . . . . . . . . . . . . . | 180 | |
| Accumulated Depreciation . . . . . . . . . . . . . . . . . . . . . . . . . . . . . . . . . | 460 | |
| Equipment . . . . . . . . . . . . . . . . . . . . . . . . . . . . . . . . . . . . . . | | 600 |
| Cash (Operations—Gain on Sale of Equipment Subtraction) . . . . . . . | | 40 |

The statement of cash flows classifies all cash proceeds as investing activities and none as operating activities. Most firms acquire and sell fixed assets with the objective of providing a capacity to carry out operations rather than as a means of generating operating income.

Fixed assets disposed of at a loss instead of a gain require an addback to net income in deriving cash flow from operations. The work sheet entry, assuming the data of the preceding entry except that Ellwood Corporation sells the equipment for $110, would be as follows:

| | | |
|---|---|---|
| Cash (Investing—Disposal of Equipment) . . . . . . . . . . . . . . . . . . . . . . . | 110 | |
| Accumulated Depreciation . . . . . . . . . . . . . . . . . . . . . . . . . . . . . . . . . | 450 | |
| Cash (Operations—Loss on Disposal of Equipment Addback) . . . . . . . . . . . | 40 | |
| Equipment . . . . . . . . . . . . . . . . . . . . . . . . . . . . . . . . . . . . . . | | 600 |

## LINE 8: EQUITY IN UNDISTRIBUTED EARNINGS OF AFFILIATE

The balance sheet indicates that Ellwood Corporation owns 40 percent of the common stock of Company B. During Year 2, Company B earned $1,200 and paid $400 of dividends. Ellwood Corporation made the following entries on its books during the year.

| | | |
|---|---|---|
| **(8)** Investment in Company B . . . . . . . . . . . . . . . . . . . . . . . . . . . . . . . | 480 | |
| Equity in Earnings of Affiliate . . . . . . . . . . . . . . . . . . . . . . . . . | | 480 |

| Assets | = | Liabilities | + | Shareholders' Equity | (Class.) |
|---|---|---|---|---|---|
| +480 | | | | +480 | IncSt → RE |

| Δ Cash | = | Δ L | + | Δ SE | − | Δ N$A |
|---|---|---|---|---|---|---|
| 0 | | 0 | | +480 | | +480 |

Records equity in earnings of $480 = 0.40 × $1,200.

| | | |
|---|---|---|
| Cash . . . . . . . . . . . . . . . . . . . . . . . . . . . . . . . . . . . . . . . . . . . . . | 160 | |
| Investment in Company B . . . . . . . . . . . . . . . . . . . . . . . . . . . | | 160 |

| Assets | = | Liabilities | + | Shareholders' Equity | (Class.) |
|---|---|---|---|---|---|
| +160 | | | | | |
| −160 | | | | | |

| Δ Cash | = | Δ L | + | Δ SE | − | Δ N$A |
|---|---|---|---|---|---|---|
| +160 (Opns.) | | 0 | | 0 | | −160 |

Records dividends received of $160 = 0.40 × $400.

Net income of Ellwood Corporation on line **1** of **Exhibit 13.3** includes $480 of equity income. It received only $160 of cash. Thus, the work sheet subtracts $320 (= $480 − $160) from net income in deriving cash from operations.

| | | |
|---|---|---|
| **(8a)** Investment in Company B . . . . . . . . . . | 320 | |
| Cash (Operations—Equity in Undistributed Earnings Subtraction) . . . . . . . . . . | | 320 |

Analytic entry recorded in T-account work sheet.

## LINE 9: DECREASE IN PREPAYMENTS

Because prepayments decreased by $200 during Year 2, the firm expensed less cash during Year 2 for new prepayments than it expensed prepayments of earlier years. Assume that all prepayments relate to selling and administrative activities. The journal entries that Ellwood Corporation made in the accounting records during the year had the following combined effect:

| | | |
|---|---|---|
| **(9)** Selling and Administrative Expenses . . . . . . . . . . | 3,550 | |
| Cash . . . . . . . . . . | | 3,350 |
| Prepayments . . . . . . . . . . | | 200 |

| Δ Cash | = | Δ L | + | Δ SE | − | Δ N$A |
|---|---|---|---|---|---|---|
| −3,350 (Opns.) | | 0 | | −3,550 | | −200 |

To explain the change in the work sheet for Prepayments, add back $200 to net income for the credit change in an operating current asset account so that cash flow from operations reports expenditures, not expenses.

| | | |
|---|---|---|
| **(9a)** Cash (Operations—Decrease in Prepayments) . . . . . . . . . . | 200 | |
| Prepayments . . . . . . . . . . | | 200 |

## LINE 10: INCREASE IN ACCOUNTS PAYABLE

An increase in accounts payable indicates that new purchases on account during Year 2 exceeded payments during Year 2 for previous purchases on account. This increase in accounts payable, a credit change in an operating current liability account, implicitly provides cash. If you think of this source of cash as financing, you have the right idea. Suppliers have provided financing so that Ellwood Corporation can acquire goods on account. You might think of it this way. Imagine a firm borrows from a supplier, debiting Cash and crediting Notes Payable. Then the firm uses the cash to acquire inventory or other items. You can see that the supplier has provided cash, and the firm increases a current liability account. A firm buying on account has achieved the same result, except that it credits Accounts Payable, not Notes Payable. Because the supplier ties the financing to the purchase of goods used in operations, accounting classifies this source of cash in the operating, not financing, section of the statement of cash flows.

| | | |
|---|---|---|
| **(10a)** Cash (Operations—Increase in Accounts Payable) . . . . . . . . . . | 780 | |
| Accounts Payable (for inventory) . . . . . . . . . . | | 780 |

We will explore the effect that the adjustment for the change in accounts payable has on the equation for the change in cash when we discuss the adjustment for inventory.

## LINE 11: INCREASE IN ADVANCES FROM CUSTOMERS

The $400 increase in customer advances means that the firm received $400 more cash during Year 2 than it recognized as revenue. The work sheet adds this excess to net income in deriving cash flow from operations.

| | | |
|---|---|---|
| **(11a)** Cash (Operations—Increase in Advances from Customers) | 400 | |
| Advances from Customers | | 400 |

We will consider the effect that the adjustment from advances for customers has on the equation for the change in cash next, when we discuss the adjustment for accounts receivable.

## LINE 12: INCREASE IN ACCOUNTS RECEIVABLE

The increase in accounts receivable indicates that the firm collected less cash from customers than the amount shown for sales on account. The work sheet subtracts the increase in accounts receivable, a debit change in an operating current asset account, in deriving cash flow from operations.

| | | |
|---|---|---|
| **(12a)** Accounts Receivable (net) | 900 | |
| Cash (Operations—Increase in Accounts Receivable) | | 900 |

Note that this entry automatically incorporates the effect of any change in the Allowance for Uncollectible Accounts. The work sheet could make separate work sheet entries for the change in gross accounts receivable and the change in allowance for uncollectible accounts.

We can now summarize the effect that changes in accounts receivable and advances from customers have on the equation for changes in cash. Ellwood Corporation made entries during the year with the following combined effect:

| | | |
|---|---|---|
| Cash | 10,900 | |
| Accounts Receivable (net) | 900 | |
| Advances from Customers | | 400 |
| Sales Revenue | | 10,500 |

| Assets | = | Liabilities | + | Shareholders' Equity | (Class.) |
|---|---|---|---|---|---|
| +10,900 | | +400 | | +10,500 | IncSt → RE |
| +900 | | | | | |

| Δ Cash | = | Δ L | + | Δ SE | − | Δ N$A |
|---|---|---|---|---|---|---|
| +10,900 (Opns.) | | +400 | | +10,500 | | +900 |

## LINE 13: INCREASE IN INVENTORIES

The increase in inventories indicates the firm purchased more merchandise than it sold during Year 2. The work sheet subtracts this debit change in inventory in deriving cash flow from operations.

| | | |
|---|---|---|
| **(13a)** Inventories | 850 | |
| Cash (Operations—Increase in Inventories) | | 850 |

We can now consider the effect on cash of the change in inventories and the change in accounts payable. Ellwood Corporation made entries during the year that had the following combined effect:

| | | |
|---|---|---|
| Cost of Goods Sold | 6,000 | |
| Inventories | 850 | |
| Accounts Payable (for inventory) | | 780 |
| Cash | | 6,070 |

| Assets | = | Liabilities | + | Shareholders' Equity | (Class.) |
|---|---|---|---|---|---|
| +850 | | +780 | | −6,000 | IncSt → RE |
| −6,070 | | | | | |

| Δ Cash | = | Δ L | + | Δ SE | − | Δ N$A |
|---|---|---|---|---|---|---|
| −6,070 (Opns.) | | +780 | | −6,000 | | +850 |

## LINE 14: DECREASE IN WARRANTIES PAYABLE

Recall that firms estimate future warranty costs on current sales using the allowance method for warranties. The Warranties Payable account increases for the estimated cost of future warranty services on products sold during the period and decreases by the actual cost of warranty services performed. During Year 2, the firm paid $200 more in warranty claims than it reported as expenses on the income statement. Ellwood Corporation includes estimated warranty expense of $920 in selling and administrative expenses in its income statement in **Exhibit 13.1**. The firm made entries during the year with the following combined effect:

| | | |
|---|---|---|
| **(9)** Selling and Administrative Expenses | 920 | |
| Warranties Payable | 300 | |
| Cash | | 1,220 |

| Δ Cash | = | Δ L | + | Δ SE | − | Δ N$A |
|---|---|---|---|---|---|---|
| −1,220 (Opns.) | | −300 | | −920 | | 0 |

The work sheet subtracts this decrease in Warranties Payable, a debit change in an operating current liability account so that cash flow from operations reports cash expenditures, not expenses.

| | | |
|---|---|---|
| **(14a)** Warranties Payable | 300 | |
| Cash (Operations—Decrease in Warranties Payable) | | 300 |

Cash flow from operations is $910 for Year 2.

## LINES 15 AND 16

See the discussion for lines **4** and **7**.

## LINE 17: ACQUISITION OF EQUIPMENT

The firm acquired equipment costing $1,300 during Year 2. The analytic entry for this investing activity is as follows:

| | | |
|---|---|---|
| **(17a)** Equipment | 1,300 | |
| Cash (Investing—Acquisition of Equipment) | | 1,300 |

Cash flow from investing for Year 2 is a net outflow of $1,070.

## LINE 18: SHORT-TERM BANK BORROWING

Ellwood Corporation borrowed $750 during Year 2 from its bank under a short-term borrowing arrangement. Even though this loan is short-term, the statement of cash flows classifies it as a financing instead of an operating activity. The analytic entry on the work sheet is as follows:

| | | |
|---|---|---|
| **(18a)** Cash (Financing—Short-Term Bank Borrowing) | 750 | |
| Bank Note Payable | | 750 |

## LINE 19: LONG-TERM BONDS ISSUED

The firm issued long-term bonds totaling $400 during Year 2.

| | | |
|---|---|---|
| **(19a)** Cash (Financing—Long-Term Bonds Issued) | 400 | |
| Bonds Payable | | 400 |

## LINE 20: PREFERRED STOCK ISSUED

The firm issued preferred stock totaling $200 during the year.

| | | |
|---|---|---|
| **(20a)** Cash (Financing—Preferred Stock Issued) | 200 | |
| Preferred Stock | | 200 |

## LINE 21: RETIREMENT OF LONG-TERM DEBT AT MATURITY

Ellwood Corporation retired $1,500 of long-term debt at maturity. The income statement in **Exhibit 13.1** shows no gain or loss on retirement of debt. Thus, Ellwood Corporation must have retired the debt at its book value. We make the following work sheet entry:

| | | |
|---|---|---|
| **(21a)** Bonds Payable | 1,500 | |
| Cash (Financing—Retirement of Long-Term Debt) | | 1,500 |

If the firm had retired the debt prior to maturity, the firm would likely have recognized a gain or loss. The work sheet would eliminate the gain or loss from net income in computing cash flow from operations and classify as a financing activity the full amount of cash used to retire the debt.

## LINE 22: ACQUISITION OF COMMON STOCK

The firm acquired common stock costing $130 during Year 2. The analytic entry is as follows:

| | | |
|---|---|---|
| **(22a)** Treasury Stock | 130 | |
| Cash (Financing—Acquisition of Common Stock) | | 130 |

## LINE 23: DIVIDENDS

Ellwood Corporation declared and paid $390 of dividends to its shareholders during Year 2. The analytic entry is as follows:

| | | |
|---|---|---|
| **(23a)** Retained Earnings . . . . . . . . . . . . . . . . . . . . . . . . . . . . . . . . . . . . | 390 | |
| Cash (Financing—Dividends) . . . . . . . . . . . . . . . . . . . . . . . . . . . . . | | 390 |

Net cash outflow for financing totaled $670 during the year.

## Noncash Investing and Financing Transactions

Some investing and financing transactions do not involve cash and therefore do not appear on the statement of cash flows. These transactions nevertheless help explain changes in balance sheet accounts. The accountant must enter these transactions in the T-account work sheet to account fully for all balance sheet changes and compute correctly the portion of the changes affecting cash.

**Write-Down of Marketable Equity Securities** During Year 2, Ellwood Corporation wrote down marketable equity securities to their market value. The journal entry made for this write-down is as follows:

| | | |
|---|---|---|
| **(24)** Unrealized Holding Loss on Marketable Equity Securities Available for Sale . . . . . . . . . . . . . . . . . . . . . . . . . . . . . . . . . . . . | 20 | |
| Marketable Equity Securities Available for Sale . . . . . . . . . . . . . . . | | 20 |

| Assets | = | Liabilities | + | Shareholders' Equity | (Class.) |
|---|---|---|---|---|---|
| −20 | | | | −20 | OCInc → AOCInc |

| Δ Cash | = | Δ L | + | Δ SE | − | Δ N$A |
|---|---|---|---|---|---|---|
| 0 | | 0 | | −20 | | −20 |

This entry does not affect cash and therefore does not appear in the statement of cash flows. It does, however, help explain the change during the year in the two marketable equity securities accounts above and requires the following entry in the T-account work sheet:

| | | |
|---|---|---|
| **(24a)** Unrealized Holding Loss on Marketable Equity Securities Available for Sale . . . . . . . . . . . . . . . . . . . . . . . . . . . . . . . . . . | 20 | |
| Marketable Equity Securities Available for Sale . . . . . . . . . . . . . . . | | 20 |

**Write-Up of Investment in Securities** During Year 2, Ellwood Corporation also wrote up its Investment in Company A, a security available for sale, to reflect market value. The journal entry for the write-up is as follows:

| | | |
|---|---|---|
| **(25)** Investment in Company A . . . . . . . . . . . . . . . . . . . . . . . . . . . . . . . . | 30 | |
| Unrealized Holding Gain on Investment in Securities . . . . . . . . . . | | 30 |

| Assets | = | Liabilities | + | Shareholders' Equity | (Class.) |
|---|---|---|---|---|---|
| +30 | | | | +30 | OCInc → AOCInc |

| Δ Cash | = | Δ L | + | Δ SE | − | Δ N$A |
|---|---|---|---|---|---|---|
| 0 | | 0 | | +30 | | +30 |

This entry does not affect cash flows but explains the change during the year in the two investment in securities accounts above and requires the following entry in the T-account work sheet:

| | | |
|---|---|---|
| **(25a)** Investment in Company A . . . . . . . . . . . . . . . . . . . . . . . . . . . . . . . . | 30 | |
| Unrealized Holding Gain on Investment in Securities . . . . . . . . . . . | | 30 |

**Capitalization of Leases** During Year 2, Ellwood Corporation signed a long-term lease for a building. It classified the lease as a capital lease and recorded it in the accounts as follows:

| | | |
|---|---|---|
| **(26)** Building . . . . . . . . . . . . . . . . . . . . . . . . . . . . . . . . . . . . . . . . . . . . | 300 | |
| Capitalized Lease Obligation . . . . . . . . . . . . . . . . . . . . . . . . . . | | 300 |

| Assets | = | Liabilities | + | Shareholders' Equity | (Class.) |
|---|---|---|---|---|---|
| +300 | | +300 | | | |

| Δ Cash | = | Δ L | + | Δ SE | – | Δ N$A |
|---|---|---|---|---|---|---|
| 0 | | +300 | | | | +300 |

Note that this entry does not affect cash. It does affect the investing and financing activities of Ellwood Corporation and requires disclosure in a supplementary schedule or notes to the financial statements. The accountant makes the following entry in the T-account work sheet:

| | | |
|---|---|---|
| **(26a)** Building . . . . . . . . . . . . . . . . . . . . . . . . . . . . . . . . . . . . . . . . . . . | 300 | |
| Capitalized Lease Obligation . . . . . . . . . . . . . . . . . . . . . . . . . . | | 300 |

The firm must disclose transactions such as this in notes, not in the statement of cash flows. Even though they involve investing activities, they do not involve cash.

**Conversion of Debt into Equity** During Year 2, investors in bonds of Ellwood Corporation exercised their option to convert their debt securities into shares of common stock. The entry made in the accounting records to record the conversion is as follows:

| | | |
|---|---|---|
| **(27)** Bonds Payable . . . . . . . . . . . . . . . . . . . . . . . . . . . . . . . . . . . . . . | 300 | |
| Common Stock . . . . . . . . . . . . . . . . . . . . . . . . . . . . . . . . . . . | | 100 |
| Additional Paid-in Capital . . . . . . . . . . . . . . . . . . . . . . . . . . . . | | 200 |

| Assets | = | Liabilities | + | Shareholders' Equity | (Class.) |
|---|---|---|---|---|---|
| | | -300 | | +100 | ContriCap |
| | | | | +200 | ContriCap |

| Δ Cash | = | Δ L | + | Δ SE | – | Δ N$A |
|---|---|---|---|---|---|---|
| 0 | | -300 | | +100 | | 0 |
| | | | | +200 | | |

The accountant reflects this financing transaction on the T-account work sheet by making the following entry:

| | | |
|---|---|---|
| **(27a)** Bonds Payable . . . . . . . . . . . . . . . . . . . . . . . . . . . . . . . . . . . . . | 300 | |
| Common Stock . . . . . . . . . . . . . . . . . . . . . . . . . . . . . . . . . . . | | 100 |
| Additional Paid-in Capital . . . . . . . . . . . . . . . . . . . . . . . . . . . . | | 200 |

**Exhibit 13.4** presents a T-account work sheet for Ellwood Corporation for Year 2.

## Illustration of the Direct Method for Cash Flows from Operations

**Exhibit 13.5** derives Cash Flows from Operations presented with the direct method for Ellwood Corporation. While the direct method's presentation of cash flow requires less understanding of the contrast between cash and accrual accounting, its derivation requires the same understanding as does the indirect method. Every addback and subtraction in the indirect presentation appears in the direct method's derivation.

To see the relation between the indirect and direct methods, consider the following contrast of the equivalent arithmetic used for the two derivations of cash flow from operations.

- The indirect method starts with the total for net income and removes the effects of gains and losses from nonoperating transactions. Then, it adds or subtracts balance sheet changes involving operating accounts. Take net income stripped of nonoperating gains and losses, then list under it, vertically, additions and subtractions for balance sheet changes.

**EXHIBIT 13.5**

**ELLWOOD CORPORATION**
**Deriving Direct Method Cash Flow from Operations Using Data from T-Account Work Sheet**

1. Copy Income Statement and Cash Flow from Operations
2. Copy Information from T-Account Work Sheet Next to Related Income Statement Item
3. Sum Across Rows to Derive Direct Receipts and Expenditures

| Operations | (a) | Indirect Method (b) | Changes in Related Balance Sheet Accounts from T-Account Work Sheet (c) | Direct Method (d) | From Operations: Receipts less Expenditures |
|---|---|---|---|---|---|
| Sales | $10,500 | $ 400 | = Advances from Customers Receivable Increase | $10,000 | Receipts from Customers |
| | | (900) | = Accounts Receivable Increase | | |
| Interest and Dividends | 320 | | | 320 | Receipts from Investments |
| Equity in Earnings of Affiliate | 480 | (320) | Dividends Received were only $160 | 160 | Receipts from Equity Method Investments |
| Gain on Disposal of Equipment | 40 | (40) | Gain Produces No Cash from Operations | — | |
| Cost of Goods Sold | (6,000) | 450 | Depreciation on Manufacturing Facilities | (5,470) | Payments for Inventory |
| | | 150 | Amortization of Patents Used in Manufacturing | | |
| | | 780 | = Accounts Payable Increase | | |
| | | (850) | = Increase in Inventories | | |
| Selling and Administrative Expenses | (3,550) | 250 | Depreciation on Administrative Buildings and Equipment | (3,400) | Payments for Selling and Administrative Services |
| | | 200 | = Prepayments Decrease | | |
| | | (300) | = Decrease in Warranties Payable | | |
| Loss on Sale of Marketable Equity Securities | (30) | 30 | Loss Uses No Cash | — | |
| Interest Expense | (450) | (50) | = Coupon Payments in Excess of Interest Expense | (500) | Payments for Interest |
| Income Tax Expense | (300) | 100 | Deferred Income Taxes Uses No Cash This Period | (200) | Payments for Income Taxes |
| Net Income | $ 1,010 | $1,010 | Totals | $ 910 | = Cash Flow from Operations Derived via Direct Method |
| | | $ 910 | = Cash Flow from Operations Derived via Indirect Method | | |

- The direct method starts with the components of income, the individual revenues and expenses, but not gains and losses, then adds or subtracts the same balance sheet changes involving the same operating accounts. Take an income statement line, then list next to it, horizontally, additions and subtractions.

The indirect method presents the net of revenues less expenses, then adds to, and subtracts from, that total. The direct method starts with a line of the income statement, then adds to, and subtracts from, that component. Because the amounts for balance sheet changes added and subtracted are the same, the final result, cash flow from operations, must be the same.

We think you will better understand cash flow from operations if you master the direct method, because its presentation, if not its derivation, will match your intuition. In addition, understanding the cause of changes from period to period in cash flow from operations comes easier from the direct method's presentation. Few firms, however, use the direct method in their public presentations.

## Problem 13.1 for Self-Study

**Effects of transactions on the statement of cash flows.** **Exhibit 4.16** in **Chapter 4** presents a simplified statement of cash flows. For each of the transactions that follow, indicate the number(s) of the line(s) in **Exhibit 4.16** affected by the transaction and the amount and direction (increase or decrease) of the effect. If the transaction affects net income, be sure to indicate whether it increases or decreases. Ignore income tax effects.

**a.** A firm sells for $12,000 equipment that originally cost $30,000 and has accumulated depreciation of $16,000 at the time of sale.
**b.** A firm owns 25 percent of the common stock of an investee acquired several years ago at book value and uses the equity method. The investee had net income of $80,000 and paid dividends of $20,000 during the period.
**c.** A firm, as lessee (tenant), records lease payments of $50,000 on capital leases for the period, of which $35,000 represents interest expense.
**d.** Income tax expense for the period totals $120,000, of which the firm pays $90,000 immediately and defers the remaining $30,000 because of temporary differences between the accounting principles used for financial reporting and those used for tax reporting.
**e.** A firm owns 10 percent of the common stock of an investee acquired at its book value several years ago and accounts for it at market value as a long-term investment. The investee had net income of $100,000 and paid dividends of $40,000 during the period. The market value at the end of the period equaled the market value at the beginning of the period.

## Interpreting the Statement of Cash Flows

**Chapter 4** points out that the proper interpretation of information in the statement of cash flows requires

- an understanding of the economic characteristics of the industries in which a firm conducts operations, and
- a multiperiod view.

This section discusses the interpretation of the statement of cash flows more fully.

### Relation between Net Income and Cash Flow from Operations

Net income and cash flow from operations differ for two principal reasons:

1. Changes in noncurrent assets and noncurrent liabilities
2. Changes in operating working capital accounts

**Changes in Noncurrent Assets and Noncurrent Liabilities** The extent to which a firm adjusts net income for changes in noncurrent assets and noncurrent liabilities in deriving cash flow from operations depends on the nature of its operations. Capital-intensive firms will likely show a substantial addback to net income for depreciation expense, whereas service firms will show a smaller amount. Rapidly growing firms usually show an addback for deferred tax expense, whereas firms that stop growing or that shrink show a subtraction. Firms that grow or diversify by acquiring minority ownership positions in other businesses will often show a subtraction from net income for equity in undistributed earnings. Firms that decrease in size will usually show additions or subtractions for losses and gains on the disposal of assets.

**Changes in Operating Working Capital Accounts** The adjustment for changes in operating working capital accounts depends in part on a firm's rate of growth. Rapidly growing firms usually experience significant increases in accounts receivable and inventories. Some firms use suppliers or other creditors to finance these working capital needs (classified as operating activities), whereas other firms use short- or long-term borrowing or equity financing (classified as financing activities).

## Relations between Cash Flows from Operating, Investing, and Financing Activities

The product life-cycle concept from microeconomics and marketing provides useful insights into the relations between cash flows from operating, investing, and financing activities.

During the introduction phase, cash outflow exceeds cash inflow from operations because operations are not yet earning profits while the firm must invest in accounts receivable and inventories. Investing activities result in a net cash outflow to build productive capacity. Firms must rely on external financing during this phase to overcome the negative cash flow from operations and investing.

The growth phase portrays cash flow characteristics similar to the introduction phase. The growth phase reflects sales of successful products, and net income turns positive. A growing firm makes more sales, but it also needs to acquire more goods to sell. Because it usually must pay for the goods it acquires before it collects for the goods it sells, the growing firm finds itself ever short of cash from operations. The faster it grows (even though profitable), the more cash it needs. Banks do not like to lend for such needs. They view such needs (even though for current assets) as a permanent part of the firm's financing needs. Thus banks want firms to use shareholders' equity or long-term debt to finance growth in nonseasonal inventories and receivables.

The maturing of a product alters these cash flow relations. Net income usually reaches a peak, and working capital stops growing. Operations generate positive cash flow, enough to finance expenditures on property, plant, and equipment. Capital expenditures usually maintain, rather than increase, productive capacity. Firms use the excess cash flow to repay borrowing from the introduction and growth phases and to begin paying dividends to shareholders.

Weakening profitability—from reduced sales or reduced profit margins on existing sales—signals the beginning of the decline phase, but ever-declining accounts receivable and inventories can produce positive cash flow from operations. In addition, sales of unneeded property, plant, and equipment can result in positive cash flow from investing activities. Firms can use the excess cash flow to repay remaining debt or diversify into other areas of business.

Biotechnology firms are in their growth phase, consumer foods companies are in their mature phase, and U.S. auto manufacturers are in the late maturity or, perhaps, the early decline phase.

# Solution to Self-Study Problem

### SUGGESTED SOLUTION TO PROBLEM 13.1 FOR SELF-STUDY

(Effects of transactions on the statement of cash flows.)

**a.** The journal entry to record this transaction is as follows:

| | | |
|---|---|---|
| Cash | 12,000 | |
| Accumulated Depreciation | 16,000 | |
| Loss on Sale of Equipment | 2,000 | |
| Equipment | | 30,000 |

*(continued)*

| Assets | = | Liabilities | + | Shareholders' Equity | (Class.) |
|---|---|---|---|---|---|
| +12,000 | | | | −2,000 | IncSt → RE |
| +16,000 | | | | | |
| −30,000 | | | | | |

| Δ Cash | = | Δ L | + | Δ SE | − | Δ N$A |
|---|---|---|---|---|---|---|
| +12,000 (Invst.) | | 0 | | −2,000 | | +16,000<br>−30,000 |

The debit to the Cash account results in an increase on line **(11)** of $12,000. Selling equipment is an investing transaction, so line **(6)** increases by $12,000. The loss on the sale reduces net income, so line **(3)** decreases by $2,000. Because the loss does not use cash, line **(4)** increases by $2,000 to add back the loss to net income when computing cash flow from operations.

**b.** The journal entry to record this transaction is as follows:

| | | |
|---|---|---|
| Cash | 5,000 | |
| Investment in Securities | 15,000 | |
| Equity in Earnings of Affiliate | | 20,000 |

| Assets | = | Liabilities | + | Shareholders' Equity | (Class.) |
|---|---|---|---|---|---|
| +5,000 | | | | +20,000 | IncSt → RE |
| +15,000 | | | | | |

| Δ Cash | = | Δ L | + | Δ SE | − | Δ N$A |
|---|---|---|---|---|---|---|
| +5,000 (Opns.) | | 0 | | +20,000 | | +15,000 |

The debit to the Cash account results in an increase on line **(11)** of $5,000. Line **(3)** increases by $20,000 for the equity in earnings. Because the firm receives only $5,000 in cash, line **(5)** must increase by $15,000 to subtract from earnings the excess of equity in earnings over the dividends received.

**c.** The journal entry to record this transaction is as follows:

| | | |
|---|---|---|
| Interest Expense | 35,000 | |
| Capitalized Lease Obligation | 15,000 | |
| Cash | | 50,000 |

| Assets | = | Liabilities | + | Shareholders' Equity | (Class.) |
|---|---|---|---|---|---|
| −50,000 | | −15,000 | | −35,000 | IncSt → RE |

| Δ Cash | = | Δ L | + | Δ SE | − | Δ N$A |
|---|---|---|---|---|---|---|
| −50,000 (Opns.) | | −15,000 | | −35,000 | | 0 |

The credit to the Cash account reduces line **(11)** by $50,000. The recognition of interest expense reduces net income on line **(3)** by $35,000. This amount represents an operating use of cash and therefore requires no addback or subtraction in computing cash flow from operations. The remaining cash payment of $15,000 is a financing use of cash, so line **(9)** increases by $15,000.

**d.** The journal entry to record this transaction is as follows:

| | | |
|---|---|---|
| Income Tax Expense | 120,000 | |
| Deferred Tax Liability | | 30,000 |
| Cash | | 90,000 |

| Assets | = | Liabilities | + | Shareholders' Equity | (Class.) |
|---|---|---|---|---|---|
| −90,000 | | +30,000 | | −120,000 | IncSt → RE |

| Δ Cash | = | Δ L | + | Δ SE | − | Δ N$A |
|---|---|---|---|---|---|---|
| −90,000 (Opns.) | | +30,000 | | −120,000 | | 0 |

The credit to the Cash account results in a reduction on line **(11)** of $90,000. The recognition of income tax expense reduces net income on line **(3)** by $120,000. Because the firm used only $90,000 in cash for income taxes this period, line **(4)** increases by $30,000 for the portion of the expense that did not use cash.

**e.** The journal entry to record this transaction is as follows:

| | | |
|---|---|---|
| Cash | 4,000 | |
| Dividend Revenue | | 4,000 |

| Assets | = | Liabilities | + | Shareholders' Equity | (Class.) |
|---|---|---|---|---|---|
| +4,000 | | | | +4,000 | IncSt → RE |

| Δ Cash | = | Δ L | + | Δ SE | − | Δ N$A |
|---|---|---|---|---|---|---|
| +4,000 (Opns.) | | 0 | | +4,000 | | 0 |

The debit to the Cash account results in an increase on line **(11)** of $4,000. The recognition of dividend revenue increases net income on line **(3)** by $4,000. Because dividends received from investments in securities are operating transactions and the amount of the dividends revenue equals the amount of cash received, the accountant makes no adjustment to net income when computing cash flow from operations.

## Problems and Cases

For additional student resources, content, and interactive quizzes for this chapter visit the FACMU website: **www.thomsonedu.com/accounting/stickney**

1. **Effects of transactions on statement of cash flows. Exhibit 4.16** in **Chapter 4** provides a simplified statement of cash flows. For each of the transactions that follow, indicate the number(s) of the line(s) in **Exhibit 4.16** affected by the transaction and the amount and direction (increase or decrease) of the effect. If the transaction affects net income on line **(3)** or cash on line **(11)**, be sure to indicate if it increases or decreases the line. Ignore income tax effects. Indicate the effects of each transaction on the Cash Change Equation.
   **a.** A firm declares cash dividends of $15,000, of which it pays $12,000 immediately to its shareholders; it will pay the remaining $3,000 early in the next accounting period.
   **b.** A firm borrows $75,000 from its bank.
   **c.** A firm sells for $20,000 machinery originally costing $40,000 and with accumulated depreciation of $35,000.

**d.** A firm as lessee records lease payments on operating leases of $28,000 for the period.
**e.** A firm acquires, with temporarily excess cash, marketable equity securities costing $39,000.
**f.** A firm writes off a fully depreciated truck originally costing $14,000.
**g.** A marketable equity security (available for sale) acquired during the current period for $90,000 has a market value of $82,000 at the end of the period. Indicate the effect of any year-end adjusting entry to apply the market value method.
**h.** A firm records interest expense of $15,000 for the period on bonds issued several years ago at a discount, comprising a $14,500 cash payment and a $500 addition to Bonds Payable.
**i.** A firm records an impairment loss of $22,000 for the period on goodwill arising from the acquisition several years ago of an 80 percent investment in a subsidiary.

**2. Effects of transactions on statement of cash flows. Exhibit 4.16** in **Chapter 4** provides a simplified statement of cash flows. For each of the transactions that follow, indicate the number(s) of the line(s) in **Exhibit 4.16** affected by the transaction and the amount and direction (increase or decrease) of the effect. If the transaction affects net income on line **(3)** or cash on line **(11)**, be sure to indicate if it increases or decreases the line. Ignore income tax effects. Indicate the effects of each transaction on the Cash Change Equation.
**a.** A firm acquires a building costing $400,000, paying $40,000 cash and signing a promissory note to the seller for $360,000.
**b.** A firm using the allowance method records $32,000 of bad debt expense for the period.
**c.** A firm using the allowance method writes off accounts totaling $28,000 as uncollectible.
**d.** A firm owns 30 percent of the common stock of an investee acquired several years ago at book value. The investee had net income of $40,000 and paid dividends of $50,000 during the period.
**e.** A firm sells for $22,000 marketable equity securities (available for sale) originally costing $25,000 and with a book value of $23,000 at the time of sale.
**f.** Holders of a firm's preferred stock with a book value of $10,000 convert their preferred shares into common stock with a par value of $2,000. Use the book value method.
**g.** A firm gives land with an acquisition cost and market value of $5,000 in settlement of the annual legal fees of its corporate attorney.
**h.** A firm reduces the liability account Rental Fees Received in Advance for $8,000 when it provides rental services.
**i.** A firm reclassifies long-term debt of $30,000, maturing within the next year, as a current liability.

**3. Effects of transactions on statement of cash flows. Exhibit 4.16** in **Chapter 4** provides a simplified statement of cash flows. For each of the transactions that follow, indicate the number(s) of the line(s) in **Exhibit 4.16** affected by the transaction and the amount and direction (increase or decrease) of the effect. If the transaction affects net income on line **(3)** or cash on line **(11)**, be sure to indicate if it increases or decreases the line. Ignore income tax effects. Indicate the effects of each transaction on the Cash Change Equation.
**a.** A firm using the percentage-of-completion method for long-term contracts recognizes $15,000 of revenue for the period.
**b.** A local government donates land with a market value of $50,000 to a firm as an inducement to locate manufacturing facilities in the area.
**c.** A firm writes down long-term investments in securities by $8,000 to reflect the market value method.
**d.** A firm records $60,000 depreciation on manufacturing facilities for the period. The firm has sold all goods it manufactured this period.
**e.** A firm using the allowance method recognizes $35,000 as warranty expense for the period.
**f.** A firm using the allowance method makes expenditures totaling $28,000 to provide warranty services during the period.
**g.** A firm recognizes income tax expense of $80,000 for the period, comprising $100,000 paid currently and a $20,000 reduction in the Deferred Income Tax Liability account.
**h.** A firm writes down inventories by $18,000 to reflect the lower-of-cost-or-market valuation.

**4. Working backward from the statement of cash flows. Exhibit 13.6** presents a statement of cash flows for Alcoa for Year 9. Give the entry made on the T-account work sheet for each of

**EXHIBIT 13.6** **ALCOA (Problem 4)**

**Income Statement for Year 9 (all dollar amounts in millions)**

| | |
|---|---|
| Sales Revenues | $20,465.0 |
| Gain on Sale of Marketable Securities | 20.8 |
| Equity in Earnings of Affiliates | 214.0 |
| Total Revenues and Gains | $20,699.8 |
| Cost of Goods Sold | $ 9,963.3 |
| General and Administrative Expenses | 5,570.2 |
| Interest Expense | 2,887.3 |
| Income Tax Expense | 911.6 |
| Total Expenses | $19,332.4 |
| Net Income | $ 1,367.4 |

**Statement of Cash Flows for Year 9 (all dollar amounts in millions)**

| | |
|---|---|
| OPERATIONS | |
| (1) Net Income | $ 1,367.4 |
| **Adjustments for Noncash Transactions:** | |
| (2) Depreciation | 664.0 |
| (3) Increase in Deferred Tax Liability | 82.0 |
| (4) Equity in Undistributed Earnings of Affiliates | (47.1) |
| (5) Gain from Sale of Marketable Securities Available for Sale | (20.8) |
| (6) (Increase) Decrease in Accounts Receivable | 74.6 |
| (7) (Increase) Decrease in Inventories | (198.9) |
| (8) (Increase) Decrease in Prepayments | (40.3) |
| (9) Increase (Decrease) in Accounts Payable for Inventory | 33.9 |
| (10) Increase (Decrease) in Other Current Liabilities | (110.8) |
| Cash Flow from Operations | $ 1,804.0 |
| INVESTING | |
| (11) Sale of Marketable Securities Available for Sale | $ 49.8 |
| (12) Acquisition of Marketable Securities Available for Sale | (73.2) |
| (13) Acquisition of Property, Plant, and Equipment | (875.7) |
| (14) Acquisition of Subsidiaries | (44.5) |
| Cash Flow from Investing | $ (943.6) |
| FINANCING | |
| (15) Common Stock Issued to Employees | $ 34.4 |
| (16) Repurchase of Common Stock | (100.9) |
| (17) Dividends Paid to Shareholders | (242.9) |
| (18) Additions to Short-Term Borrowing | 127.6 |
| (19) Additions to Long-Term Debt | 121.6 |
| (20) Payments on Long-Term Debt | (476.4) |
| Cash Flow from Financing | $ (536.6) |
| Change in Cash | $ 323.8 |
| Cash, Beginning of Year | 506.8 |
| Cash, End of Year | $ 830.6 |
| SUPPLEMENTARY INFORMATION | |
| (21) Acquisition of Property, Plant, and Equipment by Mortgaged Borrowing | $ 76.9 |
| (22) Acquisition of Property, Plant, and Equipment by Capital Leases | 98.2 |
| (23) Conversion of Debt into Common Stock | 47.8 |

(24) Other Current Liabilities represents obligations for General and Administrative Expenses.

the numbered line items. For example, the work sheet entry for line **(1)** is as follows (amounts in millions):

| | | |
|---|---|---|
| Cash (Operations—Net Income) | 1,367.4 | |
| Retained Earnings | | 1,367.4 |

5. **Deriving direct method presentation of cash flow from operations using data from the T-account work sheet.** Refer to the data in **Exhibit 13.6** for Alcoa for Year 9. Derive a presentation of cash flow from operations using the direct method.
6. **Working backward from the statement of cash flows. Exhibit 13.7** presents a statement of cash flows from Ingersoll-Rand for Year 5. Give the entry made on the T-account work sheet

**EXHIBIT 13.7**

**INGERSOLL-RAND**
**Statement of Cash Flows**
**Year 5**
**(all dollar amounts in millions)**
**(Problem 6)**

| | |
|---|---|
| OPERATIONS | |
| (1) Net Income | $ 270.3 |
| Adjustments for Noncash Transactions: | |
| (2) Depreciation | 179.4 |
| (3) Gain on Sale of Property, Plant, and Equipment | (3.6) |
| (4) Equity in Earnings of Affiliates | (41.5) |
| (5) Deferred Income Taxes | 15.1 |
| (6) (Increase) Decrease in Accounts Receivable | 50.9 |
| (7) (Increase) Decrease in Inventories | (15.2) |
| (8) (Increase) Decrease in Other Current Assets | (33.1) |
| (9) Increase (Decrease) in Accounts Payable | (37.9) |
| (10) Increase (Decrease) in Other Current Liabilities | 19.2 |
| Cash Flow from Operations | $ 403.6 |
| INVESTING | |
| (11) Capital Expenditures | $(211.7) |
| (12) Proceeds from Sale of Property, Plant, and Equipment | 26.5 |
| (13) (Increase) Decrease in Marketable Securities | (4.6) |
| (14) Advances from Equity Companies | 18.4 |
| Cash Flow from Investing | $(171.4) |
| FINANCING | |
| (15) Decrease in Short-Term Borrowing | $ (81.5) |
| (16) Issue of Long-Term Debt | 147.6 |
| (17) Payment of Long-Term Debt | (129.7) |
| (18) Proceeds from Exercise of Stock Options | 47.9 |
| (19) Proceeds from Sale of Treasury Stock | 59.3 |
| (20) Dividends Paid | (78.5) |
| Cash Flow from Financing | $ (34.9) |
| Change in Cash | $ 197.3 |
| Cash, Beginning of Year | 48.3 |
| Cash, End of Year | $ 245.6 |
| SUPPLEMENTARY INFORMATION | |
| (21) New Capital Leases Signed | $ 147.9 |
| (22) Conversion of Preferred Stock into Common Stock | 62.0 |
| (23) Issue of Common Stock to Acquire Investments in Securities | 94.3 |

for each of the numbered line items. For example, the work sheet entry for line **(1)** is as follows (amounts in millions):

| | | |
|---|---|---|
| Cash (Operations—Net Income) | 270.3 | |
| Retained Earnings | | 270.3 |

7. **Preparing a statement of cash flows.** (Adapted from CPA examination.) The management of Warren Corporation, concerned over a decrease in cash, provides you with the comparative analysis of changes in account balances between December 31, Year 4, and December 31, Year 5, appearing in **Exhibit 13.8**.

During Year 5, Warren Corporation engaged in the following transactions:

**(1)** Purchased new machinery for $463,200. In addition, it sold certain obsolete machinery, having a book value of $73,200, for $57,600. It made no other entries in Machinery and Equipment or related accounts other than provisions for depreciation.

**(2)** Paid $2,400 of legal costs in a successful defense of a new patent, which it correctly debited to the Patents account. It recorded patent amortization amounting to $5,040 during Year 5.

**(3)** Purchased 120 shares of preferred stock, par value $100, at $110 and subsequently canceled it. Warren Corporation debited the premium paid to Retained Earnings.

**(4)** On December 10, Year 5, the board of directors declared a cash dividend of $0.24 per share, payable to holders of common stock on January 10, Year 6.

**EXHIBIT 13.8**

**WARREN CORPORATION**
**Changes in Account Balances Between December 31, Year 4, and December 31, Year 5 (Problem 7)**

| | December 31 | |
|---|---|---|
| | Year 4 | Year 5 |
| DEBIT BALANCES | | |
| Cash | $ 223,200 | $ 174,000 |
| Accounts Receivable | 327,600 | 306,000 |
| Inventories | 645,600 | 579,600 |
| Securities Held for Plant Expansion Purposes | — | 180,000 |
| Machinery and Equipment | 776,400 | 1,112,400 |
| Leasehold Improvements | 104,400 | 104,400 |
| Patents | 36,000 | 33,360 |
| Totals | $2,113,200 | $2,489,760 |
| CREDIT BALANCES | | |
| Allowance for Uncollectible Accounts | $ 20,400 | $ 19,200 |
| Accumulated Depreciation of Machinery and Equipment | 446,400 | 499,200 |
| Allowance for Amortization of Leasehold Improvements | 58,800 | 69,600 |
| Accounts Payable | 126,000 | 279,360 |
| Cash Dividends Payable | — | 48,000 |
| Current Portion of 6 Percent Serial Bonds Payable | 60,000 | 60,000 |
| 6 Percent Serial Bonds Payable (noncurrent portion) | 360,000 | 300,000 |
| Preferred Stock | 120,000 | 108,000 |
| Common Stock | 600,000 | 600,000 |
| Retained Earnings | 321,600 | 506,400 |
| Totals | $2,113,200 | $2,489,760 |

(5) The following presents a comparative analysis of retained earnings as of December 31, Year 4 and Year 5:

| | December 31 | |
|---|---|---|
| | Year 4 | Year 5 |
| Balance, January 1 | $157,200 | $321,600 |
| Net Income | 206,400 | 234,000 |
| Subtotal | $363,600 | $555,600 |
| Dividends Declared | (42,000) | (48,000) |
| Premium on Preferred Stock Repurchased | — | (1,200) |
| Balance, December 31 | $321,600 | $506,400 |

(6) Warren Corporation wrote off accounts totaling $3,600 as uncollectible during Year 5.

**a.** Prepare a T-account work sheet for the preparation of a statement of cash flows.

**b.** Prepare a formal statement of cash flows for Warren Corporation for the year ending December 31, Year 5, using the indirect method.

8. **Preparing a statement of cash flows.** (Adapted from CPA examination.) Roth Company has prepared its financial statements for the year ended December 31, Year 6, and for the three months ended March 31, Year 7. You have been asked to prepare a statement of cash flows for the three months ended March 31, Year 7. **Exhibit 13.9** presents the company's balance sheet at December 31, Year 6, and March 31, Year 7, and **Exhibit 13.10** presents its income

**EXHIBIT 13.9** ROTH COMPANY Balance Sheet (Problem 8)

| | December 31, Year 6 | March 31, Year 7 |
|---|---|---|
| Cash | $ 37,950 | $131,100 |
| Marketable Securities Available for Sale | 24,000 | 10,200 |
| Accounts Receivable (net) | 36,480 | 73,980 |
| Inventory | 46,635 | 72,885 |
| Total Current Assets | $145,065 | $288,165 |
| Land | 60,000 | 28,050 |
| Building | 375,000 | 375,000 |
| Equipment | — | 122,250 |
| Accumulated Depreciation | (22,500) | (24,375) |
| Investment in 30-Percent-Owned Company (using equity method) | 91,830 | 100,470 |
| Other Assets | 22,650 | 22,650 |
| Total Assets | $672,045 | $912,210 |
| Accounts Payable | $ 31,830 | $ 25,995 |
| Dividend Payable | — | 12,000 |
| Income Taxes Payable | — | 51,924 |
| Total Current Liabilities | $ 31,830 | $ 89,919 |
| Other Liabilities | 279,000 | 279,000 |
| Bonds Payable | 71,550 | 169,275 |
| Deferred Income Tax | 765 | 1,269 |
| Preferred Stock | 45,000 | — |
| Common Stock | 120,000 | 165,000 |
| Unrealized Holding Loss on Marketable Securities | (750) | (750) |
| Retained Earnings | 124,650 | 208,497 |
| Total Equities | $672,045 | $912,210 |

**EXHIBIT 13.10**

**ROTH COMPANY**
**Income Statement Data**
**For the Three Months Ended March 31, Year 7**
**(Problem 8)**

| | |
|---|---|
| Sales | $364,212 |
| Gain on Sale of Marketable Securities | 3,600 |
| Equity in Earnings of 30-Percent-Owned Company | 8,640 |
| Gain on Condemnation of Land | 16,050 |
| Total Revenues | $392,502 |
| Cost of Sales | $207,612 |
| General and Administration Expenses | 33,015 |
| Depreciation | 1,875 |
| Interest Expense | 1,725 |
| Income Taxes | 52,428 |
| Total Expenses | $296,655 |
| Net Income | $ 95,847 |

statement for the three months ended March 31, Year 7. You are satisfied that the amounts presented are correct.

Your discussion with the company's controller and a review of the financial records reveal the following information:

**(1)** On January 8, Year 7, the company sold marketable securities for cash. The firm had purchased these securities on December 31, Year 6. The firm purchased no marketable securities during Year 7.
**(2)** The company's preferred stock is convertible into common stock at a rate of one share of preferred for two shares of common. The preferred stock and common stock have par values of $2 and $1, respectively.
**(3)** On January 17, Year 7, the local government condemned three acres of land. Roth Company received an award of $48,000 in cash on March 22, Year 7. It does not expect to purchase additional land as a replacement.
**(4)** On March 25, Year 7, the company purchased equipment for cash.
**(5)** Interest expense on bonds payable exceeded the cash coupon payments by $225 during the three-month period. On March 29, Year 7, the company issued bonds payable for cash.
**(6)** Roth Company declared $12,000 in dividends during the three months.

**a.** Prepare a T-account work sheet for the preparation of a statement of cash flows, defining funds as cash and cash equivalents.
**b.** Prepare a formal statement of cash flows for Roth Company for the three months ending March 31, Year 7. Use the indirect method.
**c.** Derive a presentation of cash flows from operations using the direct method.

**9. Preparing a statement of cash flows.** (Adapted from CPA examination.) **Exhibit 13.11** presents a comparative statement of financial position for Biddle Corporation as of December 31, Year 1 and Year 2. **Exhibit 13.12** presents an income statement for Year 2. Additional information follows:

**(1)** On February 2, Year 2, Biddle issued a 10 percent stock dividend to shareholders of record on January 15, Year 2. The market price per share of the common stock on February 2, Year 2, was $15.
**(2)** On March 1, Year 2, Biddle issued 1,900 shares of common stock for land. The common stock and land had current market values of approximately $20,000 on March 1, Year 2.
**(3)** On April 15, Year 2, Biddle repurchased long-term bonds with a face and book value of $25,000. It reported a gain of $6,000 on the income statement.
**(4)** On June 30, Year 2, Biddle sold equipment costing $26,500, with a book value of $11,500, for $9,500 cash.
**(5)** On September 30, Year 2, Biddle declared and paid a $0.04 per share cash dividend to shareholders of record on August 1, Year 2.
**(6)** On October 10, Year 2, Biddle purchased land for $42,500 cash.
**(7)** Deferred income taxes represent temporary differences relating to the use of different depreciation methods for income tax and financial statement reporting.

**EXHIBIT 13.11**

**BIDDLE CORPORATION**
**Statement of Financial Position**
**(Problem 9)**

| | December 31 | |
|---|---|---|
| | Year 1 | Year 2 |
| ASSETS | | |
| Cash | $ 45,000 | $ 50,000 |
| Accounts Receivable (net of allowance for doubtful accounts of $10,000 and $8,000, respectively) | 70,000 | 105,000 |
| Inventories | 110,000 | 130,000 |
| Total Current Assets | $225,000 | $285,000 |
| Land | 100,000 | 162,500 |
| Plant and Equipment | 316,500 | 290,000 |
| Less Accumulated Depreciation | (50,000) | (45,000) |
| Patents | 16,500 | 15,000 |
| Total Assets | $608,000 | $707,500 |
| LIABILITIES AND SHAREHOLDERS' EQUITY | | |
| **Liabilities** | | |
| Accounts Payable | $100,000 | $130,000 |
| Accrued Liabilities | 105,000 | 100,000 |
| Total Current Liabilities | $205,000 | $230,000 |
| Deferred Income Taxes | 50,000 | 70,000 |
| Long-Term Bonds (due December 15, Year 13) | 90,000 | 65,000 |
| Total Liabilities | $345,000 | $365,000 |
| **Shareholders' Equity** | | |
| Common Stock, Par Value $5, Authorized 50,000 Shares, Issued and Outstanding 21,000 and 25,000 Shares, Respectively | $105,000 | $125,000 |
| Additional Paid-in Capital | 85,000 | 116,500 |
| Retained Earnings | 73,000 | 101,000 |
| Total Shareholders' Equity | $263,000 | $342,500 |
| Total Liabilities and Shareholders' Equity | $608,000 | $707,500 |

**EXHIBIT 13.12**

**BIDDLE CORPORATION**
**Income Statement**
**For the Year Ended December 31, Year 2**
**(Problem 9)**

| | |
|---|---|
| Sales | $500,000 |
| Gain in Repurchase of Bonds | 6,000 |
| Total Revenues | $506,000 |
| Expenses: | |
| Cost of Goods Sold | $280,000 |
| Salary and Wages | 95,000 |
| Depreciation | 10,000 |
| Patent Amortization | 1,500 |
| Loss on Sale of Equipment | 2,000 |
| Interest | 8,000 |
| Miscellaneous | 4,000 |
| Total Expenses | $400,500 |
| Income before Income Taxes | $105,500 |
| Income Taxes | |
| Current | $ 25,000 |
| Deferred | 20,000 |
| Provision for Income Taxes | $ 45,000 |
| Net Income | $ 60,500 |
| Earnings per Share | $ 2.45 |

**a.** Prepare a T-account work sheet for the preparation of a statement of cash flows.
**b.** Prepare a formal statement of cash flows for Biddle Corporation for the year ended December 31, Year 2. Use the indirect method.

**10. Preparing a statement of cash flows.** (Adapted from CPA examination.) **Exhibit 13.13** presents the comparative balance sheets for Plainview Corporation for Year 4 and Year 5. The following additional information relates to Year 5 activities:

**(1)** The Retained Earnings account changed as follows:

| | | |
|---|---|---|
| Retained Earnings, December 31, Year 4 | | $758,200 |
| Add Net Income | | 236,580 |
| Subtotal | | $994,780 |
| Deduct: | | |
| Cash Dividends | $130,000 | |
| Loss on Reissue of Treasury Stock | 3,000 | |
| Stock Dividend | 100,200 | 233,200 |
| Retained Earnings, December 31, Year 5 | | $761,580 |

**(2)** On January 2, Year 5, Plainview Corporation sold for $127,000 marketable securities with an acquisition cost and a book value of $110,000. The firm used the proceeds from this sale, the funds in the bond sinking fund, and the amount received from the issuance of the 8 percent debentures to retire the 6 percent mortgage bonds.

**(3)** The firm reissued treasury stock on February 28, Year 5. It treats "losses" on the reissue of treasury stock as a charge to Retained Earnings.

**EXHIBIT 13.13**

**PLAINVIEW CORPORATION**
**Comparative Balance Sheets**
**December 31, Year 4 and Year 5**
**(Problem 10)**

| | Year 4 | Year 5 |
|---|---|---|
| ASSETS | | |
| Cash | $ 165,300 | $ 142,100 |
| Marketable Securities (at market value) | 129,200 | 122,600 |
| Accounts Receivable (net) | 371,200 | 312,200 |
| Inventories | 124,100 | 255,200 |
| Prepayments | 22,000 | 23,400 |
| Bond Sinking Fund | 63,000 | — |
| Investment in Subsidiary (at equity) | 152,000 | 134,080 |
| Plant and Equipment (net) | 1,534,600 | 1,443,700 |
| Total Assets | $2,561,400 | $2,433,280 |
| SOURCES OF FINANCING | | |
| Accounts Payable | $ 213,300 | $ 238,100 |
| Notes Payable—Current | 145,000 | — |
| Accrued Payables | 18,000 | 16,500 |
| Income Taxes Payable | 31,000 | 97,500 |
| Deferred Income Taxes (noncurrent) | 128,400 | 127,900 |
| 6 Percent Mortgage Bonds Payable (due Year 17) | 310,000 | — |
| 8 Percent Debentures Payable (due Year 25) | — | 125,000 |
| Common Stock, $10 Par Value | 950,000 | 1,033,500 |
| Additional Paid-in Capital | 51,000 | 67,700 |
| Accumulated Other Comprehensive Income | | |
| Unrealized Holding Gain on Marketable Securities | 2,500 | 2,500 |
| Retained Earnings | 755,700 | 759,080 |
| Treasury Stock—at Cost of $3 per Share | (43,500) | (34,500) |
| Total Sources of Financing | $2,561,400 | $2,433,280 |

(4) The firm declared a stock dividend on October 31, Year 5, when the market price of Plainview Corporation's stock was $12 per share.

(5) On April 30, Year 5, a fire destroyed a warehouse that cost $100,000 and on which depreciation of $65,000 had accumulated. The loss was not insured. Plainview Corporation properly included the loss in the Continuing Operations section of the income statement.

(6) Plant and equipment transactions consisted of the sale of a building at its book value of $4,000 and the purchase of machinery for $28,000.

(7) The firm wrote off accounts receivable as uncollectible totaling $16,300 in Year 4 and $18,500 in Year 5. It recognized expired insurance of $4,100 in Year 4 and $3,900 in Year 5.

(8) The subsidiary, which is 40 percent owned, reported a loss of $44,800 for Year 5.

a. Prepare a T-account work sheet for Plainview Corporation for Year 5, defining funds as cash and cash equivalents.

b. Prepare a formal statement of cash flows using the indirect method for the year ending December 31, Year 5.

11. **Preparing and interpreting the statement of cash flows. Exhibit 13.14** presents a comparative balance sheet and **Exhibit 13.15** presents a comparative income statement for UAL Corporation for Year 9 and Year 10. Expenditures on new property, plant, and equipment were $1,568 million in Year 9 and $2,821 million in Year 10. Changes in other noncurrent assets are investing activities, and changes in other noncurrent liabilities are financing activities.

a. Prepare T-account work sheets for Year 9 and Year 10 for a statement of cash flows.

b. Prepare a comparative statement of cash flows for Year 9 and Year 10 using the indirect method.

c. Comment on the relations between cash flows from operating, investing, and financing activities for Year 9 and Year 10.

12. **Preparing and interpreting the statement of cash flows.** Irish Paper Company (Irish) manufactures and markets various paper products around the world. Paper manufacturing is a capital-intensive activity. A firm that does not adequately use its manufacturing capacity will experience poor operating performance. Sales of paper products tend to be cyclical with general economic conditions, although consumer paper products are less cyclical than business paper products.

**Exhibit 13.16** on page 33 presents comparative income statements and **Exhibit 13.17** on page 34 presents comparative balance sheets for Irish Paper Company for Year 9, Year 10, and Year 11. Additional information appears below (amounts in millions).

(1)

| Cash Flow Information | Year 9 | Year 10 | Year 11 |
|---|---|---|---|
| Investments in Affiliates[a] | $ (92) | $ 86 | $ (13) |
| Expenditures on Property, Plant, and Equipment | (775) | (931) | (315) |
| Long-Term Debt Issued | 449 | 890 | 36 |

[a]Excludes earnings and dividends.

(2) Depreciation expense was $306 million in Year 9, $346 million in Year 10, and $353 million in Year 11.

(3) During Year 9, Irish purchased outstanding stock warrants for $201 million. It recorded the transaction by debiting the Common Stock account.

(4) During Year 9, Irish sold timberlands at a gain. It received cash of $5 million and a long-term note receivable for $220 million, which it includes in Other Assets on the balance sheet.

(5) In addition to the cash expenditures presented above, Irish acquired property, plant, and equipment during Year 10 costing $221 million by assuming a long-term mortgage payable.

(6) During Year 11, Irish resold treasury stock for an amount greater than its cost.

(7) Changes in Other Assets are investing activities.

a. Prepare T-account work sheets for a statement of cash flows for Irish for Year 9, Year 10, and Year 11.

b. Prepare a comparative statement of cash flows for Irish for Year 9, Year 10, and Year 11 using the indirect method.

c. Comment on the pattern of cash flows from operating, investing, and financing activities for each of the three years.

13. **Preparing a statement of cash flows.** (Adapted from a problem prepared by Stephen A. Zeff.) Selected information from the accounting records of Breda Enterprises Inc. appears

**EXHIBIT 13.14**

**UAL CORPORATION**
**Comparative Balance Sheet**
**(all dollar amounts in millions)**
**(Problem 11)**

| | December 31 | | |
|---|---|---|---|
| | **Year 8** | **Year 9** | **Year 10** |
| ASSETS | | | |
| Cash | $1,087 | $ 465 | $ 221 |
| Marketable Securities | — | 1,042 | 1,066 |
| Accounts Receivable (net) | 741 | 888 | 913 |
| Inventories | 210 | 249 | 323 |
| Prepayments | 112 | 179 | 209 |
| Total Current Assets | $2,150 | $2,823 | $2,732 |
| Property, Plant, and Equipment | 7,710 | 7,704 | 8,587 |
| Accumulated Depreciation | (3,769) | (3,805) | (3,838) |
| Other Assets | 610 | 570 | 605 |
| Total Assets | $6,701 | $7,292 | $8,086 |
| LIABILITIES AND SHAREHOLDERS' EQUITY | | | |
| Accounts Payable | $ 540 | $ 596 | $ 552 |
| Short-Term Borrowing | 121 | 446 | 447 |
| Current Portion of Long-Term Debt | 110 | 84 | 89 |
| Advances from Customers | 619 | 661 | 843 |
| Other Current Liabilities | 1,485 | 1,436 | 1,826 |
| Total Current Liabilities | $2,875 | $3,223 | $3,757 |
| Long-Term Debt | 1,418 | 1,334 | 1,475 |
| Deferred Tax Liability | 352 | 364 | 368 |
| Other Noncurrent Liabilities | 715 | 719 | 721 |
| Total Liabilities | $5,360 | $5,640 | $6,321 |
| Common Stock | $ 119 | $ 119 | $ 120 |
| Additional Paid-in Capital | 48 | 48 | 52 |
| Accumulated Other Comprehensive Income | | | |
| Unrealized Holding Gain on Marketable Securities | — | 85 | 92 |
| Retained Earnings | 1,188 | 1,512 | 1,613 |
| Treasury Stock | (14) | (112) | (112) |
| Total Shareholders' Equity | $1,341 | $1,652 | $1,765 |
| Total Liabilities and Shareholders' Equity | $6,701 | $7,292 | $8,086 |

below. The firm uses a calendar year as its reporting period. You are asked to prepare a statement of cash flows for Breda Enterprises Inc. for Year 6. Use the indirect method. Key all figures in the statement of cash flows to the numbered items below.

**(1)** Net income for Year 6 is $90,000.

**(2)** Beginning and ending balances in three accounts relating to the firm's customers were as follows:

| | December 31, Year 5 | December 31, Year 6 |
|---|---|---|
| Accounts Receivable (gross) | $41,000 | $53,000 |
| Allowance for Uncollectible Accounts | 1,800 | 3,200 |
| Advances from Customers | 3,700 | 1,000 |

**EXHIBIT 13.15**

**UAL CORPORATION**
**Comparative Income Statement**
**(all dollar amounts in millions)**
**(Problem 11)**

| | Year 9 | Year 10 |
|---|---|---|
| REVENUES | | |
| Sales | $ 9,794 | $11,037 |
| Interest Revenue | 121 | 123 |
| Gains on Dispositions of Property, Plant, and Equipment | 106 | 286 |
| Total Revenues | $10,021 | $11,446 |
| EXPENSES | | |
| Compensation | $ 3,158 | $ 3,550 |
| Fuel | 1,353 | 1,811 |
| Commissions | 1,336 | 1,719 |
| Depreciation | 517 | 560 |
| Other Operating Costs | 2,950 | 3,514 |
| Interest | 169 | 121 |
| Income Taxes | 214 | 70 |
| Total Expenses | $ 9,697 | $11,345 |
| Net Income | $ 324 | $ 101 |

**EXHIBIT 13.16**

**IRISH PAPER COMPANY**
**Comparative Income Statements**
**(all dollar amounts in millions)**
**(Problem 12)**

| | Year 9 | Year 10 | Year 11 |
|---|---|---|---|
| Sales | $5,066 | $5,356 | $4,976 |
| Equity in Earnings of Affiliates | 31 | 38 | 30 |
| Interest Income | 34 | 23 | 60 |
| Gain (Loss) on Sale of Property, Plant, and Equipment | 221 | 19 | (34) |
| Total Revenues | $5,352 | $5,436 | $5,032 |
| Cost of Goods Sold | $3,493 | $3,721 | $3,388 |
| Selling Expenses | 857 | 925 | 1,005 |
| Administrative Expenses | 303 | 414 | 581 |
| Interest Expense | 158 | 199 | 221 |
| Income Tax Expense | 165 | 8 | (21) |
| Total Expenses | $4,976 | $5,267 | $5,174 |
| Net Income | $ 376 | $ 169 | $ (142) |

On November 1, Year 6, a customer gave the firm a six-month, 8 percent, $15,000 note in satisfaction of an account receivable of $15,000. Interest is payable at maturity. This was the only note receivable held by the company during Year 6.

**EXHIBIT 13.17**

**IRISH PAPER COMPANY**
**Comparative Balance Sheets**
**(all dollar amounts in millions)**
**(Problem 12)**

| | Year 8 | Year 9 | Year 10 | Year 11 |
|---|---|---|---|---|
| ASSETS | | | | |
| Cash | $ 374 | $ 49 | $ 114 | $ 184 |
| Accounts Receivable (net) | 611 | 723 | 829 | 670 |
| Inventories | 522 | 581 | 735 | 571 |
| Prepayments | 108 | 54 | 54 | 56 |
| Total Current Assets | $1,615 | $1,407 | $1,732 | $1,481 |
| Investments in Affiliates | 254 | 375 | 322 | 333 |
| Property, Plant, and Equipment | 5,272 | 5,969 | 7,079 | 7,172 |
| Accumulated Depreciation | (2,160) | (2,392) | (2,698) | (2,977) |
| Other Assets | 175 | 387 | 465 | 484 |
| Total Assets | $5,156 | $5,746 | $6,900 | $6,493 |
| LIABILITIES AND SHAREHOLDERS' EQUITY | | | | |
| Accounts Payable | $ 920 | $ 992 | $1,178 | $1,314 |
| Current Portion of Long-Term Debt | 129 | 221 | 334 | 158 |
| Other Current Liabilities | 98 | 93 | 83 | 38 |
| Total Current Liabilities | $1,147 | $1,306 | $1,595 | $1,510 |
| Long-Term Debt | 1,450 | 1,678 | 2,455 | 2,333 |
| Deferred Income Taxes | 607 | 694 | 668 | 661 |
| Total Liabilities | $3,204 | $3,678 | $4,718 | $4,504 |
| Preferred Stock | $ 7 | $ 7 | $ 7 | $ 7 |
| Common Stock | 629 | 428 | 432 | 439 |
| Retained Earnings | 1,331 | 1,648 | 1,758 | 1,557 |
| Treasury Stock | (15) | (15) | (15) | (14) |
| Total Shareholders' Equity | $1,952 | $2,068 | $2,182 | $1,989 |
| Total Liabilities and Shareholders' Equity | $5,156 | $5,746 | $6,900 | $6,493 |

**(3)** The balances in Merchandise Inventory and Accounts Payable were as follows:

| | December 31, Year 5 | December 31, Year 6 |
|---|---|---|
| Merchandise Inventory | $47,000 | $43,000 |
| Accounts Payable | 27,000 | 39,000 |

**(4)** During Year 6 the firm sold, for $25,000 cash, equipment with a book value of $38,000. The firm also purchased equipment for cash. Depreciation expense for Year 6 was $42,000. The balance in the Equipment account at acquisition cost decreased $26,000 between the beginning and end of Year 6. The balance in the Accumulated Depreciation account increased $11,000 between the beginning and end of Year 6.

**(5)** The balances in the Leasehold Asset and Lease Liability accounts were as follows on various dates:

| | December 31, Year 4 | December 31, Year 5 | December 31, Year 6 |
|---|---|---|---|
| Leasehold Asset (net) | $0 | $76,000 | $71,000 |
| Lease Liability | 0 | 76,000 | 73,600 |

On December 31, Year 5, the firm signed a long-term lease which, by its terms, qualified as a capital lease. The firm made a payment under the lease of $10,000 on December 31, Year 6.

**(6)** The firm declared cash dividends during Year 6 of $26,000, of which $10,000 remains unpaid on December 31, Year 6. During Year 6, the firm paid $8,000 cash for dividends declared during Year 5.

**(7)** The firm classifies all marketable securities as available for sale. It purchased no marketable securities during Year 6 but sold marketable equity securities that had originally cost $4,500 for $9,100 cash in November, Year 6. The market values of marketable equity securities were $4,000 on December 31, Year 5, and $10,500 on December 31, Year 6. These amounts were also the book values of the securities on these two dates.

**(8)** Investors in $100,000 face value of convertible bonds of Breda Enterprises Inc. converted them into 8,000 shares of the firm's $12 par value common stock during Year 6. The common stock had a market value of $15 per share on the conversion date. Breda Enterprises Inc. had originally issued the bonds at a premium. Their book value on the date of the conversion was $105,000. The firm chose the generally accepted (alternative) accounting principle of recording the issuance of the common stock at market value and recognizing a loss of $15,000. The loss is not classified as an extraordinary item. The firm amortized $1,500 of the bond premium between January 1, Year 6, and the date of the conversion.

**14. Interpreting the statement of cash flows. Exhibit 13.18** presents a statement of cash flows for L.A. Gear, manufacturer of athletic shoes and sportswear, for three recent years.

**a.** What is the likely reason for the negative cash flow from operations?

**EXHIBIT 13.18**

**L.A. GEAR**
**Statement of Cash Flows**
**(all dollar amounts in thousands)**
**(Problem 14)**

| | Year 7 | Year 8 | Year 9 |
|---|---|---|---|
| OPERATIONS | | | |
| Net Income | $ 4,371 | $ 22,030 | $ 55,059 |
| Depreciation | 133 | 446 | 1,199 |
| Noncash Compensation to Employees | — | — | 558 |
| Increase in Accounts Receivable | (12,410) | (34,378) | (51,223) |
| Increase in Inventories | (1,990) | (50,743) | (72,960) |
| Increase in Prepayments | (599) | (2,432) | (8,624) |
| Increase in Accounts Payable | 1,656 | 7,197 | 17,871 |
| Increase (Decrease) in Other Current Liabilities | (537) | 11,193 | 10,587 |
| Cash Flow from Operations | $ (9,376) | $(46,687) | $(47,533) |
| INVESTING | | | |
| Sale of Marketable Securities | $ 5,661 | — | — |
| Acquisition of Property, Plant, and Equipment | (874) | $ (2,546) | $ (6,168) |
| Acquisition of Other Noncurrent Assets | (241) | (406) | (246) |
| Cash Flow from Investing | $ 4,546 | $ (2,952) | $ (6,414) |
| FINANCING | | | |
| Increase (Decrease) in Short-Term Borrowing | $ 4,566 | $ 50,104 | $(19,830) |
| Issue of Common Stock | — | 495 | 69,925 |
| Cash Flow from Financing | $ 4,566 | $ 50,599 | $ 50,095 |
| Change in Cash | $ (264) | $ 960 | $ (3,852) |
| Cash, Beginning of Year | 3,509 | 3,245 | 4,205 |
| Cash, End of Year | $ 3,245 | $ 4,205 | $ 353 |

b. How did L.A. Gear finance the negative cash flow from operations during each of the three years? Suggest reasons for L.A. Gear's choice of financing source for each year.
c. Expenditures on property, plant, and equipment substantially exceeded the addback for depreciation expense each year. What is the likely explanation for this difference in amounts?
d. The addback for depreciation expense is a relatively small proportion of net income. What is the likely explanation for this situation?
e. L.A. Gear had no long-term debt in its capital structure during Year 7 through Year 9. What is the likely explanation for such a financial structure?

15. **Interpreting the statement of cash flows. Exhibit 13.19** presents a statement of cash flows for Campbell Soup Company for three recent years. Campbell Soup Company is in the consumer foods industry, a relatively mature industry in the United States.
    a. Cash flow from operations each year approximately equals net income plus addbacks for depreciation, deferred taxes, and other. What is the likely explanation for this relation?

**EXHIBIT 13.19**

**CAMPBELL SOUP COMPANY**
**Statement of Cash Flows**
**(all dollar amounts in millions)**
**(Problem 15)**

| | Year 6 | Year 7 | Year 8 |
|---|---|---|---|
| OPERATIONS | | | |
| Net Income | $ 223 | $ 247 | $ 274 |
| Depreciation | 127 | 145 | 171 |
| Deferred Income Taxes | 29 | 46 | 31 |
| Other Addbacks | 21 | 34 | 11 |
| (Increase) in Accounts Receivable | (19) | (40) | (55) |
| (Increase) Decrease in Inventories | 13 | (13) | 6 |
| (Increase) in Prepayments | (7) | (11) | (40) |
| Increase in Accounts Payable | 27 | 53 | 72 |
| Increase (Decrease) in Other Current Liabilities | 29 | 2 | (1) |
| Cash Flow from Operations | $ 443 | $ 463 | $ 469 |
| INVESTING | | | |
| Sale of Property, Plant, and Equipment | $ 30 | $ 21 | $ 41 |
| Sale of Marketable Securities | 328 | 535 | 319 |
| Acquisition of Property, Plant, and Equipment | (275) | (250) | (245) |
| Acquisition of Marketable Securities | (472) | (680) | (70) |
| Acquisition of Investments in Securities | — | — | (472) |
| Other Investing Transactions | (5) | (34) | (48) |
| Cash Flow from Investing | $(394) | $(408) | $(475) |
| FINANCING | | | |
| Increase in Short-Term Borrowing | — | $ 5 | $ 86 |
| Increase in Long-Term Borrowing | $ 220 | 29 | 103 |
| Issue of Common Stock | 4 | 2 | — |
| Decrease in Short-Term Borrowing | (3) | — | (5) |
| Decrease in Long-Term Borrowing | (168) | (27) | (106) |
| Acquisition of Common Stock | — | — | (28) |
| Dividends | (84) | (92) | (103) |
| Cash Flow from Financing | $ (31) | $ (83) | $ (53) |
| Change in Cash | $ 18 | $ (28) | $ (59) |
| Cash, Beginning of Year | 155 | 173 | 145 |
| Cash, End of Year | $ 173 | $ 145 | $ 86 |

**b.** In the Investing section of Campbell's statement of cash flow, what are the indications that the company is in a relatively mature industry?

**c.** In the Financing section of Campbell's statement of cash flows, what are the indications that the company is in a relatively mature industry?

**16. Interpreting the statement of cash flows.** Prime Contracting Services provides various services to government agencies under multiyear contracts. In Year 6, the services primarily involved transportation services of equipment and household furniture. Beginning in Year 7, the firm began exiting these transportation services businesses and began offering more people-based services (clerical, training). Sales increased at a compounded annual rate of 28.9 percent during the five-year period. **Exhibit 13.20** presents a statement of cash flows for Prime Contracting Services for Year 6 to Year 10. Changes in Other Current Liabilities primarily represent salaries.

**a.** What evidence do you see of the strategic shift from asset-based to people-based services?

**b.** What are the likely reasons that net income decreased between Year 6 and Year 8 while cash flow from operations increased during the same period?

**EXHIBIT 13.20**

**PRIME CONTRACTING SERVICES**
**Statement of Cash Flows**
**(all dollar amounts in millions)**
**(Problem 16)**

| | Year 6 | Year 7 | Year 8 | Year 9 | Year 10 |
|---|---|---|---|---|---|
| OPERATIONS | | | | | |
| Net Income | $ 261,243 | $ 249,438 | $ 46,799 | $ 412,908 | $ 593,518 |
| Depreciation | 306,423 | 616,335 | 826,745 | 664,882 | 606,633 |
| Deferred Income Taxes | 158,966 | 179,584 | 55,000 | (110,116) | (154,000) |
| Loss (Gain) on Disposition of Assets | 20,000 | — | — | (117,804) | (35,077) |
| Other | 2,200 | (7,226) | (51,711) | (19,377) | 9,100 |
| (Increase) Decrease in Accounts Receivable | (1,420,783) | (647,087) | (263,164) | (864,555) | 175,408 |
| (Increase) Decrease in Other Current Assets | (38,031) | (25,792) | (40,067) | (9,333) | 127,548 |
| Increase (Decrease) in Accounts Payable | 507,386 | (177,031) | (32,732) | (272,121) | (166,672) |
| Increase (Decrease) in Other Current Liabilities | 266,260 | 99,417 | 422,929 | 927,478 | (416,856) |
| Cash Flow from Operations | $ 63,664 | $ 287,638 | $ 963,799 | $ 611,962 | $ 739,602 |
| INVESTING | | | | | |
| Fixed Assets Sold | $ 80,000 | — | — | $ 117,804 | $ 175,075 |
| Employee and Officer Loans | (16,960) | $ 62,894 | — | — | |
| Fixed Assets Acquired | (2,002,912) | (911,470) | $ (56,370) | (19,222) | (48,296) |
| Cash Flow from Investing | $(1,939,872) | $(848,576) | $ (56,370) | $ 98,582 | $ 126,779 |
| FINANCING | | | | | |
| Net Increase (Decrease) in Notes Payable | $ 204,817 | $ 275,475 | $(126,932) | $ 12,650 | $ 325,354 |
| Borrowings under Equipment Loans | 943,589 | 793,590 | 208,418 | — | — |
| Borrowings under Capital Leases | 915,596 | — | — | — | — |
| Borrowings from Shareholder Loans | 127,500 | 117,422 | — | — | — |
| Repayments under Equipment Loans | (236,229) | (389,268) | (564,585) | (437,660) | (736,793) |
| Repayments under Capital Leases | (124,012) | (268,556) | (296,495) | (304,054) | — |
| Repayments under Shareholder Loans | (63,077) | — | (150,000) | — | (28,710) |
| Cash Flow from Financing | $ 1,768,184 | $ 528,663 | $(929,594) | $(729,064) | $(440,149) |
| Change in Cash | $ (108,024) | $ (32,275) | $ (22,165) | $ (18,520) | $ 426,232 |
| Cash, Beginning of Year | 186,897 | 78,873 | 46,598 | 24,433 | 5,913 |
| Cash, End of Year | $ 78,873 | $ 46,598 | $ 24,433 | $ 5,913 | $ 432,145 |

c. What are the likely reasons that net income increased between Year 8 and Year 10 while cash flow from operations was less during Year 9 and Year 10 than in Year 8?

d. How has the risk of Prime Contracting Services changed during the five years?

17. **Interpreting the statement of cash flows. Exhibit 13.21** presents a statement of cash flows for Cypres Corporation.

a. What are the likely reasons that net income increased between Year 11 and Year 13 but cash flow from operations decreased?

b. What are the likely reasons for the increased cash flow from operations between Year 13 and Year 15?

c. How has the risk of Cypres Corporation changed over the five-year period?

**EXHIBIT 13.21**

**CYPRES CORPORATION**
**Statement of Cash Flows**
**(all dollar amounts in thousands)**
**(Problem 17)**

| | Year 11 | Year 12 | Year 13 | Year 14 | Year 15 |
|---|---|---|---|---|---|
| OPERATION | | | | | |
| Net Income | $ 1,045 | $ 1,733 | $ 3,716 | $ 6,583 | $ 6,602 |
| Depreciation and Amortization | 491 | 490 | 513 | 586 | 643 |
| Other Addbacks | 20 | 25 | 243 | 151 | 299 |
| Other Subtractions | 0 | 0 | 0 | 0 | (97) |
| Working Capital Provided by Operations | $1,556 | $ 2,248 | $ 4,472 | $ 7,320 | $ 7,447 |
| (Increase) Decrease in Receivables | (750) | (2,424) | (3,589) | (5,452) | 4,456 |
| (Increase) Decrease in Inventories | (1,387) | (4,111) | (7,629) | 1,867 | 1,068 |
| Increase (Decrease) Accts. Pay-Trade | 1,228 | 2,374 | 1,393 | 1,496 | (2,608) |
| Increase (Decrease) in Other Current Liabilities | 473 | 2,865 | 4,737 | 1,649 | (1,508) |
| Cash from Operations | $ 1,120 | $ 952 | $ (616) | $ 6,880 | $ 8,855 |
| INVESTING | | | | | |
| Fixed Assets Acquired (net) | $ (347) | $ (849) | $ (749) | $(1,426) | $(1,172) |
| Marketable Securities Acquired | 0 | 0 | 0 | 0 | (3,306) |
| Other Investment Transactions | 45 | 0 | 81 | (64) | 39 |
| Cash Flow from Investing | $ (302) | $ (849) | $ (668) | $(1,490) | $(4,439) |
| FINANCING | | | | | |
| Increase in Short-Term Borrowing | $ 0 | $ 700 | $ 2,800 | $ 0 | $ 0 |
| Increase in Long-Term Borrowing | 0 | 0 | 0 | 0 | 0 |
| Issue of Capital Stock | 0 | 0 | 0 | 0 | 315 |
| Decrease in Short-Term Borrowing | 0 | 0 | 0 | (3,500) | 0 |
| Decrease in Long-Term Borrowing | (170) | (170) | (170) | (170) | (170) |
| Acquisition of Capital Stock | (27) | 0 | 0 | 0 | 0 |
| Dividends | (614) | (730) | (964) | (1,427) | (2,243) |
| Other Financing Transactions | 0 | 0 | 0 | 0 | 0 |
| Cash Flow from Financing | $ (811) | $ (200) | $ 1,666 | $(5,097) | $(2,098) |
| Change in Cash | $ 7 | $ (97) | $ 382 | $ 293 | $ 2,318 |
| Cash, Beginning of Year | 955 | 962 | 865 | 1,247 | 1,540 |
| Cash, End of Year | $ 962 | $ 865 | $ 1,247 | $ 1,540 | $ 3,858 |

APPENDIX

# Compound Interest: Concepts and Applications

## Learning Objectives

1. Begin to master compound interest concepts of future value, present value, present discounted value of single sums and annuities, discount rates, interest rates, and internal rates of return on cash flows.
2. Apply those concepts to problems of finding the single payment or, in the context of annuities, the amount for a series of payments required to meet a specified objective.
3. Begin using perpetuity growth models in valuation analysis.
4. Learn how to find the interest rate to satisfy a stated set of conditions, such as to make the present value of a stream of cash flows equal a specified amount.
5. Begin to learn how to construct a problem of the types described above from a description of a business or an accounting situation.

*O*wners of cash, like owners of other scarce resources, can permit borrowers to rent the use of their cash for a period of time. Payment for the rental of cash differs little from other rental payments, such as those made to a landlord for the use of property or to a car rental agency for the use of a car. *Interest* is payment for the use of cash. Accounting must record transactions for both the lender and the borrower caused by lenders' providing cash to borrowers.

Accountants and managers deal with interest calculations for other reasons. Expenditures for an asset most often precede the receipts for services produced by that asset. Cash received later has smaller value than cash received sooner. The difference in timing affects the amount of profit from a firm's acquiring an asset. Amounts of cash received at different times have different values. Managers use interest calculations to make valid comparisons among amounts of cash paid or received at different times. Accountants and analysts use interest-related calculations involving present values to help them estimate the values of entire divisions and firms.

Contracts involving a series of cash payments over time, such as bonds, mortgages, notes, and leases, have **present values**. The present value of a series of payments represents the single amount of cash one would pay now to receive the entire series of future cash payments.

## Compound Interest Concepts

Contracts typically state interest cost as a percentage of the amount borrowed per unit of time. Examples are 12 percent per year and 1 percent per month, which differ from one another. When the statement of interest cost does not explicitly state a period, then the rate applies to a year, so that "interest at the rate of 12 percent" means 12 percent per year. Some inflation-ravaged countries, such as Brazil, quote interest rates for a month.

The amount borrowed or lent is the **principal**. **Compound interest** means that the amount of interest earned during a period increases the principal, which is thus larger for the next interest period. For example, if you deposit $1,000 in a savings account that pays compound interest at the rate of 6 percent per year, you will earn $60 by the end of one year. If you do not withdraw the $60, then $1,060 will earn interest during the second year. During the second year, your principal of $1,060 will earn $63.60 in interest: $60 on the initial deposit of $1,000 and $3.60 on the $60 earned the first year. By the end of the second year, your principal will total $1,123.60.

When only the original principal earns interest during the entire life of the loan, the interest due at the time the borrower repays the loan is called **simple interest**. Simple interest calculations ignore interest on previously earned interest. If the lender may withdraw interest earned, or the borrower must make periodic payments with further interest charges for late payments, then compound interest techniques will still apply.

**Simple Interest Rarely Applies to Economic Calculations but Accounting Often Uses It for Convenience** The use of simple interest calculations in accounting arises in the following way: if you borrow $10,000 at a rate of 12 percent per year but compute interest for any month as $100 (= $10,000 × 0.12 × 1/12), you are using a simple interest calculation. Nearly all economic calculations, however, involve compound interest. When firms use simple interest to compute amounts for periods less than a year, some distortion of periodic numbers results, but no harm. Early periods get charged "too much" interest and later periods get charged "too little," but the distortions are minor.

**Power of Compound Interest** The force, or effect, of compound interest exceeds the intuition of many people. For example, compounded annually at 8 percent, cash doubles itself in nine years. Put another way, if you invest $100 at 8 percent compounded annually, you will have $200 in nine years. If you were to invest $1 in the stock market at age 25 and leave it there for 45 years and the market increased for the next 45 years the way it has for the last 20 years, you would have over $150 by the time you reached age 70.

Problems involving compound interest generally fall into two groups with respect to time.

- First, we may want to know the future value of cash invested or loaned today, as in the two examples in the preceding paragraph.
- Second, we may want to know the present value, or today's value, of cash to be received or paid at later dates. (If I want to have $1,000,000 available at retirement, how much must I invest today?)

In addition, the accountant must sometimes find the interest rate implicit in specified payment streams. For example, assume a bank will lend you $1,000 in return for your promise to repay $91.70 per month for one year or $73.24 per month for 15 months. You might want to know that the implied rate of interest is 1.5 percent per month for the first offer and 1.2 percent per month for the second.

## Future Value

If you invest $1 today at 12 percent compounded annually, it will grow to $1.12000 at the end of one year, $1.25440 at the end of two years, $1.40493 at the end of three years, and so on, according to the following formula:

$$F_n = P(1 + r)^n,$$

where

$F_n$ = accumulation or future value
$P$ = one-time investment today
$r$ = interest rate per period
$n$ = number of periods from today

The amount $F_n$ is the **future value** of the present payment, $P$, compounded at $r$ percent per period for $n$ periods. **Table 1** at the back of the book on page **736** shows the future values of

**TABLE A.1** **(Excerpt from Table 1)**
**Future Value of $1 at 8 Percent and 12 Percent per Period**
$F_n = (1 + r)^n$

| Number of Periods = $n$ | Rate = $r$ | |
|---|---|---|
| | 8 Percent | 12 Percent |
| 1 | 1.08000 | 1.12000 |
| 2 | 1.16640 | 1.25440 |
| 3 | 1.25971 | 1.40493 |
| 10 | 2.15892 | 3.10585 |
| 20 | 4.66096 | 9.64629 |

$P$ = \$1 for various numbers of periods and for various interest rates. Extracts from that table appear here in **Table A.1**.

## Example Problems in Computing Future Value

**Example 1** How much will \$1,000 deposited today at 8 percent compounded annually grow to in 10 years?

Refer to **Table A.1**, 10-period row, 8-percent column. One dollar deposited today at 8 percent will grow to \$2.15892; therefore, \$1,000 will grow to $\$1{,}000 \times (1.08)^{10} = \$1{,}000 \times 2.15892 =$ \$2,158.92.

**Example 2** Macaulay Corporation deposits \$10,000 in an expansion fund today. The fund will earn 12 percent per year. How much will the \$10,000 grow to in 20 years if Macaulay leaves the entire fund and all interest earned on it deposited in the fund?

One dollar deposited today at 12 percent will grow to \$9.64629 in 20 years. Therefore, \$10,000 will grow to \$96,463 (= \$10,000 × 9.64629) in 20 years.

## Present Value

The preceding section developed the computation of the future value, $F_n$, of a sum of cash, $P$, deposited or invested today. You know $P$, and you calculate $F_n$. This section deals with the problems of calculating how much principal, $P$, you must invest today in order to have a specified amount, $F_n$, at the end of $n$ periods. You know the future amount, $F_n$, the interest rate, $r$, and the number of periods, $n$; you want to find $P$. In order to have \$1 one year from today when deposits earn 8 percent, you must invest $P$ of \$0.92593 today. That is, $F_1 = P(1.08)^1$ or $\$1 = \$0.92593 \times 1.08$. Because $F_n = P(1 + r)^n$, dividing both sides of the equation by $(1 + r)^n$ yields

$$\frac{F_n}{(1 + r)^n} = P,$$

or

$$P = \frac{F_n}{(1 + r)^n} = F_n (1 + r)^{-n}.$$

## Present Value Terminology

The number $(1 + r)^{-n}$ equals the present value of \$1 to be received after $n$ periods when interest accrues at $r$ percent per period. Accountants often use the words *discount* and **discounted value** in this context as follows. The discounted present value of \$1 to be received $n$ periods in the future is $(1 + r)^{-n}$ when the discount rate is $r$ percent per period for $n$ periods. The number $r$ is the **discount rate**, and the number $(1 + r)^{-n}$ is the **discount factor** for $n$ periods. A discount factor $(1 + r)^{-n}$ is the reciprocal, or inverse, of a number, $(1 + r)^n$, in **Table A.1**. Portions of **Table 2** at the back of

**TABLE A.2** (Excerpt from Table 2)
**Present Value of $1 at 8 Percent and 12 Percent per Period**
$P = F_n(1 + r)^{-n}$

| Number of Periods = $n$ | Rate = $r$ | |
|---|---|---|
| | 8 Percent | 12 Percent |
| 1 | 0.92593 | 0.89286 |
| 2 | 0.85734 | 0.79719 |
| 3 | 0.79383 | 0.71178 |
| 10 | 0.46319 | 0.32197 |
| 20 | 0.21455 | 0.10367 |

the book on page **737**, which shows discount factors or, equivalently, present values of $1 for various interest (or discount) rates for various numbers of periods, appear in **Table A.2**.

### Example Problems in Computing Present Values

**Example 3** What is the present value of $1 due 10 years from now if the interest rate (equivalently, the discount rate) $r$ is 8 percent per year?

From **Table A.2**, 8-percent column, 10-period row, the present value of $1 to be received 10 periods hence at 8 percent is $0.46319.

**Example 4** (This is **Example 14** in **Chapter 9**.) You issue a single-payment note that promises to pay $160,000 three years from today in exchange for used equipment. How much is that promise worth today if the discount rate appropriate for such notes is 12 percent per period? (An accountant needs to know the answer to the question to record the acquisition cost of the used equipment just acquired.)

One dollar paid three years hence discounted at 12 percent has a present value of $0.71178. Thus, the promise is worth $160,000 × 0.71178 = $113,885. (Record the equipment at a cost of $113,885.)

## Changing the Compounding Period: Nominal and Effective Rates

"Twelve percent, compounded annually" states the price for a loan; this means that interest increases, or converts to, principal once a year at the rate of 12 percent. Often, however, the price for a loan states that compounding is to take place more than once a year. A savings bank may advertise that it pays 6 percent, compounded quarterly. This means that at the end of each quarter the bank credits savings accounts with interest calculated at the rate of 1.5 percent (= 6 percent/4). The investor can withdraw the interest payment or leave it on deposit to earn more interest.

The sum of $10,000 invested today at 12 percent, compounded annually, grows to a future value one year later of $11,200. If the rate of interest is 12 percent compounded semiannually, the bank adds 6 percent interest to the principal every six months. At the end of the first six months, $10,000 will have grown to $10,600; that amount will grow to $10,600 × 1.06 = $11,236 by the end of the year. Twelve percent compounded semiannually is equivalent to 12.36 percent compounded annually. Suppose that the bank quotes interest as 12 percent, compounded quarterly. It will add an additional 3 percent of the principal every three months. By the end of the year, $10,000 will grow to $10,000 × $(1.03)^4$ = $10,000 × 1.12551 = $11,255. At 12 percent compounded monthly, $1 will grow to $1 × $(1.01)^{12}$ = $1.12683 and $10,000 will grow to $11,268. Thus, 12 percent compounded monthly provides the same ending amount as 12.68 percent compounded annually. Common terminology would say that *12 percent compounded monthly* has an "effective rate of 12.68 percent compounded annually" or is "equivalent to 12.68 percent compounded annually."

For a given nominal rate, such as the 12 percent in the examples above, the more often interest compounds, the higher the effective rate of interest paid. If a nominal rate, $r$, compounds $m$ times per year, the effective rate equals $(1 + r/m)^m - 1$.

In practice, to solve problems that require computation of interest quoted at a nominal rate $r$ percent per period compounded $m$ times per period for $n$ periods, use the tables for rate $r/m$ and $m \times n$ periods. For example, 12 percent compounded quarterly for five years is equivalent to the rate found in the interest tables for $r = 12/4 = 3$ percent for $m \times n = 4 \times 5 = 20$ periods.

Some savings banks advertise that they compound interest daily or even continuously. The mathematics of calculus provides a mechanism for finding the effective rate when interest is compounded continuously. If interest compounds continuously at nominal rate $r$ per year, the effective annual rate is $e^r - 1$, where $e$ is the base of the natural logarithms. Six percent per year compounded continuously is equivalent to 6.1837 percent compounded annually. Twelve percent per year compounded continuously is equivalent to 12.75 percent compounded annually. Do not confuse the compounding period with the payment period. Some banks, for example, compound interest daily but pay interest quarterly. You can be sure that these banks do not employ clerks or even computers to calculate interest every day. They derive an equivalent effective rate to apply at the end of each quarter.

## Example Problems in Changing the Compounding Period

**Example 5** What is the future value five years hence of \$600 invested at 16 percent compounded semiannually?

Sixteen percent compounded two times per year for five years is equivalent to 8 percent per period compounded for 10 periods. **Table A.1** shows the value of $F_{10} = (1.08)^{10}$ to be 2.15892. Six hundred dollars, then, would grow to \$600 × 2.15892 = \$1,295.35.

**Example 6** How much cash must you invest today at 16 percent compounded semiannually in order to yield \$10,000 in 10 years from today?

Sixteen percent compounded two times a year for 10 years is equivalent to 8 percent per period compounded for 20 periods. The present value, **Table A.2**, of \$1 received 20 periods hence at 8 percent per period is \$0.21455. That is, \$0.21455 invested today for 20 periods at an interest rate of 8 percent per period will grow to \$1. To have \$10,000 in 20 periods (10 years), you must invest \$2,146 (= \$10,000 × \$0.21455) today.

**Example 7** A local department store offers its customers credit and advertises its interest rate at 18 percent per year, compounded monthly at the rate of 1.5 percent per month. What is the effective annual interest rate?

One and one-half percent per month for 12 months is equivalent to $(1.015)^{12} - 1 = 19.562$ percent per year. See **Table 1**, 12-period row, 1.5-percent column, where the factor is 1.19562.

Under truth in lending legislation, lenders must disclose the effective annual interest rate, called the **APR** or *annual percentage rate*, to borrowers.

**Example 8** If prices increased at the rate of 6 percent during each of two consecutive six-month periods, how much did prices increase during the entire year?

If a price index is 100.00 at the start of the year, it will be $100.00 \times (1.06)^2 = 112.36$ at the end of the year. The price change for the entire year is $(112.36/100.00) - 1 = 12.36$ percent.

# Annuities

An annuity is a series of equal payments, one per period for periods equally spaced through time. Examples of annuities include monthly rental payments, semiannual corporate bond coupon payments, and annual payments to a lessor under a lease contract. Armed with an understanding of the tables for future and present values, you can solve any annuity problem. Annuities arise so often, however, and their solution is so tedious without special tables or calculator functions that annuity problems merit special study and the use of special tables or functions.

The common computer spreadsheet programs such as Microsoft Excel® include functions for annuity and other compound interest functions. Knowing which function to use to solve a given problem and which values for the variables to insert into the formula requires the same clear understanding required to use the tables. Hence, if you want to use spreadsheet functions, you must master the use of the tables in this book or gain equivalent knowledge.

## Terminology for Annuities

*An annuity involves equally spaced payments of equal amounts.* If either the time between payments or the amounts of the payments vary, then the stream is not an annuity. An annuity with payments occurring at the end of each period is an **ordinary annuity (annuity in arrears)**. Semiannual corporate bonds usually promise that debt service (coupon) payments will be paid in arrears or, equivalently, that the first payment will not occur until after the bond has been outstanding for six months. An annuity with payments occurring at the beginning of each period is an *annuity due* or an *annuity in advance*. Rent paid at the beginning of each month is an annuity due. In a **deferred annuity**, the first payment occurs sometime later than the end of the first period.

Annuity payments can go on forever. Such annuities are *perpetuities*. Bonds that promise payments forever are *consols*. The British and the Canadian governments have issued consols from time to time. A perpetuity can be in arrears or in advance. The only difference between the two is the timing of the first payment.

Annuities can be confusing. Studying them is easier with a time line such as the one shown below.

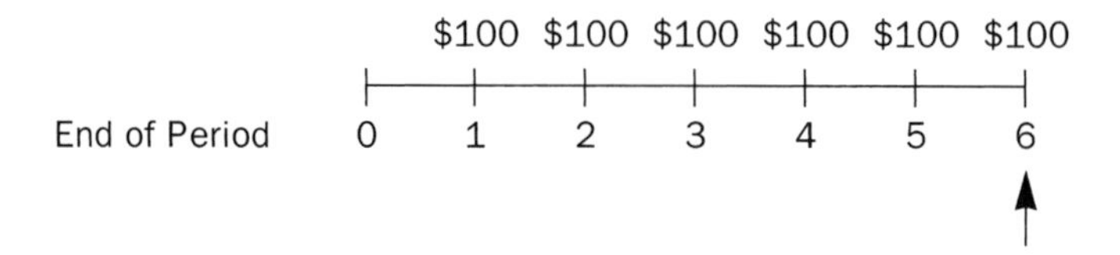


A time line marks the end of each period, numbers the period, shows the payments to be received or paid, and shows the time in which the accountant wants to value the annuity. The time line above represents an ordinary annuity (in arrears) for six periods of $100 to be valued at the end of period 6. The end of period 0 is *now*. The first payment occurs one period from now.

## Ordinary Annuities (Annuities in Arrears)

The future values of ordinary annuities appear in **Table 3** at the back of the book on page **738**, portions of which **Table A.3** reproduces.

Consider an ordinary annuity for three periods at 12 percent. The time line for the future value of such an annuity is as follows:

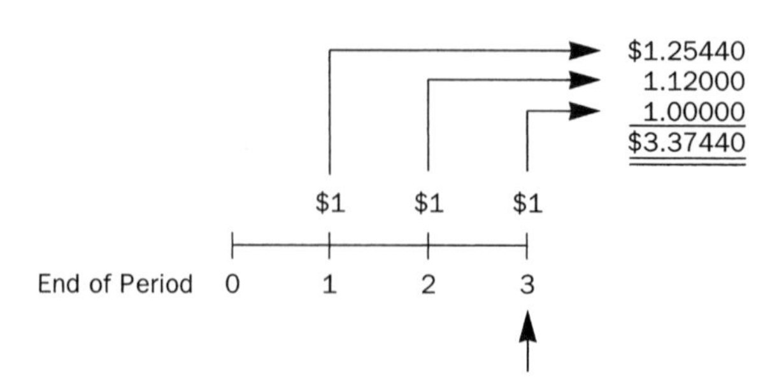


The $1 received at the end of the first period earns interest for two periods, so it grows to $1.25440 at the end of period 3 (see **Table A.1**). The $1 received at the end of the second period grows to $1.12000 by the end of period 3, and the $1 received at the end of period 3 is, of course, worth $1.00000 at the end of period 3. The entire annuity is worth $3.37440 at the end of period 3. This amount appears in **Table A.3** for the future value of an ordinary annuity for three periods at 12 percent. Factors for the future value of an annuity for a particular number of peri-

| TABLE A.3 | (Excerpt from Table 3) Future Value of an Ordinary Annuity of $1 per Period at 8 Percent and 12 Percent | |
|---|---|---|
| | Rate = $r$ | |
| Number of Periods = $n$ | 8 Percent | 12 Percent |
| 1 | 1.00000 | 1.00000 |
| 2 | 2.08000 | 2.12000 |
| 3 | 3.24640 | 3.37440 |
| 5 | 5.86660 | 6.35285 |
| 10 | 14.48656 | 17.54874 |
| 20 | 45.76196 | 72.05244 |

ods sum the factors for the future value of $1 for each of the periods. The future value of an ordinary annuity is as follows:

$$\text{Future Value of Ordinary Annuity} = \text{Periodic Payment} \times \text{Factor for the Future Value of an Ordinary Annuity}$$

Thus,

$$\$3.37440 = \$1 \times 3.37440$$

**Table 4** at the back of the book on page **739** shows the present value of ordinary annuities. **Table A.4** reproduces excerpts from **Table 4.**

The time line for the present value of an ordinary annuity of $1 per period for three periods, discounted at 12 percent, is as follows:

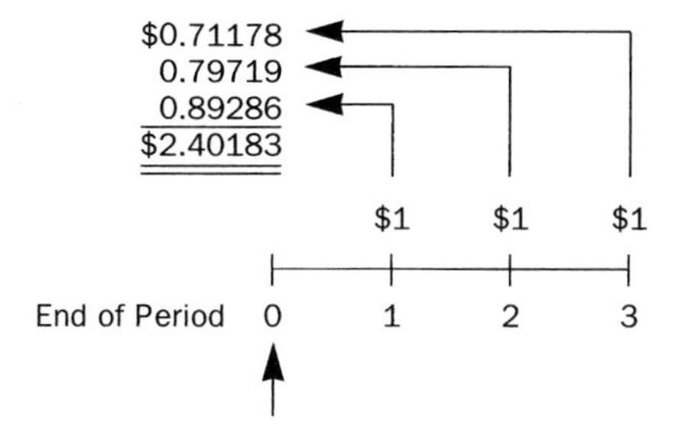


The $1 due at the end of period 1 has a present value of $0.89286, the $1 due at the end of period 2 has a present value of $0.79719, and the $1 due at the end of period 3 has a present value of $0.71178. Each of these numbers comes from **Table A.2**. The present value of the annuity sums these individual present values, $2.40183, shown in **Table A.4**.

The present value of an ordinary annuity for $n$ periods is the sum of the present value of $1 received one period from now plus the present value of $1 received two periods from now, and so on until we add on the present value of $1 received $n$ periods from now. The present value of an ordinary annuity is as follows:

$$\text{Present Value of an Ordinary Annuity} = \text{Periodic Payment} \times \text{Factor for the Present Value of an Ordinary Annuity}$$

Thus,

$$\$2.40183 = \$1 \times 2.40183$$

**TABLE A.4** (Excerpt from Table 4)
**Present Value of an Ordinary Annuity of $1 per Period at 8 Percent and 12 Percent**

| Number of Periods = $n$ | Rate = $r$ | |
|---|---|---|
| | 8 Percent | 12 Percent |
| 1 | 0.92593 | 0.89286 |
| 2 | 1.78326 | 1.69005 |
| 3 | 2.57710 | 2.40183 |
| 5 | 3.99271 | 3.60478 |
| 10 | 6.71008 | 5.65022 |
| 20 | 9.81815 | 7.46944 |

## Example Problems Involving Ordinary Annuities

**Example 9** You plan to invest $1,000 at the end of each of the next 10 years in a savings account. The savings account accumulates interest of 8 percent compounded annually. What will be the balance in the savings account at the end of 10 years?

The time line for this problem is as follows:

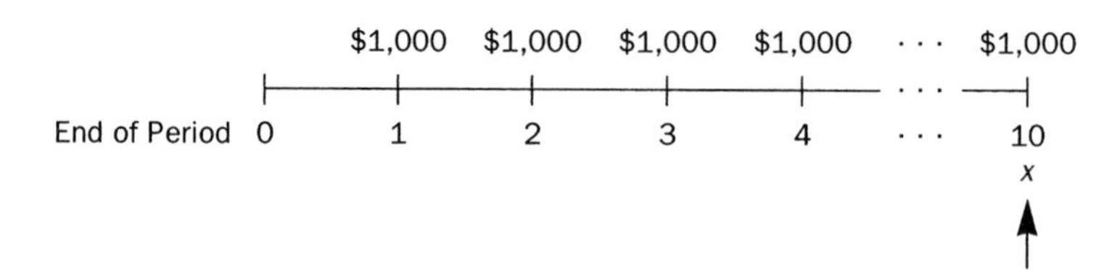


The symbol $x$ denotes the amount you must calculate. **Table A.3** indicates that the factor for the future value of an annuity at 8 percent for 10 periods is 14.48656. Thus,

| Future Value of an Ordinary Annuity | = | Periodic Payment | × | Factor for the Future Value of an Ordinary Annuity |
|---|---|---|---|---|
| $x$ | = | $1,000 | × | 14.48656 |
| $x$ | = | $14,487 | | |

**Example 10** You want to receive $600 every six months, starting six months hence, for the next five years. How much must you invest today if the funds accumulate at the rate of 8 percent compounded semiannually?

The time line is as follows:

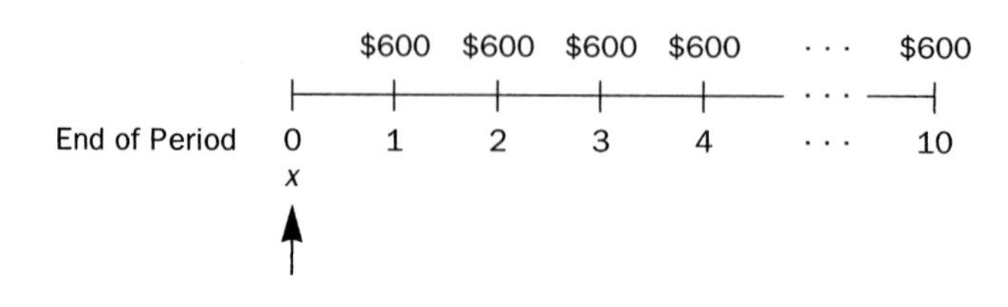

The factor from **Table 4** for the present value of an annuity at 4 percent (= 8 percent per year/2 semiannual periods per year) for 10 (= 2 periods per year × 5 years) periods is 8.11090. Thus,

$$\begin{array}{ccccc} \text{Present Value of an Ordinary Annuity} & = & \text{Periodic Payment} & \times & \text{Factor for the Present Value of an Ordinary Annuity} \\ x & = & \$600 & \times & 8.11090 \\ x & = & \$4{,}866.54 & & \end{array}$$

If you invest $4,866.54 today, the principal plus interest compounded on the principal will provide sufficient funds that you can withdraw $600 every six months for the next five years.

**Example 11** (This example also appears on page **427** in **Chapter 9** as the Western Company mortgage problem.) A company borrows $125,000 from a savings and loan association. The interest rate on the loan is 12 percent compounded semiannually. The company agrees to repay the loan in equal semiannual installments over the next five years, with the first payment six months from now. What is the required semiannual payment?

The time line is as follows:

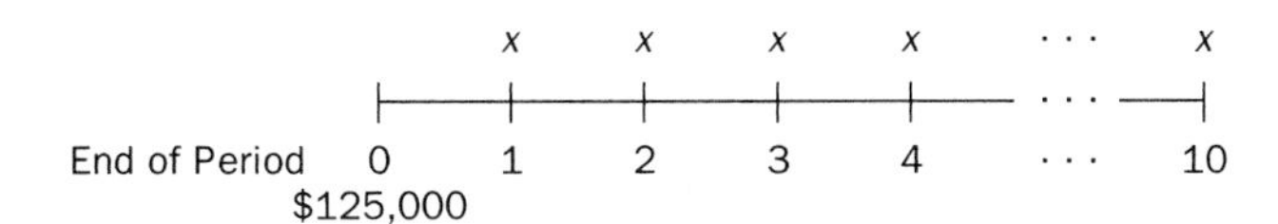


This problem resembles **Example 10** because both involve periodic future payments discounted to today. **Example 10** gives the periodic payments and asks for the present value. **Example 11** gives the present value and asks for the periodic payment. **Table 4** indicates that the present value of an annuity at 6 percent (= 12 percent per year/2 semiannual periods per year) for 10 periods (=2 periods per year × 5 years) is 7.36009. Thus,

$$\begin{array}{ccccc} \text{Present Value of an Ordinary Annuity} & = & \text{Periodic Payment} & \times & \text{Factor for the Present Value of an Ordinary Annuity} \\ \$125{,}000 & = & x & \times & 7.36009 \\ x & = & \dfrac{\$125{,}000}{7.36009} & & \\ x & = & \$16{,}983 & & \end{array}$$

To find the periodic payment, divide the present value amount of $125,000 by the present value factor. (**Exhibit 9.2** presents the amortization table for this loan. That exhibit shows the amount of each semiannual payment as $17,000, rather than $16,983, and the last payment as $16,781, less than $17,000, to compensate for the extra $17 paid in each of the preceding periods and the interest on those amounts.)

**Example 12** (This example also appears on page **475** in **Chapter 10** as the Food Barn lease problem.) A company signs a lease acquiring the right to use property for three years. The company will make lease payments of $19,709 annually at the end of this and the next two years. The discount rate is 15 percent per year. What is the present value of the lease payments?

The time line is as follows:

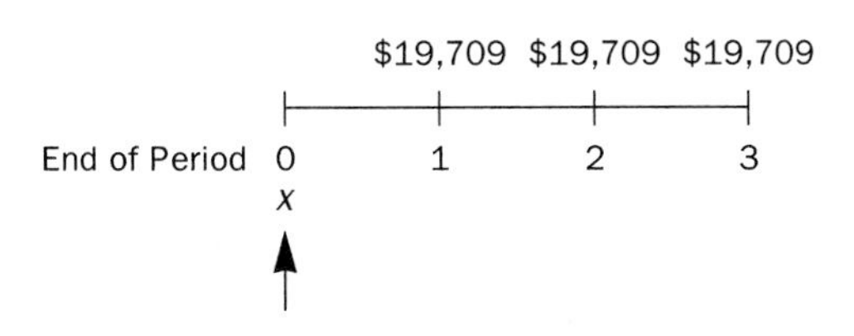

The factor from **Table 4** for the present value of an annuity at 15 percent for three periods is 2.28323. Thus,

| Present Value of an Ordinary Annuity | | Periodic Payment | | Factor for the Present Value of an Ordinary Annuity |
|---|---|---|---|---|
| $x$ | = | $19,709 | × | 2.28323 |
| $x$ | = | $45,000 | | |

The Food Barn example in **Chapter 10** gives the cost of the equipment, $45,000, and we compute the periodic rental payment with an annuity factor. Thus,

| Present Value of an Ordinary Annuity | | Periodic Payment | | Factor for the Present Value of an Ordinary Annuity |
|---|---|---|---|---|
| $45,000 | = | $x$ | × | 2.28323 |
| $x$ | = | $\frac{\$45{,}000}{2.28323}$ | | |
| $x$ | = | $19,709 | | |

**Example 13** A company promises to make annual payments to a pension fund at the end of each of the next 30 years. The payments must have a present value today of $100,000. What must the annual payment be if the fund expects to earn interest at the rate of 8 percent per year?

The time line is as follows:

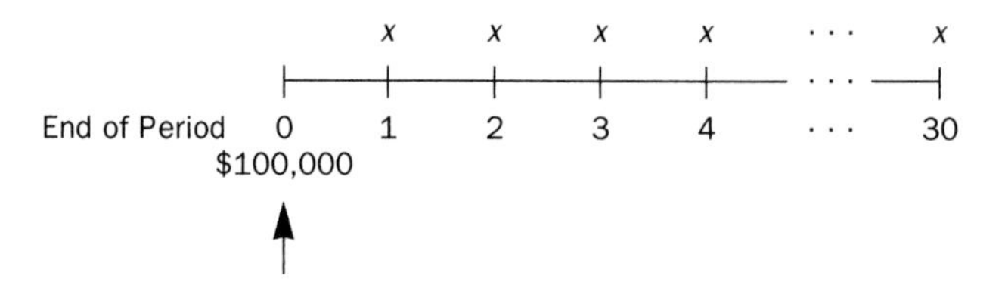


**Table 4** indicates that the factor for the present value of $1 paid at the end of the next 30 periods at 8 percent per period is 11.25778. Thus,

| Present Value of an Ordinary Annuity | | Periodic Payment | | Factor for the Present Value of an Ordinary Annuity |
|---|---|---|---|---|
| $100,000 | = | $x$ | × | 11.25778 |
| $x$ | = | $\frac{\$100{,}000}{11.25778}$ | | |
| $x$ | = | $8,883 | | |

**Example 14** Mr. Mason is 62 years old. He wants to invest equal amounts on his 63rd, 64th, and 65th birthdays so that starting on his 66th birthday he can withdraw $50,000 on each birthday for 10 years. His investments will earn 8 percent per year. How much should he invest on the 63rd through 65th birthdays?

The time line for this problem is as follows:

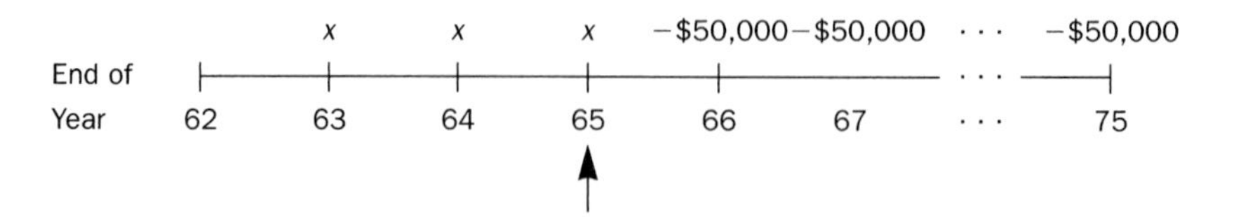


At 65, Mr. Mason needs to have accumulated a fund equal to the present value of an annuity of $50,000 per period for 10 periods, discounted at 8 percent per period. The factor from **Table A.4** for 8 percent and 10 periods is 6.71008. Thus,

| Present Value of an Ordinary Annuity | = | Periodic Payment | × | Factor for the Present Value of an Ordinary Annuity |
|---|---|---|---|---|
| $x$ | = | $50,000 | × | 6.71008 |
| $x$ | = | $335,500 | | |

The time line now appears as follows:

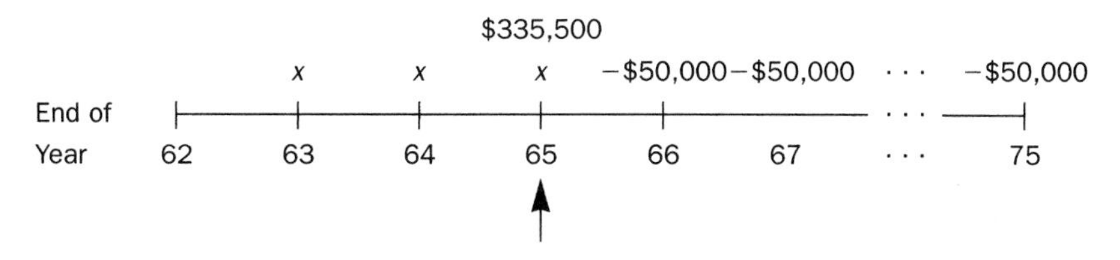


The question now becomes, how much must Mr. Mason invest on his 63rd, 64th, and 65th birthdays to accumulate to a fund of $335,500 on his 65th birthday? The factor for the future value of an annuity for three periods at 8 percent is 3.24640. Thus,

| Future Value of an Ordinary Annuity | = | Periodic Payment | × | Factor for the Future Value of an Ordinary Annuity |
|---|---|---|---|---|
| $335,500 | = | $x$ | × | 3.24640 |
| $x$ | = | $\frac{\$335,500}{3.24640}$ | | |
| $x$ | = | $103,350 | | |

The solution above expresses all calculations in terms of equivalent amounts on Mr. Mason's 65th birthday. That is, the present value of an annuity of $50,000 per period for 10 periods at 8 percent equals the future value of an annuity of $103,350 per period for three periods at 8 percent, and both of these amounts equal $335,500. You could work this problem by selecting any common time period between Mr. Mason's 62nd and 75th birthdays.

One alternative expresses all calculations in terms of equivalent amounts on Mr. Mason's 62nd birthday. To solve the problem in this way, first find the present value on Mr. Mason's 65th birthday of an annuity of $50,000 per period for 10 periods ($335,500 = $50,000 × 6.71008). Discount $335,500 back three periods using **Table 2** for present value of single payments: $266,330 = $335,500 × 0.79383. The result is the present value of the payments to be made to Mr. Mason measured as of his 62nd birthday. Then, find the amounts that Mr. Mason must invest on his 63rd, 64th, and 65th birthdays to have a present value on his 62nd birthday equal to $266,330. The calculation is as follows:

| Present Value of an Ordinary Annuity | = | Periodic Payment | × | Factor for the Present Value of an Ordinary Annuity |
|---|---|---|---|---|
| $266,330 | = | $x$ | × | 2.57710 |
| $x$ | = | $103,350 | | |

We computed the same amount, $103,350, above.

## Perpetuities

A periodic payment to be received forever is a **perpetuity**. Future values of perpetuities are undefined. One dollar to be received at the end of every period discounted at rate $r$ percent has a present value of $\$1/r$. Observe what happens in the expression for the present value of an ordinary annuity of $A per payment as $n$, the number of payments, approaches infinity:

$$P_A = \frac{A[1 - (1 + r)^{-n}]}{r}$$

As $n$ approaches infinity, $(1 + r)^{-n}$ approaches 0, so that $P_A$ approaches A(1/$r$). If the first payment of the perpetuity occurs now, the present value is A[1 + 1/$r$].

## Examples of Perpetuities

**Example 15** The Canadian government offers to pay $30 every six months forever in the form of a perpetual bond. What is that bond worth if the discount rate is 10 percent compounded semiannually?

Ten percent compounded semiannually is equivalent to 5 percent per six-month period. If the first payment occurs six months from now, the present value is $30/0.05 = $600. If the first payment occurs today, the present value is $30 + $600 = $630.

**Example 16** Every two years, the Bank of Tokyo gives ¥5 million (Japanese yen) to the university to provide a scholarship for an entering student in a two-year business administration course. If the university credits 6 percent per year to its investment accounts, how much must the bank give to the university to provide such a scholarship every two years forever, starting two years hence?

A perpetuity in arrears assumes one payment at the end of each period. Here, the period is two years. Six percent compounded once a year over two years is equivalent to a rate of $(1.06)^2 - 1 = 0.12360$ or 12.36 percent compounded once per two-year period. Consequently, the present value of the perpetuity paid in arrears every two years is ¥40.45 (= ¥5/0.1236). A gift of ¥40.45 million will provide a ¥5 million scholarship forever. If the university will award the first scholarship now, the gift must be ¥45.45 (= ¥40.45 + ¥5.00) million.

**Example 17** (This example illustrates the **relief from royalty method** for valuing trademarks.) Burns, Philp & Company (B-P) wants to value its trademark for Fleischmann's Yeast using the relief from royalty method. B-P estimates that sales of the product will continue at the rate of $100 million per year in perpetuity and that if it had to pay a royalty to another firm that owned the trademark, it would have to pay 4 percent of sales at the end of each year. It uses a discount rate of 10 percent per year in computing present values. What is the value of the trademark under these assumptions?

If B-P paid royalties of 4 percent of sales, these would total $4 (= 0.04 × $100) million per year, in perpetuity. The present value of a perpetual stream of payments of $4 million per year, discounted at 10 percent per year, is $40 (= $4/0.10) million. Under these assumptions, the trademark has a value of $40 million.

B-P, more realistically, might estimate that sales will grow each year by, say, 2 percent more than the rate of inflation. When one assumes that a perpetuity's payments, which start at $p$ per period, $4 million in this example, will grow at a constant rate $g$, 2 percent in this example, then the value of the perpetuity is $p/(r - g)$ = $4/(0.10 − 0.02) = $50 million in this example.

The next three examples all refer to the data in **Exhibit 12.2**, where we evaluate the six individual activities of Bernard Company to find the value of the entire firm.

**Example 18** Bernard Company's Activity 1 generates $100 per year, with the cash flow at the end of the year, indefinitely. The present value of this activity is $1,000 (= $100/ 0.10). Investors sometimes call this process of deriving market value from a series of future cash flows *capitalizing earnings*. The analyst might say that the earnings of $100 have a *price/earnings ratio* of 10 or that the earnings "deserve [or carry] a multiple of 10." The price/earnings ratio for a perpetuity is the reciprocal of the discount rate. In the example, where the discount rate is 10 percent per period, the price/earnings ratio is 10.

**Example 19** Bernard Company's second activity generates $30 at the end of the first year, and a cash flow that grows by 6 percent per year thereafter. The present value of this activity is $750 [= $30/(0.10 − 0.06)]: If the cash flow starts at $1 per year and grows at rate $g$ per year, then the present value of the growing cash flows is $1/($r - g$). This is the **perpetuity with growth** formula. This example and the formula for it assume the cash flows and growth last *forever*. *Forever* is a long time. Sometimes students derive silly results from assuming *forever*, particularly when the growth rate is large. We have seen analysts project growth rates so much larger than the growth of the U.S. economy that after 50 or so years, the analyzed company projects sales larger than the entire U.S. gross domestic product. Use these perpetuity-based, arithmetic tools to learn and to make approximations, but do not expect firms to grow *forever* at high growth rates.

**Example 20** Bernard Company's third activity is cyclic, generating $115 per year at the end of each odd-numbered year. The present value of this activity is $602.38, derived as follows. Think of this series of cash flows as a perpetuity with $115 per period, but each period is two years long. Then, when the discount rate is 10 percent per year, the discount rate for a two-year period must be 21 percent [= (1.10 × 1.10) − 1]. The cash flows from Year 2 onwards have present value of $547.62 (= $115/0.21) at the start of Year 2 and present value of $497.84 (= $547.62/1.10) at the start of Year 1. The $115 cash flow received at the end of Year 1 has present value of $104.54 (= $115/1.10) at the start of Year 1. The entire series has present value of $602.38 (= $497.84 + $104.54) at the start of Year 1.

## Implicit Interest Rates: Finding Internal Rates of Return

The preceding examples computed a future value or a present value given the interest rate and stated cash payment. Or, they computed the required payments given their known future value or their known present value. In other calculations, however, you know the present or the future value and the periodic payments; you must find the implicit interest rate. For example, **Chapter 9** illustrates a case in which you know that the cash price of some equipment is $10,500 and that the firm acquired the asset in exchange for a single-payment note. The note has a face value of $16,000 and matures in three years. To compute interest expense over the three-year period, you must know the **implicit interest rate (internal rate of return)**. The time line for this problem is as follows:

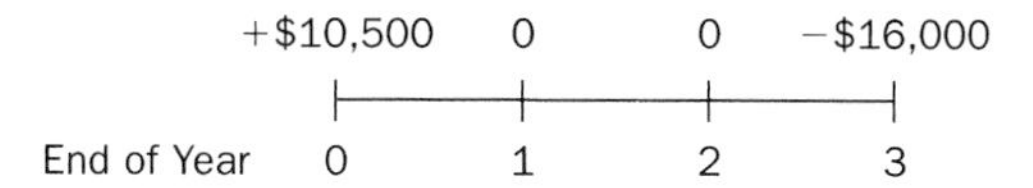


The implicit interest rate is $r$, such that

(A.1) $$0 = \$10{,}500 - \frac{\$16{,}000}{(1 + r)^3}$$

(A.2) $$\$10{,}500 = \frac{\$16{,}000}{(1 + r)^3}$$

That is, the present value of $16,000 discounted three periods at $r$ percent per period is $10,500. The present value of all current and future cash flows discounted at $r$ per period must be zero. In general, to find such an $r$ requires trial and error. In cases where $r$ appears only in one term, as here, you can find $r$ analytically. Here $r = (\$16{,}000/\$10{,}500)^{1/3} - 1 = 0.1507 = 15.1$ percent.

The general procedure is *finding the internal rate of return of a series of cash flows*. The internal rate of return of a series of cash flows is the discount rate that makes the net present values of that series of cash flows equal to zero. The steps in finding the internal rate of return are as follows:

1. Make an educated guess, called the "trial rate," at the internal rate of return. If you have no idea what to guess, try zero.
2. Calculate the present value of all the cash flows (including the one at the end of Year 0).
3. If the present value of the cash flows is zero, stop. The current trial rate is the internal rate of return.
4. If the amount found in step **2** is less than zero, try a larger interest rate as the trial rate and go back to step **2**.
5. If the amount found in step **2** is greater than zero, try a smaller interest rate as the new trial rate and go back to step **2**.

The iterations below illustrate the process for the example in **Equation A.1**.

| Iteration Number | Net Present Value: Trial Rate = $r$ | Right-Hand Side of Equation A.1 |
|---|---|---|
| 1 | 0.0% | −$5,500 |
| 2 | 10.0 | −1,521 |
| 3 | 15.0 | −20 |
| 4 | 15.5 | 116 |
| 5 | 15.2 | 34 |
| 6 | 15.1 | 7 |

With a trial rate of 15.1 percent, the right-hand side is close enough to zero that you can use 15.1 percent as the implicit interest rate in making the adjusting entries for interest expense. Continued iterations would find trial rates even closer to the true rate, which is about 15.0739 percent.

Finding the internal rate of return for a series of cash flows can be tedious; you should not attempt it unless you have at least a calculator. An exponential feature, which allows the computation of $(1 + r)$ raised to various powers, helps. Computer spreadsheets, such as Microsoft Excel®, have a built-in function to find the internal rate of return.

## Example Problem Involving Implicit Interest Rates

**Example 21** Alexis Company acquires a machine with a cash price of $10,500. It pays for the machine by giving a note for $12,000, promising to make payments equal to 7 percent of the face value, $840 (= 0.07 × $12,000), at the end of each of the next three years and a single payment of $12,000 in three years. What is the interest rate implicit in the loan?

The time line for this problem is as follows:

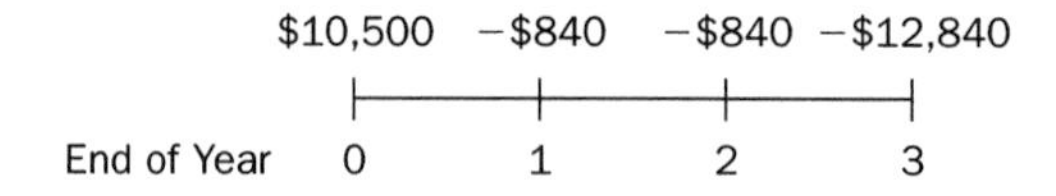


The implicit interest rate is $r$, such that

**(A.3)**
$$\$10{,}500 = \frac{\$840}{(1 + r)} + \frac{\$840}{(1 + r)^2} + \frac{\$12{,}840}{(1 + r)^3}$$

Compare this formulation to that in **Equation A.1**. Note that the left-hand side equals 0 in **Equation A.1** but not in **Equation A.3**. You may use any left-hand side that you find convenient for the particular context.

The iteration process finds an internal rate of return of 12.2 percent to the nearest tenth of 1 percent:

| Iteration Number | Trial Rate | Right-Hand Side of Equation A.3 |
|---|---|---|
| 1 | 7.0% | $12,000 |
| 2 | 15.0 | 9,808 |
| 3 | 11.0 | 10,827 |
| 4 | 13.0 | 10,300 |
| 5 | 12.0 | 10,559 |
| 6 | 12.5 | 10,428 |
| 7 | 12.3 | 10,480 |
| 8 | 12.2 | 10,506 |
| 9 | 12.1 | 10,532 |

## Summary

Accountants typically use one of four kinds of compound interest calculations: (1) the present value or (2) the future value of (3) a single payment or of (4) a series of payments. In working annuity problems, you may find a time line helpful in deciding the particular kind of annuity involved. Computer spreadsheet programs have a built-in function to perform the computations described in this appendix.

## Key Terms and Concepts

Present value
Principal
Compound interest
Simple interest
Future value
Discounted value
Discount rate
Discount factor
APR
Ordinary annuity (annuity in arrears)
Deferred annuity
Perpetuity
Perpetuity with growth
Relief from royalty method
Implicit interest rate (internal rate of return)

## Questions, Short Exercises, Exercises, Problems, and Cases

### QUESTIONS

**1.** Review the terms and concepts listed above in Key Terms and Concepts.

**2.** What is interest?

**3.** Distinguish between simple and compound interest.

**4.** Distinguish between the discounted present value of a stream of future payments and their net present value. If there is no distinction, then so state.

**5.** Distinguish between an annuity due and an ordinary annuity.

**6.** Describe the implicit interest rate for a series of cash flows and a procedure for finding it.

**7.** Does the present value of a given amount to be paid in 10 years increase or decrease if the interest rate increases? Suppose that the amount is due in 5 years? 20 years? Does the present value of an annuity to be paid for 10 years increase or decrease if the discount rate decreases? Suppose that the annuity is for 5 years? 20 years?

**8.** Rather than pay you $1,000 a month for the next 20 years, the person who injured you in an automobile accident is willing to pay a single amount now to settle your claim for injuries. Would you rather use an interest rate of 6 percent or 12 percent in computing the present value of the lump-sum settlement? Comment or explain.

**9.** The perpetuity with growth formula involves several assumptions. Which one seems least plausible?

### SHORT EXERCISES

**Short exercises 10 through 15 involve calculations of present and future value for single payments and for annuities. To make the exercises more realistic, we do not give specific guidance with each individual exercise.**

**10.** Mr. Altgeldt has $5,000 to invest. He wants to know how much it will amount to if he invests it at the following rates:

**a.** 6 percent per year for 21 years
**b.** 8 percent per year for 33 years

**11.** Mme. Barefield wishes to have $150,000 at the end of 8 years. How much must she invest today to accomplish this purpose if the interest rate is

**a.** 6 percent per year?
**b.** 8 percent per year?

**12.** Mr. Case plans to set aside $4,000 each year, the first payment to be made on January 1, 2005, and the last on January 1, 2010. How much will he have accumulated by January 1, 2010, if the interest rate is
   **a.** 6 percent per year?
   **b.** 8 percent per year?

**13.** Ms. Namura wants to have ¥45 million on her 65th birthday. She asks you to tell her how much she must deposit on each birthday from her 58th to 65th, inclusive, in order to receive this amount. Assume the following interest rates:
   **a.** 8 percent per year
   **b.** 12 percent per year

**14.** If Mr. Enmetti invests €90,000 on June 1 of each year from 2001 to 2011 inclusive, how much will he have accumulated on June 1, 2012 (note that one year elapses after last payment) if the interest rate is
   **a.** 5 percent per year?
   **b.** 10 percent per year?

**15.** Ms. Fleming has £145,000 with which she purchases an annuity on February 1, 2006. The annuity consists of six annual receipts, the first to be received on February 1, 2007. How much will she receive in each payment? Assume the following interest rates:
   **a.** 8 percent per year
   **b.** 12 percent per year

## EXERCISES

**16.** In the preceding **Short Exercises 10** through **15**, you computed a number. To do so, first you must decide on the appropriate factor from the tables, and then you use that factor in the appropriate calculation. Notice that you could omit the last step. You could write an arithmetic expression showing the factor you want to use without actually copying down the number and doing the arithmetic. For example, the notation *T(i, p, r)* means Table *i* (1, 2, 3, or 4), row *p* (periods 1 to 20, 22, 24 . . . , 40, 45, 50, 100), and column *r* (interest rates from 1/2 percent up to 20 percent). Thus, *T*(3, 16, 12) would be the factor in **Table 3** for 16 periods and an interest rate of 12 percent per period, which is 42.75328. Using this notation, you can write an expression for any compound interest problem. A clerk or a computer can evaluate the expression.

You can check that you understand this notation by observing that the following are true statements:

| | | |
|---|---|---|
| T(1, 20, 8) | = | 4.66096 |
| T(2, 12, 5) | = | 0.55684 |
| T(3, 16, 12) | = | 42.75328 |
| T(4, 10, 20) | = | 4.19247 |

In the following questions, write an expression for the answer using the notation introduced here, but do not attempt to evaluate the expression.
   **a.** Work the **a** parts of **Exercises 10** through **15**.
   **b.** How might your instructor use this notation to write examination questions on compound interest without having to supply you with tables?

**17.** **Effective interest rate.** State the rate per period and the number of periods in the following:
   **a.** 12 percent per year, for 5 years, compounded annually
   **b.** 12 percent per year, for 5 years, compounded semiannually
   **c.** 12 percent per year, for 5 years, compounded quarterly
   **d.** 12 percent per year, for 5 years, compounded monthly

**Exercises 18 through 26 involve calculations of present and future value for single payments and for annuities. To make the exercises more realistic, we do not give specific guidance with each individual exercise.**

**18.** Compute the future value of the following:
   **a.** $100 invested for 5 years at 4 percent compounded annually
   **b.** $500 invested for 15 periods at 2 percent compounded once per period
   **c.** $200 invested for 8 years at 3 percent compounded semiannually

**d.** \$2,500 invested for 14 years at 8 percent compounded quarterly
**e.** \$600 invested for 3 years at 12 percent compounded monthly

**19.** Compute the present value of the following:
**a.** \$100 due in 30 years at 4 percent compounded annually
**b.** \$250 due in 8 years at 8 percent compounded quarterly
**c.** \$1,000 due in 2 years at 12 percent compounded monthly

**20.** Compute the amount (future value) of an ordinary annuity (an annuity in arrears) of the following:
**a.** 13 rental payments of \$100 at 1 percent per period
**b.** 8 rental payments of \$850 at 6 percent per period
**c.** 28 rental payments of \$400 at 4 percent per period

**21.** Mr. Grady agrees to lease a certain property for 10 years, at the following annual rental, payable in advance:

Years 1 and 2—\$1,000 per year
Years 3 to 6—\$2,000 per year
Years 7 to 10—\$2,500 per year

What single immediate sum will pay all of these rents, discounted at
**a.** 6 percent per year?
**b.** 8 percent per year?
**c.** 10 percent per year?

**22.** To establish a fund that will provide a scholarship of \$3,000 per year indefinitely, with the first award to occur now, how much must a donor deposit if the fund earns
**a.** 6 percent per period?
**b.** 8 percent per period?

**23.** Consider the scholarship fund in the preceding question. Suppose that the first scholarship award occurs one year from now and the donor wants the scholarship to grow by 2 percent per year. How much should the donor deposit if the fund earns
**a.** 6 percent per period?
**b.** 8 percent per period?

Suppose that the first scholarship award occurs five years from now but is to grow at 2 percent per year after Year 5, the time of the first \$3,000 award. How much should the donor deposit if the fund earns
**c.** 6 percent per year?
**d.** 8 percent per year?

**24.** An old agreement obliges the state to help a rural county maintain a bridge by paying \$60,000 now and every two years thereafter forever toward the expenses. The state wants to discharge its obligation by paying a single sum to the county now for the payment due and all future payments. How much should the state pay the county if the discount rate is
**a.** 8 percent per year?
**b.** 12 percent per year?

**25.** Find the interest rate implicit in a loan of \$100,000 that the borrower discharges with two annual installments of \$55,307 each, paid at the end of Years 1 and 2.

**26.** A single-payment note promises to pay \$140,493 in three years. The issuer exchanges the note for equipment having a fair market value of \$100,000. The exchange occurs three years before the maturity date on the note. What interest rate will the accounting impute for the single-payment note?

**27.** A single-payment note promises \$67,280 at maturity. The issuer of the note exchanges it for land with a fair market value of \$50,000. The exchange occurs two years before the maturity date on the note.
**a.** What interest rate will the accounting impute for this single-payment note?
**b.** Using the imputed interest rate (sometimes called the *implicit interest* rate), construct an amortization schedule for the note. Show book value of the note at the start of each year, interest for each year, the amount reducing or increasing book value each year, and book value at the end of the year.

**28.** **Finding implicit interest rates; constructing amortization schedules.** Berman Company purchased a plot of land for possible future development. The land had fair market value of \$86,000. Berman Company gave a three-year interest-bearing note. The note had face value of $100,000 and provided for interest at a stated rate of 8 percent. The note requires payments of $8,000 at the end of each of three years, the last payment coinciding with the maturity of the note's face value of $100,000.

**a.** What is the interest rate implicit in the note, accurate to the nearest tenth of 1 percent?

**b.** Construct an amortization schedule for the note for each year. Show the book value of the note at the start of the year, interest for the year, payment for the year, amount reducing or increasing the book value of the note for each payment, and the book value of the note at the end of each year. Use the interest rate found in part **a**. See **Exhibit 10.3** for an example of an amortization schedule.

**29. Find equivalent annual rate offered for purchase discounts.** The terms of sale "2/10, net/30" mean that the buyer can take a discount of 2 percent from gross invoice price by paying the invoice within 10 days; otherwise, the buyer must pay the full amount within 30 days.

**a.** Write an expression for the implied annual rate of interest being offered by viewing the entire discount as interest for funds received sooner rather than later. (Note that by not taking the discount, the buyer borrows 98 percent of the gross invoice price for 20 days.)

**b.** The tables at the back of the book do not permit the exact evaluation of the expression derived in part **a**. The rate of interest implied is 44.59 percent per year. Use the tables to convince yourself that this astounding (to some) answer must be close to correct.

## PROBLEMS AND CASES

**Problems 30 through 44 involve using future value and present value techniques, including perpetuities, to solve a variety of realistic problems. We give no hints as to the specific calculation with the problems.**

**30.** An oil-drilling company figures that it must spend $30,000 for an initial supply of drill bits and that it must spend $10,000 every month to replace the worn-out bits. What is the present value of the cost of the bits if the company plans to be in business indefinitely and discounts payments at 1 percent per month?

**31.** If you promise to leave $35,000 on deposit at the Dime Savings Bank for four years, the bank will give you a new large, flat-screen Sony TV today and your $35,000 back at the end of four years. How much are you paying today for the TV, in effect, if the bank pays other customers 8 percent interest compounded quarterly (2 percent paid four times per year)?

**32.** When Mr. Shafer died, his estate after taxes amounted to $300,000. His will provided that Mrs. Shafer would receive $24,000 per year starting immediately from the principal of the estate and that the balance of the principal would pass to the Shafers' children upon Mrs. Shafer's death. The state law governing this estate provided for a dower option. If Mrs. Shafer elects the dower option, she renounces the will and can have one-third of the estate in cash now. The remainder will then pass immediately to their children. Mrs. Shafer wants to maximize the present value of her bequest. Should she take the annuity or elect the dower option if she will receive five payments and discounts payments at

**a.** 8 percent per year?

**b.** 12 percent per year?

(Note this problem explicitly states that Mrs. Shafer will receive five payments. In reality, life expectancy is uncertain. The correct calculation combines a mortality table with the present value tables. Actuaries deal with such calculations.)

**33.** Mrs. Heileman occasionally drinks beer. (Guess which brand.) She consumes one case in 20 weeks. She can buy beer in disposable cans for $25.20 per case or for $24.00 per case of returnable bottles if she pays a $3.00 refundable deposit at the time of purchase. If her discount rate is 1/2 percent per week, how much in present value dollars does she save by buying the returnables and thereby losing the use of the $3.00 deposit for 20 weeks?

**34.** When General Electric Company first introduced the Lucalox ceramic, screw-in light bulb, the bulb cost three and one-half times as much as an ordinary bulb but lasted five times as long. An ordinary bulb cost $1.00 and lasted about eight months. If a firm has a discount rate of 12 percent compounded three times a year, how much would it save in present-value dollars by using one Lucalox bulb?

**35.** Oberweis Dairy switched from delivery trucks with regular gasoline engines to ones with diesel engines. The diesel trucks cost $2,000 more than the ordinary gasoline trucks but $600 per year less to operate. Assume that Oberweis saves the operating costs at the end of each month. If Oberweis uses a discount rate of 1 percent per month, approximately how many months, at a minimum, must the diesel trucks remain in service for the switch to be sensible?

36. **Calculating Impairment.** On January 1, Year 1, assume that Levi Strauss opened a new textile plant to produce synthetic fabrics. The plant is on leased land; 20 years remain on the nonrenewable lease.

    The cost of the plant was $20 million. Net cash flow to be derived from the project is estimated to be $3,000,000 per year. The company does not normally invest in such projects unless the anticipated yield is at least 12 percent.

    On December 31, Year 1, the company finds cash flows from the plant to be $2,800,000 for the year. On the same day, farm experts predict cotton production will be unusually low for the next two years. Levi Strauss estimates the resulting increase in demand for synthetic fabrics will boost cash flows to $3,500,000 for each of the next two years. Subsequent years' estimates remain unchanged. Ignore tax considerations.

    **a.** Calculate the present value of the future expected cash flows from the plant when it opened.

    **b.** What is the present value of the plant on January 1, Year 2, immediately after the re-estimation of future cash flows?

    **c.** On January 2, Year 2, the day following the cotton production news release, a competitor announces plans to build a synthetic fabrics plant to open in three years. Levi Strauss keeps its Year 2 to Year 4 estimates but reduces the estimated annual cash flows for subsequent years to $2,000,000. What is the value of Levi Strauss's present plant on January 1, Year 2, after the new projections?

    **d.** On January 2, Year 2, an investor contacts Levi Strauss about purchasing a 20 percent share of the plant. If the investor expects to earn at least a 12 percent annual return on the investment, what is the maximum amount that the investor can pay? Assume that the investor and Levi Strauss both know all relevant information and use the same estimates of annual cash flows.

37. **Finding implicit interest rates (truth-in-lending laws reduce the type of deception suggested by this problem).** Friendly Loan Company advertises that it is willing to lend money for five years at the low rate of 8 percent per year. A potential borrower discovers that a five-year, $10,000 loan requires that the borrower pay the 8 percent interest in advance, with interest deducted from the loan proceeds. The borrower will collect $6,000 [= $10,000 − (= 5 × 0.08 × $10,000)] in cash and must repay the "$10,000" loan in five annual installments of $2,000, one each at the end of the next five years.

    Compute the effective interest rate implied by these loan terms.

38. **Deriving net present value of cash flows for decision to dispose of asset.** Suppose that yesterday Black & Decker Company purchased and installed a made-to-order machine tool for fabricating parts for small appliances. The machine cost $100,000. Today, Square D Company offers a machine tool that will do exactly the same work but costs only $50,000. Assume that the discount rate is 12 percent, that both machines will last for five years, that Black & Decker will depreciate both machines on a straight-line basis with no salvage value for tax purposes, that the income tax rate is and will continue to be 40 percent, and that Black & Decker earns sufficient income that it can use any loss from disposing of or depreciating the "old" machine to offset other taxable income.

    How much, at a minimum, must the "old" machine fetch on resale at this time to make purchasing the new machine worthwhile?

39. **Computation of present value of cash flows; untaxed acquisition, no change in tax basis of assets.** The balance sheet of Lynch Company shows net assets of $100,000 and owners' equity of $100,000. The assets are all depreciable assets with remaining lives of 20 years. The income statement for the year shows revenues of $700,000, depreciation of $50,000 (= $1,000,000 ÷ 20 years), no other expenses, income taxes of $260,000 (40 percent of pretax income of $650,000), and net income of $390,000.

    Bages Company is considering purchasing all of the stock of Lynch Company. It is willing to pay an amount equal to the present value of the cash flows from operations for the next 20 years discounted at a rate of 10 percent per year.

    The transaction will be a tax-free exchange; that is, after the purchase, the tax basis of the assets of Lynch Company will remain unchanged, so that depreciation charges will remain at $50,000 per year and income taxes will remain at $260,000 per year. Revenues will be $700,000 per year for the next 20 years.

    **a.** Compute the annual cash flows produced by Lynch Company.

    **b.** Compute the maximum amount Bages Company should be willing to pay.

40. **Computation of the present value of cash flows; taxable acquisition, changing tax basis of assets.** Refer to the data in the preceding problem. Assume now that the acquisition is taxable,

so that the tax basis of the assets acquired changes after the purchase. If the purchase price is $V, then depreciation charges will be $V/20 per year for 20 years. Income taxes will be 40 percent of pretax income. What is the maximum Bages Company should be willing to pay for Lynch Company?

41. **Valuation of intangibles with perpetuity formulas.** When the American Basketball Association (ABA) merged with the National Basketball Association (NBA), the owners of the ABA St. Louis Spirits agreed to dissolve their team and not enter the NBA. In return, the owners received a promise in perpetuity from the NBA that the NBA would pay to the Spirits' owners an amount each year equal to 40 percent of the TV revenues that the NBA paid to any one of its regular teams. Currently, the owners receive $4 million per year. The NBA wants to pay a single amount to the owners now and not have to pay more in the future. Of course, the owners prefer to collect more, rather than less, but here they want to know the reasonable minimum that will make them indifferent to the single payment in lieu of receiving the annual payments in perpetuity. Ignore income tax effects.
   a. Assume the owners expect the TV revenues to remain constant, so that they can expect $4 million per year in perpetuity and use an interest rate of 8 percent in their discounting calculations. What minimum price should these owners be willing to accept?
   b. Refer to the specifications for the preceding question. If the owners use a smaller interest rate for discounting, will the minimum price they are willing to accept increase, decrease, or remain unchanged?
   c. The owners use an 8-percent discount rate, and they expect TV revenues to increase by 2 percent per year in perpetuity. What minimum price should the owners be willing to accept?
   d. Refer to the specifications in **c**. If the owners use a smaller interest rate for discounting, will the minimum price they are willing to accept increase, decrease, or remain unchanged?
   e. Refer to the specifications in **c**. If the owners assume a smaller rate for growth in future receipts from the NBA, will the minimum price they are willing to accept increase, decrease, or remain unchanged?

42. **(Adapted from a problem by S. Zeff.)** Lexie T. Colleton is the chief financial officer of Ragazze, and one of her duties is to give advice on investment projects. Today's date is December 31, Year 0. Colleton requires that, to be acceptable, new investments must provide a positive net present value after discounting cash flows at 12 percent per year.

   A proposed investment is the purchase of an automatic gonculator, which involves an initial cash disbursement on December 31, Year 0. The useful life of the machine is nine years, through Year 9. Colleton expects to be able to sell the machine for cash of $30,000 on December 31, Year 9. She expects commercial production to begin on December 31, Year 1.

   Ragazze will depreciate the machine on a straight-line basis. Ignore income taxes.

   During Year 1, the break-in year, Ragazze will perform test runs in order to put the machine in proper working order. Colleton expects that the total cash outlay for this purpose will be $20,000, incurred at the end of Year 1.

   Colleton expects that the cash disbursements for regular maintenance will be $60,000 at the end of each of Years 2 through 5, inclusive, and $100,000 at the end of each of Years 6 through 8, inclusive.

   Colleton expects the cash receipts (net of all other operating expenses) from the sale of the product that the machine produces to be $130,000 at the end of each year from Year 2 through Year 9, inclusive.
   a. What is the maximum price that Ragazze can pay for the automatic gonculator on December 31, Year 0, and still earn a positive net present value of cash flows?
   b. Independent of your answer to part **a**, assume the purchase price is $250,000, which Ragazze will pay with an installment note requiring four equal annual installments starting December 31, Year 1, and an implicit interest rate of 10 percent per year. What is the amount of each payment?

43. **(Adapted from a problem by S. Zeff.)** William Marsh, CEO of Gulf Coast Manufacturing, wishes to know which of two strategies he has chosen for acquiring an automobile has lower present value of cost.

   **Strategy L.** Acquire a new Lexus, keep it for six years, then trade it in on a new car.

   **Strategy M.** Acquire a new Mercedes-Benz, trade it in after three years on a second Mercedes-Benz, keep that for another three years, then trade it in on a new car.

   Data pertinent to these choices appear below. Assume that Marsh will receive the trade-in value in cash or as a credit toward the purchase price of a new car. Ignore income taxes and use a discount rate of 10 percent per year. Gulf Coast Manufacturing depreciates automobiles on a straight-line basis over 8 years for financial reporting, assuming zero salvage value at the end of 8 years.

**a.** Which strategy has lower present value of costs?

**b.** What role, if any, do depreciation charges play in the analysis and why?

| | Lexus | Mercedes-Benz |
|---|---|---|
| Initial Cost at the Start of Year 1 | $60,000 | $ 45,000 |
| Initial Cost at the Start of Year 4 | | 48,000 |
| Trade-in Value | | |
| End of Year 3 | | 23,000 |
| End of Year 6[2] | 16,000 | 24,500 |
| Estimated Annual Cash Operating Costs, | | |
| Except Major Servicing | 4,000 | 4,500 |
| Estimated Cash Cost of Major Servicing | | |
| End of Year 4 | 6,500 | |
| End of Year 2 and End of Year 5 | | 2,500 |

[2]At this time Lexus is 6 years old; second Mercedes-Benz is 3 years old.

**44. (Wal Mart Stores; perpetuity growth model derivation of results in Chapter 5.)** Refer to the discussion on page **249** in **Chapter 5**. There, in estimating the value of a share of common stock of Wal-Mart Stores, we computed the present value of excess cash flows at the end of Year 14 (= beginning of Year 15) to be $203.6 billion. This exercise requires you to confirm that computation.

To compute the amount for the years after Year 19, note we assume that the excess cash flows are $5.962 billion at the end of Year 19 and grow at the rate of 10 percent per year thereafter. That means the cash flows for the end of Year 20 are $6.558 (= 1.10 × $5.962) billion. You can use the perpetuity growth model to verify that the present value at the end of Year 19 of that growing stream of payments is $327.9 billion. That is, if a payment (in this case $6.558 billion), grows at rate $g$ (in this case, 10 percent) per period forever, the discount rate is $r$ (in this case, 12 percent) per period, and the first payment flows at the end of the first period, then the present value of that stream is $327.9 [= $6.558/($r - g$) = $6.558/(0.12 − 0.10)] billion. Then, we discount that amount to the end of Year 14 to derive $186.1 billion. Analysts describe the $186.1 billion valuation in such computations as the *terminal value*.

(We do not expect that Wal-Mart's excess cash flows could increase forever at 10 percent per year. After a century or so, such a firm would be larger than the rest of the entire U.S. economy, combined. We use such computations to estimate values. When the discount rate (here 12 percent per year) exceeds the growth rate (here 10 percent per year) by a substantial amount (here only 2 percentage points), the present value of payments far in the future, say more than 40 years out, is negligible.)

**a.** Reproduce the numbers in Column **(6)** on page **249** using the data from Column **(5)** and the appropriate present value computations.

**b.** Re-do the valuation changing the growth rate from Year 19 from 10 percent to 9 percent.

**c.** Re-do the valuation changing the growth rate from Year 19 from 10 percent to 5 percent.

**d.** Comment on the sensitivity of this valuation modeling tool to the effect of assumed growth rates on terminal values.

**45. (Fast Growth Start-Up Company; valuation involving perpetuity growth model assumptions.)** Fast Growth Start-Up Company (FGSUC) has a new successful Internet business model. It earned $100 million of after-tax free cash flows this year. The company proposes to go public and the company's internal financial staff suggests to the board of directors that a valuation of $2.5 billion seems reasonable for the company. The investment banking firm's analyst and the financial staff at the company agree that the growth rate in free cash flows will be 25 percent per year for several years before the growth rate drops back to one more closely resembling the growth rate in the economy as a whole, which all assume to be 4 percent per year. Assume that the after-tax discount rate suitable for such a new venture is 15 percent per year.

How many years of growth in after-tax free cash flow of 25 percent per year will the FGSUC need to earn to justify a market valuation of $2.5 billion? Do not attempt to work this problem without using a spreadsheet program.

# Compound Interest, Annuity, and Bond Tables

## TABLE 1

**Future Value of $1**

$F_n = P(1 + r)^n$

$r$ = interest rate; $n$ = number of periods until valuation; $P$ = $1

| Periods = $n$ | $\frac{1}{2}$% | 1% | $1\frac{1}{2}$% | 2% | 3% | 4% | 5% | 6% | 7% | 8% | 10% | 12% | 15% | 20% | 25% |
|---|---|---|---|---|---|---|---|---|---|---|---|---|---|---|---|
| 1 ..... | 1.00500 | 1.01000 | 1.01500 | 1.02000 | 1.03000 | 1.04000 | 1.05000 | 1.06000 | 1.07000 | 1.08000 | 1.10000 | 1.12000 | 1.15000 | 1.20000 | 1.25000 |
| 2 ..... | 1.01003 | 1.02010 | 1.03023 | 1.04040 | 1.06090 | 1.08160 | 1.10250 | 1.12360 | 1.14490 | 1.16640 | 1.21000 | 1.25440 | 1.32250 | 1.44000 | 1.56250 |
| 3 ..... | 1.01508 | 1.03030 | 1.04568 | 1.06121 | 1.09273 | 1.12486 | 1.15763 | 1.19102 | 1.22504 | 1.25971 | 1.33100 | 1.40493 | 1.52088 | 1.72800 | 1.95313 |
| 4 ..... | 1.02015 | 1.04060 | 1.06136 | 1.08243 | 1.12551 | 1.16986 | 1.21551 | 1.26248 | 1.31080 | 1.36049 | 1.46410 | 1.57352 | 1.74901 | 2.07360 | 2.44141 |
| 5 ..... | 1.02525 | 1.05101 | 1.07728 | 1.10408 | 1.15927 | 1.21665 | 1.27628 | 1.33823 | 1.40255 | 1.46933 | 1.61051 | 1.76234 | 2.01136 | 2.48832 | 3.05176 |
| 6 ..... | 1.03038 | 1.06152 | 1.09344 | 1.12616 | 1.19405 | 1.26532 | 1.34010 | 1.41852 | 1.50073 | 1.58687 | 1.77156 | 1.97382 | 2.31306 | 2.98598 | 3.81470 |
| 7 ..... | 1.03553 | 1.07214 | 1.10984 | 1.14869 | 1.22987 | 1.31593 | 1.40710 | 1.50363 | 1.60578 | 1.71382 | 1.94872 | 2.21068 | 2.66002 | 3.58318 | 4.76837 |
| 8 ..... | 1.04071 | 1.08286 | 1.12649 | 1.17166 | 1.26677 | 1.36857 | 1.47746 | 1.59385 | 1.71819 | 1.85093 | 2.14359 | 2.47596 | 3.05902 | 4.29982 | 5.96046 |
| 9 ..... | 1.04591 | 1.09369 | 1.14339 | 1.19509 | 1.30477 | 1.42331 | 1.55133 | 1.68948 | 1.83846 | 1.99900 | 2.35795 | 2.77308 | 3.51788 | 5.15978 | 7.45058 |
| 10 ..... | 1.05114 | 1.10462 | 1.16054 | 1.21899 | 1.34392 | 1.48024 | 1.62889 | 1.79085 | 1.96715 | 2.15892 | 2.59374 | 3.10585 | 4.04556 | 6.19174 | 9.31323 |
| 11 ..... | 1.05640 | 1.11567 | 1.17795 | 1.24337 | 1.38423 | 1.53945 | 1.71034 | 1.89830 | 2.10485 | 2.33164 | 2.85312 | 3.47855 | 4.65239 | 7.43008 | 11.64153 |
| 12 ..... | 1.06168 | 1.12683 | 1.19562 | 1.26824 | 1.42576 | 1.60103 | 1.79586 | 2.01220 | 2.25219 | 2.51817 | 3.13843 | 3.89598 | 5.35025 | 8.91610 | 14.55192 |
| 13 ..... | 1.06699 | 1.13809 | 1.21355 | 1.29361 | 1.46853 | 1.66507 | 1.88565 | 2.13293 | 2.40985 | 2.71962 | 3.45227 | 4.36349 | 6.15279 | 10.69932 | 18.18989 |
| 14 ..... | 1.07232 | 1.14947 | 1.23176 | 1.31948 | 1.51259 | 1.73168 | 1.97993 | 2.26090 | 2.57853 | 2.93719 | 3.79750 | 4.88711 | 7.07571 | 12.83918 | 22.73737 |
| 15 ..... | 1.07768 | 1.16097 | 1.25023 | 1.34587 | 1.55797 | 1.80094 | 2.07893 | 2.39656 | 2.75903 | 3.17217 | 4.17725 | 5.47357 | 8.13706 | 15.40702 | 28.42171 |
| 16 ..... | 1.08307 | 1.17258 | 1.26899 | 1.37279 | 1.60471 | 1.87298 | 2.18287 | 2.54035 | 2.95216 | 3.42594 | 4.59497 | 6.13039 | 9.35762 | 18.48843 | 35.52714 |
| 17 ..... | 1.08849 | 1.18430 | 1.28802 | 1.40024 | 1.65285 | 1.94790 | 2.29202 | 2.69277 | 3.15882 | 3.70002 | 5.05447 | 6.86604 | 10.76126 | 22.18611 | 44.40892 |
| 18 ..... | 1.09393 | 1.19615 | 1.30734 | 1.42825 | 1.70243 | 2.02582 | 2.40662 | 2.85434 | 3.37993 | 3.99602 | 5.55992 | 7.68997 | 12.37545 | 26.62333 | 55.51115 |
| 19 ..... | 1.09940 | 1.20811 | 1.32695 | 1.45681 | 1.75351 | 2.10685 | 2.52695 | 3.02560 | 3.61653 | 4.31570 | 6.11591 | 8.61276 | 14.23177 | 31.94800 | 69.38894 |
| 20 ..... | 1.10490 | 1.22019 | 1.34686 | 1.48595 | 1.80611 | 2.19112 | 2.65330 | 3.20714 | 3.86968 | 4.66096 | 6.72750 | 9.64629 | 16.36654 | 38.33760 | 86.73617 |
| 22 ..... | 1.11597 | 1.24472 | 1.38756 | 1.54598 | 1.91610 | 2.36992 | 2.92526 | 3.60354 | 4.43040 | 5.43654 | 8.14027 | 12.10031 | 21.64475 | 55.20614 | 135.5253 |
| 24 ..... | 1.12716 | 1.26973 | 1.42950 | 1.60844 | 2.03279 | 2.56330 | 3.22510 | 4.04893 | 5.07237 | 6.34118 | 9.84973 | 15.17863 | 28.62518 | 79.49685 | 211.7582 |
| 26 ..... | 1.13846 | 1.29526 | 1.47271 | 1.67342 | 2.15659 | 2.77247 | 3.55567 | 4.54938 | 5.80735 | 7.39635 | 11.91818 | 19.04007 | 37.85680 | 114.4755 | 330.8722 |
| 28 ..... | 1.14987 | 1.32129 | 1.51722 | 1.74102 | 2.28793 | 2.99870 | 3.92013 | 5.11169 | 6.64884 | 8.62711 | 14.42099 | 23.88387 | 50.06561 | 164.8447 | 516.9879 |
| 30 ..... | 1.16140 | 1.34785 | 1.56308 | 1.81136 | 2.42726 | 3.24340 | 4.32194 | 5.74349 | 7.61226 | 10.06266 | 17.44940 | 29.95992 | 66.21177 | 237.3763 | 807.7936 |
| 32 ..... | 1.17304 | 1.37494 | 1.61032 | 1.88454 | 2.57508 | 3.50806 | 4.76494 | 6.45339 | 8.71527 | 11.73708 | 21.11378 | 37.58173 | 87.56507 | 341.8219 | 1262.177 |
| 34 ..... | 1.18480 | 1.40258 | 1.65900 | 1.96068 | 2.73191 | 3.79432 | 5.25335 | 7.25103 | 9.97811 | 13.69013 | 25.54767 | 47.14252 | 115.80480 | 492.2235 | 1972.152 |
| 36 ..... | 1.19668 | 1.43077 | 1.70914 | 2.03989 | 2.89828 | 4.10393 | 5.79182 | 8.14725 | 11.42394 | 15.96817 | 30.91268 | 59.13557 | 153.15185 | 708.8019 | 3081.488 |
| 38 ..... | 1.20868 | 1.45953 | 1.76080 | 2.12230 | 3.07478 | 4.43881 | 6.38548 | 9.15425 | 13.07927 | 18.62528 | 37.40434 | 74.17966 | 202.54332 | 1020.675 | 4814.825 |
| 40 ..... | 1.22079 | 1.48886 | 1.81402 | 2.20804 | 3.26204 | 4.80102 | 7.03999 | 10.28572 | 14.97446 | 21.72452 | 45.25926 | 93.05097 | 267.86355 | 1469.772 | 7523.164 |
| 45 ..... | 1.25162 | 1.56481 | 1.95421 | 2.43785 | 3.78160 | 5.84118 | 8.98501 | 13.76461 | 21.00245 | 31.92045 | 72.89048 | 163.9876 | 538.76927 | 3657.262 | 22958.87 |
| 50 ..... | 1.28323 | 1.64463 | 2.10524 | 2.69159 | 4.38391 | 7.10668 | 11.46740 | 18.42015 | 29.45703 | 46.90161 | 117.3909 | 289.0022 | 1083.65744 | 9100.438 | 70064.92 |
| 100 ..... | 1.64667 | 2.70481 | 4.43205 | 7.24465 | 19.21863 | 50.50495 | 131.5013 | 339.3021 | 867.7163 | 2199.761 | 13780.61 | 83522.27 | $117 \times 10^4$ | $828 \times 10^5$ | $491 \times 10^7$ |

## TABLE 2 Present Value of $1

$P = F_n(1 + r)^{-n}$

$r$ = discount rate; $n$ = number of periods until payment; $F_n$ = $1

| Periods = $n$ | ½% | 1% | 1½% | 2% | 3% | 4% | 5% | 6% | 7% | 8% | 10% | 12% | 15% | 20% | 25% |
|---|---|---|---|---|---|---|---|---|---|---|---|---|---|---|---|
| 1 | .99502 | .99010 | .98522 | .98039 | .97087 | .96154 | .95238 | .94340 | .93458 | .92593 | .90909 | .89286 | .86957 | .83333 | .80000 |
| 2 | .99007 | .98030 | .97066 | .96117 | .94260 | .92456 | .90703 | .89000 | .87344 | .85734 | .82645 | .79719 | .75614 | .69444 | .64000 |
| 3 | .98515 | .97059 | .95632 | .94232 | .91514 | .88900 | .86384 | .83962 | .81630 | .79383 | .75131 | .71178 | .65752 | .57870 | .51200 |
| 4 | .98025 | .96098 | .94218 | .92385 | .88849 | .85480 | .82270 | .79209 | .76290 | .73503 | .68301 | .63552 | .57175 | .48225 | .40960 |
| 5 | .97537 | .95147 | .92826 | .90573 | .86261 | .82193 | .78353 | .74726 | .71299 | .68058 | .62092 | .56743 | .49718 | .40188 | .32768 |
| 6 | .97052 | .94205 | .91454 | .88797 | .83748 | .79031 | .74622 | .70496 | .66634 | .63017 | .56447 | .50663 | .43233 | .33490 | .26214 |
| 7 | .96569 | .93272 | .90103 | .87056 | .81309 | .75992 | .71068 | .66506 | .62275 | .58349 | .51316 | .45235 | .37594 | .27908 | .20972 |
| 8 | .96089 | .92348 | .88771 | .85349 | .78941 | .73069 | .67684 | .62741 | .58201 | .54027 | .46651 | .40388 | .32690 | .23257 | .16777 |
| 9 | .95610 | .91434 | .87459 | .83676 | .76642 | .70259 | .64461 | .59190 | .54393 | .50025 | .42410 | .36061 | .28426 | .19381 | .13422 |
| 10 | .95135 | .90529 | .86167 | .82035 | .74409 | .67556 | .61391 | .55839 | .50835 | .46319 | .38554 | .32197 | .24718 | .16151 | .10737 |
| 11 | .94661 | .89632 | .84893 | .80426 | .72242 | .64958 | .58468 | .52679 | .47509 | .42888 | .35049 | .28748 | .21494 | .13459 | .08590 |
| 12 | .94191 | .88745 | .83639 | .78849 | .70138 | .62460 | .55684 | .49697 | .44401 | .39711 | .31863 | .25668 | .18691 | .11216 | .06872 |
| 13 | .93722 | .87866 | .82403 | .77303 | .68095 | .60057 | .53032 | .46884 | .41496 | .36770 | .28966 | .22917 | .16253 | .09346 | .05498 |
| 14 | .93256 | .86996 | .81185 | .75788 | .66112 | .57748 | .50507 | .44230 | .38782 | .34046 | .26333 | .20462 | .14133 | .07789 | .04398 |
| 15 | .92792 | .86135 | .79985 | .74301 | .64186 | .55526 | .48102 | .41727 | .36245 | .31524 | .23939 | .18270 | .12289 | .06491 | .03518 |
| 16 | .92330 | .85282 | .78803 | .72845 | .62317 | .53391 | .45811 | .39365 | .33873 | .29189 | .21763 | .16312 | .10686 | .05409 | .02815 |
| 17 | .91871 | .84438 | .77639 | .71416 | .60502 | .51337 | .43630 | .37136 | .31657 | .27027 | .19784 | .14564 | .09293 | .04507 | .02252 |
| 18 | .91414 | .83602 | .76491 | .70016 | .58739 | .49363 | .41552 | .35034 | .29586 | .25025 | .17986 | .13004 | .08081 | .03756 | .01801 |
| 19 | .90959 | .82774 | .75361 | .68643 | .57029 | .47464 | .39573 | .33051 | .27651 | .23171 | .16351 | .11611 | .07027 | .03130 | .01441 |
| 20 | .90506 | .81954 | .74247 | .67297 | .55368 | .45639 | .37689 | .31180 | .25842 | .21455 | .14864 | .10367 | .06110 | .02608 | .01153 |
| 22 | .89608 | .80340 | .72069 | .64684 | .52189 | .42196 | .34185 | .27751 | .22571 | .18394 | .12285 | .08264 | .04620 | .01811 | .00738 |
| 24 | .88719 | .78757 | .69954 | .62172 | .49193 | .39012 | .31007 | .24698 | .19715 | .15770 | .10153 | .06588 | .03493 | .01258 | .00472 |
| 26 | .87838 | .77205 | .67902 | .59758 | .46369 | .36069 | .28124 | .21981 | .17220 | .13520 | .08391 | .05252 | .02642 | .00874 | .00302 |
| 28 | .86966 | .75684 | .65910 | .57437 | .43708 | .33348 | .25509 | .19563 | .15040 | .11591 | .06934 | .04187 | .01997 | .00607 | .00193 |
| 30 | .86103 | .74192 | .63976 | .55207 | .41199 | .30832 | .23138 | .17411 | .13137 | .09938 | .05731 | .03338 | .01510 | .00421 | .00124 |
| 32 | .85248 | .72730 | .62099 | .53063 | .38834 | .28506 | .20987 | .15496 | .11474 | .08520 | .04736 | .02661 | .01142 | .00293 | .00079 |
| 34 | .84402 | .71297 | .60277 | .51003 | .36604 | .26355 | .19035 | .13791 | .10022 | .07305 | .03914 | .02121 | .00864 | .00203 | .00051 |
| 36 | .83564 | .69892 | .58509 | .49022 | .34503 | .24367 | .17266 | .12274 | .08754 | .06262 | .03235 | .01691 | .00653 | .00141 | .00032 |
| 38 | .82735 | .68515 | .56792 | .47119 | .32523 | .22529 | .15661 | .10924 | .07646 | .05369 | .02673 | .01348 | .00494 | .00098 | .00021 |
| 40 | .81914 | .67165 | .55126 | .45289 | .30656 | .20829 | .14205 | .09722 | .06678 | .04603 | .02209 | .01075 | .00373 | .00068 | .00013 |
| 45 | .79896 | .63905 | .51171 | .41020 | .26444 | .17120 | .11130 | .07265 | .04761 | .03133 | .01372 | .00610 | .00186 | .00027 | .00004 |
| 50 | .77929 | .60804 | .47500 | .37153 | .22811 | .14071 | .08720 | .05429 | .03395 | .02132 | .00852 | .00346 | .00092 | .00011 | .00001 |
| 100 | .60729 | .36971 | .22563 | .13803 | .05203 | .01980 | .00760 | .00295 | .00115 | .00045 | .00007 | .00001 | .00000 | .00000 | .00000 |

## TABLE 3 Future Value of Annuity of $1 in Arrears

$$P_F = \frac{(1 + r)^n - 1}{r}$$

$r$ = interest rate; $n$ = number of payments

| No. of Payments = $n$ | ½% | 1% | 1½% | 2% | 3% | 4% | 5% | 6% | 7% | 8% | 10% | 12% | 15% | 20% | 25% |
|---|---|---|---|---|---|---|---|---|---|---|---|---|---|---|---|
| 1 ........ | 1.00000 | 1.00000 | 1.00000 | 1.00000 | 1.00000 | 1.00000 | 1.00000 | 1.00000 | 1.00000 | 1.00000 | 1.00000 | 1.00000 | 1.00000 | 1.00000 | 1.00000 |
| 2 ........ | 2.00500 | 2.01000 | 2.01500 | 2.02000 | 2.03000 | 2.04000 | 2.05000 | 2.06000 | 2.07000 | 2.08000 | 2.10000 | 2.12000 | 2.15000 | 2.20000 | 2.25000 |
| 3 ........ | 3.01503 | 3.03010 | 3.04523 | 3.06040 | 3.09090 | 3.12160 | 3.15250 | 3.18360 | 3.21490 | 3.24640 | 3.31000 | 3.37440 | 3.47250 | 3.64000 | 3.81250 |
| 4 ........ | 4.03010 | 4.06040 | 4.09090 | 4.12161 | 4.18363 | 4.24646 | 4.31013 | 4.37462 | 4.43994 | 4.50611 | 4.64100 | 4.77933 | 4.99338 | 5.36800 | 5.76563 |
| 5 ........ | 5.05025 | 5.10101 | 5.15227 | 5.20404 | 5.30914 | 5.41632 | 5.52563 | 5.63709 | 5.75074 | 5.86660 | 6.10510 | 6.35285 | 6.74238 | 7.44160 | 8.20703 |
| 6 ........ | 6.07550 | 6.15202 | 6.22955 | 6.30812 | 6.46841 | 6.63298 | 6.80191 | 6.97532 | 7.15329 | 7.33593 | 7.71561 | 8.11519 | 8.75374 | 9.92992 | 11.25879 |
| 7 ........ | 7.10588 | 7.21354 | 7.32299 | 7.43428 | 7.66246 | 7.89829 | 8.14201 | 8.39384 | 8.65402 | 8.92280 | 9.48717 | 10.08901 | 11.06680 | 12.91590 | 15.07349 |
| 8 ........ | 8.14141 | 8.28567 | 8.43284 | 8.58297 | 8.89234 | 9.21423 | 9.54911 | 9.89747 | 10.25980 | 10.63663 | 11.43589 | 12.29969 | 13.72682 | 16.49908 | 19.84186 |
| 9 ........ | 9.18212 | 9.36853 | 9.55933 | 9.75463 | 10.15911 | 10.58280 | 11.02656 | 11.49132 | 11.97799 | 12.48756 | 13.57948 | 14.77566 | 16.78584 | 20.79890 | 25.80232 |
| 10 ........ | 10.22803 | 10.46221 | 10.70272 | 10.94972 | 11.46388 | 12.00611 | 12.57789 | 13.18079 | 13.81645 | 14.48656 | 15.93742 | 17.54874 | 20.30372 | 25.95868 | 33.25290 |
| 11 ........ | 11.27917 | 11.56683 | 11.86326 | 12.16872 | 12.80780 | 13.48635 | 14.20679 | 14.97164 | 15.78360 | 16.64549 | 18.53117 | 20.65458 | 24.34928 | 32.15042 | 42.56613 |
| 12 ........ | 12.33556 | 12.68250 | 13.04121 | 13.41209 | 14.19203 | 15.02581 | 15.91713 | 16.86994 | 17.88845 | 18.97713 | 21.38428 | 24.13313 | 29.00167 | 39.58050 | 54.20766 |
| 13 ........ | 13.39724 | 13.80933 | 14.23683 | 14.68033 | 15.61779 | 16.62684 | 17.71298 | 18.88214 | 20.14064 | 21.49530 | 24.52271 | 28.02911 | 34.35192 | 48.49660 | 68.75958 |
| 14 ........ | 14.46423 | 14.94742 | 15.45038 | 15.97394 | 17.08632 | 18.29191 | 19.59863 | 21.01507 | 22.55049 | 24.21492 | 27.97498 | 32.39260 | 40.50471 | 59.19592 | 86.94947 |
| 15 ........ | 15.53655 | 16.09690 | 16.68214 | 17.29342 | 18.59891 | 20.02359 | 21.57856 | 23.27597 | 25.12902 | 27.15211 | 31.77248 | 37.27971 | 47.58041 | 72.03511 | 109.6868 |
| 16 ........ | 16.61423 | 17.25786 | 17.93237 | 18.63929 | 20.15688 | 21.82453 | 23.65749 | 25.67253 | 27.88805 | 30.32428 | 35.94973 | 42.75328 | 55.71747 | 87.44213 | 138.1085 |
| 17 ........ | 17.69730 | 18.43044 | 19.20136 | 20.01207 | 21.76159 | 23.69751 | 25.84037 | 28.21288 | 30.84022 | 33.75023 | 40.54470 | 48.88367 | 65.07509 | 105.9306 | 173.6357 |
| 18 ........ | 18.78579 | 19.61475 | 20.48938 | 21.41231 | 23.41444 | 25.64541 | 28.13238 | 30.90565 | 33.99903 | 37.45024 | 45.59917 | 55.74971 | 75.83636 | 128.1167 | 218.0446 |
| 19 ........ | 19.87972 | 20.81090 | 21.79672 | 22.84056 | 25.11687 | 27.67123 | 30.53900 | 33.75999 | 37.37896 | 41.44626 | 51.15909 | 63.43968 | 88.21181 | 154.7400 | 273.5558 |
| 20 ........ | 20.97912 | 22.01900 | 23.12367 | 24.29737 | 26.87037 | 29.77808 | 33.06595 | 36.78559 | 40.99549 | 45.76196 | 57.27500 | 72.05244 | 102.44358 | 186.6880 | 342.9447 |
| 22 ........ | 23.19443 | 24.47159 | 25.83758 | 27.29898 | 30.53678 | 34.24797 | 38.50521 | 43.39229 | 49.00574 | 55.45676 | 71.40275 | 92.50258 | 137.63164 | 271.0307 | 538.1011 |
| 24 ........ | 25.43196 | 26.97346 | 28.63352 | 30.42186 | 34.42647 | 39.08260 | 44.50200 | 50.81558 | 58.17667 | 66.76476 | 88.49733 | 118.15524 | 184.16784 | 392.4842 | 843.0329 |
| 26 ........ | 27.69191 | 29.52563 | 31.51397 | 33.67091 | 38.55304 | 44.31174 | 51.11345 | 59.15638 | 68.67647 | 79.95442 | 109.18177 | 150.33393 | 245.71197 | 567.3773 | 1319.489 |
| 28 ........ | 29.97452 | 32.12910 | 34.48148 | 37.05121 | 42.93092 | 49.96758 | 58.40258 | 68.52811 | 80.69769 | 95.33883 | 134.20994 | 190.69889 | 327.10408 | 819.2233 | 2063.952 |
| 30 ........ | 32.28002 | 34.78489 | 37.53868 | 40.56808 | 47.57542 | 56.08494 | 66.43885 | 79.05819 | 94.46079 | 113.28321 | 164.49402 | 241.33268 | 434.74515 | 1181.881 | 3227.174 |
| 32 ........ | 34.60862 | 37.49407 | 40.68829 | 44.22703 | 52.50276 | 62.70147 | 75.29883 | 90.88978 | 110.21815 | 134.21354 | 201.13777 | 304.84772 | 577.10046 | 1704.109 | 5044.710 |
| 34 ........ | 36.96058 | 40.25770 | 43.93309 | 48.03380 | 57.73018 | 69.85791 | 85.06696 | 104.18375 | 128.25876 | 158.62667 | 245.47670 | 384.52098 | 765.36535 | 2456.118 | 7884.609 |
| 36 ........ | 39.33610 | 43.07688 | 47.27597 | 51.99437 | 63.27594 | 77.59831 | 95.83632 | 119.12087 | 148.91346 | 187.10215 | 299.12681 | 484.46312 | 1014.34568 | 3539.009 | 12321.95 |
| 38 ........ | 41.73545 | 45.95272 | 50.71989 | 56.11494 | 69.15945 | 85.97034 | 107.70955 | 135.90421 | 172.56102 | 220.31595 | 364.04343 | 609.83053 | 1343.62216 | 5098.373 | 19255.30 |
| 40 ........ | 44.15885 | 48.88637 | 54.26789 | 60.40198 | 75.40126 | 95.02552 | 120.79977 | 154.76197 | 199.63511 | 259.05652 | 442.59256 | 767.09142 | 1779.09031 | 7343.858 | 30088.66 |
| 45 ........ | 50.32416 | 56.48107 | 63.61420 | 71.89271 | 92.71986 | 121.0294 | 159.7002 | 212.7435 | 285.7493 | 386.5056 | 718.9048 | 1358.230 | 3585.12846 | 18281.31 | 91831.50 |
| 50 ........ | 56.64516 | 64.46318 | 73.68283 | 84.57940 | 112.7969 | 152.6671 | 209.3480 | 290.3359 | 406.5289 | 573.7702 | 1163.909 | 2400.018 | 7217.71628 | 45497.19 | 280255.7 |
| 100 ........ | 129.33370 | 170.4814 | 228.8030 | 312.2323 | 607.2877 | 1237.624 | 2610.025 | 5638.368 | 12381.66 | 27484.52 | 137796.1 | 696010.5 | $783 \times 10^4$ | $414 \times 10^6$ | $196 \times 10^8$ |

Note: To convert from this table to values of an annuity in advance, determine the annuity in arrears above for one more period and subtract 1.00000.

## TABLE 4

**Present Value of Annuity of $1 in Arrears**

$$P_A = \frac{1 - (1 + r)^{-n}}{r} \times \$1.00$$

**$r$ = discount rate; $n$ = number of payments**

$n$ Periods = Payments

0 $1 $1 $1 $1 $1 ← Payments in Arrears

$P_A$ $P_F$

(Value in Table 4) ← $\sum$ (Individual Values from Table 2) (Value in Table 3)

| No. of Payments = $n$ | ½% | 1% | 1½% | 2% | 3% | 4% | 5% | 6% | 7% | 8% | 10% | 12% | 15% | 20% | 25% |
|---|---|---|---|---|---|---|---|---|---|---|---|---|---|---|---|
| 1 .......... | .99502 | .99010 | .98522 | .98039 | .97087 | .96154 | .95238 | .94340 | .93458 | .92593 | .90909 | .89286 | .86957 | .83333 | .80000 |
| 2 .......... | 1.98510 | 1.97040 | 1.95588 | 1.94156 | 1.91347 | 1.88609 | 1.85941 | 1.83339 | 1.80802 | 1.78326 | 1.73554 | 1.69005 | 1.62571 | 1.52778 | 1.44000 |
| 3 .......... | 2.97025 | 2.94099 | 2.91220 | 2.88388 | 2.82861 | 2.77509 | 2.72325 | 2.67301 | 2.62432 | 2.57710 | 2.48685 | 2.40183 | 2.28323 | 2.10648 | 1.95200 |
| 4 .......... | 3.95050 | 3.90197 | 3.85438 | 3.80773 | 3.71710 | 3.62990 | 3.54595 | 3.46511 | 3.38721 | 3.31213 | 3.16987 | 3.03735 | 2.85498 | 2.58873 | 2.36160 |
| 5 .......... | 4.92587 | 4.85343 | 4.78264 | 4.71346 | 4.57971 | 4.45182 | 4.32948 | 4.21236 | 4.10020 | 3.99271 | 3.79079 | 3.60478 | 3.35216 | 2.99061 | 2.68928 |
| 6 .......... | 5.89638 | 5.79548 | 5.69719 | 5.60143 | 5.41719 | 5.24214 | 5.07569 | 4.91732 | 4.76654 | 4.62288 | 4.35526 | 4.11141 | 3.78448 | 3.32551 | 2.95142 |
| 7 .......... | 6.86207 | 6.72819 | 6.59821 | 6.47199 | 6.23028 | 6.00205 | 5.78637 | 5.58238 | 5.38929 | 5.20637 | 4.86842 | 4.56376 | 4.16042 | 3.60459 | 3.16114 |
| 8 .......... | 7.82296 | 7.65168 | 7.48593 | 7.32548 | 7.01969 | 6.73274 | 6.46321 | 6.20979 | 5.97130 | 5.74664 | 5.33493 | 4.96764 | 4.48732 | 3.83716 | 3.32891 |
| 9 .......... | 8.77906 | 8.56602 | 8.36052 | 8.16224 | 7.78611 | 7.43533 | 7.10782 | 6.80169 | 6.51523 | 6.24689 | 5.75902 | 5.32825 | 4.77158 | 4.03097 | 3.46313 |
| 10 .......... | 9.73041 | 9.47130 | 9.22218 | 8.98259 | 8.53020 | 8.11090 | 7.72173 | 7.36009 | 7.02358 | 6.71008 | 6.14457 | 5.65022 | 5.01877 | 4.19247 | 3.57050 |
| 11 .......... | 10.67703 | 10.36763 | 10.07112 | 9.78685 | 9.25262 | 8.76048 | 8.30641 | 7.88687 | 7.49867 | 7.13896 | 6.49506 | 5.93770 | 5.23371 | 4.32706 | 3.65640 |
| 12 .......... | 11.61893 | 11.25508 | 10.90751 | 10.57534 | 9.95400 | 9.38507 | 8.86325 | 8.38384 | 7.94269 | 7.53608 | 6.81369 | 6.19437 | 5.42062 | 4.43922 | 3.72512 |
| 13 .......... | 12.55615 | 12.13374 | 11.73153 | 11.34837 | 10.63496 | 9.98565 | 9.39357 | 8.85268 | 8.35765 | 7.90378 | 7.10336 | 6.42355 | 5.58315 | 4.53268 | 3.78010 |
| 14 .......... | 13.48871 | 13.00370 | 12.54338 | 12.10625 | 11.29607 | 10.56312 | 9.89864 | 9.29498 | 8.74547 | 8.24424 | 7.36669 | 6.62817 | 5.72448 | 4.61057 | 3.82408 |
| 15 .......... | 14.41662 | 13.86505 | 13.34323 | 12.84926 | 11.93794 | 11.11839 | 10.37966 | 9.71225 | 9.10791 | 8.55948 | 7.60608 | 6.81086 | 5.84737 | 4.67547 | 3.85926 |
| 16 .......... | 15.33993 | 14.71787 | 14.13126 | 13.57771 | 12.56110 | 11.65230 | 10.83777 | 10.10590 | 9.44665 | 8.85137 | 7.82371 | 6.97399 | 5.95423 | 4.72956 | 3.88741 |
| 17 .......... | 16.25863 | 15.56225 | 14.90765 | 14.29187 | 13.16612 | 12.16567 | 11.27407 | 10.47726 | 9.76322 | 9.12164 | 8.02155 | 7.11963 | 6.04716 | 4.77463 | 3.90993 |
| 18 .......... | 17.17277 | 16.39827 | 15.67256 | 14.99203 | 13.75351 | 12.65930 | 11.68959 | 10.82760 | 10.05909 | 9.37189 | 8.20141 | 7.24967 | 6.12797 | 4.81219 | 3.92794 |
| 19 .......... | 18.08236 | 17.22601 | 16.42617 | 15.67846 | 14.32380 | 13.13394 | 12.08532 | 11.15812 | 10.33560 | 9.60360 | 8.36492 | 7.36578 | 6.19823 | 4.84350 | 3.94235 |
| 20 .......... | 18.98742 | 18.04555 | 17.16864 | 16.35143 | 14.87747 | 13.59033 | 12.46221 | 11.46992 | 10.59401 | 9.81815 | 8.51356 | 7.46944 | 6.25933 | 4.86958 | 3.95388 |
| 22 .......... | 20.78406 | 19.66038 | 18.62082 | 17.65805 | 15.93692 | 14.45112 | 13.16300 | 12.04158 | 11.06124 | 10.20074 | 8.77154 | 7.64465 | 6.35866 | 4.90943 | 3.97049 |
| 24 .......... | 22.56287 | 21.24339 | 20.03041 | 18.91393 | 16.93554 | 15.24696 | 13.79864 | 12.55036 | 11.46933 | 10.52876 | 8.98474 | 7.78432 | 6.43377 | 4.93710 | 3.98111 |
| 26 .......... | 24.32402 | 22.79520 | 21.39863 | 20.12104 | 17.87684 | 15.98277 | 14.37519 | 13.00317 | 11.82578 | 10.80998 | 9.16095 | 7.89566 | 6.49056 | 4.95632 | 3.98791 |
| 28 .......... | 26.06769 | 24.31644 | 22.72672 | 21.28127 | 18.76411 | 16.66306 | 14.89813 | 13.40616 | 12.13711 | 11.05108 | 9.30657 | 7.98442 | 6.53351 | 4.96967 | 3.99226 |
| 30 .......... | 27.79405 | 25.80771 | 24.01584 | 22.39646 | 19.60044 | 17.29203 | 15.37245 | 13.76483 | 12.40904 | 11.25778 | 9.42691 | 8.05518 | 6.56598 | 4.97894 | 3.99505 |
| 32 .......... | 29.50328 | 27.26959 | 25.26714 | 23.46833 | 20.38877 | 17.87355 | 15.80268 | 14.08404 | 12.64656 | 11.43500 | 9.52638 | 8.11159 | 6.59053 | 4.98537 | 3.99683 |
| 34 .......... | 31.19555 | 28.70267 | 26.48173 | 24.49859 | 21.13184 | 18.41120 | 16.19290 | 14.36814 | 12.85401 | 11.58693 | 9.60857 | 8.15656 | 6.60910 | 4.98984 | 3.99797 |
| 36 .......... | 32.87102 | 30.10751 | 27.66068 | 25.48884 | 21.83225 | 18.90828 | 16.54685 | 14.62099 | 13.03521 | 11.71719 | 9.67651 | 8.19241 | 6.62314 | 4.99295 | 3.99870 |
| 38 .......... | 34.52985 | 31.48466 | 28.80505 | 26.44064 | 22.49246 | 19.36786 | 16.86789 | 14.84602 | 13.19347 | 11.82887 | 9.73265 | 8.22099 | 6.63375 | 4.99510 | 3.99917 |
| 40 .......... | 36.17223 | 32.83469 | 29.91585 | 27.35548 | 23.11477 | 19.79277 | 17.15909 | 15.04630 | 13.33171 | 11.92461 | 9.77905 | 8.24378 | 6.64178 | 4.99660 | 3.99947 |
| 45 .......... | 40.20720 | 36.09451 | 32.55234 | 29.49016 | 24.51871 | 20.72004 | 17.77407 | 15.45583 | 13.60552 | 12.10840 | 9.86281 | 8.28252 | 6.65429 | 4.99863 | 3.99983 |
| 50 .......... | 44.14279 | 39.19612 | 34.99969 | 31.42361 | 25.72976 | 21.48218 | 18.25593 | 15.76186 | 13.80075 | 12.23348 | 9.91481 | 8.30450 | 6.66051 | 4.99945 | 3.99994 |
| 100 .......... | 78.54264 | 63.02888 | 51.62470 | 43.09835 | 31.59891 | 24.50500 | 19.84791 | 16.61755 | 14.26925 | 12.49432 | 9.99927 | 8.33323 | 6.66666 | 5.00000 | 4.00000 |

Note: To convert from this table to values of an annuity in advance, determine the annuity in arrears above for one fewer period and add 1.00000.

**TABLE 5**

**Bond Values in Percent of Par: 10 Percent Semiannual Coupons**
**Bond value = $10/r + (100 - 10/r)(1 + r/2)^{-2n}$**
**$r$ = yield to maturity; $n$ = years to maturity**

| Years to Maturity | Market Yield Percent per Year Compounded Semiannually | | | | | | | | | | |
|---|---|---|---|---|---|---|---|---|---|---|---|
| | 8.0 | 9.0 | 9.5 | 10 | 10.5 | 11.0 | 12.0 | 13.0 | 14.0 | 15.0 | 20.0 |
| 0.5 | 100.9615 | 100.4785 | 100.2387 | 100.0 | 99.7625 | 99.5261 | 99.0566 | 98.5915 | 98.1308 | 97.6744 | 95.4545 |
| 1.0 | 101.8861 | 100.9363 | 100.4665 | 100.0 | 99.5368 | 99.0768 | 98.1666 | 97.2691 | 96.3840 | 95.5111 | 91.3223 |
| 1.5 | 102.7751 | 101.3745 | 100.6840 | 100.0 | 99.3224 | 98.6510 | 97.3270 | 96.0273 | 94.7514 | 93.4987 | 87.5657 |
| 2.0 | 103.6299 | 101.7938 | 100.8917 | 100.0 | 99.1186 | 98.2474 | 96.5349 | 94.8613 | 93.2256 | 91.6267 | 84.1507 |
| 2.5 | 104.4518 | 102.1950 | 101.0899 | 100.0 | 98.9251 | 97.8649 | 95.7876 | 93.7665 | 91.7996 | 89.8853 | 81.0461 |
| 5.0 | 108.1109 | 103.9564 | 101.9541 | 100.0 | 98.0928 | 96.2312 | 92.6399 | 89.2168 | 85.9528 | 82.8398 | 69.2772 |
| 9.0 | 112.6593 | 106.0800 | 102.9803 | 100.0 | 97.1339 | 94.3770 | 89.1724 | 84.3513 | 79.8818 | 75.7350 | 58.9929 |
| 9.5 | 113.1339 | 106.2966 | 103.0838 | 100.0 | 97.0393 | 94.1962 | 88.8419 | 83.8979 | 79.3288 | 75.1023 | 58.1754 |
| 10.0 | 113.5903 | 106.5040 | 103.1827 | 100.0 | 96.9494 | 94.0248 | 88.5301 | 83.4722 | 78.8120 | 74.5138 | 57.4322 |
| 15.0 | 117.2920 | 108.1444 | 103.9551 | 100.0 | 96.2640 | 92.7331 | 86.2352 | 80.4120 | 75.1819 | 70.4740 | 52.8654 |
| 19.0 | 119.3679 | 109.0250 | 104.3608 | 100.0 | 95.9194 | 92.0976 | 85.1540 | 79.0312 | 73.6131 | 68.8015 | 51.3367 |
| 19.5 | 119.5845 | 109.1148 | 104.4017 | 100.0 | 95.8854 | 92.0357 | 85.0509 | 78.9025 | 73.4701 | 68.6525 | 51.2152 |
| 20.0 | 119.7928 | 109.2008 | 104.4408 | 100.0 | 95.8531 | 91.9769 | 84.9537 | 78.7817 | 73.3366 | 68.5140 | 51.1047 |
| 25.0 | 121.4822 | 109.8810 | 104.7461 | 100.0 | 95.6068 | 91.5342 | 84.2381 | 77.9132 | 72.3985 | 67.5630 | 50.4259 |
| 30.0 | 122.6235 | 110.3190 | 104.9381 | 100.0 | 95.4591 | 91.2751 | 83.8386 | 77.4506 | 71.9216 | 67.1015 | 50.1642 |
| 40.0 | 123.9154 | 110.7827 | 105.1347 | 100.0 | 95.3175 | 91.0345 | 83.4909 | 77.0728 | 71.5560 | 66.7690 | 50.0244 |
| 50.0 | 124.5050 | 110.9749 | 105.2124 | 100.0 | 95.2666 | 90.9521 | 83.3825 | 76.9656 | 71.4615 | 66.6908 | 50.0036 |

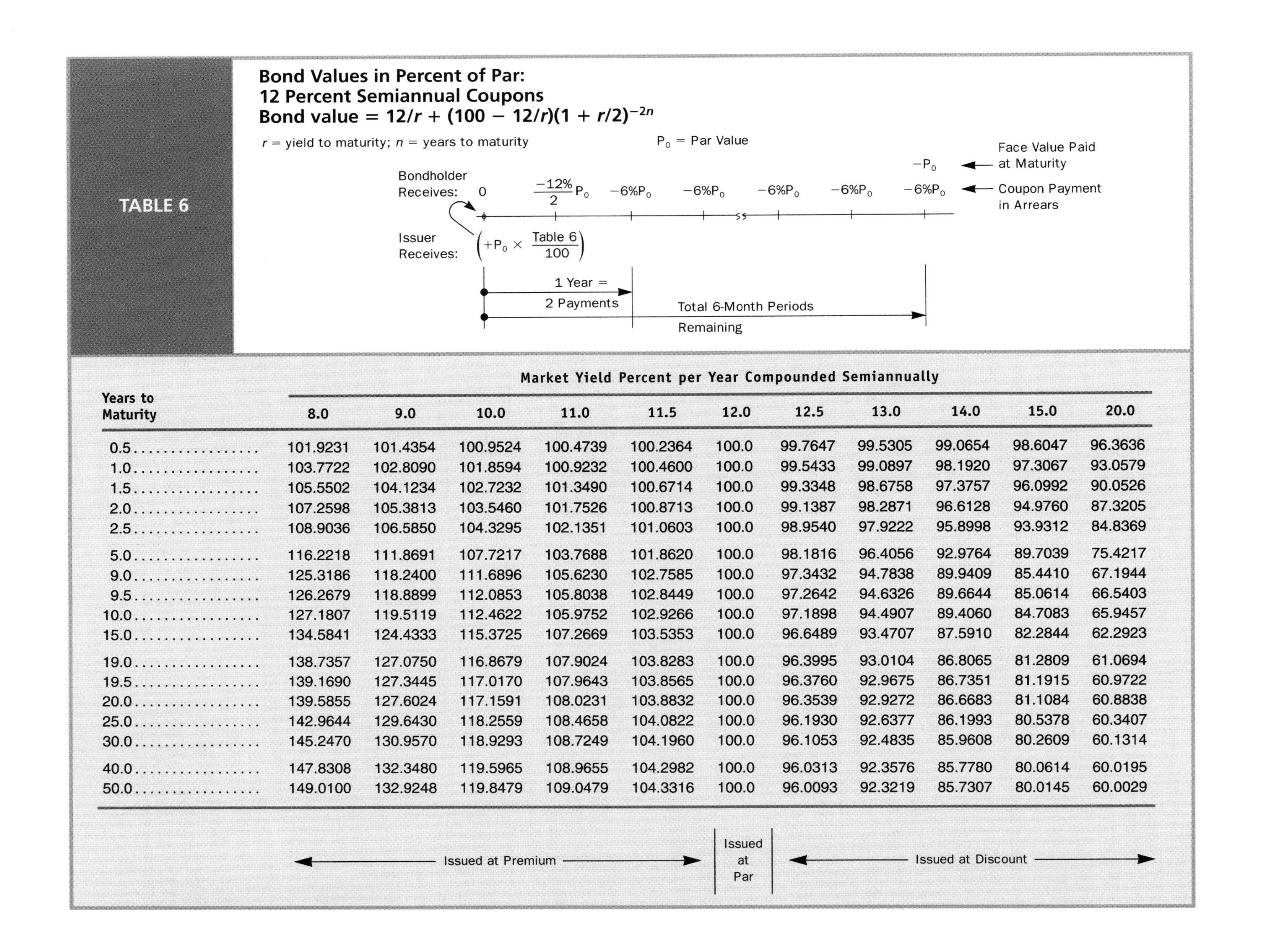

TABLE 6

**Bond Values in Percent of Par:**
**12 Percent Semiannual Coupons**
**Bond value = $12/r + (100 - 12/r)(1 + r/2)^{-2n}$**

$r$ = yield to maturity; $n$ = years to maturity



**Market Yield Percent per Year Compounded Semiannually**

| Years to Maturity | 8.0 | 9.0 | 10.0 | 11.0 | 11.5 | 12.0 | 12.5 | 13.0 | 14.0 | 15.0 | 20.0 |
|---|---|---|---|---|---|---|---|---|---|---|---|
| 0.5 | 101.9231 | 101.4354 | 100.9524 | 100.4739 | 100.2364 | 100.0 | 99.7647 | 99.5305 | 99.0654 | 98.6047 | 96.3636 |
| 1.0 | 103.7722 | 102.8090 | 101.8594 | 100.9232 | 100.4600 | 100.0 | 99.5433 | 99.0897 | 98.1920 | 97.3067 | 93.0579 |
| 1.5 | 105.5502 | 104.1234 | 102.7232 | 101.3490 | 100.6714 | 100.0 | 99.3348 | 98.6758 | 97.3757 | 96.0992 | 90.0526 |
| 2.0 | 107.2598 | 105.3813 | 103.5460 | 101.7526 | 100.8713 | 100.0 | 99.1387 | 98.2871 | 96.6128 | 94.9760 | 87.3205 |
| 2.5 | 108.9036 | 106.5850 | 104.3295 | 102.1351 | 101.0603 | 100.0 | 98.9540 | 97.9222 | 95.8998 | 93.9312 | 84.8369 |
| 5.0 | 116.2218 | 111.8691 | 107.7217 | 103.7688 | 101.8620 | 100.0 | 98.1816 | 96.4056 | 92.9764 | 89.7039 | 75.4217 |
| 9.0 | 125.3186 | 118.2400 | 111.6896 | 105.6230 | 102.7585 | 100.0 | 97.3432 | 94.7838 | 89.9409 | 85.4410 | 67.1944 |
| 9.5 | 126.2679 | 118.8899 | 112.0853 | 105.8038 | 102.8449 | 100.0 | 97.2642 | 94.6326 | 89.6644 | 85.0614 | 66.5403 |
| 10.0 | 127.1807 | 119.5119 | 112.4622 | 105.9752 | 102.9266 | 100.0 | 97.1898 | 94.4907 | 89.4060 | 84.7083 | 65.9457 |
| 15.0 | 134.5841 | 124.4333 | 115.3725 | 107.2669 | 103.5353 | 100.0 | 96.6489 | 93.4707 | 87.5910 | 82.2844 | 62.2923 |
| 19.0 | 138.7357 | 127.0750 | 116.8679 | 107.9024 | 103.8283 | 100.0 | 96.3995 | 93.0104 | 86.8065 | 81.2809 | 61.0694 |
| 19.5 | 139.1690 | 127.3445 | 117.0170 | 107.9643 | 103.8565 | 100.0 | 96.3760 | 92.9675 | 86.7351 | 81.1915 | 60.9722 |
| 20.0 | 139.5855 | 127.6024 | 117.1591 | 108.0231 | 103.8832 | 100.0 | 96.3539 | 92.9272 | 86.6683 | 81.1084 | 60.8838 |
| 25.0 | 142.9644 | 129.6430 | 118.2559 | 108.4658 | 104.0822 | 100.0 | 96.1930 | 92.6377 | 86.1993 | 80.5378 | 60.3407 |
| 30.0 | 145.2470 | 130.9570 | 118.9293 | 108.7249 | 104.1960 | 100.0 | 96.1053 | 92.4835 | 85.9608 | 80.2609 | 60.1314 |
| 40.0 | 147.8308 | 132.3480 | 119.5965 | 108.9655 | 104.2982 | 100.0 | 96.0313 | 92.3576 | 85.7780 | 80.0614 | 60.0195 |
| 50.0 | 149.0100 | 132.9248 | 119.8479 | 109.0479 | 104.3316 | 100.0 | 96.0093 | 92.3219 | 85.7307 | 80.0145 | 60.0029 |

← Issued at Premium → | Issued at Par | ← Issued at Discount →

# GLOSSARY

The definitions of many words and phrases in the glossary use other glossary terms. In a given definition, we *italicize* terms that themselves (or variants thereof) appear elsewhere under their own listings. The cross-references generally take one of two forms:

**1. absorption costing.** See *full absorption costing.*
**2. ABC.** *Activity-based costing.*

Form (1) refers you to another term for discussion of this boldfaced term. Form (2) tells you that this bold-faced term is synonymous with the *italicized* term, which you can consult for discussion if necessary.

## A

**AAA.** *American Accounting Association.*

***Abacus.*** A scholarly journal containing articles on theoretical aspects of accounting, published by Blackwell Publishers for the Accounting Foundation of the University of Sydney.

**abatement.** A complete or partial cancellation of a levy imposed by a government unit.

**ABC.** *Activity-based costing.*

**abnormal spoilage.** Actual spoilage exceeding that expected when operations are normally efficient. Usual practice treats this cost as an *expense* of the period rather than as a *product cost.* Contrast with *normal spoilage.*

**aboriginal cost.** In public utility accounting, the *acquisition cost* of an *asset* incurred by the first *entity* devoting that asset to public use; the cost basis for most public utility regulation. If regulators used a different cost basis, then public utilities could exchange assets among themselves at ever-increasing prices in order to raise the rate base and, then, prices based on them.

**absorbed overhead.** *Overhead costs* allocated to individual products at some *overhead rate;* also called *applied overhead.*

**absorption costing.** See *full absorption costing.*

***Abstracts of the EITF.*** See *Emerging Issues Task Force.*

**Accelerated Cost Recovery System (ACRS).** A form of accelerated depreciation that Congress enacted in 1981 and amended in 1986, so that now most writers refer to it as *MACRS,* or *Modified Accelerated Cost Recovery System.* The system provides percentages of the asset's cost that a firm depreciates each year for tax purposes. The percentages derive, roughly, from 150-percent *declining-balance depreciation* methods. ACRS ignores salvage value. We do not generally use these amounts for *financial accounting.*

**accelerated depreciation.** In calculating *depreciation* charges, any method in which the charges become progressively smaller each period. Examples are *double declining-balance depreciation* and *sum-of-the-years'-digits depreciation* methods.

**acceptance.** A written promise to pay; equivalent to a *promissory note.*

**account.** A device for representing the amount (*balance*) for any line (or a part of a line) in the *balance sheet* or *income statement.* Because income statement accounts explain the changes in the balance sheet account Retained Earnings, the definition does not require the last three words of the preceding sentence. An account is any device for accumulating additions and subtractions relating to a single *asset, liability,* or *owners' equity* item, including *revenues* and *expenses.*

**account analysis method.** A method of separating *fixed costs* from *variable costs* based on the analyst's judgment of whether the cost is fixed or variable. Based on their names alone, the analyst might classify *direct labor* (*materials*) *costs* as variable and *depreciation* on a factory building as fixed. In our experience, this method results in too many fixed costs and not enough variable costs—that is, analysts have insufficient information to judge management's ability to reduce costs that appear to be fixed.

**account form.** The form of *balance sheet* in which *assets* appear on the left and *equities* appear on the right. Contrast with *report form.* See *T-account.*

**accountability center.** *Responsibility center.*

**accountancy.** The British word for *accounting.* In the United States, it means the theory and practice of accounting.

**accountant's comments.** Canada: a written communication issued by a public accountant at the conclusion of a review engagement. It consists of a description of the work performed and a statement that, under the terms of the engagement, the accountant has not performed an audit and consequently expresses no opinion. (Compare *auditor's report; denial of opinion.*)

**accountant's opinion.** *Auditor's report.*

**accountant's report.** *Auditor's report.*

**accounting.** A system conveying information about a specific *entity.* The information is in financial terms and will appear in accounting statements only if the accountant can measure it with reasonable precision. The *AICPA* defines accounting as a service activity whose "function is to provide quantitative information, primarily financial in nature, about economic entities that is intended to be useful in making economic decisions."

**accounting adjustments.** *Prior-period adjustments,* changes in accounting principles accounted for on a cumulative basis, and corrections of errors. See *accounting changes.* The *FASB* indicates that it will tend to call these items "accounting adjustments," not "accounting changes," when it requires the reporting of *comprehensive income.*

***Accounting and Tax Index.*** A publication that indexes, in detail, the accounting literature of the period. Published by UMI, a subsidiary of ProQuest Company.

**accounting changes.** As defined by *APB Opinion No. 20,* a change in (1) an *accounting principle* (such as a switch from *FIFO* to *LIFO* or from *sum-of-the-years'-digits depreciation* to *straight-line depreciation*), (2) an accounting estimate (such as estimated useful lives or salvage value of depreciable assets and estimates of *warranty* costs or *uncollectible accounts*), or (3) the reporting *entity.* The firm should disclose changes of type (1). It should include in reported earnings for the period of change the cumulative effect of the change on *retained earnings* at the start of the period during which it made the change. The firm should treat changes of type (2) as affecting only the period of change and, if necessary, future periods. The firm should disclose reasons for changes of type (3) in statements reporting on operations of the period of the change, and it should show the effect of the change on all other periods, for comparative purposes. In some cases (such as a change from *LIFO* to other inventory *flow assumptions* or a change in the method of accounting for long-term construction contracts), *GAAP* treat changes of type (1) like changes of type (3). That is, for these changes the firm should restate all statements shown for prior periods to show the effect of adopting the change for those periods as well. See *all-inclusive (income) concept* and *accounting errors.*

**accounting conventions.** Methods or procedures used in accounting. Writers tend to use this term when the method or procedure has not yet received official authoritative sanction by a pronouncement of a group such as the *APB, EITF, FASB,* or *SEC.* Contrast with *accounting principles.*

**accounting cycle.** The sequence of accounting procedures starting with *journal entries* for various transactions and events and ending with the *financial statements* or, perhaps, the *post-closing trial balance.*

**accounting deficiency.** Canada: a failure to adhere to generally accepted *accounting principles* or to disclose essential information in *financial statements.*

**accounting entity.** See *entity.*

**accounting equation.** *Assets = Equities; Assets = Liabilities + Owners' Equity.*

**accounting errors.** Arithmetic errors and misapplications of *accounting principles* in previously published financial statements. The firm corrects these during the current period with direct *debits* or *credits* to *retained earnings.* In this regard, the firm treats them like *prior-period adjustments,* but technically *APB Opinion No. 9* does not classify them as prior-period adjustments. See *accounting changes,* and contrast with changes in accounting estimates as described there.

**accounting event.** Any occurrence that is recorded in the accounting records.

***Accounting Horizons.*** Quarterly journal of the *American Accounting Association.*

**accounting methods.** *Accounting principles;* procedures for carrying out accounting principles.

**accounting period.** The time period between consecutive *balance sheets;* the time period for which the firm prepares *financial statements* that measure *flows,* such as the *income statement* and the *statement of cash flows.* See *interim statements.*

**accounting policies.** *Accounting principles* adopted by a specific *entity.*

**accounting principles.** The methods or procedures used in accounting for events reported in the *financial statements.* We tend to use this term when the method or procedure has received official authoritative sanction from a pronouncement of a group such as the *APB, EITF, FASB,* or *SEC.* Contrast with *accounting conventions* and *conceptual framework.*

**Accounting Principles Board.** See *APB.*

**accounting procedures.** See *accounting principles.* However, this term usually refers to the methods for implementing accounting principles.

**accounting rate of return.** Income for a period divided by average investment during the period; based on income, rather than discounted cash flows, and hence a poor decision-making aid or tool. See *ratio.*

***Accounting Research Bulletin (ARB).*** The name of the official pronouncements of the former *Committee on Accounting Procedure (CAP)* of the AICPA. The committee issued fifty-one bulletins between 1939 and 1959. *ARB No. 43* restated and codified the parts of the first forty-two bulletins not dealing solely with definitions.

***Accounting Research Study (ARS).*** One of a series of studies published by the Director of Accounting Research of the *AICPA* and "designed to provide professional accountants and others interested in the development of accounting with a discussion and documentation of accounting problems." The AICPA published fifteen such studies in the period 1961–73.

***Accounting Review.*** Scholarly publication of the *American Accounting Association.*

***Accounting Series Release (ASR).*** See *SEC.*

**accounting standards.** *Accounting principles.*

**Accounting Standards Executive Committee (AcSEC).** The senior technical committee of the *AICPA* authorized to speak for the AICPA in the areas of *financial accounting* and reporting as well as *cost accounting.*

**accounting system.** The procedures for collecting and summarizing financial data in a firm.

***Accounting Terminology Bulletin (ATB).*** One of four releases of the Committee on Terminology of the *AICPA* issued in the period 1953–57.

***Accounting Trends and Techniques.*** An annual *AICPA* publication that surveys the reporting practices of 600 large corporations. It presents tabulations of specific practices, terminology, and disclosures along with illustrations taken from individual annual reports.

**accounts payable.** A *liability* representing an amount owed to a *creditor;* usually arising from the purchase of *merchandise* or materials and supplies, not necessarily due or past due; normally, a *current liability.*

**accounts receivable.** Claims against a *debtor;* usually arising from sales or services rendered, not necessarily due or past due; normally, a *current asset.*

**accounts receivable turnover.** Net sales on account divided by average accounts receivable. See *ratio.*

**accretion.** Occurs when a *book value* grows over time, such as a *bond* originally issued at a *discount;* the correct technical term is "accretion," not "amortization." This term also refers to an increase in economic worth through physical change caused by natural growth, usually said of a natural resource such as timber. Contrast with *appreciation.* See *amortization.*

**accrual.** Recognition of an *expense* (or *revenue*) and the related *liability* (or *asset*) resulting from an *accounting event,* frequently from the passage of time but not signaled by an explicit cash transaction; for example, the recognition of interest expense or revenue (or wages, salaries, or rent) at the end of a period even though the firm makes no explicit cash transaction at that time. Cash flow follows accounting recognition; contrast with *deferral.*

**accrual basis of accounting.** The method of recognizing *revenues* as a firm sells *goods* (or delivers them) and as it renders *services,* independent of the time when it receives cash. This system recognizes *expenses* in the period when it recognizes the related revenue, independent of the time when it pays cash. *SFAC No. 1* says, "Accrual accounting attempts to record the financial effects on an enterprise of transactions and other events and circumstances that have cash consequences for the enterprise in the periods in which those transactions, events, and circumstances occur rather than only in the periods in which cash is received or paid by the enterprise." Contrast with the *cash basis of accounting.* See *accrual* and *deferral.* We could more correctly call this "accrual/deferral" accounting.

**accrue.** See *accrued,* and contrast with *incur.*

**accrued.** Said of a *revenue* (*expense*) that the firm has earned (recognized) even though the related *receivable* (*payable*) has a future due date. We prefer not to use this adjective as part of an account title. Thus, we prefer to use Interest Receivable (Payable) as the account title rather than Accrued Interest Receivable (Payable). See *matching convention* and *accrual.* Contrast with *incur.*

**accrued depreciation.** An incorrect term for *accumulated depreciation.* Acquiring an asset with cash, capitalizing it, and then amortizing its cost over periods of use is a process of *deferral* and allocation, not of *accrual.*

**accrued payable.** A *payable* usually resulting from the passage of time. For example, *salaries* and *interest* accrue as time passes. See *accrued.*

**accrued receivable.** A *receivable* usually resulting from the passage of time. See *accrued.*

**accumulated benefit obligation.** See *projected benefit obligation* for definition and contrast.

**accumulated depreciation.** A preferred title for the asset *contra account* that shows the sum of *depreciation* charges on an asset since the time the firm acquired it. Other account titles are *allowance* for *depreciation* (acceptable term) and *reserve* for *depreciation* (unacceptable term).

**accumulated other comprehensive income.** *Balance sheet* amount in *owners' equity* showing the total of all *other comprehensive income* amounts from all prior periods.

**accurate presentation.** The qualitative accounting objective suggesting that information reported in financial statements should correspond as precisely as possible with the economic effects underlying transactions and events. See *fair presentation* and *full disclosure.*

**acid test ratio.** *Quick ratio.*

**acquisition cost.** Of an *asset,* the net *invoice* price plus all *expenditures* to place and ready the asset for its intended use. The other expenditures might include legal fees, transportation charges, and installation costs.

**ACRS.** *Accelerated Cost Recovery System.*

**AcSEC.** *Accounting Standards Executive Committee* of the *AICPA.*

**activity accounting.** *Responsibility accounting.*

**activity-based costing (ABC).** Method of assigning *indirect costs,* including nonmanufacturing *overhead costs,* to products and services. ABC assumes that almost all overhead costs associate with activities within the firm and vary with respect to the *drivers* of those activities. Some practitioners suggest that ABC attempts to find the drivers for all indirect costs; these people note that in the long run, all costs are *variable,* so *fixed* indirect costs do not occur. This method first assigns costs to activities and then to products based on the products' usage of the activities.

**activity-based depreciation.** *Production method* (*depreciation*).

**activity-based management (ABM).** Analysis and management of activities required to make a product or to produce a service. ABM focuses attention to enhance activities that add value to the customer and to reduce activities that do not. Its goal is to satisfy customer needs while making smaller demands on costly resources. Some refer to this as "activity management."

**activity basis.** *Costs* are *variable* or *fixed* (*incremental* or *unavoidable*) with respect to some activity, such as production of units (or the undertaking of some new project). Usage calls this activity the "activity basis."

**activity center.** Unit of the organization that performs a set of tasks.

**activity variance.** *Sales volume variance.*

**actual cost (basis).** *Acquisition* or *historical cost.* Also contrast with *standard cost.*

**actual costing (system).** Method of allocating costs to products using actual *direct materials,* actual *direct labor,* and actual *factory overhead.* Contrast with *normal costing* and *standard costing.*

**actuarial.** An adjective describing computations or analyses that involve both *compound interest* and probabilities, such as the computation of the *present value* of a life-contingent *annuity.* Some writers use the word even for computations involving only one of the two.

**actuarial accrued liability.** A 1981 report of the Joint Committee on Pension Terminology (of various actuarial societies) agreed to use this term rather than *prior service cost.*

**ad valorem.** A method of levying a tax or duty on goods by using their estimated value as the tax base.

**additional paid-in capital.** An alternative acceptable title for the *capital contributed in excess of par* (or *stated*) *value account.*

**additional processing cost.** *Costs* incurred in processing *joint products* after the *split-off point.*

**adequate disclosure.** An auditing standard that, to achieve *fair presentation* of *financial statements,* requires *disclosure* of *material* items. This *auditing standard* does not, however, require publicizing all information detrimental to a company. For example, the company may face a lawsuit, and disclosure might require a *debit* to a *loss* account and a *credit* to an *estimated liability.* But the court might view the making of this entry as an admission of liability, which could adversely affect the outcome of the suit. The firm should debit expense or loss for the expected loss, as required by *SFAS No. 5,* but need not use such accurate account titles that the court can spot an admission of liability.

**adjunct account.** An *account* that accumulates additions to another account. For example, Premium on Bonds Payable is adjunct to the liability Bonds Payable; the effective liability is the sum of the two account balances at a given date. Contrast with *contra account.*

**adjusted acquisition (historical) cost.** Sometimes said of the *book value* of a *plant asset,* that is, *acquisition cost* less *accumulated depreciation.* Also, cost adjusted to a *constant-dollar* amount to reflect *general price-level changes.*

**adjusted bank balance of cash.** The *balance* shown on the statement from the bank plus or minus amounts, such as for unrecorded deposits or outstanding checks, to reconcile the bank's balance with the correct cash balance. See *adjusted book balance of cash.*

**adjusted basis.** The *basis* used to compute gain or loss on the disposition of an *asset* for tax purposes. See also *book value.*

**adjusted book balance of cash.** The *balance* shown in the firm's account for cash in bank plus or minus amounts, such as for *notes* collected by the bank or bank service charges, to reconcile the account balance with the correct cash balance. See *adjusted bank balance of cash.*

**adjusted trial balance.** *Trial balance* taken after *adjusting entries* but before *closing entries.* Contrast with *pre-*and *post-closing trial balances.* See *unadjusted trial balance* and *post-closing trial balance.* See also *work sheet.*

**adjusting entry.** An entry made at the end of an *accounting period* to record a *transaction* or other *accounting event* that the firm has not yet recorded or has improperly recorded during the accounting period; an entry to update the accounts. See *work sheet.*

**adjustment.** An *account* change produced by an *adjusting entry.* Sometimes accountants use the term to refer to the process of restating *financial statement* amounts to *constant dollars.*

**administrative costs (expenses).** *Costs* (*expenses*) incurred for the firm as a whole, in contrast with specific functions such as manufacturing or selling; includes items such as salaries of top executives, general office rent, legal fees, and auditing fees.

**admission of partner.** Occurs when a new partner joins a *partnership.* Legally, the old partnership dissolves, and a new one comes into being. In practice, however, the firm may keep the old accounting records in use, and the accounting entries reflect the manner in which the new partner joined the firm. If the new partner merely purchases the interest of another partner, the accounting changes the name for one capital account. If the new partner contributes *assets* and *liabilities* to the partnership, then the firm must recognize them. See *bonus method.*

**ADR.** See *asset depreciation range.*

**advances from (by) customers.** A preferred title for the *liability* account representing *receipts* of *cash* in advance of delivering the *goods* or rendering the *service.* After the firm delivers the goods or services, it will recognize *revenue.* Some refer to this as "deferred revenue" or "deferred income," terms likely to confuse the unwary because the item is not yet *revenue* or *income.*

**advances to affiliates.** *Loans* by a parent company to a *subsidiary;* frequently combined with "investment in subsidiary" as "investments and advances to subsidiary" and shown as a *noncurrent asset* on the parent's *balance sheet.* The consolidation process eliminates these advances in *consolidated financial statements.*

**advances to suppliers.** A preferred term for the *asset* account representing *disbursements* of cash in advance of receiving *assets* or *services.*

**adverse opinion.** An *auditor's report* stating that the financial statements are not fair or are not in accord with *GAAP.*

**affiliated company.** A company controlling or controlled by another company.

**after closing.** Post-closing; a *trial balance* at the end of the period.

**after cost.** *Expenditures* to be made after *revenue* recognition. For example, *expenditures* for *repairs* under warranty are after cost. Proper recognition of after cost involves a debit to expense at the time of the sale and a credit to an *estimated liability.* When the firm discharges the liability, it debits the estimated liability and credits the assets consumed.

**AG (Aktiengesellschaft).** Germany: the form of a German company whose shares can trade on the stock exchange.

**agency fund.** An account for *assets* received by governmental units in the capacity of trustee or agent.

**agency theory.** A branch of economics relating the behavior of *principals* (such as owner nonmanagers or bosses) and that of their *agents* (such as nonowner managers or subordinates). The principal assigns responsibility and authority to the agent, but the agent's own risks and preferences differ from those of the principal. The principal cannot observe all activities of the agent. Both the principal and the agent must consider the differing risks and preferences in designing incentive contracts.

**agent.** One authorized to transact business, including executing contracts, for another.

**aging accounts receivable.** The process of classifying *accounts receivable* by the time elapsed since the claim came into existence for the purpose of estimating the amount of uncollectible accounts receivable as of a given date. See *sales contra, estimated uncollectibles,* and *allowance for uncollectibles.*

**aging schedule.** A listing of *accounts receivable,* classified by age, used in *aging accounts receivable.*

**AICPA (American Institute of Certified Public Accountants).** The national organization that represents *CPA*s. See *AcSEC.* It oversees the writing and grading of the Uniform CPA Examination. Each state sets its own requirements for becoming a CPA in that state. See *certified public accountant.* Web site: www.aicpa.org.

**all-capital earnings rate.** *Rate of return on assets.*

**all-current method.** *Foreign currency translation* in which all *financial statement* items are translated at the *current exchange rate.*

**all-inclusive (income) concept.** A concept that does not distinguish between *operating* and *nonoperating revenues* and *expenses.* Thus, the only entries to retained earnings are for *net income* and *dividends.* Under this concept, the *income statement* reports all *income, gains,* and *losses;* thus, net income includes events usually reported as *prior-period adjustments* and as *corrections of errors. GAAP* do not include this concept in its pure form, but *APB Opinions No. 9* and *No. 30* move far in this direction. They do permit retained earnings entries for prior-period adjustments and correction of errors.

**allocate.** To divide or spread a *cost* from one *account* into several accounts, to several products or activities, or to several periods.

**allocation base.** The systematic method that assigns *joint costs* to *cost objectives.* For example, a firm might assign the cost of a truck to periods based on miles driven during the period; the allocation base is miles. Or the firm might assign the cost of a factory supervisor to a product based on *direct labor* hours; the allocation base is direct labor hours.

**allocation of income taxes.** See *deferred income tax.*

**allowance.** A balance sheet *contra account* generally used for *receivables* and depreciable assets. See *sales* (or *purchase*) *allowance* for another use of this term.

**allowance for funds used during construction.** In accounting for public utilities, a *revenue* account *credited* for *implicit interest* earnings on *shareholders' equity* balances. One principle of public utility regulation and rate setting requires that customers should pay the full costs of producing the services (e.g., electricity) that they use, nothing more and nothing less. Thus, an electric utility must capitalize into an *asset* account the full costs, but no more, of producing a new electric power-generating plant. One of the costs of building a new plant is the *interest* cost on cash tied up during construction. If *funds* are explicitly borrowed by an ordinary business, the journal entry for interest of $1,000 is typically:

| | | |
|---|---|---|
| Interest Expense . . . . . . . . . . . . . . . . | 1,000 | |
| Interest Payable . . . . . . . . . . . . . | | 1,000 |
| Interest expense for the period. | | |

If the firm is constructing a new plant, then another entry would be made, capitalizing interest into the plant-under-construction account:

| | | |
|---|---|---|
| Construction Work-in-Progress . . . . . . | 750 | |
| Interest Expense . . . . . . . . . . . . . | | 750 |
| Capitalize relevant portion of interest relating to construction work in progress into the asset account. | | |

The cost of the *plant asset* increases; when the firm uses the plant, it charges *depreciation.* The interest will become an expense through the depreciation process in the later periods of use, not currently as the firm pays for interest. Thus, the firm reports the full cost of the electricity generated during a given period as expense in that period. But suppose, as is common, that the electric utility does not explicitly borrow the funds but uses some of its own funds, including funds raised from equity issues as well as from debt. Even though the firm incurs no explicit interest expense or other explicit expense for capital, the funds have an *opportunity cost.* Put another way, the plant under construction will not have lower economic cost just because the firm used its own cash rather than borrowing. The public utility using its own funds, on which it would have to pay $750 of interest if it had explicitly borrowed the funds, will make the following entry:

| | | |
|---|---|---|
| Construction Work-in-Progress . . . . . . . | 750 | |
| Allowance for Funds Used during Construction . . . . . . . . . | | 750 |
| Recognition of interest, an opportunity cost, on own funds used. | | |

The allowance account is a form of *revenue,* to appear on the income statement, and the firm will close it to Retained Earnings, increasing it. On the *statement of cash flows* it is an income or revenue item not producing funds, and so the firm must subtract it from net income in deriving *cash provided by operations*. *SFAS No. 34* specifically prohibits nonutility companies from capitalizing, into plant under construction, the opportunity cost (interest) on their own funds used.

**allowance for uncollectibles (accounts receivable).** A *contra account* that shows the estimated *accounts receivable* amount that the firm expects not to collect. When the firm uses such an allowance, the actual write-off of specific accounts receivable (*debit* allowance, *credit* specific customer's account) does not affect *revenue* or *expense* at the time of the write-off. The firm reduces revenue when it debits *bad debt expense* (or, our preference, a revenue contra account) and credits the allowance; the firm can base the amount of the credit to the allowance on a percentage of sales on account for a period of time or compute it from *aging accounts receivable*. This contra account enables the firm to show an estimated receivables amount that it expects to collect without identifying specific uncollectible accounts. See *allowance method*.

**allowance method.** A method of attempting to match all *expenses* of a transaction with their associated *revenues;* usually involves a debit to expense and a credit to an *estimated liability,* such as for estimated warranty expenditures, or a debit to a revenue (*contra*) account and a credit to an asset (*contra*) account, such as in some firms' accounting for uncollectible accounts. See *allowance for uncollectibles* for further explanation. When the firm uses the allowance method for *sales discounts,* the firm records sales at gross invoice prices (not reduced by the amounts of discounts made available). The firm *debits* an estimate of the amount of discounts to be taken to a revenue contra account and *credits* an allowance account, shown contra to *accounts receivable*.

**American Accounting Association (AAA).** An organization primarily for academic accountants but open to all interested in accounting. It publishes the *Accounting Review* and several other journals.

**American Institute of Certified Public Accountants.** See *AICPA*.

**American Stock Exchange (AMEX) (ASE).** A public market where various corporate *securities* are traded.

**AMEX.** *American Stock Exchange*.

**amortization.** Strictly speaking, the process of liquidating or extinguishing ("bringing to death") a *debt* with a series of payments to the *creditor* (or to a *sinking fund*). From that usage has evolved a related use involving the accounting for the payments themselves: "amortization schedule" for a mortgage, which is a table showing the allocation between *interest* and *principal*. The term has come to mean writing off ("liquidating") the cost of an asset. In this context it means the general process of *allocating* the *acquisition cost* of an asset either to the periods of benefit as an *expense* or to *inventory* accounts as a *product cost*. This is called *depreciation* for *plant assets, depletion* for *wasting assets* (natural resources), and "amortization" for *intangibles*. *SFAC No. 6* refers to amortization as "the accounting process of reducing an amount by periodic payments or write-downs." The expressions "unamortized debt discount or premium" and "to amortize debt discount or premium" relate to *accruals,* not to *deferrals*. The expressions "amortization of long-term assets" and "to amortize long-term assets" refer to deferrals, not accruals. Contrast with *accretion*.

**amortized cost.** A measure required by *SFAS No. 115* for *held-to-maturity securities*. This amount results from applying the method described at *effective interest method*. The firm records the security at its initial cost and computes the *effective interest rate* for the security. Whenever the firm receives cash from the issuer of the security or whenever the firm reaches the end of one of its own *accounting periods* (that is, reaches the time for its own *adjusting entries*), it takes the following steps. It multiplies the amount currently recorded on the books by the effective interest rate (which remains constant over the time the firm holds the security). It debits that amount to the debt security account and credits the amount to Interest Revenue. If the firm receives cash, it debits Cash and credits the debt security account. The firm recomputes the book value of the debt security as the book value before these entries plus the increase for the interest revenue less the decrease for the cash received. The resulting amount is the amortized cost for the end of that period.

**analysis of variances.** See *variance analysis*.

**annual report.** A report prepared once a year for shareholders and other interested parties. It includes a *balance sheet,* an *income statement,* a *statement of cash flows,* a reconciliation of changes in *owners' equity* accounts, a *summary of significant accounting principles,* other explanatory *notes,* the *auditor's report,* and comments from management about the year's events. See *10-K* and *financial statements*.

**annuitant.** One who receives an *annuity*.

**annuity.** A series of payments of equal amount, usually made at equally spaced time intervals.

**annuity certain.** An *annuity* payable for a definite number of periods. Contrast with *contingent annuity*.

**annuity due.** An *annuity* whose first payment occurs at the start of period 1 (or at the end of period 0). Contrast with *annuity in arrears*.

**annuity in advance.** An *annuity due*.

**annuity in arrears.** An *ordinary annuity* whose first payment occurs at the end of the first period.

**annuity method of depreciation.** See *compound interest depreciation*.

**antidilutive.** Said of a *potentially dilutive* security that will increase *earnings per share* if its holder *exercises* it or *converts* it into common stock. In computing *diluted earnings per share,* the firm must assume that holders of antidilutive securities will not exercise their options or convert securities into common shares. The opposite assumption would lead to increased reported earnings per share in a given period.

**APB.** Accounting Principles Board of the *AICPA*. It set *accounting principles* from 1959 through 1973, issuing 31 *APB Opinions* and 4 *APB Statements*. The *FASB* superseded it.

***APB Opinion.*** The name for the APB pronouncements that compose much of *generally accepted accounting principles;* the APB issued 31 APB Opinions from 1962 through 1973.

***APB Statement.*** The *APB* issued four *APB Statements* between 1962 and 1970. The *Statements* were approved by at least two-thirds of the board, but they state recommendations, not requirements. For example, *Statement No. 3* (1969) suggested the publication of *constant-dollar* financial statements but did not require them.

***APBs.*** An abbreviation used for *APB Opinions*.

**applied cost.** A *cost* that a firm has *allocated* to a department, product, or activity; not necessarily based on actual costs incurred.

**applied overhead.** *Overhead costs* charged to departments, products, or activities. Also called *absorbed overhead*.

**appraisal.** In valuing an *asset* or *liability,* a process that involves expert opinion rather than evaluation of explicit market transactions.

**appraisal method of depreciation.** The periodic *depreciation* charge that equals the difference between the beginning-of-period and the end-of-period appraised values of the *asset* if that difference is positive. If negative, there is no charge. Not based on *historical cost,* this method is thus not generally accepted.

**appreciation.** An increase in economic value caused by rising market prices for an *asset*. Contrast with *accretion*.

**appropriated retained earnings.** See *retained earnings, appropriated*.

**appropriation.** In governmental accounting, an *expenditure* authorized for a specified amount, purpose, and time.

**appropriation account.** In governmental accounting, an account set up to record specific authorizations to spend. The governmental unit credits this account with appropriation amounts. At the end of the period, the unit closes to (debits) this account all *expenditures* during the period and all *encumbrances* outstanding at the end of the period.

**approximate net realizable value method.** A method of assigning joint costs to *joint products* based on revenues minus *additional processing costs* of the end products.

***ARB.*** *Accounting Research Bulletin*.

**arbitrage.** Strictly speaking, the simultaneous purchase in one market and sale in another of a *security* or commodity in hope of making a *profit* on price differences in the different markets. Often writers use this term loosely when a trader sells an item that is somewhat different from the item purchased; for example, the sale of shares of common stock and the simultaneous purchase of a *convertible bond* that is convertible into identical common shares. The trader hopes that the market will soon see that the similarities of the items should make them have equal market values. When the market values converge, the trader closes the positions and profits from the original difference in prices, less trading costs.

**arbitrary.** Having no causation basis. Accounting theorists and practitioners often, properly, say, "Some cost allocations are arbitrary." In that sense, the accountant does not mean that the allocations are capricious or haphazard but does mean that theory suggests no unique solution to the allocation problem at hand. Accountants require that arbitrary allocations be systematic, rational, and consistently followed over time.

**arm's length.** A transaction negotiated by unrelated parties, both acting in their own self-interests; the basis for a *fair market value* estimation or computation.

**arrears.** *Cumulative dividends* that the firm has not yet declared. See *annuity in arrears* for another context.

***ARS.*** *Accounting Research Study*.

**articles of incorporation.** Document filed with state authorities by persons forming a corporation. When the state returns the document with a certificate of incorporation, the document becomes the corporation's *charter*.

**articulate.** The relation between any operating statement (for example, *income statement* or *statement of cash flows*) and comparative balance sheets, where the operating statement explains (or reconciles) the change in some major balance sheet category (for example, *retained earnings* or *working capital*).

**ASE.** *American Stock Exchange*.

***ASR.*** *Accounting Series Release*.

**assess.** To value property for the purpose of property taxation; to levy a charge on the owner of property for improvements thereto, such as for sewers or sidewalks. The taxing authority computes the assessment.

**assessed valuation.** For real estate or other property, a dollar amount that a government uses as a basis for levying taxes. The amount need not have some relation to *market value*.

**asset.** *SFAC No. 6* defines assets as "probable future economic benefits obtained or controlled by a particular entity as a result of past transactions. . . . An asset has three essential characteristics: (a) it embodies a probable future benefit that involves a capacity, singly or in combination with other assets, to contribute directly or indirectly to future net cash inflows, (b) a particular entity can obtain the benefit and control others' access to it, and (c) the transaction or other event giving rise to the entity's right to or control of the benefit has already occurred." A footnote points out that "probable" means that which we can reasonably expect or believe but that is not certain or proved. You may understand condition (c) better if you think of it as requiring that a future benefit cannot be an asset if it arises from an *executory contract,* a mere exchange of promises. Receiving a purchase order from a customer provides a future benefit, but it is an executory contract, so the order cannot be an asset. An asset may be *tangible* or *intangible,* short-term (current) or long-term (noncurrent).

**asset depreciation range (ADR).** The range of *depreciable lives* allowed by the *Internal Revenue Service* for a specific depreciable *asset.*

**asset turnover.** Net sales divided by average assets. See *ratio.*

**assignment of accounts receivable.** Transfer of the legal ownership of an account receivable through its sale. Contrast with *pledging* accounts receivable, where the receivables serve as *collateral* for a *loan.*

***ATB.*** *Accounting Terminology Bulletin.*

**at par.** A *bond* or *preferred shares* issued (or selling) at *face amount.*

**attachment.** The laying claim to the *assets* of a borrower (or debtor) by a lender (or creditor) when the borrower has failed to pay debts on time.

**attest.** An auditor's rendering of an *opinion* that the *financial statements* are fair. Common usage calls this procedure the "attest function" of the CPA. See *fair presentation.*

**attestor.** Typically independent *CPAs,* who *audit financial statements* prepared by management for the benefit of users. The *FASB* describes accounting's constituency as comprising preparers, attestors, and users.

**attribute measured.** The particular *cost* reported in the balance sheet. When making physical measurements, such as of a person, one needs to decide the units with which to measure, such as inches or centimeters or pounds or grams. One chooses the attribute height or weight independently of the measuring unit, English or metric. Conventional accounting uses *historical cost* as the attribute measured and *nominal dollars* as the measuring unit. Some theorists argue that accounting would better serve readers if it used *current cost* as the attribute measured. Others argue that accounting would better serve readers if it used *constant dollars* as the measuring unit. Some, including us, think accounting should change both the measuring unit and the attribute measured. One can measure the attribute historical cost in nominal dollars or in constant dollars. One can also measure the attribute current cost in nominal dollars or constant dollars. Choosing between the two attributes and the two measuring units implies four different accounting systems. Each of these four has its uses.

**attribute(s) sampling.** The use of sampling technique in which the observer assesses each item selected on the basis of whether it has a particular qualitative characteristic in order to ascertain the rate of occurrence of this characteristic in the population. See also *estimation sampling.* Compare *variables sampling.* Example of attributes sampling: take a sample population of people, note the fraction that is male (say, 40 percent), and then infer that the entire population contains 40 percent males. Example of variables sampling: take a sample population of people, observe the weight of each sample point, compute the mean of those sampled people's weights (say 160 pounds), and then infer that the mean weight of the entire population equals 160 pounds.

**audit.** Systematic inspection of accounting records involving analyses, tests, and *confirmations.* See *internal audit.*

**audit committee.** A committee of the board of directors of a *corporation,* usually comprising outside directors, who nominate the independent auditors and discuss the auditors' work with them. If the auditors believe the shareholders should know about certain matters, the auditors, in principle, first bring these matters to the attention of the audit committee; in practice, the auditors may notify management before they notify the audit committee.

***Audit Guides.*** See *Industry Audit Guides.*

**audit program.** The procedures followed by the *auditor* in carrying out the *audit.*

**audit trail.** A reference accompanying an entry, or *post,* to an underlying source record or document. Efficiently checking the accuracy of accounting entries requires an audit trail. See *cross-reference.*

***Auditing Research Monograph.*** Publication series of the *AICPA.*

**auditing standards.** Standards promulgated by the *PCAOB* for auditors to follow in carrying out their attest functions. The PCAOB began operations in earnest in 2003, and initially has said that it would use the standards originally promulgated by the *AICPA,* including general standards, standards of field work, and standards of reporting. According to the AICPA, these standards "deal with the measures of the quality of the performance and the objectives to be attained" rather than with specific auditing procedures. As time passes, the PCAOB will substitute its rules for those of the AICPA.

**Auditing Standards Board.** *AICPA* operating committee that promulgates auditing rules. The new operations of the *PCAOB,* after 2003, render uncertain what this Board will do.

**auditor.** Without a modifying adjective, usually refers to an external auditor—one who checks the accuracy, fairness, and general acceptability of accounting records and statements and then *attests* to them. See *internal auditor.*

**auditor's opinion.** *Auditor's report.*

**auditor's report.** The auditor's statement of the work done and an opinion of the *financial statements.* The auditor usually gives unqualified ("clean") opinions but may qualify them, or the auditor may disclaim an opinion in the report. Often called the "accountant's report." See *adverse opinion.*

**AudSEC.** The former Auditing Standards Executive Committee of the *AICPA,* now functioning as the *Auditing Standards Board.*

**authorized capital stock.** The number of *shares* of stock that a corporation can issue; specified by the *articles of incorporation.*

**available for sale, securities.** *Marketable securities* a firm holds that are classified as neither *trading securities* nor *held-to-maturity* (*debt*) *securities.* This classification is important in *SFAS No. 115,* which requires the owner to carry marketable equity securities on the balance sheet at market value, not at cost. Under *SFAS No. 115,* the income statement reports *holding gains and losses* on trading securities but not on securities available for sale. The required accounting *credits* (*debits*) holding gains (losses) on securities available for sale directly to an *owners' equity* account. On sale, the firm reports realized gain or loss as the difference between the selling price and the original cost, for trading securities, and as the difference between the selling price and the book value at the beginning of the period of sale, for securities available for sale and for debt securities held to maturity. By their nature, however, the firm will only rarely sell debt securities "held to maturity."

**average.** The arithmetic mean of a set of numbers; obtained by summing the items and dividing by the number of items.

**average collection period of receivables.** See *ratio.*

**average-cost flow assumption.** An inventory *flow assumption* in which the cost of units equals the *weighted average* cost of the *beginning inventory* and purchases. See *inventory equation.*

**average tax rate.** The rate found by dividing *income tax* expense by *net income* before taxes. Contrast with *marginal tax rate* and *statutory tax rate.*

**avoidable cost.** A *cost* that ceases if a firm discontinues an activity; an *incremental* or *variable cost.* See *programmed cost.*

# B

**backflush costing.** A method of *allocating indirect costs* and *overhead;* used by companies that hope to have zero or small *work-in-process inventory* at the end of the period. The method *debits* all *product costs* to *cost of goods sold* (or *finished goods inventory*) during the period. To the extent that work in process actually exists at the end of the period, the method then debits work-in-process and *credits* cost of goods sold (or finished goods inventory). This method is "backflush" in the sense that costing systems ordinarily, but not in this case, allocate first to work-in-process and then forward to cost of goods sold or to finished goods. Here, the process allocates first to cost of goods sold (or finished goods) and then, later if necessary, to work-in-process.

**backlog.** Orders for which a firm has insufficient *inventory* on hand for current delivery and will fill in a later period.

**backlog depreciation.** In *current cost accounting,* a problem arising for the *accumulated depreciation* on *plant assets.* Consider an *asset* costing $10,000 with a 10-year life depreciated with the straight-line method. Assume that a similar asset has a current cost of $10,000 at the end of the first year but $12,000 at the end of the second year. Assume that the firm bases the depreciation charge on the average current cost during the year, $10,000 for the first year and $11,000 for the second. The depreciation charge for the first year is $1,000 and for the second is $1,100 (= 0.10 × $11,000), so the *accumulated depreciation account* is $2,100 after two years. Note that at the end of the second year, the firm has used 20 percent of the asset's future benefits, so the accounting records based on current costs must show a *net book value* of $9,600 (= 0.80 × $12,000), which results only if accumulated depreciation equals $2,400, so that book value equals $9,600 (= $12,000 − $2,400). But the sum of the depreciation charges equals only $2,100 (= $1,000 + $1,100). The *journal entry* to increase the accumulated depreciation account requires a *credit* to that account of $300. The backlog depreciation question arises: what account do we debit? Some theorists would *debit* an *income* account, and others would *debit* a *balance sheet owners' equity* account without reducing current-period earnings. The answer to the question of what to debit interrelates with how the firm records the *holding gains* on the asset. When the firm debits the asset account for $2,000 to increase the recorded amount from $10,000 to $12,000, it records a holding gain of $2,000 with a credit. Many theorists believe that whatever account the firm credits for the holding gain is the same account that the firm should debit for backlog depreciation. This is sometimes called "catch-up depreciation."

**bad debt.** An *uncollectible account;* see *bad debt expense* and *sales contra, estimated uncollectibles.*

**bad debt expense.** The name for an *account debited* in both the *allowance method* for *uncollectible accounts* and the *direct write-off method.* Under the allowance method, some prefer to treat the account as a revenue contra, not as an expense, and give it an account title such as Uncollectible Accounts Adjustment.

**bad debt recovery.** Collection, perhaps partial, of a specific account receivable previously written off as uncollectible. If a firm uses the *allowance method,* it will usually *credit* the *allowance* account, assuming that it has correctly assessed the amount of bad debts but has merely misjudged the identity of one of the nonpaying customers. If the firm decides that its charges for bad debts have been too large, it will credit the Bad Debt Expense account. If the firm uses the *direct write-off* method, it will credit a *revenue account.*

**bailout period.** In a *capital budgeting* context, the total time that elapses before accumulated cash inflows from a project, including the potential *salvage value* of assets at various times, equal or exceed the accumulated cash outflows. Contrast with *payback period,* which assumes completion of the project and uses terminal salvage value. Bailout, in contrast with payback, takes into account, at least to some degree, the *present value* of the cash flows after the termination date that the analyst is considering. The potential salvage value at any time includes some estimate of the flows that can occur after that time.

**balance.** As a noun, the opening balance in an *account* plus the amounts of increases less the amounts of decreases. (In the absence of a modifying adjective, the term means closing balance, in contrast to opening balance. The closing balance for a period becomes the opening balance for the next period.) As a verb, "balance" means to find the value of the arithmetic expression described above.

**balance sheet.** Statement of financial position that shows Total *Assets* = Total *Liabilities* + *Owners' Equity*. The *balance sheet* usually classifies Total Assets as (1) *current assets,* (2) *investments,* (3) *property, plant, and equipment,* or (4) *intangible assets*. The balance sheet accounts composing Total Liabilities usually appear under the headings Current Liabilities and Long-Term Liabilities.

**balance sheet account.** An account that can appear on a balance sheet; a *permanent account.* Contrast with *temporary account*.

**balanced scorecard.** A set of performance targets, not all expressed in dollar amounts, for setting an organization's goals for its individual employees or groups or divisions. A community relations employee might, for example, set targets in terms of number of employee hours devoted to local charitable purposes.

**balloon.** Most *mortgage* and *installment loans* require relatively equal periodic payments. Sometimes the loan requires relatively equal periodic payments with a large final payment. Usage calls the large final payment a "balloon" payment and the loan, a "balloon" loan. Although a coupon bond meets this definition, usage seldom, if ever, applies this term to bond loans.

**bank balance.** The amount of the balance in a checking account shown on the *bank statement*. Compare with *adjusted bank balance of cash,* and see *bank reconciliation schedule*.

**bank prime rate.** See *prime rate*.

**bank reconciliation schedule.** A schedule that explains the difference between the book balance of the cash in a bank account and the bank's statement of that amount; takes into account the amount of items such as checks that have not cleared or deposits that have not been recorded by the bank, as well as errors made by the bank or the firm.

**bank statement.** A statement sent by the bank to a checking account customer showing deposits, checks cleared, and service charges for a period, usually one month.

**bankrupt.** Occurs when a company's *liabilities* exceed its *assets* and the firm or one of its creditors has filed a legal petition that the bankruptcy court has accepted under the bankruptcy law. A bankrupt firm is usually, but need not be, *insolvent*.

**base stock method.** A method of inventory valuation that assumes that a firm must keep on hand at all times a minimum normal, or base stock, of goods for effective continuity of operations. The firm values this base quantity at *acquisition cost* of the inventory on hand in the earliest period when inventory was on hand. Firms may not use this method, either for financial reporting or for tax reporting, but most theorists consider it to be the forerunner of the *LIFO* cost flow assumption.

**basic accounting equation.** *Accounting equation*.

**basic cost-flow equation.** *Cost-flow equation*.

**basic earnings per share (BEPS).** *Net income* to *common shareholders,* divided by the weighted average number of common shares *outstanding* during the period. Required by *SFAS No. 128* and by *IASB*.

**basis.** *Acquisition cost,* or some substitute therefor, of an *asset* or *liability* used in computing gain or loss on disposition or retirement; *attribute measured*. This term appears in both *financial* and *tax reporting,* but the basis of a given item need not be the same for both purposes.

**basis point.** One one-hundredth (=1/100). Terminology usually quotes *interest rates* in percentage terms, such as "5.60 percent" or "5.67 percent." The difference between those two interest rates is described as "7 basis points" or seven one hundredths of one percent. Financial writers often extend this usage to other contexts involving decimals. For example, if the mean grade point average in the class is 3.25 and a given student scores 3.30, we might say that the student scored "5 basis points" above the class average.

**basket purchase.** Purchase of a group of *assets* (and *liabilities*) for a single price; the acquiring firm must assign *costs* to each item so that it can record the individual items with their separate amounts in the *accounts*.

**bear.** One who believes that security prices will fall. A "bear market" refers to a time when stock prices are generally declining. Contrast with *bull*.

**bearer bond.** See *registered bond* for contrast and definition.

**beginning inventory.** Valuation of *inventory* on hand at the beginning of the *accounting period,* equals *ending inventory* from the preceding period.

**behavioral congruence.** *Goal congruence*.

**benchmarking.** Process of measuring a firm's performance, products, and services against standards based on best levels of performance achievable or, sometimes, achieved by other firms.

**benefit element (of stock options).** The amount by which the *market value* of a *share* exceeds the *exercise price* of the *stock option*.

**BEPS.** *Basic earnings per share*.

**betterment.** An *improvement,* usually *capitalized,* not *expensed*.

**bid.** An offer to purchase, or the amount of the offer.

**big bath.** A *write-off* of a substantial amount of costs previously treated as *assets;* usually occurs when a corporation drops a business line that earlier required a large investment but that proved to be unprofitable. The term is sometimes used to describe a situation in which a corporation takes a large write-off in one period in order to free later periods of gradual write-offs of those amounts. In this sense it frequently occurs when the top management of the firm changes.

**Big 4; Final 4.** The four largest U.S. *public accounting* partnerships; in alphabetical order: Andersen; Deloitte & Touche (U.S. national practice of the international firm Deloitte Touche Tohmatsu); Ernst & Young; KPMG Peat Marwick; and Pricewaterhouse Coopers.

**Big N.** The largest U.S. *public accounting* partnerships. When we first prepared this glossary, there were eight such partnerships, referred to as the "Big 8." See *Big 4.* The term "Big N" came into use when various of the *Big 8* proposed to merge with each other and the ultimate number of large partnerships was in doubt, which it still is.

**bill.** An *invoice* of charges and *terms of sale* for *goods* and *services;* also, a piece of currency.

**bill of materials.** A specification of the quantities of *direct materials* that a firm expects to use to produce a given job or quantity of output.

**blocked currency.** Currency that the holder, by law, cannot withdraw from the issuing country or exchange for the currency of another country.

**board.** *Board of directors.*

**board of directors.** The governing body of a corporation; elected by the shareholders.

**bond.** A certificate to show evidence of debt. The *par value* is the *principal* or face amount of the bond payable at maturity. The *coupon rate* is the amount of the yearly payments divided by the principal amount. Coupon bonds have attached coupons that the holder can redeem at stated dates. Increasingly, firms issue not coupon bonds but registered bonds; the firm or its agent keeps track of the owners of registered bonds. Normally, bonds call for semiannual payments.

**bond conversion.** The act of exchanging *convertible bonds* for *preferred* or *common shares.*

**bond discount.** From the standpoint of the issuer of a *bond* at the issue date, the excess of the *par value* of a bond over its initial sales price and, at later dates, the excess of par over the sum of the following two amounts: initial issue price and the portion of discount already *amortized;* from the standpoint of a bondholder, the difference between par value and selling price when the bond sells below par.

**bond indenture.** The contract between an issuer of *bonds* and the bondholders.

**bond premium.** Exactly parallel to *bond discount* except that the issue price (or current selling price) exceeds *par value.*

**bond ratings.** Corporate and *municipal bond* issue ratings, based on the issuer's existing *debt* level, its previous record of payment, the *coupon rate* on the bonds, and the safety of the *assets* or *revenues* that are committed to paying off *principal* and *interest.* Moody's Investors Service and Standard & Poor's Corporation publish bond ratings: Moody's top rating is Aaa; Standard & Poor's is AAA.

**bond redemption.** Retirement of *bonds.*

**bond refunding.** To incur *debt,* usually through the issue of new *bonds,* intending to use the proceeds to retire an *outstanding* bond issue.

**bond sinking fund.** See *sinking fund.*

**bond table.** A table showing the current price of a *bond* as a function of the *coupon rate,* current (remaining) term *maturity,* and effective *yield to maturity* (or *effective rate*).

**bonus.** Premium over normal *wage* or *salary,* paid usually for meritorious performance.

**bonus method.** One of two methods to recognize an excess, say $10,000, when a *partnership* admits a new partner and when the new partner's capital account is to show an amount larger than the amount of *tangible* assets that he or she contributes. First, the old partners may transfer $10,000 from themselves to the new partner. This is the bonus method. Second, the partnership may recognize goodwill in the amount of $10,000, with the credit to the new partner's capital account. This is the *goodwill method.* (Notice that the new partner's percentage of total ownership differs under the two methods.) If the new partner's capital account is to show an amount smaller than the tangible assets that he or she contributed, then the old partners will receive bonus or goodwill, depending on the method.

**book.** As a verb, to record a transaction; as a noun, usually plural, the *journals* and *ledgers;* as an adjective, see *book value.*

**book cost.** *Book value.*

**book inventory.** An *inventory* amount that results not from physical count but from the amount of beginning inventory plus *invoice* amounts of net purchases less invoice amounts of *requisitions* or withdrawals; implies a *perpetual inventory* method.

**book of original entry.** *Journal.*

**book value.** The amount shown in the books or in the *accounts* for an *asset, liability,* or *owners' equity* item. The term is generally used to refer to the *net* amount of an *asset* or group of assets shown in the account that records the asset and reductions, such as for *amortization,* in its cost. Of a firm, it refers to the excess of total assets over total liabilities; *net assets.*

**book value per share of common stock.** Common *shareholders' equity* divided by the number of shares of common stock outstanding. See *ratio.*

**bookkeeping.** The process of analyzing and recording transactions in the accounting records.

**boot.** The additional cash paid (or received) along with a used item in a trade-in or exchange transaction for another item. See *trade-in.*

**borrower.** See *loan.*

**branch.** A sales office or other unit of an enterprise physically separated from the home office of the enterprise but not organized as a legally separate *subsidiary.* Writers seldom use this term to refer to manufacturing units.

**branch accounting.** An accounting procedure that enables the firm to report the financial position and operations of each *branch* separately but later combine them for published statements.

**breakeven analysis.** See *breakeven chart.*

**breakeven chart.** Two kinds of breakeven charts appear here. The charts use the following information for one month. Revenue is $30 per unit.

| Cost Classification | Variable Cost, Per Unit | Fixed Cost, Per Month |
|---|---|---|
| Manufacturing costs: | | |
| Direct material | $ 4 | — |
| Direct labor | 9 | — |
| Overhead | 4 | $3,060 |
| Total manufacturing costs | $17 | $3,060 |
| Selling, general, and administrative costs | 5 | 1,740 |
| Total costs | $22 | $4,800 |

The cost-volume-profit graph presents the relation between changes in volume to the amount of *profit*, or *income*. Such a graph shows total *revenue* and total *costs* for each volume level, and the user reads profit or loss at any volume directly from the chart. The profit-volume graph does not show revenues and costs but more readily indicates profit (or loss) at various output levels. Keep in mind two caveats about these graphs:

1. Although the curve depicting *variable cost* and total cost appears as a straight line for its entire length, at low or high levels of output, variable cost will probably differ from $22 per unit. The variable cost figure usually results from studies of operations at some broad central area of production, called the *relevant range*. The chart will not usually provide accurate results for low (or high) levels of activity. For this reason, the total cost and the profit-loss curves sometimes appear as dotted lines at lower (or higher) volume levels.
2. This chart, simplistically, assumes a single-product firm. For a multiproduct firm, the horizontal axis would have to be stated in dollars rather than in physical units of output. Breakeven charts for multiproduct firms necessarily assume that the firm sells constant proportions of the several products, so that changes in this mixture, as well as in costs or selling prices, invalidate such a chart.

**(a) Cost-Volume-Profit Graph**

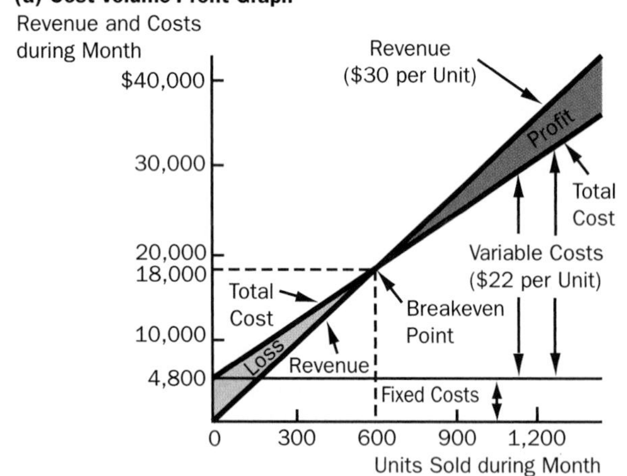


**(b) Profit-Volume Graph**

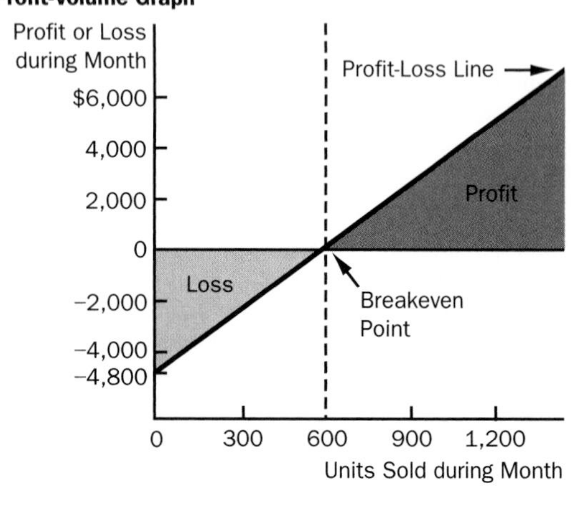


**breakeven point.** The volume of sales required so that total *revenues* equals total *costs;* may be expressed in units (*fixed costs* ÷ *contribution per unit*) or in sales dollars [selling price per unit × (fixed costs ÷ contribution per unit)].

**budget.** A financial plan that a firm uses to estimate the results of future operations; frequently used to help control future operations. In governmental operations, budgets often become the law. See *standard costs* for further elaboration and contrast.

**budgetary accounts.** In governmental accounting, the accounts that reflect estimated operations and financial condition, as affected by estimated *revenues, appropriations,* and *encumbrances*. Contrast with *proprietary accounts,* which record the transactions.

**budgetary control.** Management of governmental (nongovernmental) unit in accordance with an official (approved) *budget* in order to keep total expenditures within authorized (planned) limits.

**budgeted cost.** See *standard costs* for definition and contrast.

**budgeted statements.** *Pro forma statements* prepared before the event or period occurs.

**bull.** One who believes that security prices will rise. A "bull market" refers to a time when stock prices are generally rising. Contrast with *bear.*

**burden.** See *overhead costs.*

**burn rate.** A new business usually begins life with cash-absorbing operating losses but with a limited amount of cash. The "burn rate" measures how long the new business can survive before operating losses must stop or the firm must receive a new infusion of cash. Writers usually express the burn rate in months.

**business combination.** As defined in *APB Opinion No. 16,* the bringing together into a single accounting *entity* of two or more incorporated or unincorporated businesses. The new entity will account for the *merger* either with the *purchase method* or the *pooling-of-interests method.* See *conglomerate.*

**business entity.** *Entity; accounting entity.*

**BV (*besloten vennootschap*).** Netherlands: a private limited-liability company.

**bylaws.** The rules adopted by the shareholders of a corporation; specify the general methods for carrying out the functions of the corporation.

**by-product.** A *joint product* whose sales value is so small relative to the sales value of the other joint product(s) that it does not receive normal accounting treatment. The costs assigned to by-products reduce the costs of the main product(s). Accounting allocates by-products a share of joint costs such that the expected gain or loss at their sale is zero. Thus, by-products appear in the *accounts* at *net realizable value*.

# C

**C corporation.** In tax terminology, a corporation paying its own income taxes. Contrast with *S corporation.*

**CA.** *Chartered accountant.*

**call.** An option to buy *shares* of a publicly traded corporation at a fixed price during a fixed time span. Contrast with *put.*

**call premium.** See *callable bond.*

**call price.** See *callable bond.*

**callable bond.** A *bond* for which the issuer reserves the right to pay a specific amount, the call price, to retire the obligation before its *maturity* date. If the issuer agrees to pay more than the *face amount* of the bond when called, the excess of the payment over the face amount is the "call premium."

**called-up share capital.** UK: *common stock* at *par value.*

**Canadian Institute of Chartered Accountants.** The national organization that represents *chartered accountants* in Canada. Web site: www.cica.ca.

**cancelable lease.** See *lease.*

**CAP.** *Committee on Accounting Procedure.*

**capacity.** Stated in units of product, the amount that a firm can produce per unit of time; stated in units of input, such as *direct labor* hours, the amount of input that a firm can use in production per unit of time. A firm uses this measure of output or input in allocating *fixed costs* if the amounts producible are normal, rather than maximum, amounts.

**capacity cost.** A *fixed cost* incurred to provide a firm with the capacity to produce or to sell. Consists of *standby costs* and *enabling costs*. Contrast with *programmed costs*.

**capacity variance.** *Production volume variance.*

**capital.** *Owners' equity* in a business; often used, equally correctly, to mean the total assets of a business; sometimes used to mean *capital assets*. Sometimes used to mean *funds* raised. This word causes confusion in accounting and finance. Uninformed users mix up the funds (and their uses) with the sources of the funds. Consider the following transactions. A firm raises $100 cash from investors and uses the $100 to acquire *inventory* and *plant assets.* Did the investors "invest capital" of $100 or did the firm "invest capital" of $100? You will hear "invest capital" used for both sides of that transaction. Now focus on the firm who issued the shares and received the cash. Some would say the first transaction, the issue of shares, "raised capital." (If you ask of a person who answers this way, "What is the *capital,* the increase in owners' equity or the increased cash?" you will not get a clear answer, consistent across all such people.) Others would say only the second transaction, spending the cash, raised capital and only then for the plant assets, not the inventory. When a regulator focuses on a bank's capital ratios, it looks to the right-hand side of the balance sheet, not to how the firm has invested its funds. Sometimes bank regulators will take the owners' equity total and subtract from that amount the amount of intangible assets, resulting in a total with no clear conception, which they call "tangible capital." See *cost of capital* for further discussion of the confusion between the cost of raising funds and the return to, or *opportunity cost* of, investing funds. The confusion is so prevalent that we tend to avoid using the word.

**capital asset.** Properly used, a designation, for income tax purposes, that describes property held by a taxpayer except *cash,* inventoriable *assets,* goods held primarily for sale, most depreciable property, *real estate, receivables,* certain *intangibles,* and a few other items. Sometimes writers use this term imprecisely to describe *plant* and *equipment,* which are clearly not capital assets under the income-tax definition. Writers often use the term to refer to an *investment* in *securities*.

**capital budget.** Plan of proposed outlays for acquiring long-term *assets* and the means of *financing* the acquisition.

**capital budgeting.** The process of choosing *investment* projects for an enterprise by considering the *present value* of cash flows and deciding how to raise the funds the investment requires.

**capital consumption allowance.** The term used for *depreciation expense* in national income accounting and the reporting of funds in the economy.

**capital contributed in excess of par (or stated) value.** A preferred title for the account that shows the amount received by the issuer for *capital stock* in excess of *par* (or *stated*) *value*.

**capital expenditure (outlay).** An *expenditure* to acquire long-term *assets*.

**capital gain.** The excess of proceeds over *cost,* or other *basis,* from the sale of a *capital asset* as defined by the Internal Revenue Code. If the taxpayer has held the capital asset for a sufficiently long time before sale, then the gain is taxed at a rate lower than that used for other gains and ordinary income.

**capital lease.** A *lease* treated by the *lessee* as both the borrowing of funds and the acquisition of an *asset* to be *amortized*. The lessee (tenant) recognizes both the *liability* and the asset on its balance sheet. Expenses consist of *interest* on the *debt* and *amortization* of the asset. The *lessor* (landlord) treats the lease as the sale of the asset in return for a series of future cash receipts. Contrast with *operating lease*.

**capital loss.** A negative capital gain; see *capital gain.*

**capital rationing.** In a *capital budgeting* context, the imposition of constraints on the amounts of total capital expenditures in each period.

**capital stock.** The ownership shares of a corporation. Consists of all classes of *common* and *preferred shares.*

**capital structure.** The composition of a corporation's equities; the relative proportions of short-term debt, long-term debt, and *owners' equity*.

**capital surplus.** An inferior term for *capital contributed in excess of par* (or *stated*) *value*.

**capitalization of a corporation.** A term used by investment analysts to indicate *shareholders' equity* plus bonds outstanding.

**capitalization of earnings.** The process of estimating the *fair value* of a firm by computing the *net present value* of the predicted *net income* (not *cash flows*) of the firm for the future.

**capitalization rate.** An *interest rate* used to convert a series of payments or receipts or earnings into a single *present value*.

**capitalize.** To record an *expenditure* that may benefit a future period as an *asset* rather than to treat the expenditure as an *expense* of the period of its occurrence. Whether expenditures for advertising or for research and development should be capitalized is controversial, but *SFAS No. 2* forbids capitalizing *R&D* costs. We believe GAAP should allow firms to capitalize expenditures when they lead to future benefits and thus meet the criterion to be an asset.

**carryback, carryforward, carryover.** The use of losses or tax credits in one period to reduce income taxes payable in other periods. Two common kinds of carrybacks exist: for net operating losses and for *capital losses.* They apply against taxable income. In general, carrybacks are for three years, with the earliest year first. The taxpayer can carry forward operating losses for fifteen years. Corporate capital loss carryforwards are for five years. Individuals can carry forward capital losses indefinitely.

**carrying cost.** Costs (such as property taxes and insurance) of holding, or storing, *inventory* from the time of purchase until the time of sale or use.

**carrying value (amount).** *Book value.*

**CASB (Cost Accounting Standards Board).** A board authorized by the U.S. Congress to "promulgate cost-accounting standards designed to achieve uniformity and consistency in the cost-accounting principles followed by defense contractors and subcontractors under federal contracts." The *principles* the CASB promulgated since 1970 have considerable weight in practice wherever the *FASB* has not established a standard. Congress allowed the CASB to go out of existence in 1980 but reinstated it in 1990.

**cash.** Currency and coins, negotiable checks, and balances in bank accounts. For the *statement of cash flows,* "cash" also includes *marketable securities* held as *current assets*.

**cash basis of accounting.** In contrast to the *accrual basis of accounting,* a system of accounting in which a firm recognizes *revenues* when it receives *cash* and recognizes *expenses* as it makes *disbursements*. The firm makes no attempt to match *revenues* and *expenses* in measuring *income*. See *modified cash basis*.

**cash budget.** A schedule of expected cash *receipts* and *disbursements*.

**cash change equation.** For any *period,* the change in *cash* equals the change in *liabilities* plus the change in *owners' equity* minus the change in noncash *assets.*

**cash collection basis.** The *installment method* for recognizing *revenue*. Do not confuse with the *cash basis of accounting*.

**cash conversion cycle.** *Cash cycle.*

**cash cycle.** The period of time during which a firm converts *cash* into *inventories,* inventories into *accounts receivable,* and *receivables* back into cash. Sometimes called *earnings cycle.*

**cash disbursements journal.** A specialized *journal* used to record *expenditures* by *cash* and by *check*. If a *check register* is also used, a cash disbursements journal records only expenditures of currency and coins.

**cash discount.** A sales or purchase price reduction allowed for prompt payment.

**cash dividend.** See *dividend.*

**cash equivalent.** According to *SFAS No. 95,* "short-term, highly liquid investments that are both readily convertible to known amounts of cash [and] so near their maturity that they present insignificant risk of changes in value because of changes in interest rates. . . . Examples of items commonly considered to be cash equivalents are Treasury bills, commercial paper, [and] money market funds."

**cash equivalent value.** A term used to describe the amount for which an *asset* could be sold. Sometimes called *market value* or *fair market price* (*value*).

**cash flow.** Cash *receipts* minus *disbursements* from a given *asset,* or group of assets, for a given period. Financial analysts sometimes use this term to mean *net income + depreciation + depletion + amortization*. See also *operating cash flow* and *free cash flow.*

**cash flow from operations.** Receipts from customers and from investments less expenditures for inventory, labor, and services used in the usual activities of the firm, less interest expenditures. See *statement of cash flows* and *operations*. Same as *cash provided by operations.*

**cash-flow hedge.** A hedge of an exposure to variability in the cash flows of a recognized *asset* or *liability* or of a forecasted transaction, such as expected future foreign sales.

**cash flow statement.** *Statement of cash flows.*

**cash provided by operations.** An important subtotal in the *statement of cash flows*. This amount equals the total of revenues producing *cash* less *expenses* requiring cash. Often, the amount appears as *net income* plus expenses not requiring cash (such as depreciation charges) minus revenues not producing cash (such as revenues recognized under the *equity method* of accounting for a long-term investment). The statement of cash flows maintains the same distinctions between *continuing operations, discontinued operations,* and *income* or *loss* from *extraordinary items* as does the *income statement.*

**cash receipts journal.** A specialized *journal* used to record all *receipts* of *cash.*

**cash (surrender) value of life insurance.** An amount equal not to the face value of the policy to be paid in the event of death but to the amount that the owner could realize by immediately canceling the policy and returning it to the insurance company for cash. A firm owning a life insurance policy reports it as an asset at an amount equal to this value.

**cash yield.** See *yield.*

**cashier's check.** A bank's own *check* drawn on itself and signed by the cashier or other authorized official. It is a direct obligation of the bank. Compare with *certified check.*

**catch-up depreciation.** *Backlog depreciation.*

**CCA.** *Current cost accounting; current value accounting.*

**central corporate expenses.** General *overhead expenses* incurred in running the corporate headquarters and related supporting activities of a corporation. Accounting treats these expenses as *period expenses.* Contrast with *manufacturing overhead. Line of business reporting* must decide how to treat these expenses—whether to allocate them to the individual segments and, if so, how to allocate them.

**central processing unit (CPU).** The computer system component that carries out the arithmetic, logic, and data transfer.

**certificate.** The document that is the physical embodiment of a *bond* or a *share of stock;* a term sometimes used for the *auditor's report.*

**certificate of deposit.** A form of *deposit* in a bank or thrift institution. Federal law constrains the rate of interest that banks can pay to their depositors. Current law allows banks to pay a rate higher than the one allowed on a *time deposit* if the depositor promises to leave funds on deposit for several months or more. When the bank receives such funds, it issues a certificate of deposit. The depositor can withdraw the funds before maturity by paying a penalty.

**certified check.** The *check* of a depositor drawn on a bank. The bank inserts the words "accepted" or "certified" on the face of the check, with the date and a signature of a bank official. The check then becomes an obligation of the bank. Compare with *cashier's check.*

**certified financial statement.** A financial statement attested to by an independent *auditor* who is a *CPA.*

**certified internal auditor.** See *CIA.*

**certified management accountant.** *CMA.*

**certified public accountant (CPA).** An accountant who has satisfied the statutory and administrative requirements of his or her jurisdiction to be registered or licensed as a public accountant. In addition to passing the Uniform CPA Examination administered by the *AICPA,* the CPA must meet certain educational, experience, and moral requirements that differ from jurisdiction to jurisdiction. The jurisdictions are the 50 states, the District of Columbia, Guam, Puerto Rico, and the Virgin Islands.

**CGA (Certified General Accountant).** Canada: an accountant who has satisfied the experience, education, and examination requirements of the Certified General Accountants' Association.

**chain discount.** A series of *discount* percentages. For example, if a chain discount of 10 and 5 percent is quoted, then the actual, or *invoice,* price is the nominal, or list, price times 0.90 times 0.95, or 85.5, percent of invoice price.

**change fund.** Coins and currency issued to cashiers, delivery drivers, and so on.

**changes, accounting.** See *accounting changes.*

**changes in financial position.** See *statement of cash flows.*

**charge.** As a noun, a *debit* to an account; as a verb, to debit.

**charge off.** To treat as a *loss* or *expense* an amount originally recorded as an *asset;* use of this term implies that the charge is not in accord with original expectations.

**chart of accounts.** A list of names and numbers, systematically organized, of *accounts.*

**charter.** Document issued by a state government authorizing the creation of a corporation.

**chartered accountant(s) (CA).** The title used in Australia, Canada, and Scotland for an accountant who has satisfied the requirements of the institute of his or her jurisdiction to be qualified to serve as a *public accountant.* In the UK other than Scotland, members use the initials ACA or FCA: *A* means Associate and *F* means Fellow; the Associate has less experience than does the Fellow. A partnership of chartered accountants signs its firm name with the letters *CA.* In Canada, each provincial institute or order has the right to administer the examination and set the standards of performance and ethics for Chartered Accountants in its province. For a number of years, however, the provincial organizations have pooled their rights to qualify new members through the Interprovincial Education Committee, and the result is that there are nationally set and graded examinations given in English and French. Deviation from the pass/fail grade awarded by the Board of Examiners (a subcommittee of the Inter-provincial Education Committee) is rare.

**check.** The Federal Reserve Board defines a check as "a *draft* or order upon a bank or banking house purporting to be drawn upon a deposit of funds for the payment at all events of a certain sum of money to a certain person therein named or to him or his order or to bearer and payable instantly on demand." It must contain the phrase "pay to the order of." The amount shown on the check must be clearly readable, and the check must have the signature of the drawer. The drawer need not date the check. In the accounts, the drawer usually reduces the *balance* in the *cash account* when it issues the check, not later when the check clears the bank. See *remittance advice.*

**check register.** A *journal* to record *checks* issued.

**CIA (Certified Internal Auditor).** One who has satisfied certain requirements of the *Institute of Internal*

*Auditors* including experience, ethics, education, and passing examinations.

**CICA.** *Canadian Institute of Chartered Accountants.*

**CIF (cost, insurance, and freight).** In contracts, a term used along with the name of a given port, such as New Orleans, to indicate that the quoted price includes insurance, handling, and freight charges up to delivery by the seller at the given port.

**circulating capital.** *Working capital.*

**clean opinion.** See *auditor's report.*

**clean surplus concept.** The notion that all entries to the *retained earnings* account must record *net income* and *dividends.* See *comprehensive income.* Contrast with *current operating performance concept.* This concept, with minor exceptions, now controls *GAAP.* (See *APB Opinions No. 9* and *No. 30.*)

**clearing account.** An account containing amounts to be transferred to another account(s) before the end of the *accounting period.* Examples are the *income summary* account (whose balance transfers to *retained earnings*) and the purchases account (whose balance transfers to *inventory* or to *cost of goods sold*).

**close.** As a verb, to transfer the *balance* of a *temporary* or *contra* or *adjunct account* to the main account to which it relates; for example, to transfer *revenue* and *expense* accounts directly, or through the *income summary* account, to an *owners' equity* account or to transfer *purchase discounts* to purchases.

**closed account.** An *account* with equal *debits* and *credits,* usually as a result of a *closing entry.* See *ruling an account.*

**closing entries. closing process.** The *entries* that accomplish the transfer of balances in *temporary accounts* to the related *balance sheet accounts.* See *work sheet.*

**closing inventory.** *Ending inventory.*

**CMA (Certified Management Accountant) certificate.** Awarded by the *Institute of Certified Management Accountants* of the *Institute of Management Accountants* to those who pass a set of examinations and meet certain experience and continuing-education requirements.

**CoCoA.** *Continuously Contemporary Accounting.*

**coding of accounts.** The numbering of *accounts,* as for a *chart of accounts,* that is necessary for computerized accounting.

**coinsurance.** Common condition of insurance policies that protect against hazards such as fire or water damage. These often specify that the owner of the property may not collect the full amount of insurance for a loss unless the insurance policy covers at least some specified "coinsurance" percentage, usually about 80 percent, of the *replacement cost* of the property. Coinsurance clauses induce the owner to carry full, or nearly full, coverage.

**COLA.** Cost-of-living adjustment. See *indexation.*

**collateral.** *Assets* pledged by a *borrower* who will surrender those assets if he or she fails to repay a *loan.*

**collectible.** Capable of being converted into *cash*—now if due, later otherwise.

**collusion.** Cooperative effort by employees to commit fraud or another unethical act.

**combination.** See *business combination.*

**comfort letter.** A letter in which an auditor conveys negative assurances as to unaudited financial statements in a prospectus or draft financial statements included in a preliminary prospectus.

**commercial paper.** Short-term notes issued by corporate borrowers.

**commission.** Employee remuneration, usually expressed as a percentage, based on an activity rate, such as sales.

**committed costs.** *Capacity costs.*

**Committee on Accounting Procedure (CAP).** Predecessor of the *APB.* The *AICPA*'s principles-promulgating body from 1939 through 1959. Its 51 pronouncements are *Accounting Research Bulletins.*

**common cost.** *Cost* resulting from the use of *raw materials,* a facility (for example, plant or machines), or a service (for example, fire insurance) that benefits several products or departments. A firm must allocate this cost to those products or departments. Common costs result when two or more departments produce multiple products together even though the departments could produce them separately; *joint costs* occur when two or more departments must produce multiple products together. Many writers use "common costs" and "joint costs" synonymously. See *joint cost, indirect costs, overhead,* and *sterilized allocation.*

**common-dollar accounting.** *Constant-dollar accounting.*

**common monetary measuring unit.** For U.S. corporations, the dollar. See also *stable monetary unit assumption* and *constant-dollar accounting.*

**common shares.** *Shares* representing the class of owners who have residual claims on the *assets* and *earnings* of a *corporation* after the firm meets all *debt* and *preferred shareholders'* claims.

**common-size statement.** A *percentage statement* usually based on total *assets* or *net sales* or *revenues.*

**common-stock equivalent.** A *security* whose primary value arises from its holder's ability to exchange it for *common shares;* includes *stock options, warrants,* and also *convertible bonds* or *convertible preferred stock* whose *effective interest rate* at the time of issue is less than two-thirds the average Aa corporate bond yield. See *bond ratings.*

**company-wide control.** See *control system.*

**comparative (financial) statements.** *Financial statements* showing information for the same company for different times, usually two successive years for balance sheets and three for *income* and *cash flow statements.* Nearly all published financial statements are in this form. Contrast with *historical summary.*

**compensating balance.** The amount required to be left on deposit for a loan. When a bank lends funds to customers, it often requires that the customers keep on deposit in their checking accounts an amount equal to some percentage—say, 20 percent—of the loan. Such amounts effectively increase the *interest* *rate*. The borrower must disclose the amounts of such balances in *notes* to the *financial statements*.

**completed contract method.** Recognizing *revenues* and *expenses* for a job or order only when the firm finishes it, except that when the firm expects a loss on the contract, the firm must recognize all revenues and expenses in the period when the firm first foresees a loss. Accountants generally use this term only for long-term contracts. This method is otherwise equivalent to the *sales basis of revenue recognition*.

**completed sales basis.** See *sales basis of revenue recognition*.

**compliance audit.** Objectively obtaining and evaluating evidence regarding assertions, actions, and events to ascertain the degree of correspondence between them and established performance criteria.

**compliance procedure.** An *audit* procedure used to gain evidence as to whether the prescribed internal controls are operating effectively.

**composite cost of capital.** See *cost of capital*.

**composite depreciation** or **composite life method.** *Group depreciation* when the items are of unlike kind. The term also applies when the firm depreciates as a whole a single item (for example, a crane, which consists of separate units with differing service lives, such as the chassis, the motor, the lifting mechanism, and so on), rather than treating each of its components separately.

**compound entry.** A *journal entry* with more than one *debit* or more than one *credit* or both. See *trade-in transaction* for an example.

**compound interest.** *Interest* calculated on *principal* plus previously undistributed interest.

**compound interest depreciation.** A method designed to hold the *rate of return* on an asset constant. First find the *internal rate of return* on the cash inflows and outflows of the asset. The periodic depreciation charge equals the cash flow for the period less the internal rate of return multiplied by the asset's book value at the beginning of the period. When the cash flows from the asset are constant over time, usage sometimes refers to the method as the "annuity method" of depreciation.

**compounding period.** The time period, usually a year or a portion of a year, for which a firm calculates *interest*. At the end of the period, the borrower may pay interest to the lender or may add the interest (that is, convert it) to the principal for the next interest-earning period.

**comprehensive budget.** *Master budget*.

**comprehensive income.** Defined in *SFAC No. 3* as "the change in equity (net assets) of an entity during a period from transactions and other events and circumstances from nonowner sources. It includes all changes in equity during a period except those resulting from investments by owners and distributions to owners." In this definition, "equity" means *owners' equity* or *shareholders' equity*. *SFAS No. 130* requires firms to report comprehensive income as part of a statement showing *earnings* (primarily from realized transactions), comprehensive income (with additions for all other changes in owners' equity, primarily *holding gains and losses* and *foreign exchange gains and losses*), and comprehensive income plus *accounting adjustments*. The *FASB* encourages the discontinuation of the term "net income." The terms "earnings" and "comprehensive income" denote different concepts, with totals different from that of the old "net income." *SFAS No. 130* requires that the firm report comprehensive income in a format having the same prominence as other *financial statements*. We cannot predict which "income total"—earnings or comprehensive income—users of financial statements will focus on. See *Exhibits 12.4* and *12.6* for the three formats the FASB suggests firms use.

**comptroller.** Same meaning and pronunciation as *controller*.

**conceptual framework.** A coherent system of interrelated objectives and fundamentals, promulgated by the *FASB* primarily through its *SFAC* publications, expected to lead to consistent standards for *financial accounting* and reporting.

**confidence level.** The measure of probability that the actual characteristics of the population lie within the stated precision of the estimate derived from a sampling process. A sample estimate may be expressed in the following terms: "Based on the sample, we are 95 percent sure [confidence level] that the true population value is within the range of X to Y [precision]." See *precision*.

**confirmation.** A formal memorandum delivered by the customers or suppliers of a company to its independent *auditor* verifying the amounts shown as receivable or payable. The auditor originally sends the confirmation document to the customer. If the auditor asks that the customer return the document whether the *balance* is correct or incorrect, usage calls it a "positive confirmation." If the auditor asks that the customer return the document only if it contains an error, usage calls it a "negative confirmation."

**conglomerate.** *Holding company*. This term implies that the owned companies operate in dissimilar lines of business.

**conservatism.** A *reporting objective* that calls for anticipation of all *losses* and *expenses* but defers recognition of *gains* or *profits* until they are *realized* in *arm's-length* transactions. In the absence of certainty, report events to minimize cumulative income. Conservatism does not mean reporting low income in every *accounting period*. Over long-enough time spans, income is cash-in less cash-out. If a (conservative) reporting method shows low income in early periods, it must show higher income in some later period.

**consignee.** See *on consignment*.

**consignment.** See *on consignment*.

**consignor.** See *on consignment*.

**consistency.** Treatment of like *transactions* in the same way in consecutive periods so that financial statements will be more comparable than otherwise; the reporting policy implying that a reporting *entity,* once it adopts specified procedures, should follow them from period to period. See *accounting changes* for the treatment of inconsistencies.

**consol.** A *bond* that never matures; a *perpetuity* in the form of a bond; originally issued by Great Britain after the Napoleonic wars to consolidate debt issues of that period. The term arose as an abbreviation for "consolidated annuities."

**consolidated financial statements.** Statements that are issued by legally separate companies and that show financial position and income as they would appear if the companies were one economic *entity*.

**constant dollar.** A hypothetical unit of *general purchasing power,* denoted "C$" by the *FASB*.

**constant-dollar accounting.** Accounting that measures items in *constant dollars*. See *historical cost/constant-dollar accounting* and *current cost/nominal-dollar accounting*. Sometimes called "general price level–adjusted accounting" or "general purchasing power accounting."

**constant-dollar date.** The time at which the *general purchasing power* of one *constant dollar* exactly equals the *general purchasing power* of one *nominal dollar;* that is, the date when C$1 = $1. When the constant-dollar date is midperiod, the nominal amounts of *revenues* and *expenses* spread evenly throughout the period equal their constant-dollar amounts but end-of-period *balance sheet* amounts measured in constant midperiod dollars differ from their nominal-dollar amounts. When the constant-dollar date is at the end of the period, the constant-dollar amounts equal the nominal-dollar amounts on a balance sheet for that date.

**constrained share company.** Canada: a public company whose *charter* specifies that people who are Canadian citizens or who are corporations resident in Canada must own a prescribed percentage of the shares.

**constructive liability.** *FASB's* term for an item recorded as an accounting *liability,* which the firm has no obligation to pay but intends to pay. An example is the liability with related *expense* that management establishes for future cash payments for severance payments for employees it intends to discharge in a restructuring.

**constructive receipt.** An item included in taxable income when the taxpayer can control funds whether or not it has received cash. For example, *interest* added to *principal* in a savings account is constructively received.

**Consumer Price Index (CPI).** A *price index* computed and issued monthly by the Bureau of Labor Statistics of the U.S. Department of Labor. The index attempts to track the price level of a group of goods and services purchased by the average consumer. The CPI is used in *constant-dollar accounting*.

**contingency.** A potential *liability.* If a specified event occurs, such as a firm's losing a lawsuit, it would recognize a liability. The notes disclose the contingency, but so long as it remains contingent, it does not appear in the balance sheet. *SFAS No. 5* requires treatment as a contingency until the outcome is "probable" and the amount of payment can be reasonably estimated, perhaps within a range. When the outcome becomes probable (the future event is "likely" to occur) and the firm can reasonably estimate the amount (using the lower end of a range if it can estimate only a range), then the firm recognizes a liability in the accounts, rather than just disclosing it. A *material* contingency may lead to a qualified, "*subject to*" auditor's opinion. Firms do not record *gain* contingencies in the accounts but merely disclose them in notes.

**contingent annuity.** An *annuity* whose number of payments depends on the outcome of an event whose timing is uncertain at the time the annuity begins; for example, an annuity payable until death of the *annuitant*. Contrast with *annuity certain*.

**contingent issue (securities).** Securities issuable to specific individuals at the occurrence of some event, such as the firm's attaining a specified level of earnings.

**contingent liability.** *Contingency*. Avoid this term because it refers to something not (yet) a *liability* on the *balance sheet*.

**contingent obligation.** *Contingency.*

**continuing appropriation.** A governmental *appropriation* automatically renewed without further legislative action until altered or revoked or expended.

**continuing operations.** See *income from continuing operations*.

**continuity of operations.** The assumption in accounting that the business *entity* will continue to operate long enough to carry out its current plans. The *going-concern assumption*.

**continuous budget.** A *budget* that adds a future period as the current period ends. This budget, then, always reports on the same number of periods.

**continuous compounding.** *Compound interest* in which the *compounding period* is every instant of time. See *e* for the computation of the equivalent annual or periodic rate.

**continuous improvement.** Modern *total quality management (TQM)* practitioners believe that the process of seeking quality is never complete. This attitude reflects that assumption, seeking always to improve activities.

**continuous inventory method.** The *perpetual inventory* method.

**Continuously Contemporary Accounting (CoCoA).** A name coined by the Australian theorist Raymond J. Chambers to indicate a combination of *current value accounting* in which the *measuring unit* is *constant dollars* and the *attribute measured* is *exit value*.

**contra account.** An *account,* such as *accumulated depreciation,* that accumulates subtractions from another account, such as machinery. Contrast with *adjunct account*.

**contributed capital.** Name for the *owners' equity* account that represents amounts paid in, usually in *cash,* by owners; the sum of the balances in *capital stock* accounts plus *capital contributed in excess of par* (or *stated*) *value* accounts. Contrast with *donated capital.*

**contributed surplus.** An inferior term for *capital contributed in excess of par value.*

**contribution approach.** *Income statement* preparation method that reports *contribution margin,* by separating *variable costs* from *fixed costs,* in order to emphasize the importance of cost-behavior patterns for purposes of planning and control.

**contribution margin.** *Revenue* from *sales* less all variable *expenses.* Contrast with *gross margin.*

**contribution margin ratio.** *Contribution margin* divided by *net sales;* usually measured from the price and cost of a single unit; sometimes measured in total for companies with multiple products.

**contribution per unit.** Selling price less *variable costs* per unit.

**contributory.** Said of a *pension plan* in which employees, as well as employers, make payments to a pension *fund.* Note that the provisions for *vesting* apply only to the employer's payments. Whatever the degree of vesting of the employer's payments, employees typically get back all their payments, with interest, in case of death or other cessation of employment before retirement.

**control (controlling) account.** A summary *account* with totals equal to those of entries and balances that appear in individual accounts in a *subsidiary ledger.* Accounts Receivable is a control account backed up with an account for each customer. Do not change the balance in a control account unless you make a corresponding change in one of the subsidiary accounts.

**control system.** A device used by top management to ensure that lower-level management carries out its plans or to safeguard assets. Control designed for a single function within the firm is "operational control"; control designed for autonomous segments that generally have responsibility for both revenues and costs is "divisional control"; control designed for activities of the firm as a whole is "companywide control." Systems designed for safeguarding *assets* are "internal control" systems.

**controllable cost.** A *cost* influenced by the way a firm carries out operations. For example, marketing executives control advertising costs. These costs can be *fixed* or *variable.* See *programmed costs* and managed costs.

**controlled company.** A company in which an individual or corporation holds a majority of the voting shares. An owner can sometimes exercise effective control even though it owns less than 50 percent of the shares.

**controller.** A title for the chief accountant of an organization; often spelled *comptroller.*

**convergence.** The process by which the International Accounting Standards Board (IASB) collaborates with selected national standard setters around the world to reduce or eliminate substantive differences between the international and national standards at a high level of quality. The national standard setter with which the IASB works most proactively in the convergence process is the Financial Accounting Standards Board (FASB).

**conversion.** The act of exchanging a convertible security for another security.

**conversion audit.** An examination of changeover procedures, and new accounting procedures and files, that takes place when a significant change in the accounting system (e.g., a change from a manual to a computerized system or a change of computers) occurs.

**conversion cost.** *Direct labor* costs plus factory *overhead* costs incurred in producing a product; that is, the cost to convert raw materials to finished products. *Manufacturing cost.*

**conversion period.** *Compounding period;* also, period during which the holder of a *convertible bond* or *convertible preferred stock* can convert it into *common shares.*

**convertible bond.** A *bond* whose owner may convert it into a specified number of shares of *capital stock* during the *conversion period.*

**convertible preferred stock.** *Preferred shares* whose owner may convert them into a specified number of *common shares.*

**cookie jar accounting.** When a firm records a *loss,* such as for an *asset impairment,* it *debits* loss and *credits* an *asset account.* If management thinks the market expects a loss, then it can increase the amount of the reported loss even larger than actual. It knows that future *income* can be made to be larger by the amount of the excess loss recognized currently and management can choose the time for the future *earnings* boost. Arthur Levitt, Jr., former Chairman of the *SEC,* criticized this practice and called it "cookie jar accounting," because firms can overstate losses or expenses on one period to store, as in a jar, reportable earnings for discretionary reporting in the future.

**cooperative.** An incorporated organization formed for the benefit of its members (owners), who are either producers or consumers, in order to acquire for them profits or savings that otherwise accrue to middlemen. Members exercise control on the basis of one vote per member.

**coproduct.** A product sharing production facilities with another product. For example, if an apparel manufacturer produces shirts and jeans on the same line, these are coproducts. Distinguish coproducts from *joint products* and *by-products* that, by their very nature, a firm must produce together, such as the various grades of wood a lumber factory produces.

**copyright.** Exclusive right granted by the government to an individual author, composer, playwright, or the like for the life of the individual plus 50 years. If a firm receives the copyright, then the right extends 75 years after the original publication. The *economic life* of a copyright can be less than the legal life, such as, for example, the copyright of this book.

**core deposit intangible.** A bank borrows funds from its customers, called "depositors," who open checking and savings accounts. Those depositors can take out their funds at any time, but usually don't. The amount that depositors leave on deposit for long periods of time are called "core deposits." The bank lends those funds to other customers, called "borrowers," at *interest rates* larger than the amount it pays the depositors for the funds. (For checking accounts, the rate the bank pays depositors is often zero.) The fact that the depositors can remove their funds at any time, but, on average, leave amounts on deposit relatively permanently means that the bank can lend those funds for relatively long periods of time, usually at higher interest rates, than it can charge for shorter-term loans. (See *yield curve.*) The bank's ability to borrow from some customers at a low rate and lend to other customers at a high rate creates wealth for the bank. Bankers and banking analysts call the cause of this wealth the "core deposit intangible." It represents an *asset* not recognized in the financial statements by the bank that created with wealth, although some *SEC* commissioners have expressed the thought that accounting should recognize such items as assets. When one bank buys another in a *purchase,* however, it will pay for this asset and will record it as an asset. Usually, the acquiring bank does not use the specific account title "Core Deposit Intangible," but instead uses the account title *Goodwill.*

**corner.** The control, of a quantity of shares or a commodity, sufficiently large that the holder can control the market price.

**corporation.** A legal entity authorized by a state to operate under the rules of the entity's *charter.*

**correcting entry.** An *adjusting entry* that properly records a previously, improperly recorded *transaction.* Do not confuse with entries that correct *accounting errors.*

**correction of errors.** See *accounting errors.*

**cost.** The sacrifice, measured by the *price* paid or to be paid, to acquire *goods* or *services.* See *acquisition cost* and *replacement cost.* Terminology often uses "cost" when referring to the valuation of a good or service acquired. When writers use the word in this sense, a cost is an *asset.* When the benefits of the acquisition (the goods or services acquired) expire, the cost becomes an *expense* or *loss.* Some writers, however, use "cost" and "expense" as synonyms. Contrast with *expense.* The word "cost" appears in more than 50 accounting terms, each with sometimes subtle distinctions in meaning. See *cost terminology* for elaboration. Clarity requires that the user include with the word "cost" an adjective or phrase to be clear about intended meaning.

**cost accounting.** Classifying, summarizing, recording, reporting, and allocating current or predicted *costs;* a subset of *managerial accounting.*

**Cost Accounting Standards Board.** See *CASB.*

**cost accumulation.** Bringing together, usually in a single *account,* all *costs* of a specified activity. Contrast with *cost allocation.*

**cost allocation.** Assigning *costs* to individual products or time periods. Contrast with *cost accumulation.*

**cost-based transfer price.** A *transfer price* based on *historical costs.*

**cost behavior.** The functional relation between changes in activity and changes in *cost;* for example: *fixed* versus *variable costs; linear* versus *curvilinear cost.*

**cost/benefit criterion.** Some measure of *costs* compared with some measure of *benefits* for a proposed undertaking. If the costs exceed the benefits, then the analyst judges the undertaking not worthwhile. This criterion will not yield good decisions unless the analyst estimates all costs and benefits flowing from the undertaking.

**cost center.** A unit of activity for which a firm accumulates *expenditures* and *expenses.*

**cost driver.** A factor that causes an activity's costs. See *driver* and *activity basis.*

**cost-effective.** Among alternatives, the one whose benefit, or payoff, per unit of cost is highest; sometimes said of an action whose expected benefits exceed expected costs whether or not other alternatives exist with larger benefit-cost ratios.

**cost estimation.** The process of measuring the functional relation between changes in activity levels and changes in cost.

**cost flow assumption.** See *flow assumption.*

**cost-flow equation.** Beginning Balance + Transfers In = Transfers Out + Ending Balance; BB + TI = TO + EB.

**cost flows.** Costs passing through various classifications within an entity. See *flow of costs* for a diagram.

**cost method (for investments).** In accounting for an investment in the *capital stock* or *bonds* of another company, method in which the firm shows the investment at *acquisition cost* and treats only *dividends* declared or *interest receivable* as *revenue;* not allowed by *GAAP.*

**cost method (for treasury stock).** The method of showing *treasury stock* in a *contra account* to all other items of *shareholders' equity* in an amount equal to that paid to reacquire the stock.

**cost object(ive).** Any activity for which management desires a separate measurement of *costs.* Examples include departments, products, and territories.

**cost of capital.** *Opportunity cost* of funds invested in a business; the rate of return that rational owners require an asset to earn before they will devote that asset to a particular purpose; sometimes measured as the average annual rate that a company must pay for its *equities.* In *efficient capital markets,* this cost is the *discount rate* that equates the expected *present value* of all future cash flows to common shareholders with the market value of common stock at a given time. Analysts often measure the cost of capital by taking a *weighted average* of the firm's *debt* and various *equity securities.* We sometimes call the measurement so derived the "composite cost of capital," and some analysts confuse this measurement of the cost of capital with the cost of capital itself. For example, if the equities of a firm include substantial amounts for the *deferred income tax liability,* the composite cost of capital will underestimate the true cost of capital, the required rate of return on a firm's assets, because the deferred income tax liability has no explicit cost.

**cost of goods manufactured.** The sum of all costs allocated to products completed during a period, including materials, labor, and *overhead*.

**cost of goods purchased.** Net purchase price of goods acquired plus costs of storage and delivery to the place where the owner can productively use the items.

**cost of goods sold.** Inventoriable *costs* that firms *expense* because they sold the units; equals *beginning inventory* plus *cost of goods purchased* or *manufactured* minus *ending inventory*.

**cost of sales.** Generally refers to *cost of goods sold*, occasionally to *selling expenses*.

**cost or market, whichever is lower.** See *lower of cost or market*.

**cost percentage.** One less *markup percentage; cost* of *goods available for sale* divided by selling prices of goods available for sale (when FIFO is used); *cost* of *purchases* divided by selling prices of purchases (when LIFO is used). See *markup* for further detail on inclusions in the calculation of cost percentage.

**cost pool.** *Indirect cost pool;* groupings or aggregations of costs, usually for subsequent analysis.

**cost principle.** The *principle* that requires reporting *assets* at *historical* or *acquisition cost*, less accumulated *amortization*. This principle relies on the assumption that cost equals *fair market value* at the date of acquisition and that subsequent changes are not likely to be significant.

**cost-recovery-first method.** A method of *revenue* recognition that *credits inventory* as the firm receives cash collections and continues until the firm has collected cash equal to the sum of all costs. Only after the firm has collected cash equal to costs does it recognize *income*. A firm may not use this method in financial reporting unless the total amount of collections is highly uncertain. It is never allowed for income tax reporting. Contrast with the *installment method*, allowed for both book and tax, in which the firm credits *constant* proportions of each cash collection both to cost and to income.

**cost sheet.** Statement that shows all the elements composing the total cost of an item.

**cost structure.** For a given set of total costs, the percentages of fixed and variable costs, typically two percentages adding to 100 percent.

**cost terminology.** The word "cost" appears in many accounting terms. The accompanying exhibit classifies some of these terms according to the distinctions between the terms in accounting usage. Joel Dean was, to our knowledge, the first to attempt such distinctions; we have used some of his ideas here. We discuss some of the terms in more detail under their own listings.

**cost-to-cost.** The *percentage-of-completion method* in which the firm estimates the fraction of completion as the ratio of costs incurred to date divided by the total costs the firm expects to incur for the entire project.

## Cost Terminology: Distinctions among Terms Containing the Word "Cost"

| Terms (Synonyms Given in Parentheses) | | | Distinctions and Comments |
|---|---|---|---|
| | | | 1. The following pairs of terms distinguish the basis measured in accounting. |
| Historical Cost (Acquisition Cost) | v. | Current Cost | A distinction used in financial accounting. Current cost can be used more specifically to mean replacement cost, net realizable value, or present value of cash flows. "Current cost" is often used narrowly to mean replacement cost. |
| Historical Cost (Actual Cost) | v. | Standard Cost | The distinction between historical and standard costs arises in product costing for inventory valuation. Some systems record actual costs; others record the standard costs. |
| | | | 2. The following pairs of terms denote various distinctions among historical costs. For each pair of terms, the sum of the two kinds of costs equals total historical cost used in financial reporting. |
| Variable Cost | v. | Fixed Cost (Constant Cost) | Distinction used in breakeven analysis and in the design of cost accounting systems, particularly for product costing. See *(4)*, below, for a further subdivision of fixed costs and *(5)*, below, for the economic distinction between marginal and average cost closely paralleling this one. |
| Traceable Cost | v. | Common Cost (Joint Cost) | Distinction arises in allocating manufacturing costs to product. Common costs are allocated to product, but the allocations are more or less arbitrary. The distinction also arises in preparing segment reports and in separating manufacturing from nonmanufacturing costs. |
| Direct Cost | v. | Indirect Cost | Distinction arises in designing cost accounting systems and in product costing. Direct costs can be traced directly to a cost object (e.g., a product, a responsibility center), whereas indirect costs cannot. |

*(continued on next page)*

| Terms (Synonyms Given in Parentheses) | | | Distinctions and Comments |
|---|---|---|---|
| Out-of-Pocket Cost (Outlay Cost; Cash Cost) | v. | Book Cost | Virtually all costs recorded in financial statements require a cash outlay at one time or another. The distinction here separates expenditures to occur in the future from those already made and is used in making decisions. Book costs, such as for depreciation, reduce income without requiring a future outlay of cash. The cash has already been spent. See future cost v. past cost in (5), below. |
| Incremental Cost (Marginal Cost; Differential Cost) | v. | Unavoidable Cost (Inescapable Cost; Sunk Cost) | Distinction used in making decisions. Incremental costs will be incurred (or saved) if a decision is made to go ahead (or to stop) some activity, but not otherwise. Unavoidable costs will be reported in financial statements whether the decision is made to go ahead or not, because cash has already been spent or committed. Not all unavoidable costs are book costs, such as, for example, a salary that is promised but not yet earned and that will be paid even if a no-go decision is made.<br><br>The economist restricts the term marginal cost to the cost of producing one more unit. Thus the next unit has a marginal cost; the next week's output has an incremental cost. If a firm produces and sells a new product, the related new costs would properly be called incremental, not marginal. If a factory is closed, the costs saved are incremental, not marginal. |
| Escapable Cost | v. | Inescapable Cost (Unavoidable Cost) | Same distinction as incremental cost v. unavoidable cost, but this pair is used only when the decision maker is considering stopping something—ceasing to produce a product, closing a factory, or the like. See next pair. |
| Avoidable Cost | v. | Unavoidable Cost | A distinction sometimes used in discussing the merits of variable and absorption costing. Avoidable costs are treated as product costs and unavoidable costs are treated as period expenses under variable costing. |
| Controllable Cost | v. | Uncontrollable Cost | The distinction here is used in assigning responsibility and in setting bonus or incentive plans. All costs can be affected by someone in the entity; those who design incentive schemes attempt to hold a person responsible for a cost only if that person can influence the amount of the cost. |
| | | | 3. In each of the following pairs, used in historical cost accounting, the word "cost" appears in one of the terms where "expense" is meant. |
| Expired Cost | v. | Unexpired Cost | The distinction is between expense and asset. |
| Product Cost | v. | Period Cost | The terms distinguish product cost from period expense. When a given asset is used, is its cost converted into work-in-process and then finished goods on the balance sheet until the goods are sold, or is it an expense shown on this period's income statement? Product costs appear on the income statement as part of cost of goods sold in the period when the goods are sold. Period expenses appear on the income statement with an appropriate caption for the item in the period when the cost is incurred or recognized. |
| | | | 4. The following subdivisions of fixed (historical) costs are used in analyzing operations. The relation between the components of fixed costs is as follows:<br><br>Fixed Costs = Capacity Costs + Programmed Costs<br><br>Fixed Costs = (Semifixed Costs + "Pure" Fixed Costs) + Fixed Portions of Semivariable Costs<br><br>Capacity Costs = Standby Costs + Enabling Costs |
| Capacity Cost (Committed Cost) | v. | Programmed Cost (Managed Cost; Discretionary Cost) | Capacity costs give a firm the capability to produce or to sell. Programmed costs, such as for advertising or research and development, may not be essential, but once a decision to incur them is made, they become fixed costs. |
| Standby Cost | v. | Enabling Cost | Standby costs will be incurred whether capacity, once acquired, is used or not, such as property taxes and depreciation on a factory. Enabling costs, such as for a security force, can be avoided if the capacity is unused. |

(continued on next page)

| Terms (Synonyms Given in Parentheses) | | | Distinctions and Comments |
|---|---|---|---|
| Semifixed Cost | v. | Semivariable Cost | A cost that is fixed over a wide range but that can change at various levels is a semifixed cost or "step cost." An example is the cost of rail lines from the factory to the main rail line, where fixed cost depends on whether there are one or two parallel lines but is independent of the number of trains run per day. Semivariable costs combine a strictly fixed component cost plus a variable component. Telephone charges usually have a fixed monthly component plus a charge related to usage. |
| | | | 5. The following pairs of terms distinguish among economic uses or decision-making uses or regulatory uses of cost terms. |
| Fully Absorbed Cost | v. | Variable Cost (Direct Cost) | Fully absorbed costs refer to costs where fixed costs have been allocated to units or departments as required by generally accepted accounting principles. Variable costs, in contrast, may be more relevant for making decisions, such as setting prices. |
| Fully Absorbed Cost | v. | Full Cost | In full costing, all costs, manufacturing costs as well as central corporate expenses (including financing expenses), are allocated to products or divisions. In full absorption costing, only manufacturing costs are allocated to products. Only in full costing will revenues, expenses, and income summed over all products or divisions equal corporate revenues, expenses, and income. |
| Opportunity Cost | v. | Outlay Cost (Out-of-Pocket Cost) | Opportunity cost refers to the economic benefit forgone by using a resource for one purpose instead of for another. The outlay cost of the resource will be recorded in financial records. The distinction arises because a resource is already in the possession of the entity with a recorded historical cost. Its economic value to the firm, opportunity cost, generally differs from the historical cost; it can be either larger or smaller. |
| Future Cost | v. | Past Cost | Effective decision making analyzes only present and future outlay costs, or out-of-pocket costs. Opportunity costs are relevant for profit maximizing; past costs are used in financial reporting. |
| Short-Run Cost | v. | Long-Run Cost | Short-run costs vary as output is varied for a given configuration of plant and equipment. Long-run costs can be incurred to change that configuration. This pair of terms is the economic analog of the accounting pair, see (2) above, variable and fixed costs. The analogy is not perfect because some short-run costs are fixed, such as property taxes on the factory, from the point of view of breakeven analysis. |
| Imputed Cost | v. | Book Cost | In a regulatory setting some costs, for example the cost of owners' equity capital, are calculated and used for various purposes; these are imputed costs. Imputed costs are not recorded in the historical costs accounting records for financial reporting. Book costs are recorded. |
| Average Cost | v. | Marginal Cost | The economic distinction equivalent to fully absorbed cost of product and variable cost of product. Average cost is total cost divided by number of units. Marginal cost is the cost to produce the next unit (or the last unit). |
| Differential Cost (Incremental Cost) | v. | Variable Cost | Whether a cost changes or remains fixed depends on the activity basis being considered. Typically, but not invariably, costs are said to be variable or with respect to an activity basis such as changes in production levels. Typically, but not invariably, costs are said to be incremental or not with respect to an activity basis such as the undertaking of some new venture. For example, consider the decision to undertake the production of food processors, rather than food blenders, which the manufacturer has been making. To produce processors requires the acquisition of a new machine tool. The cost of the new machine tool is incremental with respect to a decision to produce food processors instead of food blenders but, once acquired, becomes a fixed cost of producing food processors. If costs of direct labor hours are going to be incurred for the production of food processors or food blenders, whichever is produced (in a scenario when not both are to be produced), such costs are variable with respect to production measured in units but are not incremental with respect to the decision to produce processors rather than blenders. This distinction is often blurred in practice, so a careful understanding of the activity basis being considered is necessary to understand the concepts being used in a particular application. |

**cost-volume-profit analysis.** A study of the sensitivity of *profits* to changes in units sold (or produced) or costs or prices.

**cost-volume-profit graph (chart).** A graph that shows the relation between *fixed costs, contribution per unit, break-even point,* and *sales*. See *breakeven chart.*

**costing.** The process of calculating the cost of activities, products, or services; the British word for *cost accounting.*

**counterparty.** The term refers to the opposite party in a legal contract. In accounting and finance, a frequent usage arises when an entity purchases (or sells) a *derivative* financial contract, such as an *option, forward contract,* and *futures contract.*

**coupon.** That portion of a *bond* document redeemable at a specified date for payments. Its physical form resembles a series of tickets; each coupon has a date, and the holder either deposits it at a bank, just like a check, for collection or mails it to the issuer's agent for collection.

**coupon rate.** Of a *bond,* the amount of annual coupons divided by par value. Contrast with *effective rate.*

**covenant.** A promise with legal validity. A loan covenant specifies the terms under which the lender can force the borrower to repay funds otherwise not yet due. For example, a *bond* covenant might say that the *principal* of a bond issue falls due on December 31, 2010, unless the firm's *debt-equity ratio* falls below 40 percent, in which case the amount becomes due immediately.

**CPA.** See *certified public accountant.* The *AICPA* suggests that no periods appear in the abbreviation.

**CPI.** *Consumer price index.*

**CPP.** Current purchasing power; usually used, primarily in the UK, as an adjective modifying the word "accounting" to mean the accounting that produces *constant-dollar financial statements.*

**Cr.** Abbreviation for *credit,* always with initial capital letter. Quiz: what do you suppose *Cr.* stands for? For the answer, see *Dr.*

**creative accounting.** Selection of *accounting principles* and interpretation of transactions or events designed to manipulate, typically to increase but sometimes merely to smooth, reported *income from continuing operations;* one form of *fraudulent financial reporting.* Many attempts at creative accounting involve premature *revenue recognition.*

**credit.** As a noun, an entry on the right-hand side of an *account;* as a verb, to make an entry on the right-hand side of an account; records increases in *liabilities, owners' equity, revenues,* and *gains;* records decreases in *assets* and *expenses.* See *debit and credit conventions.* This term also refers to the ability or right to buy or borrow in return for a promise to pay later.

**credit bureau.** An organization that gathers and evaluates data on the ability of a person to meet financial obligations and sells this information to its clients.

**credit loss.** The amount of accounts receivable that the firm finds, or expects to find, *uncollectible.*

**credit memorandum.** A document used by a seller to inform a buyer that the seller is crediting (reducing) the buyer's account receivable because of *errors, returns,* or *allowances;* also, the document provided by a bank to a depositor to indicate that the bank is increasing the depositor's balance because of some event other than a deposit, such as the collection by the bank of the depositor's *note receivable.*

**creditor.** One who lends. In the UK, *account payable.*

**critical path method (CPM).** A method of *network analysis* in which the analyst estimates normal duration time for each activity within a project. The critical path identifies the shortest completion period based on the most time-consuming sequence of activities from the beginning to the end of the network. Compare *PERT.*

**cross-reference (index).** A number placed beside each *account* in a *journal entry* indicating the *ledger* account to which the record keeper posted the entry and placing in the ledger the page number of the journal where the record keeper first recorded the journal entry; used to link the *debit* and *credit* parts of an entry in the ledger accounts back to the original entry in the journal. See *audit trail.*

**cross-section analysis.** Analysis of *financial statements* of various firms for a single period of time; contrast with *time-series analysis,* in which analysts examine statements of a given firm for several periods of time.

**Crown corporation.** Canada and UK: a corporation that is ultimately accountable, through a minister of the Crown, to Parliament or a legislature for the conduct of its affairs.

**cum div. (dividend).** The condition of shares whose quoted market price includes a declared but unpaid dividend. This condition pertains between the declaration date of the dividend and the record date. Compare *ex div. (dividend).*

**cum rights.** The condition of securities whose quoted market price includes the right to purchase new securities. Compare *ex rights.*

**cumulative dividend.** Preferred stock *dividends* that, if not paid, accrue as a commitment that the firm must pay before it can declare dividends to common shareholders.

**cumulative preferred shares.** *Preferred* shares with *cumulative dividend* rights.

**current assets.** *Cash* and other *assets* that a firm expects to turn into cash, sell, or exchange within the normal operating cycle of the firm or one year, whichever is longer. One year is the usual period for classifying asset balances on the balance sheet. Current assets include *cash, marketable securities, receivables, inventory,* and *current prepayments.*

**current cost.** *Cost* stated in terms of current values (of *productive capacity*) rather than in terms of *acquisition cost.* See *net realizable value* and *current selling price.*

**current cost accounting.** The *FASB's* term for *financial statements* in which the *attribute measured* is *current cost.*

**current cost/nominal-dollar accounting.** Accounting based on *current cost* valuations measured in *nominal dollars.* Components of *income* include an *operating margin* and *holding gains and losses.*

**current exchange rate.** The rate at which the holder of one unit of currency can convert it into another at the end of the *accounting period* being reported on or, for *revenues, expenses, gains,* and *losses,* the date of recognition of the transaction.

**current exit value.** *Exit value.*

**current fund.** In governmental accounting, a synonym for *general fund.*

**current funds.** *Cash* and other assets readily convertible into cash; in governmental accounting, funds spent for operating purposes during the current period; includes *general,* special revenue, *debt service,* and *enterprise funds.*

**current (gross) margin.** See *operating margin based on current costs.*

**current liability.** A debt or other obligation that a firm must discharge within a short time, usually the *earnings cycle* or one year, normally by expending *current assets.*

**current operating performance concept.** The notion that reported *income* for a period ought to reflect only ordinary, normal, and recurring operations of that period. A consequence is that *extraordinary* and nonrecurring items are entered directly in the Retained Earnings account. Contrast with *clean surplus concept.* This concept is no longer acceptable. (See *APB Opinion No. 9* and *No. 30.*)

**current ratio.** Sum of *current assets* divided by sum of *current liabilities.* See *ratio.*

**current realizable value.** *Realizable value.*

**current replacement cost.** Of an *asset,* the amount currently required to acquire an identical asset (in the same condition and with the same service potential) or an asset capable of rendering the same service at a current *fair market price.* If these two amounts differ, use the lower. Contrast with *reproduction cost.*

**current selling price.** The amount for which an *asset* could be sold as of a given time in an *arm's-length* transaction rather than in a forced sale.

**current service costs.** *Service costs* of a *pension plan.*

**current value accounting.** The form of accounting in which all assets appear at *current replacement cost* (*entry value*) or *current selling price* or *net realizable value* (*exit value*) and all *liabilities* appear at *present value.* Entry and exit values may differ from each other, so theorists have not agreed on the precise meaning of "current value accounting."

**current yield.** Of a *bond,* the annual amount of *coupons* divided by the current market price of the bond. Contrast with *yield to maturity.*

**currently attainable standard cost.** *Normal standard cost.*

**curvilinear (variable) cost.** A continuous, but not necessarily linear (straight-line), functional relation between activity levels and *costs.*

**customers' ledger.** The *ledger* that shows *accounts receivable* of individual customers. It is the *subsidiary ledger* for the *control account* Accounts Receivable.

**cutoff rate.** *Hurdle rate.*

# D

**data bank.** An organized file of information, such as a customer name and address file, used in and kept up-to-date by a processing system.

**database.** A comprehensive collection of interrelated information stored together in computerized form to serve several applications.

**database management system.** Generalized software programs used to handle physical storage and manipulation of databases.

**days of average inventory on hand.** See *ratio.*

**days of grace.** The days allowed by law or contract for payment of a debt after its due date.

**DCF.** *Discounted cash flow.*

**DDB.** *Double declining-balance depreciation.*

**debenture bond.** A *bond* not secured with *collateral.*

**debit.** As a noun, an entry on the left-hand side of an *account;* as a verb, to make an entry on the left-hand side of an account; records increases in *assets* and *expenses;* records decreases in *liabilities, owners' equity,* and *revenues.* See *debit and credit conventions.*

**debit and credit conventions.** The conventional use of the *T-account* form and the rules for debit and credit in *balance sheet accounts* (see below). The equality of the two sides of the *accounting equation* results from recording equal amounts of *debits* and *credits* for each *transaction.*

**Typical Asset Account**

| | |
|---|---|
| Opening Balance | |
| Increase | Decrease |
| + | – |
| Dr. | Cr. |
| Ending Balance | |

**Typical Liability Account**

| | |
|---|---|
| | Opening Balance |
| Decrease | Increase |
| – | + |
| Dr. | Cr. |
| | Ending Balance |

**Typical Owners' Equity Account**

| | |
|---|---|
| | Opening Balance |
| Decrease | Increase |
| – | + |
| Dr. | Cr. |
| | Ending Balance |

Revenue and expense accounts belong to the owners' equity group. The relation and the rules for debit and credit in these accounts take the following form:

**Owners' Equity**

| | |
|---|---|
| Decrease | Increase |
| – | + |
| Dr. | Cr. |

**Expenses**

| | |
|---|---|
| Dr. | Cr. |
| + | – |
| * | |

**Revenues**

| | |
|---|---|
| Dr. | Cr. |
| – | + |
| | * |

*Normal balance before closing

**debit memorandum.** A document used by a seller to inform a buyer that the seller is debiting (increasing) the amount of the buyer's *accounts receivable.* Also, the document provided by a bank to a depositor to indicate that the bank is decreasing the depositor's *balance* because of some event other than payment for a *check,* such as monthly service charges or the printing of checks.

**debt.** An amount owed. The general name for *notes, bonds, mortgages,* and the like that provide evidence of amounts owed and have definite payment dates.

**debt capital.** *Noncurrent liabilities.* See *debt financing,* and contrast with *equity financing.*

**debt-equity ratio.** Total *liabilities* divided by total equities. See *ratio.* Some analysts put only total shareholders' equity in the denominator. Some analysts restrict the numerator to *long-term debt.*

**debt financing.** *Leverage.* Raising *funds* by issuing *bonds, mortgages,* or *notes.* Contrast with *equity financing.*

**debt guarantee.** See *guarantee.*

**debt ratio.** *Debt-equity ratio.*

**debt service fund.** In governmental accounting, a *fund* established to account for payment of *interest* and *principal* on all general-obligation *debt* other than that payable from special *assessments.*

**debt service payment.** The payment required by a lending agreement, such as periodic coupon payment on a bond or installment payment on a loan or a lease payment. It is sometimes called "interest payment," but this term will mislead the unwary. Only rarely will the amount of a debt service payment equal the interest expense for the period preceding the payment. A debt service payment will always include some amount for interest, but the payment will usually differ from the interest expense.

**debt service requirement.** The amount of cash required for payments of *interest,* current maturities of *principal* on outstanding *debt,* and payments to *sinking funds* (corporations) or to the debt service fund (governmental).

**debtor.** One who borrows; in the UK, *account receivable.*

**decentralized decision making.** Management practice in which a firm gives a manager of a business unit responsibility for that unit's *revenues* and *costs,* freeing the manager to make decisions about prices, sources of supply, and the like, as though the unit were a separate business that the manager owns. See *responsibility accounting* and *transfer price.*

**declaration date.** Time when the *board of directors* declares a *dividend.*

**declining-balance depreciation.** The method of calculating the periodic *depreciation* charge by multiplying the *book value* at the start of the period by a constant percentage. In pure declining-balance depreciation, the constant percentage is $1 - \sqrt[n]{s/c}$, where *n* is the *depreciable life, s* is *salvage value,* and *c* is *acquisition cost.* See *double declining-balance depreciation.*

**deep discount bonds.** Said of *bonds* selling much below (exactly how much is not clear) *par value.*

**defalcation.** Embezzlement.

**default.** Failure to pay *interest* or *principal* on a *debt* when due.

**defeasance.** Transaction with the economic effect of *debt retirement* that does not retire the debt. When *interest rates* increase, many firms find that the *market value* of their outstanding *debt* has dropped substantially below its *book value.* In *historical cost accounting* for debt retirements, retiring debt with a *cash* payment less than the book value of the debt results in a gain (generally, an *extraordinary item*). Many firms would like to retire the outstanding debt issues and report the gain. Two factors impede doing so: (1) the gain can be a taxable event generating adverse *income tax* consequences; and (2) the transaction costs in retiring all the debt can be large, in part because the firm cannot easily locate all the debt holders or persuade them to sell back their bonds to the issuer. The process of "defeasance" serves as the economic equivalent to retiring a debt issue while it saves the issuer from experiencing adverse tax consequences and from actually having to locate and retire the bonds. The process works as follows. The debt-issuing firm turns over to an independent trustee, such as a bank, amounts of cash or low-risk government bonds sufficient to make all debt service payments on the outstanding debt, including bond retirements, in return for the trustee's commitment to make all debt service payments. The debt issuer effectively retires the outstanding debt. It debits the liability account, credits Cash or Marketable Securities as appropriate, and credits Extraordinary Gain on Debt Retirement. The trustee can retire debt or make debt service payments, whichever it chooses. For income tax purposes, however, the firm's debt remains outstanding. The firm will have taxable interest *deductions* for its still-outstanding debt and taxable interest *revenue* on the investments held by the trustee for debt service. In law, the term "defeasance" means "a rendering null and void." This process renders the outstanding debt economically null and void, without causing a taxable event.

**defensive interval.** A financial *ratio* equal to the number of days of normal cash *expenditures* covered by *quick assets.* It is defined as follows:

$$\frac{\text{Quick Assets}}{\text{(All Expenses Except Amortization and Others Not Using Funds} \div 365)}$$

The denominator of the ratio is the cash expenditure per day. Analysts have found this ratio useful in predicting *bankruptcy.*

**deferral.** The accounting process concerned with past *cash receipts* and *payments;* in contrast to *accrual;* recognizing a liability resulting from a current cash receipt (as for magazines to be delivered) or recognizing an asset from a current cash payment (as for prepaid insurance or a long-term depreciable asset).

**deferral method.** See *flow-through method* (of accounting for the *investment credit*) for definition and contrast.

**deferred annuity.** An *annuity* whose first payment occurs sometime after the end of the first period.
**deferred asset.** *Deferred charge.*
**deferred charge.** *Expenditure* not recognized as an *expense* of the period when made but carried forward as an *asset* to be *written off* in future periods, such as for advance rent payments or insurance premiums. See *deferral.*
**deferred cost.** *Deferred charge.*
**deferred credit.** Sometimes used to indicate *advances from customers.*
**deferred debit.** *Deferred charge.*
**deferred expense.** *Deferred charge.*
**deferred gross margin.** *Unrealized gross margin.*
**deferred income.** *Advances from customers.*
**deferred income tax (liability).** An *indeterminate-term liability* that arises when the pretax income shown on the tax return is less than what it would have been had the firm used the same *accounting principles* and *cost basis* for *assets* and *liabilities* in tax returns as it used for financial reporting. *SFAS No. 109* requires that the firm debit income tax *expense* and credit deferred income tax with the amount of the taxes delayed by using accounting principles in tax returns different from those used in financial reports. See *temporary difference, timing difference, permanent difference,* and *installment sales.* If, as a result of temporary differences, cumulative taxable income exceeds cumulative reported income before taxes, the deferred income tax account will have a *debit* balance, which the firm will report as a *deferred charge.*
**deferred revenue.** Sometimes used to indicate *advances from customers.*
**deferred tax.** See *deferred income tax.*
**deficit.** A *debit balance* in the Retained Earnings account; presented on the balance sheet in a *contra account* to shareholders' equity; sometimes used to mean negative *net income* for a period.
**defined-benefit plan.** A *pension plan* in which the employer promises specific dollar amounts to each eligible employee; the amounts usually depend on a formula that takes into account such things as the employee's earnings, years of employment, and age. The employer adjusts its cash contributions and pension expense to *actuarial* experience in the eligible employee group and investment performance of the pension *fund.* This is sometimes called a "fixed-benefit" pension plan. Contrast with *money purchase plan.*
**defined-contribution plan.** A *money purchase* (*pension*) *plan* or other arrangement, based on formula or discretion, in which the employer makes cash contributions to eligible individual employee *accounts* under the terms of a written plan document. The trustee of the funds in the account manages the funds, and the employee-beneficiary receives at retirement (or at some other agreed time) the amount in the fund. The employer makes no promise about that amount. Profit-sharing pension plans are of this type.
**deflation.** A period of declining *general price-level changes.*
**demand deposit.** *Funds* in a *checking account* at a bank.
**demand loan.** See *term loan* for definition and contrast.
**denial of opinion.** Canada: the statement that an *auditor,* for reasons arising in the *audit,* is unable to express an opinion on whether the *financial statements* provide *fair presentation.*
**denominator volume.** Capacity measured in the number of units the firm expects to produce this period; when divided into *budgeted fixed costs,* results in fixed costs applied per unit of product.
**department(al) allocation.** Obtained by first accumulating *costs* in *cost pools* for each department and then, using separate rates, or sets of rates, for each department, allocating from each cost pool to products produced in that department.
**dependent variable.** See *regression analysis.*
**depletion.** Exhaustion or *amortization* of a *wasting asset* or *natural resource.* Also see *percentage depletion.*
**depletion allowance.** See *percentage depletion.*
**deposit intangible.** See *core deposit intangible.*
**deposit, sinking fund.** Payments made to a *sinking fund.*
**deposit method (of revenue recognition).** A method of *revenue* recognition that is the same as the *completed sale* or *completed contract method.* In some contexts, such as when the customer has the right to return goods for a full refund or in retail land sales, the customer must make substantial payments while still having the right to back out of the deal and receive a refund. When the seller cannot predict with reasonable precision the amount of cash it will ultimately collect and when it will receive cash, the seller must *credit* Deposits, a *liability account,* rather than *revenue.* (In this regard, the accounting differs from that in the completed contract method, in which the account credited offsets the *Work-in-Process* inventory account.) When the *sale* becomes complete, the firm credits a revenue account and *debits* the Deposits account.
**deposits (by customers).** A *liability* that the firm *credits* when receiving *cash* (as in a bank, or in a grocery store when the customer pays for soda-pop bottles with cash to be repaid when the customer returns the bottles) and when the firm intends to discharge the liability by returning the cash. Contrast with the liability account *Advances from Customers,* which the firm credits on receipt of cash, expecting later to discharge the liability by delivering goods or services. When the firm delivers the goods or services, it credits a *revenue* account.
**deposits in transit.** Deposits made by a firm but not yet reflected on the *bank statement.*
**depreciable cost.** That part of the *cost* of an asset, usually *acquisition cost* less *salvage value,* that the firm will charge off over the life of the asset through the process of *depreciation.*
**depreciable life.** For an *asset,* the time period or units of activity (such as miles driven for a truck) over which the firm allocates the *depreciable cost.* For tax returns, depreciable life may be shorter than estimated *service life.*

**depreciation.** *Amortization of plant assets;* the process of allocating the cost of an asset to the periods of benefit—the *depreciable life;* classified as a *production cost* or a *period expense,* depending on the asset and whether the firm uses *full absorption* or *variable costing.* Depreciation methods described in this glossary include the *annuity method, appraisal method, composite method, compound interest method, declining-balance method, production method, replacement method, retirement method, straight-line method, sinking fund method,* and *sum-of-the-years'-digits method.*

**depreciation reserve.** An inferior term for *accumulated depreciation.* See *reserve.* Do not confuse with a replacement *fund.*

**derivative (financial instrument).** A financial instrument, such as an option to purchase a share of stock, created from another, such as a share of stock; an instrument, such as a *swap,* whose value depends on the value of another asset called the "underlying"—for example, the right to receive the difference between the interest payments on a fixed-rate five-year loan for $1 million and the interest payments on a floating-rate five-year loan for $1 million. To qualify as a derivative under *FASB* rules, *SFAS No. 133,* the instrument has one or more underlyings, and one or more notional amounts or payment provisions or both, it either does not require an initial net investment or it requires one smaller than would be required for other types of contracts expected to have a similar response to changes in market factors, and its terms permit settlement for cash in lieu of physical delivery or the instrument itself trades on an exchange. See also *forward contract* and *futures contract.*

**detective controls.** *Internal controls* designed to detect, or maximize the chance of detection of, errors and other irregularities.

**determination.** See *determine.*

**determine.** A term often used (in our opinion, overused) by accountants and those who describe the accounting process. A leading dictionary associates the following meanings with the verb "determine": settle, decide, conclude, ascertain, cause, affect, control, impel, terminate, and decide upon. In addition, accounting writers can mean any one of the following: measure, allocate, report, calculate, compute, observe, choose, and legislate. In accounting, there are two distinct sets of meanings: those encompassed by the synonym "cause or legislate" and those encompassed by the synonym "measure." The first set of uses conveys the active notion of causing something to happen, and the second set of uses conveys the more passive notion of observing something that someone else has caused to happen. An accountant who speaks of cost or income "determination" generally means measurement or observation, not causation; management and economic conditions cause costs and income to be what they are. One who speaks of accounting principles "determination" can mean choosing or applying (as in "determining depreciation charges" from an allowable set) or causing to be acceptable (as in the *FASB's* "determining" the accounting for *leases*). In the long run, income is cash-in less cash-out, so management and economic conditions "determine" (cause) income to be what it is. In the short run, reported income is a function of accounting principles chosen and applied, so the accountant "determines" (measures) income. A question such as "Who determines income?" has, therefore, no unambiguous answer. The meaning of "an accountant determining acceptable accounting principles" is also vague. Does the clause mean merely choosing one principle from the set of generally acceptable principles, or does it mean using professional judgment to decide that some of the generally accepted principles are not correct under the current circumstances? We try never to use "determine" unless we mean "cause." Otherwise we use "measure," "report," "calculate," "compute," or whatever specific verb seems appropriate. We suggest that careful writers will always "determine" to use the most specific verb to convey meaning. "Determine" seldom best describes a process in which those who make decisions often differ from those who apply technique. The term *predetermined (factory) overhead rate* contains an appropriate use of the word.

**development stage enterprise.** As defined in *SFAS No. 7,* a firm whose planned principal *operations* have not commenced or, having commenced, have not generated significant *revenue.* The financial statements should identify such enterprises, but no special *accounting principles* apply to them.

**diagnostic signal.** See *warning signal* for definition and contrast.

**differentiable cost.** The cost increments associated with infinitesimal changes in volume. If a total cost curve is smooth (in mathematical terms, differentiable), then we say that the curve graphing the derivative of the total cost curve shows differentiable costs.

**differential.** An adjective used to describe the change (increase or decrease) in a *cost, expense, investment, cash flow, revenue, profit,* and the like as the firm produces or sells one or more additional (or fewer) units or undertakes (or ceases) an activity. This term has virtually the same meaning as *incremental,* but if the item declines, "decremental" better describes the change. Contrast with *marginal,* which means the change in cost or other item for a small (one unit or even less) change in number of units produced or sold.

**differential analysis.** Analysis of *differential costs, revenues, profits, investment, cash flow,* and the like.

**differential cost.** See *differential.*

**dilution.** A potential reduction in *earnings per share* or *book value* per share by the potential *conversion* of securities or by the potential exercise of *warrants* or *options.*

**diluted earnings per share.** For *common stock,* smallest *earnings per share* figure that one can obtain by computing an earnings per share for all possible combinations of assumed *exercise* or *conversion* of *potentially dilutive securities.*

**dilutive.** Said of a *security* that will reduce *earnings per share* if it is exchanged for *common shares*.

**dip(ping) into LIFO layers.** See *LIFO inventory layer*.

**direct access.** Access to computer storage where information can be located directly, regardless of its position in the storage file. Compare *sequential access*.

**direct cost.** Cost of *direct material* and *direct labor* incurred in producing a product. See *prime cost*. In some accounting literature, writers use this term to mean the same thing as *variable cost*.

**direct costing.** Another, less-preferred, term for *variable costing*.

**direct-financing (capital) lease.** See *sales-type (capital) lease* for definition and contrast.

**direct labor (material) cost.** Cost of labor (material) applied and assigned directly to a product; contrast with *indirect labor (material)*.

**direct labor variance.** Difference between actual and *standard direct labor* allowed.

**direct method.** See *statement of cash flows*.

**direct posting.** A method of bookkeeping in which the firm makes *entries* directly in *ledger accounts*, without using a *journal*.

**direct write-off method.** See *write-off method*.

**disbursement.** Payment by *cash* or by *check*. See *expenditure*.

**DISC (domestic international sales corporation).** A U.S. *corporation*, usually a *subsidiary*, whose *income* results primarily from exports. The parent firm usually defers paying *income tax* on 50 percent of a DISC's income for a long period. Generally, this results in a lower overall corporate tax for the *parent* than would otherwise be incurred.

**disclaimer of opinion.** An *auditor's report* stating that the auditor cannot give an opinion on the *financial statements*. Usually results from *material* restrictions on the scope of the audit or from material uncertainties, which the firm has been unable to resolve by the time of the audit, about the accounts.

**disclosure.** The showing of facts in *financial statements*, *notes* thereto, or the *auditor's report*.

**discontinued operations.** See *income from discontinued operations*.

**discount.** In the context of *compound interest*, *bonds*, and *notes*, the difference between *face amount* (or *future value*) and *present value* of a payment; in the context of *sales* and *purchases*, a reduction in price granted for prompt payment. See also *chain discount*, *quantity discount*, and *trade discount*.

**discount factor.** The reciprocal of one plus the *discount rate*. If the discount rate is 10 percent per period, the discount factor for three periods is $1/(1.10)^3 = (1.10)^{-3} = 0.75131$.

**discount rate.** *Interest rate* used to convert future payments to *present values*.

**discounted bailout period.** In a *capital budgeting* context, the total time that must elapse before discounted value of net accumulated cash flows from a project, including potential *salvage value* at various times of assets, equals or exceeds the *present value* of net accumulated cash outflows. Contrast with *discounted payback period*.

**discounted cash flow (DCF).** Using either the *net present value* or the *internal rate of return* in an analysis to measure the value of future expected cash *expenditures* and *receipts* at a common date. In discounted cash flow analysis, choosing the alternative with the largest *internal rate of return* may yield wrong answers given *mutually exclusive projects* with differing amounts of initial investment for two of the projects. Consider, to take an unrealistic example, a project involving an initial investment of $1, with an *IRR* of 60 percent, and another project involving an initial investment of $1 million, with an IRR of 40 percent. Under most conditions, most firms will prefer the second project to the first, but choosing the project with the larger IRR will lead to undertaking the first, not the second. Usage calls this shortcoming of choosing between alternatives based on the magnitude of the internal rate of return, rather than based on the magnitude of the *net present value* of the cash flows, the "scale effect."

**discounted payback period.** The shortest amount of time that must elapse before the discounted *present value* of cash inflows from a project, excluding potential *salvage value*, equals the discounted present value of the cash outflows.

**discounting a note.** See *note receivable discounted* and *factoring*.

**discounts lapsed (lost).** The sum of *discounts* offered for prompt payment that the purchaser did not take because the discount period expired. See *terms of sale*.

**discovery sampling.** Acceptance sampling in which the analyst accepts an entire population if and only if the sample contains no disparities.

**discovery value accounting.** See *reserve recognition accounting*.

**discretionary cost center.** See *engineered cost center* for definition and contrast.

**discretionary costs.** *Programmed costs*.

***Discussion Memorandum.*** A neutral discussion of all the issues concerning an accounting problem of current concern to the *FASB*. The publication of such a document usually signals that the FASB will consider issuing an *SFAS* or *SFAC* on this particular problem. The discussion memorandum brings together material about the particular problem to facilitate interaction and comment by those interested in the matter. A public hearing follows before the FASB will issue an *Exposure Draft*.

**dishonored note.** A *promissory note* whose maker does not repay the loan at *maturity*, for a *term loan*, or on demand, for a *demand loan*.

**disintermediation.** Moving funds from one interest-earning account to another, typically one promising a higher rate. Federal law regulates the maximum *interest rate* that both banks and savings-and-loan associations can pay for *time deposits*. When free-market interest rates exceed the regulated interest ceiling for such time deposits, some depositors withdraw their funds and invest them elsewhere at a higher interest rate. This process is known as "disintermediation."

**distributable income.** The portion of conventional accounting net income that the firm can distribute to owners (usually in the form of *dividends*) without impairing the physical capacity of the firm to continue operations at current levels. Pretax distributable income is conventional pretax income less the excess of *current cost* of goods sold and *depreciation* charges based on the replacement cost of *productive capacity* over cost of goods sold and depreciation on an *acquisition cost basis*. Contrast with *sustainable income*. See *inventory profit*.

**distributable surplus.** Canada and UK: the statutory designation to describe the portion of the proceeds of the issue of shares without *par value* not allocated to share capital.

**distributed processing.** Processing in a computer information network in which an individual location processes data relevant to it while the operating system transmits information required elsewhere, either to the central computer or to another local computer for further processing.

**distribution expense.** *Expense* of selling, advertising, and delivery activities.

**dividend.** A distribution of assets generated from *earnings* to owners of a corporation. The firm may distribute cash (cash dividend), stock (stock dividend), property, or other securities (dividend in kind). Dividends, except stock dividends, become a legal liability of the corporation when the corporation's board declares them. Hence, the owner of stock ordinarily recognizes *revenue* when the board of the corporation declares the dividend, except for stock dividends. See also *liquidating dividend* and *stock dividend*.

**dividend yield.** *Dividends* declared for the year divided by market price of the stock as of the time for which the analyst computes the yield.

**dividends in arrears.** Dividends on *cumulative preferred stock* that the corporation's board has not yet declared in accordance with the preferred stock contract. The corporation must usually clear such arrearages before it can declare dividends on *common shares*.

**dividends in kind.** See *dividend*.

**division.** A more or less self-contained business unit that is part of a larger family of business units under common control.

**divisional control.** See *control system*.

**divisional reporting.** See *segment reporting*.

**dollar sign rules.** In accounting statements or schedules, place a dollar sign beside the first figure in each column and beside any figure below a horizontal line drawn under the preceding figure.

**dollar-value LIFO method.** A form of *LIFO* inventory accounting with inventory quantities (*layers*) measured in dollar, rather than physical, terms. The method adjusts for changing prices by using specific price indexes appropriate for the kinds of items in the inventory.

**domestic international sales corporation.** See *DISC*.

**donated capital.** A *shareholders' equity* account credited when the company receives gifts, such as land or buildings, without issuing shares or other owners' equity interest in return. A city might donate a plant site hoping the firm will build a factory and employ local residents. Do not confuse with *contributed capital*.

**double declining-balance depreciation (DDB).** *Declining-balance depreciation* in which the constant percentage used to multiply by book value in computing the depreciation charge for the year is $2/n$, where $n$ is the *depreciable life* in periods. Omit *salvage value* from the depreciable amount. Thus if the asset cost \$100 and has a depreciable life of five years, the depreciation in the first year would be \$40 = 2/5 × \$100, in the second year would be \$24 = 2/5 × (\$100 − \$40), and in the third year would be \$14.40 = 2/5 × (\$100 − \$40 − \$24). By the fourth year, the remaining undepreciated cost could be depreciated under the straight-line method at \$10.80 = 1/2 × (\$100 − \$40 − \$24 − \$14.40) per year for tax purposes. Note that salvage value does not affect these computations except that the method will not depreciate the book value below salvage value.

**double entry.** In recording transactions, a system that maintains the equality of the accounting equation or the balance sheet. Each entry results in recording equal amounts of *debits* and *credits*.

**double T-account.** *T-account* with an extra horizontal line showing a change in the account balance to be explained by the subsequent entries into the account.

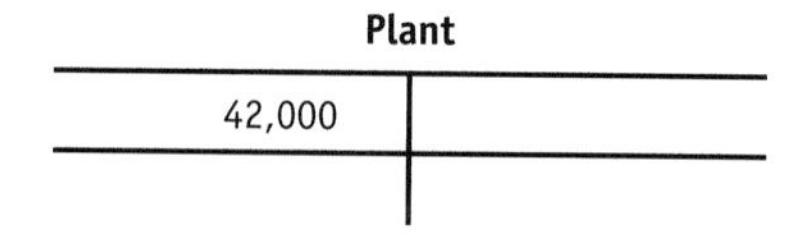


This account shows an increase in the asset account, plant, of \$42,000 to be explained. Such accounts are useful in preparing the *statement of cash flows;* they are not a part of the formal record-keeping process.

**double taxation.** Occurs when the taxing authority (U.S. or state) taxes corporate income as earned (first tax) and then the same taxing authority taxes the aftertax income, distributed to owners as dividends, again as personal income tax (second tax).

**doubtful accounts.** *Accounts receivable* that the firm estimates to be *uncollectible*.

**Dr.** The abbreviation for *debit*, always with the initial capital letter. *Dr.* is a shortened form of the word *debitor*, and *Cr.* comes from the word *creditor*. In the early days of double-entry record keeping in the UK, the major asset was accounts receivable, called *creditors*, and the major liability was accounts payable, called *debitors*. Thus the *r* in *Cr.* does not refer to the *r* in *credit* but to the second *r* in *creditor*.

**draft.** A written order by the first party, called the drawer, instructing a second party, called the drawee (such as a bank) to pay a third party, called the payee. See also *check, cashier's check, certified check, NOW account, sight draft,* and *trade acceptance*.

**drawee.** See *draft.*

**drawer.** See *draft.*

**drawing account.** A *temporary account* used in *sole proprietorships* and *partnerships* to record payments to owners or partners during a period. At the end of the period, the firm closes the drawing account by crediting it and debiting the owner's or partner's share of income or, perhaps, his or her capital account.

**drawings.** Payments made to a *sole proprietor* or to a *partner* during a period. See *drawing account.*

**driver, cost driver.** A cause of costs incurred. Examples include processing orders, issuing an engineering change order, changing the production schedule, and stopping production to change machine settings. The notion arises primarily in product costing, particularly *activity-based costing.*

**drop ship(ment).** Occurs when a distributor asks a manufacturer to send an order directly to the customer (ordinarily a manufacturer sends goods to a distributor, who sends the goods to its customer). Usage calls the shipment a "drop shipment" and refers to the goods as "drop shipped."

**dry-hole accounting.** See *reserve recognition accounting* for definition and contrast.

**dual-transactions assumption (fiction).** Occurs when an analyst, in understanding cash flows, views transactions not involving *cash* as though the firm first generated cash and then used it. For example, the analyst might view the issue of *capital stock* in return for the *asset* land as though the firm issued stock for *cash* and then used cash to acquire the land. Other examples of transactions that could involve the dual-transaction assumption are the issue of a *mortgage* in return for a noncurrent asset and the issue of stock to bondholders on *conversion* of their *convertible bonds.*

**dual transfer prices.** Occurs when the *transfer price charged* to the buying *division* differs from that *credited* to the selling division. Such prices make sense when the selling division has excess capacity and, as usual, the *fair market value* exceeds the *incremental cost* to produce the goods or services being transferred.

**duality.** The *double entry* record-keeping axiom that every *transaction* must result in equal *debit* and *credit* amounts.

**dumping.** A foreign firm's selling a good or service in the United States at a price below market price at home or, in some contexts, below some measure of cost (which concept is not clearly defined). The practice is illegal in the United States if it harms (or threatens to harm) a U.S. industry.

# E

**e.** The base of natural logarithms; 2.71828. . . . If *interest* compounds continuously during a period at stated rate of $r$ per period, then the effective *interest rate* is equivalent to interest compounded once per period at rate $i$ where $i = e^r - 1$. Tables of $e^r$ are widely available. If 12 percent annual interest compounds continuously, the effective annual rate is $e^{.12} - 1 = 12.75$ percent. Interest compounded continuously at rate $r$ for $d$ days is $e^{rd/365} - 1$. For example, interest compounded continuously for 92 days at 12 percent is $e^{.12 \times 92/365} - 1 = 3.07$ percent.

**earn-out.** For two merging firms, an agreement in which the amount paid by the acquiring firm to the acquired firm's shareholders depends on the future earnings of the acquired firm or, perhaps, of the *consolidated entity.*

**earned surplus.** A term that writers once used, but no longer use, for *retained earnings.*

**earnings.** A term with no precise meaning but used to mean *income* or sometimes *profit.* The *FASB,* in requiring that firms report *comprehensive income,* encouraged firms to use the term "earnings" for the total formerly reported as *net income.* Firms will likely only slowly change from using the term "net income" to the term "earnings."

**earnings, retained.** See *retained earnings.*

**earnings cycle.** The period of time, or the series of transactions, during which a given firm converts *cash* into *goods* and *services,* then sells goods and services to customers, and finally collects cash from customers. *Cash cycle.*

**earnings per share (of common stock).** *Net income* to common shareholders (net income minus *preferred dividends*) divided by the average number of *common shares* outstanding; see also *basic earnings per share* and *diluted earnings per share.* See *ratio.*

**earnings per share (of preferred stock).** *Net income* divided by the average number of *preferred shares* outstanding during the period. This ratio indicates how well income covers (or protects) the preferred dividends; it does not indicate a legal share of *earnings.* See *ratio.*

**earnings statement.** *Income statement.*

**easement.** The acquired right or privilege of one person to use, or have access to, certain property of another. For example, a public utility's right to lay pipes or lines under the property of another and to service those facilities.

**EBIT.** *Earnings* before *interest and (income) taxes;* acronym used by analysts.

**EBITDA.** *Earnings* before *interest, (income) taxes, depreciation,* and *amortization;* acronym used by analysts to focus on a particular measure of *cash flow* used in valuation. This is not the same as, but is similar in concept to, *cash flow from operations.* Some analysts exclude *nonrecurring* items from this total.

**economic consequences.** The *FASB* says that in setting *accounting principles,* it should take into account the real effects on various participants in the business world. It calls these effects "economic consequences."

**economic depreciation.** Decline in *current cost* (or *fair value*) of an *asset* during a period.

**economic entity.** See *entity.*

**economic life.** The time span over which the firm expects to receive the benefits of an *asset.* The economic life of a *patent, copyright,* or *franchise* may be less than the legal life. *Service life.*

**economic order quantity (EOQ).** In mathematical *inventory* analysis, the optimal amount of stock to order when demand reduces inventory to a level called the "reorder point." If $A$ represents the *incremental cost* of placing a single order, $D$ represents the total demand for a period of time in units, and $H$ represents the incremental holding cost during the period per unit of inventory, then the economic order quantity is $EOQ = \sqrt{2AD/H}$. Usage sometimes calls *EOQ* the "optimal lot size."

***ED.*** *Exposure Draft.*

**EDGAR.** Electronic Data, Gathering, Analysis, and Retrieval system; rules and systems adopted by the *SEC* in 1993 to ensure that all the paperwork involved in the filings submitted by more than 15,000 public companies are electronically submitted.

**EDP.** *Electronic data processing.*

**effective interest method.** In computing *interest expense* (or *revenue*), a systematic method that makes the interest expense (revenue) for each period divided by the amount of the net *liability* (*asset*) at the beginning of the period equal to the *yield rate* on the liability (asset) at the time of issue (acquisition). Interest for a period is the yield rate (at time of issue) multiplied by the net liability (asset) at the start of the period. The *amortization* of discount or premium is the *plug* to give equal *debits* and *credits*. (Interest expense is a debit, and the amount of debt service payment is a credit.)

**effective (interest) rate.** Of a liability such as a bond, the *internal rate of return* or *yield to maturity* at the time of issue. Contrast with *coupon rate*. If the borrower issues the bond for a price below *par,* the effective rate is higher than the coupon rate; if it issues the bond for a price greater than par, the effective rate is lower than the coupon rate. In the context of *compound interest,* the effective rate occurs when the *compounding period* on a *loan* differs from one year, such as a nominal interest rate of 12 percent compounded monthly. The effective interest is the single rate that one could use at the end of the year to multiply the *principal* at the beginning of the year and give the same amount as results from compounding interest each period during the year. For example, if 12 percent per year compounds monthly, the effective annual interest rate is 12.683 percent. That is, if you compound $100 each month at 1 percent per month, the $100 will grow to $112.68 at the end of the year. In general, if the nominal rate of $r$ percent per year compounds $m$ times per year, then the effective rate is $(1 + r/m)^m - 1$.

**efficiency variance.** A term used for the *quantity variance* for materials or labor or *variable overhead* in a *standard costing system.*

**efficient capital market.** A market in which security prices reflect all available information and react nearly instantaneously and in an unbiased fashion to new information.

**efficient market hypothesis.** The finance supposition that security prices trade in *efficient capital markets.*

**EITF.** *Emerging Issues Task Force.*

**electronic data processing.** Performing computations and other data-organizing steps in a computer, in contrast to doing these steps by hand or with mechanical calculators.

**eligible.** Under income tax legislation, a term that restricts or otherwise alters the meaning of another tax or accounting term, generally to signify that the related assets or operations may receive a specified tax treatment.

**eliminations.** In preparing *consolidated statements, work sheet* entries made to avoid duplicating the amounts of *assets, liabilities, owners' equity, revenues,* and *expenses* of the consolidated *entity* when the firm sums the accounts of the *parent* and *subsidiaries.*

**Emerging Issues Task Force (EITF).** A group convened by the *FASB* to deal more rapidly with accounting issues than the FASB's due-process procedures can allow. The task force comprises about 20 members from public accounting, industry, and several trade associations. It meets every six weeks. Several FASB board members usually attend and participate. The chief accountant of the *SEC* has indicated that the SEC will require that published financial statements follow guidelines set by a consensus of the EITF. The EITF requires that nearly all its members agree on a position before that position receives the label of "consensus." Such positions appear in *Abstracts of the EITF,* published by the FASB. Since 1984, the EITF has become one of the promulgators of *GAAP.*

**employee stock option.** See *stock option.*

**Employee Stock Ownership Trust (or Plan).** See *ESOT.*

**employer, employee payroll taxes.** See *payroll taxes.*

**enabling costs.** A type of *capacity cost* that a firm will stop incurring if it shuts down operations completely but will incur in full if it carries out operations at any level. Examples include costs of a security force or of a quality-control inspector for an assembly line. Contrast with *standby costs.*

**encumbrance.** In governmental accounting, an anticipated *expenditure* or *funds* restricted for an anticipated expenditure, such as for outstanding purchase orders. *Appropriations* less expenditures less outstanding encumbrances yields unencumbered balance.

**ending inventory.** The *cost* of *inventory* on hand at the end of the *accounting period;* often called "closing inventory." Ending inventory from the end of one period becomes the *beginning inventory* for the next period.

**endorsee.** See *endorser.*

**endorsement.** See *draft.* The *payee* signs the draft and transfers it to a fourth party, such as the payee's bank.

**endorser.** A *note* or *draft payee,* who signs the note after writing "Pay to the order of X," transfers the note to person X, and presumably receives some benefit, such as cash, in return. Usage refers to person X as the "endorsee." The endorsee then has the rights of the payee and may in turn become an endorser by endorsing the note to another endorsee.

**engineered cost center.** Responsibility center with sufficiently well-established relations between inputs and outputs that the analyst, given data on inputs, can predict the outputs or, conversely, given the outputs, can estimate the amounts of inputs that the process should have used. Consider the relation between pounds of flour (input) and loaves of bread (output). Contrast discretionary cost center, where such relations are so imprecise that analysts have no reliable way to relate inputs to outputs. Consider the relation between advertising the corporate logo or trademark (input) and future revenues (output).

**engineering method (of cost estimation).** To estimate unit cost of product from study of the materials, labor, and *overhead* components of the production process.

**enterprise.** Any business organization, usually defining the accounting *entity*.

**enterprise fund.** A *fund* that a governmental unit establishes to account for acquisition, operation, and maintenance of governmental services that the government intends to be self-supporting from user charges, such as for water or airports and some toll roads.

**entity.** A person, *partnership, corporation,* or other organization. The *accounting entity* that issues accounting statements may not be the same as the entity defined by law. For example, a *sole proprietorship* is an accounting entity, but the individual's combined business and personal assets are the legal entity in most jurisdictions. Several affiliated corporations may be separate legal entities but issue *consolidated financial statements* for the group of companies operating as a single economic entity.

**entity theory.** The corporation view that emphasizes the form of the *accounting equation* that says *assets* = *equities*. Contrast with *proprietorship theory*. The entity theory focuses less on the distinction between *liabilities* and *shareholders' equity* than does the proprietorship theory. The entity theory views all equities as coming to the corporation from outsiders who have claims of differing legal standings. The entity theory implies using a *multiple-step* income statement.

**entry value.** The *current cost* of acquiring an asset or service at a *fair market price. Replacement cost.*

**EOQ.** *Economic order quantity.*

**EPS.** *Earnings per share.*

**EPVI.** *Excess present value index.*

**equalization reserve.** An inferior title for the allowance or *estimated liability* account when the firm uses the *allowance method* for such things as maintenance expenses. Periodically, the accountant will debit maintenance *expense* and credit the allowance. As the firm makes *expenditures* for maintenance, it will debit the allowance and credit cash or the other asset used in maintenance.

**equities.** *Liabilities* plus *owners' equity*. See *equity*.

**equity.** A claim to *assets;* a source of assets. *SFAC No. 3* defines equity as "the residual interest in the assets of an entity that remains after deducting its liabilities." Thus, many knowledgeable people use "equity" to exclude liabilities and count only owners' equities. We prefer to use the term to mean all liabilities plus all owners' equity because there is no other single word that serves this useful purpose. We fight a losing battle.

**equity financing.** Raising *funds* by issuing *capital stock*. Contrast with *debt financing*.

**equity method.** In accounting for an *investment* in the stock of another company, a method that debits the proportionate share of the earnings of the other company to the investment account and credits that amount to a *revenue* account as earned. When the investor receives *dividends,* it debits *cash* and credits the investment account. An investor who owns sufficient shares of stock of an unconsolidated company to exercise significant control over the actions of that company must use the equity method. It is one of the few instances in which the firm recognizes revenue without an increase in *working capital*.

**equity ratio.** *Shareholders' equity* divided by total *assets*. See *ratio*.

**equivalent production.** *Equivalent units*.

**equivalent units (of work).** The number of units of completed output that would require the same costs that a firm would actually incur for the production of completed and partially completed units during a period. For example, if at the beginning of a period the firm starts 100 units and by the end of the period has incurred costs for each of these equal to 75 percent of total costs to complete the units, then the equivalent units of work for the period would be 75. This is used primarily in *process costing* calculations to measure in uniform terms the output of a continuous process.

**ERISA (Employee Retirement Income Security Act of 1974).** The federal law that sets most *pension plan* requirements.

**error accounting.** See *accounting errors*.

**escalator clause.** Inserted in a purchase or rental contract, a clause that permits, under specified conditions, upward adjustments of price.

**escapable cost.** *Avoidable cost*.

**ESOP** (Employee Stock Ownership Plan). See *ESOT*.

**ESOT** (Employee Stock Ownership Trust). A trust *fund* that is created by a corporate employer and that can provide certain tax benefits to the corporation while providing for employee stock ownership. The corporate employer can contribute up to 25 percent of its payroll per year to the trust. The corporation may deduct the amount of the contribution from otherwise taxable income for federal *income tax* purposes. The trustee of the assets must use them for the benefit of employees—for example, to fund death or retirement benefits. The assets of the trust are usually the *common shares,* sometimes nonvoting, of the corporate employer. For an example of the potential *tax shelter,* consider the case of a corporation with $1 million of *debt* outstanding, which it wants to retire, and an annual payroll of $2 million. The corporation sells $1 million of common stock to the ESOT. The ESOT borrows $1 million with the loan guaranteed by, and therefore a *contingency* of, the corporation. The corporation uses the $1 million proceeds of the stock issue to retire its outstanding debt. (The debt of the corporation has been replaced with the debt of the ESOT.) The corporation can contribute $500,000 (= 0.25 × $2 million payroll) to the ESOT each year and treat the contribution as a deduction for tax purposes. After a little more than two years, the ESOT has received sufficient funds to retire its loan. The corporation has effectively repaid its original $1 million debt with pretax dollars. Assuming an income tax rate of 40 percent, it has saved $400,000 (= 0.40 × $1 million) of aftertax dollars *if* the $500,000 expense for the contribution to the ESOT for the pension benefits of employees would have been made, in one form or another, anyway. Observe that the corporation could use the proceeds ($1 million in the example) of the stock issued to the ESOT for any of several different purposes: financing expansion, replacing plant assets, or acquiring another company. Basically this same form of pretax-dollar financing through pensions is available with almost any corporate pension plan, with one important exception. The trustees of an ordinary pension trust must invest the assets prudently, and if they do not, they are personally liable to the employees. Current judgment about prudent investment requires diversification—trustees should invest pension trust assets in a wide variety of investment opportunities. (The trustee may not ordinarily invest more than 10 percent of a pension trust's assets in the parent's common stock.) Thus the ordinary pension trust cannot, in practice, invest all, or even most, of its assets in the parent corporation's stock. This constraint does not apply to the investments of an ESOT. The trustee may invest all ESOT assets in the parent company's stock. The ESOT also provides a means for closely held corporations to achieve wider ownership of shares without *going public*. The laws enabling ESOTs provide for the independent professional appraisal of shares not traded in public markets and for transactions between the corporation and the ESOT or between the ESOT and the employees to be based on the appraised values of the shares.

**estate planning.** The arrangement of an individual's affairs to facilitate the passage of assets to beneficiaries and to minimize taxes at death.

**estimated expenses.** See *after cost*.

**estimated liability.** The preferred terminology for estimated costs the firm will incur for such uncertain things as repairs under *warranty*. An estimated liability appears on the *balance sheet*. Contrast with *contingency*.

**estimated revenue.** A term used in governmental accounting to designate revenue expected to accrue during a period independent of whether the government will collect it during the period. The governmental unit usually establishes a *budgetary account* at the beginning of the budget period.

**estimated salvage value.** Synonymous with *salvage value* of an *asset* before its retirement.

**estimates, changes in.** See *accounting changes*.

**estimation sampling.** The use of sampling technique in which the sampler infers a qualitative (e.g., fraction female) or quantitative (e.g., mean weight) characteristic of the population from the occurrence of that characteristic in the sample drawn. See *attribute(s) sampling; variables sampling*.

**EURL (entreprise unipersonnelle à responsabilité limitée).** France: similar to *SARL* but having only one shareholder.

**ex div. (dividend).** Said of *shares* whose market price quoted in the market has been reduced by a *dividend* already declared but not yet paid. The *corporation* will send the dividend to the person who owned the share on the *record date*. One who buys the share ex dividend will not receive the dividend although the corporation has not yet paid it.

**ex rights.** The condition of securities whose quoted market price no longer includes the right to purchase new securities, such rights having expired or been retained by the seller. Compare *cum rights*.

**except for.** Qualification in *auditor's report*, usually caused by a change, approved by the auditor, from one acceptable accounting principle or procedure to another.

**excess present value.** In a *capital budgeting* context, *present value* (of anticipated net cash inflows minus cash outflows including initial cash outflow) for a project. The analyst uses the *cost of capital* as the *discount rate*.

**excess present value index.** *Present value* of future *cash* inflows divided by initial cash outlay.

**exchange.** The generic term for a transaction (or, more technically, a reciprocal transfer) between one entity and another; in another context, the name for a market, such as the New York Stock Exchange.

**exchange gain or loss.** The phrase used by the *FASB* for *foreign exchange gain or loss*.

**exchange rate.** The *price* of one country's currency in terms of another country's currency. For example, the British pound sterling might be worth U.S. $1.60 at a given time. The exchange rate would be stated as "one pound is worth one dollar and sixty cents" or "one dollar is worth £.625" (= £1/$1.60).

**excise tax.** Tax on the manufacture, sale, or consumption of a commodity.

**executory contract.** A mere exchange of promises; an agreement providing for payment by a payor to a payee on the performance of an act or service by the payee, such as a labor contract. Accounting does not recognize benefits arising from executory contracts as *assets*, nor does it recognize obligations arising from such contracts as *liabilities*. See *partially executory contract*.

**exemption.** A term used for various amounts subtracted from gross income in computing taxable income. Usage does not call all such subtractions "exemptions." See *tax deduction*.

**exercise.** Occurs when owners of an *option* or *warrant* purchase the security that the option entitles them to purchase.

**exercise price.** See *option*.

**exit value.** The proceeds that would be received if assets were disposed of in an *arm's-length transaction. Current selling price; net realizable value.*

**expected value.** The mean or arithmetic *average* of a statistical distribution or series of numbers.

**expected value of (perfect) information.** Expected *net benefits* from an undertaking with (perfect) information minus expected net benefits of the undertaking without (perfect) information.

**expendable fund.** In governmental accounting, a *fund* whose resources, *principal,* and earnings the governmental unit may distribute.

**expenditure.** Payment of *cash* for goods or services received. Payment may occur at the time the purchaser receives the goods or services or at a later time. Virtually synonymous with *disbursement* except that disbursement is a broader term and includes all payments for goods or services. Contrast with *expense.*

**expense.** As a noun, a decrease in *owners' equity* accompanying the decrease in *net assets* caused by selling goods or rendering services or by the passage of time; a "gone" (net) asset; an expired cost. Measure expense as the *cost* of the (net) assets used. Do not confuse with *expenditure* or *disbursement,* which may occur before, when, or after the firm recognizes the related expense. Use the word "cost" to refer to an item that still has service potential and is an asset. Use the word "expense" after the firm has used the asset's service potential. As a verb, "expense" means to designate an expenditure—past, current, or future—as a current expense.

**expense account.** An *account* to accumulate *expenses; closed* to *retained earnings* at the end of the accounting period; a *temporary owners' equity* account; also used to describe a listing of expenses that an employee submits to the employer for reimbursement.

**experience rating.** A term used in insurance, particularly unemployment insurance, to denote changes from ordinary rates to reflect extraordinarily large or small amounts of claims over time by the insured.

**expired cost.** An *expense* or a *loss.*

***Exposure Draft* (*ED*).** A preliminary statement of the *FASB* (or the *APB* between 1962 and 1973) showing the contents of a pronouncement being considered for enactment by the board.

**external reporting.** Reporting to shareholders and the public, as opposed to internal reporting for management's benefit. See *financial accounting,* and contrast with *managerial accounting.*

**extraordinary item.** A *material expense* or *revenue* item characterized both by its unusual nature and by its infrequency of occurrence; appears along with its income tax effects separately from ordinary income and *income from discontinued operations* on the *income statement.* Accountants would probably classify a *loss* from an earthquake as an extraordinary item. Accountants treat gain (or loss) on the retirement of *bonds* as an extraordinary item under the terms of *SFAS No. 4.*

# F

**face amount (value).** The nominal amount due at *maturity* from a *bond* or *note* not including the contractual periodic payment that may also come due on the same date. Good usage calls the corresponding amount of a stock certificate the *par* or *stated value,* whichever applies.

**factoring.** The process of buying *notes* or *accounts receivable* at a *discount* from the holder owed the debt; from the holder's point of view, the selling of such notes or accounts. When the transaction involves a single note, usage calls the process "discounting a note."

**factory.** Used synonymously with *manufacturing* as an adjective.

**factory burden.** *Manufacturing overhead.*

**factory cost.** *Manufacturing cost.*

**factory expense.** *Manufacturing overhead. Expense* is a poor term in this context because the item is a *product cost.*

**factory overhead.** Usually an item of *manufacturing cost* other than *direct labor* or *direct materials.*

**fair market price (value).** See *fair value.*

**fair presentation (fairness).** One of the qualitative standards of financial reporting. When the *auditor's report* says that the *financial statements* "present fairly . . . ," the auditor means that the accounting alternatives used by the entity all comply with *GAAP.* In recent years, however, courts have ruled that conformity with *generally accepted accounting principles* may be insufficient grounds for an opinion that the statements are fair. *SAS No. 5* requires that the auditor judge the accounting principles used in the statements to be "appropriate in the circumstances" before attesting to fair presentation.

**fair value, fair market price (value).** Price (value) negotiated at *arm's length* between a willing buyer and a willing seller, each acting rationally in his or her own self-interest. The accountant may estimate this amount in the absence of a monetary transaction. This is sometimes measured as the present value of expected cash flows.

**fair-value hedge.** A hedge of an exposure to changes in the *fair value* of a recognized *asset* or *liability* or of an unrecognized firm commitment.

**FASAC.** *Financial Accounting Standards Advisory Council.*

**FASB (Financial Accounting Standards Board).** An independent board responsible, since 1973, for establishing *generally accepted accounting principles.* Its official pronouncements are *Statements of Financial Accounting Concepts* (*SFAC*), *Statements of Financial Accounting Standards* (*SFAS*), and *FASB Interpretations.* See also *Discussion Memorandum* and *Technical Bulletin.* Web site: www.fasb.org.

***FASB Interpretation.*** An official *FASB* statement interpreting the meaning of *Accounting Research Bulletins, APB Opinions,* and *Statements of Financial Accounting Standards.*

***FASB Technical Bulletin.*** See *Technical Bulletin.*

**favorable variance.** An excess of actual *revenues* over expected revenues; an excess of *standard cost* over actual cost.

**federal income tax.** *Income tax* levied by the U.S. government on individuals and corporations.

**Federal Insurance Contributions Act.** See *FICA*.

**Federal Unemployment Tax Act.** See *FUTA*.

**Fédération des Experts Comptables Européens (FEE).** This body, the European Federation of Accountants, which is based in Brussels, is the representative organization of the European accounting profession on matters relating to the initiatives of the European Union, including especially its administrative wing, the European Commission. See www.fee.org.

**feedback.** The process of informing employees about how their actual performance compares with the expected or desired level of performance, in the hope that the information will reinforce desired behavior and reduce unproductive behavior.

**FEI.** *Financial Executives Institute.*

**FICA (Federal Insurance Contributions Act).** The law that sets *Social Security taxes* and benefits.

**fiduciary.** Someone responsible for the custody or administration of property belonging to another; for example, an executor (of an estate), agent, receiver (in *bankruptcy*), or trustee (of a trust).

**FIFO** (first-in, first-out). The *inventory flow assumption* that firms use to compute *ending inventory* cost from most recent purchases and *cost of goods sold* from oldest purchases including beginning inventory. FIFO describes cost flow from the viewpoint of the income statement. From the balance sheet perspective, *LISH* (last-in, still-here) describes this same cost flow. Contrast with *LIFO*.

**finance.** As a verb, to supply with *funds* through the *issue* of stocks, bonds, notes, or mortgages or through the retention of earnings.

**financial accounting.** The accounting for *assets, equities, revenues,* and *expenses* of a business; primarily concerned with the historical reporting, to external users, of the *financial position* and operations of an *entity* on a regular, periodic basis. Contrast with *managerial accounting*.

**Financial Accounting Foundation.** The independent foundation (committee), governed by a board of trustees, that raises funds to support the *FASB* and *GASB*.

**Financial Accounting Standards Advisory Council (FASAC).** A committee of academics, preparers, attestors, and users giving advice to the *FASB* on matters of strategy and emerging issues. The council spends much of each meeting learning about current developments in standard-setting from the FASB staff.

**Financial Accounting Standards Board.** *FASB*.

**Financial Executives Institute (FEI).** An organization of financial executives, such as chief accountants, *controllers,* and treasurers, of large businesses. In recent years, the FEI has been a critic of the FASB because it views many of the FASB requirements as burdensome while not *cost-effective*.

**financial expense.** An *expense* incurred in raising or managing *funds*.

**financial flexibility.** As defined by *SFAC No. 5,* "the ability of an entity to take effective actions to alter amounts and timing of cash flows so it can respond to unexpected needs and opportunities."

**financial forecast.** See *financial projection* for definition and contrast.

**financial instrument.** The *FASB* defines this term as follows: "Cash, evidence of an ownership interest in an entity, or a contract that both:

[a] imposes on one entity a contractual obligation (1) to deliver cash or another financial instrument to a second entity or (2) to exchange financial instruments on potentially unfavorable terms with the second entity, and

[b] conveys to that second entity a contractual right (1) to receive cash or another financial instrument from the first entity or (2) to exchange other financial instruments on potentially favorable terms with the first entity."

**financial leverage.** See *leverage*.

**financial position (condition).** Statement of the *assets* and *equities* of a firm; displayed as a *balance sheet*.

**financial projection.** An estimate of *financial position,* results of *operations*, and changes in cash flows for one or more future periods based on a set of assumptions. If the assumptions do not represent the most likely outcomes, then auditors call the estimate a "projection." If the assumptions represent the most probable outcomes, then auditors call the estimate a "forecast." "Most probable" means that management has evaluated the assumptions and that they are management's judgment of the most likely set of conditions and most likely outcomes.

**financial ratio.** See *ratio*.

**financial reporting objectives.** Broad objectives that are intended to guide the development of specific *accounting standards;* set out by *FASB SFAC No. 1.*

***Financial Reporting Release.*** Series of releases, issued by the SEC since 1982; replaces the *Accounting Series Release*. See *SEC*.

**financial statements.** The *balance sheet, income statement, statement of retained earnings, statement of cash flows,* statement of changes in *owners' equity accounts,* statement of *comprehensive income,* and *notes* thereto.

**financial structure.** *Capital structure.*

**financial vice-president.** Person in charge of the entire accounting and finance function; typically one of the three most influential people in the company.

**financial year.** Australia and UK: term for *fiscal year.*

**financing activities.** Obtaining resources from (a) owners and providing them with a return on and a return of their *investment* and (b) *creditors* and repaying amounts borrowed (or otherwise settling the obligation). See *statement of cash flows*.

**financing lease.** *Capital lease.*

**finished goods (inventory account).** Manufactured product ready for sale; a *current asset* (inventory) account.

**firm.** Informally, any business entity. (Strictly speaking, a firm is a *partnership*.)

**firm commitment.** The *FASB,* in *SFAS No. 133,* defines this as "an agreement with an unrelated party, binding on both parties and usually legally enforceable," which requires that the firm promise to pay a specified amount of a currency and that the firm has sufficient disincentives for nonpayment that the firm will probably make the payment. A firm commitment resembles a *liability,* but it is an *executory contract,* so is not a liability. *SFAS No. 133* allows the firm to recognize certain financial *hedges* in the balance sheet if they hedge firm commitments. The *FASB* first used the term in *SFAS No. 52* and *No. 80* but made the term more definite and more important in *SFAS No. 133.* This is an early, perhaps the first, step in changing the recognition criteria for assets and liabilities to exclude the test that the future benefit (asset) or obligation (liability) not arise from an executory contract.

**first-in, first-out.** See *FIFO.*

**fiscal year.** A period of 12 consecutive months chosen by a business as the *accounting period* for *annual reports,* not necessarily a *natural business year* or a calendar year.

**FISH.** An acronym, conceived by George H. Sorter, for *f*irst-*i*n, *s*till-*h*ere. FISH is the same cost flow assumption as *LIFO.* Many readers of accounting statements find it easier to think about inventory questions in terms of items still on hand. Think of LIFO in connection with *cost of goods sold* but of FISH in connection with *ending inventory.* See *LISH.*

**fixed assets.** *Plant assets.*

**fixed assets turnover.** *Sales* divided by average total *fixed assets.*

**fixed benefit plan.** A *defined-benefit plan.*

**fixed budget.** A plan that provides for specified amounts of *expenditures* and *receipts* that do not vary with activity levels; sometimes called a "static budget." Contrast with *flexible budget.*

**fixed charges earned (coverage) ratio.** *Income* before *interest expense* and *income tax expense* divided by interest expense.

**fixed cost (expense).** An *expenditure* or *expense* that does not vary with volume of activity, at least in the short run. See *capacity costs,* which include *enabling costs* and *standby costs,* and *programmed costs* for various subdivisions of fixed costs. See *cost terminology.*

**fixed interval sampling.** A method of choosing a sample: the analyst selects the first item from the population randomly, drawing the remaining sample items at equally spaced intervals.

**fixed liability.** *Long-term* liability.

**fixed manufacturing overhead applied.** The portion of *fixed manufacturing overhead cost* allocated to units produced during a period.

**fixed overhead variance.** Difference between *actual fixed manufacturing costs* and fixed manufacturing costs applied to production in a *standard costing system.*

**flexible budget.** *Budget* that projects receipts and expenditures as a function of activity levels. Contrast with *fixed budget.*

**flexible budget allowance.** With respect to manufacturing overhead, the total cost that a firm should have incurred at the level of activity actually experienced during the period.

**float.** *Checks* whose amounts the bank has *added* to the depositor's bank account but whose amounts the bank has not yet reduced from the *drawer's* bank account.

**flow.** The change in the amount of an item over time. Contrast with *stock.*

**flow assumption.** An assumption used when the firm makes a *withdrawal* from *inventory.* The firm must compute the cost of the withdrawal by a flow assumption if the firm does not use the *specific identification* method. The usual flow assumptions are *FIFO, LIFO,* and *weighted average.*

**flow of costs.** *Costs* passing through various classifications within an *entity* engaging, at least in part, in manufacturing activities. See the diagram on the next page for a summary of *product* and *period cost* flows.

**flow-through method.** Accounting for the *investment credit* to show all income statement benefits of the credit in the year of acquisition rather than spreading them over the life of the asset acquired (called the "deferral method"). The *APB* preferred the deferral method in *Opinion No. 2* (1962) but accepted the flow-through method in *Opinion No. 4* (1964). The term also applies to *depreciation* accounting in which the firm uses the *straight-line method* for financial reporting and an *accelerated depreciation* method for tax reporting. Followers of the flow-through method would not recognize a *deferred tax liability. APB Opinion No. 11* prohibits the use of the flow-through approach in financial reporting, although some regulatory commissions have used it.

**FOB.** Free on board some location (for example, FOB shipping point, FOB destination). The *invoice* price includes delivery at seller's expense to that location. Title to goods usually passes from seller to buyer at the FOB location.

**folio.** A page number or other identifying reference used in posting to indicate the source of entry.

**footing.** Adding a column of figures.

**footnotes.** More detailed information than that provided in the *income statement, balance sheet, statement of retained earnings,* and *statement of cash flows.* These are an integral part of the statements, and the *auditor's report* covers them. They are sometimes called "notes."

**forecast.** See *financial projection* for definition and contrast.

**foreclosure.** Occurs when a lender takes possession of property for his or her own use or sale after the borrower fails to make a required payment on a *mortgage.* Assume that the lender sells the property but that the proceeds of the sale are too small to cover the outstanding balance on the loan at the time of foreclosure. Under the terms of most mortgages, the lender becomes an unsecured creditor of the borrower for the still-unrecovered balance of the loan.

**foreign currency.** For *financial statements* prepared in a given currency, any other currency.

Flow of Costs (and Sales Revenue)

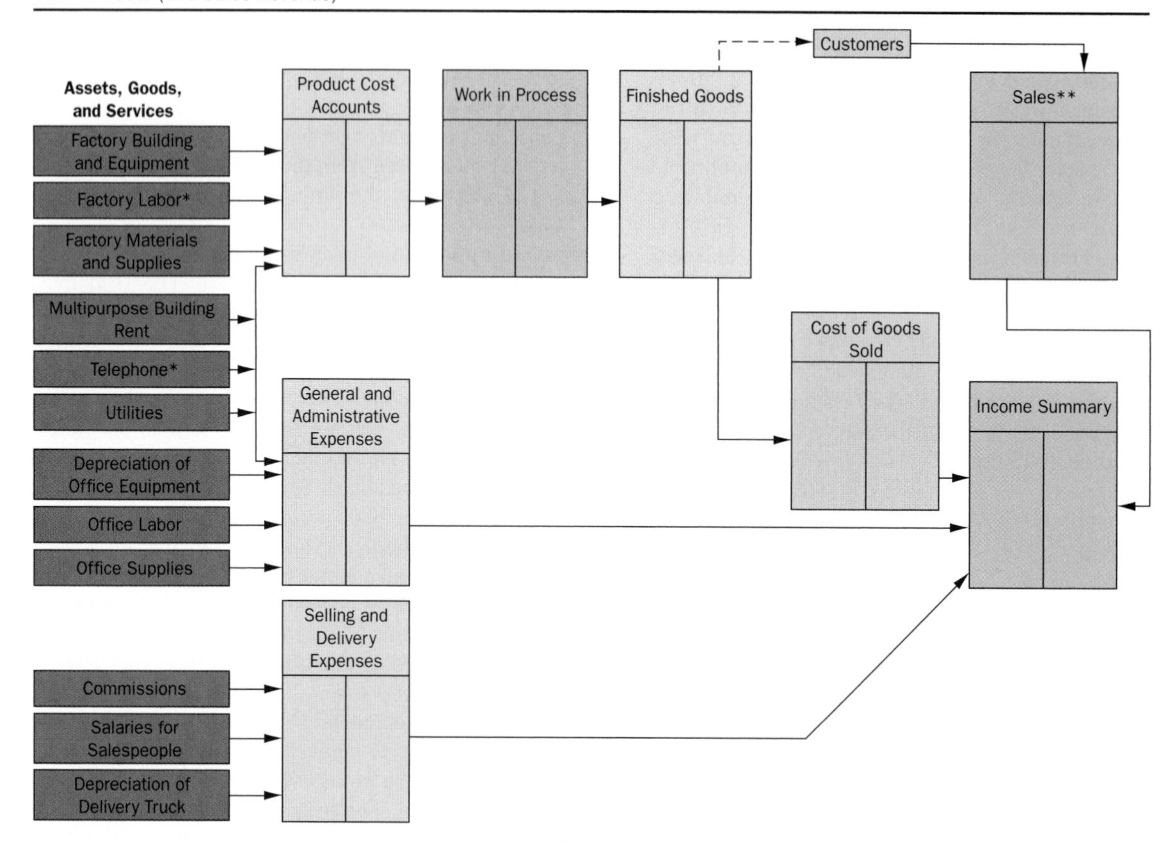


*The credit in the entry to record these items is usually to a payable; for all others, the credit is usually to an asset, or to an asset contra account.

**When the firm records sales to customers, it credits the Sales account. The debit is usually to Cash or Accounts Receivable.

**foreign currency translation.** Reporting in the currency used in financial statements the amounts denominated or measured in a different currency.

**foreign exchange gain or loss.** Gain or loss from holding *net* foreign *monetary items* during a period when the *exchange rate* changes.

**foreign sales corporation.** See *FSC.*

**forfeited share.** A share to which a subscriber has lost title because of nonpayment of a *call.*

**Form 10-K.** See *10-K.*

**Form 20-F.** See *20-F.*

**forward contract.** An agreement to purchase or sell a specific commodity or financial instrument for a specified price, the *forward price,* at a specified date. Contrast with *futures contract.* Typically, forward contracts are not traded on organized exchanges (unlike *futures contract*), so the parties to the agreement sacrifice liquidity but gain flexibility in setting contract quantities, qualities, and settlement dates.

**forward-exchange contract.** An agreement to exchange at a specified future date currencies of different countries at a specified rate called the "forward rate."

**forward price.** The price of a commodity for delivery at a specified future date; in contrast to the "spot price," the price of that commodity on the day of the price quotation.

**franchise.** A privilege granted or sold, such as to use a name or to sell products or services.

**fraudulent conveyance.** A transfer of goods or cash that a court finds illegal. *Creditors* of a *bankrupt* firm usually receive less than the firm owed them. For example, a creditor of a bankrupt firm might collect from the trustee of the bankrupt firm only $0.60 for every dollar the bankrupt firm owed. Creditors, anticipating bankruptcy, sometimes attempt to persuade the firm to pay the debt in full before the firm declares bankruptcy, reducing the net assets available to other creditors. Bankruptcy laws have rules forbidding such transfers from a near-bankrupt firm to some of its creditors. Such a transfer is called a "fraudulent conveyance." Courts sometimes ask accountants to judge whether a firm had liabilities exceeding assets even before the firm went into bankruptcy. When the court can find that economic bankruptcy occurred before legal bankruptcy, it will declare transfers of assets to creditors after economic bankruptcy to be fraudulent conveyances and have the assets returned to the trustees (or to a legal entity called the "bankrupt's estate") for redistribution to all creditors.

**fraudulent financial reporting.** Intentional or reckless conduct that results in materially misleading *financial statements.* See *creative accounting.*

**free cash flow.** This term has no standard meaning. Some financial statement analysts use it to mean *cash flow from operations + interest expense + income tax expense*. Others mean the excess of cash flow from operations over cash flow for investing. Usage varies so much that you should ascertain the meaning intended in context by this phrase.

**free on board.** *FOB*.

**freight-in.** The *cost* of freight or shipping incurred in acquiring *inventory,* preferably treated as a part of the cost of *inventory;* often shown temporarily in an *adjunct account* that the acquirer closes at the end of the period with other purchase accounts to the inventory account.

**freight-out.** The *cost* of freight or shipping incurred in selling *inventory,* treated by the seller as a selling *expense* in the period of sale.

**FSC (foreign sales corporation).** A foreign *corporation* engaging in certain export activities, some of whose *income* the United States exempts from federal *income tax*. A U.S. corporation need pay no income taxes on *dividends* distributed by an FSC out of *earnings* attributable to certain foreign income.

**full absorption costing.** The *costing* method that assigns all types of manufacturing costs (*direct material, direct labor, fixed* and *variable overhead*) to units produced; required by *GAAP;* also called "absorption costing." Contrast with *variable costing*.

**full costing, full costs.** The total cost of producing and selling a unit; often used in *long-term* profitability and pricing decisions. Full cost per unit equals *full absorption cost* per unit plus *marketing, administrative, interest,* and other *central corporate expenses,* per unit. The sum of full costs for all units equals total costs of the firm.

**full disclosure.** The reporting policy requiring that all significant or *material* information appear in the financial statements. See *fair presentation*.

**fully vested.** Said of a *pension plan* when an employee (or his or her estate) has rights to all the benefits purchased with the employer's contributions to the plan even if the employee does not work for this employer at the time of death or retirement.

**function.** In governmental accounting, said of a group of related activities for accomplishing a service or regulatory program for which the governmental unit has responsibility; in mathematics, a rule for associating a number, called the dependent variable, with another number (or numbers), called independent variable(s).

**functional classification.** *Income statement* reporting form that classifies *expenses* by function, that is, cost of goods sold, administrative expenses, financing expenses, selling expenses. Contrast with *natural classification*.

**functional currency.** Currency in which an entity carries out its principal economic activity.

**fund.** An *asset* or group of assets set aside for a specific purpose. See also *fund accounting*.

**fund accounting.** The accounting for resources, obligations, and *capital* balances, usually of a not-for-profit or governmental *entity,* which the entity has segregated into *accounts* representing logical groupings based on legal, donor, or administrative restrictions or requirements. The groupings are "funds." The accounts of each fund are *self-balancing,* and from them one can prepare a *balance sheet* and an operating statement for each fund. See *fund* and *fund balance*.

**fund balance.** In governmental accounting, the excess of assets of a *fund* over its liabilities and reserves; the not-for-profit equivalent of *owners' equity*.

**funded.** Said of a *pension plan* or other obligation when the firm has set aside *funds* for meeting the obligation when it comes due. The federal law for pension plans requires that the firm fund all *normal costs* when it recognizes them as expenses. In addition, the firm must fund *prior service cost* of pension plans over 30 or over 40 years, depending on the circumstances.

**funding.** Replacing *short-term* liabilities with *long-term* debt.

**funds.** Generally *working capital;* current assets less current liabilities; sometimes used to refer to *cash* or to cash and *marketable securities*.

**funds provided by operations.** See *cash provided by operations*.

**funds statement.** An informal name often used for the *statement of cash flows*.

**funny money.** Said of securities, such as *convertible preferred stock, convertible bonds, options,* and *warrants,* that have aspects of *common shares* but that did not reduce reported *earnings per share* before the issuance of *APB Opinion No. 9* in 1966 and *No. 15* in 1969.

**FUTA (Federal Unemployment Tax Act).** Provides for taxes to be collected at the federal level, to help subsidize the individual states' administration of their unemployment compensation programs.

**future value.** Value at a specified future date of a sum increased at a specified *interest rate*.

**futures contract.** An agreement to purchase or sell a specific commodity or financial instrument for a specified price, at a specific future time or during a specified future period. Contrast with *forward contract.* When traded on an organized exchange, the exchange sets the minimum contract size and expiration date(s). The exchange requires that the holder of the contract settle in cash each day the fluctuations in the value of the contract. That is, each day, the exchange marks the contract to market value, called the "(daily) settlement price." A contract holder who has lost during the day must put up more cash, and a holder who has gained receives cash.

# G

**GAAP.** *Generally accepted accounting principles;* a plural noun. In the UK and elsewhere, this means "generally accepted accounting practices."

**GAAS.** *Generally accepted auditing standards;* a plural noun. Do not confuse with *GAS*.

**gain.** In *financial accounting* contexts, the increase in *owners' equity* caused by a transaction that is not part of a firm's typical, day-to-day operations and not part of owners' *investment* or *withdrawals*. Accounting distinguishes the meaning of the term "gain" (or *loss*) from that of related terms. First, gains (and losses) generally refer to nonoperating, incidental, peripheral, or nonroutine transactions: gain on sale of land in contrast to *gross margin* on *sale* of *inventory*. Second, gains and losses are *net* concepts, not gross concepts: gain or loss results from subtracting some measure of *cost* from the measure of inflow. *Revenues* and *expenses*, on the other hand, are gross concepts; their difference is a net concept. Gain is non-routine and net, *profit* or *margin* is routine and net; revenue from *continuing operations* is routine and gross; revenue from *discontinued operations* is nonroutine and gross. Loss is net but can be either routine ("loss on sale of inventory") or not ("loss on disposal of segment of business").

In *managerial accounting* and lay contexts, the difference between some measure of *revenue* or *receipts* or *proceeds* and some measure of costs, such as direct costs or variable costs or fully absorbed costs or full costs (see *cost terminology*). Because the word can have so many different meanings, careful writers should be explicit to designate one.

**gain contingency.** See *contingency*.

**GAS.** *Goods available for sale*. Do not confuse with *GAAS*.

**GASB (Governmental Accounting Standards Board).** An independent body responsible, since 1984, for establishing accounting standards for state and local government units. It is part of the *Financial Accounting Foundation*, parallel to the *FASB*, and currently consists of five members.

**GbR (Gesellschaft des bürgerlichen Rechtes).** Germany: a *partnership* whose members agree to share in specific aspects of their own separate business pursuits, such as an office. This partnership has no legal form and is not a separate accounting *entity*.

**GDP Implicit Price Deflator (index).** A *price index* issued quarterly by the Office of Business Economics of the U.S. Department of Commerce. This index attempts to trace the price level of all *goods and services* composing the *gross domestic product*. Contrast with *Consumer Price Index*.

**gearing.** UK: *financial leverage*.

**gearing adjustment.** A *revenue* representing part of a *holding gain*. Consider a firm that has part of its assets financed by *noncurrent liabilities* and that has experienced *holding gains* on its *assets* during a period. All the increase in wealth caused by the holding gains belongs to the owners; none typically belongs to the lenders. Some British accounting authorities believe that published *income statements* should show part of the holding gain in *income* for the period. The part they would report in income is the fraction of the gain equal to the fraction that debt composes of total financing; for example, if debt equals 40 percent of total equities and the holding gain equals $100 for the period, the amount to appear in income for the period would be $40. Usage calls that part the "gearing adjustment."

**general debt.** A governmental unit's debt legally payable from general revenues and backed by the full faith and credit of the governmental unit.

**general expenses.** *Operating expenses* other than those specifically identified as cost of goods sold, selling, and administration.

**general fixed asset (group of accounts).** Accounts showing a governmental unit's long-term assets that are not accounted for in *enterprise, trust*, or intragovernmental service funds.

**general fund.** A nonprofit entity's assets and liabilities not specifically earmarked for other purposes; the primary operating fund of a governmental unit.

**general journal.** The formal record in which the firm records transactions, or summaries of similar transactions, in *journal entry* form as they occur. Use of the adjective "general" usually implies that the journal has only two columns for cash amounts or that the firm also uses various *special journals*, such as a *check register* or *sales journal*.

**general ledger.** The name for the formal *ledger* containing all the financial statement accounts. It has equal debits and credits, as evidenced by the *trial balance*. Some of the accounts in the general ledger may be *control accounts*, supported by details contained in *subsidiary ledgers*.

**general partner.** *Partnership* member who is personally liable for all debts of the partnership; contrast with *limited partner*.

**general price index.** A measure of the aggregate prices of a wide range of goods and services in the economy at one time relative to the prices during a base period. See *Consumer Price Index* and *GDP Implicit Price Deflator*. Contrast with *specific price index*.

**general price level–adjusted statements.** See *constant-dollar accounting*.

**general price-level changes.** Changes in the aggregate prices of a wide range of goods and services in the economy. These price measurements result from using a *general price index*. Contrast with *specific price changes*.

**general purchasing power.** The command of the dollar over a wide range of goods and services in the economy. The general purchasing power of the dollar is inversely related to changes in a general price index. See *general price index*.

**general purchasing-power accounting.** See *constant-dollar accounting*.

**generally accepted accounting principles (GAAP).** As previously defined by the *CAP, APB*, and now the *FASB*, the conventions, rules, and procedures necessary to define accepted accounting practice at a particular time; includes both broad guidelines and relatively detailed practices and procedures. In the United States the FASB defines GAAP to include accounting pronouncements of the *SEC* and other government agencies as well as a variety of authoritative sources, such as this book.

**generally accepted auditing standards (GAAS).** The *PCAOB* has explicitly stated that it began compiling its auditing promulgations with *GAAS*, as issued by the *AICPA*, but "a reference to generally accepted auditing standards in auditors' reports is no longer appropriate or necessary." The phrase has referred to the standards, as opposed to particular procedures, that the AICPA promulgated (in *Statements on Auditing Standards*) and that concern "the auditor's professional quantities" and "the judgment exercised by him in the performance of his examination and in his report." Currently, there have been ten such standards: three general ones (concerned with proficiency, independence, and degree of care to be exercised), three standards of field work, and four standards of reporting. The first standard of reporting requires that the *auditor's report* state whether the firm prepared the *financial statements* in accordance with *generally accepted accounting principles*. Thus, before the PCAOB became the auditing rulemaker, the typical auditor's report says that the auditor conducted the examination in accordance with generally accepted auditing standards and that the firm prepared the statements in accordance with generally accepted accounting principles. The report will not refer to the standards of the Public Company Accounting Oversight Board (United States). See *auditor's report*.

**geographic segment.** A single operation or a group of operations that are located in a particular geographic area and that generate revenue, incur costs, and have assets used in or associated with generating such revenue.

**G4+1.** A group concerned with unifying accounting standards across countries. It originally comprised the *FASB*, *CICA* (Canada), the Accounting Standards Board (UK), and the Australian Accounting Standards Board, plus the *IASB*. Hence, the name: a group of four national standard-setters plus the *IASB*. The group now includes participants from New Zealand.

**GIE (groupement d'intérêt économique).** France: a joint venture, normally used for exports and research-and-development pooling.

**GmbH (Gesellschaft mit beschränkter Haftung).** Germany: a private company with an unlimited number of shareholders. Transfer of ownership can take place only with the consent of other shareholders. Contrast with *AG*.

**goal congruence.** The idea that all members of an organization have incentives to perform for a common interest, such as *shareholder* wealth maximization for a *corporation*.

**going-concern assumption.** For accounting purposes, accountants' assumption that a business will remain in operation long enough to carry out all its current plans. This assumption partially justifies the *acquisition cost* basis, rather than a *liquidation* or *exit value* basis, of accounting.

**going public.** Said of a business when its *shares* become widely traded rather than being closely held by relatively few shareholders; issuing shares to the general investing public.

**goods.** Items of merchandise, supplies, raw materials, or finished goods. Sometimes the meaning of "goods" is extended to include all *tangible* items, as in the phrase "goods and services."

**goods available for sale.** The sum of *beginning inventory* plus all acquisitions of merchandise or finished goods during an *accounting period*.

**goods-in-process.** *Work-in-process*.

**goodwill.** The excess of cost of an acquired firm (or operating unit) over the current *fair market value* of the separately identifiable *net assets* of the acquired unit. Before the acquiring firm can recognize goodwill, it must assign a *fair market value* to all identifiable assets, even when not recorded on the books of the acquired unit. For example, if a firm has developed a *patent* that does not appear on its books because of *SFAS No. 2*, if another company acquires the firm, the acquirer will recognize the patent at an amount equal to its estimated fair market value. The acquirer will compute the amount of goodwill only after assigning values to all assets it can identify. Informally, the term indicates the value of good customer relations, high employee morale, a well-respected business name, and so on, all of which the firm or analyst expects to result in greater-than-normal earning power.

**goodwill method.** A method of accounting for the *admission* of a new partner to a *partnership* when the new partner will receive a portion of capital different from the value of the *tangible* assets contributed as a fraction of tangible assets of the partnership. See *bonus method* for a description and contrast.

**Governmental Accounting Standards Advisory Council.** A group that consults with the *GASB* on agenda, technical issues, and the assignment of priorities to projects. It comprises more than a dozen members representing various areas of expertise.

**Governmental Accounting Standards Board.** *GASB*.

**GPL (general price level).** Usually used as an adjective modifying the word "accounting" to mean *constant-dollar accounting*.

**GPLA (general price level–adjusted accounting).** *Constant-dollar accounting*.

**GPP (general purchasing power).** Usually used as an adjective modifying the word "accounting" to mean *constant-dollar accounting*.

**graded vesting.** Said of a *pension plan* in which not all employees currently have fully *vested* benefits. By law, the benefits must vest according to one of several formulas as time passes.

**grandfather clause.** An exemption in new accounting *pronouncements* exempting transactions that occurred before a given date from the new accounting treatment. For example, *APB Opinion No. 17*, adopted in 1970, exempted *goodwill* acquired before 1970 from required *amortization*. The term "grandfather" appears in the title to *SFAS No. 10*.

**gross.** Not adjusted or reduced by deductions or subtractions. Contrast with *net*, and see *gain* for a description of how the difference between net and gross affects usage of the terms *revenue, gain, expense,* and *loss*.

**gross domestic product (GDP).** The market value of all goods and services produced by capital or labor within a country, regardless of who owns the capital or of the nationality of the labor; most widely used measure of production within a country. Contrast with gross national product (GNP), which measures the market value of all goods and services produced with capital owned by, and labor services supplied by, the residents of that country regardless of where they work or where they own capital. In the United States in recent years, the difference between GDP and GNP equals about two-tenths of 1 percent of GDP.

**gross margin.** *Net sales* minus *cost of goods sold.*

**gross margin percent.** 100 × (1 − *cost of goods sold/net sales*) = 100 × (*gross margin/net sales*).

**gross national product (GNP).** See *gross domestic product* for definition and contrast.

**gross price method (of recording purchase or sales discounts).** The firm records the *purchase* (or *sale*) at the *invoice price,* not deducting the amounts of *discounts* available. Later, it uses a *contra* account to purchases (or sales) to record the amounts of discounts taken. Since information on discounts lapsed will not emerge from this system, most firms should prefer the *net price method* of recording purchase discounts.

**gross profit.** *Gross margin.*

**gross profit method.** A method of estimating *ending inventory* amounts. First, the firm measures *cost of goods sold* as some fraction of sales; then, it uses the *inventory equation* to value *ending inventory.*

**gross profit ratio.** *Gross margin* divided by *net sales.*

**gross sales.** All *sales* at *invoice* prices, not reduced by *discounts, allowances, returns,* or other adjustments.

**group depreciation.** In calculating *depreciation* charges, a method that combines similar assets rather than depreciating them separately. It does not recognize gain or loss on retirement of items from the group until the firm sells or retires the last item in the group. See *composite life method.*

**Group of 4 Plus 1.** See *G4+1.*

**guarantee.** A promise to answer for payment of debt or performance of some obligation if the person liable for the debt or obligation fails to perform. A guarantee is a *contingency* of the *entity* making the promise. Often, writers use the words "guarantee" and "warranty" to mean the same thing. In precise usage, however, "guarantee" means some person's promise to perform a contractual obligation such as to pay a sum of cash, whereas "warranty" refers to promises about pieces of machinery or other products. See *warranty.*

# H

**half-year convention.** In *tax accounting* under *ACRS,* and sometimes in *financial accounting,* an assumption that the firm acquired *depreciable assets* at midyear of the year of acquisition. When the firm uses this convention, it computes the *depreciation charge* for the year as one-half the charge that it would have used if it had acquired the assets at the beginning of the year.

**hardware.** The physical equipment or devices forming a computer and peripheral equipment.

**hash total.** Used to establish accuracy of data processing; a control that takes the sum of data items not normally added together (e.g., the sum of a list of part numbers) and subsequently compares that sum with a computer-generated total of the same values. If the two sums are identical, then the analyst takes some comfort that the two lists are identical.

***Hasselback.*** An annual directory of accounting faculty at colleges and universities; gives information about the faculty's training and fields of specialization. James R. Hasselback, of Florida State University, has compiled the directory since the 1970s; Prentice-Hall distributes it. On-line, you can find it at the Rutgers University accounting Web site: www.rutgers.edu/Accounting/.

**health-care benefits obligation.** At any time, the present value of the non-pension benefits promised by an employer to employees during their retirement years.

**hedge.** To reduce, perhaps cancel altogether, one risk the entity already bears, by purchasing a security or other financial instrument. For example, a farmer growing corn runs the risk that corn prices may decline before the corn matures and can be brought to market. Such a farmer can arrange to sell the corn now for future delivery, hedging the risk of corn price changes. A firm may have a *receivable* denominated in German marks due in six months. It runs the risk that the exchange rate between the dollar and the mark will change and the firm will receive a smaller number of dollars in the future than it would receive from the same number of marks received today. Such a firm may hedge its exposure to risk of changes in the exchange rate between dollars and German marks in a variety of ways.

**held-to-maturity securities.** *Marketable debt securities* that a firm expects to, and has the ability to, hold to *maturity;* a classification important in *SFAS No. 115,* which generally requires the owner to carry marketable securities on the balance sheet at market value, not at cost. Under *SFAS No. 115,* the firm may show held-to-maturity debt securities at *amortized cost.* If the firm lacks either the expectation or the intent to hold the debt security to its maturity, then the firm will show that security at market value as a security *available for sale.*

**hidden reserve.** An amount by which a firm has understated *owners' equity,* perhaps deliberately. The understatement arises from an undervaluation of *assets* or overvaluation of *liabilities.* By undervaluing assets on this period's *balance sheet,* the firm can overstate *net income* in some future period by disposing of the asset: actual *revenues* less artificially low cost of assets sold yields artificially high net income. No *account* in the *ledger* has this title.

**hire-purchase agreement (contract).** UK: a *lease* containing a purchase *option.*

**historical cost.** *Acquisition cost; original cost;* a *sunk cost.*

**historical cost/constant-dollar accounting.** Accounting based on *historical cost* valuations measured in *constant dollars.* The method restates *nonmonetary items* to reflect changes in the *general purchasing power* of the dollar since the time the firm acquired specific *assets* or incurred specific *liabilities*. The method recognizes a *gain* or *loss* on *monetary items* as the firm holds them over time periods when the general purchasing power of the dollar changes.

**historical exchange rate.** The rate at which one currency converts into another at the date a transaction took place. Contrast with *current exchange rate*.

**historical summary.** A part of the *annual report* that shows items, such as *net income, revenues, expenses, asset* and *equity* totals, *earnings per share,* and the like, for five or ten periods including the current one. Usually not as much detail appears in the historical summary as in *comparative statements,* which typically report as much detail for the two preceding years as for the current year. Annual reports may contain both comparative statements and a historical summary.

**historical valuation.** Showing balance sheet amounts at acquisition cost, sometimes reduced for *accumulated amortization;* sometimes reduced to *lower of cost or market.*

**holdback.** Under the terms of a contract, a portion of the progress payments that the customer need not pay until the contractor has fulfilled the contract or satisfied financial obligations to subcontractors.

**holding company.** A company that confines its activities to owning *stock* in, and supervising management of, other companies. A holding company usually owns a controlling interest in—that is, more than 50 percent of the voting stock of—the companies whose stock it holds. Contrast with *mutual fund*. See *conglomerate*. In British usage, the term refers to any company with controlling interest in another company.

**holding gain or loss.** Difference between end-of-period price and beginning-of-period price of an asset held during the period. The financial statements ordinarily do not separately report realized holding gains and losses. Income does not usually report unrealized gains at all, except on *trading securities*. See *lower of cost or market*. See *inventory profit* for further refinement, including *gains* on *assets* sold during the period.

**holding gain or loss net of inflation.** Increase or decrease in the *current cost* of an asset while it is held; measured in units of *constant dollars*.

**horizontal analysis.** *Time-series analysis*.

**horizontal integration.** An organization's extension of activity in the same general line of business or its expansion into supplementary, complementary, or compatible products. Compare *vertical integration*.

**house account.** An account with a customer who does not pay sales commissions.

**human resource accounting.** A term used to describe a variety of proposals that seek to report the importance of human resources—knowledgeable, trained, and loyal employees—in a company's earning process and total assets.

**hurdle rate.** Required rate of return in a *discounted cash flow* analysis.

**hybrid security.** *Security,* such as a *convertible bond,* containing elements of both *debt* and *owners' equity*.

**hypothecation.** The *pledging* of property, without transfer of title or possession, to secure a loan.

# I

**IAA.** *Interamerican Accounting Association*.

**IASB.** *International Accounting Standards Board*.

**ICMA (Institute of Certified Management Accountants).** See *CMA* and *Institute of Management Accountants*.

**ideal standard costs.** *Standard costs* set equal to those that a firm would incur under the best-possible conditions.

**IIA.** *Institute of Internal Auditors*.

**IMA.** *Institute of Management Accountants*.

**impairment.** Reduction in *market value* of an *asset*. When the firm has information indicating that its long-lived *assets,* such as *plant,* identifiable *intangibles,* and *goodwill,* have declined in *market value* or will provide a smaller future benefit that originally anticipated, it tests to see if the decline in value is so drastic that the expected future cash flows from the asset have declined below *book value*. If then-current book value exceeds the sum of expected cash flows, an asset impairment has occurred. At the time the firm judges that an impairment has occurred, the firm writes down the book value of the asset to its then-current *fair value,* which is the market value of the asset or, if the firm cannot assess the market value, the expected *net present value* of the future cash flows.

**impairment loss.** See *impairment*. This term refers to the amount by which the firm writes down the asset.

**implicit interest.** *Interest* not paid or received. See *interest, imputed*. All transactions involving the deferred payment or receipt of cash involve interest, whether explicitly stated or not. The implicit interest on a single-payment *note* equals the difference between the amount collected at maturity and the amount lent at the start of the loan. One can compute the implicit *interest rate* per year for loans with a single cash inflow and a single cash outflow from the following equation:

$$\left[\frac{\text{Cash Received at Maturity}}{\text{Cash Lent}}\right]^{(1/t)} - 1$$

where $t$ is the term of the loan in years; $t$ need not be an integer.

**imprest fund.** *Petty cash fund*.

**improvement.** An *expenditure* to extend the useful life of an *asset* or to improve its performance (rate of output, cost) over that of the original asset; sometimes called "betterment." The firm capitalizes such expenditures as part of the asset's cost. Contrast with *maintenance* and *repair*.

**imputed cost.** A cost that does not appear in accounting records, such as the *interest* that a firm could earn on cash spent to acquire inventories rather than, say, government bonds. Or, consider a firm that owns the buildings it occupies. This firm has an imputed cost for rent in an amount equal to what it would have to pay to use similar buildings owned by another or equal to the amount it could collect from someone renting the premises from the firm. *Opportunity cost.*

**imputed interest.** See *interest, imputed.*

**in the black (red).** Operating at a profit (loss).

**in-process R&D.** When one firm acquires another, the acquired firm will often have *research and development* activities under way that, following *GAAP,* it has *expensed.* The acquiring firm will pay for these activities to the extent they have value and will then, following GAAP, write off the activities. For each dollar of in-process R&D the acquiring firm identities and immediately *expenses,* it will have one less dollar of *goodwill* or other assets to *amortize.* Some acquirers have overstated the valuations of acquired in-process R&D in order to increase immediate *write-offs* and subsequent, recurring *income.*

**incentive compatible compensation.** Said of a compensation plan that induces managers to act for the interests of owners while acting also in their own interests. For example, consider that a time of rising prices and increasing inventories when using a *LIFO* cost flow assumption implies paying lower *income taxes* than using *FIFO*. A bonus scheme for managers based on accounting *net income* is not incentive-compatible because owners likely benefit more under LIFO, whereas managers benefit more if they report using FIFO. See *LIFO conformity rule* and *goal congruence*.

**income.** *Excess of revenues* and *gains* over *expenses* and *losses* for a period; *net income*. The term is sometimes used with an appropriate modifier to refer to the various intermediate amounts shown in a *multiple-step income statement* or to refer to revenues, as in "rental income." See *comprehensive income.*

**income accounts.** *Revenue* and *expense accounts.*

**income before taxes.** On the *income statement,* the difference between all *revenues* and *expenses* except *income tax* expense. Contrast with *net income.*

**income determination.** See *determine*.

**income distribution account.** *Temporary account* sometimes debited when the firm declares *dividends;* closed to *retained earnings.*

**income from continuing operations.** As defined by *APB Opinion No. 30,* all *revenues* less all *expenses* except for the following: results of operations (including *income tax* effects) that a firm has discontinued or will discontinue; *gains* or *losses,* including income tax effects, on disposal of segments of the business; gains or losses, including income tax effects, from *extraordinary items;* and the cumulative effect of *accounting changes.*

**income from discontinued operations.** *Income,* net of tax effects, from parts of the business that the firm has discontinued during the period or will discontinue in the near future. Accountants report such items on separate lines of the *income statement,* after *income from continuing operations* but before *extraordinary items.*

**income (revenue) bond.** See *special revenue debt.*

**income smoothing.** A method of timing business *transactions* or choosing *accounting principles* so that the firm reports smaller variations in *income* from year to year than it otherwise would. Although some managements set income smoothing as an objective, no standard-setter does.

**income statement.** The statement of *revenues, expenses, gains,* and *losses* for the period, ending with *net income* for the period. Accountants usually show the *earnings-per-share* amount on the income statement; the *reconciliation* of beginning and ending balances of *retained earnings* may also appear in a combined statement of income and retained earnings. See *income from continuing operations, income from discontinued operations, extraordinary items, multiple-step,* and *single-step.*

**income summary.** In problem solving, an *account* that serves as a surrogate for the *income statement.* In using an income summary, close all *revenue* accounts to the Income Summary as *credits* and all *expense* accounts as *debits*. The *balance* in the account, after you make all these *closing entries,* represents income or loss for the period. Then, close the income summary balance to retained earnings.

**income tax.** An annual tax levied by the federal and other governments on the income of an entity.

**income tax allocation.** See *deferred income tax* (*liability*) and *tax allocation: intra-statement*.

**incremental.** An adjective used to describe the increase in *cost, expense, investment, cash flow, revenue, profit,* and the like if the firm produces or sells one or more units or if it undertakes an activity. See *differential.*

**incremental cost.** See *incremental.*

**incur.** Said of an obligation of a firm, whether or not that obligation is *accrued.* For example, a firm incurs interest expense on a loan as time passes but accrues that interest only on payment dates or when it makes an *adjusting entry*.

**indenture.** See *bond indenture.*

**independence.** The mental attitude required of the *CPA* in performing the *attest* function. It implies that the CPA is impartial and that the members of the auditing CPA firm own no stock in the corporation being audited.

**independent accountant.** The *CPA* who performs the *attest* function for a firm.

**independent variable.** See *regression analysis*.

**indeterminate-term liability.** A *liability* lacking the criterion of being due at a definite time. This term is our own coinage to encompass the *minority interest.*

**indexation.** An attempt by lawmakers or parties to a contract to cope with the effects of *inflation*. Amounts fixed in law or contracts are "indexed" when these amounts change as a given measure of price changes. For example, a so-called escalator clause (COLA—cost of living allowance or adjustment) in a labor contract might provide that hourly wages will be increased as the *Consumer Price Index* increases. Many economists have suggested the indexation of numbers fixed in the *income tax* laws. If, for example, the personal *exemption* is $2,500 at the start of the period, if prices rise by 10 percent during the period, and if the personal exemption is indexed, then the personal exemption would automatically rise to $2,750 (= $2,500 + 0.10 × $2,500) at the end of the period.

**indirect cost pool.** Any grouping of individual costs that a firm does not identify with a *cost objective.*

**indirect costs.** Production costs not easily associated with the production of specific goods and services; *overhead costs.* Accountants may *allocate* them on some *arbitrary* basis to specific products or departments.

**indirect labor (material) cost.** An *indirect cost* for labor (material), such as for supervisors (supplies).

**indirect method.** See *statement of cash flows.*

**individual proprietorship.** *Sole proprietorship.*

***Industry Audit Guides.*** A series of *AICPA* publications providing specific accounting and *auditing principles* for specialized situations. Audit guides have been issued covering government contractors, state and local government units, investment companies, finance companies, brokers and dealers in securities, and many other subjects.

**inescapable cost.** A *cost* that the firm or manager cannot avoid (see *avoidable*) because of an action. For example, if management shuts down two operating rooms in a hospital but still must employ security guards in unreduced numbers, the security costs are "inescapable" with respect to the decision to close the operating rooms.

**inflation.** A time of generally rising prices.

**inflation accounting.** Strictly speaking, *constant-dollar accounting.* Some writers incorrectly use the term to mean *current cost accounting.*

**information circular.** Canada: a document, accompanying the notice of a shareholders' meeting, prepared in connection with the solicitation of proxies by or on behalf of the management of the corporation. It contains information concerning the people making the solicitation, election of directors, appointment of auditors, and other matters to be acted on at the meeting.

**information system.** A system, sometimes formal and sometimes informal, for collecting, processing, and communicating data that are useful for the managerial functions of decision making, planning, and control and for financial reporting under the *attest* requirement.

**inherent interest rate.** *Implicit interest* rate.

**insolvent.** Unable to pay debts when due; said of a company even though *assets* exceed *liabilities.*

**installment.** Partial payment of a debt or partial collection of a receivable, usually according to a contract.

**installment contracts receivable.** The name used for *accounts receivable* when the firm uses the *installment method* of recognizing revenue. Its *contra account, unrealized gross margin,* appears on the balance sheet as a subtraction from the amount receivable.

**installment sales.** Sales on account when the buyer promises to pay in several separate payments, called *installments.* The seller may, but need not, account for such sales using the *installment method.* If the seller accounts for installment sales with the *sales basis of revenue recognition* for financial reporting but with the installment method for income tax returns, then it will have *deferred income tax* (*liability*).

**installment (sales) method.** Recognizing *revenue* and *expense* (or *gross margin*) from a sales transaction in proportion to the fraction of the selling price collected during a period; allowed by the *IRS* for income tax reporting but acceptable in *GAAP* (*APB Opinion No. 10*) only when the firm cannot estimate cash collections with reasonable precision. See *realized* (and *unrealized*) *gross margin.*

**Institute of Certified Management Accountants (ICMA).** See *CMA* and *Institute of Management Accountants.*

**Institute of Internal Auditors (IIA).** The national association of accountants who are engaged in internal auditing and are employed by business firms; administers a comprehensive professional examination. Those who pass the exam qualify to be designated *CIA* (Certified Internal Auditor).

**Institute of Management Accountants (IMA).** Formerly, the National Association of Accountants, NAA; a society open to those engaged in management accounting; parent organization of the *ICMA,* which oversees the *CMA* program.

**insurance.** A contract for reimbursement of specific losses; purchased with insurance premiums. "Self-insurance" is not insurance but is merely the noninsured's willingness to assume the risk of incurring losses while saving the premium.

**intangible asset.** A nonphysical right that gives a firm an exclusive or preferred position in the marketplace. Examples are *copyright, patent, trademark, goodwill, organization costs, capitalized* advertising cost, computer programs, licenses for any of the preceding, government licenses (e.g., broadcasting or the right to sell liquor), *leases,* franchises, mailing lists, exploration permits, import and export permits, construction permits, and marketing quotas. Invariably, accountants define "intangible" using a "for example" list, as we have just done, because accounting has been unable to devise a definition of "intangible" that will include items such as those listed above but exclude stock and bond certificates. Accountants classify these items as tangibles, even though they give their holders a preferred position in receiving dividends and interest payments.

**Interamerican Accounting Association (IAA).** An organization, headquartered in Miami, devoted to facilitating interaction between accounting practitioners in the Americas.

**intercompany elimination.** See *eliminations.*

**intercompany profit.** Profit within an organization. If one *affiliated company* sells to another, and the goods remain in the second company's *inventory* at the end of the period, then the first company has not yet realized a *profit* by a sale to an outsider. The profit is "intercompany profit," and the accountant eliminates it from net *income* when preparing *consolidated income statements* or when the firm uses the *equity method.*

**intercompany transaction.** *Transaction* between a *parent company* and a *subsidiary* or between subsidiaries in a *consolidated entity;* the accountant must eliminate the effects of such a transaction when preparing *consolidated financial statements.* See *intercompany profit.*

**intercorporate investment.** Occurs when a given *corporation* owns *shares* or *debt* issued by another.

**interdepartment monitoring.** An *internal control* device. The advantage of allocating *service department costs* to *production departments* stems from the incentives that this gives those charged with the costs to control the costs incurred in the service department. That process of having one group monitor the performance of another is interdepartment monitoring.

**interest.** The charge or cost for using cash, usually borrowed funds. Interest on one's own cash used is an *opportunity cost, imputed interest.* The amount of interest for a loan is the total amount paid by a borrower to a lender less the amount paid by the lender to the borrower. Accounting seeks to allocate that interest over the time of the loan so that the interest rate (= interest charge/amount borrowed) stays constant each period. See *interest rate* for discussion of the quoted amount. See *effective interest rate* and *nominal interest rate.*

**interest, imputed.** The difference between the face amount and the present value of a promise. If a borrower merely promises to pay a single amount, sometime later than the present, then the face amount the borrower will repay at *maturity* will exceed the present value (computed at a *fair market* interest rate, called the "imputed interest rate") of the promise. See also *imputed cost.*

**interest factor.** One plus the *interest* rate.

**interest method.** See *effective interest method.*

**interest rate.** A basis used for computing the cost of borrowing funds; usually expressed as a ratio between the number of currency units (e.g., dollars) charged for a period of time and the number of currency units borrowed for that same period of time. When the writers and speakers do not state a period, they almost always mean a period of one year. See *interest, simple interest, compound interest, effective (interest) rate,* and *nominal interest rate.*

**interest rate swap.** See *swap.*

**interfund accounts.** In governmental accounting, the accounts that show transactions between funds, especially interfund receivables and payables.

**interim statements.** Statements issued for periods less than the regular, annual *accounting period.* The *SEC* requires most corporations to issue interim statements on a quarterly basis. In preparing interim reports, a problem arises that the accountant can resolve only by understanding whether interim reports should report on the interim period (1) as a self-contained accounting period or (2) as an integral part of the year so that analysts can make forecasts of annual performance. For example, assume that at the end of the first quarter, a retailer has dipped into old LIFO layers, depleting its *inventory,* so that it computes *LIFO cost of goods sold* artificially low and *net income* artificially high, relative to the amounts the firm would have computed if it had made the "normal" purchases, equal to or greater than sales. The retailer expects to purchase inventory sufficiently large so that when it computes cost of goods sold for the year, there will be no *dips into old LIFO layers* and income will not be artificially high. The first approach will compute the quarterly income from low cost of goods sold using data for the dips that have actually occurred by the end of the quarter. The second approach will compute quarterly income from cost of goods sold assuming that purchases were equal to "normal" amounts and that the firm did not dip into old LIFO layers. *APB Opinion No. 28* and the *SEC* require that interim reports be constructed largely to satisfy the second purpose.

**internal audit, internal auditor.** An *audit* conducted by the firm's own employees, called "internal auditors," to ascertain whether the firm's *internal control* procedures work as planned. Contrast with an external audit conducted by a *CPA.*

**internal controls.** Policies and procedures designed to provide management with reasonable assurances that employees behave in a way that enables the firm to meet its organizational goals. See *control system.*

**internal rate of return (IRR).** The discount rate that equates the net *present value* of a stream of cash outflows and inflows to zero.

**internal reporting.** Reporting for management's use in planning and control. Contrast with *external reporting* for financial statement users.

**Internal Revenue Service (IRS).** Agency of the U.S. Treasury Department responsible for administering the Internal Revenue Code and collecting income and certain other taxes.

**International Accounting Standards (IAS).** Standards set between 1973 and 2000 by the International Accounting Standards Committee (IASC), which was succeeded in 2001 by the International Accounting Standards Board (IASB). The IASC's standards were known as International Accounting Standards (IAS) and are still cited by this name today even if they have subsequently been revised by the IASB.

**International Accounting Standards Board (IASB).** The independent body, based in London, that sets International Financial Reporting Standards (IFRS) and revises International Accounting Standards (IAS), which are required to be used, or may be used, in more the 100 countries. It succeeded the International Accounting Standards Committee (IASC) in 2001. Since 2002, the IASB and the Financial Accounting Standards Board (FASB) have been collaborating on the mutual convergence of their respective standards at a high level of quality. See www.iasb.org. Developments concerning IFRS may be monitored on the Deloitte web site, www.iasplus.com.

**International Accounting Standards Committee (IASC).** The body that set International Accounting Standards (IAS) between 1973 and 2000, which was succeeded in 2001 by the International Accounting Standards Board (IASB).

**International Federation of Accountants (IFAC).** A federation, based in New York, of virtually all national professional accounting bodies around the world. Its committees develop standards in auditing, education, ethics and public sector financial reporting. See www.ifac.com.

**International Financial Reporting Interpretations Committee (IFRIC).** This committee, based in London, is the interpretive body of the International Accounting Standards Board (IASB). Its recommended interpretations of International Accounting Standards (IAS) and International Financial Reporting Standards (IFRS) must be approved by the IASB before they become official.

**International Financial Reporting Standards (IFRS).** Standards set by the International Accounting Standards Board (IASB). Some of the standards previously issued by the International Accounting Standards Committee (IASC), even if revised by the IASB since 2001, are still referred to as International Accounting Standards (IAS).

**International Organization of Securities Commissions.** *IOSCO.*

**interperiod tax allocation.** See *deferred income tax (liability).*

**interpolation.** The estimation of an unknown number intermediate between two (or more) known numbers.

***Interpretations.*** See *FASB Interpretation.*

**intrastatement tax allocation.** See *tax allocation: intrastatement.*

**inventoriable costs.** *Costs* incurred that the firm adds to the cost of manufactured products; *product costs (assets)* as opposed to *period expenses.*

**inventory.** As a noun, the *balance* in an asset *account,* such as raw materials, supplies, work-in-process, and finished goods; as a verb, to calculate the *cost* of goods on hand at a given time or to count items on hand physically.

**inventory equation.** *Beginning inventory* + net additions − withdrawals = *ending inventory.* Ordinarily, additions are net purchases, and withdrawals are *cost of goods sold.* Notice that ending inventory, appearing on the balance sheet, and cost of goods sold, appearing on the income statement, must add to a fixed sum. The larger is one; the smaller must be the other. In valuing inventories, the firm usually knows beginning inventory and net purchases. Some inventory methods (for example, some applications of the *retail inventory method*) measure costs of goods sold and use the equation to find the cost of ending inventory. Most methods measure cost of ending inventory and use the equation to find the cost of goods sold (withdrawals). In *current cost* (in contrast to *historical cost*) *accounting,* additions (in the equation) include holding gains, whether realized or not. Thus the current cost inventory equation is as follows: Beginning Inventory (at Current Cost) + Purchases (where Current Cost is Historical Cost) + Holding Gains (whether Realized or Not) − Ending Inventory (at Current Cost) = Cost of Goods Sold (Current Cost).

**inventory holding gains.** See *inventory profit.*

**inventory layer.** See *LIFO inventory layer.*

**inventory profit.** A term with several possible meanings. Consider the data in the accompanying illustration. The firm uses a *FIFO cost flow assumption* and derives its *historical cost* data. The assumed *current cost* data resemble those that the FASB suggested in *SFAS No. 89.* The term *income from continuing operations* refers to revenues less expenses based on current, rather than historical, costs. To that subtotal, add realized holding gains to arrive at realized (conventional) income. To that, add unrealized holding gains to arrive at economic income. The term "inventory profit" often refers (for example, in some *SEC* releases) to the realized holding gain, $110 in the illustration. The amount of inventory profit will usually be material when the firm uses FIFO and when prices rise. Other analysts, including us, prefer to use the term "inventory profit" to refer to the total *holding gain,* $300 (= $110 + $190, both realized and unrealized), but writers use this meaning less often. In periods of rising prices and increasing inventories, the realized holding gains under a FIFO cost flow assumption will exceed those under LIFO. In the illustration, for example, assume under LIFO that the historical cost of goods sold is $4,800, that historical LIFO cost of beginning inventory is $600, and that historical LIFO cost of ending inventory is $800. Then income from continuing operations, based on

Inventory Profit Illustration

| | (Historical) Acquisition Cost Assuming FIFO | Current Cost |
|---|---|---|
| ASSUMED DATA | | |
| Inventory, 1/1 | $ 900 | $1,100 |
| Inventory, 12/31 | 1,160 | 1,550 |
| Cost of Goods Sold for the Year | 4,740 | 4,850 |
| Sales for the Year | $5,200 | $5,200 |
| INCOME STATEMENT FOR THE YEAR | | |
| Sales | $5,200 | $5,200 |
| Cost of Goods Sold | 4,740 | 4,850 |
| (1) Income from Continuing Operations | | $ 350 |
| Realized Holding Gains | | 110[a] |
| (2) Realized Income = Conventional Net Income (under FIFO) | $ 460 | $ 460 |
| Unrealized Holding Gain | | 190[b] |
| (3) Economic Income | | $ 650 |

[a]Realized holding gain during a period is current cost of goods sold less historical cost of goods sold; for the year the realized holding gain under FIFO is $110 = $4,850 − $4,740. Some refer to this as "inventory profit."

[b]The total unrealized holding gain at any time is current cost of inventory on hand at that time less historical cost of that inventory. The unrealized holding gain during a period is unrealized holding gain at the end of the period less the unrealized holding gain prior to this year: $200 = $1,100 − $900. Unrealized holding gain during the year = ($1,550 − $1,160) − ($1,100 − $900) = $390 − $200 = $190.

current costs, remains \$350 (= \$5,200 − \$4,850), realized holding gains are \$50 (= \$4,850 − \$4,800), realized income is \$400 (= \$350 + \$50), the unrealized holding gain for the year is \$250 [= (\$1,550 − \$800) − (\$1,100 − \$600)], and economic income is \$650 (= \$350 + \$50 + \$250). The cost flow assumption has only one real effect on this series of calculations: the split of the total holding gain into realized and unrealized portions. Thus, economic income does not depend on the cost flow assumption. Holding gains total \$300 in the illustration. The choice of cost flow assumption determines the portion reported as realized.

**inventory turnover.** Number of times the firm sells the average *inventory* during a period; *cost of goods sold* for a period divided by average inventory for the period. See *ratio*.

**inventory valuation allowance.** A preferred term for the difference between the *FIFO* (or *current cost*) of *inventory* and its *LIFO* valuation. Many in business refer to this as the "LIFO reserve." See *reserve* for an explanation of why we dislike that term.

**invested capital.** *Contributed capital.*

**investee.** A company in which another entity, the "investor," owns stock.

**investing activities.** Acquiring and selling *securities* or productive *assets* expected to produce *revenue* over several *periods*.

**investment.** An *expenditure* to acquire property or other *assets* in order to produce *revenue;* the asset so acquired; hence a *current* expenditure made in anticipation of future income; said of other companies' *securities* held for the long term and appearing in a separate section of the *balance sheet;* in this context, contrast with *marketable securities*.

**investment center.** A *responsibility center,* with control over *revenues, costs,* and *assets*.

**investment credit.** A reduction in income tax liability sometimes granted by the federal government to firms that buy new equipment. This item is a credit in that the taxpayer deducts it from the tax bill, not from pretax income. The tax credit has been a given percentage of the purchase price of the assets purchased. The government has changed the actual rules and rates over the years. As of 1999, there is no investment credit. See *flow-through method* and *carryforward*.

**investment decision.** The decision whether to undertake an action involving production of goods or services; contrast with financing decision.

**investment tax credit.** *Investment credit.*

**investment turnover ratio.** A term that means the same thing as *total assets turnover ratio*.

**investments.** A balance sheet heading for tangible assets held for periods longer than the operating cycle and not used in revenue production (assets not meeting the definitions of *current assets* or *property, plant, and equipment*).

**invoice.** A document showing the details of a sale or purchase *transaction*.

**IOSCO (International Organization of Securities Commissions).** The name, since 1983, of a confederation of regulators of securities and futures markets. Members come from over 80 countries. The IOSCO encourages the *IASB* to eliminate accounting alternatives and to ensure that accounting standards are detailed and complete, with adequate disclosure requirements, and that financial statements are user-friendly.

**I.O.U.** An informal document acknowledging a debt, setting out the amount of the debt and signed by the debtor.

**IRR.** *Internal rate of return.*

**IRS.** *Internal Revenue Service.*

**isoprofit line.** On a graph showing feasible production possibilities of two products that require the use of the same, limited resources, a line showing all feasible production possibility combinations with the same *profit* or, perhaps, *contribution margin*.

**issue.** A corporation exchange of its stock (or *bonds*) for cash or other *assets*. Terminology says the corporation "issues," not "sells," that stock (or bonds). Also used in the context of withdrawing supplies or materials from inventory for use in operations and of drawing a *check*.

**issued shares.** Those shares of *authorized capital stock* that a *corporation* has distributed to the shareholders. See *issue.* Shares of *treasury stock* are legally issued but are not *outstanding* for the purpose of voting, *dividend declarations,* and *earnings-per-share* calculations.

# J

**JIT.** See *just-in-time inventory.*

**job cost sheet.** A schedule showing actual or budgeted inputs for a special order.

**job development credit.** The name used for the *investment credit* in the 1971 tax law, since repealed, on this subject.

**job (-order) costing.** Accumulation of *costs* for a particular identifiable batch of product, known as a job, as it moves through production.

**joint cost.** Cost of simultaneously producing or otherwise acquiring two or more products, called joint products, that a firm must, by the nature of the process, produce or acquire together, such as the cost of beef and hides of cattle. Generally, accounting allocates the joint costs of production to the individual products in proportion to their respective sales value (or, sometimes and usually not preferred, their respective physical quantities) at the *split-off* point. Other examples include *central corporate expenses* and *overhead* of a department when it manufactures several products. See *common cost* and *sterilized allocation*.

**joint cost allocation.** See *joint cost.*

**joint product.** One of two or more outputs with significant value that a firm must produce or acquire simultaneously. See *by-product* and *joint cost*.

**journal.** The place where the firm records transactions as they occur; the book of original entry.

**journal entry.** A dated *journal* recording, showing the accounts affected, of equal *debits* and *credits,* with an explanation of the *transaction,* if necessary.

***Journal of Accountancy.*** A monthly publication of the *AICPA.*

***Journal of Accounting and Economics.*** Scholarly journal published by the William E. Simon Graduate School of Business Administration of the University of Rochester.

***Journal of Accounting Research.*** Scholarly journal containing articles on theoretical and empirical aspects of accounting; published by the Graduate School of Business of the University of Chicago.

**journal voucher.** A *voucher* documenting (and sometimes authorizing) a *transaction,* leading to an entry in the *journal.*

**journalize.** To make an entry in a *journal.*

**judgment(al) sampling.** A method of choosing a sample in which the analyst subjectively selects items for examination, in contrast to selecting them by statistical methods. Compare *random sampling.*

**junk bond.** A low-rated *bond* that lacks the merit and characteristics of an investment-grade bond. It offers high yields, typically in excess of 15 percent per year, but also possesses high risk of default. Sometimes writers, less pejoratively, call these "high-yield bonds." No clear line separates junk from nonjunk bonds.

**just-in-time inventory (production) (JIT).** In managing *inventory* for manufacturing, system in which a firm purchases or manufactures each component just before the firm uses it. Contrast with systems in which firms acquire or manufacture many parts in advance of needs. JIT systems have much smaller carrying costs for inventory, ideally none, but run higher risks of incurring *stockout* costs.

# K

**k.** Two to the tenth power ($2^{10}$ or 1,024), when referring to computer storage capacity. The one-letter abbreviation derives from the first letter of the prefix "kilo-" (which means 1,000 in decimal notation).

**Kaizen costing.** A management concept that seeks continuous improvements, likely occurring in small incremental amounts, by refinements of all components of a production process.

**KG (Kommanditgesellschaft).** Germany: similar to a general partnership (*OHG*) except that some of its members may limit their liability. One of the partners must be a *general partner* with unlimited liability.

**kiting.** A term with slightly different meanings in banking and auditing contexts. In both, however, it refers to the wrongful practice of taking advantage of the *float,* the time that elapses between the deposit of a *check* in one bank and its collection at another. In the banking context, an individual deposits in Bank A a check written on Bank B. He (or she) then writes checks against the deposit created in Bank A. Several days later, he deposits in Bank B a check written on Bank A, to cover the original check written on Bank B. Still later, he deposits in Bank A a check written on Bank B. The process of covering the deposit in Bank A with a check written on Bank B and vice versa continues until the person can arrange an actual deposit of cash. In the auditing context, kiting refers to a form of *window dressing* in which the firm makes the amount of the account Cash in Bank appear larger than it actually is by depositing in Bank A a check written on Bank B without recording the check written on Bank B in the *check register* until after the close of the *accounting period.*

**know-how.** Technical or business information that is of the type defined under *trade secret* but that a firm does not maintain as a secret. The rules of accounting for this *asset* are the same as for other *intangibles.*

# L

**labor variances.** The *price* (or *rate*) and *quantity* (or *usage*) *variances* for *direct labor* inputs in a *standard costing system.*

**laid-down cost.** Canada and UK: the sum of all direct costs incurred for procurement of goods up to the time of physical receipt, such as invoice cost plus customs and excise duties, freight, and cartage.

**land.** An *asset* shown at *acquisition cost* plus the *cost* of any nondepreciable *improvements;* in accounting, implies use as a plant or office site rather than as a *natural resource,* such as timberland or farmland.

**lapping (accounts receivable).** The theft, by an employee, of cash sent in by a customer to discharge the latter's *payable.* The employee conceals the theft from the first customer by using cash received from a second customer. The employee conceals the theft from the second customer by using cash received from a third customer, and so on. The process continues until the thief returns the funds or can make the theft permanent by creating a fictitious *expense* or receivable write-off or until someone discovers the fraud.

**lapse.** To expire; said of, for example, an insurance policy or discounts that are made available for prompt payment and that the purchaser does not take.

**last-in, first-out.** See *LIFO.*

**layer.** See *LIFO inventory layer.*

**lead time.** The time that elapses between placing an order and receiving the *goods* or *services* ordered.

**learning curve.** A mathematical expression of the phenomenon that incremental unit costs to produce decrease as managers and labor gain experience from practice.

**lease.** A contract calling for the lessee (user) to pay the lessor (owner) for the use of an asset. A cancelable lease allows the lessee to cancel at any time. A noncancelable lease requires payments from the lessee for the life of the lease and usually shares many of the economic characteristics of *debt financing.* Most long-term noncancelable leases meet the usual criteria for classifying them as *liabilities,* and GAAP require the firm to show them as liabilities. *SFAS No. 13* and the *SEC* require disclosure, in notes to the financial statements, of the commitments for long-term noncancelable leases. See *capital lease* and *operating lease*.

**leasehold.** The *asset* representing the right of the lessee to use leased property. See *lease* and *leasehold improvement*.

**leasehold improvement.** An *improvement* to leased property. The firm should *amortize* it over the *service life* or the life of the lease, whichever is shorter.

**least and latest rule.** Paying the least amount of taxes as late as possible within the law to minimize the *present value* of tax payments for a given set of operations. Sensible taxpayers will follow this rule. When a taxpayer knows that tax rates will increase later, the taxpayer may reduce the present value of the tax burden by paying smaller taxes sooner. Each set of circumstances requires its own computations.

**ledger.** A book of accounts; book of final entry. See *general ledger* and *subsidiary ledger.* Contrast with *journal.*

**legal capital.** The amount of *contributed capital* that, according to state law, the firm must keep permanently in the firm as protection for creditors.

**legal entity.** See *entity*.

**lender.** See *loan*.

**lessee.** See *lease*.

**lessor.** See *lease*.

**letter stock.** Privately placed *common shares;* so called because the *SEC* requires the purchaser to sign a letter of intent not to resell the shares.

**leverage.** More than proportional result from extra effort or financing. Some measure of output increases faster than the measure of input. "Operating leverage" refers to the tendency of *net income* to rise at a faster rate than sales in the presence of *fixed costs*. A doubling of sales, for example, usually implies a more than doubling of net income. "Financial leverage" (or "capital leverage") refers to an increase in rate of return larger than the increase in explicit financing costs—the increased rate of return on *owners' equity* (see *ratio*) when an *investment* earns a return larger than the after-tax *interest rate* paid for *debt* financing. Because the interest charges on debt usually do not change, any *incremental* income benefits owners and none benefits debtors. When writers use the term "leverage" without a qualifying adjective, the term usually refers to financial leverage, the use of *long-term* debt in securing *funds* for the *entity*.

**leveraged lease.** A special form of lease involving three parties: a *lender,* a *lessor,* and a *lessee*. The lender, such as a bank or insurance company, lends a portion, say 80 percent, of the cash required for acquiring the *asset*. The lessor puts up the remainder, 20 percent, of the cash required. The lessor acquires the asset with the cash, using the asset as security for the loan, and leases it to the lessee on a *noncancelable* basis. The lessee makes periodic lease payments to the lessor, who in turn makes payments on the loan to the lender. Typically, the lessor has no obligation for the debt to the lender other than transferring a portion of the receipts from the lessee. If the lessee should default on the required lease payments, then the lender can repossess the leased asset. The lessor usually has the right to benefit from the tax deductions for *depreciation* on the asset, for *interest expense* on the loan from the lender, and for any *investment credit*. The lease is leveraged in the sense that the lessor, who takes most of the risks and enjoys most of the rewards of ownership, usually borrows most of the funds needed to acquire the asset. See *leverage.*

**liability.** An obligation to pay a definite (or reasonably definite) amount at a definite (or reasonably definite) time in return for a past or current benefit (that is, the obligation arises from a transaction that is not an *executory contract*); a probable future sacrifice of economic benefits arising from present obligations of a particular *entity* to *transfer assets* or to provide services to other entities in the future as a result of past *transactions* or events. *SFAC No. 6* says that "probable" refers to that which we can reasonably expect or believe but that is neither certain nor proved. A liability has three essential characteristics: (1) the obligation to transfer assets or services has a specified or knowable date, (2) the entity has little or no discretion to avoid the transfer, and (3) the event causing the obligation has already happened, that is, it is not executory.

**lien.** The right of person A to satisfy a claim against person B by holding B's property as security or by seizing B's property.

**life annuity.** A *contingent annuity* in which payments cease at the death of a specified person(s), usually the *annuitant(s)*.

**LIFO (last-in, first-out).** An *inventory* flow assumption in which the *cost of goods sold* equals the cost of the most recently acquired units and a firm computes the *ending inventory cost* from the costs of the oldest units. In periods of rising prices and increasing inventories, LIFO leads to higher reported expenses and therefore lower reported income and lower balance sheet inventories than does FIFO. Contrast with *FIFO*. See *FISH* and *inventory profit*.

**LIFO conformity rule.** The *IRS* rule requiring that companies that use a *LIFO cost flow assumption* for *income taxes* must also use LIFO in computing *income* reported in *financial statements* and forbidding the disclosure of pro forma results from using any other cost flow assumption.

**LIFO, dollar-value method.** See *dollar-value LIFO method*.

**LIFO inventory layer.** A portion of LIFO inventory cost on the *balance sheet*. The *ending inventory* in physical quantity will usually exceed the *beginning inventory*. The *LIFO cost flow assumption* assigns to this increase in physical quantities a cost computed from the prices of the earliest purchases during the year. The LIFO inventory then consists of layers, sometimes called "slices," which typically consist of relatively small amounts of physical quantities from each of the past years when purchases in physical units exceeded sales in units. Each layer carries the prices from near the beginning of the period when the firm acquired it. The earliest layers will typically (in periods of rising prices) have prices much less than current prices. If inventory quantities should decline in a subsequent period—a "dip into old LIFO layers"—the latest layers enter cost of goods sold first.

**LIFO reserve.** *Unrealized holding gain* in *ending inventory*: current or *FIFO historical* cost of ending inventory less LIFO *historical cost*. A better term for this concept is "excess of current cost over LIFO historical cost." See *reserve*.

**limited liability.** The legal concept that shareholders of corporations are not personally liable for debts of the company.

**limited partner.** A *partnership* member who is not personally liable for debts of the partnership. Every partnership must have at least one *general partner*, who is fully liable.

**line-of-business reporting.** See *segment reporting*.

**line of credit.** An agreement with a bank or set of banks for short-term borrowings on demand.

**linear programming.** A mathematical tool for finding profit-maximizing (or cost-minimizing) combinations of products to produce when a firm has several products that it can produce but faces linear constraints on the resources available in the production processes or on maximum and minimum production requirements.

**liquid.** Said of a business with a substantial amount (the amount is unspecified) of *working capital*, especially *quick assets*.

**liquid assets.** *Cash, current marketable securities*, and sometimes, *current receivables*.

**liquidating dividend.** A *dividend* that a firm declares in the winding up of a business to distribute its assets to the shareholders. Usually the recipient treats this as a return of *investment*, not as *revenue*.

**liquidation.** Payment of a debt; sale of assets in closing down a business or a segment thereof.

**liquidation value per share.** The amount each *share* of stock will receive if the *board* dissolves a corporation; for *preferred stock* with a liquidation preference, a stated amount per share.

**liquidity.** Refers to the availability of *cash*, or near-cash resources, for meeting a firm's obligations.

**LISH.** An acronym, conceived by George H. Sorter, for *last-in, still-here*. LISH is the same cost flow assumption as *FIFO*. Many readers of accounting statements find it easier to think about inventory questions in terms of items still on hand. Think of FIFO in connection with *cost of goods sold* but of LISH in connection with *ending inventory*. See *FISH*.

**list price.** The published or nominally quoted price for goods.

**list price method.** See *trade-in transaction*.

**loan.** An arrangement in which the owner of property, called the lender, allows someone else, called the borrower, the use of the property for a period of time, which the agreement setting up the loan usually specifies. The borrower promises to return the property to the lender and, often, to make a payment for the use of the property. This term is generally used when the property is *cash* and the payment for its use is *interest*.

**LOCOM.** *Lower of cost or market*.

**long-lived (term) asset.** An asset whose benefits the firm expects to receive over several years; a *noncurrent* asset, usually includes *investments, plant assets*, and *intangibles*.

**long-term (construction) contract accounting.** The *percentage-of-completion method* of *revenue* recognition; sometimes used to mean the *completed contract method*.

**long-term debt ratio.** *Noncurrent liabilities* divided by total *assets*.

**long-term liability (debt).** *Noncurrent liability*.

**long-term, long-run.** A term denoting a time or time periods in the future. How far in the future depends on context. For some securities traders, "long-term" can mean anything beyond the next hour or two. For most managers, it means anything beyond the next year or two. For government policymakers, it can mean anything beyond the next decade or two. For geologists, it can mean millions of years.

**long-term solvency risk.** The risk that a firm will not have sufficient *cash* to pay its *debts* sometime in the *long run*.

**loophole.** Imprecise term meaning a technicality allowing a taxpayer (or *financial statements*) to circumvent the intent, without violating the letter, of the law (or *GAAP*).

**loss.** Excess of *cost* over net proceeds for a single transaction; negative *income* for a period; a cost expiration that produced no *revenue*. See *gain* for a discussion of related and contrasting terms and how to distinguish loss from *expense*.

**loss contingency.** See *contingency*.

**lower of cost or market (LOCOM).** A basis for valuation of *inventory*. This basis sets inventory value at the lower of *acquisition cost* or *current replacement cost* (market), subject to the following constraints. First, the market value of an item used in the computation cannot exceed its *net realizable value*—an amount equal to selling price less reasonable costs to complete production and to sell the item. Second, the market value of an item used in the computation cannot be less than the net realizable value minus the normal *profit* ordinarily realized on disposition of completed items of this type. The basis chooses the lower-of-cost-or-market valuation as the lower of acquisition *cost* or replacement cost (*market*) subject to the upper and lower bounds on replacement cost established in the first two steps. Thus,

| | | |
|---|---|---|
| Market Value | = | Midvalue of (Replacement Cost, Net Realizable Value, Net Realizable Value Less Normal Profit Margin) |
| Lower of Cost or Market Valuation | = | Minimum (Acquisition Cost, Market Value) |

The accompanying exhibit illustrates the calculation of the lower-of-cost-or-market valuation for four inventory items. Notice that each of the four possible outcomes occurs once in measuring lower of cost or market. Item 1 uses acquisition cost; item 2 uses net realizable value; item 3 uses replacement cost; and item 4 uses net realizable value less normal profit margin.

| | Item | | | |
|---|---|---|---|---|
| | 1 | 2 | 3 | 4 |
| *Calculation of Market Value* | | | | |
| (a) Replacement Cost | $92 | $96 | $92 | $96 |
| (b) Net Realizable Value | 95 | 95 | 95 | 95 |
| (c) Net Realizable Value Less Normal Profit Margin [= (b) − $9] | 86 | 86 | 86 | 86 |
| (d) Market = Midvalue [(a), (b), (c)] | 92 | 95 | 92 | 95 |
| *Calculation of Lower of Cost or Market* | | | | |
| (e) Acquisition Cost | 90 | 97 | 96 | 90 |
| (f) Market [= (d)] | 92 | 95 | 92 | 95 |
| (g) Lower of Cost or Market = Minimum [(e), (f)] | 90 | 95 | 92 | 90 |

A taxpayer may not use the lower-of-cost-or-market basis for inventory on tax returns in combination with a *LIFO cost flow assumption.* In the context of inventory, once the firm writes down the asset, it establishes a new "original cost" basis and ignores subsequent increases in market value in the accounts.

The firm may apply lower of cost or market to individual items of inventory or to groups (usually called *pools*) of items. The smaller the group, the more *conservative* the resulting valuation.

Omit hyphens when you use the term as a noun, but use them when you use the term as an adjectival phrase.

**Ltd., Limited.** UK: a private limited corporation. The name of a private limited company must include the word "Limited" or its abbreviation "Ltd."

**lump-sum acquisition.** *Basket purchase.*

# M

**MACRS.** *Modified Accelerated Cost Recovery System.* See *Accelerated Cost Recovery System.* Since 1986, MACRS has been the accelerated depreciation method required for U.S. income tax purposes.

**maintenance.** *Expenditures* undertaken to preserve an *asset's* service potential for its originally intended life. These expenditures are *period expenses* or *product costs.* Contrast with *improvement,* and see *repair.*

**make-or-buy decision.** A managerial decision about whether the firm should produce a product internally or purchase it from others. Proper make-or-buy decisions in the short run result only when a firm considers *incremental costs* in the analysis.

**maker (of note) (of check).** One who signs a *note* to borrow; one who signs a *check;* in the latter context, synonymous with "drawer." See *draft.*

**management.** Executive authority that operates a business.

**management accounting.** See *managerial accounting.*

***Management Accounting.*** Monthly publication of the *IMA.*

**management audit.** An audit conducted to ascertain whether a firm or one of its operating units properly carries out its objectives, policies, and procedures; generally applies only to activities for which accountants can specify qualitative standards. See *audit* and *internal audit.*

**management by exception.** A principle of management in which managers focus attention on performance only if it differs significantly from that expected.

**management by objective (MBO).** A management approach designed to focus on the definition and attainment of overall and individual objectives with the participation of all levels of management.

**management information system (MIS).** A system designed to provide all levels of management with timely and reliable information required for planning, control, and evaluation of performance.

**management's discussion and analysis (MD&A).** A discussion of management's views of the company's performance; required by the *SEC* to be included in the *10-K* and in the *annual report* to shareholders. The information typically contains discussion of such items as liquidity, results of *operations, segments,* and the effects of *inflation.*

**managerial (management) accounting.** Reporting designed to enhance the ability of management to do its job of decision making, planning, and control. Contrast with *financial accounting.*

**manufacturing cost.** Cost of producing goods, usually in a factory.

**manufacturing expense.** An imprecise, and generally incorrect, alternative title for *manufacturing overhead.* The term is generally incorrect because these costs are usually *product costs,* not expenses.

**manufacturing overhead.** General manufacturing *costs* that are not directly associated with identifiable units of product and that the firm incurs in providing a capacity to carry on productive activities. Accounting treats *fixed* manufacturing overhead cost as a *product cost* under *full absorption costing* but as an *expense* of the period under *variable costing.*

**margin.** *Revenue* less specified expenses. See *contribution margin, gross margin,* and *current margin.*

**margin of safety.** Excess of actual, or budgeted, sales over *breakeven* sales; usually expressed in dollars but may be expressed in units of product.

**marginal cost.** The *incremental cost* or *differential cost* of the last unit added to production or the first unit subtracted from production. See *cost terminology* and *differential* for contrast.

**marginal costing.** *Variable costing.*

**marginal revenue.** The increment in *revenue* from the sale of one additional unit of product.

**marginal tax rate.** The amount, expressed as a percentage, by which income taxes increase when taxable income increases by one dollar. Contrast with *average tax rate*.

**markdown.** See *markup* for definition and contrast.

**markdown cancellation.** See *markup* for definition and contrast.

**market-based transfer price.** A *transfer price* based on external market data rather than internal company data.

**market price.** See *fair value.*

**market rate.** The rate of *interest* a company must pay to borrow *funds* currently. See *effective rate.*

**market value.** *Fair market value.*

**marketable equity securities.** *Marketable securities* representing *owners' equity* interest in other companies, rather than *loans* to them.

**marketable securities.** Other companies' stocks and *bonds* held that can be readily sold on stock exchanges or over-the-counter markets and that the company plans to sell as cash is needed; classified as *current assets* and as part of "cash" in preparing the *statement of cash flows*. If the firm holds these same securities for *long-term* purposes, it will classify them as *noncurrent assets*. *SFAS No. 115* requires that all marketable equity and all debt securities (except those debt securities the holder has the ability and intent to hold to maturity) appear at market value on the balance sheet. The firm reports changes in market value in income for *trading securities* but debits holding losses (or credits holding gains) directly to owners' equity accounts for *securities available for sale.*

**marketing costs.** Costs incurred to sell; includes locating customers, persuading them to buy, delivering the goods or services, and collecting the sales proceeds.

**markon.** See *markup* for definition and contrast.

**markup.** The difference between the original selling price of items acquired for *inventory* and the cost. Precise usage calls this "markon," although many business people use the term "markup." Because of confusion of this use of "markup" with its precise definition (see below), terminology sometimes uses "original markup." If the originally established retail price increases, the precise term for the amount of price increase is "markup," although terminology sometimes uses "additional markup." If a firm reduces selling price, terminology uses the terms "markdown" and "markup cancellation." "Markup cancellation" refers to reduction in price following "additional markups" and can, by definition, be no more than the amount of the additional markup; "cancellation of additional markup," although not used, is descriptive. "Markdown" refers to price reductions from the original retail price. A price increase after a markdown is a "markdown cancellation." If original cost is $12 and original selling price is $20, then markon (original markup) is $8; if the firm later increases the price to $24, the $4 increase is markup (additional markup); if the firm later lowers the price to $21, the $3 reduction is markup cancellation; if the firm further lowers the price to $17, the $4 reduction comprises $1 markup cancellation and $3 markdown; if the firm later increases the price to $22, the $5 increase comprises $3 of markdown cancellation and $2 of markup (additional markup). Accountants track markup cancellations and markdowns separately because they deduct the former (but not the latter) in computing the selling prices of goods available for sale for the denominator of the *cost percentage* used in the conventional *retail inventory method.*

**markup cancellation.** See *markup* for definition and contrast.

**markup percentage.** *Markup* divided by (acquisition cost plus *markup*).

**master budget.** A *budget* projecting all *financial statements* and their components.

**matching convention. matching principle.** The concept of recognizing cost expirations (*expenses*) in the same accounting period during which the firm recognizes related *revenues;* combining or simultaneously recognizing the revenues and expenses that jointly result from the same *transactions* or other events.

**material.** As an adjective, it means relatively important, capable of influencing a decision (see *materiality*); as a noun, *raw material.*

**materiality.** The concept that accounting should disclose separately only those events that are relatively important (no operable definition yet exists) for the business or for understanding its statements. *SFAC No. 2* suggests that accounting information is material if "the judgment of a reasonable person relying on the information would have been changed or influenced by the omission or misstatement."

**materials variances.** *Price* and *quantity variances* for *direct materials* in *standard costing systems;* difference between actual cost and standard cost.

**matrix.** A rectangular array of numbers or mathematical symbols.

**matrix inverse.** For a given square *matrix* **A**, the matrix, $\mathbf{A}^{-1}$ such that $\mathbf{AA}^{-1} = \mathbf{A}^{-1}\mathbf{A} = \mathbf{I}$, the identity matrix. Not all square matrices have inverses. Those that do not are "singular"; those that do are nonsingular.

**maturity.** The date at which an obligation, such as the *principal* of a *bond* or a *note,* becomes due.

**maturity value.** The amount expected to be collected when a loan reaches *maturity*. Depending on the context, the amount may be *principal* or principal and *interest.*

**MBO.** *Management by objective.*

**MD&A.** *Management's discussion and analysis* section of the *annual report.*

**measuring unit.** See *attribute measured* for definition and contrast.

**merchandise.** *Finished goods* bought by a retailer or wholesaler for resale; contrast with finished goods of a manufacturing business.

**merchandise costs.** Costs incurred to sell a product, such as commissions and advertising.

**merchandise turnover.** *Inventory turnover* for merchandise. See *ratio.*

**merchandising business.** As opposed to a manufacturing or service business, one that purchases (rather than manufactures) *finished goods* for resale.

**merger.** The joining of two or more businesses into a single *economic entity*. See *holding company*.

**minority interest.** A *balance sheet account* on *consolidated statements* showing the *equity* in a less-than-100-percent-owned *subsidiary* company; equity allocable to those who are not part of the controlling (majority) interest; may be classified either as shareholders' equity or as a liability of *indeterminate term* on the consolidated balance sheet. The *income statement* must subtract the minority interest in the current period's income of the less-than-100-percent-owned subsidiary to arrive at consolidated *net income* for the period.

**minority investment.** A holding of less than 50 percent of the *voting stock* in another corporation; accounted for with the *equity method* when the investor owns sufficient shares that it can exercise "significant influence" and as *marketable securities* otherwise. See *mutual fund*.

**minutes book.** A record of all actions authorized at corporate *board of directors* or shareholders' meetings.

**MIS.** *Management information system.*

**mix variance.** One of the *manufacturing variances*. Many *standard cost* systems specify combinations of inputs—for example, labor of a certain skill and materials of a certain quality grade. Sometimes combinations of inputs used differ from those contemplated by the standard. The mix variance attempts to report the cost difference caused by those changes in the combination of inputs.

**mixed cost.** A *semifixed* or a *semivariable* cost.

**Modified Accelerated Cost Recovery System (MACRS).** Name used for the *Accelerated Cost Recovery System*, originally passed by Congress in 1981 and amended by Congress in 1986.

**modified cash basis.** The *cash basis of accounting* with long-term assets accounted for using the *accrual basis of accounting*. Most users of the term "cash basis of accounting" actually mean "modified cash basis."

**monetary assets and liabilities.** See *monetary items*.

**monetary gain or loss.** The firm's *gain* or *loss* in *general purchasing power* as a result of its holding *monetary assets* or liabilities during a period when the *general purchasing power of the dollar* changes; explicitly reported in *constant-dollar accounting*. During periods of *inflation*, holders of net monetary assets lose, and holders of net monetary liabilities gain, general purchasing power. During periods of *deflation*, holders of net monetary assets gain, and holders of net monetary liabilities lose, general purchasing power.

**monetary items.** Amounts fixed in terms of dollars by statute or contract; *cash, accounts receivable, accounts payable,* and *debt*. The distinction between monetary and nonmonetary items is important for *constant-dollar accounting* and for *foreign exchange gain or loss* computations. In the foreign exchange context, account amounts denominated in dollars are not monetary items, whereas amounts denominated in any other currency are monetary.

**monetary-nonmonetary method.** *Foreign currency translation* that translates all *monetary items* at the *current exchange rate* and translates all *nonmonetary items* at the *historical rate*.

**money.** A word seldom used with precision in accounting, at least in part because economists have not yet agreed on its definition. Economists use the term to refer to both a medium of exchange and a store of value. See *cash* and *monetary items*. Consider a different set of issues concerning the phrase, "making money." Lay terminology uses this to mean "earning income" whether, as a result, the firm increased its cash balances or other net assets. The user does not typically mean that the firm has increased cash equal to the amount of net income, although the unaware listeners often think the phrase means this. Given that usage equates "making money" with "earning income," in this sense "money" has a credit balance not a debit balance. Since cash typically has a debit balance, the phrase "making money" is even more troublesome. Consider the following language from the U.S. statutes on forfeitures required of some who commit illegal acts: ". . . the amount of money acquired through illegal transactions . . ." Does the law mean the cash left over after the lawbreaker has completed the illegal transactions, the income earned from the transactions, or something else? Sometimes "making money" means avoiding a cost, not recognized in financial accounting. Consider the following sets of questions and see how you have to think to decide whether, in a given question, "money" refers to a debit or a credit. Assume I start with $10 in cash.

1. I took a cab and it cost $10; I spent money. Did the cabbie make money? Does the cabbie have money?
2. I decided to walk, so I didn't spend $10. Did I make money?
3. I canceled the trip. Did I make money?

"Money" sometimes refers to debits and sometimes to credits; "making money" sometimes means earning accounting income and sometimes avoiding a cost, not reported in accounting, so careful writing about accounting avoids the word.

**money purchase plan.** A *pension plan* in which the employer contributes a specified amount of cash each year to each employee's pension fund; sometimes called a *defined-contribution plan;* contrast with *defined-benefit plan*. The plan does not specify the benefits ultimately received by the employee, since these benefits depend on the rate of return on the cash invested. As of the mid-1990's, most corporate pension plans were defined-benefit plans because both the law and *generally accepted accounting principles* for pensions made defined-benefit plans more attractive than money purchase plans. *ERISA* makes money purchase plans relatively more attractive than they had been. We expect the relative number of money purchase plans to continue to increase.

**mortality table.** Data of life expectancies or probabilities of death for persons of specified age and sex.

**mortgage.** A claim given by the borrower (mortgagor) to the lender (mortgagee) against the borrower's property in return for a loan.

**moving average.** An *average* computed on observations over time. As a new observation becomes available, analysts drop the oldest one so that they always compute the average for the same number of observations and use only the most recent ones.

**moving average method.** *Weighted-average inventory method.*

**multiple-step.** Said of an *income statement* that shows various subtotals of *expenses* and *losses* subtracted from *revenues* to show intermediate items such as *operating income,* income of the enterprise (operating income plus *interest* income), income to investors (income of the enterprise less *income taxes*), net income to shareholders (income to investors less interest charges), and income retained (net income to shareholders less dividends). See *entity theory.*

**municipal bond.** A *bond* issued by a village, town, or city. *Interest* on such bonds is generally exempt from federal *income taxes* and from some state income taxes. Because bonds issued by state and county governments often have these characteristics, terminology often calls such bonds "municipals" as well. These are also sometimes called "taxexempts."

**mutual fund.** An investment company that issues its own stock to the public and uses the proceeds to invest in securities of other companies. A mutual fund usually owns less than 5 or 10 percent of the stock of any one company and accounts for its investments using current *market values.* Contrast with *holding company.*

**mutually exclusive (investment) projects.** Competing investment projects in which accepting one project eliminates the possibility of undertaking the remaining projects.

# N

**NAARS.** *National Automated Accounting Research System.*

**NASDAQ (National Association of Securities Dealers Automated Quotation System).** A computerized system to provide brokers and dealers with price quotations for securities traded *over the counter* as well as for some *NYSE* securities.

**National Association of Accountants (NAA).** Former name for the *Institute of Management Accountants (IMA).*

**National Automated Accounting Research System (NAARS).** A computer-based information-retrieval system containing, among other things, the complete text of most public corporate annual reports and *Forms 10-K.* Users may access the system through the *AICPA.*

**natural business year.** A 12-month period chosen as the reporting period so that the end of the period coincides with a low point in activity or inventories. See *ratio* for a discussion of analyses of financial statements of companies using a natural business year.

**natural classification.** *Income statement* reporting form that classifies *expenses* by nature of items acquired, that is, materials, wages, salaries, insurance, and taxes, as well as depreciation. Contrast with *functional classification.*

**natural resources.** Timberland, oil and gas wells, ore deposits, and other products of nature that have economic value. Terminology uses the term *depletion* to refer to the process of *amortizing* the cost of natural resources. Natural resources are "nonrenewable" (for example, oil, coal, gas, ore deposits) or "renewable" (timberland, sod fields); terminology often calls the former "wasting assets." See also *reserve recognition accounting* and *percentage depletion.*

**negative confirmation.** See *confirmation.*

**negative goodwill.** See *goodwill.* When a firm acquires another company, and the *fair market value* of the *net assets* acquired exceeds the purchase price, *APB Opinion No. 16* requires that the acquiring company reduce the valuation of noncurrent assets (except *investments* in *marketable securities*) until the purchase price equals the adjusted valuation of the fair market value of net assets acquired. If, after the acquiring company reduces the valuation of noncurrent assets to zero, the valuation of the remaining net assets acquired still exceeds the purchase price, then the difference appears as a credit balance on the balance sheet as negative goodwill. For negative goodwill to exist, someone must be willing to sell a company for less than the fair market value of net current assets and marketable securities. Because such bargain purchases are rare, one seldom sees negative goodwill in the financial statements. When it does appear, it generally signals unrecorded obligations, such as a contingency related to a pending lawsuit.

**negotiable.** Legally capable of being transferred by *endorsement.* Usually said of *checks* and *notes* and sometimes of *stocks* and *bearer bonds.*

**negotiated transfer price.** A *transfer price* set jointly by the buying and the selling divisions.

**net.** Reduced by all relevant deductions.

**net assets.** Total *assets* minus total *liabilities;* equals the amount of *owners' equity.* Often, we find it useful to split the balance sheet into two parts: owners' equity and all the rest. The "rest" is total assets less total liabilities. To take an example, consider one definition of *revenue*: the increase in owners' equity accompanying the net assets increase caused by selling goods or rendering services. An alternative, more cumbersome way to say the same thing is: the increase in owners' equity accompanying the assets increase or the liabilities decrease, or both, caused by selling goods or rendering services. Consider the definition of *goodwill*: the excess of purchase price over the fair market value of identifiable net assets acquired in a purchase transaction. Without the phrase "net assets," the definition might be as follows: the excess of purchase price over the fair market value of identifiable assets reduced by the fair market value of identifiable liabilities acquired in a purchase transaction.

**net bank position.** From a firm's point of view, *cash* in a specific bank less *loans* payable to that bank.

**net book value.** *Book value.*

**net current asset value (per share).** *Working capital* divided by the number of common shares outstanding. Some analysts think that when a common share trades in the market for an amount less than net current asset value, the shares are undervalued and investors should purchase them. We find this view naive because it ignores, generally, the efficiency of capital markets and, specifically, unrecorded obligations, such as for executory contracts and contingencies, not currently reported as *liabilities* in the *balance sheet* under *GAAP*.

**net current assets.** *Working capital* = *current assets* − *current liabilities*.

**net income.** The excess of all *revenues* and *gains* for a period over all *expenses* and *losses* of the period. The FASB is proposing to discontinue use of this term and substitute *earnings*. See *comprehensive income.*

**net loss.** The excess of all *expenses* and *losses* for a period over all *revenues* and *gains* of the period; negative *net income*.

**net markup.** In the context of *retail inventory methods, markups* less markup cancellations; a figure that usually ignores *markdowns* and markdown cancellations.

**net of tax method.** A nonsanctioned method for dealing with the problem of *income tax allocation;* described in *APB Opinion No. 11*. The method subtracts deferred tax items from specific *asset* amounts rather than showing them as a deferred credit or *liability.*

**net of tax reporting.** Reporting, such as for *income from discontinued operations, extraordinary items,* and *prior-period adjustments,* in which the firm adjusts the amounts presented in the *financial statements* for all income tax effects. For example, if an extraordinary loss amounted to $10,000, and the marginal tax rate was 40 percent, then the extraordinary item would appear "net of taxes" as a $6,000 loss. Hence, not all a firm's income taxes necessarily appear on one line of the income statement. The reporting allocates the total taxes among *income from continuing operations, income from discontinued operations, extraordinary items,* cumulative effects of *accounting changes,* and *prior-period adjustments*.

**net operating profit.** *Income from continuing operations.*

**net present value.** Discounted or *present value* of all cash inflows and outflows of a project or of an *investment* at a given *discount rate*.

**net price method (of recording purchase or sales discounts).** Method that records a *purchase* (or *sale*) at its *invoice* price less all *discounts* made available, under the assumption that the firm will take nearly all discounts offered. The purchaser debits, to an *expense* account, discounts lapsed through failure to pay promptly. For purchases, management usually prefers to know about the amount of discounts lost because of inefficient operations, not the amounts taken, so that most managers prefer the net price method to the *gross price method*.

**net realizable (sales) value.** Current selling price less reasonable costs to complete production and to sell the item. Also, a method for *allocating joint costs* in proportion to *realizable values* of the joint products. For example, joint products A and B together cost $100; A sells for $60, whereas B sells for $90. Then a firm would allocate to A ($60/$150) × $100 = 0.40 × $100 = $40 of cost while it would allocate to B ($90/$150) × $100 = $60 of cost.

**net sales.** Sales (at gross invoice amount) less *returns, allowances,* freight paid for customers, and *discounts* taken.

**net working capital.** *Working capital;* the term "net" is redundant in accounting. Financial analysts sometimes mean *current assets* when they speak of working capital, so for them the "net" is not redundant.

**net worth.** A misleading term with the same meaning as *owners' equity*. Avoid using this term; accounting valuations at historical cost do not show economic worth.

**network analysis.** A project planning and scheduling method, usually displayed in a diagram, that enables management to identify the interrelated sequences that it must accomplish to complete the project.

**New York Stock Exchange (NYSE).** A public market in which those who own seats (a seat is the right to participate) trade various corporate *securities*.

**next-in, first-out.** See *NIFO*.

**NIFO (next-in, first-out).** A *cost flow assumption,* one not allowed by GAAP. In making decisions, many managers consider *replacement costs* (rather than *historical costs*) and refer to them as NIFO costs.

**no par.** Said of *stock* without a *par value*.

**nominal accounts.** *Temporary accounts,* such as *revenue* and *expense* accounts; contrast with *balance sheet accounts*. The firm *closes* all nominal accounts at the end of each *accounting period*.

**nominal amount (value).** An amount stated in dollars, in contrast to an amount stated in *constant dollars*. Contrast with *real amount* (*value*).

**nominal dollars.** The measuring unit giving no consideration to differences in the *general purchasing power of the dollar* over time. The face amount of currency or coin, a *bond,* an *invoice,* or a *receivable* is a nominal-dollar amount. When the analyst adjusts that amount for changes in *general purchasing power,* it becomes a *constant-dollar* amount.

**nominal interest rate.** A rate specified on a *debt* instrument; usually differs from the market or *effective rate;* also, a rate of *interest* quoted for a year. If the interest compounds more often than annually, then the *effective interest rate* exceeds the nominal rate.

**noncancelable.** See *lease*.

**nonconsolidated subsidiary.** An *intercorporate investment* in which the parent owns more than 50 percent of the shares of the *subsidiary* but accounts for the investment with the *cost method*.

**noncontributory.** Said of a *pension plan* in which only the employer makes payments to a pension *fund.* Contrast with *contributory.*

**noncontrollable cost.** A cost that a particular manager cannot *control.*

**noncurrent.** Of a *liability,* due in more than one year (or more than one *operating cycle*); of an *asset,* the firm will enjoy the future benefit in more than one year (or more than one operating cycle).

**nonexpendable fund.** A governmental fund whose *principal,* and sometimes earnings, the entity may not spend.

**noninterest-bearing note.** A *note* that does not specify explicit interest. The *face value* of such a note will exceed its *present value* at any time before *maturity* value so long as *interest rates* are positive. *APB Opinion No. 21* requires that firms report the present value, not face value, of long-term noninterest-bearing notes as the *asset* or *liability* amount in financial statements. For this purpose, the firm uses the *historical interest rate.* See *interest, imputed.*

**nonmanufacturing costs.** All *costs* incurred other than those necessary to produce goods. Typically, only manufacturing firms use this designation.

**non-monetary items.** All items that are not monetary. See *monetary items.*

**nonoperating.** In the *income statement* context, said of *revenues* and *expenses* arising from *transactions* incidental to the company's main line(s) of business; in the *statement of cash flows* context, said of all financing and investing sources or uses of cash in contrast to cash provided by operations. See *operations.*

**nonprofit corporation.** An incorporated *entity,* such as a hospital, with owners who do not share in the earnings. It usually emphasizes providing services rather than maximizing income.

**nonrecurring.** Said of an event that is not expected to happen often for a given firm. *APB Opinion No. 30* requires firms to disclose separately the effects of such events as part of *ordinary* items unless the event is also unusual. See *extraordinary* item.

**nonvalue-added activity.** An activity that causes costs without increasing a product's or service's value to the customer.

**normal cost.** Former name for *service cost* in accounting for pensions and other postemployment benefits.

**normal costing.** Method of charging costs to products using actual *direct materials,* actual *direct labor,* and predetermined *factory overhead* rates.

**normal costing system.** *Costing* based on *actual material* and *labor* costs but using *predetermined overhead* rates per unit of some *activity* basis (such as *direct labor hours* or machine hours) to apply overhead to production. Management decides the rate to charge to production for overhead at the start of the period. At the end of the period the accounting multiplies this rate by the actual number of units of the base activity (such as actual direct labor hours worked or actual machine hours used during the period) to apply overhead to production.

**normal spoilage.** Costs incurred because of ordinary amounts of spoilage. Accounting prorates such costs to units produced as *product costs.* Contrast with *abnormal spoilage.*

**normal standard cost, normal standards.** The *cost* a firm expects to incur under reasonably efficient operating conditions with adequate provision for an average amount of rework, spoilage, and the like.

**normal volume.** The level of production that will, over a time span, usually one year, satisfy purchasers' demands and provide for reasonable *inventory* levels.

**note.** An unconditional written promise by the maker (borrower) to pay a certain amount on demand or at a certain future time.

**note receivable discounted.** A *note* assigned by the holder to another. The new holder of the note typically pays the old holder an amount less than the *face value* of the note, hence the word "discounted." If the old holder assigns the note to the new holder with recourse, the old holder has a *contingent liability* until the maker of the note pays the debt. See *factoring.*

**notes.** Some use this word instead of *footnotes* when referring to the detailed information included by management as an integral part of the *financial statements* and covered by the *auditor's report.*

**NOW (negotiable order of withdrawal) account.** Negotiable order of withdrawal. A *savings account* whose owner can draw an order to pay, much like a *check* but technically not a check, and give it to others, who can redeem the order at the savings institution.

**number of days sales in inventory (or receivables).** Days of average inventory on hand (or average collection period for receivables). See *ratio.*

**NV (naamloze vennootschap).** Netherlands: a public limited liability company.

**NYSE.** *New York Stock Exchange.*

# O

**OASD(H)I.** *Old Age, Survivors, Disability, and (Hospital) Insurance.*

**objective.** See *reporting objectives* and *objectivity.*

**objective function.** In *linear programming,* the name of the profit (or cost) criterion the analyst wants to maximize (or minimize).

**objectivity.** The reporting policy implying that the firm will not give formal recognition to an event in financial statements until the firm can measure the magnitude of the events with reasonable accuracy and check that amount with independent verification.

**obsolescence.** An asset's *market value* decline caused by improved alternatives becoming available that will be more *cost-effective.* The decline in market value does not relate to physical changes in the asset itself. For example, computers become obsolete long before they wear out. See *partial obsolescence.*

**Occupational Safety and Health Act.** *OSHA.*

**off-balance-sheet financing.** A description often used for an obligation that meets all the tests to be classified a liability except that the obligation arises from an *executory contract* and, hence, is not a *liability*. Consider the following example. Miller Corporation desires to acquire land costing \$25 million, on which it will build a shopping center. It could borrow the \$25 million from its bank, paying interest at 12 percent, and buy the land outright from the seller. If so, both an asset and a liability will appear on the balance sheet. Instead, it borrows \$5 million and purchases for \$5 million from the seller an *option* to buy the land from the seller at any time within the next six years for a price of \$20 million. The option costs Miller Corporation \$5 million immediately and provides for continuing "option" payments of \$2.4 million per year, which precisely equal Miller Corporation's borrowing rate multiplied by the remaining purchase price of the land: \$2.4 million = 0.12 × \$20 million. Although Miller Corporation need not continue payments and can let the option lapse at any time, it also has an obligation to begin developing on the site immediately. Because Miller Corporation has invested a substantial sum in the option, will invest more, and will begin immediately developing the land, Miller Corporation will almost certainly exercise its option before expiration. The seller of the land can take the option contract to the bank and borrow \$20 million, paying interest at Miller Corporation's borrowing rate, 12 percent per year. The continuing option payments from Miller Corporation will be sufficient to enable the seller to make its payments to the bank. *Generally accepted accounting principles* view Miller Corporation as having acquired an option for \$5 million rather than having acquired land costing \$25 million in return for \$25 million of debt. The firm will not recognize debt on the balance sheet until it borrows more funds to exercise the option.

**off-balance-sheet risk.** A contract that exposes an entity to the possibility of loss but that does not appear in the financial statements. For example, a *forward-exchange contract* generally does not appear on the balance sheet because it is an *executory contract*. The contract may reduce or increase the entity's exposure to foreign-exchange risk (the chance of loss due to unfavorable changes in the foreign-exchange rate). It may also expose the entity to credit risk (the chance of loss that occurs when the *counterparty* to the contract cannot fulfill the contract terms). *SFAS No. 105* requires entities to describe contracts with off-balance-sheet risk.

**OHG (Offene Handelsgesellschaft).** Germany: a general *partnership*. The partners have unlimited *liability*.

**Old Age, Survivors, Disability, and (Hospital) Insurance, or OASD(H)I.** The technical name for Social Security under the Federal Insurance Contributions Act (*FICA*).

**on consignment.** Said of goods delivered by the owner (the consignor) to another (the consignee) to be sold by the consignee. On delivery of the goods from the consignor to the consignee, the consignor can, but need not, make an entry transferring the goods at cost from Finished Goods Inventory to another *inventory account*, such as Goods out on Consignment. The consignor recognizes *revenue* only when the consignee has sold the goods to customers. Under such an arrangement, the owner of the goods bears the inventory holding costs until the ultimate seller (consignee) sells them. The owner also bears the risk that the items will never sell to final customers, but manufacturers or distributors who provide generous return options to their customers can achieve this aspect of consignment sales in an outright sale. The consignment protects the consignor from the consignee's bankruptcy, as the arrangement entitles the owner either to the return of the property or to payment of a specified amount. The goods are *assets* of the consignor. Such arrangements provide the consignor with better protection than an outright *sale on account* to the consignee in *bankruptcy*. In event of bankruptcy, the ordinary seller, holding an account receivable, had no special claim to the return of the goods, whereas a consignor can reclaim the goods without going through bankruptcy proceedings, from which the consignor might recover only a fraction of the amounts owed to it.

**on (open) account.** Said of a *purchase* (or *sale*) when the seller expects payment sometime after delivery and the purchaser does not give a *note* evidencing the *debt*. The purchaser has generally signed an agreement sometime in the past promising to pay for such purchases according to an agreed time schedule. When the firm sells (purchases) on open account, it *debits* (*credits*) *Accounts Receivable* (*Payable*).

**one-line consolidation.** Said of an *intercorporate investment* accounted for with the *equity method*. With this method, the *income* and *balance sheet* total *assets* and *equities* amounts are identical to those that would appear if the parent consolidated the investee firm, even though the income from the investment appears on a single line of the income statement and the net investment appears on a single line in the Assets section of the balance sheet.

**one-write system.** A system of bookkeeping that produces several records, including original documents, in one operation by the use of reproductive paper and equipment that provides for the proper alignment of the documents.

**open account.** Any *account* with a nonzero *debit* or *credit balance*. See *on (open) account*.

**operating.** An adjective used to refer to *revenue* and *expense* items relating to the company's main line(s) of business. See *operations*.

**operating accounts.** *Revenue, expense,* and *production cost accounts*. Contrast with *balance sheet accounts*.

**operating activities.** For purposes of the *statement of cash flows*, all *transactions* and *events* that are neither *financing activities* nor *investing activities*. See *operations*.

**operating budget.** A formal *budget* for the *operating cycle* or for a year.

**operating cash flow.** *Cash flow from operations*. Financial statement analysts sometimes use this term to mean *cash flow from operations* − *capital expendi-*

*tures – dividends*. This usage leads to such ambiguity that the reader should always confirm the definition that the writer uses before drawing inferences from the reported data.

**operating cycle.** *Earnings cycle*.

**operating expenses.** *Expenses* incurred in the course of *ordinary* activities of an *entity;* frequently, a classification including only *selling, general,* and *administrative expenses,* thereby excluding *cost of goods sold, interest,* and *income tax* expenses. See *operations*.

**operating lease.** A *lease* accounted for by the *lessee* without showing an *asset* for the lease rights (*leasehold*) or a *liability* for the lease payment obligations. The lessee reports only rental payments during the period, as *expenses* of the period. The asset remains on the lessor's *books,* where rental collections appear as *revenues*. Contrast with *capital lease*.

**operating leverage.** Usually said of a firm with a large proportion of *fixed costs* in its *total costs*. Consider a book publisher or a railroad: such a firm has large costs to produce the first unit of service; then, the *incremental costs* of producing another book or transporting another freight car are much less than the *average cost,* so the *gross margin* on the sale of the subsequent units is relatively large. Contrast this situation with that, for example, of a grocery store, where the *contribution margin* equals less than 5 percent of the selling price. For firms with equal profitability, however defined, we say that the one with the larger percentage increase in income from a given percentage increase in dollar sales has the larger operating leverage. See *leverage* for contrast of this term with "financial leverage." See *cost terminology* for definitions of terms involving the word "cost."

**operating margin.** *Revenues* from *sales* minus *cost of goods sold* and *operating expenses*.

**operating margin based on current costs.** *Revenues* from *sales* minus *current cost* of goods sold; a measure of operating efficiency that does not depend on the *cost flow assumption* for *inventory;* sometimes called "current (gross) margin." See *inventory profit* for illustrative computations.

**operating ratio.** See *ratio*.

**operational control.** See *control system*.

**operations.** A word not precisely defined in *accounting*. Generally, analysts distinguish operating activities (producing and selling *goods* or *services*) from financing activities (raising funds) and *investing activities*. Acquiring goods on account and then paying for them one month later, though generally classified as an operating activity, has the characteristics of a financing activity. Or consider the transaction of selling plant assets for a price in excess of book value. On the *income statement,* the gain appears as part of income from operations ("continuing operations" or "discontinued" operations, depending on the circumstances), but the *statement of cash flows* reports all the funds received below the Cash from Operations section, as a nonoperating source of cash, "disposition of noncurrent assets." In income tax accounting, an "operating loss" results whenever deductions exceed taxable revenues.

**opinion.** The *auditor's report* containing an attestation or lack thereof; also, *APB Opinion*.

**opinion paragraph.** Section of *auditor's report,* generally following the *scope paragraph* and giving the auditor's conclusion that the *financial statements* are (rarely, are not) in accordance with *GAAP* and present fairly the *financial position,* changes in financial position, and the results of *operations*.

**opportunity cost.** The *present value* of the *income* (or *costs*) that a firm could earn (or save) from using an *asset* in its best alternative use to the one under consideration.

**opportunity cost of capital.** *Cost of capital*.

**option.** The legal right to buy or sell something during a specified period at a specified price, called the *exercise* price. If the right exists during a specified time interval, it is known as an "American option." If it exists for only one specific day, it is known as a "European option." Do not confuse employee stock options with *put* and *call* options, traded in various public markets.

**ordinary annuity.** An *annuity in arrears*.

**ordinary income.** For income tax purposes, reportable *income* not qualifying as *capital gains*.

**organization costs.** The *costs* incurred in planning and establishing an *entity;* example of an *intangible* asset. The firm must treat these costs as *expenses* of the period, even though the *expenditures* clearly provide future benefits and meet the test to be *assets*.

**original cost.** *Acquisition cost;* in public utility accounting, the acquisition cost of the *entity* first devoting the *asset* to public use. See *aboriginal cost*.

**original entry.** Entry in a *journal*.

**OSHA (Occupational Safety and Health Act).** The federal law that governs working conditions in commerce and industry.

**other comprehensive income.** According to the FASB, *comprehensive income* items that are not themselves part of earnings. See *comprehensive income*. To define comprehensive income does not convey its essence. To understand comprehensive income, you need to understand how it differs from *earnings* (or *net income*), the concept measured in the *earnings* (*income*) *statement*. The term *earnings* (or *net income*) refers to the sum of all components of comprehensive income *minus* the components of other comprehensive income.

**outlay.** The amount of an *expenditure*.

**outlier.** Said of an observation (or data point) that appears to differ significantly in some regard from other observations (or data points) of supposedly the same phenomenon; in a *regression analysis,* often used to describe an observation that falls far from the fitted regression equation (in two dimensions, line).

**out-of-pocket.** Said of an *expenditure* usually paid for with cash; an *incremental* cost.

**out-of-stock cost.** The estimated decrease in future *profit* as a result of losing customers because a firm has insufficient quantities of *inventory* currently on hand to meet customers' demands.

**output.** Physical quantity or monetary measurement of *goods* and *services* produced.

**outside director.** A corporate board of directors member who is not a company officer and does not participate in the corporation's day-to-day management.

**outstanding.** Unpaid or uncollected; when said of *stock,* refers to the shares issued less *treasury stock;* when said of *checks,* refers to a check issued that did not clear the *drawer's* bank prior to the *bank statement* date.

**over-and-short.** Title for an *expense account* used to account for small differences between book balances of cash and actual cash and vouchers or receipts in *petty cash* or *change funds.*

**overapplied (overabsorbed) overhead.** Costs applied, or *charged,* to product and exceeding actual *overhead costs* during the period; a *credit balance* in an overhead account after overhead is assigned to product.

**overdraft.** A *check* written on a checking account that contains funds less than the amount of the check.

**overhead costs.** Any *cost* not directly associated with the production or sale of identifiable goods and services; sometimes called "burden" or "indirect costs" and, in the UK, "oncosts"; frequently limited to manufacturing overhead. See *central corporate expenses* and *manufacturing overhead.*

**overhead rate.** Standard, or other predetermined rate, at which a firm applies *overhead costs* to products or to services.

**over-the-counter.** Said of a *security* traded in a negotiated transaction, as on *NASDAQ,* rather than in an auctioned one on an organized stock exchange, such as the *New York Stock Exchange.*

**owners' equity.** *Proprietorship; assets* minus *liabilities; paid-in capital* plus *retained earnings* of a corporation; partners' capital accounts in a *partnership;* owner's capital account in a *sole proprietorship.*

# P

**paid-in capital.** Sum of balances in *capital stock* and *capital contributed in excess of par* (or *stated*) *value* accounts; same as *contributed capital* (minus *donated capital*). Some use the term to mean only *capital contributed in excess of par* (or *stated*) *value.*

**paid-in surplus.** See *surplus.*

**P&L.** Profit-and-loss statement; *income statement.*

**paper profit.** A *gain* not yet realized through a *transaction;* an *unrealized holding gain.*

**par.** See *at par* and *face amount.*

**par value.** *Face amount* of a *security.*

**par value method.** In accounting for *treasury stock,* method that *debits* a common stock account with the *par value* of the shares required and allocates the remaining debits between the *Additional Paid-in Capital* and *Retained Earnings* accounts. Contrast with *cost method.*

**parent company.** Company owning more than 50 percent of the voting shares of another company, called the *subsidiary.*

**Pareto chart.** A graph of a skewed statistical distribution. In many business settings, a relatively small percentage of the potential population causes a relatively large percentage of the business activity. For example, some businesses find that the top 20 percent of the customers buy 80 percent of the goods sold. Or, the top 10 percent of products account for 60 percent of the revenues or 70 percent of the profits. The statistical distribution known as the Pareto distribution has this property of skewness, so a graph of a phenomenon with such skewness has come to be known as a Pareto chart, even if the underlying data do not actually well fit the Pareto distribution. Practitioners of *total quality management* find that in many businesses, a small number of processes account for a large fraction of the quality problems, so they advocate charting potential problems and actual occurrences of problems to identify the relatively small number of sources of trouble. They call such a chart a "Pareto chart."

**partial obsolescence.** One cause of decline in *market value* of an *asset.* As technology improves, the economic value of existing *assets* declines. In many cases, however, it will not pay a firm to replace the existing asset with a new one, even though it would acquire the new type rather than the old if it did make a new acquisition currently. In these cases, the accountant should theoretically recognize a loss from partial obsolescence from the firm's owning an old, out-of-date asset, but *GAAP* do not permit recognition of partial obsolescence until the sum of future cash flows from the asset total less than book value; see *impairment.* The firm will carry the old asset at *cost* less *accumulated depreciation* until the firm retires it from service so long as the *undiscounted* future *cash flows* from the asset exceed its book value. Thus management that uses an asset subject to partial obsolescence reports results inferior to those reported by a similar management that uses a new asset. See *obsolescence.*

**partially executory contract.** *Executory contract* in which one or both parties have done something other than merely promise.

**partially funded.** Said of a *pension plan* in which the firm has not funded all earned benefits. See *funded* for funding requirements.

**partially vested.** Said of a *pension plan* in which not all employee benefits have *vested.* See *graded vesting.*

**participating dividend.** *Dividend* paid to preferred shareholders in addition to the minimum preferred dividends when the *preferred stock* contract provides for such sharing in earnings. Usually the contract specifies that dividends on *common shares* must reach a specified level before the preferred shares receive the participating dividend.

**participating preferred stock.** *Preferred stock* with rights to *participating dividends.*

**partner's drawing.** A payment made to a partner and debited against his or her share of income or capital. The name of a *temporary account,* closed to the partner's capital account, to record the debits when the partner receives such payments.

**partnership.** Contractual arrangement between individuals to share resources and operations in a jointly run business. See *general* and *limited partner* and *Uniform Partnership Act.*

**patent.** A right granted for up to 20 years by the federal government to exclude others from manufacturing, using, or selling a claimed design, product, or plant (e.g., a new breed of rose) or from using a claimed process or method of manufacture; an *asset* if the firm acquires it by purchase. If the firm develops it internally, current *GAAP* require the firm to *expense* the development costs when incurred.

**payable.** Unpaid but not necessarily due or past due.

**pay-as-you-go.** Said of an *income tax* scheme in which the taxpayer makes periodic payments of income taxes during the period when it earns the income to be taxed; in contrast to a scheme in which the taxpayer owes no payments until the end of, or after, the period when it earned the income being taxed (called PAYE—pay-as-you-earn—in the UK). The phrase is sometimes used to describe an *unfunded pension plan,* or retirement benefit plan, in which the firm makes payments to pension plan beneficiaries from general corporate funds, not from cash previously contributed to a fund. Under this method, the firm debits expense as it makes payments, not as it incurs the obligations. This is not acceptable as a method of accounting for pension plans, under *SFAS No. 87,* or as a method of *funding,* under *ERISA.*

**payback period.** Amount of time that must elapse before the cash inflows from a project equal the cash outflows.

**payback reciprocal.** One divided by the *payback period.* This number approximates the *internal rate of return* on a project when the project life exceeds twice the payback period and the cash inflows are identical in every period after the initial period.

**PAYE (pay-as-you-earn).** See *pay-as-you-go* for contrast.

**payee.** The person or entity who receives a cash payment or who will receive the stated amount of cash on a *check.* See *draft.*

**payout ratio.** *Common stock dividends* declared for a year divided by net *income* to common stock for the year; a term used by financial analysts. Contrast with *dividend yield.*

**payroll taxes.** Taxes levied because the taxpayer pays salaries or wages; for example, *FICA* and unemployment compensation insurance taxes. Typically, the employer pays a portion and withholds part of the employee's wages.

**P/E ratio.** *Price-earnings ratio.*

**Pension Benefit Guarantee Corporation (PBGC).** A federal corporation established under *ERISA* to guarantee basic pension benefits in covered pension plans by administering terminated pension plans and placing *liens* on corporate assets for certain unfunded pension liabilities.

**pension fund.** *Fund,* the assets of which the trustee will pay to retired ex-employees, usually as a *life annuity;* generally held by an independent trustee and thus not an *asset* of the employer.

**pension plan.** Details or provisions of employer's contract with employees for paying retirement *annuities* or other benefits. See *funded, vested, service cost, prior service cost, money purchase plan,* and *defined-benefit plan.*

**per books.** An expression used to refer to the *book value* of an item at a specific time.

**percent.** Any number, expressed as a decimal, multiplied by 100.

**percentage depletion (allowance).** Deductible *expense* allowed in some cases by the federal *income tax* regulations; computed as a percentage of gross income from a *natural resource* independent of the unamortized cost of the *asset.* Because the amount of the total deductions for tax purposes usually exceeds the cost of the asset being *depleted,* many people think the deduction is an unfair tax advantage or *loophole.*

**percentage-of-completion method.** Recognizing *revenues* and *expenses* on a job, order, or contract (1) in proportion to the *costs* incurred for the period divided by total costs expected to be incurred for the job or order ("cost to cost") or (2) in proportion to engineers' or architects' estimates of the incremental degree of completion of the job, order, or contract during the period. Contrast with *completed contract method.*

**percentage of sales method.** Measuring *bad debt expense* as a fraction of *sales revenue.* When the firm uses this method, it must periodically check the adequacy of the *allowance for uncollectibles* by *aging accounts receivable.* See *allowance method* for contrast.

**percentage statement.** A statement containing, in addition to (or instead of) dollar amounts, ratios of dollar amounts to some base. In a percentage *income statement,* the base is usually either *net sales* or total *revenues,* and in a percentage *balance sheet,* the base is usually total *assets.*

**period.** *Accounting period.*

**period cost.** An inferior term for *period expense.*

**period expense (charge).** *Expenditure,* usually based on the passage of time, charged to operations of the accounting period rather than *capitalized* as an asset. Contrast with *product cost.*

**periodic inventory.** In recording *inventory,* a method that uses data on beginning inventory, additions to inventories, and ending inventory to find the cost of withdrawals from inventory. Contrast with *perpetual inventory.*

**periodic procedures.** The process of making *adjusting entries* and *closing entries* and preparing the *financial statements,* usually by use of *trial balances* and *work sheets.*

**permanent account.** An account that appears on the *balance sheet.* Contrast with *temporary account.*

**permanent difference.** Difference between reported income and taxable income that will never reverse and, hence, requires no entry in the *deferred income tax* (*liability*) account; for example, nontaxable state and municipal *bond* interest that will appear on the financial statements. Contrast with *temporary difference.* See *deferred income tax liability.*

**permanent file.** The file of working papers that are prepared by a public accountant and that contain the information required for reference in successive professional engagements for a particular organization, as distinguished from working papers applicable only to a particular engagement.

**perpetual annuity.** *Perpetuity.*

**perpetual inventory.** *Inventory* quantity and amount records that the firm changes and makes current with each physical addition to or withdrawal from the stock of goods; an inventory so recorded. The records will show the physical quantities and, frequently, the dollar valuations that should be on hand at any time. Because the firm explicitly computes *cost of goods sold,* it can use the *inventory equation* to compute an amount for what *ending inventory* should be. It can then compare the computed amount of ending inventory with the actual amount of ending inventory as a *control* device to measure the amount of *shrinkages.* Contrast with *periodic inventory.*

**perpetuity.** An *annuity* whose payments continue forever. The *present value* of a perpetuity in *arrears* is $p/r$ where $p$ is the periodic payment and $r$ is the *interest rate* per period. If a perpetuity promises $100 each year, in arrears, forever and the interest rate is 8 percent per year, then the perpetuity has a value of $1,250 = $100/0.08.

**perpetuity (with) growth model.** See *perpetuity.* A *perpetuity* whose cash flows grow at the rate $g$ per period and thus has *present value* of $1/(r - g)$. Some call this the "Gordon Growth Model" because Myron Gordon wrote about applications of this formula and its variants in the 1950s. John Burr Williams wrote about them in the 1930s.

**personal account.** *Drawing account.*

**PERT (Program Evaluation and Review Technique).** A method of *network analysis* in which the analyst makes three time estimates for each activity—the optimistic time, the most likely time, and the pessimistic time—and gives an expected completion date for the project within a probability range.

**petty cash fund.** Currency and coins maintained for expenditures that the firm makes with cash on hand.

**physical units method.** A method of allocating a *joint cost* to the *joint products* based on a physical measure of the joint products; for example, allocating the cost of a cow to sirloin steak and to hamburger, based on the weight of the meat. This method usually provides nonsensical (see *sterilized allocation*) results unless the physical units of the joint products tend to have the same value.

**physical verification.** *Verification,* by an *auditor,* performed by actually inspecting items in *inventory, plant assets,* and the like, in contrast to merely checking the written records. The auditor may use statistical sampling procedures.

**planning and control process.** General name for the management techniques comprising the setting of organizational goals and *strategic plans, capital budgeting, operations* budgeting, comparison of plans with actual results, performance evaluation and corrective action, and revisions of goals, plans, and budgets.

**plant.** *Plant assets.*

**plant asset turnover.** Number of dollars of *sales* generated per dollar of *plant assets;* equal to sales divided by average *plant assets.*

**plant assets.** *Assets* used in the revenue-production process. Plant assets include buildings, machinery, equipment, land, and natural resources. The phrase "property, plant, and equipment" (though often appearing on balance sheets) is therefore a redundancy. In this context, "plant" used alone means buildings.

**plantwide allocation method.** A method for *allocating overhead costs* to product. First, use one *cost pool* for the entire plant. Then, allocate all costs from that pool to products using a single overhead *allocation* rate, or one set of rates, for all the products of the plant, independent of the number of departments in the plant.

**PLC (public limited company).** UK: a publicly held *corporation.* Contrast with *Ltd.*

**pledging.** The borrower assigns *assets* as security or *collateral* for repayment of a loan.

**pledging of receivables.** The process of using expected collections on *accounts receivable* as *collateral* for a loan. The borrower remains responsible for collecting the receivable but promises to use the proceeds for repaying the debt.

**plow back.** To retain *assets* generated by earnings for continued investment in the business.

**plug.** Process for finding an unknown amount. For any *account,* beginning balance + additions − deductions = ending balance; if you know any three of the four items, you can find the fourth with simple arithmetic, called "plugging." In making a *journal entry,* often you know all *debits* and all but one of the *credits* (or vice versa). Because *double-entry* bookkeeping requires equal debits and credits, you can compute the unknown quantity by subtracting the sum of the known credits from the sum of all the debits (or vice versa), also called "plugging." Accountants often call the unknown the "plug." For example, in amortizing a *discount* on *bonds payable* with the *straight-line depreciation* method, *interest expense* is a plug: interest expense = interest payable + *discount amortization.* See *trade-in transaction* for an example. The term sometimes has a bad connotation for accountants because plugging can occur in a slightly different context. During the process of preparing a *preclosing trial balance* (or *balance sheet*), often the sum of the debits does not equal the sum of the credits. Rather than find the error, some accountants are tempted to force equality by changing one of the amounts, with a plugged debit or credit to an account such as Other Expenses. No harm results from this procedure if the amount of the error is small compared with asset totals, since spending tens or hundreds of dollars in a bookkeeper's or accountant's time to find an error of a few dollars will not be *cost-effective.* Still, most accounting teachers rightly disallow this use of plugging because exercises and problems set for students provide enough information not to require it.

**point of sale.** The time, not the location, at which a *sale* occurs.

**pooling-of-interests method.** Accounting for a *business combination* by adding together the *book value* of the *assets* and *equities* of the combined firms; generally leads to a higher reported *net income* for the combined firms than results when the firm accounts for the business combination as a purchase because the *market values* of the merged assets generally exceed their book values. *GAAP* no longer allows this treatment in the U.S. Contrast with *purchase method.* Called *uniting-of-interests method* by the *IASB* and allowed by it under some conditions.

**population.** The entire set of numbers or items from which the analyst samples or performs some other analysis.

**positive confirmation.** See *confirmation.*

**post.** To record entries in an *account* to a *ledger,* usually as transfers from a *journal.*

**post-closing trial balance.** *Trial balance* taken after the accountant has *closed* all *temporary accounts.*

**post-statement events.** Events that have *material* impact and that occur between the end of the *accounting period* and the formal publication of the *financial statements.* Even though the events occur after the end of the period being reported on, the firm must disclose such events in notes if the auditor is to give a *clean opinion.*

**potentially dilutive.** A *security* that its holder may convert into, or exchange for, common stock and thereby reduce reported *earnings per share; options, warrants, convertible bonds,* and *convertible preferred stock.*

**PPB.** *Program budgeting.* The second "P" stands for "plan."

**practical capacity.** Maximum level at which a plant or department can operate efficiently.

**precision.** The degree of accuracy for an estimate derived from a sampling process, usually expressed as a range of values around the estimate. The analyst might express a sample estimate in the following terms: "Based on the sample, we are 95 percent sure [confidence level] that the true population value is within the range of X to Y [precision]." See *confidence level.*

**preclosing trial balance.** *Trial balance* taken at the end of the period before *closing entries;* in this sense, an *adjusted trial balance;* sometimes taken before *adjusting entries* and then synonymous with *unadjusted trial balance.*

**predatory prices.** Setting prices below some measure of cost in an effort to drive out competitors with the hope of recouping losses later by charging monopoly prices. Illegal in the United States if the prices set are below long-run variable costs. We know of no empirical evidence that firms are successful at recoupment.

**predetermined (factory) overhead rate.** Rate used in applying *overhead costs* to products or departments developed at the start of a period. Compute the rate as estimated overhead cost divided by the estimated number of units of the overhead allocation base (or *denominator volume*) activity. See *normal costing.*

**preemptive right.** The privilege of a *shareholder* to maintain a proportionate share of ownership by purchasing a proportionate share of any new stock issues. Most state corporation laws allow corporations to pay shareholders to waive their preemptive rights or state that preemptive rights exist only if the *corporation charter* explicitly grants them. In practice, then, preemptive rights are the exception rather than the rule.

**preference as to assets.** The rights of *preferred shareholders* to receive certain payments before common shareholders receive payments in case the board dissolves the corporation.

**preferred shares.** *Capital stock* with a claim to *income* or *assets* after *bondholders* but before *common shares. Dividends* on preferred shares are *income distributions,* not *expenses.* See *cumulative preferred stock.*

**premium.** The excess of issue (or market) price over *par value.* For a different context, see *insurance.*

**premium on capital stock.** Alternative but inferior title for *capital contributed in excess of par* (or *stated*) *value.*

**prepaid expense.** An *expenditure* that leads to a *deferred charge* or *prepayment.* Strictly speaking, this is a contradiction in terms because an *expense* is a gone asset, and this title refers to past *expenditures,* such as for rent or insurance premiums, that still have future benefits and thus are *assets.* We try to avoid this term and use "prepayment" instead.

**prepaid income.** An inferior alternative title for *advances from customers.* Do not call an item *revenue* or *income* until the firm earns it by delivering goods or rendering services.

**prepayments.** *Deferred charges; assets* representing *expenditures* for future benefits. Rent and insurance premiums paid in advance are usually current prepayments.

**present value.** Value today (or at some specific date) of an amount or amounts to be paid or received later (or at other, different dates), discounted at some *interest* or *discount rate;* an amount that, if invested today at the specified rate, will grow to the amount to be paid or received in the future.

**price.** The quantity of one *good* or *service,* usually *cash,* asked in return for a unit of another good or service. See *fair value.*

**price-earnings (P/E) ratio.** At a given time, the market value of a company's *common share,* per share, divided by the *earnings per* common *share* for the past year. The analyst usually bases the denominator on *income from continuing operations* or, if the analyst thinks the current figure for that amount does not represent a usual situation—such as when the number is negative or, if positive, close to zero—on some estimate of the number. See *ratio.*

**price index.** A series of numbers, one for each period, that purports to represent some *average* of prices for a series of periods, relative to a base period.

**price level.** The number from a *price index* series for a given period or date.

**price level-adjusted statements.** *Financial statements* expressed in terms of dollars of uniform purchasing power. The statements restate *nonmonetary* items to reflect changes in general *price levels* since the time the firm acquired specific *assets* and incurred *liabilities*. The statements recognize a *gain* or *loss* on *monetary items* as the firm holds them over time periods when the general *price level* changes. Conventional financial statements show *historical costs* and ignore differences in purchasing power in different periods.

**price variance.** In accounting for *standard costs,* an amount equal to (actual cost per unit – standard cost per unit) times actual quantity.

**prime cost.** Sum of *direct materials* plus *direct labor* costs assigned to product.

**prime rate.** The loan rate charged by commercial banks to their creditworthy customers. Some customers pay even less than the prime rate and others, more. The *Federal Reserve Bulletin* is the authoritative source of information about historical prime rates.

**principal.** An amount on which *interest* accrues, either as *expense* (for the borrower) or as *revenue* (for the lender); the *face amount* of a *loan;* also, the absent owner (principal) who hires the manager (agent) in a "principal-agent" relationship.

**principle.** See *generally accepted accounting principles*.

**prior-period adjustment.** A *debit* or *credit* that is made directly to *retained earnings* (and that does not affect *income* for the period) to adjust earnings as calculated for prior periods. Such adjustments are now rare. Theory suggests that accounting should correct for errors in accounting estimates (such as the *depreciable life* or *salvage value* of an asset) by adjusting retained earnings so that statements for future periods will show correct amounts. But *GAAP* require that corrections of such estimates flow through current, and perhaps future, *income statements*. See *accounting changes* and *accounting errors*.

**prior service cost.** *Present value* at a given time of a *pension plan's* retroactive *benefits*. "Unrecognized prior service cost" refers to that portion of prior service cost not yet *debited* to *expense*. See *actuarial accrued liability* and *funded.* Contrast with *normal cost*.

**pro forma statements.** Hypothetical statements; financial statements as they would appear if some event, such as a *merger* or increased production and sales, had occurred or were to occur; sometimes spelled as one word, "proforma."

**probable.** In many of its definitions, the *FASB* uses the term "probable." See, for example, *asset, firm commitment, liability.* A survey of practicing accountants revealed that the average of the probabilities that those surveyed had in mind when they used the term "probable" was 85 percent. Some accountants think that any event whose outcome is greater than 50 percent should be called "probable." The FASB uses the phrase "more likely than not" when it means greater than 50 percent.

**proceeds.** The *funds* received from the disposition of assets or from the issue of securities.

**process costing.** A method of *cost accounting* based on average costs (total cost divided by the *equivalent units* of work done in a period); typically used for assembly lines or for products that the firm produces in a series of steps that are more continuous than discrete.

**product.** *Goods* or *services* produced.

**product cost.** Any *manufacturing cost* that the firm can—or, in some contexts, should—debit to an *inventory* account. See *flow of costs,* for example. Contrast with *period expenses*.

**product life cycle.** Time span between initial concept (typically starting with research and development) of a good or service and the time when the firm ceases to support customers who have purchased the good or service.

**production cost.** *Manufacturing cost*.

**production cost account.** A *temporary account* for accumulating *manufacturing costs* during a period.

**production department.** A department producing salable *goods* or *services;* contrast with *service department*.

**production method (depreciation).** One form of *straight-line depreciation.* The firm assigns to the depreciable asset (e.g., a truck) a *depreciable life* measured not in elapsed time but in units of output (e.g., miles) or perhaps in units of time of expected use. Then the *depreciation* charge for a period is a portion of depreciable cost equal to a fraction computed as the actual output produced during the period divided by the expected total output to be produced over the life of the asset. This method is sometimes called the "units-of-production (or output) method."

**production method (revenue recognition).** *Percentage-of-completion method* for recognizing *revenue*.

**production volume variance.** Standard fixed *overhead* rate per unit of normal *capacity* (or base activity) times (units of base activity budgeted or planned for a period minus actual units of base activity worked or assigned to product during the period); often called a "volume variance."

**productive capacity.** One *attribute measured* for *assets*. The *current cost* of *long-term assets* means the cost of reproducing the productive capacity (for example, the ability to manufacture one million units a year), not the cost of reproducing the actual physical assets currently used (see *reproduction cost*). *Replacement cost* of productive capacity will be the same as reproduction cost of assets only in the unusual case when no technological improvement in production processes has occurred and the relative prices of goods and services used in production have remained approximately the same as when the firm acquired the currently used goods and services.

**profit.** Excess of *revenues* over *expenses* for a *transaction;* sometimes used synonymously with *net income* for the period.

**profit and loss account.** UK: *retained earnings*.

**profit-and-loss sharing ratio.** The fraction of *net income* or loss allocable to a partner in a *partnership;* need not be the same fraction as the partner's share of capital.

**profit-and-loss statement.** *Income statement.*

**profit center.** A *responsibility center* for which a firm accumulates both *revenues* and *expenses.* Contrast with *cost center.*

**profit margin.** *Sales* minus all *expenses.*

**profit margin percentage.** *Profit margin* divided by *net sales.*

**profit maximization.** The doctrine that the firm should account for a given set of operations so as to make reported *net income* as large as possible; contrast with *conservatism.* This concept in accounting differs from the profit-maximizing concept in economics, which states that the firm should manage operations to maximize the present value of the firm's wealth, generally by equating *marginal costs* and *marginal revenues.*

**profit-sharing plan.** A *defined-contribution plan* in which the employer contributes amounts based on *net income.*

**profit variance analysis.** Analysis of the causes of the difference between budgeted profit in the *master budget* and the profits earned.

**profit-volume analysis (equation).** Analysis of effects, on *profits,* caused by changes in volume or *contribution margin* per unit or *fixed costs.* See *breakeven chart.*

**profit-volume graph.** See *breakeven chart.*

**profit-volume ratio.** *Net income* divided by net sales in dollars.

**profitability.** Some measure of the firm's ability to earn *income* relative to the size of the firm or its *revenues.*

**profitability accounting.** *Responsibility accounting.*

**program budgeting (PPB).** Specification and analysis of inputs, outputs, costs, and alternatives that link plans to *budgets.*

**programmed cost.** A *fixed cost* not essential for carrying out operations. For example, a firm can control costs for research and development and advertising designed to generate new business, but once it commits to incur them, they become fixed costs. These costs are sometimes called managed costs or *discretionary costs.* Contrast with *capacity costs.*

**progressive tax.** Tax for which the rate increases as the taxed base, such as income, increases. Contrast with *regressive tax.*

**project financing arrangement.** As defined by *SFAS No. 47,* the financing of an investment project in which the lender looks principally to the *cash flows* and *earnings* of the project as the source of funds for repayment and to the *assets* of the project as *collateral* for the loan. The general *credit* of the project entity usually does not affect the terms of the financing either because the borrowing entity is a *corporation* without other assets or because the financing provides that the lender has no direct *recourse* to the entity's owners.

**projected benefit obligation.** The *actuarial present value* at a given date of all pension benefits attributed by a *defined-benefit pension* formula to employee service rendered before that date. The analyst measures the obligation using assumptions as to future compensation levels if the formula incorporates future compensation, as happens, for example, when the plan bases the eventual pension benefit on wages of the last several years of employees' work lives. Contrast to "accumulated benefit obligation," where the analyst measures the obligation using employee compensation levels at the time of the measurement date.

**projected financial statement.** *Pro forma* financial statement.

**projection.** See *financial projection* for definition and contrast.

**promissory note.** An unconditional written promise to pay a specified sum of cash on demand or at a specified date.

**proof of journal.** The process of checking the arithmetic accuracy of *journal entries* by testing for the equality of all *debits* and all *credits* since the last previous proof.

**property dividend.** A *dividend in kind.*

**property, plant, and equipment.** See *plant assets.*

**proportionate consolidation.** Canada: a presentation of the *financial statements* of any investor-investment relationship, whereby the investor's pro rata share of each *asset, liability, income* item, and *expense* item appears in the *financial statements* of the investor under the various *balance sheet* and *income statement* headings.

**proprietary accounts.** See *budgetary accounts* for definition and contrast in the context of governmental accounting.

**proprietorship.** *Assets* minus *liabilities* of an *entity;* equals *contributed capital* plus *retained earnings.*

**proprietorship theory.** The corporation view that emphasizes the form of the *accounting equation* that says *assets* − *liabilities* = *owners' equity;* contrast with *entity theory.* The major implication of a choice between these theories deals with the treatment of *subsidiaries.* For example, the proprietorship theory views *minority interest* as an *indeterminate-term liability.* The proprietorship theory implies using a *single-step income statement.*

**prorate.** To *allocate* in proportion to some base; for example, to allocate *service department* costs in proportion to hours of service used by the benefited department or to allocate *manufacturing variances* to product sold and to product added to *ending inventory.*

**prorating variances.** See *prorate.*

**prospectus.** Formal written document describing *securities* a firm will issue. See *proxy.*

**protest fee.** Fee charged by banks or other financial agencies when the bank cannot collect items (such as *checks*) presented for collection.

**provision.** Part of an *account* title. Often the firm must recognize an *expense* even though it cannot be sure of the exact amount. The entry for the estimated expense, such as for *income taxes* or expected costs under *warranty,* is as follows:

(*continued*)

| | | |
|---|---|---|
| Expense (Estimated) . . . . . . . . . . . . . | X | |
| Liability (Estimated) . . . . . . . . . . | | X |

American terminology often uses "provision" in the expense account title of the above entry. Thus, Provision for Income Taxes means the estimate of income tax expense. (British terminology uses "provision" in the title for the estimated liability of the above entry, so that Provision for Income Taxes is a balance sheet account.)

**proxy.** Written authorization given by one person to another so that the second person can act for the first, such as to vote shares of stock; of particular significance to accountants because the *SEC* presumes that management distributes financial information along with its proxy solicitations.

**public accountant.** Generally, this term is synonymous with *certified public accountant.* Some jurisdictions, however, license individuals who are not CPAs as public accountants.

**public accounting.** That portion of accounting primarily involving the *attest* function, culminating in the *auditor's report.*

**PuPU.** Acronym for *pu*rchasing *p*ower *u*nit; conceived by John C. Burton, former chief accountant of the *SEC.* Those who think that *constant-dollar accounting* is not particularly useful poke fun at it by calling it "PuPU accounting."

**purchase allowance.** A reduction in sales *invoice price* usually granted because the purchaser received *goods* not exactly as ordered. The purchaser does not return the goods but agrees to keep them for a price lower than originally agreed upon.

**purchase discount.** A reduction in purchase *invoice price* granted for prompt payment. See *sales discount* and *terms of sale.*

**purchase investigation.** An investigation of the financial affairs of a company for the purpose of disclosing matters that may influence the terms or conclusion of a potential acquisition.

**purchase method.** Accounting for a *business combination* by adding the acquired company's assets at the price paid for them to the acquiring company's assets. Contrast with *pooling-of-interests method.* The firm adds the acquired assets to the books at current values rather than original costs; the subsequent *amortization expenses* usually exceed those (and reported income is smaller than that) for the same business combination accounted for as a pooling of interests. *GAAP* in the U.S. require that the acquirer use the purchase method but other countries still allow the pooling-of-interests method.

**purchase order.** Document issued by a buyer authorizing a seller to deliver goods, with the buyer to make payment later.

**purchasing power gain or loss.** *Monetary gain or loss.*

**push-down accounting.** An accounting method used in some *purchase transactions.* Assume that Company A purchases substantially all the *common shares* of Company B but that Company B must still issue its own *financial statements.* The question arises, shall Company B change the *basis* for its *assets* and *equities* on its own books to the same updated amounts at which they appear on Company A's *consolidated financial statements*? Company B uses "push-down accounting" when it shows the new asset and equity bases reflecting Company A's purchase, because the method "pushes down" the new bases from Company A (where *GAAP* require them) to Company B (where the new bases would not appear in *historical cost accounting*). Since 1983, the *SEC* has required push-down accounting under some circumstances.

**put.** An option to sell *shares* of a publicly traded corporation at a fixed price during a fixed time span. Contrast with *call.*

# Q

**qualified report (opinion).** *Auditor's report* containing a statement that the auditor was unable to complete a satisfactory examination of all things considered relevant or that the auditor has doubts about the financial impact of some *material* item reported in the financial statements. See *except for* and *subject to.*

**quality.** In modern usage, a product or service has quality to the extent it conforms to specifications or provides customers the characteristics promised them.

**quality of earnings.** A phrase with no single, agreed-upon meaning. Some who use the phrase use it with different meanings on different occasions. "Quality of earnings" has an accounting aspect and a business cycle aspect.

In its accounting aspect, managers have choices in measuring and reporting *earnings.* This discretion can involve any of the following: selecting *accounting principles* or standards when *GAAP* allow a choice; making estimates in the application of accounting principles; and timing transactions to allow recognizing *nonrecurring* items in earnings. In some instances the range of choices has a large impact on reported earnings and in others, small. (1) Some use the phrase "quality of earnings" to mean the degree to which management can affect reported income by its choices of accounting estimates even though the choices recur every period. These users judge, for example, insurance companies to have low-quality earnings. Insurance company management must reestimate its liabilities for future payments to the insured each period, thereby having an opportunity to report periodic earnings within a wide range. (2) Others use the phrase to mean the degree to which management actually takes advantage of its flexibility. For them, an insurance company that does not vary its methods and estimating techniques, even though it has the opportunity to do so, has high-quality earnings. (3) Some have in mind the proximity in time between *revenue* recognition and cash collection. For them, the smaller the time delay, the higher will be the quality. (4) Still others use the phrase to mean the degree to which managers who have a choice among the items with large influence on earnings choose the ones that result in income measures that are more likely to recur. For them, the more likely an item of earnings is to recur, the higher will be its quality. Often these last two groups trade off with each other. Consider a dealer leasing a car on a long-term *lease,* receiving monthly collections. The dealer who uses *sales-type lease* accounting scores low on proximity of revenue recognition (all at the time of signing the lease) to cash collection but highlights the nonrepetitive nature of the transaction. The leasing dealer who uses *operating lease* accounting has perfectly matching revenue recognition and cash collection, but the *recurring* nature of the revenue gives a misleading picture of a repetitive transaction. The phrase "item of earnings" in (4) is ambiguous. The writer could mean the underlying economic event (which occurs when the lease for the car is signed) or the revenue recognition (which occurs every time the dealer using operating lease accounting receives cash). Hence, you should try to understand what other speakers and writers mean by "quality of earnings" when you interpret what they say and write. Some who refer to "earnings quality" suspect that managers will usually make choices that enhance current earnings and present the firm in the best light, independent of the ability of the firm to generate similar earnings in the future. In this book, particularly in *Chapter 14*, we use the term in the first sense—opportunity to manage earnings, not necessarily exploitation of that opportunity.

In the business cycle aspect, management's action often has no impact on the stability and recurrence of earnings. Compare a company that sells consumer products and likely has sales repeating every week with a construction company that builds to order. Companies in noncyclical businesses, such as some public utilities, likely have more stable earnings than ones in cyclical businesses, such as steel. Some use "quality of earnings" to refer to the stability and recurrence of basic revenue-generating activities. Those who use the phrase this way rarely associate earnings quality with accounting issues.

**quality of financial position.** Because of the *articulation* of the *income statement* with the *balance sheet,* the factors that imply a high (or low) *quality of earnings* also affect the balance sheet. Users of this phrase have in mind the same accounting issues as they have in mind when they use the phrase "quality of earnings."

**quantitative performance measure.** A measure of output based on an objectively observable quantity, such as units produced or *direct costs* incurred, rather than on an unobservable quantity or a quantity observable only nonobjectively, like quality of service provided.

**quantity discount.** A reduction in purchase price as quantity purchased increases. The Robinson-Patman Act constrains the amount of the discount. Do not confuse with *purchase discount*.

**quantity variance.** *Efficiency variance;* in *standard cost* systems, the standard price per unit times (actual quantity used minus standard quantity that should be used).

**quasi-reorganization.** A *reorganization* in which no new company emerges or no court has intervened, as would happen in *bankruptcy*. The primary purpose is to rid the balance sheet of a *deficit* (negative *retained earnings*) and give the firm a "fresh start."

**quick assets.** *Assets* readily convertible into *cash;* includes cash, current marketable securities, and current receivables.

**quick ratio.** Sum of (cash, current marketable securities, and current receivables) divided by *current liabilities;* often called the "acid test ratio." The analyst may exclude some nonliquid receivables from the numerator. See *ratio*.

# R

**$R^2$.** The proportion of the statistical variance of a *dependent variable* explained by the equation fit to *independent variable(s)* in a *regression analysis*.

**Railroad Accounting Principles Board (RAPB).** A board brought into existence by the Staggers Rail Act of 1980 to advise the Interstate Commerce Commission on accounting matters affecting railroads. The RAPB was the only cost-accounting body authorized by the government during the decade of the 1980s (because Congress ceased funding the CASB during the 1980s). The RAPB incorporated the pronouncements of the CASB and became the government's authority on cost accounting principles.

**R&D.** See *research and development*.

**random number sampling.** For choosing a sample, a method in which the analyst selects items from the *population* by using a random number table or generator.

**random sampling.** For choosing a sample, a method in which all items in the population have an equal chance of being selected. Compare *judgment(al) sampling*.

**RAPB.** *Railroad Accounting Principles Board*.

**rate of return on assets.** *Return on assets*.

**rate of return on common stock equity.** See *ratio*.

**rate of return on shareholders' (owners') equity.** See *ratio*.

**rate of return (on total capital).** See *ratio* and *return on assets*.

**rate variance.** *Price variance,* usually for *direct labor costs*.

**ratio.** The number resulting when one number divides another. Analysts generally use ratios to assess aspects of profitability, solvency, and liquidity. The commonly used financial ratios fall into three categories: (1) those that summarize some aspect of *operations* for a period, usually a year, (2) those that summarize some aspect of *financial position* at a given moment—the moment for which a balance sheet reports, and (3) those that relate some aspect of operations to some aspect of financial position. *Exhibit 5.12* lists the most common financial ratios and shows separately both the numerator and the denominator for each ratio.

For all ratios that require an average balance during the period, the analyst often derives the average as one half the sum of the beginning and the ending balances. Sophisticated analysts recognize, however, that particularly when companies use a fiscal year different from the calendar year, this averaging of beginning and ending balances may mislead. Consider, for example, the rate of *return on assets* of Sears, Roebuck & Company, whose fiscal year ends on January 31. Sears chooses a January 31 closing date at least in part because inventories are at a low level and are therefore easy to count—it has sold the Christmas merchandise, and the Easter merchandise has not yet all arrived. Furthermore, by January 31, Sears has collected for most Christmas sales, so receivable amounts are not unusually large. Thus at January 31, the amount of total assets is lower than at many other times during the year. Consequently, the denominator of the rate of return on assets, total assets, for Sears more likely represents the smallest amount of total assets on hand during the year rather than the average amount. The return on assets rate for Sears and other companies that choose a fiscal year end to coincide with low points in the inventory cycle is likely to exceed the ratio measured with a more accurate estimate of the average amounts of total assets.

**raw material.** Goods purchased for use in manufacturing a product.

**reacquired stock.** *Treasury shares.*

**real accounts.** *Balance sheet accounts,* as opposed to *nominal accounts.* See *permanent accounts.*

**real amount (value).** An amount stated in *constant dollars.* For example, if the firm sells an investment costing $100 for $130 after a period of 10 percent general *inflation,* the *nominal amount* of *gain* is $30 (= $130 − $100) but the real amount of gain is C$20 (= $130 − 1.10 × $100), where "C$" denotes constant dollars of purchasing power on the date of sale.

**real estate.** *Land* and its *improvements,* such as landscaping and roads, but not buildings.

**real interest rate.** Interest rate reflecting the productivity of capital, not including a premium for inflation anticipated over the life of the loan.

**realizable value.** *Fair value* or, sometimes, *net realizable (sales) value.*

**realization convention.** The accounting practice of delaying the recognition of *gains* and *losses* from changes in the market price of *assets* until the firm sells the assets. However, the firm recognizes unrealized losses on *inventory* (or *marketable securities* classified as *trading securities*) prior to sale when the firm uses the *lower-of-cost-or-market* valuation basis for inventory (or the *fair value* basis for marketable securities).

**realize.** To convert into *funds;* when applied to a *gain* or *loss,* implies that an *arm's-length transaction* has taken place. Contrast with *recognize;* the firm may recognize a loss (as, for example, on *marketable equity securities*) in the financial statements even though it has not yet realized the loss via a transaction.

**realized gain (or loss) on marketable equity securities.** An income statement account title for the difference between the proceeds of disposition and the *original cost* of *marketable equity securities.*

**realized holding gain.** See *inventory profit* for definition and an example.

**rearrangement costs.** Costs of reinstalling assets, perhaps in a different location. The firm may, but need not, *capitalize* them as part of the assets cost, just as is done with original installation cost. The firm will *expense* these costs if they merely maintain the asset's future benefits at their originally intended level before the relocation.

**recapitalization.** *Reorganization.*

**recapture.** Name for one kind of tax payment. Various provisions of the *income tax* rules require a refund by the taxpayer (recapture by the government) of various tax advantages under certain conditions. For example, the taxpayer must repay tax savings provided by *accelerated depreciation* if the taxpayer prematurely retires the item providing the tax savings.

**receipt.** Acquisition of *cash.*

**receivable.** Any *collectible,* whether or not it is currently due.

**receivable turnover.** See *ratio.*

**reciprocal holdings.** Company A owns stock of Company B, and Company B owns stock of Company A; or Company B owns stock of Company C, which owns stock of Company A.

**recognize.** To enter a transaction in the accounts; not synonymous with *realize.*

**reconciliation.** A calculation that shows how one balance or figure derives from another, such as a reconciliation of retained earnings or a *bank reconciliation schedule.* See *articulate.*

**record date.** The date at which the firm pays *dividends* on payment date to those who own the stock.

**recourse.** The rights of the lender if a borrower does not repay as promised. A recourse loan gives the lender the right to take any of the borrower's assets not exempted from such taking by the contract. See also *note receivable discounted.*

**recovery of unrealized loss on trading securities.** An *income statement account title* for the *gain* during the current period on *trading securities.*

**recurring.** Occurring again; occurring repetitively; in accounting, an adjective often used in describing *revenue* or *earnings.* In some contexts, the term "recurring revenue" is ambiguous. Consider a construction contractor who accounts for a single long-term project with the *installment method,* with revenue recognized at the time of each cash collection from the customer. The recognized revenue is recurring, but the transaction leading to the revenue is not. See *quality of earnings.*

**redemption.** Retirement by the issuer, usually by a purchase or *call,* of *stocks* or *bonds.*

**redemption premium.** *Call premium.*

**redemption value.** The price a corporation will pay to retire *bonds* or *preferred stock* if it calls them before *maturity.*

**refinancing.** An adjustment in the *capital structure* of a *corporation,* involving changes in the nature and amounts of the various classes of *debt* and, in some cases, *capital* as well as other components of *shareholders' equity. Asset* carrying values in the accounts remain unchanged.

**refunding bond issue.** Said of a *bond* issue whose proceeds the firm uses to retire bonds already *outstanding.*

**register.** A collection of consecutive entries, or other information, in chronological order, such as a check register or an insurance register that lists all insurance policies owned. If the firm records entries in the register, it can serve as a *journal.*

**registered bond.** A bond for which the issuer will pay the *principal* and *interest,* if registered as to interest, to the owner listed on the books of the issuer; as opposed to a bearer bond, in which the issuer must pay the possessor of the bond.

**registrar.** An *agent,* usually a bank or trust company, appointed by a corporation to keep track of the names of shareholders and distributions to them.

**registration statement.** Required by the Securities Act of 1933, statement of most companies that want to have owners of their securities trade the securities in public markets. The statement discloses financial data and other items of interest to potential investors.

**regression analysis.** A method of *cost estimation* based on statistical techniques for fitting a line (or its equivalent in higher mathematical dimensions) to an observed series of data points, usually by minimizing the sum of squared deviations of the observed data from the fitted line. Common usage calls the cost that the analysis explains the "dependent variable"; it calls the variable(s) we use to estimate cost behavior "independent variable(s)." If we use more than one independent variable, the term for the analysis is "multiple regression analysis." See $R^2$, *standard error,* and *t-value.*

**regressive tax.** Tax for which the rate decreases as the taxed base, such as income, increases. Contrast with *progressive tax.*

**Regulation S-K.** The *SEC*'s standardization of nonfinancial statement disclosure requirements for documents filed with the SEC.

**Regulation S-T.** The *SEC*'s regulations specifying formats for electronic filing and the *EDGAR* system.

**Regulation S-X.** The *SEC*'s principal accounting regulation, which specifies the form and content of financial reports to the SEC.

**rehabilitation.** The improving of a used *asset* via an extensive repair. Ordinary *repairs* and *maintenance* restore or maintain expected *service potential* of an asset, and the firm treats them as *expenses.* A rehabilitation improves the asset beyond its current service potential, enhancing the service potential to a significantly higher level than before the rehabilitation. Once rehabilitated, the asset may be better, but need not be, than it was when new. The firm will *capitalize expenditures* for rehabilitation, like those for *betterments* and *improvements.*

**reinvestment rate.** In a *capital budgeting* context, the rate at which the firm invests cash inflows from a project occurring before the project's completion. Once the analyst assumes such a rate, no project can ever have multiple *internal rates of return.* See *Descartes' rule of signs.*

**relative performance evaluation.** Setting performance targets and, sometimes, compensation in relation to the performance of others, perhaps in different firms or divisions, who face a similar environment.

**relative sales value method.** See *net realizable (sales) value.*

**relevant cost.** Cost used by an analyst in making a decision. *Incremental cost; opportunity cost.*

**relevant range.** Activity levels over which costs are linear or for which *flexible budget* estimates and *breakeven charts* will remain valid.

**remit earnings.** An expression likely to confuse a reader without a firm understanding of accounting basics. A firm generates *net assets* by earning *income* and retains net assets if it does not declare *dividends* in the amount of net income. When a firm declares dividends and pays the cash (or other net assets), some writers would say the firm "remits earnings." We think the student learns better by conceiving earnings as a *credit balance.* When a firm pays dividends it sends net assets, things with debit balances, not something with a credit balance, to the recipient. When writers say firms "remit earnings," they mean the firms send assets (or net assets) that previous earnings have generated and reduce *retained earnings.*

**remittance advice.** Information on a *check stub,* or on a document attached to a check by the *drawer,* that tells the *payee* why a payment is being made.

**rent.** A charge for use of land, buildings, or other assets.

**reorganization.** In the *capital structure* of a corporation, a major change that leads to changes in the rights, interests, and implied ownership of the various security owners; usually results from a *merger* or an agreement by senior security holders to take action to forestall *bankruptcy.*

**repair.** An *expenditure* to restore an *asset's* service potential after damage or after prolonged use. In the second sense, after prolonged use, the difference between repairs and maintenance is one of degree and not of kind. A repair is treated as an *expense* of the period when incurred. Because the firm treats repairs and maintenance similarly in this regard, the distinction is not important. A repair helps to maintain capacity at the levels planned when the firm acquired the *asset.* Contrast with *improvement.*

**replacement cost.** For an asset, the current fair market price to purchase another, similar asset (with the same future benefit or service potential). *Current cost.* See *reproduction cost* and *productive capacity.* See also *distributable income* and *inventory profit.*

**replacement cost method of depreciation.** Method in which the analyst augments the original-cost *depreciation* charge with an amount based on a portion of the difference between the *current replacement cost* of the asset and its *original cost.*

**replacement system of depreciation.** See *retirement method of depreciation* for definition and contrast.

**report.** *Financial statement; auditor's report.*

**report form.** *Balance sheet* form that typically shows *assets* minus *liabilities* as one total. Then, below that total appears the components of *owners' equity* summing to the same total. Often, the top section shows *current* assets less current liabilities before *noncurrent assets* less noncurrent liabilities. Contrast with *account form.*

**reporting objectives (policies).** The general purposes for which the firm prepares *financial statements.* The *FASB* has discussed these in *SFAC No. 1.*

**representative item sampling.** Sampling in which the analyst believes the sample selected is typical of the entire population from which it comes. Compare *specific item sampling.*

**reproduction cost.** The *cost* necessary to acquire an *asset* similar in all physical respects to another asset for which the analyst requires a *current value.* See *replacement cost* and *productive capacity* for contrast.

**required rate of return (RRR).** *Cost of capital.*

**requisition.** A formal written order or request, such as for withdrawal of supplies from the storeroom.

**resale value.** *Exit value; net realizable value.*

**research and development (R&D).** A form of economic activity with special accounting rules. Firms engage in research in hopes of discovering new knowledge that will create a new product, process, or service or of improving a present product, process, or service. Development translates research findings or other knowledge into a new or improved product, process, or service. *SFAS No. 2* requires that firms expense costs of such activities as incurred on the grounds that the future benefits are too uncertain to warrant *capitalization* as an asset. This treatment seems questionable to us because we wonder why firms would continue to undertake R&D if there was no expectation of future benefit; if future benefits exist, then R&D *costs* should be assets that appear, like other assets, at *historical cost.*

**reserve.** The worst word in accounting because almost everyone not trained in accounting, and some who are, misunderstand it. The common confusion is that "reserves" represent a pool of *cash* or other *assets* available when the firm needs them. Wrong. Cash always has a *debit balance.* Reserves always have a *credit* balance. When properly used in accounting, "reserves" refer to an account that appropriates *retained earnings* and restricts dividend declarations. Appropriating retained earnings is itself a poor and vanishing practice, so the word should seldom appear in accounting. In addition, "reserve" was used in the past to indicate an asset *contra account* (for example, "reserve for depreciation") or an *estimated liability* (for example, "reserve for warranty costs"). In any case, reserve accounts have *credit* balances and are not pools of *funds,* as the unwary reader might infer. If a company has set aside a pool of *cash* (or *marketable securities*) to serve some specific purpose such as paying for a new factory, then it will call that cash a *fund.* No other word in accounting causes so much misunderstanding by nonexperts as well as by "experts" who should know better. A leading unabridged dictionary defines "reserve" as "cash, or assets readily convertible into cash, held aside, as by a corporation, bank, state or national government, etc. to meet expected or unexpected demands." This definition is absolutely wrong in accounting. Reserves are not funds. For example, the firm creates a contingency fund of $10,000 by depositing cash in a fund and makes the following entry:

| | | |
|---|---|---|
| Dr. Contingency Fund . . . . . . . . . . . | 10,000 | |
| Cr. Cash . . . . . . . . . . . . . . . . . . | | 10,000 |

The following entry may accompany the previous entry, if the firm wants to appropriate retained earnings:

| | | |
|---|---|---|
| Dr. Retained Earnings . . . . . . . . . . . | 10,000 | |
| Cr. Reserve for Contingencies . . . . | | 10,000 |

The transaction leading to the first entry has economic significance. The second entry has little economic impact for most firms. The problem with the word "reserve" arises because the firm can make the second entry without the first—a company can create a reserve, that is, appropriate retained earnings, without creating a fund. The problem results, at least in part, from the fact that in common usage, "reserve" means a pool of assets, as in the phrase "oil reserves." The *Internal Revenue Service* does not help in dispelling confusion about the term "reserves." The federal *income tax* return for corporations uses the title "Reserve for Bad Debts" to mean "Allowance for Uncollectible Accounts" and speaks of the "Reserve Method" in referring to the *allowance method* for estimating *revenue* or *income* reductions from estimated *uncollectibles.*

**reserve recognition accounting (RRA).** One form of *accounting* for natural resources. In exploration for natural resources, the problem arises of how to treat the expenditures for exploration, both before the firm knows the outcome of the efforts and after it knows the outcome. Suppose that the firm spends $10 million to drill 10 holes ($1 million each) and that nine of them are dry whereas one is a gusher containing oil with a *net realizable value* of $40 million. Dry hole, or *successful efforts,* accounting would expense $9 million and *capitalize* $1 million, which the firm will *deplete* as it lifts the oil from the ground. *SFAS No. 19,* now suspended, required *successful efforts costing.* Full costing would expense nothing but would capitalize the $10 million of drilling costs that the firm will deplete as it lifts the oil from the single productive well. Reserve recognition accounting would capitalize $40 million, which the firm will deplete as it lifts the oil, with a $30 million *credit* to *income* or *contributed capital.* The *balance sheet* shows the *net realizable value* of proven oil and gas reserves. The *income statement* has three sorts of items: (1) current income resulting from production or "lifting profit," which is the *revenue* from sales of oil and gas less the expense based on the current valuation amount at which these items have appeared on the balance sheet, (2) profit or loss from exploration efforts in which the current value of new discoveries is revenue and all the exploration cost is expense, and (3) gain or loss on changes in current value during the year, which accountants in other contexts call a *holding gain or loss*.

**reset bond.** A bond, typically a *junk bond,* that specifies that periodically the issuer will reset the coupon rate so that the bond sells at *par* in the market. Investment bankers created this type of instrument to help ensure the purchasers of such bonds of getting a fair rate of return, given the riskiness of the issuer. If the issuer gets into financial trouble, its bonds will trade for less than par in the market. The issuer of a reset bond promises to raise the interest rate and preserve the value of the bond. Ironically, the reset feature has often had just the opposite effect. The default risk of many issuers of reset bonds has deteriorated so much that the bonds have dropped to less than 50 percent of par. To raise the value to par, the issuer would have to raise the interest rate to more than 25 percent per year. That rate is so large that issuers have declared bankruptcy rather than attempt to make the new large interest payments; this then reduces the market value of the bonds rather than increases them.

**residual income.** In an external reporting context, a term that refers to *net income* to *common shares* (= net income less *preferred stock dividends*). In *managerial accounting,* this term refers to the excess of income for a *division* or *segment* of a company over the product of the *cost of capital* for the company multiplied by the average amount of capital invested in the division during the period over which the division earned the income.

**residual security.** A *potentially dilutive security*. *Options, warrants, convertible bonds,* and *convertible preferred stock.*

**residual value.** At any time, the estimated or actual *net realizable value* (that is, proceeds less removal costs) of an *asset,* usually a depreciable *plant asset*. In the context of depreciation accounting, this term is equivalent to *salvage value* and is preferred to *scrap value* because the firm need not scrap the asset. It is sometimes used to mean net *book value*. In the context of a *noncancelable* lease, it is the estimated value of the leased asset at the end of the lease period. See *lease*.

**resources supplied.** *Expenditures* made for an activity.

**resources used.** *Cost driver* rate times cost driver volume.

**responsibility accounting.** Accounting for a business by considering various units as separate entities, or *profit centers,* giving management of each unit responsibility for the unit's *revenues* and *expenses*. See *transfer price*.

**responsibility center.** An organization part or *segment* that top management holds accountable for a specified set of activities. Also called "accountability center." See *cost center, investment center, profit center,* and *revenue center*.

**restricted assets.** Governmental resources restricted by legal or contractual requirements for a specific purpose.

**restricted retained earnings.** That part of *retained earnings* not legally available for *dividends*. See *retained earnings, appropriated*. Bond indentures and other loan contracts can curtail the legal ability of the corporation to declare dividends without formally requiring a retained earnings appropriation, but the firm must disclose such restrictions.

**retail inventory method.** Ascertaining cost amounts of *ending inventory* as follows (assuming *FIFO*): cost of ending inventory = (selling price of *goods available for sale* − sales) × *cost percentage*. The analyst then computes cost of goods sold from the inventory equation; costs of beginning inventory, purchases, and ending inventory are all known. (When the firm uses *LIFO,* the method resembles the *dollar-value LIFO method.*) See *markup*.

**retail terminology.** See *markup*.

**retained earnings.** Net *income* over the life of a corporation less all *dividends* (including capitalization through *stock dividends*); *owners' equity* less *contributed capital.*

**retained earnings, appropriated.** An *account* set up by crediting it and debiting *retained earnings;* used to indicate that a portion of retained earnings is not available for dividends. The practice of appropriating retained earnings is misleading unless the firm marks all capital with its use, which is not practicable, nor sensible, since capital is fungible—all the *equities* jointly fund all the *assets*. The use of formal retained earnings appropriations is declining.

**retained earnings statement.** A *reconciliation* of the beginning and the ending balances in the *retained earnings account;* required by *generally accepted accounting principles* whenever the firm presents *comparative balance sheets* and an *income statement.* This reconciliation can appear in a separate statement, in a combined statement of income and retained earnings, or in the balance sheet.

**retirement method of depreciation.** A method in which the firm records no entry for *depreciation expense* until it retires an *asset* from service. Then, it makes an entry *debiting* depreciation expense and *crediting* the asset account for the cost of the asset retired. If the retired asset has a *salvage value,* the firm reduces the amount of the debit to depreciation expense by the amount of salvage value with a corresponding debit to cash, receivables, or salvaged materials. The "replacement system of depreciation" is similar, except that the debit to depreciation expense equals the cost of the new asset less the salvage value, if any, of the old asset. Some public utilities used these methods. For example, if the firm acquired ten telephone poles in Year 1 for $60 each and replaces them in Year 10 for $100 each when the salvage value of the old poles is $5 each, the accounting would be as follows:

(continued)

**Retirement Method**

| | | |
|---|---|---|
| Plant Assets . . . . . . . . . . . . . . . . . . . | 600 | |
| Cash . . . . . . . . . . . . . . . . . . . . . | | 600 |
| To acquire assets in Year 1. | | |
| Depreciation Expense . . . . . . . . . . . . . | 550 | |
| Salvage Receivable . . . . . . . . . . . . . . . | 50 | |
| Plant Assets . . . . . . . . . . . . . . . . | | 600 |
| To record retirement and depreciation in Year 10. | | |
| Plant Assets . . . . . . . . . . . . . . . . . . . | 1,000 | |
| Cash . . . . . . . . . . . . . . . . . . . . . . | | 1,000 |
| To record acquisition of new assets in Year 10. | | |

**Replacement Method**

| | | |
|---|---|---|
| Plant Assets . . . . . . . . . . . . . . . . . . . | 600 | |
| Cash . . . . . . . . . . . . . . . . . . . . . . | | 600 |
| To acquire assets in Year 1. | | |
| Depreciation Expense . . . . . . . . . . . . . | 950 | |
| Salvage Receivable . . . . . . . . . . . . . . . | 50 | |
| Cash . . . . . . . . . . . . . . . . . . . . . | | 1,000 |
| To record depreciation on old asset in amount quantified by net cost of replacement asset in Year 10. | | |

The retirement method is like *FIFO* in that it records the cost of the first assets as depreciation and puts the cost of the second assets on the balance sheet. The replacement method is like *LIFO* in that it records the cost of the second assets as depreciation expense and leaves the cost of the first assets on the balance sheet.

**retirement plan.** *Pension plan.*

**retroactive benefits.** In initiating or amending a *defined-benefit pension plan,* benefits that the benefit formula attributes to employee services rendered in periods prior to the initiation or amendment. See *prior service costs.*

**return.** A schedule of information required by governmental bodies, such as the tax return required by the *Internal Revenue Service;* also the physical return of merchandise. See also *return on investment.*

**return on assets (ROA).** *Net income* plus after-tax *interest charges* plus *minority interest* in income divided by average total *assets;* perhaps the single most useful ratio for assessing management's overall operating performance. Most financial economists would subtract average noninterest-bearing *liabilities* from the denominator. Economists realize that when liabilities do not provide for explicit interest charges, the creditor adjusts the terms of contract, such as setting a higher selling price or lower discount, to those who do not pay cash immediately. (To take an extreme example, consider how much higher salary a worker who receives a salary once per year, rather than once per month, would demand.) This ratio requires in the numerator the income amount before the firm accrues any charges to suppliers of funds. We cannot measure the interest charges implicit in the noninterest-bearing liabilities because they cause items such as cost of goods sold and salary expense to be somewhat larger, since the interest is implicit. Subtracting their amounts from the denominator adjusts for their implicit cost. Such subtraction assumes that assets financed with noninterest-bearing liabilities have the same rate of return as all the other assets.

**return on investment (ROI), return on capital.** *Income* (before distributions to suppliers of capital) for a period; as a rate, this amount divided by average total assets. The analyst should add back *interest,* net of tax effects, to *net income* for the numerator. See *ratio.*

**revenue.** The *owners' equity* increase accompanying the *net assets* increase caused by selling goods or rendering services; in short, a service rendered; *sales* of products, merchandise, and services and earnings from *interest, dividends, rents,* and the like. Measure revenue as the expected *net present value* of the net assets the firm will receive. Do not confuse with *receipt* of *funds,* which may occur before, when, or after revenue is recognized. Contrast with *gain* and *income.* See also *holding gain.* Some writers use the term *gross income* synonymously with *revenue;* avoid such usage.

**revenue center.** Within a firm, a *responsibility center* that has control only over revenues generated. Contrast with *cost center.* See *profit center.*

**revenue expenditure.** A term sometimes used to mean an *expense,* in contrast to a capital *expenditure* to acquire an *asset* or to discharge a *liability.* Avoid using this term; use *period expense* instead.

**revenue received in advance.** An inferior term for *advances from customers.*

**reversal (reversing) entry.** An *entry* in which all *debits* and *credits* are the credits and debits, respectively, of another entry, and in the same amounts. The accountant usually records a reversal entry on the first day of an *accounting period* to reverse a previous *adjusting entry,* usually an *accrual.* The purpose of such entries is to make the bookkeeper's tasks easier. Suppose that the firm pays salaries every other Friday, with paychecks compensating employees for the two weeks just ended. Total salaries accrue at the rate of $5,000 per five-day workweek. The bookkeeper is accustomed to making the following entry every other Friday:

| | | |
|---|---|---|
| **(1)** Salary Expense . . . . . . . . . . . . . . . . | 10,000 | |
| Cash . . . . . . . . . . . . . . . . . . . | | 10,000 |
| To record salary expense and salary payments. | | |

If the firm delivers paychecks to employees on Friday, November 25, then the *adjusting entry* made on November 30 (or perhaps later) to record accrued salaries for November 28, 29, and 30 would be as follows:

| | | |
|---|---|---|
| **(2)** Salary Expense . . . . . . . . . . . . . . . . | 3,000 | |
| Salaries Payable . . . . . . . . . . . | | 3,000 |
| To charge November operations with all salaries earned in November. | | |

The firm would close the Salary Expense account as part of the November 30 closing entries. On the next payday, December 9, the salary entry would be as follows:

| | | |
|---|---|---|
| **(3)** Salary Expense . . . . . . . . . . . . . . . . | 7,000 | |
| Salaries Payable . . . . . . . . . . . . . . . . . | 3,000 | |
| Cash . . . . . . . . . . . . . . . . . . . | | 10,000 |
| To record salary payments split between expense for December (seven days) and liability carried over from November. | | |

To make entry *(3)*, the bookkeeper must look back into the records to see how much of the debit is to Salaries Payable accrued from the previous month in order to split the total debits between December expense and the liability carried over from November. Notice that this entry forces the bookkeeper both (a) to refer to balances in old accounts and (b) to make an entry different from the one customarily made, entry *(1)*. The reversing entry, made just after the books have been closed for the second quarter, makes the salary entry for December 9 the same as that made on all other Friday paydays. The reversing entry merely *reverses* the adjusting entry *(2)*:

| | | |
|---|---|---|
| **(4)** Salaries Payable . . . . . . . . . . . . . . . . . | 3,000 | |
| Salary Expense . . . . . . . . . . . . | | 3,000 |
| To reverse the adjusting entry. | | |

This entry results in a zero balance in the Salaries Payable account and a credit balance in the Salary Expense account. If the firm makes entry *(4)* just after it closes the books for November, then the entry on December 9 will be the customary entry *(1)*. Entries *(4)* and *(1)* together have exactly the same effect as entry *(3)*.

The procedure for using reversal entries is as follows: the firm makes the required adjustment to record an accrual (*payable* or *receivable*) at the end of an *accounting period;* it makes the closing entry as usual; as of the first day of the following period, it makes an entry reversing the adjusting entry; when the firm makes (or receives) a payment, it records the entry as though it had not recorded an adjusting entry at the end of the preceding period. Whether a firm uses reversal entries affects the record-keeping procedures but not the financial statements.

This term is also used to describe the entry reversing an incorrect entry before recording the correct entry.

**reverse stock split.** A stock split in which the firm decreases the number of shares *outstanding*. See *stock split*.

**revolving fund.** A fund whose amounts the firm continually spends and replenishes; for example, a *petty cash fund*.

**revolving loan.** A *loan* that both the borrower and the lender expect to renew at *maturity*.

**right.** The privilege to subscribe to new *stock* issues or to purchase stock. Usually, securities called *warrants* contain the rights, and the owner of the warrants may sell them. See also *preemptive right.*

**risk.** A measure of the variability of the *return on investment*. For a given expected amount of return, most people prefer less risk to more risk. Therefore, in rational markets, investments with more risk usually promise, or investors expect to receive, a higher rate of return than investments with lower risk. Most people use "risk" and "uncertainty" as synonyms. In technical language, however, these terms have different meanings. We use "risk" when we know the probabilities attached to the various outcomes, such as the probabilities of heads or tails in the flip of a fair coin. "Uncertainty" refers to an event for which we can only estimate the probabilities of the outcomes, such as winning or losing a lawsuit.

**risk-adjusted discount rate.** Rate used in discounting cash flows for projects more or less risky than the firm's average. In a *capital budgeting* context, a decision analyst compares projects by comparing their net *present values* for a given *interest* rate, usually the cost of capital. If the analyst considers a given project's outcome to be much more or much less risky than the normal undertakings of the company, then the analyst will use a larger interest rate (if the project is riskier) or a smaller interest rate (if less risky) in discounting, and the rate used is "risk-adjusted."

**risk-free rate.** An interest rate reflecting only the pure interest rate plus an amount to compensate for inflation anticipated over the life of a loan, excluding a premium for the risk of default by the borrower. Financial economists usually measure the risk-free rate in the United States from U.S. government securities, such as Treasury bills and notes.

**risk premium.** Extra compensation paid to employees or extra *interest* paid to lenders, over amounts usually considered normal, in return for their undertaking to engage in activities riskier than normal.

**ROA.** *Return on assets.*

**ROI.** *Return on investment;* usually used to refer to a single project and expressed as a ratio: *income* divided by average *cost* of *assets* devoted to the project.

**royalty.** Compensation for the use of property, usually a patent, copyrighted material, or natural resources. The amount is often expressed as a percentage of receipts from using the property or as an amount per unit produced.

**RRA.** *Reserve recognition accounting.*

**RRR.** Required rate of return. See *cost of capital.*

**rule of 69.** Rule stating that an amount of cash invested at $r$ percent per period will double in $69/r + 0.35$ periods. This approximation is accurate to one-tenth of a period for interest rates between 1/4 and 100 percent per period. For example, at 10 percent per period, the rule says that a given sum will double in $69/10 + 0.35 = 7.25$ periods. At 10 percent per period, a given sum actually doubles in 7.27+ periods.

**rule of 72.** Rule stating that an amount of cash invested at $r$ percent per period will double in $72/r$ periods. A reasonable approximation for interest rates between 4 and 10 percent but not nearly as accurate as the *rule of 69* for interest rates outside that range. For example, at 10 percent per period, the rule says that a given sum will double in 72/10 = 7.2 periods.

**rule of 78.** The rule followed by many finance companies for allocating earnings on *loans* among the months of a year on the sum-of-the-months'-digits basis when the borrower makes equal monthly payments to the lender. The sum of the digits from 1 through 12 is 78, so the rule allocates 12/78 of the year's earnings to the first month, 11/78 to the second month, and so on. This approximation allocates more of the early payments to interest and less to principal than does the correct, compound-interest method. Hence, lenders still use this method even though present-day computers can make the compound-interest computation as easily as they can carry out the approximation. See *sum-of-the-years'-digits depreciation*.

**ruling (and balancing) an account.** The process of summarizing a series of entries in an *account* by computing a new *balance* and drawing double lines to indicate that the new balance summarizes the information above the double lines. An illustration appears below. The steps are as follows: (1) Compute the sum of all *debit* entries including opening debit balance, if any— $1,464.16. (2) Compute the sum of all credit entries including opening credit balance, if any— $413.57. (3) If the amount in (1) exceeds the amount in (2), then write the excess as a credit with a checkmark— $1,464.16 − $413.57 = $1,050.59. (4) Add both debit and credit columns, which should both now sum to the same amount, and show that identical total at the foot of both columns. (5) Draw double lines under those numbers and write the excess of debits over credits as the new debit balance with a checkmark. (6) If the amount in (2) exceeds the amount in (1), then write the excess as a debit with a checkmark. (7) Do steps (4) and (5) except that the excess becomes the new credit balance. (8) If the amount in (1) equals the amount in (2), then the balance is zero, and only the totals with the double lines beneath them need appear.

**Rutgers Accounting Web site.** See http://www.rutgers.edu/Accounting/ for a useful compendium of accounting information.

# S

**S corporation.** A corporation taxed like a *partnership*. Corporation (or partnership) agreements allocate the periodic *income* to the individual shareholders (or partners) who report these amounts on their individual *income tax* returns. Contrast with *C corporation*.

**SA (société anonyme).** France: A *corporation*.

***SAB***. *Staff Accounting Bulletin* of the *SEC*.

**safe-harbor lease.** A form of *tax-transfer lease*.

**safety stock.** Extra items of *inventory* kept on hand to protect against running out.

**salary.** Compensation earned by managers, administrators, and professionals, not based on an hourly rate. Contrast with *wage*.

**sale.** A *revenue* transaction in which the firm delivers *goods* or *services* to a customer in return for cash or a contractual obligation to pay.

**sale and leaseback.** A *financing* transaction in which the firm sells improved property but takes it back for use on a long-term *lease*. Such transactions often have advantageous income tax effects but usually have no effect on *financial statement income*.

**sales activity variance.** *Sales volume variance*.

**sales allowance.** A sales *invoice* price reduction that a seller grants to a buyer because the seller delivered *goods* different from, perhaps because of damage, those the buyer ordered. The seller often accumulates amounts of such adjustments in a temporary *revenue contra account* having this, or a similar, title. See *sales discount*.

**sales basis of revenue recognition.** Recognition of *revenue* not when a firm produces goods or when it receives orders but only when it has completed the sale by delivering the goods or services and has received cash or a claim to cash. Most firms recognize revenue on this basis. Compare with the *percentage-*

An Open Account, Ruled and Balanced
(Steps indicated in parentheses correspond to steps described in "ruling an account.")

| | Date 2007 | Explanation | Ref. | Debit (1) | Date 2007 | Explanation | Ref. | Credit (2) | |
|---|---|---|---|---|---|---|---|---|---|
| | Jan. 2 | Balance | ✓ | 100.00 | | | | | |
| | Jan. 13 | | VR | 121.37 | Sept. 15 | | J | .42 | |
| | Mar. 20 | | VR | 56.42 | Nov. 12 | | J | 413.15 | |
| | June 5 | | J | 1,138.09 | Dec. 31 | Balance | ✓ | 1,050.59 | (3) |
| | Aug. 18 | | J | 1.21 | | | | | |
| | Nov. 20 | | VR | 38.43 | | | | | |
| | Dec. 7 | | VR | 8.64 | | | | | |
| (4) | 2001 | | | 1,464.16 | 2001 | | | 1,464.16 | (4) |
| (5) | Jan. 1 | Balance | ✓ | 1,050.59 | | | | | |

*of-completion method* and the *installment method.* This is identical with the *completed contract method,* but the latter term ordinarily applies only to *long-term* construction projects.

**sales contra, estimated uncollectibles.** A title for the contra-revenue account to recognize estimated reductions in income caused by *accounts receivable* that will not be collected. See *bad debt expense, allowance for uncollectibles,* and *allowance method.*

**sales discount.** A sales *invoice* price reduction usually offered for prompt payment. See *terms of sale* and *2/10, n/30.*

**sales return.** The physical return of merchandise. The seller often accumulates amounts of such returns in a temporary revenue contra account.

**sales-type (capital) lease.** A form of *lease.* See *capital lease.* When a manufacturer (or other firm) that ordinarily sells goods enters a capital lease as *lessor,* the lease is a "sales-type lease." When a financial firm, such as a bank or insurance company or leasing company, acquires the asset from the manufacturer and then enters a capital lease as lessor, the lease is a "direct-financing-type lease." The manufacturer recognizes its ordinary profit (sales price less *cost of goods sold,* where sales price is the *present value* of the contractual lease payments plus any down payment) on executing the sales-type capital lease, but the financial firm does not recognize profit on executing a capital lease of the direct-financing type.

**sales value method.** *Relative sales value method.* See *net realizable value method.*

**sales volume variance.** Budgeted *contribution margin* per unit times (planned sales volume minus actual sales volume).

**salvage value.** Actual or estimated selling price, net of removal or disposal costs, of a used *plant asset* that the firm expects to sell or otherwise retire. See *residual value.*

**SAR.** *Summary annual report.*

**SARL (société à responsabilité limitée).** France: a *corporation* with limited liability and a life of no more than 99 years; must have at least two and no more than 50 *shareholders.*

***SAS.*** *Statement on Auditing Standards* of the *AICPA.*

**scale effect.** See *discounted cash flow.*

**scatter diagram.** A graphic representation of the relation between two or more variables within a population.

**schedule.** A supporting set of calculations, with explanations, that show how to derive figures in a *financial statement* or tax return.

**scientific method.** *Effective interest method* of *amortizing bond discount* or *premium.*

**scrap value.** *Salvage value* assuming the owner intends to junk the item. A *net realizable value. Residual value.*

**SEC (Securities and Exchange Commission).** An agency authorized by the U.S. Congress to regulate, among other things, the financial reporting practices of most public corporations. The SEC has indicated that it will usually allow the *FASB* to set accounting principles, but it often requires more disclosure than the FASB requires. The SEC states its accounting requirements in its *Accounting Series Releases* (*ASR*), *Financial Reporting Releases, Accounting and Auditing Enforcement Releases, Staff Accounting Bulletins* (these are, strictly speaking, interpretations by the accounting staff, not rules of the commissioners themselves), and *Regulations S-X.* See also *registration statement, 10-K,* and *20-F.*

**secret reserve.** *Hidden reserve.*

**Securities and Exchange Commission.** *SEC.*

**security.** Document that indicates ownership, such as a *share* of *stock,* or indebtedness, such as a *bond,* or potential ownership, such as an *option* or *warrant.*

**security available for sale.** According to *SFAS No. 115* (1993), a *debt* or *equity security* that is not a *trading security,* or a debt security that is not a *security held to maturity.*

**security held to maturity.** According to *SFAS No. 115* (1993), a *debt security* the holder has both the ability and the intent to hold to *maturity;* valued in the *balance sheet* at amortized acquisition cost: the book value of the security at the end of each period is the book value at the beginning of the period multiplied by the historical *yield* on the security (measured as of the time of purchase) less any cash the holder receives at the end of this period from the security.

**segment (of a business).** As defined by *APB Opinion No. 30,* "a component of an *entity* whose activities represent a separate major line of business or class of customer. . . . [It may be] a *subsidiary,* a division, or a department, . . . provided that its *assets,* results of *operations,* and activities can be clearly distinguished, physically and operationally for financial reporting purposes, from the other assets, results of operations, and activities of the entity." In *SFAS No. 14,* a segment is defined as a "component of an enterprise engaged in promoting a product or service or a group of related products and services primarily to unaffiliated customers . . . for a profit." *SFAS No. 131* defines operating segments using the "management approach" as components of the enterprise engaging in revenue- and expense-generating business activities "whose operating results are regularly reviewed by the enterprise's chief operating decision maker to make decisions about resources . . . and asset performance."

**segment reporting.** Reporting of *sales, income,* and *assets* by *segments of a business,* usually classified by nature of products sold but sometimes by geographical area where the firm produces or sells goods or by type of customers; sometimes called "line of business reporting." The accounting for segment income does not allocate *central corporate expenses* to the segments.

**self-balancing.** A set of records with equal *debits* and *credits* such as the *ledger* (but not individual accounts), the *balance sheet,* and a *fund* in nonprofit accounting.

**self-check(ing) digit.** A digit forming part of an account or code number, normally the last digit of the number, which is arithmetically derived from the other numbers of the code and is used to detect errors in transcribing the code number. For example, assume the last digit of the account number is the remainder after summing the preceding digits and dividing that sum by nine. Suppose the computer encounters the account numbers 7027261-7 and 9445229-7. The program can tell that something has gone wrong with the encoding of the second account number because the sum of the first seven digits is 35, whose remainder on division by 9 is 8, not 7. The first account number does not show such an error because the sum of the first seven digits is 25, whose remainder on division by 9 is, indeed, 7. The first account number may be in error, but the second surely is.

**self-insurance.** See *insurance*.

**self-sustaining foreign operation.** A foreign operation both financially and operationally independent of the reporting enterprise (owner) so that the owner's exposure to exchange-rate changes results only from the owner's net investment in the foreign entity.

**selling and administrative expenses.** *Expenses* not specifically identifiable with, or assigned to, production.

**semifixed costs.** *Costs* that increase with activity as a step function.

**semivariable costs.** *Costs* that increase strictly linearly with activity but that are positive at zero activity level. Royalty fees of 2 percent of sales are variable; royalty fees of $1,000 per year plus 2 percent of sales are semivariable.

**senior securities.** *Bonds* as opposed to *preferred stock*; *preferred stock* as opposed to *common stock*. The firm must meet the senior security claim against *earnings* or *assets* before meeting the claims of less-senior securities.

**sensitivity analysis.** A study of how the outcome of a decision-making process changes as one or more of the assumptions change.

**sequential access.** Computer-storage access in which the analyst can locate information only by a sequential search of the storage file. Compare *direct access*.

**serial bonds.** An *issue* of *bonds* that mature in part at one date, another part on another date, and so on. The various maturity dates usually occur at equally spaced intervals. Contrast with *term bonds*.

**service basis of depreciation.** *Production method*.

**service bureau.** A commercial data-processing center providing service to various customers.

**service cost, (current) service cost.** *Pension plan expenses incurred* during an *accounting period* for employment services performed during that period. Contrast with *prior service cost*. See *funded*.

**service department.** A department, such as the personnel or computer department, that provides services to other departments rather than direct work on a salable product. Contrast with *production department*. A firm must allocate costs of service departments whose services benefit manufacturing operations to *product costs* under *full absorption costing*.

**service life.** Period of expected usefulness of an asset; may differ from *depreciable life* for income tax purposes.

**service potential.** The future benefits that cause an item to be classified as an *asset*. Without service potential, an item has no future benefits, and accounting will not classify the item as an asset. *SFAC No. 6* suggests that the primary characteristic of service potential is the ability to generate future net cash inflows.

**services.** Useful work done by a person, a machine, or an organization. See *goods*.

**setup.** The time or costs required to prepare production equipment for doing a job.

***SFAC.*** *Statement of Financial Accounting Concepts* of the *FASB*.

***SFAS.*** *Statement of Financial Accounting Standards*. See *FASB*.

**shadow price.** An opportunity cost. A *linear programming* analysis provides as one of its outputs the potential value of having available more of the scarce resources that constrain the production process, for example, the value of having more time available on a machine tool critical to the production of two products. Common terminology refers to this value as the "shadow price" or the "dual value" of the scarce resource.

**share.** A unit of *stock* representing ownership in a corporation.

**share premium.** UK: *additional paid-in capital* or *capital contributed in excess of par value*.

**shareholders' equity.** *Proprietorship* or *owners' equity* of a corporation. Because *stock* means inventory in Australia, the UK, and Canada, their writers use the term "shareholders' equity" rather than the term "stockholders' equity."

**short-run.** The opposite of *long-run* or *long-term*.

**short-term.** Current; ordinarily, due within one year.

**short-term liquidity risk.** The risk that an *entity* will not have enough *cash* in the *short run* to pay its *debts*.

**shrinkage.** An excess of *inventory* shown on the *books* over actual physical quantities on hand; can result from theft or shoplifting as well as from evaporation or general wear and tear. Some accountants, in an attempt to downplay their own errors, use the term to mean record-keeping mistakes that they later must correct, with some embarrassment, and that result in material changes in reported income. One should not use the term "shrinkage" for the correction of mistakes because adequate terminology exists for describing mistakes.

**shutdown cost.** Those fixed costs that the firm continues to incur after it has ceased production; the costs of closing down a particular production facility.

**sight draft.** A demand for payment drawn by Person A to whom Person B owes cash. Person A presents the *draft* to Person B's (the debtor's) bank in expectation that Person B will authorize his or her bank to disburse the funds. Sellers often use such drafts when selling goods to a new customer in a different city. The seller is uncertain whether the buyer will pay the bill. The seller sends the *bill* of lading, or other evidence of ownership of the goods, along with a sight draft to the buyer's bank. Before the warehouse holding the goods can release them to the buyer, the buyer must instruct its bank to honor the sight draft by withdrawing funds from the buyer's account. Once the bank honors the sight draft, it hands to the buyer the bill of lading or other document evidencing ownership, and the goods become the property of the buyer.

**significant influence.** A firm will use the *equity method* of accounting for its *investment* in another when it has significant influence over the *investee company. GAAP* do not define this term, but give examples of it: representation on the *board of directors,* participation in policy making processes, *material* intercompany transactions, interchange of managerial personnel, and technological dependency. Ultimately, the accountant must judge whether the investor's influence is significant.

**simple interest.** *Interest* calculated on *principal* where interest earned during periods before maturity of the loan does not increase the principal amount earning interest for the subsequent periods and the lender cannot withdraw the funds before maturity. Interest = principal × interest rate × time, where the rate is a rate per period (typically a year) and time is expressed in units of that period. For example, if the *rate* is annual and the time is two months, then in the formula, use 2/12 for *time.* Simple interest is seldom used in economic calculations except for periods of less than one year and then only for computational convenience. Contrast with *compound interest.*

**single-entry accounting.** Accounting that is neither *self-balancing* nor *articulated.* That is, it does not rely on equal *debits* and *credits.* The firm makes no *journal entries* and must *plug* to derive *owners' equity* for the *balance sheet.*

**single proprietorship.** *Sole proprietorship.*

**single-step.** Said of an *income statement* in which *ordinary revenue* and *gain* items appear first, with their total. Then come all ordinary *expenses* and *losses,* with their total. The difference between these two totals, plus the effect of *income from discontinued operations* and *extraordinary items,* appears as *net income.* Contrast with *multiple-step* and see *proprietorship theory.*

**sinking fund.** *Assets* and their earnings earmarked for the retirement of bonds or other long-term obligations. Earnings of sinking fund investments become taxable income of the company.

**sinking fund method of depreciation.** Method in which the periodic charge is an equal amount each period so that the *future value* of the charges, considered as an *annuity,* will accumulate at the end of the depreciable life to an amount equal to the *acquisition cost* of the asset. The firm does not necessarily, or even usually, accumulate a *fund* of cash. Firms rarely use this method.

**skeleton account.** *T-account.*

**slide.** The name of the error made by a bookkeeper in recording the digits of a number correctly with the decimal point misplaced; for example, recording $123.40 as $1,234.00 or as $12.34. If the only errors in a *trial balance* result from one or more slides, then the difference between the sum of the *debits* and the sum of the *credits* will be divisible by nine. Not all such differences divisible by nine result from slides. See *transposition error.*

**SMAC (Society of Management Accountants of Canada).** The national association of accountants whose provincial associations engage in industrial and governmental accounting. The association undertakes research and administers an educational program and comprehensive examinations; those who pass qualify to be designated CMA (Certified Management Accountants), formerly called RIA (Registered Industrial Accountant).

**SNC (société en nom collectif).** France: a *partnership.*

**soak-up method.** The *equity method.*

**Social Security taxes.** Taxes levied by the federal government on both employers and employees to provide *funds* to pay retired persons (or their survivors) who are entitled to receive such payments, either because they paid Social Security taxes themselves or because Congress has declared them eligible. Unlike a *pension plan,* the Social Security system does not collect funds and invest them for many years. The tax collections in a given year pay primarily for benefits distributed that year. At any given time the system has a multitrillion-dollar unfunded obligation to current workers for their eventual retirement benefits. See *Old Age, Survivors, Disability, and (Hospital) Insurance.*

**software.** The programming aids, such as compilers, sort and report programs, and generators, that extend the capabilities of and simplify the use of the computer, as well as certain operating systems and other control programs. Compare *hardware.*

**sole proprietorship.** A firm in which all *owners' equity* belongs to one person.

**solvent.** Able to meet debts when due.

**SOP.** *Statement of Position* (of the *AcSEC* of the *AICPA*).

**sound value.** A phrase used mainly in appraisals of *fixed assets* to mean *fair market price* (*value*) or *replacement cost* in present condition.

**source of funds.** Any *transaction* that increases *cash* and *marketable securities* held as *current assets.*

**sources and uses statement.** *Statement of cash flows.*

**SOYD.** *Sum-of-the years'-digits depreciation.*

**SP (société en participation).** France: a silent *partnership* in which the managing partner acts for the partnership as an individual in transacting with others who need not know that the person represents a partnership.

**special assessment.** A compulsory levy made by a governmental unit on property to pay the costs of a specific improvement or service presumed not to benefit the general public but only the owners of the property so assessed; accounted for in a special assessment fund.

**special journal.** A *journal,* such as a sales journal or cash disbursements journal, to record *transactions* of a similar nature that occur frequently.

**special purpose entity (vehicle). SPE (SPV).** A term unknown by laymen until the Enron shambles of

2001. A legal entity—corporation or partnership or trust—with substantial owners' equity provided by investors other than the firm organizing the SPE. The independent owners must have significant risks and rewards of ownership. A parent firm, such as Enron, establishes the SPE to transfer assets and debt off its balance sheet. Some of Enron's SPEs appear not to have satisfied all the GAAP tests to be kept off the balance sheet. Many SPEs have a legitimate purpose. Consider, for example, a firm that wishes to sell its accounts receivable. It sells the receivables to a corporation in which others have provided owners' equity funding. The corporation runs the risks (or will reap the rewards) of collecting less (or more) from the receivables than the organizers of the SPE forecast.

**special revenue debt.** A governmental unit's debt backed only by revenues from specific sources, such as tolls from a bridge.

**specific identification method.** Method for valuing *ending inventory* and *cost of goods sold* by identifying actual units sold and remaining in inventory and summing the actual costs of those individual units; usually used for items with large unit values, such as precious jewelry, automobiles, and fur coats.

**specific item sampling.** Sampling in which the analyst selects particular items because of their nature, value, or method of recording. Compare *representative item sampling.*

**specific price changes.** Changes in the market prices of specific *goods* and *services.* Contrast with *general price-level changes.*

**specific price index.** A measure of the price of a specific good or service, or a small group of similar goods or services, at one time relative to the price during a base period. Contrast with *general price index.* See *dollar-value LIFO method.*

**spending variance.** In *standard cost systems,* the *rate* or *price variance* for *overhead costs.*

**split.** *Stock split.* Sometimes called "split-up."

**split-off point.** In accumulating and allocating costs for *joint products,* the point at which all costs are no longer *joint costs* but at which an analyst can identify costs associated with individual products or perhaps with a smaller number of *joint products.*

**spoilage.** See *abnormal spoilage* and *normal spoilage.*

**spot price.** The price of a commodity for delivery on the day of the price quotation. See *forward price* for contrast.

**spreadsheet.** For many years, a term that referred specifically to a *work sheet* organized like a *matrix* that provides a two-way classification of accounting data. The rows and columns both have labels, which are *account* titles. An entry in a row represents a *debit,* whereas an entry in a column represents a *credit.* Thus, the number "100" in the "cash" row and the "accounts receivable" column records an entry debiting cash and crediting accounts receivable for $100. A given row total indicates all debit entries to the account represented by that row, and a given column total indicates the sum of all credit entries to the account represented by the column. Since personal-computer software has become widespread, this term has come to refer to any file created by programs such as Lotus 1-2-3® and Microsoft Excel®. Such files have rows and columns, but they need not represent debits and credits. Moreover, they can have more than two dimensions.

**squeeze.** A term sometimes used for *plug.*

***SSARS.*** *Statement on Standards for Accounting and Review Services.*

**stabilized accounting.** *Constant-dollar accounting.*

**stable monetary unit assumption.** In spite of *inflation,* which appears to be a way of life, the assumption that underlies historical cost/nominal-dollar accounting—namely that one can meaningfully add together current dollars and dollars of previous years. The assumption gives no specific recognition to changing values of the dollar in the usual *financial statements.* See *constant-dollar accounting.*

***Staff Accounting Bulletin.*** An interpretation issued by the staff of the Chief Accountant of the *SEC* "suggesting" how the accountants should apply various *Accounting Series Releases* in practice. The suggestions are part of *GAAP.*

**stakeholder.** An individual or group, such as employees, suppliers, customers, and shareholders, who have an interest in the corporation's activities and outcomes.

**standard cost.** Anticipated *cost* of producing a unit of output; a predetermined cost to be assigned to products produced. Standard cost implies a norm—what costs should be. Budgeted cost implies a forecast—something likely, but not necessarily, a "should," as implied by a norm. Firms use standard costs as the benchmark for gauging good and bad performance. Although a firm may similarly use a budget, it need not. A budget may be a planning document, subject to changes whenever plans change, whereas standard costs usually change annually or when technology significantly changes or when costs of labor and materials significantly change.

**standard costing.** *Costing* based on *standard costs.*

**standard costing system.** *Product costing* using *standard costs* rather than actual costs. The firm may use either *full absorption* or *variable costing* principles.

**standard error (of regression coefficients).** A measure of the uncertainty about the magnitude of the estimated parameters of an equation fit with a *regression analysis.*

**standard manufacturing overhead.** *Overhead costs* expected to be incurred per unit of time and per unit produced.

**standard price (rate).** Unit price established for materials or labor used in *standard cost systems.*

**standard quantity allowed.** The direct material or direct labor (inputs) quantity that production should have used if it had produced the units of output in accordance with preset *standards.*

**standby costs.** A type of *capacity cost,* such as property taxes, incurred even if a firm shuts down operations completely. Contrast with *enabling costs.*

**stated capital.** Amount of capital contributed by shareholders; sometimes used to mean *legal capital.*

**stated value.** A term sometimes used for the *face amount of capital stock,* when the *board* has not designated a *par value.* Where there is stated value per share, capital *contributed in excess of stated value* may come into being.

**statement of affairs.** A *balance sheet* showing immediate *liquidation* amounts rather than *historical costs,* usually prepared when *insolvency* or *bankruptcy* is imminent. Such a statement specifically does not use the *going-concern assumption.*

**statement of cash flows.** A schedule of *cash receipts* and *payments,* classified by *investing, financing,* and *operating activities;* required by the *FASB* for all for-profit companies. Companies may report operating activities with either the direct method (which shows only receipts and payments of cash) or the indirect method (which starts with *net income* and shows adjustments for *revenues* not currently producing cash and for *expenses* not currently using cash). "Cash" includes cash equivalents such as Treasury bills, commercial paper, and *marketable securities* held as *current assets.* This is sometimes called the "funds statement." Before 1987, the FASB required the presentation of a similar statement called the *statement of changes in financial position,* which tended to emphasize *working capital,* not cash.

**statement of changes in financial position.** As defined by *APB Opinion No. 19,* a statement that explains the changes in *working capital* (or cash) balances during a period and shows the changes in the working capital (or cash) accounts themselves. The *statement of cash flows* has replaced this statement.

**statement of charge and discharge.** A financial statement, showing *net assets* or *income,* drawn up by an executor or administrator, to account for receipts and dispositions of cash or other assets in an estate or trust.

***Statement of Financial Accounting Concepts* (*SFAC*).** One of a series of *FASB* publications in its *conceptual framework* for *financial accounting* and reporting. Such statements set forth objectives and fundamentals to be the basis for specific financial accounting and reporting standards.

***Statement of Financial Accounting Standards* (*SFAS*).** See *FASB.*

**statement of financial position.** *Balance sheet.*

***Statement of Position* (*SOP*).** A recommendation, on an emerging accounting problem, issued by the *AcSEC* of the *AICPA.* The AICPA's Code of Professional Ethics specifically states that *CPAs* need not treat *SOPs* as they do rules from the *FASB,* but a CPA would be wary of departing from the recommendations of an *SOP.*

**statement of retained earnings (income).** A statement that reconciles the beginning-of-period and the end-of-period balances in the *retained earnings* account. It shows the effects of *earnings, dividend declarations,* and *prior-period adjustments.*

**statement of significant accounting policies (principles).** A summary of the significant *accounting principles* used in compiling an *annual report;* required by *APB Opinion No. 22.* This summary may be a separate exhibit or the first *note* to the financial statements.

***Statement on Auditing Standards* (*SAS*).** A series addressing specific auditing standards and procedures. *No. 1* (1973) of this series codifies all statements on auditing standards previously promulgated by the *AICPA.*

***Statement on Standards for Accounting and Review Services* (*SSARS*).** Pronouncements issued by the *AICPA* on unaudited *financial statements* and unaudited financial information of nonpublic entities.

**static budget.** *Fixed budget.* Budget developed for a set level of the driving variable, such as production or sales, which the analyst does not change if the actual level deviates from the level set at the outset of the analysis.

**status quo.** Events or cost incurrences that will happen or that a firm expects to happen in the absence of taking some contemplated action.

**statutory tax rate.** The tax rate specified in the *income tax* law for each type of income (for example, *ordinary income, capital gain or loss*).

**step allocation method.** *Step-down method.*

**step cost.** *Semifixed cost.*

**step-down method.** In *allocating service department* costs, a method that starts by allocating one service department's costs to *production departments* and to all other service departments. Then the firm allocates a second service department's costs, including costs allocated from the first, to production departments and to all other service departments except the first one. In this fashion, a firm may allocate all service departments costs, including previous allocations, to production departments and to those service departments whose costs it has not yet allocated.

**step(ped) cost.** *Semifixed cost.*

**sterilized allocation.** Desirable characteristics of cost allocation methods. Optimal decisions result from considering *incremental costs* only. Optimal decisions never require *allocations* of *joint* or *common costs.* A "sterilized allocation" causes the optimal decision choice not to differ from the one that occurs when the accountant does not allocate joint or common costs "sterilized" with respect to that decision. Arthur L. Thomas first used the term in this context. Because *absorption costing* requires that product costs absorb all manufacturing costs and because some allocations can lead to bad decisions, Thomas (and we) advocate that the analyst choose a sterilized allocation scheme that will not alter the otherwise optimal decision. No single allocation scheme is always sterilized with respect to all decisions. Thus, Thomas (and we) advocate that decisions be made on the basis of incremental costs before any allocations.

**stewardship.** Principle by which management is accountable for an *entity's* resources, for their efficient use, and for protecting them from adverse impact. Some theorists believe that accounting has as a primary goal aiding users of *financial statements* in their assessment of management's performance in stewardship.

**stock.** A measure of the amount of something on hand at a specific time. In this sense, contrast with *flow.* See *inventory* and *capital stock.*

**stock appreciation rights.** An employer's promise to pay to the employee an amount of *cash* on a certain future date, with the amount of cash being the difference between the *market value* of a specified number of *shares* of *stock* in the employer's company on the given future date and some base price set on the date the rights are granted. Firms sometimes use this form of compensation because changes in tax laws in recent years have made *stock options* relatively less attractive. *GAAP* compute compensation based on the difference between the market value of the shares and the base price set at the time of the grant.

**stock dividend.** A so-called *dividend* in which the firm distributes additional *shares* of *capital stock* without cash payments to existing shareholders. It results in a *debit* to *retained earnings* in the amount of the market value of the shares issued and a *credit* to *capital stock* accounts. Firms ordinarily use stock dividends to indicate that they have permanently reinvested earnings in the business. Contrast with a *stock split,* which requires no entry in the capital stock accounts other than a notation that the *par* or *stated value* per share has changed.

**stock option.** The right to purchase or sell a specified number of shares of *stock* for a specified price at specified times. Employee stock options are purchase rights granted by a corporation to employees, a form of compensation. Traded stock options are *derivative* securities, rights created and traded by investors, independent of the corporation whose stock is optioned. Contrast with *warrant.*

**stock right.** See *right.*

**stock split(-up).** Increase in the number of common shares outstanding resulting from the issuance of additional shares to existing shareholders without additional capital contributions by them. Does not increase the total *value* (or *stated value*) of *common shares* outstanding because the *board* reduces the par (or stated) value per share in inverse proportion. A three-for-one stock split reduces par (or stated) value per share to one-third of its former amount. A stock split usually implies a distribution that increases the number of shares outstanding by 20 percent or more. Compare with *stock dividend.*

**stock subscriptions.** See *subscription* and *subscribed stock.*

**stock warrant.** See *warrant.*

**stockholders' equity.** See *shareholders' equity.*

**stockout.** Occurs when a firm needs a unit of *inventory* to use in production or to sell to a customer but has none available.

**stockout costs.** *Contribution margin* or other measure of *profits* not earned because a seller has run out of *inventory* and cannot fill a customer's order. A firm may incur an extra cost because of delay in filling an order.

**stores.** *Raw materials,* parts, and supplies.

**straight-debt value.** An estimate of the *market value* of a *convertible bond* if the bond did not contain a conversion privilege.

**straight-line depreciation.** Method in which, if the *depreciable life* is *n* periods, the periodic *depreciation* charge is $1/n$ of the *depreciable cost;* results in equal periodic charges. Accountants sometimes call it "straight-time depreciation."

**strategic plan.** A statement of the method for achieving an organization's goals.

**stratified sampling.** In choosing a *sample,* a method in which the investigator first divides the entire *population* into relatively homogeneous subgroups (strata) and then selects random samples from these subgroups.

**street security.** A stock certificate in immediately transferable form, most commonly because the issuing firm has registered it in the name of the broker, who has endorsed it with "payee" left blank.

**Subchapter S corporation.** A firm legally organized as a *corporation* but taxed as if it were a *partnership.* Tax terminology calls the corporations paying their own income taxes *C corporations.*

**subject to.** In an *auditor's report,* qualifications usually caused by a *material* uncertainty in the valuation of an item, such as future promised payments from a foreign government or outcome of pending litigation.

**subordinated.** *Debt* whose claim on income or assets has lower priority than claims of other debt.

**subscribed stock.** A *shareholders' equity* account showing the capital that the firm will receive as soon as the share-purchaser pays the subscription price. A subscription is a legal contract, so once the share-purchaser signs it, the firm makes an entry *debiting* an *owners' equity contra account* and *crediting* subscribed stock.

**subscription.** Agreement to buy a *security* or to purchase periodicals, such as magazines.

**subsequent events.** *Poststatement events.*

**subsidiary.** A company in which another company owns more than 50 percent of the voting shares.

**subsidiary ledger.** The *ledger* that contains the detailed accounts whose total appears in a *controlling account* of the *general ledger.*

**subsidiary (ledger) accounts.** The *accounts* in a *subsidiary ledger.*

**successful efforts costing.** In petroleum accounting, the *capitalization* of the drilling costs of only those wells that contain gas or oil. See *reserve recognition accounting* for an example.

**summary annual report (SAR).** Condensed financial statements distributed in lieu of the usual *annual report.* Since 1987, the *SEC* has allowed firms to include such statements in the annual report to shareholders as long as the firm includes full, detailed statements in SEC filings and in *proxy* materials sent to shareholders.

**summary of significant accounting principles.** *Statement of significant accounting policies (principles).*

**sum-of-the-years'-digits depreciation (SYD, SOYD).** An *accelerated depreciation* method for an asset with *depreciable life* of *n* years where the charge in period *i* ($i = 1, \ldots, n$) is the fraction $(n + 1 - i)/[n(n + 1)/2]$ of the *depreciable cost.* If an asset has a depreciable cost of $15,000 and a five-year depreciable life, for example, the depreciation charges would be $5,000 (= 5/15 × $15,000) in the first year, $4,000 in the second, $3,000 in the third, $2,000 in the fourth, and $1,000 in the fifth. The name derives from the fact that the denominator in the fraction is the sum of the digits 1 through *n*.

**sunk cost.** Past *costs* that current and future decisions cannot affect and, hence, that are irrelevant for decision making aside from *income tax* effects. Contrast with *incremental costs* and *imputed costs*. For example, the *acquisition cost* of machinery is irrelevant to a decision of whether to scrap the machinery. The current *exit value* of the machinery is the *opportunity cost* of continuing to own it, and the cost of, say, the electricity to run the machinery is an incremental cost of its operation. Sunk costs become relevant for decision making when the analysis requires taking *income taxes* (*gain* or *loss* on disposal of asset) into account, since the cash payment for income taxes depends on the tax basis of the asset. Avoid this term in careful writing because it is ambiguous. Consider, for example, a machine costing $100,000 with current *salvage* value of $20,000. Some (including us) would say that $100,000 (the *gross* amount) is "sunk"; others would say that only $80,000 (the *net* amount) is "sunk."

**supplementary statements (schedules).** Statements (schedules) in addition to the four basic *financial statements* (*balance sheet, income statement, statement of cash flows,* and the *statement of retained earnings*).

**surplus.** A word once used but now considered poor terminology; prefaced by "earned" to mean *retained earnings* and prefaced by "capital" to mean *capital contributed in excess of par* (or *stated*) *value.*

**surplus reserves.** *Appropriated retained earnings.* A phrase with nothing to recommend it: of all the words in accounting, *reserve* is the most objectionable, and *surplus* is the second-most objectionable.

**suspense account.** A *temporary account* used to record part of a transaction before final analysis of that transaction. For example, if a business regularly classifies all sales into a dozen or more different categories but wants to deposit the proceeds of cash sales every day, it may credit a sales suspense account pending detailed classification of all sales into Durable Goods Sales, Women's Clothing Sales, Men's Clothing Sales, Housewares Sales, and so on.

**sustainable income.** The part of *distributable income* (computed from *current cost* data) that the firm can expect to earn in the next accounting period if it continues operations at the same levels as were maintained during the current period. *Income from discontinued operations,* for example, may be distributable but not sustainable.

**swap.** A currency swap is a financial instrument in which the holder promises to pay to (or receive from) the *counterparty* the difference between *debt* denominated in one currency (such as U.S. dollars) and the payments on debt denominated in another currency (such as German marks). An interest-rate swap typically obligates the party and counterparty to exchange the difference between fixed- and floating-rate interest payments on otherwise similar loans.

**S-X.** See *Regulation S-X.*

**SYD.** *Sum-of-the-years'-digits depreciation.*

# T

**T-account.** Account form shaped like the letter T with the title above the horizontal line. *Debits* appear on the left of the vertical line, *credits* on the right.

**take-home pay.** The amount of a paycheck; earned wages or *salary* reduced by deductions for *income taxes, Social Security taxes,* contributions to fringe-benefit plans, union dues, and so on. Take-home pay might be as little as half of earned compensation.

**take-or-pay contract.** As defined by *SFAS No. 47,* a purchaser-seller agreement that provides for the purchaser to pay specified amounts periodically in return for products or services. The purchaser must make specified minimum payments even if it does not take delivery of the contracted products or services.

**taking a bath.** To incur a large loss. See *big bath.*

**tangible.** Having physical form. Accounting has never satisfactorily defined the distinction between tangible and intangible assets. Typically, accountants define intangibles by giving an exhaustive list, and everything not on the list is defined as tangible. See *intangible asset* for such a list.

**target cost.** *Standard cost.* Sometimes, target price less expected profit margin.

**target price.** Selling price based on customers' value in use of a good or service, constrained by competitors' prices of similar items.

**tax.** A nonpenal, but compulsory, charge levied by a government on income, consumption, wealth, or other basis, for the benefit of all those governed. The term does not include fines or specific charges for benefits accruing only to those paying the charges, such as licenses, permits, special assessments, admission fees, and tolls.

**tax allocation: interperiod.** See *deferred income tax liability.*

**tax allocation: intrastatement.** The showing of income tax effects on *extraordinary items, income from discontinued operations,* and *prior-period adjustments,* along with these items, separately from income taxes on other income. See *net-of-tax reporting.*

**tax avoidance.** See *tax shelter* and *loophole.*

**tax basis of assets and liabilities.** A concept important for applying *SFAS No. 109* on *deferred income taxes.* Two *assets* will generally have different *book values* if the firm paid different amounts for them, *amortizes* them on a different schedule, or both. Similarly a single asset will generally have a book value different from what it will have for tax purposes if the firm recorded different *acquisition* amounts for the asset for book and for tax purposes, amortizes it differently for book and for tax purposes, or both. The difference between financial book value and income tax basis becomes important in computing deferred income tax amounts. The adjusted cost in the financial records is the "book basis," and the adjusted amount in the tax records is the "tax basis." Differences between book and tax basis can arise for *liabilities* as well as for assets.

**tax credit.** A subtraction from taxes otherwise payable. Contrast with *tax deduction*.

**tax deduction.** A subtraction from *revenues* and *gains* to arrive at taxable income. Tax deductions differ technically from tax *exemptions,* but both reduce gross income in computing taxable income. Both differ from *tax credits,* which reduce the computed tax itself in computing taxes payable. If the tax rate is the fraction $t$ of pretax income, then a *tax credit* of \$1 is worth \$1/$t$ of *tax deductions*. Deductions appear on tax returns, *expenses* appear on *income statements*.

**tax evasion.** The fraudulent understatement of taxable revenues or overstatement of deductions and expenses or both. Contrast with *tax shelter* and *loophole*.

**tax-exempts.** See *municipal bonds*.

**tax shelter.** The legal avoidance of, or reduction in, *income taxes* resulting from a careful reading of the complex income-tax regulations and the subsequent rearrangement of financial affairs to take advantage of the regulations. Often writers use the term pejoratively, but the courts have long held that a taxpayer has no obligation to pay taxes any larger than the legal minimum. If the public concludes that a given tax shelter is "unfair," then Congress can, and has, changed the laws and regulations. The term is sometimes used to refer to the investment that permits tax avoidance. See *loophole*.

**tax shield.** The amount of an *expense,* such as *depreciation,* that reduces taxable income but does not require *working capital*. Sometimes this term includes expenses that reduce taxable income and use working capital. A depreciation deduction (or *R&D expense* in the expanded sense) of \$10,000 provides a tax shield of \$3,700 when the marginal tax rate is 37 percent.

**taxable income.** *Income* computed according to *IRS* regulations and subject to *income taxes*. Contrast with income, net income, income before taxes (in the *income statement*), and *comprehensive income* (a *financial reporting* concept). Use the term "pretax income" to refer to income before taxes on the income statement in financial reports.

**tax-transfer lease.** One form of *capital lease*. Congress has in the past provided business with an incentive to invest in qualifying *plant and equipment* by granting an *investment credit,* which, though it occurs as a reduction in *income taxes* otherwise payable, effectively reduces the purchase price of the assets. Similarly, Congress continues to grant an incentive to acquire such assets by allowing the *Modified Accelerated Cost Recovery System* (*MACRS,* form of unusually *accelerated depreciation*). Accelerated depreciation for tax purposes allows a reduction of taxes paid in the early years of an asset's life, providing the firm with an increased *net present value* of *cash flows*. The *IRS* administers both of these incentives through the income tax laws, rather than paying an outright cash payment. A business with no taxable income in many cases had difficulty reaping the benefits of the investment credit or of accelerated depreciation because Congress had not provided for tax refunds to those who acquire qualifying assets but who have no taxable income. In principle, a company without taxable income could lease from another firm with taxable income an asset that it would otherwise purchase. The second firm acquires the asset, gets the tax-reduction benefits from the acquisition, and becomes a lessor, leasing the asset (presumably at a lower price reflecting its own costs lowered by the tax reductions) to the unprofitable company. Before 1981, tax laws discouraged such leases. That is, although firms could enter into such leases, they could not legally transfer the tax benefits. Under certain restrictive conditions, the tax law now allows a profitable firm to earn tax credits and take deductions while leasing to the firm without tax liability in such leases. These are sometimes called "safe-harbor leases."

***Technical Bulletin.*** The *FASB* has authorized its staff to issue bulletins to provide guidance on financial accounting and reporting problems. Although the FASB does not formally approve the contents of the bulletins, their contents are part of *GAAP.*

**technology.** The sum of a firm's technical *trade secrets* and *know-how,* as distinct from its *patents*.

**temporary account.** *Account* that does not appear on the *balance sheet; revenue* and *expense* accounts, their *adjuncts* and *contras, production cost accounts, dividend distribution accounts,* and purchases-related accounts (which close to the various inventories); sometimes called a "nominal account."

**temporary difference.** According to the *SFAS No. 109* (1992) definition: "A difference between the tax basis of an asset or liability and its reported amount in the financial statements that will result in taxable or deductible amounts in future years." Temporary differences include *timing differences* and differences between *taxable income* and pretax income caused by different cost bases for assets. For example, a plant asset might have a cost of \$10,000 for financial reporting but a basis of \$7,000 for income tax purposes. This temporary difference might arise because the firm has used an accelerated depreciation method for tax but straight-line for book, or the firm may have purchased the asset in a transaction in which the fair value of the asset exceeded its tax basis. Both situations create a temporary difference.

**temporary investments.** Investments in *marketable securities* that the owner intends to sell within a short time, usually one year, and hence classifies as *current assets*.

**10-K.** The name of the annual report that the *SEC* requires of nearly all publicly held corporations.

**term bonds.** A *bond issue* whose component bonds all mature at the same time. Contrast with *serial bonds*.

**term loan.** A loan with a *maturity* date, as opposed to a demand loan, which is due whenever the lender requests payment. In practice, bankers and auditors use this phrase only for loans for a year or more.

**term structure.** A phrase with different meanings in *accounting* and *financial economics*. In accounting, it refers to the pattern of times that must elapse before *assets* turn into, or produce, *cash* and the pattern of times that must elapse before *liabilities* require cash. In financial economics, the phrase refers to the pattern of interest rates as a function of the time that elapses for loans to come due. For example, if six-month loans cost 6 percent per year and 10-year loans cost 9 percent per year, this is called a "normal" term structure because the longer-term loan carries a higher rate. If the six-month loan costs 9 percent per year and the 10-year loan costs 6 percent per year, the term structure is said to be "inverted." See *yield curve*.

**terms of sale.** The conditions governing payment for a sale. For example, the terms *2/10, n(et)/30* mean that if the purchaser makes payment within 10 days of the invoice date, it can take a *discount* of 2 percent from *invoice* price; the purchaser must pay the invoice amount, in any event, within 30 days, or it becomes overdue.

**theory of constraints (TOC).** Concept of improving operations by identifying and reducing bottlenecks in process flows.

**thin capitalization.** A state of having a high *debt-equity ratio*. Under income tax legislation, the term has a special meaning.

**throughput contract.** As defined by *SFAS No. 47*, an agreement that is signed by a shipper (processor) and by the owner of a transportation facility (such as an oil or natural gas pipeline or a ship) or a manufacturing facility and that provides for the shipper (processor) to pay specified amounts periodically in return for the transportation (processing) of a product. The shipper (processor) must make cash payments even if it does not ship (process) the contracted quantities.

**throughput contribution.** Sales dollars minus the sum of all short-run variable costs.

**tickler file.** A collection of *vouchers* or other memoranda arranged chronologically to remind the person in charge of certain duties to make payments (or to do other tasks) as scheduled.

**time-adjusted rate of return.** *Internal rate of return*.

**time cost.** *Period cost*.

**time deposit.** Cash in bank earning interest. Contrast with *demand deposit*.

**time value element (of stock options).** All else equal, the longer one has to exercise an *option*, the more valuable that option. This term refers to the part of an option's value on a date caused by the option's not expiring until sometime after that date. All else equal, the more volatile the *market price* of the *security* receivable on exercise, the more valuable is this element.

**time-series analysis.** See *cross-section analysis* for definition and contrast.

**times-interest (charges) earned.** Ratio of pretax *income* plus *interest* charges to interest charges. See *ratio*.

**timing difference.** The major type of *temporary difference* between taxable income and pretax income reported to shareholders; reverses in a subsequent period and requires an entry in the *deferred income tax* account; for example, the use of *accelerated depreciation* for tax returns and *straight-line depreciation* for financial reporting. Contrast with *permanent difference*.

**Toronto Stock Exchange (TSE).** A public market where various corporate securities trade.

**total assets turnover.** *Sales* divided by average total *assets*.

**total quality management (TQM).** Concept of organizing a company to excel in all its activities in order to increase the quality of products and services.

**traceable cost.** A *cost* that a firm can identify with or assign to a specific product. Contrast with a *joint cost*.

**trade acceptance.** A *draft* that a seller presents for signature (acceptance) to the buyer at the time it sells goods. The draft then becomes the equivalent of a *note receivable* of the seller and a *note payable* of the buyer.

**trade credit.** Occurs when one business allows another to buy from it in return for a promise to pay later. Contrast with "consumer credit," which occurs when a business extends a retail customer the privilege of paying later.

**trade discount.** A *list price discount* offered to all customers of a given type. Contrast with a *discount* offered for prompt payment and with *quantity discount*.

**trade-in.** Acquiring a new *asset* in exchange for a used one and perhaps additional cash. See *boot* and *trade-in transaction*.

**trade-in transaction.** The accounting for a trade-in; depends on whether the firm receives an asset "similar" to (and used in the same line of business as) the asset traded in and whether the accounting is for *financial statements* or for *income tax* returns. Assume that an old asset cost $5,000, has $3,000 of *accumulated depreciation* (after recording depreciation to the date of the trade-in), and hence has a *book value* of $2,000. The old asset appears to have a market value of $1,500, according to price quotations in used asset markets. The firm trades in the old asset on a new asset with a list price of $10,000. The firm gives up the old asset and $5,500 cash (*boot*) for the new asset. The generic entry for the trade-in transaction is as follows:

| | | | |
|---|---|---|---|
| New Asset .................... | A | | |
| Accumulated Depreciation (Old Asset) .................. | 3,000 | | |
| Adjustment on Exchange of Asset .................... | B | or | B |
| Old Asset .................. | | | 5,000 |
| Cash ...................... | | | 5,500 |

(1) The *list price* method of accounting for trade-ins rests on the assumption that the list price of the new asset closely approximates its market value. The firm records the new asset at its list price (A = $10,000 in the example); B is a *plug* (= $2,500 credit in the example). If B requires a *debit* plug, the Adjustment on Exchange of Asset is a *loss;* if B requires a *credit* plug (as in the example), the adjustment is a *gain.*

(2) Another theoretically sound method of accounting for trade-ins rests on the assumption that the price quotation from used-asset markets gives a market value of the old asset that is a more reliable measure than the market value of the new asset determined by list price. This method uses the *fair market price* (*value*) of the old asset, $1,500 in the example, to determine B (= $2,000 book value − $1,500 assumed proceeds on disposition = $500 debit or loss). The exchange results in a loss if the book value of the old asset exceeds its market value and in a gain if the market value exceeds the book value. The firm records the new asset on the books by plugging for A (= $7,000 in the example).

(3) For income tax reporting, the taxpayer must recognize neither gain nor loss on the trade-in. Thus the taxpayer records the new asset for tax purposes by assuming B is zero and plugging for A (= $7,500 in the example). In practice, firms that want to recognize the loss currently will sell the old asset directly, rather than trading it in, and acquire the new asset entirely for cash.

(4) *Generally accepted accounting principles* (*APB Opinion No. 29*) require a variant of these methods. The basic method is (1) or (2), depending on whether the list price of the new asset (1) or the quotation of the old asset's market value (2) provides the more reliable indication of market value. If the basic method requires a debit entry, or loss, for the Adjustment on Exchange of Asset, then the firm records the trade-in as in (1) or (2) and recognizes the full amount of the loss currently. If, however, the basic method requires a credit entry, or gain, for the Adjustment on Exchange of Asset, then the firm recognizes the gain currently if the old asset and the new asset are not "similar." If the assets are similar and the party trading in receives no cash, then it recognizes no gain and the treatment resembles that in (3); that is B = 0, plug for A. If the assets are similar and the firm trading in receives cash—a rare case—then it recognizes a portion of the gain currently. The portion of the gain recognized currently is the fraction cash received/fair market value of total consideration received. (When the firm uses the list price method, (1), it assumes that the market value of the old asset is the list price of the new asset plus the amount of cash received by the party trading in.)

A summary of the results of applying *GAAP* to the example follows.

| More Reliable Information as to Fair Market Value | Old Asset Compared with New Asset: Similar | Not Similar |
|---|---|---|
| New Asset List Price ....... | A = $7,500<br>B = 0 | A = $10,000<br>B = 2,500 gain |
| Old Asset Market Price ..... | A = $7,000<br>B = 500 loss | A = $ 7,000<br>B = 500 loss |

**trade payables (receivables).** *Payables* (*receivables*) arising in the ordinary course of business transactions. Most *accounts payable* (*receivable*) are of this kind.

**trade secret.** Technical or business information such as formulas, recipes, computer programs, and marketing data not generally known by competitors and maintained by the firm as a secret; theoretically capable of having an indefinite, finite life. A famous example is the secret process for Coca-Cola® (a registered *trademark* of the company). Compare with *know-how*. The firm will capitalize this intangible asset only if purchased and then will amortize it over a period not to exceed 40 years. If the firm develops the intangible internally, the firm will *expense* the costs as incurred and show no asset.

**trademark.** A distinctive word or symbol that is affixed to a product, its package, or its dispenser and that uniquely identifies the firm's products and services. See *trademark right*.

**trademark right.** The right to exclude competitors in sales or advertising from using words or symbols that are so similar to the firm's *trademarks* as possibly to confuse consumers. Trademark rights last as long as the firm continues to use the trademarks in question. In the United States, trademark rights arise from use and not from government registration. They therefore have a legal life independent of the life of a registration. Registrations last 20 years, and the holder may renew them as long as the holder uses the trademark. Although a trademark right might have an indefinite life, *GAAP* require amortization over some estimate of its life, not to exceed 40 years. Under *SFAS No. 2,* the firm must *expense* internally developed trademark rights.

**trading on the equity.** Said of a firm engaging in *debt financing;* frequently said of a firm doing so to a degree considered abnormal for a firm of its kind. *Leverage*.

**trading securities.** *Marketable securities* that a firm holds and expects to sell within a relatively short time; a classification important in *SFAS No. 115,* which requires the owner to carry marketable equity securities on the balance sheet at market value, not at cost. Contrast with *available for sale, securities* and *held-to-maturity securities*. Under *SFAS No. 115,* the balance sheet reports trading securities at market value on the balance sheet date, and the income statement reports *holding gains and losses* on trading securities. When the firm sells the securities, it reports realized gain or loss as the difference between the selling price and the market value at the last balance sheet date.

**transaction.** A *transfer* (of more than promises—see *executory contract*) between the accounting *entity* and another party or parties.

**transfer.** Under *SFAC No. 6,* consists of two types: "reciprocal" and "nonreciprocal." In a reciprocal transfer, or "exchange," the entity both receives and sacrifices. In a nonreciprocal transfer, the entity sacrifices but does not receive (examples include gifts, distributions to owners) or receives but does not sacrifice (investment by owner in entity). *SFAC No. 6* suggests that the term "internal transfer" is self-contradictory and that writers should use the term "internal event" instead.

**transfer agent.** Usually a bank or trust company designated by a corporation to make legal transfers of *stock* (*bonds*) and, perhaps, to pay *dividends* (*coupons*).

**transfer price.** A substitute for a *market,* or *arm's-length, price* used in *profit,* or *responsibility center, accounting* when one segment of the business "sells" to another segment. Incentives of profit center managers will not coincide with the best interests of the entire business unless a firm sets transfer prices properly.

**transfer-pricing problem.** The problem of setting *transfer prices* so that both buyer and seller have *goal congruence* with respect to the parent organization's goals.

**translation adjustment.** The effect of *exchange-rate* changes caused by converting the value of a net investment denominated in a *foreign currency* to the entity's reporting currency. *SFAS No. 52* requires firms to translate their net investment in relatively self-contained foreign operations at the *balance sheet* date. Year-to-year changes in value caused by exchange-rate changes accumulate in an *owners' equity* account, sometimes called the "cumulative translation adjustment."

**translation gain (or loss).** *Foreign exchange gain* (*or loss*).

**transportation-in.** *Freight-in.*

**transposition error.** An error in record keeping resulting from reversing the order of digits in a number, such as recording "32" for "23." If the only errors in a *trial balance* result from one or more transposition errors, then the difference between the sum of the *debits* and the sum of the *credits* will be divisible by nine. Not all such differences result from transposition errors. See *slide.*

**treasurer.** The financial officer responsible for managing cash and raising funds.

**treasury bond.** A bond issued by a corporation and then reacquired. Such bonds are treated as retired when reacquired, and an *extraordinary gain or loss* on reacquisition is recognized. This term also refers to a *bond* issued by the U.S. Treasury Department.

**treasury shares.** *Capital stock* issued and then reacquired by the corporation. Such reacquisitions result in a reduction of *shareholders' equity* and usually appear on the balance sheet as contra to shareholders' equity. Accounting recognizes neither *gain* nor *loss* on transactions involving treasury stock. The accounting debits (if positive) or credits (if negative) any difference between the amounts paid and received for treasury stock transactions to *additional paid-in capital.* See *cost method* and *par value method.*

**treasury stock.** *Treasury shares.*

**trial balance.** A two-column listing of *account balances.* The left-hand column shows all accounts with *debit* balances and their total. The right-hand column shows all accounts with *credit* balances and their total. The two totals should be equal. Accountants compute trial balances as a partial check of the arithmetic accuracy of the entries previously made. See *adjusted, preclosing, post-closing, unadjusted trial balance, plug, slide,* and *transposition error.*

**troubled debt restructuring.** As defined in *SFAS No. 15,* a concession (changing of the terms of a *debt*) that is granted by a *creditor* for economic or legal reasons related to the *debtor's* financial difficulty and that the creditor would not otherwise consider.

**TSE.** *Toronto Stock Exchange.*

**t-statistic.** For an estimated *regression* coefficient, the estimated coefficient divided by the *standard error* of the estimate.

**turnover.** The number of times that *assets,* such as *inventory* or *accounts receivable,* are replaced on average during the period. Accounts receivable turnover, for example, is total sales on account for a period divided by the average accounts receivable balance for the period. See *ratio.* In the UK, "turnover" means *sales.*

**turnover of plant and equipment.** See *ratio.*

**t-value.** In *regression analysis,* the ratio of an estimated regression coefficient divided by its *standard error.*

**20-F.** Form required by the *SEC* for foreign companies issuing or trading their securities in the United States. This form reconciles the foreign accounting amounts resulting from using foreign *GAAP* to amounts resulting from using U.S. GAAP.

**two T-account method.** A method for computing either (1) *foreign-exchange gains and losses* or (2) *monetary gains* or *losses* for *constant-dollar accounting statements.* The left-hand *T-account* shows actual net balances of *monetary items,* and the right-hand T-account shows implied (common) dollar amounts.

**2/10, n(et)/30.** See *terms of sale.*

# U

**unadjusted trial balance.** *Trial balance* taken before the accountant makes *adjusting* and *closing entries* at the end of the period.

**unappropriated retained earnings.** *Retained earnings* not appropriated and therefore against which the *board* can declare *dividends* in the absence of retained earnings restrictions. See *restricted retained earnings*.

**unavoidable cost.** A *cost* that is not an *avoidable cost*.

**uncertainty.** See *risk* for definition and contrast.

**uncollectible account.** An *account receivable* that the *debtor* will not pay. If the firm uses the preferable *allowance method,* the entry on judging a specific account to be uncollectible *debits* the allowance for uncollectible accounts and *credits* the specific account receivable. See *bad debt expense* and *sales contra, estimated uncollectibles*.

**unconsolidated subsidiary.** A *subsidiary* not consolidated and, hence, not accounted for in the *equity method*.

**uncontrollable cost.** The opposite of *controllable cost*.

**underapplied (underabsorbed) overhead.** An excess of actual *overhead costs* for a period over costs applied, or charged, to products produced during the period; a *debit balance* remaining in an overhead account after the accounting assigns overhead to product.

**underlying document.** The record, memorandum, *voucher,* or other signal that is the authority for making an *entry* into a *journal*.

**underwriter.** One who agrees to purchase an entire *security issue* for a specified price, usually for resale to others.

**undistributed earnings.** *Retained earnings*. Typically, this term refers to that amount retained for a given year.

**unearned income (revenue).** *Advances from customers;* strictly speaking, a contradiction in terms because the terms "income" and "revenue" mean earned.

**unemployment tax.** See *FUTA*.

**unencumbered appropriation.** In governmental accounting, portion of an *appropriation* not yet spent or encumbered.

**unexpired cost.** An *asset*.

**unfavorable variance.** In *standard cost* accounting, an excess of expected revenue over actual revenue or an excess of actual cost over standard cost.

**unfunded.** Not *funded*. An obligation or *liability,* usually for *pension costs,* exists, but no *funds* have been set aside to discharge the obligation or liability.

**Uniform Partnership Act.** A model law, enacted by many states, to govern the relations between partners when the *partnership* agreement fails to specify the agreed-upon treatment.

**unissued capital stock.** *Stock* authorized but not yet issued.

**uniting-of-interests method.** The IASB's term for the *pooling-of-interests method.* The IASB allows uniting of interests only when the merging firms are roughly equal in size and the shareholders retain substantially the same, relative to each other, voting rights and interests in the combined entity after the combination as before.

**units-of-production method.** The *production method of depreciation*.

**unlimited liability.** The legal obligation of *general partners* or the sole proprietor for all debts of the *partnership* or *sole proprietorship*.

**unqualified opinion.** See *auditor's report*.

**unrealized appreciation.** An *unrealized holding gain;* frequently used in the context of *marketable securities*.

**unrealized gain (loss) on marketable securities.** An *income statement account* title for the amount of *gain (loss)* during the current period on the portfolio of *marketable securities* held as *trading securities*. *SFAS No. 115* requires the firm to recognize, in the income statement, gains and losses caused by changes in market values, even though the firm has not yet *realized* them.

**unrealized gross margin (profit).** A *contra account* to *installment accounts receivable* used with the *installment method* of revenue recognition; shows the amount of profit that the firm will eventually realize when it collects the receivable. Some accountants show this account as a *liability*.

**unrealized holding gain.** See *inventory profit* for the definition and an example.

**unrecovered cost.** *Book value* of an *asset*.

**unused capacity.** The difference between resources supplied and resources used.

**usage variance.** *Efficiency variance*.

**use of funds.** Any transaction that reduces funds (however "funds" is defined).

**useful life.** *Service life*.

# V

**valuation account.** A *contra account* or *adjunct account*. When the firm reports *accounts receivable* at expected collectible amounts, it will credit any expected uncollectible amounts to the *allowance for uncollectibles,* a valuation account. In this way, the firm can show both the gross receivables amount and the amount it expects to collect. *SFAC No. 6* says a valuation account is "a separate item that reduces and increases the carrying amount" of an asset (or liability). The accounts are part of the related assets (or liabilities) and are not assets (or liabilities) in their own right.

**value.** Monetary worth. This term is usually so vague that you should not use it without a modifying adjective unless most people would agree on the amount. Do not confuse with cost. See *fair market price (value), entry value,* and *exit value*.

**value added.** *Cost* of a product or *work-in-process* minus the cost of the material purchased for the product or work-in-process.

**value-added activity.** Any activity that increases the usefulness to a customer of a product or service.

**value chain.** The set of business functions that increase the usefulness to the customer of a product or service; typically including research and development, design of products and services, production, marketing, distribution, and customer service.

**value engineering.** An evaluation of the activities in the value chain to reduce costs.

**value variance.** *Price variance*.

**variable annuity.** An *annuity* whose periodic payments depend on some uncertain outcome, such as stock market prices.

**variable budget.** *Flexible budget.*

**variable costing.** In allocating costs, a method that assigns only *variable manufacturing costs* to products and treats *fixed manufacturing costs* as *period expenses.* Contrast with *full absorption costing.*

**variable costs.** *Costs* that change as activity levels change. Strictly speaking, variable costs are zero when the activity level is zero. See *semivariable costs.* In accounting, this term most often means the sum of *direct costs* and variable *overhead.*

**variable overhead variance.** Difference between actual and *standard variable overhead costs.*

**variable rate debt.** *Debt* whose interest rate results from the periodic application of a formula, such as "three-month LIBOR [London Interbank Offered Rate] plus 1 percent [one hundred basis points] set on the 8th day of each February, May, August, and November."

**variables sampling.** The use of a sampling technique in which the sampler infers a particular quantitative characteristic of an entire population from a sample (e.g., mean amount of accounts receivable). See also *estimation sampling.* See *attribute(s) sampling* for contrast and further examples.

**variance.** Difference between actual and *standard costs* or between *budgeted* and actual *expenditures* or, sometimes, *expenses.* The word has completely different meanings in accounting and in statistics, where it means a measure of dispersion of a distribution.

**variance analysis.** *Variance investigation.* This term's meaning differs in statistics.

**variance investigation.** A step in managerial control processes. *Standard costing systems* produce *variance* numbers of various sorts. These numbers seldom exactly equal to zero. Management must decide when a variance differs sufficiently from zero to study its cause. This term refers both to the decision about when to study the cause and to the study itself.

**variation analysis.** Analysis of the causes of changes in financial statement items of interest such as *net income* or *gross margin.*

**VAT (Value-added tax).** A tax levied on the market value of a firm's outputs less the market value of its purchased inputs.

**vendor.** A seller; sometimes spelled "vender."

**verifiable.** A qualitative *objective* of financial reporting specifying that accountants can trace items in *financial statements* back to *underlying documents*—supporting *invoices,* canceled *checks,* and other physical pieces of evidence.

**verification.** The auditor's act of reviewing or checking items in *financial statements* by tracing back to *underlying documents*—supporting *invoices,* canceled *checks,* and other business documents—or sending out *confirmations* to be returned. Compare with *physical verification.*

**vertical analysis.** Analysis of the financial statements of a single firm or across several firms for a particular time, as opposed to *horizontal* or *time-series analysis,* in which the analyst compares items over time for a single firm or across firms.

**vertical integration.** The extension of activity by an organization into business directly related to the production or distribution of the organization's end products. Although a firm may sell products to others at various stages, a vertically integrated firm devotes the substantial portion of the output at each stage to the production of the next stage or to end products. Compare *horizontal integration.*

**vested.** An employee's *pension plan* benefits that are not contingent on the employee's continuing to work for the employer.

**visual curve fitting method.** One crude form of cost *estimation.* Sometimes, when a firm needs only rough approximations of the amounts of *fixed* and *variable costs,* management need not perform a formal *regression analysis* but can plot the data and draw a line that seems to fit the data. Then it can use the parameters of that line for the rough approximations.

**volume variance.** *Production volume variance;* less often, used to mean *sales volume variance.*

**voucher.** A document that signals recognition of a *liability* and authorizes the disbursement of cash; sometimes used to refer to the written evidence documenting an *accounting entry,* as in the term *journal voucher.*

**voucher system.** In controlling *cash,* a method that requires someone in the firm to authorize each *check* with an approved *voucher.* The firm makes no *disbursements* of currency or coins except from *petty cash funds.*

**vouching.** The function performed by an *auditor* to ascertain that underlying data or documents support a *journal entry.*

# W

**wage.** Compensation of employees based on time worked or output of product for manual labor. But see *take-home pay.*

**warning signal.** Tool used to identify quality-control problems; only signals a problem. Contrast with *diagnostic signal,* which both signals a problem and suggests its cause.

**warrant.** A certificate entitling the owner to buy a specified number of shares at a specified time(s) for a specified price; differs from a *stock option* only in that the firm grants options to employees and issues warrants to the public. See *right.*

**warranty.** A promise by a seller to correct deficiencies in products sold. When the seller gives warranties, proper accounting practice recognizes an estimate of warranty *expense* and an *estimated liability* at the time of sale. See *guarantee* for contrast in proper usage.

**wash sale.** The sale and purchase of the same or similar *asset* within a short time period. For *income tax* purposes, the taxpayer may not recognize *losses* on a sale of stock if the taxpayer purchases equivalent stock within 30 days before or after the date of sale.

**waste.** Material that is a residue from manufacturing operations and that has no sale value. Frequently, this has negative value because a firm must incur additional costs for disposal.

**wasting asset.** A *natural resource* that has a limited *useful life* and, hence, is subject to *amortization,* called *depletion.* Examples are timberland, oil and gas wells, and ore deposits.

**watered stock.** Shares issued for *assets* with *fair market price* (*value*) less than *par* or *stated value.* The firm records the assets on the books at the overstated values. In the law, for shares to be considered watered, the *board of directors* must have acted in bad faith or fraudulently in issuing the shares under these circumstances. The term originated from a former practice of cattle owners who fed cattle ("stock") large quantities of salt to make them thirsty. The cattle then drank much water before their owner took them to market. The owners did this to make the cattle appear heavier and more valuable than otherwise.

**weighted average.** An average computed by counting each occurrence of each value, not merely a single occurrence of each value. For example, if a firm purchases one unit for $1 and two units for $2 each, then the simple average of the purchase prices is $1.50, but the weighted average price per unit is $5/3 = $1.67. Contrast with *moving average.*

**weighted-average inventory method.** Valuing either *withdrawals* or *ending inventory* at the *weighted-average* purchase price of all units on hand at the time of withdrawal or of computation of ending inventory. The firm uses the *inventory equation* to calculate the other quantity. If a firm uses the *perpetual inventory* method, accountants often call it the *moving average method.*

**where-got, where-gone statement.** A term allegedly used in the 1920s by W. M. Cole for a statement much like the *statement of cash flows.* Noted accounting historian S. Zeff reports that Cole actually used the term "where-got-gone" statement.

**wind up.** To bring to an end, such as the life of a corporation. The *board* winds up the life of a corporation by following the winding-up provisions of applicable statutes, by surrendering the charter, or by following *bankruptcy* proceedings. See also *liquidation.*

**window dressing.** The attempt to make financial statements show *operating* results, or a *financial position,* more favorable than they would otherwise show.

**with recourse.** See *note receivable discounted.*

**withdrawals.** *Assets* distributed to an owner. *Partner's drawings.* See *inventory equation* for another context.

**withholding.** Deductions that are taken from *salaries* or *wages,* usually for *income taxes,* and that the employer remits, in the employee's name, to the taxing authority.

**without recourse.** See *note receivable discounted.*

**work sheet (program).** (1) A computer program designed to combine explanations and calculations. This type of program helps in preparing *financial statements* and *schedules.* (2) A tabular schedule for convenient summary of *adjusting* and *closing entries.* The work sheet usually begins with an *unadjusted trial balance.* Adjusting entries appear in the next two columns, one for *debits* and one for *credits.* The work sheet carries the horizontal sum of each line to the right into either the *income statement* or the *balance sheet* column, as appropriate. The *plug* to equate the income statement column totals is, if a debit, the income or, if a credit, a loss for the period. That income will close retained earnings on the balance sheet. The income statement credit columns are the revenues for the period, and the debit columns are the expenses (and revenue contras) that appear on the income statement. "Work sheet" also refers to *schedules* for ascertaining other items that appear on the *financial statements* and that require adjustment or compilation.

**working capital.** *Current assets* minus *current liabilities;* sometimes called "net working capital" or "net current assets."

**work(ing) papers.** The schedules and analyses prepared by the *auditor* in carrying out investigations before issuing an *opinion* on *financial statements.*

**work-in-process (inventory account).** Partially completed product; appears on the balance sheet as *inventory.*

**worth.** *Value.* See *net worth.*

**worth-debt ratio.** Reciprocal of the *debt-equity ratio.* See *ratio.*

**write down.** To *write off,* except that the firm does not charge all the *asset's* cost to *expense* or *loss;* generally used for nonrecurring items.

**write off.** To *charge* an *asset* to *expense* or *loss;* that is, to *debit* expense (or loss) and *credit* the asset.

**write-off method.** For treating *uncollectible accounts,* a method that *debits bad debt expense* and *credits* accounts receivable of specific customers as the firm identifies specific accounts as uncollectible. The firm cannot use this method when it can estimate uncollectible amounts and they are significant. See *bad debt expense, sales contra, estimated uncollectibles,* and the *allowance method* for contrast.

**write up.** To increase the recorded *cost* of an *asset* with no corresponding *disbursement* of *funds;* that is, to *debit* asset and *credit revenue* or, perhaps, *owners' equity;* seldom done in the United States because currently accepted accounting principles await actual transactions before recording asset increases. An exception occurs in accounting for *marketable equity securities.*

**XBRL. eXtensible Business Reporting Language.** A language created by over thirty partners, including the *AICPA,* to promote automated processing of business information by software on a computer. The main idea is that financial data get coded lables, called "tags," not locations in a financial statement, so that the user can access only the data needed for a particular use, without downloading and extracting the needed data from a *balance sheet.* For example, if you download a company's *annual report,* you can go to the balance sheet and copy out the amounts for *current assets* and *current liabilities,* then, divide the first by the second to get the *current ratio.* Using XBRL, you'd write an arithmetic expression such as:

> tag for current assets[particular company]/tag for current liabilities[same company]

The XBRL would extract from the data just the two numbers, corresponding to the two tags you programmed.

The initiative to construct XBRL began around 2000; Microsoft was the first to prepare, in 2002, financial statement data available for public use in XBRL; and the *SEC,* in 2005, has allowed companies to file financial data at the SEC using XBRL. In 2005, the *PCAOB* issued guidelines for the audit of such XBRL filings at the SEC. We guess that use of XBRL will mushroom over the next decade. It serves users' needs in a way no other widely available system does.

# Y

**yield.** *Internal rate of return* of a stream of cash flows. Cash yield is cash flow divided by book value. See also *dividend yield.*

**yield curve.** The relation between *interest rates* and the term to maturity of loans. Ordinarily, longer-term loans have higher interest rates than shorter-term loans. This is called a "normal" yield curve. Sometimes long-term and short-term rates are approximately the same—a "flat" yield curve. Sometimes short-term loans have a higher rate than long-term ones—an "inverted" yield curve. *Term structure* of interest rates.

**yield to maturity.** At a given time, the *internal rate of return* of a series of cash flows; usually said of a *bond;* sometimes called the "effective rate."

**yield variance.** Measures the input-output relation while holding the standard mix of inputs constant: (standard price multiplied by actual amount of input used in the standard mix)—(standard price multiplied by standard quantity allowed for the actual output). It is the part of the *efficiency variance* not called the *mix variance.*

# Z

**zero-base(d) budgeting (ZBB).** One philosophy for setting budgets. In preparing an ordinary *budget* for the next period, a manager starts with the budget for the current period and makes adjustments as seem necessary because of changed conditions for the next period. Since most managers like to increase the scope of the activities managed and since most prices increase most of the time, amounts in budgets prepared in the ordinary, incremental way seem to increase period after period. The authority approving the budget assumes that managers will carry out operations in the same way as in the past and that next period's expenditures will have to be at least as large as those of the current period. Thus, this authority tends to study only the increments to the current period's budget. In ZBB, the authority questions the process for carrying out a program and the entire budget for the next period. The authority studies every dollar in the budget, not just the dollars incremental to the previous period's amounts. The advocates of ZBB claim that in this way, (1) management will more likely delete programs or divisions of marginal benefit to the business or governmental unit, rather than continuing with costs at least as large as the present ones, and (2) management may discover and implement alternative, more cost-effective ways of carrying out programs. ZBB implies questioning the existence of programs and the fundamental nature of the way that firms carry them out, not merely the amounts used to fund them. Experts appear to divide evenly as to whether the middle word should be "base" or "based."

# INDEX

# B

# C

# D

## G

# S

# T



# U

# Present Value of $1

## (Table 2 Excerpt, p. 737)

| Periods = $n$ | $\frac{1}{2}$% | 1% | $1\frac{1}{2}$% | 2% | 3% | 4% | 5% | 6% | 7% | 8% | 10% | 12% |
|---|---|---|---|---|---|---|---|---|---|---|---|---|
| 1 | .99502 | .99010 | .98522 | .98039 | .97087 | .96154 | .95238 | .94340 | .93458 | .92593 | .90909 | .89286 |
| 2 | .99007 | .98030 | .97066 | .96117 | .94260 | .92456 | .90703 | .89000 | .87344 | .85734 | .82645 | .79719 |
| 3 | .98515 | .97059 | .95632 | .94232 | .91514 | .88900 | .86384 | .83962 | .81630 | .79383 | .75131 | .71178 |
| 4 | .98025 | .96098 | .94218 | .92385 | .88849 | .85480 | .82270 | .79209 | .76290 | .73503 | .68301 | .63552 |
| 5 | .97537 | .95147 | .92826 | .90573 | .86261 | .82193 | .78353 | .74726 | .71299 | .68058 | .62092 | .56743 |
| 6 | .97052 | .94205 | .91454 | .88797 | .83748 | .79031 | .74622 | .70496 | .66634 | .63017 | .56447 | .50663 |
| 7 | .96569 | .93272 | .90103 | .87056 | .81309 | .75992 | .71068 | .66506 | .62275 | .58349 | .51316 | .45235 |
| 8 | .96089 | .92348 | .88771 | .85349 | .78941 | .73069 | .67684 | .62741 | .58201 | .54027 | .46651 | .40388 |
| 9 | .95610 | .91434 | .87459 | .83676 | .76642 | .70259 | .64461 | .59190 | .54393 | .50025 | .42410 | .36061 |
| 10 | .95135 | .90529 | .86167 | .82035 | .74409 | .67556 | .61391 | .55839 | .50835 | .46319 | .38554 | .32197 |
| 11 | .94661 | .89632 | .84893 | .80426 | .72242 | .64958 | .58468 | .52679 | .47509 | .42888 | .35049 | .28748 |
| 12 | .94191 | .88745 | .83639 | .78849 | .70138 | .62460 | .55684 | .49697 | .44401 | .39711 | .31863 | .25668 |
| 13 | .93722 | .87866 | .82403 | .77303 | .68095 | .60057 | .53032 | .46884 | .41496 | .36770 | .28966 | .22917 |
| 14 | .93256 | .86996 | .81185 | .75788 | .66112 | .57748 | .50507 | .44230 | .38782 | .34046 | .26333 | .20462 |
| 15 | .92792 | .86135 | .79985 | .74301 | .64186 | .55526 | .48102 | .41727 | .36245 | .31524 | .23939 | .18270 |
| 16 | .92330 | .85282 | .78803 | .72845 | .62317 | .53391 | .45811 | .39365 | .33873 | .29189 | .21763 | .16312 |
| 17 | .91871 | .84438 | .77639 | .71416 | .60502 | .51337 | .43630 | .37136 | .31657 | .27027 | .19784 | .14564 |
| 18 | .91414 | .83602 | .76491 | .70016 | .58739 | .49363 | .41552 | .35034 | .29586 | .25025 | .17986 | .13004 |
| 19 | .90959 | .82774 | .75361 | .68643 | .57029 | .47464 | .39573 | .33051 | .27651 | .23171 | .16351 | .11611 |
| 20 | .90506 | .81954 | .74247 | .67297 | .55368 | .45639 | .37689 | .31180 | .25842 | .21455 | .14864 | .10367 |
| 22 | .89608 | .80340 | .72069 | .64684 | .52189 | .42196 | .34185 | .27751 | .22571 | .18394 | .12285 | .08264 |
| 24 | .88719 | .78757 | .69954 | .62172 | .49193 | .39012 | .31007 | .24698 | .19715 | .15770 | .10153 | .06588 |
| 26 | .87838 | .77205 | .67902 | .59758 | .46369 | .36069 | .28124 | .21981 | .17220 | .13520 | .08391 | .05252 |
| 28 | .86966 | .75684 | .65910 | .57437 | .43708 | .33348 | .25509 | .19563 | .15040 | .11591 | .06934 | .04187 |
| 30 | .86103 | .74192 | .63976 | .55207 | .41199 | .30832 | .23138 | .17411 | .13137 | .09938 | .05731 | .03338 |
| 32 | .85248 | .72730 | .62099 | .53063 | .38834 | .28506 | .20987 | .15496 | .11474 | .08520 | .04736 | .02661 |
| 34 | .84402 | .71297 | .60277 | .51003 | .36604 | .26355 | .19035 | .13791 | .10022 | .07305 | .03914 | .02121 |
| 36 | .83564 | .69892 | .58509 | .49022 | .34503 | .24367 | .17266 | .12274 | .08754 | .06262 | .03235 | .01691 |
| 38 | .82735 | .68515 | .56792 | .47119 | .32523 | .22529 | .15661 | .10924 | .07646 | .05369 | .02673 | .01348 |
| 40 | .81914 | .67165 | .55126 | .45289 | .30656 | .20829 | .14205 | .09722 | .06678 | .04603 | .02209 | .01075 |
| 45 | .79896 | .63905 | .51171 | .41020 | .26444 | .17120 | .11130 | .07265 | .04761 | .03133 | .01372 | .00610 |
| 50 | .77929 | .60804 | .47500 | .37153 | .22811 | .14071 | .08720 | .05429 | .03395 | .02132 | .00852 | .00346 |
| 100 | .60729 | .36971 | .22563 | .13803 | .05203 | .01980 | .00760 | .00295 | .00115 | .00045 | .00007 | .00001 |

# Present Value of Annuity of $1 in Arrears

## (Table 4 Excerpt, p. 739)

| No. of Payments = $n$ | $\frac{1}{2}$% | 1% | $1\frac{1}{2}$% | 2% | 3% | 4% | 5% | 6% | 7% | 8% | 10% | 12% |
|---|---|---|---|---|---|---|---|---|---|---|---|---|
| 1 | .99502 | .99010 | .98522 | .98039 | .97087 | .96154 | .95238 | .94340 | .93458 | .92593 | .90909 | .89286 |
| 2 | 1.98510 | 1.97040 | 1.95588 | 1.94156 | 1.91347 | 1.88609 | 1.85941 | 1.83339 | 1.80802 | 1.78326 | 1.73554 | 1.69005 |
| 3 | 2.97025 | 2.94099 | 2.91220 | 2.88388 | 2.82861 | 2.77509 | 2.72325 | 2.67301 | 2.62432 | 2.57710 | 2.48685 | 2.40183 |
| 4 | 3.95050 | 3.90197 | 3.85438 | 3.80773 | 3.71710 | 3.62990 | 3.54595 | 3.46511 | 3.38721 | 3.31213 | 3.16987 | 3.03735 |
| 5 | 4.92587 | 4.85343 | 4.78264 | 4.71346 | 4.57971 | 4.45182 | 4.32948 | 4.21236 | 4.10020 | 3.99271 | 3.79079 | 3.60478 |
| 6 | 5.89638 | 5.79548 | 5.69719 | 5.60143 | 5.41719 | 5.24214 | 5.07569 | 4.91732 | 4.76654 | 4.62288 | 4.35526 | 4.11141 |
| 7 | 6.86207 | 6.72819 | 6.59821 | 6.47199 | 6.23028 | 6.00205 | 5.78637 | 5.58238 | 5.38929 | 5.20637 | 4.86842 | 4.56376 |
| 8 | 7.82296 | 7.65168 | 7.48593 | 7.32548 | 7.01969 | 6.73274 | 6.46321 | 6.20979 | 5.97130 | 5.74664 | 5.33493 | 4.96764 |
| 9 | 8.77906 | 8.56602 | 8.36052 | 8.16224 | 7.78611 | 7.43533 | 7.10782 | 6.80169 | 6.51523 | 6.24689 | 5.75902 | 5.32825 |
| 10 | 9.73041 | 9.47130 | 9.22218 | 8.98259 | 8.53020 | 8.11090 | 7.72173 | 7.36009 | 7.02358 | 6.71008 | 6.14457 | 5.65022 |
| 11 | 10.67703 | 10.36763 | 10.07112 | 9.78685 | 9.25262 | 8.76048 | 8.30641 | 7.88687 | 7.49867 | 7.13896 | 6.49506 | 5.93770 |
| 12 | 11.61893 | 11.25508 | 10.90751 | 10.57534 | 9.95400 | 9.38507 | 8.86325 | 8.38384 | 7.94269 | 7.53608 | 6.81369 | 6.19437 |
| 13 | 12.55615 | 12.13374 | 11.73153 | 11.34837 | 10.63496 | 9.98565 | 9.39357 | 8.85268 | 8.35765 | 7.90378 | 7.10336 | 6.42355 |
| 14 | 13.48871 | 13.00370 | 12.54338 | 12.10625 | 11.29607 | 10.56312 | 9.89864 | 9.29498 | 8.74547 | 8.24424 | 7.36669 | 6.62817 |
| 15 | 14.41662 | 13.86505 | 13.34323 | 12.84926 | 11.93794 | 11.11839 | 10.37966 | 9.71225 | 9.10791 | 8.55948 | 7.60608 | 6.81086 |
| 16 | 15.33993 | 14.71787 | 14.13126 | 13.57771 | 12.56110 | 11.65230 | 10.83777 | 10.10590 | 9.44665 | 8.85137 | 7.82371 | 6.97399 |
| 17 | 16.25863 | 15.56225 | 14.90765 | 14.29187 | 13.16612 | 12.16567 | 11.27407 | 10.47726 | 9.76322 | 9.12164 | 8.02155 | 7.11963 |
| 18 | 17.17277 | 16.39827 | 15.67256 | 14.99203 | 13.75351 | 12.65930 | 11.68959 | 10.82760 | 10.05909 | 9.37189 | 8.20141 | 7.24967 |
| 19 | 18.08236 | 17.22601 | 16.42617 | 15.67846 | 14.32380 | 13.13394 | 12.08532 | 11.15812 | 10.33560 | 9.60360 | 8.36492 | 7.36578 |
| 20 | 18.98742 | 18.04555 | 17.16864 | 16.35143 | 14.87747 | 13.59033 | 12.46221 | 11.46992 | 10.59401 | 9.81815 | 8.51356 | 7.46944 |
| 22 | 20.78406 | 19.66038 | 18.62082 | 17.65805 | 15.93692 | 14.45112 | 13.16300 | 12.04158 | 11.06124 | 10.20074 | 8.77154 | 7.64465 |
| 24 | 22.56287 | 21.24339 | 20.03041 | 18.91393 | 16.93554 | 15.24696 | 13.79864 | 12.55036 | 11.46933 | 10.52876 | 8.98474 | 7.78432 |
| 26 | 24.32402 | 22.79520 | 21.39863 | 20.12104 | 17.87684 | 15.98277 | 14.37519 | 13.00317 | 11.82578 | 10.80998 | 9.16095 | 7.89566 |
| 28 | 26.06769 | 24.31644 | 22.72672 | 21.28127 | 18.76411 | 16.66306 | 14.89813 | 13.40616 | 12.13711 | 11.05108 | 9.30657 | 7.98442 |
| 30 | 27.79405 | 25.80771 | 24.01584 | 22.39646 | 19.60044 | 17.29203 | 15.37245 | 13.76483 | 12.40904 | 11.25778 | 9.42691 | 8.05518 |
| 32 | 29.50328 | 27.26959 | 25.26714 | 23.46833 | 20.38877 | 17.87355 | 15.80268 | 14.08404 | 12.64656 | 11.43500 | 9.52638 | 8.11159 |
| 34 | 31.19555 | 28.70267 | 26.48173 | 24.49859 | 21.13184 | 18.41120 | 16.19290 | 14.36814 | 12.85401 | 11.58693 | 9.60857 | 8.15656 |
| 36 | 32.87102 | 30.10751 | 27.66068 | 25.48884 | 21.83225 | 18.90828 | 16.54685 | 14.62099 | 13.03521 | 11.71719 | 9.67651 | 8.19241 |
| 38 | 34.52985 | 31.48466 | 28.80505 | 26.44064 | 22.49246 | 19.36786 | 16.86789 | 14.84602 | 13.19347 | 11.82887 | 9.73265 | 8.22099 |
| 40 | 36.17223 | 32.83469 | 29.91585 | 27.35548 | 23.11477 | 19.79277 | 17.15909 | 15.04630 | 13.33171 | 11.92461 | 9.77905 | 8.24378 |
| 45 | 40.20720 | 36.09451 | 32.55234 | 29.49016 | 24.51871 | 20.72004 | 17.77407 | 15.45583 | 13.60552 | 12.10840 | 9.86281 | 8.28252 |
| 50 | 44.14279 | 39.19612 | 34.99969 | 31.42361 | 25.72976 | 21.48218 | 18.25593 | 15.76186 | 13.80075 | 12.23348 | 9.91481 | 8.30450 |
| 100 | 78.54264 | 63.02888 | 51.62470 | 43.09835 | 31.59891 | 24.50500 | 19.84791 | 16.61755 | 14.26925 | 12.49432 | 9.99927 | 8.33323 |

Note: To convert from this table to values of an annuity in advance, determine the annuity in arrears above for one fewer period and add 1.00000.